Springer Collected Works in Mathematics

More information about this series at http://www.springer.com/series/11104

David Mumford

Selected Papers I

On the Classification of Varieties and Moduli Spaces

Reprint of the 2004 Edition

Springer

David Mumford
Applied Mathematics
Brown University
Providence, RI, USA

ISSN 2194-9875
Springer Collected Works in Mathematics
ISBN 978-1-4939-9535-6

This Springer imprint is published by the registered company Springer Science+Business Media, LLC, part of Springer Nature
The registered company address is: 233 Spring Street, New York, NY 10013, U.S.A.

David Mumford

Selected Papers

On the Classification of Varieties and Moduli Spaces

Springer

David Mumford
Division of Applied Mathematics
Brown University
Providence, RI 02912
david_mumford@brown.edu

Mathematics Subject Classification (2000): 01A75, 14Lxx, 14Kxx

Library of Congress Cataloging-in-Publication Data
Mumford, David, 1937–
Selected papers on the classifications of varieties and moduli spaces / David Mumford.
p. cm.
Includes bibliographical references and index.
ISBN 0-387-21092-X (acid-free paper)
1. Algebraic varieties--Classification theory. 2. Moduli theory. I. Title.
QA564.M858 2004
516.3'53–dc22 2004045274

ISBN 978-0-387-21092-6 Printed on acid-free paper.

9 8 7 6 5 4 3 2 1 SPIN 10953020

Springer-Verlag is a part of *Springer Science+Business Media*

springeronline.com

Preface

Firstly, I want to thank David Gieseker, George Kempf, Herbert Lange, Ina Lindemann, and Eckart Viehweg for their tremendous efforts in bringing this volume out, in writing three very excellent and thorough commentaries, and in pulling the whole project together. I am very grateful to all of you. I also want to express my great sorrow at the early decease of George Kempf, who had been a great partner in developing some of these ideas back in the 1960s and a good friend over many years.

It is a strange sensation to read the commentaries in this book and look back through a mist (a cliché but quite true), now only half knowing things that were once so vivid and so central a part of my life. One of the best parts is to remember all the collaborators, students, and colleagues I was blessed with and what fun it was to see the ball passed back and forth, ideas come to fruition, and totally unexpected connections emerge, such as those with physics. So often, new ideas come out of a casual conversation, only indirectly related to the application they find. I've been working for 20 years now in another field—stochastic methods and their applications to thought—but, boy, is algebraic geometry a beautiful field!

By my count, this volume contains about three-fifths of my papers in algebraic geometry. But it's fair to say that these were the papers where my heart was. The point is, I love maps, that is "maps" in the sense of "maps of the world," "charts of the ocean," "atlases of the sky"! I think one of the key things that attracted me to this group of problems was the hope of making a *map* of some parts of the world of algebraic varieties. An algebraic variety felt like a tangible thing in the lectures of Oscar Zariski, so why shouldn't you venture out, like Magellan, and uncover their geography? The sought-for map involves finding first the different continents, something like the 3-way elliptic/parabolic/hyperbolic curvature trichotomy, Enriques's classification of surfaces, and its generalizations to higher dimensions with Kodaira dimension and being uniruled. And it involves the interiors of the continents, that is, constructing the continuous moduli spaces and finding their properties and invariants. And how about abelian varieties? Well, these are the "group objects in the category of projective varieties" (as Grothendieck would put it), and the astonishing thing is that while being far from simple, they carry a magical algebraic machinery, that of theta functions, which gives both them and their moduli spaces a very explicit structure.

Besides being a form of cartography, the theory of moduli spaces has the wonderful feature of having many doors, many techniques by which this theory can be developed. Of course, there is traditional algebraic geometry, but there is also invariant theory, complex-analytic techniques such as Teichmüller theory, global topological techniques, and purely characteristic p methods such as counting objects over finite fields. This is another part of its charm.

In reading the commentaries in this book, I see that one fact stands out: the story is not finished, and many people have become interested. When I was a student in the 1950s, I doubt there were more than a dozen or so active researchers in the field. Now it is a river in full flood. I am sure that there are marvelous things waiting to be uncovered. To everyone in the field, good luck in this exciting enterprise.

David Mumford
Tenants Harbor
September 2003

Contents

Part III. The Classification of Surfaces and Other Varieties

List of Publications

I. Research in Algebraic Geometry: 1959–1982

Books:

1. *Lectures on Curves on Surfaces*, (with G. Bergman), Princeton University Press, 1964.
2. *Geometric Invariant Theory*, Springer-Verlag, 1965; 2nd enlarged edition, (with J. Fogarty), 1982; 3rd enlarged edition, (with F. Kirwan and J. Fogarty), 1994.
3. *The Red Book of Varieties and Schemes*, mimeographed notes from the Harvard Mathematics Department 1967, reprinted in Lecture Notes in Mathematics 1348, Springer-Verlag 1988.
4. *Abelian Varieties*, Oxford University Press, 1st edition 1970, 2nd edition 1974.
5. 6 Appendices to *Algebraic Surfaces*, by O. Zariski, 2nd edition, Springer-Verlag, 1971.
6. *Toroidal Embeddings I* (with G. Kempf, F. Knudsen and B. Saint-Donat), Lecture Notes in Mathematics 339, Springer-Verlag 1973.
7. *Curves and their Jacobians*, University of Michigan Press, 1975.
8. *Smooth Compactification of Locally Symmetric Varieties* (with A. Ash, M. Rapoport, Y. Tai), Lie Groups: History Frontiers and Applications, Vol. 4, Math. Sci. Press 1975.
9. *Algebraic Geometry I: Complex Projective Varieties*, Springer-Verlag, New York, 1976.
10. *Tata Lectures on Theta* (with C. Musili, M. Nori, P. Norman, E. Previato and M. Stillman), Birkhauser-Boston, Part I, 1982, Part II, 1983, Part III, 1991.

Papers:

1. Topology of Normal Singularities and a Criterion for Simplicity, *Publ. de l'Institut des Hautes Etudes Scientifiques*, 1961, pp. 1–22.
2. Pathologies of Modular Geometry, *Amer. J. of Math.*, 1961, pp. 339–342.

3. An Elementary theorem in Geometric Invariant Theory, *Bull. Amer. Math. Soc.*, 1961, pp. 483–487.
4. Projective Invariants of Projective Structures, *International Congress of Mathematicians*, Stockholm 1962, pp. 526–530.
5. Further Pathologies in Algebraic Geometry, *Amer. J. of Math.*, 1962, pp. 642–648.
6. The Canonical Ring of an Algebraic Surface, an appendix to a paper by Oscar Zariski, *Ann. Of Math.*, **76**, 1962, pp. 612–615.
7. Topics in the Theory of Moduli, (in Japanese), *Sugaku*, 1963.
8. Two Fundamental Theorem on Deformations of Polarized Varieties (with T. Matsusaka), *Amer. J. of Math.*, 1964, pp. 668–684.
9. A Remark on Mordell's Conjecture, *Amer. J. of Math.*, 1965, pp. 1007–1016.
10. Picard Groups of Moduli Problems, in *Proc. of a Conference in Arithmetic Algebraic Geometry*, Harper and Row, 1965.
11. On the Equations Defining Abelian Varieties I, II, III, *Inventiones Mathematicae*, 1966, **1**, pp. 287–384; 1967, **3**, pp. 75–135 and pp. 215–244.
12. Families of Abelian Varieties, in *Proc. of Symposium in Pure Math.*, **9**, Amer. Math. Soc., 1966.
13. Pathologies III, *Amer. J. of Math.*, 1967, **89**, pp. 94–104.
14. Abelian Quotients of the Teichmuller Modular Group, *Journal d'Analyse*, 1967, **28**, pp. 227–244.
15. Deformations and Liftings of Finite Commutative Group Schemes (with F. Oort), *Inventiones Mathematicae*, 1968, **5**, pp. 317–334.
16. Periods of Moduli Spaces of Bundles on Curves (with P. Newstead), *Amer. J. of Math.*, 1968, **90**, pp. 1200–1208.
17. Enriques' Classification of Surfaces in Char. p, I, in *Global Analysis (papers in honor of K. Kodaira)*, Spencer and Iyanaga editors, U. of Tokyo Press, 1969, pp. 325–339.
18. Bi-extensions of Formal Groups, in *Algebraic Geometry*, Oxford University Press, 1969, pp. 307–322.
19. A Note on Shimura's paper "Discontinuous Groups and Abelian Varieties," *Math. Annalen*, 1969, **181**, pp. 345–351.
20. Rational Equivalences of 0-cycles on Surfaces, *J. of Math. of Kyoto Univ.*, 1969, **9**, pp. 195–204.
21. The Irreducibility of the Space of Curves of Given Genus (with P. Deligne), *Publ. Math. Dfe l'I.H.E.S.*, 1969, **36**, pp. 75–109.
22. Varieties Defined by Quadratic Equations, in *Questions on Algebraic Varieties*, C.I.M.E., 1969, publ. by Editioni Cremonese, 1970.
23. Theta Characteristics of an Algebraic Curve, *Annales de l'Ecole Norm. Sup.*, 1971, pp. 181–192.

24. A Remark on Mahler's Compactness Theorem, *Proc. Amer. Math. Soc.*, 1971., **28**, pp. 289–194.
25. The Structure of the Moduli Spaces of Curves and Abelian Varieties, *Actes du Congress Int. du Math.*, Nice, 1970; publ. by Gauthier-Villars, 1971.
26. An Analytic Construction of Degenerating Curves Over Complete Local Rings, *Composito Math.*, 1972, **24**, pp. 129–174.
27. An Analytic Construction of Degenerating Abelian Varieties over Complete Rings, *Composito Math.*, 1972, **24**, pp. 239–272.
28. Some Elementary Examples of Unirational Varieties which are not Rational (with M. Artin), *J. London Math. Soc.*, 1972, **25**, pp. 75–95.
29. Introduction to the Theory of Moduli (with K. Suominen), in *Algebraic Geometry, Oslo 1970*, F. Oort editor, Wolters-Noordhoff, 1971.
30. A Rank 2 Vector Bundle on **P**4 with 15,000 Symmetries (with G. Horrocks), *Topology*, 1973, **12**, pp. 63–81.
31. An Example of a Unirational 3-fold which is not Rational, *Accad. Naz. dei Lincei*, 1973.
32. A Remark on the Paper of M. Schlessinger, in *Complex Analysis, 1972*, Rice University Studies, **59**, 1973, pp. 113–117.
33. Prym Varictics I, in *Contributions to Analysis*, Academic Press, 1974, pp. 325–350.
34. A New Approach to Compactifying Locally Symmetric Varieties, in *Discrete Subgroups of Lie Groups*, Oxford University Press, 1975, pp. 211–224.
35. Matsusaka's Big Theorem (with D. Lieberman), in *Algebraic Geometry, Arcata 1974*, AMS Proc. of Symposia in Pure Math., **29**, 1975, pp. 513–530.
36. The Self-Intersection Formula and the "Formule-Clef" (with A.T. Lascu and D.B. Scott), *Math. Proc. Camb. Phil. Soc.*, 1975, **78**, pp. 117–123.
37. Hilbert's 14th Problem - The Finite Generation of Subgroups such as Rings of Invariants, in *Proc. of a Conference on Hilbert's Problems*, Dekalb, 1974, Amer. Math. Soc.
38. Pathologies IV, Amer. J. of Math., 1975, **97**, pp. 847–849.
39. Enriques Classification of Surfaces in Char. p, II and III (with E. Bombieri), in *Complex Analysis and Algebraic Geometry*, Baily and Shioda editors, Cambridge Univ. Press, 1977, pp. 23–42, and *Invent. Math.*, 1976, **35**, pp. 197–232.
40. Stability of Projective Varieties, Monographie No. 24 de *L'Enseignement Math.*, **23**, 1977, pp. 39–110.
41. Hirzebruch's Proportionality Theorem in the non-compact case, *Invent. Math.*, 1977, **42**., pp. 239–272.
42. An Algebro-Geometric Construction of Commuting Operators and of Solutions to the Toda Lattice Equation, Korteweg de Vries Equation and Related Non-Linear Equations, in *Proc. of the Int. Symp. on Alg. Geom.*, Kinokuniya, Tokyo 1977.

43. Some Footnotes to the Work of C.P. Ramanujam and On Ramanujam's Work in Algebraic Geometry, in *C.P. Ramanujam–A Tribute*, Studies in Math. No 8, Tata Institute of Fundamental Research, 1978.
44. An Instinct for the Key Idea (with J. Tate), *Science*, 1978, **202**, pp. 737–739.
45. An Algebraic Surface with K ample, $(K^2) = 9$, $p_g = q = 0$, *Amer. J. of Math.*, 1979, **101**, pp. 233–244.
46. The Spectrum of Difference Operators and Algebraic Curves (with P. Van Moerbeke), *Acta Math.*, 1979, **143**, pp. 93–154.
47. On the Kodaira Dimension of the Moduli Space of Curves (with J. Harris), *Inv. Math.*, 1982.
48. On the Kodaira Dimension of the Siegel Modular Variety, in *Algebraic Geometry - Open Problems, Lecture Notes in Mathematics 997*, Springer-Verlag 1983, pp. 348–375.
49. Towards an Enumerative Geometry of Moduli Space of Curves, in *Arithmetic and Geometry*, edited by M. Artin, J. Tate, Birkhauser-Boston, 1983, pp. 271–326.
50. Foreword for non-mathematicians, in *The Unreal Life of Oscar Zariski*, by Carol Parikh, Academic Press, 1990.
51. What can be Computed in Algebraic Geometry? (with Dave Bayer), in *Computational Algebraic Geometry and Commutative Algebra*, ed. D. Eisenbud & L. Robbiano, Camb Univ. Press, 1993, pp. 1–48.

II. Research in Vision: 1983-2001

Books:

1. *Filtering, Segmentation and Depth*, (with M. Nitzberg and T. Shiota), Springer Lecture Notes in Computer Science **662**, 1993.
2. *Two and Three dimensional Patterns of the Face*, (with P. Giblin, G. Gordon, P. Hallinan and A. Yuille), AK Peters, 1999.
3. *Pattern Theory through Examples*, (with A. Desolneux), in preparation.

Papers:

1. The Representation of Shape (with A. Latto and J. Shah), in *Proceedings of the 1984 IEEE Workshop on Computer Vision*, pp. 183–191, 1984.
2. Boundary Detection by Minimizing Functionals I (with J. Shah), in *Image Understanding 1989*, Ablex Press, preliminary version in *1985 IEEE Conference on Computer Vision and Pattern Recognition* (CVPR), 1985.
3. The Problem of Robust Shape Descriptors, in *Proc of 1st IEEE International Conference on Computer Vision* (ICCV), 1987, pp. 602–606.
4. Discriminating Figure from Ground: the role of edge detection and region growing (with S. Kosslyn, L. Hillger and R. Herrnstein), *Proc. Nat. Acad. Sci.*, 1987, **84**, pp. 7354–7358.

5. Optimal Approximations of Piecewise Smooth Functions and Associated Variational Problems (with J. Shah), *Comm. in Pure and Appl. Math.*, 1989, **42**, pp. 577–685.
6. Teaching Pigeons an Abstract Relational Rule: Insideness (with R. Herrnstein, W. Vaughan and S. Kosslyn), *Perception and Psychophysics*, 1989, **46**, pp. 56–64. 2
7. The 2.1D Sketch (with M. Nitzberg), in *Proc. of 3rd IEEE International Conference on Computer Vision* (ICCV), 1990, pp. 138–144.
8. Parametrizing Exemplars of Categories, *J. Cognitive Neuroscience*, 1991, **3**, pp. 87–88.
9. On the Computational Architecture of the Neocortex, I: The role of the thalamo-cortical loop, *Biological Cybernetics*, 1991, **65**, pp. 135–145; II: The role of cortico-cortical loops, *Biological Cybernetics*, **66**, pp. 241–251.
10. Mathematical Theories of Shape: do they model perception?, in *Proc. Conference 1570, Soc. Photo-optical & Ind. Engineers*, 1991, pp. 2–10.
11. Texture Segmentation by Minimizing Vector-Valued Energy Functionals: the coupled membrane model (with Tai Sing Lee and Alan Yuille), *Proc. European Conf. Comp. Vision, 1992*, Lecture Notes in Computer Science 588, pp. 165–173.
12. A Bayesian Treatment of the Stereo Correspondence Problem Using Half-Occluded Regions, (with P. Belhumeur), *Proc. IEEE Conf. Comp. Vision and Pattern Recognition, 1992* (CVPR), pp. 506–512.
13. Elastica and Computer Vision, in *Algebraic Geometry and its Applications*, ed. C. Bajaj, Springer-Verlag, 1993, pp. 507–518.
14. Commentary on Grenander & Miller "Representations of Knowledge in Complex Systems", *Proc. Royal Stat. Soc.*, 1994.
15. Pattern Theory: a Unifying Perspective, in *Proceedings 1st European Congress of Mathematics*, Birkhauser-Boston, 1994. Revised version in *Perception as Bayesian inference*, ed. D. Knill and W. Richards, Cambridge Univ. Press, 1996, pp. 25–62.
16. Neuronal Architectures for Pattern-theoretic Problems, in *Large Scale Neuronal Theories of the Brain*, MIT Press, 1994, pp. 125–152.
17. Chordal completions of planar graphs (with F.R.K. Chung), *J. of Combinatorics*, **62**, 1994, pp. 96–106.
18. The Bayesian Rationale for Energy Functionals, in *Geometry-Driven Diffusion in Computer Vision*, Bart Romeny editor, Kluwer Academic, 1994, pp. 141–153.
19. Thalamus, in *The Handbook of Brain Theory and Neural Networks*, M. Arbib editor, MIT Press, 1995.
20. Neural correlates of boundary and medial axis representations in primate striate cortex, (with T.S. Lee, K. Zipser & P.H. Schiller), ARVO abstract, 1995.
21. The Statistical Description of Visual Signals, *ICIAM 95* ed. K. Kirshgassner, O. Mahrenholtz & R. Mennicken, Akademie Verlag, 1996.

22. Review of *Variational Methods in image segmentation*, by J-M Morel & S. Solimini, *Bull. Amer. Math. Soc.*, **33**, 1996, 211–216.
23. Issues in the mathematical modeling of cortical functioning and thought, in *The Legacy of Norbert Wiener: A Centennial Symposium*, ed. D. Jerison et al, Amer. Math. Society, 1997, pp. 235–260.
24. Visual Search and Shape from Shading Modulate Contextual Processing in Macaque Early Visual Cortices, (with T.S. Lee, R. Romero, A. Tobias & T. Moore), *Neuroscience Abstract*, 1997.
25. FRAME: Filters, Random Field and Maximum Entropy, (with S.C. Zhu and Y. Wu), *Int. J. Comp. Vis.*, **27**, 1998.
26. Minimax Entropy Principle and its Application to Texture Modeling (with S.C. Zhu and Y.N. Wu), *Neural Computation*, **9**, 1997, 1627-60.
27. Prior Learning and Gibbs Reaction-Diffusion (with Song Chun Zhu), *IEEE Trans. Patt. Anal. and Mach. Int.*, **19**, 1997, 1236-50.
28. The Role of V1 in Shape Representation (with Tai Sing Lee, Song Chun Zhu & Victor Lamme), *Computational Neuroscience*, ed. Bower, Plenum Press, 1997.
29. The Role of Primary Visual Cortex in Higher Level Vision (with T.S. Lee, R. Romero and V. Lamme), *Vision Research*, **38**, 1998, 2429-2454.
30. Thalamus, in *MIT Encyclopedia of the Cognitive Sciences*, MIT Press, 1999.
31. The Statistics of Natural Images and Models (with J. Huang), *Proc. IEEE Conf. Comp. Vision and Pattern Rec.* 1999, pp. 541–547, Comp Sci Press.
32. Statistics of range images (with Jinggang Huang and Ann Lee), *Proc. IEEE Conf. Comp. Vision and Pattern Rec.* 2000, pp. 324–331, Comp Sci Press.
33. Stochastic Models for Generic Images (with Basilis Gidas), *Quarterly Appl. Math.*, **59**, 2001, pp. 85–111.
34. Occlusion models for natural images: A statistical study of a scale-invariant dead-leaves model, (with Ann Lee and Jinggang Huang), *Int. J. Computer Vision*, **41**, 2001, pp. 35–59.
35. Surface evolution under curvature flow (with Conglin Lu and Yan Cao), to appear, Special Issue on Partial Differential equations in Image Proc. Comp. Vision and Comp. Graphics, *Journal of Visual Communication and Image Representation*, 2001.
36. Neural activity in early visual cortex reflects behavioral experience and higher-order perceptual saliency (with Tai Sing Lee, C. Yang, R. Romero), *Nature Neuroscience*, **5**, 2002, 589–597.
37. Geometric Structure Estimation of Axially Symmetric Pots from Small Fragments (with Yan Cao), in *Proc. of Int. Conf. on Signal Processing, Pattern Recognition, and Applications*, Crete, 2002.
38. Pattern Theory: The Mathematics of Perception, in *Proceedings of ICM 2002, Beijing*, **vol. 1**, 2002, 401–422.
39. The Nonlinear Statistics of High-contrast Patches in Natural Images (with Ann Lee and Kim Pedersen), *Int. J. Comp. Vision*, **54**, 2003, 83–103.

40. Hierarchical Bayesian Inference in the Visual Cortex, (with Tau Sing Lee), *Journal of the Optical Society of America*, **20**, 2003, 1434–1448.
41. Riemannian Geometries on Spaces of Plane Curves (with Peter Michor), submitted to J. of the European Math. Society, 2003.

III. Other Work

1. Contributor to *Multi-variable Calculus*, The Calculus Consortium based at Harvard, Wiley, 1995.
2. Calculus Reform – For the Millions, *Notices Amer. Math. Soc.*, May 1997, pp. 559–563.
3. Trends in the Profession of Mathematics, *Mitteilungen der Deutsche Mathemtiker Verein* REF, 1998.
4. The Dawning of the Age of Stochasticity, in *Mathematics: Frontiers and Perspectives*, edited by V. Arnold, M. Atiyah, P. Lax and B. Mazur, AMS, 2000.
5. *Indra's Pearls* (with C. Series and D. Wright), Cambridge University Press, 2002.

Mumford's work on the moduli of curves and vector bundles

David Gieseker

Much of Mumford's work has been devoted to the study of the moduli spaces of curves. Let g be an integer larger than one. Riemann knew that a Riemann surface of genus g depended on $3g-3$ complex parameters, suggesting that the set of all curves of genus g should form a $3g-3$ dimensional space $\mathcal{M}_g$. With some historical hindsight, the main problems were to establish the existence of $\mathcal{M}_g$, to compactify $\mathcal{M}_g$ and to establish properties of $\mathcal{M}_g$. Mumford began by studying the existence of $\mathcal{M}_g$. Prior to Mumford's work on the existence of $\mathcal{M}_g$, Ahlfors and Bers had shown that $\mathcal{M}_g$ existed as an analytic space and Bailey had shown that $\mathcal{M}_g$ was a quasi-projective variety over $\mathbf{C}$[2]. Mumford was the first to formulate a purely algebraic approach to the problem of the existence of $\mathcal{M}_g$ as a quasi-projective variety valid in all characteristics. Mumford actually used two quite different approaches to this existence problem, both using Geometric Invariant Theory (GIT).

Invariant theory was one of the major areas of nineteenth century mathematics. The prototypical problem in invariant theory is to write down quantities depending on n unordered points in $\mathbf{P}^1$ which are invariant under the group of linear fractional transformations. In modern language and some historical hindsight, one would construct an explicit injective map ϕ from the quotient Q of $S^n(\mathbf{P}^1)$ by the group G of automorphisms of $\mathbf{P}^1$ to some projective space. However, it turns out that the quotient Q cannot exist as a separated variety, so it is impossible to construct such a map ϕ. An important goal in the nineteenth century was to explicitly find a basis of these invariants as an algebra. Hilbert solved this problem, although by an indirect argument which did not yield an explicit invariants. In addition, Hilbert also developed a method to pick out a large open subset of points of $S^n(\mathbf{P}^1)$ so that one could define invariants for this set of points [46]. He gave a criterion to determine if a given set of points was in this set or not using one parameter subgroups (in modern language). Hilbert gave a definitive solution to the nineteenth century problems about invariants; the subject vanished from the consciousness of the mathematical community and [46] was largely forgotten.

Unaware of Hilbert's work, Mumford developed GIT in the case of a reductive group G acting on a polarized variety Y and realized the resulting theory could be used as a tool for studying the existence of moduli spaces in algebraic geometry [M8]. For most applications, one may assume that $G = SL(N)$, which we will do in this article. Mumford defined an open set $Y_s \subset Y$ of 'stable' points so that the quotient of Y by G exists as a separated quasi-projective variety. Further, there is a related idea of a point being semi-stable and the quotient by G of the semi-stable points modulo a certain equivalence relation is a projective variety compactifying the quotient of the stable points by G. For many practical purposes, one may assume that Y is just a projective space by embedding Y equivariantly in a projective space on which G acts. In this case, a point is semi-stable if there is an invariant homogeneous polynomial not vanishing at the point. The quotient is then formed by using all the invariant homogeneous polynomials of some fixed large degree to map the set of semi-stable points to a projective space.

It is very difficult to check whether a given point is stable or not using this definition, since a direct application would involve calculating the invariant polynomials. Mumford gave a criterion, called the Hilbert-Mumford criterion, to determine if a given point is stable by examining the behavior of the point under all one parameter subgroups of the reductive group. Thus a point P in projective space $\mathbf{P}^n$ is not semi-stable iff there is a homomorphism λ from the multiplicative group to G so that the limit of $\lambda(\alpha)(Q)$ is zero, where Q is a lift of P to affine $N+1$ space.

The applications of GIT to constructing moduli spaces often proceed in several steps. One starts with a set S of objects one is trying to construct the moduli space for, e.g. S could be the set of isomorphism classes of curves of genus g. One then defines a scheme Y and an action of a group G on Y so that there is a natural map Φ from S to quotient of Y by G. In our example, Y might be the Chow points of curves in some projective space, G would be the automorphisms of the projective space, and Φ would take the Chow point of some n canonical image of $X \in S$. The Hilbert-Mumford criterion allows one to translate the problem of showing $\Phi(X)$ is stable into a problem about the geometry of X. If a moduli problem can be attacked with GIT, then not only can existence of the moduli space be established, but one also gets a compactification of the moduli space and an ample line bundle on the moduli space. At the time of [M8], Mumford's theory was confined to characteristic zero. However, since then the theory has been extended by Haboush, Procesi and Formanek, and Seshadri to all characteristics. [79, 80, 28, 40] Over $\mathbf{C}$, a beautiful connection has been made by Kempf, Ness and Kirwan to symplectic geometry and the moment map [52, 53, 71, 51, 12, 76] . Kirwan's work also enables one to relate the cohomology of the moduli space to the cohomology of Y. Another recent development is the consideration of variable polarizations. [88, 89, 14]

Mumford's first method for the construction of the moduli space of curves was to construct the moduli of principally polarized abelian varieties and to use Torelli's theorem to obtain the existence of $\mathcal{M}_g$. The idea here was to start with an abelian variety A of dimension g in $\mathbf{P}^k$ and to pick a large integer n and an ordering on the set D of n division points of A. Motivated by the Hilbert-Mumford criterion, Mumford defined a notion of a set of points in $\mathbf{P}^k$ being stable. He was able to show that D was stable. Mumford reduced the problem of finding a moduli space for abelian varieties to finding a quotient of the set of stable points in $\mathbf{P}^k$ modulo the automorphism group of $\mathbf{P}^k$. GIT then provided such a quotient. However, at the time GIT was not known to be valid in general, so Mumford actually constructed the quotient by hand [12].

Mumford's second method is to study the stability of the Chow point of the n-canonical embedding of a curve C for $n > 2$. Given an abstract curve C, the Chow point of the n canonical embedding of C is determined up to projective equivalence. Mumford established that the Chow point of this curve was stable using the Hilbert-Mumford criterion and so established the existence of $\mathcal{M}_g$ in characteristic zero. A variant of this approach has been used by the author to study surfaces of general type [30] and by Morrison to study certain classes of ruled surfaces [68]

Mumford also dealt with the moduli of vector bundles of rank r on a curve C in [M8]. One of the main problems here is that the moduli space of all bundles of a fixed degree and rank is non-separated. Mumford was able to construct a quasi-projective moduli space for a large class of 'stable' bundles. If we fix a large number N of points P_i on C and consider a vector bundle E of rank r generated by global sections and an isomorphism ϕ of $\mathbf{C}^n$ with $H^0(E)$, we can define N codimension r subspaces V_i of $\mathbf{C}^n$ as the kernel of the evaluation map of $H^0(E)$ to E_{P_i}. Mumford showed that stability of the set of V_i in the sense of GIT is a property of the vector bundle. He called such a bundle a stable bundle and was able to show that there was a moduli space for stable bundles. Narasimhan and Seshadri had independently found the same definition of stable bundle while investigating bundles arising from the unitary representations of the fundamental group of the curve [62] and the full details of the GIT theory were worked out in [78].

Mumford and Newstead in [M7] examined the structure of some of the moduli spaces of stable bundles on a smooth curve C. Let L a line bundle of degree one on C. Let M be the space of stable bundles whose determinant bundle is isomorphic to L on C. M is a smooth projective variety. Mumford and Newstead established the existence of a universal bundle E on $C \times M$ and used the second Chern class of E to get a map ϕ between $H^1(X)$ and $H^3(M)$. They then showed that ϕ is an isomorphism of Hodge structures. Thus the intermediate Jacobian of M is isomorphic to the Jacobian of C.

In [M6], Deligne and Mumford developed the theory of stable curves, which had been earlier developed in characteristic zero by Meyer and Mumford. A stable curve is a connected curve with only nodes and with no smooth rational components which meet the rest of the curve in fewer than three points. The main properties of stable curves is that they form a good separated compactification $\bar{\mathcal{M}}_g$ of $\mathcal{M}_g$ and that they have a smooth local deformation space. In fact, $\bar{\mathcal{M}}_g - \mathcal{M}_g$ is a divisor with normal crossings. It is quite surprising that $\mathcal{M}_g$ has such a simple and elegant compactification. The components of $\bar{\mathcal{M}}_g - \mathcal{M}_g$ are denoted Δ_i for i from 1 to $[\frac{n}{2}]$. The generic point of Δ_0 is an irreducible curve with one node, and the generic point of Δ_i for $i > 0$ consists of two smooth curves C_1 of genus i and C_2 of genus $g-i$ glued together at one point. (One technical problem which I have suppressed is that $\bar{\mathcal{M}}_g$ is not quite smooth at the curves with non-trivial automorphisms.) The main application of stable curves in [M6] was to show that $\mathcal{M}_g$ is irreducible in all characteristics. The main idea of [M6] is that there is a space $\bar{\mathcal{M}}_{g,\mathbf{Z}}$ over $Spec(\mathbf{Z})$, whose geometric fibers are moduli spaces of stable curves over the various characteristics. Then one can use Zariski's main theorem to deduce the connectedness of the special fibers from the known connectedness of the general fiber in characteristic zero. At the time of [M6], it was not known that $\bar{\mathcal{M}}_g$ was projective, so two arguments were given to deduce the irreducibility of $\bar{\mathcal{M}}_g$, the first elementary and the second using stacks. An elegant approach to this problem is in [29]. The concept of stable curves has been generalized by Kontsevich to the concept of stable maps [58, 57]. This approach has since become one of the most important tools in enumerative geometry, providing a mathematical context for mirror symmetry and quantum cohomology. This in turn has led to the resolution of the mirror conjecture for a wide variety of Calabi-Yau manifolds. [38, 60, 6]. The idea of stable curves has also been essential in arithmetic applications, e.g. Faltings' proof of the Mordell conjecture [24] and GIT has proved useful in other arithmetic applications [27].

Mumford used the theory of stable curves in a striking way to study the problem of uniformization of curves in p-adic analysis. Let K be a field complete with respect to a discrete valuation. Let $\mathcal{O}$ be the ring of integers of K. Tate had discovered a uniformization of elliptic curves over K with non-integral j-invariant. Tate associated to any $q \in \mathcal{O}$ which is not a unit in $\mathcal{O}$ an elliptic curve C defined over K with non-integral j invariant and a map $\pi : K^* \to C_K$. Further, any elliptic curve with non-integral j invariant and stable reduction occurs as such a C. Tate's map π is a non-Archimedian analogue of the classical map from $\mathbf{C}^*$ to an elliptic curve X obtained by taking the usual isomorphism of $\mathbf{C}/\Lambda$ with X and dividing $\mathbf{C}$ by a cyclic subgroup H of Λ to get a map $\mathbf{C}/H = \mathbf{C}^* \to X$. To make sense of his map as an analytic map, Tate developed the theory of rigid analytic spaces. This theory gives a good theory of analytic continuation on p-adic spaces, which the most naive theory of p-adic spaces does not. Rigid analytic spaces are intimately connected with the theory of formal schemes over $\mathcal{O}$. The idea is that a formal scheme flat over $\mathcal{O}$ gives a rigid analytic space and that two such schemes give the same analytic space if one can get from one to the other by blowing up and down along centers situated on the special fiber. Mumford in [M5] produced a striking generalization of Tate's example. A curve over K is k-split degenerate if it can be extended to a stable curve over $\mathcal{O}$ so that the special fiber consists of rational curves and the double points are all rational over k, the residue field of $\mathcal{O}$. These curves are now known as Mumford curves. Mumford showed that there is an analogue of Schottky's uniformization for any k-split degenerate curve. $\Gamma \subset PGL(2, K)$ is called a Schottky group if Γ is finitely generated and every non-trivial element is hyperbolic. To each k-split degenerate curve, Mumford assigned a Schottky group and vice-versa. Given a Schottky group Γ, he defined a formal scheme $\mathcal{X}$ over $\mathcal{O}$ obtained by blowing up the formal completion of $\mathbf{P}^1$ over $\mathcal{O}$ and an action of Γ on $\mathcal{X}$. The formal completion of the curve associated to Γ is just the quotient of $\mathcal{X}$ by action of Γ. Conversely, given a k-split degenerate curve, one can take the formal scheme associated to that curve. This formal scheme actually has a universal covering space in the étale topology and Mumford showed this universal cover arises from a Schottky group. A nice exposition of these results is in [75]. These ideas have been useful in number theory, e.g. [18]. Faltings [25] obtained a beautiful analogue of the Narasimhan-Seshadri's construction of semi-stable bundles from unitary representations in the context of Mumford curves.

[M3] gave a construction of projective curves which admit no smooth deformations by constructing a family of singular curves of arithmetic genus g which have more than $3g - 3$ dimensional moduli. Thus generically such curves cannot be deformed to smooth curves. This then shows there are curve singularities which cannot even be locally smoothed. Following Iarabino, Mumford produces such curves by looking at subspaces V of $m^{\nu}/m^{2\nu}$, where m is a maximal ideal of some point on a smooth curve and ν is some large integer.

In [M4], Mumford returned to Geometric Invariant Theory as it applies to the moduli space of curves. If a moduli space can be constructed by GIT, then there is automatically a compactification of this moduli space. Thus constructing $\mathcal{M}_g$ by GIT by using the n-canonical Chow point produces a compactification $\bar{\mathcal{M}}_g^n$ of $\mathcal{M}_g$. For some small value of n, this compactification is not the Deligne-Mumford compactification by stable curves. However, for $n \geq 5$, the compactification just turns out to the Deligne-Mumford compactification. In fact, one can construct the moduli of stable curves over $Spec\,\mathbf{Z}$ and use Zariski's theorem to establish the irreducibility of $\mathcal{M}_g$. (The projectivity of $\bar{\mathcal{M}}_g$ had earlier been established by Knudsen in characteristic zero [54]. [32] gives treatment parallel to Mumford's.) To show that $\bar{\mathcal{M}}_g^n = \bar{\mathcal{M}}_g$, Mumford did not directly prove that the Chow point of the n-canonical embedding of a stable curve is GIT stable. Rather he ruled out anything but the Chow points of n-canonically embedded Deligne-Mumford curves being in $\bar{\mathcal{M}}_g^n$. For instance, he shows that if a curve X has a cusp, then the Chow point of the n-canonical embedding of X cannot be GIT stable. It is often easy to show something is unstable in GIT, since all one has to do is to construct a single one parameter subgroup which destabilizes the object. In practice, such a one parameter subgroup is often quite simple and directly related to the geometry of the situation. Kempf and Rousseau showed that a non-semi-stable point always defines a canonical worst one parameter subgroup [50, 77]. An alternate proof of the projectivity of $\mathcal{M}_g$ has been given by [55].

[M4] also began the study of line bundles on $\bar{\mathcal{M}}_g$. There is a sequence of cohomology classes λ_n defined on the $\bar{\mathcal{M}}_g$ which are defined as the Chern classes of the highest exterior powers of the pushdowns of the nth power of the relative canonical bundle of the universal curve over $\bar{\mathcal{M}}_g$. λ_2 restricted to $\mathcal{M}_g$ can be identified with the canonical bundle of $\mathcal{M}_g$, since we can identify $H^1(X, \Omega^2)$ with the cotangent space of the point X in $\mathcal{M}_g$. Mumford established a formula for λ_n in terms of $\lambda = \lambda_1$ and the δ_i, the Chern classes of the components Δ_i of complement of $\mathcal{M}_g$ in $\bar{\mathcal{M}}_g$. In particular, on $\mathcal{M}_g$, we have

$$\lambda_2 = \lambda_1^{\otimes 13}.$$

Mumford proves these formulas by applying Grothendieck's Riemann-Roch Theorem. He also investigates which line bundles on $\bar{\mathcal{M}}_g$ are ample.

[M2] represents the culmination of Mumford's work on $\mathcal{M}_g$. For low values of g, $\mathcal{M}_g$ is unirational and it was conjectured that $\mathcal{M}_g$ was unirational for all g. [11, 10] It came as a great surprise when Harris and Mumford established that if $g \geq 25$ and is odd, $\mathcal{M}_g$ is of general type. Harris and Mumford introduced a divisor $D_k \subset \mathcal{M}_g$ which consists of all the curves with are k-fold covers of $\mathbf{P}^1$, where $k = \frac{g+1}{2}$. They calculated the Chern class of canonical bundle of $\bar{\mathcal{M}}_g$ and $\bar{D}_k$ in terms of λ_1 and the δ_i. From these calculations, it is easy to see that for $k > 12$ canonical bundle of $\bar{\mathcal{M}}_g$ is a positive multiple of λ_1 plus an effective divisor. Since high multiples of λ_1 have a lot of sections, so does the canonical bundle of $\bar{\mathcal{M}}_g$. The main problem is to calculate $\bar{D}_k$ in terms of λ_1 and the δ_i. Harris and Mumford first showed that $\bar{D}_k$ is a linear combination of λ_1 and the δ_i. They then identified $\bar{D}_k \cap \Delta_i$. The coefficients of the above linear combination can then be calculated by using certain test curves in the Δ_i. The problem ultimately reduced to using techniques of Griffiths and Harris to calculate the number of special divisors of certain curves. The identification of $\bar{D}_k \cap \Delta_i$ was accomplished by studying a compactification of D_k. The points of this compactification consist of maps from semi-stable curves to a tree of $\mathbf{P}^1$'s, generalizing a construction of Beauville. A technical difficulty is that $\bar{\mathcal{M}}_g$ is not smooth, but rather is locally a quotient of a manifold by a finite group. This difficulty is overcome by using results of Reid and Tai [86]. This work has been extended by several authors [20, 43, 44]. Teichmüller theory and other analytic techniques have proved effective in studying the properties of these spaces. [41, 42, 94, 56]

In [M1], Mumford develops an enumerative geometry of $\bar{\mathcal{M}}_g$. First, he considers the problem of defining the Chow ring of $\bar{\mathcal{M}}_g$. The main difficulty is that $\bar{\mathcal{M}}_g$ is not smooth, although it is not too far away from being smooth as it is locally the quotient of a smooth variety by a finite group. This concept is formalized into the idea of a Q-variety. Thus Mumford defines a good Chow ring of $\bar{\mathcal{M}}_g$ together with a theory of Chern classes of vector bundles. He further introduces some elements κ_i in the Chow ring of $\bar{\mathcal{M}}_g$, also introduced independently by Miller. [66, 67] These classes are obtained by pushing down products of the first Chern class of the relative dualizing sheaf of the universal curve over $\bar{\mathcal{M}}_g$. Mumford conjectured that these κ_i are generators for the stable cohomology ring of $\mathcal{M}_g$. [22, 49, 21, 23, 61] represent more recent work on [M1].Mumford's conjecture have been established in generality by Madsen and Weiss [63] using topological methods. [92] gives an excellent survey of recent results. These classes are important in topological field theory [19].

Much of the existence theory of moduli spaces for both varieties and vector bundles has been generalized to higher dimensions using GIT. Viehweg has used GIT to construct quasi-projective moduli spaces for canonically polarized varieties in characteristic 0 [93] Rather than apply GIT to the relevant Hilbert scheme H with its usual embedding as was carried out by the author in [30], Viehweg constructed a new line bundle $\mathcal{L}$ on H, so that it is easy to check that a Hilbert points are stable with respect to $\mathcal{L}$. He also obtained some results for polarized varieties. Also Alexeev, Kollár and Shepard-Barron have shown that the moduli space of surfaces of general type has a good projective compactification [93]. Moduli spaces for semi-stable sheaves on an algebraic surface were first constructed in [31]. It is necessary to add certain non-locally free sheaves as the space of semi-stable vector bundles is not compact (Gieseker compactification). Maruyama has extended this work to show that moduli spaces for semi-stable torsion free coherent sheaves with fixed numerical data exist using GIT, except that it is unknown if these spaces are of finite type in non-zero characteristic [64]. There has been an extensive investigations into the properties of these moduli spaces, especially if the base variety is a curve or a surface. [3, 13, 7, 81, 36, 65, 83, 59, 73, 82, 17, 34, 9, 74, 37, 72, 69, 35, 47, 87, 48]. GIT has also proved useful in Arakelov theory [84, 8, 95]. The theory of stable curves has become an important tool in many areas of algebraic geometry. For instance, this theory has been useful in the theory of special divisors as developed by the author in [33], which in turn was based on work of Griffiths and Harris [39] and then further developed by Eisenbud and Harris [45].

One of the most interesting recent developments has been the interplay between physics and the theory of stable bundles and curves. In general, many of the objects physicists consider can be transformed into algebro-geometric objects. In particular, the theory of stable curves has found an important place in the theory of the bosonic string. The amplitudes for Polyokov's string theory can be reduced to holomorphic integrals over $\mathcal{M}_g$. Thus the algebraic geometry of $\bar{\mathcal{M}}_g$ can be used to study string theory [70]. Of crucial importance for this transformation is Mumford's formula $\lambda_2 = \lambda_1^{\otimes 13}$ on $\mathcal{M}_g$. String theory has also led to much activity in the theory of stable bundles on curves. Verlinde gave formulas to calculate the Euler characteristics of some line bundles on the space of semi-stable bundles on an algebraic curve. The topology and geometry of the moduli space of semi-stable bundles has been investigated by numerous mathematicians.[17, 5, 26, 88, 85, 4, 90]

Another connection between physics and stability is Yang-Mills theory. The earliest manifestation of Yang-Mills theory in the theory of bundles of higher rank is the result of Narasimhan and Seshadri mentioned above. Penrose discovered that the theory of instantons could be translated into the theory of stable bundles on $\mathbf{P}^3$. Atiyah and Bott [1] then realized that Yang-Mills theory could be used to investigate the topology of the moduli space of semi-stable bundles on curves, at least in principle. Donaldson established that a stable bundle on a surface X admits a unique irreducible Hermitian-Einstein metric. His method can be interpreted as an infinite dimensional analogue of GIT[16]. This work has been generalized by Yau and Uhlenbeck [91]. Donaldson has been able to define a sequence of invariants of X, which only depend on differential structure of X, and which in principle can be computed by studying the moduli space of semi-stable bundles on X. [15] For instance, the proof of the non-triviality of these invariants depends on the projectivity of the moduli space for semi-stable bundles on curves.

Mumford's ideas have been greatly influential in algebraic geometry. On a personal note, I feel that virtually all my work has come from a close reading of Mumford's papers, sometimes in non-obvious ways. For instance, my work on special divisors [33] was largely influenced by a close study of [M5]. Many other algebraic geometers have also been strongly influenced by Mumford's papers. Yet the ideas in these papers are far from exhausted. Young algebraic geometers would do well to study these papers. They contain a wealth of ideas waiting to be developed.

References

[1] M. Atiyah and R. Bott. The Yang-Mills equations over Riemann surfaces. *Trans. Roy. Soc. London*, 308, 1982.

[2] W. Baily. On the moduli of Jacobian varieties. *Annals of Mathematics*, 71, 1960.

[3] W. Barth. Moduli of vector bundles on the projective plane. *Inv. Math*, 42, 1977.

[4] A. Beauville. Conformal blocks, fusion rules and the Verlinde formula. In *Proceedings of the Hirzebruch 65 Conference on Algebraic Geometry (Ramat Gan 1993)*, volume 9 of *Israel Math. Conf. Proc.*

[5] A. Bertram and A. Szenes. Hilbert polynomials of moduli spaces of rank 2 vector bundles I. *Topology*, 32, 1993.

[6] Aaron Bertram. Another way to enumerate rational curves with torus actions. *Invent. Math.*, 142(3):487–512, 2000.

[7] F. Bogomolov. Holomorphic tensors and vector bundles on projective varieties. *Izv. Akad. Nauk SSR*, 42, 1978.

[8] J.-B. Bost. Semi-stability and heights of cycles. *Inv. Math.*, 118, 1994.

[9] L. Caparoso. A compactification of the universal Picard variety over the moduli space of stable curves. *JAMS*, 7, 1994.

[10] M. Chang and Z. Ran. The kodaira dimension of the moduli spaces of curves of genus 15. *J. Diff. Geom.*, 24, 1986.

[11] M. Chang and Z. Ran. Unirationality of the moduli space of curves of genus 11, 13 (and 12). *Inv. Math.*, 76, 1986.

[12] F. Kirwan D. Mumford J. Fogarty. *Geometric invariant theory (third edition)*. Springer Verlag, 1994.

[13] U. V. Desale and S. Ramanan. Classification of vector bundles over hyperelliptic curves. *Inv. Math.*, 38, 1976.

[14] Igor V. Dolgachev and Yi Hu. Variation of geometric invariant theory quotients,. *Inst. Hautes Études Sci. Publ. Math.*,, (87,):5–56,, 1998,. With an appendix by Nicolas Ressayre,.

[15] S. Donaldson and P. Kronheimer. *The geometry of four manifolds*. Oxford Univ. Press, 1990.

[16] S. K. Donaldson. Anti self-dual Yang-Mills connections over complex algebraic surfaces and stable vector bundles. *Proc. London Math. Soc. (3)*, 50, 1985.

[17] J.-M. Drezet and M. Narasimhan. Groupe de picard des variétés de modules de fibrés. *Inv. Math.*, 97, 1989.

[18] V. G. Drinfeld. Elliptic Modules I. *Math. USSR*, Sb23, 1974.

[19] B. Dubrovin. Geometry of 2D topological field theories. In *Integrable Systems and Quantum groups*, volume 1620 of *Lecture Notes in Math.* Springer.

[20] D. Eisenbud and J. Harris. Limit linear series, the irrationality of $\mathcal{M}_g$, and other applications. *Bull. Amer. Math. Soc.*, 10, 1984.

[21] C. Faber and R. Pandharipande. Logarithmic series and Hodge integrals in the tautological ring. *Michigan Math. J.*, 48:215–252, 2000. With an appendix by Don Zagier, Dedicated to William Fulton on the occasion of his 60th birthday.

[22] C. F. Faber. Chow rings of the moduli spaces of curves. I the chow ring of $\bar{\mathcal{M}}_3$. *Ann. of Math.*, 132, 1990.

[23] Carel Faber. A conjectural description of the tautological ring of the moduli space of curves. In *Moduli of curves and abelian varieties*, Aspects Math., E33, pages 109–129. Vieweg, 1999.

[24] G. Faltings. Endlichkeitssätze für abelsche Varietäten über Zahlkörpern. *Invent. Math.*, 73(3):349–366, 1983.

[25] G. Faltings. Semi-stable vector bundles on Mumford curves. *Inv. Math.*, 73, 1983.

[26] G. Faltings. A proof of the Verlinde formula. *J. Algebraic Geom*, 3, 1994.

[27] Gerd Faltings. Mumford-Stabilität in der algebraischen Geometrie. In *Proceedings of the International Congress of Mathematicians, Vol. 1, 2 (Zürich, 1994)*, pages 648–655, Basel, 1995. Birkhäuser.

[28] E. Formanek and C. Procesi. Mumford's conjecture for the general linear group. *Adv. in Math.*, 19, 1976.

[29] W. Fulton. On the irreducibility of the moduli space of curves. *Inv. Math.*, 67, 1982.

[30] D. Gieseker. Global moduli for surfaces of general type. *Invent. Math.*, 43(3):233–282, 1977.

[31] D. Gieseker. On the moduli of vector bundles on an algebraic surface. *Ann. of Math. (2)*, 106(1):45–60, 1977.

[32] D. Gieseker. *Lectures on moduli of curves*, volume 69 of *Tata Institute of Fundamental Research Lectures on Mathematics and Physics*. Published for the Tata Institute of Fundamental Research, Bombay, 1982.

[33] D. Gieseker. Stable curves and special divisors: Petri's conjecture. *Invent. Math.*, 66(2):251–275, 1982.

[34] D. Gieseker. A degeneration of the moduli space of stable bundles. *J. Diff. Geom.*, 19, 1984.

[35] D. Gieseker and J. Li. Irreducibility of moduli of rank-2 vector bundles on algebraic surfaces. *J. Diff. Geom.*, 40, 1994.

[36] D. Gieseker and J. Li. Moduli of high rank vector bundles over surfaces. *JAMS*, 9, 1996.

[37] D. Gieseker and I. Morrison. Hilbert stability of rank two bundles on curves. *J. Diff. Geom.*, 19, 1984.

[38] Alexander Givental. The mirror formula for quintic threefolds. In *Northern California Symplectic Geometry Seminar*, volume 196 of *Amer. Math. Soc. Transl. Ser. 2*, pages 49–62. Amer. Math. Soc., Providence, RI, 1999.

[39] Phillip Griffiths and Joseph Harris. On the variety of special linear systems on a general algebraic curve. *Duke Math. J.*, 47(1):233–272, 1980.

[40] W. Haboush. Reductive groups are geometrically reductive. *Ann. of Math.*, 102, 1975.

[41] J. Harer. The second homology group of the mapping class group of an orientable surface. *Inv. Math.*, 72, 1983.

[42] J. Harer and D. Zagier. The Euler characteristic of the moduli space of curves. *Inv. Math.*, 85, 1986.

[43] J. Harris. On the Kodaira dimension of the moduli space of curves II. *Inv. Math.*, 75, 1984.

[44] J. Harris and I. Morrison. Slopes of effective divisors on the moduli space of curves. *Inv. Math.*, 99, 1990.

[45] Joe Harris and Ian Morrison. *Moduli of curves.* Springer-Verlag, New York, 1998.

[46] D. Hilbert. Über die vollen invariantensysteme. *Math. Annalen*, 42, 1893.

[47] N. Hitchin. Stable bundles and integrable systems. *Duke Math. J.*, 54, 1987.

[48] Daniel Huybrechts and Manfred Lehn. *The geometry of moduli spaces of sheaves.* Friedr. Vieweg & Sohn, Braunschweig, 1997.

[49] E. Izadi. The Chow group of the moduli space of curves of genus 5. In *The moduli space of curves (Texel Island 1994)*, volume 129 of *Progr. Math.* Birkhäuser Boston.

[50] G. Kempf. Instability in invariant theory. *Ann. of Math.*, 108, 1978.

[51] G. Kempf and L. Ness. On the lengths of vectors in representation spaces. 732, 1972.

[52] F. Kirwan. *Cohomology of quotients in symplectic and algebraic geometry*, volume 31 of *Math. Notes.* Princeton University Press, 1984.

[53] F. Kirwan. The cohomology rings of moduli spaces over Riemann surfaces. *JAMS*, 5, 1992.

[54] F. Knudsen. The projectivity of the moduli space of stable curves III. *Math. Scand.*, 200, 1983.

[55] J. Kollár. Projectivity of complete moduli. *J. Diff. Geom*, 32, 1990.

[56] M. Kontsevich. Intersection theory on the moduli space of curves and the matrix Airy function. *Comm. Math. Phys.*, 147, 1992.

[57] M. Kontsevich and Yu. Manin. Gromov-Witten classes, quantum cohomology, and enumerative geometry. *Comm. Math. Phys.*, 164(3):525–562, 1994.

[58] Maxim Kontsevich. Enumeration of rational curves via torus actions. In *The moduli space of curves (Texel Island, 1994)*, pages 335–368. Birkhäuser Boston, Boston, MA, 1995.

[59] J. Li. Kodaira dimension of moduli space of vector bundles on surfaces. *Inv. Math.*, 115, 1994.

[60] Bong H. Lian, Kefeng Liu, and Shing-Tung Yau. Mirror principle, a survey. In *Current developments in mathematics, 1998 (Cambridge, MA)*, pages 35–82. Int. Press, Somerville, MA, 1999.

[61] Eduard Looijenga. On the tautological ring of $\mathcal{M}_g$. *Invent. Math.*, 121(2):411–419, 1995.

[62] C. S. Seshadri M. S. Narasimhan. Stable and unitary bundles on compact Riemann surfaces. *Annals of Math.*, 82, 1965.

[63] I. Madsen and M. Weiss. The stable moduli space of riemann surfaces: Mumford's conjecture. *arXiv:math.AT/0212321.*

[64] M. Maruyama. Moduli of stable sheaves I. *J. Math. Kyoto*, 17, 1977.

[65] V. Mehta and A. Ramanathan. Semistable sheaves on projective varieties and their restrictions to curves. *Math. Ann.*, 89, 1984.

[66] E. Miller. The homology of the mapping class group. *J. Diff. Geom.*, 24, 1986.

[67] S. Morita. Characteristic classes of surface bundles. *Invent. Math.*, 90, 1987.

[68] I. Morrison. Projective stability of ruled surfaces. *Inv. Math.*, 56, 1980.

[69] S. Mukai. Symplectic structure on the moduli space of sheaves on an Abelian or $K3$ surface. *Inv. Math.*, 77.

[70] Subhashis Nag. Mathematics in and out of string theory,. In *Topology and Teichmüller spaces (Katinkulta, 1995)*,, pages 187–220,. World Sci. Publishing,, River Edge, NJ,, 1996,.

[71] L. Ness. A stratification of the null cone via the moment map. *Amer. J. Math.*, 106, 1984.

[72] K. O'Grady. Moduli of vector bundles on projective surfaces: some basic results. *Inv. Math.*, 123, 1996.

[73] Christian Okonek, Michael Schneider, and Heinz Spindler. *Vector bundles on complex projective spaces.* Birkhäuser Boston, Mass., 1980.

[74] R. Pandharipande. A compactification over $\bar{\mathcal{M}}_g$ of the universal moduli of slope-semistable vector bundles. *JAMS*, 9, 1996.

[75] M. Raynaud. Construction analytique de courbes en géometrie non archimédienne (d'aprés David Mumford). In *Séminaire Bourbaki, 25e anné (1972-3) Exp. No. 427*, volume 383 of *Lecture Notes in Math.* Springer, 1974.

[76] Z. Reichstein. Stability and equivariant maps. *Inv. Math.*, 96, 1989.

[77] G. Rousseau. Immeubles sphériques et théorie des invariants. *C. R. Acad. Sci. Paris*, 286, 1978.

[78] C. S. Seshadri. Space of unitary vector bundles on an Riemann surface. *Ann. of Math.*, 85, 1967.

[79] C. S. Seshadri. Quotient spaces modulo reductive algebraic groups. *Ann. of Math. (2)*, 95:511–556; errata, ibid. (2) **96 (1972), 599**,, 1972.

[80] C. S. Seshadri. Geometric reductivity over arbitrary base. *Advances in Math.*, 26(3):225–274, 1977.

[81] C. S. Seshadri. *Fibrés vectoriels sur les courbes algébriques,.* Société Mathématique de France,, Paris,, 1982,. Notes written by J.-M. Drezet from a course at the École Normale Supérieure, June 1980,.

[82] T. Oda C. S. Seshadri. Compactification of the generalized Jacobian variety. *Trans. AMS*, 253, 1979.

[83] C. Simpson. Higgs bundles and local systems. *Publ. Math. IHES*, 74, 1992.

[84] C. Soulé. Successive minima on arithmetic varieties. *Compositio Math.*, 96, 1995.

[85] A. Szenes. Verification of Verlinde's formulas for $SU(2)$. *Internat. Math. Res. Notices*, 93, 1991.

[86] Y. Tai. On the Kodaira dimension of the moduli space of abelian varieties. *Inv. Math.*, 68, 1982.

[87] M. Thaddeus. Conformal field theory and the cohomology of the moduli space of stable bundles. *J. Diff. Geom.*, 35, 1992.

[88] M. Thaddeus. Stable pairs, linear systems and the Verlinde formula. *Inv. Math.*, 117, 1994.

[89] Michael Thaddeus. Geometric invariant theory and flips. *J. Amer. Math. Soc.*,, 9,(3,):691–723,, 1996,.

[90] Akihiro Tsuchiya, Kenji Ueno, and Yasuhiko Yamada. Conformal field theory on universal family of stable curves with gauge symmetries. In *Integrable systems in quantum field theory and statistical mechanics*, pages 459–566. Academic Press, Boston, MA, 1989.

[91] K. Uhlenbeck and S.-T. Yau. On the existence of hermitian Yang-Mills connections on stable bundles over compact Kahler manifolds. *Comm. Pure Appl. Math.*, 39, 1989.

[92] Ravi Vakil. The moduli space of curves and its tautological ring. *Notices of the AMS*, 50, 2003.

[93] E. Viehweg. *Quasi-projective Moduli for Polarized Manifolds*. Ergebnisse. Springer, 1995.

[94] S. Wolpert. The geometry of the moduli space of Riemann surfaces. *Bull. Amer. Math. Soc.*, 11, 1984.

[95] Shouwu Zhang. Heights and reductions of semi-stable varieties,. *Compositio Math.*,, 104,(1,):77–105,, 1996,.

References to Mumford's papers in this volume

[M1] Towards an Enumerative Geometry of Moduli Space of Curves, in *Arithmetic and Geometry*, edited by M. Artin, J. Tate, Birkhauser-Boston, 1983, pp. 271–326.

[M2] On the Kodaira Dimension of the Siegel Modular Variety, in *Algebraic Geometry - Open Problems, Lecture Notes in Mathematics 997*, Springer-Verlag 1983, pp. 348–375.

[M3] Pathologies IV, Amer. J. of Math., 1975, **97**, pp. 847–849.

[M4] Stability of Projective Varieties, Monographie No. 24 de *L'Enseignement Math.*, **23**, 1977, pp. 39–110.

[M5] An Analytic Construction of Degenerating Curves Over Complete Local Kings, *Composito Math.*, 1972, **24**, pp. 129–174.

[M6] The Irreducibility of the Space of Curves of Given Genus (with P. Deligne), *Publ. Math. Dfe l'I.H.E.S.*, 1969, **36**, pp. 75–109.

[M7] Periods of Moduli Spaces of Bundles on Curves (with P. Newstead), *Amer. J of Math.*, 1968, **90**, pp. 1200–1208.

[M8] Projective Invariants of Projective Structures, *International Congress of Mathematicians*, Stockholm 1962, pp. 526–530.

AN ELEMENTARY THEOREM IN GEOMETRIC INVARIANT THEORY

BY DAVID MUMFORD

Communicated by Raoul Bott, May 18, 1961

The purpose of this note is to prove the key theorem in a construction of the arithmetic scheme of moduli M of curves of any genus. This construction, which relies heavily on Grothendieck's whole theory of schemes, may be briefly outlined as follows: first one defines the family K of tri-canonical models of curves C of genus g, any characteristic, in $\mathbf{P}^{5g-6}$, as a sub-scheme of one of Grothendieck's Hilbert schemes [**3**]. Second, one maps $K \to A$, where A (a sub-scheme of another Hilbert scheme) parametrizes the full projective family of the polarized Jacobians $J \subset \mathbf{P}^N$ of these curves. It may then be shown that M should be the orbit space $K/\mathrm{PGL}(5g-6)$; and it can also be shown that this orbit space exists if the orbit space $A/\mathrm{PGL}(N)$ exists. But, in general, given any family A of polarized abelian varieties $V \subset \mathbf{P}^N$ invariant under projective transformations, $A/\mathrm{PGL}(N)$ does exist; for simplicity assume that a section serving as an identity is rationally defined in the whole family A. Then A may be identified with the family of 0-cycles $\mathfrak{A} \subset \mathbf{P}^N$ which are the points of order m (suitable m) on the abelian varieties V. Now neglecting for simplicity the group of permutations of the m^{2g} points of these 0-cycles, this reduces the problem to constructing the orbit space $(\mathbf{P}^N)^{m^{2g}}/\mathrm{PGL}(N)$. In this paper, an apparently very natural open sub-scheme[1] $(\mathbf{P}^n)^m_0 \subset (\mathbf{P}^n)^m$, any n, m, is constructed such that in fact $(\mathbf{P}^n)^m_0$ is a principal fibre bundle over its quotient by $\mathrm{PGL}(n)$. This result is apparently new even over the complex numbers. The methods used are entirely elementary, and no special techniques are used to deal with the generalization from varieties to schemes. Moreover, except for the replacement of $\mathbf{Z}$ by k, no changes are necessary or appear possible should the reader wish to consider the objects as varieties rather than schemes.

1. Let $\mathbf{P}^n$ be projective n-space over $\mathbf{Z}$, $(\mathbf{P}^n)^m$ the m-fold product with itself. Let homogeneous coordinates in the ith factor be $X_0^{(i)}$, $X_1^{(i)}, \cdots, X_n^{(i)}$, and for all $(n+1)$-tuples $i_0, i_1, \cdots, i_n$ let

$$D_{i_0, i_1, \cdots, i_n} = \operatorname*{Det}_{0 \leq k, l \leq n} \left(X_l^{(i_k)}\right).$$

[1] See also [**2**] for a slightly larger open sub-scheme.

Define $(\mathbf{P}^n)_0^m \subset (\mathbf{P}^n)^m$ to be the "set of m-tuples" $(P_1, P_2, \cdots, P_m)$, $P_i \in \mathbf{P}^n$, such that less than $m/(n+1) P_i$ are contained in any single hyperplane of $\mathbf{P}^n$. This is immediately seen to be an open sub-scheme.

PROPOSITION. *Let $P = (P_1, P_2, \cdots, P_m)$ be a geometric point in $(\mathbf{P}^n)_0^m$.*

(a) *There are positive integers N and N_0 such that for every i, a monomial $\Pi^{(i)}$ in the D's exists such that $\Pi^{(i)}(P) \neq 0$, and the degree of $\Pi^{(i)}$ in $X_*^{(j)}$ is N if $j \neq i$, and $N - N_0$ if $j = i$.*

(b) *There is a positive integer N such that for every $i \neq j$, a monomial $\Pi^{(i,j)}$ exists such that $\Pi^{(i,j)}(P) \neq 0$, and the degree of $\Pi^{(i,j)}$ in $X_*^{(k)}$ is N if $k \neq i, j$, and $N+1$ if $k = i$, and $N-1$ if $k = j$.*

PROOF. Let E be the real vector space of dimension m, and let H be the convex cone spanned by the points

$$P_{i_0, i_1, \cdots, i_n} = (x_1, x_2, \cdots, x_m),$$

where $x_i = 0$ if $i \neq i_k$, any k, $x_i = 1$ if $i = i_k$, some k, $D_{i_0, i_1, \cdots, i_n}(P) \neq 0$. By means of assigning to each monomial in the D's the corresponding additive expression in the P's, the result (a) can readily be translated to the assertion that $(1, 1, \cdots, 1) \in \operatorname{Int}(H)$. But if this is false, there is a linear functional on E, zero at $(1, 1, \cdots, 1)$ and negative on H, i.e. there exist $\alpha_1, \cdots, \alpha_m$ such that

(i) $$\sum_{k=0}^{n} \alpha_{i_k} \leqq 0 \qquad \text{if } D_{i_0, i_1, \cdots, i_n}(P) \neq 0,$$

(ii) $$\sum_{i=1}^{m} \alpha_i = 0.$$

Say without loss of generality that $\alpha_1 \geqq \alpha_2 \geqq \cdots \geqq \alpha_m$. Pick the sequence $i_0, i_1, \cdots, i_n$ as follows: $i_0 = 1$; $i_1 =$ smallest i such that $P_1 \neq P_i$; $i_2 =$ smallest i such that P_1, P_{i_1}, P_i do not lie on a line; etc. It is easy to see that the hypothesis on P implies $i_n - 1 < m/(n+1)$, hence $i_n \leqq [m/(n+1)] + 1 = \mu + 1$, for instance. Now

$$\begin{aligned} 0 &\geqq \mu \sum_{k=0}^{n} \alpha_{i_k} \\ &\geqq \mu\alpha_1 + \mu n \alpha_{i_n} \\ &\geqq \mu\alpha_1 + \mu n \alpha_{\mu+1} \\ &\geqq \mu\alpha_1 + (m - \mu)\alpha_{\mu+1} \qquad (\text{since } \alpha_{\mu+1} \leqq 0) \\ &\geqq \sum_{i=1}^{\alpha} \alpha_i + \sum_{i=\mu+1}^{m} \alpha_i = 0. \end{aligned}$$

Hence all equality signs hold, and one sees easily that all $\alpha_i=0$. To obtain (b), given i, j, note first that there exist $i_1, \cdots, i_n$ such that $D_{i,i_1,\cdots,i_n}(P)\neq 0$, and $D_{j,i_1,\cdots,i_n}(P)\neq 0$ which follows easily using only that $<m/2$ points P_i lie in any one hyperplane. Then set:

$$\Pi^{(i,j)} = D_{i,i_1,\cdots,i_n}\cdot(D_{j,i_1,\cdots,i_n})^{N_0-1}\cdot\Pi^{(i_1)}\cdot \cdots \cdot\Pi^{(i_n)}\cdot\Pi^{(j)}. \quad \text{Q.E.D.}$$

2. THEOREM. *There exist a quasi-projective scheme M, and a morphism ψ: $(\mathbf{P}^n)_0^m\rightarrow M$ such that $(\mathbf{P}^n)_0^m$ is a principal fibre bundle over M, with group* PGL(n).

PROOF. Let

$$X_{i_1,\cdots,i_m} = X_{i_1}^{(1)}\cdot X_{i_2}^{(2)}\cdots X_{i_m}^{(m)}, \qquad S = \mathbf{Z}[\cdots, X_j^{(i)}, \cdots],$$

$$R = \mathbf{Z}[\cdots, X_{i_1,\cdots,i_m}, \cdots], \qquad S_0 = \mathbf{Z}[\cdots, D_{i_0,\cdots,i_n}, \cdots],$$

and

$$R_0 = S_0 \cap R.$$

The first step is:

(A) *R_0 is finitely generated.* First notice that R_0 is generated by the monomials in $D_{i_0,\cdots,i_n}$'s that are homogeneous of the same degree in each set of variables $X_*^{(i)}$. Following Hilbert [1], use:

LEMMA OF GORDAN. *A finite system of homogeneous linear Diophantine equations has a finite set of positive integral solutions so that every other positive integral solution is a positive integral combination of them.*

If the $D_{i_0,\cdots,i_n}$ are listed as $D^{(1)}, D^{(2)}, \cdots, D^{(N)}$, then the monomials $D^{(1)^{r_1}}\cdot D^{(2)^{r_2}}\cdots D^{(N)^{r_N}}$ in R_0 are those satisfying:

$$\sum_{\substack{[1 \text{ occurs among} \\ \text{subscripts of } D^{(i)}]}} r_i = \sum_{\substack{[2 \text{ occurs among} \\ \text{subscripts of } D^{(i)}]}} r_i = \text{etc.}$$

and the lemma applies. Q.E.D.

Now recall the well-known lemma [2]:

LEMMA. *$R=\sum_n R_n$ a graded ring, finitely generated over R_0. There exists an N such that if $R(N)=\sum_n R_{Nn}$, then $R(N)$ is finitely generated over R_0 by elements of $R(N)_1=R_N$.*

Pick such an N for R_0 and consider the inclusion $R_0(N)\subset R(N)$. This corresponds to a rational map of projective schemes:

$$\bar{\psi}: (\mathbf{P}^n)^m \rightarrow \overline{M}.$$

(B) *$\bar{\psi}$ is defined on $(\mathbf{P}^n)_0^m$.* By the usual translation into algebra, this means that if P is a geometric point in $(\mathbf{P}^n)_0^m$, there is an $\alpha\in R_{0,Nk}$, some positive k, such that $\alpha(P)\neq 0$. But by part (a) of the proposition, there is indeed a monomial $\pi\in R_{0,k}$, $\pi(P)\neq 0$. Q.E.D. Let $\psi=\bar{\psi}|(\mathbf{P}^n)_0^m$. It is immediate that, if π_i: $\mathrm{PGL}(n)\times(\mathbf{P}^n)_0^m\to(\mathbf{P}^n)_0^m$ are the morphisms of the action of the group $\mathrm{PGL}(n)$ and the projection, for $i=1, 2$ resp., then $\psi\circ\pi_1=\psi\circ\pi_2$. We wish to show that given any geometric point $P\in(\mathbf{P}^n)_0^m$, there is an open sub-scheme $U\subset\overline{M}$, $\psi(P)\in U$, and a morphism s: $U\to(\mathbf{P}^n)_0^m$ such that

(i) $\psi\circ s=$identity,

(ii) $U\times\mathrm{PGL}(n)\to^A\psi^{-1}(U)$ given by $(x, \sigma)\to\pi_1(\sigma, s(x))$ on geometric points, is an isomorphism.

When this is shown, it follows that $M=\psi((\mathbf{P}^n)_0^m)$ is open in $\overline{M}$ and the theorem is fully proven. s will be constructed by the help of a very simple "*Typische Darstellung*" (as such identities are called in classical invariant theory). Namely, seek

(i) $I_{i,j}\in R_{0,Nk}$, $0\leqq i\leqq n$, and $1\leqq j\leqq m$,

(ii) $\alpha_j\in S_0$, $1\leqq j\leqq m$, homogeneous in $X_*^{(k)}$, all k,

(iii) $\sigma_{i,j}\in S$, $0\leqq i$, $j\leqq n$, homogeneous in $X_*^{(k)}$, all k with degree independent of i and j, but dependent on k, such that

$$(*)\qquad \alpha_j X_i^{(j)}=\sum_{k=0}^{n}\sigma_{i,k}I_{k,j},\qquad 0\leqq i\leqq n,\ 1\leqq j\leqq m,$$

$$(**)\qquad \alpha_j(P)\neq 0,\qquad 1\leqq j\leqq m.$$

Then $\{\sigma_{i,j}\}$ define a rational map ϕ: $(\mathbf{P}^n)^m\to\mathrm{PGL}(n)$, while for each j, $\{I_{i,j}\}$ define a rational map $\overline{M}\to\mathbf{P}^n$, hence together a rational map $\overline{M}\to^S(\mathbf{P}^n)^m$. Define $U\subset\overline{M}$ as the "set of points" $Q\in\overline{M}$ such that for all j, $\alpha_j(Q)\neq 0$. Clearly by (*), s is a morphism when restricted to U, and ϕ is a morphism when restricted to $\psi^{-1}(U)$. Moreover (*) then translates to the statement that the composed morphism below is the identity:

$$\psi^{-1}(u)\xrightarrow{\phi\times(s\circ\psi)}\mathrm{PGL}(n)\times(\mathbf{P}^n)^m\xrightarrow{\pi_1}(\mathbf{P}^n)^m.$$

Then (i) follows formally, and to show (ii), define $B:\psi^{-1}(U)\to U\times\mathrm{PGL}(n)$ by $\psi\times\phi$, then $A\circ B=$identity also follows formally. To see that $B\circ A=$identity, note first that for any geometric points P and σ in $(\mathbf{P}^n)_0^m$ and $\mathrm{PGL}(n)$ resp., $\pi_1(\sigma, P)=P$ implies $\sigma=e$; for there are clearly $n+1$ independent points P_i' among $P=(P_1, \cdots, P_m)$, and since $<m/(n+1)$ of the P_i are in any one hyperplane, all P_i could not lie in one of the $n+1$ hyperplanes spanned by $(P_0', \cdots, \hat{P}_i', \cdots, P_n')$. Consequently $B\circ A$ is the identity on geometric points, and since U and $\mathrm{PGL}(n)$ are reduced, $B\circ A$ is the identity.

Finally, construct the Typische Darstellung as follows. Given P, obviously there exist $i_0, \cdots, i_n$ such that $D_{i_0,\cdots,i_n}(P) \neq 0$. Simply multiply through by a suitable function the standard identity:

$$(*)_0 \qquad D_{i_0,\cdots,i_n} X_i^{(j)} = \sum_{k=0}^{n} D_{i_0,\cdots,\hat{i}_k,j,\cdots,i_n} \cdot X_i^{(i_k)} .$$

For some arbitrary α, set

$$\alpha_j = (D_{i_0,\cdots,i_n})^{N_0} \cdot \Pi^{(\alpha,j)} \cdot \Pi^{(i_0,\alpha)} \cdot \ \cdots \ \cdot \Pi^{(i_n,\alpha)} \cdot \Pi^{(i_0)} \cdot \ \cdots \ \cdot \Pi^{(i_n)},$$

$$I_{k,j} = (D_{i_0,\cdots,i_n})^{N_0-1} \cdot D_{i_0,\cdots,\hat{i}_k,j,\cdots,i_n} \cdot \Pi^{(\alpha,j)} \cdot \Pi^{(i_k,\alpha)} \cdot \Pi^{(i_0)} \cdot \ \cdots \ \cdot \Pi^{(i_n)},$$

$$\sigma_{i,k} = \Pi^{(i_0,\alpha)} \cdot \ \cdots \ \cdot \hat{\Pi}^{(i_k,\alpha)} \cdot \ \cdots \ \cdot \Pi^{(i_n,\alpha)} \cdot X_i^{(i_k)}$$

and (*) and (**) follow. Q.E.D.

References

1. D. Hilbert, *Über die Endlichkeit des Invariantensystems für binäre Grundformen*, Collected Works, vol. 2, Berlin, 1933.

2. D. Mumford, *Orbit spaces of varieties by linear groups in char. O*, forthcoming.

3. A. Grothendieck, *Théorie de descente*. IV, Séminaire Bourbaki, mai, 1961.

Harvard University

PROJECTIVE INVARIANTS OF PROJECTIVE STRUCTURES AND APPLICATIONS

By DAVID MUMFORD

The basic problem that I wish to discuss is this: if V is a variety, or scheme, parametrizing the set, or functor, of all structures of some type in projective n-space $\mathbf{P}_n$, then the group $PGL(n)$ of automorphisms of $\mathbf{P}_n$ acts on V. Then under what conditions does there exist a quotient or orbit space $V/PGL(n)$, i.e. when can we construct enough "projective invariants" for these structures? For example, let V parametrize the set of hypersurfaces of degree m, with certain types of singularities; or let V parametrize the set of tri-canonical space curves of given genus, or even n-canonical surfaces with at most "negligible singularities" [1]; or let V parametrize the set of 0-cycles of degree m in $\mathbf{P}_n$; or let V parametrize the set of all morphisms of a fixed scheme X into $\mathbf{P}_n$. Moreover, I wish to illustrate how such questions are one essential step in several basic existence and construction problems of algebraic geometry.

One approach to this problem is afforded by the invariant theory of the representations of reductive groups. Here you generalize the problem first: consider an arbitrary action of an algebraic group G on a variety (or scheme) V and seek an orbit space V/G. Then you specialize the problem by (*a*) assuming G is reductive, (*b*) restricting attention to quasi-projective orbit spaces V/G. Of course, if, in particular, the characteristic is 0, $PGL(n)$ is reductive. Now suppose you have a projective embedding $V \subset \mathbf{P}_N$. If V is a normal variety, and if this embedding is defined by a complete linear system on V, it is possible to prove that the action of G on V extends to an action of G on $\mathbf{P}_N$. In any case, if this occurs, I say that *G acts linearly* on $V \subset \mathbf{P}_N$.

Now, if the action of G is linear, its action is induced by a linear and unimodular representation of some finite covering G^* of G on the affine cone $\mathbf{A}^{N+1}$ over the ambient $\mathbf{P}_N$. Then I make the definition:

A point $x \in V$ is *stable* for the action of G on V, relative to the embedding $V \subset \mathbf{P}_N$, if for one (and hence all) homogeneous points $x^* \in \mathbf{A}^{N+1}$ over x, (i) the stabilizer of x^* is a finite subgroup of G^*, and (ii) the orbit of x^* under G^* is closed in $\mathbf{A}^{N+1}$.

Now assume that a reductive algebraic group G acts linearly on $V \subset \mathbf{P}_N$. The fundamental theorem is this:

THEOREM. *The set of stable points forms an open set U in V, and a quasi-projective orbit space U/G exists.*

In case G is semi-simple and of characteristic 0, I can say more. First, let us call the action $\alpha: G \times V \to V$ of G on V a *proper action* if the morphism $\alpha \times Pr_2: G \times V \to V \times V$ is proper (see [2]). Then:

THEOREM. (1) *A point $x \in V$ is stable if and only if there is a hypersurface section H of V which is invariant under G, and such that $x \notin H$ and the orbit of x is closed in $V - H$.*

(2) *The action of G on the set of stable points U is proper.*

(3) *If G acts properly on V, and if a quasi-projective orbit space V/G exists, then for some projective embedding $V \subset \mathbf{P}_N$, every point of V is stable.*

Finally, if G is any reductive algebraic group in characteristic O, then I can analyze the manner in which stability breaks down in the following way:

THEOREM. *If $x \in V$ is not stable for the action of G, then there is a Borel subgroup $B \subset G$ and a 1-parameter subgroup $\mathbf{G}_m \subset B$ such that, if $\mathbf{G}_m$ is any conjugate of $\mathbf{G}_m$ in B, then x is not stable for the action of $\mathbf{G}_m$ on V.*

These results are not definitive: I conjecture that they all are valid for semi-simple groups of any characteristic. However, they suggest the kind of answer that should be found for the original question: to every type of structure in $\mathbf{P}_n$, there is a stability condition which is sufficient and, in general, necessary for the existence of enough projective invariants to classify these structures; and, moreover, this stability condition is always of the form: there is no flag in $\mathbf{P}_n$ which has "too high an order of contact" with the given structure.

The following case has been worked out exhaustively by myself and J. Tate in all characteristics, and even over the scheme of integers: let $V = (\mathbf{P}_n)^m$, i.e. V parametrizes ordered 0-cycles of degree m in $\mathbf{P}_n$. Then relative to the Segre embedding of V, a point $x = (x_1, x_2, \ldots, x_m)$ is stable if and only if:

For all linear subspaces $\mathrm{L} \subset \mathbf{P}_n$, then:

$$\frac{(\text{number of points } x_i \text{ in subspace } L)}{(\text{total number of points } x_i)} < \frac{\dim (L) + 1}{n+1}.$$

Then in all characteristics, or even over the scheme of integers, the set of stable 0-cycles forms an open set $U \subset (\mathbf{P}_n)^m$, and a quasi-projective orbit space $U/PGL(n)$ exists. In fact, *U is a principle fibre bundle over $U/PGL(n)$*. Our techniques are entirely elementary (see [3] and [6]).

I would like to illustrate how this simple result can be used to prove the existence of (i) a moduli scheme for polarized abelian varieties, and (ii) (according to a suggestion of Grothendieck) the Picard scheme of any variety.[1] First, let me fix some notations: if $U \subset (\mathbf{P}_n)^m$ is as above, let $Q_{n,m} = U/PGL(n)$. Second, since U is a principal fibre bundle over $Q_{n,m}$ with group $PGL(n)$, if $PGL(n)$ acts on any scheme X, we can form an associated fibre bundle with fibre X; in particular, if $X = \mathbf{P}_n$, denote the associated fibre bundle by $P_{n,m}$. Let $\pi: P_{n,m} \to Q_{n,m}$ be the bundle morphism. It is easy to see that the bundle $P_{n,m}/Q_{n,m}$ has m distinguished sections $s_i: Q_{n,m} \to P_{n,m}$ associated to the maps which take the m-tuple $(x_1, x_2, \ldots, x_m)$ to its ith factor x_i. Intuitively, regard $P_{n,m}$ plus the collection of sections (s_i) as the universal family of projective n-spaces with given stable 0-cycle of degree m.

To simplify the treatment of the moduli problem, let us consider only the question of finding a quasi-projective variety $M_{g,d}$ whose points parametrize "naturally" the set of all abelian varieties A of dimension g, plus a very ample[2] divisor class D such that $(D^g) = d \cdot g$! For the question of

[1] In fact, it can be used to prove much stronger results on the existence of relative Picard schemes: see [4].

[2] in the sense of Grothendieck, i.e. induced via a projective embedding.

making explicit the sense of the word "naturally", and the question of replacing D by its numerical equivalence class, see [6]. Now pick any n such that the characteristic does not divide n, and $n > d \cdot \sqrt{g!}$. Let $\nu = n^{2g}$. Then consider the following set:

Subvarieties $A \subset P_{d-1,\nu}$ such that

(a) $\pi(A)$ is a single point $q \in Q_{d-1,\nu}$,

(b) the degree of A in $\pi^{-1}(q)$ equals $d \cdot g!$, and the dimension of A is g,

(c) A admits a structure of abelian variety such that $\{s_i(q)\}$, for $i = 1, 2, \ldots, \nu$ is the set of points of order n.

Then, on the one hand, this set is parametrized naturally by a locally closed subset of a suitable Chow variety. But, on the other hand, it is isomorphic to the set of abelian varieties A of dimension g, plus a very ample divisor class D such that $(D^g) = d \cdot g!$, *plus* an ordering of the set of points of order n. [Namely, one can prove that the inequality $n > d \cdot \sqrt{g!}$ insures that the set of points of order n form a stable 0-cycle in $\mathbf{P}_{d-1}$ via the embedding defined by the complete linear system $|D|$. Let q be the equivalence class of this 0-cycle mod $PGL(d-1)$; then there is a unique identification of $\pi^{-1}(q)$ with the $\mathbf{P}_{d-1}$ ambient to A, under which $s_i(q)$ corresponds to the ith point of order n. Then A is embedded in $\pi^{-1}(q) \subset P_{d-1,\nu}$ via this identification.] Therefore, the set of all A, plus D alone, is parametrized by the quotient of this set by the group of permutations of ν letters. As such a quotient of a quasi-projective variety is well-known to exist, this solves the problem.

To simplify the treatment of the Picard scheme of a projective variety V, let us consider only the construction of the reduced and connected Picard scheme. The question then may be loosely described as that of finding a variety P whose points parametrize "naturally" the set of all Cartier divisor classes D on V which are algebraically equivalent to 0. But this is equivalent to doing the same for the set of all Cartier divisor classes D on V which are algebraically equivalent to some fixed very ample divisor D_0. But if D_0 is chosen sufficiently ample, and if we associate to each divisor class D the morphism $\phi: V \to \mathbf{P}_n$ defined by the complete linear system $|D|$, then this set will be isomorphic to the set of orbits under the group $PGL(n)$ in the set of morphisms $\phi: V \to \mathbf{P}_n$ which are algebraically equivalent to a fixed $\phi_0: V \to \mathbf{P}_n$. This is a problem of the type originally posed.

It can be reduced to the 0-cycle problem already solved as follows: pick a large number of sufficiently generic points $x_1, x_2, \ldots x_N$ in V, i.e. such that: For all $\phi: V \to \mathbf{P}_n$ algebraically equivalent to ϕ_0, the set of points $\phi x_1, \phi x_2, \ldots, \phi x_N$ is a stable 0-cycle in $\mathbf{P}_n$.

Then consider the following set:

Morphisms $\phi: V \to P_{n,N}$ such that

(a) $\pi \circ \phi$ maps V to a single point $q \in Q_{n,N}$,

(b) $s_i(q) = \phi(x_i)$,

(c) ϕ, as a morphism from V to $\pi^{-1}(q) \cong \mathbf{P}_n$ is algebraically equivalent to ϕ_0.

Then on the one hand this set is parametrized by a locally closed subset of a suitable Chow variety (i.e. via the graph of ϕ), and on the other hand it is isomorphic to the set of orbits described above.

Here is another example of a stability condition and the resulting quotient theorem. Assume the characteristic is 0, and consider, instead of sequences of points in $\mathbf{P}_n$, sequences of linear subspaces of any dimension. Thus, if $Grass_{k,n}$ stands for the Grassmannian of k-dimensional linear subspaces of $\mathbf{P}_n$, I ask for orbit spaces of the type $(Grass_{k,n})^m/PGL(n)$. Then, in fact, relative to the usual Plücker embedding of the Grassmannian, and to the Segre embedding of this product, it turns out that a point $x=(L_1,L_2,\ldots,L_m)$ of $(Grass_{k,n})^m$ is stable if and only if:

For all linear subspaces $L\subset P_n$, then:

$$\frac{\sum_i (\dim (L\cap L_i)+1)}{\sum_i (\dim L_i+1)}<\frac{(\dim L+1)}{(n+1)}.$$

Then, by the fundamental theorem, the set of stable points forms an open set U, and an orbit space $U/PGL(n)$ exists.

This result can be applied to the problem of classifying vector bundles over a variety in exactly the same way as the result on 0-cycles was applied to the problem of classifying line bundles, i.e. Cartier divisor classes. Of course, it is well-known that the set of vector bundles even over an algebraic curve is not a separated space; in fact, it is not even locally separated, because of the "jump" phenomenon noted by Kodaira and Spencer [5]. However, again, a basic stability condition avoids all these difficulties. For simplicity, let us consider only vector bundles over a fixed-curve C.

Definition. A vector bundle E is *stable* if for all sub-bundles F,

$$\operatorname{Deg} c_1(F)<\operatorname{Deg} c_1(\mathrm{E})\cdot\frac{\operatorname{rank} F}{\operatorname{rank} E},$$

where c_1 denotes the first chern class.

In other words, a vector bundle is stable if all its subbundles are "less ample" than itself. To illustrate the stability condition, let me mention its simplest properties:

(i) If L is a line bundle, then E is stable if and only if $E\otimes L$ is stable; moreover, E is stable if and only if $\check{E}$ is stable.
(ii) If E_1 and E_2 are two vector bundles, $E_1\oplus E_2$ is never stable.
(iii) A line bundle is always stable.
(iv) If a vector bundle E of rank 2 is not stable, then either E is isomorphic to $L_1\oplus L_2$, *or* there is a unique sub-bundle L for which $\geqslant$ holds in the definition and E can be canonically described as an extension.

Then I can prove the following theorem:

Theorem. *The set of all stable vector bundles of rank* r *over a fixed curve* C *in characteristic* 0 *is "naturally" isomorphic to the set of points of a non-singular quasi-projective variety* $V_r(\mathrm{C})$.

A more complicated example of a stability condition is given by the action of $PGL(2)$ on the variety of plane curves of degree n. There seems to be no simple general rule describing when a plane curve is stable; however, I can prove that if $n\geqslant 3$, then at least every non-singular curve is stable. For low values of n, the precise answer is given by:

n	Stable curves
1, 2	None.
3	Non-singular.
4	No triple points and no tacnodes.
5	No triple points with 3 coincident tangents, or with 2 tangents forming a tacnode.

To conclude, I want to pose a question that seems to me to be the most interesting problem in extending the results discussed above. Moreover, I think this problem is the central one on the road to solving the general problem of algebraic moduli of polarized non-singular varieties. The question is: *when is the Chow point of a non-singular subvariety* $V \subset \mathbf{P}_n$ *stable (in the usual projective embedding of the Chow variety)*? Perhaps more reasonable is the stabilized form of this problem: Given a subvariety $V \subset \mathbf{P}_n$, when is there an n_0 such that if $n \geqslant n_0$, the Chow point of V in the n-ple embedding $V \subset \mathbf{P}_N$ is stable? I have no conjecture to make.

REFERENCES

[1]. ARTIN, M., Some numerical criteria for the contractability of curves on an algebraic surface. *Amer. J. Math.*, 84 (1962), 485.

[2]. PALAIS, R., On the existence of slices for actions of non-compact Lie groups. *Ann. Math.*, 73 (1961), 295.

[3]. MUMFORD, D., An elementary theorem in geometric invariant theory. *Bull. Amer. Math. Soc.*, 67 (1961), 483.

[4]. GROTHENDIECK, A., *Séminaire Bourbaki*, exp. 232 and 236, 1962.

[5]. KODAIRA, K. & SPENCER, D. C., On deformations of complex analytic structures. *Ann. Math.*, 67 (1958), 328.

Most of the material discussed above will be published in:

[6]. MUMFORD, D., Geometric invariant theory. (Forthcoming.)

PERIODS OF A MODULI SPACE OF BUNDLES ON CURVES.

By D. Mumford and P. Newstead.

We will work over the complex numbers in this paper. For all curves C, and for all integers (n, d), the problem arises of determining the structure of the "space" of all vector bundles E, with rank n and degree $(= \deg c_1(E))d$. The problem has been considerably clarified recently by the introduction of the concept of stable and semi-stable bundles: [4], [6], [10]. It has been proven, in particular, that for each n and each line bundle L on C such that n and $\deg L$ are relatively prime, then the set:

$$S_{n,L}(C) = \begin{cases} \text{set of all stable vector bundles } E \text{ on } C \text{ of} \\ \text{rank } n \text{ such that } \Lambda^n E \cong L \end{cases}$$

has a natural structure of a non-singular projective variety of dimension $(n^2-1)\cdot(g-1)$, where $g = \text{genus } (C)$. It is important to note that the map

$$E \mapsto E \otimes M$$

for a line bundle M induces an isomorphism

$$S_{n,L}(C) \xrightarrow{\approx} S_{n,L\otimes M^n}(C)$$

hence the variety $S_{n,L}(C)$ depends essentially only on the residue class of $\deg L \bmod n$.

We wish to look at the case $g \geqq 2$, $n = 2$, $\deg L$ odd. In this case, we may assume for simplicity that a base point $x_0 \in C$ has been chosen that L is taken to be the line bundle whose sections form the sheaf $O_{-C}(x_0)$. We abbreviate $S_{2,L}(C)$ now to $S_2^-(C)$. The topology of these varieties has been described in [7] and when the genus of C is 2, their complete structure is described in [8]. $S_2^-(C)$ has dimension $3g-3$ and is known to be birationally equivalent to $\mathbf{P}_{3g-3}$. In particular, it is simply connected and the invariants $h^{0,p} = h^{p,0}$ are all 0, ([9]). In [7], it is also proven that $B_2 = 1$, $B_3 = 2g$. Now for non-singular projective varieties X with $h^{0,3} = h^{3,0} = 0$, a very interesting invariant is Weil's "intermediate jacobian" attached to $H^3(X)$. This is an abelian variety, which we shall denote $J^2(X)$, which is by definition:

Received June 12, 1967.

1200

$$J^2(X) \cong H^3(X, \mathbf{R})/\mathrm{Image}[H^3(X, \mathbf{Z})]$$

where $H^3(X, \mathbf{R})$ is given a complex structure via the decomposition

$$H^3(X, \mathbf{R}) \otimes \mathbf{C} \cong H^{2,1} \oplus H^{1,2}$$

since this induces an isomorphism

$$H^3(X, \mathbf{R}) \cong H^{1,2} = H^2(X, \Omega^1).$$

cf. [11], [1], [3]. Weil also showed that a polarization on X induces a polarization on $J^2(X)$ in a canonical way.

If $\mathrm{Alb}(C)$ denotes the albanese, or jacobian, variety of C, then our main result is:

THEOREM. $J^2[S_2^-(C)] \cong \mathrm{Alb}(C)$.

Note that $S_2^-(C)$ has a *unique* polarization since $B_2 = 1$, hence $J^2(S_2^-(C))$ has a *canonical* polarization, just as $\mathrm{Alb}(C)$ does. It is easy to check that our isomorphism is compatible with these canonical polarizations, hence by Torelli's theorem, we conclude:

COROLLARY. *If* $S_2^-(C_1) \cong S_2^-(C_2)$, *then* $C_1 \cong C_2$.

Before beginning the proof, we must recall Weil's map relating $J^2(X)$ to codimension 2 cycles on X:

let Y be an non-singular parameter space,
let W be an algebraic cycle on $X \times Y$ of codimension 2.

Then we get

$w \in H^4(X \times Y, \mathbf{Z})$, the fundamental class of W
esp. $w_{3,1} \in (H^3(X, \mathbf{Z})/\text{torsion}) \otimes H^1(Y, \mathbf{Z})$, the (3,1)-component of w.

Then $w_{3,1}$ defines a map

$$\begin{array}{ccc} \phi_W : \widehat{H^1(Y, \mathbf{R})} \Big/ \begin{matrix}\text{linear maps} \\ \text{which are} \\ \text{integral on} \\ H^1(Y, \mathbf{Z})\end{matrix} & \longrightarrow & H^3(X, \mathbf{R})/\mathrm{Image}\, H^3(X, \mathbf{Z}) \\ \| & & \| \\ \mathrm{Alb}(Y) & & J^2(X) \end{array}$$

12

which is easily seen to be complex-analytic using the fact that w is of type $(2,2)$ in the Hodge decomposition of H. Note the obvious fact:

LEMMA 1. *ϕ_W is an isomorphism if and only if $w_{3,1}$ is "unimodular," (i.e., written out as a matrix in terms of bases of $H^3(X,\mathbf{Z})/torsion$, $H^1(Y,\mathbf{Z})$, it is a square matrix with* $\det = \pm 1$).

1. In the sequel, we abbreviate $S_2^-(C)$ by S. The first step in our proof is to construct a universal vector bundle E on $S \times C$, i.e., one whose restriction to $\{t\} \times C$, for any $t \in S$, is exactly the vector bundle E_t on C corresponding to the point $t \in S$. This is a problem in descent theory. In fact, S can be described as a quotient $R/PGL(\nu)$, where R is a non-singular quasi-projective variety, and $PGL(\nu)$ acts freely on R; and where there is a vector bundle F on $R \times C$ whose restriction to $\{t\} \times C$, any $t \in R$, is the vector bundle on C corresponding to the image of t in S: cf. [10], p. 321. However, the difficulty is that the action of $PGL(\nu)$ on R does not, a priori, lift to an action on F. Instead, $GL(\nu)$ acts on F satisfying

1) $\mathbf{G}_m = \text{center } (GL(\nu))$ acts on F by homotheties
2) if $\pi\colon GL(\nu) \to PGL(\nu)$ is the canonical map, and T_g represents the action of an element g, then the diagram

$$\begin{array}{ccc} F & \xrightarrow{\quad T_g \quad} & F \\ \downarrow & & \downarrow \\ R \times C & \xrightarrow[\quad T_{\pi(g)} \times 1_C \quad]{} & R \times C \end{array} \qquad \text{commutes.}$$

The way out of this type of impasse is to find a "functorial" way of associating to every vector bundle E on C (of the type being considered) a 1-dimensional vector space $\lambda(E)$ such that multiplication by α in E induces multiplication by α in $\lambda(E)$. By functorial we mean that the procedure extends to families of such vector bundles: if E is a vector bundle on $T \times C$ (for any algebraic scheme T) whose restriction to $\{t\} \times C$ is of the type under consideration, then we should get a line bundle $\lambda(E)$ on T. Moreover, for any diagram of vector bundles

$$\begin{array}{ccc} E_1 & \xrightarrow{\quad g \quad} & E_2 \\ \downarrow & & \downarrow \\ T_1 \times C & \xrightarrow[\quad f \times 1_C \quad]{} & T_2 \times C \end{array}$$

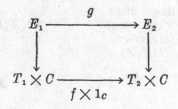

making E_1 into a fibre product of E_2 and $T_1 \times C$ over $T_2 \times C$, we should *be given* a definite isomorphism of $\lambda(E_1)$ with $f^*(\lambda(E_2))$. For example, if $T_1 = T_2 = \mathrm{Spec}(\boldsymbol{C})$, $E_1 = E_2 = E$, and g is multiplication by a scalar $\alpha \neq 0$, we are then given an induced automorphism of $\lambda(E)$: we want this automorphism to be multiplication by α too (it might turn out to be multiplication by α^n instead). All this data is subject to an obvious co-cycle condition: compare [5], p. 64. If we can find such data, we get as a consequence a line bundle $\lambda(F)$ on R, plus an action of $GL(\nu)$ on $\lambda(F)$ in which the center acts by homotheties. If we then define

$$F' = F \otimes p_1^*(\lambda(F)^{-1}),$$

we get a new vector bundle on $R \times C$ with the same restrictions to the fibres $\{t\} \times C$ as before; but where in the natural action of $GL(\nu)$ on F', the action of the center $\boldsymbol{G}_m$ on F and $p_1^*(\lambda(F)^{-1})$ cancel each other out, i.e., $PGL(\nu)$ acts on F. Then $F/PGL(\nu)$ is the sought-for universal vector bundle on $S \times C$.

Here's how to construct λ. We limit ourselves to the case $T = \mathrm{Spec}(\boldsymbol{C})$, E a vector bundle on C, since the generalization of λ to an arbitrary base will be clear. Recall E has rank 2, degree 1, and is stable:

a) $H^1(E \otimes (\Omega_C{}^1)^k) = (0)$, if $k \geqq 1$.

Proof. This group is dual to $H^0(\hat{E} \otimes (\Omega_C{}^1)^{1-k})$ and if this were non-zero, we would get a non-zero homomorphism

$$(\Omega_C{}^1)^{k-1} \to \hat{E}$$

hence a sub-line-bundle $G \subset \hat{E}$ of degree $\geqq 2(k-1)(g-1) \geqq 0$. This contradicts the stability of E.

b) If $V_k(E) = H^0(E \otimes (\Omega_C{}^1)^k)$, then

$$\mathrm{Dim}\, V_k(E) = (2g-2)(2k-1) + 1.$$

Proof. Riemann-Roch.

c) Set $\lambda(E) = [\Lambda^{2g-1} V_1(E)]^{\otimes(3g-1)} \otimes [\Lambda^{6g-5} V_2(E)]^{\otimes(-g)}$.
Then multiplication by α in E induces the endomorphism, multiplication by α, in each $V_k(E)$, hence it induces multiplcation by α to the power

$$(3g-1)(2g-1)+(6g-5)(-g)$$

in $\lambda(E)$. This number happens to be 1!

We now know that E exists. Next consider the chern classes of E. We have

$$c_2(E)\in H^4(S\times C,\mathbf{Z})\cong(H^2(S,\mathbf{Z})\otimes H^2(C,\mathbf{Z}))$$
$$c_1(E)\in H^2(S\times C,\mathbf{Z})\cong H^2(C,\mathbf{Z})\otimes H^2(S,\mathbf{Z})$$
$$\otimes(H^3(S,\mathbf{Z})\otimes H^1(C,\mathbf{Z}))$$
$$\otimes H^4(S,\mathbf{Z}).$$

Note that any bundle $E\otimes p_1{}^*M$, M a line bundle on S, would have the same universal property that E does, so $c_1(E)$ is not very interesting. However, let

$$\alpha=(c_2(E))_{3,1}=[\text{component of } c_2(E) \text{ in } H^3(S,\mathbf{Z})\otimes H^1(C,\mathbf{Z})].$$

A simple computation of chern classes shows that α is independent of this modification of E. According to [7], $H^3(S,\mathbf{Z})$ and $H^1(C,\mathbf{Z})$ have the same rank. In fact:

PROPOSITION 1. *α is unimodular.*

This will be proven in § 2. Assuming this, it follows from Lemma 1 that if $W=$ the *algebraic* 2nd chern class of E, then Weil's map ϕ_W: Alb$(C)\to J^2(S)$ is an isomorphism, as required. Although it is not essential, it will be convenient in § 2 to know that $H^3(S,\mathbf{Z})$ is torsion-free. In fact, the torsion subgroup of $H^3(X,\mathbf{Z})$—for any non-singular complete variety X over $\mathbf{C}$—is a *birational invariant* of X known as the "topological Brauer group" (cf. [12], Cor. (7.3) and equation (8.9), p. 59). And S is birationally equivalent to $\mathbf{P}_{3g-3}$ which has no H^3 at all!

2. We start by recalling the results of [6]. In fact, let S_0 be the subset of $SU(2)^{2g}$ consisting of points $(A_1,\cdots,A_{2g})$ such that

$$\prod_{i=1}^{g}A_{2i-1}A_{2i}A_{2i-1}{}^{-1}A_{2i}{}^{-1}=-I.$$

Then S_0 is an orientable submanifold of $SU(2)^{2g}$ and there is a natural map

$$p\colon S_0\to S,$$

which is a principal fibration with group $PU(2)$. The map p may be determined as follows. Let $\tilde{C}$ be the simply-connected covering of C which is ramified over x_0 with ramification index 2. The group π of this covering is generated by elements $a_1, \cdots, a_{2g}$ subject to the single relation

$$[\prod_{i=1}^{g} a_{2i-1}a_{2i}a_{2i-1}^{-1}a_{2i}^{-1}]^2 = e.$$

Thus a point of S_0 may be regarded as a representation of π, and this representation defines a stable bundle E over C of rank 2, with $\Lambda^2 E \cong L$, and hence a point of S. So we get a map $p\colon S_0 \to S$. Notice that the a_i determine elements of $\pi_1(C)$ and hence of $H_1(C;\mathbf{Z})$, and that these elements form a basis for $H_1(C;\mathbf{Z})$; let $\{\alpha_i\}$ be the dual basis of $H^1(C;\mathbf{Z})$.

LEMMA 2. *$p^*\colon H^3(S;\mathbf{Z}) \to H^3(S_0;\mathbf{Z})$ is an isomorphism.*

Proof. Since $H^1(PU(2);\mathbf{Z}) = 0$ and $H^1(S;\mathbf{Z}) = 0$ (S is simply-connected), the spectral sequence of the fibration p gives rise to an exact sequence

$$H^0(S;H^2(PU(2);\mathbf{Z})) \to H^3(S;\mathbf{Z}) \xrightarrow{p^*} H^3(S_0;\mathbf{Z}) \to H^0(S;H^3(PU(2);\mathbf{Z})).$$

Now the first group in this sequence is $\mathbf{Z}_2$ and the last is $\mathbf{Z}$. Moreover $H^3(S;\mathbf{Z})$ is torsion-free (see §1) and has the same rank as $H^3(S_0;\mathbf{Z})$ by the results of [7]. The lemma now follows.

LEMMA 3. *The homomorphism $H^3(SU(2)^{2g},\mathbf{Z}) \to H^3(S_0;\mathbf{Z})$ induced by the inclusion of S_0 in $SU(2)^{2g}$ is an isomorphism.*

Proof. Lemma 3 of [7] shows that the homomorphism

$$H_3(S_0;\mathbf{Z}) \to H_3(SU(2)^{2g},\mathbf{Z})$$

is surjective, except possibly for some 2-primary torsion. However, in this simple case, the same argument can be used to prove that the homomorphism is really surjective. It follows at once that $H^3(SU(2)^{2g};\mathbf{Z})$ is contained in $H^3(S_0;\mathbf{Z})$ as a direct summand. The lemma now follows from the fact that the ranks of these two groups are equal (see [7]) and that $H^3(S_0;\mathbf{Z})$ is torsion-free by Lemma 2.

Now let $p_i\colon S_0 \to SU(2)$ denote the projection on the i-th factor and let

$$\beta_i = p_i^*[\text{generator of } H^3(SU(2);\mathbf{Z})].$$

Then by Lemma 3 the β_i form a basis for $H^3(S_0;\mathbf{Z})$. In view of Lemma 2, it is now sufficient to prove:

PROPOSITION 2. *$c_2[(p \times 1_C)^*E]_{3,1} = \sum_{i=1}^{2g} \beta_i \otimes \alpha_i$.*

Now choose embedding $s_i\colon S^1 \to C - x_0$ which represent the generators a_i of π. Then

$$\begin{aligned} s_i^*(\alpha_j) &= 0 & i \neq j \\ &= \text{generator of } H^1(S^1;\mathbf{Z}) & i = j. \end{aligned}$$

Hence Proposition 2 will follow at once from

PROPOSITION 3.

$$c_2[(1_{S_0} \times s_i)^*(p \times 1_C)^*E]_{3,1} = \beta_i \otimes [\textit{generator of } H^1(S^1;\mathbf{Z})].$$

We now need to recall a few more details from [6]. Let E_ρ be the bundle over C corresponding to the representation $\rho \in S_0$. Then ([6] Remark 6.2) we can write down coordinate transformations for E_ρ as follows. Choose a finite open covering $\{U_i\}$ $(i = 0, 1, \cdots, m)$ of C such that every non-empty intersection of the sets U_i is contractible. Assume $x_0 \in U_0$, $x_0 \notin U_i$ for $i \neq 0$. Assume moreover that there exist discs D_i in $\tilde{C}$ such that U_0 is the quotient of D_0 by $\mathbf{Z}_2$ and that for $i \neq 0$, D_i maps homeomorphically onto U_i. For every i, j, k, where $k = i$ or j, let $W_{ij,k}$ be a connected component of $v^{-1}(U_i \cap U_j) \cap D_k$ (where $v\colon \tilde{C} \to C$ is the covering map). If $U_i \cap U_j = \emptyset$, $i \neq j$, $W_{ij,k}$ maps homeomorphically onto $U_i \cap U_j$; let γ_{ij} be the element of π such that $\gamma_{ij} W_{ij,j} = W_{ji,i}$. Then a set of coordinate transformations g_{ij} for E_ρ is given by

$$\begin{aligned} g_{ij} &= \rho(\gamma_{ij}) \text{ on } U_i \cap U_j,\ i \neq 0,\ j \neq 0 \\ &= f_i \cdot \rho(\gamma_{0i}) \text{ on } U_0 \cap U_i,\ i \neq 0, \end{aligned}$$

where f_i is an analytic scalar function on $U_0 \cap U_i$ which is independent of ρ. Note that the coordinate transformations depend differentially on ρ, so that the same g_{ij} (now regarded as functions on $S_0 \times U_i \cap S_0 \times U_j$) define a differentiable bundle E' over $S_0 \times C$ which is a differentiable family of analytic bundles over C.

Now $E' \mid \{\rho\} \times C \cong E_\rho \cong (p \times 1_C)^*E \mid \{\rho\} \times C$ for all $\rho \in S_0$. Since E is stable, it follows that

$$\dim H^0(C; \underset{\text{Anal.}}{\mathrm{Hom}}(E', (p \times 1_C)^*E) \mid \{\rho\} \times C) = 1$$

for all ρ. So by Proposition 2.7 of [2],

$$\bigcup_{\rho \in S_0} H^0(C; \underset{\text{Anal.}}{\mathrm{Hom}}(E', (p \times 1_C)^*E) \mid \{\rho\} \times C)$$

has a natural structure of differentiable line bundle over S_0. Let L be the induced line bundle over $S_0 \times C$. There is then an obvious isomorphism

$$E' \otimes L \cong (p \times 1_C)^* E.$$

So

$$c_2[(p \times 1_C)^* E]_{3,1} = c_2(E')_{3,1}.$$

Using the above explicit description of the bundle E', we see that for any continuous map $s\colon S^1 \to C - x_0$, $(1_{S_0} \times s)^* E'$ can be described as follows: take a trivial bundle of rank 2 over $S_0 \times [0,1]$ and glue its ends together by means of the map

$$S_0 \to SU(2),$$

defined by

$$\rho \mapsto \rho(a), \text{ where } a \in \pi \text{ corresponds to } s.$$

Apply this when $s = s_i$, $a = a_i$, and $\rho(a_i) = p_i(\rho)$; so Proposition 3 will follow at once from

LEMMA 4. *Let W be a space and let F be the bundle over $W \times S^1$ obtained by glueing together the two ends of the trivial bundle of rank 2 over $W \times [0,1]$ by means of the map $f\colon W \to SU(2)$. Then*

$$c_2(F) = f^*[\text{generator of } H^3(SU(2);\mathbf{Z})] \otimes [\text{generator of } H^1(S^1;\mathbf{Z})].$$

Proof. F is the bundle induced by f from the bundle obtained by taking $W = SU(2)$, $f = 1_{SU(2)}$ in the construction. Hence it is sufficient to prove the lemma for this special case. But then it follows from the fact that $H^*(BSU(2))$ is generated by c_2.

This completes the proof of Proposition 3 and hence of our theorem.

REFERENCES.

[1] P. Griffiths, "Periods of integrals on algebraic manifolds," I, II, III, forthcoming.
[2] K. Kodaira and D. C. Spencer, "On deformations of complex analytic structures," *Annals of Mathematics*, vol. 67 (1958), p. 328.
[3] D. Lieberman, "Higher Picard varieties," forthcoming.

[4] D. Mumford, "Projective invariants of projective structures," *International Mathematical Congress*, Stockholm, 1962, p. 526.

[5] ———, "Picard groups of moduli problems," *Proc. Conf. Arith. Alg. Geom.*, Harper and Row, 1966.

[6] M. S. Narasimhan and C. S. Seshadri, "Stable and unitary vector bundles on a compact Riemann surface," *Annals of Mathematics*, vol. 82 (1965), p. 540.

[7] P. E. Newstead, "Topological properties of some spaces of stable bundles," *Topology*, vol. 6 (1967), p. 241.

[8] ———, "Stable bundles of rank 2 and odd degree over a curve of genus 2," forthcoming.

[9] J. P. Serre, "On the fundamental group of a unirational variety," *Journal of the London Mathematical Society*, vol. 34 (1959), p. 481.

[10] C. S. Seshadri, "Space of unitary vector bundles on a compact Riemann surface," *Annals of Mathematics*, vol. 84 (1967), p. 303.

[11] A. Weil, "Sur la variété de Picard," *American Journal of Mathematics*, vol. 74 (1952), p. 865.

[12] A. Grothendieck, *Le groupe de Brauer III*, mimeographed notes from Institut Hautes Etudes Sciences.

Actes, Congrès intern. math., 1970. Tome 1, p. 457 à 465.

THE STRUCTURE OF THE MODULI SPACES OF CURVES AND ABELIAN VARIETIES

by DAVID MUMFORD

§ 1. The purpose of this talk is to collect together what seem to me to be the most basic moduli spaces (for curves and abelian varieties) and to indicate some of their most important interrelations and the key features of their internal structure, in particular those that come from the theta functions. We start with abelian varieties. Fix an integer $g \geqslant 1$. To classify g-dimensional abelian varieties, the natural moduli spaces are:

$$\mathscr{A}^{(n)} = \left\{ \begin{array}{l} \text{moduli space of pairs } (X, \lambda),\ X \text{ a } g\text{-dimensional} \\ \text{abelian variety, } \lambda : X \to \hat{X} \text{ a polarization such} \\ \text{that } \deg(\lambda) = n^2 \end{array} \right\}$$

Here and below when we talk of a moduli space, we mean a coarse moduli space in the sense of [11], p. 99 and in all cases these moduli spaces will actually exist as schemes of finite type over Spec (Z). This can be proven by the methods of [11], Ch. 7, for instance, which also shows that all the moduli spaces used are quasi-projective at least over every open set Spec $Z\left[\frac{1}{p}\right]$.

The local structure of $\mathscr{A}^{(n)}$ seems quite difficult to work out at some points. However, for every sequence $\delta_1, \ldots, \delta_g$ such that $\delta_1 \mid \ldots \mid \delta_g$, $\prod_{i=1}^{g} \delta_i = n$, let

$$\mathscr{A}^{(\delta)} = \left\{ \begin{array}{l} \text{the open subscheme of } \mathscr{A}^{(n)} \text{ of pairs } (X, \lambda) \\ \text{such that} \\ \qquad \ker(\lambda) \cong \prod_1^g Z/\delta_i Z \times \prod_1^g \mu_{\delta_i} \end{array} \right\}$$

The $\mathscr{A}^{(\delta)}$'s are disjoint and exhaust all of $\mathscr{A}^{(n)}$ except for (X, λ)'s such that char$|$n and ker (λ) contains a subgroup isomorphic to α_p. The local structure of $\mathscr{A}^{(\delta)}$ is not hard to work out (using results of Serre-Tate [20], and Grothendieck and myself on the formal deformation theory of abelian varieties and p-divisible groups, see Oort [17]). In particular all components of $\mathscr{A}^{(\delta)}$ dominate Spec (Z). Now I have proven that for all n and all p, the open subset of $\mathscr{A}^{(n)} \times$ Spec Z/pZ of (X, λ)'s such that X is ordinary (*) is *dense* (cf. [14] for a sketch of the proof). Therefore $\bigcup_\delta \mathscr{A}^{(\delta)}$ is dense in $\mathscr{A}^{(n)}$. Since $\mathscr{A}^{(\delta)} \times$ Spec (C) is irreducible (see below), it follows that the components of $\mathscr{A}^{(n)}$ are the closures $\overline{\mathscr{A}^{(\delta)}}$ of the $\mathscr{A}^{(\delta)}$ and that all of them dominate Spec (Z). It is *not*

(*) i. e. has the maximal number p^g of points of order p.

known, however, whether the geometric fibres of $\mathscr{A}^{(\delta)}$ over finite primes are irreducible or not.

Now these various schemes $\mathscr{A}^{(n)}$ are all related by the isogeny correspondences:

$$Z_{n_1,n_2,k} = \left\{ (X,\lambda)\varepsilon\mathscr{A}^{(n_1)},\ (Y,\mu)\varepsilon\mathscr{A}^{(n_2)} \;\middle|\; \begin{array}{l} \exists \text{ an isogeny } \pi : X \to Y \\ \text{of degree } k \text{ such that} \\ n_1{}^2\hat{\pi}\circ\mu\circ\pi = (kn_2)^2\lambda \end{array} \right\}$$

To uniformize all of these, one introduces a second more convenient sequence of moduli spaces. Firstly, over the base scheme $Z[\zeta_n]$, ζ_n a primitive n-th root of 1, let

$$\mathscr{A}_n^* = \left\{ \begin{array}{l} \text{moduli space of triples } (X,\lambda,\alpha),\ X \text{ a } g\text{-dimensional} \\ \text{abelian variety, } \lambda : X \overset{\approx}{\rightarrow} \hat{X} \text{ a principal polarization,} \\ \text{and } \alpha : X_n \overset{\approx}{\rightarrow} (Z/nZ)^g \times \mu_n^g \text{ a symplectic isomorphism} \end{array} \right\}$$

These spaces are normal and irreducible and form a tower with respect to the natural quasi-finite morphisms $\mathscr{A}_{nm}^* \to \mathscr{A}_n^*$ given by $(X,\lambda,\alpha) \mapsto (X,\lambda,\operatorname{res}_{X_n}\alpha)$. Secondly, we enlarge these schemes somewhat by letting $\mathscr{A}_n$ be the normalization of $\mathscr{A}_1$ in the field of rational functions $Q(\mathscr{A}_n^*, \zeta_n)$. Then $\mathscr{A}_n$ is a normal irreducible scheme in which $\mathscr{A}_n^*$ is an open subscheme, and the $\mathscr{A}_n$'s form a tower with respect to finite morphisms $\mathscr{A}_{nm} \to \mathscr{A}_n$. Note that $\mathscr{A}_n = \mathscr{A}_n^*$ except over primes dividing n. Moreover, if $n \geqslant 3$, $\mathscr{A}_n$ is smooth over Z except at non-ordinary abelian schemes in characteristics dividing n. Next, we can uniformize very nearly all of $\mathscr{A}^{(\delta)}$ by the natural morphism:

$$\begin{array}{c} \mathscr{A}_{\delta_g}^* \to \mathscr{A}^{(\delta)} \\ (X,\lambda,\alpha) \mapsto (Y,\mu) \end{array}$$

where Y is the etale covering of X defined by requiring its dual to be the quotient:

$$\hat{Y} = \hat{X}/\alpha^{-1}[(0) \times \Pi\mu_{\delta_i}],$$

and μ is the polarization on Y induced by λ. In the tower $\{\mathscr{A}_n\}$ one now has the Hecke ring of correspondences instead of the isogeny correspondences. These come essentially from 2 types of morphisms:

(*a*) $Q(\mathscr{A}_n, \zeta_n)$ is a Galois extension of $Q(\mathscr{A}_1, \zeta_n)$ with Galois group $\mathrm{Sp}(2g, Z/nZ)$, hence $\mathrm{Sp}(2g, Z/nZ)$ acts as a group of automorphisms of $\mathscr{A}_n$;

(*b*) the morphism

$$\begin{array}{c} \mathscr{A}_n^* \to \mathscr{A}_1 \\ (X,\lambda,\alpha) \mapsto (Y,\mu) \end{array}$$

(where $Y = X/\alpha^{-1}[(Z/nZ)^g \times (0)]$, and if $\pi : X \to Y$ is the natural map, then $\mu : Y \overset{\approx}{\rightarrow} \hat{Y}$ is determined by the requirement $\hat{\pi}\circ\mu\circ\pi = n\lambda$). For a discussion on Hecke operators in the classical case, see Shimura [21]. The picture is even clearer when you pass to an inverse limit: e. g. for all n,

$$\varprojlim_k \mathscr{A}_{n^k} \times \operatorname{Spec} R_n$$

where

$$R_n = Z\left[\frac{1}{n}, \zeta_n, \zeta_{n^2}, \ldots\right]$$

exists as a scheme and $\prod_{p|n} \mathrm{Sp}(2g, Q_p)$ acts on it (See [12], § 9 for the case $n = 2$).

Over the complex ground field, these moduli spaces have well-known analytic uniformizations coming from the theory of Siegel modular forms:

$$\mathscr{A}^{(n)} \times \mathrm{Spec}\,(C) = \coprod_{\delta} \mathscr{A}^{(\delta)} \times \mathrm{Spec}\,(C)$$
$$\mathscr{A}^{(\delta)} \times \mathrm{Spec}\,(C) = \mathfrak{H}/\Gamma_\delta$$
$$\mathscr{A}_n \times \mathrm{Spec}\,(C) = \mathscr{A}_n^* \times \mathrm{Spec}\,(C) = \mathfrak{H}/\Gamma(n)$$

where

$$\mathfrak{H} = \text{Siegel upper } \frac{1}{2} - \text{plane} = \left\{ Z \,\middle|\, \begin{array}{l} Z = g \times g \text{ complex matrix} \\ {}^tZ = Z,\ \mathrm{Im}\, Z > 0 \end{array} \right\}$$
$$\Gamma(n) = \{ A \in \mathrm{Sp}\,(2g, Z)/(\pm I) \mid A = I_{2g} \bmod n \}$$
$$\Gamma_\delta = \{ A \in GL(2g, Z)/(\pm I) \mid {}^tA . J_\delta . A = J_\delta \},$$

where

$$J_\delta = \left(\begin{array}{c|c} 0 & \begin{matrix} \delta_1 & & \\ & \ddots & \\ & & \delta_g \end{matrix} \\ \hline \begin{matrix} -\delta_1 & & \\ & \ddots & \\ & & -\delta_g \end{matrix} & 0 \end{array} \right)$$

Thus $\mathfrak{H}$ is the analytic " inverse limit " of the $\mathscr{A}_n$'s over Spec C. All the irreducibility assertions made so far are proven by these analytic uniformizations.

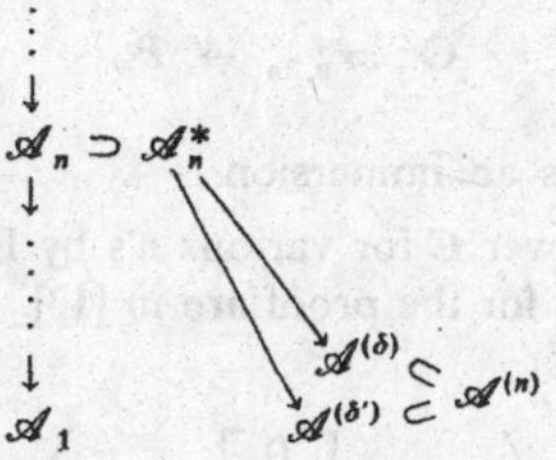

Summary of moduli spaces.

§ 2. The next point is that there is a moduli space intermediate in the tower between $\mathscr{A}_n^*$ and $\mathscr{A}_{2n}^*$ on which there are *canonical coordinates*. Following Igusa, we christen this $\mathscr{A}_{n,2n}^*$ and it is defined as follows in char. $\neq 2$:

$\mathscr{A}_{n,2n}^* = \Big\{$ moduli space of triples (X, L, α), X an abelian variety of dimension g, L an ample symmetric invertible sheaf, α a symmetric isomorphism:

$$\alpha : \mathscr{G}(L) \xrightarrow{\sim} G_m \times (Z/nZ)^g \times \mu_n^g$$

such that

i) if n even, $e_*^L \equiv 1$ on X_2,
ii) if n odd, e_*^L takes the value $+1$ more often than the value -1. $\Big\}$

For definitions of $\mathcal{G}(L)$, e_*^L, etc., see [12], § 1, 2 and [15], § 23. There is an obvious map

$$\mathcal{A}^*_{n,2n} \to \mathcal{A}^*_n$$
$$(X, L, \alpha) \mapsto \left(X, \frac{1}{n}\varphi_L, \bar{\alpha}\right)$$

where $\bar{\alpha}$ is the induced map from $\mathcal{G}(L)/G_m \cong X_n$ to $(Z/nZ)^g \times \mu_n^g$. There is a not so obvious map $\mathcal{A}^*_{2n} \times \operatorname{Spec} Z\left[\frac{1}{2}\right] \to \mathcal{A}^*_{n,2n}$ (see [12], § 2). Over C, $\mathcal{A}^*_{n,2n}$ is simply the quotient $\mathfrak{H}/\Gamma(n, 2n)$, where $\Gamma(n, 2n)$ is the subgroup between $\Gamma(n)$ and $\Gamma(2n)$ described by Igusa [9]. Canonical coordinates on $\mathcal{A}^*_{n,2n}$ (where $n \geqslant 2$) are defined as follows:

i) $\mathcal{G}(L)$ and hence $G_m \times (Z/nZ)^g \times \mu_n^g$ acts on $H^0(X, L)$. Write this action as $U_{(\lambda,a,b)}: H^0(X, L) \to H^0(X, L)$,

ii) there is a section $\sigma \in H^0(X, L)$ unique up to scalars such that $U_{(1,0,c)}\sigma = \sigma$, all $c \in \mu_n^g$,

iii) let $\sigma \to \sigma(0)$ denote evaluation of sections at $0 \in X$. We obtain a function:

$$(Z/nZ)^g \to K$$
$$\alpha \mapsto (U_{(1,\alpha,0)}\sigma)(0)$$

unique up to multiplication by a constant, which is never identically zero.

iv) If $N = n^g - 1$, and the homogeneous coordinates of P_N are put in one-one correspondence with the elements of $(Z/nZ)^g$, this defines a morphism:

$$\Theta: \mathcal{A}^*_{n,2n} \to P_N$$

THEOREM. — If $n \geqslant 4$, Θ is an immersion.

This theorem was proven over C for various n's by Baily [4] and Igusa [9]; in the general case, all the essentials for the proof are in [13]. Over C, Θ is the morphism defined by

$$Z \mapsto \left(\ldots, \theta_{nZ}\begin{bmatrix} 0 \\ \alpha/n \end{bmatrix}(0), \ldots\right)_{\alpha \in (Z/nZ)^g}$$

where $Z \in \mathfrak{H}$, and θ is Riemann's theta function. If $8 \mid n$, one can even find a finite set of homogeneous quartic polynominals—Riemann's theta relations—such that the image of Θ is an open part of the subscheme of P_N defined by these quartics (see [12], § 6).

Even in the char. p case, it is possible to reformulate these canonical coordinates as values of a type of theta function. These theta functions are not functions on the universal covering space of X, but rather on the Tate group.

If p = char. of ground field,
$V(X)$ = group of sequences $\{x_i\}$, $i \geqslant 1$ but $p \nmid i$, where $x_i \in X$, $nx_{in} = x_i$ and x_1 has finite order k prime to p.
Let $T(X) = \{(x_i) \in V(X)$ such that $x_1 = 0\}$.
We get an exact sequence:

$$0 \to T(X) \to V(X) \to (\text{torsion on } X \text{ prime to } p) \to 0$$

We use the result:

THEOREM. — Let L be an ample symmetric invertible sheaf of degree 1 on an abelian variety X of char. p. If $p \nmid 2n$, then for all $x \in X_n$ for every choice of a point $y \in X$ such that $2ny = x$, there is a canonical isomorphism:

$$L \otimes_{\mathcal{O}_X} k(x) \cong L \otimes_{\mathcal{O}_X} k(0)$$

COROLLARY. — If $\sigma \in \Gamma(L)$, then evaluating σ via the above isomorphisms defines a function

$$\theta : V(X) \rightarrow L \otimes_{\mathcal{O}_X} k(0) \cong k$$

such that if $x \in \frac{1}{n} T(X)$, then $\theta(x + y) = \theta(x)$ if $y \in 2nT(X)$.

In fact the functions that we obtain in this way have the following properties:

a) $$\theta(x + a) = e_*\left(\frac{a}{2}\right)e\left(\frac{a}{2}, x\right)\theta(x), \; x \in V(X), \; a \in T(X)$$

where

$$e_* : \frac{1}{2} T(X) \rightarrow \{\pm 1\} \quad \text{and} \quad e : V(X) \times V(X) \rightarrow k^*$$

are the functions induced by e_*^L and e_n on $V(X)$.

b) $\theta(-x) = \pm\, \theta(x)$, the sign depending on the Arf invariant of e_*^L.

c) $$\prod_{i=1}^{4} \theta(x_i) = 2^{-g} \sum_{\eta \in \frac{1}{2}T(X)/T(X)} e(y, \eta)\left(\prod_{i=1}^{4} \theta(x_i + y + \eta)\right),$$

where

$$y = -\frac{1}{2}\Sigma x_i$$

d) $\forall x \in V(X), \quad \exists \eta \in \frac{1}{2} T(X)$ such that $\theta(x + \eta) \neq 0$.

e) Up to an elementary linear transformation whose coefficients are roots of 1, the set of values of θ on $\frac{1}{n} T(X)$ is equal to the set of values of the canonical coordinates Θ on the triple (X, L^{n^2}, α) (for any symmetric α).

f) Over C, if Z is a period matrix for X, θ is essentially the function $a \mapsto \theta_Z[a](0)$, $a \in Q^{2g}$.

g) Moreover, if we restrict the domain to $V_2(X)$, this 2-adic Tate group, all functions $\psi : V_2(X) \rightarrow k$ satisfying *a)*, *b)*, *c)* and *d)* arise from a unique principally polarized abelian variety.

(Cf. [12], § 8 through § 12).

§ 3. We turn next to curves. Fix $g \geq 2$. Let

$$\mathcal{M} = \left\{ \begin{array}{l} \text{moduli space of non-singular} \\ \text{complete curves } C \text{ of genus } g \end{array} \right\}$$

$\mathcal{M}$ is not only irreducible, but it has irreducible geometric fibres over Spec $(\mathbf{Z})$, cf. [5]. This is proven by introducing a compacification $\overline{\mathcal{M}}$ of $\mathcal{M}$, where

$$\overline{\mathcal{M}} = \left\{ \begin{array}{l} \text{moduli space of stable complete curves } C \\ \text{such that } \dim H^1(o_C) = g \end{array} \right\}$$

and where a *stable curve* is one with at most ordinary double points and such that every non-singular rational component has at least 3 double points on it.

$\overline{\mathcal{M}}$ has recently been proven by F. Knudsen, Seshadri and my self to be a scheme projective over $\mathbf{Z}$

Define:

$$\begin{aligned} t: \mathcal{M} &\to \mathcal{A}_1 \\ C &\mapsto (\mathrm{Pic}^0 (C), \lambda) \end{aligned}$$

where λ is the theta polarization, viz: fixing a base point x_0 on C, we obtain a morphism:

$$\begin{aligned} \phi: C &\to \mathrm{Pic}^0 C \\ x &\mapsto \text{class of } o_C(x - x_0) \end{aligned}$$

hence

$$\widehat{\mathrm{Pic}^0 C} = \mathrm{Pic}^0 (\mathrm{Pic}^0 C) \overset{\phi^*}{\underset{\sim}{\to}} \mathrm{Pic}^0 (C)$$

and $\lambda = -(\phi^*)^{-1}$.

According to Torelli's theorem (cf. [1], [10]), t is injective on geometric points. Its image however is not closed since t extends to a morphism on $\tilde{\mathcal{M}}$:

$$\left\{ \begin{array}{l} \text{Stable curves made from} \\ \text{non-singular components} \\ \text{connected together like a} \\ \text{tree} \end{array} \right\} = \tilde{\mathcal{M}} \qquad \begin{array}{c} \overline{\mathcal{M}} \\ \cup \\ \tilde{\mathcal{M}} \;\overset{\tilde{t}}{\searrow} \\ \cup \\ \mathcal{M} \underset{t}{\to} \mathcal{A}_1 \end{array}$$

and $\tilde{t}$ can be shown to be a proper morphism taking each stable curve C in $\tilde{\mathcal{M}}$ to $\mathrm{Pic}^0 C$ with a suitable polarization (cf. Hoyt [8]). Let $\mathcal{T} = \overline{t(\mathcal{M})} = \tilde{t}(\tilde{\mathcal{M}})$: this is called the *Torelli locus*. A famous classical problem is to describe $\mathcal{T}$, or its inverse image in some $\mathcal{A}_n$, by explicit equations, e. g. polynomials in the theta-nulls. Partial results on this were obtained in characteristic zero by Riemann [18], Schottky, and Schottky-Jung [19]. Their results have been rigorously established recently by Farkas and Rauch [6], and some interesting generalizations have been given by Fay [7]. A completely different approach to this problem is given in the beautiful paper of Andreotti and Mayer [2]. I want to finish by sketching the key point in Schottky's theory and stating a theorem on what his equations do characterize. We assume char. $\neq 2$.

Let $\Pi: \hat{C} \to C$ be an etale double covering, and let $\iota: \hat{C} \to \hat{C}$ be the corresponding involution. If g = genus of C, then $2g - 1$ = genus of $\hat{C}$. The Jacobians $J = \mathrm{Pic}^0 C$, $\hat{J} = \mathrm{Pic}^0 \hat{C}$ are related by 2 homeomorphisms:

$$\hat{J} \underset{\Pi^*}{\overset{Nm}{\rightleftarrows}} J$$

such that $Nm \circ \Pi^* = 2_J$. ι acts on $\hat{J}$ also. Define:

$$P = \text{locus of points } \{\iota x - x\} \text{ in } \hat{J}.$$

We get isogenies:

$$\hat{J} \underset{\alpha}{\overset{\beta}{\rightleftarrows}} J \times P$$

$$\alpha(x, y) = \Pi^* x + y$$
$$\beta(z) = (Nmz, z - \iota z)$$

such that $\alpha \circ \beta = \beta \circ \alpha =$ mult. by 2. Next fix γ, a division class on C such that $2\gamma \equiv K_C$, the canonical divisor class, and such that dim $H^0(o(\gamma))$ is even (cf. [16] for this). We get symmetric divisors:

$$\Theta = \{ \text{locus of div. classes } \sum_1^{g-1} P_i - \gamma \} \subset J$$

$$\hat{\Theta} = \{ \text{locus of div. classes } \sum_1^{2g-2} P_i - \Pi^*\gamma \} \subset \hat{J}$$

representing the standard polarizations of J and $\hat{J}$.

LEMMA

a) $\Pi^{*-1}(\hat{\Theta}) = \Theta + \Theta_\kappa$, where $\{0, \kappa\} = \text{Ker}(\Pi^*)$.
b) $\exists$ a symmetric divisor Ξ on P such that $\hat{\Theta}.P = 2\Xi$.
c) $\alpha^{-1}(\hat{\Theta}) \equiv \Theta \times P + \Theta_\kappa \times P + 2J \times \Xi$.

In particular, Ξ has degree 1 and defines a principal polarization on P. Abstractly put now we have a situation with.

i) 3 abelian varieties X, Y, Z of dimensions g, $g-1$, $2g-1$ resp.,
ii) 3 ample degree 1 symmetric divisors $\Theta_X \subset X$, $\Theta_Y \subset Y$, $\Theta_Z \subset Z$, which define as in § 2 theta-functions θ_X on $V(X)$, θ_Y on $V(Y)$ and θ_Z on $V(Z)$,
iii) isogenies $Z \underset{\alpha}{\overset{\beta}{\rightleftarrows}} X \times Y$ such that $\alpha \circ \beta = \beta \circ \alpha =$ mult. by 2. In such a case, $Z \cong X \times Y/H$, where H is a so-called Göpel group, and $V(Z) \cong V(X) \times V(Y)$. Moreover θ_Z can be computed from θ_X and θ_Y by one of the basic theta formulas. But then the lemma, esp. part *a)*, implies non-trivial identities on θ_X and θ_Y. In fact, it follows that for a suitable $\eta \in \frac{1}{2} T(X)$ with image κ in X and a suitable homeomorphism:

$$\varphi : \left\{ x \in \frac{1}{2} T(X) \mid e(x, \eta) = 1 \right\} \to \frac{1}{2} T(Y)$$

$$(*) \quad \begin{cases} \dfrac{\theta_X(x).\theta_X(x+\eta)}{\theta_Y(\varphi x)^2} = \dfrac{\theta_X(y).\theta_X(y+\eta)}{\theta_Y(\varphi y)^2} \\ \text{all} \quad x, y \in \frac{1}{2} T(X) \\ \text{with} \quad e(x, \eta) = e(y, \eta) = 1 \end{cases}$$

If we globalize this set-up, we get the following moduli situation: $\mathcal{M}_*$ is to be the normalization of $\mathcal{M}$ is a suitable finite algebraic extension of its function field such that for every point of $\mathcal{M}_*$ there is given rationally not only a curve C of genus g, but

(a) a (4,8)-structure on J (i. e. a point of $\mathscr{A}_{(4,8)}$ lying over J in $\mathscr{A}_1$), (b) a double covering $\Pi : \hat{C} \to C$, (c) a (4,8)-structure on P (cf. [5], pp. 104-108 for a precise discussion of such " non-abelian levels "). Thus, if we let $\mathscr{A}$'s (resp. $\mathscr{B}$'s) represent moduli spaces for abelian varieties of dim. g (resp. dim. $g - 1$), we have morphisms:

$$\begin{array}{ccc} & & \mathscr{A}_{4,8} \\ & \overset{t_*}{\nearrow} & \\ \mathscr{M}^* & & \\ & \underset{s_*}{\searrow} & \\ & & \mathscr{B}_{4,8} \end{array}$$

and since θ_J on $\frac{1}{2}T(J)$ $\left(\text{resp. } \theta_P \text{ on } \frac{1}{2}T(P)\right)$ are coordinates on $\mathscr{A}_{4,8}$ (resp. $\mathscr{B}_{4,8}$), the identities (*) define a locus $\mathscr{C} \subset \mathscr{A}_{4,8} \times \mathscr{B}_{4,8}$ (the η and φ must be independent of the curve you start with). We find:

THEOREM. — Im (t_*, s_*) is an open subset of one of the components of locus $\mathscr{C}$ of solutions of the Schottky-Jung identites (*) inside the moduli space $\mathscr{A}_{4,8} \times \mathscr{B}_{4,8}$.

REFERENCES

[1] A. Andreotti. — On a theorem of Torelli, *Am. J. Math.*, 80 (1958).

[2] — and A. Mayer. — On period relations for abelian integrals on algebraic curves, *Annali di Scuolo Norm. di Pisa* (1967).

[3] M. Artin. — The implicit function theorem in algebraic geometry, in *Algebraic Geometry*, Oxford Univ. Press (1969).

[4] W. Baily. — On the moduli of abelian varieties with multiplications, *J. Math. Soc. Japan*, 15 (1963).

[5] P. Deligne and D. Mumford. — The irreducibility of the space of curves of given genus, *Pub. I. H. E. S.*, 36 (1969).

[6] H. Farkas and Rauch. — Two kinds of theta constants and period relations on a Riemann Surface, *Proc. Nat. Acad. U. S.*, 62 (1969).

[7] J. Fay. — *Special moduli and theta relations*, Ph. D. Thesis, Harvard (1970).

[8] W. Hoyt. — On products and algebraic families of jacobian varieties, *Annals of Math.*, 77 (1963).

[9] J. I. Igusa. — On the graded ring of theta-constants, *Am. J. Math.*, 86 (1964) and 88 (1966).

[10] T. Matsusaka. — On a theorem of Torelli, *Am. J. Math.*, 80 (1958).

[11] D. Mumford. — *Geometric Invariant theory*, Springer-Verlag, Heidelberg (1965).

[12] —. — On the equations defining abelian varieties, *Inv. Math.*, Part I: vol. 1 (1966), Parts II and III: vol. 3 (1967).

[13] —. — Varieties defined by quadratic equations, in Questioni sulle varieta algebriche, *Corsi dal C. I. M. E.*, 1969, Edizioni Cremonese, Roma.

[14] —. — *Bi-extensions of formal groups*, *in Algebraic Geometry*, Oxford Univ. Press (1969).

[15] —. — *Abelian varieties*, Oxford Univ. Press (1970).

[16] —. — Theta characteristics of an algebraic curve, to appear in *Annales de l'École Normale Supérieure* (1971).

[17] F. Oort. — Finite group schemes, local moduli for abelian varieties and lifting problems, to appear in *Proc. Nordic Summer school of 1970*, Wolters-Noordhoff, Groningen.

[18] B. Riemann. — *Collected works*, Nachtrag IV, Dover edition (1953).

[19] F. Schottky and H. Jung. — Neue Sätze über symmetralfunctionen und die Abelschen funktionen der Riemann'sche theorie, *Sitzungber, Berlin Akad. Wissensch.*, vol. I (1909).

[20] J. P. Serre and J. Tate. — *Proc. of seminar on formal groups at Woods Hole Summer Institute* (1964), mimeographed notes.

[21] G. Shimura. — Moduli of abelian varieties and number theory, in algebraic groups and discontinuous subgroups, *Am. Math. Soc.* (1966).

Harvard University,
Department of Mathematics,
2, Divinity Avenue,
Cambridge, Mass. 02138
(U. S. A.)

COMPOSITIO MATHEMATICA, Vol. 24, Fasc. 2, 1972, pag. 129–174
Wolters-Noordhoff Publishing
Printed in the Netherlands

AN ANALYTIC CONSTRUCTION OF DEGENERATING CURVES OVER COMPLETE LOCAL RINGS

by

David Mumford

This is the first half of a 2 part paper, the first of which deals with the construction of curves and the second with abelian varieties. The idea of investigating the p-adic analogs of classical uniformizations of curves and abelian varieties is due to John Tate. In a very beautiful and influential piece of unpublished work, he showed that if K is a complete non-Archimedean valued field, and E is an elliptic curve over K *whose j-invariant is not an integer*, then E can be analytically uniformized. This uniformization is *not* a holomorphic map:

$$\pi : A^1_K \to E$$

generalizing the universal covering space

$$\pi : \boldsymbol{C} \to E \text{ (= closed points of an elliptic curve over } \boldsymbol{C}),$$

but instead is a holomorphic map:

$$\pi_2 : A^1_K - \{0\} \to E$$

generalizing an infinite cyclic covering π_2 over $\boldsymbol{C}$:

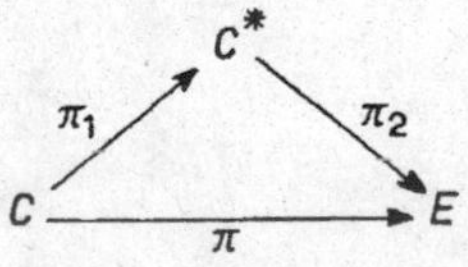

$$\pi_1(z) = e^{2\pi i (z/\omega_1)}$$

ω_1 one of the 2 periods of E. Here you can take holomorphic map to mean holomorphic in the sense of the non-Archimedean function theory of Grauert and Remmert [G-R]. But the uniformization π_2 is more simply expressed by embedding E in $\boldsymbol{P}^2_K$ and defining the three homogeneous coordinates of $\pi(z)$ by three everywhere convergent Laurent series.

The purpose of my work is 2-fold: The first is to generalize Tate's results both to curves of higher genus and to abelian varieties. This gives a very useful tool for investigating the structures at infinity of the moduli spaces. It gives for instance an abstract analog of the Fourier series development of modular forms. Our work here overlaps to some extent with the work of Morikawa [Mo] and McCabe [Mc] generalizing Tate's uniformization to higher-dimensional abelian varieties. The second purpose is to understand the *algebraic* meaning of these uniformizations. For instance, in Tate's example, π defines not only a holomorphic map but also a formal morphism from the Néron model of G_m to the Néron model of E over the ring of integers $A \subset K$. And from an algebraic point of view, it is very unnatural to uniformize only curves over the quotient fields K of complete one-dimensional rings A: one wants to allow A to be a higher-dimensional local ring as well, (this is essential in the applications to moduli for instance). But when dim $A > 1$, there is no longer any satisfactory theory of holomorphic functions and spaces over K.

In this introduction, I would first like to explain (in the case K is a discretely-valued complete local field) what to expect for curves of higher genus. We can do this by carrying a bit further the interesting analogies between the real, complex and p-adic structures of $PGL(2)$ as developed recently by Bruhat, Tits and Serre:

(A) *real case*: $PSL(2, \boldsymbol{R})$ acts *isometrically* and transitively on the upper $\frac{1}{2}$-plane and the boundary can be identified with $\boldsymbol{RP}^1$ (the real line, plus ∞):

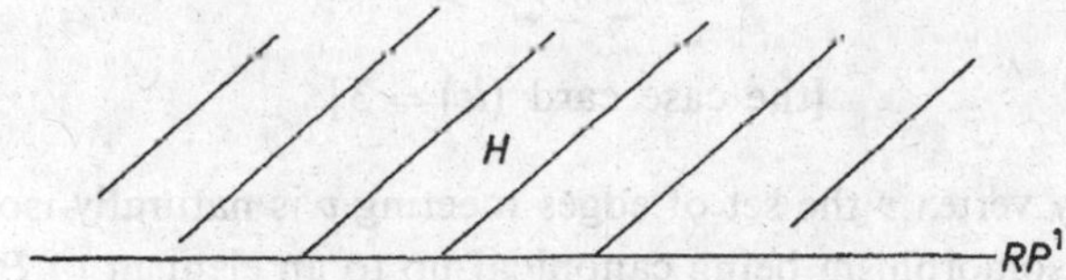

$$z \mapsto \frac{az+b}{cz+d} \qquad ds^2 = \frac{1}{y^2}(dx^2+dy^2)$$

(B) *complex case*: $PGL(2, \boldsymbol{C})$ acts isometrically and transitively on the upper $\frac{1}{2}$-space [1] H' and the boundary can be identified with $\boldsymbol{CP}^1$:

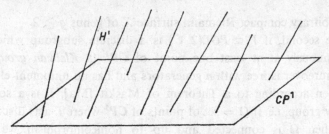

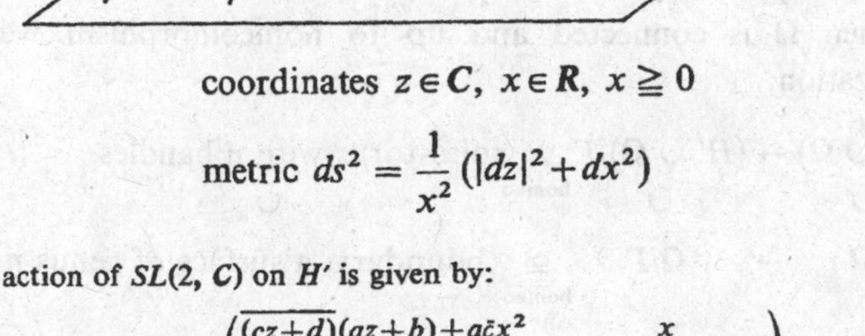

coordinates $z \in \boldsymbol{C}$, $x \in \boldsymbol{R}$, $x \geqq 0$

$$\text{metric } ds^2 = \frac{1}{x^2}(|dz|^2+dx^2)$$

[1] The action of $SL(2, \boldsymbol{C})$ on H' is given by:

$$(z, x) \mapsto \left(\frac{\overline{(cz+d)}(az+b)+a\bar{c}x^2}{|cz+d|^2+|c|^2x^2}, \frac{x}{|cz+d|^2+|c|^2x^2}\right)$$

($z \in \boldsymbol{C}$, $x \in \boldsymbol{R}^+$).

(C) *p-adic case*: $PGL(2, K)$ acts isometrically and transitively on the tree Δ of Bruhat-Tits, (whose vertices correspond to the subgroups $gPGL(2, A)g^{-1}$, and whose edges have length 1 and correspond to the subgroups gBg^{-1}, $B = \left\{\begin{pmatrix} a & b \\ c & d \end{pmatrix} | a, b, c, d \in A, c \in m, ad \notin m\right\}$ modulo A^*) and the set of whose ends can be identified with KP^1 (for details, cf. § 1, below and Serre [S]):

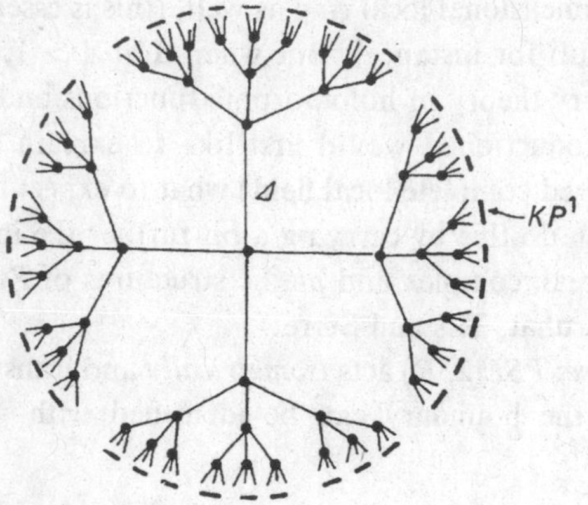

[the case card $(k) = 3$].

In Δ, for any vertex v the set of edges meeting v is naturally isomorphic to kP^1, the isomorphism being canonical up to an element of $PGL(2, k)$ (where $k = A/m$).

In the first case, if $\Gamma \subset PSL(2, \mathbf{R})$ is a discrete subgroup with no elements of finite order such that $PSL(2, \mathbf{R})/\Gamma$ is compact, we obtain Koebe's uniformization

$$H \to H/\Gamma = X$$

of an arbitrary compact Riemann surface X of genus $g \geqq 2$.

In the second, if $\Gamma \subset PGL(2, \mathbf{C})$ is a discrete subgroup which acts discontinuously at at least one point of $\mathbf{C}P^1$ (a *Kleinian group*) and which moreover is free with n generators and has no unipotent elements in it, then according to a Theorem of Maskit [Ma], Γ is a so-called Schottky group, i.e. if Ω = set of points of $\mathbf{C}P^1$ where Γ acts discontinuously, then Ω is connected and up to homeomorphism we get a uniformization:

$$\begin{array}{ccccc} (H' \cup \Omega) & \to & (H' \cup \Omega)/\Gamma & \underset{\text{homeo}}{\cong} & \text{solid torus with } n \text{ handles} \\ \cup & & \cup & & \cup \\ \Omega & \xrightarrow{\pi} & \Omega/\Gamma & \underset{\text{homeo}}{\cong} & \{\text{boundary, a surface of genus } n\} \end{array}$$

In particular Ω/Γ is a compact Riemann surface of genus n and for a suitable standard basis $a_1, \cdots, a_n, b_1, \cdots, b_n$ of $\pi_1(\Omega/\Gamma)$, π is the partial covering corresponding to the subgroup

$$N \subset \pi_1(\Omega/\Gamma)$$
$$N = \text{least normal subgroup containing } a_1, \cdots, a_n.$$

The uniformization π is what is called in the classical literature the Schottky uniformization. It is the one which has a p-adic analog.

In the third case, let $\Gamma \subset PGL(2, K)$ be any discrete subgroup consisting entirely of hyperbolic elements [2]. Then Ihara [I] proved that Γ is free: let Γ have n generators. Again, let Ω = set of closed points of $\boldsymbol{P}^1_K$ where Γ acts discontinuously (equivalently, Ω is the set of points which are not limits of fixed points of elements of Γ).

Then I claim that there is a curve C of genus n and a holomorphic isomorphism:

$$\pi : \Omega/\Gamma \approxeq C.$$

Moreover Δ/Γ has a very nice interpretation as a graph of the specialization of C over the ring A. In fact

a) there will be a smallest subgraph $(\Delta/\Gamma)_0 \subset \Delta/\Gamma$ such that

$$\pi_1((\Delta/\Gamma)_0) \simeq \pi_1(\Delta/\Gamma) \text{ and } (\Delta/\Gamma)_0$$

will be finite:

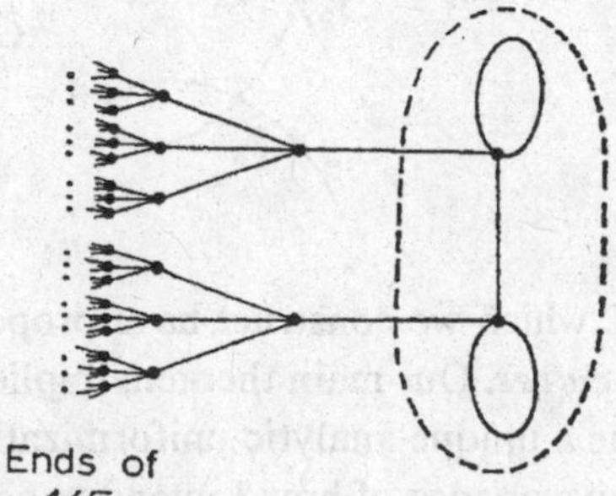

b) C will have a canonical specialization $\bar{C}$ over A, where $\bar{C}$ is a singular curve of arithmetic genus n made up from copies of $\boldsymbol{P}^1_K$ with a finite number of distinct pairs of k-rational points identified to form ordinary double points. Such a curve $\bar{C}$ will be called a *k-split degenerate curve of genus n.*

c) $C(K)$, the set of K-rational points of C, will be naturally isomorphic to the set of ends of Δ/Γ; $\bar{C}(k)$, the set of k-rational points of $\bar{C}$, will be naturally isomorphic to the set of edges of Δ/Γ that meet vertices of

[2] If K is locally compact, this is equivalent to asking simply that Γ has no elements of finite order, since suitable powers of a non-hyperbolic element not of finite order must converge to the identity.

$(\Delta/\Gamma)_0$ (so that the components of $\bar{C}$ correspond to the edges of Δ/Γ meeting a fixed vertex of $(\Delta/\Gamma)_0$ and the double points of $\bar{C}$ correspond to the edges of $(\Delta/\Gamma)_0$); and finally the specialization map

$$C(K) \to \bar{C}(k)$$

is equal, under the above identifications, to the map:

$$(\text{Ends of } \Delta/\Gamma) \to \begin{pmatrix} \text{edges of } \Delta/\Gamma \\ \text{meeting } (\Delta/\Gamma)_0 \end{pmatrix}$$

which takes an end to the last edge in the shortest path from that end to $(\Delta/\Gamma)_0$.

Examples. We have illustrated a case where the genus is 2, $\bar{C}$ has 2 components, each with one double point and meeting each other once:

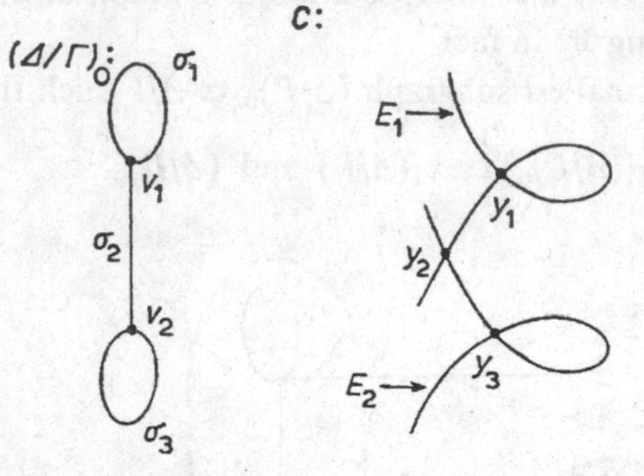

$E_i \leftrightarrow v_i$

$y_i \leftrightarrow \sigma_i$

$\bar{C} = E_1 \cup E_2$

Because all the curves C which we construct have property (b), we refer to them as *degenerating curves*. Our main theorem implies that every such degenerating curve C has a unique analytic uniformization $\pi : \Omega/\Gamma \simeq C$.

Next I would like to give an idea of how I intend to construct *algebraic* objects which imply the existence of the analytic uniformization π, which express the way the analytic map specializes over A, and which will generalize to the case $\dim A > 1$. Given a discretely-valued complete local field K, one has:

a) the category of holomorphic spaces X over K, in the sense e.g. of Grauert and Remmert [G-R],

b) the category of formal schemes $\mathfrak{X}$ over A, locally of topological finite type over A, with $m\mathcal{O}_{\mathfrak{X}}$ as a defining sheaf of ideals.

There is a functor:

$$\mathfrak{X} \to \mathfrak{X}_{an}$$

from the category of formal schemes to the category of holomorphic spaces given as follows:

i) as a point set $\mathscr{X}_{an} \cong$ set of reduced irreducible formal subschemes $Z \subset \mathscr{X}$ such that Z is finite over A, but $Z \not\subset$ the closed fibre $\mathscr{X}_0$;

ii) if $\emptyset : \mathscr{X}_{an} \to \text{Max}(\mathscr{X}) =$ (closed points of $\mathscr{X}$) is the specialization map $Z \mapsto Z \cap \mathscr{X}_0$, then for all $U \subset \mathscr{X}$ affine open, $\emptyset^{-1}(\text{Max}\, U) = V$ is an affinoid subdomain of $\mathscr{X}_{an}$ with affinoid ring $\Gamma(U, \mathcal{O}_{\mathscr{X}})$.

According to results of Raynaud, the category of holomorphic spaces looks like a kind of localization of the category of formal schemes with respect to blowings-up of subschemes concentrated in the closed fibre. In fact, he has proven that the category of holomorphic spaces admitting a finite covering by affinoids is equivalent to this localization of the category of formal schemes of finite type. What happens in our concrete situation is that the holomorphic spaces Ω and C both have *canonical* liftings into the category of formal schemes and the analytic map $\pi : \Omega \to C$ is induced by a formal morphism. For the uniformization of abelian varieties, discussed in the 2nd paper of this series, the lifting turns out not to be canonical; however, a whole class of such liftings can be singled out, which is non-empty and for which π lifts too. Thus the whole situation is lifted into the category of formal schemes where it can be generalized to higher-dimensional base rings A. Let me illustrate this lifting in Tate's original case of an elliptic curve. First of all, what formal scheme over A gives rise to the holomorphic space $A_K^1 - \{0\} = G_{m,K}$? If we take the formal completion of the algebraic group G_m over Spec (A), the holomorphic space that we get is only the unit circle:

$$T \subset G_{m,K}$$
$$T = \{z \mid |z| = 1\}.$$

If we take formal completion of Raynaud's 'Néron model' of G_m over Spec (A) (cf. [R]), we get the subgroup:

$$\bigcup_{n=-\infty}^{+\infty} \pi^n T \subset G_{m,K}$$
$$(\pi) = \text{max. ideal of } A.$$

To get the full $G_{m,K}$ start with $P^1 \times \text{Spec}\,(A)$. Blow up (0), (∞) in the closed fibre P_K^1; then blow up again the points where the 0-section and ∞-section meet the closed fibre; repeat infinitely often. (See figure on next page.)

The result is a scheme P_∞, only locally of finite type over Spec (A). If we omit the double points of the closed fibre, we get Raynaud's 'Néron model'. However, if we take the whole affair, the holomorphic space associated to its formal completion is $G_{m,K}$. On the other hand, the Néron model of E over Spec (A) will have a canonical 'compactifica-

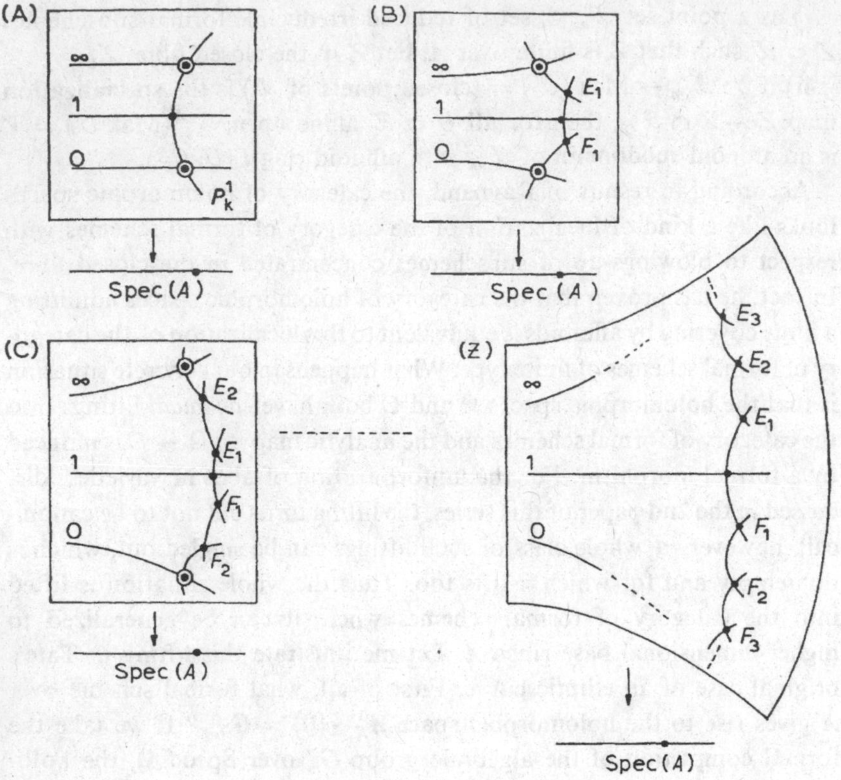

tion': it can be embedded in a unique normal scheme $\mathscr{E}$ proper over Spec (A) by adding a finite set of points. It will look like this (possibly, after replacing K by a suitable quadratic extension):

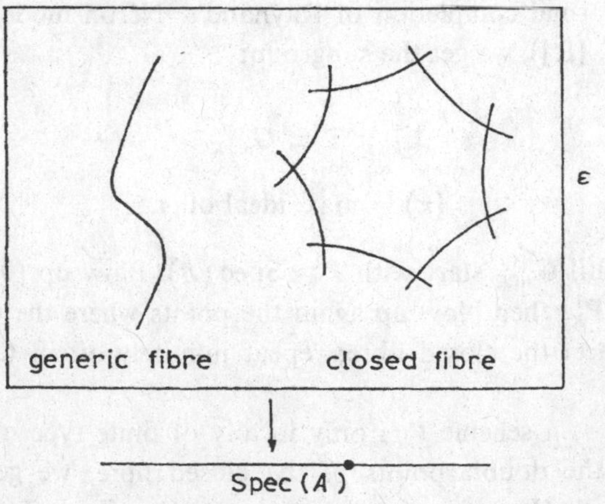

$\mathscr{E}$ will in fact be regular. Finally, the analytic uniformization which we denoted π_2 will come from a formal étale morphism from P_∞ to $\mathscr{E}$

which simply wraps the infinite chain in the closed fibre of P_∞ around and around the polygon which is the closed fibre of $\mathscr{C}$.

We can now state fully our main result:

For every Schottky group $\Gamma \subset PGL(2, K)$, there is a canonical formal scheme $\mathscr{P}$ over A on which Γ acts freely and whose associated holomorphic space is the open set $\Omega \subset \boldsymbol{P}^1_K$. There is a one-one correspondence between a) conjugacy classes of Schottky groups Γ, and b) isomorphism classes of curves C over K which are the generic fibres of normal schemes $\mathscr{C}$ over A whose closed fibre $\bar{C}$ is a k-split degenerate curve, set up by requiring that $\mathscr{P}/\Gamma$ is formally isomorphic to $\mathscr{C}$.

Some notation

$\boldsymbol{P}^1_K$ = projective line over K

$K\boldsymbol{P}^1$ = K-rational points of $\boldsymbol{P}^1_K = K+K-(0,0)/K^*$

$\langle u, v\rangle$ = module generated by u, v

$R(X)$ = field of rational functions on an integral scheme X.

$\hat{X}$ = formal completion of a scheme X over a complete local ring A, along its closed fibre.

1. Trees

Let A be a complete integrally closed noetherian local ring, with quotient field K, maximal ideal m and residue field $k = A/m$. Let $S =$ Spec (A), $S_\eta =$ Spec (K) and $S_0 =$ Spec (k):

$$S_\eta \hookrightarrow S \hookleftarrow S_0.$$

We are interested in certain finitely generated subgroups of $PGL(2, K)$ that we will call Schottky groups. First of all define a morphism:

$$t : PGL(2) \to \boldsymbol{A}^1$$

by

$$t\begin{pmatrix} a & b \\ c & d\end{pmatrix} = \frac{(a+d)^2}{ad-bc}$$

Here and below we will describe elements of $PGL(2)$ by 2×2 matrices, considered modulo multiplication by a scalar, without further comment.

Lemma (1.1). *Let $\gamma \in PGL(2, K)$. Then $t^{-1}(\gamma) \in m$ if and only if γ can be represented:*

$$A\cdot\begin{pmatrix} \mu & 0 \\ 0 & 1\end{pmatrix}\cdot A^{-1}, \quad \mu \in m.$$

Proof. On the one hand:

$$t^{-1}\left(A\cdot\begin{pmatrix}\mu & 0\\ 0 & 1\end{pmatrix}\cdot A^{-1}\right)=\frac{\mu}{(\mu+1)^2}\in m.$$

Conversely, suppose $t^{-1}(\gamma)=\nu\in m$, and let the matrix C represent γ. Then $\det C=\nu\cdot(\operatorname{Tr}C)^2$ and the characteristic polynomial of C is

$$X^2-(\operatorname{Tr}C)\cdot X+\nu\cdot(\operatorname{Tr}C)^2=0.$$

By Hensel's lemma, this has 2 distinct roots in K of the form

$$X_1=u\cdot\operatorname{Tr}C$$
$$X_2=\nu\cdot u^{-1}\cdot\operatorname{Tr}C$$

where u = a unit in A.

Then γ is also represented by the matrix $C'=C/u\cdot\operatorname{Tr}C$ with eigenvalues 1 and $\nu/u^2\in m$. Therefore C' has the required form.

Q.E.D.

DEFINITION (1.2). The elements $\gamma\in PGL(2,K)$ such that $t^{-1}(\gamma)\in m$ will be called *hyperbolic*.

From the lemma it follows immediately that if γ is hyperbolic, then γ as an automorphism of $\boldsymbol{P}_K^1$ has 2 distinct fixed points P and Q, both rational over K, and such that the differential $d\gamma|_P$ = mult. by μ in T_P (the tangent space to $\boldsymbol{P}_K^1$ at P), $\mu\in m$, while $d\gamma|_Q$ = mult. by μ^{-1}; P is called the *attractive fixed point* of γ and Q the *repulsive fixed point*.

DEFINITION (1.3). A *Schottky group* $\Gamma\subset PGL(2,K)$ is a finitely generated subgroup such that every $\gamma\in\Gamma$, $\gamma\neq e$, is hyperbolic.

These are probably the most natural class of groups to look at. However, there is a particular type which is easier to prove theorems about and which include *all* Schottky groups in the case $\dim A=1$:

DEFINITION (1.4). *A flat Schottky group* $\Gamma\subset PGL(2,K)$ has the extra property that if $\Sigma\subset K\boldsymbol{P}^1$ is the set of fixed points of the elements $\gamma\in\Gamma$, then for any $P_1,P_2,P_3,P_4\in\Sigma$, $R(P_1,P_2;P_3,P_4)$ or its inverse is in A, i.e. the cross-ratio R defines a morphism from S to $\boldsymbol{P}^1$.

The construction of flat Schottky groups is not so easy and we postpone this until § 4. For the time being, we simply assume that one is given.

The structure of $PGL(2,K)$ and of Γ is best displayed, following the method of Bruhat and Tits [B-T] by introducing:

$\Delta^{(0)}\cong$ {set of sub A-modules $M\subset K+K$, M free of rank 2, modulo the identification $M\sim\lambda\cdot M$, $\lambda\in K^*$, (the image $\{M\}$ in $\Delta^{(0)}$ of a module M will be called the *class* of M)}

$\cong$ {set of schemes P/S with generic fibre $\boldsymbol{P}_K^1$, such that $P\cong\boldsymbol{P}_S^1$, modulo isomorphism}.

These sets will be identified by the map

$$M \mapsto P = \text{Proj (Symmetric } A\text{-algebra on Hom } (M, A)).$$

This is easily seen to be a bijection under which the set of A-valued points of P equals the set of elements $x \in M - mM$, modulo A^*. Intuitively, P is the scheme of one-dimensional subspaces of the rank 2 vector bundle M.

DEFINITION (1.5). If $\{M\} \in \Delta^{(0)}$, we denote the corresponding scheme P/S by $\mathbf{P}(M)$.

Note that $PGL(2, K)$ acts on $\Delta^{(0)}$:

$\forall X \in GL(2, K), \forall M \subset K+K$,
let $X(M) = \{X \cdot x | x \in M\}$.
Then the class $\{X(M)\}$ depends only on the image $\{X\}$ of X in $PGL(2, K)$ and on the class $\{M\}$.

The stabilizer of the module $A+A$ is:

$PGL(2, A)$ = elements of $PGL(2, K)$ represented by matrices $\begin{pmatrix} a & b \\ c & d \end{pmatrix}$, $a, b, c, d \in A$, $ad - bc \in A^*$,

and the stabilizers of the other modules M are conjugates of $x \cdot PGL(2, A) \cdot x^{-1}$ in $PGL(2, K)$.

Moreover $PGL(2, K)$ acts transitively on $\Delta^{(0)}$, so $\Delta^{(0)}$ can be naturally identified with the coset space $PGL(2, K)/PGL(2, A)$.

Less obvious is the fact that any 3 distinct points $x_1, x_2, x_3 \in KP^1$ determined canonically an element of $\Delta^{(0)}$: let $w_1, w_2, w_3 \in K+K$ be homogeneous coordinates for x_1, x_2, x_3. Then there is a linear equation: $a_1 w_1 + a_2 w_2 + a_3 w_3 = 0$, unique up to scalar. Let $M = \sum_{i=1}^{3} A \cdot a_i w_i$. The class of multiples $\{M\}$ of M is determined by the x_i alone. We will write this class as $\{M(x_1, x_2, x_3)\}$.

Unlike the case where dim $A = 1$, the full set $\Delta^{(0)}$ is rather unmanageable. We need to introduce the concept:

DEFINITION (1.6). $\{M_1\}, \{M_2\} \in \Delta^{(0)}$ are *compatible* if there exists a basis u, v of M_1, and elements $\lambda \in K^*$, $\alpha \in A$ such that $\lambda u, \lambda\alpha v$ is a basis of M_2, (M_i representatives of $\{M_i\}$).

It is easy to check that this definition is symmetric and that the principal ideal (α) is uniquely determined by $\{M_1\}$ and $\{M_2\}$. Since (α) measures the 'distance' of $\{M_1\}$ from $\{M_2\}$, we write:

$$(\alpha) = \rho(\{M_1\}, \{M_2\}).$$

Moreover, when dim $A = 1$, every pair $\{M_1\}, \{M_2\}$ is compatible. If M_i' are representatives of the classes $\{M_i\}$ such that

$$M'_1 \supset M'_2 \supset \alpha \cdot M'_1,$$

we call M'_1, M'_2 *representatives in standard position.*

Now then, let

$$\Delta_\Gamma^{(0)} = \text{set of classes } \{M(x_1, x_2, x_3)\}, \text{ where } x_1, x_2, x_3 \in \Sigma,$$
$$\Sigma = \text{set of fixed points of elements of } \Gamma.$$

The flatness of the Schottky group Γ is obviously equivalent to the property:

$$* \begin{cases} \forall x_1, x_2, x_3, x_4 \in \Sigma, \text{ these points have homogeneous coordinates} \\ w_1, w_2, w_3, w_4 \in K+K \text{ such that} \\ \quad w_3 = w_1 + w_2 \\ \quad w_4 = a_1 w_1 + a_2 w_2, \quad a_i \in A \\ \text{and either } a_1 \text{ or } a_2 \text{ is not in } m. \end{cases}$$

This now gives us:

PROPOSITION (1.7). *Any 2 classes* $\{M_1\}, \{M_2\} \in \Delta_\Gamma^{(0)}$ *are compatible.*

PROOF. First note the

LEMMA (1.8). *If* $x_1, x_2, x_3, x_4 \in P_K^1$ *have property* $(*)$, *then for some* $i, j \in \{1, 2, 3\}$, *with* $i \neq j$, $\{M(x_1, x_2, x_3)\} = \{M(x_i, x_j, x_4)\}$.

PROOF. Let w_i be coordinates for x_i as in $(*)$. Then if a_1 and $a_2 \notin m$, one checks that $M(x_1, x_2, x_3) = M(x_1, x_2, x_4)$; if $a_1 \notin m$, $a_2 \in m$, then $M(x_1, x_2, x_3) = M(x_2, x_3, x_4)$; and if $a_1 \in m$, $a_2 \notin m$, then

$$M(x_1, x_2, x_3) = M(x_1, x_3, x_4).$$

Now let $\{M_1\} = M(x_1, x_2, x_3)$, $\{M_2\} = M(y_1, y_2, y_3)$. Choose coordinates w_i for x_i and u_i for y_i such that

a) $M_1 = A \cdot w_1 + A \cdot w_2$, $w_3 = w_1 + w_2$,

b) $u_i = a_i w_1 + b_i w_2$, where $a_i, b_i \in A$ but if $a_i \in m$, then $b_i \notin m$.

Next, if the ratios $a_i : b_i \bmod m$ in kP^1 are all distinct, one checks immediately that the u_i are related by $\lambda_1 u_1 + \lambda_2 u_2 + \lambda_3 u_3 = 0$ where $\lambda_i \in A$, $\lambda_i \notin m$. This implies that

$$M_2 = (\text{module generated by the } u_i) = M_1,$$

hence M_1 and M_2 are obviously compatible. Ncw if the ratios $a_i : b_i \bmod m$ are not all distinct, then at least one of the triples $(0:1)$, $(1:0)$, $(1:1)$ is different from all three ratios $a_i : b_i \bmod m$. Permuting the three w_i's, we may as well assume that $(1:0)$ does not cocur, i.e. $b_i \notin m$ for all i. Multiplying u_i by a unit, we can normalize it so that now:

b′) $u_i = a_i w_1 + w_2$.

Now by the lemma, $\{M_2\} = \{M(y_i, y_j, x_1)\}$ for some i and j. The linear equation relating u_i, u_j and w_1 is:

$$u_i - a_i w_1 = u_j - a_j w_1.$$

Therefore $M(y_i, y_j, x_1)$ is the module generated by u_i, u_j and $(a_i - a_j)w_1$. Thus u_i and w_1 are a basis of M_1 and u_i and $(a_i - a_j)w_1$ are a basis of M_2. Hence $\{M_1\}$ and $\{M_2\}$ are compatible.

Q.E.D.

I claim that for any 3 compatible classes of modules, there is a multiplicative triangle 'inequality' relating their 'distances' from each other:

PROPOSITION (1.9). *Let $\{M_1\}, \{M_2\}, \{M_3\} \in \Delta^{(0)}$ be distinct but compatible. Let $(\alpha_{ij}) = \rho(\{M_i\}, \{M_j\})$ and let*

$$M_1 \supset M_2 \supset \alpha_{12} M_1$$
$$M_1 \supset M_3 \supset \alpha_{13} M_1$$

be representatives in standard position. Then if $N = M_2 + M_3$, there exist $u, v \in M_1$ and $\lambda_1, \lambda_2, \lambda_3 \in A$ such that:

$$M_1 = \langle u, v \rangle$$
$$N = \langle u, \lambda_1 v \rangle$$
$$M_2 = \langle u, \lambda_1 \lambda_2 v \rangle$$
$$M_3 = \langle u + \lambda_1 v, \lambda_1 \lambda_3 v \rangle$$
$$(\alpha_{ij}) = (\lambda_i \lambda_j).$$

In particular, for all permutations i, j, k of 1, 2, 3, $\alpha_{ij} | \alpha_{ik} \alpha_{jk}$.

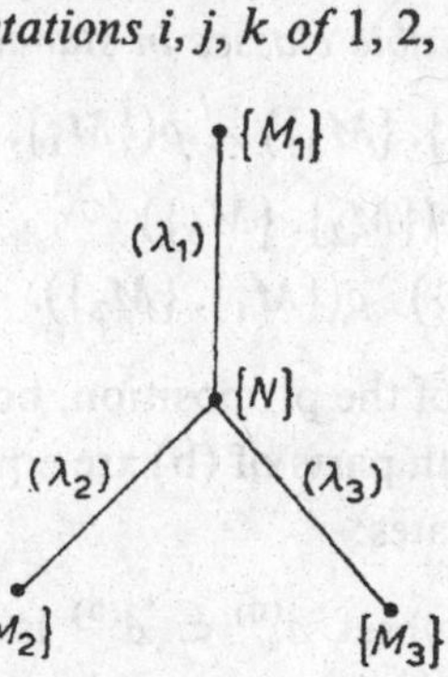

CLUMSY PROOF. First choose $u \in M_2$ such that $u \notin m \cdot M_1$. Secondly choose $\bar{v} \in M_1/mM_1$ such that $\bar{v}$ is not in either of the 1-dimensional subspaces $M_2/M_2 \cap mM_1$ or $M_3/M_3 \cap mM_1$ of M_1/mM_1. Lift $\bar{v}$ to

$v \in M_1$. Then first of all $\bar{u}$ and $\bar{v}$ generate M_1/mM_1, hence u and v generate M_1. Secondly u and $\alpha_{12}v$ lie in M_2 and since M_1 and M_2 are in standard position, it is easy to see that they must generate M_2. Thirdly, for some $\lambda_1 \in A$, $u+\lambda_1 v \in M_3$. Then since M_1 and M_3 are in standard position, $u+\lambda_1 v$ and $\alpha_{13}v$ must generate M_3. Now use the fact that M_2 and M_3 are compatible: for some $\xi \in K^*$, $M_2 \supset \xi \cdot M_3$ and this pair is in standard position. Then

$$\xi \cdot (u+\lambda_1 v) \in M_2$$
$$\xi \cdot \alpha_{13} v \in M_2$$

and one of these is not in $m \cdot M_2$. This implies that $\xi \in A$, $\xi\lambda_1 = \zeta \cdot \alpha_{12}$, $\xi\alpha_{13} = \eta \cdot \alpha_{12}$, (where ζ and $\eta \in A$), and furthermore that either ξ, ζ or η is a unit. Firstly, suppose ξ is a unit but ζ is not. Note that we may replace u by $u' = u+\alpha_{12}v$ then u' and $\alpha_{12}v$ still generate M_2 and $u'+\lambda_1' v$ and $\alpha_{13}v$ generate M_3 where $\lambda_1' = \lambda_1 - \alpha_{12}$. But then $\xi\lambda_1' = \zeta'\alpha_{12}$, where $\zeta' = \zeta - 1$ is a unit. Therefore by suitable choice of u, we can assume that ζ is a unit. Secondly, suppose η is a unit but ζ is not. In this case, note that $u+\lambda_1' v$ and $\alpha_{13}v$ still generate M_3 where $\lambda_1' = \lambda_1 + \alpha_{13}$. And then $\xi\lambda_1' = \xi\lambda_1 + \xi\alpha_{13} = (\zeta+\eta)\alpha_{12} = \zeta'\alpha_{12}$ where ζ' is a unit. Thus we can always assume that ζ is a unit. Then if $\lambda_2 = \xi \cdot \zeta^{-1}$ and $\lambda_3 = \eta \cdot \zeta^{-1}$ it follows that $\alpha_{12} = \lambda_1\lambda_2$ and $\alpha_{13} = \lambda_1\lambda_3$ hence M_2 and M_3 are generated as required. It follows immediately that M_2+M_3 is generated by u and $\lambda_1 v$. To evaluate α_{23}, note that the 2 modules $M_2 \supset \lambda_2 M_3$ are in standard position (since $\lambda_2 u+\lambda_1\lambda_2 v$ is in $\lambda_2 M_3$ but not in mM_2) and that $\lambda_2 M_3$ is generated by $\lambda_2 u+\lambda_1\lambda_2 v$ and by $\lambda_2\lambda_3 u$ hence $(\lambda_2\lambda_3) = \rho(\{M_2\}, \{M_3\})$.

Q.E.D.

COROLLARY (1.10). *If* $M_1 \supset M_2 \supset \alpha_{12}M_1$, $M_1 \supset M_3 \supset \alpha_{13}M_1$ *are representatives of* 3 *compatible classes in standard position, then*

a) $M_2 \supset M_3 \Leftrightarrow \rho(\{M_1\}, \{M_3\}) = \rho(\{M_1\}, \{M_2\}) \cdot \rho(\{M_2\}, \{M_3\})$,

b) $M_1 = M_2+M_3 \Leftrightarrow \rho(\{M_2\}, \{M_3\})$
$= \rho(\{M_2\}, \{M_1\}) \cdot \rho(\{M_1\}, \{M_3\})$.

PROOF. In the notation of the proposition, both parts of (a) are equivalent to λ_2 being a unit; both parts of (b) are equivalent to λ_1 being a unit.

This Proposition motivates:

DEFINITION (1.11). A subset $\Delta_*^{(0)} \subset \Delta^{(0)}$ is *linked* if a) every pair of elements $\{M_1\}, \{M_2\} \in \Delta_*^{(0)}$ is compatible, b) for every triple $\{M_1\}$, $\{M_2\}, \{M_3\} \in \Delta_*^{(0)}$, if we pick representatives $M_1 \supset M_2$, $M_1 \supset M_3$ in standard position, then M_2+M_3 (which is a free A-module by the proposition) defines a class $\{M_2+M_3\}$ in $\Delta_*^{(0)}$.

We must check that $\Delta_\Gamma^{(0)}$ has both these fine properties:

Theorem (1.12). *$\Delta_\Gamma^{(0)}$ is linked.*

Proof. Suppose $\{M_i\} = \{M(x_i, y_i, z_i)\}$, $i = 1, 2, 3$, where all these points come from Σ. We saw above that all these classes are compatible. Choose representatives $M_1 \supset M_2, M_1 \supset M_3$ in standard position, and choose homogeneous coordinates $u_i, v_i, w_i \in M_i$ for x_i, y_i and z_i such that the linear relation $\alpha_i u_i + \beta_i v_i + \gamma_i w_i = 0$ has the property $\alpha_i, \beta_i, \gamma_i \in A^*$. Since $M_2 \not\subset mM_1$, one of u_2, v_2, or w_2 is in the set $M_1 - mM_1$. Renaming, we can assume $u_2 \notin mM_1$. Similarly, we can assume $u_3 \notin mM_1$. Next, the images $\bar{u}_1, \bar{v}_1, \bar{w}_1$ of u_1, v_1, w_1 in M_1/mM_1 are all distinct, so one of them is different from both $\bar{u}_2$ and $\bar{u}_3$. Renaming, we can assume $\bar{u}_1 \neq \bar{u}_2$ or $\bar{u}_3$. Let us construct a module in the class $\{M(x_1, x_2, x_3)\} \in \Delta_\Gamma^{(0)}$. We must find the linear equation relating u_1, u_2, u_3: since $\bar{u}_1, \bar{u}_2, \bar{u}_3 \in M_1/mM_1$ are related by an equation $\bar{\alpha}\bar{u}_1 + \bar{\beta}\bar{u}_2 + \bar{u}_3 = 0$, where $\bar{\alpha}, \bar{\beta} \in A/m$, and $\bar{\beta} \neq 0$, it follows that u_1, u_2, u_3 are related by an equation $\alpha u_1 + \beta u_2 + u_3 = 0$, where $\alpha, \beta \in A$, $\beta \notin m$. Therefore

$$\langle u_2, u_3 \rangle \in \{M(x_1, x_2, x_3)\}.$$

On the other hand, if we choose generators $u, v \in M_1$ as in the previous Proposition, it follows that

$$u_2 = \sigma_2 u + \tau_2(\lambda_1 \lambda_2 v), \qquad \sigma_2, \tau_2 \in A, \sigma \notin m$$
$$u_3 = \sigma_3(u + \lambda_1 v) + \tau_3(\lambda_1 \lambda_3 v), \qquad \sigma_3, \tau_3 \in A, \sigma_3 \notin m.$$

If $\lambda_2, \lambda_3 \in m$, then since $M_2 + M_3 = \langle u, \lambda_1 v \rangle$, it follows that u_2 and u_3 have distinct images $\bar{u}_2, \bar{u}_3 \in M_2 + M_3/m \cdot (M_2 + M_3)$. Therefore $M_2 + M_3 = \langle u_2, u_3 \rangle$ whose class is in $\Delta_\Gamma^{(0)}$. If either λ_2 or λ_3 is in A^*, then $M_2 \supset M_3$ or $M_3 \supset M_2$ and $M_2 + M_3$ equals either M_2 or M_3, whose class is in $\Delta_\Gamma^{(0)}$. *Q.E.D.*

Linked subsets $\Delta_*^{(0)} \subset \Delta^{(0)}$ are very nice objects. They can be fitted together in a natural way into a *tree*.

Tree theorem (1.13). *If $\Delta_*^{(0)}$ is a linked subset of $\Delta^{(0)}$, then $\Delta_*^{(0)}$ is the set of a vertices of a connected tree Δ_* in which a principal ideal (α_σ) is associated to each edge σ and such that for every pair of classes $\{P\}, \{Q\} \in \Delta_*^{(0)}$, if they are linked in the tree as follows:*

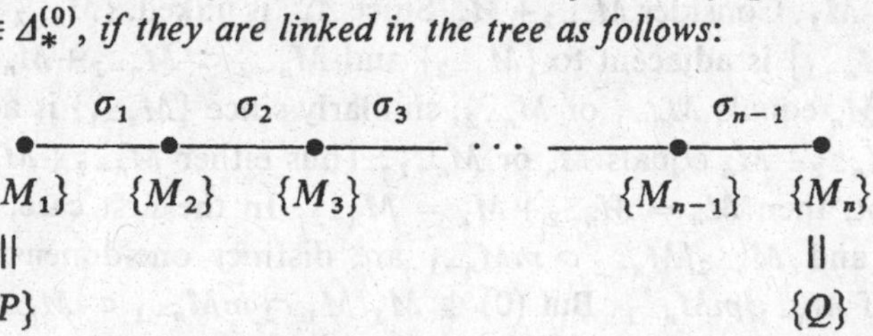

then

$$(*)\qquad \rho(\{P\},\{Q\}) = \prod_{i=1}^{n-1}(\alpha_{\sigma_i}).$$

Proof. First, let us call $\{P\}, \{Q\} \in \Delta_*^{(0)}$ *adjacent* if there is no $\{R\} \in \Delta_*^{(0)}$ such that:

$$\rho(\{P\},\{Q\}) = \rho(\{P\},\{R\})\cdot\rho(\{R\},\{Q\}),$$
$$\{R\} \neq \{P\} \text{ or } \{Q\}.$$

Join 2 adjacent classes by an edge σ and set $(\alpha_\sigma) = \rho(\{P\},\{Q\})$. This gives us a graph in any case. Starting now with any $\{P\}, \{Q\}$, consider all sequences

$$\{P\} = \{M_1\}, \{M_2\}, \cdots, \{M_n\} = \{Q\} \quad \text{in } \Delta_*^{(0)}$$

such that

$$\rho(\{P\},\{Q\}) = \prod_{i=1}^{n-1}\rho(\{M_i\},\{M_{i+1}\}).$$

By the noetherian assumption on A, there is a maximal sequence of this type. Then each pair $\{M_i\}, \{M_{i+1}\}$ must be adjacent and this proves that $\{P\}$ and $\{Q\}$ are joined in our graph by a sequence of edges. Therefore the graph is connected.

To prove that our graph is a tree and to prove $(*)$, it suffices, by an obvious induction, to prove:

Lemma (1.14). *Let $\{M_1\}, \{M_2\}, \cdots, \{M_n\} \in \Delta_*^{(0)}$ such that $\{M_i\}$, and $\{M_{i+1}\}$ are adjacent $(1 \leqq i \leqq n-1)$. Assume*

$$\rho(\{M_1\},\{M_{n-1}\}) = \prod_{i=1}^{n-2}\rho(\{M_i\},\{M_{i+1}\}).$$

Then either

$$\rho(\{M_1\},\{M_n\}) = \prod_{i=1}^{n-1}\rho(\{M_i\},\{M_{i+1}\})$$

or

$$\{M_n\} = \{M_{n-2}\}.$$

Proof of lemma. Let $M_{n-1} \supset M_n$, $M_{n-1} \supset M_{n-2}$, $M_{n-1} \supset M_1$ be representatives in standard position. By the Corollary (1.10) $M_{n-1} \supset M_{n-2} \supset M_1$. Consider $M_{n-2}+M_n$. Since $\Delta_*^{(0)}$ is linked, $\{M_{n-2}+M_n\} \in T$. Since $\{M_{n-1}\}$ is adjacent to $\{M_{n-2}\}$ and $M_{n-2} \subset M_{n-2}+M_n \subset M_{n-1}$, $M_{n-2}+M_n$ equals M_{n-1} or M_{n-2}; similarly since $\{M_{n-1}\}$ is adjacent to $\{M_n\}$, $M_{n-2}+M_n$ equals M_n or M_{n-1}. Thus either $M_{n-2}+M_n = M_{n-1}$ or, if not, then $M_n = M_{n-2}+M_2 = M_{n-2}$. In the first case, $M_n/M_n \cap mM_{n-1}$ and $M_{n-2}/M_{n-2} \cap mM_{n-1}$ are distinct one-dimensional subspaces of M_{n-1}/mM_{n-1}. But $(0) \subsetneqq M_1/M_1 \cap mM_{n-1} \subset M_{n-2}/M_{n-2} \cap$

mM_{n-1}, so $M_1/M_1 \cap mM_{n-1}$ must be the same subspace as $M_{n-2}/M_{n-2} \cap m \cdot M_{n-1}$. Therefore M_1 and M_n together generate M_{n-1}/mM_{n-1}, hence they generate M_{n-1}. By Cor. (1.10) applied to the triple M_1, M_{n-1}, M_n this means that

$$\begin{aligned} \rho(\{M_1\}, \{M_n\}) &= \rho(\{M_1\}, \{M_{n-1}\}) \cdot \rho(\{M_{n-1}\}, \{M_n\}) \\ &= \prod_{i=1}^{n-1} \rho(\{M_i\}, \{M_{i+1}\}). \end{aligned}$$ *Q.E.D.*

COROLLARY (1.15). *In a canonical way, $\Delta_\Gamma^{(0)}$ is the set of vertices of a tree Δ_Γ on which Γ acts.*

COROLLARY (1.16) (Ihara). *Γ is a free group.*

PROOF. I claim that Γ acts freely on Δ_Γ. In fact, if $\gamma \in \Gamma$, $\gamma \neq e$, has a fixed point P, then P is either a vertex or the midpoint of an edge. In the latter case, γ^2 fixes the 2 endpoints of this edge. But the stabilizers in $PGL(2, K)$ of the elements of $\Delta^{(0)}$ are the various subgroups

$$xPGL(2, A)x^{-1} \subset PGL(2, K).$$

Since every $\gamma \in \Gamma$ is hyperbolic, neither γ nor γ^2 can belong to any such subgroup. Thus Γ acts freely on a tree, hence Γ itself must be free. *Q.E.D.*

COROLLARY (1.17) (Bruhat-Tits). *If* $\dim A = 1$, *the whole of $\Delta^{(0)}$ is, in a canonical way, the set of vertices of a tree Δ on which the whole group $PGL(2, K)$ acts.*

It can be shown further when $\dim A = 1$ that for all $\gamma \in PGL(2, K)$, either γ is hyperbolic and has no fixed point on Δ; or γ is not hyperbolic and γ has a fixed point on Δ in which case then γ^2 is in some subgroup $g \cdot PGL(2, A) \cdot g^{-1}$.

For any linked subset $\Delta_*^{(0)} \subset \Delta^{(0)}$, let Δ_* be the associated tree. We can add a boundary to Δ_* that has an interesting interpretation: let

$$\partial\Delta_* = \text{the set of ends of } \Delta_*$$

[Where an *end* is an equivalence class of subtrees of Δ_* isomorphic to:

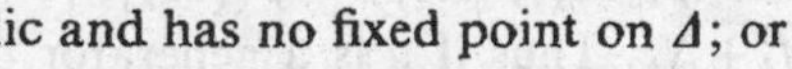

two such being 'equivalent' if they differ only in a finite set of vertices]. Let $\bar{\Delta}_* = \Delta_* \cup \partial\Delta_*$: this is a topological space if an open set is a subset $U \cup V$, where $U \subset \Delta_*$ is open and $V \subset \partial\Delta_*$ is the set of ends represented by subtrees in U.

PROPOSITION (1.18). a) *There is a natural injection*

$$i : \partial\Delta_* \hookrightarrow KP^1.$$

b) *If* $\Delta_* = \Delta_\Gamma$, $\Sigma \subset i(\partial\Delta_\Gamma)$.

c) *If* $\dim A = 1$, $\Delta_* = \Delta$, *then i is a bijection of* $\partial\Delta$ *and* KP^1.

PROOF. Let $\{M_1\}, \{M_2\}, \cdots$ be an infinite sequence of adjacent vertices of Δ_* which defines an end $e \in \partial\Delta_*$. Represent these by modules in standard position:

$$M_1 \supset M_2 \supset M_3 \supset \cdots.$$

Then it is easy to check that $\bigcap_{n=1}^{\infty} M_n$ is a free A-module of rank 1 in $K+K$. If $u \in \cap M_n$ is a generator, u defines the point $i(e) \in KP^1$. Note that $u \notin mM_1$ and if $v \in M_1$ is such that $\bar{u} \neq \bar{v}$ in M_1/mM_1, then

$$M_1 = \langle u, v\rangle, M_2 = \langle u, \alpha_2 v\rangle, M_3 = \langle u, \alpha_3 v\rangle, \cdots,$$

where $(\alpha_n) = \rho(\{M_1\}, \{M_n\})$. Next, let e' be a 2nd end and assume $e \neq e'$. From general properties of trees, it follows that there is a unique subtree of Δ_* isomorphic to:

(we call such a subtree a *line*) defining e at one end, e' at the other. Pick a base point $\{P\}$ on this, and let its vertices in the 2 directions be $\{M_1\}, \{M_2\}, \cdots$ and $\{N_1\}, \{N_2\}, \cdots$. Represent these by modules in standard positions:

$$P \supset M_1 \supset M_2 \supset \cdots$$
$$P \supset N_1 \supset N_2 \supset \cdots$$

By Cor. (1.10), since $\rho(\{M_k\}, \{N_k\}) = \rho(\{M_k\}, \{P\}) \cdot \rho(\{P\}, \{N_k\})$, it follows that $P = M_k + N_k$. Thus if u is a generator of $\cap M_n$ and v is a generator of $\cap N_n$, then $P = \langle u, v\rangle$. In particular $K \cdot u$ and $K \cdot v$ are distinct subspaces of $K+K$. But $K \cdot u$ represents $i(e)$, $K \cdot v$ represents $i(e')$. Therefore $i(e) \neq i(e')$. To prove (b), note that when $\Delta_* = \Delta_\Gamma$, Γ acts on Δ_Γ, on $\partial\Delta_\Gamma$ and on KP^1 and i is Γ-linear. On the other hand, any fixed point free automorphism γ of a tree leaves invariant a unique line (its *axis*) and it acts on its axis by a translation. Thus γ fixes 2 distinct ends of the tree. In our case, every $\gamma \in \Gamma$, $\gamma \neq e$, has therefore 2 fixed points in $\partial\Delta_\Gamma$ hence in $i(\partial\Delta_\Gamma)$, and hence $i(\partial\Delta_\Gamma)$ contains the 2 fixed points of γ in KP^1. To prove (c), let $x \in KP^1$ and let $u \in K+K$ be homogeneous coordinates for x. Let $v \in K+K$ be any vector that is not a multiple of u and let $M_n = \langle u, \pi^n v\rangle$, where $(\pi) = m$, the maximal ideal in A. Then $\{M_n\}_{n \geq 0}$ is a half-line in the full tree Δ whose end is mapped by i to x. *Q.E.D.*

DEFINITION (1.19). $i(\partial\Delta_\Gamma) \subset KP^1$ will be denoted $\bar{\Sigma}$ and called the *limit point set* of Γ.

In fact, when $\dim A = 1$, it is easy to check that $\bar{\Sigma}$ is precisely the closure of Σ in the natural topology on KP^1.

We can link together via the map i several of the ideas we have been working with:

PROPOSITION (1.20). *Let $\Delta_*^{(0)}$ be any linked subset of $\Delta^{(0)}$, and let Δ_* be the corresponding tree. For any $x, y, z \in \partial\Delta_*$, the class $\{M(ix, iy, iz)\}$ is in $\Delta_*^{(0)}$ and equals to unique vertex of Δ_* such that the paths from v to the 3 ends x, y, z all start off on different edges.*

PROOF. In fact, let the module M represent v, and let $M \supset M_x$, $M \supset M_y$, $M \supset M_z$ be representatives in standard positions of M and the module classes next on the path from v to x, y and z respectively. Then ix, iy, iz are represented by homogeneous coordinates $u, v, w \in K+K$ such that

$$\begin{aligned} u &\in M_x - mM, \\ v &\in M_y - mM, \\ w &\in M_z - mM. \end{aligned}$$

Since $\{M\}$ is between $\{M_x\}$ and $\{M_y\}$ in Δ_*, it follows from (1.10) that $M_x + M_y = M$ hence $\bar{u}, \bar{v} \in M/mM$ are distinct. Similarly $\bar{w}$ is distinct from $\bar{u}$ and $\bar{v}$. Therefore, if $\alpha u + \beta v + \gamma w = 0$ is the linear relation on u, v, w, it follows that $\alpha, \beta, \gamma \in A - m$. Therefore

$$\{M(x, y, z)\} = \{A \cdot u + A \cdot v + A \cdot w\} = \{M\} = v.$$

COROLLARY (1.21). *Every vertex of Δ_Γ meets at least 3 distinct edges.*

PROPOSITION (1.22). *Let $\Delta_*^{(0)}$ be any linked subset of $\Delta^{(0)}$ and let Δ_* be the corresponding tree. Then for any $x_1, x_2, x_3, x_4 \in \partial\Delta_*$, $R(ix_1, ix_2, ix_3, ix_4) \in A$ or $R(ix_1, ix_2, ix_3, ix_4)^{-1} \in A$.*

PROOF. Let v be the vertex of Δ_* joined to x_1, x_2, x_3 by paths starting on different edges. Represent v by M and ix_1, ix_2, ix_3 by coordinates $u_1, u_2, u_3 \in M - mM$ as in the proof of the previous Proposition.

Moreover we can represent x_4 by coordinates $u_4 \in M - mM$ too. Then

$$\begin{aligned} u_3 &= \alpha u_1 + \beta u_2, && \alpha, \beta \in A - m \\ u_4 &= \gamma u_1 + \delta u_2, && \gamma, \delta \in A \text{ and } \gamma \text{ or } \delta \notin m. \end{aligned}$$

In this case,

$$R(ix_1, ix_2, ix_3, ix_4) = \frac{\delta\alpha}{\beta\gamma} \begin{cases} \in A & \text{if } \gamma \notin m, \\ \in A^{-1} & \text{if } \delta \notin m. \end{cases} \qquad Q.E.D.$$

We can use Proposition (1.20) to prove the important:

THEOREM (1.23). *For all flat Schottky groups Γ, Δ_Γ/Γ is a finite graph.*

PROOF. Let $\gamma_1, \cdots, \gamma_n$ be free generators of Γ and let v be any vertex of Δ_Γ. Let σ_i be the path in Δ_Γ from v to $\gamma_i(v)$ and let $S = \sigma_1 \cup \cdots \cup \sigma_n$. S is a finite tree and I claim that S maps *onto* Δ_Γ/Γ. This is equivalent to saying that

$$\tilde{S} = \bigcup_{\gamma \in \Gamma} \gamma(S) = \bigcup_{\gamma \in \Gamma} \bigcup_{i=1}^{n} (\text{path from } \gamma(v) \text{ to } \gamma\gamma_i(v))$$

is equal to Δ_Γ. Note first that $\tilde{S}$ is connected. In fact, if $\gamma \in \Gamma$, write γ as a word:

$$\gamma = \gamma_{i_1}^{\varepsilon_1}\gamma_{i_2}^{\varepsilon_2}\cdots\gamma_{i_N}^{\varepsilon_N}, \quad \varepsilon_i = \pm 1, \ 1 \leqq i_l \leqq n, \text{ all } l.$$

Then a typical point of $\tilde{S}$ is on the path from $\gamma(v)$ to $\gamma\gamma_i(v)$. It is joined to v by the sequence of paths:

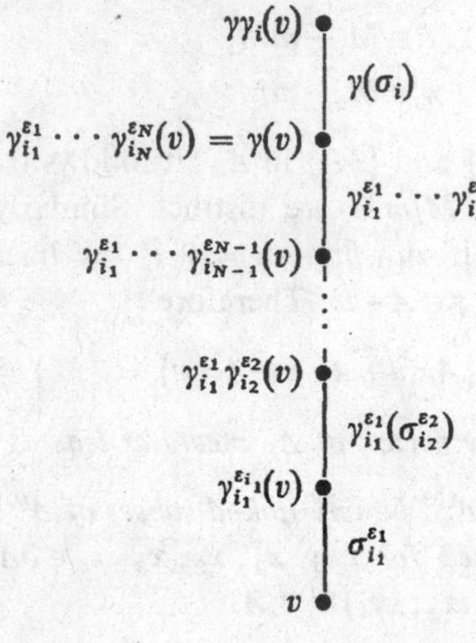

where $\sigma_i^1 = \sigma_i$ and $\sigma_i^{-1} = \gamma_i^{-1}(\sigma_i) = (\text{path from } v \text{ to } \gamma_i^{-1}(v))$. Note secondly that for every $x \in \Sigma$, the end $i^{-1}x$ of Δ_Γ is actually an end of the subtree $\tilde{S}$. In fact, if x is a fixed point of $\gamma \in \Gamma$, it suffices to join the points

$$\cdots, \gamma^{-2}v, \gamma^{-1}v, v, \gamma v, \gamma^2 v, \cdots$$

by paths in $\tilde{S}$. The result is a line in $\tilde{S}$, invariant under γ, plus an infinite set of spurs one leading to each of the points $\gamma^n v$. The 2 ends of this line are the 2 ends of Δ_Γ fixed by γ and one of these is $i^{-1}x$. Thus $i^{-1}x$ is, in fact, an end of $\tilde{S}$. Finally, suppose w is a vertex of $\Delta_\Gamma - \tilde{S}$. Then $w = \{M(x, y, z)\}$ for some $x, y, z \in \Sigma$. Since $w \notin \tilde{S}$, w is connected to $\tilde{S}$ by a unique path τ:

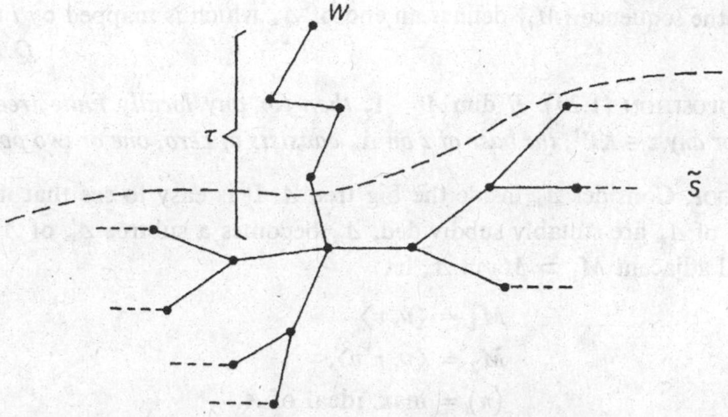

Thus all the paths from w to all ends of $\tilde{S}$ start with the same edge. Since $i^{-1}x$, $i^{-1}y$, $i^{-1}z$ are all ends of $\tilde{S}$, this contradicts Prop. (1.20). Hence $\Delta_\Gamma = \tilde{S}$. *Q.E.D.*

COROLLARY (1.24). *Δ_Γ is a locally finite tree.*

DEFINITION (1.25). Let $\{M\}, \{N\} \in \Delta^{(0)}$ be compatible, and let $z \in KP^1$. Then $\{M\}$ *separates* $\{N\}$ from z if there are representatives $N \supset M$ in standard positions and homogeneous coordinates $z^* \in K+K$ of z such that:

$$z^* \in M, z^* \notin m \cdot N.$$

The following is almost immediate:

PROPOSITION (1.26). *Let $\Delta_*^{(0)} \subset \Delta^{(0)}$ be a linked subset and let e be an end of $\partial\Delta_*$. Let $\{M\}, \{N\} \in \Delta_*^{(0)}$. Then $\{M\}$ separates $\{N\}$ from $i(e)$ if and only if the line in Δ_* from $\{N\}$ to the end e passes through $\{M\}$.*

DEFINITION (1.27). Let $\Delta_*^{(0)} \subset \Delta^{(0)}$ be any linked subset and let $z \in KP^1$. The *base* of z of $\Delta_*^{(0)}$ is the set of all $\{M\} \in \Delta_*^{(0)}$ which are not separated from z by any $\{N\} \in \Delta_*^{(0)}$.

PROPOSITION (1.28). *The base of z on $\Delta_*^{(0)}$ is empty if and only if $z \in i(\partial\Delta_*)$.*

PROOF. If $z \in i(\partial\Delta_*)$, the base of z is empty by (1.26). Conversely, suppose the base is empty. Choose homogeneous coordinates $z^* \in K+K$ for z. Then for all $\{M\} \in \Delta_*^{(0)}$, there is a representative $M \in \{M\}$ such that $z^* \notin m \cdot M$ and an $\{N\} \in \Delta_*^{(0)}$ such that

$$(*) \qquad M \not\supseteq N \ni z^*.$$

Start with any $\{M_1\} \in \Delta_*^{(0)}$. Call the N satisfying $(*)$ M_2. With M as M_2, call the N satisfying $(*)$ M_3. Continuing in this way, we get an infinite sequence:

$$M_1 \not\supseteq M_2 \not\supseteq M_3 \not\supseteq \cdots \ni z^*.$$

Then the sequence $\{M_i\}$ defines an end of Δ_* which is mapped by i to z.
Q.E.D.

PROPOSITION (1.29). *If* $\dim A = 1$, *then for any locally finite tree* Δ_* *and for any* $z \in KP^1$, *the base of* z *on* Δ_* *consists of zero, one or two points.*

PROOF. Consider Δ_* inside the big tree Δ. It is easy to see that if the edges of Δ_* are suitably subdivided, Δ_* becomes a subtree Δ'_* of Δ; i.e. for all adjacent $M_1 \supset M_2$ in Δ_* let

$$M_1 = \langle u, v \rangle$$
$$M_2 = \langle u, \pi^r v \rangle,$$
$$(\pi) = \text{max. ideal of } A.$$

Then adding the intermediate classes

$$\{\langle u, \pi^i v \rangle\}, \qquad 1 \leqq i \leqq r-1.$$

has the effect of subdividing the edge in Δ_* between $\{M_1\}$ and $\{M_2\}$ so that it becomes a path in Δ. Now every $z \in KP^1$ is an end of Δ, and every end of Δ which is not an end of Δ_* can be joined to the subtree Δ'_* by a unique shortest path. If $z \in KP^1 - i(\partial\Delta_*)$, let $v(z)$ be the vertex of Δ'_* where this path meets Δ'_*. Then it follows from (1.26) that the base of z on Δ_* is $\{v(z)\}$ if $v(z)$ is a vertex of Δ_*; or it equals the 2 endpoints of the edge of Δ_* containing $v(z)$ if $v(z)$ is not a vertex of Δ_*. *Q.E.D.*

In case $\dim A > 1$, the points $z \in KP^1 - i(\partial\Delta_*)$ can have wildly diverse kinds of bases on Δ_*. Take the case $\Delta_* = \Delta_\Gamma$. Then heuristically, Γ does not act equally discontinuously at all the points of $KP^1 - \bar{\Sigma}$. An important definition is this:

DEFINITION (1.30). If Γ is a flat Schottky group, then $\Omega_\Gamma = \{z \in KP^1 \mid$ the base of z on Δ_Γ is finite and non-empty$\}$; Ω_Γ is called the set of *strict discontinuity*, or the set of points where Γ acts *strictly discontinuously*. (Note that if $\dim A = 1$, Ω_Γ is simply $KP^1 - \bar{\Sigma}$, the usual set of discontinuity.)

2. From trees to schemes

We turn now from the construction of trees to the construction of actual or formal schemes over S[3]. Consider the set of all reduced and

[3] It is interesting that the trees which Bruhat and Tits associated to $PGL(2, K)$ are, in fact, a highly developed special case of the graphs that have been used for a long time in the theory of algebraic surfaces to plot the configurations of intersecting curves. To be precise, if $\dim A = 1$, we consider the inverse system of surfaces obtained by blowing up closed points on the 2-dimensional scheme $P^1 \times S$. The graph which plots the components of their closed fibres over S and their intersection relations is canonically isomorphic to Δ.

irreducible schemes Z over S whose generic fibre is $\boldsymbol{P}^1_K$. These form a partially ordered set if $Z_1 > Z_2$ means that there is an S-morphism from Z_1 to Z_2 which restricts to the identity on the generic fibre. In this partially ordered set, any 2 elements have a least upper bound, called their *join*. We want to study

1. the joins of finite sets of schemes $\boldsymbol{P}(M)$, $\{M\} \in \Delta^{(0)}$,
2. certain special infinite joins that exist as formal schemes over S but not as actual schemes.

PROPOSITION (2.1). *Let* $\{M_1\}, \{M_2\} \in \Delta^{(0)}$. *Let* P_{12} *be the join of* $\boldsymbol{P}(M_1)$ *and* $\boldsymbol{P}(M_2)$. *Then*

(a) *if* $\{M_1\}, \{M_2\}$ *are not compatible, the closed fibre of* P_{12} *is isomorphic to* $\boldsymbol{P}(M_1)_0 \times \boldsymbol{P}(M_2)_0$; *in particular, the fibres of* P_{12} *over S do not all have the same dimension hence* P_{12} *is not flat over S.*

(b) *if* $\{M_1\}, \{M_2\}$ *are compatible and* $(\alpha) = \rho(\{M_1\}, \{M_2\})$, *then* P_{12} *is a normal scheme, flat over S, and its fibre over* $s \in S$ *is:*

(b$_1$) $\boldsymbol{P}(M_i)_s$ *if* $\alpha(s) \neq 0$ $\quad (i = 1$ *or* $2)$,

(b$_2$) $\left\{\begin{array}{l} \boldsymbol{P}(M_1)_s \cup \boldsymbol{P}(M_2)_s \\ \textit{meeting transversely in one} \\ k(s)\textit{-rational point} \end{array}\right\}$ *if* $\alpha(s) = 0$.

PROOF. Let u_1, v_1 be a basis of M_1, and let $u_2 = au_1 + bv_1$, $v_2 = cu_1 + dv_1$ be a basis of M_2. Define map s:

$$\left.\begin{array}{r} X_i, Y_i : M_i \to A \\ X_i(u_i) = Y_i(v_i) = 1 \\ X_i(v_i) = Y_i(u_i) = 0 \end{array}\right\} \quad i = 1, 2.$$

Then $\boldsymbol{P}(M_i) = \operatorname{Proj} A[X_i, Y_i]$. But

$$X_1 = aX_2 + cY_2$$
$$Y_1 = bX_2 + dY_2$$

and these equations define the generic isomorphism of $\boldsymbol{P}(M_1)$ and $\boldsymbol{P}(M_2)$. Now P_{12} is just the closure in $\boldsymbol{P}(M_1) \times_S \boldsymbol{P}(M_2)$ of the graph of this generic isomorphism, i.e. P_{12} is the closure in

$$\operatorname{Proj} A[X_1 X_2, X_1 Y_2, Y_1 X_2, Y_1 Y_2]$$

of the curve in the generic fibre defined by

$$(*) \qquad aX_2 Y_1 - bX_1 X_2 + cY_1 Y_2 - dY_2 X_1 = 0.$$

According to the lemma of Ramanujam-Samuel (EGA IV.21) this closure either contains the whole closed fibre of $\boldsymbol{P}(M_1) \times_S \boldsymbol{P}(M_2)$ or else is a relative Cartier divisor over S. If the latter is true, then the closure must be defined as a subscheme of $\mathbf{P}(M_1)\times_S\mathbf{P}(M_2)$ by a suitable multiple of equation (*) with all coefficients in A, and not all coefficients in m. In particular P_{12} is then flat over S and its fibres are curves in $\mathbf{P}(M_1)_s\times\mathbf{P}(M_2)_s$ defined by equations of type (*). But over any field L, an equation of type (*) which is not identically zero defines a curve in $\mathbf{P}^1_L\times\mathbf{P}^1_L$ which is (i) a graph of an isomorphism of the 2 factors if $ad-bc\neq 0$ or (ii) equal to $\mathbf{P}^1_L\times(\alpha)\cup(\beta)\times\mathbf{P}^1_L$ for some $\alpha,\beta\in L\mathbf{P}^1$ if $ad-bc=0$.

To tie these possibilities up with compatibility of the $\{M_i\}$, note that on the one hand if $\lambda a,\lambda b,\lambda c,\lambda d\in A$ and not all are in m, then $M_1\supset\lambda M_2$ and $mM_1\not\supset\lambda\cdot M_2$: thus $\{M_1\}$ and $\{M_2\}$ are compatible and $M_1,\lambda M_2$ are representatives in standard position. It is easy to check that

$$(\lambda a\cdot\lambda d-\lambda\beta\cdot\lambda c)=\rho(\{M_1\},\{M_2\}).$$

On the other hand, if $\{M_1\}$ and $\{M_2\}$ are compatible, choose $M_1\supset M_2$ to be representatives in standard position. Then $a,b,c,d\in A$, and not all are in m.

Finally, the normality of P_{12} in the 2nd clase is a formal consequence of the rest: since S is normal and P_{12} is flat and generically smooth over S, it is certainly non-singular in codimension one. And if $f\in m$, then the ideal $f\cdot A\subset A$ has no embedded components, and since none of the fibres of P_{12}/S have embedded components, $f\cdot\mathcal{O}_{P_{12}}$ has no embedded components either. Thus P_{12} is normal. *Q.E.D.*

PROPOSITION (2.2). *Let $\{M_1\},\{M_2\}\in\Delta^{(0)}$ be compatible and let $z\in K\mathbf{P}^1$. Then $\{M_1\}$ separates $\{M_2\}$ from z if and only if*

$$\mathrm{cl}_{P_{12}}(z)\cap\mathbf{P}(M_2)_0=\emptyset,$$

where $\mathrm{cl}_{P_{12}}(z)$ is the closure of $\{z\}$ in P_{12} and $\mathbf{P}(M_2)_0$ is the component of the closed fibre of P_{12} isomorphic to the closed fibre of $\mathbf{P}(M_2)$.

PROOF. Since the closed fibre of P_{12} minus $\mathbf{P}(M_2)_0$ is isomorphic to $\mathbf{P}(M_1)_0$ minus a point, $\mathrm{cl}_{P_{12}}(z)\cap\mathbf{P}(M_2)_0=\emptyset$ implies that $\mathrm{cl}_{P_{12}}(z)$ meets the closed fibre in P_{12} in a finite set of points where P_{12} is smooth over S. Therefore by the lemma of Ramanujam-Samuel, $\mathrm{cl}_{P_{12}}(z)$ will be a relative Cartier divisor in this case; hence $\mathrm{cl}_{P_{12}}(z)$ will be the image of a section of P_{12} over S. Thus $\mathrm{cl}_{P_{12}}(z)\cap\mathbf{P}(M_2)_0=\emptyset$ is equivalent to z extending to a section of P_{12} not meeting the component $\mathbf{P}(M_2)_0$ of the closed fibre. But if

$$M_2=\langle u,v\rangle,$$
$$M_1=\langle u,\alpha v\rangle$$

are representatives in standard position, then for z to define a section of $\boldsymbol{P}(M_1)$ means that z has homogeneous coordinates:

$$z^* = \beta u + \gamma(\alpha v), \qquad \beta, \gamma \in A, \qquad \beta \text{ or } \gamma \notin m.$$

In this case $\mathrm{cl}_{P_{12}}(z)$ meets $\boldsymbol{P}(M_1)_0$ in a point other than the one where $\boldsymbol{P}(M_2)_0$ meets $\boldsymbol{P}(M_1)_0$ if and only if $\beta \notin m$. Thus $\{M_1\}$ separates $\{M_2\}$ and z if and only if z defines a section of $\boldsymbol{P}(M_1)$ passing through a closed point other than the point where $\boldsymbol{P}(M_2)_0$ meets it in P_{12}; which is the same as saying that z defines a section of P_{12} not meeting $\boldsymbol{P}(M_2)_0$. *Q.E.D.*

PROPOSITION (2.3): *Let $\{M_1\}, \cdots, \{M_k\} \in \Delta^{(0)}$ be pairwise compatible. Let Z be the join of $\boldsymbol{P}(M_1), \cdots, \boldsymbol{P}(M_k)$. Then Z is a normal scheme, proper and flat over S and its closed fibre Z_0 satisfies:*

i) *it is reduced, connected and 1-dimensional,*

ii) *its components are naturally isomorphic to the schemes $\boldsymbol{P}(M_1)_0$, $\cdots \boldsymbol{P}(M_k)_0$ respectively,*

iii) *2 components meet in at most one point and no set of components meets to form a loop,*

(iv) *every singular point z is locally isomorphic over k to the union W_l of the coordinate axes in A^l.*

PROOF. By definition, Z is the closure in $\boldsymbol{P}(M_1) \times_S \cdots \times_S \boldsymbol{P}(M_k)$ of the graph of the generic isomorphism of all the factors. Therefore Z_0 is connected by Zariski's connectedness theorem (EGA, III. 4.3.1). For every i and $j (1 \leqq i, j \leqq k)$, let p_{ij} be the projection onto $\boldsymbol{P}(M_i) \times_S \boldsymbol{P}(M_j)$. Since p_{ij} is proper, it follows from the proof of the previous proposition that:

$$(*) \qquad p_{ij}(Z_0) = \boldsymbol{P}(M_i) \times (a) + (b) \times \boldsymbol{P}(M_j) \qquad \text{some } a, b \in k\boldsymbol{P}^1.$$

Therefore each component of Z_0 is 'parallel to one of the coordinate axes', i.e. has the form:

$$(a_1) \times (a_2) \times \cdots \times \boldsymbol{P}(M_i)_0 \times \cdots \times (a_k)$$

for some i, and is naturally isomorphic to $\boldsymbol{P}(M_i)_0$ for this i. Moreover it follows immediately from $(*)$ that for each i, exactly one of the components of Z_0 is parallel to the ith coordinate axis. This proves (ii). For any union Z_0 of coordinate axes, $(Z_0)_{\text{red}}$ is locally isomorphic to the scheme W_l in (iv). Moreover, if Z_0 had a loop, then for some i there would have to be 2 or more components parallel to the ith axis. The only point which is not very clear is that Z_0 is reduced at its singular points. But note that the ideal of W_l in A^l is generated by the monomials $(X_i X_j)$ which are the defining equations of $p_{ij}(W_l)$; therefore the scheme-theoretic intersection $\bigcap_{i,j} p_{ij}^{-1}[\mathbf{P}(M_i)_0 \times (a) + (b) \times \mathbf{P}(M_n)_0]$ is already reduced, so *a fortiori* Z_0 is reduced. Finally, Z is flat over S by EGA 4.15.2, and normal by the same argument used in Prop. (2.1). *Q.E.D.*

Finally, we have:

PROPOSITION (2.4): *Let $\Delta_*^{(0)} \subset \Delta^{(0)}$ be a finite linked subset. Let $\mathbf{P}(\Delta_*)$ be the join of all the schemes $\mathbf{P}(M_i)$, $\{M_i\} \in \Delta_*^{(0)}$. Then in addition to the above properties, we have also:*

i) *Z_0 has only double points,*

ii) *in the one-one correspondence between the components of the closed fibre Z_0 and the elements of $\Delta_*^{(0)}$ 2 components meet if and only if the corresponding elements of $\Delta_*^{(0)}$ are adjacent in the tree Δ_*.*

In other words, if we make a tree out of Z_0 by taking a vertex for each component and an edge for each point of intersection, we obtain geometrically the tree Δ_*.

PROOF. In fact, if Z_0 has a point of multiplicity $\geqq 3$, this would mean that $\geqq 3$ components of Z_0 all met each other. Since we know Δ_* is a tree, this would contradict (ii). Thus it suffices to prove (ii). Suppose $\{M_1\}$, $\{M_2\} \in \Delta_*^{(0)}$ are *not* adjacent. This means there is an $\{M_3\} \in \Delta_*^{(0)}$ different from $\{M_1\}$ and $\{M_2\}$ such that

$$\rho(\{M_1\}, \{M_2\}) = \rho(\{M_1\}, \{M_3\}) \cdot \rho(\{M_3\}, \{M_2\}).$$

Therefore we can find representatives of these classes in standard position:

$$\begin{aligned} M_1 &= \langle u, v\rangle, \\ M_3 &= \langle u, \alpha v\rangle, (\alpha) = \rho((M_1\}, (M_3\}) \subset m, \\ M_2 &= \langle u, \alpha \cdot \beta v\rangle, (\beta) = \rho((M_3\}, (M_2\}) \subset m. \end{aligned}$$

Let X, Y be defined by

$$\begin{aligned} X(u) &= Y(v) = 1, \\ X(v) &= Y(u) = 0. \end{aligned}$$

Then:

$$\begin{aligned} \mathbf{P}(M_1) &= \operatorname{Proj} A[X, Y], \\ \mathbf{P}(M_3) &= \operatorname{Proj} A[X, Y/\alpha], \\ \mathbf{P}(M_2) &= \operatorname{Proj} A[X, Y/\alpha\beta]. \end{aligned}$$

Now form the join Z_{123} of $\mathbf{P}(M_1), \mathbf{P}(M_2), \mathbf{P}(M_3)$:

$$\begin{aligned} Z_{123} &= \operatorname{Proj} A[X^3, X^2Y/\alpha\beta, XY^2/\alpha^2\beta, Y^3/\alpha^2\beta] \\ &\cong \operatorname{Proj} A[u_1, u_2, u_3, u_4]/(\beta u_2^2 - u_1u_3, \alpha u_3^2 - u_2u_4, \alpha\beta u_2u_3 - u_1u_4). \end{aligned}$$

From this it follows easily that the closed fibre of Z_{123} has 3 components, connected like this:

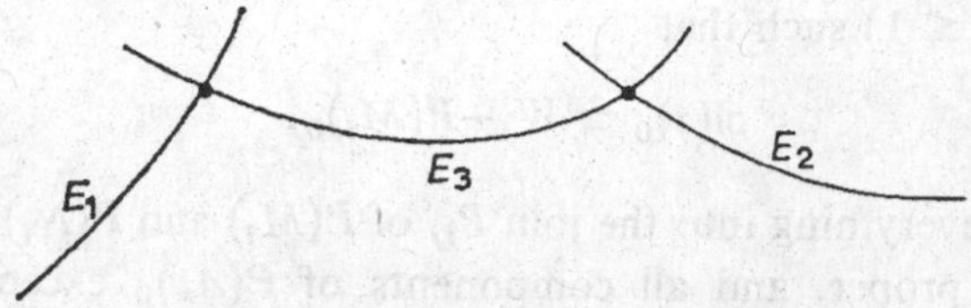

where E_i maps onto $\boldsymbol{P}(M_i)_0$. But now there is a natural map of Z onto Z_{123}, and the component of Z_0 corresponding to $\{M_i\}$ must map onto $E_i \subset (Z_{123})_0$. Since $E_1 \cap E_2 = \emptyset$, it follows that in Z_0 the components corresponding to $\{M_1\}$ and $\{M_2\}$ do not meet! This proves that the only components of Z_0 that can meet are those corresponding to adjacent vertices of Δ_*. But Δ_* is a tree so it is disconnected by leaving out any edge. Thus if any pair of components of Z_0 corresponding to adjacent vertices of Δ_0 did not actually meet, Z_0 would be disconnected. Thus (ii) is completely proven. *Q.E.D.*

PROPOSITION (2.5): *Let $\Delta_*^{(0)} \subset \Delta^{(0)}$ be a finite linked subset, and let $\boldsymbol{P}(\Delta_*)$ be the join of the $\boldsymbol{P}(M_i)$, $\{M_i\} \in \Delta_*^{(0)}$. Let $z \in K\boldsymbol{P}^1$, and let* cl(z) *denote the closure of $\{z\}$ in $\boldsymbol{P}(\Delta_*)$. Then for all $\{M_i\} \in \Delta_*^{(0)}$*

$$\mathrm{cl}(z) \cap \boldsymbol{P}(M_i)_0 \neq \emptyset$$

if and only if $\{M_i\}$ is in the base of z on $\Delta_*^{(0)}$.

PROOF. The statement is equivalent to: $[\mathrm{cl}(z) \cap \boldsymbol{P}(M_i)_0 = \emptyset] \Leftrightarrow [\exists \{M_j\} \in \Delta_*^{(0)}$ separating $\{M_i\}$ and $z]$. The implication $\Leftarrow$ is an immediate consequence of (2.2). Conversely, suppose $\mathrm{cl}(z) \cap \boldsymbol{P}(M_i)_0 = \emptyset$. Let $x_1, \cdots, x_l$ be the double points on $\boldsymbol{P}(M_i)_0$, let $\boldsymbol{P}(N_1)_0, \cdots, \boldsymbol{P}(N_l)_0$ be the other components of $\boldsymbol{P}(\Delta_*)_0$ through $x_1, \cdots, x_l$ respectively, and let $W_t \subset \boldsymbol{P}(\Delta_*)_0$ be the union of the components of $\boldsymbol{P}(\Delta_*)_0$ which are connected to $\boldsymbol{P}(N_t)_0$ without passing through $\boldsymbol{P}(M_i)_0$:

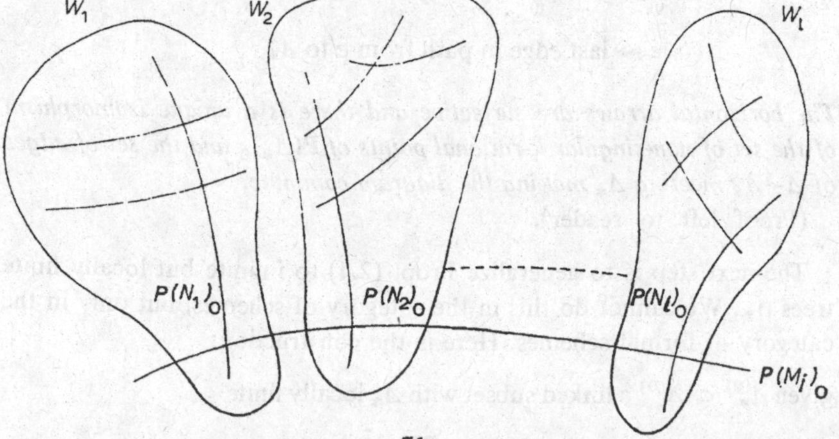

By Zariski's connectedness theorem, $\mathrm{cl}(z)_0$ is connected, hence there is some j $(1 \leqq j \leqq 1)$ such that

$$\mathrm{cl}(z)_0 \subset W_j - \boldsymbol{P}(M_i)_0.$$

Now project everything into the join P_{ij} of $\boldsymbol{P}(M_i)$ and $\boldsymbol{P}(N_j)$. Since the projection is proper, and all components of $\boldsymbol{P}(\Delta_*)_0$ except $\boldsymbol{P}(M_i)_0$, $\boldsymbol{P}(N_j)_0$ are mapped to one point, it follows that $p_{ij}(\mathrm{cl}\,(z)_0) =$ one point, and it is still disjoint from $\boldsymbol{P}(M_i)_0$. Therefore by (2.2), $\{N_j\}$ separates z from $\{M_i\}$.

In the case $\dim A = 1$, an important class of trees Δ_* are the subtrees of the full Δ, i.e. linked sets $\Delta_*^{(0)}$ such that if $\{M_1\}, \{M_2\} \in \Delta_*^{(0)}$ are adjacent, then equivalently $\rho(\{M_1\}, \{M_2\}) =$ max. ideal of A, or $\{M_1\}$, $\{M_2\}$ are adjacent in Δ. These have the following easy characterization:

PROPOSITION (2.6). *If* $\dim A = 1$, *and* $\Delta_*^{(0)} \subset \Delta^{(0)}$ *is a finite linked set then* $\boldsymbol{P}(\Delta_*)$ *is regular if and only if* Δ_* *is a subtree of* Δ.

We omit the proof, which is easy. When $\dim A = 1$, $\mathrm{cl}(z)$ is necessarily isomorphic to S, i.e. it is the image of a section of $\boldsymbol{P}(\Delta_*)$ over S. Therefore $\mathrm{cl}(z) \cap \boldsymbol{P}(\Delta_*)_0$ is a single k-rational point of $\boldsymbol{P}(\Delta_*)_0$. If, moreover, $\boldsymbol{P}(\Delta_*)$ is regular, it must be a non-singular point of $\boldsymbol{P}(\Delta_*)_0$ and we have the following nice interpretation of the map $z \mapsto \mathrm{cl}\,(z) \cap \boldsymbol{P}(\Delta_*)_0$.

PROPOSITION (2.7): *Let* $\dim A = 1$ *and let* $\Delta_* \subset \Delta$ *be a finite subtree Consider the maps*:

$$\begin{array}{ccc}
z & \mapsto & \mathrm{cl}(z) \cap \boldsymbol{P}(\Delta_*)_0 \\
\Cap & & \Cap \\
KP^1 & \mapsto & [\text{non-singular } k\text{-rational points of } \boldsymbol{P}(\Delta_*)_0] \\
\wr\| & & \\
\text{ends of } \Delta & \to & [\text{edges of } \Delta - \Delta_* \text{ that meet } \Delta_*] \\
\Cup & & \Cup \\
e & \mapsto & \text{last edge in path from } e \text{ to } \Delta_*
\end{array}$$

The horizontal arrows are surjective and there is a unique isomorphism of the set of non-singular k-rational points of $\boldsymbol{P}(\Delta_*)_0$ *and the set of edges of* $\Delta - \Delta_*$ *meeting* Δ_* *making the diagram commute.*

(Proof left to reader).

The next step is to generalize Prop. (2.4) to infinite but locally finite trees Δ_*. We cannot do this in the category of schemes, but only in the category of formal schemes. Here is the construction:

given $\Delta_*^{(0)} \subset \Delta^{(0)}$ a linked subset with Δ_* locally finite

I) for all finite subtrees $S \subset \Delta_*$ let $\boldsymbol{P}(S)$ be the finite join as above,

II) when $S_1 \subset S_2$, there is a natural morphism

$$p : \boldsymbol{P}(S_2) \to \boldsymbol{P}(S_1)$$

giving us an inverse system,

III) let $\mathscr{P}(S) =$ formal completion of $\boldsymbol{P}(S)$ along its closed fibre $\boldsymbol{P}(S)_0$. We get again an inverse system:

$$p : \mathscr{P}(S_2) \to \mathscr{P}(S_1).$$

IV) For all S, let $\mathscr{P}(S)'$ be the maximal open subset $U \subset \mathscr{P}(S)$ such that for all finite subtrees $S \subset T \subset \Delta_*$, the morphism $\mathrm{res}_U\, p$ is an isomorphism:

$$\begin{array}{ccc} p^{-1}(U) & \xrightarrow[\mathrm{res}_U p]{} & U \\ \cap & & \cap \\ \mathscr{P}(T) & \xrightarrow[p]{} & \mathscr{P}(S). \end{array}$$

V) Then the inverse system of $\mathscr{P}(S)$'s become a direct system of $\mathscr{P}(S)'$'s in which all morphisms:

$$\mathscr{P}(S_2)' \to \mathscr{P}(S_1)'$$

are open immersions. Let

$$\mathscr{P}(\Delta_*) = \varinjlim_{S} \mathscr{P}(S)'.$$

PROPOSITION (2.8): *$\mathscr{P}(\Delta_*)$ is a normal formal scheme flat over S, such that $m \cdot \mathcal{O}_{\mathscr{P}(\Delta)}$ is a defining sheaf of ideals. The closed fibre has the properties:*

i) it is reduced, connected, 1-dimensional and locally of finite type over k,

ii) it has at most ordinary double points and these are k-rational,

iii) its components are all isomorphic to $\boldsymbol{P}^1_k$ and are in one-one correspondence with the elements of $\Delta_*^{(0)}$,

iv) 2 components meet if and only if the corresponding vertices of $\Delta_*^{(0)}$ are adjacent and then in exactly one point.

PROOF. This follows immediately from the previous Propositions once we have proven the following lemma.

LEMMA (2.9). Let $S \subset \Delta_*$ *be a finite subtree. If a vertex v of S is such that all edges of Δ_* which meet v lie in S, then the component E_v of $\boldsymbol{P}(S)_0$ corresponding to v lies entirely in the open set $\mathscr{P}(S)'$.*

PROOF OF LEMMA. Let $S \subset T \subset \Delta_*$, where T is another finite subtree containing all edges of Δ_* meeting edges meeting v, i.e.

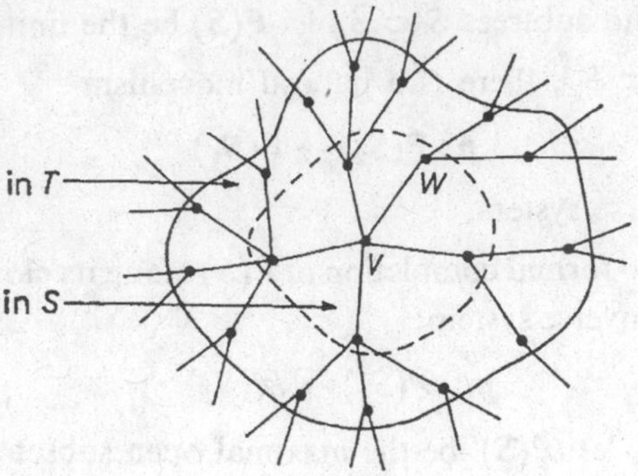

Let $U \subset \boldsymbol{P}(S)_0$ be the open set consisting of E_v, plus those points x of the components E_w which meet E_v such that $p^{-1}(x)$ does not meet any other component of $\boldsymbol{P}(T)_0$:

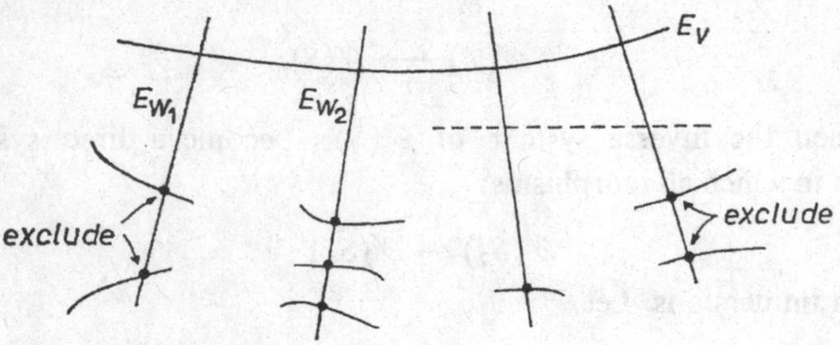

Then even if $T' \supset T$ is a bigger finite subtree, it follows from Prop. (2.3) that in the diagram:

$$\begin{array}{ccc} p^{-1}(U) & \xrightarrow{\text{res}_U\, p_0} & U \\ \cap & & \cap \\ \boldsymbol{P}(T')_0 & \xrightarrow[p_0]{} & \boldsymbol{P}(S)_0 \end{array}$$

$\text{res}_U\, p_0$ is an isomorphism. Since $p: \boldsymbol{P}(T') \to \boldsymbol{P}(S)$ is proper and surjective and $\boldsymbol{P}(S)$ is reduced, there is an open set $V \subset \boldsymbol{P}(S)$ such that $V \supset U$ and $p^{-1}(V) \to V$ is an isomorphism. Therefore the inverse image of U in the formal scheme $\mathscr{P}(T')$ is mapped isomorphically to the open subscheme U of $\mathscr{P}(S)$. Therefore $U \subset \mathscr{P}(S)'$. *Q.E.D.*

Speaking heuristically, $\mathscr{P}(\Delta_*)$ is the *infinite formal join* of the schemes $\boldsymbol{P}(M)$, for all $\{M\} \in \Delta_*^{(0)}$.

For every $z \in K\boldsymbol{P}^1$, we can talk about the closure of z, $\text{cl}(z)\hat{}$, in $\mathscr{P}(\Delta_*)$. In fact, we can form $\text{cl}_S(z)$ in $\boldsymbol{P}(S)$, take its formal completion $\text{cl}_S(z)\hat{}$ in $\mathscr{P}(S)$, and restrict it to $\mathscr{P}(S)'$. Then if $S_1 \subset S_2$, $\text{cl}_{S_1}(z)\hat{}$ is just the restriction of $\text{cl}_{S_2}(z)\hat{}$ to $\mathscr{P}(S_1)$, hence there is a unique formal subscheme:

$$\text{cl}(z)\hat{} \subset \mathscr{P}(\Delta_*)$$

such that

$$\mathrm{cl}(z)^\wedge \cap \mathscr{P}(S)' = \mathrm{cl}_S(z)^\wedge \cap \mathscr{P}(S)',$$

all finite subtrees S.

PROPOSITION (2.10). *$\mathrm{cl}(z)^\wedge$ is proper over S if and only if the base of S on Δ_* is finite. If $\mathrm{cl}(z)^\wedge$ is non-empty as well then it is the formal completion of the proper scheme $\mathrm{cl}_S(z)$ for all sufficiently large finite subtrees $S \subset \Delta_*$.*

This is an immediate consequence of (2.5).

COROLLARY (2.11). *If Γ is a flat Schottky group and $z \in KP^1$, then Γ acts strictly discontinuously at z if and only if $\mathrm{cl}(z)^\wedge \subset \mathscr{P}(\Delta_\Gamma)$ is non-empty and proper over S.*

In case $\dim A = 1$, we have the infinite generalization of (2.7):

PROPOSITION (2.12). *Let $\dim A = 1$ and let Δ_* be a locally finite subtree of Δ. Consider the maps*

$$\begin{array}{ccc} z & \mapsto & \mathrm{cl}(z) \cap (\Delta_*)_0 \\ \Cap & & \Cap \\ KP^1 - i(\partial\Delta_*) & \rightarrow & [\textit{non-singular k-rational points of } (\Delta_*)_0] \\ & & \wr\| \\ \partial\Delta - \partial\Delta_* & \rightarrow & [\textit{edges of } \Delta - \Delta_* \textit{ meeting } \Delta_*] \\ \Cup & & \Cup \\ e & \mapsto & \textit{last edge in path from } e \textit{ to } \Delta_* \end{array}$$

The horizontal maps are surjective and there is a unique isomorphism of the set of non-singular k-rational points of $\mathscr{P}(\Delta_*)_0$ and the set of edges of $\Delta - \Delta_*$ meeting Δ_* making the diagram commute.

(Proof left to reader).

3. The construction of the quotient

We now restrict ourselves to the case $\Delta_* = \Delta_\Gamma$. Then the group Γ acts on $\mathscr{P}(\Delta_\Gamma)$. The final step in our construction is to form a quotient $\mathscr{P}(\Delta_\Gamma)/\Gamma$ and to algebrize it.

THEOREM (3.1). *There exists a unique pair $(\mathscr{X}, \pi)$ consisting of a formal scheme $\mathscr{X}$ proper over S and a surjective étale S-morphism*

$$\pi : \mathscr{P}(\Delta_\Gamma) \rightarrow \mathscr{X}$$

such that

a) *$\forall \gamma \in \Gamma$, if $[\gamma]$ represents the induced automorphism of $\mathscr{P}(\Delta_\Gamma)$, then $\pi \circ [\gamma] = \pi$,*

b) $\forall x, y \in \mathscr{P}(\Delta_\Gamma)$, $\pi(x) = \pi(y) \Leftrightarrow x = [\gamma]y$, some $\gamma \in \Gamma$.

Moreover $\mathscr{X}$ is normal, is flat and projective over S, and is algebraizable. *$\mathscr{X}$ will be written $\mathscr{P}(\Delta_\Gamma)/\Gamma$.*

PROOF. As a topological space, $\mathscr{X}$ must equal the quotient of the underlying topological space to $\mathscr{P}(\Delta_\Gamma)$ by Γ, and its structure sheaf must be the subsheaf of $\pi_*(\mathcal{O}_{\mathscr{P}(\Delta_\Gamma)})$ of Γ-invariants. Therefore $\mathscr{X}$ is unique. To construct $\mathscr{X}$, we proceed in two stages:

(i) prove the results for a suitable $\Gamma_0 \subset \Gamma$ of finite index,

(ii) prove them for Γ.

The point is that since Γ acts freely on the tree Δ_Γ, no $\gamma \in \Gamma$ $(\gamma \neq e)$ takes any component of $\mathscr{P}(\Delta_\Gamma)_0$ into itself. Even better, there is a normal subgroup $\Gamma_0 \subset \Gamma$ of finite index such that no $\gamma \in \Gamma_0 (\gamma \neq e)$ takes a vertex of Δ_Γ into itself or to an adjacent vertex. Therefore no $\gamma \in \Gamma_0(\gamma \neq e)$ takes a component of $\mathscr{P}(\Delta_\Gamma)_0$ into itself or into a second component meeting the fiirst one. From this it follows that Γ_0 acts on $\mathscr{P}(\Delta_\Gamma)_0$ discontinuously in the Zariski topology: i.e. every $x \in \mathscr{P}(\Delta_\Gamma)_0$ has an open neighbourhood U such that $U \cap \gamma U = \emptyset$, all $\gamma \in \Gamma_0(\gamma \neq e)$. But in this case a quotient $\mathscr{P}(\Delta_\Gamma)/\Gamma_0$ can be constructed simply by re-glueing! To be precise cover $\mathscr{P}(\Delta_\Gamma)$ by affine open subschemes $\mathrm{Spf}(A_i)$ whose underlying open subsets have the above property. Then for every i, j, there is at most one element $\gamma_{ij} \in \Gamma_0$ such that

$$\gamma_{ij}(\mathrm{Spf}\,(A_i)) \cap \mathrm{Spf}\,(A_j) \neq \emptyset.$$

Glue $\mathrm{Spf}(A_i)$ to $\mathrm{Spf}(A_j)$ on this overlap via the map $[\gamma_{ij}]$. This gives a formal scheme $\mathscr{Y}$ and a morphism

$$\pi_0 : \mathscr{P}(\Delta_\Gamma) \to \mathscr{Y}$$

which is surjective and locally an isomorphism such that

(a) $\pi \circ [\gamma] = \pi$, all $\gamma \in \Gamma_0$, and

(b) $\pi(x) = \pi(y)$ implies $x = [\gamma]y$, some $\gamma \in \Gamma_0$.

Note that since Δ_Γ/Γ is a finite graph, so is Δ_Γ/Γ_0. Therefore $\mathscr{Y}_0$ has only a finite number of components and is proper over k. Therefore $\mathscr{Y}$ is proper over S. Obviously $\mathscr{Y}$ is normal and flat over S too since it is locally isomorphic to $\mathscr{P}(\Delta_\Gamma)$.

Now for each component E of $\mathscr{Y}_0$, choose a point $x \in E$ which is not in any other component of $\mathscr{Y}_0$. Let $\bar{d}_E \in \mathcal{O}_{x,\mathscr{Y}_0} = \mathcal{O}_{x,\mathscr{Y}}/m \cdot \mathcal{O}_{x,\mathscr{Y}}$ be a generator of the maximal ideal and let $d_E \in \mathcal{O}_{x,\mathscr{Y}}$ lift $\bar{d}_E$. Let $d_E = 0$ define the relative Cartier divisor $D_E \subset \mathscr{Y}$, and let $D = \Sigma D_E$. Then D is relatively ample on $\mathscr{Y}$ over S, hence $\mathscr{Y}$ is projective. Now we can apply Grothendieck's algebraizability theorem (EGA III.5) to conclude that

$\mathscr{Y}$ is the formal completion of a unique scheme Y, projective over S, along its closed fibre Y_0.

Finally, Y is a projective scheme, hence any finite subset of Y is contained in an affine. Therefore its Γ/Γ_0-orbits are contained in affines and there exists a quotient $Y/(\Gamma/\Gamma_0)$(cf. [M1], § 7). Let $\mathscr{X}$ be the formal completion of $Y/(\Gamma/\Gamma_0)$ along its closed fibre. This $\mathscr{X}$ has all the required properties. *Q.E.D.*

DEFINITION (3.2). P_Γ is the scheme, projective over S, whose formal completion is $\mathscr{P}(\Delta_\Gamma)/\Gamma$.

We recall the concept of a *stable curve* over S in the sense of Deligne and Mumford: this is a scheme C, proper and flat over S, whose geometric fibres are reduced connected and 1-dimensional; have at most ordinary double points; and such that their non-singular rational components, if any, meet the remaining components in at least 3 points. Moreover a stable curve C over a field k will be call *degenerate* if Pic^0_C is a torus, or equivalently if the normalizations of all the components of $C \times_k \bar{k}$ ($\bar{k}$ an algebraic closure of k) are rational curves. C is called *k-split degenerate* if the normalizations of all the components of C are isomorphic to $\mathbf{P}^1_k$, and if all the double points are k-rational with 2 k-rational branches. This means that C is gotten by identifying in pairs a finite set of distinct k-rational points of a finite union of copies of $\mathbf{P}^1_k$.

THEOREM (3.3). *If n is the number of generators of Γ, then P_Γ is a stable curve over S of genus n, whose generic fibre is smooth over K and whose special fibre is k-split degenerate. Moreover its special fibre $(P_\Gamma)_0$ has the property*:

(*) *there is a $1-1$ correspondence between components of $(P_\Gamma)_0$ and vertices of Δ_Γ/Γ, and between double points of $(P_\Gamma)_0$ and edges of Δ_Γ/Γ, such that a component contains a double point if and only if the corresponding vertex is an endpoint of the corresponding edge.*

PROOF. Since $(P_\Gamma)_0$ is the quotient of $\mathscr{P}(\Delta_\Gamma)_0$ by Γ, the asserted properties of a stable curve are clear except for the requirement that every non-singular rational component meets the other components in $\geqq 3$ points. But by Prop. (1.20), every vertex of Δ_Γ/Γ is met by at least 3 edges so this is *OK*. Now a deformation of a stable curve is stable, so P_Γ is stable. Finally, the formal completion of P_Γ is normal, so P_Γ is normal. Therefore, its generic fibre is regular. Since it is also a stable curve, it is smooth over K too. *Q.E.D.*

Let $P_\Gamma(K)$ denote the set of K-rational points of P_Γ. We can now construct a map

$$\pi : \Omega_\Gamma \to P_\Gamma(K)$$

from the set of strict discontinuity to the set of rational points of the smooth curve $(P_\Gamma)_\eta$. In fact if $z \in \Omega_\Gamma$, we have seen that we can form

$$\mathrm{cl}(z)^\wedge \subset \mathscr{P}(\Delta_\Gamma)$$

and that $\mathrm{cl}(z)^\wedge$ is the formal completion of scheme $\mathrm{cl}(z)$ proper over S, and birational to it. This gives us a formal morphism.

$$\hat{p}_z : \mathrm{cl}\,(z)^\wedge \to \mathscr{P}(\Delta_\Gamma)/\Gamma$$

which by Grothendieck's theorem (EGA III.5) comes from a morphism:

$$p_z : \mathrm{cl}(z) \to P_\Gamma$$

Let $\pi(z)$ be the image of the generic point under this map.

PROPOSITION (3.4). *If $z_1, z_2 \in \Omega_\Gamma$, then*

$$[\pi(z_1) = \pi(z_2)] \Leftrightarrow [\exists \gamma \in \Gamma, \text{ such that } \gamma(z_1) = z_2].$$

PROOF. '$\Leftarrow$' is obvious. Conversely, say $\pi(z_1) = \pi(z_2)$. Let Z be the join of $\mathrm{cl}(z_1)$, $\mathrm{cl}(z_2)$: Z is proper over S and birational to it. In particular, Z_0 is connected. By assumption the 2 morphisms:

$$\begin{array}{ccccc} & \nearrow & \mathrm{cl}(z_1) & \overset{p_{z_1}}{\searrow} & \\ Z & & & & P_\Gamma \\ & \searrow & \mathrm{cl}(z_2) & \underset{p_{z_2}}{\nearrow} & \end{array}$$

are equal. Therefore the 2 formal morphisms:

$$\begin{array}{ccl} & \nearrow & \mathrm{cl}(z_1)^\wedge \subset \mathscr{P}(\Delta_\Gamma) \\ \hat{Z} & & \\ & \searrow & \mathrm{cl}(z_2)^\wedge \subset \mathscr{P}(\Delta_\Gamma) \end{array}$$

both lift the same formal morphism to $\mathscr{P}(\Delta_\Gamma)/\Gamma$. Since Z_0 is connected and $\mathscr{P}(\Delta_\Gamma)$ is étale over $\mathscr{P}(\Delta_\Gamma)/\Gamma$, these 2 differ by the action of some $\gamma \in \Gamma$. Therefore for a large finite subtree $S \subset \Delta_\Gamma$, we get a commutative diagram:

$$\begin{array}{ccl} & \nearrow & \mathrm{cl}(z_1) \subset P(S) \\ Z & & \qquad\quad \downarrow [\gamma] \\ & \searrow & \mathrm{cl}(x_2) \subset P(\gamma(S)). \end{array}$$

Evaluating these on the generic point, it follows that $\gamma(z_1) = z_2$. *Q.E.D.*

THEOREM (3.5). *If A is a regular local ring, then π is surjective.*

Proof. Note that if $\Gamma_1 \subset \Gamma$ is a subgroup of finite index, then

i) the set of fixed points Σ_1 of Γ_1 and Σ of Γ are the same (since $\forall \gamma \in \Gamma$, γ^n is in Γ_1 for some n) hence

ii) $\Delta_{\Gamma_1} = \Delta_\Gamma$ and

iii) $\mathscr{P}(\Delta_{\Gamma_1}) = \mathscr{P}(\Delta_\Gamma)$, hence

iv) P_{Γ_1} is a finite étale covering of P_Γ.

The first step in our proof is that the map

$$P_{\Gamma_1}(K) \to P_\Gamma(K)$$

is surjective. To see this, take some $z \in P_\Gamma(K)$ and let $\mathrm{cl}(z)$ be the closure of z in P_Γ, and let $\mathrm{cl}(z)'$ be the normalization of $\mathrm{cl}(z)$. Taking the fibre product:

$$\begin{array}{ccc} W & \longrightarrow & P_{\Gamma_1} \\ \downarrow & & \downarrow \\ \mathrm{cl}(z)' & \xrightarrow{f} & P_\Gamma \end{array}$$

we get an induced finite étale covering W of $\mathrm{cl}(z)'$. Since $\mathrm{cl}(z)'$ is normal, so is W and hence the components W_i of W are disjoint and are each finite and étale over $\mathrm{cl}(z)'$. Let K_i be the function field of W_i and let A_i be the normalization of A in K_i. Since the projection

$$\mathrm{cl}(z)' \to \mathrm{Spec}\, A$$

is birational, $\mathrm{cl}(z)'$ and Spec A are isomorphic outside a closed set Z of codimension two in $\mathrm{Spec}(A)$. Therefore over $\mathrm{Spec}(A)-Z$, W_i is isomorphic to $\mathrm{Spec}(A_i)$. In particular, $\mathrm{Spec}(A_i)$ is unramified over $\mathrm{Spec}(A)$ in codimension one. But by the theorem of the purity of the branch locus, which applies since A is regular, this proves that A_i is unramified everywhere over A; hence by Hensel's lemma, A_i is isomorphic to A. Therefore $K_i = K$ and $W_i \cong \mathrm{cl}(z)'$. This moves that not only is

$$P_{\Gamma_1}(K) \to P_\Gamma(K)$$

surjective, but also that there is a lifting f_1:

$$\begin{array}{ccc} & \overset{f_1}{\nearrow} & P_{\Gamma_1} \\ & & \downarrow \\ \mathrm{cl}(z)' & \xrightarrow{f} & P_\Gamma \end{array}$$

for every $\Gamma_1 \subset \Gamma$ if finite index.

The second step is that by passing to the formal completion of $\mathrm{cl}(z)'$ along the closed fibre, there exists a lifting $\hat{g}$:

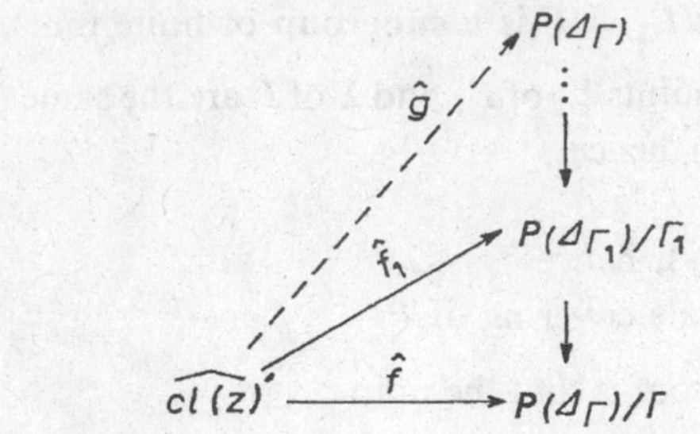

In fact let k be the maximum number of edges meeting any vertex of Δ_Γ and let n be the number of components of the closed fibre of $\mathrm{cl}(z)'$. Then if Γ_1 is sufficiently small, every $\gamma \in \Gamma_1$ ($\gamma \neq e$) maps every vertex v of Δ_Γ to a vertex $\gamma(v)$ joined to v by a line with more than $(k+1)n$ edges. Let $U \subset (P_{\Gamma_1})_0$ be the open subset consisting of all points of all the components that meet $f_1(\mathrm{cl}(z)'_0)$ except those that lie also on components disjoint from $f_1(\mathrm{cl}(z)'_0)$. Then U has at most $(k+1)n$ components. Let $\hat{U}$ be the corresponding open sub-formal scheme of $\mathscr{P}(\Delta_{\Gamma_1})/\Gamma_1 = P_{\Gamma_1}^{\wedge}$. Let $\hat{V}$ be the inverse image of $\hat{U}$ in $\mathscr{P}(\Delta_\Gamma)$. By our assumption on the way Γ_1 operates in Δ_Γ, no 2 components E and $\gamma(E)$ of $\hat{V}$ ($\gamma \in \Gamma_1, \gamma \neq e$) can be joined by a line of components of $\hat{V}$: hence $\hat{V}$ is the disjoint union of copies of $\hat{U}$. Choosing one of these, we can lift $\hat{f}_1$ uniquely to a morphism $\hat{g}$ of $\widehat{\mathrm{cl}(z)}'$ into this component.

Thirdly, $\hat{g}(\mathrm{cl}(z)'_0)$ is proper over k, hence it lies in one of the approximating pieces:

$$\mathscr{P}(S)' \subset \mathscr{P}(\Delta_\Gamma)$$
$$S \subset \Delta_\Gamma \text{ finite subtree.}$$

Thus $\hat{g}$ can be algebraized to a true morphism:

$$g : \mathrm{cl}(z)' \to P(S).$$

The image of the generic point here is a point of KP^1 which clearly lies in Ω_Γ and is mapped by π to z. *Q.E.D.*

Before ending this section, I would like to discuss briefly the special features of the dim $A = 1$ case and indicate how the somewhat more precise and elegant formulation given in the Introduction can be worked out. When dim $A = 1$, one has the big tree Δ and it usually is more convenient to replace the tree Δ_Γ by the tree Δ'_Γ where vertices are:

a) the vertices of Δ_Γ
b) the vertices of Δ intermediate between 2 vertices of Δ_Γ.

Then Δ'_Γ is a subtree of Δ hence $\mathscr{P}(\Delta'_\Gamma)$ is regular by (2.6): in fact, $\mathscr{P}(\Delta'_\Gamma)$ is just the minimal resolution of the normal surface $\mathscr{P}(\Delta_\Gamma)$. Let $\mathscr{P}(\Delta'_\Gamma)/\Gamma$ be the formal completion of P'_Γ. Then P'_Γ is regular and is the minimal resolution of the normal surface P_Γ. Generically, $P'_\Gamma = P_\Gamma$, but the closed fibre is now only a semi-stable curve – i.e. a reduced connected 1-dimensional scheme with at most ordinary double points and such that every non-singular rational component meets the other components in at least 2 points. P'_Γ is the so-called *minimal model* of the curve $(P_\Gamma)_\eta$ over A (cf. [$D-M$], p. 87, [L] and [$\check{S}$]). Moreover, the closed fibre $(P'_\Gamma)_0$ has only rational components one for each vertex of the graph Δ'_Γ/Γ; and one double point for each edge of the graph Δ'_Γ/Γ. Δ'_Γ/Γ is the graph referred to as $(\Delta/\Gamma)_0$ in the Introduction.

EXAMPLE.

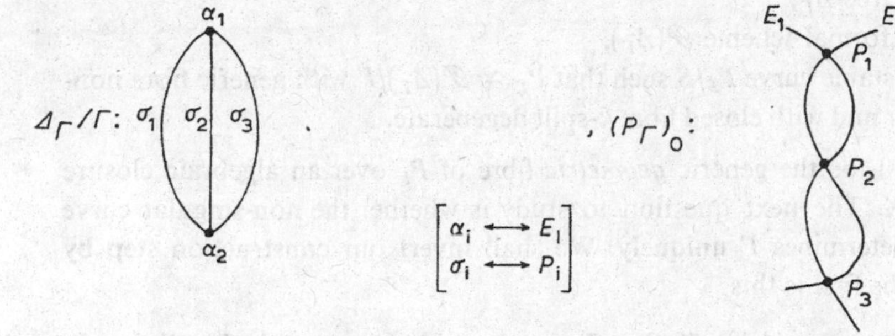

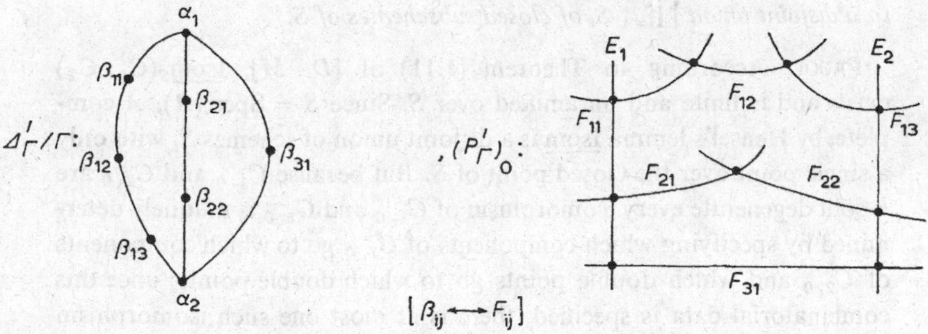

All this is an immediate generalization of (3.3) and is proven in exactly the same way. Finally applying Prop. (2.12) to Δ'_Γ, we get a commutative diagram on the upper-left.

(*See figure on next page.*)

Then the commutative diagram on the lower-right is deduced by taking the quotient of each set by Γ.

$$\Omega_\Gamma \xrightarrow{\ z \mapsto \text{cl}(z)_0\ } [\text{n.s. } k\text{-rat. pts. of } \mathscr{P}(\Delta'_\Gamma)_0]$$

$$\Big\| \Big\| \ \searrow^{\pi}\ P_\Gamma(K) \xrightarrow{\ z \mapsto \text{cl}(z)_0\ } \Big\| \Big\| \xrightarrow{\ \ \ } [\text{n.s. } k\text{-rat. pts. of } (P'_\Gamma)_0]$$

$$\partial\Delta - \partial\Delta'_\Gamma \Big| \Big|\!-\!\!\longrightarrow [\text{edges of } \Delta - \Delta'_\Gamma \text{ meeting } \Delta'_\Gamma] \searrow \ \Big\| \Big\|$$

$$\searrow \partial(\Delta/\Gamma) \xrightarrow{\ \ \ \ \ } [\text{edges of } \Delta_\Gamma/\Gamma - \Delta'_\Gamma/\Gamma \text{ meeting } \tilde\Delta'_\Gamma/\Gamma]$$

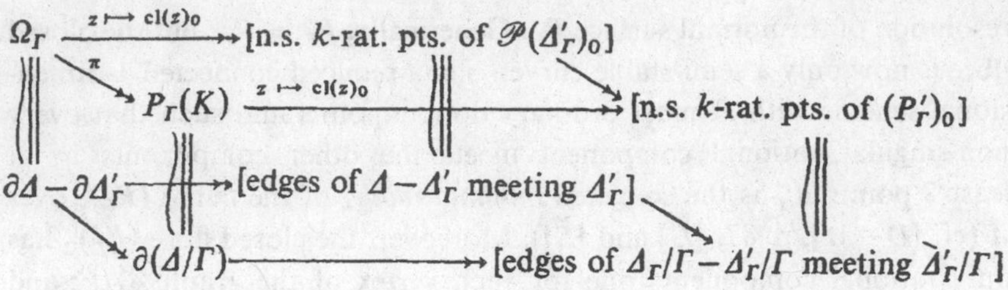

4. Existence and uniqueness

Summarising the discussion up to this point, we have started with a flat Schottky group $\Gamma \subset PGL(2, K)$, and have then constructed:

a) a tree Δ_Γ,

b) a formal scheme $\mathscr{P}(\Delta_\Gamma)$,

c) a stable curve P_Γ/S such that $\hat{P}_\Gamma \cong \mathscr{P}(\Delta_\Gamma)/\Gamma$ with generic fibre non-singular and with closed fibre k-split degenerate.

Let $(P_\Gamma)_{\bar\eta}$ be the generic *geometric* fibre of P_Γ over an algebraic closure $\bar{K}$ of K. The next question to study is whether the non-singular curve $(P_\Gamma)_{\bar\eta}$ determines Γ uniquely. We shall invert our construction step by step. The first is this:

PROPOSITION (4.1): *If C_1, C_2 are 2 stable curves over S with k-split degenerate fibres, then the scheme $\underline{\text{Isom}}_S(C_1, C_2)$ exists and is isomorphic to a disjoint union $\coprod_{i=1}^N S_i$ of closed subschemes of S.*

PROOF. According to Theorem (1.11) of $[D-M]$, $\underline{\text{Isom}}_S(C_1, C_2)$ exists and is finite and unramified over S. Since $S = \text{Spec }(A)$, A complete, by Hensel's lemma $\underline{\text{Isom}}$ is a disjoint union of schemes S_i with only a single point over the closed point of S. But because $C_{1,0}$ and $C_{2,0}$ are k-split degenerate every isomorphism of $C_{1,0}$ and $C_{2,0}$ is uniquely determined by specifying which components of $C_{1,0}$ go to which components of $C_{2,0}$ and which double points go to which double points; once this combinatorial data is specified, there is at most one such isomorphism of $C_{1,0}$ and $C_{2,0}$ and if it exists at all, it is rational over k. Therefore the closed points $s_i \in S_i$ are k-rational. Since S_i is finite and unramified over S, $S_i \to S$ must in fact be a closed immersion.   Q.E.D.

COROLLARY (4.2): *Every isomorphism of the generic geometric fibres $(P_{\Gamma_1})_{\bar\eta}$, $(P_{\Gamma_2})_{\bar\eta}$ extends uniquely to an isomorphism of P_{Γ_1} and P_{Γ_2}.*

PROOF. The hypothesis means that one of the components S_i of $\underline{\text{Isom}}_S(P_{\Gamma_1}, P_{\Gamma_2})$ has a point over the generic point of S, hence S_i is isomorphic to S and defines an S-isomorphism of P_{Γ_1}, P_{Γ_2}.

Proposition (4.3). *The morphism $\mathscr{P}(\Delta_\Gamma) \to \hat{P}_\Gamma$ makes $\mathscr{P}(\Delta_\Gamma)$ into the universal covering space of $\hat{P}_\Gamma$.*

Proof. In fact, the category of formal étale coverings of $\hat{P}_\Gamma$ is isomorphic to the category of étale coverings of $(P_\Gamma)_0$. Since the closed fibre $\mathscr{P}(\Delta_\Gamma)_0$ is connected and is a tree-like union of copies of $\boldsymbol{P}_k^1$, it is simply connected and must be the universal covering space of $(P_\Gamma)_0$. *Q.E.D.*

Definition (4.4). An *exterior isomorphism* of 2 groups G_1, G_2 is the set of isomorphisms $\alpha\varphi\alpha^{-1}$ conjugate to an ordinary isomorphism φ.

When we talk of $\pi_1(X)$, for a connected scheme X, in order to have a well-defined group depending functorially on X we have to fix a geometric base point $x : \operatorname{Spec}(\Omega) \to X$ (Ω an algebraically closed field), and then π_1 should be written $\pi_1(X, x)$. But up to exterior isomorphism, π_1 is independent of x, hence so long as we only talk of exterior isomorphisms, we can write $\pi_1(X)$.

Corollary (4.5). *Γ, as an abstract group, is canonically exterior-isomorphic to $\pi_1(\hat{P}_\Gamma)$(or $\pi_1((P_\Gamma)_0)$).*

Corollary (4.6). *Starting with an isomorphism*

$$\bar{\varphi} : (P_{\Gamma_1})_{\bar{\eta}} \simeq (P_{\Gamma_2})_{\bar{\eta}},$$

$\bar{\varphi}$ first extends uniquely to an isomorphism

$$\varphi : P_{\Gamma_1} \to P_{\Gamma_2}$$

hence to an isomorphism

$$\hat{\varphi} : \hat{P}_{\Gamma_1} \to P_{\Gamma_2},$$

hence to a pair consisting of an isomorphism

$$\alpha : \Gamma_1 \to \Gamma_2$$

and an α-equivariant isomorphism

$$\tilde{\varphi} : \mathscr{P}(\Delta_{\Gamma_1}) \to \mathscr{P}(\Delta_{\Gamma_2}).$$

Then $(\alpha, \tilde{\varphi})$ is unique up to a change $(\alpha', \tilde{\varphi}') = (\gamma\alpha\gamma^{-1}, \gamma \circ \tilde{\varphi})(\gamma \in \Gamma_2)$.

The last step is to show how the function field $R(\boldsymbol{P}_K^1)$ can be identified inside the field of meromorphic functions on $\mathscr{P}(\Delta_\Gamma)$, so that Γ as *a subgroup of* PGL(2, K) ($= \operatorname{Aut}_K R(\boldsymbol{P}_K^1)$) can be recovered from Γ as a group of automorphisms of $\mathscr{P}(\Delta_\Gamma)$.

Proposition (4.7): *Let $\mathscr{D} \subset \mathscr{P}(\Delta_\Gamma)$ be a positive relative Cartier divisor such that $\mathscr{D}_0$ meets only one component of the closed fibre $\mathscr{P}(\Delta_\Gamma)_0$. Then $R(\boldsymbol{P}_K^1)$, as a field of meromorphic functions on $\mathscr{P}(\Delta_\Gamma)$ is the quotient field of the A-algebra:*

$$\bigcup_{n=1}^{\infty} \Gamma(\mathscr{P}(\Delta_\Gamma), \mathcal{O}_{\mathscr{P}(\Delta_\Gamma)}(n\mathscr{D})).$$

PROOF. Let $\mathscr{D}_0$ meet the component of $\mathscr{P}(\Delta_\Gamma)_0$ corresponding to the vertex $\{M\} \in \Delta_\Gamma^{(0)}$. Consider the projection:

$$p : \mathscr{P}(\Delta_\Gamma) \to \boldsymbol{P}(M)^\wedge.$$

Then $\mathscr{D}$ is the inverse image of a relative Cartier divisor in $\boldsymbol{P}(M)^\wedge$. By Grothendieck's existence Theorem (EGA III.5), this divisor is the formal completion of a relative Cartier divisor $D \subset \boldsymbol{P}(M)$. We have homomorphisms:

$$\Gamma(\boldsymbol{P}(M), \mathcal{O}_{\boldsymbol{P}(M)}(nD)) \approx \Gamma(\boldsymbol{P}(M)^\wedge, \mathcal{O}_{\boldsymbol{P}(M)^\wedge}(n\hat{D})) \underset{p^*}{\hookrightarrow} \Gamma(\mathscr{P}(\Delta_\Gamma), \mathcal{O}_{\mathscr{P}(\Delta_\Gamma)}(n\mathscr{D})),$$

the first being an isomorphism by EGA III.4.1. Since $R(\boldsymbol{P}_K^1)$ is the quotient field of

$$\bigcup_{n=1}^{\infty} \Gamma(\boldsymbol{P}(M), \mathcal{O}_{\boldsymbol{P}(M)}(nD)),$$

the proposition will follow if we show that the second of these maps is an isomorphism too. This follows from:

LEMMA (4.8). *Let $\Delta_* \subset \Delta_\Gamma$ be any finite subtree. Let $p : \mathscr{P}(\Delta_\Gamma) \to \boldsymbol{P}(\Delta_*)^\wedge$ be the projection. Then*

$$p_*(\mathcal{O}_{\mathscr{P}(\Delta_\Gamma)}) = \mathcal{O}_{\boldsymbol{P}(\Delta_*)^\wedge}.$$

PROOF. It suffices to prove that for affine open affine $U \subset \boldsymbol{P}(\Delta_*)_0$ and every ideal $I \subset A$ of finite codimension, that

$$\Gamma(U, \mathcal{O}_{\boldsymbol{P}(\Delta_*)}/I \cdot \mathcal{O}_{\boldsymbol{P}(\Delta_*)}) \to \Gamma(p^{-1}(U), \mathcal{O}_{\mathscr{P}(\Delta_\Gamma)}/I \cdot \mathcal{O}_{\mathscr{P}(\Delta_\Gamma)})$$

is an isomorphism. We check this by induction on $\dim_k(A/I)$. If $I = m$, note that the open subscheme $p^{-1}(U)$ of the closed fibre $\mathscr{P}(\Delta_\Gamma)_0$ is obtained from U by adding infinite trees of $\boldsymbol{P}_k^1$'s at a finite set of points of U. Since global sections of $\mathcal{O}_{\boldsymbol{P}_k^1}$ are constants in k, a section on $p^{-1}(U)$ of $\mathcal{O}_{\mathscr{P}(\Delta_\Gamma)_0}$ is just a section on U of $\mathcal{O}_{\boldsymbol{P}(\Delta_*)_0}$ extended as a constant to each of these trees. The result is true when $I = m$. In general, if $I_0 = I + A \cdot \eta$, where $m \cdot \eta \subset I$, then by flatness of $\mathscr{P}(\Delta_\Gamma)$ and $\boldsymbol{P}(M)$ over S, we get a diagram:

$$\begin{array}{ccccccc}
0 \to \mathcal{O}_{\boldsymbol{P}(\Delta_*)}/m \cdot \mathcal{O}_{\boldsymbol{P}(\Delta_*)} & \xrightarrow{\eta} & \mathcal{O}_{\boldsymbol{P}(\Delta_*)}/I \cdot \mathcal{O}_{\boldsymbol{P}(\Delta_*)} & \to & \mathcal{O}_{\boldsymbol{P}(\Delta_*)}/I_0\, \mathcal{O}_{\boldsymbol{P}(\Delta_*)} \to 0 \\
\downarrow \alpha & & \downarrow \beta & & \downarrow \gamma \\
0 \to p_*\mathcal{O}_{\mathscr{P}(\Delta_\Gamma)}/m \cdot \mathcal{O}_{\mathscr{P}(\Delta_\Gamma)} & \xrightarrow{\eta} & p_*\mathcal{O}_{\mathscr{P}(\Delta_\Gamma)}/I \cdot \mathcal{O}_{\mathscr{P}(\Delta_\Gamma)} & \to & p_*\mathcal{O}_{\mathscr{P}(\Delta_\Gamma)}/I_0\, \mathcal{O}_{\mathscr{P}(\Delta_\Gamma)}.
\end{array}$$

Then α and γ are isomorphisms by induction, hence so is β.

Note that plenty of such $\mathscr{D}$'s exist: just take a closed point x of the closed fibre $\mathscr{P}(\Delta_\Gamma)_0$ where $\mathscr{P}(\Delta_\Gamma)$ is smooth over S: let $f \in \mathcal{O}_{x,\mathscr{P}(\Delta_\Gamma)}$ be an element such that

$$m_{x,\mathscr{P}(\Delta_\Gamma)} = m \cdot \mathcal{O}_{x,\mathscr{P}(\Delta_\Gamma)} + f \cdot \mathcal{O}_{x,\mathscr{P}(\Delta_\Gamma)}.$$

Then $f = 0$ defines such a $\mathscr{D}$. Therefore we have:

COROLLARY (4.9). *In the situation of Corollary (4.6), the map induced by $\tilde{\varphi}$ on the meromorphic functions restricts to an isomorphism:*

$$\varphi^* : R(\boldsymbol{P}_K^1) \to R(\boldsymbol{P}_K^1).$$

If φ^* is given by an element $g \in \mathrm{PGL}(2, K)$. then $\alpha : \Gamma_1 \to \Gamma_2$ is given by

$$\alpha(\gamma) = g\gamma g^{-1}$$

PROOF. The first part follows immediately from (4.8) since such $\mathscr{D}$'s exist and $\tilde{\varphi}$ takes such a $\mathscr{D}$ to another such $\mathscr{D}$. Since $\tilde{\varphi}$ is α-equivariant, so is φ^*, i.e. acting on $R(\boldsymbol{P}_K^1)$, we find

$$\varphi^* \circ \gamma = \alpha(\gamma) \circ \varphi^*, \text{ all } \gamma \in \Gamma_1.$$

Hence if φ^* is given by the action of $g \in PGL(2, K)$ on $R(\boldsymbol{P}_K^1)$, $g\gamma = \alpha(\gamma)g$, all $\gamma \in \Gamma_1$. *Q.E.D.*

COROLLARY (4.10).

$$\begin{aligned}\mathrm{Isom}_{\bar{K}}((P_\Gamma)_{\bar{\eta}}, (P_{\Gamma_2})_{\bar{\eta}}) &= \mathrm{Isom}_S(P_{\Gamma_1}, P_{\Gamma_2}) \\ &= \{g \in PGL(2, K) | g\Gamma_1 g^{-1} = \Gamma_2\}/\text{modulo } g \sim g',\end{aligned}$$

where $g \sim g'$ if $\exists \gamma_2 \in \Gamma_2$ such that $g' = \gamma_2 g$, or equivalently $\exists \gamma_1 \in \Gamma_1$ such that $g' = g\gamma_1$.

PROOF. It suffices to note that everything can be reversed: given $g \in PGL(2, K)$ such that $g\Gamma_1 g^{-1} = \Gamma_2$, then g defines a $\tilde{\varphi}$ and an α, hence a $\hat{\varphi}$, hence a φ. *Q.E.D.*

COROLLARY (4.11). $(P_{\Gamma_1})_{\bar{\eta}}$ *is isomorphic to* $(P_{\Gamma_2})_{\bar{\eta}}$ *if and only if Γ_1 is conjugate to Γ_2 in $PGL(2, K)$. All isomorphisms are, moreover, rational over K.*

COROLLARY (4.12). Aut $((P_\Gamma)_{\bar{\eta}})$ *is isomorphic to $N(\Gamma)/\Gamma$, where $N(\Gamma)$ is the normalizer of Γ in $PGL(2, K)$. All automorphisms are, moreover, rational K.*

We turn finally to the question of the existence of these uniformizations. We wish to start only with a stable curve C/S with non-singular generic fibre and k-split degenerate closed fibre and reverse the construction step-by-step.

Step I: Let $\mathscr{C}$ be the formal completion of C along its closed fibre. Note that C must be normal, hence so is $\mathscr{C}$.

Step II: Let $p_0 : P_0 \to C_0$ be the universal covering scheme of C_0 and let Γ be the group of cover transformations. It is clear that P_0 is an infinite union of copies of $\mathbf{P}^1_k$, each one joined to a finite number of others (at least 3 others) at k-rational double points, but the whole being connected as a tree. More precisely, if we make a graph Δ (resp. G) to illustrate P_0 (resp. C_0) in the usual way (one vertex for each component, one edge for each double point), then Δ is a tree, Γ acts freely on Δ and $\Delta/\Gamma \cong G$.

Step III: Since the category of étale coverings of $\mathscr{C}$ and of C_0 are equivalent, there is a unique formal scheme $\mathscr{P}$ with closed fibre P_0, and formal étale morphism

$$p : \mathscr{P} \to \mathscr{C}$$

extending p_0. Moreover, Γ acts on $\mathscr{P}$ so that $\mathscr{C} \cong \mathscr{P}/\Gamma$.

Step IV: For each component M of $\mathscr{P}_0$, let $\mathscr{D} \subset \mathscr{P}$ be a positive relative Cartier divisor such that $\mathscr{D}_0$ meets only the component M of $\mathscr{P}_0$. Let:

$$\mathbf{P}(M) = \text{Proj} \sum_{n=0}^{\infty} \Gamma(\mathscr{P}, \mathcal{O}_{\mathscr{P}}(n\mathscr{D})).$$

PROPOSITION (4.13). *$\mathbf{P}(M) \cong \mathbf{P}^1 \times S$, and there is a canonical formal morphism*:

$$p_M : \mathscr{P} \to \mathbf{P}(M)^\wedge$$

which on the closed fibre $\mathscr{P}_0$ maps every component $M' \neq M$ to a point and maps M isomorphically onto $\mathbf{P}(M)_0$.

PROOF. First we need:

LEMMA (4.14). $H^1(\mathscr{P}_0, \mathcal{O}_{\mathscr{P}_0}(n\mathscr{D})) = (0)$, all $n \geqq 0$.

PROOF. Let $\mathscr{P}'_0$ be the disjoint union of the components of $\mathscr{P}_0$, and let $q : \mathscr{P}'_0 \to \mathscr{P}_0$ be the obvious morphism. We have an exact sequence:

$$0 \to \mathcal{O}_{\mathscr{P}_0} \to q_*(\mathcal{O}_{\mathscr{P}'_0}) \to \bigoplus_{\substack{\text{double}\\ \text{points } x}} k(x) \to 0.$$

It is obvious that

$$H^1(q_*(\mathcal{O}_{\mathscr{P}'_0}) \cong H^1(\mathcal{O}_{\mathscr{P}'_0}) \cong \prod_{\substack{\text{components}\\ M}} [H^1(\mathcal{O}_M)] = (0),$$

and that

$$(\prod_{\substack{\text{components}\\ M}} [H^0(\mathcal{O}_M)]) = H^0(\mathcal{O}_{\mathscr{P}'_0}) \to \prod_{\substack{\text{double}\\ \text{points } x}} k(x)$$

is surjective (since $\mathscr{P}_0$ is connected together like a tree!). Therefore $H^1(\mathcal{O}_{\mathscr{P}_0}) = (0)$. Moreover, if $n > 0$, use

$$0 \to \mathcal{O}_{\mathscr{P}_0} \to \mathcal{O}_{\mathscr{P}_0}(n\mathscr{D}_0) \to \mathscr{F} \to 0$$

where $\dim(\operatorname{Supp}\mathscr{F}) = 0$. From this it follows that $H^1(\mathcal{O}_{\mathscr{P}_0}(n\mathscr{D}_0)) = (0)$ too.

LEMMA (4.15). *For all $n \geqq 0$, $\Gamma(\mathscr{P}, \mathcal{O}_{\mathscr{P}}(n\mathscr{D}))$ is a finitely generated free A-module such that*

$$\Gamma(\mathscr{P}, \mathcal{O}_{\mathscr{P}}(n\mathscr{D})) \otimes_A k \cong \Gamma(\mathscr{P}_0, \mathcal{O}_{\mathscr{P}_0}(n\mathscr{D}_0)) \cong \Gamma(M, \mathcal{O}_M(n\mathscr{D}_0)).$$

PROOF. It suffices to prove that for all $I \subset A$ of finite codimension, $\Gamma(\mathscr{P}, \mathcal{O}_{\mathscr{P}}(n)/I \cdot \mathcal{O}_{\mathscr{P}}(n\mathscr{D}))$ is a finitely generated free A/I-module such that

$$\Gamma(\mathscr{P}, \mathcal{O}_{\mathscr{P}}(n\mathscr{D})/I \cdot \mathcal{O}_{\mathscr{P}}(n\mathscr{D})) \otimes_A A/m = \Gamma(\mathscr{P}_0, \mathcal{O}_{\mathscr{P}_0}(n\mathscr{D}_0)).$$

If $I = m$, note that every section of $\mathcal{O}_{\mathscr{P}_0}(n\mathscr{D}_0)$ is just a section of $\mathcal{O}_M(n\mathscr{D}_0)$ extended as a constant to all other components of $\mathscr{P}_0$, hence

$$\text{res}\colon \Gamma(\mathscr{P}_0, \mathcal{O}_{\mathscr{P}_0}(n\mathscr{D}_0)) \overset{\approx}{\to} \Gamma(M, \mathcal{O}_{\mathscr{P}_0}(n\mathscr{D}_0))$$

is an isomorphism. In general, use induction of $\dim A/I$. If $I_0 = I + A \cdot \eta$, where $m \cdot \eta \subset I$, we get a diagram (because of *flatness* of $\mathscr{P}$ over S):

$$0 \to H^0(\mathcal{O}_{\mathscr{P}_0}(n\mathscr{D}_0)) \xrightarrow{\eta} H^0(\mathcal{O}_{\mathscr{P}}(n\mathscr{D})/I \cdot \mathcal{O}_{\mathscr{P}}(n\mathscr{D}))$$
$$\to H^0(\mathcal{O}_{\mathscr{P}}(n\mathscr{D})/I_0 \cdot \mathcal{O}_{\mathscr{P}}(n\mathscr{D})) \to H^1(\mathcal{O}_{\mathscr{P}_0}(n\mathscr{D}_0)) = (0).$$

Using this, the assertion for I_0 implies immediately the assertion for I. *Q.E.D.*

It follows from (4.15) that $\sum_{n=0}^{\infty} \Gamma(\mathscr{P}, \mathcal{O}_{\mathscr{P}}(n\mathscr{D}))$ is a free finitely generated graded A-algebra, hence its Proj is a flat and proper scheme over S; moreover, its closed fibre is just

$$\operatorname{Proj}\{\sum_{n=0}^{\infty} \Gamma(M, \mathcal{O}_M(n\mathscr{D}_0))\},$$

which is M itself. Since $M \cong \boldsymbol{P}_k^1$ and all deformations of $\boldsymbol{P}^1$ are trivial, this proves that $\boldsymbol{P}(M) \cong \boldsymbol{P}^1 \times S$. Finally, $\mathcal{O}_{\mathscr{P}_0}(\mathscr{D}_0)$ is generated by its global sections, hence by (4.15), $\mathcal{O}_{\mathscr{P}}(\mathscr{D})$ is generated by its sections. Therefore there is a formal morphism from $\mathscr{P}$ to $\boldsymbol{P}(M)\hat{}$. Since $\mathscr{D}_0$ is very ample on M and since all sections of $\mathcal{O}_{\mathscr{P}_0}(n\mathscr{D}_0)$ (any $n \geqq 0$) are constant on all other components of $\mathscr{P}_0$, the last assertions of the Proposition are obvious. *Q.E.D.*

PROPOSITION (4.16). *For any 2 components M_1, M_2 of $\mathscr{P}_0$, consider the morphism $p_{M_1,M_2}\colon \mathscr{P} \to (\boldsymbol{P}(M_1) \times_S \boldsymbol{P}(M_2))\hat{}$. There is a unique* *relative Carter divisor* $Z \subset \mathbf{P}(M_1)\times_S \mathbf{P}(M_2)$ *defined by an equation*:

$$ax_1x_2+bx_1y_2+cy_1x_2+dy_1y_2 = 0, \quad a, b, c, d \in A$$

(where x_i, y_i *are homogeneous coordinates on* $\mathbf{P}(M_i)$*) such that* p_{M_1,M_2} *factors through Z. Moreover* $ad - bc \neq 0$ *but* $ad-bc \in m$.

PROOF. Via the isomorphism $\mathbf{P}(M_i) \cong \mathbf{P}^1 \times S$, let the sheaf $\mathcal{O}(1) \otimes \mathcal{O}_S$ go over to the sheaf L_i on $\mathbf{P}(M_i)$. Then $\Gamma(\mathbf{P}(M_i), L_i) \cong A\cdot x_i \oplus A\cdot y_i$. Let $K = p_1^*L_1 \otimes p_2^*L_2$ on $\mathbf{P}(M_1)\times_S\mathbf{P}(M_2)$. Then:

$$\Gamma(\mathbf{P}(M_1)\times_S\mathbf{P}(M_2), K) \cong Ax_1x_2 \oplus Ay_1x_2 \oplus Ax_1y_2 \oplus Ay_1y_2.$$

Let $\hat{L}_i = p_{M_i}^*(L_i)$, $\hat{K} = p_{M_1,M_2}^*(K)$ be the induced sheaves on $\mathscr{P}$, and let $L_{i,0}$, K_0 be the induced sheaves on $\mathscr{P}_0$. The first step is to check that:

a) $H^0(\mathscr{P}_0, K_0) \cong H^0(\mathbf{P}(M_1)_0\times\mathbf{P}(M_2)_0, K_0)$/modulo 1-dimensional subspace of form $\lambda(\alpha x_1+\beta y_1)\cdot(\gamma x_2+\delta y_2)$,

b) $H^1(\mathscr{P}_0, K_0) = (0)$.

In fact, $L_{i,0}$ is a trivial invertible sheaf on all components of $\mathscr{P}_0$ except M_i; it follows easily that if we pick arbitrary sections in the 2-dimensional spaces:

$$H^0(M_1, K_0 \otimes \mathcal{O}_{M_1}), \quad H^0(M_2, K \otimes \mathcal{O}_{M_2}),$$

they extend to at most one section of $\mathscr{P}_0$, and that there is one condition for them to do so, namely that they induce the same constant section in the link between M_1 and M_2:

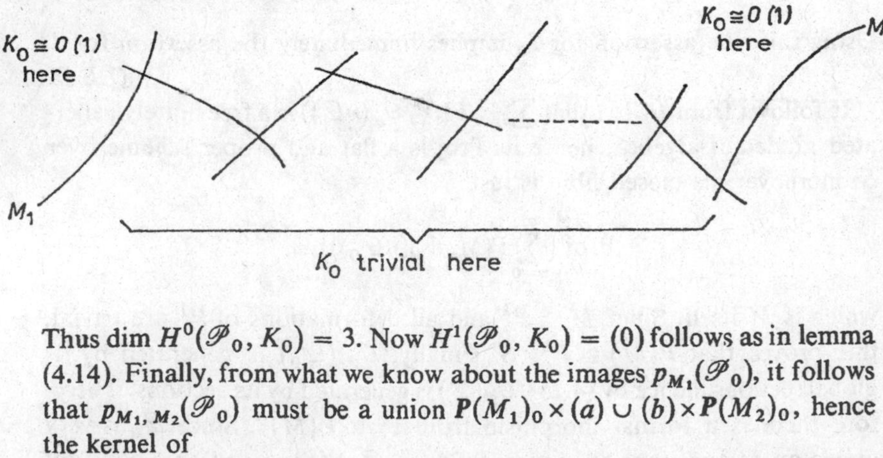

Thus dim $H^0(\mathscr{P}_0, K_0) = 3$. Now $H^1(\mathscr{P}_0, K_0) = (0)$ follows as in lemma (4.14). Finally, from what we know about the images $p_{M_1}(\mathscr{P}_0)$, it follows that $p_{M_1,M_2}(\mathscr{P}_0)$ must be a union $\mathbf{P}(M_1)_0\times(a) \cup (b)\times\mathbf{P}(M_2)_0$, hence the kernel of

$$H^0(\mathbf{P}(M_1)_0\times\mathbf{P}(M_2)_0, K_0) \xrightarrow{p_{M_1,M_2}} H^0(\mathscr{P}_0, K_0)$$

is 1-dimensional and generated by an element $(\alpha x_1+\beta y_1)(\gamma x_2+\delta y_2)$.

The second step is that $H^0(\mathscr{P}, K)$ is a free A-module of rank 3 such that

$$H^0(\mathscr{P}, \hat{K}) \otimes_A k \cong H^0(\mathscr{P}_0, K_0).$$

This follows from (a) and (b) just as in lemma (4.15). Therefore

$$H^0(\boldsymbol{P}(M_1) \times_S \boldsymbol{P}(M_2), K) \to H^0(\mathscr{P}, \hat{K})$$

is a homomorphism of a free rank 4 module to a free rank 3 module; after $\otimes_A k$, it becomes surjective, so it is already surjective. Therefore $H^0(\boldsymbol{P}(M_1) \times_S \boldsymbol{P}(M_2), K) \cong H^0(\mathscr{P}, \hat{K}) \oplus A \cdot f$, where $f = ax_1x_2 + \cdots + dy_1y_2$ and $f \equiv (\alpha x_1 + \beta y_1)(\gamma x_2 + \delta y_2) \bmod m$. Thus $f = 0$ defines a relative Cartier divisor Z through which p_{M_1, M_2} factors. Finally, if $ad - bc = 0$, then f splits into a product over A as well as over k; then Z is reducible: say $Z = Z_1 \cup Z_2$. If $\mathscr{P}_i = p^{-1}_{M_1, M_2}(Z_i)$, then $\mathscr{P} = \mathscr{P}_1 \cup \mathscr{P}_2$. But $\mathscr{P}$ is normal and connected so this is absurd. *Q.E.D.*

COROLLARY (4.17). *Let $R(\boldsymbol{P}(M))$ be the field of rational functions on $\boldsymbol{P}(M)$ and let $p_M^* R(\boldsymbol{P}(M))$ be the induced field of meromorphic functions on $\mathscr{P}$. Then*

$$p_{M_1}^* R(\boldsymbol{P}(M_1)) = p_{M_2}^* R(\boldsymbol{P}(M_2))$$

for any 2 components M_1, M_2 of $\mathscr{P}_0$.

PROOF. In fact, p_{M_i} factors:

$$\mathscr{P} \xrightarrow{p_{M_1, M_2}} Z \xrightarrow{q_i} \boldsymbol{P}(M_i)$$

and since $ad - bc \neq 0$, Z is irreducible and $q_i^* R(\boldsymbol{P}(M_i)) = R(Z)$. Thus $p_{M_1}^* R(\boldsymbol{P}(M_1)) = p_{M_1, M_2}^* R(Z) = p_{M_2}^* R(\boldsymbol{P}(M_2))$. *Q.E.D.*

Step V. Choose once and for all an isomorphism:

(*) $\quad p_M^* R(\boldsymbol{P}(M)) \cong R(\boldsymbol{P}_K^1)$, the field of rational functions on $\boldsymbol{P}_K^1$.

The isomorphism of $p_M^* R(\boldsymbol{P}(M))$ with $R(\boldsymbol{P}_K^1)$ induces an isomorphism of the generic fibre $\boldsymbol{P}(M)_\eta$ with $\boldsymbol{P}_K^1$. Thus $\boldsymbol{P}(M)$ becomes a $\boldsymbol{P}^1$-bundle over S with generic fibre $\boldsymbol{P}_K^1$, i.e. $\boldsymbol{P}(M)$ define an element $\{M\} \in \Delta^{(0)}$. Thus we have associated an element of $\Delta^{(0)}$ to each component of $\mathscr{P}_0$. Let $\Delta_*^{(0)}$ be the set of elements of $\Delta^{(0)}$ that we get. By Prop. (4.16), the join Z of 2 $\boldsymbol{P}(M)$'s is flat over S with reducible closed fibre: therefore if $M_1 \neq M_2$ are 2 components of $\mathscr{P}_0$, the corresponding elements of $\Delta^{(0)}$ are distinct and compatible by (2.1). Since $\mathscr{P}_0$ has only double points, it follows from (2.3) that $\Delta_*^{(0)}$ is a linked subset. I claim that

$$\mathscr{P} \cong \mathscr{P}(\Delta_*).$$

In fact, it is easy to see that there is a formal morphism $\pi : \mathscr{P} \to \mathscr{P}(\Delta_*)$ which is an isomorphism on the closed fibre. Then apply the easy:

LEMMA (4.18). *Let $f : \mathscr{X} \to \mathscr{Y}$ be a formal morphism of formal schemes over S, whose topologies are defined by the ideals $m \cdot \mathcal{O}_{\mathscr{X}}$, $m \cdot \mathcal{O}_{\mathscr{Y}}$ respectively. If $f_0 : \mathscr{X}_0 \to \mathscr{Y}_0$ is an isomorphism and $\mathscr{X}$ is flat over S, then f is an isomorphism.*

Step VI. By Cor. (4.17), Γ leaves invariant the field of meromorphic functions $p_M^* R(P(M))$ on $\mathscr{P}$, hence by our basic identification of this field with $R(P)_K^1$, Γ acts faithfully on $R(P_K^1)$. This induces an embedding

$$\Gamma \hookrightarrow PGL(2, K).$$

Identifying Γ with its image here, then Γ maps $\Delta_*^{(0)}$ into itself and this induces an action of Γ on $\mathscr{P}(\Delta_*)$, and hence induces an action of Γ on $\mathscr{P}$ equal to the one we started with. It remains only to prove:

PROPOSITION (4.19). *Γ is a flat Schottky group and $\Delta_* = \Delta_\Gamma$.*

PROOF. Let $\gamma \in \Gamma$, $\gamma \neq e$. Since γ acts freely on the tree Δ_*, it leaves fixed 2 ends $x, y \in \partial\Delta_*$. Therefore γ leaves fixed $ix, iy \in KP^1$. Let $\{M\}$ be a vertex on the line in Δ_* joining the end x to the end y. Then ix and iy are represented by homogeneous coordinates $u, v \in K+K$ such that $M = A \cdot u + A \cdot v$. Reordering x and y if necessary, we can assume that $\gamma\{M\}$ separates $\{M\}$ from x. Then $\gamma\{M\}$ is represented by a module N such that

$$M \not\supseteq N \ni u,$$
$$u \notin mM.$$

But if we lift γ to an element $\tilde{\gamma} \in GL(2, K)$, then

$$\tilde{\gamma}(M) = \lambda \cdot N, \text{ some } \lambda \in K^*,$$
$$\tilde{\gamma}(u) = \sigma \cdot u,$$
$$\tilde{\gamma}(v) = \tau \cdot v.$$

Then $N = \langle \sigma/\lambda u, \tau/\lambda v \rangle$, so $\sigma/\lambda \in A^*$, $\tau/\lambda \in m$. Then

$$t^{-1}(\gamma) = \frac{\sigma\tau}{(\sigma+\tau)^2} = \frac{\sigma/\lambda \cdot \tau/\lambda}{(\sigma/\lambda+\tau/\lambda)^2} = \text{unit} \cdot \frac{\tau}{\lambda} \in m.$$

Therefore γ is hyperbolic.

Next, the fixed points of the elements of Γ are contained in the set $i(\partial\Delta_*)$, hence by Prop. (1.22), any 4 of them have a cross-ratio in A or A^{-1}. Thus Γ is a flat Schottky group. Moreover, $\Delta_\Gamma^{(0)} \subset \Delta_*^{(0)}$ since by Prop. (1.20) $\{M(x, y, z)\} \in \Delta_*^{(0)}$ for any 3 fixed points x, y, z of Γ. Conversely, say v is a vertex of Δ_*. Since C_0 was a stable curve, $\mathscr{P}_0$ has the property that every component has at least 3 double points on it. Therefore every vertex of Δ_* is an endpoint of at least 3 edges. Take 3 edges meeting v. In Δ_*/Γ, choose 3 loops starting and ending at the image of v that start off on these edges. Let these loops define $\gamma_1, \gamma_2, \gamma_3 \in \Gamma \cong \pi_1(\Delta_*/\Gamma)$. Then the paths from v to $\gamma_i v$ start on these 3 edges. Let $\gamma_i^n v$ tend to an end $x_i \in \partial\Delta_*$. Then $i(x_i)$ is a fixed point of γ_i, hence

$$v = \{M(ix_1, ix_2, ix_3)\} \in \Delta_\Gamma^{(0)}. \qquad Q.E.D.$$

This completes the proof of:

THEOREM (4.20): *Every stable curve over S with non-singular generic fibre and k-split degenerate closed fibre is isomorphic to P_Γ for a unique flat Schottky group $\Gamma \subset PGL(2, K)$.*

BIBLIOGRAPHY

F. BRUHAT AND J. TITS

[B-T] Groupes algébriques semi-simple sur un corps local, To appear in Publ. I.H.E.S.

P. DELIGNE AND D. MUMFORD

[D-M] The irreducibility of the space of curves of given genus. Publ. I.H.E.S. 36 (1969) 75–109.

A. GROTHENDIECK AND J. DIEUDONNÉ

[EGA] Eléments de la geométrie algébrique, Publ. I.H.E.S., 4, 8, 11, etc.

H. GRAUERT AND R. REMMERT

[G-R] Forthcoming book on the foundations of global p-adic function theory.

Y. IHARA

[I] On discrete subgroups of the 2×2 projective linear group over p-adic fields, J. Math. Soc. Japan 18 (1966) 219–235.

S. LICHTENBAUM

[L] Curves over discrete valuation rings, Amer. J. Math. 90 (1968) 380–000.

D. MUMFORD

[M1] Abelian varieties, Oxford Univ. Press, 1970.

B. MASKIT

[Ma] A characterization of Schottky groups, J. d'Analyse 19 (1967) 227–230.

J. MCCABE

[Mc] Harvard Univ. thesis on P-adic theta functions, (1968) unpublished.

H. MORIKAWA

[Mo] Theta functions and abelian varieties over valuation rings, Nagoya Math. J. 20 (1962).

M. RAYNAUD

[R] Modèles de Néron, C.R. Acad. Sci., Paris 262 (1966) 413–414.

J.-P. SERRE

[S] Groupes Discrets, Mimeo. notes from course at College de France, 1968–69.

I. ŠAFAREVITCH

[Š] Lectures on minimal models, Tata Institute Lecture Notes, Bombay, 1966.

(Oblatum 16–III–71)

Mathematics Institute
University of Warwick
Coventry, Warks
England

PATHOLOGIES IV.

By David Mumford.

In this note I would like to use the beautifully simple method introduced by Tony Iarrobino [1]—when he proved that there are 0-dimensional subschemes of $\mathbf{P}^3$ which are not specializations of reduced subschemes—to prove here that there are also reduced and irreducible complete curves which are not specializations of non-singular curves. Since there are no global obstructions in deforming reduced curves, this also shows that there are complete reduced 1-dimensional local rings with no flat deformation which is generically smooth.

Start with a complete non-singular curve C of genus g with no automorphisms over an algebraically closed ground field k. Choose a point $x \in C$ and a large even integer ν. Note that if V is any k-vector space where

$$m_{x,C}^{2\nu} \subset V \subset m_{x,C}^{\nu}$$

then $k+V$ is a subring of $\mathcal{O}_{x,C}$. For each such V, define a new curve:

$$\pi : C \to C(V)$$

by:

(a) π is a bijection, and an isomorphism

$$\operatorname{res} \pi : C - \{x\} \xrightarrow{\approx} C(V) - \{\pi x\}.$$

(b) $\mathcal{O}_{\pi x, C(V)} = k + V.$

Note that if V_1, V_2 are two such vector spaces, then

$$C(V_1) \approx C(V_2) \Rightarrow V_1 = V_2.$$

(In fact, C is the normalization of each $C(V)$; hence any $o : C(V_1) \xrightarrow{\approx} C(V_2)$ lifts to $o' : C \to C$ which must be the identity, hence $k + V_1 = \mathcal{O}_{\pi(x), C(V_1)} = \mathcal{O}_{\pi(x), C(V_2)} = k + V_2$.) Moreover, the curves $C(V)$ can all be fitted together into

Manuscript received December 14, 1973.

American Journal of Mathematics, Vol. 97, No. 3, pp. 847–849

a family if we fix the integer $\dim_k V/m_{x,C}^{2\nu}$: for all k, $0 \leqslant k \leqslant \nu$,

let G = Grassmanian of k-dimensional subspaces of $m_{x,C}^{\nu}/m_{x,C}^{2\nu}$,

let $\mathcal{V} \subset (m_{x,C}^{\nu}/m_{x,C}^{2\nu}) \otimes_k \mathcal{O}_G$ be the universal family,

let $C(\mathcal{V})$ be the scheme equal to $C \times G$ as topological space, with structure sheaf defined by:

$$\begin{array}{ccc}
\mathcal{O}_{C\times G} & \xrightarrow{\alpha} & (\mathcal{O}_{x,C}/m_{x,C}^{2\nu}) \otimes_k \mathcal{O}_G \longrightarrow 0 \\
 & & \cup \\
\cup & & [k + (m_{x,C}^{\nu}/m_{x,C}^{2\nu})] \otimes_k \mathcal{O}_G \\
 & & \cup \\
\alpha^{-1}(\mathcal{V} + \mathcal{O}_G) & - - - \rightarrow & \mathcal{V} + \mathcal{O}_G \\
\| \text{def.} & & \\
\mathcal{O}_{C(\mathcal{V})} & &
\end{array}$$

Since $[(\mathcal{O}_{x,C}/m_{x,C}^{2\nu}) \otimes_k \mathcal{O}_G]/\mathcal{V}$ is a locally free $\mathcal{O}_G$-sheaf, $\mathcal{O}_{C(\mathcal{V})}$ is flat over $\mathcal{O}_G$, i.e., $C(\mathcal{V})$ is flat over G.

Now choose $k = \nu/2$ and calculate:

(i) $\dim G = k(\nu - k) = \nu^2/4$

(ii) $p_a(C(V)) = g + \dim_k [\mathcal{O}_{x,C}/\mathcal{O}_{\pi(x),C(V)}]$

$$= g + \left(\frac{3\nu}{2} - 1\right)$$

Therefore if $\nu \gg 0, \dim G \geqslant 3p_a(C(V)) - 3$! I claim that this implies that almost all the curves $C(V)$ are not specializations of non-singular curves, because of:

Lemma. *Let $p: \mathcal{C} \to S$ be a flat and proper family of reduced and irreducible singular curves $C_s = p^{-1}(s)$ such that*

(a) *$\forall s \in S$, $\{s' | C_{s'} \approx C_s\}$ is finite*

(b) *$p_a(C_s) \geqslant 2$, S is irreducible and $\dim S \geqslant 3p_a(C_s) - 3$,*

then almost all curves C_s are not specializations of non-singular curves.

Proof. If the conclusion is false, then after replacing S by a Zariski open subset we can extend the family $\mathcal{C}/S$ like this:

$$\begin{array}{ccc}
\mathcal{C} & \longrightarrow & \mathcal{C}^* \\
\downarrow & & \downarrow \\
S & \longrightarrow & S^*
\end{array} \quad ; \quad \begin{array}{l} S^* \text{ irreducible,} \\ \dim S^* = \dim S + 1 \end{array}$$

so that $\mathcal{C}^*$ is generically smooth over S^*. In fact $\mathcal{C}$ will carry a relatively ample L, so we may use $p_*L^{\otimes n}$ $(n \gg 0)$ to embed $\mathcal{C}$ in some $\mathbf{P}^N$-bundle $\mathcal{P}$ over S. Moreover, if a C_S $(s \in S_0)$ is abstractly a specialization of a non-singular curve, so is the embedded curve $C_S \subset \mathbf{P}^N$. So take S^* to be a suitable subvariety of the Hilbert scheme of $\mathcal{P}$ over S. Once we have $\mathcal{C}^*/S^*$, consider the two induced families:

$$\mathcal{C}_i^* = \mathcal{C}^* \times_{S^*} (S^* \times S^*) \qquad \text{(formed via } p_i : S^* \times S^* \rightarrow S^*, i = 1,2)$$

and the scheme

$$I = \mathbf{Isom}_{S^* \times S^*}(\mathcal{C}_1^*, \mathcal{C}_2^*)$$

whose points over $(s_1, s_2) \in S^* \times S^*$ are isomorphisms $o: C_{s_1} \rightarrow C_{s_2}$. Look at the morphisms:

$$q \; \begin{array}{c} I \\ \downarrow \!\! \big) \\ S^* \end{array} \; \delta \qquad \begin{array}{l} q(o: C_{s_1} \rightarrow C_{s_2}) = s_1 \\ \delta(s) = [\text{id.}: C_s \rightarrow C_s] \end{array}$$

Since $\dim S^* = \dim S + 1 > 3p_a(C_s) - 3$, whenever C_s is non-singular, the same non-singular curve must occur in the family $\mathcal{C}^*/S^*$ infinitely often; thus when C_s is non-singular, some component of $q^{-1}(s)$ through $\delta(s)$ is positive-dimensional. Now by upper semi-continuity of dimensions of fibres of a morphism, it follows that for every s, $q^{-1}(s)$ has a positive-dimensional component through $\delta(s)$. Now let $D_1 = \text{Im}(S \rightarrow S^*)$, $D_2 = \{s | C_s \text{ is singular}\}$; then $\overline{D}_1$ is a component of D_2 and let $D_1^0 = D_1$-(closure of $D_2 - D_1$). Choose $s \in D_1^0$ and consider how $q^{-1}(s)$ can have a positive-dimensional component γ through $\delta(s)$. By (b), Aut(C_s) is finite; by (a), there are only finitely many $s' \in D_1$ with $C_s \approx C_{s'}$; certainly $C_s \not\approx C_{s'}$ if $s' \in S^* - D_2$ because C_s is singular while $C_{s'}$ is non-singular; and since $s \not\in$ (closure $D_2 - D_1$), γ cannot lie over $(s) \times \overline{D_2 - D_1}$ in $S^* \times S^*$. Thus there is nowhere for γ to go! Contradiction.

REFERENCES.

[1] A. Iarrobino, "Reducibility of the families of 0-dimensional schemes on a variety," *Inv. Math.* 15 (1972), pp. 72–77.

STABILITY OF PROJECTIVE VARIETIES

by David MUMFORD

LECTURES GIVEN AT THE "INSTITUT DES HAUTES ÉTUDES SCIENTIFIQUES", BURES-SUR-YVETTE (FRANCE), MARCH-APRIL 1976, UNDER THE SPONSORSHIP OF THE INTERNATIONAL MATHEMATICAL UNION. NOTES BY IAN MORRISON.

Extrait de *L'Enseignement Mathématique*
T. XXIII (1977) pp. 39–110

STABILITY OF PROJECTIVE VARIETIES

by David MUNFORD

CONTENTS

The most direct approach to the construction of moduli spaces of algebraic varieties is via the theory of invariants: one describes the varieties by some sort of numerical projective data, canonically up to the action of some algebraic group, and then seeks to make these numbers canonical by applying invariant polynomials to the data, or equivalently by forming a quotient of the data by the group action. The main difficulty in this approach is to prove that "enough invariants exist": their values on the projective data must distinguish non-isomorphic varieties.

Take as an example the moduli space $\mathcal{M}_g$ of curves of genus $g \geqq 2$ over some algebraically closed field k. Given C, such a curve, we obtain by choosing a basis B of $\Gamma(C, (\Omega_c^1)^{\otimes l})$, an embedding $\Phi: C \to \mathbf{P}^{(2l-1)(g-1)-1} = \mathbf{P}^N$. Let F be the Chow form of $\Phi(C)$ (cf. 1.16). Changing the basis B subjects $\Phi(C)$ to a projective transformation and F to the corresponding contragradient transformation. So if we could find "enough" polynomials I_λ in the coefficients of F which are invariant under this action of $SL(N+1)$ then the image of the map given by $C \mapsto (..., I_\lambda(F), ...)$ would be $\mathcal{M}_g$.

As of two years ago, this process could be carried out only when char $k = 0$ and C was smooth; and moduli spaces in characteristic p had to be constructed via the much more explicit theory of moduli of abelian varieties (cf. [14] and [15]). Since then, however, two very nice things have been proven:

a) W. Haboush [10] by making a systematic use of Steinberg representations has shown that all reductive groups are geometrically reductive (cf. Remark 1.2. vi). This was independently shown for $SL(n)$, by Processi and Formanek [25], using the idea that the group ring of an infinite permutation group has "radical" zero: i.e. for each $x \in R$, $x \neq 0$, there exists $y \in R$ such that xy

is not nilpotent. For a complete treatment of the new situation in characteristic p moduli problems see Seshadri [20].

b) D. Gieseker [9] using the concept of asymptotic stability (cf. 1.17) has established the numerical criterion for stability (c_s of 1.1) for surfaces of general type. Inspired by Gieseker's ideas, the author has extended this method to the "stable" curves of Deligne and Mumford [6]. (These are curves C with dim $H^1(C, \mathcal{O}_C) = g$, ordinary double points but no worse singularities and no smooth rational components meeting the remainder of the curve in fewer than three points; they are important because the most natural compactification $\overline{\mathcal{M}}_g$ of $\mathcal{M}_g$ is the moduli space for stable curves of genus g.) The power of the ideas of Gieseker is by no means exhausted. It looks like nice results may be possible for other surfaces, perhaps even for singular surfaces and the technique suggests several nice problems: in particular, it may lead to a proof of the surjectivity of the period map for K3 surfaces. The new ideas and results of these lectures are largely inspired by Gieseker's results (cf. especially corollary 3.2 below).

My goal is to outline this method and its applications, especially to the completed moduli spaces of curves $\overline{\mathcal{M}}_g$, indicating open problems. The field is moving ahead rapidly and may be greatly simplified in the near future.

We will work in general over an arbitrary ground field k.

§ 1. Stable points of representation, examples and Chow forms

For more details on the notations, definitions and properties which follow see Mumford [14], which we will call G.I.T. or Seshadri [20].

Fix k an algebraically closed field,

G a reductive algebraic group over k (i.e. $G =$ [semi-simple group $\times$ $\mathbf{G}_m^n$]/finite central subgroup),

V an n-dimensional representation of G,

$x \in V$.

There are three possibilities for x whose equivalent formulations are summarized in table 1.1 below.

1.1.

x unstable	x semi-stable	x stable
(a_u) $0 \in \overline{O^G(x)}$	(a_{ss}) $0 \notin \overline{O^G(x)}$	(a_s) i) $O^G(x)$ is closed in V ii) stab (x) is finite
(b_u) $\forall$ non-constant G-invariant homogeneous polynomials f $f(x) = 0$	(b_{ss}) $\exists$ a non-constant G-invariant homogeneous polynomial f s.t. $f(x) \neq 0$	(b_s) i) $\forall y \in V - O^G(x)$, $\exists$ a G-invariant polynomial f s.t. $f(x) \neq f(y)$ ii) tr $\deg_k k(V)^G = \dim V\text{-}\dim G$
(c_u) $\exists$ a 1-PS λ of G s.t. the weights of x with respect to λ are all positive	(c_{ss}) $\forall$ 1-PS's λ of G the weights of x with respect to λ are not all positive	(c_s) for all non-trivial 1-PS's λ of G, x has both positive and negative weights with respect to λ

1.2. Remarks. i) Recall that a 1-PS (one parameter subgroup) λ of G is just a homomorphism $\lambda\colon \mathbf{G}_m \to G$. Such λ can always be diagonalized in a suitable basis:

$$\lambda(t) = \begin{bmatrix} t^{r_1} & & & 0 \\ & \cdot & & \\ & & \cdot & \\ & & & \cdot \\ 0 & & & t^{r_n} \end{bmatrix}$$

If in this basis $x = (x_1, \ldots, x_n)$, the set of weights of x with respect to λ is the set of r_i for which $x_i \neq 0$.

ii) Unstable is *not* the opposite of stable, but of semi-stable. We will use non-stable as the opposite of stable.

iii) The important part of stability is the condition: $O^G(x)$ closed in V. In virtually all the cases that will interest us the finiteness of stab (x) will be automatic (but cf. the remark following 1.15).

iv) A point x is stable if it merely has negative weights with respect to every non-trivial 1-PS λ, for then it also has positive weights with respect to λ, namely, its negative weights with respect to λ^{-1}.

v) The proofs of $c_u \Rightarrow a_u \Rightarrow b_u$ and of $b_s \Rightarrow a_s \Rightarrow c_s$ are obvious: for example, if λ is a 1-PS for which all weights of x are positive, then $\lambda(t)\,x \to 0$ at $t \to 0$; i.e. $c_u \Rightarrow a_u$.

vi) The proofs of $a_s \Rightarrow b_s$ and $b_u \Rightarrow a_u$ are achieved by reduction to the special case called geometric reducivity of G. A group G is called geometrically reductive if

a) whenever V_0 is an invariant codimension-1 subspace of a vector space V in which G is represented, there exists an n for which the codimension-1 invariant subspace $V^0 \cdot \mathrm{Symm}^{n-1} V \subset \mathrm{Symm}^n V$ has an invariant 1-dimensional complement.

   But notice that this is the same as saying that

b) whenever $x \neq 0$ is a G-invariant point, then there exists a G-invariant polynomial f such that $f(x) \neq 0$ and $f(0) = 0$. (Just consider x as a functional on the dual $\hat{V}$ and apply a) to its kernel there).

And b) is a special case of $a_s \Rightarrow b_s$. When char $k = 0$ we can take the polynomial f to be linear, for by complete reducibility the invariant subspace generated by x is invariantly complemented. A simple example shows this does not happen in char p. Take $p = 2$, $G = SL(2)$, $V =$ the space of symmetric bilinear functions on k^2, and x a non-degenerate skew-symmetric form ($x \in V$ because $p = 2$!). Then x is $SL(2)$-invariant and there are no G-invariant non-zero linear functionals on V. A quadratic f which does work is the determinant.

vii) The remaining implications $c_s \Rightarrow a_s$ and $a_u \Rightarrow c_u$ are essentially consequences of the surjectivity of the natural map

$$\left\{\begin{matrix} \text{1PS's } \lambda \text{ of } G \\ \lambda: \mathbf{G}_m \to G \end{matrix}\right\} \longrightarrow G(k[[t]]) \diagdown G(k((t))) \diagup G(k[[t]])$$

where λ is considered as a $k((t))$-valued point of G by composition with the canonical map

$$\operatorname{Spec} k((t)) \to \operatorname{Spec} k[t, t^{-1}] = \mathbf{G}_m$$

1.3. Let V_{ss} (resp. V_s) denote the Zariski-open cones of semi-stable (resp. stable) points. $V - V_{ss}$ is the Zariski-closed cone of unstable points. The conditions b of 1.1 tell us that if we try to map $\mathbf{P}(V)$ to a projective space by invariant polynomials, we can only hope to achieve a well-defined map on $\mathbf{P}(V)_{ss}$ and an embedding on $\mathbf{P}(V)_s$. From the point of view of quotients this can be expressed by:

PROPOSITION 1.3. *Let* $X = \operatorname{Proj} k[V]^G$. *Then there is a diagram*

$$\begin{array}{ccccc} \mathbf{P}(V) \supset & \mathbf{P}(V)_{ss} & \supset & \mathbf{P}(V)_s \\ & \pi \big\downarrow & & \big\downarrow \pi_s \\ & X & \supset & X_s \end{array}$$

such that i) *if* $x, y \in \mathbf{P}(V)_s$, $\pi_s(x) = \pi_s(y) \Leftrightarrow \exists\, g \in G$ s.t. $x = gy$

ii) *if* $x, y \in \mathbf{P}(V)_{ss}$, $\pi(x) = \pi(y) \Leftrightarrow \overline{O^G(x)} \cap \overline{O^G(y)} \cap \mathbf{P}(V)_{ss} \neq \varnothing$.

We now want to look at some examples to illustrate the application of these ideas.

1.4. "BAD" ACTIONS. Using results of T. Kimura and M. Sato [11] [1]), we can give a list of all representations of simple algebraic groups in charac-

[1]) Plus help given by J. Tits.

teristic 0 in which all vectors are unstable. The point is that there are *very* few such representations.

G	V
$SL(W)$	$W^i, \hat{W}^i, 1 \leqslant k < \dim W$
$SL(W)$	$\Lambda^2 W, \Lambda^2 \hat{W}$
$\dim W$ odd	$\Lambda^2 W \oplus \hat{W}, \Lambda^2 \hat{W} \oplus W$, $\Lambda^2 W \oplus \Lambda^2 \hat{W}$, $\Lambda^2 \hat{W} \oplus \Lambda^2 W$
Sp (W)	W
Spin (10)	W [redacted] where W is a 16-dimensional half-spin representation

Corrections due to V. Kac

1.5. Discriminant. If G is semi-simple and char $k = 0$ then any irreducible representation V has the form $V = \Gamma(G/B, L)$ for a suitable line bundle L on G/B (B is a Borel subgroup of G). To a point x in V associate the divisor H_x on G/B which is the zero set of the corresponding section. Except in the extremely unusual case that the set of singular H_x is of co-dimension > 1, there is an irreducible invariant polynomial δ, the discriminant, such that

1) $\delta(x) = 0 \Leftrightarrow H_x$ is singular

2) $V - (\delta = 0)$ consists of semi-stable points.

An interesting case is

Lemma 1.6. *Let $G = SL(n)$, $V = \Lambda^l(k^n)$. If $W \subset k^n$ is a subspace of codimension l then let Φ_W denote the natural map $\Lambda^2 W \otimes \Lambda^{l-2}(k^n) \to \Lambda^l(k^n)$. If $2 < l < n - 2$ or n is even $l = 2$ or $n - 2$, then there is a G-invariant δ such that $\delta(x) = 0 \Leftrightarrow x \in \operatorname{Im}(\Phi_W)$ for some W.*

When $l = 2$ and $n = 2m + 1$ we have seen that there are no invariants; corresponding to these cases the Grassmanian of lines in $\mathbf{P}^{2m}$ in its Plücker embedding in projective space has the unusual property that the singular hyperplane sections are of codimension $\geqq 2$ in the set of all such sections.

Question: if not every point of V is unstable, then is the set of singular hyperplane sections H_x of codimension 1?

For $l = 2$ and n even or $l = 3$, $n \leqq 8$, one can check that x is unstable $\Leftrightarrow \delta(x) = 0$, hence δ generates the ring of invariants. It would be nice to have a necessary and sufficient condition for a 3-form to be unstable for higher n as well.

1.7. 0-CYCLES. For $G = SL(W)$, $\dim W = 2$,

$$\begin{aligned} V_n &= \mathrm{Symm}^n(\hat{W}) \\ &= \text{vector space of homogeneous polynomials } f \text{ of degree } n \text{ on } W, \\ \mathbf{P}(V_n) &= \text{space of 0-cycles of } n \text{ unordered points on the projective line } \mathbf{P}(W)\text{, the roots of an } f \text{ determining the cycle.} \end{aligned}$$

If $f = \sum_{i=0}^{n} a_i x^{n-i} y^i$ and λ is the one-parameter subgroup given by $t \mapsto \begin{pmatrix} t & 0 \\ 0 & t^{-1} \end{pmatrix}$ in these coordinates, then $\lambda(t) f = \sum_{i=0}^{n} a_i t^{n-2i} x^{n-i} y^i$. For f to be stable, the weights $(n-2i)$ associated to the non-zero coefficients of f must lie on both sides of 0: i.e. if $j \geqq n/2$, neither x^j nor y^j divide f.

a_n a_{n-1} a_1 a_0 coefficient
$-n$ $-n+2$ 0 $n-2$ n weight

In fact, the stability of f is equivalent to the same condition with respect to all linear forms l: $l^j \nmid f$ if $j \geqq n/2$.

Thus $\mathbf{P}(V_n)_s = \{\text{0-cycles with no points of multiplicity} \geqq n/2\}$

$\mathbf{P}(V_n)_{ss} = \{\text{0-cycles with no points of multiplicity} > n/2\}$.

1.8. REMARK. In the example above we can also prove that semi-stability is a purely topological character. I claim that if n is odd and f is unstable then the action of G near $\bar{f} \in \mathbf{P}(V_n)$ is bad: on all open neighbourhoods of the orbit of $\bar{f}$, G acts non-properly and the orbit space is non-Haussdorf. Let's see this for $n = 7$. Consider the following deformations of a 7-point cycle.

(Subscripts indicate multiplicities)

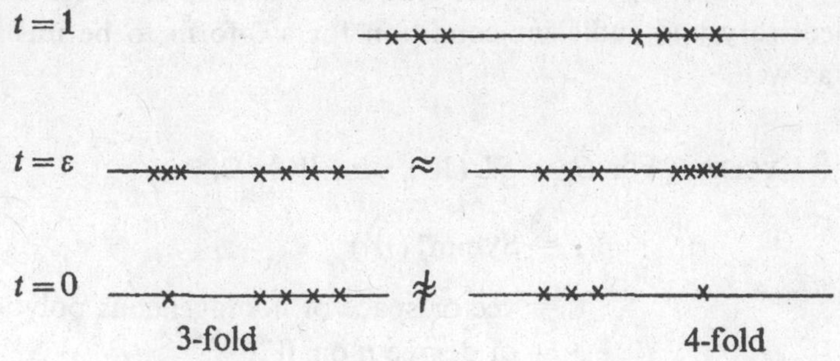

At each intermediate stage the two cycles are projectively equivalent, but the unstable limiting cycle in the right is clearly not equivalent to the limit on the left. In fact, any pair of cycles with the multiplicities indicated on the line $t = 0$ arise in this way as simultaneous limits of projectively equivalent 0-cycles. Moreover, there are cycles of the same type as the left hand limit in any neighbourhood of the orbit of the right limit—just bring a multiplicity one point in towards the triple point; so the orbit space cannot be Hausdorff near the right limit.

1.9. CURVES. Here $G = SL(W)$, dim $W = 3$, $V_n = \mathrm{Symm}^n(\hat{W})$, as before, and a point $f \in V_n$ defines a plane curve of degree n. There is a very simple way to decide the stability of f. Represent f as below by a triangle of coefficients, T.

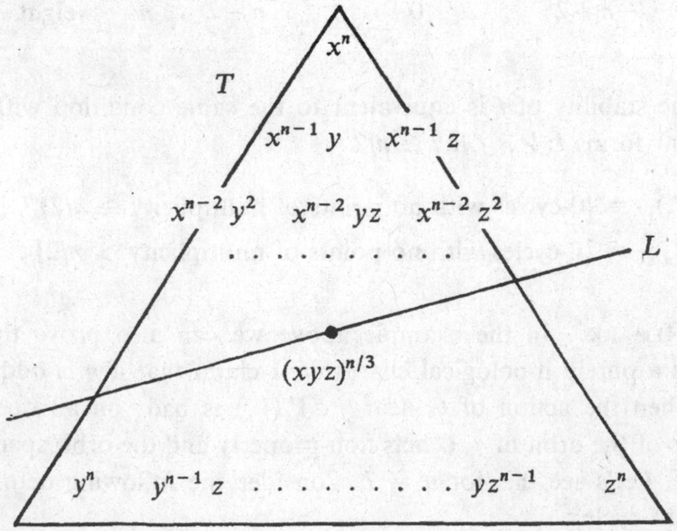

We can coordinatize this triangle by 3 coordinates i_x, i_y, i_z (the exponents of x, y and z respectively) related by $i_x + i_y + i_z = n$. The condition that a line L with equation $ai_x + bi_y + ci_z = 0$, $(a, b, c) \neq (0, 0, 0)$, should pass through the centre of this triangle is just $a + b + c = 0$; if L also passes through a point with integral coordinates then a, b and c can be chosen integral. It is now easy to check that the weights of the 1-PS

$$t \longmapsto \begin{pmatrix} t^a & & 0 \\ & t^b & \\ 0 & & t^c \end{pmatrix}$$

at f are just the values of the form defining L at the non-zero coefficients of f. In suitable coordinates every 1-PS is of this form so:

f is unstable $\Leftrightarrow$ in some coordinates, all non-zero coefficients of f lie to one side of some L

f is stable (resp. semi-stable) $\Leftrightarrow$ for all choices of coordinates and all L, f has non-zero coordinates on both sides of L (resp. f has non-zero coordinates on both sides of L or has non-zero coefficients on L).

Roughly speaking, a stable f can only have certain restricted singularities. We summarize what happens for small n, showing the "worst" triangle T for f with given singularities, and the associated L when f is not stable.

1.10. $n = 2$: We can achieve the diagram below for a non-singular quadric f by choosing coordinates so that $(1, 0, 0) \in f$ and $z = 0$ is the tangent line there, so f is never stable. We cannot make the xz coefficient of f zero without making f singular so f is always semi-stable; indeed, we know f always has non-zero discriminant. A singular quadric always has a diagram like that on the right: make $(1, 0, 0)$ the double point. Henceforth, we leave the checking of the diagrams to the reader.

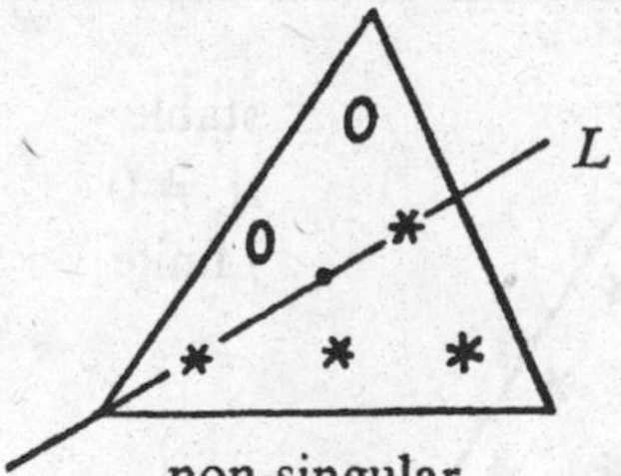

non-singular
quadric

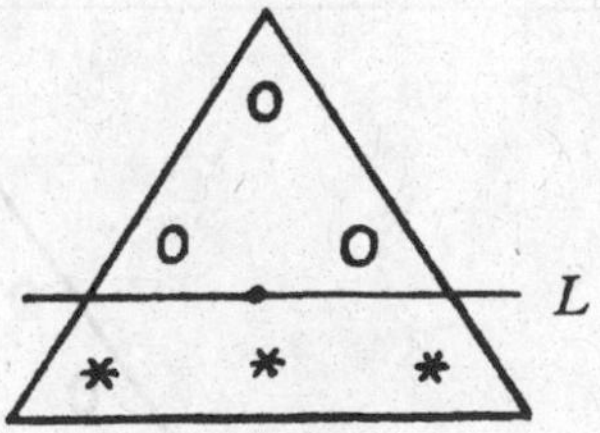

singular
quadric

1.11. $n = 3$: It is well known that in this case the ring of invariants is generated by two invariants, A of degree 4 and B of degree 6. If we set $\Delta = 27A^3 + 4B^2$, then up to a constant the classical j-invariant is just A^3/Δ. The possibilities are:

SINGULARITIES OF f	"WORST" TRIANGLE	STABILITY AND INVARIANTS
f has triple point	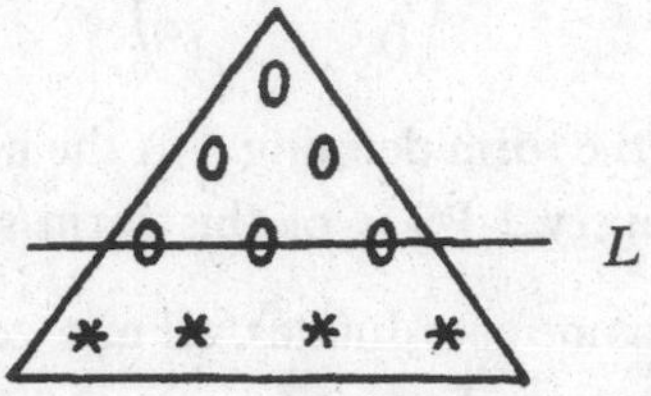	unstable $A = B = 0$ j undefined
f has a cusp or two components tangent at a point.	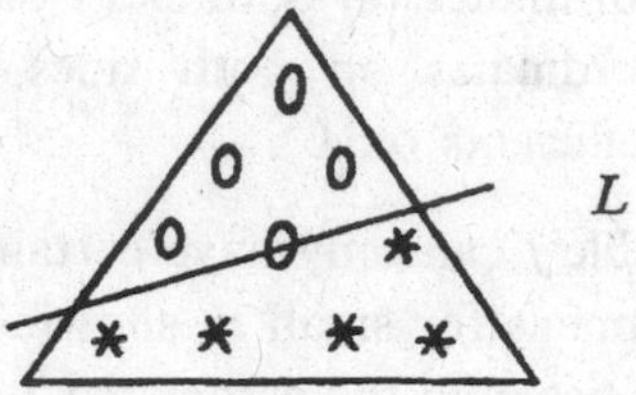	unstable $A = B = 0$ j undefined
f has ordinary double points (this includes the reducible cases: f is a conic and a transversal line, f is a triangle)	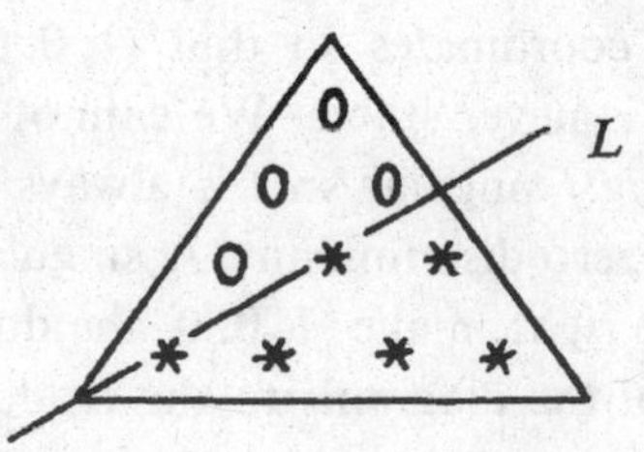	semi-stable and not stable $\Delta = 0$ but $A, B \neq 0$ hence $j = \infty$
f smooth	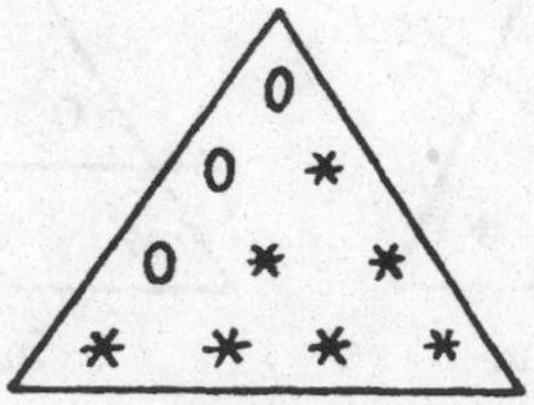	stable $\Delta \neq 0$ j finite

We remark that in this case, we have

$$\begin{array}{ccc} \mathcal{M}_1 & \cong & \mathbf{A}^1 \\ \cap & & \cap \\ \overline{\mathcal{M}_1} & = & \mathbf{P}^2 \end{array}$$

and that the j-invariant is a true modulus. Note that from a moduli point of view all three semi-stable types are equivalent.

1.12. $n = 4$: There are already quite a few diagram types here. Their enumeration can be summarized by saying that f is unstable if and only if f has a triple point or consists of a cubic and an inflectional tangent line; f is stable if and only if f has only ordinary double points or ordinary cusps (i.e. singularities with local equation $y^2 = x^3 +$ higher terms). The remaining f's with a tacnode (a double point with local equation $y^2 = x^4 +$ higher terms) are strictly semi-stable.

1.13. Remark. The fact that for $n \geqq 4$ curves with sufficiently tame cusps are semi-stable (or even stable!) is a definite problem because

i) such curves do not appear in the good compactification $\overline{\mathcal{M}}_g$ of the moduli space of non-singular curves of genus g. But

ii) if we wish to obtain a compactification of $\mathcal{M}_g$ as the quotient space of some subset of $\mathbf{P}(V_n)$ by G, the natural candidate is $\mathbf{P}(V_n)_{ss}$; so these curves must be let in.

For example, when $n = 4$, we have

$$\begin{array}{ccc} \mathcal{M}_{3,\text{ non-hyperelliptic}} & \cong & [\mathbf{P}(V_4) - (\delta = 0)]/G \\ \cap & & \cap \\ \mathcal{M}_3 & & \mathbf{P}(V_4)_s/G \\ \cap & & \cap \\ \overline{\mathcal{M}}_3 & \overset{\alpha}{\underset{\beta}{\leftrightarrows}} & \mathbf{P}(V_4)_{ss}/G \end{array}$$

$\overline{\mathcal{M}}_3$ is the moduli space for "stable" curves of genus 3: (see introduction). Recall from Proposition 1.3 that $\mathbf{P}(V_4)_{ss}/G$ is just the projectivization of the full rings of invariants of $\mathbf{P}(V_4)$. The rational maps α and β induced by the top isomorphism enable us to make a topological comparison of these two compactifications. Let's see geometrically how cuspidal curves in $\mathbf{P}(V_4)_{ss}$ prevent α and β from being continuous.

First α: the diagram below shows on the left a deformation on $\mathcal{M}_3$ with limit in $\overline{\mathcal{M}}_3$, and on the right the same deformation followed to its limit in $\mathbf{P}(V_4)_{ss}/G$.

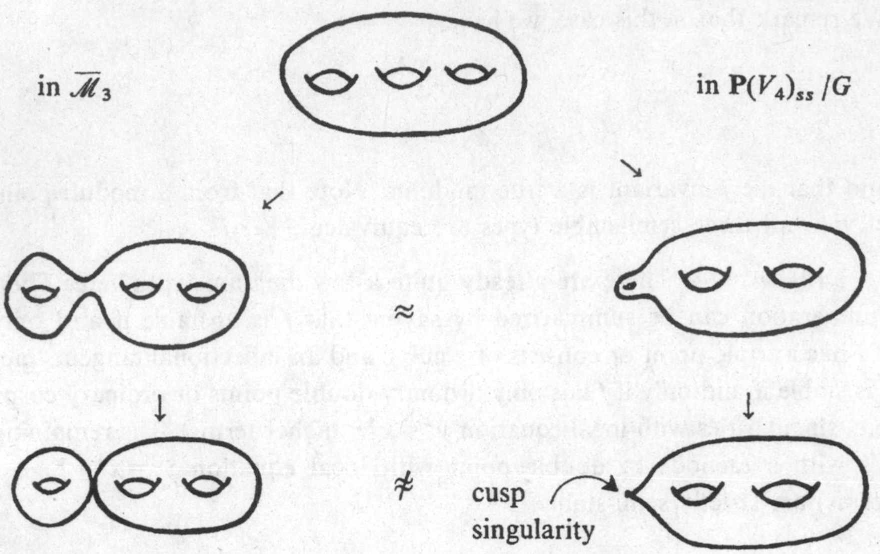

In the limit on the right, the value of the j-invariant of the shrinking elliptic curve has been lost! So α blows up a point representing a curve C with a cusp to the set of points representing joins of an arbitrary elliptic curve with the desingularization $\tilde{C}$ of C. α also blows up the point representing a double conic to the family of all hyperelliptic curves.

As for β, look at the double pinching below:

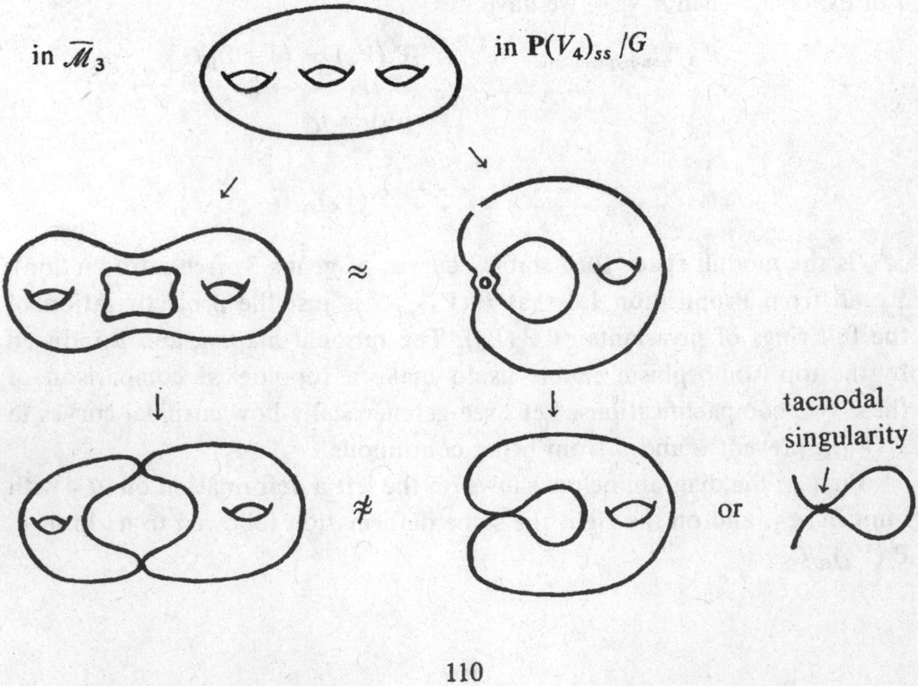

Here it is the manner in which the tangent spaces of the two branches have been glued at the tacnodal point which has been lost in the limiting curve on the left: this glueing corresponds on the left to the relative rate at which the two pinches are made. Thus β has blown up the point corresponding to the double join of two elliptic curves to a family of tacnodal quartics.

1.14. SURFACES. Here $G = SL(W)$, $\dim W = 4$ and $V_n = \mathrm{Symm}^n(\hat{W})$ as before. The technique for determining stability here is essentially that given for curves in 1.9 except that one has a tetrahedron T of coefficients and 1-PS's determine central planes, L: and, of course, the computations required to apply the technique are much more complicated (cf. the case $n = 4$ below). For small n, the situation is summarized below.

n	TYPE OF SINGULARITIES	STABILITY
$n = 2$	non-singular	semi-stable, not stable
	singular	unstable
$n = 3$	non-singular or with ordinary double points of type A1	stable
	ordinary double points of type A2	semi-stable, not stable
	triple points, double curve, higher double points	unstable
$n = 4$ (due to Jayant Shah [26])	singularities at most rational double points, or ordinary double curves possibly with pinch points, but no double *line*, and if reducible then no component a plane, no multiple components	stable
	A triple point whose tangent cone has only ordinary double points; or a double line not as below; or an irrational double point not as below; or a plane plus a cubic meeting in a plane cubic curve with only ordinary double points; or a non-singular quadric counted twice	semi-stable — but not stable
	a) quadruple point, or triple point whose tangent cone has cusp, b) $x = y = 0$ is double line and $f \in (x^2, xyz^2, xy^2, y^3)$ c) a higher double point of form: $f \in (x^2, xy^2, xyz^2, xz^3, y^3z, y^4)$	unstable

2

1.15. ADJOINT STABILITY.

PROPOSITION 1.15. *Let G be any semi-simple group with Lie algebra* $\mathfrak{g}$. *Then $X \in \mathfrak{g}$ is unstable* $\Leftrightarrow$ ad X *is nilpotent.*

Proof: ($\Rightarrow$) From the formula $\mathrm{ad}\,(\mathrm{Ad}\, g\,(x)) = \mathrm{Ad}\, g \circ \mathrm{ad}\, x \circ \mathrm{Ad} g^{-1}$ it is immediate that the characteristic polynomial $\det(tI - \mathrm{ad}\, x)$ is G-invariant, hence that is coefficients are invariant functions. If x is unstable, these all vanish so ad$x$ is nilpotent.
($\Leftarrow$) If ad$x$ is nilpotent then the $\{\exp t\,(x) \mid t \in k\}$ is a unipotent subgroup of G which must be contained in the unipotent radical $R_u\,(B)$ of some Borel subgroup B of G. Fix a maximal torus $T \subset B$, so $B = R_u\,.\,T$. Then by the structure theorem of semi-simple groups we can write $\mathfrak{g} = t + \left(\sum_{\alpha>0} \mathfrak{g}_\alpha\right) + \left(\sum_{\alpha<0} \mathfrak{g}_\alpha\right)$ where $t = \mathrm{Lie}\,(T)$ and $\left(\sum_{\alpha>0} \mathfrak{g}_\alpha\right) = \mathrm{Lie}\,(R_u\,(B))$. Let χ_α be the character of T, which is associated to $\alpha = (\alpha_i)$ (i.e. if $w \in T$, $y \in \mathfrak{g}_\alpha$ then $\mathrm{Ad}\,(w)\,(y) = \chi_\alpha\,(w)\,y$), and let l be a linear functional on the group of characters of T defining the given ordering: i.e.,

$$l\,(\chi_\alpha) = \sum_i c_i\,\alpha_i > 0 \quad \text{if} \quad \alpha > 0 \quad \text{and} \quad l\,(\chi_\alpha) < 0 \quad \text{if} \quad \alpha < 0\,.$$

We can always choose l so that all the c_i are integers. If we define a 1-PS $\lambda: \mathbf{G}_m \to T$ by $\lambda\,(t) = (\ldots, t^{c_i}, \ldots)$, then the weights of X with respect to λ are some subset of $\{\, l\,(\alpha) \mid \alpha > 0\,\}$, hence are positive. Thus X is unstable.

REMARK. There are no stable points. One can show that the regular semi-simple elements of $\mathfrak{g}$ have closed orbits of maximal dimension but their stabilizers will be their centralizers, i.e. maximal tori of G, and hence far from finite.

1.16. CHOW FORM. The Chow form is the answer to the problem of describing by an explicit set of numbers a general subvariety $V^r \subset \mathbf{P}^n$. In two cases, the problem has a very easy answer: a hypersurface has its equation F and a linear space L^r has its Plücker coordinates. The Chow form is just a clever combination of these two special cases. Suppose V^r has degree d. There are two ways to proceed

i) If $u = (u_i) \in \mathbf{P}^n$ write H_u for the hyperplance $\sum u_i X_i = 0$. One shows that there is an irreducible polynomial Φ_V such that

$$[V \cap H_u^{(0)} \cap \ldots \cap H_u^{(r)} \neq \varnothing] \Leftrightarrow [\Phi_V\,(u_i^{(0)}, \ldots, u_i^{(r)}) = 0]$$

Moreover Φ_V is multihomogeneous of degree d in each of the sets of variables $(u_0^{(j)}, \ldots, u_n^{(j)})$, Φ_V is unique up to a scalar, and Φ_V determines V.

ii) If G = Grassmanian of L^{n-r-1}'s in $\mathbf{P}^n$ and $\mathcal{O}_G(1)$ is the ample line bundle on G defined by its Plücker embedding, then the set of $L \in G$ such that $L \cap V \neq \varnothing$ is the divisor D_V of zeroes of some section if $\mathcal{O}_G(d)$ and V and D_V determine each other. (Unfortunately, D_V is almost always a singular divisor.)

These methods give the same result via the identification:

$$\bigoplus_{d=0}^{\infty} \Gamma(G, \mathcal{O}_G(d)) = \left\{\begin{array}{l}\text{Homogeneous}\\ \text{coordinate}\\ \text{ring of } G\end{array}\right\}$$

$$= \left\{\begin{array}{l}\text{Subring } W \text{ of } \mathbf{C}[\ldots, U_i^{(j)}, \ldots] \text{ generated}\\ \text{by the Plücker coordinates}\\ P_{i_0, \ldots, i_r} = \det_{(r+1, r+1)}(U_{i_l}^{(j)}),\ i_0 < i_1 < \ldots < i_r\end{array}\right\}$$

Letting W_d be the dth graded piece of W, the identification furnishes an irreducible representation

$$\mathrm{Symm}^d(\Lambda^{r+1}(\mathbf{C}^{n+1})) \twoheadrightarrow W_d \hookrightarrow \bigotimes^{r+1} \mathrm{Symm}^d(\mathbf{C}^{n+1})$$

Thus, although we will usually consider the Chow form as a point of the $SL(n+1)$ representation $\otimes^{r+1} \mathrm{Symm}^d(\mathbf{C}^{n+1})$ this form lies in the irreducible piece W_d and can be thought of as defining a divisor on the Grassmanian. For more details on Chow forms, see Samuel [17, Ch. 1 § 9].

1.17. Asymptotic stability. We will say that a variety $V^r \subset \mathbf{P}^n$ is *Chow stable* or simply *stable* if its Chow form is stable for the natural $SL(n+1)$-action. If L is an ample line bundle on V, we say that (V, L) is *asymptotically stable* if

$$\exists\, n_0 \text{ s.t. } \forall n \geq n_0,\ \Phi_{\Gamma(L^n)}(V) \subset \mathbf{P}^{h^0(L^n)-1} \text{ is stable.}$$

Attention: a stable variety need not be asymptotically stable (nor, of course, vice versa). Indeed, one of the main goals of this exposition is to show that the asymptotically stable curves are exactly the "stable" curves of Deligne and Mumford, and that by using asymptotic stability we can construct $\bar{\mathcal{M}}_g$ as a "quotient" moduli space for these curves.

§ 2. A criterion for $X^r \subset \mathbf{P}^n$ to be stable

If $f(a)$ is an integer-valued function which is represented by a rational polynomial of degree at most r in n for large n, we will denote by n.l.c. (f) (the normalized leading coefficient of f) the integer e for which $f(n) = e\dfrac{n^r}{r!}$ + lower order terms. (What r is to be taken, will always be clear from the context.)

Proposition 2.1 [1]. *(The "Hilbert-Hilbert-Samuel" Polynomial). Suppose X is a k-variety (not necessarily complete), L is an invertible sheaf on X and $\mathscr{I} \subset \mathcal{O}_X$ is an ideal sheaf such that $Z = \operatorname{Supp} \mathcal{O}_X/\mathscr{I}$ is proper over k. Then there is a polynomial $P(n, m)$ of total degree $\leqq r$, such that, for large m*

$$\chi(L^n/\mathscr{I}^m L^n) = P(n, m).$$

Proof. We can compactify X and extend L to a line bundle on this compactification, without altering the validity of the theorem so we may as well assume X proper over k. Let $\pi\colon B \to X$ be the blow-up of X along $\mathscr{I}$ (i.e. $B = B_{\mathscr{I}}(X) = \operatorname{Proj}(\mathcal{O}_X \oplus \mathscr{I} \oplus \mathscr{I}^2 \oplus \ldots)$) and let E be the exceptional divisor on B so that $\mathscr{I} . \mathcal{O}_B = \mathcal{O}(-E)$. The well-known theorems of F.A.C. (Serre [18]) for the vanishing of higher cohomology in the relative case imply that when $m \gg 0$:

i) $\pi_*(\mathcal{O}(-mE)) = \mathscr{I}^m$

ii) $R^i\pi_*(\mathcal{O}(-mE)) = (0), \ i > 0$

Now examine the exact sequence:

$$0 \longrightarrow \mathscr{I}^m L^n \longrightarrow L^n \longrightarrow L^n/\mathscr{I}^m L^n \longrightarrow 0$$

The Hilbert polynomial for $\chi(L^n)$ certainly satisfies the conditions on P. Moreover, in view of i) and ii); we have for $m \gg 0$:

$$\chi(X, \mathscr{I}^m L^n) = \chi(B, \pi^* L^n(-mE)) = \chi(B, (\pi^* L)^{\otimes n} \otimes \mathcal{O}(-E)^{\otimes m})$$

so, a theorem of Snapper [5, 21] guarantees that this last Euler characteristic is also a polynomial of the required type for large m and n. By the additivity of χ we are done.

[1]) This result and its geometric interpretation are essentially due to C. P. Ramanujam [16].

Definition 2.2. *In the situation of Proposition* 2.1, *we denote by* $e_L(\mathcal{I})$ *(the multiplicity of* $\mathcal{I}$ *measured via* L*) the integer* n.l.c. $(\chi(L^n/\mathcal{I}^n L^n))$.

Examples. i) If $\mathcal{I} = 0$ and X is complete, P is the Hilbert polynomial of L. ii) If Z is set-theoretically a point x then P is the Hilbert-Samuel polynomial of $\mathcal{I}$ as an ideal of $\mathcal{O}_{x,X}$ and $e(\mathcal{I})$ is its multiplicity there: in particular, it is independent of L. Note that, in general, $e_L(\mathcal{I})$ depends on the formal completion of X along Z and the pull-backs of $\mathcal{I}, L$ to this formal completion.

2.3. **Classical geometric interpretation.** Let $X^r \subset \mathbf{P}^n$ be a projective variety, $L = \mathcal{O}_X(1)$, and Λ be a subspace of $\Gamma(\mathbf{P}^n, \mathcal{O}(1))$. Define L_Λ to be the linear subspace of $\mathbf{P}^n$ given by $s = 0$, $s \in \Lambda$. Define $\mathcal{I}_\Lambda$ to be the ideal sheaf generated by the sections $s \in \Lambda$, i.e. $\mathcal{I}_\Lambda . L$ is the subsheaf of L generated by those sections and $Z = \operatorname{Supp}(\mathcal{O}_X/\mathcal{I}_\Lambda) = X \cap L_\Lambda$ is the set of their base points.

If $p_\Lambda : \mathbf{P}^n - L_\Lambda \to \mathbf{P}(\Lambda) = \mathbf{P}^m$ is the canonical projection, and π is the blow-up of X along $\mathcal{I}_\Lambda$ then there is a unique map q making the following diagram commute:

$$\begin{array}{ccc} X - Z & \xrightarrow{\ \text{res}\ p_\Lambda\ } & \mathbf{P}^m \\ \cap & \cap & \nearrow{\scriptstyle q} \\ X & \xleftarrow{\ \pi\ } & B = B_{\mathcal{I}_\Lambda}(X) \end{array}$$

Moreover, because sections of $\mathcal{O}_{\mathbf{P}^m}(1)$ pull back to sections of $\mathcal{I}_\Lambda . L$ on X and are blown-up to sections of L twisted by minus the exceptional divisor E,

$$q^*(\mathcal{O}_{\mathbf{P}^m}(1)) = (\pi^* L)(-E). \tag{2.4}$$

Define $p_\Lambda(X)$, the image of X by the projection p_Λ, to be [cycle $(q(B))$]: that is, $q(B)$ with multiplicity equal to the degree of B over $q(B)$ if these have the same dimension and 0 otherwise. I claim

Proposition 2.5. $e_L(\mathcal{I}_\Lambda) = \deg X - \deg p_\Lambda(X)$.

Proof. If H is the divisor class of a hyperplane section on X, then

$$\deg X = (H^r) = \text{n.l.c.}\,(\chi(\mathcal{O}_X(n))).$$

By 2.4, q is defined by the linear system of divisors of the form $\pi^{-1}(H) - E$, hence

$$\deg p_\Lambda(x) = ((\pi^{-1}(H) - E)^r) = \text{n.l.c. } \chi(\pi^*(\mathcal{O}(n)(-nE)).$$

Finally, from its definition

$$\begin{aligned} e_L(\mathcal{I}_\Lambda) &= \text{n.l.c. } \chi(\mathcal{O}_X(n)/\mathcal{I}^n\mathcal{O}_X(n)) \\ &= \text{n.l.c. } \chi(\mathcal{O}_X(n)) - \text{n.l.c. } \chi(\mathcal{I}^n\mathcal{O}_X(n)) \\ &= \deg X - \deg p_\Lambda(X) \end{aligned}$$

This proof brings out the geometry even more clearly. If $H_1, \ldots, H_r$ are generic hyperplanes in $\mathbf{P}^r$ then

$$\deg(X) = \#(X \cap H_1 \cap \ldots \cap H_r), \quad (\# \text{ denoting cardinality}).$$

As the H_i specialize to hyperplanes H_i' of the form $s = 0$, $s \in \Lambda$ (remaining otherwise generic) the points in this intersection specialize to either:

i) points outside Z: these points correspond to points in the intersection of $\mathrm{Im}(q)$ with r generic hyperplanes on $\mathbf{P}^n$, and each of these is the specialization of $\deg q$ of the original points i.e. $\deg p_\Lambda(X)$ points specialize in this way

ii) points in Z: $e_L(\mathcal{I}_\Lambda)$ measures the number of points which specialize in this way.

For example, if $X^1 \subset \mathbf{P}^2$ is a curve of degree d, $y = (0, 0, 1)$ is on X and $\Lambda = kX_0 + kX_1$, then $|Z| = \{y\}$, $p_\Lambda(x_0, x_1, x_2) = (x_0, x_1)$ and the picture is:

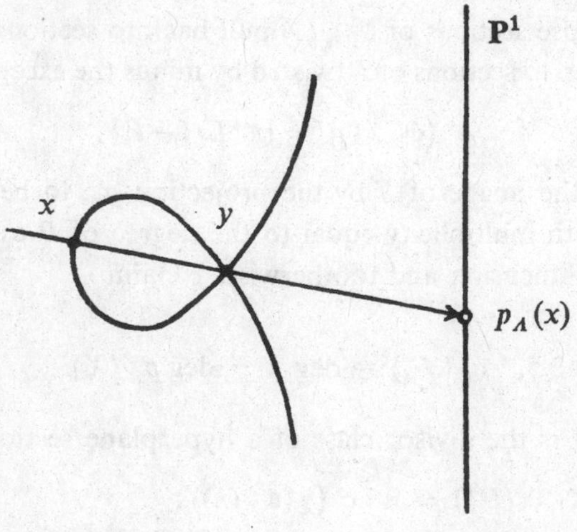

Thus $p_A(X) = (a\mathbf{P}^1)$, where a is the degree of the covering p; a generic line meets X in d points and as this line specializes to a non-tangent line through y it meets X at y on $\text{mult}_y(X) = e_L(\mathscr{I}_A)$ points and meets X away from y in $d - e_L(\mathscr{I}_A) = a$ points.

The following technical facts will be useful in calculating the the invariants $e_L(\mathscr{I})$.

Proposition 2.6. a) *If (in the situation of Proposition 2.1) L and $\mathscr{I} . L$ are generated by their sections then* $\left| h^0(L^n/\mathscr{I}^n L^n) - e_L(\mathscr{I})\dfrac{n^r}{r!}\right| = O(n^{r-1})$. *(Thus we can calculate $e_L(\mathscr{I})$ from the dimensions of spaces of sections.)*

b) *Suppose, in addition, we are given a diagram*

$$\begin{array}{ccc} X & \supsetneqq & X_0 = f^{-1}(0) \\ f\big\downarrow & & \big\downarrow \\ \text{Spec}(A) & \ni & 0 \end{array}$$

where f is proper, and a finite dimensional vector space $W \subset \Gamma(X, \mathscr{I} L)$ which

i) *generates $\mathscr{I} . L$*

ii) *defines a closed immersion $X - X_0 \hookrightarrow \mathbf{P}(\hat{W})$*

Then the dimensions of the kernel and cokernel of the map

$(\Gamma(X, L^n)/A$-submodule generated by the image of $W^{\otimes n} \to \Gamma(L^n/\mathscr{I}^n L^n)$

are both $O(n^{r-1})$.

Proof. The idea in a) is to show that $h^i(L^n/\mathscr{I}^n . L^n) = O(n^{r-1})$, $i \geq 1$. We first remark that is a compactification $\bar{X}$ of X over which L extends to a line bundle $\bar{L}$ such that

i) $\bar{L}$ is generated by its sections

ii) some $W \subset \Gamma(X, L)$ which generates $\mathscr{I} . L$ extends to a $\bar{W} \subset \Gamma(\bar{X}, \bar{L})$.

Indeed, on any compactification $\bar{X}$, there exists a coherent sheaf $\bar{\mathscr{F}}$ such that $\bar{\mathscr{F}}|_X \cong L$ and $\bar{\mathscr{F}}$ has properties i) and ii), and the pullback of $\bar{\mathscr{F}}$ to the blow-up $B_{\bar{\mathscr{F}}_1}(\bar{X})$ is a line bundle with these properties: so we might as well replace $\bar{X}$ by $B_{\bar{\mathscr{F}}}(\bar{X})$. Then if we take an ideal sheaf $\bar{\mathscr{I}}$ such that $\bar{W}$ generates $\bar{\mathscr{I}} . \bar{L}$, $\bar{\mathscr{I}} = \mathscr{I} . \mathscr{I}'$ where $\mathscr{I}'$ is supported on $\bar{X} - X$ only, and it suffices to show $h^i(\bar{L}^n/\bar{\mathcal{J}}^n\bar{L}^n) = O(n^{r-1})$ $i \geq 1$ since $\bar{L}^n/\bar{\mathcal{J}}^n\bar{L}^n \cong \bar{L}^n/\mathcal{J}^n\bar{L}^n \oplus \bar{L}^n/\mathcal{J}'^n.\bar{L}^n$ so this bounds $h^i(L^n/\mathcal{J}^nL^n)$. To do this, it suffices, in turn, to bound $h^i(\bar{X}, \bar{L}^n)$ and $h^i(\bar{X}, \bar{\mathcal{J}}^n.\bar{L}^n) = h^i(B_{\bar{\mathcal{J}}}(\bar{X}), \bar{L}(-\bar{E})^{\otimes n})$ (where E is the exceptional divisor on $B_{\bar{\mathcal{J}}}(\bar{X})$). These bounds follow from:

LEMMA 2.7. *If X^r is proper over k and L is a line bundle on X generated by its sections, then* $h^i(L^{\otimes n}) = O(n^{r-1})$, $i \geq 1$.

Proof. Let X_0 be the image of X in $\mathbf{P}^n$ under the map given by the sections of L. Then $L = \pi^*(\mathcal{O}_{X_0}(1))$ and

$$\begin{aligned} H^i(X, L^{\otimes n}) &= H^i(X, \pi^*(\mathcal{O}_{X_0}(n))) \\ &\cong H^0(X_0, (R^i\pi_*\mathcal{O}_{X_0}) \otimes \mathcal{O}_{X_0}(n)) \\ &\quad \text{for } n \text{ large.} \end{aligned}$$

The last isomorphism follows from first applying the Leray spectral sequence, and then noting that all the terms involving higher cohomology groups vanish for large n, by the ampleness of $\mathcal{O}_{X_0}(1)$. But if $p \in \operatorname{Supp} R^i\pi_*\mathcal{O}_{X_0}$ for $i \geq 1$, the fibre $\pi^{-1}(p)$ has positive dimension, hence $\dim \operatorname{Supp} R^i\pi_*\mathcal{O}_{X_0} \leq r-1$ which gives the desired $O(n^{r-1})$ bound on the dimension of the last space.

A suitable compactification and an argument like that in the proof of a), reduce the part of the statement of b) about the cokernel to bounding an $h^1(\mathcal{J}^n.L^n)$ and this is accompanied as in a) by a blow-up and the lemma. The procedure for dealing with the kernel is somewhat different: What we want to control is the dimension

$$(H^0(\mathcal{J}^nL^n)/A\text{-submodule generated by the image of } W^{\otimes n})$$

That is to say, for $n \gg 0$, the dimension of:

$$(H^0(B(X), \pi^*L^n(-nE))/A\text{-submodule generated by image of } W^{\otimes n})$$

Let $B = B_{\mathcal{J}}(X)$ and q be the proper, birational map $B \xrightarrow{q} B' \subset \mathbf{P}^n \times \operatorname{Spec} A$ induced by W. Then $q^*(\mathcal{O}_{B'}(1)) = \pi^*L(-E)$ and for large n, we have

$$\begin{array}{ccc} H^0(B, L^n(-nE)) & \cong & H^0(B', q_*(\mathcal{O}_B) \otimes \mathcal{O}_{B'}(n)) \\ \cup\!\!\uparrow & & \uparrow\cup \\ \left[\begin{array}{l} A\text{-submodule} \\ \text{generated by} \\ \text{the image of } W^{\otimes n} \end{array}\right] & \cong & H^0(B', \mathcal{O}_{B'}(n)) \end{array}$$

The cokernel of the inclusion on the right is just $H^0(B', q_*(\mathcal{O}_B)/\mathcal{O}_{B'}(n))$. But the support of this last sheaf is proper over $0 \in \operatorname{Spec} A$, hence of dimension less than r, so a final application of the lemma completes the proof.

2.8. Fix : $X^r \subset \mathbf{P}^n$ a projective variety,

$X_0, \ldots, X_n$ coordinates on $\mathbf{P}^n$,

Φ_X the Chow form of X,

$$\lambda(t) = \begin{bmatrix} t^{\rho_0} & & 0 \\ & \ddots & \\ 0 & & t^{\rho_n} \end{bmatrix} \cdot t^{-k}, \quad \rho_0 \geqq \rho_1 \geqq \ldots \geqq \rho_n \geqq 0,$$

k chosen so that this is a 1-PS of $SL(n+1)$, i.e. $k = -\sum \rho_i/n+1$.

We define an ideal sheaf $\mathscr{I} \subset \mathcal{O}_{X \times A^1}$ by

$\mathscr{I} \cdot [\mathcal{O}_X(1) \otimes \mathcal{O}_{\mathbf{A}^1}] =$ subsheaf generated by $\{ t^{\rho_i} X_i \}$, $i = 0, \ldots, n$.

Remarks. i) From an examination of the generators of $\mathscr{I}$, one sees that the support of the subscheme $Z = \mathcal{O}_{X \times \mathbf{A}^1}/\mathscr{I}$ is concentrated over $0 \in \mathbf{A}^1$; if we normalize the ρ_i so that $\rho_n = 0$ then the support of $\mathscr{I}$ also lies over the section $X_n = 0$ in X.

ii) Consider the weighted flag:

$$\begin{array}{ccccc} (X_1 = \ldots = X_n = 0) & \subset & (X_2 = \ldots = X_n = 0) \subset \ldots & \subset & (X_n = 0) \\ \| & & \| & & \| \\ L_0 & & L_1 & & L_{n-1} \\ \text{weight } \rho_0 & & \text{weight } \rho_1 & & \text{weight } \rho_{n-1} \end{array}$$

The subscheme Z looks roughly like a union of ρ_i^{th}-order normal neighborhoods of $L_i \cap X$. It is easily seen to depend only on the weighted flag and not on the splitting defined by λ.

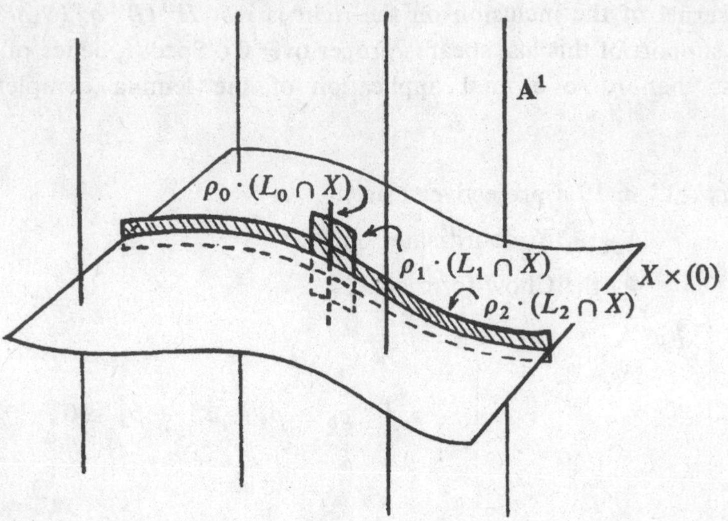

iii) Roughly speaking, $e_{\mathcal{O}_{\mathbf{A}^1} \otimes \mathcal{O}_X(1)}(\mathcal{I})$, which we will denote $e(\mathcal{I})$ measures the degree of contact of this weighted flag with X[1]. The multiplicity of $\mathcal{I}$ can be expected to get bigger, for example, if L_0 becomes a more singular point of X or if L_{n-1} oscillates to X to higher degree. The main theorem of this chapter makes this more precise:

THEOREM 2.9. *In the situation of* 2.8, Φ_X *is stable (resp.: semi-stable) with respect to* λ *if and only if:*

$$e(\mathcal{I}) < \frac{(r+1)\deg X}{n+1} \cdot \sum_{i=0}^{n} \rho_i$$

$$\left(\text{resp.: } e(\mathcal{I}) \leq \frac{(r+1)\deg X}{n+1} \cdot \sum_{i=0}^{n} \rho_i\right)$$

Proof. We begin with a definition.

DEFINITION 2.10. *If* $\mu\colon \mathbf{G}_m \to GL(W)$ *is a representation of* $\mathbf{G}_m$ *and* W_i *is the eigenspace where* $\mathbf{G}_m$ *acts by the character* t^i, *then the* μ*-weight of* W *is* $\sum_{i=-\infty}^{\infty} i \,.\, \dim W_i$. *If* $w \in W_i$ *then we say* i *is the* μ*-weight of* w.

[1]) It seems to be a general fact of life that one must go up to some $(r+1)$ dimensional variety—here $X \times \mathbf{A}^1$—to measure such a contact on an r-dimensional variety.

1) THE LIMIT CYCLE. If $X^{\lambda(t)}$ is the image of X by $\lambda(t)$, then taking $\lim_{t\to 0} X^{\lambda(t)}$ gives a scheme $X^{\lambda(0)}$ and an underlying cycle $\tilde{X}$, both of which are fixed by λ. Moreover, $\Phi_{X^{\lambda(t)}} = (\Phi_X)^{\lambda(t)}$ so if $\Phi_X = \sum_{i=a}^{b} \Phi_{X,i}$ where $\Phi_{X,i}$ is the component of Φ_X in the ith weight space; then

$$\begin{aligned}\Phi_{X^{\lambda(t)}} &= \sum_{i=a}^{b} t^i\, \Phi_{X,i} \\ &= t^a\,[\Phi_{X,a} + t\,(\text{other terms})]\end{aligned}$$

Hence, $\Phi_{\tilde{X}} = \Phi_{X,a}$ and a is the λ-weight of $\Phi_{\tilde{X}}$. By definition, Φ_X is stable (resp: semi-stable) with respect to λ if and only if $a < 0$ (resp: $a \leqq 0$) or equivalently if and only if the λ-weight of $\Phi_{\tilde{X}}$ is < 0 (resp: $\leqq 0$).

2) The next step is to connect this weight with a Hilbert polynomial; this is done by:

PROPOSITION 2.11. *Let $V^r \subset \mathbf{P}$ be fixed by a* 1-PS λ *of* $SL(n+1)$, *let I be the homogeneous ideal of V and let* $R_n = (k\,[x_0, \ldots, X_n]/I)_n$ (i.e. $V = \operatorname{Proj}\,(\bigoplus_{n=0}^{\infty} R_n)$). *Let a_V be the λ-weight of Φ_V and r_n^V be the λ-weight of R_n. Then for large n, r_n^V is represented by a polynomial in n of degree at most* $(r+1)$ *with* n.l.c. a_V.

Proof. a) Assume V is linear. In suitable coordinates, we can write $V = V(X_{r+1}, \ldots, X_n)$ and $\lambda(t) = \begin{bmatrix} t^{a_0} & & 0 \\ & \ddots & \\ 0 & & t^{a_n} \end{bmatrix}$. Then in the notation of 1.16, the Chow form of V is the monomial

$$\Phi_V = \det(U_i^{(j)}),\ i, j = 0, \ldots, n\,.$$

Hence $\Phi_{\tilde{V}} = \Phi_V$ and has weight $\sum_{i=0}^{r} a_i$. On the other hand the λ-weight of R_n depends only on $a_0 \ldots a_r$, is symmetric in these weights, and is linear in the vector $(a_0, \ldots, a_r)$, hence depends only on $\sum_{i=0}^{r} a_i$. By considering the case $a_0 = \ldots = a_r$ we see that

$$r_n^V = \frac{n}{r+1} \left(\sum_{i=0}^{r} a_i \right) \dim R_n = a_V \cdot \frac{n}{r+1} \cdot \binom{n}{r}$$

which is certainly of the form claimed.

b) V is a positive cycle of linear spaces. Here it is more convenient to consider the ideal I instead of V. By noetherian induction, we can suppose the claim proven for all λ-fixed ideals $I' \supsetneq I$. Then if $V = \sum a_i L_i$, let J_1 be the ideal of L_1, and choose an $a \in k[X] - I$ which is a λ-eigenvector of weight, say, w and such that $J_1 a \subset I$. Now look at the exact sequence:

$$0 \to a + I/I \to k[x]/I \to k[x]/I + a \to 0$$

The claim is true for $I + a$ by the noetherian induction. If $I' = \{ f \mid af \in I \}$ $\supset J_1 \supsetneq I$, then via the shift of weights by w, $a + I/I \cong k[x]/I'$; but this shift changes the λ-weight by an amount $w \cdot \dim [(k[x]/I')_n]) = O(n^r)$, hence does not affect the leading coefficient of the λ-weight. The claim for I', which also follows from the noetherian induction, thus proves the claim for I.

c) Reduction to case b). Recall the Borel fixed point theorem: if G is a connected solvable algebraic group acting on a projective variety W, then there is a fixed point on $\overline{O^G(y)}$ for every $y \in W$. Let $[V]$ be the associated point of V in $\mathrm{Hilb}_{\mathbf{P}^n}$ and consider the orbit of $[V]$ under the action of a maximal torus $T \subset SL(n+1)$ containing $\lambda(t)$. Let $[V_0]$ be a T-invariant point in $\overline{O^T([V])}$. Then V_0 is a sum of linear spaces, since these are the only T-invariant subvarieties of $\mathbf{P}^n$. If we decompose Φ_V by $\Phi_V = \sum_\alpha \Phi_V^\alpha$, where α runs over the characters of T and Φ_V^α is the part of Φ_V on which T acts with weight α, then for any $\tau \in T$, $\Phi_V^\tau = \sum_\alpha c_\alpha^\tau \Phi_V^\alpha$ for suitable constants c_α^τ. Since Φ_{V_0} is both T-invariant and a limit of forms Φ_V^τ, $\tau \in T$, $\Phi_{V_0} = \Phi^\alpha$ for some α. Moreover since V is a λ-invariant point, all the characters α appearing in the decomposition of Φ_V must have the same value on λ, hence the λ-weight of Φ_{V_0} is the λ-weight of Φ_V.

It remains only to compare the homogeneous coordinate rings. Now V and V_0 are members of a flat family V_t, $t \in S$ for some connected parameter space S, so that if $n \gg 0$, $H^0(V_t, \mathcal{O}_{V_t}(n))$ are the fibres of a vector bundle over S. This means that the λ-action on these fibres varies continuously, hence that the λ-weights of all the fibres are equal. Now the claim for V follows from b).

REMARK. The relation between Chow forms and Hilbert points in c) is really much more general: in fact, Knudsen [12] has shown that there is a canonical isomorphism of 1-dimensional vector spaces $k \,.\, \Phi_V \cong [(r+1)^{\text{st}}$ "differences"—formed via $\otimes$—of successive spaces in the sequence $\Lambda^{\dim R_n} R_n]$, and it is possible to base the whole proof of 2.11 on this.

3) Next we will see how to obtain $X^{\lambda(0)}$ by blowing up $\mathscr{I}$. Consider the map

$$\Lambda_1 \;:\; \mathbf{G}_m \times X \to \mathbf{P}$$
$$(t, X) \mapsto \lambda(t)(x)\,.$$

If the embedding of X is defined by $s_0, \ldots, s_n \in \Gamma\,[X, \mathcal{O}_X(1)]$ and the action of $\lambda(t)$ is by $(a_0, \ldots, a_n) \mapsto (t^{r_0}a_0, \ldots, t^{r_n}a_n)$ with $r_0 \geqq r_1 \geqq \ldots \geqq r_n$ and $\sum_{i=0}^{n} r_i = 0$ (i.e. $(0, \ldots, 0, 1)$ is an attractive fixed point and $(1, 0, \ldots, 0)$ is a repulsive fixed point), then $\Lambda^*_1(X_1) = t^{r_i}s_i$. Now $t^{-\gamma}$ is a unit on $\mathbf{G}_m \times X$, so changing the identification $\Lambda_1^*(\mathcal{O}_{\mathbf{P}^n}(1)) \cong \mathcal{O}_{\mathbf{G}_m} \otimes \mathcal{O}_X(1)$ by this unit we can assume $\Lambda_1^*(X_1) = t^{\rho_i}s_i$ where $\rho_i = r_i - \gamma$ is normalized as in 2.8 so that $\rho_n \geqq 0$. Then Λ_1 "extends" to a rational map $\mathbf{A}^1 \times X \to \mathbf{P}^n$ which is defined by the section $\{\, t^{\rho_i}s_i \,\} \in \Gamma\,(\mathbf{A}^1 \times X, p_2^*\mathcal{O}_X(1)$. $\mathscr{I}$ is just the ideal sheaf these generate in $\mathcal{O}_{\mathbf{A}1 \times X}$ and Z is just the set of base points of the rational map. Blowing up along $\mathscr{I}$ gives the picture

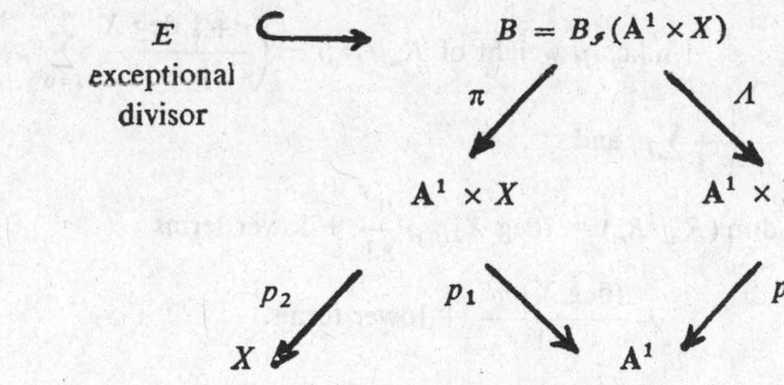

where the morphism Λ is defined by the sections $\{\, t^{\rho_i}s_i \}$ in $\Gamma\,[B, (p_2\pi)^*(\mathcal{O}\,(1))\,(-E)]$. Now Im (Λ) is the closed subscheme of $\mathbf{A}^1 \times \mathbf{P}^n$ given by Proj $(\bigoplus_{m=0}^{m} R_m)$ where

(2.12) $$R_m = \begin{bmatrix} k\,[t]\text{-submodule of } \Gamma(X, \mathcal{O}(m)) \otimes_k k\,[t] \\ \text{generated by } m^{\text{th}} \text{ degree monomials in } \{ t^{\rho_i} s_i \} \end{bmatrix}$$

In fact, Im Λ is *flat* over $\mathbf{A}^1$, because of:

LEMMA 2.13. *Let S be a non-singular curve, X flat over S and $f: X \to Y$ be a proper map over S. Then the scheme $(f(X), \mathcal{O}_Y/\ker f^*)$ is flat over S.*

Proof. We may as well suppose $S = \operatorname{Spec} R$; and then this amounts to showing the $\mathcal{O}_Y/\ker f^*$ has no R-torsion: if $a \in \mathcal{O}_Y/\ker f^*$, $r \in R$, then $r \,.\, a = 0 \Rightarrow r \,.\, f^* a = 0 \Rightarrow f^* a = 0 \Rightarrow a = 0$.

In particular, we see that $X^{\lambda(0)}$ is the fibre of Im Λ over $t = 0$, i.e. $X^{\lambda(0)} = \operatorname{Proj}\left(\bigoplus_{m=0}^{m} R_m/tR_m \right)$.

4) The proof is completed by making precise the relation between $\mathcal{I}$ and the λ-weight of $\Phi_{\tilde{X}}$. One must be careful however because there are two $\mathbf{G}_m$-actions on R_m/tR_m, that given by the identification $R_1/tR_1 = \oplus\, (t^{r_i} s_i)\, k$, which is just λ, and that given by the identification $R_1/tR_1 = \oplus\, (t^{\rho_i} s_i)\, k$; call this action μ. The weights of μ on R_m/tR_m are just those of λ translated by $m\gamma$. By Proposition 2.11

$$\begin{aligned} \lambda\text{-weight of } \Phi_{\tilde{X}} &= \text{n.l.c. } (\lambda\text{-weight of } R_m/tR_m) \\ &= \text{n.l.c. } (\mu\text{-weight of } R_m/tR_m + \gamma m \dim (R_m/tR_m)) \\ &= \text{n.l.c. } (\mu\text{-weight of } R_m/tR_m) - \left(\frac{r+1 \deg X}{n+1} \sum_{i=0}^{n} \rho_i \right) \end{aligned}$$

using $\gamma = -\dfrac{1}{n+1} \sum \rho_i$ and

$$\begin{aligned} \dim (R_m/tR_m) &= (\deg X_{\lambda(0)}) \frac{m^r}{r!} + \text{lower terms} \\ &= \frac{(\deg X)\, m^r}{r!} + \text{lower terms.} \end{aligned}$$

A droll lemma allows us to re-express the μ-weight of R_m/tR_m.

LEMMA 2.14. *Let W be a k-vector space and let $\mathbf{G}_m$ act by μ on W with weights $\rho_n \geqq \rho_{n-1} \dots \geqq \rho_0 = 0$. Let W_i be the eigenspace of weight ρ_i and let W^* be the $k\,[t]$-submodule of $W \otimes k\,[t]$ generated by $\oplus\, t^{\rho_i} W_i$. Then* $\dim (k\,[t] \otimes W/W^*) = $ *μ-weight of W^*/tW^*.*

Proof by Diagram:

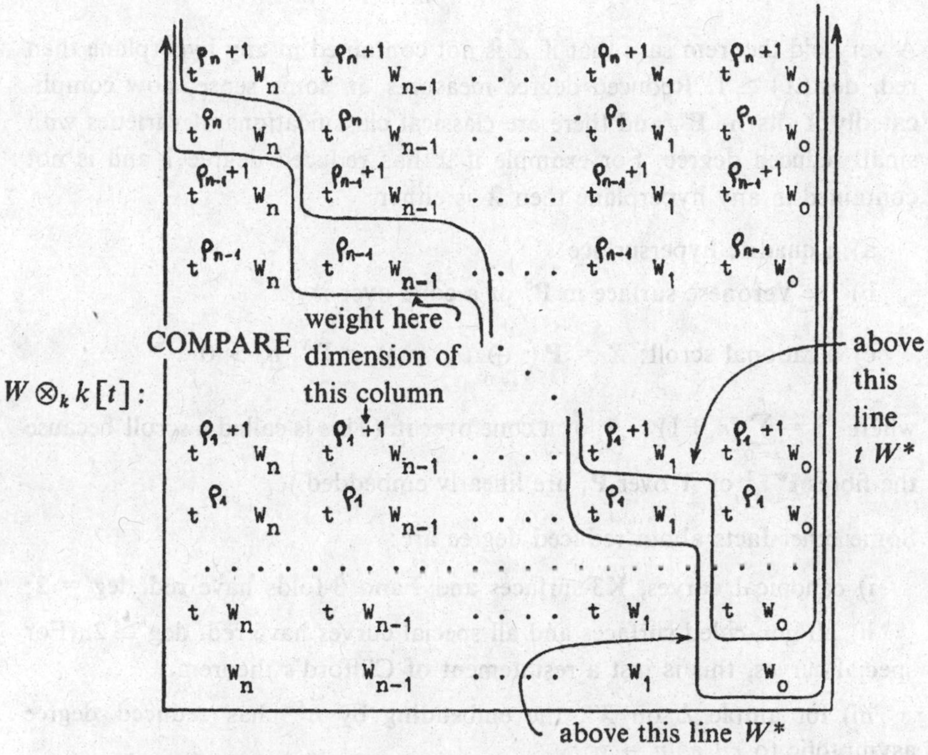

Recalling the definition of R_m (2.12), and applying this to the μ-action on R_m/tR_m, we see that the μ-weight of R_m/tR_m is just: $\dim(\Gamma(X, \mathcal{O}(m)) \otimes_k k[t]/R_m)$. But the sections $\{t^{\rho_i}s_i\}$ whose m^{th} tensor powers generate R_m, also generate $\mathcal{I} . p_2^*(\mathcal{O}_{X(1)})$ so by a) and b) of Proposition 2.6, this last dimension can be used to calculate $e(\mathcal{I})$. Putting all this together, we see that:

$$\Phi_X \text{ is stable with respect to } \lambda$$

$$\Leftrightarrow \lambda\text{-weight of } \Phi_X < 0$$

$$\Leftrightarrow e_L(\mathcal{I}) - \frac{(r+1)}{(n+1)} \deg X \sum_{i=0}^{n} \rho_i < 0$$

which, with the analogous statement for semi-stability, is our theorem.

2.15. Interpretation via reduced degree. If $X^r \subset \mathbf{P}^n$ is a variety, its reduced degree is defined to be:

$$\text{red. deg } (X) = \frac{\deg X}{n + 1 - r}$$

A very old theorem says that if X is not contained in any hyperplane then red. deg $(X) \geqq 1$. Reduced degree measures, in some sense, how complicatedly X sits in $\mathbf{P}^n$, and there are classical classifications of varieties with small reduced degree. For example if X has reduced degree 1 and is not contained in any hyperplane then X is either

a) a quadric hypersurface

b) the Veronese surface in $\mathbf{P}^5$ or a cone over it

c) a rational scroll: $X = \mathbf{P}\left(\bigoplus_{i=0}^{r} \mathcal{O}_{\mathbf{P}^1}(n_i)\right) \subset \mathbf{P}^N,\ n_i > 0$

where $N = \sum_{i=0}^{r} (n_i+1) - 1$, or a cone over it. (This is called a scroll because the fibres $\mathbf{P}^{r-1}$ of X over $\mathbf{P}_1$ are linearly embedded.)

Some other facts about reduced degree are:

i) canonical curves, K3-surfaces and Fano 3-folds have red. deg $= 2$;

ii) all non-ruled surfaces and all special curves have red. deg $\geqq 2$. (For special curves, this is just a restatement of Clifford's theorem.)

iii) for ample L on X^r, the embedding by $L^{\otimes r}$ has reduced degree asymptotic to $r\,!$ as $n \to \infty$;

iv) red-deg is preserved under taking of proper hyperplane sections.

It would be very interesting to know whether almost all 3-folds (in a sense similar to that of ii) for surfaces) have red. deg $\geqq 2 + \varepsilon$. The following definition is introduced only tentatively as a means of linking the present ideas to older ideas (e.g. Albanese's method to simplify singularities of varieties):

2.16. Definition. *A variety* $X^r \subset \mathbf{P}^n$ *is linearly stable (resp. linearly semi-stable) if, whenever* $L^{n-m-1} \subset \mathbf{P}^n$ *is a linear space such that the image cycle* $p_L(X)$ *of* X *under the projection* $p_L : \mathbf{P}^n - L \to \mathbf{P}^m$ *has dimension* r, *then* red deg $p_L(X) >$ red deg X (*resp.* red-deg $p_L(X) \geqq$ red deg X).

Attention: p_L is allowed to be finite to 1, and which case $p_L(X)$ must be taken to be the image cycle. Linear stability is a property of the linear system embedding X; if $X^r \subset \mathbf{P}^n$ is embedded by $\Gamma(X, L)$, then X linearly stable means that for all subspaces $\Lambda \subset \Gamma(X, L)$

$$\frac{\deg p_L(X)}{\dim \Lambda - r} > \frac{\deg X}{n+1-r}$$

or equivalently, by applying Proposition 2.5,

$$e(\mathcal{I}_\Lambda) < \frac{\deg X}{n+1-r} (\text{codim } \Lambda)$$

EXAMPLES. i) when X is a curve of genus 0, it is linearly semi-stable but not stable. When $g \geqq 1$, Clifford's theorem shows that X is linearly stable whenever it is embedded by a complete non-special linear system (see § 4 below).

ii) $\mathbf{P}^2$ is linearly unstable when embedded by $\mathcal{O}(n)$, $n \geqq 3$ because it projects to the Veronese surface. In view of the next proposition, a very interesting problem is that of finding large classes of linearly (semi)-stable surfaces.
(It may, however, turn out that linear stability is really too strong, or unpredictable, a property for surfaces in which case this Proposition is not very interesting !)

PROPOSITION 2.17. *Fix $X^r \subset \mathbf{P}^n$, let C be any smooth curve and let L be an ample line bundle on C. Let $\Phi_i : C \times X \to \mathbf{P}^{N(i)}$ be the embedding defined by $\{S_j \otimes X_l\}$ where $\{S_j\}$ is a basis of $\Gamma(L^{\otimes i})$ and $X_l \in \Gamma(X, \mathcal{O}_X(1))$ are the homogeneous coordinates. If $\Phi_i(C \times X)$ is linearly semi-stable for all large i, then X^r is Chow-semi-stable.*

Proof. Choose a 1-PS: $\lambda(t) = \begin{bmatrix} t^{\rho_0} & & 0 \\ & \cdot & \\ & & \cdot \\ 0 & & t^{\rho_n} \end{bmatrix} t^{-\frac{\Sigma \rho_i}{n+1}}$

as in (2.8).
Choose a point $p \in C$ an isomorphism $L_p \cong \mathcal{O}_p$ and an i large enough that $L^{\otimes i}$ is very ample and $L^{\otimes i}(-\rho_0 p)$ is non-special. Then the map

$$\bigoplus_{l=1}^{n} \Gamma(C, L^{\otimes i}) \,.\, X_l \xrightarrow{\ \Phi_i\ } \bigoplus_{l=0}^{n} [\mathcal{O}_{p,C}/\mathcal{M}^{\rho_0}_{p,C}] \cdot X_i$$

is surjective. Let Λ^i be the inverse image of $\bigoplus_{l=0}^{n} [(\mathcal{M}^{\rho_l}_{p,C}/\mathcal{M}^{\rho_0}_{p,C}) \cdot X_l]$ under this map and let $\mathcal{I}^i_\Lambda \subset \mathcal{O}_{C \times X}$ be the induced ideal. Since all the $L^{\otimes i}$ are trivial near p and $\mathcal{I}^i_\Lambda$ has support on the fibre of $X \times C$ over P, the ideals

3

$\mathscr{I}^i_\Lambda$ are independent of i; we denote this ideal by $\mathscr{I}_\Lambda$. The hypothesis says that for large i

$$e(\mathscr{I}_\Lambda) \leqq \frac{\deg(C \times X)}{(n+1)(h^0(L^i) - r - 1)} \operatorname{codim} \Lambda$$

$$= \frac{(r+1) \deg X \deg L^{\otimes i}}{(n+1)(\deg L^{\otimes i} - g + 1) - r - 1} \cdot \sum_{l=0}^{n} \rho_l$$

and letting $i \to \infty$,

$$e(\mathscr{I}_\Lambda) \leqq \frac{(r+1) \deg X}{n+1} \sum_{l=0}^{n} \rho_l$$

But $C \times X$ along $p \times X$ is formally isomorphic to $\mathbf{A}^1 \times X$ along $0 \times X$ with corresponding $\mathscr{I}'_\Lambda s$, so by Theorem 2.9., X is Chow-semi-stable.

§ 3. Effect of Singular Points on Stability

We begin with an application of Theorem 2.9.

PROPOSITION 3.1. *Let $X^1 \subset \mathbf{P}^n$ be a curve with no embedded components such that* $\deg X/n+1 < 8/7$. *If X is Chow-semi-stable, then X has at most ordinary double points.*

REMARKS. i) When $n = 2$, $\deg X/n+1 < 8/7 \Leftrightarrow \deg X < 4$ and the proposition confirms what we have seen in 1.10 and 1.11

ii) Suppose L is ample on X^1 and $X_m \subset \mathbf{P}^{N(m)}$ is the embedding of X defined by $\Gamma(X, L^{\otimes m})$. By Riemann-Roch, $\deg X_m/N(m) \to 1$ as $m \to \infty$, hence:

COROLLARY 3.2. *An asymptotically stable curve X has at most ordinary double points.*

In particular, if $X \subset \mathbf{P}^2$ has degree $\geqq 4$ and has one ordinary cusp, then, in $\mathbf{P}^2$, X is stable but when re-embedded in high enough space, X is unstable! The fact that this surprising flip happens was discovered by D. Gieseker and came as an amazing revelation to me, as I had previously assumed without proof the opposite.

iii) We will see in Proposition 3.14 that the constant 8/7 is best possible.

Proof of 3.1. We note first that a semi-stable X of any dimension cannot be contained in a hyperplane: if $X \subset V(X_0)$, then X has only positive weights with respect to the 1-PS

$$\lambda(t) = \begin{bmatrix} t^{-n} & & & & 0 \\ & t & & & \\ & & \cdot & & \\ & & & \cdot & \\ & & & & \cdot \\ 0 & & & & t \end{bmatrix}$$

The plan is clear: by Theorem 2.9, it suffices to show that if x is a bad singularity of X, then there is a 1-PS.

$$\lambda(t) = \begin{bmatrix} t^{\rho_0} & & & & 0 \\ & \cdot & & & \\ & & \cdot & & \\ & & & \cdot & \\ 0 & & & & t^{\rho_n} \end{bmatrix}$$

such that

$$e(\mathscr{I}) \geqq \frac{16}{7} \sum_{i=0}^{n} \rho_i > \frac{\deg X \,.\, (r+1)}{(n+1)} \sum_{i=0}^{n} \rho_i \,.$$

First, if $x \in X$ has multiplicity at least three, then take coordinates $(X_0, \ldots, X_n)$ so that $x = (1, 0, \ldots, 0)$ and let $\lambda(t) = \begin{bmatrix} t & & & & \\ & 1 & & & \\ & & \cdot & & \\ & & & \cdot & \\ & & & & 1 \end{bmatrix}$ Then $\mathscr{I} \,.\, \mathcal{O}_{A1 \times X}(1)$ is generated by $\{ tX_0, X_1, \ldots, X_n \}$. Since $\{ X_1, \ldots, X_n \}$ generate $\mathscr{M}_{x,X}$ and X_0 is a unit at x, $\mathscr{I} = (t, \mathscr{M}_x)\, \mathcal{O}_{A1 \times X}$, i.e. $\mathscr{I}$ is the maximal ideal of $(0, x)$ on $\mathbf{A}^1 \times X$. Therefore, $e(\mathscr{I}) = \text{mult}_{(0,x)} (\mathbf{A}^1 \times X) = \text{mult}_x X \geqq 3$, which does what we want since $16/7 \sum_{i=0}^{n} \rho_i = 16/7 < 3$.

Now if $x \in V$ is a non-ordinary double point—i.e. a double point whose tangent cone is reduced to a single line—then $\dim (\mathscr{M}_{x,X}/\mathscr{M}^2_{x,X}) = 2$ and $\mathscr{M}_{x,X} \underset{\neq}{\supset} I \underset{\neq}{\supset} \mathscr{M}^2_{x,X}$ where I is the ideal of the tangent cone at x. Choose coordinates $(X_0, \ldots, X_n)$ such that

i) $X_0(x) \neq 0$

ii) $v = X_1/X_0$ and $u = X_2/X_0$ span $\mathscr{M}_{x,X}/\mathscr{M}^2_{x,X}$

iii) $u \in I$ so that $u^2 \in \mathscr{M}^3_{x,X}$.

iv) $X_3/X_0, \ldots, X_n/X_0 \in \mathscr{M}^2_{x,X}$

Then if $\lambda(t) = \begin{bmatrix} t^4 & & & & & & & 0 \\ & t^2 & & & & & & \\ & & t & & & & & \\ & & & 1 & & & & \\ & & & & \cdot & & & \\ & & & & & \cdot & & \\ & & & & & & \cdot & \\ 0 & & & & & & & 1 \end{bmatrix}$ the associated ideal is

$\mathcal{I} = (t^4 X_0, t^2 X_1, t X_2, X_3, \ldots, X_n)$. But $\mathcal{O}_{A^1 \times X}/\mathcal{I}$ is supported only at the point $(0, x)$ hence $e(\mathcal{I})$ is again Hilbert-Samuel multiplicity and is at least equal to the multiplicity of the possibly larger ideal $\mathcal{I}' = (t^4, t^2 v, tu, \mathcal{M}^2_{x,X})$. If I is the ideal $(t^4, \mathcal{M}^2_{x,X})$, then since

$$(t^2 v)^2 = t^4 v^2 \in I^2$$
$$(tu)^4 = t^4 (u^2)^2 \in t^4 (\mathcal{M}^3_{x,X})^2 \subset I^4 \quad \text{by iii)}$$

$\mathcal{I}'$ is integral over I. Hence

$$e(\mathcal{I}) \geqq e(\mathcal{I}') = e(I) = (4) . (2) . e(\mathcal{M}_{x,X}) = 16 = \frac{16}{7} \sum_{i=0}^{n} \rho_i$$

as required.

The attempt to systematize this theorem leads to a numerical measure of the degree of singularity of a point. The results that follow are part of a joint investigation of this concept by D. Eisenbud and myself. Full proofs will appear later. Many of these results have also been discovered independently by Jayant Shah.

DEFINITION 3.3. *If $\mathcal{O}$ is an equi-characteristic* [1]) *local ring of dimension r, and $k \geqq 0$ is an integer, then we define $e_k(\mathcal{O})$, the kth flat multiplicity of $\mathcal{O}$, by*

$$e_0(\mathcal{O}) = \sup \left\{ \frac{e(I)}{r!\, \text{col}(I)} \;\middle|\; I \text{ of finite colength in } \mathcal{O} \right\}$$
$$e_k(\mathcal{O}) = e_0(\mathcal{O}[[t_1, \ldots, t_k]])$$

It is obvious that if $\hat{\mathcal{O}}$ is the completion of $\mathcal{O}$, then $e_k(\hat{\mathcal{O}}) = e_k(\mathcal{O})$.

PROPOSITION 3.4. $e_k(\mathcal{O}) \geqq \max(1, e(\mathcal{O})/(r+k)!)$.

[1]) The hypothesis on $\mathcal{O}$ can be avoided, and the proof simplified, by a use of the associated graded ring instead of the Borel fixed point theorem (D. Eisenbud).

Proof. The second bound is obvious. To get the first note that if J is any ideal of finite colength then $e(J^n) = n^r e(J)$ and $\mathrm{col}(J^n) = \dfrac{e(J)\, n^r}{r!} + O(n^{r-1})$, hence

$$\frac{e(J^n)}{r!\,\mathrm{col}(J^n)} \to 1 \quad \text{as} \quad n \to \infty .$$

To get an upper bound on e_k we first obtain another lower bound!

PROPOSITION 3.5. *$e_0(\mathcal{O}) \geqq e_0(\mathcal{O}[[t]])$; moreover if $r = \dim \mathcal{O} > 0$ and there is equality, then the sup defining $e_0(\mathcal{O}[[t]])$ is not attained. Hence*

$$e_0(\mathcal{O}) \geqq e_1(\mathcal{O}) \geqq e_2(\mathcal{O}) \geqq \ldots\ldots\ldots \geqq 1 .$$

Proof. We begin by giving a lemma which is useful in the applications of e_0 as well.

LEMMA 3.6. *Let $\mathcal{J}$ be the set of ideals of $\mathcal{O}[[t]]$ of the form $I = \bigoplus_{i=0}^{\infty} I_i t^i$, where I_i is an increasing sequence of ideals of finite colength in $\mathcal{O}$ such that $I_N = \mathcal{O}$ for some N. Then*

$$e_0(\mathcal{O}[[t]]) = \sup_{I \in \mathcal{J}} \frac{e(I)}{r!\,\mathrm{col}(I)}$$

Proof. For any equi-characteristic local ring R, let Hilb^n_R be the subscheme of the Grassmanian of codimension n subspaces of $R/\mathcal{M}^n_R$ parametrizing those subspaces which are ideals: since any ideal in R of colength n contains $\mathcal{M}^n_R$, Hilb^n_R parameterizes these ideals. Let $e\colon \mathrm{Hilb}^n_R \to \mathbf{Z}$ be the map assigning to an ideal its multiplicity. By results of Teissier and Lejeune [23], e is upper-semi-continuous.

The natural $\mathbf{G}_m$-action on $\mathcal{O}[[t]]$ by $t \to \lambda t$ induces a $\mathbf{G}_m$-action on $\mathrm{Hilb}^n_{\mathcal{O}[[t]]}$. By the Borel fixed point theorem, there is, for every I, an ideal fixed by this action in $O^{\mathbf{G}_m}(I)$. Such an ideal must, by the upper-semi-continuity of multiplicity have multiplicity at least as large as $e(I)$. Thus, to compute $e_0(\mathcal{O}[[t]])$ it suffices to look at $\mathbf{G}_m$-fixed ideals of finite colength and $\mathcal{J}$ is just the set of such ideals.

Fix $I = \bigoplus_{i=0}^{\infty} I_i t^i$, where $I_0 \subset I_1 \subset \ldots \subset I_N = \mathcal{O}$ is an increasing sequence of ideals in $\mathcal{O}$. Clearly $\mathrm{col}(I) = \sum_{i=0}^{N-1} \mathrm{col}(I_i)$. To bound $e(I)$ we note that

$$I^n \supset (I_0^n) \oplus (I_0^{n-1} I_1 t) \oplus (I_0^{n-2} I_1^2 t^2) \oplus \ldots \oplus (I_0 I_1^{n-1} t^{n-1}) \oplus$$
$$\oplus (I_1^n t^n) \oplus (I_1^{n-1} I_2 t^{n+1}) \oplus \ldots\ldots\ldots \oplus (I_{N-2} I_{N-1}^{n-1} t^{(N-1)n-1}) \oplus$$
$$\oplus (I_{N-1}^n t^{(N-1)n}) \oplus (I_{N-1}^{n-1} t^{(N-1)n+1}) \oplus \ldots\ldots \oplus (I_{N-1} t^{Nn-1}) \tag{3.7}$$
$$\oplus (\mathcal{O} t^{Nn}) \oplus \ldots$$
$$\Rightarrow I^n \supset (I_0^n \oplus I_0^n t \oplus \ldots \oplus I_0^n t^{n-1}) \oplus (I_1^n t^n \oplus \ldots \oplus I_1^n t^{2n-1}) \oplus \ldots\ldots$$
$$\oplus (I_{N-1}^n t^{(N-1)n} \oplus I_{N-1}^{n-1} t^{(N-1)n+1} \oplus \ldots. \oplus I_{N-1} t^{Nn-1})$$
$$\oplus \mathcal{O} t^{Nn} \oplus \ldots$$

Hence,

$$\operatorname{col}(I^n) \leqq \sum_{i=0}^{N-2} n \operatorname{col}(I_i^n) + \sum_{j=1}^{n} \operatorname{col}(I_{n-1}^j)$$
$$= \frac{n^{r+1}}{r!} \sum_{i=0}^{N-2} e(I_i) + \frac{n^{r+1}}{(r+1)!} e(I_{N-1}) + O(n^r)$$

(We have evaluated the second sum by "integration"!)
Finally

$$\frac{e(I)}{(r+1)!\operatorname{col}(I)} \leqq \frac{(r+1)\sum_{i=0}^{N-2} e(I_i) + e(I_{N-1})}{(r+1)!\sum_{i=0}^{N-1}\operatorname{col}(I_i)} \leqq \frac{\sum_{i=0}^{N-1} e(I_i)}{r!\sum_{i=0}^{N-1}\operatorname{col}(I_i)},$$

with strict inequality if $r > 0$

$$\leqq \max_i \frac{e(I_i)}{r!\operatorname{col}(I_i)} \leqq e_0(\mathcal{O}).$$

Corollary 3.8. *If $\mathcal{O}$ is regular, $e_0(\mathcal{O}) = 1$ and if $r > 1$, the defining sup is not attained.*

Corollary 3.9. (Lech [1]). *For all $\mathcal{O}$ and all $I \subset \mathcal{O}$, $e(I) \leqq r!\, e(\mathcal{O}) \operatorname{col}(I)$, hence $e_0(\mathcal{O}) \leqq e(\mathcal{O})$.*

Proof. None of the quantities involved change if we complete $\mathcal{O}$. But after doing this, we can write $\mathcal{O}$ as a finite module over $\mathcal{O}_0 = k[[t_1, \ldots, t_r]]$ so that:

(*) There is a sub $\mathcal{O}_0$-module $\mathcal{O}_0^{e(\mathcal{O})} \subset \mathcal{O}$ such that the quotient $\mathcal{O}/\mathcal{O}_0$ is an $\mathcal{O}_0$-torsion module M.

[1]) Cf. [13], Theorem 3.

Let $I_0 = I \cap \mathcal{O}_0$. Then $\mathrm{col}(I) \geqq \mathrm{col}(I_0)$ and

$$\begin{aligned}\dim(\mathcal{O}/I^n) &\leqq \dim \mathcal{O}/I_0^n\,\mathcal{O}\\ &\leqq \dim(M/I_0^n M) + \dim(\mathcal{O}_0^{e(\mathcal{O})}/I_0^n\,\mathcal{O}^{e(\mathcal{O})})\end{aligned}$$

Condition (*) implies that $\dim(M/I_0^n M)$ is represented by a polynomial of degree less than r, hence

$$\begin{aligned}e(I) &\leqq e(\mathcal{O})\, e(I_0)\\ &\leqq r\,!\, e(\mathcal{O})\,\mathrm{col}(I_0) \text{ by Corollary 3.8}\\ &\leqq r\,!\, e(\mathcal{O})\,\mathrm{col}(I)\end{aligned}$$

We state two other useful properties of e_k:

PROPOSITION 3.10. i) *If* $\mathcal{O}$ *and* $\mathcal{O}'$ *are local domains with the same fraction field and* $\mathcal{O}'$ *is integral over* $\mathcal{O}$, *then* $e_k(\mathcal{O}') \leqq e_k(\mathcal{O})$.

ii) *If* $\mathcal{O} = (k[[t]] + \mathscr{P})$ *is an augmented* $k[[t]]$*-algebra, let* $\mathcal{O}_\eta = \mathcal{O}_{\mathscr{P}}$, *a local ring with residue field* $k((t))$ *and let* $\mathcal{O}_s = \mathcal{O}/t\mathcal{O}$ *be its specialization over* k; *then* $e_k(\mathcal{O}_\eta) \leqq e_k(\mathcal{O}_s)$.

We come now to the main definitions.

DEFINITION 3.11. $\mathcal{O}$ *is semi-stable if* $e_1(\mathcal{O}) = 1$; $\mathcal{O}$ *is stable if, in addition, the defining sup is not attained.*

This terminology is justified by the following proposition which shows that the semi-stability of the local rings on a variety X is just the local impact of the global condition of asymptotic semi-stability for X.

PROPOSITION 3.12. *Fix a variety* X^r, *an ample line bundle* $L = \mathcal{O}_X(D)$ *on* X, *and* $p \in X$. *Then if* $\mathcal{O}_{p,X}$ *is unstable,* (X, L) *is asymptotically unstable.*

Proof. Choose an ideal $I \subset \mathcal{O}_{p,X}[[t]]$ such that

i) $e(I) = (1+\varepsilon)(r+1)\,!\,\mathrm{col}(I)$, $\varepsilon > 0$

ii) $I = \bigoplus_{i=0}^{\infty} I_i t^i$, $I_0 \subset I_1 \subset \ldots \subset I_N = \mathcal{O}_{p,X}$ a sequence of ideals of finite colength. (This is possible because of Lemma 3.6).

Let Φ_m denote the projective embedding of X by $\Gamma(X, L^{\otimes m})$. Choose m large enough that

a) for all $Q \in X, \Gamma(X^r, L^m) \xrightarrow{\psi} \Gamma(X, L^m / I_0 \mathcal{M}_{Q,X} \cdot L^m)$ is surjective

b) L^m is very ample

c) $h^0(X, L^m) > \dfrac{1}{1+\varepsilon} \dfrac{m^r(D^r)}{r!} = \dfrac{1}{1+\varepsilon} \dfrac{\deg \Phi_m(X)}{r!}$

(That the last condition can always be realized is a consequence of Riemann-Roch for X.)

Next choose a basis $X_{i,j}$, $0 \leqq i \leqq N$, of $\Gamma(X, L^m)$ such that

$$X_{0,j} \text{ is a basis of } \psi^{-1}(I_0),$$
$$X_{1,j} \text{ is a basis of } \psi^{-1}(I_1)/\psi^{-1}(I_0),$$
$$\cdots\cdots\cdots$$
$$X_{N,j} \text{ is a basis of } \Gamma(X, L^m)/\psi^{-1}(I_{N-1}),$$

Finally, let λ be the 1-PS which multiplies $X_{i,j}$ by t^i : i.e. in the form of (2.8) $\rho^{(i,j)} = i$; then by assumption (a) the ideal $\mathscr{I}$ corresponding to λ in (2.8) is just I and is supported at the single point $(0, p) \in \mathbf{A}^1 \times X$. Moreover, by condition a)

$$\sum_{i,j} \rho^{(i,j)} = N \dim(\mathcal{O}/I_{N-1}) + (N-1)\dim(I_{N-1}/I_{N-2}) + \ldots + 2\dim(I_2/I_1) + \dim I_1/I_0 = \operatorname{col}(I)$$

(This is Lemma 2.14 again). Hence,

$$\begin{aligned} e(\mathscr{I}) &= e(I) \\ &= (1+\varepsilon) \, . \, (r+1)! \operatorname{col}(I) \\ &> (1+\varepsilon) \, . \, (r+1) \, . \, \frac{\deg \Phi_m(X)}{(1+\varepsilon) h^0(L^m)} \cdot \sum_{i,j} \rho^{(i,j)} \\ &= \frac{(r+1)\deg \Phi_m(X)}{h^0(L^m)} \cdot \sum_{i,j} \rho^{(i,j)} \end{aligned}$$

By Theorem 2.9, $\Phi_m(X)$ is unstable.

Restating Corollary 3.7 gives us a trivial class of stable points:

PROPOSITION 3.13. *If $\mathcal{O}$ is regular and of positive dimension it is stable.*

The next step is to pindown the meaning of semi-stability for small dimensional local rings. For dimension 1, we can be quite explicit:

Proposition 3.14. *If* $\dim \mathcal{O} = 1$ *and* $\mathcal{O}$ *is Cohen-Macauley (i.e.* Spec $\mathcal{O}$ *has no embedded components), then:*

i) $\mathcal{O}$ *stable* $\Leftrightarrow \mathcal{O}$ *regular* $\Leftrightarrow e(\mathcal{O}) = e_0(\mathcal{O}) = e_1(\mathcal{O}) = \ldots = 1$.

ii) $\mathcal{O}$ *semi-stable but not stable* $\Leftrightarrow \mathcal{O}$ *an ordinary double point* $\Leftrightarrow e(\mathcal{O}) = e_0(\mathcal{O}) = 2,\ e_1(\mathcal{O}) = e_2(\mathcal{O}) = \ldots = 1$.

iii) $\mathcal{O}$ *a higher double point* $\Rightarrow e_1(\mathcal{O}) \geqq 8/7$.

iv) $\mathcal{O}$ *a triple point or higher multiplicity* $\Rightarrow e_1(\mathcal{O}) \geqq 3/2$.

Proof. If $\mathcal{O}$ is a triple or higher point, so is $\mathcal{O}[[t]]$, hence $e(\mathcal{O}[[t]]) \geqq 3$, and by Proposition 3.4, $e_1(\mathcal{O}) = e_0(\mathcal{O}[[t]]) \geqq 3/2$.

As for Cohen-Macaulay double points, when char. $\neq 2$ these are all of the form $\hat{\mathcal{O}} = k[[x, y]]/(x^2 - y^n)$, $2 \leqq n \leqq \infty$. (Think of $\hat{\mathcal{O}}$ as a quadratic free $k[[y]]$-algebra; the argument can be readily adapted to char. 2 also). If $n \geqq 3$, then in $k[[x, y, t]]/(x^2 - y^n)$, take $I = (x^2, xy, y^2, xt, yt^2, t^4)$. (This, of course, is the ideal of Proposition 3.1 again). I has complementary basis $(1, x, y, t, yt, t^2, t^3)$, hence $\operatorname{col}(I) = 7$. I claim $e(I) = 16$, which will imply iii). We first note that I is integral over (y^2, t^4). We compute the multiplicity of (y^2, t^4) as

$$\textit{intersection-multiplicity at } \mathcal{M}\ ((\mathrm{Spec}\ \mathcal{O}) \,.\, (y^2 = 0) \,.\, (t^4 = 0))$$
$$= 8 \,.\, \textit{intersection-multiplicity}\ ((\mathrm{Spec}\ \mathcal{O}) \,.\, (y = 0) \,.\, (t = 0))$$
$$= 16$$

since $\mathcal{O}$ is a double point.

When $\mathcal{O}$ is an ordinary double point, I claim $e_0(\mathcal{O}[[t]]) = 1$. Since this value is attained by the maximal ideal $\mathcal{M}$: $\dfrac{e(\mathcal{M})}{2!\operatorname{col}(\mathcal{M})} = \dfrac{2}{2} = 1$, this will prove ii), hence i) in view of Proposition 3.13.

In general, if $\mathcal{O} = k[[x, y]]/(x \,.\, y)$, an ideal $I \subset \mathcal{O}[[t]]$ corresponds to a pair of ideals $J \subset k[[x, t]]$ and $K \subset k[[y, t]]$ such that $J + (x)/(x)$ and $K + (y)/(y)$ have the same image, say (t^n), in $k[[t]]$. A rough picture is given below: the condition on the two ideals ensures that they glue along the intersection of the two planes.

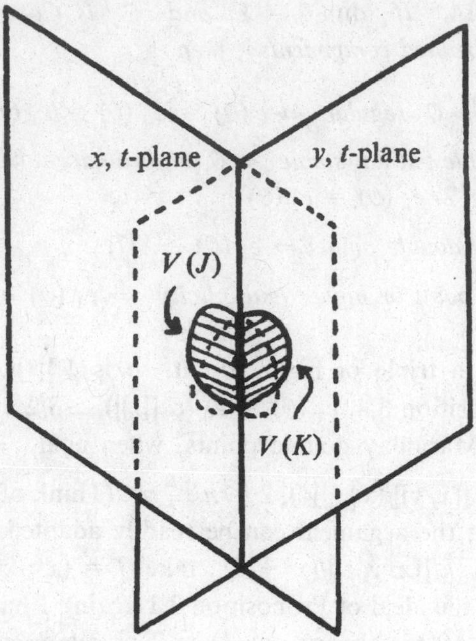

In this situation, col (I) = col (J) + col $(K) - n$, and $e(I) = e(J) + e(K)$, so the inequality $e(I)/2 \,.\, \text{col}(I) \leqq 1$ follows from:

LEMMA 3.15. *If* $I \subset k[[x, y]]$ *and* $I + (x) = (x, y^a)$, *then* $e(I) \leqq 2\, \text{col}(I) - a$.

Proof. By applying Lemma 3.6, we can reduce to the case where I is generated by monomials:

$$I = \bigoplus_{l=0}^{\infty} (y^{r_l} . x^l) . k[[y]], \text{ with } a = r_0 \geqq r_1 \geqq \ldots \geqq r_N = 0.$$

Then as in (3.7):

$$I^n \supset (y^{nr_0})\, k \oplus (y^{n-1)r_0+r_1}x)\, k \oplus (y^{(n-2)r_0+2r_1}x^2)\, k \oplus \ldots$$
$$\oplus (y^{nr_1}x^n)\, k \oplus (y^{(n-1)r_1+r_2}x^{n+1})\, k \oplus \ldots\ldots \oplus (y^{nr_2}x^{2n})\, k \oplus \ldots\ldots\ldots\ldots$$

$$\Rightarrow \text{col}(I^n) \leqq \frac{n(n+1)}{2} r_0 + n^2 r_1 + n^2 r_2 + \ldots\ldots + n^2 r_{N-1}$$

$$\Rightarrow \frac{e(I)}{2} \leqq \frac{r_0}{2} + r_1 + \ldots\ldots + r_{N-1} = \text{col}(I) - \frac{a}{2}.$$

REMARK. If $I \subset \mathcal{O}[[t]]$ is of the form of Lemma 3.6, the expansion (3.7) for I^n, which we have used again here, can be used to give even better bounds for $e(I)$. To get these however, requires the more involved theory of mixed multiplicities which will be discussed in § 4.

The meaning of semi-stability for two dimensional singularities is not yet completely worked out, but what follows gives a good overview of the situation.

DEFINITION 3.16. *If* $\mathcal{O}$ *is a normal 2-dimensional local ring,* x *is the closed point of* Spec $\mathcal{O}$, *and* $X^* \xrightarrow{\pi}$ Spec $\mathcal{O}$ *is a resolution of* $\mathcal{O}$ *(i.e.* π *is proper and birational), then we define*

i) *big genus of* $\mathcal{O} = \dim R^1 \pi_* (\mathcal{O}_{X^*})$
($R^1\pi_*$ *is a torsion* $\mathcal{O}$*-module supported at* x)

ii) *little genus of* $\mathcal{O} = \sup_Z (p_a(\mathcal{O}_Z))$, *where* Z *runs over the effective cycles on* $\pi^{-1}(x)$.

Wagreich [24] has shown that big genus $\geqq$ little genus—hence the names—and Artin [3] has shown that if the little genus is zero then so is the big genus. (But when little genus $= 1$, big genus may be > 1). We call $\mathcal{O}$: rational (resp. strongly elliptic) if its big genus is 0 (resp. 1), and weakly elliptic if its little genus is 1.

If there is to be any hope of constructing compact moduli spaces for semi-stable surfaces, the non-normal singularity $xyz = 0$ must be semi-stable—in fact, it is. But $xyz = 0$ is the cone over a plane triangle so the triple point on it is really a degenerate "elliptic" singularity. In fact, $xyz = 0$ is a limit of the family of non-singular cubics $xyz + t(x^3+y^3+z^3) = 0$. Similarly, the standard singularities A_{n-1}: $xy = z^n$ and D_n: $x^2 = y^2 z + z^n$ have non-normal limits $xy = 0$ and $x^2 = y^2 z$ respectively as $n \to \infty$. We can summarize these considerations in the heuristic conjecture: the semi-stable singularities of surfaces will be a limited class of rational and strongly elliptic normal singularities and their non-normal limits.

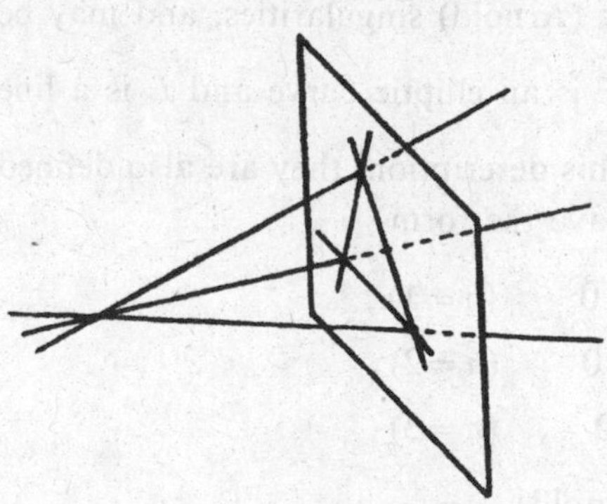

We now list without proof some classes of semi-stable singularities.

3.17. Elliptic polygonal cones. In $\mathbf{P}^{n-1}$ take a generic n-gon $\bigcup_{i=0}^{n} \overline{p_i p_{i+1}}$ ($p_0 = p_{n+1}$) and take the cone in $\mathbf{C}^n$ over it. This is a union of n-planes crossing normally in pairs and meeting at an n-fold point at the origin. We also allow the degenerate cases $n = 2$ (local equation $x^2 = y^2 z^2$) and $n = 1$ (local equation $x^2 = y^2(y + z^2)$) which correspond respectively, to glueing two planes to each other along a pair of transversal lines, and to glueing a pair of transversal lines in a plane together as shown below.

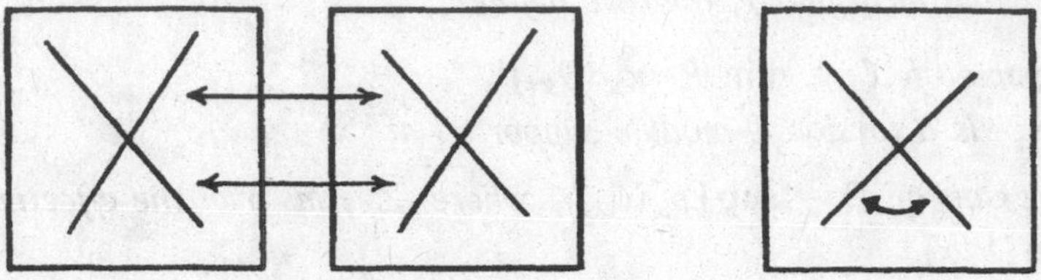

Proposition 3.18. *Elliptic polygonal n-cones are semi-stable if and only if* $1 \leqq n \leqq 6$. *Moreover, all small deformations of these singularities are semi-stable.*

Examples of such singularities are:

i) Cone over a smooth elliptic curve with generic j in $\mathbf{P}^n$, $3 \leqq n \leqq 5$. (In fact, I expect this holds for arbitrary j). These are also called the simple elliptic (Saito) or parabolic (Arnold) singularities, and may be described as $\bigoplus_{m=0}^{\infty} \Gamma(E, L^m)$ where E is an elliptic curve and L is a line bundle of positive degree n: with this description, they are also defined for $n = 1, 2$. For small n, these have the form

$$x^2 + y^3 + z^6 + a(y^2 z^2) = 0 \qquad (n=1),$$
$$x^2 + y^4 + z^4 + a(y^2 z^2) = 0 \qquad (n=2),$$
$$x^3 + y^3 + z^3 + a(xyz) = 0 \qquad (n=3).$$

ii) The hyperbolic singularities of Arnold:

$$xyz + x^n + y^m + z^p = 0 \qquad \frac{1}{n} + \frac{1}{m} + \frac{1}{p} < 1.$$

iii) Rational double points.

iv) Pinch points: these have local equation $x^2 = y^2 z$.

3.18. RATIONAL POLYGONAL CONES. In $\mathbf{P}^{n-1}$ take $(n-1)$ generic line segments $P_0 P_1 \cup P_1 P_2 \ldots \cup P_{n-1} P_n$ and in $\mathbf{C}^n$ take the cone over them: one obtains $(n-2)$ planes crossing normally in $(n-1)$ lines.

PROPOSITION 3.19. *Rational polygonal n-cones are semi-stable if and only if* $2 \leqq n \leqq 6$. *Hence, all small deformations of these singularities are semi-stable.*

A typical singularity which arises in this way is the cone over a rational normal curve in $\mathbf{P}^{n-1}$, $2 \leqq n \leqq 6$.

By applying the semi-stability condition to the ideal $I = \bigoplus_{j=0}^{i} t^{i-j} . (\widetilde{I^j}) \subset \mathcal{O}[[t]]$, where I is an ideal in $\mathcal{O}$ and $\sim$ denotes integral closure in $\mathcal{O}$, one can prove the following necessary condition for semi-stability:

PROPOSITION 3.19. *If* $\mathcal{O}^r$ *is semi-stable,* $I \subset \mathcal{O}$ *and* $P(i) = \dim(\mathcal{O}/(\widetilde{I^i}))$, *then*

$$P(1) + \ldots\ldots + P(i) \geqq \frac{e(I)\, i^{r+1}}{(r+1)!} .$$

When $r = 2$, and $\mathcal{O}$ is Cohen-Macaulay this reduces us to *ten* basic types of singularities. In the first few cases we have listed the singularities of this type which are actually semi-stable.

1) Regular points: always stable.

2) Double coverings of $\mathbf{C}^2$ with branch curve of multiplicity $\leqq 4$: semi-stable here are,

   a) rational double points and their non-normal limits $xy = 0$, $x = y^2 z$,

   b) hyperbolic double points,

   c) parabolic double points.

3) Triple points in $\mathbf{C}^3$: Semi-stable are,

   a) cones over non-singular elliptic curves,

   b) hyperbolic triple points.

4-5) Triple and quadruple points in $\mathbf{C}^4$.

6-7) Quadruple and quintuple points in $\mathbf{C}^5$.

8-9) Quintuple and sextuple points in $\mathbf{C}^6$.

10) Sextuple points in $\mathbf{C}^7$.

REMARK. With Eisenbud, we made some computations by computor to eliminate cases; the computer came up with some amusing examples. For instance it found an ideal I in $k[[x, y, z, t]]/(x^2+y^3+z^7)$ with col (I) = 63,398, mult (I) = 381,024, showing that $e_0 \geqq 1.000167$, hence that the singularity $x^2 + y^3 + z^7 = 0$ is unstable.

Further restrictions, confirming the heuristic conjecture, on what singularities are semi-stable are provided by:

PROPOSITION 3.20. *If $\mathcal{O}$ is normal and semi-stable then $\mathcal{O}$ is rational or weakly elliptic. Moreover, there are no cuspidal curves, i.e. generically all singular curves are ordinary.*

We omit the proof except to note that the last statement comes from the observation that for large n the choices $I_n = (T^9, u^{9n}, v^{9n})^{\sim}$ show that $e_2(k[[T^2, T^3]]) \geqq 1 + 22/221$!

Now suppose $\mathcal{O}$ is not Cohen-Macaulay. We can create a slew of stable $\mathcal{O}$'s using i) of Proposition 3.10. For example if $k[[x, y]] \supset \mathcal{O} \supset k[[x, xy, y^2]]$, then $\mathcal{O}$ is semi-stable since the ring on the right which is the pinch point is semi-stable; a typical example is $\mathcal{O} = k[[x, xy, y^2, y^3]]$, a very partial pinch in which only the y-tangent has been removed. Fortunately most of these points cannot appear as singularities of varieties on boundary of moduli spaces as they have no smooth deformations. More precisely, (cf. [27]):

THEOREM 3.21. *If $\mathcal{O}$ is a 2-dimensional local ring which is not Cohen-Macauley such that $\mathcal{O} = \mathcal{O}'/t\mathcal{O}'$ where $\mathcal{O}'$ is a normal 3-dimensional local ring; let $\mathcal{O}_{\text{norm}}$ be its normalization and $\tilde{\mathcal{O}} = \{ a \in \mathcal{O}_{\text{norm}} \mid$ for some $n, \mathcal{M}_{\mathcal{O}}^n a \subset \mathcal{O} \}$.*

Then i) *$\tilde{\mathcal{O}}$ is a local ring*

ii) *If in addition $\mathcal{O}$ has characteristic* 0, *then*

$$\dim(\tilde{\mathcal{O}}/\mathcal{O}) \leqq \text{big genus of } \tilde{\mathcal{O}}.$$

REMARK. If, as seems likely, in view of Proposition 3.20 the big genus of the Cohen-Macaulay ring $\tilde{\mathcal{O}}$ is 0 or 1, this means that $\mathcal{O}$ must be nearly Cohen-Macauley.

We conclude this section by outlining an as yet completely uninvestigated approach to deciding which singularities should be allowed on the objects of a moduli space.

DEFINITION 3.22. *$\mathcal{O}^r$ is an insignificant limit singularity if, whenever $\mathcal{O}'$ is an $(r+1)$ dimensional local ring such that $\mathcal{O} = \mathcal{O}'/t\mathcal{O}'$ for some $t \in \mathcal{O}'$, $\pi: X \to \operatorname{Spec} \mathcal{O}'$ is a resolution of $\operatorname{Spec} \mathcal{O}'$ and $E \subset X$ is an exceptional divisor (i.e. $\dim \pi(E) < \dim E$), then E is birationally ruled, that is, the function field of E is a purely transcendental extension of a proper subfield. Equivalently, setting $\mathcal{O}/\mathcal{M}_{\mathcal{O}} = k$, this says that whenever R is a discrete rank 1 valuation ring containing $\mathcal{O}'$ with tr. $\text{deg.}_k R/\mathcal{M}_R = r$, then $R/\mathcal{M}_R = K(t)$, for some K such that tr. $\text{deg.}_k K = r - 1$.*

EXAMPLES. 1) $xy = 0$ is insignificant because on deforming this only A_n singularities arise.

2) $x^2 + y^3 = 0$ is significant because the deformation $t^6 = x^2 + y^3$ blows up to a non-singular elliptic curve with $(E^2) = -1$. Similarly I can show that all higher plane curve singularities are significant.

3) $x^3 + y^3 + y^4 = 0$ is significant because $t^{12} = x^3 + y^3 + y^4$ blows up to a 3-fold containing a K3 surface.

4) Jayant Shah [26] has proven that rational double points and Arnold's parabolic and hyperbolic singularities are insignificant. As a limiting case, normal crossings $xyz = 0$ is insignificant.

REMARKS. 1) Why should birational ruling of exceptional divisors be the right criterion for insignificance? The reason is that all exceptional divisors which arise from blow-ups of non-singular points are birationally ruled and all birationally ruled varieties arise in this way. So on the one hand, such exceptional divisors must be permitted, and on the other, the examples suggest that sufficiently tame singularities cannot "swallow" anything else.

2) The examples suggest that $\mathcal{O}$ semi-stable and $\mathcal{O}$ insignificant are closely related. For instance, perhaps these are the same when embedding-dim $\mathcal{O} = 1$. In dim 2 for example, after hyperbolic and parabolic singularities in the Dolgacev-Arnold list [2, 7] of 2-dimensional singularities come 31 special singularities. These are all unstable and in a recent letter to me Dolgacev remarks that all of these have deformations which blow up to K3 surfaces as in Example 3. If semi-stability and insignificance turn out to be roughly the same in arbitrary dimension, we would have a very powerful tool to apply to moduli problems.

§ 4. Asymptotic Stability of Canonically Polarized Curves

The chief difficulty of using the numerical criterion of Theorem 2.9 to prove the stability of a projective variety is that it is necessary to look inside $\mathcal{O}_{X \times \mathbf{A}^1}$ to compute the multiplicity $e_L(\mathcal{I})$. To circumvent this difficulty, we will construct an upper bound on $e_L(\mathcal{I})$ in terms of data on X alone. For curves, this bound involves only the multiplicities of ideals $\mathcal{I} \subset \mathcal{O}_X$, but for higher dimensional varieties—in particular, surfaces—it requires a theory of mixed multiplicities, i.e. multiplicities for several ideals simultaneously. To motivate the global theory, we will first describe what happens in the local case. Here the basic ideas were introduced by Teissier and Rissler [22]. Recall that if $\mathcal{O}$ is a local ring of dimension r with infinite residue field and I is an ideal of finite colength in it then whenever $f_1, \ldots f_r$ are sufficiently generic elements of I, $e(I) = e((f_1, \ldots, f_r))$. This suggests

Definition 4.1. *If $\mathcal{O}^r$ is a local ring and $I_1, \ldots, I_r$ are ideals of finite colength in $\mathcal{O}$, the mixed multiplicity of the I_i is defined by*

$$e(I_1, \ldots, I_r) = e((f_1, \ldots, f_r))$$

where $f_i \in I_i$ is a sufficiently generic element. (The set of integers $e((f_1, \ldots, f_r))$ has some minimal element and a choice $(f_1, \ldots, f_r)$ is sufficiently generic if the minimum is attained for these f_i.)

The basic property of these multiplicities is:

Proposition 4.2. *Let $I_1, \ldots, I_k$ be ideals of finite colength of a local ring $\mathcal{O}^r$ and let*

$$P_r(m_1, \ldots, m_k) = \sum_{\substack{\Sigma r_i = r \\ r_i \geq 0}} \frac{1}{\prod (r_i !)} \cdot e(I_1^{[r_1]}, \ldots, I_k^{[r_k]}) \cdot m_1^{r_1} \ldots m_k^{r_k}$$

where $I_i^{[r_i]}$ indicates that I_i appears r_i times. Then

i) $$\left| \dim (\mathcal{O} / \prod_{i=1}^{k} I_i^{m_i}) - P_r(m_1, \ldots, m_k) \right| = 0((\textstyle\sum m_i)^{r-1})$$

ii) *There exists a polynomial of total degree r*

$$P(m_1, \ldots, m_k) = P_r(m_1, \ldots, m_k) + \text{lower order terms}$$

and an N_0 such that if $m_i \geq N_0$ for all i, then

$$\dim (\mathcal{O} / \prod I_i^{m_i}) = P(m_1, \ldots, m_k).$$

Proof. See Teissier and Rissler [22].
Using this we obtain the estimate:

PROPOSITION 4.3. *Let* $I \subset \mathcal{O}[[t]]$ *be an ideal of finite codimension and let* $I_k = \{ a \in \mathcal{O} \mid at^k \in I \}$; *then* $I_0 \subseteq I_1 \subseteq \ldots \subseteq I_N = \mathcal{O}, N \geqslant 0$. *Then for all sequences* $0 = r_0 < r_1 < \ldots < r_l = N$,

$$e(I) \leqslant \sum_{k=0}^{l-1} (r_{k+1} - r_k) \sum_{j=0}^{r} e(I_{r_k}^{[j]}, I_{r_{k+1}}^{[r-j]}) .$$

Proof. Since $I \supset \oplus\, t^{r_i} I_{r_i}$

$$\begin{aligned} I^n \supset\; & I_{r_0}^n (\mathcal{O} + t\mathcal{O} + \ldots + t^{r_1 - 1}\mathcal{O}) + I_{r_0}^{n-1} I_{r_1} (t^{r_1}\mathcal{O} + t^{r_1+1}\mathcal{O} + \ldots + t^{2r_1 - 1}\mathcal{O}) \\ & + \ldots + I_{r_0} I_{r_1}^{n-1} (t^{(n-1)r_1}\mathcal{O} + \ldots + t^{nr_1 - 1}\mathcal{O}) \\ & + I_{r_1}^n (t^{nr_1}\mathcal{O} + \ldots + t^{(n-1)r_1 + r_2 - 1}\mathcal{O}) + I_{r_1}^{n-1} I_{r_2} (t^{(n-1)r_1 + r_2}\mathcal{O} + \ldots) \\ & + \ldots + I_{r_{l-1}}^n (t^{nr_l - 1}\mathcal{O} + \ldots) + I_{r_{l-1}}^{n-1} (t^{(n-1)r_l - 1 + r_l}\mathcal{O} + \ldots) \\ & + \ldots + t^{nr_l}\mathcal{O}[[t]] . \end{aligned}$$

whence

$$\dim(\mathcal{O}[[t]]/I^n) \leqslant \sum_{k=0}^{l} (r_{k+1} - r_k) \sum_{i=0}^{n-1} \dim(\mathcal{O}/(I_{r_k}^{n-i} \,.\, I_{r_{k+1}}^{i}))$$

$$(4.4) \quad = \sum_{k=0}^{l} (r_{k+1} - r_k) \sum_{i=0}^{n-1} \left[\sum_{j=0}^{r} \frac{1}{j!\,(r-j)!} e(I_{r_k}^{[r-j]}, I_{r_{k+1}}^{[j]})\, (n-i)^{r-j} i^j + R_i \right]$$

By Proposition 4.2 i) each remainder terms R_i is $O(n^{r-1})$. Indeed, ii) of 4.2 says that except when i or $n - i < N_0$, the R_i are all represented by a polynomial of degree $r - 1$ so that we can obtain a uniform $O(n^{r-1})$ estimate for the R_i; hence $\sum_{i=0}^{n-1} R_i = O(n^r)$.

But the n.l.c. of the $(r+1)^{\text{st}}$ degree polynomial representing $\dim(\mathcal{O}[[t]]/I^n)$ is by definition $e(I)$; so evaluating the n.l.c. of the sum in (4.4) using the lemma below, gives the proposition.

LEMMA 4.5. $$\frac{j!\,(r-j)!}{(r+1)!} n^{r+1} = \sum_{i=0}^{n-1} (n-i)^{r-j} i^j + O(n^r)$$

Proof. We can reexpress the left hand side in terms of the β-function as

$$\frac{j!\,(r-j)!}{(r+1)!} n^{r+1} = \beta(j, r-j)\, n^{r+1} = \left(\int_0^1 t^j (1-t)^{r-j}\, dt \right) n^{r+1} ,$$

and the right hand side is just another expression for n^{r+1} times this integral as a Riemann sum plus error term.

To globalize these ideas we combine them with some results of Snapper [5, 21].

DEFINITION 4.6. *Let X^r be a variety, L be a line bundle on X and $\mathscr{I}_1, \ldots, \mathscr{I}_r$ be ideals on $\mathcal{O}_X$ such that* $\operatorname{supp}(\mathcal{O}_X/\mathscr{I}_i)$ *is proper. Choose a compactification $\bar{X}$ of X on which L extends to a line bundle $\bar{L}$ and let $\pi: \bar{B} \to \bar{X}$ be the blowing up of $\bar{X}$ along $\prod \mathscr{I}_i$ so that $\pi^{-1}(\mathscr{I}_i) = \mathcal{O}_{\bar{B}}(-E_i)$. Let $\pi^* L = \mathcal{O}_{\bar{B}}(D)$. We define*

$$e_L(\mathscr{I}_1, \ldots, \mathscr{I}_r) = (D^r) - ((D - E_1) \cdot \cdots\cdots \cdot (D - E_r)).$$

We omit the check that this definition is independent of the choice of $\bar{X}$ and $\bar{L}$.

4.7. CLASSICAL GEOMETRIC INTERPRETATION. Suppose X is a projective variety, $L = \mathcal{O}_X(1)$ and $\mathscr{I}_i \cdot L$ is generated by a space of sections $W_i \subset \Gamma(\mathbf{P}^n, \mathcal{O}(1))$. If $H_1, \ldots, H_r$ are generic hyperplanes of $\mathbf{P}^n$, then $\#(H_1 \cap \ldots \cap H_r \cap X) = \deg X$. One sees by an argument like that of Proposition 2.5, that as the H_i specialize to hyperplanes defined by elements of W_i but otherwise generic, the number of points in $H_1 \cap \ldots \cap H_r \cap X$ which specialize to a point in one of the W_i's is just $e_L(\mathscr{I}_1, \ldots, \mathscr{I}_r)$.

We can globalize Proposition 4.2 to give an interpretation of the mixed multiplicity by Hilbert polynomials.

PROPOSITION 4.8. i) *Let X^r be a variety, $L_1, \ldots, L_n$ be line bundles on X and $\mathscr{I}_1, \ldots, \mathscr{I}_l$ be ideals in $\mathcal{O}_X$ such that* $\operatorname{supp}(\mathcal{O}_X/\mathscr{I}_i)$ *is proper for all i. Then there is a polynomial $P(n, m)$ of total degree r and an M_0 such that if $m_j \geq M_0$ for all j then*

$$\chi\left(X, \bigotimes_{i=1}^{k} L_i^{n_i} \Big/ \prod_{j=1}^{l} \mathscr{I}_j^{m_j} \cdot \bigotimes_{i=1}^{k} L_i^{n_i}\right) = P(n, m).$$

Now suppose all the line bundles are the same, say L and let

$$P_r(m_1, \ldots, m_l) = \sum_{\substack{\Sigma r_i = r \\ r_i \geq 0}} \frac{1}{\prod (r_i!)} e_L(\mathscr{I}_1^{[r_1]}, \ldots, \mathscr{I}_l^{[r_l]}) m_1^{r_1} \ldots m_l^{r_l}$$

Then

ii) $P(\sum m_i; m_1, \ldots, m_l) = P_r(m_1, \ldots, m_l) +$ lower order terms

iii) $| \chi (X, L^{\Sigma m_i} / \prod \mathscr{I}_j^{m_j} \otimes L^{\Sigma m_i}) - P_r(m_1, \ldots, m_l) | = O((\sum_{j=1}^{l} m_j)^{r-1})$

(i.e. *we retain an estimate assuming only* $\sum m_j$ *is large*).

Proof. Making a suitable compactification of X will not alter the Euler characteristics so we may assume X is compact.

Before proceeding we recall certain facts: If $R = \bigoplus_{n_i \geq 0} R_{n_1, \ldots, n_l}$ is a multigraded ring we can form a scheme Proj (R) in the obvious way from multi-homogeneous prime ideals. Quasi-coherent sheaves $\mathscr{F}$ on Proj (R) correspond to multigraded R-modules $M = \bigoplus M_{n_1, \ldots, n_l}$. Suppose $R_{0, \ldots, 0} = k$ a field and that R is generated by the homogeneous pieces $R_{0, \ldots, 0, 1, 0, \ldots, 0}$. Then we get invertible sheaves $L_1, \ldots, L_l$ on Proj (R) from the modules M_i, where M_i = (R with i^{th}-grading shifted by 1), and the multigraded variant of the F.A.C. vanishing theorem for higher cohomology says that if $\mathscr{F}$ is a coherent sheaf on Proj (R) then

$$H^i(\mathscr{F} \otimes (\otimes L_j^{n_j})) = \begin{cases} M_{n_1, \ldots, n_l}, & i = 0 \\ (0), & i > 0 \end{cases} \qquad \text{if } n_j \gg 0, \text{ all } j$$

Now if $\mathscr{I}_1, \ldots, \mathscr{I}_k$ are ideal sheaves on X such that supp $(\mathcal{O}_X/\mathscr{I}_j)$ is proper for all i, let $\mathscr{A} = \bigoplus_{m_j \geq 0} \mathscr{I}_1^{m_1} \ldots \mathscr{I}_l^{m_l}$. Then $\mathscr{A}$ is a multigraded sheaf of $\mathcal{O}_X$-algebras. Let B = Proj $(\mathscr{A})$; the blow up of X along $\prod \mathscr{I}_j$ is just $\pi: B \to X$. If E_j is the exceptional divisor corresponding to $\mathscr{I}_j$, then when $\mathcal{O}_B(-\sum m_j E_j)$ is coherent and when all the m_j are large the relative versions of the vanishing theorems say:

a) $R^i \pi_* (\mathcal{O}(-\sum m_j E_j)) = 0,\ i > 0$

b) $\pi_* \mathcal{O}(-\sum m_j E_j) = \prod_{j=1}^{l} \mathscr{I}_j^{m_j}$

In any case,

c) supp $R^i \pi_* (\mathcal{O}(-\sum m_j E_j))$ has dimension less than $r, i > 0$,

d) $\pi_* (\mathcal{O}(-\sum m_j E_j)) = \prod_i \mathscr{I}_i^{m_i}$ except on a set of dimension less than r.

From a) and b) we deduce that when all the m_j are large, $\chi(\prod \mathscr{I}_j^{m_j}) = \chi(\pi^* \mathcal{O}(-\sum m_j E_j))$. Thus, $\chi(X, \otimes L_i^{n_i} / \prod \mathscr{I}_j^{m_j} L_i^{n_i}) = \chi(X, \otimes L_i^{n_i}) - \chi(B, \otimes L_i^{n_i}(-\sum m_j E_j))$ and both of these last Euler characteristics polynomials of degree $\leq r$ by Snapper [5, 21]. Now if $\pi^* L = \mathcal{O}_B(D)$, his result also says,

$$r!\,.\,\text{n.l.c.}\,(\chi(X, L^{\Sigma m_j}/\prod \mathscr{I}_j^{m_j} \otimes L^{\Sigma m_j}) = (\sum m_j)^r (D^r) - ((\sum m_j (D - E_j))^r)$$

$$= \sum_{\substack{\Sigma r_j = r \\ r_j \geq 0}} \frac{r!}{\prod (r_j!)} \prod (m_j (D - E_j))^{r_j}$$

$$= \sum_{\substack{\Sigma r_j = r \\ r_j \geq 0}} \frac{r!}{\prod (r_j!)} e_L(\mathscr{I}_1^{[r_1]}, \ldots, \mathscr{I}_l^{[r_l]}) \,.\, m_1^{r_1} \ldots m_l^{r_l}$$

which is ii). Fix an N such that ii) holds when all $m_j \geq N$.

Now suppose I is a proper subset of $\{1, \ldots, l\}$, J is its complement and that values $m_i < N$ are fixed for all $i \in I$. Let $\pi_J : B_J \to X$ be the blow up of X along $\prod_{j \in J} \mathscr{I}_j$. As above we deduce that $\exists N'$ depending on I and the m_i, $i \in I$ such that if $m_j > N'$, $\forall j \in J$, then

$$\chi(X, \mathscr{I}_1^{m_1} \ldots \mathscr{I}_k^{m_k}) = \chi(B_J, \prod_{i \in I} \mathscr{I}_i^{m_i}(-\sum_{j \in J} m_j E_j)).$$

Then applying c) and d) we see that for some C, also depending on I and the m_i, $i \in I$,

$$|\chi(B, \mathcal{O}(-\sum m_i E_i)) - \chi(B_J, \prod_{i \in I} \mathscr{I}_i^{m_i}(-\sum_{j \in J} m_j E_j))| \leq C(\sum_{j \in J} m_j)^{r-1}.$$

Combining this with the argument used in the proof of i) and ii) shows that for some C' (depending on I and the m_i, $i \in I$)

$$|\chi(X, L^{\Sigma m_j}/\prod \mathscr{I}_j^{m_j} L^{\Sigma m_j}) - P_r(m_1 \ldots m_l)| \leq C'(\sum_{j \in J} m_j)^{r-1}.$$

From ii), we get an estimate of this type with a uniform constant C', when all the $m_j \geq N$. Since there are only finitely many sets I and for each of these only finitely many choices for the m_i, $i \in I$ with $m_i < N$ we can combine all these estimates to show: there exists M and C'' such that if any $m_i > M$, then

$$|\chi(X, L^{\Sigma m_j}/\prod_j \mathscr{I}_j^{m_j} L^{\Sigma m_j}) - P_r(m_1, \ldots, m_l)| \leq C''((\sum_j m_j)^{r-1})$$

which is iii).

The following analogue of Proposition 2.6 allows us to calculate mixed multiplicities in terms of the dimensions of spaces of sections.

PROPOSITION 4.9. *If $L, \mathscr{I}_1 L, \ldots, \mathscr{I}_l L$ are generated by their sections, then*

$$|\chi(X, L^{\Sigma m_j}/(\prod \mathscr{I}_j^{m_j}) L^{\Sigma m_j}) - \dim(\Gamma(X, L^{\Sigma m_j})/\Gamma(X, \prod \mathscr{I}^{m_j} L^{\Sigma m_j}))|$$
$$= O((\sum m_j)^{r-1})$$

Proof. We give only a sketch of the proof which is very similar to that of Proposition 2.6. One first shows as in the proof of 2.6a), that for $i > 0$, $h^i(L^{\Sigma m_j}/\prod \mathscr{I}^{m_j} L^{\Sigma m_j}) = O((\sum m_j)^{r-1})$, hence that

$$|\chi(X, L^{\Sigma m_j}/\prod \mathscr{I}_j^{m_j} L^{\Sigma m_j}) - \dim \Gamma(X, L^{\Sigma m_j}/\prod \mathscr{I}_j^{m_j} L^{\Sigma m_j})| = O((\textstyle\sum m_j)^{r-1})$$

Using the long exact sequence

$$0 \to \Gamma(X, \prod \mathscr{I}_j^{m_j} L^{\Sigma m_j}) \to \Gamma(X, L^{\Sigma m_j}) \to \Gamma(X, L^{\Sigma m_j}/\prod \mathscr{I}_j^{m_j} L^{\Sigma m_j}) \to \dots$$

this reduces the proposition to showing that

$$\dim\left(\operatorname{coker}\left(\Gamma(X, L^{\Sigma m_j}) \to \Gamma(X, L^{\Sigma m_j}/\prod \mathscr{I}_j^{m_j} L^{\Sigma m_j})\right)\right) = O((\textstyle\sum m_j)^{r-1})$$

and this is done exactly as in the proof of 2.6b). (Note that the extra hypotheses of 2.6b) were not used in this part of the proof.)

The global form of Proposition 4.3 is:

PROPOSITION 4.10. *Given a variety X, a line bundle L on X and an ideal $\mathscr{I} \subset \mathcal{O}_{X \times \mathbf{A}^1}$ with $\operatorname{supp}(\mathcal{O}_{X\times \mathbf{A}^1}/\mathscr{I})$ proper in $X \times (0)$, let $\mathscr{I}_k = \{a \in \mathcal{O}_X \mid t^k a \in \mathscr{I}\}$ so that $\mathscr{I}_0 \subseteq \mathscr{I}_1 \subseteq \dots \subseteq \mathscr{I}_N = \mathcal{O}_X$ and let $L_1 = L \otimes \mathcal{O}_{\mathbf{A}^1}$. Suppose that L, $\mathscr{I}_k L$ and $\mathscr{I} L_1$ are generated by their sections. Then for all sequences $0 = r_0 < r_1 < \dots < r_l = N$,*

$$e_{L_1}(\mathscr{I}) \leq \sum_{k=0}^{l} (r_{k+1} - r_k) \sum_{j=0}^{r} e_L(\mathscr{I}_{r_k}^{[j]}, \mathscr{I}_{r_{k+1}}^{[r-j]}).$$

Proof. By Proposition 4.9, $e_{L_1}(\mathscr{I})$ is calculated by the order of growth of

$$\dim\left[H^0(X \times \mathbf{A}^1, L_1^n)/H^0(X \times \mathbf{A}^1, \mathscr{I}^n \cdot L_1^n)\right].$$

Exactly as in Proposition 4.3, for each n, we introduce using the r_i's an approximating ideal sheaf $\mathscr{I}'_n$:

$$\mathscr{I}^n \supset \mathscr{I}'_n = \bigoplus_{k=0}^{\infty} t^k \cdot \mathscr{I}_{n,k}$$

where $\mathscr{I}_{n,0} \subset \mathscr{I}_{n,1} \subset \dots \subset \mathscr{I}_{n,N} = \mathcal{O}_X$ for $N \gg 0$. Since

$$H^0(X \times \mathbf{A}^1, \mathscr{I}^n \cdot L_1^n) \supset H^0(X \times \mathbf{A}^1, \mathscr{I}'_n \cdot L_1^n) = \bigoplus_{k=0}^{\infty} H^0(X, \mathscr{I}_{n,k} \cdot L^n),$$

it follows that

$$\dim (H^0(X \times \mathbf{A}^1, L_1^n)/H^0(X \times \mathbf{A}^1, \mathscr{I}^n . L_1^n)$$
$$\leq \sum_{k=0}^{\infty} \dim (H^0(X, L^n)/H^0(X, \mathscr{I}_{n,k} . L^n))$$

The rest of the proof follows Proposition 4.3 exactly, using 4.9 again to get the estimate

$$\dim (H^0(X, L^n)/H^0(X, \mathscr{I}^i_{r_k} . \mathscr{I}^{n-i}_{r_{k+1}} . L^n))$$

for $\chi(L^n/\mathscr{I}^i_{r_k} . \mathscr{I}^{n-i}_{r_{k+1}} . L^n)$.

COROLLARY 4.11. *If in Proposition* 4.10, X *is a curve*

$$e_{L_1}(\mathscr{I}) \leq \min_{0=r_0<r_1\ldots<r_l=N} \left[\sum_{k=0}^{e} (r_{k+1}-r_k) . (e_L(\mathscr{I}_{r_k}) + e_L(\mathscr{I}_{r_{k+1}}))\right]$$

If X *is a surface,*

$$e_{L_1}(\mathscr{I})$$
$$\leq \min_{0=r_0<r_1\ldots<r_l=N} \left[\sum_{k=0}^{l} (r_{k+1}-r_k) . (e_L(\mathscr{I}_{r_k}) + e_L(\mathscr{I}_{r_k}, \mathscr{I}_{r_{k+1}}) + e_L(\mathscr{I}_{r_k}))\right]$$

We now show how this upper bound proves the asymptotic stability of non-singular curves. It turns out that the estimate is, however, *not* sufficiently sharp to prove the asymptotic stability of curves with ordinary double points: more precisely, if $\mathscr{I}$ is the ideal associated to a 1-PS λ with normalized weights ρ_i then the estimate of the corollary may be greater than $\frac{2 \deg X}{n+1} . \sum \rho_i$ (cf. Theorem 2.9)

THEOREM 4.12. *If* $C^1 \subset \mathbf{P}^N$ *is a linearly stable (resp.: semi-stable) curve, then* C *is Chow stable (resp.: semi-stable).*

Proof. We prove the stable case; the semi-stable case follows by replacing the strict inequalities in the proof by inequalities.

Fix coordinates $X_0, \ldots, X_N$ on $\mathbf{P}^N$ and a 1-PS

$$\lambda(t) = \begin{bmatrix} t^{\rho_0} & & 0 \\ & \ddots & \\ 0 & & t^{\rho_N} \end{bmatrix}, \quad \rho_0 \geq \rho_1 \geq \ldots \geq \rho_N = 0$$

Let $\mathscr{I}$ be the associated ideal on $\mathcal{O}_{C\times\mathbf{A}^1}$ and let $\mathscr{I}_k \subset \mathcal{O}_C$ be the ideal defined by $\mathscr{I}_k . L =$ [sheaf generated by $X_k, \ldots, X_N$]; thus $\mathscr{I} = \sum_{k=0}^{N} t^{\rho_k} \mathscr{I}_k$. The linear stability of X implies (cf. 2.16), $e(\mathscr{I}_k) < \frac{\deg C}{N} . \operatorname{codim} < X_k, \ldots, X_N >$ $= \frac{\deg C . k}{N}$. So using Corollary 4.11,

$$e_L(\mathscr{I}) \leqq \min_{0=s_0<\ldots<s_k=N} [\sum(\rho_{s_k}-\rho_{s_{k+1}})(e_L(\mathscr{I}_{s_k}) + e_L(\mathscr{I}_{s_{k+1}}))]$$

$$< \min_{0=s_0<\ldots<s_k=N} \left[\sum(\rho_{s_k}-\rho_{s_{k+1}})(s_k+s_{k+1})\frac{\deg C}{N}\right]$$

In view of the Lemma below this implies $e_L(\mathscr{I}) < \frac{2\deg C}{N+1}\sum_{i=0}^{N}\rho_i$ which in turn implies C is stable by Theorem 2.9.

Lemma 4.13. *If* $\rho_0 \geqq \ldots \geqq \rho_n = 0$, *then*

$$\min_{0=s_0<\ldots<s_l=n}\left[\sum(\rho_{s_k}-\rho_{s_{k+1}}) . \left(\frac{s_k+s_{k+1}}{2}\right)\right] \leqq \frac{n}{n+1}\sum_{k=0}^{n}\rho_k$$

Proof. Draw the Newton polygon of the points (k, ρ_k) as shown below

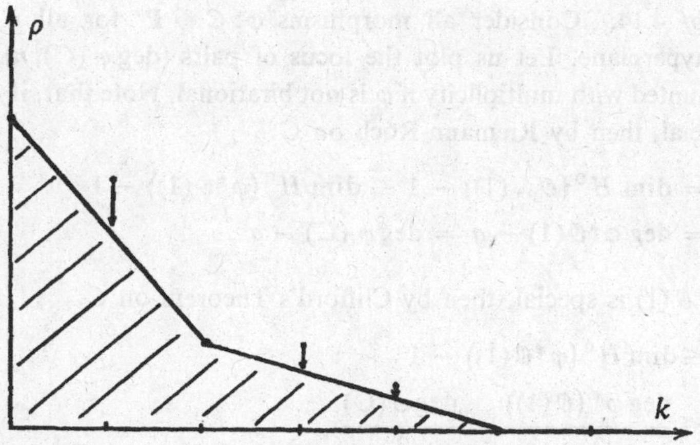

The left hand side is just the area under this polygon so moving the points above the polygon down onto it as shown, does not affect this expression. Since this can only decrease the right hand side we may assume all the ρ_i are on this polygon. Then the left hand expression can be calculated with $s_k = k$ and it becomes

$$\frac{1}{2}\rho_0 + \rho_1 + \dots + \rho_{n-1} + \frac{1}{2}\rho_n = \rho_0 + \dots + \rho_n - \frac{1}{2}(\rho_0 + \rho_n)$$

$$\leqq \rho_0 + \dots + \rho_n - \frac{1}{n+1}(\rho_0 + \dots + \rho_n)$$

since the Newton polygon is convex. But the last expression is just $\frac{n}{n+1}(\rho_0 + \dots + \rho_n)$, hence the lemma.

THEOREM 4.14. *If $C \subset \mathbf{P}^N$ is a smooth curve embedded by $\Gamma(C, L)$ where L is a line bundle of degree d, then*

i) $d > 2g > 0 \Rightarrow C$ *linearly stable,*

ii) $d \geqq 2g \geqq 0 \Rightarrow C$ *linearly semi-stable.*

Combining this result with Theorem 4.13 gives the main theorem of this section:

THEOREM 4.15. *If C is a smooth curve of genus $g \geqq 1$ embedded by a complete linear system of degree $d > 2g$ then C is Chow-stable.*

Proof of 4.14. Consider all morphisms $\varphi: C \to \mathbf{P}^n$ for all n, where $\varphi(C) \not\subset$ hyperplane. Let us plot the locus of pairs $(\deg \varphi(C), n)$, where $\varphi(C)$ is counted with multiplicity if φ is not birational. Note that, if $\varphi^*\mathcal{O}(1)$ is non-special, then by Riemann-Roch on C:

$$\begin{aligned} n &= \dim H^0(\mathcal{O}_{\mathbf{P}^n}(1)) - 1 \leqq \dim H^0(\varphi^*\mathcal{O}(1)) - 1 \\ &= \deg \varphi^*\mathcal{O}(1) - g = \deg \varphi(C) - g \end{aligned}$$

while if $\varphi^*\mathcal{O}(1)$ is special, then by Clifford's Theorem on C:

$$\begin{aligned} n &\leqq \dim H^0(\varphi^*\mathcal{O}(1)) - 1 \\ &\leqq \frac{\deg \varphi^*(\mathcal{O}(1))}{2} = \frac{\deg \varphi(C)}{2} \end{aligned}$$

This gives us the diagram

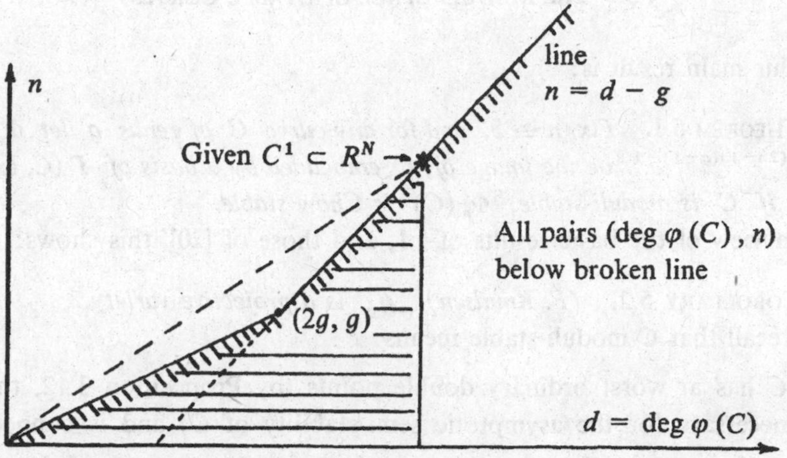

The reduced degree of $\varphi(C)$ is just d/n, the inverse of the slope of the joining (0, 0) to the plotted point (n, d). In case (i), by assumption, the given curve $C^1 \subset \mathbf{P}^N$ corresponds to a point on the upper bounding segment, such as * in our picture. Any projection of C corresponds to a point (n', d') in the shaded area with $d' \leqq d, n' < n$. From the diagram it is clear that the slope decreases, or the reduced degree increases: this is exactly what linear stability means. In case (ii), we allow the given curve C to correspond to the vertex $(2g, g)$ of the boundary, or allow $g = 0$, when the boundary line is just $n = d$. In these cases, the slope at least cannot increase, or the reduced degree cannot decrease under projection.

Remark. Curves with ordinary double points are *not*, in general, linearly stable since projecting from a double point lowers the degree by 2, but decreases the dimension of the ambient space by only 1. In fact, linear stability is somewhat too strong a condition for most moduli problems: Chow stability for varieties of dimension r apparently allows points of multiplicity up to $(r+1)$! while linear stability allows only points of multiplicity up to r !

§ 5. The Moduli Space of Stable Curves

Our main result is:

THEOREM 5.1. *Fix* $n \geqq 5$, *and for any curve* C *of genus* g *let* $\Phi_n(C) \subset \mathbf{P}^{(2n-1)(g-1)-1}$ *be the image of* C *embedded by a basis of* $\Gamma(C, \omega_c^{\otimes n})$. *Then if* C *is moduli-stable,* $\Phi_n(C)$ *is Chow stable.*

In view of the basic results of § 1, and those of [20], this shows:

COROLLARY 5.2. *(F. Knudsen)* $\overline{\mathcal{M}}_g$ *is a projective variety.*

Recall that C moduli-stable means

(1) C has at worst ordinary double points (by Proposition 3.12, this is necessary for the asymptotic semi-stability of C) and is connected,

(2) C has no smooth rational components meeting the rest of the curve in fewer than three points:
this condition is necessary to ensure that C has only finitely many automorphisms.

We will call C moduli semi-stable if it satisfies (1) and

(2′) C has no smooth rational components meeting the rest of the curve in only one point.

Note that if C is moduli semi-stable, then the set of its smooth rational components meeting the rest of the curve in exactly 2 points form a finite set of chains and if each of these is replaced by a point, we get a moduli stable curve:

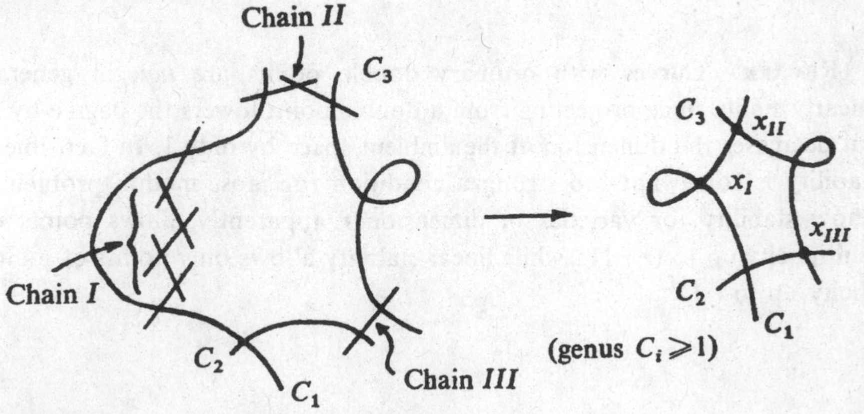

We will case these the rational chains of C.

It would be more satisfactory to have a direct proof of Theorem 5.1 similar to the proof of the stability of smooth curves given in § 4. But curves with double points are not usually linearly stable (cf. the remark following Theorem 4.14) and, in fact, the estimates in Corollary 4.11 do not suffice to prove stability for such curves. We will therefore take an indirect approach.

Proof of 5.1. We begin by recalling the useful valuative criterion:

LEMMA 5.3. *Suppose a reductive group* G *acts on a k-vector space* V. *Let* $K = k((t))$ *and suppose* $x \in V_K$ *is G-stable. Then there is a finite extension* $K' = k'((t')) \supset K$, *and elements* $g \in G_{K'}$, $\lambda \in (K')^*$ *such that the point* $\lambda g(x) \in V \otimes_k K'$ *lies in* $V \otimes_k k'[[t']]$ *and specializes as* $t \to 0$ *to a point* $\overline{\lambda g(x)}$ *with closed orbit. Thus* $\overline{\lambda g(x)}$ *is either stable or semi-stable with a positive dimensional stabilizer.*

Proof. The diagram below is defined over k:

$$\begin{array}{l} \mathbf{P}(V) \supset \mathbf{P}(V)_{ss} \\ \qquad\qquad \Big\downarrow \pi \\ \qquad\qquad X = \text{Proj (graded ring of invariants on } V) \end{array}$$

The point $\pi(x) \in X_K$ specializes to a point $\overline{\pi(x)} \in X_k$. Let $\bar{y}$ be a lifting of this point to V_{ss} with $O^G(\bar{y})$ closed. In the scheme $V \times \operatorname{Spec} k[[t]]$ form the closure Z of $\mathbf{G}_m \,.\, O^G(x)$. The lemma follows if we prove that $\bar{y} \in Z$. If $\bar{y} \notin Z$, then Z and $O^G(\bar{y})$ are closed disjoint G invariant subsets of $V \times \operatorname{Spec} k[[t]]$, hence there exists a homogeneous G-invariant f such that $f(x) = 0$ but $f(\bar{y}) \neq 0$. Then for some n, $f^{\otimes n}$ descends to a section of some line bundle on $X \times \operatorname{Spec} k[[t]]$. But then $f(\pi(x)) = 0$ and $f(\overline{\pi(x)}) \neq 0$ are contradictory.

Now suppose that C is a moduli stable curve of genus g over k. Let $\mathscr{C}/k[[t]]$ be a family of curves with fibre C_0 over $t = 0$ equal to C and generic fibre C_η smooth. At the double points of C_0, $\mathscr{C}$ looks formally like $xy = t^n$, that is has only A_{n-1}-type singularities and hence is normal. Embed C_η in $\mathbf{P}^N$ $(N = (2n-1)(g-1)-1)$ by $\Gamma(C_\eta\, \omega_{C_\eta}{}^{\otimes n})$ and let $\Phi(C_\eta)$ denote its image there. Then Lemma 5.3 says that by replacing $k[[t]]$ with some finite extension and choosing a suitable basis of $\Gamma(C_\eta, \omega_{C_\eta}{}^{\otimes n})$—this corresponds to choosing g, λ—we may assume that the closure $\mathscr{D}$ in $\mathbf{P}^N \times \operatorname{Spec} k[[t]]$ of $\Phi(C_\eta)$ satisfies

i) $D_\eta = C_\eta$

ii) D_0 Chow-stable or Chow semi-stable with positive dimensional stabilizer.

I now claim:

(5.4) $\mathscr{D} = \Phi(\mathscr{C})$, the image of $\mathscr{C}$ under a $k[[t]]$ basis of

$$\Gamma(\mathscr{C}, \omega_{\mathscr{C}}{}^n{}_{/k[[t]]})$$

In particular this implies $D_0 = C_0 = C$ and since C has finite stabilizer this means D_0, hence C, is Chow stable.

The main step in the proof of (5.4) is to show that D_0 is moduli semi-stable as a scheme, and the key difficulty in doing this is to show that D_0 has only ordinary double points. At first glance, this seems rather obvious, since from Proposition 3.12 it follows easily that as a cycle D_0 has no multiplicities and has only ordinary double points. But ordinary double points on a limit cycle arise in two ways:

i)

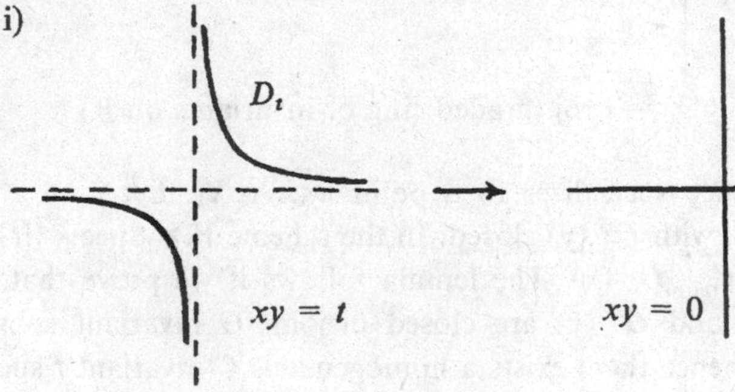

ii)

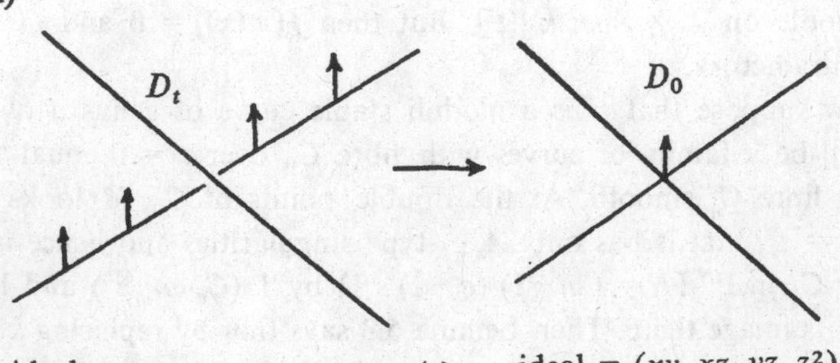

ideal $= \langle x, z\rangle . \langle y, z-t\rangle$ ideal $= (xy, xz, yz, z^2)$

In the second case the scheme D_0 has an embedded component (the first order normal neighbourhood in the z-direction) at the double point so in the limit scheme the double point is not ordinary. If case (ii) occurred for D_0, then since D_0 is Chow semi-stable, it must span $\mathbf{P}^N$ set-theoretically. But $\Gamma(D_0, \mathcal{O}_{D_0}(1))$ has a torsion section supported at the double point: so D_0 would have to be embedded by a non-complete linear system $\sum \subset \Gamma(D_0, \mathcal{O}_{D_0}(1))$ of torsion-free sections, $\dim \sum = \dim H^0(D_\eta, \mathcal{O}_{D_\eta}(1))$. Consequently $H^1(D_0, \mathcal{O}_{D_0}(1)) \neq (0)$ too. That this cannot happen in the situation of (5.4) follows from:

PROPOSITION 5.5. *Let* $C \subset \mathbf{P}^n$ *be a* 1*-dimensional scheme such that*

a) $n + 1 = \deg C + \chi(\mathcal{O}_C), \chi(\mathcal{O}_C) < 0$,

b) C *is Chow semi-stable,*

c) $\dfrac{\deg C}{n+1} < \dfrac{8}{7}$.

Then i) C *is embedded by a complete non-special* [1]) *linear system,*

ii) C *is a moduli semi-stable curve with rational chains of length at most one consisting of straight lines.*

Moreover if $v = \dfrac{\deg C}{\deg \omega_C}$ *(where* ω_C *is the Grothendieck dualizing sheaf) and* $C = C_1 \cup C_2$ *is a decomposition of* C *into two sets of components such that* $\mathscr{W} = C_1 \cap C_2$ *and* $w = \# \mathscr{W}$ *then*

iii) $$|\deg C_1 - v \deg_{C_1}(\omega_C)| \leq \frac{w}{2}$$

REMARKS. 1) It is clear that D_0 satisfies the hypotheses of the lemma. Indeed a) is satisfied by D_η and is preserved under specialization. The key point of the Proposition to replace this by the stronger condition i)

2) Roughly, iii) says that the degrees of the components of C are roughly in proposition to their "natural" degrees. We will see later on that this is enough to force $\mathscr{D} = \mathscr{C}$.

Proof. From b), c) and Proposition 3.1 we know that the cycle of C has no multiplicity and only ordinary double points. Hence C_{red} is a scheme

---

[1]) Non-special means $H^1(C, \mathcal{O}_C(1)) = (0)$.

5

having only ordinary double points and differing from C only by embedded components.

Suppose we are given a decomposition $C_{\text{red}} = C_1 \cup C_2$; let $\mathscr{W} = C_1 \cap C_2$, $w = \# \mathscr{W}$, L_i be the smallest linear subspace containing C_i and $n_i = \dim L_i$. We can assume $L_1 = V(X_{n_1+1} \ldots X_n)$. For the 1-PS λ given by

$$t \mapsto \left[\begin{array}{ccc|ccc} t & & & & & \\ & \ddots & & & 0 & \\ & & t & & & \\ \hline & & & 1 & & \\ & 0 & & & \ddots & \\ & & & & & 1 \end{array}\right] \begin{array}{l} n_1+1 \\ \\ \\ \\ n-n_1 \\ \\ \end{array} \qquad \sum \rho_i = n_1 + 1$$

the associated ideal $\mathscr{I}$ in $\mathcal{O}_{C_{\text{red}} \times \mathbf{A}^1}$ is given by $\mathscr{I} = (t, I(L_1))$. To evaluate $e(\mathscr{I})$ we use an easy lemma whose proof is left to the reader

LEMMA 5.6. *If $X' \xrightarrow{f} X$ is a proper morphism of r-dimensional, possibly reducible "varieties", birational on each component, L is a line bundle on X, and $\mathscr{I}$ is an ideal sheaf on X such that* $\operatorname{supp}(\mathcal{O}_X/\mathscr{I})$ *is proper, then* $e_{f^*(L)}(f^*(\mathscr{I})) = e_L(\mathscr{I})$.

Letting $\mathscr{I}_i$ be the pullback of $\mathscr{I}$ to C_i, the lemma says $e_L(\mathscr{I}) = e_{L_1}(\mathscr{I}_1) + e_{L_2}(\mathscr{I}_2)$. But $\mathscr{I}_1 = t \,.\, \mathcal{O}_{C_1 \times \mathbf{A}^1}$ and support $\mathscr{I}_2$ contains $(0) \times \mathscr{W}$ so this implies[1] $e_L(\mathscr{I}) \geqq 2 \deg C_1 + w$. Using b) and Theorem 2.8 this gives

$$(5.7) \qquad w + 2 \deg C_1 \leqq \frac{\deg C}{n+1} \,.\, 2 \,.\, (n_1+1) \leqq \frac{16}{7}(n_1+1)$$

If C_1 as any component of C_{red}, then this implies:

a) $H^1(C_1, \mathcal{O}_{C_1}(1)) = 0$: if not, then by Clifford's theorem

$$h^0(C_1, \mathcal{O}_{C_1}(1)) \leqq \frac{\deg C_1}{2} + 1$$

[1]) This argument has a gap: see Appendix, p. 72

so by (5.7)

$$\deg C_1 \leqq \frac{8}{7} h^0(C_1, \mathcal{O}_{C_1}(1)) \leqq \frac{8}{14} \deg C_1 + \frac{8}{7},$$

which implies $\deg C_1 \leqq 2$, hence C_1 is rational and then $H^1(C_1, \mathcal{O}_{C_1}(1)) = (0)$ anyway.

b) $H^1(C_1, \mathcal{O}_{C_1}(1)(-\mathscr{W})) = (0)$: indeed from (5.7) and Riemann-Roch,

$$\deg C_1 + \frac{1}{2} w \leqq \frac{8}{7}(\deg C_1 - g_1 + 1), \text{ whence}$$

$$\deg \mathcal{O}_{C_1}(1)(-\mathscr{W}) = \deg C_1 - w \geqq 8(g_1 - 1) + \frac{5}{2} w.$$

The last expression is greater than $2g_1 - 2$ unless $w = 0$, when b) reduces to a), or $g_1 = 0$ and $w = 1$ or 2. But in this case $\mathcal{O}_{C_1}(1)(-\mathscr{W}) = \mathcal{O}_{\mathbf{P}1}(e)$, with $e \geqq 1 - 2 = -1$.

Together a) and b) imply $H^1(C, \mathcal{O}_C(1)) = 0$. In fact, if C_{red} has components C_i, then there is an exact sequence

$$0 \to \oplus \mathcal{O}_{C_i}(1)(-\mathscr{W}_i) \to \mathcal{O}_{C_{\text{red}}}(1) \to \mathscr{M} \to 0$$

where $\mathscr{M}$ has 0-dimensional support, hence $H^1(C_{\text{red}}, \mathcal{O}_{C_{\text{red}}}(1)) = 0$, and if $\mathscr{N}$ is the sheaf of nilpotents in $\mathcal{O}_C$, then $\mathscr{N}$ has 0-dimensional support and the conclusion follows from an examination of the exact sequence

$$0 \to \mathscr{N} \to \mathcal{O}_C \to \mathcal{O}_{C\text{red}} \to 0.$$

Therefore hypothesis (a) can be rewritten $n + 1 = h^0(\mathcal{O}_C(1))$. Since C is not contained in a hyperplane, C is embedded by a complete linear system. But now if $\mathscr{N} \neq (0)$, then set-theoretically C will still be contained in a hyperplane, contradicting its Chow semi-stability; so $C = C_{\text{red}}$ and all that we have said about C_{red} above is true of C.

Using the fact that

$$\chi(\mathcal{O}_C) = -\chi(\omega_C) = -(\deg \omega_C + \chi(\mathcal{O}_C))$$

it follows that $\deg C/n + 1 = 2v/2v - 1$ and we can rewrite (5.7) in terms of v as

$$\frac{w}{2} + \deg C_1 \leqq \left(\frac{2v}{2v-1}\right)(\deg C_1 - g_1 + 1)$$

or equivalently

$$\frac{w}{2} \geqq v(2g_1 - 2 + w) - \deg C_1 = v \deg_{C_1}(\omega_C) - \deg C_1 .$$

Then since

$$\begin{aligned} 0 &= v(\deg(\omega_C)) - \deg C \\ &= v \deg_{C_1}(\omega_C) + v \deg_{C_2}(\omega_C) - \deg C_1 - \deg C_2 , \end{aligned}$$

we obtain iii): $\frac{w}{2} \geqq \left| v \deg_{C_1}(\omega_C) - \deg C_1 \right|$.

Now suppose C has a smooth rational component C_1 meeting the rest of the curve in w points $P_1, \ldots, P_w$. Then $\omega_C \mid C_1$ is just the sheaf of differentials on C_1 with poles at $P_1, \ldots, P_w$, so if $w \leqq 2$, $\deg_{C_1}(\omega_C) \leqq 0$. Using iii) this shows $\deg C_1 \leqq \frac{1}{2}$ if $w = 1$, absurd, and $\deg C_1 \leqq 1$ if $w = 2$. Moreover, if, in this last case one of the P_1 lies on a smooth rational curve C_2 meeting the rest of C in only 1 other point, as in the diagram below

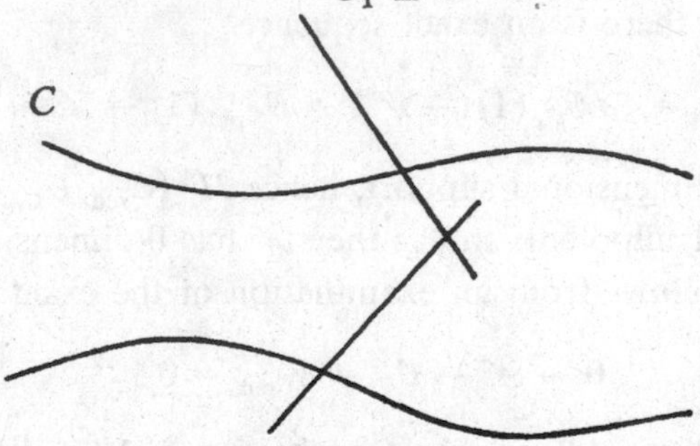

then $\omega_C \mid_{C_1} \cong \mathcal{O}_{C_1}$ and $\omega_C \mid_{C_1} \cong \mathcal{O}_{C_2}$ so $\deg_{C_1 \cup C_2}(\omega_C) = 0$. Using iii) again, we find $\deg(C_1 \cup C_2) \leqq \frac{1}{2}\, 2 = 1$, and as this is absurd, we have proved all parts of the Proposition.

We are now ready to show that $\mathscr{D} = \mathscr{C}$. Since D_0 is moduli semi-stable, it follows that $\mathscr{D}$ is a normal two-dimensional scheme with only type A_n singularities. Moreover $\omega^{\otimes n}_{\mathscr{D}/k[[t]]}$ is generated by its sections if $n \geqq 3$ and defines a morphism from $\mathscr{D}$ to a scheme $\mathscr{D}'/k\,[[t]]$, where $D'_\eta = D_\eta$, $D'_0 = D_0$ with rational chains blown down to points. Thus $\mathscr{D}'$ is a family of moduli-stable curves over $k\,[[t]]$ with generic fibre $\mathscr{C}_\eta$. Since there is only one such (cf. [6]), it follows that $\mathscr{D}' = \mathscr{C}$. Thus we have a diagram:

$$\begin{array}{ccccc} C_\eta & \xleftarrow{\approx} & D_\eta & \xrightarrow{\Phi\eta} & \mathbf{P}^N \times \operatorname{Spec} k((t)) \\ \cap & & \cap & & \cap \\ \mathscr{C} & \longleftarrow & \mathscr{D} & \xrightarrow[\Phi]{} & \mathbf{P}^N \times \operatorname{Spec} k[[t]] \end{array}$$

$$\Phi_\eta^*(\mathcal{O}_{\mathbf{P}^N}(1)) = \omega_{D_\eta/k((t))}^{\otimes n}.$$

Let $L = \mathcal{O}_{\mathscr{D}}(1)$. It follows that $L \cong \omega_{\mathscr{D}/k[[t]]}^{\otimes n}(-\sum r_i D_i)$, where D_i are the components of D_0. Multiplying the isomorphism by $t^{\min(r_i)}$, we can assume $r_i \geqq 0$, $\min r_i = 0$. Let $D_1 = \bigcup_{r_i=0} D_i$, $D_2 = \bigcup_{r_i>0} D_i$. If f is a local equation of $\sum r_i D_i$, then $f \not\equiv 0$ in any component of D_1 since $r_i = 0$ on all these while $f(x) = 0$, all $x \in D_1 \cap D_2$, so

$$\#(D_1 \cap D_2) \leqq \deg_{D_1}(\mathcal{O}_{\mathscr{D}_0}(\textstyle\sum r_i D_i)).$$

But this last degree equals $(\deg D_1 - n \deg_{D_1}(\omega_{D_0}))$ which contradicts iii) of Proposition 5.5 unless all r_i are zero. Hence $L = \omega_{\mathscr{C}}^{\otimes n}$ which shows $\mathscr{D} = \mathscr{C}$.

Line bundles on the moduli space

For the remainder of this section we examine Pic $(\overline{\mathscr{M}}_g)$. We fix a genus $g \geqq 2$ and an $e \geqq 3$. Then for all stable C, $\omega_C^{\otimes e}$ is very ample and in this embedding C has degree $d = 2e(g-1)$, the ambient space has dimension $\nu - 1$ where $\nu = (2e-1)(g-1)$ and C has Hilbert polynomial $P(X) = dX - (g-1)$. Let $H \subset \operatorname{Hilb}^P_{\mathbf{P}^{\nu-1}}$ be the locally closed smooth subscheme of e-canonical stable curves C, let $C \subset H \times \mathbf{P}^{\nu-1}$ be the universal curve and let

$$\mathrm{ch}: H \to \mathrm{Div} = \mathrm{Div}^{d,d} = \left\{\begin{array}{l}\text{projective space of bihomogeneous forms}\\ \text{of bidegree } (d,d) \text{ in dual coordinates}\\ u, v \text{ (cf. § 1).}\end{array}\right\}$$

be the Chow map. These are related by the diagram

$$\begin{array}{ccccc} & & C & & \\ & & \downarrow{\scriptstyle \pi} & & \\ \mathrm{Div} & \xleftarrow{\mathrm{ch}} & H & \xrightarrow{\rho} & \overline{\mathscr{M}}_g = H/PGL(\nu) \end{array}$$

If Pic $(H, PGL(\nu))$ is the Picard group of invertible sheaves on H with $PGL(\nu)$-action, we have a diagram

$$\mathrm{Pic}(\overline{\mathcal{M}}_g) \xrightarrow{\rho^*} \mathrm{Pic}(H, PGL(v)) \xrightarrow{\alpha} \mathrm{Pic}(H)^{PGL(v)} \subset \mathrm{Pic}(H).$$

In this situation, we have:

LEMMA 5.8. *In the sequence above, ρ^* is injective with torsion cokernel and α is an isomorphism.*

Proof. α is an isomorphism by Prop. 1.4 [14]; ρ^* injective is easy; coker ρ^* torsion can be proved, for instance, using Seshadri's construction, Th. 6.1 [19].
This lemma allows us to examine Pic $(\overline{\mathcal{M}}_g)$ by looking inside Pic $(H)^{PGL(v)}$ which is a much easier group to come to grips with.

DEFINITION 5.9. *Let $\Delta \subset H$ be the divisor of singular curves, $\delta = \mathcal{O}_H(\Delta)$ and $\lambda_n = \Lambda^{\max}(\pi_*(\omega_{C/H}{}^{\otimes n}))$, $(n \geqq 1)$. We write λ for λ_1.*

The sheaves λ_n and δ are the most obviously interesting invertible sheaves on H from a moduli point of view. The next theorem expresses all of these in terms just involving λ and δ.

THEOREM 5.10. $\lambda_n = \mu^{\binom{n}{2}} \otimes \lambda$ *where* $\mu = \lambda^{12} \otimes \delta^{-1}$.

Proof. The proof is based on Grothendieck's relative Riemann-Roch theorem (see Borel-Serre [4]), which we will briefly recall.

Let X and Y be complete smooth varieties over k, $A(X)$ be the Chow ring of X and $\mathcal{F}$ be a coherent sheaf on X. Let $c_i(\mathcal{F}) \in A(X)$ denote the ith Chern class of $\mathcal{F}$, Chern $(\mathcal{F}) \in A(X) \otimes \mathbf{Q}$ its Chern character and $\mathcal{T}(\mathcal{F}) \in A(X) \otimes \mathbf{Q}$ its Todd genus. These are related by:

(5.11) $$\mathrm{Chern}(\mathcal{F}) = rk\,\mathcal{F} + c_1(\mathcal{F}) + \frac{c_1(\mathcal{F})^2}{2} - c_2(\mathcal{F})$$

+ terms of higher codimension,

$$\mathcal{T}(\mathcal{F}) = 1 - \frac{c_1(\mathcal{F})}{2} + \frac{c_1(\mathcal{F})^2 + c_2(\mathcal{F})}{12}$$

+ terms of higher codimension.

Let $K(Y)$ be the Grothendieck group of $Y, f: X \to Y$ be a proper map' and $f_!(\mathcal{F}) = \sum (-1)^i [\mathbf{R}^i f_* \mathcal{F}] \in K(Y)$. The relative Riemann-Roch theorem expresses the Chern character of $f_!(\mathcal{F})$, *modulo torsion* as

$$\mathrm{Chern}(f_! \mathcal{F}) = f_*(\mathrm{Chern}\,\mathcal{F} \cdot \mathcal{T}(\Omega^1_{X/Y}))$$

which using (5.11) gives:

$$(5.12)\quad rk f_! \mathscr{F} + c_1(f_! \mathscr{F}) + \ldots\ldots\ldots$$

$$= f_* \left[\left(rk(\mathscr{F}) + c_1(\mathscr{F}) + \frac{c_1(\mathscr{F})^2}{2} - c_2(\mathscr{F}) \right) \cdot \left(1 - \frac{c_1(\Omega^1_{X/Y})}{2} + \frac{c_1(\Omega^1_{X/Y})^2 + c_2(\Omega^1_{X/Y})}{12} \right) \right]$$

For the time being, we work implicitly modulo torsion.

Now suppose $\mathscr{F}$ is a line bundle such that $R^i f_* (\mathscr{F}) = 0$, $i > 0$ and suppose $\dim X = \dim Y + 1$. Then the codimension 1 term on the left of (5.12) (i.e. on Y) correspônds to the codimension two term on the right (i.e. on X). Since $c_2(\mathscr{F}) = 0$, this gives

$$(5.13)\quad c_1(f_* \mathscr{F}) = c_1(f_! \mathscr{F})$$

$$= f_* \left[\frac{c_1(\Omega^1_{X/Y})^2 + c_2(\Omega^1_{X/Y})}{12} - \frac{c_1(\mathscr{F})\, c_1(\Omega^1_{X/Y})}{2} + \frac{c_1(\mathscr{F})^2}{2} \right]$$

In case $f: C \to S$ is a moduli-stable curve over S, $X = C$ and $Y = S$, we can simplify this. Indeed I claim that if Sing C is the singular set on C and I_{sing} is its ideal, then

i) codim Sing $C = 2$

ii) the canonical homomorphism $\Omega^1_{C/S} \to \omega_{C/S}$ induces an isomorphism $\Omega^1_{C/S} = I_{\text{sing}} \cdot \omega_{C/S}$.

We certainly have the isomorphism of ii) off Sing C. At a singular point C has a local equation of the form $xy = t^n$, where t is a parameter on S, x and y are affine coordinates on the fibre. Moreover locally C is singular only at the points $(0, 0)$ in the fibres where $t = 0$, so Sing C has codimension 2. Near the singular point

$$\Omega^1_{C/S} = (\mathcal{O}_C dx + \mathcal{O}_C dy)/(x dy + y dx)\, \mathcal{O}_C$$

while $\omega_{C/S}$ is the invertible sheaf generated by the differential ζ which is given by dx/x outside $x = 0$ and by $-dy/y$ outside $y = 0$. Thus

$$\Omega^1_{C/S} = \mathscr{M}_{(0,0),C} \cdot \zeta = \mathscr{M}_{(0,0),C} \cdot \omega_{C/S} \cdot$$

Recall the following corollary to Riemann-Roch: if X is a smooth variety, $Y \subset X$ a subvariety of codim r and $\mathscr{F}$ is coherent on Y, then considering $\mathscr{F}$ as a sheaf on X

$$c_i(\mathscr{F}) = \begin{cases} 0, 1 \leqq i \leqq r-1 \\ ((-1)^{r-1}(r-1)!\, rk\, \mathscr{F})\, Y, i = r \end{cases}$$

Set $X = C$, $Y = \text{Sing } C$ and $\mathscr{F} = \Omega^1_{C/S}$. The Whitney product formula applied to the chern classes of the exact sequence

$$0 \to \Omega^1_{C/S} \to \omega_{C/S} \to \omega_{C/S} \otimes \mathcal{O}_{\text{Sing } C} \to 0$$

gives, taking account of the corollary

$$1 + c_1(\omega_{C/S})$$
$$= (1 + c_1(\Omega^1_{C/S}) + c_2(\Omega^1_{C/S}) + \ldots..) . (1 + 0 - [\text{Sing } C] + \ldots..)$$

Equating terms of equal codimension, we see that $c_1(\Omega^1_{C/S}) = c_1(\omega)$ and $c_2(\Omega^1_{C/S}) = [\text{Sing } C]$ so that (5.13) becomes

$$c_1(f_* \mathscr{F}) = f_* \left[\frac{c_1(\omega_{C/S})^2 + [\text{Sing } C]}{12} - \frac{c_1(\mathscr{F})\, c_1(\omega_{C/S})}{2} + \frac{c_1(\mathscr{F})^2}{2} \right]$$

Applying this to the map $\pi: C \to H$, when $\mathscr{F} = \omega^{\otimes n}_{C/H}$ gives

$$\lambda_n = \Lambda^{\max}(\pi_* \omega^{\otimes n}_{C/H}) = c_1(\pi_* \omega^{\otimes n}_{C/H})$$
$$= \pi_* \left[\frac{c_1(\omega_{C/H})^2 + [\text{Sing } C]}{12} - \frac{c_1(\omega^{\otimes n}_{C/H})\, c_1(\omega_{C/H})}{2} + \frac{c_1(\omega^{\otimes n}_{C/H})^2}{2} \right]$$
$$= \binom{n}{2} \pi_*(c_1(\omega_{C/H})^2) + \frac{\pi_*(c_1(\omega_{C/H})^2) + [\Delta]}{12}$$

Setting[1]) $n = 1$, we see that $\lambda = \left[\frac{\pi_*(c_1(\omega_{C/H})^2) + [\Delta]}{12} \right]$ and $\pi_*(c_1(\omega_{C/H})^2) = 12\lambda - [\Delta]$. Plugging these values back in gives us the theorem up to torsion. But in fact:

LEMMA 5.14. *Over* **C**, Pic $(H, PGL(\nu))$ *is torsion free.*

Note that this will prove what we want because the invertible sheaves that we are trying to show are isomorphic all "live" on the full scheme $H_{\mathbf{Z}}$ over Spec **Z** of stable $\mathscr{C}$-canonical curves. If they are isomorphic on $H_{\mathbf{Z}}$, they are isomorphic after any base change. But on the other hand, I claim that Pic $(H, PGL(\nu))$ injects into Pic $(H_{\mathbf{C}}, PGL_{\mathbf{C}}(\nu))$:

---

[1]) For $n = 1$, $R^1\pi_*(\omega_{C/H})$ is not zero, but it is the trivial line bundle, hence doesn't affect $\pi_!$.

If L is a line bundle on H with $PGL(v)$ action such that $L \otimes \mathbf{C}$ is trivial over $H_{\mathbf{C}}$, then

$$H^0(H, L)^{PGL(v)} \otimes \mathbf{C} = H^0(H_{\mathbf{C}}, L \otimes \mathbf{C})^{PGL(v)}$$

$$\wr\!\wr \Big\downarrow \alpha$$

$$H^0(H_{\mathbf{C}}, \mathcal{O}_{H\mathbf{C}})^{PGL(v)} = \mathbf{C}$$

since $H_{\mathbf{C}}/_{PGL(v)}$ is compact. Thus we can find a non-zero section $s \in H^0(H, L)^{PGL(v)}$, which over $\mathbf{C}$ can be used to give the trivialization α. Over $\mathbf{C}$, s has no zeros so the divisor $(s)_0$ of the zeros of s on H, has support only over the closed fibres of Spec ($\mathbf{Z}$). Mumford and Deligne [6] have shown that $H \to \operatorname{Spec} \mathbf{Z}$ is smooth with irreducible fibres, hence $(s)_0 = \sum r_i \pi^{-1}(p)$, $r_i \geq 0$ i.e. $(s)_0 = (n)$ for some integer n. Then $\left(\frac{s}{n}\right)$ is a global section of L with no zeros so L is trivial.

Proof of Lemma. Over $\mathbf{C}$, we have Teichmüller theory at our disposal. Let Π be a standard model of a group with generators $\{a_i, b_i \mid 1 \leq i \leq g\}$ mod the relation $\prod_{i=1}^{g} (a_i b_i a_i^{-1} b_i^{-1}) = 1$. Then the Teichmüller modular group Γ is

$$\Gamma = \{\alpha \mid \alpha : \Pi \to \Pi \text{ is an orientation preserving isomorphism}\} / \text{inner automorphisms}$$

The Teichmüller space $\mathcal{T}_g$ is given by

$$\mathcal{T}_g = \left\{ (C, \alpha) \;\middle|\; \begin{array}{l} C \text{ a smooth curve of genus } g \text{ and } \alpha\colon \pi_1(C) \to \Pi \text{ an} \\ \text{orientation preserving isomorphism given up to inner} \\ \text{automorphism} \end{array} \right\}$$

Fix a model M_g of the real surface of genus g, and identify $\pi_1(M_g)$ and Π. Then Γ is generated by the maps which are induced by certain automorphisms of M_g, called Dehn twists. The Dehn twist h_γ corresponding to a loop $\gamma\colon [0, 1] \to M_g$ on M_g is given by taking an ε-collar $\gamma \times [-\varepsilon, \varepsilon]$ about γ, letting h = identify off the collar and letting $h(\gamma(t), \eta - \varepsilon) = \left(\gamma\left(t + \frac{\eta}{2\varepsilon}\right), \eta - \varepsilon\right)$ as shown below.

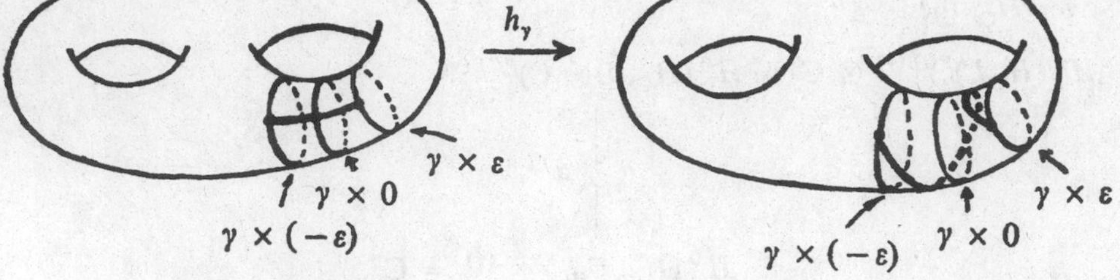

Up to inner automorphism h_γ is determined by which of the pictures below results from cutting open M_g along γ. We have name these elements of Γ in the diagrams:

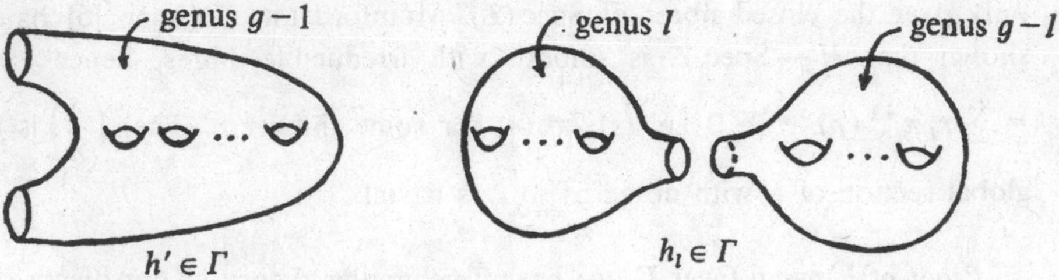

The Dehn twist h_γ can also be described as the monodromy map obtained by going around a curve C_0 with one double point for which γ is the vanishing cycle.

The components of $\Delta \subset H$ correspond to the different ways of putting a stable double point on a smooth moduli stable curve C. They are the closures of the sets of curves of the forms shown below: again, we name these components in the diagram:

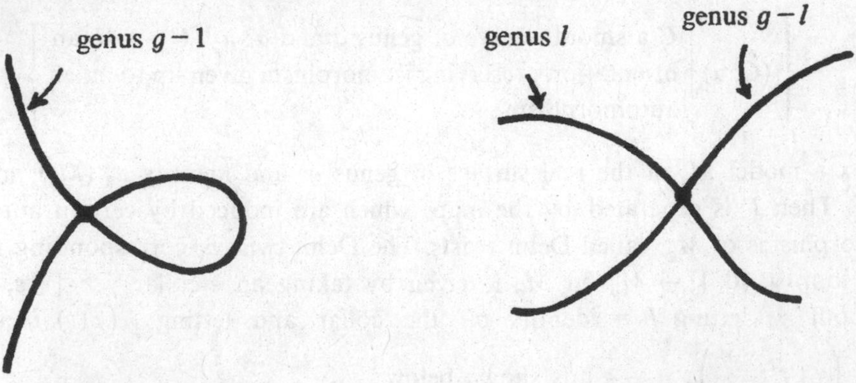

$$\Delta' = \left(\begin{array}{l} \text{closure of set} \\ \text{formed by curves} \\ \text{like this} \end{array} \right) \qquad \Delta_l = \left(\begin{array}{l} \text{closure of set} \\ \text{formed by curves} \\ \text{like this} \end{array} \right)$$

Let $\tilde{H} = \left\{ (C, \alpha, B) \,\middle|\, \begin{array}{l} (C, \alpha) \in \mathcal{T}_g, B \text{ a basis of the } e\text{-tuple dif-} \\ \text{ferentials on } C \text{ given up to a scalar} \end{array} \right\}$

Suppose we are given a line bundle L on H with $PGL(\nu)$-action such that $L^n \cong \mathcal{O}_H$. L induces a cyclic covering H' of H plus a lifting of the $PGL(\nu)$-action to H'. If we choose n minimal this covering is not split: we denote its structure group by Γ_L. Let $\tilde{H}'$ be the pullback of covering over $\tilde{H}$, and let $\mathcal{T}'_g$ denote the quotient of $\tilde{H}'$ by $PGL(\nu)$—this is a covering of $\mathcal{T}_g$. These coverings are related by

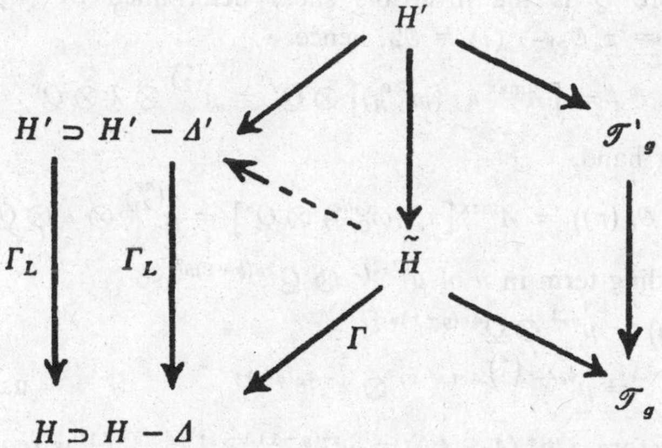

$\mathcal{T}_g$ is simply connected so the cover $\mathcal{T}'_g \to \mathcal{T}_g$ splits, hence so does $\tilde{H}' \to \tilde{H}$. A section of this last cover gives a map from $\tilde{H}$ to $H' - \Delta'$ (shown dashed in the diagram), so Γ_L is a quotient of Γ, of finite order.

Let γ' [resp. γ_e] be a loop at a fixed base point $P_0 \in H - \Delta$ going around Δ' [resp.: Δ_e] but homotopic to 0 in H. Fix a point $\tilde{P}_0 \in \tilde{H}$ over P_0. The monodromy characterization of the Dehn twists implies that γ' [resp.: γ_e] lifted to $\tilde{H}$ goes from $\tilde{P}_0$ to $h'(\tilde{P}_0)$ [resp.: to $h_e(\tilde{P}_0)$]. Since γ' [resp.: γ_e] are homotopic to 0 in H, and the covering $H' - \Delta'$ extends over H, this implies that the image of h' [resp.: h_l] in Γ_L is 0. But these elements and their conjugates generate Γ_L, so $\Gamma_L = \{1\}$, hence $L \cong \mathcal{O}_H$, proving the lemma and the theorem.

In order to describe the ample cone on Pic $(\bar{\mathcal{M}}_g)$ we prove:

THEOREM 5.15. $\mathrm{Ch}^*(\mathcal{O}_{\mathrm{Div}}(\nu)) = (\mu^e \otimes \lambda^{-4})^{e(g-1)}$

Proof. The proof depends on a result which we simply quote from Fogarty [8] or Knudsen [12]:

PROPOSITION 5.16. *Let* S *be a locally closed subscheme of a Hilbert scheme* $\mathrm{Hilb}^P_{\mathbf{P}^{\nu-1}}$, Ch *be the associated Chow map* $\mathrm{Ch}: S \to \mathrm{Div}$ *and* $Z \subset \mathbf{P}^\nu \times S$ *have relative dimension r over S. Then if* $n \gg 0$, $\Lambda^{\max} p_{2,*}(\mathcal{O}_Z(n))$ $= \bigotimes_{i=0}^{r+1} \mu_i^{\binom{n}{i}}$ *and* $\mathrm{Ch}^*(\mathcal{O}_{\mathrm{Div}}(1)) = \mu_{r+1}$, *where* μ_i *are suitable invertible sheaves on* S.

In the situation of our theorem, with $S = H$ and $Z = C$, $\mathcal{O}_C(1) = \omega_{C/H}^{\otimes e} \otimes \pi^* Q$ where Q is the invertible sheaf determined by $(\pi_* \omega_{C/H}^{\otimes e}) \otimes Q = \pi_* \mathcal{O}_C(1) = \pi_* \mathcal{O}_{\mathbf{P}^{\nu-1}}(1) = \mathcal{O}_H^\nu$, hence

$$\mathcal{O}_H = [\Lambda^{\max} \pi_* (\omega_{C/H}^{\otimes e})] \otimes Q^\nu = \mu^{\binom{e}{2}} \otimes \lambda \otimes Q^\nu . \tag{5.17}$$

On the other hand,

$$\Lambda^{\max}(\pi_* \mathcal{O}_C(n)) = \Lambda^{\max}[\pi_* (\omega_{C/H}^{\otimes ne}) \otimes Q^n] = \mu^{\binom{ne}{2}} \otimes \lambda \otimes Q^{P(n).n} .$$

This has leading term in n of $\mu^{n^2e^2/2} \otimes Q^{2e(g-1)n^2}$ so

$$\begin{aligned} Ch^*(\mathcal{O}_{\mathrm{Div}}(\nu)) &= \mu^{\nu e^2} \otimes Q^{4e(g-1)\nu} \\ &= \mu^{\nu e^2 - \binom{e}{2}.4e(g-1)} \otimes \lambda^{-4e(g-1)} \qquad \text{using (5.17)}. \end{aligned}$$

Finally, therefore, $\mathrm{Ch}^*(\mathcal{O}_{\mathrm{Div}}(\nu)) = \mu^{e^2(g-1)} \otimes \lambda^{-4e(g-1)}$ as required.

COROLLARY 5.18. *If* $e \geq 5$, $\mu^e \otimes \lambda^{-4} (= \lambda^{12e-4} \otimes \delta^{-e})$ *is "ample on* $\overline{\mathcal{M}}_g$", *i.e. those positive powers of this bundle which are pull-backs of bundles on* $\overline{\mathcal{M}}_g$ *are ample on* $\overline{\mathcal{M}}_g$.

Proof. This is an immediate consequence of the Theorem and our main result: that $PGL(\nu)$-invariant sections of $\mathrm{Ch}^*(\mathcal{O}_{\mathrm{Div}}(1))$ define a projective embedding of $\overline{\mathcal{M}}_g$.

REMARK 5.19. A similar argument using the facts that

(1) $\omega^{\otimes e}$ is base point free for all canonical curves when $e \geq 2$,

(2) smooth curves are stable if $d > 2g$,

shows that if $e \geq 2$, the sections of $\lambda^{12e-4} \otimes \delta^{-e}$ on $\overline{\mathcal{M}}_g$ separate points on $\mathcal{M}_g$.

To get a good picture of the ample cone on $\overline{\mathcal{M}}_g$ we need to use the realization via Θ functions $\mathcal{A}_{g,1} \xrightarrow{\Theta} \mathbf{P}^N$ of the moduli scheme $\mathcal{A}_{g,1}$ of principally polarized abelian varieties. More precisely, let $J : \mathcal{M}_g \to \mathcal{A}_{g,1}$ be the map taking a curve C to its Jacobian. Then we have:

THEOREM 5.20. *In characteristic* 0, *the morphism* $\mathcal{M}_g \xrightarrow{J} \mathcal{A}_{g,1} \xrightarrow{\Theta} \mathbf{P}^N$ *extends to a morphism* $\overline{\mathcal{M}}_g \xrightarrow{\theta} \mathbf{P}^N$ *so that for some* m, $\theta^*(\mathcal{O}_{\mathbf{P}^N}(1)) = \lambda^m$.

Proof. See Arakelov [1] or Knudsen [12].

REMARK. This should also hold in characteristic p, but it seems to be a rather messy problem there.

Putting together 5.18 and 5.20, we get a whole sector in the (a, b)-plane such that $\lambda^b \otimes \delta^{-a}$ is ample for (a, b) in this sector. This is depicted in the diagram below:

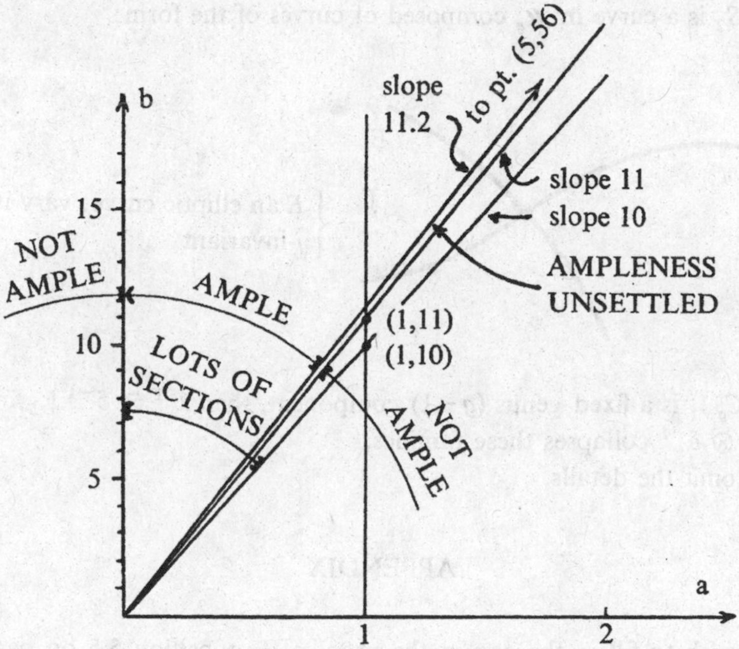

The fact that λ and $\lambda^{11} \otimes \delta^{-1}$ are not ample can be seen by examining the following 2 curves in $\overline{\mathcal{M}}_g$:

(1) If S_1 is a curve in $\overline{\mathcal{M}}_g$ composed of curves of the form:

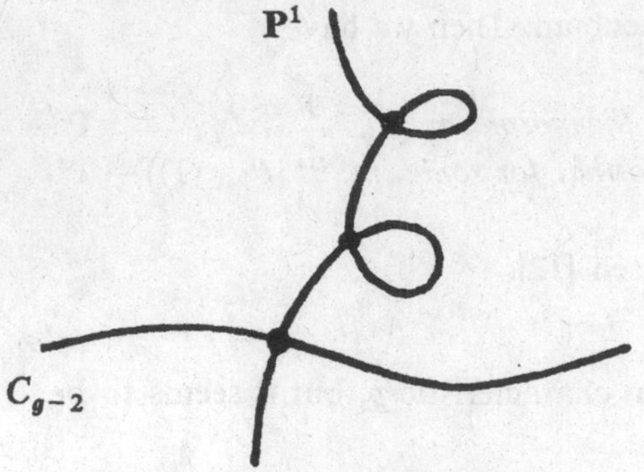

{ vary cross ratio of the 4 points in $\mathbf{P}^1$ making up the 2 double points

where C_{g-2} is a fixed genus $(g-2)$ component, then $\lambda \mid_{S_1} = \mathcal{O}_{S_1}$, hence sections of λ always collapse such families.

(2) If S_2 is a curve in $\overline{\mathcal{M}}_g$ composed of curves of the form:

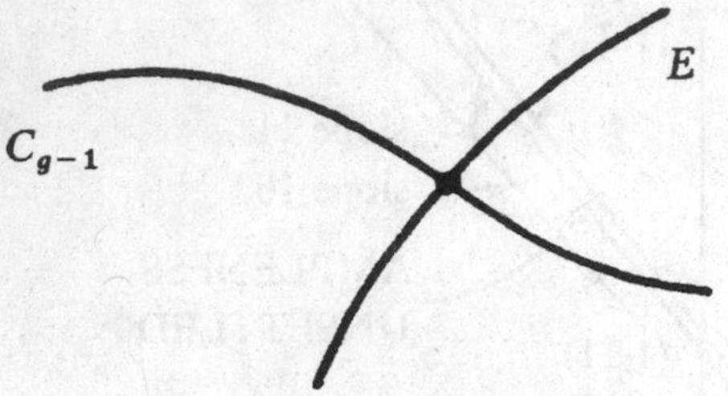

{ E an elliptic curve: vary its j-invariant

where C_{g-1} is a fixed genus $(g-1)$ component, then $\lambda^{11} \otimes \delta^{-1} \mid_{S_2} = \mathcal{O}_{S_2}$ i.e. $\lambda^{11} \otimes \delta^{-1}$ collapses these families.

We omit the details.

APPENDIX

We wish to fill in the gap in the proof of Proposition 5.5 on page 59. The difficulty occurs if the support of $\mathcal{I}$, i.e. $(0) \times L_1$, contains some of the components of C_2 meeting C_1. In this case, the inequality

$$e_L(\mathcal{I}_2) \geqslant w$$

is not clear. Indeed, if $D_1, \ldots, D_k$ are the components of C_2 meeting C_1, $w_i = \#(D_i \cap C_1)$, and $\mathcal{K}_i$ is the pull-back of $\mathcal{I}_2$ to D_i, then

$$\begin{aligned} e_L(\mathscr{I}_2) &= \sum e_L(\mathscr{K}_i) \\ e_L(\mathscr{K}_i) &\geqslant w_i \text{ if } L_1 \not\supseteq D_i \\ &= 2 \deg D_i \text{ if } L_1 \supseteq D_i . \end{aligned}$$

Now suppose C_1 is *irreducible* and $D_i \subseteq L_1$. Then (5.7) is modified to:

$$2 \deg D_i + 2 \deg C_1 \leqslant \frac{16}{7}(n_1+1) .$$

Since C_1 spans L_1, $n_1 \leqslant \deg C_1$. Substituting this, we find

$$\deg D_i \leqslant \frac{\deg C_1}{7} + \frac{8}{7}$$

hence $\deg D_i \leqslant \deg C_1$ (except in the lowest case $\deg C_1 = 1$; in this case, C_1 is a line, so $C_1 = L_1$ and $\operatorname{Supp} \mathscr{K}_i = D_i \cap L_1 \subsetneqq D_i$). Now the reverse of this inequality cannot be true too. This means that if we apply the same arguement to

$$C_{\text{red}} = D_i \cup \overline{(C - D_i)}$$

then the linear span M of D_i cannot contain C_1. Therefore

$$\begin{aligned} & w_i + 2 \deg D_i \leqslant \frac{16}{7}(\dim M_i + 1) \leqslant \frac{16}{7}(\deg D_i + 1) \\ & \therefore\ w_i \leqslant 2 \deg D_i \\ & \therefore\ e_L(\mathscr{K}_i) \geqslant w_i \quad \text{in all cases} \\ & \therefore\ e_L(\mathscr{I}_2) \geqslant w \quad \text{as required.} \end{aligned}$$

This proves (5.7) if C_1 is irreducible, hence (a) and (b) that follow are correct. In particular, (b) shows that $\mathcal{O}_{C_1}(1)(-\mathscr{W})$ always has sections, unless C_1 is a line and $\# \mathscr{W} = 2$. The next paragraph shows that C is embedded by a complete linear system. So when $\Gamma(\mathcal{O}_{C_1}(1)(-\mathscr{W})) \neq (0)$, there is a hyperplane containing all components of C except C_1. Returning to the general case of (5.7) where C_1 is any subset of the components of C, it follows that the linear span L_1 of C_1 contains only C_1 and the *lines* D_i which meet C_1 in 2 points. For these, $\# (D_i \cap C_1) = 2 \deg D_i$, so in all cases it is true that $e_L(\mathscr{I}_2) \geqslant w$ as required.

BIBLIOGRAPHY

[1] Arakelov, S. Families of algebraic curves with fixed degeneracies. *Izvest. Akad. Nauk. 35* (1971).

[2] Arnold, V. I. Critical points of functions. *Proc. Int. Congress at Vancouver*, vol. I (1974).

[3] Artin, M. On isolated rational singularities of surfaces. *Am. J. Math. 88* (1966), p. 129.

[4] Borel, A. and J.-P. Serre. Le théorème de Riemann-Roch (d'après Grothendieck). *Bull. Soc. Math. France 36* (1958).

[5] Cartier, P. Sur un théorème de Snapper. *Bull. Soc. Math. France 88* (1960), p. 333.

[6] Deligne, P. and D. Mumford. The irreducibility of the space of curves of given genus. *Publ. I.H.E.S. 36* (1969), p. 75.

[7] Dolgacev, I. V. Factor-conical singularities of complex surfaces. *Funkc. Anal. i Pril. 8* (1974), pp. 75-76.

[8] Fogarty, J. Truncated Hilbert Functors. *J. Reine u. Ang. Math. 234* (1969), p. 65.

[9] Gieseker, D. Global moduli for surfaces of general type. *To appear.*

[10] Habousch, W. Reductive groups are geometrically reductive. *Annals of Math. 102* (1975).

[11] Kimura, T. and M. Sato. A classification of irreducible prehomogeneous Vector Spaces and their relative invariants. *Nagoya J. Math.* (1976).

[12] Knudsen, F. Projectivity of the moduli space of curves. Part I: Div and Det and Part II. *To appear Math. Scand.*

[13] Lech, C. Note on multiplicity of ideals. *Arkiv for Matematik 4* (1960).

[14] Mumford, D. *Geometric Invariant Theory.* Springer-Verlag, 1965.

[15] —— On the equations defining abelian varieties. *Inv. Math., vol. 1 and 3* (1967).

[16] Ramanujam, C. P. On a geometric interpretation of multiplicity. *Inv. Math. 22* (1973).

[17] Samuel, P. *Méthodes d'algèbre abstraite en géométrie algébrique.* Springer-Verlag, 1955.

[18] Serre, J.-P. Faisceaux algébriques cohérents. *Annals of Math. 61* (1955).

[19] Seshadri, C. S. Quotient spaces modulo reductive algebraic groups. *Annals of Math. 95* (1972), p. 511.

[20] —— Geometric reductivity over arbitrary base. *Preprint.*

[21] Snapper, E. Polynomials associated with divisors. *J. Math. Mech. 9* (1960).

[22] Teissier, B. Cycles évanescents, sections planes et conditions de Whitney. *Astérisque 7-8* (1973), pp. 285-362.

[23] —— and M. Lejeune. Normal cones and sheaves of relative jets. *Comp. Math. 28* (1974), p. 305.

[24] Wagreich, P. Elliptic singularities of surfaces. *Am. J. Math. 92* (1970), p. 419.

[25] Formanek, E. and C. Procesi. Mumford's conjecture for the general linear group. *Adv. in Math.* (1976).

[26] Shah, J. Monodromy of semi-stable quartic surfaces and sextic double planes. Ph. D. thesis, *M.I.T.*, 1974.

[27] Munford, D. Some footnotes to the work of C.P. Ramanujam, *to appear in volume published by Tata Institute, Bombay.*

Invent. math. 67, 23–86 (1982)

Inventiones mathematicae

On the Kodaira Dimension of the Moduli Space of Curves

Joe Harris and David Mumford

The purpose of this paper is to prove that the moduli space $\mathcal{M}_g$ of curves of genus g over $\mathbf{C}$ is of general type if g is odd and $g \geqq 25$. Moreover, the Kodaira dimension is at least 0 if $g=23$. It appears that a variant of our technique, which is technically more difficult, will prove that $\mathcal{M}_g$ is of general type for all sufficiently large g. In order to keep this paper to a reasonable length, we are treating only the odd genus case here, postponing the even genus case to a later paper.

The result of this paper should be contrasted with an earlier weaker result [12] to the effect that $\mathcal{M}_g$ is of "log general type" in the sense that there are enough n-canonical forms [1] ω on $\mathcal{M}_g$-Sing$(\mathcal{M}_g)$ with logarithmic poles at infinity (i.e., if $\overline{\mathcal{M}}_g - \mathcal{M}_g = V(x_1)$, then $\omega = a(x)(dx_1 \wedge \ldots \wedge dx_{3g-3}/x_1)^{\otimes n}$) to separate points generically. By the results of Brylinski [3], this implies that for *all* g, moduli spaces $\mathcal{M}_g^{(\alpha)}$ of sufficiently high level α are of general type (the levels in question are non-abelian levels).

The proof is based on the analysis of a special divisor $D_k \subset \mathcal{M}_g$ with a very natural *geometric* meaning:

$$D_k = \left(\text{locus of curves } C \text{ which are } k\text{-fold covers of } \mathbf{P}^1 \text{ where } k = \frac{g+1}{2}\right).$$

We consider the closure $\overline{D}_k$ of D_k in $\overline{\mathcal{M}}_g$ and compute the divisor class of $\overline{D}_k$ in terms of the basic divisor classes

$$\lambda, \delta_0, \delta_1, \ldots, \delta_{[g/2]} \in \operatorname{Pic}(\mathcal{M}_g) \otimes \mathbf{Q}$$

introduced in [13]. Here λ is c_1 of the "Hodge bundle" L_λ, where L_λ has fibre $\Lambda^g H^0(\omega_C)$ over a curve C, and δ_i are the divisor classes of components Δ_i of $\overline{\mathcal{M}}_g - \mathcal{M}_g$. In particular, Δ_0 is the closure of the locus of *irreducible* singular curves. The final result of § 5 will be:

[1] By an n-canonical form on a smooth variety X^r, we mean a differential form locally expressed as

$$a(x_1, \ldots, x_r)(dx_1 \wedge \ldots \wedge dx_r)^{\otimes n}$$

$$\text{Class of } \overline{D}_k \equiv \frac{(2k-4)!}{k!(k-2)!}\left\{6(k+1)\lambda - k\delta_0 - \sum_{\alpha=1}^{k-1} 3\alpha(2k-1-\alpha)\delta_\alpha\right\}.$$

On the other hand, using Grothendieck's Riemann-Roch formula, we calculate in § 2 the canonical class of $\overline{\mathcal{M}}_g$:

$$K_{\overline{\mathcal{M}}_g} \equiv 13\lambda - 2\delta_0 - 3\delta_1 - 2\delta_2 - \ldots - 2\delta_{[\frac{g}{2}]}.$$

Putting these together proves

$$c_k K_{\overline{\mathcal{M}}_g} \equiv 2\overline{D}_k + c_k\left(1-\frac{12}{k}\right)\lambda + \begin{pmatrix}\text{comb. of } \delta_1, \ldots, \delta_{[\frac{g}{2}]} \\ \text{with positive coefficients}\end{pmatrix}$$

if $c_k = (2k-4)!/(k-1)!(k-2)!$. It turns out also that the singularities of $\overline{\mathcal{M}}_g$ are sufficiently mild so that all divisors in $|nK_{\overline{\mathcal{M}}_g}|$ define n-canonical differential forms on $\overline{\mathcal{M}}_g$ without poles on the resolution of $\overline{\mathcal{M}}_g$. Thus if $|nK|_{\text{bir}}$ denotes the linear system of birationally holomorphic forms, we have

$$|nc_k K_{\overline{\mathcal{M}}_g}|_{\text{bir}} \supset (\text{eff. div.}) + \left|nc_k\left(1-\frac{12}{k}\right)\lambda\right|.$$

Since λ is ample on Satake's compactification $\overline{\mathcal{A}}_g$, the sections of $L_\lambda^{\otimes n}$ on $\overline{\mathcal{M}}_g$ define a birational morphism to $\mathbf{P}^N$. Therefore if $k=12$, $|c_k K_{\overline{\mathcal{M}}_g}|_{\text{bir}} \neq \phi$, and if $k>12$, $|nc_k K_{\overline{\mathcal{M}}_g}|_{\text{bir}}$ defines a birational map for $n \gg 0$.

The idea of this proof owes a great deal to E. Freitag and Y.-S. Tai, although most of the links to their work are not apparent. First and foremost, it was Freitag who saw beyond the classical picture according to which geometrically natural moduli spaces all seemed to be unirational, although their "higher level" variants were of general type: a picture largely based on the heavily studied case of $\mathcal{M}_1$, the moduli space of elliptic curves. Freitag showed that the moduli space $\mathcal{A}_g$ of principally polarized abelian varieties is not unirational for $g \equiv 1\ (8)$, $g \geqq 17$ [4], and for $24|g$ [5]. Tai then showed that, in fact, $\mathcal{A}_g$ is of general type for all $g \geqq 9$ ([17] to be published). He introduced 2 important new techniques: the first was the proof that the singularities of $\mathcal{A}_g$ "didn't matter", except for the one along the 1st boundary component $\mathcal{A}_{g-1}$. In particular, Tai and, independently, Reid [15] found an important hypothesis on the action of a finite group G on $\mathbf{C}^r$ which implies that n-canonical differential forms on $\mathbf{C}^r/G$ have no poles on the resolution of $\mathbf{C}^r/G$. We use this criterion again here. Secondly, by means of the Hirzebruch Proportionality theorem, Tai proved that if g is large enough, there are Siegel modular forms of weight $n(g)$, vanishing at $\mathcal{A}_{g-1}$ to order $m(g)$ with $m(g)/n(g)$ *arbitrarily large.* Subsequent to this, E. Freitag and one of the authors discussed at length how these techniques could be extended to $\mathcal{M}_g$. They sought to use mixed holomorphic tensors on $\mathcal{A}_g$, i.e., differential forms which are not exterior forms, or symmetric forms, but sections of the full tensor power $(\Omega^1_{\mathcal{A}_g})^{\otimes n}$, so as to produce from theta series n-canonical differential forms on $\mathcal{M}_g$. This still looks quite hopeful. But what led directly to the present paper was the discovery of a particular holomorphic tensor on $\mathcal{A}_g$ which defines a map from the ring of

Siegel modular forms vanishing at $\mathcal{A}_{g-1}$ to sufficiently large order to the pluricanonical ring of $\mathcal{M}_g$. (It is still not clear whether or not all Siegel modular forms with this vanishing at $\mathcal{A}_{g-1}$ do not, by some fluke, also vanish on the locus of Jacobians, in which case this map is zero.) But the forms defined by this map must at least vanish on a certain divisor $D' \subset \mathcal{M}_g$:

$$D' = \left\{ \begin{matrix} \text{locus of curves } C \text{ with a line bundle } L \text{ such that} \\ h^0(L) \geqq 2,\ h^0(\Omega^1_C \otimes L^{-2}) \geqq 1 \end{matrix} \right\}.$$

If g is odd, the D above is one of the components of D'. In this roundabout way, Freitag's techniques pointed directly to the method of the present proof!

This paper suggests that there is much interest in a systematic investigation of the relations between the fundamental classes of the many subvarieties of $\mathcal{M}_g$ defined by geometric conditions, with the hope of establishing a calculus of such cycles. Such a study was in fact first undertaken by Steven Diaz, who is investigating the global geometry of the locus in $\overline{\mathcal{M}}_g$ of curves with abnormal Weierstrass points; his work has been extremely valuable in developing the techniques of this paper. Hopefully, a more complete picture of the geometry of $\mathcal{M}_g$ will emerge from such investigations; a great deal of work remains to be done. Another important question is whether $\mathcal{M}_g$ actually carries non-zero holomorphic *exterior* p-forms for some p. The analogous theory for $\mathcal{A}_g$ suggests that this *might* happen for $p=g$, $2g-1$ or $3g-3$ (cf. Anderson [1], Stillman [16]).

The paper is organized as follows. In §1, we use the Reid-Tai criterion to prove that n-canonical forms on the open set $\overline{\mathcal{M}}{}^0_g$ parametrizing curves without automorphisms automatically are holomorphic on a resolution of $\overline{\mathcal{M}}_g$. In §2, we compute $K_{\overline{\mathcal{M}}_g}$. In §3 and §6, we calculate the class of $\overline{D}_k$, using a compactification of the Hurwitz moduli scheme of k-fold coverings, developed in §4, and using counts of the number of pencils of certain types on a generic curve of genus g, due to Griffiths and Harris, developed in §5.

In order to indicate the significance of this result we would like to point out the following easy result:

Proposition. *Assume for some g that the Kodaira dimension of* $\mathcal{M}_g$ *is at least* 0. *Then if C is a generic curve of genus g* (i.e. *the corresponding point* $[C] \in \mathcal{M}_g$ *lies in no subvariety defined over* $\mathbf{Q}$), *and F is an algebraic surface containing C on which C moves in a non-trivial linear system, then F is birational to* $C \times \mathbf{P}^1$.

Proof. The Kodaira dimension being at least 0 means that $\mathcal{M}_g$ carries a n-canonical differential $a(x)\,(dx_1 \wedge \ldots \wedge dx_{3g-3})^{\otimes n}$ with no poles on a compact smooth model of $\mathcal{M}_g$. This implies that $\mathcal{M}_g$ is not "uniruled", i.e. there is no dominant rational map

$$\mathbf{P}^1 \times W^{3g-4} \to \mathcal{M}_g.$$

Therefore the images of all non-constant maps

$$\mathbf{P}^1 \to \mathcal{M}_g$$

lie on subvarieties of $\mathcal{M}_g$ defined over $\mathbf{Q}$. Therefore, given $C\in|C|$ on F, a pencil in $|C|$ defines a map from $\mathbf{P}^1$ to $\mathcal{M}_g$ through $[C]$, which must be constant, i.e. all $C'\in|C|$ are isomorphic to C. Since C has no automorphism, this means that F is birational to $C\times\mathbf{P}^1$. Q.E.D.

Corollary. *If g is odd, $g\geqq 23$, then a generic curve of genus g does not occur in a non-trivial linear system on any non-ruled surface.*

§ 1. Pluri-canonical Forms on $\overline{\mathcal{M}}_g$

If V^n is any quasi-projective variety, by k-canonical forms on V we understand holomorphic tensors ω given on the open set V_{reg} of smooth points of V by

$$\omega=a(x_1,\ldots,x_n)(dx_1\wedge\ldots\wedge dx_n)^{\otimes k}$$

such that for one and hence all desingularizations

$$\pi\colon \tilde{V}\to V$$

of V, ω extends to a holomorphic tensor of this type on all of $\tilde{V}$. Pluri-canonical forms[2] refers to k-canonical forms for all k. If V is smooth, then $o_V(nK)$ is the sheaf of n-canonical forms on V.

As in the introduction, $\overline{\mathcal{M}}_g$ stands for the coarse moduli space of stable curves of genus g: it is a normal projective variety. Related to $\overline{\mathcal{M}}_g$, we introduce the following further varieties:

$$\begin{array}{ll} & \overline{\mathcal{M}}_g\leftarrow\tilde{\mathcal{M}}_g=\text{desingularization of }\overline{\mathcal{M}}_g\\ & \quad\;\cup\\ \text{open set of smooth pts. of }\overline{\mathcal{M}}_g & =\overline{\mathcal{M}}_{g,\text{reg}}\\ & \quad\;\cup\\ \text{open set of curves } C \text{ w/o automorphisms} & =\overline{\mathcal{M}}^0_g\end{array}$$

The purpose of this section is to prove the following theorem:

Theorem 1. *If $g\geqq 4$, then for all n, every n-canonical form on $\overline{\mathcal{M}}^0_g$ extends to an n-canonical form on $\tilde{\mathcal{M}}_g$. More precisely:*

$$\Gamma(\overline{\mathcal{M}}^0_g, o_{\overline{\mathcal{M}}_g}(nK))=\Gamma(\tilde{\mathcal{M}}_g, o_{\tilde{\mathcal{M}}_g}(nK)).$$

Proof. Recall that locally $\overline{\mathcal{M}}_g$ can be described as follows: let C be a stable curve of genus g and let

$$\pi\colon \mathscr{C}\to\Delta^{3g-3}$$

[2] There is a general confusion of terminology for referring to these and other tensor forms which are not *exterior* forms. We suggest k-canonical forms as a good phrase to distinguish these from exterior k-forms or symmetric k-forms

be its local universal deformation space. Aut(C) is a finite group which operates on $\mathscr{C}$ and Δ^{3g-3}. Then a neighborhood of the point $[C]\in\overline{\mathcal{M}}_g$ defined by C is isomorphic as analytic space to

$$\Delta^{3g-3}/\mathrm{Aut}(C).$$

Moreover the action of a finite group on a smooth space can always be made linear in suitable coordinates, hence if $T_{0,\Delta}$ is the tangent space to Δ^{3g-3} at 0,

$$\Delta^{3g-3}/\mathrm{Aut}(C)\cong \text{neigh. of } 0 \text{ in } [T_{0,\Delta}/\mathrm{Aut}(C)].$$

On the other hand, $T_{0,\Delta}$ is the space of infinitesimal deformations of C which is well known to be

$$\mathrm{Ext}^1(\Omega^1_C, \mathcal{O}_C)$$

where Ω^1_C is the sheaf of Kähler differentials on C. By Serre duality,

$$\mathrm{Ext}^1(\Omega^1_C, \mathcal{O}_C)\cong H^0(\Omega^1_C\otimes\omega_C)^*$$

where ω_C is the dualizing sheaf on C[3]. Thus finally

$$(\text{neigh. of } [C] \text{ in } \overline{\mathcal{M}}_g)\cong(\text{neigh. of } 0 \text{ in } H^0(\Omega^1_C\otimes\omega_C)^*/\mathrm{Aut}(C)).$$

Note that in this description, the open set of points on the left which are in $\mathcal{M}^0_g$ is equal to the open set of points on the right where Aut(C) acts freely.

We have therefore particular cases of the problem:

V a vector space of dimension d
$G\subset GL(V)$ a finite group
$V_0\subset V$ open set where G acts freely
When do n-canonical forms on V_0/G extend holomorphically to a resolution $\widetilde{V/G}$ of V/G?

The following criterion is due independently to M. Reid [15] and Y.-S. Tai [17]:

Reid-Tai Criterion: In the above situation, for all $g\in G$, let g be conjugate to

$$\begin{pmatrix} \zeta^{a_1} & & 0 \\ & \ddots & \\ 0 & & \zeta^{a_d} \end{pmatrix}$$

where ζ is a primitive m^{th} root of 1, and $0\leqq a_i<m$. If for all g and ζ,

[3] If C is smooth, $\omega_C\cong\Omega^1_C$. But at double points given by $xy=0$, Ω^1_C is generated by dx, dy mod one relation $x\,dy+y\,dx=0$, and has the torsion submodule $\mathbb{C}\cdot(x\,dy)$, while ω_C is free on one generator given by the differentials $\frac{dx}{x}$ on $y=0$, $-\frac{dy}{y}$ on $x=0$

$$\sum_{i=1}^{d} a_i/m \geqq 1$$

then any n-canonical form on V_0/G extends holomorphically to $\widetilde{V/G}$.

In fact, we need a slight generalization of the Reid-Tai criterion. This says that for *any* G-action, certain *n*-canonical forms ω extend to $\widetilde{V/G}$, namely those which are holomorphic on V_0/G *and*, for all $g \in G$ for which $\sum a_i/m < 1$ (for some ζ) – call these bad g's – ω is also holomorphic on the divisors $E \subset \widetilde{V/G}$ mapping onto

$$\operatorname{Im}(\{x \in V \mid gx = x\} \to V/G).$$

Equivalently, we must assume ω to be holomorphic on the resolution of some open set $U \subset V/G$, where U contains V_0/G and U contains the image in V/G of the generic point η_g of $\{x \in V \mid gx = x\}$, for all *bad* g. We will give a proof in appendix 1 to this section.

Our main task is therefore to investigate, case-by-case, what are the eigenvalues of an automorphism ϕ of a stable curve C acting on $H^0(\Omega^1_C \otimes \omega_C)$. We take the case of a smooth C first.

Proposition. *Let C be a smooth curve over* **C**, *and let ϕ be an automorphism of C of order n. Let ζ be any primitive n^{th} root of* 1, *and let the action ϕ on $H^0(o_C(2K))^*$ be given by*

$$\begin{pmatrix} \zeta^{a_1} & & 0 \\ & \ddots & \\ 0 & & \zeta^{a_{3g-3}} \end{pmatrix} \quad \textit{where } 0 \leqq a_i < n.$$

Then either

$$\sum_{i=1}^{3g-3} (a_i/n) \geqq 1$$

or else (C, ϕ) is one of the following cases:

i) genus $C = 0$

ii) genus $C = 1$

iii) genus $C = 2$, $n = 2$, *ϕ is the hyperelliptic involution*

iv) genus $C = 2$, *C is a double cover of an elliptic curve and ϕ is the associated involution*

v) genus $C = 3$, *C is hyperelliptic and ϕ is the hyperelliptic involution.*

Proof. Let C_0 be the quotient of C by ϕ, let g_0 be the genus of C_0 and let $P_1, \ldots, P_\beta \in C_0$ be the branch points of the cover

$$\pi \colon C \to C_0.$$

Let m_i = order of branching of π over P_i: i.e., there are n/m_i points $Q_{i,\alpha} \in C$ over P_i and π is given locally at each $Q_{i,\alpha}$ by the m_i^{th} root of a suitable local parameter at P_i.

Consider the family of cyclic branched coverings

$$\pi' : C' \to C_0'$$

which you get by varying the moduli of C_0 and varying the branch points $P_i \in C_0$. The dimension of this family is $3g_0 - 3 + \beta$. This defines a subvariety $W \subset \mathcal{M}_g$ of codimension $3(g-g_0)-\beta$ containing C such that all $C' \in W$ admit automorphisms ϕ' deforming ϕ. On the other hand, if the local deformation space of C is $p: \mathcal{C} \to S$, S a germ of smooth $(3g-3)$-dimensional manifold, then ϕ acts naturally on $\mathcal{C}$ and S:

$$\begin{array}{ccc} \mathcal{C} & \xrightarrow{\ \phi_2\ } & \mathcal{C} \\ {\scriptstyle p}\downarrow & & \downarrow{\scriptstyle p} \\ S & \xrightarrow[\ \phi_1\ ]{} & S \end{array}$$

In suitable coordinates on S, ϕ_1 acts linearly. The subvariety W of $\mathcal{M}_g$ is just the image of the fixed point set of ϕ_1 on S. Therefore, the dimension of W is the dimension of the subspace of $T_{0,S} \cong H^0(o_C(2K))$ fixed by ϕ, i.e., the number of a_i equal to 0. This proves:

$$\#\{i \mid a_i = 0\} = 3g_0 - 3 + \beta,$$

$$\#\{i \mid a_i \geqq 1\} = 3(g - g_0) - \beta.$$

Therefore

$$\sum_{i=1}^{3g-3} (a_i/n) \geqq \frac{3(g-g_0)-\beta}{n}.$$

But by Hurwitz's formula:

$$2g - 2 = n(2g_0 - 2) + \sum_{i=1}^{\beta} \frac{n}{m_i}(m_i - 1).$$

Let's now assume $\sum (a_i/n) < 1$, hence in particular

$$n > 3(g - g_0) - \beta.$$

Combining this with Hurwitz's formula, you easily check:

$$\frac{2}{3} > \frac{2n-2}{n}(g_0 - 1) + \sum_{i=1}^{\beta} \left(1 - \frac{2}{3n} - \frac{1}{m_i}\right). \qquad (*)$$

Note that as $n \geqq 2$, $m_i \geqq 2$,

$$1 - \frac{2}{3n} - \frac{1}{m_i} \geqq \frac{1}{6} > 0,$$

hence $(*)$ implies immediately that $g_0 \leqq 1$.

Moreover, the m_i cannot be chosen as arbitrary divisors of n because of the following:

Lemma. *Let $M = \text{l.c.m.}\,(m_i)$. For all primes p dividing M, let $p^r | M$, $p^{r+1} \nmid M$, and let*

$$I = \{i | p^r | m_i\}.$$

Then I has at least 2 elements in it, and if $p=2$, I consists of an even number of i's. Moreover, if $g_0=0$, $n=M$.

Proof. The covering C of C_0 is defined by a surjective homomorphism:

$$\pi_1(C_0 - \{P_1, \ldots, P_\beta\}) \twoheadrightarrow \mathbf{Z}/n\mathbf{Z}.$$

This factors through $H_1(C_0 - \{P_1, \ldots, P_\beta\})$ because $\mathbf{Z}/n\mathbf{Z}$ is abelian. Let e_i be a small loop around P_i. Then $H_1(C_0 - \{P_1, \ldots, P_\beta\})$ is the direct sum of $H_1(C_0)$ and $\oplus \mathbf{Z} e_i$ modulo the one relation $\sum e_i = 0$. Let e_i be mapped to $\bar{e}_i \in \mathbf{Z}/n\mathbf{Z}$. It's easy to see that m_i is exactly the order of $\bar{e}_i$. But

a) $\sum_{i=1} \bar{e}_i = 0$ in $\mathbf{Z}/n\mathbf{Z}$

and

b) if $g_0 = 0$, hence $H_1(C_0) = (0)$, $\{\bar{e}_i\}$ generate $\mathbf{Z}/n\mathbf{Z}$.

With the notation in the lemma, if $m = M/p$

$$\sum_{i=1}^{\beta} m\bar{e}_i = 0 \quad \text{and} \quad m\bar{e}_i = 0 \text{ if } i \notin I.$$

Therefore $\# I \geqq 2$. And if $p=2$, $m\bar{e}_i$ is in the subgroup $\mathbf{Z}/2\mathbf{Z}$ of $\mathbf{Z}/n\mathbf{Z}$, hence $\# I$ is even. Also, if $g_0 = 0$, then $M = n$ by (b).

In particular, this shows that $\beta = 0$ or $\beta \geqq 2$, and if $n=2$, then β is even. Using these restrictions on the m_i, a lengthy but straightforward calculation allows one to *list all the solutions* to (*). The result is:

a) $g_0 = 1$, $\beta = 0$

b) $g_0 = 1$, $n = \beta = 2$, all $m_i = 2$

c) $g_0 = 0$, $n = 2$, $\beta = 2$, 4, 6 or 8, all $m_i = 2$

d) $g_0 = 0$, $\beta = 2$, any n

e) $g_0 = 0$, $\beta = 3$ with one of the following triples (m_1, m_2, m_3):

$(2, 4k, 4k)$ $\quad n = 4k$

$(2, 2k, k)$, k odd $\quad n = 2k$

$(3, 3k, 3k)$ $\quad n = 3k$

$(3, 3k, k)$, $3 \nmid k$ $\quad n = 3k$

$(4, 4k, k)$, k odd, $5 \leqq k \leqq 13$

$(4, 4k, 2k)$, $k = 3, 5, 7$

$(4, 8k, 8k)$, $k = 1, 2$

$(5, 5k, 5k)$, $k = 1, 2$

$(5, 5k, k)$, $k = 6, 7, 8$

$(7, 7, 7)$

f) $g_0=0$, $\beta=4$ with one of the following quadruples (m_1, m_2, m_3, m_4):

$$\begin{array}{lll}(2,2,k,k), & k=3,5, & n=2k\\ (2,2,2k,2k), & k=2,3, & n=2k\\ (3,3,3,3), & & n=3\\ (2,6,3,3), & & n=6.\end{array}$$

The Proposition asserts that a), b), c) and d) are in fact the only cases where $\sum a_i/n<1$. In cases e) and f) we need to look more closely and evaluate the a_i's.

For (f), we can describe the curves C which occur as follows:

for $(2,2,k,k)$, k odd, C and ϕ are

$$y^2=(x^k-1)(x^k-a)$$

$$\phi(x,y)=(\zeta_k x, -y),\ \zeta_k \text{ a primitive } k^{\text{th}} \text{ root of } 1$$

for $(2,2,2k,2k)$, C and ϕ are

$$y^2=x(x^k-1)(x^k-a)$$

$$\phi(x,y)=(\zeta_k x, \zeta_{2k} y),\ \zeta_{2k}^2=\zeta_k$$

for $(3,3,3,3)$, C is

$$y^2=(x^3-1)(x^3-a) \text{ again}$$

but ϕ is

$$\phi(x,y)=(\zeta_3 x, y),\ \zeta_3 \text{ a primitive } 3^{\text{rd}} \text{ root of } 1,$$

for $(2,6,3,3)$, C is the non-hyperelliptic curve of genus 3

$$y^3=(x^2-1)(x^2-a)$$

$$\phi(x,y)=(-x, \zeta_3 y).$$

From this description, one calculates the a_i's in this table:

(m_1, m_2, m_3, m_4)	(a_i/n)
(2, 2, 3, 3)	(0/6, 2/6, 4/6)
(2, 2, 4, 4)	(0/4, 2/4, 2/4)
(2, 2, 5, 5)	$(\frac{0}{10}, \frac{1}{10}, \frac{2}{10}, \frac{4}{10}, \frac{4}{10}, \frac{6}{10}, \frac{6}{10}, \frac{8}{10}, \frac{9}{10})$
(2, 2, 6, 6)	$(\frac{0}{6}, \frac{2}{6}, \frac{2}{6}, \frac{3}{6}, \frac{4}{6}, \frac{4}{6})$
(3, 3, 3, 3)	(0/3, 1/3, 2/3)
(2, 6, 3, 3)	(0/6, 1/6, 2/6, 3/6, 4/6, 4/6)

which confirms the Proposition in this case. (Note that the ζ_k and ζ_3 used to describe ϕ may be *any* primitive k^{th} or 3rd root of 1, hence one must not merely check

$$\sum a_i/n \geq 1$$

but also check that for all $j \in \{1, \ldots, n-1\}$ relatively prime to n,

$$\sum \mathrm{res}_n(j \cdot a_i)/n \geqq 1$$

where $\mathrm{res}_n(k)$ is the residue in $\{0, 1, \ldots, n-1\}$ of $k \bmod n$.)

For (e), a similar check could be carried out, but because of its tediousness and the possibility of error, it seemed easier *and* much more convincing to write a computer program to calculate $\sum a_i/n$ one at a time for all cyclic covers of $\mathbf{P}^1$ with 3 branch points. This is reproduced in appendix 2 where a table of the minimum values of $\sum(a_i/n)$ for each $g \leqq 18$ is also given. On the other hand, the computer only checks finitely many cases and the 1st 4 types in (e) are infinite families. These are the curves

$$\begin{aligned}
y^2 &= x^k - 1, & \phi(x, y) &= (\zeta_k x, -y),\ k \text{ odd},\\
y^2 &= x(x^{2k} - 1), & \phi(x, y) &= (\zeta_{2k} x, \zeta_{4k} y)\\
y^3 &= x(x^k - 1), & \phi(x, y) &= (\zeta_k x, \zeta_{3k} y)\\
y^3 &= x^2(x^k - 1), & \phi(x, y) &= (\zeta_k x, \zeta_{3k}^2 y)
\end{aligned}$$

for which the a_i are readily worked out explicitly. We omit this.

Using this analysis as a building block, we look next at automorphism of singular stable curves:

Theorem 2. *Let C be a stable (possibly singular) curve of arithmetic genus g and let ϕ be an automorphism of C of order n. Let ζ be any primitive n^{th} root of 1 and let the action of ϕ on $H^0(\Omega_C^1 \otimes \omega_C)$ be given by*

$$\begin{pmatrix} \zeta^{a_1} & & 0 \\ & \ddots & \\ 0 & & \zeta^{a_{3g-3}} \end{pmatrix}, \quad \textit{where} \quad 0 \leqq a_i < n.$$

We assume $g \geqq 4$.

Then either

$$\sum_{i=1}^{3g-3} a_i/n \geqq 1$$

or else (C, ϕ) belongs to one of the following cases:

(i) $C = C_1 \cup C_2$ *where* $\mathrm{genus}(C_1) = g - 1$,

C_2 is either elliptic or rational with one node, $C_1 \cap C_2 = \{P\}$, $n = 2$ and

$$\phi|_{C_1} = \textit{identity}$$

$$\phi|_{C_2} = \textit{inverse with respect to origin } P$$

(ii) $C = C_1 \cup C_2$ *as above, but C_2 is elliptic with $j(C_2) = 0$, $n = 6$ and*

$\phi|_{C_1} = \textit{identity}$

$\phi|_{C_2} = $ *one of the two automorphisms of C_2 of order 6 fixing P.*

(iii) $C=C_1\cup C_2$ *as above, but* C_2 *is elliptic with* $j(C_2)=12^3$, $n=4$ *and*

$\phi|_{C_1}=$ *identity*

$\phi|_{C_2}=$ *one of the two automorphisms of* C_2 *of order 4 fixing P.*

Proof. Note that we have proved the theorem for smooth curves C. Also note that the set of exceptions (i), (ii) and (iii) forms a closed set in the moduli space $\bar{\mathcal{M}}_g$. Therefore we can prove the theorem by induction on the number of double points $P\in C$, checking for each (C,ϕ) that either

a) $\sum(a_i/n)\geqq 1$ or

b) (C,ϕ) has a deformation (C',ϕ') with fewer double points.

This is because in any family $\mathscr{C}\to S$ with connected base and with an automorphism $\phi\colon\mathscr{C}\to\mathscr{C}$ over S of order n, the eigenvalues of ϕ on $H^0(\Omega^1_{C_s}\otimes\omega_{C_s})$ vary continuously and are n^{th} roots of 1, hence are constant.

Note next $\Omega^1_C\otimes\omega_C$ has the following local description: at smooth points $P\in C$ with local coordinate x, it is an invertible sheaf with generator $dx^{\otimes 2}$; at ordinary double points $P\in C$, if C is given locally by $x\cdot y=0$, it is generated by the differentials:

$$\omega_1=dx^{\otimes 2}/x,$$
$$\omega_2=dy^{\otimes 2}/y$$

subject to the relation

$$y\omega_1=x\omega_2.$$

Thus

$$y\omega_1=\frac{y\,dx^{\otimes 2}}{x}=\frac{x\,dy^{\otimes 2}}{y}$$

generates a submodule of dimension 1 over k (because $x(y\omega_1)=0$ and $y(y\omega_1)=y(x\omega_2)=0$), and mod this we have the direct sum of the sheaves of quadratic differentials on the 2 branches at P, with simple poles at P. This gives rise to the exact sequence:

$$0\to\bigoplus_{\substack{\text{double pts.}\\ P\in C}}(\text{tor}_P)\to\Omega^1_C\otimes\omega_C\to\bigoplus_{\alpha}o_{C_\alpha}\left(2K_{C_\alpha}+\sum P_\beta\right)\to 0$$

where C_α are the normalizations of the components of C, and for each α, the $P_\beta\in C_\alpha$ are those points of C_α whose image in C are double points of C. Therefore:

$$0\to\bigoplus_{\substack{\text{double}\\ \text{pts.}\,P}}(\text{tor}_P)\to H^0(\Omega^1_C\otimes\omega_C)\to\bigoplus_{\alpha}H^0\left(o_{C_\alpha}\left(2K_\alpha+\sum_\beta P_\beta\right)\right)\to 0.$$

The first step is to analyze the eigenvalues of ϕ on $\bigoplus(\text{tor}_P)$. Say P is a double point of C and $\{P,\phi P,\phi^2P,\ldots,\phi^{m-1}P\}$ are distinct, $\phi^m P=P$, where $m|n$. Then ϕ^m acts on tor_P and if this action is trivial, take $\bar{e}\in(\text{tor}_P)$ and consider the element $e\in H^0(\Omega^1_C\otimes\omega_C)$ given by

$$e\in\bigoplus_Q(\text{tor}_Q),\qquad e_Q=\begin{cases}\phi^i\bar{e} & \text{if } Q=\phi^i P\\ 0 & \text{if } Q\notin\{P,\phi P,\ldots,\phi^{n-1}P\}.\end{cases}$$

Then $\phi e = e$. Thus dually, e defines a deformation of C to which ϕ lifts and in which the double points $\phi^i P$ disappear. By induction on the number of double points, this case is taken care of. Next suppose ϕ^m acts non-trivially on (tor_P). We then calculate all the eigenvalues of ϕ on

$$(\mathrm{tor}_P) \oplus \ldots \oplus (\mathrm{tor}_{\phi^{m-1}P}).$$

Let ζ be a primitive n^{th} root of 1. We must have

$$\phi^m \bar{e} = \zeta^{m\ell} \bar{e}, \quad 1 \leqq \ell < \frac{n}{m}.$$

Let

$$e_a = \sum_{i=0}^{m-1} \zeta^{ia} \cdot (\phi^i \bar{e}).$$

Then

$$\begin{aligned}
\phi(e_a) &= \sum_{i=0}^{m-1} \zeta^{ia} \cdot (\phi^{i+1} \bar{e}) \\
&= \sum_{i=1}^{m} \zeta^{(i-1)a} (\phi^i \bar{e}) \\
&= \zeta^{-a} \sum_{i=1}^{m-1} \zeta^{ia} (\phi^i \bar{e}) + \zeta^{(m-1)a} \cdot \zeta^{m\ell} \cdot \bar{e} \\
&= \zeta^{-a} \cdot e_a
\end{aligned}$$

provided $m(a+l) \equiv 0 \pmod{n}$, i.e., $a \equiv -l \pmod{n/m}$. Therefore

$$\zeta^{\ell}, \zeta^{\ell + \frac{n}{m}}, \ldots, \zeta^{\ell + (m-1)\frac{n}{m}}$$

are the eigenvalues of ϕ on

$$(\mathrm{tor}_P) \oplus \ldots \oplus (\mathrm{tor}_{\phi^{m-1}P}).$$

The corresponding part of the sum $\sum a_i/n$ works out to be

$$\frac{\ell \cdot m}{n} + \frac{1}{m}(1 + 2 + \ldots + (m-1)) = \ell \cdot \frac{m}{n} + \frac{m-1}{2}.$$

In particular, if $m \geqq 3$, it follows that $\sum a_i/n \geqq 1$ already. If $\sum (a_i/n) < 1$, it follows that either

a) ϕ fixes all double points $P \in C$ or

b) ϕ fixes all but one pair $\{P, \phi P\}$ which are interchanged, and $n \geqq 6$.

Moreover, if C has δ double points, the torsion eigenvalues contribute at least

$$\delta/n, \qquad \text{resp. } \delta/n + 1/2$$

to $\sum a_i/n$, in case a), resp. b).

The second step is to analyze the action of ϕ on the set of components of C. We shall prove that if for any α, $\phi C_\alpha \neq C_\alpha$ then either $\sum a_i/n \geqq 1$ or (C, ϕ)

has a deformation with fewer double points. To see this, consider that the image C'_α of C_α in C looks like:

(a) $C_\alpha = \mathbf{P}^1$, C'_α 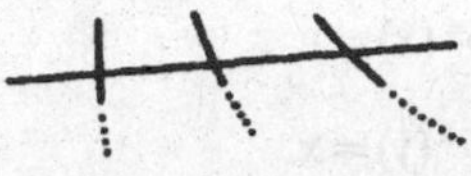

(b) $C_\alpha \cong \mathbf{P}^1$, C'_α 

(c) $C_\alpha \cong \mathbf{P}^1$, C'_α 

(d) $C_\alpha \cong \mathbf{P}^1$, C'_α 

(e) C_α elliptic, C'_α 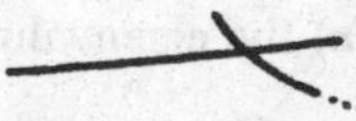

(f) If $g_\alpha = \text{genus}(C_\alpha)$, and $\delta_\alpha = \#\{\text{pts. } P_\beta \in C_\alpha \text{ mapping to double pts. of } C\}$, then

$$3g_\alpha - 3 + \delta_\alpha \geqq 2.$$

As in the analysis of double points, suppose $C_\alpha, \phi C_\alpha, \ldots, \phi^{m-1} C_\alpha$ are distinct and $\phi^m C_\alpha = C_\alpha$. Look at the action of ϕ on

$$W = \bigoplus_{k=0}^{m-1} H^0(o_{\phi^k C_\alpha}(2K_{(\phi^k C_\alpha)} + \sum P_\beta)).$$

The same calculation given for double points shows that if

$$k = \dim H^0(o_{C_\alpha}(2K_\alpha + \sum P_\beta))$$

and if the eigenvalues of ϕ^m here are $\zeta^{m\ell_1}, \ldots, \zeta^{m\ell_k}$, then the eigenvalues of ϕ on W are

$$\zeta^{\ell_i + j\left(\frac{n}{m}\right)}, \quad 1 \leqq i \leqq k, \quad 0 \leqq j < \frac{n}{m},$$

hence W gives *a* contribution to $\sum(a_i/n)$ at least equal to

$$k \cdot \frac{m-1}{2}.$$

Therefore if $\sum a_i/n < 1$, either $k=0$ or $k=1$ and $m=2$. In cases a), b), $k=0$, in cases c), d), e), $k=1$ and in case f), $k \geqq 2$. On the other hand, in cases c), d), e), if any double point moves, it also gives a contribution of 1/2 to $\sum a_i/n$, hence together with the contribution from W, $\sum a_i/n \geqq 1$. But if all double points are fixed, $\phi(C'_\alpha)$ must be the second component through all double points P where C'_α meets another component of C, i.e., $C = C'_\alpha \cup \phi(C'_\alpha)$. In this case, one sees immediately that $g \leqq 3$. Turning to case a), at least one of the three double points on C'_α must be fixed, call it P_0, hence $\phi(C_\alpha)$ must be the second component of C through P_0. Therefore ϕ interchanges these 2 components and ϕ^2 fixes C_α. But ϕ^2 also fixes all double points, hence ϕ^2 is an automorphism of $C'_\alpha \cong \mathbf{P}^1$ with 3 fixed points, hence $\phi^2|_{C_\alpha}$ = identity. At P_0, let $x \cdot y = 0$ be a local equation of C. Then $\phi^2|_{C_\alpha}$ = identity implies that ϕ acts by:

$$\phi^*(x) = y,$$

$$\phi^*(y) = x.$$

But then ϕ fixes the torsion differential $y\,dx^{\otimes 2}/x$ (or alternately, one can argue that ϕ lifts to the universal deformation $xy = t$ of P_0 by $\phi^* x = y$, $\phi^* y = x$, $\phi^* t = t$). Therefore (C, ϕ) has a deformation in which P_0 disappears. Finally, in case (b), the double point of C'_α must move, hence the other double point of C on C'_α does *not* move, and the whole of C will be nothing but $C'_\alpha \cup \phi(C'_\alpha)$. Then $g = 2$ which was excluded.

This reduces us to the case where ϕ fixes every component of C. But then every component C_α contributes to $\sum(a_i/n)$ the eigenvalues of ϕ on

$$H^0(o_{C_\alpha}(2K_{C_\alpha} + \sum P_\beta)).$$

In most cases, by the Proposition, the eigenvalues of ϕ on $H^0(o_{C_\alpha}(2K_{C_\alpha}))$ already give us $\sum(a_i/n) \geqq 1$. In particular, it follows that for all α, one of the following is true:

a) $\phi|_{C_\alpha}$ is the identity,

b) $C_\alpha \cong \mathbf{P}^1$,

c) C_α is elliptic,

d) C_α is hyperelliptic of genus 2 or 3, $\phi|_{C_\alpha}$ = hyperelliptic involution or

e) C_α has genus 2, and is a double cover of an elliptic curve with $\phi|_{C_\alpha}$ = sheet interchange.

We can now argue that ϕ fixes all double points of C too. This goes as follows: note that ϕ^2 fixes all double points of C, hence ϕ^4 fixes all pairs consisting of a double point of C and a branch of C at P, i.e., all P_β in all C_α. Since in case (b), $\mathbf{P}^1$ has at least three P_β's on it, $\phi^4|_{C_\alpha}$ = identity in this case. Moreover in case (c), $\phi|_{C_\alpha}$ either has a fixed point, hence has order $\leqq 6$, or is a translation. And if it is a translation, then $\phi^4|_{C_\alpha}$ is still a translation and also fixes the points P_β on C_α, i.e., $\phi^4|_{C_\alpha}$ = identity. Therefore in all cases the order of ϕ on each C_α is 1, 2, 3, 4 or 6. Now suppose ϕ moves a double point. This can happen in 2 ways:

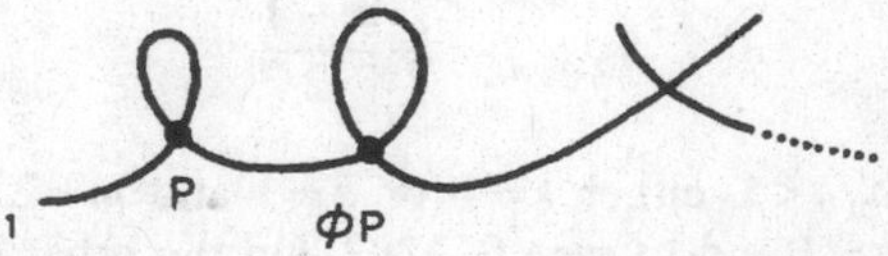

or

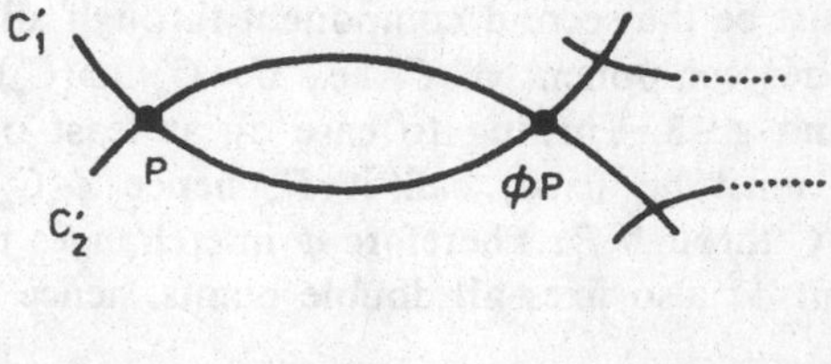

In the first case, $\mathrm{tor}_P \oplus \mathrm{tor}_{\phi P}$ contributes at least $1/2+2/n_1$ to $\sum a_i/n$ where n_1 $=$order of $\phi|_{C_1}$. If $n_1 \leqq 4$, we are done. If $n_1=6$, then C_1 is elliptic with $j(C_1)$ $=0$. Then we get an extra term from the action of ϕ on

$$H^0(\mathcal{o}_{C_1}(2K_{C_1}))\subset H^0(\mathcal{o}_{C_1}(2K_{C_1}+\sum P_\beta)).$$

If dz is the translation-invariant differential on C_1, then $dz^{\otimes 2}$ generates $\mathcal{o}(2K_{C_1})$ and ϕ acts by $z_1 \mapsto \zeta_6 z_1$. Thus we have a term of 1/3 and

$$1/2+2/6+1/3>1.$$

In the second case, if $n_1=\mathrm{lcm}$ (order of ϕ on C_1, C_2), then again $\mathrm{tor}_P \oplus \mathrm{tor}_{\phi P}$ contributes at least $1/2+2/n_1$ to $\sum a_i/n$. If $n_1 \leqq 4$, we are done. If $n_1 \geqq 6$, then one of the C_i, say C_1, is again elliptic with $j(C_1)=0$. If $n_1=6$, we conclude as before. And if $n_1>6$, the order of ϕ on C_2 must be 4. But $P\in C_2$ satisfies $\phi P \neq P$, $\phi^2 P=P$, so C_2 cannot be $\mathbf{P}^1$ (because the map $z\to\sqrt{-1}z$ on $\mathbf{P}^1$ has no points of order exactly 2). Thus C_2 must be elliptic with $j(C_2)=12^3$. Then $dz^{\otimes 2}\in H^0(\mathcal{o}(2K_{C_2}))$ contributes 1/2 to $\sum a_i/n$ and we are done. This completes the proof that ϕ fixes all double points of C.

We now go back to the list a)–e) and enumerate the possible components that C can have.

Case a). Any C_α with $\phi|_{C_\alpha}=$identity.

Case b). $C_\alpha=\mathbf{P}^1$. In suitable coordinates on C_α, $\phi|_{C_\alpha}$ is the map $z\mapsto\zeta z$. But all $P_\beta\in C_\alpha$ are either fixed by ϕ or of order 2. ϕ has only 2 fixed points and there are at least 3 P_β's on C_α. Thus at least one of the P_β, call it P_1, has order 2. But ϕ has a point of order exactly 2 only if ϕ itself has order 2: $\phi(z)=-z$. Now P_1 and ϕP_1 are mapped to the same double point P of C and because ϕ^2 is the identity on C_α, we see that (C,ϕ) near P is given by $xy=0$, $\phi^* x=y$, $\phi^* y=x$. As above, this means that (C,ϕ) can be deformed to eliminate P, so this case is taken care of.

Case c). C_α is elliptic.

Case c1). $\phi|_{C_\alpha}$ is a translation. Since $\phi^2 P_\beta=P_\beta$ for all β, $\phi|_{C_\alpha}$ has order 2 and $\phi^2=$identity. Then as in case (b), the double points of C which are images of these P_β can be deformed away and this case is taken care of.

Case c2). $\phi|_{C_\alpha}$ fixes $0\in C_\alpha$ and has order 2, i.e., it is the inverse in the group structure on C_α. As $\phi^2|_{C_\alpha}=$identity, we are through as above if $\phi P_\beta \neq P_\beta$ for some β. Therefore we may assume all P_β are points of order 2. Now for any 2 points P_1, $P_2\in C_\alpha$ of order 2,

$$H^0(\mathcal{o}_{C_\alpha}(P_1+P_2))$$

contains one even function, namely 1, and one odd function, in fact one with a simple pole at each P_i. Thus

$$H^0(\mathcal{o}_{C_\alpha}(2K_{C_\alpha}+\sum P_\beta))$$

contributes eigenvalues

$$(0, 1/2, \ldots, 1/2)$$

to $\sum a_i/n$. Thus if there are 3 or 4 P_β's, we are through, and we are left with 2 cases for C'_α:

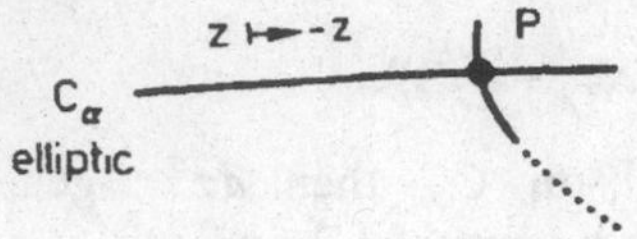

P fixed, C_α contributes 0 to $\sum a_i/n$

or

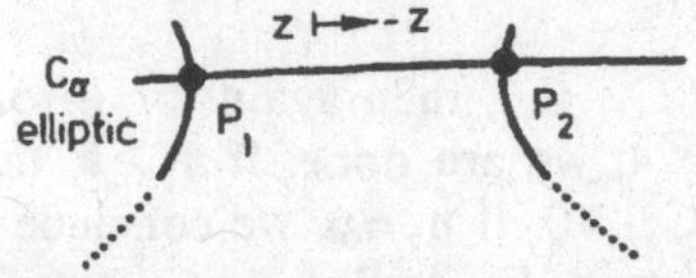

P_1, P_2 fixed, C_α contributes 1/2 to $\sum a_i/n$.

Case c3). $\phi|_{C_\alpha}$ fixes $0 \in C_\alpha$ and has order 3. Then $j(\alpha)=0$ and each P_β is a fixed point. There are 3 of these. If 2 or 3 of these occur as P_β, we are done by considering the eigenvalues of ϕ on $H^0(o_{C_\alpha}(2K_\alpha+P_1+P_2))$. In fact, if dz is the translation-invariant differential and $P_1=0$, then locally the 2 sections here look like

$$dz^{\otimes 2}, z\,dz^{\otimes 2}.$$

Under $z \mapsto \zeta_3 z$, these give a contribution $\frac{1}{3}+\frac{2}{3}$ to $\sum a_i/n$. This leaves the case of one P_β, or C'_α looking like:

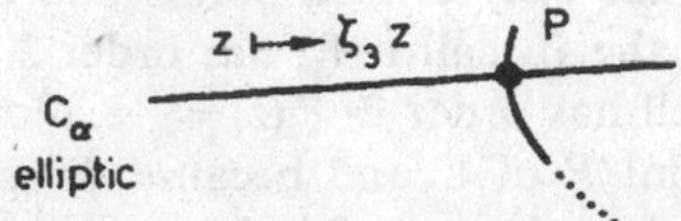

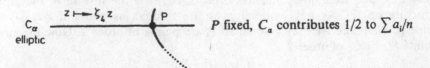

P fixed, C_α contributes 1/3 to $\sum a_i/n$.

Case c4). $\phi|_{C_\alpha}$ fixes $0 \in C_\alpha$ and has order 4. Then $j(C_\alpha)=12^3$, ϕ has 2 fixed points $0=P_0$ and P_1 and $\phi^2=$inverse has 2 more fixed points P_2 and ϕP_2. Thus $\{P_\beta\}$ could consist in $\{P_0\}$, $\{P_0, P_1\}$, $\{P_0, P_2, \phi P_2\}$, $\{P_0, P_1, P_2, \phi P_2\}$. ($\{P_2, \phi P_2\}$ is impossible because C'_α must meet the rest of C somewhere). The eigenvalues on $H^0(o_{C_\alpha}(2K_{C_\alpha}+\sum P_\beta))$ are respectively (1/2), (1/2, 1/4), (1/2, 1/4, 3/4) in the first 3 cases, so the only cases where $\sum a_i/n<1$ could occur are the first 2 cases:

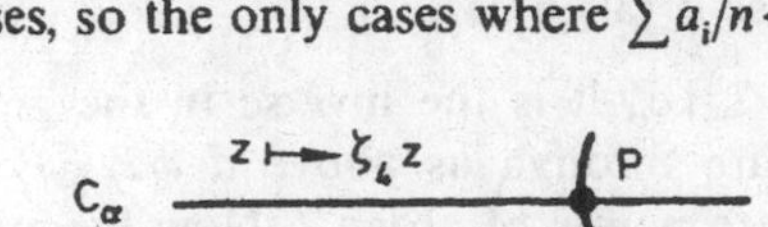

P fixed, C_α contributes 1/2 to $\sum a_i/n$

or

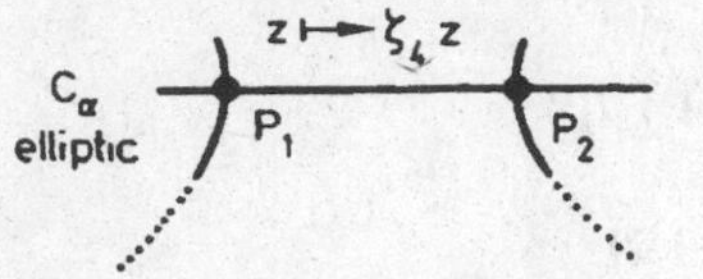

P_1, P_2 fixed, C_α contributes 3/4 to $\sum a_i/n$

Case c5). $\phi|_{C_\alpha}$ fixes $0 \in C_\alpha$ and has order 6. Then $j(C_\alpha)=0$, ϕ has only $0=P_0$ as fixed point, but ϕ^2 has 2 further fixed points P_1, ϕP_1. The set $\{P_\beta\}$ is either $\{P_0\}$ or $\{P_0, P_1, \phi P_1\}$ and the eigenvalues for these are $(\frac{1}{3})$, $(\frac{1}{3}, \frac{2}{3}, \frac{1}{6})$. (In fact, $dz^{\otimes 2} \in H^0(o_{C_\alpha}(2K_{C_\alpha}))$ is transformed under $z \mapsto \zeta_6 z$ by ζ_3; and a basis of $H^0(o_{C_\alpha}(2K_{C_\alpha}+P_0+P_1+\phi P_1))$ looks locally like $dz^{\otimes 2}$, $\dfrac{dz^{\otimes 2}}{z}$ and $z^2 dz^{\otimes 2}$.) The only case to consider is therefore

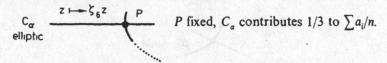

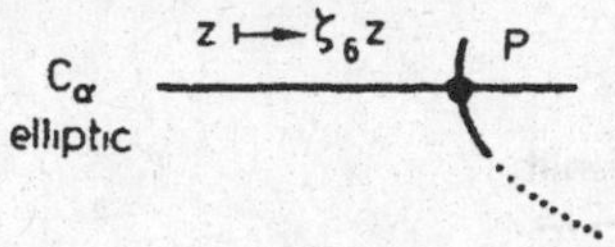

P fixed, C_α contributes 1/3 to $\sum a_i/n$.

Case d). C_α is hyperelliptic of genus 2 or 3, $\phi|_{C_\alpha}$ = hyperelliptic involution. The set $\{P_\beta\}$ must consist in at least one fixed point of ϕ, i.e., a Weierstrass point of C_α and possibly further Weierstrass points or pairs P, ϕP. But the dimension of the (-1)-eigenspace of ϕ on $H^0(o(2K_C+\sum P_\beta))$ is always $\geqq 2$ if $g=3$ and P_1 is Weierstrass point. And it is 1 if $g=2$ and $\{P_\beta\}$ consists in a single Weierstrass point, $\geqq 2$ if $g=2$ and there are further P. So the only case not eliminated is

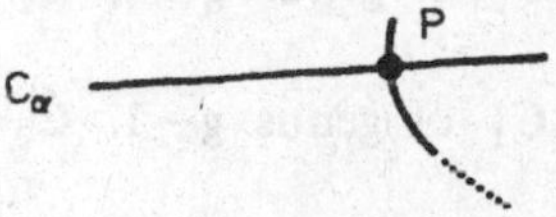

C_α of genus 2, P fixed, C_α contributes 1/2 to $\sum a_i/n$.

Case e). C of genus 2, double cover of elliptic curve, ϕ = sheet interchange. If P is a fixed point of ϕ, then the dimension of the (-1)-eigenspace of ϕ on $H^0(o(2K_C+P))$ is 2, giving $(\frac{1}{2}, \frac{1}{2})$ to $\{a_i/n\}$. So this case is eliminated.

We are left with a problem of patching together the few curves of genus 1 and 2 above with a lot of C_α's where ϕ acts identically. In each curve so obtained, we must add up all the contributions to $\sum a_i/n$. One notes that the following lead to $\sum a_i/n \geqq 1$:

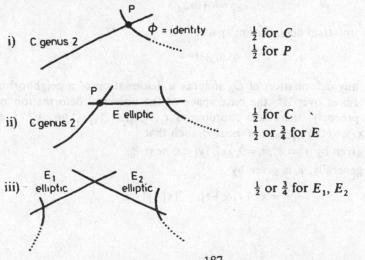

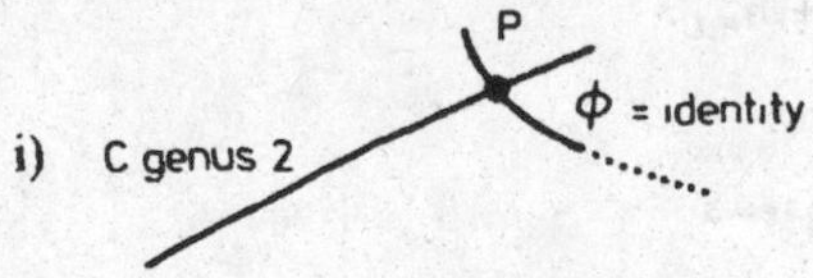

$\frac{1}{2}$ for C
$\frac{1}{2}$ for P

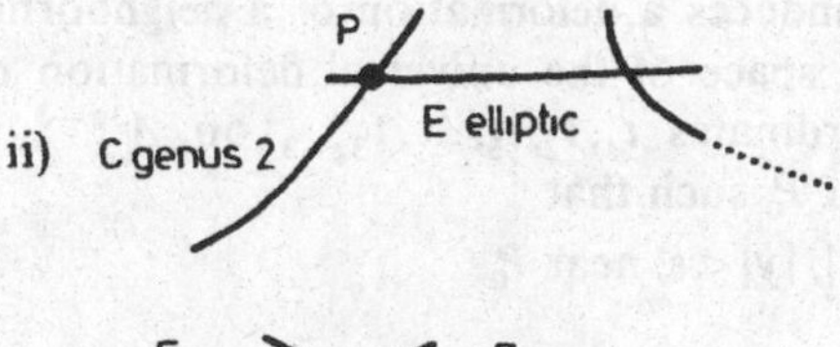

$\frac{1}{2}$ for C
$\frac{1}{2}$ or $\frac{3}{4}$ for E

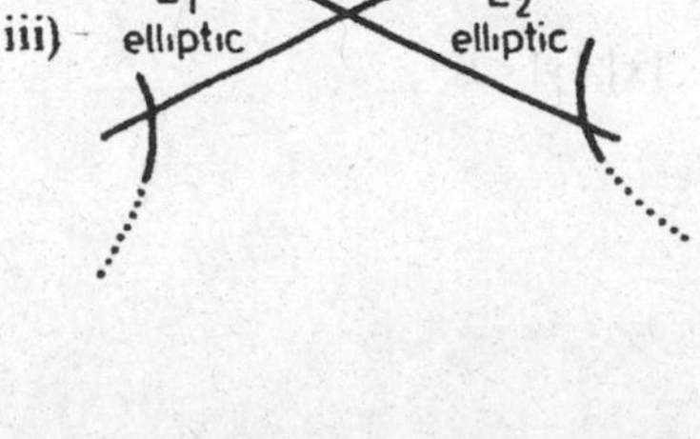

$\frac{1}{2}$ or $\frac{3}{4}$ for E_1, E_2

iv) 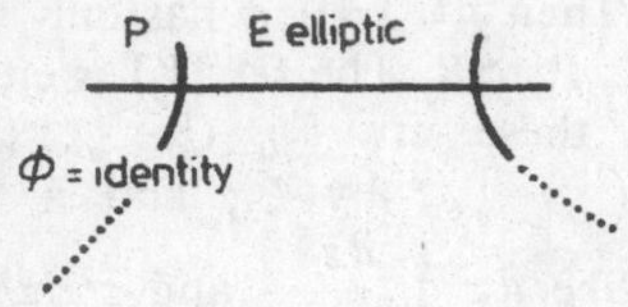

$\frac{1}{2}$ for E or $\frac{3}{4}$ for E
$\frac{1}{2}$ for P $\geqq\frac{1}{4}$ for P

Excluding cases where $g \leqq 3$, it's easy to see that we are left with "the elliptic tails":

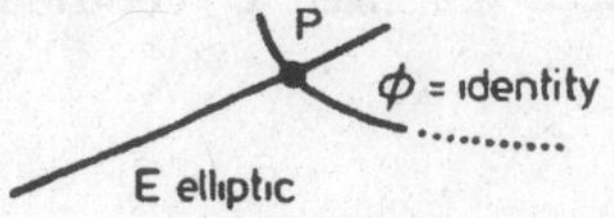

Finally, if $\phi|_E$ has order 3, one finds that E contributes 1/3, P 2/3 so this gives $\sum a_i/n \geqq 1$ too. Q.E.D.

By Reid-Tai's criterion, it now follows that an n-canonical form ω on $\overline{\mathcal{M}}_g^0$ is holomorphic on $\tilde{\mathcal{M}}_g$ provided it is holomorphic on the resolution of a neighborhood of the following 3 points in $\overline{\mathcal{M}}_g$:

a) the generic curve $C = C_1 \cup C_2$, $C_1 \cap C_2 = \{P\}$, C_1 of genus $g-1$, C_2 elliptic,

b) the generic curve $C = C_1 \cup C_2$, $C_1 \cap C_2 = \{P\}$, C_1 of genus $g-1$, C_2 elliptic with $j(C_2)=0$,

c) the generic curve $C = C_1 \cup C_2$, $C_1 \cap C_2 = \{P\}$, C_1 of genus $g-1$, C_2 elliptic with $j(C_2)=12^3$.

In fact, we now take any smooth curve C_1 without automorphisms and a point $P \in C_1$ and consider the curve in $\overline{\mathcal{M}}_g$ parametrizing the stable curves $C_1 \cup C_2$, where $C_1 \cap C_2 = \{P\}$ and C_2 is *any* elliptic curve or is a rational curve with node. We shall show that any ω holomorphic on $\overline{\mathcal{M}}_g^0$ is also holomorphic on the resolution of a neighborhood of this curve.

To do this, let C_0 be the curve obtained from C_1 by making P into an ordinary cusp $P_0 \in C_0$, i.e.,

$$o_{C_0} = k + m_{P,C}^2.$$

Consider the universal deformation space:

$$\mathscr{C} \to \Delta^{3g-3}$$

of C_0. Since any deformation of C_0 induces a deformation of a neighborhood of P_0, Δ^{3g-3} fibres over Δ^2, the base space of the universal deformation of a cusp. More precisely, there are coordinates $t_1, t_2, t_3, \ldots, t_{3g-3}$ on Δ^{3g-3} and coordinates $x, y, t_1, \ldots, t_{3g-3}$ on $\mathscr{C}$ near P_0 such that

i) C_0 is given by $y^2 = x^3$, $t_i = 0$, $|x|, |y| < \varepsilon$, near P_0,

ii) more generally, $\mathscr{C}$ is given by

$$y^2 = x^3 + t_1 x + t_2, \quad |x|, |y| < \varepsilon.$$

Outside of the codimension 2 set $V(t_1, t_2)$ (i.e., $t_1 = t_2 = 0$) $\mathscr{C}$ is a family of stable curves of genus g. Moreover, the fibres of $\mathscr{C}$ are all irreducible curves which either have one cusp, one node, or are smooth. Considering

$$\mathrm{Isom}_{\Delta^{3g-3}}(\mathscr{C},\mathscr{C}) = \left\{ \begin{array}{c} \text{scheme of triples } s_1, s_2, \phi, \\ s_1, s_2 \in \Delta^{3g-3}, \\ \phi: C_{s_1} \xrightarrow{\ \simeq\ } C_{s_2} \text{ an isomorphism} \end{array} \right\}$$

we see that because C_0 has no automorphisms, all nearby curves in the family $\mathscr{C}$ have no automorphisms and occur only once in the family. Therefore we have a holomorphic map

$$f: \Delta^{3g-3} - V(t_1, t_2) \to \overline{\mathscr{M}}_g^0$$

which is injective, and therefore an isomorphism of $\Delta^{3g-3} - V(t_1, t_2)$ with an open subset of $\overline{\mathscr{M}}_g^0$. We want to study the singularity of f at $V(t_1, t_2)$. To do this, we want to convert $\mathscr{C}/\Delta^{3g-3}$ into a family of stable curves.

Let

$$S \to \Delta^{3g-3}$$

be the normalization of the blow up of the ideal (t_1^3, t_2^2). We shall show that f extends to an isomorphism $\tilde{f}$ of S with a suitable open set in $\mathscr{M}_g$. S is covered by 2 charts:

$$S_1 \text{ with coordinates } t_1, t_2, \frac{t_1^2}{t_2}, \frac{t_1^3}{t_2^2},$$

$$S_2 \text{ with coordinates } t_1, t_2, \frac{t_2}{t_1}, \frac{t_2^2}{t_1^3}.$$

Let $p_1: \tilde{S}_1 \to S_1$ be the normalization of S_1 in the 6-cyclic covering $u_2 = t_2^{1/6}$. Then $\tilde{S}_1$ is smooth with coordinates $u_1, u_2, t_3, \ldots, t_{3g-3}$, p_1 being given by

$$t_1 = u_1 u_2^4,$$

$$t_2 = u_2^6,$$

$$(t_1^2/t_2) = u_1^2 u_2^2,$$

$$(t_1^3/t_2^2) = u_1^3.$$

Moreover, the group μ_6 of 6th roots of 1 acts on $\tilde{S}_1$ by

$$(u_1, u_2) \mapsto (\zeta^2 u_1, \zeta u_2)$$

so that $S_1 \cong \tilde{S}_1/\mu_6$. Let p_2: $\tilde{S}_2 \to S_2$ be the normalization of S_2 in the 4-cyclic covering $u_1 = t_1^{1/4}$. Then $\tilde{S}_2$ is smooth with coordinates $u_1, u_2, t_3, \ldots, t_{3g-3}$, p_2 being given by

$$\begin{aligned} t_1 &= u_1^4, \\ t_2 &= u_2 u_1^6, \\ t_2/t_1 &= u_2 u_1^2, \\ t_2^2/t_1^3 &= u_2^2. \end{aligned}$$

Moreover, the group μ_4 of 4th roots of 1 acts on $\tilde{S}_2$ by

$$(u_1, u_2) \mapsto (\zeta u_1, \zeta^2 u_2)$$

so that $S_2 = \tilde{S}_2/\mu_4$.

Now pull the family $\mathscr{C} \to \Delta^{3g-3}$ back to $\tilde{S}_1$ and $\tilde{S}_2$. Over $\tilde{S}_1$, the family is given near P_0 by

$$y^2 = x^3 + u_1 u_2^4 x + u_2^6.$$

Let $\mathscr{C}_1$ be the normalization of the blow-up of $\mathscr{C} \times_{(\Delta^{3g-3})} \tilde{S}_1$ in the ideal (x, u_2^2). It is covered by the 2 charts

$$\mathscr{C}_{1,a}\colon\ y'^2 = x'^3 + u_1 x' + 1, \quad \text{where} \quad x' = x/u_2^2,\ y' = y/u_2^3$$

$$\mathscr{C}_{1,b}\colon\ x'' \cdot y'' = u_2, \quad \text{where} \quad x'' = u_2 x/y,\ |x''|, |y''| < \varepsilon,\ y'' = y/x$$

$$\left(\text{note that: } y''^2 = x\left(1 + u_1\left(\frac{u_2^2}{x}\right)^2 + \left(\frac{u_2^2}{x}\right)^3\right)\right.$$

$$x''^2 = \frac{(u_2^2/x)}{(1 + u_1(u_2^2/x)^2 + (u_2^2/x)^3)}$$

$$\left.\text{so that } x'', y'' \text{ are integrally dependent on } x, u_1, \frac{u_2^2}{x} \text{ when } \left|\frac{u_2^2}{x}\right| < \varepsilon.\right)$$

In particular, the fibre of $\mathscr{C}_1$ over points where $u_2 = 0$ looks like:

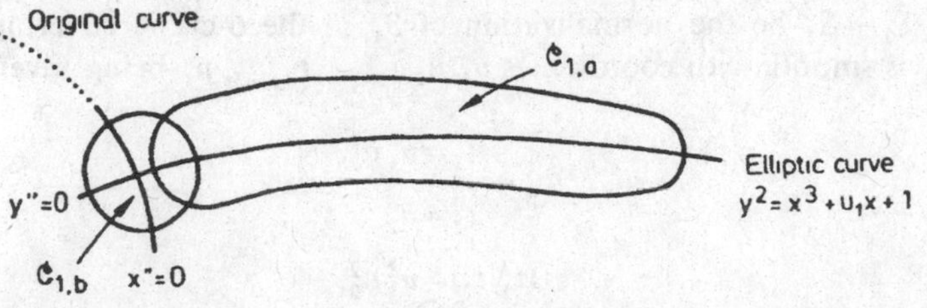

Moreover, $\mathscr{C}_1$ is a smooth variety, fibred in stable curves over $\tilde{S}_1$. The same works for the pull-back of $\mathscr{C}$ to $\tilde{S}_2$, which is given near P_0 by

$$y^2 = x^3 + u_1^4 x + u_2 u_1^6.$$

Let $\mathscr{C}_2$ be the normalization of the blow-up of $\mathscr{C} \times_{(\Delta^{3g-3})} \tilde{S}_2$ in the ideal (x, u_1^2). It is covered by the 2 charts:

$$\mathscr{C}_{2,a}\colon\ y'^2 = x_1'^3 + x' + u_2 \quad \text{where } y' = y/u_1^3,\ x' = x/u_1^2,$$

$$\mathscr{C}_{2,b}\colon\ x''y'' = u_1, \quad \text{where } x'' = \frac{xu_1}{y},\ y'' = \frac{y}{x},\ |x''|, |y''| < \varepsilon$$

$$\left(\text{note that } y''^2 = x\left(1 + \left(\frac{u_1^2}{x}\right)^2 + u_2\left(\frac{u_1^2}{x}\right)^3\right)\right.$$

$$x''^2 = \frac{u_1^2/x}{1 + (u_1^2/x)^2 + u_2(u^2/x)^3}$$

so that x'', y'' are integrally dependent on x, u_2, u_1^2/x when $|u_1^2/x| < \varepsilon$.)

Thus $\mathscr{C}_2$ has the same form as $\mathscr{C}_1$, except that the elliptic "tail" is now:

$$y^2 = x^3 + x + u_2.$$

It follows that the map f extends to holomorphic maps $\tilde{f}_1\colon \tilde{S}_1 \to \overline{\mathscr{M}}_g$, $\tilde{f}_2\colon \tilde{S}_2 \to \overline{\mathscr{M}}_g$ in the diagram:

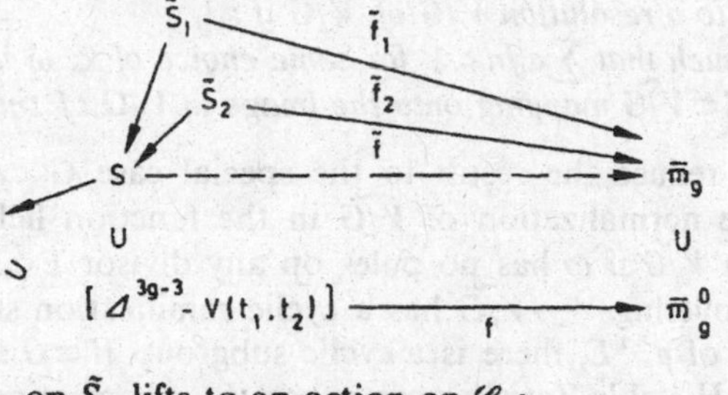

But the action of μ_6 on $\tilde{S}_1$ lifts to an action on $\mathscr{C}_1$:

$$(x', y') \mapsto (\zeta^4 x', \zeta^3 y'),$$

$$(x'', y'') \mapsto (\zeta x'', y'')$$

and the action of μ_4 on $\tilde{S}_2$ lifts to an action on $\mathscr{C}_2$:

$$(x', y') \mapsto (\zeta^2 x', \zeta y'),$$

$$(x'', y'') \mapsto (\zeta x'', y'').$$

Thus $\tilde{f}_1$, $\tilde{f}_2$ factor through

$$\tilde{f}\colon S \to \overline{\mathscr{M}}_g.$$

Examining the fibres, we see that $\tilde{f}$ is injective, which proves:

Lemma. *S is isomorphic to a neighborhood in $\overline{\mathscr{M}}_g$ of the curve parametrizing the set of stable curves:*

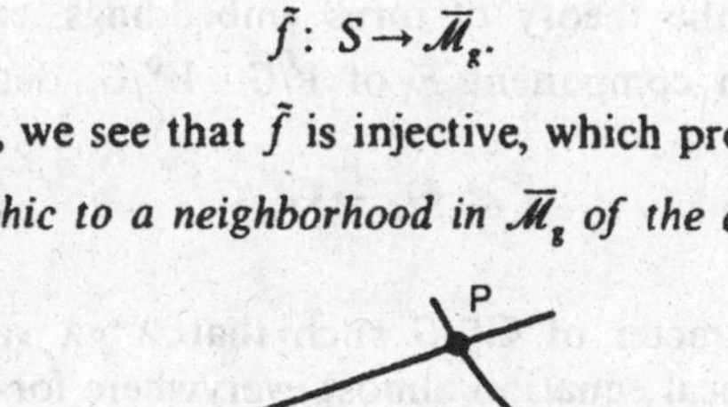

where E is an arbitrary elliptic curve or rational curve.

It now follows from this lemma that any holomorphic tensor on $\overline{\mathcal{M}}_g^0$

a) restricts to $\Delta^{3g-3} - V(t_1, t_2)$

b) extends holomorphically to Δ^{3g-3}, since $V(t_1, t_2)$ has codimension 2

c) pulls back holomorphically to a resolution of S, hence

d) extends holomorphically to a resolution of $\overline{\mathcal{M}}_g$ over all points made up of a smooth curve C_0 without automorphisms joined at one point to an elliptic curve or nodal rational curve.

Together with the Reid-Tai criterion, this completes the proof of Theorem 1.

Appendix 1 to §1

Reid-Tai's Criterion. Let V be a vector space of dimension v, $G \subset GL(V)$ a finite group. For all $g \in G$ of order n, let the eigenvalues of g be $\zeta^{a_1}, \ldots, \zeta^{a_v}$, where ζ is a primitive n^{th} root of 1 and $0 \leqq a_i < n$. Moreover, let $V^0 \subset V$ be the open set where G acts freely, and let ω be an m-canonical differential on V^0/G. Then ω extends holomorphically to a resolution $\widetilde{V/G}$ of V/G if:

for all $g \in G$ such that $\sum a_i/n < 1$ for some choice of ζ, ω is holomorphic along all divisors $E \subset \widetilde{V/G}$ mapping onto the image in V/G of the fixed point set V^g.

Proof. We first reduce the result to the special case G cyclic. In fact, let p: $\tilde{V} \to \widetilde{V/G}$ be the normalization of $\widetilde{V/G}$ in the function field $\mathbf{C}(V)$. Then ω is holomorphic on $\widetilde{V/G}$ if ω has no poles on any divisor $E \subset \widetilde{V/G}$. But over each divisor E, the covering $\tilde{V} \to \widetilde{V/G}$ has a cyclic ramification subgroup, i.e., for all components E_1 of $p^{-1}E$, there is a cyclic subgroup $H \subset G$ fixing E_1 identically such that in an H-stable Zariski-open subset $U_E \subset \tilde{V}$ meeting E_1, p factors:

$$U_E \longrightarrow U_E/H \xrightarrow[p_E]{\text{étale}} \widetilde{V/G}.$$

In particular, ω is regular on E if $p_E^*\omega$ is regular on E_1. But if the criterion holds for G, it also holds for H, hence ω is regular on U_E/H, hence ω is regular on E.

Now assume G is cyclic, $g \in G$ is a generator and that $X_1, \ldots, X_v$ are coordinates such that $g^* X_i = \zeta^{a_i} X_i$. Then G is contained in the torus $\mathbf{G}_m^v$ of all diagonal automorphisms of V, and we may assume that the resolution $\widetilde{V/G}$ is $\mathbf{G}_m^v$-equivariant. Now $\widetilde{V/G}$ is a smooth equivariant partial compactification of the torus $\mathbf{G}_m^v/G$. By the theory of torus embeddings, each divisor E of $\widetilde{V/G} - (\mathbf{G}_m^v/G)$, hence each component E of $\widetilde{V/G} - V^0/G$, determines a monomorphism:

$$\lambda\colon \mathbf{G}_m \to \mathbf{G}_m^v/G$$

such that

a) if X^α is a character of $\mathbf{G}_m^v/G$ such that $X^\alpha \circ \lambda$ vanishes to order 1 at $0 \in \overline{\mathbf{G}}_m$, then X^α is a local equation almost everywhere for E,

b) the function field $\mathbf{C}(E)$ of E is generated by the restrictions of the characters X^α such that $X^\alpha \circ \lambda \equiv 1$.

c) $\lim_{t\to 0} \lambda(t)(a) \in E$ for all $a \in \mathbf{G}_m^v/G$.

The homomorphisms λ may be described by

$$\lambda^*(X_i) = t^{\ell_i}, \quad 1 \leqq i \leqq v$$

where

$$(\ell_1, \ldots, \ell_v) \in \mathbf{Z}^v + \mathbf{Z}\left(\frac{a_1}{n}, \ldots, \frac{a_v}{n}\right),$$

and if λ is associated to $E \subset \widetilde{V/G}$, then $\ell_i \geqq 0$, $1 \leqq i \leqq v$.

Now let

$$\omega = \left(\sum_{\alpha \geqq 0} c_\alpha X^\alpha\right)(dX_1 \wedge \ldots \wedge dX_v)^{\otimes m}$$

be the given m-canonical differential. Write

$$\omega = \left(\sum_{\alpha \geqq 0} c_\alpha X^{\alpha + m\varepsilon}\right)\left(\frac{dX_1}{X_1} \wedge \ldots \wedge \frac{dX_v}{X_v}\right)^{\otimes m},$$

$$\varepsilon = (1, \ldots, 1).$$

Let $Y_i = X^{\beta_i}$, $1 \leqq i \leqq v$, be a basis of the character group of $\mathbf{G}_m^v/G$ such that $Y_1 \circ \lambda$ vanishes to order 1 at $t = 0$, $Y_i \circ \lambda \equiv 1$, $2 \leqq i \leqq v$. Then

$$\frac{dY_1}{Y_1} \wedge \ldots \wedge \frac{dY_v}{Y_v} = (\text{const.}) \frac{dX_1}{X_1} \wedge \ldots \wedge \frac{dX_v}{X_v},$$

hence ω is holomorphic along E if and only if

$$c_\alpha \neq 0 \Rightarrow X^{\alpha + m\varepsilon} = Y_1^{r_1} \ldots Y_v^{r_v} \quad \text{where } r_1 \geqq m.$$

But r_1 is just the order of vanishing of $X^{\alpha + m\varepsilon} \circ \lambda$ at $t = 0$, i.e.,

$$r_1 = \langle \alpha + m\varepsilon, (\ell_1, \ldots, \ell_v) \rangle.$$

But then as $\alpha_i \geqq 0$, $\ell_i \geqq 0$, we get

$$r_1 \geqq m\left(\sum \ell_i\right).$$

Now

$$\ell_i = k_i + k a_i/n, \quad \text{some } k, k_i \in \mathbf{Z}.$$

Note that

$$(g^k)^* X_i = \zeta^{k a_i} X_i = \zeta^{n k_i + k a_i} X_i = \zeta^{n \ell_i} X_i.$$

Let

$$n\ell_i \equiv a_i' (\text{mod } n), \quad 0 \leqq a_i' < n$$

so that

$$(g^k)^* X_i = \zeta^{a_i'} X_i.$$

If

$$\sum a_i'/n \geqq 1,$$

it follows that

$$\sum \ell_i \geqq \sum \frac{a_i'}{n} \geqq 1,$$

hence $r_1 \geqq m$ as required. We are also done if $\ell_i \geqq 1$ for any i. Now assume $\sum \frac{a_i'}{n} < 1$ and $\ell_i < 1$, all i. Then g^k is one of the "bad" elements of G, so that by the assumption, ω is holomorphic on those E's over the fixed point set of $V^{(g^k)}$. But

$$\begin{aligned} V^{(g^k)} &= \{(X_1, \dots, X_v) \mid X_i \neq 0 \Rightarrow a_i' = 0\} \\ &= \{(X_1, \dots, X_v) \mid X_i \neq 0 \Rightarrow \ell_i \in \mathbf{Z}\} \\ &= \{(X_1, \dots, X_v) \mid X_i \neq 0 \Rightarrow \ell_i = 0\}. \end{aligned}$$

By property (c) above, E maps onto $V^{(g^k)}$, so we are done for this E too. Q.E.D.

Appendix 2 to § 1

The following BASIC program which was run on the second author's Apple II looks one at a time at all cyclic covers of $\mathbf{P}^1$ branched in $0, 1, \infty$. These are all given by

$$y^n = x^a (x-1)^b.$$

We may assume, permuting the branch points if necessary, that

$$1 \leqq a \leqq \frac{n}{3}, \qquad a \mid n,$$

$$a \leqq b \leqq \frac{n-a}{2}, \qquad (a, b) = 1.$$

The program calculates the genus g which by Hurwitz' formula is:

$$g = \frac{n - (n, a) - (n, b) - (n, a+b)}{2} + 1.$$

It can be checked that $n \leqq 4g+2$. If $H^0(o_C(2K))$ is decomposed into eigenspaces V_i, $0 \leqq i < n$, under ϕ: $(x, y) \rightarrow (x, e^{2\pi i/n} y)$, then the program calculates $e(i) = \dim V_i$ and for all j, $1 \leqq j \leqq n-1$, $(j, n) = 1$, it calculates $\sum_1^{3g-3} a_i$ for the automorphism ϕ^j: this is the variable z. The max and min over all j are called za, zb, and the max and min of $\sum a_i/n$ for each g for curves studied so far are kept in the arrays min(g), max(g).

```
]LIST

 1  G0=18: REM G0 IS THE MAXIMUM GENUS WHICH WILL
    BE CONSIDERED
 10 DIM E(500): DIM MIN(100): DIM MAX(100)
 20 FOR G=2 TO G0
 30 MIN(G)=1000000
```

```
35  MAX(G)=0
40  NEXT G
100 FOR N=3 TO (4*G0)+2: REM N IS THE ORDER OF THE
    CYCLIC COVER
105 REM WE LOOK AT THE N-TH ROOT OF (X**A)*((X-1)**B)
110 FOR A=1 TO INT(N/3)
115 X=N: Y=A: GOSUB 1000: D1=X
120 IF (D1<A) THEN GOTO 510: REM WE WANT A TO DIVIDE N
130 FOR B=A TO INT((N-A)/2)
150 X=D1: Y=B: GOSUB 1000
160 IF X>1 THEN GOTO 500: REM WE WANT G.C.D. (A, B)=1
170 X=N: Y=B: GOSUB 1000: D2=X
180 X=N: Y=A+B: GOSUB 1000: D3=X
190 G=(N-D1-D2-D3)/2+1: REM G IS THE GENUS
    OF THE COVER
192 IF G<2 THEN GOTO 500
193 IF G>G0 THEN GOTO 500
200 FOR I=1 TO N-1
210 E(I)=INT((I*A-2*D1)/N)+INT((I*B-2*D2)/N)
    +INT((-I*(A+B)-2*D3)/N)+3
215 IF E(I)<0 THEN E(I)=0
230 NEXT I
245 REM E(I) IS THE DIMENSION OF THE I-TH EIGENSPACE
    IN THE QUADRATIC DIFFERENTIALS
250 FOR J=1 TO N-1: REM WE LOOK AT THE J-TH POWER
    OF THE AUTOMORPHISM
255 X=N: Y=J: GOSUB 1000
256 IF X>1 GOTO 320: REM WE WANT G.C.D. (N, J)=1
260 Z=0
270 FOR I=1 TO N-1
280 K=J*I-N*INT(J*I/N)
290 Z=Z+K*E(I): REM Z IS WHAT WE WROTE ABOVE AS
    THE SUM OF ASUBI
295 NEXT I
298 IF J>1 GOTO 300
299 ZA=Z: ZB=Z: GOTO 320
300 IF (ZA<Z) THEN ZA=Z: IF (ZB>Z) THEN ZB=Z
320 NEXT J
330 IF (ZB/N)<MIN(G) THEN MIN(G)=ZB/N
340 IF (ZA/N)>MAX(G)THEN MAX(G)=ZA/N
500 NEXT B
510 NEXT A
520 NEXT N
600 FOR G=2 TO G0
610 PRINT G; "MIN="; MIN(G); "MAX="; MAX(G)
620 NEXT G
630 STOP
```

```
1000  REM THIS SUBROUTINE STARTS WITH TWO INTEGERS X, Y
      AND CHANGES X TO THEIR G.C.D.
1005  IF Y<=X GOTO 1020
1010  Z=X: X=Y: Y=Z
1020  IF Y=0 THEN RETURN
1030  Z=X-Y*INT(X/Y)
1040  X=Y: Y=Z
1050  GOTO 1020
```

The program gave the following output; where the first number on each line is the genus, the second the min of $\sum a_i/n$, the third the max of $\sum a_i/n$:

```
 2  MIN=1.2 MAX=1.8
 3  MIN=2.35714285 MAX=3.64285715
 4  MIN=3.55555556 MAX=5.44444444
 5  MIN=4.77272727 MAX=7.22727273
 6  MIN=6 MAX=9
 7  MIN=7.23333333 MAX=10.7666667
 8  MIN=8.47058824 MAX=12.5294118
 9  MIN=9.71052632 MAX=14.2894737
10  MIN=10.952381 MAX=16.047619
11  MIN=12.1956522 MAX=17.8043478
12  MIN=13.44 MAX=19.56
13  MIN=14.6851852 MAX=21.3148148
14  MIN=15.9310345 MAX=23.0689655
15  MIN=17.1774194 MAX=24.8225806
16  MIN=18.4242424 MAX=26.5757576
17  MIN=19.6714286 MAX=28.3285714
18  MIN=20.9189189 MAX=30.0810811

Break in 630
```

The regular growth of min $\sum a_i/n$ and max $\sum a_i/n$ indicates that there is some simple proposition at work, but we have not investigated this. All we care about here is that min $\sum a_i/n \geqq 1$.

§2. The Canonical Divisor Class on $\bar{\mathcal{M}}_g$

As in the previous section, $\bar{\mathcal{M}}_g^0$ is the open set of $\bar{\mathcal{M}}_g$ parametrizing curves without automorphisms. The components of $\bar{\mathcal{M}}_g - \bar{\mathcal{M}}_g^0$ of codimension 2 or more are simply the components of the singular locus of $\bar{\mathcal{M}}_g$. But $\mathcal{M}_g - \bar{\mathcal{M}}_g^0$ has one component of codimension 1: namely, the locus Δ_1 of curves with "elliptic tails" encountered in the previous section. We shall first calculate the canonical divisor $K_{\mathcal{M}_g}$ on the open set $\bar{\mathcal{M}}_g^0$, and afterwards, indicate how to modify the calculation to give the canonical divisor $K_{\mathcal{M}_g}$ on the full open set of smooth points of $\bar{\mathcal{M}}_g$.

As before, let

$$\pi\colon \bar{\mathscr{C}}_g^0 \to \bar{\mathscr{M}}_g^0$$

be the universal family of stable, automorphism-free curves. We follow the technique in [13], pp. 99–102 and apply Grothendieck's relative Riemann-Roch theorem to the morphism π. For all coherent sheaves $\mathscr{F}$ on $\bar{\mathscr{C}}_g^0$,

$$ch(\pi_!\mathscr{F}) = \pi_*(ch(\mathscr{F})\cdot T(\Omega^1_{\mathscr{C}/\mathscr{M}})) \quad \text{in } A(\bar{\mathscr{M}}_g^0)\otimes\mathbf{Q}.$$

We wish to apply this for

$$\mathscr{F} = \Omega^1_{\mathscr{C}/\mathscr{M}} \otimes \omega_{\mathscr{C}/\mathscr{M}}.$$

In this case,

$$\pi_*\mathscr{F} \cong T^*_{\bar{\mathscr{M}}_g^0}, \qquad R^1\pi_*\mathscr{F} = (0)$$

because the cotangent space to $\bar{\mathscr{M}}_g^0$ at every point x is canonically isomorphic to $H^0(C_x, \Omega^1_{C_x}\otimes\omega_{C_x}) = H^0(\pi^{-1}(x), \mathscr{F}\otimes o_{\pi^{-1}_x})$, and because $H^1(C_x, \Omega^1_{C_x}\otimes\omega_{C_x}) = (0)$. Therefore

$$K_{\bar{\mathscr{M}}_g^0} \equiv [ch(\pi_!\mathscr{F})]_1.$$

We now follow closely the calculations of [13], pp. 99–102:

$$\begin{aligned} K_{\bar{\mathscr{M}}_g^0} &= \pi_*(ch(\mathscr{F})\cdot T(\Omega^1_{\mathscr{C}/\mathscr{M}}))_1 \\ &= \pi_*\left(\left(1 + c_1(\mathscr{F}) + \frac{c_1(\mathscr{F})^2}{2} - c_2(\mathscr{F})\right)\cdot\left(1 - \frac{c_1(\Omega^1)}{2} + \frac{c_1(\Omega^1)^2 + c_2(\Omega^1)}{12}\right)\right)_1 \\ &= \pi_*\left(\frac{c_1(\mathscr{F})^2}{2} - c_2(\mathscr{F}) - \frac{c_1(\mathscr{F})\cdot c_1(\Omega^1)}{2} + \frac{c_1(\Omega^1)^2 + c_2(\Omega^1)}{12}\right). \end{aligned}$$

Moreover

$$\Omega^1_{\mathscr{C}/\mathscr{M}} = I_{\text{sing}}\cdot\omega_{\mathscr{C}/\mathscr{M}}$$

where I_{sing} is the ideal of the singular locus, hence

$$\begin{aligned} c_1(\Omega^1_{\mathscr{C}/\mathscr{M}}) &= c_1(\omega), \\ c_1(\mathscr{F}) &= 2c_1(\omega), \\ c_2(\Omega^1_{\mathscr{C}/\mathscr{M}}) &= [\operatorname{sing}\mathscr{C}], \\ c_2(\mathscr{F}) &= [\operatorname{sing}\mathscr{C}]. \end{aligned}$$

Therefore

$$(*) \qquad \begin{aligned} K_{\bar{\mathscr{M}}_g^0} &\equiv \pi_*\left(2c_1(\omega)^2 - [\operatorname{Sing}\mathscr{C}] - c_1(\omega)^2 + \frac{c_1(\omega)^2 + [\operatorname{Sing}\mathscr{C}]}{12}\right) \\ &= \frac{13}{12}\pi_*(c_1(\omega)^2) - \frac{11}{12}\pi_*([\operatorname{Sing}\mathscr{C}]). \end{aligned}$$

On the other hand, let

$$\lambda = c_1(\pi_*\omega_{\mathscr{C}/\mathscr{M}})$$

be the so-called Hodge divisor class on $\bar{\mathscr{M}}_g^0$. Let

$$\delta = \pi_*([\operatorname{Sing}\mathscr{C}])$$

be the divisor class of singular curves $\bar{\mathcal{M}}_g^0 - \mathcal{M}_g$. Then we proved in [13] that

$$\pi_*(c_1(\omega)^2) = 12\lambda - \delta.$$

Combining this with (*) and recalling that $\operatorname{Pic}(\bar{\mathcal{M}}_g^0)$ is torsion-free ([13], p. 102) we have proven.

Theorem 2. *On the smooth variety* $\bar{\mathcal{M}}_g^0$, $K_{\mathcal{M}_g} \equiv 13\lambda - 2\delta$.

Does this continue to hold on the bigger open set $\bar{\mathcal{M}}_{g,\mathrm{reg}}$ of all smooth points? The answer depends on how λ and δ are defined on $\bar{\mathcal{M}}_{g,\mathrm{reg}}$. Recall from [13] that to deal with the problems posed by curves with automorphisms, one has 2 approaches:

a) one can introduce the group of line bundles on the *moduli functor*, i.e., for all flat proper families $\pi\colon C \to S$ of stable curves, a line bundle $L(\pi)$ on S and for all Cartesian diagrams

$$\begin{array}{ccc} C_1 & \xrightarrow{\ f\ } & C_2 \\ {\scriptstyle \pi_1}\downarrow & & \downarrow{\scriptstyle \pi_2} \\ S_1 & \xrightarrow[\ g\ ]{} & S_2 \end{array}$$

between such families, an isomorphism

$$L(\pi_1) \cong g^* L(\pi_2)$$

with obvious compatibility requirements. We call this $\operatorname{Pic}_{\mathrm{fun}}(\bar{\mathcal{M}}_g)$

b) one can introduce the locally closed subscheme H_g of a suitable Hilbert scheme parametrizing stable curves in a fixed projective space $\mathbf{P}^{\nu-1}$. With H_g, one has

$$p\colon Z_g \to H_g\text{: a universal family of curves}$$
$$(PGL(\nu) \text{ acting on } H_g).$$

Then one considers the group $\operatorname{Pic}(H_g)^{PGL(\nu)}$ of isomorphism classes of line bundles on H_g invariant under $PGL(\nu)$.

As shown in [13], $\operatorname{Pic}_{\mathrm{fun}}(\bar{\mathcal{M}}_g) \cong \operatorname{Pic}(H_g)^{PGL(\nu)}$ and $\operatorname{Pic}(\bar{\mathcal{M}}_g)$ itself is a subgroup of these of finite index. Moreover, $\operatorname{Pic}_{\mathrm{fun}}(\bar{\mathcal{M}}_g)$ is torsion free, so the convenient way to relate these groups is to think of them all as lattices in the same $\mathbf{Q}$-vector space

$$\operatorname{Pic}(\bar{\mathcal{M}}_g) \otimes \mathbf{Q} \cong \operatorname{Pic}_{\mathrm{fun}}(\bar{\mathcal{M}}_g) \otimes \mathbf{Q} \cong \operatorname{Pic}(H_g)^{PGL(\nu)} \otimes \mathbf{Q}.$$

In the big group $\operatorname{Pic}(H_g)^{PGL(\nu)}$, we *define* λ to be $c_1(p_*\omega_{Z/H})$ and δ to be the class of the divisor on H_g of singular curves. Now if L is a line bundle on H_g with $PGL(\nu)$ acting equivariantly on it, then for every curve C with automorphism ϕ of order n, ϕ is induced by $\phi' \in PGL(\nu)$ fixing the point $[C] \in H_g$ defined by C. Then ϕ' acts on the fibre $L_{[C]}$ of the line bundle by an nth root of 1. If all these roots of 1 are trivial, L descends to a line bundle on $\bar{\mathcal{M}}_g = H_g/PGL(\nu)$. For example, if C is

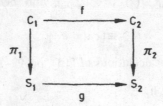

C_1 of genus $g-1$, E elliptic, let $g\colon C \to C$ be defined by $g|_{C_1} = \text{id.}$, $g|_E = -1$. For a line bundle L to descend to $\operatorname{Pic}(\bar{\mathscr{M}}_{g,\text{reg}})$, we only need that this g acts by $+1$ on $L_{[C]}$.

As an example, take the Hodge line bundle

$$L_\lambda = \Lambda^g p_*(\omega_{Z_g/H_g}).$$

g acts on $\Lambda^g H^0(\omega_C)$ by -1, hence L_λ and the Hodge divisor class $\lambda = c_1(\Lambda^g p_* \omega_{Z/H})$ do *not* descend to $\bar{\mathscr{M}}_{g,\text{reg}}$, but the square of the bundle, or twice the divisor class do. Thus the divisor class λ lies in

$$\tfrac{1}{2}\operatorname{Pic}(\bar{\mathscr{M}}_{g,\text{reg}}).$$

Next, if L_{δ_0} is the line bundle on H_g defined by the divisor of irreducible singular curves, and L_{δ_i}, $1 \leqq i \leqq \left[\frac{g}{2}\right]$ are the line bundles on H_g defined by the divisors of curves with double points separating them into pieces of genus i, $g-i$, then the line bundle L_δ of all singular curves breaks up

$$L_\delta = \bigotimes_{i=0}^{[\frac{g}{2}]} L_{\delta_i}.$$

The fibre of L_{δ_i} over a point $[C]$ is

$$\bigotimes_P \Lambda^2(m_p/m_p^2)^*$$

where P runs over all double points of C of type i (i.e., non-separating double points for $i=0$, separating with pieces of genus i, $g-i$ for $i \geqq 1$). One checks that $L_{\delta_0}, L_{\delta_2}, \ldots, L_{\delta_{[g/2]}}$ all descend to $\bar{\mathscr{M}}_{g,\text{reg}}$ and are, in fact, the line bundles defined by the divisors $\Delta_0, \Delta_2, \ldots, \Delta_{[\frac{g}{2}]} \subset \bar{\mathscr{M}}_{g,\text{reg}}$ of singular curves of various types. But the automorphism g of the curve C with elliptic tail acts by (-1) on the fibre of L_{δ_1} over $[C]$: hence $L_{\delta_1}^{\otimes 2}$ descends to $\bar{\mathscr{M}}_{g,\text{reg}}$ and is the line bundle defined by the divisor $\Delta_1 \subset \bar{\mathscr{M}}_{g,\text{reg}}$ of curves with elliptic tails. Thus in terms of divisor classes on $\bar{\mathscr{M}}_{g,\text{reg}}$:

$$\delta_1 = \tfrac{1}{2}\Delta_1; \quad \delta_i = \Delta_i, \quad i \neq 1$$

$$\delta = \Delta_0 + \tfrac{1}{2}\Delta_1 + (\Delta_2 + \ldots + \Delta_{[\frac{g}{2}]}).$$

Note incidentally that $L_\lambda \otimes L_{\delta_1}$ also descends to $\bar{\mathscr{M}}_{g,\text{reg}}$, hence $\lambda + \frac{1}{2}\Delta_1 \in \operatorname{Pic}(\bar{\mathscr{M}}_{g,\text{reg}})$. On the other hand, the equality of divisor classes on H_g:

$$c_1(p_*(\Omega^1_{Z/H} \otimes \omega_{Z/H})) \equiv 13\lambda - 2\delta$$

is proven by exactly the same proof used to prove the same equality in $\text{Pic}(\bar{\mathscr{M}}_g^0)$. The last step is the claim:

Lemma. $K_{(\mathscr{M}_g, \text{reg})} \equiv c_1(p_*(\Omega^1_{Z/H} \otimes \omega_{Z/H})) - \delta_1$.

Proof. If $f: Z_g \to \bar{\mathscr{M}}_g$ is the canonical map, then on the open set $f^{-1}(\bar{\mathscr{M}}_g^0)$, there is a canonical isomorphism:

$$\alpha: f^*(\Omega^{3g-3}_{\bar{\mathscr{M}}_g}|_{\mathscr{M}_g^0}) \xrightarrow{\approx} \Lambda^{3g-3} p_*(\Omega^1_{Z/H} \otimes \omega_{Z/H})|_{f^{-1}\mathscr{M}_g^0}.$$

If $\Delta_1' \subset H_g$ is the divisor of curves with elliptic tails, *a priori* α has a zero or pole on Δ_1' of some order ℓ, and on all of $f^{-1}(\bar{\mathscr{M}}_{g,\text{reg}})$

$$f^*(\Omega^{3g-3}_{\bar{\mathscr{M}}_g}), \quad \Lambda^{3g-3} p_*(\Omega^1_{Z/H} \otimes \omega_{Z/H})(\ell \Delta_1')$$

are isomorphic. To compute ℓ, let $C = C_1 \cup E$ be a curve with elliptic tail such that C_1 has no automorphisms, $j(E) \neq 0, 12^3, \infty$. Let (Δ_t^{3g-3}) be the base space of the universal deformation of C, with coordinates $t_1, t_2, \ldots, t_{3g-3}$ where the automorphisms $g: C \to C$ acts by

$$g^* t_1 = -t_1, \quad g^* t_i = t_i, \quad 2 \leqq i \leqq 3g-3.$$

($t_1 = 0$ being the locus of singular curves). Then $\bar{\mathscr{M}}_g$ near $[C]$ is just $\Delta_t^{3g-3}/\{e, g\}$, which is the polycylinder (Δ_s^{3g-3}) with coordinates $s_1 = t_1^2$, $s_2 = t_2, \ldots, s_{3g-3} = t_{3g-3}$. Then $ds_1 \wedge \ldots \wedge ds_{3g-3}$ is a local basis of $\Omega^{3g-3}_{\bar{\mathscr{M}}_g}$, while $\Omega^{3g-3}_{\Delta_t}$ is locally the same as $\Lambda^{3g-3} p_*(\Omega^1_{Z/H} \otimes \omega_{Z/H})$: thus $dt_1 \wedge \ldots \wedge dt_{3g-3}$ is a local basis of the latter. But

$$ds_1 \wedge \ldots \wedge ds_{3g-3} = 2t_1(dt_1 \wedge \ldots \wedge dt_{3g-3})$$

hence the map from $\Omega^{3g-3}_{\bar{\mathscr{M}}_g}$ to $\Lambda^{3g-3} p_*(\Omega^1_{Z/H} \otimes \omega_{Z/H})$ has a simple zero along the locus $t_1 = 0$. This proves that

$$f^*(\Omega^{3g-3}_{\bar{\mathscr{M}}_g}|_{\mathscr{M}_{g,\text{reg}}}) \cong \Lambda^{3g-3} p_*(\Omega^1_{Z/H} \otimes \omega_{Z/H})(-\Delta_1')|_{f^{-1}(\mathscr{M}_{g,\text{reg}})}.$$

Taking chern classes, this proves the lemma.

This proves:

Theorem 2 *bis. On the smooth variety* $\bar{\mathscr{M}}_{g,\text{reg}}$,

$$K_{\mathscr{M}_g} \equiv 13\lambda - 2\delta_0 - 3\delta_1 - 2\delta_2 - \ldots - 2\delta_{[\frac{g}{2}]}$$

$$\equiv 13\lambda - 2\Delta_0 - \tfrac{3}{2}\Delta_1 - 2\Delta_2 - \ldots - 2\Delta_{[\frac{g}{2}]}.$$

§3. The Class of the Divisor D_k, I

For the rest of this article, we assume that the genus g under consideration is odd, and let

$$g = 2k - 1.$$

As in the Introduction, we introduce as the fundamental point of this proof the divisor:

$$D_k \subset \mathcal{M}_g,$$
$$D_k = \{\text{locus of curves } C \text{ which are } k\text{-fold covers of } \mathbf{P}^1\}.$$

Thus if $g=3$, $k=2$, $D_2 \subset \mathcal{M}_3$ is the hyperelliptic locus. And if $g=5$, $k=3$, $D_3 \subset \mathcal{M}_5$ is the trigonal locus. It is well known that $\dim D_k = 3g-4$, hence D_k is a divisor. (This can be checked by the usual dimension count, considering the number of branch points for a generic covering $\pi: C \to \mathbf{P}^1$.) The purpose of this section is to prove*:

Theorem 3. *For all $g=2k-1$, there is a rational number a_k such that on $\mathcal{M}_g^0$:*

$$[D_k] \equiv a_k \lambda.$$

Corollary. *If $\bar{D}_k \subset \overline{\mathcal{M}}_g$ is the closure of D_k, then there are also integers $n_{k,\ell}$, $0 \leqq l \leqq [g/2]$ such that on $\overline{\mathcal{M}}_{g,\mathrm{reg}}$*

$$[\bar{D}_k] \equiv a_k(\lambda + \tfrac{1}{2}[\Delta_1]) + \sum_{\ell=0}^{[\frac{g}{2}]} n_{k,\ell}[\Delta_\ell].$$

The Corollary follows because $[\bar{D}_k] - a_k(\lambda + \frac{1}{2}[\Delta_1])$ is an integral divisor class on $\overline{\mathcal{M}}_{g,\mathrm{reg}}$, trivial on the open set $\mathcal{M}_{g,\mathrm{reg}}$, hence is an integral combination of the components Δ_ℓ of $\overline{\mathcal{M}}_g - \mathcal{M}_g$.

To prove the theorem, we shall apply Porteous' formula ([2] or [9]) and the Riemann-Roch theorem again. As above, let

$$\pi: \mathcal{C}_g^0 \to \mathcal{M}_g^0$$

be the universal family of curves. Let

$$\mathcal{C}_g^{0,n} = \mathcal{C}_g^0 \times_{\mathcal{M}_g^0} \cdots \times_{\mathcal{M}_g^0} \mathcal{C}_g^0 \quad (n \text{ factors}).$$

Let

$p_i: \mathcal{C}_g^{0,n} \to \mathcal{C}_g^0$ be the ith projection,

$\pi_n: \mathcal{C}_g^{0,n} \to \mathcal{M}_g^0$ be the canonical map.

Let

$$\Delta_{ij} \subset \mathcal{C}_g^{0,n}$$

be the $(i,j)^{\text{th}}$ diagonal, and let

$$K_i \in \operatorname{Pic}(\mathcal{C}_g^{0,n})$$

stand for the divisor class

$$p_i^*(c_1(\Omega^1_{\mathcal{C}_g^0/\mathcal{M}_g^0})).$$

We are interested in the Zariski-closed subset $Z \subset \mathcal{C}_g^{0,k}$ defined by:

$$Z = \left\{(P_1, \ldots, P_k) \in \mathcal{C}_g^{0,k} \,\middle|\, h^0\left(o_C\left(\sum_1^k P_i\right)\right) \geqq 2\right\}.$$

---

* See note at end of article.

D_k is, by definition, $\pi_k(Z) \subset \mathcal{M}_g$. For simplicity in what follows, we drop the g and 0 in $\mathscr{C}_g^{0,k}$. To compute Z, consider

$$R^0 p_* o_{\mathscr{C}^{k+1}}\left(\sum_{j=1}^{k} \Delta_{j,k+1}\right)$$

and

$$R^1 p_* o_{\mathscr{C}^{k+1}}\left(\sum_{j=1}^{k} \Delta_{j,k+1}\right)$$

where

$$p: \mathscr{C}^{k+1} \to \mathscr{C}^k$$

is the projection onto the 1st k factors. Since $\mathscr{C}^k$ is an integral scheme and for generic $P_1, \ldots, P_k$, $H^0\left(o_C\left(\sum_1^k P_i\right)\right)$ consists only in constants, the 0th direct image above is just $o_{\mathscr{C}^k}$.

Consider the exact sequence on $\mathscr{C}_g^{0,k+1}$:

$$0 \to o_{\mathscr{C}^{k+1}} \to o_{\mathscr{C}^{k+1}}(\textstyle\sum \Delta_{j,k+1}) \to o_{\mathscr{C}^{k+1}}(\sum \Delta_{j,k+1})/o_{\mathscr{C}^{k+1}} \to 0.$$

Taking higher direct images, we find:

$$0 \to \underbrace{p_* o_{\mathscr{C}^{k+1}}(\textstyle\sum \Delta_{j,k+1})/o_{\mathscr{C}^{k+1}}}_{\text{loc. free } rk\, k} \xrightarrow{\alpha} \underbrace{R^1 p_* o_{\mathscr{C}^{k+1}}}_{\text{loc. free } rk\, g} \to R^1 p_* o_{\mathscr{C}^{k+1}}(\sum \Delta_{j,k+1}) \to 0$$

Since the 2 locally free sheaves, after $\otimes \mathbf{k}(z)$, $z \in \mathscr{C}^k$, give

$$H^0(o_C(\textstyle\sum P_j)/o_C) \xrightarrow{\alpha(z)} H^1(o_C)$$

with kernel $H^0(o_C(\sum P_j))/\mathbf{C}$, it follows that

$$Z = \{z \in \mathscr{C}^k \mid rk\, \alpha(z) \leq k-1\}.$$

By Porteous' formula, this implies

$$[Z] = c_{g-k+1}(R^1 p_* o_{\mathscr{C}^{k+1}}(\textstyle\sum \Delta_{j,k+1})).$$

Here $[Z]$ will be the class of Z counted with some multiplicity: since the result we seek is just that $[D_k] \in \mathbf{Q} \cdot \lambda$, this does not matter**. Since $p_* o_{\mathscr{C}^{k+1}}(\sum \Delta_{j,k+1}) = o_{\mathscr{C}^k}$,

$$[Z] = c_{g-k+1}(-p_! o_{\mathscr{C}^{k+1}}(\textstyle\sum \Delta_{j,k+1})).$$

By Grothendieck's Riemann-Roch:

$$\begin{aligned}[Z] &= \text{polyn. in } ch_\ell(p_! o_{\mathscr{C}^{k+1}}(\textstyle\sum \Delta_{j,k+1}))\\ &= \text{polyn. in } p_*(ch(o_{\mathscr{C}^{k+1}}(\textstyle\sum \Delta_{j,k+1})) \cdot Td(\Omega^1_{\mathscr{C}^{k+1}/\mathscr{C}^k}))_\ell\\ &= \text{polyn. in classes of the form}\\ &\qquad p_*\left(\text{polyn. in } \sum_{j=1}^{k} [\Delta_{j,k+1}] \text{ and } K_{k+1}\right).\end{aligned}$$

** In fact, it is not hard to see that the multiplicity is one

To see what this can be, note the easy identities:

$$[\Delta_{j_1,k+1}]\cdot[\Delta_{j_2,k+1}]=[\Delta_{j_1,k+1}]\cdot p^*[\Delta_{j_1,j_2}]$$
$$[\Delta_{j,k+1}]^2=-[\Delta_{j,k+1}]\cdot p^*(K_j)$$
$$[\Delta_{j,k+1}]\cdot K_{k+1}=[\Delta_{j,k+1}]\cdot p^*(K_j)$$
$$p_*([\Delta_{j,k+1}]\cdot p^*(a))=a, \quad \text{any } a\in A(\mathscr{C}^{k+1})$$
$$p_*(K_{k+1}^\ell)=\pi_k^*(\pi_{1*}(K_{\mathscr{C}/\mathscr{M}}^\ell)).$$

The second of these is because the self-intersection of $\Delta_{j,k+1}$ is c_1 of its normal bundle, and the normal bundle is the restriction to $\Delta_{j,k+1}$ of $p_j^*\Omega^1_{\mathscr{C}/\mathscr{M}}$. The last comes from the Cartesian diagram

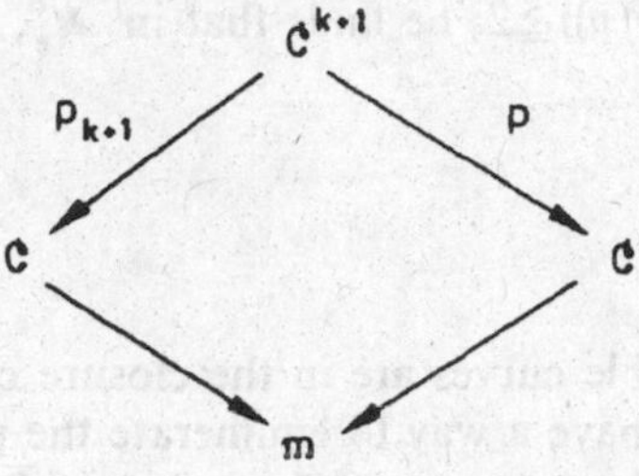

It follows that

$$[Z]=\text{polyn. in } [\Delta_{j_1,j_2}]\text{'s}, [K_j]\text{'s}, \pi_k^*(\pi_{1,*}(K_{\mathscr{C}/\mathscr{M}}^\ell)).$$

Now $[D_k]$ is *not* $\pi_{k,*}([Z])$. Indeed $\pi_{k,*}([Z])=0$ because if a curve has one $\sum P_i$ moving in a pencil, it of course has ∞^1 such cycles. To chop $[Z]$ down in dimension, ask that the first point P_1 in the cycle $\sum P_i$ be a member of a fixed canonical divisor. This gives us

$$(2g-2)(k-1)!\,[D_k]=\pi_{k,*}([Z]\cdot K_1).$$

Therefore

$$[D_k]=\text{polyn. in } \pi_{k,*}(\text{polyn. in } [\Delta_{j_1,j_2}]\text{'s}, [K_j]\text{'s}, \pi_k^*(\pi_{1,*}(K^\ell))).$$

Now factor π_k:

$$\mathscr{C}^k\to\mathscr{C}^{k-1}\to\mathscr{C}^{k-2}\to\ldots\to\mathscr{C}\to\mathscr{M},$$

and take images of the above polynomial one at a time. Using the previous identities, it is clear that under each projection

$$p_\ell\colon \mathscr{C}^\ell\to\mathscr{C}^{\ell-1}$$

$p_{\ell,*}$ carries any polyn. in $[\Delta_{j_1,j_2}]$'s, $[K_j]$'s and $\pi_\ell^*(\pi_{1,*}(K^\ell))$ into a polynomial of the same type. Finally, projecting to $\mathscr{M}$ itself we deduce

$$[D_k]=\text{polyn. in } \pi_{1,*}(K^\ell)\text{'s}.$$

But $[D_k]$ is a divisor, so this just means

$$[D_k]=\text{multiple of } \pi_{1,*}(K^2).$$

By [13], however,

$$\pi_{1,*}(K^2)=12\lambda. \quad \text{Q.E.D.}$$

It should be pointed out that this argument establishes more generally that for any integers $k, a_1, \ldots, a_n \geqq 2g^{n\geqq 1}$ and $g=2k+n-\Sigma a_i$, the divisor $\bar{D}_{a,k}$ in $\mathcal{M}_g$ defined as the closure of

$$D_{a,k}=\left\{C\in\mathcal{M}_g \,\middle|\, \begin{matrix}\exists \text{ divisor } D=\Sigma a_i p_i+D_0\in C_k \\ h^0(O_C(D))\geqq 2\end{matrix}\right\}$$

is similarly linearly equivalent to a linear combination of λ and the $[\Delta_i]$. Indeed, Diaz has used this set-up to explicitly calculate the coefficient of λ in the expression for the class of the divisor $D_{g-1,g-1}$ of curves with a Weierstrass point p with $h^0(C, O_C((g-1)p))\geqq 2$; he finds that in $\mathcal{M}_g^0$, this coefficient equals $g^2(3g-1)(g-1)/2$.***

§ 4. Parametrization of $\bar{D}_k$

It is not obvious which stable curves are in the closure of D_k, especially which *reducible* stable curves. To have a way to enumerate the points of $\bar{D}_k$ as well as to determine the tangent plane to smooth branches of $\bar{D}_k$, we introduce a new moduli space, which will be a compactification of what is usually called the Hurwitz scheme. Recall that the Hurwitz scheme in its simplest form parametrizes the family of k-sheeted coverings of $\mathbf{P}^1$ with b ordinary branch points:

$$H_{k,b}=\left\{\begin{array}{l}\text{moduli space of the data}\\ \qquad \pi\colon C\to\mathbf{P}^1 \text{ of degree } k\\ \qquad P_1,\ldots,P_b\in\mathbf{P}^1 \text{ distinct}\\ C \text{ smooth curve, } \pi \text{ with one ordinary branch}\\ \text{point over each } P_i\text{, otherwise unbranched}\end{array}\right\}.$$

(It is usual to treat the $\{P_i\}$ as a cycle $\sum P_i$, but for our purposes we wish to order them.) By the usual theory (see [6]), $H_{k,b}$ is itself a finite étale cover of the space of sequences $\{P_i\}$:

$$H_{k,b}\to[(\mathbf{P}^1)^b-\bigcup_{i<j}\Delta_{ij}]/PGL(2)\cong(\mathbf{P}^1)^{b-3}-S$$

where $S=(\bigcup_{i<j}\Delta_{ij})\cup\bigcup_i p_i^{-1}(\{0,1,\infty\})$ (normalizing $P_{b-2}=0$, $P_{b-1}=1$, $P_b=\infty$). By Hurwitz's formula, the genus g of C is given by

$$2g-2=-2k+b$$

and we have a diagram

*** See note at end of article

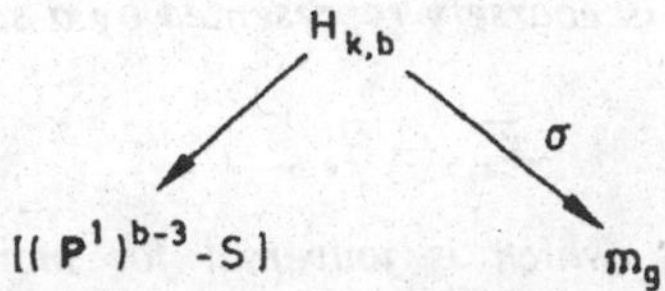

In particular, if $g=2k-1$, $D_k=\sigma(H_{k,b})$. We want to compactify $H_{k,b}$ in such a way that σ extends to a morphism

$$\sigma\colon \overline{H}_{k,b}\to\overline{\mathcal{M}}_g.$$

We do this by means of the theory of F. Knudsen [10] of b-pointed curves with fine structure. Knudsen has introduced a smooth projective compactification

$$P_b\supset[(\mathbf{P}^1)^{b-3}-S]$$

which as a moduli space can be described as follows:

i) a stable b-pointed curve is a reduced, connected curve C with at most ordinary double points, plus b smooth distinct points $P_1,\ldots,P_b\in C$ such that every smooth rational component E of C contains at least 3 points which are either P_i's or double points of C

ii) $P_b=\left\{\begin{matrix}\text{set of } b\text{-pointed stable curves } C \text{ with } p_a(C)=0\\ \text{up to isomorphism}\end{matrix}\right\}$

iii) the open dense set $(\mathbf{P}^1)^{b-3}-S$ is the set of $(C,P_1,\ldots,P_b)$ in P_b where C is irreducible.

In fact, Knudsen describes P_b as an explicit blow-up of $(\mathbf{P}^1)^{b-3}$ along an ideal sheaf with support S.

We next describe a functor that will be coarsely represented by the sought-for scheme $\overline{H}_{k,b}$:

Definition. $\overline{\mathcal{H}}_{k,b}$ is the functor which associates to a scheme S the set of isomorphism classes of the following data:

i) a stable b-pointed curve $(D;P_1,\ldots,P_b)$ of genus 0 over S,

ii) an admissible covering $\pi\colon C\to D$.

By an *admissible covering*, we mean that C/S itself is a proper flat family of reduced connected curves with at most ordinary double points, that π is étale except at unique smooth points $Q_i\colon S\to C$, one over each $P_i\colon S\to D$, where it has ordinary branching (i.e., $\pi\colon C\to D$ is analytically just $u=x^2$, x coordinate on C over S, u coordinate on D over S), and except over the double points of D/S. For each $s\in S$ and each point x of C_s over a double point y of a fibre D_s, C_s has an ordinary double point and locally C, D and π are described by:

$$\begin{aligned}&C\colon\ xy=a,\quad a\in\hat{o}_s,\ x,y \text{ generate } \hat{m}_{x,C}\\ &D\colon\ uv=a^p,\qquad u,v \text{ generate } \hat{m}_{y,D}\\ &\pi\colon\ u=x^p,\ v=y^p\end{aligned}$$

for some p. (This definition generalizes Beauville's admissible double coverings, used to compactify the space of double coverings of a curve of genus g.)

Theorem 4. *The functor $\overline{\mathscr{H}}_{k,b}$ is coarsely represented by a scheme $\overline{H}_{k,b}$ finite over P_b, i.e., there is a morphism*

$$\overline{\mathscr{H}}_{k,b} \to \overline{H}_{k,b}$$

bijective on **C**-*valued points, which is universal for morphisms from $\overline{\mathscr{H}}_{k,b}$ to schemes. Moreover, $\overline{H}_{k,b}$ represents $\overline{\mathscr{H}}_{k,b}$ on the open set of coverings $\pi: C \to D$ such that C has no automorphism $\alpha: C \to C$ with $\pi \circ \alpha = \pi$ except the identity.*

Proof. The functor $\overline{\mathscr{H}}_{k,b}$ and hence the scheme $\overline{H}_{k,b}$, if it exists, both lie over P_b, so the problem is local over P_b. Therefore we may cover P_b by suitable open sets and make the construction separately over each. Take a point $[D_0] \in P_b$. It may happen that every component of D_0 has at least one of the b-points P_i on it. But if not, choose further points $P_{b+1}, \dots, P_c \in D_0$ so we have one P_i in each component and in some neighborhood $U \subset P_b$ of $[D_0]$, choose smooth disjoint sections of the universal curve $\mathscr{D} \to P_b$ through these points. (This is possible because there are birational morphisms $\mathscr{D} \to \mathbf{P}^1 \times P_b$ which over $[D_0]$ take any one of the components of $\mathscr{D}$ isomorphically to $\mathbf{P}^1$, collapsing the rest to points; and $\mathbf{P}^1 \times P_b$ has a section through any point of $\mathbf{P}^1 \times [D_0]$ - see Knudsen [10].) Choose U small enough so that for all $[D] \in U$, the sections $P_1, \dots, P_c$ meet every component of D. Moreover, choose U's well enough so we may find a local coordinate t_i on the fibres of $\mathscr{D}$ over U, $t_i = 0$ on the section $P_i: U \to \mathscr{D}$ to 1st order. Then for all $[D] \in U$, if $\pi: C \to D$ is any admissible cover, $o_C\left(\sum_1^c \pi^{-1} P_i\right)$ is ample on C. For some n, $o_C\left(n \sum_1^c \pi^{-1} P_i\right)$ is very ample for all $\pi: C \to D$.

Next define a "rigidified" version of $\overline{\mathscr{H}}_{k,b}$ over U:

$\overline{\mathscr{H}}^U_{k,b}$ = functor of families of admissible coverings $\pi: C \to D$, $[D] \in U$ *plus* orderings of $\pi^{-1}(P_i)$:

$P_{1,1}, \dots, P_{1,k-1} \in C$, the points over $P_1 \in D$, $P_{1,1}$ ramified,
... ...
$P_{b,1}, \dots, P_{b,k-1} \in C$, the points over $P_b \in D$, $P_{b,1}$ ramified,
$P_{b+1,1}, \dots, P_{b+1,k} \in C$, the points over $P_{b+1} \in D$
... ...
$P_{c,1}, \dots, P_{c,k} \in C$, the points over $P_c \in D$

plus a choice of square root

$$\sqrt{t_i} \in \hat{o}_{P_{i,1}, C}, \qquad 1 \leq i \leq b.$$

Note that by changing these choices in the obvious way, the finite group

$$G = (\mathbf{Z}/2\mathbf{Z} \times \Sigma_{k-2})^b \times (\Sigma_k)^{c-b}$$

acts on $\overline{\mathscr{H}}^U_{k,b}$, where Σ_ℓ = permutations of ℓ letters. Via this action $\overline{\mathscr{H}}_{k,b}(\operatorname{Spec} \mathbf{C}) \supset \overline{\mathscr{H}}^U_{k,b}(\operatorname{Spec} \mathbf{C})/G$. We shall show that $\overline{\mathscr{H}}^U_{k,b}$ is representable. To do this, note that the projection

$$\pi: C \to D$$

and the choice of $t_i \in m_{P_i,D}$, $t \notin m^2_{P_i,D}$, defines isomorphisms:

$$\hat{o}_{P_{i,1},C} \xrightarrow{\approx} \hat{o}_{P_i,D}[\sqrt{t_i}] \xrightarrow{\approx} \mathbf{C}[[\sqrt{t_i}]] \quad \text{if } 1 \leqq i \leqq b$$

or

$$\hat{o}_{P_{i,j},C} \xrightarrow{\approx} \hat{o}_{P_i,D} \xrightarrow{\approx} \mathbf{C}[[t_i]] \quad \text{if } j>1 \text{ or } i>b.$$

Therefore, if N is large enough, we get an injection:

$$H^0\left(C, o_C\left(n\sum_1^c \pi^{-1}P_i\right)\right) \longrightarrow \mathbf{C}^{k(N+1)c}$$

$$f \longmapsto (\ldots, a_{ijk}, \ldots)$$

if, near P_{ij}, f is expanded:

$$f = \sum_{k=0}^{N} a_{ijk} t_i^{-n+k} + \ldots$$

or

$$f = \sum_{k=0}^{2N+1} a_{i1k} t_i^{-n+k/2} + \ldots \quad \text{if } j=1,\ i \leqq b.$$

In other words, for each admissible π: $C \to D$, we get canonically both a subspace $V \subset \mathbf{C}^{k(N+1)c}$ and an embedding

$$C \subset \mathbf{P}(V).$$

The dimension of V is given by the Riemann-Roch theorem on C as $nkc-g+1$. Reversing this process, let G be the Grassmannian of $(nkc-g+1)$-dimensional subspaces of $\mathbf{C}^{k(N+1)c}$, let $\mathscr{V} \to G$ be the universal vector bundle, and let H be the Hilbert scheme of $\mathbf{P}(\mathscr{V})$ over G of curves in $\mathbf{P}(\mathscr{V})$ with Hilbert polynomial $nkcX-g+1$. By the given ordering of the coordinates in $\mathbf{C}^{k(N+1)c}$, we get maps:

$$\lambda_{ij}\colon V \subset \mathbf{C}^{k(N+1)c} \xrightarrow[\substack{\text{via}\\ \text{coord}\\ a_{ijk}}]{} \left\{\begin{matrix}\text{V. Sp. of series}\\ \sum_{k=0}^{N} a_k t_i^{-n+k}\end{matrix}\right\} \xrightarrow[\substack{\text{mult. by}\\ t_i^n}]{} \mathbf{C}[[t_i]]/(t_i^{N+1})$$

hence over a suitable open subset of G, we get canonical embeddings

$$\phi_{ij}\colon \operatorname{Spec} \mathbf{C}[[t_i]]/(t_i^{N+1}) \hookrightarrow \mathbf{P}(V).$$

(Replace t_i by $\sqrt{t_i}$ if $j=1$, $i \leqq b$.) Let $H_1 \subset H$ be the locally closed subscheme of points z where ϕ_{ij} exist and $\operatorname{Im}\phi_{ij}$ is a subscheme of the curve C_z in $\mathbf{P}(\mathscr{V})$ defined by z. Let $H_2 \subset H_1$ be the locally closed subset of connected reduced curves C_z with at most ordinary double points with $\operatorname{Im}\phi_{ij}$ being disjoint smooth points $P_{i,j}$ of C_z, with C_z embedded by a complete linear system and with $o_{C_z}(1) \cong o_{C_z}\left(2n\sum_1^b P_{i1} + n\sum_{\text{rest}} P_{ij}\right)$. Let $H_3 \to H_2 \times U$ be the Hilbert scheme representing morphisms π: $C_z \to D$, $[D] \in U$, and let $H_4 \subset H_3$ be the locally closed set where π is finite of degree k over each component and

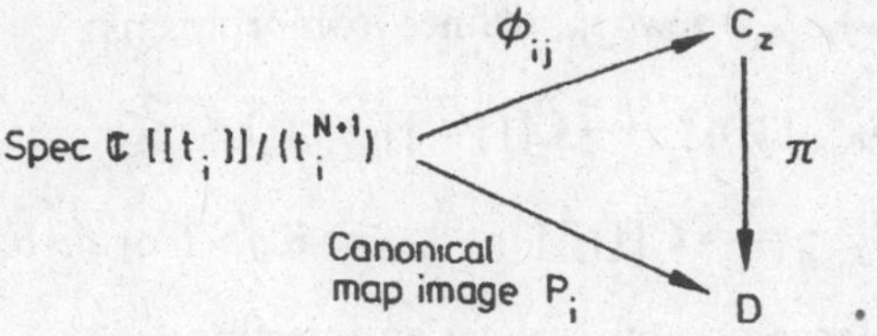

commutes (modify this for $\phi_{i1}, \ldots, \phi_{ib}$ in the obvious way). Then H_4 will represent $\overline{\mathscr{H}}_{k,b}^{U}$!

It follows that H_4/G coarsely represents the open subfunctor of $\overline{\mathscr{H}}_{k,b}$ of π: $C \to D$ with $[D] \in U$, and represents it where G acts freely. But a fixed point of $g \in G$ means a covering π: $C \to D$ such that there is an isomorphism:

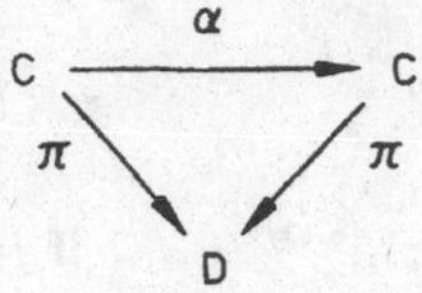

which permutes the finite sets $\pi^{-1}(Q_i)$ and/or acts by $\sqrt{t_i} \to -\sqrt{t_i}$ at the ramified points $P_{i,1}$. Glueing together the schemes H_4/G over various open sets $U \subset P_b$, we construct $\overline{H}_{k,b}$.

It is clear that the morphism $\overline{H}_{k,b} \to P_b$ is finite to one. To see that it is finite, we check that the functor $\overline{\mathscr{H}}_{k,b}$ satisfies the weak valuative criterion for properness, hence $\overline{H}_{k,b}$ is proper over $\mathbf{C}$. This means that given

$$\pi\colon C \to D$$

over the field $\mathbf{C}((t))$, then taking a suitable root $t^{1/n}$ of t, π extends to a family of admissible coverings:

$$\pi_n\colon \mathscr{C}_n \to \mathscr{D}_n$$

over the ring $\mathbf{C}[[t^{1/n}]]$. In fact, by Knudsen's results, P_b is complete, hence $D \subset \mathscr{D}_1$, $\mathscr{D}_1$ a b-pointed stable curve over $\mathbf{C}[[t]]$. Let $\mathscr{C}_1$ be the partial normalization of $\mathscr{D}_1$ in the fraction field of C defined by

$$o_{\mathscr{C}_1} = (\text{functions integral over } o_{\mathscr{D}}, \text{ generically in } o_C).$$

$\mathscr{C}_1$ may be ramified over one of the components of the fibre $(\mathscr{D}_1)_0$ over $t=0$. But replacing t by $t^{1/n}$, by Abhyankar's lemma, this no longer happens. This gives us $\mathscr{C}_n \to \mathscr{D}_n$ which, by the purity of the branch locus, is ramified only over the sections P_i: Spec $k[[t]] \to \mathscr{D}_n$ and the singular points of the fibres of $\mathscr{D}_n$. I claim that $\mathscr{C}_n$ must be an admissible covering. There are 2 points to check: what happens over an ordinary double point of $(\mathscr{D}_n)_0$ which lifts to the generic fibre $(\mathscr{D}_n)_\eta$, and what happens over those that don't. In the first case, $\mathscr{D}_n$ has the local equation

$$u \cdot v = 0$$

and $\mathscr{C}_n$ is a covering of the smooth branch $u=0$ plus a covering of the smooth branch $v=0$. So if $(\mathscr{C}_n)_\eta$ has p-fold branching on the curve $u=v=0$ on the surface

$u=0$ and on the curve $u=v=0$ on the surface $v=0$, $\mathscr{C}_n$ must do the same. In the second case, $\mathscr{D}_n$ has the local equation

$$u \cdot v = s^m \qquad (s = t^{1/n})$$

for some m, i.e., $\mathscr{D}_n$ has a singularity of type A_{m-1}. The universal cover of $\mathscr{D}_n$, ramified only at the origin, is then

$$x' \cdot y' = s, \qquad x' = u^{1/m}, \qquad y' = v^{1/m}.$$

Thus $\mathscr{C}_n$ must be given by

$$x \cdot y = x^k, \qquad x^\ell = u, \qquad y^\ell = v$$

for some factorization $k\ell = m$, and this is admissible. Q.E.D.

The scheme $\overline{H}_{k,b}$ maps on the one hand to P_b by considering only D in the cover $\pi\colon C \to D$, and on the other hand to $\overline{\mathscr{M}}_g$ by considering only C:

$$\begin{array}{ccc} & \overline{H}_{k,b} & \\ \swarrow & & \searrow \\ P_b & & \overline{m}_g \end{array}$$

(C is not necessarily stable, but one may contract unnecessary smooth rational components with only 1 or 2 double points, obtaining a stable curve - see Knudsen [10].) If $g = 2k-1$, it is clear that the divisor $\overline{D}_k \subset \overline{\mathscr{M}}_g$ is the image of the morphism $\overline{H}_{k,b} \to \overline{\mathscr{M}}_g$.

We want to discuss the local structure of $\overline{H}_{k,b}$. More precisely, what family pro-represents the infinitesimal deformations of a particular admissible covering $\pi\colon C \to D$? Let $\{Q_i\}_{1 \le i \le d}$ be the double points of D and let $\pi^{-1}(Q_i) = \{Q_{i,1}, \ldots, Q_{i,r(i)}\}$. Say the covering π is ramified with order $p(i,j)$ at $Q_{i,j}$. Now let

$$\mathscr{D} \longrightarrow \operatorname{Spec} \mathbb{C}[[t_1, \ldots, t_{b-3}]], \qquad p_i \text{ (sections)}$$

be the universal deformation of the b-pointed curve D. If u_i, v_i are coordinates at $Q_i \in D$ so that D is given by $u_i v_i = 0$, then if the t_i are suitably chosen, we may assume that $\mathscr{D}$ is given locally near Q_i by the equation

$$u_i v_i = t_i.$$

In order to lift the covering $\pi\colon C \to D$ to a covering $\pi\colon \mathscr{C} \to \mathscr{D}$, Grothendieck's theory ([8] or [14]) tells us that we must merely give $\mathscr{C}$ locally near all non-étale points of $\pi\colon C \to D$. Over each of the b sections P_i of $\mathscr{D}/\mathbb{C}[[t_1, \ldots, t_{b-3}]]$, let s_i be a function on $\mathscr{D}$ vanishing to 1st order on the image of P_i. Then near

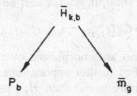

P_i, the covering $\pi: \mathscr{C} \to \mathscr{D}$ is given by adjoining $\sqrt{s_i}$. At each Q_{ij}, however, we must define $\mathscr{C}$ by

$$x_{ij} y_{ij} = t_{ij}$$

and π by

$$u_i = x_{ij}^{p(i,j)}, \quad v_i = y_{ij}^{p(i,j)}.$$

In order to do this, t_i must equal $t_{ij}^{p(i,j)}$. The universal way to do this is to take our parameter space to be:

$$\operatorname{Spec} \mathcal{O}_{[\pi: C \to D]},$$

where

$$\mathcal{O}_{[\pi: C \to D]} = \mathbf{C}[[t_1, \ldots, t_{b-3}, t_{1,1}, \ldots, t_{i,j}, \ldots, t_{d,r(d)}]]/(t_{ij}^{p(i,j)} - t_i, \text{ all } i,j).$$

Over this base, we have now defined a covering $\pi: \mathscr{C} \to \mathscr{D}$. The complete local rings of the scheme $\bar{H}_{k,b}$ are, as usual, the rings of invariants of $\mathcal{O}_{[\pi: C \to D]}$ under the finite group of automorphisms $\alpha: C \to C$ such that $\pi \circ \alpha = \pi$.

Unfortunately, some of the rings $\mathcal{O}_{[\pi: C \to D]}$ are rather messy, esp. not integrally closed. Suppose, however, that for all double points $Q_i \in D$, there is *at most one* $Q_{ij} \in C$ over Q_i such that $p(i,j) > 1$. Let it be the point $Q_{i,1}$. Then

$$\mathcal{O}_{[\pi: C \to D]} = \mathbf{C}[[t_{1,1}, \ldots, t_{d,1}, t_{d+1}, \ldots, t_{b-3}]]$$

and if, furthermore, C has no automorphisms over D, $\bar{H}_{k,b}$ is smooth at the point defined by $\pi: C \to D$. For further applications it seems that the normalization of $\bar{H}_{k,b}$ is probably more useful than $\bar{H}_{k,b}$ itself. Note that the integral closure of $\mathcal{O}_{[\pi: C \to D]}$ is a semi-local ring whose local rings are all regular.

As an application of the surjective map

$$\bar{H}_{k,b} \xrightarrow{\lambda} \bar{D}_k \subset \bar{\mathscr{M}}_g, \quad \begin{array}{l} g = 2k-1 \\ b = 2g + 2k - 2 \end{array}$$

we can describe set-theoretically at least part of $\bar{D}_k$ fairly easily. The first part of the following theorem can be proven easily without use of $\bar{H}_{k,b}$, but the second is harder:

Theorem 5. *Let C be a stable curve of genus $g = 2k-1$, $[C] \in \bar{\mathscr{M}}_g$ the corresponding point. Then*

a) *if C is irreducible, $[C] \in \bar{D}_k$ if and only if there exists a torsion-free rank 1 sheaf $\mathscr{F}$ on C such that*

$$h^0(\mathscr{F}) \geqq 2$$
$$\chi(\mathscr{F}) = 2 - k.$$

b) *if $C = C_1 \cup C_2$, C_i irreducible, $C_1 \cap C_2 = \{p\}$, then $[C] \in \bar{D}_k$ if and only if there are torsion-free rank 1 sheaves $\mathscr{F}_1$ on C_1, $\mathscr{F}_2$ on C_2 and an integer ℓ such that:*

$$h^0(\mathscr{F}_1) \geqq 2, \quad h^0(\mathscr{F}_2) \geqq 2,$$
$$h^0(\mathscr{F}_1(-\ell p)) \geqq 1, \quad h^0(\mathscr{F}_2(-\ell p)) \geqq 1,$$
$$\chi(\mathscr{F}_1) + \chi(\mathscr{F}_2) = 3 - k + \ell.$$

Proof. We first prove (a). By the theory of the compactification of Pic(C), the space of pairs $(C, \mathscr{F})$, C irreducible, $\mathscr{F}$ torsion-free rank 1 is proper over the moduli space of such C's, and the conditions in (a) define a closed subset of this space. Therefore the set of irreducible C such that such an $\mathscr{F}$ exists is closed in the space of such C's. Moreover, if C is smooth and $[C] \in D_k$, let π: $C \to \mathbf{P}^1$ be the given covering of degree k, $\mathscr{F} = \pi^* \mathcal{O}(1)$. Then

$$\chi(\mathscr{F}) = \deg \pi - g + 1 = k - (2k-1) + 1 = 2 - k.$$

This proves that the C's satisfying the conditions in (a) contains the set of irreducible C's such that $[C] \in \bar{D}_k$. Conversely, let C satisfy the conditions of (a). We will check that $[C]$ lifts to a point of $\bar{H}_{k,b}$, hence $[C] \in \operatorname{Im}(\bar{H}_{k,b} \to \bar{\mathscr{M}}_g) = \bar{D}_k$. To see this, let C' be the desingularization of C. We shall embed C' in a curve C'' with ordinary double points so that collapsing rational curves of C'' we get back to C and we construct at the same time an admissible covering

$$\pi\colon C'' \to P, \qquad P \text{ a } b\text{-pointed stable rational curve.}$$

We start by choosing $s_1, s_2 \in \Gamma(\mathscr{F})$ and letting $\mathscr{F}' \subset \mathscr{F}$ be the subsheaf generated by s_1, s_2. Let A be the set of double points of C. At each $z \in A$, $\mathscr{F}'_z$ is isomorphic either to $\mathcal{O}_{z,C}$ or to $m_{z,C}$. Call these subsets $A_1, A_2 \subset A$. Let C^* be the partial desingularization of C obtained by separating the branches of C at the $z \in A_2$ only. Then there is an invertible sheaf $\mathscr{F}^*$ on C^* such that if

$$C' \xrightarrow{f} C^* \xrightarrow{g} C$$

are canonical maps, $\mathscr{F}' = g_*(\mathscr{F}^*)$. The function s_1/s_2 defines a morphism

$$\pi^*\colon C^* \to \mathbf{P}^1$$

hence

$$\pi = \pi^* \circ f\colon C' \to \mathbf{P}^1.$$

Note that

$$\begin{aligned}
\# A_2 + \deg(\pi) &= \# A_2 + \deg(\pi^*) \\
&= \# A_2 + \deg(\mathscr{F}^*) \\
&= \# A_2 + \chi(\mathscr{F}^*) + p_a(C^*) - 1 \\
&= \chi(\mathscr{F}') + (\# A_2 + p_2(C^*)) - 1 \\
&= \chi(\mathscr{F}') + p_a(C) - 1 \\
&\leq \chi(\mathscr{F}) + 2k - 2 \\
&= k.
\end{aligned}$$

Now $\pi\colon C'' \to P$ will be built up starting from $\pi\colon C' \to \mathbf{P}^1$. Let $S \subset \mathbf{P}^1$ be

a) the multiple branch points of π

b) the images of the points $z_{1,i}, z_{2,i} \in C'$ over the double points $z_i \in C$

c) the images of $k - \# A_2 - \deg(\pi)$ further generic points $w_i \in C'$.

Then P is the original $\mathbf{P}^1$, henceforth called $(\mathbf{P}^1)_0$, with a "tail" $\mathbf{P}^1$ glued at each point of S.

Firstly, if $x \in (\mathbf{P}^1)_0$ is a multiple branch point, let $\pi^{-1}(x) = \sum n_i y_i$. For each i, add to C' a copy of $\mathbf{P}^1$ glued to C' at y_i and mapped to the tail at x by a generic map $\mathbf{P}^1 \to \mathbf{P}^1$ of degree n_i, with n_i-fold branching at y_i:

Examples: $n_1 = 2$, $n_2 = 1$, $n_3 = 3$

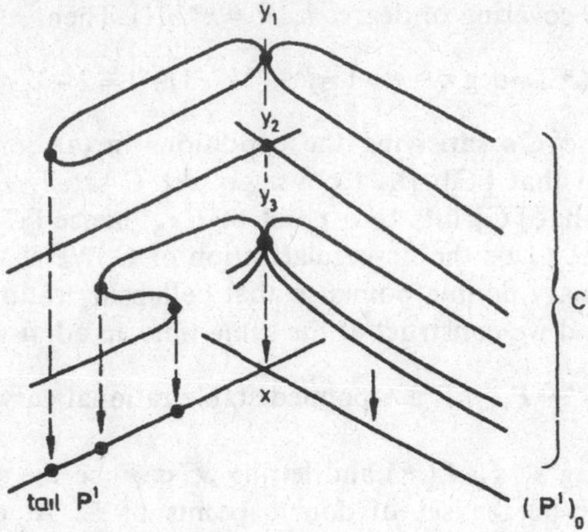

Secondly, if $x \in (\mathbf{P}^1)_0$ is the image of $z_{1,i}$ *and* $z_{2,i}$ (this happens if $z_i \in A_1$), let π be ramified to order ℓ_1 at $z_{1,i}$, ℓ_2 at $z_{2,i}$. Then add one copy of $\mathbf{P}^1$ to C' meeting C' at $z_{1,i}$ and $z_{2,i}$, and lying over the tail at x with degree $\ell_1 + \ell_2$ by a generic map $\mathbf{P}^1 \to \mathbf{P}^1$ ramified to order ℓ_1 at $z_{1,i}$, ℓ_2 at $z_{2,i}$:

Example: $\ell_1 = 2$, $\ell_2 = 1$

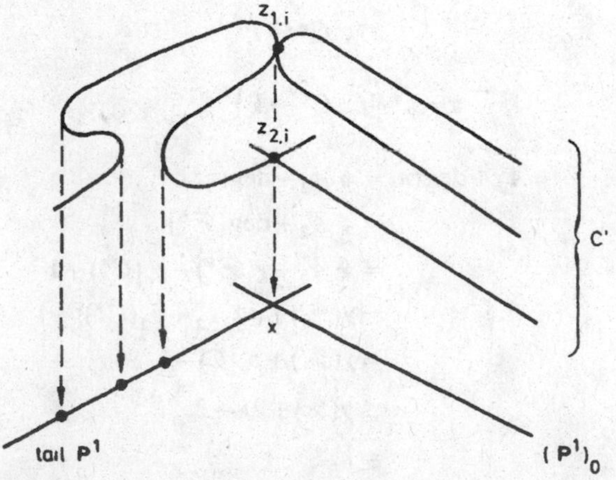

(At other points of C' in $\pi^{-1}(x)$, add further $\mathbf{P}^1$'s as in the 1st step.) Thirdly, if $x_1, x_2 \in (\mathbf{P}^1)_0$, $x_1 \neq x_2$ are the images of $z_{1,i}$ and $z_{2,i}$ (this happens if $z_i \in A_2$), let π be ramified to order ℓ_1 at $z_{1,i}$, ℓ_2 at $z_{2,i}$ again. This time, however, add 3 copies of $\mathbf{P}^1$ to C', copy A over the tail at x_1, copy B over the tail at x_2 and copy C over $(\mathbf{P}^1)_0$ as follows:

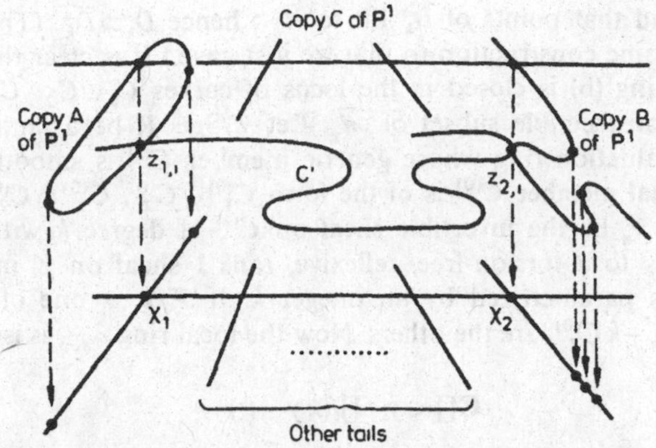

$\ell_1=1$, $\ell_2=2$ is illustrated. Note that copy C goes isomorphically to $(\mathbf{P}^1)_0$, copy B goes to the tail at x_1 with degree ℓ_1+1, generic except for ℓ_1-fold ramification at $z_{1,i}$, copy C goes to the tail at x_2 with degree ℓ_2+1, generic except for ℓ_2-fold ramification at $z_{2,i}$. Both copy A and B meet copy C at their remaining (unramified) point over x_1 (resp. x_2). Finally, more $\mathbf{P}^1$'s are joined to copy C to cover all the other tails added to $(\mathbf{P}^1)_0$. Fourthly, at the $k-\#A_2-\deg\pi$ generic points x added to S, the following "plumbing fixture" is thrown in to soup up the degree of π to k:

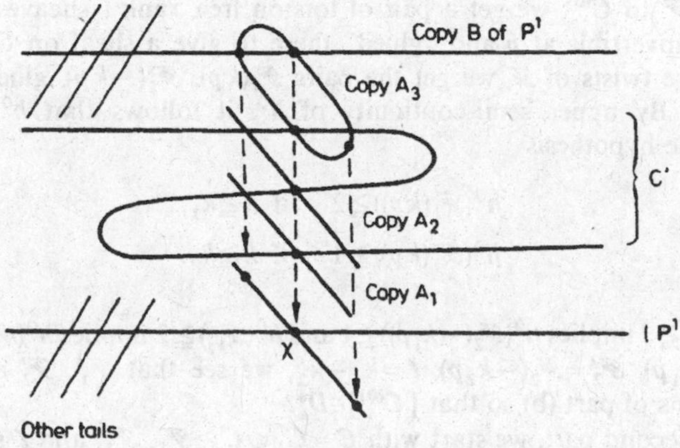

Thus $\deg\pi$ $\mathbf{P}^1$'s are added over the tail at x, all but one isomorphic to it, one of degree 2 over it. To the new point over x introduced by this last $\mathbf{P}^1$, a copy B of $\mathbf{P}^1$ isomorphic to $(\mathbf{P}^1)_0$ is joined, and to it further $\mathbf{P}^1$'s over the other tails.

After some reflection, the reader will see that the resulting C'' is an admissible cover of P of degree k, and arithmetic genus $g=2k-1$, which blows down to the original C when the extra $\mathbf{P}^1$'s are collapsed to points. This completes the proof of part (a).

To check part (b), let $\bar{D}_k^*$ be the locus of $[C]$'s satisfying the conditions of parts (a) or (b). We show first that $\bar{D}_k^*$ is closed in the open set of $\bar{\mathcal{M}}_g$ of curves with at most 2 components, meeting at most once, hence in this open set, $\bar{D}_k$

$\subset \bar{D}_k^*$, and second that points of $\bar{D}_k^*$ lift to $\bar{H}_{k,b}$ hence $\bar{D}_k \supset \bar{D}_k^*$. (The latter will use almost the same construction to that we just gave.) It is clear that the locus of curves satisfying (b) is closed in the locus of curves $C_1 \cup C_2$, $C_1 \cap C_2 = \{p\}$. Thus $\bar{D}_k^*$ is a constructible subset of $\bar{\mathcal{M}}_g$. Let $\mathscr{C}/\operatorname{Spec} R$ be a family of stable curves over a valuation ring whose generic member $C^{(\eta)}$ is smooth and in D_k and whose special member $C^{(0)}$ is of the form $C_1^{(0)} \cup C_2^{(0)}$, $C_1^{(0)} \cap C_2^{(0)} = \{p\}$, $C_i^{(0)}$ irreducible. Let $\mathscr{F}_\eta$ be the invertible sheaf on $C^{(\eta)}$ of degree k with $h^0(\mathscr{F}_\eta) \geqq 2$. Then $\mathscr{F}_\eta$ extends to a torsion-free, reflexive, rank 1 sheaf on $\mathscr{C}$ in an infinite number of ways parametrized by an integer k. If $\mathscr{F}/\mathscr{C}$ is one of these, then $\mathscr{F}(-k C_1^{(0)}) \cong \mathscr{F}(+k C_2^{(0)})$ are the others. Now the local ring $\hat{o}_{\mathscr{C},p}$ is isomorphic to

$$\mathbf{C}[[x, y, t]]/(x\,y - t^n)$$

for some n, where tR is the maximal ideal of R and $(x, t) =$ ideal of $C_1^{(0)}$, (y, t) $=$ ideal of $C_2^{(0)}$. Then the group $\operatorname{Pic}(\operatorname{Spec} \hat{o}_{\mathscr{C},p} - \{m\})$ is cyclic of order n, being given by the restrictions to the punctured spectrum $\operatorname{Spec} \hat{o}_{\mathscr{C},p} - \{m\}$ of the ideal sheaves (x, t^k), $0 \leqq k \leqq n-1$. If $\mathscr{F}$ is isomorphic, as invertible sheaf on $\operatorname{Spec} \hat{o}_{\mathscr{C},p} - \{m\}$ to (x, t^k), then $\mathscr{F}(k C_1^{(0)})$ is invertible at p. Replacing $\mathscr{F}$ by this, we may as well assume that $\mathscr{F}$ itself is invertible at p, hence so are all the sheaves in the sequence

$$\ldots, \mathscr{F}(-n C_1^{(0)}), \mathscr{F}, \mathscr{F}(n C_1^{(0)}), \mathscr{F}(2n C_1^{(0)}), \ldots.$$

Restricting $\mathscr{F}$ to $C^{(0)}$, we get a pair of torsion free, rank 1 sheaves $\mathscr{F}_1$ on $C_1^{(0)}$, $\mathscr{F}_2$ on $C_2^{(0)}$, invertible at p and "glued" there to give a sheaf on $C^{(0)}$. Restricting the above twists of $\mathscr{F}$, we get the pairs $\mathscr{F}_1(kp)$, $\mathscr{F}_2(-kp)$, glued to a sheaf $\mathscr{F}_k$ on $C^{(0)}$. By upper semi-continuity of h^0, it follows that $h^0(\mathscr{F}_k) \geqq 2$. Fix $k_1 > k_2$ by the hypothesis

$$h^0(\mathscr{F}_1(kp)) \geqq 2 \quad \text{if } k \geqq k_1,$$
$$h^0(\mathscr{F}_1(kp)) \geqq 1 \quad \text{if } k \geqq k_2.$$

Then $h^0(\mathscr{F}_{k_1}) \geqq 2$ implies $h^0(\mathscr{F}_2(-k_1 p)) \geqq 1$ and $h^0(\mathscr{F}_{k_2}) \geqq 2$ implies $h^0(\mathscr{F}_2(-k_2 p)) \geqq 2$. If $\mathscr{F}_1' = \mathscr{F}_1(k_1 p)$, $\mathscr{F}_2' = \mathscr{F}_2(-k_2 p)$, $\ell = k_1 - k_2$, we see that $\mathscr{F}_1'$, $\mathscr{F}_2'$ and ℓ satisfy the conditions of part (b) so that $[C^{(0)}] \in \bar{D}_k^*$.

For the second part, we start with $C = C_1 \cup C_2$, $\mathscr{F}_1$, $\mathscr{F}_2$, and ℓ satisfying (b). Take 2 sections in $\mathscr{F}_1$, one a section of $\mathscr{F}_1(-\ell p)$, spanning a subsheaf $\mathscr{F}_1'$, and 2 sections in $\mathscr{F}_2$, one a section of $\mathscr{F}_2(-\ell p)$, spanning $\mathscr{F}_2'$ and let their ratios define morphisms of the desingularizations $C_i^{(0)\prime}$ of $C_i^{(0)}$:

$$\pi_1\colon C_1^{(0)\prime} \to \mathbf{P}^1 \text{ of degree } k_1, \text{ ramified to order } \ell_1 \text{ at } p$$
$$\pi_2\colon C_2^{(0)\prime} \to \mathbf{P}^1 \text{ of degree } k_2, \text{ ramified to order } \ell_2 \text{ at } p.$$

If it happens that $\ell_1 = \ell_2$, we can join them into an admissible cover by identifying p in $C_1^{(0)\prime}$, $C_2^{(0)\prime}$:

Example: $\ell_1=\ell_2=2,\ n=4$

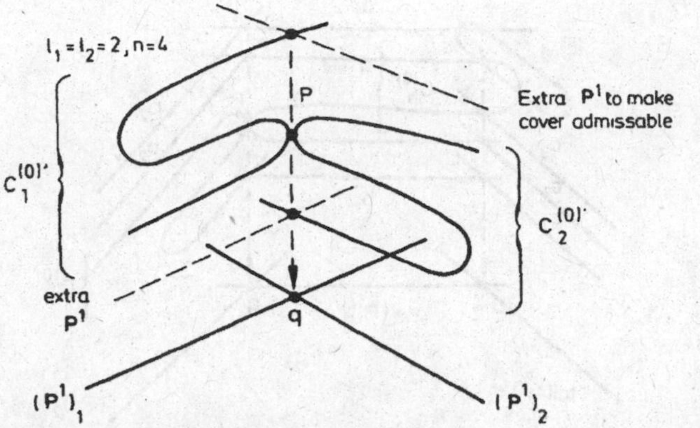

Otherwise, we need an extra $\mathbf{P}^1$ to join them:

Example. $\ell_1=3,\ \ell_2=2,\ n=5$

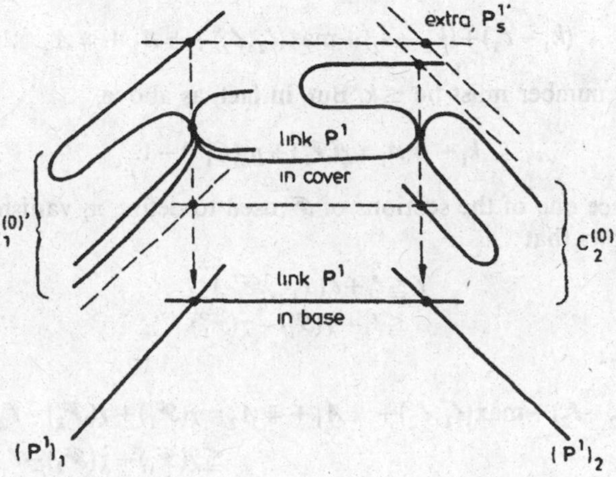

If $\ell_1>\ell_2$, add a $\mathbf{P}^1$ to the cover of degree ℓ_1 over the link $\mathbf{P}^1$ in the base ramified to order ℓ_1 at the point P in $C_1^{(0),'}$ and to order ℓ_2 at the point $p\in C_2^{(0),'}$. At this point, the covering is "filled in", exactly as in the case of an irreducible C. To make it admissible, extra $\mathbf{P}^1$'s are needed at multiple branch points of π_1 and π_2. To join the two points of $C_1^{(0),'}$ (resp. $C_2^{(0),'}$) over double points of $C_1^{(0)}$ (resp. $C_2^{(0)}$), linking $\mathbf{P}^1$'s are needed. Finally the degree of the whole cover must be augmented to k. Thus for example, in the case $\ell_1=\ell_2=2$, $k_1=k_2=3$, $k=5$, one would add $\mathbf{P}^1$'s as follows:

Example:

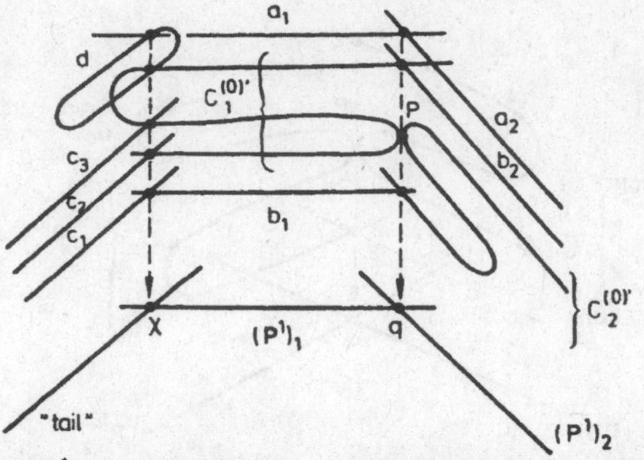

All a_i, b_i, c_i, d are $\mathbf{P}^1$'s. In general, $k_1-\ell_1$ $\mathbf{P}^1$'s (like b_2) are added to $C_1^{(0)}$ to extend its other points over q across $(\mathbf{P}^1)_2$, and $k_2-\ell_2$ $\mathbf{P}^1$'s (like b_1) are added to $C_2^{(0)}$ to extend its other points over q across $(\mathbf{P}^1)_1$.

The main point to be checked is that k is large enough to accommodate all these curves. Thus if

A_i = set of double points of $C_i^{(0)}$ where $\mathscr{F}_i'$ is *not* invertible,

to carry out our constructions, we need

$$(k_1-\ell_1)+(k_2-\ell_2)+\max(\ell_1,\ell_2)+\#A_1+\#A_2$$

sheets, so this number must be $\leqq k$. But, in fact, as above,

$$k_i+\#A_i=\chi(\mathscr{F}_i')+p_a(C_i^{(0)})-1.$$

Moreover, since one of the sections of $\mathscr{F}_i$ used to define π_i vanishes to order at least ℓ, it follows that

$$\begin{aligned}\ell &\leqq \ell_i+\ell(\mathscr{F}_{i,p}/\mathscr{F}_{i,p}')\\ &\leqq \ell_i+\chi(\mathscr{F}_i)-\chi(\mathscr{F}_i').\end{aligned}$$

Thus if $\ell_1\geqq\ell_2$:

$$\begin{aligned}(k_1-\ell_1)+(k_2-\ell_2)+\max(\ell_1,\ell_2)+\#A_1+\#A_2&=\chi(\mathscr{F}_1')+\chi(\mathscr{F}_2')-\ell_2+p_a(C^{(0)})-2\\ &\leqq\chi(\mathscr{F}_1)+\chi(\mathscr{F}_2)-\ell+2k-3\\ &=(3-k+\ell)-\ell+2k-3\\ &=k. \quad \text{Q.E.D.}\end{aligned}$$

Working this out more explicitly, we can draw a series of Corollaries from this theorem:

Corollary 1. *Let* Int Δ_0 *be the locus of irreducible curves C with one double point p. Writing these curves as the quotient of a curve C' of genus $g-1$ by identifying 2 distinct points p_1, p_2, we have:*

$$(\operatorname{Int}\Delta_0)\cap\bar{D}_k=\left\{(C',p_1,p_2)\,\middle|\,\begin{array}{l}\exists \text{ a line bundle } L \text{ on } C' \text{ of degree } k\\ \text{s.t. } h^0(L)\geqq 2,\ h^0(L(-p_1-p_2))\geqq 1\end{array}\right\}.$$

The map $(C', p_1, p_2) \mapsto C'$ carries all components of this intersection onto $\mathcal{M}_{g-1}$.

Proof. This comes from part (a) of the theorem. The condition above comes from the case where $\mathcal{F}$ is invertible. The other possibility is that $\mathcal{F}_p \cong m_p$. Such an $\mathcal{F}$ is the image of a line bundle M on C' of degree $k-1$ such that $h^0(M) \geqq 2$. Let $L = M(p_1)$ and the condition above is satisfied.

Corollary 2. *Let* $\operatorname{Int} \Delta_{0,1}$ *be the locus of curves of the type*

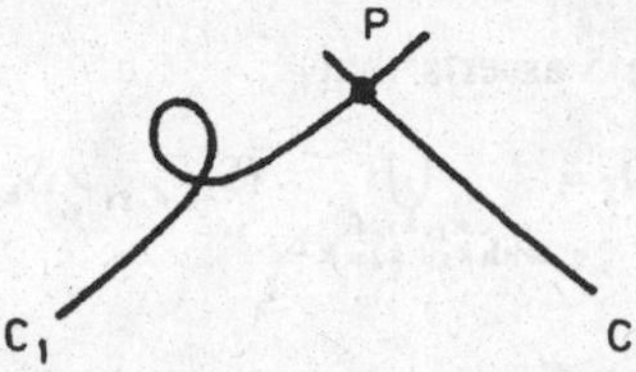

where C_2 is smooth of genus $g-1$, and C_1 is $\mathbf{P}^1$ with $0, \infty$ identified. Describing these curves by the pair (C_2, p), we have:

$$(\operatorname{Int} \Delta_{0,1}) \cap \bar{D}_k = \left\{ (C_2, p) \,\middle|\, \begin{array}{c} \exists \text{ a line bundle } L \text{ on } C_2 \text{ of degree } k \\ \text{s.t. } h^0(L) \geqq 2, h^0(L(-2p)) \geqq 1 \end{array} \right\}.$$

Proof. We use part (b) of the theorem. For $\mathcal{F}_1$ to exist, it is necessary and sufficient that $\chi(\mathcal{F}_1)$, ℓ satisfy:

$$\ell = 1, \chi(\mathcal{F}_1) \geqq 2 \quad \text{or} \quad \ell \geqq 2, \chi(\mathcal{F}_1) \geqq \ell.$$

But this means that $\mathcal{F}_2$ is a line bundle on C_2 such that either

$$\deg \mathcal{F}_2 \leqq k, h^0(\mathcal{F}_2) \geqq 2, h^0(\mathcal{F}_2(-\ell p)) \geqq 1, \ell \geqq 2$$

or

$$\deg \mathcal{F}_2 \leqq k-1, h^0(\mathcal{F}_2) \geqq 2, h^0(\mathcal{F}_2(-p)) \geqq 1.$$

All of these imply the existence of $\mathcal{F}_2$ such that

$$\deg \mathcal{F}_2 = k, h^0(\mathcal{F}_2) \geqq 2, h^0(\mathcal{F}_2(-2p)) \geqq 1.$$

For the next Corollary, we define:

$\mathcal{M}_{g,1}$ = moduli space of pairs (C, p), C smooth of genus g, $p \in C$
$S_{k,\ell,g}$ = subset of $\mathcal{M}_{g,1}$ of pairs C, p such that there exists a line bundle L on C of degree k with $h^0(L) \geqq 2$, $h^0(L(-\ell p)) \geqq 1$.

Corollary 3. *If $1 \leqq g_1 \leqq k-1$ and $g_2 = g - g_1$, let $\operatorname{Int} \Delta_{g_1}$ be the locus of curves $C = C_1 \cup C_2$ where C_1, C_2 are smooth with genus g_1, g_2 and $C_1 \cap C_2 = \{p\}$. Note that:*

$$\operatorname{Int} \Delta_{g_1} \cong \mathcal{M}_{g_1,1} \times \mathcal{M}_{g_2,1}.$$

Then

$$(\operatorname{Int} \Delta_{g_1}) \cap \bar{D}_k = \bigcup_{\frac{g+1}{2} \leqq k_1 \leqq \min(k, g_1)} (S_{k_1, 2k_1 - g_1, g_1} \times \mathcal{M}_{g_2,1})$$

$$\cup \bigcup_{\frac{g_2+1}{2} \leqq k_2 \leqq \min(k, g_2)} (\mathcal{M}_{g_1,1} \times S_{k_2, 2k_2 - g_2, g_2}).$$

Proof. We apply (b). Note that as the C_i are smooth, the $\mathcal{F}_i$ are line bundles. Let $k_i = \deg \mathcal{F}_i$. Now k_1, k_2 and ℓ satisfy:

$$(k_1+1-g_1)+(k_2+1-g_2)=\chi(\mathcal{F}_1)+\chi(\mathcal{F}_2)=3-k+\ell$$

hence

$$\begin{aligned} k_1+k_2 &= 1-k+\ell+(g_1+g_2) \\ &= k+\ell \end{aligned}$$

Now part (b) of Theorem 5 asserts that

$$(*) \qquad (\operatorname{Int} \Delta_{g_1}) \cap \bar{D}_k = \bigcup_{\substack{k_1, k_2, \ell \\ \text{with } k_1+k_2=k+\ell}} (S_{k_1,\ell,g_1} \times S_{k_2,\ell,g_2}).$$

But note that:

i) if $2k-\ell-1 \geqq g$, $S_{k,\ell,g} = \mathcal{M}_{g,1}$

ii) $S_{k,\ell,g} \subset S_{k+1,\ell+1,g}$.

For (i), see [7]; for (ii), if L has degree k and puts (C,p) in $S_{k,\ell,g}$, then $L(p)$ has degree $k+1$ and puts (C,p) in $S_{k+1,\ell+1,g}$. Now if k_1, k_2, ℓ satisfy

$$k_1+k_2=k+\ell,$$

note that:

$$\text{either } 2k_1-\ell-1 \geqq g_1, \quad \text{or} \quad 2k_2-\ell-1 \geqq g_2.$$

If not,

$$g_i \geqq 2k_i-\ell, \quad i=1,2$$

hence

$$2k-1=g_1+g_2 \geqq 2k_1+2k_2-2\ell=2k$$

a contradiction.

Therefore, in the union in (*), one of the 2 factors is always $\mathcal{M}_{g_1,1}$ or $\mathcal{M}_{g_2,1}$. But if

$$2k_1-\ell-1>g_1,$$

then if we replace k_2 by k_2+1 and ℓ by $\ell+1$, then

$$\begin{aligned} S_{k_1,\ell,g_1} \times S_{k_2,\ell,g_2} &= \mathcal{M}_{g_1,1} \times S_{k_2,\ell,g_2} \\ &\subset \mathcal{M}_{g_1,1} \times S_{k_2+1,\ell,g_2+1} \\ &= S_{k_1,\ell+1,g_1} \times S_{k_2+1,\ell,g_2+1}. \end{aligned}$$

Thus in the union in (*), we need only consider the terms with

$$2k_1-\ell-1=g_1, \; 2k_2-\ell=g_2$$

or

$$2k_1-\ell=g_1, \; 2k_2-\ell-1=g_2.$$

This is exactly the Corollary.

Note that in the special cases $g_1=1$ or 2, the Corollary reduces to:

$$\begin{aligned} (\operatorname{Int} \Delta_1) \cap \bar{D}_k &= \mathcal{M}_{1,1} \times S_{k,2,g-1}, \\ (\operatorname{Int} \Delta_2) \cap \bar{D}_k &= (S_{2,2,2} \times \mathcal{M}_{g-2,1}) \cup (\mathcal{M}_{2,1} \times S_{k,3,g-2}) \\ &\quad \cup (\mathcal{M}_{2,1} \times S_{k-1,1,g-2}). \end{aligned}$$

Only in these two cases do we have components of $\operatorname{Int} \Delta_{g_1} \cap \bar{D}_k$ consisting entirely of curves with automorphisms: viz. $\mathcal{M}_{1,1} \times S_{k,2,g-1}$ and $S_{2,2,2} \times \mathcal{M}_{g-2,1}$ (the latter means a curve C_1 of genus 2 meeting a curve C_2 of genus $g-2$ at a point p, where $p \in C_1$ is a Weierstrass point).

Corollary 2 plus the case $g_1 = 1$ combine to say:

Corollary 4.

$$\left(\begin{matrix}\textit{locus of curves } C_1 \cup C_2,\ C_2 \textit{ smooth of genus } g-1 \\ C_1 \textit{ smooth or singular of genus } 1,\ C_1 \cap C_2 = \{p\}\end{matrix}\right) \cap \bar{D}_k = \overline{\mathcal{M}}_{1,1} \times S_{k,2,g-1}.$$

§5. Counting Pencils on the Generic Curve

We need to refine the results in Corollaries 1 and 3 of §4 by determining the intersection multiplicities in the intersections $\bar{D}_k \cap \operatorname{Int} \Delta_i$, and we will need to count the intersections of these cycles with the curves in Δ_i obtained by varying the double points used to get a singular stable curve in Δ_i. In order to do both, we need two fundamental results counting pencils on the generic curve of genus g, which are essentially but not completely contained in [7]. The results are:

Theorem A. *For all $g \geqq 1$ and all d such that*

$$\frac{g}{2} + 1 \leqq d \leqq g + 1,$$

let

$$a(d,g) = (2d-g-1)\frac{g!}{d!(g-d+1)!}.$$

Then for almost all pairs (C,p), C a curve of genus g, $p \in C$, there is a finite number $a(d,g)$ of line bundles L on C of degree d such that

$$(*) \qquad h^0(L) \geqq 2,\ h^0(L(-2d-g-1)\,p)) \geqq 1.$$

Moreover, for each L, $h^0(L) = 2$, L is generated by $H^0(L)$, and $H^0(L)$ defines a covering

$$\pi\colon C \to \mathbf{P}^1$$

of degree d with all ordinary branch points except for one $(2d-g-1)$-fold branch point at p, all lying over distinct points of $\mathbf{P}^1$.

Theorem B. *For all $g \geqq 1$ and all d such that*

$$\frac{g}{2} + 1 \leqq d \leqq g$$

let

$$b(d,g) = (2d-g-1)(2d-g)(2d-g+1)\frac{g!}{d!(g-d)!}.$$

Then for almost all curves C of genus g, there is a finite number $b(d,g)$ of pairs (L,p), L a line bundle on C of degree d and points $p \in C$ such that

$$(**) \qquad h^0(L) \geqq 2, h^0(L(-(2d-g)p)) \geqq 1.$$

Moreover, for each L, $h^0(L)=2$, L is generated by $H^0(L)$ and $H^0(L)$ defines a covering

$$\pi: C \to \mathbf{P}^1$$

of degree d with all ordinary branch points and, if $2d-g \geqq 3$, one $(2d-g)$-fold branch point, all lying over distinct points of $\mathbf{P}^1$.

Proof. Theorem A follows directly from results of [7]. To see this, note first that if C is a general curve, $p \in C$ a general point, then by a naive dimension count there will exist no map

$$\pi: C \to \mathbf{P}^1,$$

ramified at p, whose branch divisor B on $\mathbf{P}^1$ is supported on fewer than $3g+1$ points; or any such map π of degree $d-k$ with a $(2d-g-1-k)$-fold branch point at p for $k>0$; or any such map of degree d with a $(2d-g-1)$-fold branch point at p such that

$$h^0(C, \pi^* \mathcal{O}_{\mathbf{P}^1}(1)) \geqq 3.$$

Thus, if (C,p) is general, any line bundle L of degree d on C satisfying $(*)$ must satisfy as well the rest of the conditions of the theorem. It then remains only to count the number of points of intersection in the dth symmetric product C_d of C of the cycles

$$C_d^1 = \{D \mid h^0(C, \mathcal{O}_C(D)) \geqq 2\}$$

and

$$X_p^{2d-g-1} = \{D \mid D-(2d-g-1)p \geqq 0\}.$$

This is readily done: the class of the cycle C_d^1 is found to be

$$c_d^1 = \frac{\theta^{g-d+1}}{(g-d+1)!} - \frac{x\theta^{g-d}}{(g-d)!}$$

where x is the class of the divisor $X_p = \{D \mid D-p \geqq 0\}$ and θ the pullback from the Jacobian of C of the class of the θ-divisor. By Poincaré's formula, the intersection numbers

$$x^{d-\alpha}\theta^{\alpha} = \frac{g!}{(g-\alpha)!}$$

and hence the intersection number

$$(c_d^1 \cdot x^{2d-g-1}) = a(d,g).$$

Finally, in [7] it is shown that if (C, p) is general, then C^1_d is reduced, and C^1_d intersects X^{2d-g-1}_p transversely in exactly $a(d, g)$ points, proving Theorem A.[3]

The proof of Theorem B, by contrast, requires a little more care. As in the first case, naive dimension counts show us that, for general C, no line bundle L of degree d on C can satisfy (**) and violate any of the remaining conditions. The problem is thus reduced to computing the intersection number in C_d of the cycle C^1_d with the diagonal

$$\Delta^{2d-g} = \{D \mid D-(2d-g)p \geqq 0 \quad \textit{for some } p \in C\}$$

and showing that this intersection is transverse.

The intersection number is readily computed. The class of C^1_d is as before; and for $p+q=g-d+1$ the degree of the pullback, via the diagonal map

$$\Delta\colon C \times C_{g-d} \to C_d$$
$$(p, E) \mapsto E+(2d-g)p$$

of the class $x^p \cdot \theta^q$ is given by

$$\Delta^*(x^p\, \theta^q) = \frac{g!}{(g-p)!}\,(2d-g)^2\, p + (2d-g)\, q$$

(cf. [11, 2]). Combining these yields the intersection number

$$(C^1_d \cdot \Delta^{2d-g}) = b(d, g).$$

It remains to check that for general C this intersection is transverse. To begin with, we may identify the tangent spaces involved as follows:

i) At any point $D \in C_d$, we have natural identifications

$$T_D(C_d) = \Gamma(\mathcal{O}_C(D)/\mathcal{O}_C)$$

and

$$T^*_D(C_d) = \Gamma(K_C/K_C(-D)),$$

with the pairing of tangent and cotangent spaces given by the residue.

ii) If $D = m \cdot p + q_1 + \ldots + q_{d-m}$ with $p, q, \ldots, q_{d-m}$ all distinct, then Δ^m is smooth at D with tangent space equal to the subspace of $\Gamma(\mathcal{O}_C(D)/\mathcal{O}_C)$:

$$T_D(\Delta^m) = \text{Annihilator}\,[\Gamma(K_C((m-1)p-D)/K_C(-D))]$$

and

iii) The tangent space to C^1_d at D with $h^0(C, \mathcal{O}_C(D)) = 2$ is given as the annihilator of the image of the map

$$\tilde{\mu}_0 = r \circ \mu_0$$

[3] In point of fact, the statement made in [7] is that C^1_d intersects the cycle $x_{p_1} \cap \ldots \cap x_{p_{2d-g-1}}$ transversely for $p_1, \ldots, p_{2d-g-1}$ general points of C; but the argument applies equally in this case

where

$$\mu_0 : H^0(C, \mathcal{O}(D)) \otimes H^0(C, K_C(-D)) \to H^0(C, K_C)$$

is multiplication, and

$$r: H^0(C, K_C) \to H^0(C, K_C/K_C(-D))$$

is evaluation at D.

Statements i) and iii) are standard (cf. [2]), and ii) is elementary. Combining them, we see that a divisor D on C, whose associated line bundle $L = \mathcal{O}_C(D)$ satisfies the conditions of Theorem B, is a transverse point of intersection of C_d^1 and Δ^{2d-g} if and only if

$$H^0(C, K_C(-2D + (2d-g-1)p)) = 0.$$

To see that this in fact holds, let D be such a divisor, and

$$\pi: C \to \mathbf{P}^1$$

be the corresponding map; let

$$R_\pi = (2d-g-1)p + q_1 + \ldots + q_{3g-1}$$

be the ramification divisor of π. Let $\mathcal{H}$ be the versal deformation space for the map π, and let $\mathcal{H}' \subset \mathcal{H}$ be the subvariety of $\mathcal{H}$ of maps with a $(2d-g)$-fold branch point over $\pi(p)$. Then the tangent space to $\mathcal{H}$ at π is given by

$$T_\pi(\mathcal{H}) = H^0(C, \eta)$$

where η is the normal sheaf of π, defined by the exact sequence

$$0 \longrightarrow \theta_C \longrightarrow \pi^* \theta_{\mathbf{P}^1} \longrightarrow \eta \longrightarrow 0,$$

and the differential of the map

$$\phi: \mathcal{H} \to \mathcal{M}_g$$

is given by the coboundary map

$$H^0(C, \eta) \to H^1(C, \theta_C)$$

in this sequence. Finally, we may identify the tangent space to $\mathcal{H}'$ at π with the subspace of $T_\pi(\mathcal{H}) = H^0(C, \eta)$ of sections of η vanishing in a neighborhood of p, i.e., the sections of η':

$$0 \longrightarrow \theta_C \longrightarrow \pi^* \theta_{\mathbf{P}}(-(m-1)p) \longrightarrow \eta' \longrightarrow 0.$$

Now, we observe that since π is a point in a general fiber of $\phi|_{\mathcal{H}'}$, by Sard's theorem the differential ϕ_*, restricted to $T_\pi(\mathcal{H}')$, must be surjective. This means that the map d below is surjective:

$$H^0(\eta') \xrightarrow{\ d\ } H^1(\theta_C) \longrightarrow H^1(\pi^* \theta_{\mathbf{P}^1}(-(m-1)P)) \longrightarrow 0.$$

But $\theta_{\mathbf{P}^1} \cong o_{\mathbf{P}^1}(2)$ and $\pi^* o_{\mathbf{P}^1}(1) = o_C(D)$, hence

$$(0) = H^1(o_C(2D-(m-1)P)) \\ = H^0(K_C(-2D+(m-1)P))^*,$$

as required.

Using Theorems A and B, we can work out the intersection multiplicities in $(\operatorname{Int}\Delta_i) \cap \bar{D}_k$:

Theorem 6. a) *In the notation of Corollary 1, Δ_0 and $\bar{D}_k$ intersect generically transversely:*

$$(\operatorname{Int}\Delta_0)\cdot \bar{D}_k = \left\{ (C', p_1, p_2) \,\middle|\, \begin{matrix} \exists\, L \text{ on } C' \text{ of degree } k \\ h^0(L) \geqq 2,\ h^0(L(-p_1-p_2)) \geqq 1 \end{matrix} \right\} \text{ w. mult. one}$$

b) *In the notation of Corollary 3, if $g_1 \geqq 3$*

$$(\operatorname{Int}\Delta_{g_1})\cdot \bar{D}_k = \sum_{\frac{g_1+1}{2} \leqq k_1 \leqq \min(k, g_1)} a(k_1+g_2-k+1, g_2)\cdot S_{k_1, 2k_1-g_1, g_1} \times \mathscr{M}_{g_2,1} \\ + \sum_{\frac{g_2+1}{2} \leqq k_2 \leqq \min(k, g_2)} a(k_2+g_1-k+1, g_1)\, \mathscr{M}_{g_1,1} \times S_{k_2, 2k_2-g_2, g_2}.$$

This is still true for $g_1 = 1$ and 2 if the intersection is taken, not in $\bar{\mathscr{M}}_g$, but in the universal deformation space of a curve in $\operatorname{Int}\Delta_{g_1}$ (n.b. $\operatorname{Int}\Delta_1$ and $\operatorname{Int}\Delta_2$ have divisors in the singular locus of $\bar{\mathscr{M}}_g$, so the intersection product is not well-defined).

Proof of a). In the set-theoretic intersection $\operatorname{Int}\Delta_0 \cap \bar{D}_k$, there is an open dense set consisting of those $C = C'/(p_1 \sim p_2)$ for which

0) C' has no automorphisms

i) there is a *unique* line bundle L on C' of degree k with $h^0(L) \geqq 2$, $h^0(L(-p_1-p_2)) \geqq 1$.

ii) for this L, $h^0(L) = 2$ and $H^0(L)$ generates L

iii) if π: $C' \to \mathbf{P}^1$ is the covering of degree k defined by $H^0(L)$, then π has only ordinary branch points q_i and $\pi(q_i)$, $\pi(p_i)$ are all distinct points of $\mathbf{P}^1$.

In fact, by Theorem B, for almost all C' of genus $g-1 = 2k-2$, there are $b(k, 2k-2)$ pairs (L, q) of line bundles L such that $\deg L = k$, $h^0(L) \geqq 2$ and points $q \in C'$ such that

$$h^0(L(-2q)) \geqq 1.$$

Each distinct L_k defines a covering π_k: $C' \to \mathbf{P}^1$, hence a curve

$$\Gamma_k \subset C' \times C' \\ \Gamma_k = \{(x, y) \mid \pi_k x = \pi_k y\}.$$

Since almost all $(p_1, p_2) \in C' \times C'$ are on a unique curve Γ_k if they are on any of them, this proves i), ii) and iii).

Now for such $C = C'/(p_1 \sim p_2)$, $[C]$ is the image of *a unique* point of $\bar{H}_{k,b}$, namely the admissible cover π: $C'' \to P$ of degree k:

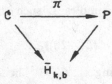

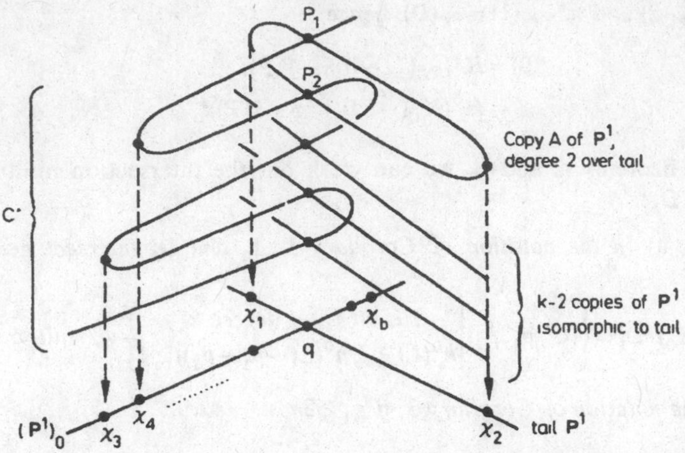

More precisely, if Σ_b is the permutation group on b letters, Σ_b acts on $\bar{H}_{k,b}$ by permuting the labelling of the branch points $\{x_i\}$ and this covering is the unique point of $\bar{H}_{k,b}/\Sigma_b$ over $[C]$. Note that this admissible covering has an automorphism ϕ of order 2, however:

$\phi = \mathrm{id}$, on C'

$\phi|_{\text{tail}\,\mathbf{P}^1}$ = automorphism fixing q, interchanging the two branch points x_1, x_2 on the tail

$\phi|_{\text{copy}\,A}$ = automorphism fixing p_1, p_2, interchanging the two ramified points.

However, C'' has *no* automorphisms over P fixing all the points of ramification. Therefore, if z is the point of $\bar{H}_{k,b}$ representing π: $C'' \to P$, z is a smooth point but if $\sigma \in \Sigma_b$ is the permutation interchanging x_1, x_2 and fixing the other branch points, it follows that $\sigma z = z$. In fact, σ fixes the smooth divisor $\Delta \subset \bar{H}_{k,b}$ of all admissible coverings like π: $C'' \to P$ where $P = (\mathbf{P}^1)_0 \cup (\text{tail}\,\mathbf{P}^1)$ but where $x_3, \ldots, x_b$ are allowed to vary in $(\mathbf{P}^1)_0$, and C' varies accordingly. There is a set $t_1, \ldots, t_{b-3}$ of local coordinates on $\bar{H}_{k,b}$ near z such that at all points of $\pi^{-1}(q)$, the universal family

$$\begin{array}{ccc} \mathcal{C} & \xrightarrow{\ \pi\ } & \mathbf{P} \\ & \searrow \quad \swarrow & \\ & \bar{H}_{k,b} & \end{array}$$

has local equations

$$xy = t_1,$$

$t_1 = 0$ is the local equation of Δ, and

$$\sigma^*(t_1) = -t_1$$
$$\sigma^*(t_i) = t_i, \quad 2 \leq i \leq b-3.$$

Then $t_1^2, t_2, \ldots, t_{b-3}$ are local coordinates on $\bar{H}_{k,b}/\{e,\sigma\}$ near the image of z.

Now consider the local analytic curve $\gamma\colon \Delta_s \to \bar{H}_{k,b}/\{e,\sigma\}$ given by $t_1^2 = s$, $t_2 = \ldots = t_{b-3} = 0$. Consider the diagram:

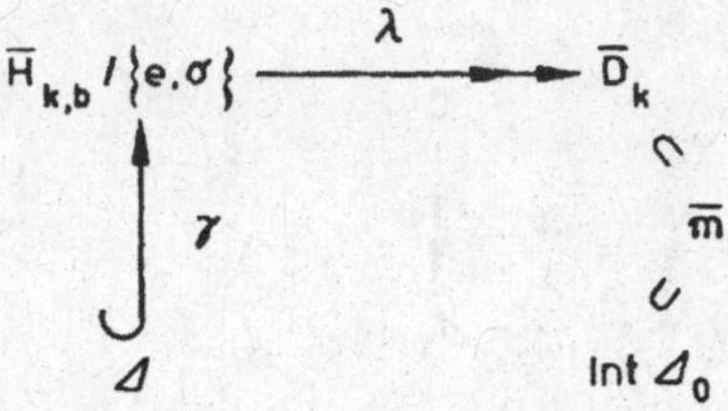

We shall check that the curve $\lambda\circ\gamma$ in $\bar{\mathscr{M}}_g$ is transverse to Int Δ_0 at the point $[C]$. This implies that the divisor $\lambda^*(\mathrm{Int}\,\Delta_0)$ on $\bar{H}_{k,n}/\{e,\sigma\}$ is Δ with multiplicity one, hence $\bar{D}_k$ and Int Δ_0 meet transversely. To see that $\lambda\circ\gamma\colon \Delta \to \bar{\mathscr{M}}_g$ is transverse to Δ_0, consider the restriction of $\mathscr{C}$ to $t_2 = \ldots = t_{b-3} = 0$. It is a family of curves with only ordinary double points:

whose fibres are smooth if $t_1 \neq 0$, and is

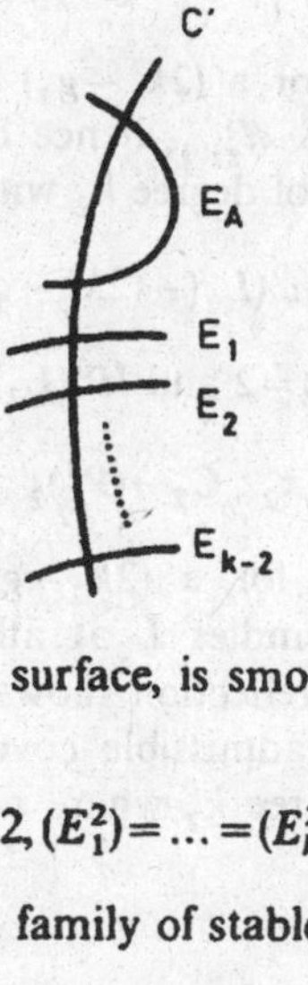

over $t_1 = 0$. Moreover, $\mathscr{C}_1$, as a surface, is smooth because its local equation is $xy = t_1$. It follows that on $\mathscr{C}_1$,

$$(E_A^2) = -2, (E_1^2) = \ldots = (E_{k-2}^2) = -1.$$

Blowing down the E's, we get a family of stable curves

with the same curves when $t_1 \neq 0$, and with fibre C over $t_1=0$. Since E_A blows down to an ordinary surface double point, $\mathscr{C}_2$ has equation $xy=t_1^2$ at the image of E_A. But $\mathscr{C}_2/\Delta_{t_1}$ is just the pull-back to Δ_{t_1} of the family of curves over Δ_s given by the morphism $\lambda \circ \gamma \colon \Delta_s \to \overline{\mathscr{M}}_g$. Thus it is induced by a family

$$\begin{array}{c} \mathfrak{C}_3 \\ \big\downarrow \\ \Delta_s \end{array}$$

where $\mathscr{C}_3$ has local equation $x \cdot y = s$ at the image of E_A and is again smooth. This means that $\mathscr{C}_3$ restricts to the universal deformation of the singular point of C, hence is a curve on $\overline{\mathscr{M}}_g$ transverse to the locus Δ_0 of singular curves.

Proof of b). The situation is similar with the two sums, so let's take a sufficiently general point $[C] \in S_{k_1, 2k_1-g_1, g_1} \times \mathscr{M}_{g_2,1}$ and compute the intersection multiplicity here. Now $C = C_1 \cup C_2$, C_1 of genus g_1, C_2 of genus g_2, $C_1 \cap C_2 = \{p\}$. Since by Theorem B for almost all C_1 of genus g_1, there are $b(k_1, g_1)$ pairs (L_1, p) such that $\deg L_1 = k_1$, $h^0(L_1) \geqq 2$, $h^0(L_1(-(2k_1-g_1)p)) \geqq 1$, and since for almost all C_1, the points p are distinct, it follows that for almost all $(C_1, p) \in S_{k_1, 2k_1-g_1, g_1}$, there is a unique L_1 of degree k_1 with $h^0(L_1) \geqq 2$, $h^0(L_1(-(2k_1-g_1)p)) \geqq 1$. Moreover, for this L_1, $h^0(L_1)=2$ and $H^0(L_1)$ defines a covering

$$\pi_1 \colon C_1 \to (\mathbf{P}^1)_1$$

with general branching except for a $(2k_1-g_1)$-fold branch point at p. As for (C_2, p), this is a general point of $\mathscr{M}_{g_2,1}$, hence by Theorem A for all k_2 it has exactly $a(k_2, g_2)$ line bundles L_2 of degree k_2 with

$$h^0(L_2) \geqq 2, \quad h^0(L_2(-(2k_2-g_2-1)p)) \geqq 1.$$

Moreover, for all of these, $h^0(L_2)=2$ and $H^0(L_2)$ defines

$$\pi_2 \colon C_2 \to (\mathbf{P}^1)_2$$

with general branching except for a $(2k_2-g_2)$-fold branch point at p. In particular, there are no line bundles L at all of degree k_2 with $h^0(L_2) \geqq 2$, $h^0(L_2(-(2k-g_2)p)) \geqq 1$. A little reflection shows that the only ways to lift C to a point of $\bar{H}_{k,b}$ are to use the admissible coverings defined by L_1 on C_1 of degree k_1 and L_2 on C_2 of degree k_2 where both have ℓ-fold branching at p and $\ell = 2k_1 - g_1 = 2k_2 - g_2 - 1$:

Example. $\ell = 2$,
$k_1 = 4$,
$k_2 = 3$,
$k = 5$.

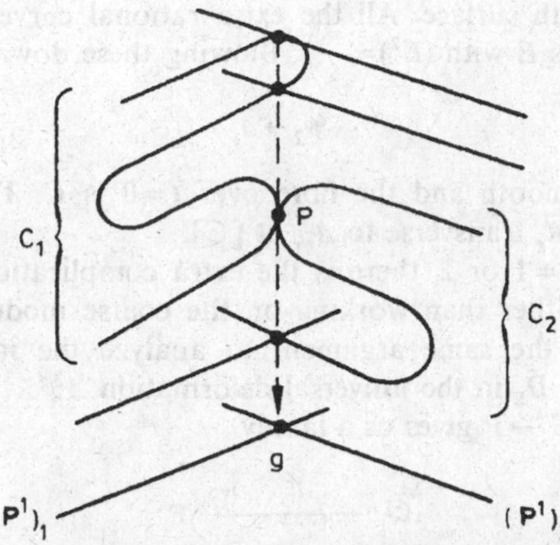

Here the degree of π is k and over $(\mathbf{P}^1)_1$ we have C_1 of degree k_1 and $k-k_1$ copies of $\mathbf{P}^1$ mapping isomorphically to $(\mathbf{P}^1)_1$; and over $(\mathbf{P}^1)_2$, we have C_2 of degree k_2 and $k-k_2$ copies of $\mathbf{P}^1$ mapping isomorphically to $(\mathbf{P}^1)_2$.

In order to do this, we must choose k_2 by the equation:

$$2k_2-g_2-1=2k_1-g_1,$$

i.e.,

$$\begin{aligned} k_2 &= k_1+\frac{g_2+1-g_1}{2} \\ &= k_1+g_2+\frac{1-g_1-g_2}{2} \\ &= k_1+g_2-k+1. \end{aligned}$$

Now, whereas L_1 is unique, there are $a(k_2,g_2)$ choices for L_2, so this gives exactly $a(k_2,g_2)$ points of $\bar{H}_{k,b}/\Sigma_b$ over $[C]\in\bar{D}_k$. Thus $\bar{D}_k$ is a divisor with $a(k_2,g_2)$ branches at $[C]$, and what the theorem says is that each of these branches meets Int Δ_{g_1} transversely. Assuming to begin with that $g_1\geqq 3$, the argument is similar to case (a) except simpler because the covering $\pi\colon C''\to P=(\mathbf{P}^1)_1\cup(\mathbf{P}^1)_2$ has no automorphisms. Thus $\bar{H}_{k,b}$ is smooth at the point z representing this covering and z is not fixed by any $\sigma\in\Sigma_b$. If we embed $\pi\colon C''\to P$ in a one-dimensional family of admissible coverings, we get

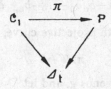

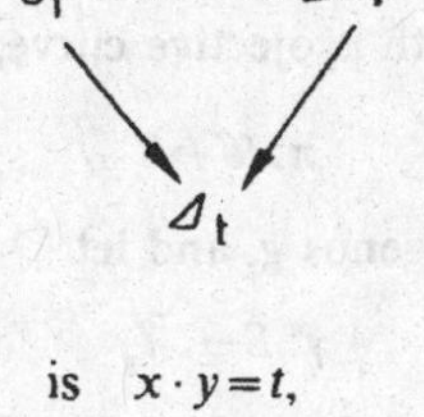

where locally near p

$$\begin{aligned} \mathscr{C}_1 &\quad \text{is} \quad x\cdot y=t, \\ \mathscr{P} &\quad \text{is} \quad u\cdot v=t^e, \\ \pi &\quad \text{is} \quad u=x^e,\ v=y^e. \end{aligned}$$

Thus $\mathscr{C}_1$ is a smooth surface. All the extra rational curves in the fibre of $\mathscr{C}_1$ over $t=0$ are curves E with $(E^2)=-1$. Blowing these down, we get a family of stable curves

$$\mathscr{C}_2 \to \Delta_t$$

where $\mathscr{C}_2$ is still smooth and the fibre over $t=0$ is C. Therefore this family defines a curve in $\bar{\mathscr{M}}_g$ transverse to Δ_{g_1} at $[C]$.

Now in case $g_1=1$ or 2, there is the extra complication that C may have automorphisms. Rather than working on the coarse moduli space, it is more convenient to use the same argument to analyze the intersection of the 2 divisors $\operatorname{Int}\Delta_{g_1}$ and $\bar{D}_k$ in the universal deformation Δ^{3g-3} of C. The universal deformation of $\pi\colon C'' \to P$ gives us a family

$$\begin{array}{ccc} \mathscr{C} & \xrightarrow{\ \pi\ } & P \\ & \searrow \quad \swarrow & \\ & \Delta^{b-3} & \end{array}$$

plus a uniformization

$$\Delta^{b-3} \to \bar{D}_k \subset \Delta^{3g-3}$$

of the branch of $\bar{D}_k$ near $[C]$ defined by $\pi\colon C'' \to P$. Restricting this deformation to a one-dimensional family as above, we get a curve on Δ^{b-3} mapped to a curve in Δ^{3g-3} transverse to the boundary component. Hence these divisors meet transversely in Δ^{3g-3}.

§ 6. The Class of the Divisor $\bar{D}_k$, II

In this section we finally compute the divisor class of $\bar{D}_k$ in $\operatorname{Pic}(\bar{\mathscr{M}}_{g,\mathrm{reg}})$ as a combination of $\lambda, \delta_0, \ldots, \delta_{k-1}$. We start with the relation

$$\bar{D}_k \equiv a\lambda - b_0\delta_0 - b_1\delta_1 - \ldots - b_{k-1}\delta_{k-1}$$

proved in § 3, but with unknown constants a, b_i, and we determine the constants by restricting this relation to suitable curves in $\bar{\mathscr{M}}_g$.

To describe the general method, suppose D is any divisor on $\bar{\mathscr{M}}_g$, i.e., a combination of codimension 1 subvarieties, but not necessarily a Cartier divisor (= one with a single local equation $f \in \mathbb{C}(\mathscr{M}_g)^*$ everywhere). Assume

$$D \equiv a\lambda - b_0\delta_0 - \ldots - b_{k-1}\delta_{k-1}$$

in $\operatorname{Pic}(\bar{\mathscr{M}}_{g,\mathrm{reg}})$. Let S be a smooth projective curve, let

$$\pi\colon \mathscr{C} \to S$$

be a family of stable curves of genus g, and let

$$\gamma\colon S \to \bar{\mathscr{M}}_g$$

be the induced morphism. Then first of all if $\gamma(S) \not\subset \operatorname{Supp}(D)$ we can define the divisor $\gamma^* D$ in a canonical way (even though $\gamma(S)$ may meet $\operatorname{Sing}(\bar{\mathscr{M}}_g) \cap \operatorname{Supp} D$). From general principles, we can apply the homomorphism:

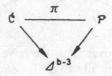

$$\mathrm{Pic}(\bar{\mathcal{M}}_{g,\mathrm{reg}}) \to \mathrm{Pic}_{\mathrm{fun}}(\bar{\mathcal{M}}_g)$$

to $\mathcal{O}_{\bar{\mathcal{M}}_{g,\mathrm{reg}}}(D)$ and "evaluate" the image on the family $\pi: \mathscr{C} \to S$, to get a line bundle on S. But even more, for every $s \in S$, let

$$\pi_s: \mathscr{C}_s \to \Delta^{3g-3}$$

be the universal deformation of the curve C_s which is the fibre of π over s. Then the morphism γ factors:

$$\begin{pmatrix}\text{neigh of}\\ s \in S\end{pmatrix} \xrightarrow[\gamma_1]{} \Delta^{3g-3} \xrightarrow[\gamma_2]{} \bar{\mathcal{M}}_g.$$

$\gamma_2^*(D)$ is a divisor on Δ^{3g-3}, hence a Cartier divisor (as Δ^{3g-3} is smooth), and we then pull it back by γ_1 as Cartier divisor:

$$\gamma^*(D) \underset{\mathrm{def}}{=} \gamma_1^*(\gamma_2^* D).$$

On the other hand, $\gamma^*(\lambda)$ is defined directly as c_1 of the line bundle $\Lambda^g \pi_* \omega_{\mathscr{C}/S}$, and $\gamma^* \delta_i$ are defined directly too as in Knudsen [10]. That is to say, if $\gamma(S) \not\subset \Delta_i$, $\gamma^* \delta_i$ is represented by the Cartier divisor $\gamma^* \Delta_i$ $(i \neq 1)$ or $\frac{1}{2}\gamma^* \Delta_1$ $(i=1)$. And if $\gamma(S) \subset \Delta_i$ and the curves C_s have exactly one double point x_s of type i, then $\gamma^* \delta_i$ is c_1 of the line boundle:

$$s \mapsto \Lambda^2(\Omega^1_{C_s} \otimes \mathbf{k}(x_s))^*.$$

For our first family, choose general curves C_1, C_2 of genera $\alpha \leqq k-1$ and $2k-1-\alpha$ respectively. Choose $p \in C_2$ a general point and let $S_2 = C_1 \times \{p\} \subset C_1 \times C_2$; denote by S_1 the diagonal $\Delta \subset C_1 \times C_1$; and let

$$\pi: T \to C_1$$

be the curve over C_1 obtained by identifying S_1 and S_2 in $C_1 \times C_1 \cup C_1 \times C_2$ over C_2. This is, the family whose fiber over $q \in C_1$ is the reducible curve obtained by identifying $q \in C_1$ with $p \in C_2$.

The degrees of the divisors λ and δ_i on T are readily calculated: first of all, we see that $\pi_* \omega_{T/C_1}$ is the trivial bundle $(H^0(C_1, \omega_{C_1}) \oplus H^0(C_2, \omega_{C_2})) \otimes \mathcal{O}_{C_1}$, so that $\deg \lambda = 0$. Clearly, $\deg \delta_i = 0$ on T for $i \neq \alpha$; and since the normal space to Δ_α in $\bar{\mathcal{M}}_{2k-1}$ at the point T_q is the tensor product $T_q(C_1) \otimes T_p(C_2)$, we have

$$\begin{aligned}\deg \delta_\alpha &= \deg(N_{\Delta_\alpha/\mathcal{M}} \otimes \mathcal{O}_T)\\ &= \deg(N_{S_1/C_1 \times C_2} \otimes N_{S_2/C_1 \times C_2})\\ &= 2 - 2\alpha.\end{aligned}$$

It remains to calculate the degree of the divisor $\bar{D}_k$ on T. By § 4, Theorem 5, Corollary 3, set-theoretically

$$\bar{D}_k \cap C_1 = \left\{ q \in C_1 \;\middle|\; \begin{array}{l}\text{For some } i, 0 \leqq i \leqq \frac{\alpha-1}{2}\\ \exists \text{ line bundle } L \text{ on } C_1 \text{ of degree } \alpha - i\\ \text{with } h^0(L) \geqq 2, h^0(L(-(\alpha-2i)q)) \geqq 1\end{array} \right\}$$

which by Theorem B consists in $b(\alpha-i,\alpha)$ points. But by Theorem 6, the multiplicity of each point in $a(k-i,2k-1-\alpha)$. (Note that for generic C_1, the divisor $S_{\alpha-i,\alpha-2i,\alpha}$ on $\mathcal{M}_{\alpha,1}$ and the curve $\{(C_1,q)|q\in C_1\}$ on $\mathcal{M}_{\alpha,1}$ must meet transversely.) Therefore

$$\begin{aligned}\deg_{C_1}\bar{D}_k &= \sum_{i=0}^{\frac{\alpha-1}{2}} b(\alpha-i,\alpha)\,a(k-i,2k-1-\alpha)\\ &= \sum_{i=0}^{\alpha/2} \frac{\alpha!(2k-1-\alpha)!}{(\alpha-i)!\,i!(k-i)!\,(k-\alpha+i)!}(\alpha-2i-1)(\alpha-2i)^2(\alpha-2i+1)\\ &= \frac{1}{2}\sum_{i=0}^{\alpha} \frac{\alpha!(2k-1-\alpha)!}{(\alpha-i)!\,i!(k-i)!\,(k-\alpha+i)!}(\alpha-2i-1)(\alpha-2i)^2(\alpha-2i+1).\end{aligned}$$

(The last equality coming from the fact that the sum is unaltered under the substitution $i\rightsquigarrow\alpha-i$.) Now, writing

$$\begin{aligned}&(\alpha-2i-1)(\alpha-2i)^2(\alpha-2i+1)\\ &\quad=[(k-i)-(k-\alpha+i)]\times[(\alpha-i)(\alpha-i-1)(\alpha-i-2)-3(\alpha-i)(\alpha-i-1)i\\ &\qquad+3(\alpha-i)i(i-1)-i(i-1)(i-2)+3(\alpha-i)(\alpha-i-1)-3i(i-1)],\end{aligned}$$

this sum becomes

$$\begin{aligned}&\frac{\alpha(\alpha-1)(\alpha-2)}{2}\sum_{i=0}^{\alpha}\left[\binom{\alpha-3}{i}\binom{2k-1-\alpha}{k-i-1}-3\binom{\alpha-3}{i-1}\binom{2k-1-\alpha}{k-i-1}\right.\\ &\qquad+3\binom{\alpha-3}{i-2}\binom{2k-1-\alpha}{k-i-1}-\binom{\alpha-3}{i-3}\binom{2k-1-\alpha}{k-i-1}\\ &\qquad-\binom{\alpha-3}{i}\binom{2k-1-\alpha}{k-i}+3\binom{\alpha-3}{i-1}\binom{2k-1-\alpha}{k-i}\\ &\qquad\left.-3\binom{\alpha-3}{i-2}\binom{2k-1-\alpha}{k-i}+\binom{\alpha-3}{i-3}\binom{2k-1-\alpha}{k-i}\right]\\ &+\frac{\alpha(\alpha-1)}{2}\sum_{i=0}^{\alpha}\left[3\binom{\alpha-2}{i}\binom{2k-1-\alpha}{k-i-1}-3\binom{\alpha-2}{i-2}\binom{2k-1-\alpha}{k-i-1}\right.\\ &\qquad\left.-3\binom{\alpha-2}{i}\binom{2k-1-\alpha}{k-i}+3\binom{\alpha-2}{i-2}\binom{2k-1-\alpha}{k-i}\right]\\ &=\frac{\alpha(\alpha-1)(\alpha-2)}{2}\left[-6\binom{2k-4}{k-2}+8\binom{2k-4}{k-3}-2\binom{2k-4}{k-4}\right]\\ &\qquad+\frac{\alpha(\alpha-1)}{2}\left[3\binom{2k-3}{k-2}-3\binom{2k-3}{k-3}\right]\\ &=-6\alpha(\alpha-1)(\alpha-2)\cdot\frac{(2k-4)!}{k!(k-2)!}+6\alpha(\alpha-1)\frac{(2k-3)!}{k!(k-2)!}\\ &=6\alpha(\alpha-1)(2k-1-\alpha)\frac{(2k-4)!}{k!(k-2)!}.\end{aligned}$$

We conclude, then, that

(3)
$$b_\alpha = \frac{\deg D_k}{2\alpha - 2}$$
$$= \frac{3\alpha(2k-1-\alpha)}{k} \cdot \frac{(2k-4)!}{(k-1)!\,(k-2)!}$$

for $\alpha \geqq 2$.

Note that if we choose to vary the point $p \in C_2$ and fix q to be a general point of C_1, the resulting family $\pi\colon T \to C_2$ has intersection numbers

$$\deg \lambda = 0,$$
$$\deg \delta_i = 0, \quad i \neq \alpha,$$
$$\deg \delta_\alpha = 2 - 2(2k-1-\alpha),$$
$$= -2(2k-2-\alpha)$$

and, as before,

$$\deg D_k = \sum_{i=0}^{\frac{\alpha+1}{2}} a(\alpha - i + 1, \alpha) \cdot b(k-i, 2k-1-\alpha)$$
$$= \sum_{i=0}^{\frac{\alpha+1}{2}} \frac{\alpha!\,(2k-1-\alpha)!}{(\alpha-i+1)!\,i!\,(k-i)!\,(k+i-\alpha-1)!}(\alpha-2i)(\alpha-2i+1)^2(\alpha-2i+2).$$

Setting $\beta = \alpha + 1$, this is

$$\sum_{i=0}^{\beta/2} \frac{(\beta-1)!\,(2k-\beta)!}{(\beta-i)!\,i!\,(k-i)!\,(k+i-\beta)!}(\beta-2i-1)(\beta-2i)^2(\beta-2i+1)$$
$$= \frac{2k-\beta}{\beta} \cdot 6\beta(\beta-1)(2k-\beta-1)\frac{(2k-4)!}{k!\,(k-2)!}$$
$$= 6\alpha(2k-\alpha-1)(2k-\alpha-2)\frac{(2k-4)!}{k!\,(k-2)!}.$$

As before, then, we conclude that

$$b_\alpha = \frac{\deg D_k}{2(2k-2-\alpha)} = \frac{3\alpha(2k-\alpha-1)}{k} \frac{(2k-4)!}{(k-1)!\,(k-2)!}$$

The difference here is that this formula is now established as well for $\alpha = 1$, i.e.,

(4)
$$b_1 = \frac{6(k-1)}{k} \cdot \frac{(2k-4)!}{(k-1)!\,(k-2)!}.$$

Our second family of curves lies entirely in Δ_1, consisting of a fixed curve C_2 of genus $g-1$, plus a variable elliptic curve E attached at a constant point of C_2. To construct it, let $\pi_1: X \to B$ be a map from a smooth surface X to a curve B, whose fibers are all stable curves of genus 1; let $S_1 \subset X$ be a section of the map. On the other hand, let C_2 be a general curve of genus $g-1$, $p \in C_2$ a general point, and $S_2 = B \times \{p\} \subset B \times C_2$. Finally, we let

$$\pi\colon U \to B$$

be the curve over B obtained by identifying $S_1 \subset X$ and $S_2 \subset B \times C_2$.

To compute the degrees of the various divisors λ, δ_α, D_k on U, note that if we set

$$d = -\deg N_{S_1/X}$$

then the degree of the j-function associated to π_1 is just $12d$; accordingly

$$\begin{aligned}\deg \delta_0 &= 12d,\\ \deg \lambda &= \deg \pi_* \omega_{U/B}\\ &= \deg(\pi_1)_* \omega_{X/B}\\ &= d,\end{aligned}$$

and

$$\begin{aligned}\deg \delta_1 &= \deg(N_{S_1/X} \otimes N_{S_2/B\times C_2})\\ &= \deg N_{S_1/X}\\ &= -d.\end{aligned}$$

Of course, $\deg \delta_\alpha = 0$ for $\alpha \geqq 2$. On the other hand, we see from Corollary 4, § 4, that U *is disjoint from* $\bar{D}_k$, since C_2, being general, possesses no line bundles L of degree $k-1$ with $h^0(L)=2$ and only finitely many of degree k; and p, being a general point of C_2, will not be a branch point of any of the associated coverings $\pi\colon C_2 \to \mathbf{P}^1$. We conclude, then, immediately

$$a \cdot \deg \lambda - b_0 \deg \delta_0 - b_1 \deg \delta_1 = 0,$$

i.e.,

$$a - 12b_0 + b_1 = 0. \tag{5}$$

For our last family, take C a general curve of genus $2k-2$, $p \in C$ a general point. Let S_1 and S_2 be the proper transforms of the diagonal Δ and the cross-section $C \times \{p\}$ in the blow-up $\widetilde{C \times C}$ of $C \times C$ at the point (p, p). S_1 and S_2 being disjoint, we may identify them in $\widetilde{C \times C}$ over C to obtain a family

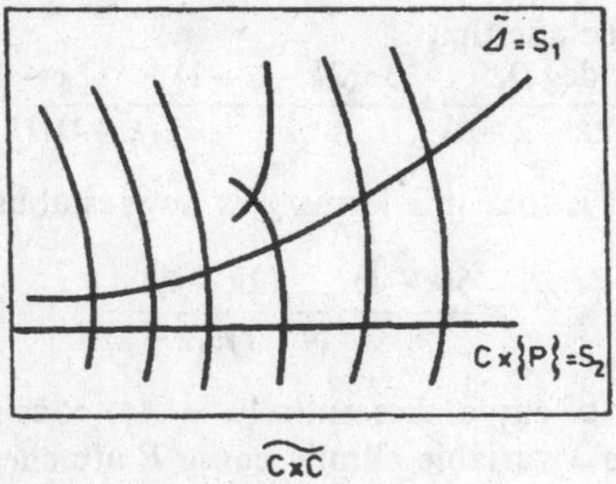

$W \xrightarrow{\pi} C$ of stable curves of genus g over C.

For this family, we have clearly

$$\begin{aligned}\deg \delta_1 &= 1,\\ \deg \delta_\alpha &= 0, \quad \alpha > 1\end{aligned}$$

and

$$\begin{aligned}\deg \delta_0 &= \deg(N_{S_1/\widetilde{C\times C}} \otimes N_{S_2/\widetilde{C\times C}})\\ &= (2-2(2k-2)-1)-1\\ &= 4-4k.\end{aligned}$$

We also have the sequence

$$0 \to H^0(C, \omega_C) \otimes \mathcal{O}_C \to \pi_* \omega_{W/C} \to \mathcal{O}_C \to 0$$

where the right-hand map is given by taking the residue at S_1, from which we conclude that

$$\deg \lambda = c_1 \pi_* \omega_{W/C} = 0.$$

Finally, since C and $p \in C$ are general, C will possess exactly $\frac{(2k-2)!}{(k-1)!(k)!}$ line bundles L of degree k with $h^0(L)=2$, each of which will have a unique section zero at p; and none of these sections will have a multiple zero. There are thus a total of $\frac{(2k-2)!}{(k-1)!(k)!}(k-1)$ points $q \in C$ such that $h^0(L(-p-q))=1$ for one of these L's. By Theorem 6a, these points occur with multiplicity one in the divisor induced by $\overline{D}_k$ on C, i.e.,

$$\deg_C \overline{D}_k = \frac{(2k-2)!}{k!(k-2)!}$$

and we conclude that

$$4(k-1)b_0 - b_1 = \frac{(2k-2)!}{k!(k-2)!}. \tag{6}$$

Since from (4) we have

$$b_1 = \frac{6(k-1)}{k} \frac{(2k-4)!}{(k-1)!(k-2)!},$$

this yields

$$\begin{aligned}b_0 &= \frac{1}{4(k-1)}\left(\frac{(2k-2)!}{k!(k-2)!} + \frac{6(k-1)}{k}\frac{(2k-4)!}{(k-1)!(k-2)!}\right)\\ &= \frac{1}{4}\cdot\frac{(2k-4)!}{k!(k-2)!}(2(2k-3)+6)\\ &= \frac{(2k-4)!}{(k-1)!(k-2)!}\end{aligned}$$

and applying the relation (5), we have in turn

$$\begin{aligned}a &= 12b_0 - b_1\\ &= \left(12 - \frac{6(k-1)}{k}\frac{(2k-4)!}{(k-1)!(k-2)!}\right)\\ &= \left(6+\frac{6}{k}\right)\cdot\frac{(2k-4)!}{(k-1)!(k-2)!}.\end{aligned}$$

All in all, then, the coefficients appearing in the expression (1) are

$$a=6(k+1)c,$$

$$b_0=kc, \quad b_\alpha=3\alpha(2k-\alpha-1)c$$

where

$$c=\frac{(2k-4)!}{k!(k-2)!}.$$

References

1. Anderson, G.: Theta functions and holomorphic differential forms on compact quotients of bounded symmetric domains. Ph.D. Thesis, Princeton, 1980
2. Arbarello, E., Cornalba, M., Griffiths, P., Harris, J.: Topics in the theory of algebraic curves. In Princeton Univ. Press (1982 in press)
3. Brylinski, J.-L.: Propriétés de ramification à l'infini du groupe modulaire de Teichmüller. Ann. Ec. Norm. Sup., **12**, 295 (1979)
4. Freitag, E.: Der Körper der Siegelschen Modulfunktionen. Abh. Math. Sem. Univ. Hamburg,
5. Freitag, E.: Die Kodairadimension von Körpern automorpher Funktionen. Crelle, **296**, 162 (1977)
6. Fulton, W.: Hurwitz schemes and the irreducibility of moduli of algebraic curves, Annals of Math., **90**, 542 (1969)
7. Griffiths, P., Harris, J.: On the variety of special linear systems on a general algebraic curve. Duke Math. J., **47**, 233 (1980)
8. Grothendieck, A., Murre, J.P.: The tame fundamental group of a formal neighborhood of a divisor with normal crossings on a scheme. Lecture Notes Berlin-Heidelberg-New York: Springer Vol. 208. 1971
9. Kleiman, S., Laksov, D.: Another proof of the existence of special divisors. Acta Math., **132**, 163 (1974)
10. Knudsen, F.: The projectivity of the moduli space of stable curves. Math. Scand. (in press 1982)
11. MacDonald, I.G.: Symmetric products of an algebraic curve. Topology, **1**, 319 (1962)
12. Mumford, D.: Hirzebruch's Proportionality Principle in the non-compact case. Inv. Math., **42**, 239 (1977)
13. Mumford, D.: Stability of projective varieties. L'Ens. Math., **23**, 39 (1977)
14. Murre, J.P.: An introduction to Grothendieck's of the fundamental group, Tata Institute Lecture Notes; Bombay (1967)
15. Reid, M.: Canonical 3-folds. Les Journées de Géometrie Algébrique d'Angers, 1979, A. Beauville, ed.
16. Stillman, M.: PhD Thesis, Harvard, 1983
17. Tai, Y.-S.: Pluri-canonical differentials on the Siegel modular variety. Invent. Math. (in press 1982)

Oblatum 3-XI-1981

Added in Proof

Since this article was written, 2 improvements have been made in Theorem 3. John Harer has proven that $\mathrm{Pic}(M_g^0)$ is infinite cyclic, hence *any* divisor D on M_g^0 satisfies

$$[D]\equiv a\cdot\lambda, \quad a\in\mathbf{Q}.$$

Secondly George Kempf has been able to carry through the calculation of the a_k of Theorem 3 via Porteous' formula, confirming that

$$a_k=6(2k-4)!\,(k+1)/k!\,(k-2)!.$$

Towards an Enumerative Geometry of the Moduli Space of Curves

David Mumford

Dedicated to Igor Shafarevitch on his 60th birthday

Introduction

The goal of this paper is to formulate and to begin an exploration of the enumerative geometry of the set of all curves of arbitrary genus g. By this we mean setting up a Chow ring for the moduli space $\mathcal{M}_g$ of curves of genus g and its compactification $\overline{\mathcal{M}}_g$, defining what seem to be the most important classes in this ring and calculating the class of some geometrically important loci in $\overline{\mathcal{M}}_g$ in terms of these classes. We take as a model for this the enumerative geometry of the Grassmannians. Here the basic classes are the Chern classes of the tautological or universal bundle that lives over the Grassmannian, and the most basic cycles are the loci of linear spaces satisfying various Schubert conditions: the so-called Schubert cycles. However, since Harris and I have shown that for g large, $\mathcal{M}_g$ is not unirational [H-M] it is not possible to expect that $\mathcal{M}_g$ has a decomposition into elementary cells or that the Chow ring of $\mathcal{M}_g$ is as simple as that of the Grassmannian. But in the other direction, J. Harer [Ha] and E. Miller [Mi] have strong results indicating that at least the low dimensional homology groups of $\mathcal{M}_g$ behave nicely. Moreover, it appers that many geometrically natural cycles are all expressible in terms of a small number of basic classes.

More specifically, the paper is divided into 3 parts. The goal of the first part is to define an intersection product in the Chow group of $\overline{\mathcal{M}}_g$. The problem is that due to curves with automorphisms, $\overline{\mathcal{M}}_g$ is singular, but in a mild way. In fact it is a "Q-variety", locally the quotient of a smooth variety by a finite group. If it were *globally* the quotient of a smooth variety by a finite group, it would be easy to define a product in $A \cdot (\overline{\mathcal{M}}_g) \otimes \mathbb{Q}$. Instead we have used the fact that $\overline{\mathcal{M}}_g$ is globally the quotient of a Cohen-Macaulay variety by a finite group, plus many of the ideas of Fulton and MacPherson, and especially a strong use of both the Grothendieck and Baum-Fulton-MacPherson forms of the Riemann-Roch theorem to achieve this goal. To handle an arbitrary Q-variety, Gillet has proposed using higher K-theory ($H^n(K^n) \cong A^n$) and this may well be the right technique.

The goal of the second part is to introduce a sequence of "tautological" classes $\kappa_i \in A^i(\overline{\mathcal{M}}_g) \otimes \mathbb{Q}$, derive some relations between them, and calculate the fundamental class of certain subvarieties, such as the hyperelliptic locus, in terms of them. Again the Grothendieck Riemann-Roch theorem is one of the main tools. Some of these results have been found independently by E. Miller [Mi], and it seems reasonable to guess, in view of the results of Harer and Miller (op. cit.), that in low codimensions $H^i(\mathcal{M}_g) \otimes \mathbb{Q}$ is a polynomial ring in the κ_i.

Finally, to make the whole theory concrete, we work out $A^{\cdot}(\overline{\mathcal{M}}_2)$ completely in Part III. An interesting corollary is the proof, as a consequence of general results only, that $\mathcal{M}_2$ is affine. It seems very worthwhile to work out $A^{\cdot}(\overline{\mathcal{M}}_g)$ or $H^{\cdot}(\mathcal{M}_g)$ for other small values of g, in order to get some feeling for the properties of these rings and their relation to the geometry of $\overline{\mathcal{M}}_g$. The techniques of Atiyah-Bott [A-B] may be very useful in doing this.

Part I: Defining a Chow Ring of the Moduli Space

§1. Fulton's Operational Chow Ring

If X is any quasi-projective variety, Fulton and Fulton-MacPherson have defined in two papers ([F1], p. 157, [F-M], p. 92) two procedures to attach to X a kind of Chow cohomology theory: a ring-valued contravariant functor. We combine here the 2 definitions taking the simplest parts of them in a way that is adequate for our applications. The theory also becomes substantially simpler if we take the *coefficients for our cycles to be* $\mathbb{Q}$, and we assume that *char* $k = 0$. This is the case we are interested in, *so we will restrict ourselves to this case henceforth.* We call the resulting ring $opA^{\cdot}(X)$. To form this ring, take:

generators : *elts*(f, α),

$f \colon X \longrightarrow Y$ *a morphism*

Y *smooth, quasi-projective*

α *a cycle on* Y

relations : $(f,\alpha) \sim 0$ if, for all $g\colon Z \longrightarrow X$, we have:
$(f \circ g)^* \alpha$ rationally equivalent to 0 on Z
via the induced map

$$A^k(Y) \xrightarrow{(f\circ g)^*} A_{n-k}(Z)$$

$(k = \text{codim of } \alpha, n = \dim \text{ of } Z)$

Equivalently, we may define

$$opA^{\cdot}(X) = \text{Image}\left\{ \varinjlim_{(X \xrightarrow{f} Y)} A^{\cdot}(Y) \longrightarrow \prod_{(Z \xrightarrow{g} X)} \text{End}\big(A_{\cdot}(Z)\big)\right\}$$

where the map is given by cap product

$$A^k(Y) \times A_l(Z) \xrightarrow{\cap} A_{l-k}(Z)$$

(cf. [F1], §2). This makes it clear that $op\,A^{\cdot}$ is a ring and a contravarient functor and that $opA^{\cdot}(X)$ acts on $A_{\cdot}(X)$ by cap product. If X is smooth, then $opA^{\cdot}(X) \cong A^{\cdot}(X)$.

Moreover, as in Fulton [F1], §3.2, for all coherent sheaves $\mathcal{F}$ with finite projective resolutions, we can define the Chern classes $c_k(\mathcal{F}) \in opA^k(X)$, (by resolution of $\mathcal{F}$, twisting and pull-back of Schubert cycles from maps of X to Grassmannians).

Using a resolution of X, we can give a very simple description of the relations in $opA^{\cdot}(X)$:

Proposition 1.1. *If $\pi\colon \tilde{X} \longrightarrow X$ is a resolution of X, then*

$$(f,\alpha) \sim 0 \Leftrightarrow (f \circ \pi)^*(\alpha) = 0 \quad \text{in } A^k(\tilde{X}),$$

i.e.,

$$opA^{\cdot}(X) \subset A^{\cdot}(\tilde{X}).$$

Proof. We must show that if

$$g\colon Z \longrightarrow X$$

is any test morphism, then $l(f \circ \pi)^* \alpha$ rationally equivalent to 0 on $\tilde{X}$ implies $l'(f \circ g)^* \alpha$ rationally equivalent to 0 on Z for some l'. But by taking a suitable subvariety of $Z \times_X \tilde{X}$ we get a diagram

$$\begin{array}{ccccc} \tilde{Z} & \xrightarrow{\tilde{g}} & \tilde{X} & & \\ p\downarrow & & \downarrow\pi & & \\ Z & \xrightarrow{g} & X & \xrightarrow{f} & Y \end{array}$$

where p is proper, surjective, generically finite of degree l''. Therefore

$$\begin{aligned} l''(f\circ g)^*\alpha &= p_*\big((f\circ g\circ p)^*\alpha\big) \\ &= p_*\big(\tilde{g}^*\big((f\circ\pi)^*\alpha\big)\big) \end{aligned}$$

hence

$$l.l''(f\circ g)^*\alpha = p_*\big(\tilde{g}^*\big(l(f\circ\pi)^*\alpha\big)\big) \sim 0.$$

This uses the formula:

$$(*)\ \text{For all}\ \begin{array}{ccc} \tilde{Z} & \xrightarrow{\tilde{h}} & Y \\ p\downarrow & & \| \\ Z & \xrightarrow[h]{} & Y \end{array} \quad \begin{array}{l} Y \text{ smooth, } \alpha \text{ a cycle on } Y, \\ p \text{ proper, surjective, generically finite of degree } d \end{array}$$

$$p_*\tilde{h}^*(\alpha) \sim dh^*(\alpha).$$

(See [F1], §2.2, part 2 of lemma). Q.E.D.

In fact, we can say more:

Proposition 1.2. *In the situation of Prop. 1.1, the image of $opA^k(X)$ in $A^k(\tilde{X})$ is contained in the subgroup of $A^k(\tilde{X})$ generated by irreducible subvarieties W of $\tilde{X}$ such that $W = \pi^{-1}(\pi(W))$.*

Proof. Let (f, α) be a generator of $opA^k(X)$, where $f\colon X \longrightarrow Y$ is a morphism and Y is smooth, quasi-projective. Let $p = f\circ\pi$, and

$$Y_k = \{y \in Y \mid \dim p^{-1}(y) \geq k\}.$$

Then use the moving lemma on Y to represent α as a cycle $\sum n_i W_i$, whose components W_i meet properly all the Y_k. Then each component W_{ij} of $p^{-1}(W_i)$ meets the open set of $\tilde{X}$ where π is an isomorphism and

$p \colon \tilde{X} \longrightarrow p(\tilde{X})$ is equidimensional. Therefore, $p^*\alpha$ is represented by a combination of the W_{ij} with suitable multiplicities and each W_{ij} satisfies $W_{ij} = \pi^{-1}(\pi(W_{ij}))$. Q.E.D.

There is also another natural way to give generators for $opA^{\cdot}(X)$:

Proposition 1.3. *Using rational coefficients in $K(X)$ too, the homomorphism*

$$ch : K(X) \longrightarrow opA^{\cdot}(X)$$

is surjective, hence $opA^{\cdot}(X)$ can be defined to be

$$\begin{array}{rcl} \text{Image}\,[K(X) & \rightarrow & A^{\cdot}(\tilde{X})] \\ \mathcal{E} & \mapsto & ch(\pi^*\mathcal{E}) \end{array}$$

($\pi \colon \tilde{X} \longrightarrow X$ a resolution of X).

Proof. It is well known that for any smooth quasi-projective Z, $ch \colon gr\, K(Z) \longrightarrow A^{\cdot}(Z)$ is a graded isomorphism, hence taking the total Chern characters, $ch \colon K(Z) \longrightarrow A^{\cdot}(Z)$ is also an isomorphism[1]. Therefore $ch \colon K(X) \longrightarrow opA^{\cdot}(X)$ is surjective.

$opA^{\cdot}(X)$ has a much subtler *covariance* for certain morphisms f, whose existence is tied up with the version of the Grothendieck-Riemann-Roch theorems for $opA^{\cdot}$. The result is this:

Theorem 1.4 (Fulton). *Let $f \colon X \longrightarrow Y$ be a projective local-complete-intersection morphism. Define $Td_f \in opA^{\cdot}(X)$ in the usual way. Then there is a homomorphism*

$$f_* \colon opA^{\cdot}(X) \longrightarrow opA^{\cdot}(Y)$$

such that

---

[1]This sounds a bid odd, but it is perhaps clarified by the observation that for any rank r and dimension n, there are universal polynomials P_k such that for all vector bundles $\mathcal{E}$ of rank r on n-dimensional varieties,

$$c_k(\mathcal{E}) = P_k(ch\,\mathcal{E}, ch\,\Lambda^2\mathcal{E}, \ldots, ch\,\Lambda^r\mathcal{E})$$

where $ch\,\mathcal{E}$ is the *total* Chern character, and these elements lie in any cohomology ring with the usual Chern formalism and rational coeffcients.

1) *for all cartesian diagrams*

$$\begin{array}{ccc} X' & \xrightarrow{h} & X \\ {\scriptstyle f'}\downarrow & & \downarrow{\scriptstyle f} \\ Y' & \xrightarrow[g]{} & Y \end{array}$$

and all $\alpha \in A.(Y')$, $\beta \in opA^{\cdot}(X)$

$$f_*\beta \cap \alpha = f'_*(\beta \cap [f]_{Y'}(\alpha)) \quad \text{in } A.(Y')$$

($[f]_{Y'}$ defined as in Fulton-MacPherson [F-M], p. 95).

2) *for all locally free sheaves* $\mathcal{E}$ *on* X,

$$ch(f_!\mathcal{E}) = f_*(ch\,\mathcal{E}.Td_f).$$

This is proven in Fulton [F2], Ch.18: here $opA^{\cdot}$ is a possibly larger Chow Ring in which (1) is the *definition* of f_*. In this ring (2) is proven, and (2) shows that $f_*\beta$ actually lies in the subring $opA^{\cdot}$ considered here.

§2. Q-Varieties and $\overline{\mathcal{M}}_g$

The moduli space $\overline{\mathcal{M}}_g$ of stable curves is an example of a variety which is locally in the étale topology a quotient of a smooth variety by a finite group. The approach we take to defining a Chow ring for $\overline{\mathcal{M}}_g$ is best studied in this more general context. Because these varieties are quite close to the objects introduced by Matsusaka [Ma], we shall call them quasi-projective Q-varieties. We define a quasi-projective Q-variety to be:

1) a quasi-projective variety X,
2) a finite atlas of charts:

$$\begin{array}{c} X_\alpha \\ \downarrow \\ X_\alpha/G_\alpha \\ \downarrow {\scriptstyle p'_\alpha} \\ X \end{array} \qquad p_\alpha = p'_\alpha \circ (X_\alpha \to X_\alpha/G_\alpha)$$

where p'_α is étale, G_α is a finite group acting, faithfully on a quasi-projective *smooth* X_α and

$$X = \bigcup_\alpha (\mathrm{Im}\ p_\alpha),$$

3) The charts should be compatible in the sense that for all α, β, let

$$X_{\alpha\beta} = \text{normalization of } X_\alpha \times_X X_\beta.$$

Then the projections

$$\begin{array}{ccc} & X_{\alpha\beta} & \\ \swarrow & & \searrow \\ X_\alpha & & X_\beta \end{array}$$

should be étale.

Here, of course, a new chart can be added if it satisfies the compatibility conditions (3) with the old ones. For any such Q-variety, we can normalize X in a Galois extension of its function field $k(X)$ containing copies of the field extensions $k(X_\alpha)$ for all α. This leads to a covering $p: \tilde{X} \longrightarrow X$ with a group G acting faithfully on $\tilde{X}$ and $X = \tilde{X}/G$. The fact that $k(\tilde{X}) \supset k(X_\alpha)$ leads to a factorization of p locally:

$$\begin{array}{ccc} \tilde{X} & \supset & \tilde{X}_\alpha \\ {\scriptstyle p}\Big\downarrow & & \downarrow \\ & & \tilde{X}_\alpha/H_\alpha \cong X_\alpha \\ X & \xleftarrow{\quad p_\alpha \quad} & \end{array}$$

$\tilde{X}_\alpha$ open in $\tilde{X}$ stabilized by $H'_\alpha \subset G$,
$H_\alpha =$ normal subgroup of H'_α with $G_\alpha \cong H'_\alpha/H_\alpha$.

Q-varieties come in various different grades. The best ones are those such that for some atlas, $\tilde{X}$ can be chosen to be itself smooth. Not so nice, but still amendable to the techniques we shall use are those where $\tilde{X}$ can be chosen to be Cohen-Macaulay. We call these Q-varieties *with a smooth global cover* and with a *Cohen-Macaulay global cover* respectively.

Another important concept is that of a *Q-sheaf on a Q-variety X*. By this, we do not mean a coherent sheaf on X, but rather a family of coherent sheaves $\mathcal{F}_\alpha$ on X_α, plus isomorphisms

$$\mathcal{F}_\alpha \otimes_{O_{X_\alpha}} O_{X_{\alpha\beta}} \cong \mathcal{F}_\beta \otimes_{O_{X_\beta}} O_{X_{\alpha\beta}}$$

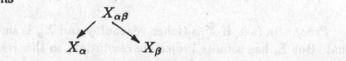

compatible on the triple overlaps. Note that such a coherent sheaf pulls back by tensor product to a family of coherent sheaves on $\tilde{X}_\alpha$, which glue together to one coherent sheaf $\tilde{\mathcal{F}}$ on $\tilde{X}$ on which G acts. Therefore, equivalently we can define a coherent sheaf on the Q-variety X to be a coherent G-sheaf $\tilde{\mathcal{F}}$ on $\tilde{X}$ such that for all α, $\tilde{\mathcal{F}}|_{\tilde{X}_\alpha}$ with its H'_α-action is the pull-back of a coherent sheaf on X_α. The importance of $\tilde{X}$ being Cohen-Macaulay is illustrated by the simple fact:

Proposition 2.1. *If $\tilde{X}$ is Cohen-Macaulay, then for any coherent sheaf $\mathcal{F}$ on the Q-variety X, $\tilde{\mathcal{F}}$ has a finite projective resolution.*

Proof. In fact, if $\tilde{X}$ is Cohen-Macaulay and X_α is smooth, $\tilde{X}_\alpha \to X_\alpha$ is flat. But $\mathcal{F}_\alpha$ has a finite projective resolution, so this resolution pulls back to such a resolution for $\tilde{\mathcal{F}}|_{\tilde{X}_\alpha}$. Q.E.D.

Consider the case of the moduli space $\overline{\mathcal{M}}_g$. Choose an integer $n \geq 3$ prime to the characteristic. Fix a free $\mathbb{Z}/n\mathbb{Z}$-module V of rank $2g$ with an alternating non-degenerate form

$$e: V \times V \longrightarrow \mu_n.$$

Fix, moreover, a flag of isotropic free submodules:

$$(0) \subset V_1 \subset V_2 \subset \cdots \subset V_g = V_g^\perp \subset \cdots \subset V_2^\perp \subset V_1^\perp \subset V$$

where $rk(V_i) = i$. Then for all stable curves C of genus g, let h be the sum of the genera of the components of its normalization. Then there is an isomorphism

$$H^1(C, \mathbb{Z}/n\mathbb{Z}) \cong V_{g-h}^\perp$$

such that the form

$$H^1 \times H^1 \xrightarrow{\cup} H^2 \xrightarrow{\text{f'al class}} \mu_n$$

corresponds to e. We may consider the auxiliary moduli space

$$\left(\overline{\mathcal{M}}_g^{(n,h)}\right)' = \text{set of pairs } (C, \phi),\ C \text{ stable curve,}$$

$$\phi : V_{g-h}^\perp \xrightarrow[\subset]{\phi\ \text{injective}} H^1(C, \mathbb{Z}/n\mathbb{Z}) \text{ a sympl. map}$$

which can be constructed by standard arguments. Inside this space, define an *open* subset by:

$$\begin{aligned}\overline{\mathcal{M}}_g^{(n,h)} = {} & \text{those pairs } (C,\phi) \text{ such that every automorphism} \\ & \alpha\colon C \longrightarrow C \text{ fixes the submodule } \operatorname{Im}\phi \subset H^1(C,\mathbb{Z}/n\mathbb{Z}) \\ & \text{and, if } \alpha \neq 1_C, \text{ then } \alpha \text{ acts non-trivially on } \operatorname{Im}\phi.\end{aligned}$$

Since the pairs (C,ϕ) in this subset have no automorphisms, $\overline{\mathcal{M}}_g^{(n,h)}$ is smooth and represents the universal deformation space of any curve occuring in it. Note that every curve C occurs in the space $\overline{\mathcal{M}}_g^{(n,h)}$ such that $g+h = rk\, H^1(C)$ (see [D-M], Th. 1.13). Next consider the finite groups

$$\begin{aligned}
G &= Sp(V,\mathbb{Z}/n\mathbb{Z}) \\
& \quad \cup \\
H'_h &= (\text{stabilizer of } V_{g-h}) \\
& \quad \cup \\
H_h &= \begin{pmatrix}\text{elements which} \\ \text{act identically} \\ \text{on } V_{g-h}^{\perp}\end{pmatrix} \\
G_h &= H'_h/H_h = (\text{induced group of automorphisms of } V_{g-h}^{\perp}).
\end{aligned}$$

Then G_h acts on $\overline{\mathcal{M}}_g^{(n,h)}$ and we have canonical morphisms

$$\begin{array}{c}
\overline{\mathcal{M}}_g^{(n,h)} \\
\downarrow \\
\overline{\mathcal{M}}_g^{(n,h)}/G_h \\
\downarrow{\scriptstyle p'_h} \\
\overline{\mathcal{M}}_g
\end{array}
\qquad p_h\colon \overline{\mathcal{M}}_g^{(n,h)} \longrightarrow \overline{\mathcal{M}}_g$$

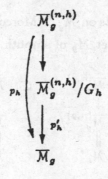

I claim that p'_h is étale. In fact, if $(C,\phi) \in \overline{\mathcal{M}}_g^{(n,h)}$, then $\mathrm{Aut}(C)$ can be identified with a subgroup of G_h and *formally* near $[C,\phi]$, the isomorphism

$$\overline{\mathcal{M}}_g^{(n,h)} \xleftarrow[\text{formal isom.}]{} \mathrm{Def}(C)$$

commutes with the action of $\mathrm{Aut}(C)$.

Therefore we have a diagram

$$\begin{array}{ccc}
\overline{\mathcal{M}}_g^{(n,h)} & \longleftarrow & \mathrm{Def}(C) \\
\downarrow & & \downarrow \\
\overline{\mathcal{M}}_g^{(n,h)}/\mathrm{Aut}(C) & \longleftarrow & \mathrm{Def}(C)/\mathrm{Aut}(C) \\
p\downarrow & & \Big\downarrow \\
\overline{\mathcal{M}}_g^{(n,h)}/G_h & & \\
q\downarrow & & \\
\overline{\mathcal{M}}_g & \longleftarrow & \mathrm{Spec}\, \hat{\mathcal{O}}_{\overline{\mathcal{M}}_g,[C]}
\end{array}$$

where the horizontal arrows are formal isomorphisms. Now $\mathrm{Def}(C)/\mathrm{Aut}(C)$ i.e., Spec of the $\mathrm{Aut}(C)$-invariants in the complete local ring representing the deformations of C, is isomorphic to Spec of the complete local ring of $\overline{\mathcal{M}}_g$ at [C]. Thus the morphism indicated by $q \circ p$ in the diagram is étale at $[C,\phi]$. Therefore so is q.

This proves that the atlas $\{p_h\colon\ \overline{\mathcal{M}}_g^{(n,h)} \longrightarrow \overline{\mathcal{M}}_g\}$ puts a structure of Q-variety on $\overline{\mathcal{M}}_g$. In this setting, what is the variety $\tilde{X}$ dominating all the charts? This is

$$\overline{\mathcal{M}}_g^{(n)} = (\text{normalization of } \overline{\mathcal{M}}_g \text{ in this field extension } \overline{\mathcal{M}}_g^{(n,g)}).$$

Note that the full group G acts on $\overline{\mathcal{M}}_g^{(n)}$. Moreover, $\overline{\mathcal{M}}_g^{(n,g)}$ is the open subset of $\overline{\mathcal{M}}_g^{(n)}$ lying over the open set $\mathcal{M}_g$ of smooth curves.

$$\begin{array}{rcccl}
p^{-1}(\mathcal{M}_g) = & \overline{\mathcal{M}}_g^{(n,g)} & \subset & \overline{\mathcal{M}}_g^{(n)} & \\
& \downarrow & & \downarrow & \\
& \overline{\mathcal{M}}_g^{(n,g)}/G & \subset & \overline{\mathcal{M}}_g^{(n)}/G & \Big) \, p \\
& \| & & \| & \\
& \mathcal{M}_g & \subset & \overline{\mathcal{M}}_g &
\end{array}$$

What we call $\overline{\mathcal{M}}_g^{(n,h)}$, for $h < g$, however, is recovered by dividing an open subset of $\overline{\mathcal{M}}_g^{(n)}$ by H_h.

Life would be particularly simple if $\overline{\mathcal{M}}_g^{(n)}$ were smooth. However, it is not, nor is it known whether the normalization of $\overline{\mathcal{M}}_g^{(n)}$ in any finite extension field is smooth. However, fortunately $\overline{\mathcal{M}}_g$ does have a Cohen-Macaulay global cover, because:

Proposition 2.2. $\overline{\mathcal{M}}_g^{(n,g)} \subset \overline{\mathcal{M}}_g^{(n)}$ *is a toroidal embedding, i.e.,* $\overline{\mathcal{M}}_g^{(n)}$ *is formally isomorphic to* $\mathbb{A}^{3g-3}$ *modulo an abelian group acting diagonally, and with* $\overline{\mathcal{M}}_g^{(n)} - \overline{\mathcal{M}}_g^{(n,g)}$ *isomorphic to the image of a union of coordinate hyperplanes in* $\mathbb{A}^{3g-3}$.

Proof. At every point of $\overline{\mathcal{M}}_g^{(n)}$, $\overline{\mathcal{M}}_g^{(n)}$ is a Galois covering of one of the smooth varieties $\overline{\mathcal{M}}_g^{(n,h)}$ with group H_h. Note that H_h is abelian of order prime to char (k). The covering is ramified only on $\overline{\mathcal{M}}_g^{(n,h)} - p_h^{-1}(\mathcal{M}_g)$. Since $\overline{\mathcal{M}}_g^{(n,h)}$ is formally the universal deformation space of some curve C, this is formally a ramified cover of $\mathbb{A}^{3g-3}$, ramified only in coordinate hyperplanes. But if char $(k) \not| \, n$, the n-cyclic extensions $k[[x_1, \ldots, x_{3g-3}]]$ ramified only over the ideals (x_i) are all given by

$$\left(\prod_{i \in I} x_i^{a_i}\right)^{1/n}, \quad I \subset \{1, \ldots, 3g-3\}.$$

Thus the covering in hand is sandwiched between $k[[x_1, \ldots, x_{3g-3}]]$ and $k[[x_1^{1/n}, \ldots, x_{3g-3}^{1/n}]]$, hence, by Galois theory, is as described.

Corollary 2.3. $\overline{\mathcal{M}}_g^{(n)}$ *is Cohen-Macaulay.*

§2*b*. *Q*-Stacks

Unfortunately, the concept of *Q*-variety, although adequate to deal with $\overline{\mathcal{M}}_g$, $g \geq 3$, or with any moduli variety whose general object has no automorphisms, breaks down for $\overline{\mathcal{M}}_2$ and $\overline{\mathcal{M}}_{1,1}$ where the general object has automorphism group $\mathbb{Z}/2\mathbb{Z}$. Consider, for instance, $\overline{\mathcal{M}}_2$. Let $\mathcal{M}_2^o \subset \overline{\mathcal{M}}_2$

be the open set of smooth curves C such that $\mathrm{Aut}(C) \cong \mathbb{Z}/2\mathbb{Z}$. Then, although $\mathcal{M}_2^o$ gives a local deformation space for its curves, it does not carry a universal family of curves. And if $\mathcal{M}_2'$ is an étale cover of $\mathcal{M}_2^o$ carrying some family

$$p\colon C' \longrightarrow \mathcal{M}_2'$$

the sheaf $\mathbb{E} = p_*\Omega^1_{C'/\mathcal{M}'}$ on $\mathcal{M}_2'$ will not be a Q-sheaf. In fact, to compare $p_1^*\mathbb{E}$, $p_2^*\mathbb{E}$ on $\mathcal{M}_2' \times_{\mathcal{M}_2} \mathcal{M}_2'$, we want an isomorphism of the 2 families

$$\begin{array}{ccc} C' \times_{\mathcal{M}_2} \mathcal{M}_2' & \cong & \mathcal{M}_2' \times_{\mathcal{M}_2} C' \\ & \searrow \quad \swarrow & \\ & \mathcal{M}_2' \times_{\mathcal{M}_2} \mathcal{M}_2' & \end{array}$$

and although these families are fibrewise isomorphic, the isomorphism is not unique and may not globalize.

To deal with this, one must use some variant of the ideal of stack (cf. [D-M], §4). The most natural thing is to replace the normalization of $X_\alpha \times_X X_\beta$ by a scheme $X_{\alpha\beta}$ which must be given as part of the data and cannot be derived from the rest. $X_{\alpha\beta}$ should map to $X_\alpha \times_X X_\beta$ and given $x \in X_{\alpha\beta}$, $y \in X_{\beta\gamma}$ with the same projection to X_β, a "composition" $x \circ y \in X_{\alpha\gamma}$ should be defined. A point $x \in X_{\alpha\beta}$ lying over $u \in X_\alpha$, $v \in X_\beta$ should be thought of as meaning an isomorphism from the object C_u corresponding to u to the object C_v corresponding to v.

Definition 2.4. A *Q-stack* is a collection of quasi-projective varieties and morphisms:

$$\begin{array}{ccccc} \coprod X_{\alpha\beta} & \overset{p_1}{\underset{p_2}{\rightrightarrows}} & \coprod X_\alpha & \xrightarrow{p} & X \\ & \curvearrowleft & & & \end{array}$$

X_α, $X_{\alpha\beta}$ smooth, X normal, p_1, p_2 étale

$$\begin{gathered} \coprod X_\alpha \to X \text{ surjective} \\ X_{\alpha\beta} \to X_\alpha \times_X X_\beta \text{ surjective, finite} \\ p_i \circ \epsilon = \text{identity} \end{gathered}$$

plus morphisms[2]:

$$\begin{aligned} X_{\alpha\beta} \times_{X_\beta} X_{\beta\gamma} &\xrightarrow{\circ} X_{\alpha\gamma} \\ X_{\alpha\beta} &\xrightarrow{-1} X_{\beta\alpha} \end{aligned}$$

making $\coprod X_{\alpha\beta}$ into a pseudo-group (i.e., $\circ$ is associative where defined, $^{-1}$ is its inverse and ϵ is an identity).

It is an interesting exercise in categorical style constructions to show that this collection of data can be derived from a finite group G acting on a normal variety $\tilde{X}$, plus open sets $\tilde{X}_\alpha \subset \tilde{X}$ stabilized by $H_\alpha \subset G$, very much as above:

$$\begin{array}{ccc} \tilde{X} & \supset & \tilde{X}_\alpha \\ & & \downarrow \\ p \downarrow & & \tilde{X}_\alpha/H_\alpha = X_\alpha \\ & \swarrow p_\alpha & \\ X = \tilde{X}/G & & \end{array}$$

and satisfying:

(2.5) a) *For all* $x \in \tilde{X}_\alpha \cap g(\tilde{X}_\beta)$ *and all* $h \in H_\alpha$ *such that* $h(x) = x$ *then* $g^{-1}hg \in H_\beta$.
b) H_α *acts faithfully on* X_α.

Then define

$$X_{\alpha\beta} = \coprod_{\substack{g = \text{repres. of double} \\ \text{cosets } H_\alpha \backslash G / H_\beta}} \tilde{X}_\alpha \cap g(\tilde{X}_\beta)/H_\alpha \cap gH_\beta g^{-1}$$

$$p_1 = \text{natural map } \tilde{X}_\alpha \cap g(\tilde{X}_\beta)/H_\alpha \cap gH_\beta g^{-1} \to \tilde{X}_\alpha/H_\alpha$$

$$p_2 = \text{the map induced by } g^{-1}$$

$$\tilde{X}_\alpha \cap g(\tilde{X}_\beta)/H_\alpha \cap gH_\beta g^{-1} \to \tilde{X}_\beta/H_\beta$$

and if

$$x \in \tilde{X}_\alpha \cap g(\tilde{X}_\beta) \text{ maps to } \overline{x} \in X_{\alpha\beta}$$

$$y \in \tilde{X}_\beta \cap g'(\tilde{X}_\gamma) \text{ maps to } \overline{y} \in X_{\beta\gamma}$$

---

[2]The maps $\circ$ can also be introduced by giving as extra data more of a semi-simplicial variety:

$$\coprod X_{\alpha\beta\gamma} \;\substack{\longrightarrow \\ \longrightarrow \\ \longrightarrow \\ \longleftarrow \\ \longleftarrow}\; \coprod X_{\alpha\beta}$$

in the usual way.

so that

$$g^{-1}x = hy, \qquad h \in H_\beta$$

then let

$$\bar{x} \circ \bar{y} = (\text{image of } x \in \tilde{X}_\alpha \cap ghg'(\tilde{X}_\gamma) \text{ in } X_{\alpha\gamma}).$$

The object so constructed is a Q-variety *if G acts faithfully on $\tilde{X}$*, but in general only a Q-stack.

Note that $\overline{\mathcal{M}}_2$ and $\overline{\mathcal{M}}_{1,1}$ are in a natural way Q-stacks. We let $\tilde{\mathcal{M}}_2, \tilde{\mathcal{M}}_{1,1}$ be the normalization of $\overline{\mathcal{M}}_2, \overline{\mathcal{M}}_{1,1}$ in the level n covering, some $n \geq 3$ and let $G = Sp(4, \mathbb{Z}/n\mathbb{Z})$, $SL(2, \mathbb{Z}/n\mathbb{Z})$ resp. The open sets X_α and subgroups H_α are defined exactly as in the case $g \geq 3$ treated above. The most general chart is any $X_\alpha \to \overline{\mathcal{M}}_2$ (resp. $\overline{\mathcal{M}}_{1,1}$) such that X_α comes with a family of the corresponding curves over it which represents locally everywhere the universal deformation space. Given 2 charts X_α, X_β, with families C_α, C_β, then $X_{\alpha\beta}$ is by definition:

$$\text{Isom}\,(C_\alpha, C_\beta) = \{(x, y, \phi) \mid x \in X_\alpha, y \in X_\beta, \phi \text{ an isom. of } C_{\alpha,x} \text{ with } C_{\beta,y}\}$$

Finally morphisms between Q-stacks X, Y are given by sets of morphisms and commuting diagrams:

$$\begin{array}{ccccc}
\coprod X_{\alpha\beta} & \rightrightarrows & \coprod X_\alpha & \to & X \\
\downarrow f_{\alpha\beta} & & \downarrow f_\alpha & & \downarrow f \\
\coprod Y_{\alpha\beta} & \rightrightarrows & \coprod Y_\alpha & \to & Y
\end{array}$$

provided the atlas for X is suitably refined. For suitable $\tilde{X}$ and $\tilde{Y}$ the morphism will be induced by a morhism

$$\tilde{f} : \tilde{X} \longrightarrow \tilde{Y}$$

which is equivariant with respect to a homomorphism $G_X \to G_Y$ of the finite groups acting on $\tilde{X}, \tilde{Y}$. However, if the Q-stack X is already presented as $\tilde{X}/G_X$ for one $\tilde{X}$, one may have to pass to a bigger covering before $\tilde{f}$ will be defined. This gives a diagram

$$\begin{array}{ccc}
\tilde{X}' & \overset{\tilde{f}}{\dashrightarrow} & \\
\downarrow & & \tilde{Y} \\
\tilde{X} & & \downarrow \\
\downarrow & & \\
\tilde{X}'/G'_X = \tilde{X}/G_X = X & \xrightarrow{f} & Y = \tilde{Y}/G_Y.
\end{array}$$

Among morphisms of Q-stacks, the simplest class consists of those that satisfy:

$$(2.6)\quad \left| \begin{array}{l} \forall \alpha, \text{ let } X'_{\alpha\alpha} = \{x \in X_{\alpha\alpha} \mid f_{\alpha\alpha}(x) = \epsilon_Y(f_\alpha(p_1(x)))\}. \\ \text{Then } X'_{\alpha\alpha} \text{ acts freely on } X_\alpha \end{array} \right.$$

These are the morphisms whose fibres are bona fide varieties, not just Q-varieties or Q-stacks. For such morphisms, it is possible to choose $\tilde{X}, \tilde{Y}$ with the same finite group G acting and $\tilde{f}$ G-equivariant:

$$\begin{array}{ccc} \tilde{X} & \xrightarrow{\tilde{f}} & \tilde{Y} \\ \downarrow & & \downarrow \\ \tilde{X}/G = X & \rightarrow & Y = \tilde{Y}/G \end{array}$$

such that, moreover, locally, $\tilde{X}_\alpha \cong (\tilde{X}_\alpha/H_\alpha) \times_{(\tilde{Y}_\alpha/H_\alpha)} \tilde{Y}_\alpha$. The fibres of $\tilde{f}$ are then the "true fibres" of f. The typical example of this is

$$\begin{array}{ccc} \overline{\mathcal{C}}_g^{(n)} & \rightarrow & \overline{\mathcal{M}}_g^{(n)} \\ \downarrow & & \downarrow \\ \overline{\mathcal{C}}_g & \rightarrow & \overline{\mathcal{M}}_g \end{array}$$

where

$\overline{\mathcal{C}}_g =$ moduli space of pairs (C, x), C a stable curve, $x \in C$

$\cong$ moduli space of pairs (C, x), a 1-pointed stable curve[3]

$\overline{\mathcal{C}}_g^{(n)} =$ normalization of $\overline{\mathcal{C}}_g$ in the covering defined by the moduli space of triples (C, x, ϕ), C a smooth curve, $x \in C$ and ϕ a level n structure on C.

Such morphisms may be called *representable* morphisms of Q-stacks. We do not need to develop the theory of Q-stacks for our applications, so we stop at merely these definitions.

[3]An n-pointed stable curve $(C, x_1, \ldots, x_n)$ is a reduced, connected curve C with at most ordinary double points plus n distinct smooth points $x_i \in C$ such that every smooth rational component E of C contains at least 3 points which are either x_i's or double points of C.

§3. The Chow Group for Q-Varieties with Cohen-Macaulay Global Covers

We want to study a quasi-projective Q-variety $\{p_\alpha: X_\alpha \longrightarrow X\}$ of dimension n such that the big global cover $\tilde{X}$ is Cohen-Macaulay. We use the notation of §2, esp. $X = \tilde{X}/G$. We also choose a resolution of singularities

$$\pi: \tilde{X}^* \longrightarrow \tilde{X}.$$

Then we assert:

Theorem 3.1. *With the above hypotheses, there is a canonical isomorphism γ between the Chow group of X and G-invariants in the operational Chow ring of $\tilde{X}$, (as usual after extending the coefficients to $\mathbb{Q}$):*

$$\gamma: A_{n-k}(X) \stackrel{\approx}{\rightarrow} opA^k(\tilde{X})^G, \quad 0 \leq k \leq n.$$

This key result does two important things for us:

a) it defines a ring structure on $A_.(X)$,
b) for all Q-sheaves $\mathcal{F}$ on the Q-variety X, we can define Chern classes

$$c_k(\mathcal{F}) \in A_.(X).$$

I don't know if these things can be done if we drop the hypothesis that $\tilde{X}$ is Cohen-Macaulay. My guess is that this hypothesis can be dropped, but more powerful tools seem to be needed to treat this case.

Proof. The first step is to define, for all subvarieties $Z \subset X$, an element $\gamma(Z) \in opA^{\cdot}(\tilde{X})^G \otimes \mathbb{Q}$. We use the local covers $p_\alpha: X_\alpha \longrightarrow X$ and let $p_\alpha^{-1}(Z)$ be the *reduced* subscheme of X_α with support $p_\alpha^{-1}(Z)$. Then $\{O_{p_\alpha^{-1}(Z)}\}$ is a Q-sheaf on the Q-variety X. As above, let p factor locally

$$\tilde{X}_\alpha \xrightarrow{q_\alpha} \tilde{X}_\alpha/H_\alpha \cong X_\alpha \xrightarrow{p_\alpha} X$$

and lift $p_\alpha^{-1}(Z)$ to its *scheme-theoretic* inverse image $q_\alpha^*(p_\alpha^{-1}(Z))$. Define

$$\tilde{Z} = \left\{ \begin{array}{l} G\text{-invariant subscheme of } \tilde{X} \text{ supported} \\ \text{on } p^{-1}(Z) \text{ such that } \tilde{Z}|_{\tilde{X}_\alpha} = q_\alpha^*(p_\alpha^{-1}(Z)) \end{array} \right\}.$$

Note that because $\tilde{X}$ is Cohen-Macaulay, and X_α is smooth, q_α is *flat*. Since $O_{p_\alpha^{-1}(Z)}$ has a finite projective resolution on X_α, this implies that O_Z has a finite projective resolution on $\tilde{X}$. Therefore, by Fulton's theory [F1], the Chern classes $c_k(O_{\tilde{Z}})$ are defined in $opA^k(\tilde{X})$. Next define the "ramification index":

$$e(Z) = \text{order of the stabilizer in } G_\alpha \text{ of almost all points of } p_\alpha^{-1}(Z).$$

Here α is any index such that $p_\alpha^{-1}(Z) \neq \phi$: the definition does not depend on α. Then let

$$\begin{aligned} \gamma(Z) &= \frac{(-1)^{k-1}e(Z)}{(k-1)!} \cdot c_k(O_{\tilde{Z}}), \quad \text{if } k = \operatorname{codim} Z, \\ (\text{or} \quad &= e(Z) ch_k(O_{\tilde{Z}}), \text{ using lemma 3.3 below}). \end{aligned}$$

An important point in the study of $\tilde{Z}$ is that the family of subschemes $\{p_\alpha^{-1}(Z)\}$ can be simultaneously resolved:

Theorem 3.2 (Hironaka). *For all subvarieties $Z \subset X$, there is a birational map π: $Z^* \longrightarrow Z, Z^*$ normal, such that for all α*

$$\left(p_\alpha^{-1}(Z) \times_Z Z^*\right)_{nor}$$

is smooth.

This is a Corollary of Hironaka's strong resolution theorem, giving a resolution compatible with the pseudo-group of all local analytic isomorphisms between open sets in the original variety: see [H], Theorem 7.1, p. 164. One may proceed as follows: first resolve $p_{\alpha_1}^{-1}(Z)$ as in[H]. Since $((\tilde{X}_{\alpha_1}/H_\alpha) \times_X (\tilde{X}_{\alpha_1}/H_{\alpha_1}))_{nor}$ is étale over $\tilde{X}_{\alpha_1}/H_{\alpha_1}$ this equivalence relation extends to one on the resolution $p_{\alpha_1}^{-1}(Z)^*$, hence there is a blow-up π_1: $Z_1 \longrightarrow Z$ such that $p_{\alpha_1}^{-1}(Z)^* \cong \left(p_{\alpha_1}^{-1}(Z) \times_Z Z_1\right)_{nor}$. Secondly, resolve $\left(p_{\alpha_2}^{-1}(Z) \times_Z Z_1\right)_{nor}$ by a blow-up over $X - p(X_{\alpha_1})$ so as not to affect the first step. Again, descend this to a further blow-up (π_2: $Z_2 \rightarrow Z_1 \rightarrow Z$) of Z. Eventually, we get the needed resolution.

Lemma 3.3. *For all Q-sheaves $\mathcal{F}$ on the Q-variety X, let $S \subset X$ be the support of $\mathcal{F}$. Then in $opA^{\cdot}(\tilde{X})$,*

a) $c_k(\tilde{\mathcal{F}}) = 0$ *if* $k < \operatorname{codim} S$

b) *if* $k = \operatorname{codim} S$, *and* $S_1, \ldots, S_n$ *are the codimension* k *components of* S, *then*

$$c_k(\tilde{\mathcal{F}}) = \sum_{i=1}^{n} l_i c_k(\mathcal{O}_{\tilde{S}_i})$$

where l_i *is the length of the stalk of* $\mathcal{F}_\alpha$ *at the generic point of* $p_\alpha^{-1}(S_i)$ *when* S_i *meets* $p(\tilde{X}_\alpha)$.

Proof. This results from an application of Fulton's Grothendieck-Riemann-Roch Theorem 1.4 and Hironaka's resolution 3.2. By the usual dévissage, reduce the lemma to the case where S is irreducible of codimension k, and $\tilde{\mathcal{F}}$ is an $\mathcal{O}_{\tilde{S}}$-module. Let $S^* \to S$ be a "resolution" as in 3.2. This gives a family of resolutions

$$S_\alpha^* = \left(p_\alpha^{-1}(S) \times_S S^*\right)_{nor} \xrightarrow{\pi_\alpha} p_\alpha^{-1}(S) \subset \tilde{X}_\alpha / H_\alpha$$

which are local complete intersection (or l.c.i.) morphisms. Therefore, by fibre product with the flat morphism $\tilde{X}_\alpha \to \tilde{X}_\alpha / H_\alpha$,

$$S_\alpha^* \times_{(\tilde{X}_\alpha / H_\alpha)} \tilde{X}_\alpha \to \tilde{X}_\alpha$$

are l.c.i. morphisms. These glue together to an l.c.i. morphism

$$\tilde{S}^* \xrightarrow{\pi} \tilde{X}$$

such that $\tilde{\mathcal{F}}$ is a $\pi_* \mathcal{O}_{\tilde{S}^*}$-module. Let $\tilde{\mathcal{F}}^* = \pi^*(\tilde{\mathcal{F}})$. Then by 1.4:

$$ch\, \pi_!(\tilde{\mathcal{F}}^*) = \pi_*(ch\, \tilde{\mathcal{F}}^* \cdot Td_\pi).$$

Now if $i > 0$, $R^i \pi_*(\tilde{\mathcal{F}}^*)$ are Q-sheaves on the Q-variety X with supports properly contained in S, so by induction we can assume

$$c_l(R^i \pi_* \tilde{\mathcal{F}}^*) = 0 \quad \text{if} \quad l \leq k, i > 0.$$

Therefore

$$ch_l(\pi_! \tilde{\mathcal{F}}^*) = ch_l(\pi_* \tilde{\mathcal{F}}^*) = ch_l(\tilde{\mathcal{F}}) \quad \text{if} \quad l \leq k.$$

But $\pi_*: opA^{\cdot}(\tilde{S}^*) \to opA^{\cdot}(\tilde{X})$ raises the codimension of a cycle by k. So

$$ch_l(\tilde{\mathcal{F}}) = 0 \quad \text{if} \quad l < k$$

and

$$\begin{aligned} ch_k(\tilde{\mathcal{F}}) &= (\text{generic rank of } \tilde{\mathcal{F}}^* \text{ as free } O_{\tilde{S}^*}\text{-module}) \cdot \pi_*(1) \\ &= \big(\text{length of } \mathcal{F}_\alpha \text{ at generic point of } p_\alpha^{-1}(S)\big) \cdot ch_k(O_{\tilde{S}}). \end{aligned}$$

Q.E.D.

Lemma 3.4. *If two cycles $\sum n_i Z_i$, $\sum m_i W_i$ on X are rationally equivalent, then*

$$\sum n_i \gamma(Z_i) = \sum m_i \gamma(W_i) \quad \textit{in } opA^{\cdot}(\tilde{X}).$$

Proof. Let L be an ample line bundle on X. Rational equivalence on X may be defined by requiring that for all subvarieties $Y \subset X$ and all $s_1, s_2 \in \Gamma(Y, L^n \otimes O_Y)$, if D_i is the divisor of zeroes of s_i on X, then

$$D_1 \underset{rat}{\sim} D_2.$$

So to prove the lemma, it will suffice to prove that for all $s \in \Gamma(Y, L^n \otimes O_Y)$, if $D = \sum n_i Z_i$ is the divisor of zeroes of s, then

$$\sum n_i \gamma(Z_i) = e(Y) \cdot [ch_k(O_{\tilde{Y}}) - ch_k(L^{-n} \otimes O_Y)].$$

To see this, use the exact sequence of Q-sheaves

$$0 \to L^{-n} \otimes O_{\tilde{Y}} \overset{\otimes s}{\to} O_{\tilde{Y}} \to O_{\tilde{D}} \to 0$$

where $\tilde{D} \subset \tilde{X}$ is the scheme of zeroes of s in $\tilde{Y}$ and the local calculation[4] that on $p^{-1}(Y)$:

$$(*) \qquad \big(\text{Divisor of zeroes of } p_\alpha^*(s) \text{ on } p_\alpha^{-1}(Y)\big) = \sum n_i \frac{e(Z_i)}{e(Y)} p_\alpha^{-1}(W_i)$$

[4]We use the lemma that if a finite group G acts faithfully on a variety Y and ϕ is a G-invariant function on Y, zero on a subvariety $W \subset Y$ of codimension 1, and $\overline{\phi}$ is the induced function on Y/G, then:

$$\text{ord}_W(\phi) = \#\{g \in G \mid g = id. \text{ on } W\} \cdot \text{ord}_{W/G}(\overline{\phi}).$$

(n.b., $p_\alpha^{-1}(Y)$, $p_\alpha^{-1}(W_i)$ are the *reduced* inverse images). Therefore

$$\begin{aligned} ch_k O_{\tilde{Y}} - ch_k L^{-n} \otimes O_{\tilde{Y}} &= ch_k O_{\tilde{D}} \\ &= \frac{(-1)^{k-1}}{(k-1)!} c_k O_{\tilde{D}} \\ &= \sum_i \begin{pmatrix} \text{length of } O_{D_\alpha} \text{ at} \\ \text{gen.pt. of } p_\alpha^{-1}(W_i) \end{pmatrix} \cdot \frac{(-1)^{k-1}}{(k-1)!} c_k O_{\tilde{W}_i} \quad \text{by (3.3)} \\ &= \sum_i \operatorname{ord}_{p_\alpha^{-1}(W_i)}(p_\alpha^* s) \cdot \frac{\gamma(W_i)}{e(W_i)} \quad \text{by def}^{\underline{n}} \text{ of } \gamma \\ &= \frac{1}{e(Y)} \sum n_i \gamma(W_i) \quad \text{by } (*). \end{aligned}$$

This proves (3.4), which shows that γ factors:

$$\gamma \colon A.(X) \longrightarrow opA^{\cdot}(\tilde{X})^G.$$

Lemma 3.5. *The composition of maps*

$$A.(X) \xrightarrow{\gamma} opA^{\cdot}(\tilde{X})^G \xrightarrow{\cap[\tilde{X}]} A.(\tilde{X})^G \xrightarrow{p_*} A.(X)$$

is multiplication by n, the degree of p.

Proof. To prove this we use another Riemann-Roch theorem: the version of Baum-Fulton-MacPherson [BFM]. This says that there is a natural transformation $\tau\colon K_o(Z) \longrightarrow A.(Z)$ for all varieties Z such that

$$\begin{array}{ccc} K^o(Z) \otimes K_o(Z) & \xrightarrow{\otimes} & K_o(Z) \\ \downarrow ch \otimes \tau & & \downarrow \tau \\ opA^{\cdot}(Z) \otimes A.(Z) & \xrightarrow{\cap} & A.(Z) \end{array}$$

commutes. By the lemma, p. 129 of [BFM] and dévissage, τ satisfies:

> *for all $\mathcal{F}$ with support $\cup Z_i$ of codimension k, $\tau(\mathcal{F})$ has codimension k and*
>
> $$\tau(\mathcal{F})_k = \text{class of } \sum \begin{pmatrix} \text{length of } \mathcal{F} \text{ at} \\ \text{gen. pt. of } Z_i \end{pmatrix} [Z_i].$$

We apply this to $Z = \tilde{X}$ and $\mathcal{F} = \mathcal{O}_{\tilde{Z}}$ where $Z \subset X$ is a subvariety. It follows that

$$ch(\mathcal{O}_{\tilde{Z}}) \cap \tau(\mathcal{O}_{\tilde{X}}) = \tau(\mathcal{O}_{\tilde{Z}}).$$

Therefore if $k =$ codimension Z,

$$\begin{aligned} p_*\big(\gamma(Z) \cap [\tilde{X}]\big) &= e(Z) \cdot p_*\big(ch_k(\mathcal{O}_{\tilde{Z}}) \cap [\tilde{X}]\big) \\ &= e(Z) \cdot p_*([\tilde{Z}]) \\ &= e(Z) \cdot [\tilde{Z} : Z] \cdot \text{class of } Z \\ &= n \cdot \text{class of } Z. \end{aligned}$$

Q.E.D.

Lemma 3.6. *If $\pi\colon \tilde{X}^* \longrightarrow \tilde{X}$ is a resolution, and $Z \subset X$ is a subvariety of codimension k such that for all components Z_i of $p^{-1}(Z)$, $\pi^{-1}(Z_i)$ is irreducible of codimension k, then $\pi^*\big(\gamma(Z)\big)$ is represented by a cycle*

$$c \cdot \sum \pi^{-1}(Z_i), \quad \text{some } c \in \mathbb{Q}, c > 0.$$

Proof. Let $U \subset \tilde{X}$ be the open set over which π is an isomorphism. Then $Z_i \cap U \neq \phi$, all i. Now

$$\begin{aligned} \pi^*\big(\gamma(Z)\big) &= e(Z) \cdot \pi^*\big(ch_k(\mathcal{O}_{\tilde{Z}})\big) \\ &= e(Z) \cdot \Big(\sum (-1)^l ch_k\big(\text{tor}_l(\mathcal{O}_{\tilde{Z}}, \mathcal{O}_{\tilde{X}^*})\big)\Big). \end{aligned}$$

But these tor_l are supported on proper subsets of $\pi^{-1}(Z_i)$, hence have no k^{th} Chern character. Therefore:

$$\begin{aligned} \pi^*\big(\gamma(Z)\big) &= e(Z) \cdot ch_k(\mathcal{O}_{\tilde{Z}} \otimes \mathcal{O}_{\tilde{X}^*}) \\ &= e(Z) \cdot \text{class of } \pi^{-1}(\tilde{Z}) \end{aligned}$$

by the Riemann-Roch theorem on $\tilde{X}^*$. Q.E.D.

Corollary 3.7. *γ is bijective.*

Proof. 3.5, 3.6 and 1.2.

This proves the theorem. A few comments can be made on the ring structure that this introduces in $A.(X)$. First of all, suppose W_1, W_2 are two cycles on X that intersect properly. Then the product $[W_1], [W_2]$ in the above ring structure can also be defined directly by assigning suitable multiplicities to the components of $\operatorname{Supp} W_1 \cap \operatorname{Supp} W_2$. In fact, define:

$$W_1 \cdot W_2 = \sum_{\substack{\text{comp.} U \text{ of} \\ \operatorname{Supp} W_1 \cap \operatorname{Supp} W_2}} i(W_1 \cap W_2; U) \cdot U$$

where if $p_\alpha^{-1}(U) \neq \phi$, then

$$i(W_1 \cap W_2; U) = \frac{e(W_1) \cdot e(W_2)}{e(U)} \cdot i(p_\alpha^{-1}(W_1) \cap p_\alpha^{-1}(W_2); p_\alpha^{-1} U).$$

Note that the intersection multiplicity on the right is taken on the smooth ambient variety $\tilde{X}_\alpha / H_\alpha$, hence is defined, e.g., by

$$\sum_l (-1)^l \begin{pmatrix} \text{length at gen. pt.} \\ \text{of } p_\alpha^{-1} U \end{pmatrix} (\operatorname{tor}_l(O_{p_\alpha^{-1} W_1}, O_{p_\alpha^{-1} W_2})).$$

The proof that this is the same as the product in $opA^{\cdot}(\tilde{X})$ is straightforward, i.e.,

$$\begin{aligned}
\gamma(Z_1) \cdot \gamma(Z_2) &= e(Z_1) e(Z_2) ch_{k_1}(O_{\tilde{Z}_1}) \cdot ch_{k_2}(O_{\tilde{Z}_2}) \\
&= e(Z_1) e(Z_2) ch_{k_1 + k_2}(O_{\tilde{Z}_1} \overset{\mathbf{L}}{\otimes} O_{\tilde{Z}_2}) \\
&\quad (\overset{\mathbf{L}}{\otimes} \text{ means take tensor product of projective resol.}) \\
&= \sum_U e(Z_1) e(Z_2) i(p_\alpha^{-1}(W_1) \cap p_\alpha^{-1}(W_2); p_\alpha^{-1} U) \cdot ch_{k_1 + k_2}(O_{\tilde{U}}) \\
&= \sum \frac{e(Z_1) e(Z_2)}{e(U)} i(p_\alpha^{-1}(W_1) \cap p_\alpha^{-1}(W_2); p_\alpha^{-1} U) \cdot \gamma(U).
\end{aligned}$$

This product could be introduced directly without relating it to the product in $opA^{\cdot}(\tilde{X})$. This has been done by Matsusaka in his book "Theory of Q-varieties", [Ma], where associativity and other standard formulae are proven. The missing ingredient, however, is the moving lemma. This follows as a Corollary of the isomorphism of $A.(X)$ with $opA^{\cdot}(\tilde{X})$, i.e., by representing a cycle on X as the projection from $\tilde{X}$ of the Chern class of a sheaf with finite resolution. In particular, I don't know any way to get a moving lemma unless some $\tilde{X}$ is Cohen-Macaulay.

Henceforth, in the study of the Chow rings of Q-varieties we shall identify $A_{n-k}(X)$ and $opA^k(\tilde{X})^G$ via the map γ, and write this as $A^k(X)$ just like the k-codimension piece of the Chow ring of an ordinary non-singular variety. This does not usually lead to any confusion, except with regard to the concept of the *fundamental class* of a subvariety $Y \subset X$. The important thing to realize here is that there are really two different notions of fundamental class, differing by a rational number, and both are important. Thus for all Y of codimension k, we will write

$$[Y] = \text{class of the cycle } Y \text{ in the Chow group}$$
$$A_{n-k}(X) = A^k(X)$$

and

$$[Y]_Q = \text{the class } ch_k(\mathcal{O}_{\tilde{Y}}) \text{ in } opA^k(\tilde{X})^G = A^k(X).$$

Since we are using the identification γ, we have:

$$[Y]_Q = \frac{1}{e(Y)} \cdot [Y].$$

When one makes calculations of intersections in local charts $\tilde{X}_\alpha / H_\alpha$, then one is verifying an identity between classes $[Y]_Q$. But when one has a rational equivalence between cycles on X, one has an identity between $[Y]$'s: e.g., if X is unirational, then for all points $P_1, P_2 \in X$,

$$[P_1] = [P_2],$$

but the point classes $[P]_Q$ are fractions $1/e(P)$ of the basic point class $[P] \in A^{\cdot}(X)$.

If X is a Q-stack, exactly the same theorem holds and we have an isomorphism

$$\gamma: A_{\cdot}(X) \xrightarrow{\approx} opA^{\cdot}(\tilde{X})^G.$$

The only difference is that a subgroup $Z \subset G$ acts identically on $\tilde{X}$. If $\#Z = z$, then the effect of this is merely to modify the ring structure on $A^{\cdot}(X)$ as follows. Let W_1, W_2 be cycles on X and consider:

i) the *Q-variety* structure on X given by the action of G/Z on $\tilde{X}$, and the multiplication $W_1 \cdot_{var} W_2$,

ii) the Q-*stack* structure on X given by the action of G on $\tilde{X}$, and the multiplication $W_1 \cdot_{st} W_2$.

Then by the moving lemma plus the formula above for proper intersections, it follows:

$$W_1 \cdot_{st} W_2 = z.W_1 \cdot_{var} W_2.$$

In particular, the identity in the Chow ring of a Q-stack X is $[X]_Q$, not $[X]$.

The Chow ring for Q-varieties, or more generally Q-stacks, has good contravariant functorial properties. We consider morphisms of Q-stacks with global Cohen-Macaulay covers:

$$X \xrightarrow{f} Y$$

as defined in §2. Then I claim:

Proposition 3.8. *There is a canonical ring homomorphism*

$$f^* : A^{\cdot}(Y) \to A^{\cdot}(X)$$

satisfying:

i) $f_*(a.f^*b) = f_*a.b$ (f_* *being defined from* $A.(X)$ *to* $A.(Y)$ *as usual),*
ii) $f^*(c_k\mathcal{E}) = c_k(f^*\mathcal{E})$ *for all* Q-*vector bundles* $\mathcal{E}$ *on* Y
iii) *if* W *is a subvariety of* Y *such that codim* $f^{-1}(W) =$ codim W *then*

$$f^*([W]) = \text{class of} \sum_{\substack{\text{comp. } V_k \\ \text{of } f^{-1}(W)}} i_k \cdot [V_k]$$

where i_k *is calculated on suitable charts* $f_\alpha : X_\alpha \to Y_\alpha$ *by pull-backs in the smooth case adjusted by* $e(W)/e(V_k)$.

Proof. Although the moving lemma plus (iii) provides us with the simplest formula for f^*, to see that f^* is well-defined, we use $opA^{\cdot}$. There is one complication. X and Y have global Cohen-Macaulay covers $\tilde{X}, \tilde{Y}$ but f may not lift to $\tilde f: \tilde X \longrightarrow \tilde Y$. Instead, we may have to 'refine' $\tilde X$:

$$\begin{array}{ccc} \tilde X' & \searrow & \\ {\scriptstyle q}\downarrow & & \\ \tilde X & & \tilde Y \\ {\scriptstyle p}\downarrow & & \downarrow \\ X & \xrightarrow{f} & Y \end{array}$$

and then $\tilde X'$ may not longer be Cohen-Macaulay. Still, once one has one Cohen-Macaulay $\tilde X$ with which to set up the theory, one proves that

$$(3.9) \qquad opA^{\cdot}(X)^{G_X} \xrightarrow{q^*} opA^{\cdot}(\tilde X')^{G'_X}$$

is an isomorphism, hence f^* may be defined by:

$$A_{\cdot}(Y) \underset{\gamma_Y}{\approxeq} opA^{\cdot}(\tilde Y)^{G_Y} \underset{\tilde f_*}{\rightarrow} opA^{\cdot}(\tilde X')^{G'_X} \approxeq opA^{\cdot}(\tilde X)^{G_X} \approxeq A_{\cdot}(X).$$

To check (3.9), use

$$\begin{array}{ccccc} A_{\cdot}(X) & \underset{\gamma_X}{\approxeq} opA^{\cdot}(\tilde X)^{G_X} & \underset{q^*}{\rightarrow} & opA^{\cdot}(\tilde X')^{G'_X} \hookrightarrow & A^{\cdot}(\tilde X'^*) \\ & \nwarrow & & & \downarrow \cap[\tilde X'] \\ & & {\scriptstyle (p\circ q)_*} & & \\ & & & & A_{\cdot}(\tilde X') \end{array}$$

($\tilde X'^* =$ resolution of $\tilde X'$) and argue as in lemmas 3.5 and 3.6. There is one hitch: namely, in 3.5, we get

$$\begin{aligned}(p\circ q)_*\big(q^*\big(\gamma(Z)\big)\cap[\tilde X']\big) &= e(Z)(p\circ q)_*\big(q^*\big(ch_k(O_{\tilde Z})\big)\cap[\tilde X']\big)\\ &= e(Z)(p\circ q)_*\big(ch_k(O_{\tilde Z}\overset{L}{\otimes} O_{\tilde X'})\cap[X']\big)\end{aligned}$$

where $\overset{L}{\otimes}$ means take a resolution of $O_{\tilde Z}$ and tensor it with $O_{\tilde X'}$. But now if $\tilde X'$ is not Cohen-Macaulay, $O_{\tilde Z}\overset{L}{\otimes} O_{\tilde X'}$ will not be a resolution of some $O_{\tilde Z'}$ and we get instead:

$$\begin{aligned}&= e(Z)\sum(-1)^l(p\circ q)_*\big(ch_k\, \mathrm{tor}_l(O_{\tilde Z}, O_{X'})\cap[\tilde X']\big)\\ &= e(Z)\cdot(p\circ q)_*\Big(\sum_n i_n\cdot[\tilde Z_n]\Big)\end{aligned}$$

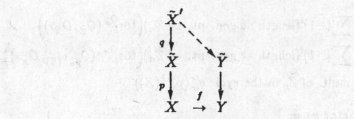

where $\tilde{Z}_l$ are the components of $(p \circ q)^{-1} Z$ and

$$\begin{aligned} i_n &= \sum (-1)^l (\text{length at gen. pt. of } \tilde{Z}_n)\Big(\operatorname{tor}_l^{O_{\tilde{X}}}(O_{\tilde{Z}}, O_{\tilde{X}'})\Big) \\ &= \sum (-1)^l (\text{length at gen. pt. of } \tilde{Z}_n)\Big(\operatorname{tor}_l^{O_{X_\alpha}}(O_{p_\alpha^{-1}(Z)}, O_{\tilde{X}'})\Big) \\ &= \text{mult. of } \tilde{Z}_n \text{ in the cycle } q_\alpha'^{*}(p_\alpha^{-1}(Z)) \end{aligned}$$

where we factor $p \circ q$:

$$\begin{array}{ccc} \tilde{X}' & \supset & \tilde{X}'_\alpha \\ \Big\downarrow & & \Big\downarrow q'_\alpha \\ & & X_\alpha \\ \Big\downarrow & \swarrow p_\alpha & \\ X & & \end{array}$$

Thus

$$\begin{aligned} e(Z)(p \circ q)_* \Big(\sum_n i_n [\tilde{Z}_n]\Big) & \\ &= e(Z) p_{\alpha,*}(\text{class of } q'_{\alpha,*}(q_\alpha'^{*}(p_\alpha^{-1} Z))) \\ &= \deg q'_\alpha \cdot e(Z) \cdot p_{\alpha,*}(p_\alpha^{-1} Z) \\ &= n.\text{class of } Z. \end{aligned}$$

f^* being defined, the rest of the proof is straightforward.

For representable morphisms $f\colon X \longrightarrow Y$ of Q-stacks, there is a further important compatibility. For such f, let

$$\tilde{f}\colon \tilde{X} \longrightarrow \tilde{Y}$$

be a G-equivariant morphism such that $X = \tilde{X}/G$, $Y = \tilde{Y}/G$ and

$$\tilde{X}_\alpha \cong X_\alpha \times_{Y_\alpha} \tilde{Y}_\alpha \tag{3.9}$$

as in §2*b*. Then we have:

Proposition 3.10. *If the morphism f on local charts*

$$f_\alpha\colon X_\alpha \longrightarrow Y_\alpha$$

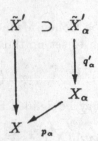

is a local complete intersection and $\tilde{Y}$ is Cohen-Macaulay, then $\tilde{f}$ is l.c.i. and the diagram

$$\begin{array}{ccc} opA^*(\tilde{X})^G & \xrightarrow{\tilde{f}_*} & opA^{\cdot}(\tilde{Y})^G \\ \gamma\uparrow & & \gamma\uparrow \\ A_{\cdot}(X) & \xrightarrow{f_*} & A_{\cdot}(Y) \end{array}$$

commutes.

Proof. Let W be a codimension k subvariety of X such that W is generically finite over $f(W)$. Then we must check

$$\tilde{f}_*(\gamma W) = [W : fW] \cdot \gamma(fW),$$

i.e.,

$$e(X) \cdot \tilde{f}_*(ch_k(\mathcal{O}_{\tilde{W}})) = [W : fW] \cdot e(fW) \cdot ch_k(\mathcal{O}_{f\tilde{W}}).$$

But by Riemann-Roch for $\tilde{f}$,

$$\begin{aligned} \tilde{f}_*(ch_k(\mathcal{O}_{\tilde{W}})) &= ch_k(\tilde{f}_*\mathcal{O}_{\tilde{W}}) \\ &= [W_\alpha : f_\alpha(W_\alpha)] \cdot ch_k(\mathcal{O}_{f\tilde{W}}) \end{aligned}$$

because $\tilde{f}_*(\mathcal{O}_{\tilde{W}})$ is generically a locally free $\mathcal{O}_{f\tilde{W}}$-algebra of length $[W_\alpha : f_\alpha W_\alpha]$ by (3.9). But now use:

$$[W_\alpha : W].[W : fW] = [W_\alpha : f_\alpha W_\alpha] \cdot [f_\alpha W_\alpha : fW]$$

and

$$\begin{aligned} [W_\alpha : W] \cdot e(W) &= [X_\alpha : X] \cdot e(X) \\ &= [G : H_\alpha] \\ &= [Y_\alpha : Y] \cdot e(Y) \\ &= [f(W)_\alpha : fW] \cdot e(fW) \end{aligned}$$

and the equality of the coefficients follows.

Part II: Basic Classes in the Chow Ring of the Moduli Space

§4. Tautological Classes

Whenever a variety or topological space is defined by some universal property, one expects that by virtue of its defining property, it possesses certain cohomology classes called tautological classes. The standard example is a Grassmannian, e.g., the Grassmannian Grass of k-planes in $\mathbb{C}^n$. By its very definition, there is a universal bundle E on Grass of rank k, and this induces Chern classes $c_l(E)$, $1 \leq l \leq k$, in both the cohomology ring of Grass and the Chow ring of Grass. These two rings are, in fact, isomorphic and generated as rings by $\{c_l(E)\}$. Moreover, one gets tautological relations from the fact that E is a sub-bundle of the trivial bundle $\mathbb{C}^n \times$ Grass. This gives an exact sequence:

$$0 \to E \to \mathcal{O}^n \to F \to 0, \quad F \text{ a bundle of rank } n-k,$$

hence

$$\left(1 + c_1(E) + \cdots + c_k(E)\right)_l^{-1} = 0, \quad l > n-k.$$

As is well known, these are a complete set of relations for the cohomology and Chow rings of Grass.

We shall begin a program of the same sort for the Chow ring (or cohomology ring) of $\overline{\mathcal{M}}_g$. Our purpose is merely to identify a natural set of tautological classes and some tautological relations. To what extent these lead to a presentation of either ring is totally unclear at the moment.

The natural place to start is with the universal curve over $\overline{\mathcal{M}}_g$. This is the same as the coarse moduli space of 1-pointed stable curves (C, P) (see Knudsen [K], Harris-Mumford [H-M]), which we call $\overline{\mathcal{M}}_{g,1}$ or $\overline{\mathcal{C}}_g$ alternatively. $\overline{\mathcal{C}}_g$ is a Q-variety, too, and everything we have said about $\overline{\mathcal{M}}_g$ applies to $\overline{\mathcal{C}}_g$ too. The morphism $\overline{\mathcal{C}}_g \to \overline{\mathcal{M}}_g$ is a representable morphism, and via level n structures, we have a covering $\tilde{\mathcal{C}}_g$ of $\overline{\mathcal{C}}_g$ and a morphism:

$$\pi: \tilde{\mathcal{C}}_g \longrightarrow \tilde{\mathcal{M}}_g$$

which is a flat, proper family of stable curves, with a finite group G acting on both, and $\overline{\mathcal{C}}_g = \tilde{\mathcal{C}}_g/G$, $\overline{\mathcal{M}}_g = \tilde{\mathcal{M}}_g/G$. If $g \geq 3$, then

$$\tilde{\mathcal{C}}_g = (\text{normalization of } \overline{\mathcal{C}}_g \times_{\overline{\mathcal{M}}_g} \tilde{\mathcal{M}}_g),$$

but if $g = 2$, the generic curve has automorphisms, $Sp(4, \mathbb{Z}/n\mathbb{Z})$ does not act faithfully on $\tilde{\mathcal{M}}_2$, and $\tilde{\mathcal{C}}_2$ is a double cover of this normalization. In any case, $\overline{\mathcal{C}}_g$ has a Q-sheaf $\omega_{\overline{\mathcal{C}}_g/\overline{\mathcal{M}}_g}$ represented by the invertible sheaf $\omega_{\tilde{\mathcal{C}}_g/\tilde{\mathcal{M}}_g}$ on $\tilde{\mathcal{C}}_g$. Henceforth, whenever we talk of sheaves on $\overline{\mathcal{C}}_g$ or $\overline{\mathcal{M}}_g$ we shall mean *Q-sheaves* and they are always represented by usual coherent sheaves on $\tilde{\mathcal{C}}_g$ and $\tilde{\mathcal{M}}_g$ with G-action. Furthermore, we shall make calculations in $A^{\cdot}(\overline{\mathcal{M}}_g)$ and $A^{\cdot}(\overline{\mathcal{C}}_g)$ by implicitly identifying these with $opA^{\cdot}(\tilde{\mathcal{M}}_g)^G$ and $opA^{\cdot}(\tilde{\mathcal{C}}_g)^G$.

Now define the *tautological classes:*

$$\begin{aligned} K_{\overline{\mathcal{C}}_g/\overline{\mathcal{M}}_g} &= c_1(\omega_{\overline{\mathcal{C}}_g/\overline{\mathcal{M}}_g}) \in A^1(\overline{\mathcal{C}}_g) \\ \kappa_l &= \left(\pi_* K^{l+1}_{\overline{\mathcal{C}}_g/\overline{\mathcal{M}}_g}\right) \in A^l(\overline{\mathcal{M}}_g) \\ \mathbb{E} &= \pi_*(\omega_{\overline{\mathcal{M}}_g/\overline{\mathcal{M}}_g}) : \text{a locally free } Q\text{-sheaf of rank } g \text{ on } \overline{\mathcal{M}}_g \\ \lambda_l &= c_l(\mathbb{E}), \quad 1 \leq l \leq g. \end{aligned}$$

I believe that the κ_l are the natural tautological classes to consider on $\overline{\mathcal{M}}_g$. On the other hand, the λ_l are the natural classes for abelian varieties. Let me sketch this link, which will not be used subsequently. In fact, if

$$\mathcal{A}_g^* = \begin{pmatrix} \text{Satake's compactification of the moduli} \\ \text{space of principally polarized abelian varieties} \end{pmatrix}$$

then there is a natural morphism

$$t\colon \overline{\mathcal{M}}_g \longrightarrow \mathcal{A}_g^*$$

carrying the point $[C]$ to the point of $\mathcal{A}_g^*$ defined by the Jacobian of C. This morphism lifts to a G-equivariant morphism

$$\tilde{t}\colon \tilde{\mathcal{M}}_g \longrightarrow \tilde{\mathcal{A}}_g$$

where $\tilde{\mathcal{A}}_g$ is a suitable toroidal compactification of the level n covering of $\mathcal{A}_g$: see Namikawa [N]. Moreover, $\tilde{\mathcal{A}}_g$ carries a universal family

$$\pi\colon \tilde{\mathcal{G}}_g \longrightarrow \tilde{\mathcal{A}}_g$$

of semi-abelian group schemes, i.e., $\tilde{\mathcal{G}}/\tilde{\mathcal{A}}$ is a group scheme whose fibres are extensions of abelian varieties by algebraic tori $(\mathbb{C}^*)^h$. The family $\tilde{\mathcal{G}}_g$ pulls back on $\tilde{\mathcal{M}}_g$ to the family of Jacobians and generalized Jacobians of $\tilde{\mathcal{C}}_g$. Over $\tilde{\mathcal{A}}_g$, define

$$\mathsf{E}' = \Omega^1_{\tilde{\mathcal{C}}_g/\tilde{\mathcal{A}}_g}\big|_{\text{o-section}}, \text{ a locally free sheaf of rank } g$$
$$\lambda'_l = c_k(\mathsf{E}'), \quad 1 \le l \le g.$$

Then it follows that

$$\tilde{t}^* \mathsf{E}' \cong \mathsf{E}$$

and

$$\tilde{t}^* \lambda'_l = \lambda_l.$$

The class $K_{\overline{\mathcal{C}}_g/\overline{\mathcal{M}}_g}$ played a central role in the basic paper [A] of Arakelov, who proved the essential case of:

Theorem 4.1 (Arakelov). *The divisor $K_{\overline{\mathcal{C}}_g/\overline{\mathcal{M}}_g}$ is numerically effective on $\overline{\mathcal{C}}_g$, i.e., for all curves $C \subset \overline{\mathcal{C}}_g$,*

$$\deg_C K_{\overline{\mathcal{C}}_g/\overline{\mathcal{M}}_g} \ge 0.$$

Proof. In fact, Arakelov proved that for all normal surfaces F fibred in stable curves over a smooth curve C, $\omega_{F/C}$ is *ample* on F. This implies that for all curves $C \subset \tilde{\mathcal{C}}_g$ such that $\pi(C) \cap \mathcal{M}_g \ne \phi$,

$$\deg_C K_{\tilde{\mathcal{C}}_g/\tilde{\mathcal{M}}_g} > 0.$$

Now suppose $C \subset \tilde{\mathcal{C}}_g$ and $\pi(C) \subset \overline{\mathcal{M}}_g - \mathcal{M}_g$.

Case 1: $\pi(C) =$ one pt. Then $\deg_C K > 0$ because ω is ample on all fibres of $\tilde{\mathcal{C}}_g \to \tilde{\mathcal{M}}_g$.

Case 2: $d\pi|_C \equiv 0$, i.e., C is in the locus Sing $\mathcal{C}$ of double points of the fibres. But Sing $\mathcal{C}$ has an étale double cover Sing$'$ $\mathcal{C}$ parametrizing pairs consisting of a double point of a fibre of π and a branch through this point. By residue

$$\omega_{\tilde{\mathcal{C}}_g|\tilde{\mathcal{M}}_g} \otimes \mathcal{O}_{\text{Sing}'\mathcal{C}} \cong \mathcal{O}_{\text{Sing}'\mathcal{C}}$$

so $\deg_C K = 0$.

Case 3: Other. After a suitable case change

$$C' \to \pi(C) \subset \tilde{M}_g$$

we can assume that the pull-back family $\tilde{C}_g \times_{\tilde{M}_g} C'$ is obtained by glueing several generically smooth stable families $Y_\alpha \to C'$ along a set of sections $t_{\alpha\beta}: C' \longrightarrow Y_\alpha$. Lying over C there will be a curve C'' contained in one of the Y_α's, say Y_{α_0}, mapping onto C' and not equal to $t_{\alpha_0\beta}(C')$, any β. The pull-back of $\omega_{\tilde{C}_g/\tilde{M}_g}$ to Y_α will be equal to $\omega_{Y_\alpha/C'}(\sum_\beta t_{\alpha_0,\beta}(C'))$ and, by Arakelov, this will have non-negative degree on C' if genus $C' \geq 2$. If genus $C' = 0$ or 1, it is easy to check that this is still the case. Q.E.D.

Corollary 4.2. *The classes κ_l are numerically effective, i.e., for all subvarieties $W \subset \overline{M}_g$ of dimension l,*

$$(W.\kappa_l) \geq 0.$$

Proof. $K_{\overline{C}/\overline{M}}$ numerically effective implies $K^{l+1}_{\overline{C}/\overline{M}}$ numerically effecitve (see [K1]), hence $\pi_*(K^{l+1}_{\overline{C}/\overline{M}})$ is numerically effective.

In fact, κ_1 is ample, see [M], §5.

§5. Tautological Relations via Grothendieck-Riemann-Roch

Grothendieck's Riemann-Roch theorem (G-R-R) is, in many cases, tailor-made to find relations among tautological classes. For example, see Atiyah-Bott [A-B], §9. We can compute the classes λ_k in terms of the classes κ_k. To do this, we apply the G-R-R to the morphism

$$\pi: \tilde{C}_g \longrightarrow \tilde{M}_g.$$

This gives us

$$ch\, \pi_!\, \omega_{\overline{C}/\overline{M}} = \pi_*\big(ch\, \omega_{\overline{C}/\overline{M}} Td^\vee(\Omega^1_{\overline{C}/\overline{M}})\big).$$

Here we use the notaion $Td^\vee(\mathcal{E})$ to write the universal multiplicative polynomial in the Chern classes of $\mathcal{E}$ such that for line bundles L,

$$\begin{aligned} Td^\vee(L) &= \frac{\lambda}{e^\lambda - 1}, \quad \lambda = c_1(L) \\ &= 1 - \frac{1}{2}\lambda - \sum_{k=1}^{\infty}(-1)^{k-1}\frac{B_k}{(2k)!}\lambda^{2k}, \end{aligned}$$

(i.e., the usual $Td(L)$ is $\lambda/1-e^{-\lambda}$ or $1+\frac{1}{2}\lambda+\cdots$). Since $R^1\pi_*\omega_{\overline{\mathcal{C}}/\overline{\mathcal{M}}}\cong\mathcal{O}_{\overline{\mathcal{M}}}$, this means:

$$ch\,\mathbb{E}=1+\pi_*(e^K\,Td^\vee(\Omega^1_{\overline{\mathcal{C}}/\overline{\mathcal{M}}})).$$

Now use the exact sequence:

$$0\to\Omega^1_{\overline{\mathcal{C}}/\overline{\mathcal{M}}}\to\omega_{\overline{\mathcal{C}}/\overline{\mathcal{M}}}\to\omega_{\overline{\mathcal{C}}/\overline{\mathcal{M}}}\otimes\mathcal{O}_{\mathrm{Sing}\,\overline{\mathcal{C}}}\to 0$$

(compare [M], pf. of 5.10). Let $\mathrm{Sing}'\,\overline{\mathcal{C}}$ be the double cover of $\mathrm{Sing}\,C$ consisting of singular points plus branches: as a Q-variety, it is an étale double cover, i.e., the map between the charts

$$(\mathrm{Sing}'\,\overline{\mathcal{C}})_\alpha\to(\mathrm{Sing}\,\overline{\mathcal{C}})_\alpha$$

which are local universal deformation spaces, is étale. Then via residue

$$\omega_{\overline{\mathcal{C}}/\overline{\mathcal{M}}}\otimes\mathcal{O}_{\mathrm{Sing}'\overline{\mathcal{C}}}\cong\mathcal{O}_{\mathrm{Sing}'\overline{\mathcal{C}}}.$$

Therefore:

$$\begin{aligned}ch\,\mathbb{E}&=1+\pi_*(e^K\cdot Td^\vee(\omega_{\overline{\mathcal{C}}/\overline{\mathcal{M}}})\cdot Td^\vee(\mathcal{O}_{\mathrm{Sing}\,\overline{\mathcal{C}}})^{-1})\\&=1+\pi_*(e^K\cdot\frac{K}{e^K-1}+[Td^\vee(\mathcal{O}_{\mathrm{Sing}\,\overline{\mathcal{C}}})^{-1}-1])\end{aligned}$$

since K intersects any cycle on $\mathrm{Sing}\,\overline{\mathcal{C}}$ in zero. Now use the lemma:

Lemma 5.1. *There is a universal power series P such that for all $i\colon Z\longrightarrow X$, an inclusion of a smooth codimension two subvariety in a smooth variety,*

$$(Td^\vee\mathcal{O}_Z)^{-1}-1=i_*[P(c_1N,c_2N)]$$

where N is the normal bundle I_Z/I_Z^2.

Proof. In fact

$$(Td^\vee\mathcal{O}_Z)^{-1}=1+(\text{polyn. in }ch_k(\mathcal{O}_Z),\quad k\geq 1)$$

and by G-R-R for i, $ch_k(\mathcal{O}_Z)$ is i_* of a polynomial in c_1N, c_2N.

To compute this polynomial P, say $Z=D_1\cdot D_2$. Then use

$$0\to\mathcal{O}_X(-D_1-D_2)\to\mathcal{O}_X(-D_1)\oplus\mathcal{O}_X(-D_2)\to\mathcal{O}_X\to\mathcal{O}_Z\to 0.$$

This gives us

$$Td^\vee O_Z = \left(Td^\vee O_X(-D_1)\right)^{-1} \cdot \left(Td^\vee O_X(-D_2)\right)^{-1} \cdot Td^\vee O_X(-D_1 - D_2)$$
$$= \left(\frac{-D_1}{e^{-D_1}-1}\right)^{-1} \cdot \left(\frac{-D_2}{e^{-D_2}-1}\right)^{-1} \cdot \left(\frac{-D_1-D_2}{e^{-D_1-D_2}-1}\right).$$

Thus

$$D_1 D_2 \cdot P(D_1 + D_2, D_1 \cdot D_2) = Td^\vee(O_Z)^{-1} - 1$$
$$= \frac{D_1}{1-e^{-D_1}} \cdot \frac{D_2}{1-e^{-D_2}} \cdot \frac{1-e^{-D_1-D_2}}{D_1+D_2} - 1$$
$$= \frac{1}{D_1+D_2} \cdot \left[D_1 \cdot \left(\frac{D_2}{1-e^{-D_2}} - 1\right) + D_2 \cdot \left(\frac{D_1}{1-e^{-D_1}} - 1\right) - D_1 \cdot D_2\right]$$
$$= \frac{D_1 D_2}{D_1+D_2} \cdot \sum_{k=1} \frac{(-1)^{k-1} B_k}{(2k)!}\left(D_1^{2k-1} + D_2^{2k-1}\right)$$

So

$$P(D_1 + D_2, D_1 \cdot D_2) = \sum_{k=1}^{\infty} \frac{(-1)^{k-1} B_k}{(2k)!}\left(\frac{D_1^{2k-1} + D_2^{2k-1}}{D_1+D_2}\right)$$
$$= \frac{1}{12} - \frac{1}{720}\left((D_1+D_2)^2 - 3D_1D_2\right) + \frac{1}{30,240}\left((D_1+D_2)^4 - 5D_1D_2(D_1+D_2)^2 + 5D_1^2D_2^2\right) + \cdots$$

Therefore

$$ch\,\mathbb{E} = 1 + \pi_*\left(\frac{K}{1-e^{-K}}\right) + (\pi \circ i)_* P(c_1 N, c_2 N).$$

Now Sing $\overline{\mathcal{C}}$ breaks up into pieces depending on whether the double point disconnects the fibre in which it lies or not, and if it does, what the genera are of the two pieces. Thus:

$$\operatorname{Sing} \overline{\mathcal{C}} = \coprod_{0 \le h \le [g/2]} \Delta_h^*$$

where Δ_0^* are the non-disconnecting double points and if $h \geq 1$, Δ_h^* are the points for which one piece has genus h. Moreover, looking at the two pieces, one sees that

$$\Delta_h^* \cong \overline{\mathcal{C}}_h \times \overline{\mathcal{C}}_{g-h} \quad \text{if} \quad 1 < h < g/2$$

while

$$\begin{aligned} \Delta_{g/2}^* &\cong \overline{\mathcal{C}}_{g/2} \times \overline{\mathcal{C}}_{g/2}/(\mathbb{Z}/2\mathbb{Z}) \text{ if } \quad g \text{ is even} \\ \Delta_0^* &\cong \overline{\mathcal{M}}_{g-1,2}/(\mathbb{Z}/2\mathbb{Z}) \end{aligned}$$

where $\overline{\mathcal{M}}_{g-1,2}$ is the space of stable curves with two ordered points P_1, P_2 and $\mathbb{Z}/2\mathbb{Z}$ permutes either the two factors or the two points. In fact, specifying a branch too, we get:

$$\begin{aligned} \operatorname{Sing}' \overline{\mathcal{C}} &= \coprod_{0 \leq h \leq [g/2]} \Delta_h' \\ \Delta_h' &\cong 2 \text{ copies of } \overline{\mathcal{C}}_h \times \overline{\mathcal{C}}_{g-h} \quad 1 \leq h \leq g/2 \\ &\cong \overline{\mathcal{C}}_{g/2} \times \overline{\mathcal{C}}_{g/2} \quad \text{if } h = g/2 \\ &\cong \overline{\mathcal{M}}_{g-1,2} \quad \text{if } h = 0. \end{aligned}$$

Let K_1, K_2 be the divisor classes defined

a) on $\overline{\mathcal{C}}_h \times \overline{\mathcal{C}}_{g-h}$ by $K_1 = p_1^* K_{\overline{\mathcal{C}}_h / \overline{\mathcal{M}}_h}$, $K_2 = p_2^* K_{\overline{\mathcal{C}}_{g-h} / \mathcal{M}_{g-h}}$

b) on $\overline{\mathcal{M}}_{g-1,2}$ by $K_i =$ conormal bundle at the ith point.

Writing out $ch\,\mathbb{E}$ finally we get

$$(5.2) \qquad \begin{aligned} ch\,\mathbb{E} = g + \sum_{l=1}^{\infty} \frac{(-1)^{l+1} \cdot B_l}{(2l)!} \cdot \\ \left[\kappa_{2l+1} + \frac{1}{2} \sum_{h=0}^{g-1} i_{h,*}(K_1^{2l-2} - K_1^{2l-3} \cdot K_2 + \cdots + K_2^{2l-2}) \right]. \end{aligned}$$

Here we have expanded $K/1 - e^{-K}$ and used the fact that $\pi_* K$ is $(2g-2)$ times the fundamental class of $\overline{\mathcal{M}}_g$. The morphism i_h is

$$\begin{aligned} i_0 &: \overline{\mathcal{M}}_{g-1,2} \to \operatorname{Sing} \overline{\mathcal{C}} \to \overline{\mathcal{M}}_g \\ i_h &: \overline{\mathcal{C}}_h \times \overline{\mathcal{C}}_{g-h} \to \operatorname{Sing} \overline{\mathcal{C}} \to \overline{\mathcal{M}}_g, \quad 1 \leq h \leq g-1. \end{aligned}$$

Note that i_0 and $i_{g/2}$ have degree 2 and the other i_h's are repeated twice in the sum: hence the factor 1/2. Moreover, we have evaluated the normal bundle to $\operatorname{Sing}\overline{\mathcal{C}}$ in $\overline{\mathcal{C}}_g$ as the direct sum of the tangent bundle to the two branches of the curve at the singular point:

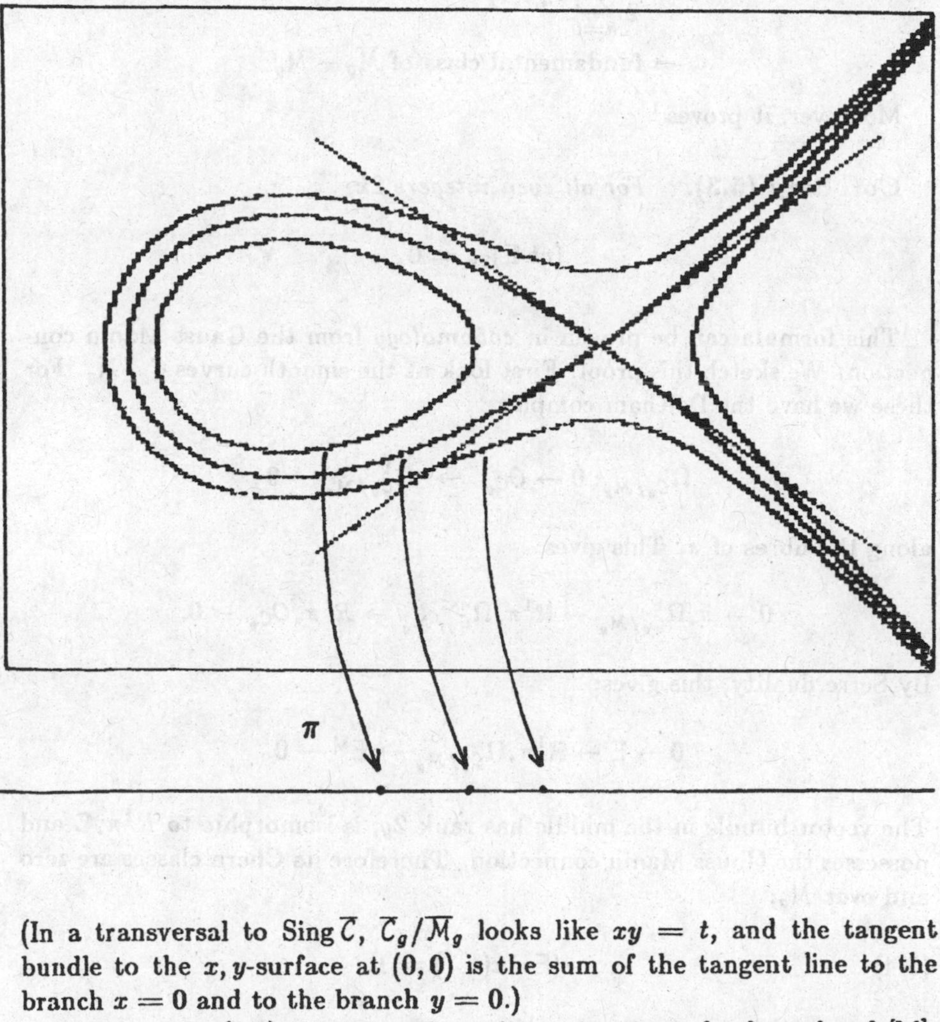

(In a transversal to $\operatorname{Sing}\overline{\mathcal{C}}$, $\overline{\mathcal{C}}_g/\overline{\mathcal{M}}_g$ looks like $xy = t$, and the tangent bundle to the x, y-surface at $(0,0)$ is the sum of the tangent line to the branch $x = 0$ and to the branch $y = 0$.)

The formula (5.2) specializes in codimension 1 to the formula of [M], p. 102:

$$\lambda_1 = c_1(\mathbb{E}) = \frac{1}{12}(\kappa_1 + \delta)$$

where

$$\delta = \frac{1}{2}\sum_{h=0}^{g-1} i_{h,*}(1)$$
$$= \text{fundamental class of } \overline{\mathcal{M}}_g - \mathcal{M}_g.$$

Moreover, it proves

Corollary (5.3). *For all even integers 2k,*

$$(ch\,\mathbb{E})_{2k} = 0.$$

This formula can be proven in *cohomology* from the Gauss-Manin connection. We sketch this proof. First look at the smooth curves $\mathcal{C}_g/\mathcal{M}_g$. For these we have the DeRham complex

$$\Omega^{\cdot}_{\mathcal{C}_g/\mathcal{M}_g}: 0 \to \mathcal{O}_{\mathcal{C}_g} \xrightarrow{d} \Omega^1_{\mathcal{C}_g/\mathcal{M}_g} \to 0$$

along the fibres of π. This gives:

$$0 \to \pi_*\Omega^1_{\mathcal{C}_g/\mathcal{M}_g} \to \mathbb{R}^1\pi_*\Omega^{\cdot}_{\mathcal{C}_g/\mathcal{M}_g} \to R^1\pi_*\mathcal{O}_{\mathcal{C}_g} \to 0.$$

By Serre duality, this gives:

$$0 \to \mathbb{E} \to \mathbb{R}^1\pi_*\Omega^{\cdot}_{\mathcal{C}_g/\mathcal{M}_g} \to \mathbb{E}^\vee \to 0.$$

The vector bundle in the middle has rank $2g$, is isomorphic to $R^1\pi_*\mathbb{C}$ and possesses the Gauss-Manin connection. Therefore its Chern classes are zero and over $\mathcal{M}_g$:

$$(5.4) \qquad c(\mathbb{E}) \cdot c(\mathbb{E}^\vee) = 1.$$

This identity can be extended to $\overline{\mathcal{M}}_g$ if we use the complex

$$\omega^{\cdot}_{\overline{\mathcal{C}}_g/\overline{\mathcal{M}}_g}: 0 \to \mathcal{O}_{\overline{\mathcal{C}}_g} \xrightarrow{d} \omega_{\overline{\mathcal{C}}_g/\overline{\mathcal{M}}_g} \to 0$$

from which we get the sequence:

$$0 \to \mathbb{E} \to \mathbb{R}^1\pi_*\omega^{\cdot}_{\overline{\mathcal{C}}_g/\overline{\mathcal{M}}_g} \to \mathbb{E}^\vee \to 0. \tag{5.5}$$

Although the Gauss-Manin connection does not extend regularly to $\mathbb{R}^1\pi_*\omega^{\cdot}_{\overline{\mathcal{C}}_g/\overline{\mathcal{M}}_g}$, it has regular singularities with a polar part which is nilpotent. This is enough to conclude that its Chern classes zero, extending (5.5) to $\overline{\mathcal{M}}_g$. This means equivalently that

$$ch(\mathbb{E}) + ch(\mathbb{E}^\vee) = 0$$

or

$$ch(\mathbb{E})_{2k} = 0, \quad k \geq 1.$$

This identity in fact holds on $\tilde{\mathcal{A}}_g$, the toroidal compactification of $\mathcal{A}_g$. It can be deduced, for instance, from the extension of Hirzebruch's proportionality theorem to $\tilde{\mathcal{A}}_g$ (see [M2]).

The conclusion to be drawn from (5.2) and (5.3) is that the even λ_k's are polynomials in the odd ones, and that all the λ_k's are polynomials in the κ_k's and in boundary cycles. Moreover, applying (5.2) in odd degree above g, we can express κ_k for k odd, $k > g$, in terms of lower κ_l's and boundary cycles. We shall strengthen this in the next section, where we find a simpler way to get identities on the κ_k's.

The exact sequence (5.4) is remarkable in another way that reveals something of the nature of $\overline{\mathcal{M}}_g$. Note that $\mathbb{E}$ tends to be a positive bundle: at least $c_1(\mathbb{E})$ is the pull-back of an ample line bundle by a birational map. But it is also a sub-bundle of a bundle with connection, i.e., the DeRham bundle $\mathbb{R}^1\pi_*\omega^{\cdot}$ is unstable yet has a connection.

§6. Tautological Relations via the Canonical Linear System

There is another very different way to get relations on the λ_i and κ_i. For this, we will not try to get the full relations in $A^{\cdot}(\overline{\mathcal{M}}_g)$ as the boundary terms seem to be a bit involved, but instead get the relations in $A^{\cdot}(\mathcal{M}_g)$. Because of the exact sequence:

$$A^{\cdot}(\overline{\mathcal{M}}_g - \mathcal{M}_g) \to A^{\cdot}(\overline{\mathcal{M}}_g) \to A^{\cdot}(\mathcal{M}_g) \to 0$$

this is the same as a relation in $A^{\cdot}(\overline{\mathcal{M}}_g)$ with an undetermined boundary term.

The method is based on the fact that for all smooth curves C, the sheaf ω_C is generated by its global sections.[5]

Now if we let $\bar{\mathcal{C}}_g/\bar{\mathcal{M}}_g$ temporarily stand for the family of *smooth* stable curves, i.e., replace $\bar{\mathcal{C}}_g$ by $\pi^{-1}(\mathcal{M}_g)$, then we have an exact sequence:

$$0 \to \mathcal{F} \to \pi^*\pi_*\omega_{\bar{\mathcal{C}}_g/\bar{\mathcal{M}}_g} \to \omega_{\bar{\mathcal{C}}_g/\bar{\mathcal{M}}_g} \to 0$$

where all these sheaves are Q-sheaves and $\mathcal{F}$ is locally free of rank $g-1$. Taking Chern classes, we get:

$$c(\mathcal{F}) = \pi^*(1+\lambda_1+\cdots+\lambda_g)\cdot(1+K_{\bar{\mathcal{C}}_g/\bar{\mathcal{M}}_g})^{-1}.$$

Using the fact that $c_n(\mathcal{F}) = 0$ if $n \geq g$, this says:

$$(K^n_{\bar{\mathcal{C}}/\bar{\mathcal{M}}}) - \pi^*(\lambda_1)\cdot(K^{n-1}_{\bar{\mathcal{C}}/\bar{\mathcal{M}}}) + \cdots + (-1)^g\pi^*(\lambda_g)\cdot(K^{n-g}_{\bar{\mathcal{C}}/\bar{\mathcal{M}}}) = 0$$

for all $n \geq g$. Taking π_*, this means

$$\begin{aligned}
\kappa_{n-1} - \lambda_1\cdot\kappa_{n-2} + \cdots + (-1)^g\lambda_g\cdot\kappa_{n-g-1} &= 0 \quad \text{if } n \geq g+2\\
\kappa_g - \lambda_1\cdot\kappa_{g-1} + \cdots + (-1)^g\lambda_g\cdot(2g-2) &= 0 \quad \text{if } n = g+1\\
\kappa_{g-1} - \lambda_1\cdot\kappa_{g-2} + \cdots + (-1)^{g-1}\lambda_{g-1}\cdot(2g-2) &= 0 \quad \text{if } n = g.
\end{aligned}$$

---

[5]If C is a singular stable curve, then one can show that $\Gamma(\omega_C)$ generates the subsheaf of ω_C of sections which are zero

i) at all double points P for which $C - P$ is disconnectd,

ii) on all components E_0 of C which are isomorphic to $\mathbf{P}^1$ and such that all double

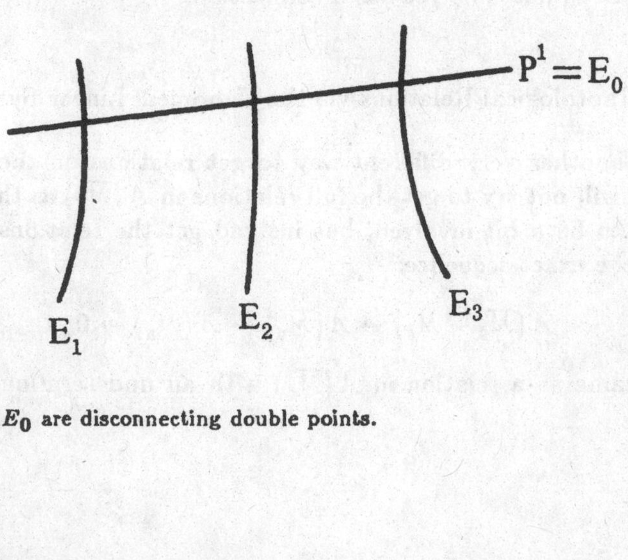

points P on E_0 are disconnecting double points.

Corollary 6.2. *For all g, all the classes λ_i, κ_i restricted to $A^{\cdot}(\mathcal{M}_g)$ are polynomials in $\kappa_1, \kappa_2, \ldots, \kappa_{g-2}$.*

Proof. This is clear except for κ_g, κ_{g-1}. Here we must combine the above relations with (5.2). There are two cases depending on whether g is even or odd. Recalling that

$$ch_n \mathbb{E} = \frac{(-1)^{n-1} \cdot c_n(\mathbb{E})}{(n-1)!} + \text{polyn. in lower Chern class}$$

it follows that if $g = 2k$ or $2k-1$, then

$$\frac{(-1)^{k+1} \cdot B_k}{(2k)!} \kappa_{2k-1} = ch_{2k-1}\mathbb{E} = \frac{\lambda_{2k-1}}{(2k-2)!} + (\text{polyn. in lower } \lambda\text{'s}).$$

If $g = 2k$, we want to show that the 2 equations

$$\begin{cases} (-1)^{k+1} B_k \cdot \kappa_{2k-1} = 2k \cdot (2k-1)\lambda_{2k-1} + \text{lower} \\ \kappa_{2k-1} = 2 \cdot (2k-1)\lambda_{2k-1} + \text{lower} \end{cases}$$

have independent leading terms, and if $g = 2k-1$, then we want to do the same with

$$\begin{cases} (-1)^{k+1} B_k \cdot \kappa_{2k-1} = 2k \cdot (2k-1)\lambda_{2k-1} + \text{lower} \\ \kappa_{2k-1} = 4(k-1) \cdot \lambda_{2k-1} + \text{lower.} \end{cases}$$

This follows, however, by inspection if $k \leq 10$ and for larger k by the estimate:

$$B_k = \frac{2 \cdot (2k)!}{(2\pi)^{2k}} \zeta(2k) > \frac{2 \cdot (2k/e)^{2k}}{(2\pi)^{2k}} = 2 \cdot \left(\frac{k}{e \cdot \pi}\right)^{2k} > 2 \cdot k \quad \text{if } k \geq 11.$$

Q.E.D.

With this approach, the first relation between $\kappa_1, \ldots, \kappa_{g-2}$ that we get occurs in codimension $g+1$ or $g+2$. One should, however, get the relation $\kappa_1^2 = 0$ in $A^{\cdot}(\mathcal{M}_3)$ so we clearly do not have all the relations on the κ_i's and λ_i's yet. It does seem reasonable to conjecture, however, that $\kappa_1, \ldots, \kappa_{g-2}$ have no relations up to something *like* codimension g, e.g., $g -$(small constant).

§7. The Tautological Classes via Arbarello's Flag of Subvarieties of $\mathcal{M}_g$

We want to consider the following subsets of $\mathcal{C}_g$ and $\mathcal{M}_g$:

$$W_l^* = \{C, x \in \mathcal{C}_g \mid h^o(\mathcal{O}_C(l\cdot x)) \geq 2\}$$
$$= \left\{ \begin{array}{l} C, x \in \mathcal{C}_g \mid \exists \text{ a morphism } \pi\colon c \to \mathbb{P}^1 \text{ of degree} \\ \qquad\qquad d \leq l \text{ with } \pi^{-1}(\infty) = d.x \end{array} \right\}$$
$$W_l = \pi(W_l^*) \subset \mathcal{M}_g$$

where $2 \leq l \leq g$. Thus W_g^* =locus of Weierstrass points in $\mathcal{C}_g$, W_{g-1} = curves with an exceptional Weierstrass of one of the two simplest types, and W_2 = hyperelliptic curves. Note that:

$$\mathcal{C}_g \supset W_g^* \supset W_{g-1}^* \supset \cdots \supset W_2^*$$
$$\mathcal{M}_g = W_g \supset W_{g-1} \supset \cdots \supset W_2.$$

I first heard of this flag from E. Arbarello who proposed (see [Arb][6]) that they might be used as a ladder to climb from the reasonably well-known space W_2 to the still mysterious $\mathcal{M}_g$.

Let me first recall and sketch the proof of the following well-known facts:

Proposition 7.1. *W_l^* is irreducible of codimension $g - l + 1$ and $W_l^* - W_{l-1}^*$ is an open dense subset smooth in the local charts for $\mathcal{C}_g$, i.e., the local deformation space for the pairs (C, x).*

Sketch of proof. Firstly, W_l^* is a determinantal subvariety of $\tilde{\mathcal{C}}_g$. In fact, consider

$$\begin{array}{ccccc} & & \tilde{\mathcal{C}}_g \times_{\tilde{\mathcal{M}}_g} \tilde{\mathcal{C}}_g \supset \Delta, \text{ diagonal} & & \\ & \swarrow^{p_1} & & \searrow^{p_2} & \\ \tilde{\mathcal{C}}_g & & & & \tilde{\mathcal{C}}_g \\ & \searrow_{\pi_1} & & \swarrow_{\pi_2} & \\ & & \tilde{\mathcal{M}}_g & & \end{array}$$

Let

$$\mathcal{F}_l = R^1 p_{2,*}(\mathcal{O}_{\tilde{\mathcal{C}}\times\tilde{\mathcal{C}}}(l\Delta)).$$

Then over a point $[C, x] \in \tilde{\mathcal{C}}_g$,

$$\mathcal{F}_l \otimes \mathbf{k}([C, x]) \cong H^1(C, \mathcal{O}_C(l.x)).$$

---

[6]Unfortunately, the proof of Theorem 3.27 in [Arb] is incomplete as it stands.

Now if $[C,x] \notin W_l^*$, $h^0(\mathcal{O}_C(lx)) = 1$ and $h^1(\mathcal{O}_C(lx)) = g - l$, while if $[C,x] \in W_l^*$, both numbers are bigger. Thus $\mathcal{F}_l$ is locally free of rank $g-l$ on $\tilde{C}_g - W_l^*$ and not locally free anywhere on W_l^*. But look at the sequence:

$$0 \to \mathcal{O}_{\tilde{C}\times\tilde{C}} \to \mathcal{O}_{\tilde{C}\times\tilde{C}}(l\Delta) \to \mathcal{O}_{\tilde{C}\times\tilde{C}}(l\Delta)/\mathcal{O}_{\tilde{C}\times\tilde{C}} \to 0$$

which gives us:

$$(7.2) \qquad 0 \to p_{2,*}(\mathcal{O}_{\tilde{C}\times\tilde{C}}(l\Delta)/\mathcal{O}_{\tilde{C}\times\tilde{C}}) \xrightarrow{\alpha} R^1p_{2,*}(\mathcal{O}_{\tilde{C}\times\tilde{C}}) \to \mathcal{F}_l \to 0.$$

The first sheaf is locally free of rank l, the second locally free of rank g, hence

$$W_l^* = \{[C,x] \in \tilde{C}_g \mid rk_{[C,x]}(\alpha) < l\}.$$

Thus the codimension of W_l^* is at most $g-l+1$. But describing $W_l^* - W_{l-1}^*$ as the set of l-fold covers of $\mathbb{P}^1$, totally ramified at ∞, one gets the *upper bound* $2g+l_1-3$ on $\dim W_{l_1}^* - W_{l_1-1}^*$ for all l_1, hence the same upper bound on $\dim W_l^*$. Comparing the two, it follows that codim W_l^* is *exactly* $g-l+1$ and W_l^* is determinantal as well as that $W_l^* - W_{l-1}^*$ is dense in W_l^*. The irreducibility of $W_l^* - W_{l-1}^*$ is a classical result of Lüroth, describing all l-fold covers of $\mathbb{P}^1$ as branched covers with a standard set of transpositions.

The smoothness of $W_l^* - W_{l-1}^*$ in the universal deformation space may be checked by the following calculation: let f have an l-fold pole at $x \in C$ and make an infinitesimal deformation C' of C over $\mathbb{C}[\epsilon]$ by glueing open sets $U_\alpha \times \operatorname{Spec} \mathbb{C}[\epsilon]$ via a 1-cocycle $D_{\alpha\beta}$ of derivations zero at x. Then f lifts to a rational function on C with l-fold pole at x if there are functions g_α with l-fold poles at x and:

$$(1+\epsilon D_{\alpha\beta})(f+\epsilon g_\alpha) = f + \epsilon g_\beta.$$

This means that $\{D_{\alpha\beta}f\} \in H^1(C, \mathcal{O}(lx))$ is zero. But $D_{\alpha\beta}f = \langle D_{\alpha\beta}, df\rangle$ is the image:

$$(7.3) \qquad \{D_{\alpha\beta}\} \in H^1(C, T_C(-x)) \xrightarrow{\langle\ ,df\rangle} H^1(C, \mathcal{O}_C(lx)).$$

Note that $H^1(C, T_C(-x))$ is the tangent space to the universal deformation space of (C,x). Moreover (7.3) is dual to the injective map

$$H^0(C, \mathcal{O}(2K_C+x)) \xleftarrow{\otimes df} H^0(C, \mathcal{O}(K_C - lx))$$

hence (7.3) is ***surjective***, i.e., the subscheme of the universal deformation space where f lifts is smooth of codimension $h^1(\mathcal{O}_C(lx)) = g - l$.

In order to work out the fundamental class of W_l^*, it is convenient to split up (7.2) into pieces as follows. Starting with

$$0 \to \mathcal{O}_{\tilde{C}\times\tilde{C}}((l-1)\Delta) \to \mathcal{O}_{\tilde{C}\times\tilde{C}}(l\Delta) \to \mathcal{O}_{\tilde{C}\times\tilde{C}}(l\Delta)/\mathcal{O}_{\tilde{C}\times\tilde{C}}((l-1)\Delta) \to 0$$

$$\mathcal{O}_{\tilde{C}\times\tilde{C}}(l\Delta)/\mathcal{O}_{\tilde{C}\times\tilde{C}}((l-1)\Delta) = \mathcal{O}_\Delta \otimes p_2^*\mathcal{O}_{\tilde{C}}(-lK_{\tilde{C}/\tilde{\mathcal{M}}})$$

we get via $R^{\cdot}p_{2,*}$:

$$0 \to \mathcal{O}_{\tilde{C}}(-lK_{\tilde{C}/\tilde{\mathcal{M}}}) \xrightarrow{\beta} \mathcal{F}_{l-1} \to \mathcal{F}_l \to 0. \tag{7.4}$$

It follows that on $\tilde{C}_g - W_{l-1}^*$ where $\mathcal{F}_{l-1}$ is locally free:

$$\begin{aligned} W_l^* &= \{[C,x] \in \tilde{C}_g \mid \beta_{[C,x]} = 0\} \\ &= \text{zeroes of the section } \beta' \in \Gamma(\tilde{C}_g, \mathcal{F}_{l-1}(lK_{\tilde{C}/\tilde{\mathcal{M}}})). \end{aligned}$$

Moreover, on the universal deformation space of $[C,x]$, this section β' vanishes to 1st order along W_l^*: in fact, the differential of β' at a point of W_l^* is a map

$$\begin{array}{ccc} T_{[C,x],\mathcal{C}_\alpha} & \longrightarrow & \mathcal{F}_{l-1}(lk)\otimes \mathrm{k}([C,x]) \\ \| & & \| \\ H^1(C, T_C(-x)) & & H^1(C, \mathcal{O}_C(lx))\otimes(\mathfrak{m}_x/\mathfrak{m}_x^2)^{\otimes l} \end{array}$$

which is readily seen to be the surjective map (7.3) (the factor $(\mathcal{M}_x/\mathcal{M}_x^2)^l$ is hidden in (7.3) in the choice of f). Thus

$$[W_l^*]_Q = c_{g-l+1}(\mathcal{F}_{l-1}(lK_{\tilde{C}/\tilde{\mathcal{M}}})), \quad \text{on } \tilde{C}_g - W_{l-1}^*. \tag{7.5}$$

But W_{l-1}^* has codimension $g - l + 2$ so

$$A^{g-l+1}(\mathcal{C}_g) \cong A^{g-l+1}(\mathcal{C}_g - W_{l-1}^*).$$

Thus (7.5) holds as an equation in $A^{g-l+1}(\mathcal{C}_g)$, hence in $opA^{g-l+1}(\tilde{C}_g)$. Now let's calculate the fundamental class of W_l^*:

$$\begin{aligned} [W_l^*]_Q &= c_{g-l+1}(\mathcal{F}_{l-1}(lK)) \\ &= c_{g-l+1}(\mathcal{F}_l(lK)) \\ &= c_{g-l+1}(\mathcal{F}_l). \end{aligned}$$

Here we have abbreviated $K_{\bar{\mathcal{C}}/\bar{\mathcal{M}}}$ to K, and the last equality follows from the general fact

$$c_n\big(\mathcal{G}(D)\big) =$$

$$c_n(\mathcal{G}) + (r-n+1)D.c_{n-1}(\mathcal{G}) + \binom{r-n+2}{2} D^2.c_{n-2}(\mathcal{G}) + \cdots + \binom{r}{n} \cdot D^n$$

($r =$ generic rank $\mathcal{G}$), whence

$$c_{r+1}\big(\mathcal{G}(D)\big) = c_{r+1}(\mathcal{G}), \text{ all divisors } D.$$

But now

$$\begin{aligned} c(\mathcal{F}_l) &= c(\mathcal{F}_{l-1}).(1-lK)^{-1} \\ &= c(\mathcal{F}_{l-2}).(1-(l-1)K)^{-1}(1-lK)^{-1} \\ &\dots \\ &= c(\mathcal{F}_0).(1-K)^{-1}.(1-2K)^{-1}.\cdots.(l-lK)^{-1} \\ &= \pi_2^*\big(c(R^1\pi_1, *\mathcal{O}_{\bar{\mathcal{C}}/\bar{\mathcal{M}}})\big).(1-K)^{-1}.\cdots.(1-lK)^{-1} \\ &= \pi_2^*\big(1-\lambda_1+\lambda_2-\cdots+(-1)^g\lambda_g\big).(1-K)^{-1}.\cdots.(1-lK)^{-1}. \end{aligned}$$

Thus

(7.6) $$\begin{aligned}[W_l^*]_Q &= (g-l+1)^{st} \text{ component of} \\ &\quad \pi^*\big(1-\lambda_1+\lambda_2-\cdots+(-1)^g\lambda_g\big).(1-K)^{-1}.\cdots.(1-lK)^{-1}.\end{aligned}$$

If we define W_l *as a cycle* as $\pi_*(W_l^*)$, we get also

(7.7) $$\begin{aligned}[W_l]_Q &= (g-l)^{th}\text{-component of} \\ &\quad \big(1-\lambda_1+\lambda_2-\cdots+(-1)^g\lambda_g\big).\pi_*[(1-K)^{-1}.\cdots.(1-lK)^{-1}].\end{aligned}$$

This shows that $[W_l]_Q$ is a polynomial in the tautological classes κ_j. Presumably the coefficient of κ_l is always non-zero and hence we can solve for the κ_l's in terms of the classes $[W_j]$, but this looks like a messy calculation.

Let's work out the hyperelliptic locus $\mathcal{H}$ as an example. Note that

$$W_2^* \to W_2$$

is a covering of degree $2g+2$ because for all hyperelliptic curves C, there are exactly $2g+2$ points x such that $h^0\big(\mathcal{O}_C(2x)\big) \geq 2$ — namely the Weierstrass points. Thus

$$[\mathcal{H}]_Q = \frac{1}{2g+2}[W_2]_Q$$

$$= \frac{1}{2g+2}\left\{\begin{matrix}(g-2)^{nd} \text{ component of} \\ (1-\lambda_1+\cdots+(-1)^g\lambda_g)\cdot\pi_*((1-K)^{-1}\cdot(1-2K)^{-1})\end{matrix}\right\}$$

$$= \frac{1}{2g+2}\{(2^g-1)\kappa_{g-2}-(2^{g-1}-1)\lambda_1\cdot\kappa_{g-3}+\cdots+$$

$$(-1)^{g-3}\cdot 7\cdot\lambda_{g-3}\cdot\kappa_3+(-1)^{g-2}\cdot(6g-6)\lambda_{g-2}\}.$$

Finally every hyperelliptic curve has an automorphism of order 2, so

$$[\mathcal{H}] = 2\cdot[\mathcal{H}]_Q$$

$$= \frac{1}{g+1}\{(2^g-1)\kappa_{g-2}-\cdots+(-1)^{g-2}(6g-2)\lambda_{g-2}\}.$$

Part III: The Case g = 2

§8. Tautological Relations in Genus 2

First of all, let's specialize the calculations of Part II to the case $g=2$ and see what we have. From $\mathbb{E}$, we get 2 elements

$$\lambda_1 \in A^1(\overline{\mathcal{M}}_2)$$
$$\lambda_2 \in A^2(\overline{\mathcal{M}}_2)$$

and because $ch_2(\mathbb{E})=0$, we get

$$\lambda_2 = \lambda_1^2/2.$$

From K on $\overline{\mathcal{C}}_2$, we get

$$\kappa_i \in A^i(\overline{\mathcal{M}}_2), \qquad i=1,2,3.$$

The calculations of §5 give us the relation:

$$\lambda_1 = \frac{1}{12}(\kappa_1+\delta). \tag{8.1}$$

Here $\overline{\mathcal{M}}_2 - \mathcal{M}_2$ has 2 components Δ_0 and Δ_1, Δ_0 the closure of the locus of irreducible singular curves, Δ_1 the locus of singular curves $C_1 \cup C_2$, $C_1 \cap C_2 =$ one pt., $p_a(C_1) = p_1(C_2) = 1$. By definition

$$\delta = [\Delta_0]_Q + [\Delta_1]_Q.$$

We shall write δ_0 for $[\Delta_0]_Q$ and δ_1 for $[\Delta_1]_Q$. We don't need (5.2) in codimension 3 but it provides an interesting check on the calculations later. It gives:

$$\frac{1}{6}\lambda_1^3 - \frac{1}{2}\lambda_1\lambda_2 = ch_3\mathrm{E} = -\frac{1}{720}\left[\kappa_3 + \frac{1}{2}\cdot\sum_{h=0}^{1} i_{h,*}((K_1+K_2)^2 - 3K_1\cdot K_2)\right]$$

or

$$60\lambda_1^3 = \kappa_3 + \frac{1}{2}\sum_{h=0}^{1} i_{h,*}(K_1^2 - K_1K_2 + K_2^2).$$

In §10*b* we shall work out these terms numerically and check this.

We can refine the calculations of §6 by working out the boundary term too. It is easy to see that if C is a stable curve of genus 2, ω_C is generated by its global sections, unless $C = C_1 \cup C_2$, $C_1 \cap C_2 = \{P\}$, in which case $\Gamma(\omega_C)$ generates $m_P \cdot \omega_C$. Therefore, working over the whole of $\bar{\mathcal{C}}_2$ we get:

$$0 \to \mathcal{F} \to \pi^*\pi_*\omega_{\bar{\mathcal{C}}_2/\bar{\mathcal{M}}_2} \to I_{\Delta_1^*}\cdot\omega_{\bar{\mathcal{C}}_2/\bar{\mathcal{M}}_2} \to 0$$

(following the notation of §5). $I_{\Delta_1^*}$ has two generators at every point, so its projective dimension is 1, i.e., $\mathcal{F}$ is locally free, hence invertible. Now use:

$$0 \to I_{\Delta_1^*}\cdot\omega \to \omega \to \omega \otimes \mathcal{O}_{\Delta_1^*} \to 0$$

and the fact that via residue, ω is trivial on the double cover Δ_1' of Δ_1^*, hence ω^2 is trivial on Δ_1^*. It follows that

$$c(\mathcal{F}) = \pi^*(1+\lambda_1+\lambda_2)\cdot(1+K_{\bar{\mathcal{C}}_2/\bar{\mathcal{M}}_2})^{-1}\cdot c(\mathcal{O}_{\Delta_1^*}).$$

A useful lemma that we can use here is:

Lemma 8.2. *If $Y \subset X$ is a local complete intersection of codimension 2, and $i\colon Y \to X$ is the inclusion, then*

$$c(\mathcal{O}_Y) = 1 - i_*(c(I_Y/I_Y^2)^{-1}).$$

Proof. If $Y = D_1.D_2$ globally, then this formula is easily checked. But by the G-R-R,

$$c(\mathcal{O}_Y) = 1 + i_*\,(\text{univ. polyn. in } c_1(I/I^2), c_2(I/I^2))$$

and the universal polynomial must be $c(I/I^2)^{-1}$ because the two are equal whenever $Y = D_1.D_2$.

As in §5, $\Delta_1' \cong \overline{M}_{1,1} \times \overline{M}_{1,1}$, i.e., there is a 2-1 map:

$$i_1 \colon \overline{M}_{1,1} \times \overline{M}_{1,1} \longrightarrow \Delta_1^*$$

and $i_1^*(I/I^2) = K_1 + K_2$. Thus

$$c(\mathcal{O}_{\Delta_1^*}) = 1 - \frac{1}{2}\cdot i_{1,*}((1 - K_1 - K_2)^{-1}).$$

Thus

$$\begin{aligned} c(\mathcal{F}) =& \pi^*(1 + \lambda_1 + \lambda_2)\cdot \\ & (1 - K + K^2 - K^3 + K^4)\cdot\left(1 - \frac{1}{2}\cdot i_{1,*}(1 + K_1 + K_2 + K_1\cdot K_2)\right). \end{aligned}$$

In particular,

$$0 = c_2(\mathcal{F}) = \pi^*\lambda_2 - K.\pi^*\lambda_1 + K^2 - [\Delta_1^*]_Q.$$

Multiplying this by K and K^2, we get even simpler formulae:

$$\begin{aligned} 0 &= K\cdot\pi^*\lambda_2 - K^2\cdot\pi^*\lambda_1 + K^3 \\ 0 &= K^2\cdot\pi^*\lambda_2 - K^3\cdot\pi^*\lambda_1 + K^4. \end{aligned}$$

Taking π_*, this gives

$$\begin{aligned} \kappa_1 &= 2\lambda_1 + \delta_1 \\ \kappa_2 &= \kappa_1\cdot\lambda_1 - 2\lambda_2 = \lambda_1\cdot(\lambda_1 + \delta_1) \\ \kappa_3 &= \kappa_2\lambda_1 - \kappa_1\lambda_2 = \frac{1}{2}\lambda_1^2(\delta_1). \end{aligned} \tag{8.3}$$

Combining (8.1) and (8.3), we see that both κ_1 and λ_1 are expressible in terms of δ_0, δ_1:

$$10\lambda_1 = \delta_0 + 2\delta_1 \tag{8.4}$$

$$5\kappa_1 = \delta_0 + 7\delta_1. \tag{8.5}$$

As κ_1 is ample, this implies the well-known fact that $\mathcal{M}_2$ is affine!

This relation (8.4) has a very simple analytic proof. Consider the modular form of weight 10 on Siegel's space $\mathcal{H}_2$ given by

$$f(Z) = \left[\prod_{a,b \text{ even}} \theta\begin{bmatrix} a \\ b \end{bmatrix}(0, Z) \right]^2$$

(Each θ has weight 1/2 and there are ten even a, b's.) It vanishes on $\mathcal{H}_2$ precisely when

$$\gamma Z = \begin{pmatrix} Z_1 & 0 \\ 0 & Z_2 \end{pmatrix}, \quad \text{some } \gamma \in Sp(4, \mathbb{Z})$$

and then to order 2. At the principal cusp

$$\begin{pmatrix} i\infty & w \\ w & z \end{pmatrix}$$

it has the form

$$\begin{aligned} &(\text{unit}) \cdot \theta\begin{bmatrix} 1 & 0 \\ 0 & 0 \end{bmatrix}^2 \cdot \theta\begin{bmatrix} 1 & 1 \\ 0 & 0 \end{bmatrix}^2 \cdot \theta\begin{bmatrix} 1 & 0 \\ 0 & 1 \end{bmatrix}^2 \cdot \theta\begin{bmatrix} 1 & 1 \\ 1 & 1 \end{bmatrix}^2 \\ &= (\text{unit}) \cdot (e^{\pi i (1/2)\Omega_{11}})^8 \\ &= \text{unit} \cdot e^{2\pi i \Omega_{11}} \end{aligned}$$

i.e., it vanishes to order 1. Thus f defines a section of $(\Lambda^2\mathbb{E})^{\otimes 10}$ whose zeroes in $\tilde{\mathcal{M}}_2$ are $2\tilde{\Delta}_1 + \tilde{\Delta}_0$. This reproves (8.4).

§9. Generators of $A.(\overline{\mathcal{M}}_2)$.

We use the exact sequence:

$$A.(Y) \to A.(X) \to A.(X - Y) \to 0$$

($Y \subset X$ closed subvariety) to get generators of $A_\cdot(\overline{\mathcal{M}}_2)$. Recall that $\mathcal{M}_2$ is known from Igusa's results [I] to be isomorphic to $\mathbb{C}^3$ modulo $\mathbb{Z}/5\mathbb{Z}$ acting by

$$(x, y, z) \mapsto (\zeta x, \zeta^2 x, \zeta^3 y).$$

Then

$$A_\cdot(\mathbb{C}^3) \to A_\cdot(\mathcal{M}_2)$$

is surjective, hence $A_k(\mathcal{M}_2) = (0)$, if $k < 3$. Thus

$$A_k(\Delta_0) \oplus A_k(\Delta_1) \to A_k(\overline{\mathcal{M}}_2)$$

is surjective if $k < 3$. In particular, $A_2(\overline{\mathcal{M}}_2)$ is generated by δ_0 and δ_1.

Define the dimension 1 subsets:

$$\Delta_{00} = \left\{ \begin{array}{l} \text{closure of curve in } \overline{\mathcal{M}}_2 \text{ parametrizing} \\ \text{irreducible rational curves with 2 nodes} \end{array} \right\}$$

$\Delta_{01} = \Delta_0 \cap \Delta_1$
$= $ Curve in $\overline{\mathcal{M}}_2$ parametrizing curves $C_0 \cup C_2$, where $C_1 \cap C_2 = \{x\}$, C_1 is elliptic or rational with one node and C_2 is rational with one node

curves in Δ_{00}

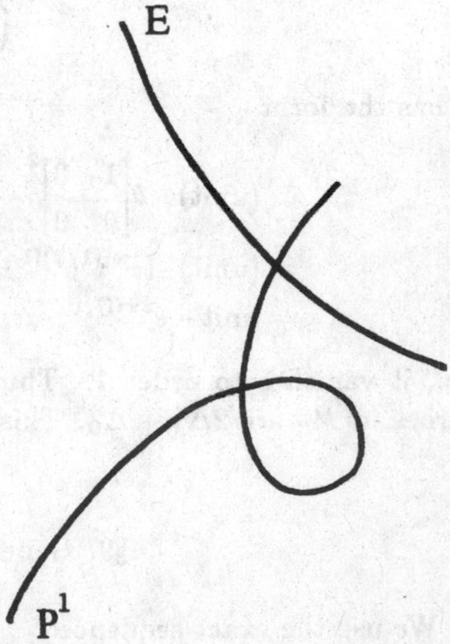

curves in Δ_{01}

Note that Δ_{00} contains, besides the irreducible curves illustrated, the two reducible curves

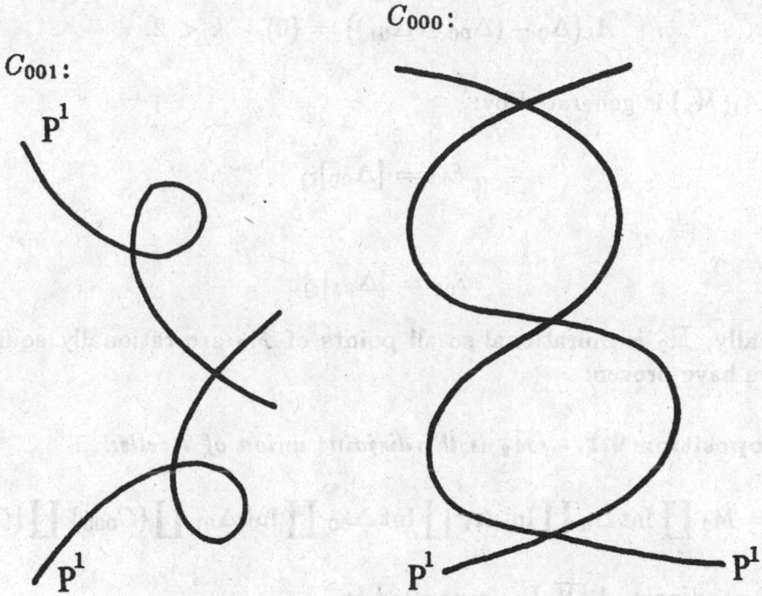

$\text{Int}\,\Delta_1 = (\Delta_1 - \Delta_{01})$ is the locus of curves $C_1 \cup C_2$ where $C_1 \cap C_2 = \{x\}$, C_1, C_2 smooth elliptic. It is isomorphic to $\text{Symm}^2 \mathcal{M}_{1,1}$, i.e., to the product of the affine j-line by itself mod the involution interchanging the factors. Therefore it is coordinatized by $j(C_1) + j(C_2)$, $j(C_1).j(C_2)$:

$$\text{Int}\,\Delta_1 \cong \mathbb{C}^2.$$

Moreover, $\text{Int}\,\Delta_0 = \Delta_0 - (\Delta_{00} \cup \Delta_{01})$ is the locus of irreducible elliptic curves with one node, i.e., the space $\mathcal{M}_{1,2}$ of triples (E, x_1, x_2), $x_1, x_2 \in E$, $x_1 \neq x_2$ mod the involution interchanging the 2 points. Write all elliptic curves as

$$y^2 = x(x-1)(x-\lambda), \quad \lambda \neq 0, 1$$

and take $x_1 =$ pt. at ∞, $x_2 = (x, y)$. Interchanging x_1, x_2 carries (x, y) to $(x, -y)$. So we get a surjective map

$$\{(x, \lambda) \mid \lambda \neq 0, 1\} \to \text{Int}\,\Delta_0.$$

Putting this together

$$A_k(\Delta_1 - \Delta_{01}) = (0), \quad k < 2$$
$$A_k(\Delta_0 - (\Delta_{00} \cup \Delta_{01})) = (0), \quad k < 2.$$

Thus $A_1(\overline{\mathcal{M}}_2)$ is generated by:

$$\delta_{00} = [\Delta_{00}]_Q$$

and

$$\delta_{01} = [\Delta_{01}]_Q.$$

Finally, $\overline{\mathcal{M}}_2$ is unirational so all points of $\overline{\mathcal{M}}_2$ are rationally equivalent and we have proven:

Proposition 9.1. *$\overline{\mathcal{M}}_2$ is the disjoint union of 7 cells:*

$$\overline{\mathcal{M}}_2 = \mathcal{M}_2 \coprod \operatorname{Int}\Delta_0 \coprod \operatorname{Int}\Delta_1 \coprod \operatorname{Int}\Delta_{00} \coprod \operatorname{Int}\Delta_{01} \coprod \{C_{000}\} \coprod \{C_{001}\}.$$

Correspondingly, $A^{\cdot}(\overline{\mathcal{M}}_2)$ is generated by

a) *1 in codimension 0,*
b) *δ_0, δ_1 in codimension 1,*
c) *δ_{00}, δ_{01} in codimension 2,*
d) *the class $[x]$ of a point in codimension 3: call this p.*

Note that by the results of §8, λ_1 and κ_1 are also generators in codimension 1. We shall see that all the above cycles are independent. This will follow as a Corollary once we work out the multiplication table for these cycles.

§10. Multiplication in $A^{\cdot}(\overline{\mathcal{M}}_2)$

We shall prove:

Theorem 10.1. *The ring $A^{\cdot}(\overline{\mathcal{M}}_2)$ has a $\mathbb{Q}$-basis consisting of* 1, δ_0,

δ_1, δ_{00}, δ_{01}, p and multiplication table:

$$\delta_0^2 = \frac{5}{3}\delta_{00} - 2\delta_{01}$$

$$\delta_0 \cdot \delta_1 = \delta_{01}$$

$$\delta_1^2 = -\frac{1}{12}\delta_{01}$$

$$\delta_0 \cdot \delta_{00} = -\frac{1}{4}p$$

$$\delta_0 \cdot \delta_{01} = \frac{1}{4}p$$

$$\delta_1 \cdot \delta_{00} = -\frac{1}{8}p$$

$$\delta_1 \cdot \delta_{01} = -\frac{1}{48}p$$

An easier way to describe the ring structure is via λ_1. Using the identities $10\lambda_1 = \delta_0 + 2\delta_1$, we can describe the multiplication by:

$$a)\ \delta_0 \cdot \delta_1 = \delta_{01}$$

$$b)\ \delta_{00} \cdot \delta_1 = \frac{1}{8}p$$

$$c)\ \delta_{00} \cdot \lambda_1 = 0$$

$$d)\ \delta_1 \cdot \lambda_1 = \frac{1}{12}\delta_{01}$$

$$e)\ \delta_0 \cdot \lambda_1 = \frac{1}{6}\delta_{00}.$$

The reader can check that these are equivalent to the relations of the Theorem.

Relations (a) and (b) are proper intersections of cycles and are proved by the explicit formula of §3: thus (a) follows because the lifts of δ_0, δ_1 to the universal deformation space of a curve $C \in \Delta_{01}$ are smooth divisors meeting transversely in the smooth curve lifting Δ_{01}. And for (b), $\Delta_{00} \cap \Delta_1$ is the one curve C_{001} whose automorphism group has order 8. In the universal deformation space of C_{001}, Δ_{00} and Δ_1 lift to a smooth curve and surface meeting transversely, so

$$\delta_{00}.\delta_1 = [C_{001}]_Q = \frac{1}{8}p.$$

c) is an immediate consequence of the general theory of Knudsen [K] or of the fact that δ_{00} is blown down to a point in the Satake compactification $\mathcal{A}_2^*$ of $\mathcal{A}_2$. To prove (d), consider

$$i_1\colon \overline{\mathcal{M}}_{1,1} \times \overline{\mathcal{M}}_{1,1} \longrightarrow \Delta_1 \subset \overline{\mathcal{M}}_2.$$

We check that $i_1^*(\lambda_1 - \frac{1}{12}\delta_0) = 0$, hence $(\lambda_1 - \frac{1}{12}\delta_0).\delta_1 = 0$. But

$$i_1^*(\lambda_1^{(2)}) = p_1^*(\lambda_1^{(1)}) + p_2^*(\lambda_1^{(1)})$$

where, for the sake of clarity, we write

$$\begin{aligned} \lambda_1^{(2)} &= \text{the class } \lambda_1 \text{ in } A^{\cdot}(\overline{\mathcal{M}}_2) \\ \lambda_1^{(1)} &= \text{the class } \lambda_1 \text{ in } A^{\cdot}(\overline{\mathcal{M}}_{1,1}). \end{aligned}$$

This is simply because on curves $E_1 \cup E_2$, E_i elliptic, $E_1 \cap E_2 =$ one point,

$$\Gamma(E_1 \cup E_2, \omega_{E_1 \cup E_2}) \cong \Gamma(E_1, \omega_{E_1}) \oplus \Gamma(E_2, \omega_{E_2}).$$

Moreover,

$$i_1^*(\delta_0^{(2)}) = p_1^*(\delta^{(1)}) + p_1^*(\delta^{(1)}).$$

But in $A^{\cdot}(\overline{\mathcal{M}}_{1,1})$, the relation

$$\lambda_1 = \frac{1}{12}\delta$$

holds. This is well known and is just the specialization to genus 1 of the theory of §5. Or else it may be seen using the elliptic modular form Δ of weight 12 with a simple pole at the cusp. Finally, to prove (*e*), consider:

$$i_o\colon \overline{\mathcal{M}}_{1,2} \longrightarrow \Delta_0 \subset \overline{\mathcal{M}}_2.$$

Then $i_{0,*}(1_{\overline{\mathcal{M}}_{1,2}}) = 2\delta_0$. One should be careful here to note that the presence of automorphisms generically on $\overline{\mathcal{M}}_2$ does not affect this: in fact

$$\begin{aligned} i_{0,*}(1_{\overline{\mathcal{M}}_{1,2}}) &= i_{0,*}([\overline{\mathcal{M}}_{1,2}]) \\ &= [\Delta_0] \quad (\text{because } \overline{\mathcal{M}}_{1,2} \to \Delta_0 \text{ is birational}) \\ &= 2\delta_0 \quad (\text{because } \mathrm{Aut}(C) = \mathbb{Z}/2\mathbb{Z}, [C] \in \Delta_0 \text{ generic}). \end{aligned}$$

Therefore

$$\begin{aligned} \lambda_1.\delta_0 &= \frac{1}{2}\lambda_1 \cdot i_{0,*}(1_{\overline{\mathcal{M}}_{2,1}}) \\ &= \frac{1}{2} i_{0,*}(i_0^*(\lambda_1)). \end{aligned}$$

Now let

$$\pi\colon \overline{\mathcal{M}}_{1,2} \longrightarrow \overline{\mathcal{M}}_{1,1}$$

be the natural projection. Note that $\overline{\mathcal{M}}_{1,1}$ is the j-line and $\overline{\mathcal{M}}_{1,2}$ is the universal family over the j-line of elliptic curves mod automorphisms. Then

$$i_0^*(\lambda_1^{(2)}) = \pi^*(\lambda_1^{(1)})$$

by Knudsen's theory. This corresponds to the fact that if E' is elliptic with one node P, and E is the normalization of E', then there is a canonical sequence

$$0 \to \Gamma(\omega_E) \to \Gamma(\omega_{E'}) \overset{res}{\to} \mathrm{k}(P) \to 0$$

hence $\Lambda^2(\Gamma(\omega_{E'})) \cong \Gamma(\omega_E)$. Therefore

$$\begin{aligned} \lambda_1.\delta_0 &= \frac{1}{2} i_{0,*}\left(\pi^*(\lambda_1^{(1)})\right) \\ &= \frac{1}{24} i_{0,*}\left(\pi^*(\delta^{(1)})\right). \end{aligned}$$

But $\pi^*(\delta^{(1)}) = [\tilde{\Delta}], [\tilde{\Delta}] \subset \overline{\mathcal{M}}_{1,2}$ being the closure of the locus of triples (C, x_1, x_2), C a rational curve with a node, x_1, x_2 distinct smooth points of C. $\tilde{\Delta}$ maps birationally to Δ_{00} in $\overline{\mathcal{M}}_2$, and the automorphism group of the generic rational curve with 2 nodes is $(\mathbb{Z}/2\mathbb{Z})^2$, hence:

$$\begin{aligned}\lambda_1.\delta_0 &= \frac{1}{24}[\Delta_{00}]\\ &= \frac{1}{6}\delta_{00}.\end{aligned}$$

Q.E.D.

§10*b*. A Check

An interesting check that these Q-stack-theoretic calculations are OK is to evaluate all terms in the identity

$$60\,\lambda_1^3 = \kappa_3 + \frac{1}{2}\sum_{h=0}^{1} i_{h,*}(K_1^2 - K_1K_2 + K_2^2)$$

obtained in §8. Using Theorem 10.1, one finds

$$60\,\lambda_1^3 = \frac{1}{48}p.$$

Using (8.3) plus theorem 10.1, one finds

$$\kappa_3 = \frac{1}{1152}p.$$

To calculate

$$i_{0,*}(K_1^2 - K_1K_2 + K_2^2)$$

let π: $\overline{\mathcal{M}}_{1,2} \longrightarrow \overline{\mathcal{M}}_{1,1}$ be the natural map and in $\tilde{\mathcal{M}}_{1,2}$ consider the points $\pi^{-1}([E])$, i.e., representing (E, x_1, x_2), with E fixed. Up to automorphisms of E, x_1 can be normalized to be the identity. Letting x_2 vary, we parametrize this subset of $\tilde{\mathcal{M}}_{1,2}$ by E itself, and describe the universal family of triples (E, x_1, x_2) as $E \times E$ over E with x_1 being given by $s_1(x) = e(x)$, x_2 by the diagonal $s_2(x) = (x, x)$:

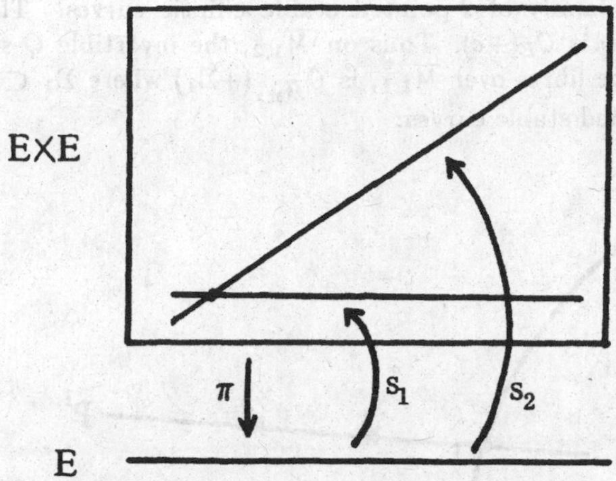

However, this allows $x_1 = x_2$ over $e \in E$, where we should have instead $E \cup \mathbb{P}^1$, $x_1, x_2 \in \mathbb{P}^1 - E \cap \mathbb{P}^1$, $x_1 \neq x_2$. Thus we must blow up $(e, e) \in E \times E$, getting:

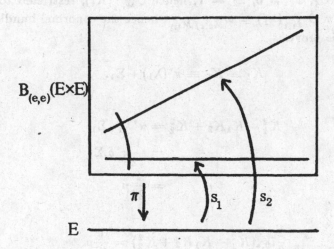

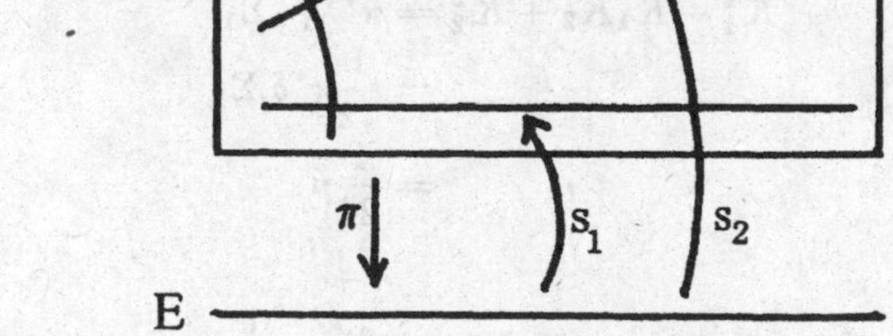

This is now a family of 2-pointed stable elliptic curves. The conormal bundle K_i to s_i is $\mathcal{O}_E(+e)$. Thus on $\overline{\mathcal{M}}_{1,2}$, the invertible Q-sheaf $\mathcal{O}(K_i)$, restricted to the fibres over $\overline{\mathcal{M}}_{1,1}$, is $\mathcal{O}_{\overline{\mathcal{M}}_{1,2}}(+\Sigma_1)$ where $\Sigma_1 \subset \overline{\mathcal{M}}_{1,2}$ is the locus of 2-pointed stable curves:

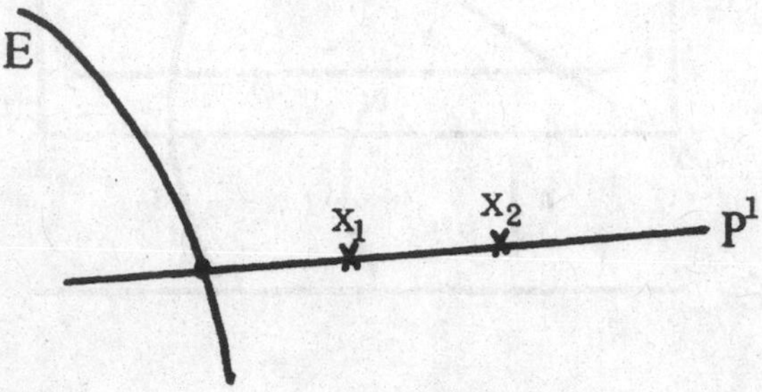

Therefore $K_1 \equiv K_2 \equiv \pi^*(A) + \Sigma_1$, for some Q-divisor class A on $\overline{\mathcal{M}}_{1,1}$. But *along* $\sum_1$, a canonical coordinate can be put on $\mathbb{P}^1$ making $E \cap \mathbb{P}^1 = \{\infty\}$, $x_1 = 0$, $x_2 = 1$, hence $\mathcal{O}_{\overline{\mathcal{M}}_{1,2}}(K_i)$, restricted to Σ_1, is trivial. Now $\pi^* \mathcal{O}_{\overline{\mathcal{M}}_{1,1}}(\lambda_1) \cong \omega_{\overline{\mathcal{M}}_{2,1}/\overline{\mathcal{M}}_{1,1}}$, hence the conormal bundle to Σ_1 is $\mathcal{O}(\lambda_1)$. This proves:

$$K_1 \equiv K_2 \equiv \pi^*(\lambda_1) + \Sigma_1.$$

Therefore

$$\begin{aligned} K_1^2 - K_1K_2 + K_2^2 &= \pi^*\lambda_1 \cdot \Sigma_1 \\ &= \frac{1}{12}\pi^*\delta.\Sigma_1 \\ &= \frac{1}{24}p \end{aligned}$$

hence

$$\frac{1}{2}i_{0,*}(K_1^2 - K_1K_2 + K_2^2) = \frac{1}{48}p.$$

Finally to calculate

$$i_{1,*}(K_1^2 - K_1K_2 + K_2^2)$$

note that on $\overline{\mathcal{M}}_{1,1} \times \overline{\mathcal{M}}_{1,1}$, $K_1^2 = K_2^2 = 0$, and K_i is the pull back from $\overline{\mathcal{M}}_{1,1}$ of λ_1. Since $\lambda_1 = \frac{1}{24}p$,

$$\frac{1}{2}i_{1,*}(K_1^2 - K_1K_2 + K_2^2) = \frac{1}{2}i_{1,*}(-\frac{1}{24} \cdot \frac{1}{24}p)$$
$$= -\frac{1}{1152}p.$$

This checks!

References

[A] Arakelov, S., *Families of algebraic curves with fixed degeneracies*, Izv. Akad. Nauk, *35* (1971).

[Arb] Arbarello, E., *Weierstrass points and moduli of curves*, Comp. Math., *29* (1974), pp. 325–342.

[A-B] Atiyah, M., and Bott, R., *The Yang-Mills equations over Riemann surfaces*, to appear.

[B-F-M] Baum, P., Fulton, W., and MacPherson, R., *Riemann-Roch for singular varieties*, Publ. I.H.E.S., *45* (1975), pp. 101–145.

[D-M] Deligne, P., and Mumford, D., *The irreducibility of the space of curves of given genus*, Publ. I.H.E.S. *36* (1969) pp. 75–109.

[F1] Fulton, W., *Rational equivalence on singular varieties*, Publ. I.H.E.S. *45* (1975), pp. 147–167.

[F2] Fulton, W., *Intersection Theory*, Springer-Verlag, 1983.

[F-M] Fulton, W. and MacPherson, R., *Categorical framework for the study of singular spaces*, Memoirs A.M.S. 243, (1981).

[Ha] Harer, J., The second homology group of the mapping class group of an orientable surface, to appear.

[H-M] Harris, J., and Mumford, D., *On the Kodaira dimension of the moduli space of curves*, Inv. Math., *67* (1982), pp. 23–86.

[H] Hironaka, H., *Bimeromorphic smoothing of a complex-analytic space*, Acta. Math. Vietnamica, *2* (1977), pp. 103–168.

[I] Igusa, J.I., *Arithmetic theory of moduli for genus two*, Annals of Math., *72* (1960), pp. 612–649.

[K] Knudsen, F., *The projectivity of the moduli space of stable curves*, Math. Scand. to appear.

[Kl] Kleiman, S., *Towards a numerical theory of ampleness*, Annals of Math., *84* (1966), pp. 293–344.

[Ma] Matsusaka, T., *Theory of Q-varieties*, Publ. Math. Soc. Japan, *8* (1964)

[Mi] Miller, E., *The homology of the mapping class group of surfaces*, in preparation.

[M] Mumford, D., *Stability of projective varieties*, L'Ens, Math., *24* (1977), pp. 39-110.

[M2] Mumford, D., *Hirzebruch's proportionality principle in the non-compact case*, Inv. math., *42* (1977), pp. 239–272.

[N] Namikawa, Y., *A new compactification of the Siegel space and degeneration of abelian varieties I and II*, Math. Annalen, *221* (1976), pp. 97–141, 201–241.

Received July 1, 1982

Professor David Mumford
Department of Mathematics
Harvard University
Cambridge, Massachusetts 02138

Theta Functions and the Moduli of Abelian Varieties

George R. Kempf † and Herbert Lange *

1 Finite theta functions and equations of abelian varieties

Let L be an ample invertible sheaf on an abelian variety X over an algebraically closed field k and let $\Lambda(L) : X \to \hat{X}$ be the usual isogeny associated to L where $\hat{X}$ denotes the dual abelian variety. The line bundle L defines a separable polarization of X if $\deg \Lambda(L)$ is prime to the characteristic of k. Let $H(L)$ be the kernel of $\Lambda(L)$. In [M1] Mumford defines a natural central extension $G(L)$ of $H(L)$ by $\mathbb{G}_m$, called the theta (or Heisenberg) group of L. The group $G(L)$ acts on $\Gamma(X, L)$ with $\mathbb{G}_m$ acting naturally. The first theorem is that, up to isomorphisms, $\Gamma(X, L)$ is the unique irreducible such representation of $G(L)$. This result has been applied to abelian varieties in papers too numerous to mention. Mumford's papers [M1] detail only separable polarizations. We will hence first discuss the separable case.

Let $\delta = (d_1, \dots, d_g)$ be a sequence of positive integers such that $d_i | d_{i+1}$. A theta structure of the polarized abelian variety (X, L) is an isomorphism $G(L) \xrightarrow{\sim} G(\delta)$ respecting $\mathbb{G}_m$. Here $G(\delta)$ denotes the standard Heisenberg group $\mathbb{G}_m \rtimes (K(\delta) \times \widehat{K(\delta)})$ with the usual group structure, where $K(\delta) = \oplus_{i=1}^{g} \mathbb{Z}/d_i\mathbb{Z}$ and $\widehat{K(\delta)}$ is the dual abelian group. Let $V(\delta)$ denote the vector space of k-valued functions on the finite group $K(\delta)$. Then $V(\delta)$ is an irreducible representation of $G(\delta)$ in which $\mathbb{G}_m$ acts naturally. Thus there is a canonical isomorphism $\beta : \Gamma(X, L) \to V(\delta)$ (determined up to a scalar) which is compatible with the group actions. Therefore, given a theta structure of (X, L), we have a canonical up to scalar basis $\{\beta^{-1}(\delta_x)\}_{x \in K(\delta)}$, where δ_x denotes the delta function. This is the basis of Mumford's algebraic theory of theta functions. The elements of $V(\delta)$ are called finite or algebraic theta functions and for many problems they are more convenient to work with than the usual analytic theta functions. The basic result is the isogeny

† George Kempf died on July 16, 2002.

* A first draft of this article was written by the first author at the beginning of 1991. The final version was written by the second author in 2003.

theorem, which describes the pull-back map $\Gamma(X, L) \to \Gamma(Y, f^*L)$ in terms of compatible theta structures for any separable isogeny $f : Y \to X$. In the third volume of [81] Mumford (in collaboration with M. Nori and P. Norman) gave an alternative treatment of finite theta functions and moreover extended these definitions to more general classes of algebraic theta functions, such as those associated to positive definite quadratic forms and those depending not only on a quadratic form but also on a spherical harmonic polynomial.

According to a theorem of Lefschetz the line bundle L gives a projective embedding $X \hookrightarrow \mathbb{P}(H^0(L))$, if L is an n-th power of an ample line bundle for some $n \geq 3$. Mumford showed in [M1,I] that X is projectively normal in $\mathbb{P}(H^0(L))$ and moreover X is cut out by quadrics, if n is a multiple of 4 or if $n \geq dim(X) + 2$. X is scheme-theoretically the zero set of explicit quadratic equations in $\mathbb{P}(H^0(L))$. We will discuss generalizations of these facts.

Mumford was the first to generalize these results to the inseparably polarized case except for the explicit forms of the equations in [79]. Using classical theta functions Koizumi showed projective normality for $n \geq 3$ in [73]. Then Sekiguchi [102] generalized Koizumi's result to characteristic p. Ohbuchi [87] gave a criterion for projective normality in case $n = 2$. In [79] Mumford generalized some of these results even to arbitrary projective varieties using a method, which is now called Castelnuovo-Mumford regularity. For a good survey on this subject see Lazarsfeld's forthcoming book [75].

Theta relations and thus equations for an abelian variety in projective space are derived with the help of an addition formula. This is a formula for the pullback $\Gamma(Y^{\times n}, L) \to \Gamma(X^{\times n}, f^*L)$, where f is an isogeny of the form $f : X^{\times n} \to Y^{\times n}$ such that L and f^*L define product polarizations of a special type. Apart from finding such a formula the problem is to give compatible theta structures of the sheaves. Mumford solved this problem in [M1] for $n = 2$, which gives Riemann's theta relations. Koizumi [73] generalized these relations over $\mathbb{C}$. In the algebraic case the general form of these results may be found in [69]. Cubic theta relations were found using an addition formula for $n = 3$ in [18]. The book [71] has an analytic discussion of the addition formula using Mumford's formalism.

Mumford's methods have also been used to study the syzygies of X in $\mathbb{P}(H^0(L))$. For this the following definition, due to Green, turned out to be useful: A very ample line bundle L is called of type N_1 if X is projectively normal and its homogeneous ideal is generated by quadrics. It is called of type N_p for some $p \geq 2$ if it is of type N_{p-1} and the p-th syzygies of the minimal resolution of X in $\mathbb{P}(H^0(L))$ are as simple as possible, i.e. generated by linear forms. Lazarsfeld conjectured that L is of type N_p if it is an n-th power with $n \geq p+3$. In [68] this is proven for $p = 1$ and moreover for $p > 1$ with a stronger condition on n. The full conjecture was proved by Pareschi [88]. In the series of papers [89] Pareschi and Popa showed some further results by introducing the notion of Mukai regularity, which is defined via the Fourier-Mukai transform and parallels and strengthens the usual Castelnuovo-Mumford regularity with respect to polarizations on abelian varieties. They have a large number of applications, e.g. to the syzygies of abelian varieties, to equations for the ideals of subvarieties of abelian varieties or to Albanese maps. Mumford's methods have also been applied to investigate ideals and moduli of polarized abelian surfaces of type $(1, d)$ as well as some abelian threefolds of type $(1, d_1, d_2)$.

The second part of Mumford's work [M1, II] deals with the moduli space M_δ of abelian varieties (X, L) with totally symmetric L of type $\delta = (d_1, \ldots, d_g)$ together with some additional structure. Influenced by Igusa [61] Mumford gives an explicit projective embedding $M_\delta \subset \mathbb{P}(V(\delta))$ defined by the theta-null values, when $d_1 = 2m$ for some $m \geq 2$. He also defines an explicit closed subscheme $\overline{M}_\delta$ of $\mathbb{P}(V(\delta))$ containing M_δ. In case d_1 is a multiple of 4, he shows that M_δ is an open subscheme of $\overline{M}_\delta$. In [68] it is shown that this is true when $m \geq 2$. The principle of this proof is very different from Mumford's and uses syzygies. It still remains an open problem to give explicit equations of the closure of M_δ in $\mathbb{P}(V(\delta))$. Mumford's tower method is linked to an infinite number of implicit such relations.

In the third paper [M1, III] Mumford used his 2-adic theta functions to construct a Satake-type compatification of the tower of moduli. Chai [23] has used Mumford's theta function technique to construct a toroidal type compactification of the moduli space in any chacteristic $\neq 2$. This is proven in any characteristic in [42] using a stack of toroidal embeddings.

2 Families of abelian varieties

Let X denote a complex smooth projective variety of dimension n. The ring $H^*(X) = \oplus_{p=0}^{n} H^{p,p}(X) \cap H^{2p}(X, \mathbb{Q})$ is called the Hodge ring of X. It is easy to see that the classes of algebraic cycles are contained in $H^*(X)$. The Hodge conjecture asserts that each of its homogeneous parts is generated by algebraic cycles as a $\mathbb{Q}$-vector space. The proof of this conjecture for the variety X consists in two steps which are both difficult in general: First one has to determine $H^*(X)$ and secondly one has to find a set of generating algebraic cycles.

In [M2] Mumford (in part jointly with Tate) gave a general method for attacking the first step in the case of abelian varieties. He introduced a group $Hg(X)$, acting on the cohomology ring, which he called the Hodge group of X (nowadays it is mostly called the special Mumford-Tate group) and which has the property that its ring of invariants is just $H^*(X)$. Given a complex polarized abelian variety (X, L) and denoting by A the alternating form associated to L, one can interpret the complex structure of X as a homomorphism of real algebraic groups $\varphi : \mathbb{S}^1 \to Sp(\mathbb{C}^g, A)$, where $\mathbb{S}^1$ denotes the circle group regarded as a one-dimensional algebraic group over the reals. $Hg(X)$ is defined as the smallest algebraic subgroup of $Sp(\mathbb{C}^g, A)$ defined over $\mathbb{Q}$ and containing $\varphi(\mathbb{S}^1)$. It can also be characterized as the largest algebraic subgroup which is defined over $\mathbb{Q}$ and which leaves invariant the elements of the Hodge rings of all powers of X.

Mumford showed in [M2] that the Hodge group is reductive. Hence one can apply classical invariant theory in order to determine the Hodge ring of X. This idea has been applied in numerous papers for many abelian varieties of some special type, too many to be quoted here.

Let $D^*(X)$ denote the subring of $H^*(X)$ generated by the classes of divisors of X. For most abelian varieties $D^*(X) = H^*(X)$, which implies that the Hodge conjecture is valid in these cases. The elements of $H^*(X) \setminus D^*(X)$ are called exceptional Hodge classes. Perhaps the first example of an abelian variety admitting exceptional Hodge classes was Mumford's example of an abelian fourfold of a particular CM-type (see [90]). Weil generalized this example in [116] to compute the Hodge ring for a class of abelian varieties which are now called of Weil type and which are defined as pairs (X, K) consisting of an abelian variety X of dimension $g = 2n$ and an imaginary quadratic field K together with an embedding of K into $End_0(X)$ such that the action of $\alpha \in K$ has eigenvalues α and $\overline{\alpha}$ with equal multiplicity n. Under some generality assumption each of these varieties admits exceptional Hodge classes in dimension n. Weil introduced these varieties as candidates where the Hodge conjecture might possibly fail (see [116], p. 429). The only cases in which the Hodge conjecture has been tested for abelian varieties of Weil type are apparently [100] (general abelian fourfolds X with $K = \mathbb{Q}(i)$ or $\mathbb{Q}(\sqrt{-3})$) and [113] (general abelian sixfolds X with $K = \mathbb{Q}(\sqrt{-3})$). Moonen and Zarhin [76] have considered the extent to which Weil's construction of exceptional Hodge classes can be generalized. Other examples of exceptional Hodge classes were given by Shioda [105], Tankeev [110] and Murty [84].

In [35] Deligne developped the theory of the Hodge group further. Moreover he showed that every Hodge cycle, i.e. element of the Hodge ring, is an absolute Hodge cycle, i.e. behaves well under the action of the absolute Galois group.

The purpose of the papers [M2] and [M3] is however not the Hodge conjecture itself. Mumford's aim was to construct families of abelian varieties parametrized by the quotients of bounded symmetric domains by arithmetic subgroups, which are characterized by some intrinsic properties of the abelian varieties occurring in them. This generalizes constructions of Kuga, Shimura and Satake, who constructed such families with various endomorphism structures. The notion characterizing Mumford's families is the Hodge type of an abelian variety X, which is defined by fixing the ways in which $\mathbb{S}^1$ is embedded in $Hg(X)$ and $Hg(X)$ is embedded in $GL(2g)$. The paper [M2] gives the construction of these families. In [M3] the relationship of these families to the families of Kuga and Shimura is clarified. Moreover an example is given of a family of abelian fourfolds of Hodge type not characterized by its ring of endomorphisms.

Deligne generalized these families of abelian varieties further by introducing the notion of a Shimura variety [33], [34]. These varieties are defined by a set of axioms which are satisfied by most of the above mentioned families of abelian varieties. Deligne also defined the notion of a canonical model of a Shimura variety. It is known that in many cases canonical models exist. In fact, the paper [M3] is applied to construct a canonical model when the Shimura variety is a modular variety (see [33]).

3 Theta characteristics

In [M4] Mumford gives direct algebraic proofs of generalizations of classical results concerning theta characteristics (i.e. line bundles on a curve whose square is the canonical bundle) proved by Riemann ([94], pp. 212 and 487) and Wirtinger [119] in the language of theta functions. Let E be a vector bundle of arbitrary rank on a smooth projective curve X over an algebraically closed field of characteristic $\neq 2$. The main theorem says that if $Q : E \to \omega_X$ is a nondegenerate quadratic form on E with values in the canonical line bundle, then the number $\dim H^0(E) \bmod 2$ is invariant under deformations. This has several consequences, the most important perhaps being an algebraic proof of a theorem of Wirtinger, which says that if $\pi : X \to Y$ is an étale double covering and L_0 a fixed theta characteristic on Y, then $\dim H^0(L \otimes \pi^* L_0) \bmod 2$ distinguishes the two components of the kernel of the norm map from the Jacobian of X to the Jacobian of Y and thus characterizes the Prym variety of π. Another consequence is that the function $e_* : T \to \mathbb{Z}/2\mathbb{Z}$ on the $\mathbb{Z}/2\mathbb{Z}$-vector space T of theta characteristics of X, defined by $e_*(L) = \dim H^0(E) \bmod 2$ is quadratic with the usual Weil pairing as associated bilinear form.

In [6] Atiyah gave analytic proofs of these results using the theory of elliptic operators. Atiyah and Rees generalized in [7] the first statement in a different direction: If E is a vector bundle on an odd-dimensional compact complex manifold X admitting a non-degenerate holomorphic quadratic form with values in the canonical line bundle of X, then under some additional assumption the holomorphic semi-characteristic $\sum_q \dim H^{2q}(X, E) \bmod 2$ is a deformation invariant.

4 Degeneration of abelian varieties

Tate discovered a uniformization of every elliptic curve E over a p-adic field K, whose j-invariant is not integral. It has the important property that a quotient of an extension of it to the ring of integers of K is a Néron model of E. Now elliptic curves can be generalized in two ways, either to curves of arbitrary genus or to abelian varieties of arbitrary dimension. Mumford gave generalizations of Tate's construction in both cases. For a discussion of the curve-case see Gieseker's commentary [49] in this volume. The paper [M5] contains the generalization to abelian varieties.

Let $S = Spec(A)$, where A is a complete and normal local domain and let G be a semi-abelian group scheme over S (i.e. a commutative group scheme over S each of whose fibres is an extension of an abelian variety by a torus) whose generic fibre is an abelian variety. Mumford shows roughly speaking that G is the quotient of an abelian group scheme $\tilde{G}$ of constant torus-rank over S by a group of periods, i.e. a free abelian subgroup of $\tilde{G}(K)$. To be more precise, he starts with a set of periods $Y \subset \tilde{G}(K)$ together with some sort of polarization, constructs a compactification $\tilde{P}$ of $\tilde{G}$ such that the action of Y extends suitably to $\tilde{P}$. Then the quotient of the formal completion of $\tilde{P}$ by the action of Y exists and can be algebraized to give a projective scheme P admitting G as an open subset, which is the wanted semi-abelian scheme over S. Moreover Mumford showed that the uniformization $\tilde{G}$ of G is uniquely determined by G, that is, does not depend on the choice of the periods. For simplicity Mumford considered only the case where $\tilde{G}$ is a split torus. The general case works in the same way (see [42]).

In 1962 Morikawa [77] had already generalized Tate's curve to higher dimensional abelian varieties in the case of base dimension 1. Chai generalized Mumford's construction in [23] and Brylinski has some additional results [22]. Faltings and Chai proved in [42] that this construction is complete, in other words gives all semi-abelian degenerations of abelian varieties. Further progress in understanding Mumford's construction of degenerating abelian varieties was achieved by Alexeev and Nakamura [3].

The main application of degeneration results is of course, that they allow good compactifications of the moduli spaces of abelian varieties (see Section 7 below). It has various other applications in arithmetic and geometry. As an example let us mention only [30], where it is shown via degeneration that the general abelian variety of type $(1, \ldots, 1, d)$ is very ample for $d > 2^g$.

5 The Horrocks-Mumford bundle

Whereas it is easy to classify vector bundles on $\mathbb{P}^1$ and the moduli spaces of vector bundles on $\mathbb{P}^2$ and $\mathbb{P}^3$ are reasonably well understood, this is not the case for vector bundles on higher dimensional projective spaces. In [M6] Horrocks and Mumford constructed a stable rank-two vector bundle F on $\mathbb{P}^4$ with an extremely rich geometry, which has attracted the interest of many algebraic geometers. Although they constructed the bundle by means of a monad, it is closely related to abelian surfaces in $\mathbb{P}^4$. In fact, every smooth zero set of a section of F is an abelian surface. Horrocks and Mumford showed that the open set $\mathbb{P}(H^0(F))_{sm}$ of smooth zero sets of sections is isomorphic to the moduli space $\mathcal{A}^0(1,5)$ of abelian surfaces with a very ample polarization of type (1,5) and a canonical level structure. The existence of abelian surfaces in $\mathbb{P}^4$ gives via Serre's method another possibility to construct the bundle F (see [55] or [13]). Comessatti showed already in 1919 [27] that certain Jacobian surfaces can be embedded in $\mathbb{P}^4$ which gives the simplest construction of the bundle F (see [74]). Ramanan proved a criterion for an abelian surface of type $(1, n), n \geq 5$ to be very ample (see [91]). Nowadays this is a consequence of Reider's criterion (see [93] or [19]). Other constructions of the bundle F were given by Sasakura [96], [97] using a generalization of Weil's matrix divisors and Sumihiro [107] using a generalized version of elementary transformations.

One of the remarkable facts of the Horrocks-Mumford bundle is its symmetry group. It is of order 15000 and can be considered as a generalization to $\mathbb{P}^4$ of the well known symmetry group of the nine flexes of an elliptic curve in $\mathbb{P}^2$.

The Horrocks-Mumford bundle F is, up to coordinate changes and pull-backs via ramified coverings $\mathbb{P}^4 \to \mathbb{P}^4$, the only known indecomposable rank-2 vector bundle on $\mathbb{P}^4$. Decker and Schreyer [32] showed that up to a change of coordinates every stable rank-2 bundle on $\mathbb{P}^4$ with Chern classes $(c_1, c_2) = (5, 10)$ is isomorphic to F.

The zero sets of sections of F are called Horrocks-Mumford surfaces. They are reasonably well understood by now, due to the work of Barth, Hulek, Moore and van de Ven (see [9], [10], [11], [12] and [60]). Every non-smooth Horrocks-Mumford surface is one of the following: a translation scroll or a tangent scroll of an elliptic normal quintic curve, an elliptic quintic scroll with a double structure, a union of five quadrics, or a union of five planes with a double structure. The moduli spaces of these surfaces have been worked out. Moreover it is shown that the above mentioned Horrocks-Mumford isomorphism $\mathcal{A}^0(1,5) \to \mathbb{P}(H^0(F))_{sm}$ can be extended to the Igusa compactification of the moduli space $\mathcal{A}(1,5)$ of abelian varieties of type $(1,5)$ with canonical level structure [57].

6 Prym varieties

Prym varieties were first introduced in Wirtinger's monograph [119] using the language of theta functions. They played a key role in the proof of the Schottky-Jung identities [101]. In [M7] Mumford gave an algebraic version of them, valid over any algebraically closed field of characteristic $\neq 2$.

Let $\pi : \tilde{C} \to C$ be an étale double covering of a curve C of genus $g+1$. The complementary abelian subvariety of $\pi^*(Jac(C))$ in the polarized Jacobian $(Jac(\tilde{C}), \tilde{\Theta})$ is a principally polarized abelian variety of dimension g, the Prym variety $(P(\pi), \Xi)$ of π. After some preliminaries, Mumford gives a purely algebraic proof of the Schottky-Jung identities which are based on the relations of the theta functions of $(Jac(C), \Theta)$ with those of $(P(\pi), \Xi)$. In the final part of the paper he computes the dimension of the singular locus of the theta divisor in a Prym variety and obtains conditions for the polarized Prym not to be a Jacobian. This result is applied to the intermediate Jacobians of certain cubic threefolds in order to show their non-rationality (see e.g. [26], [24], [14]). Since [M7] appeared, there has been extensive progress on the Schottky problem as well as on the investigation of Prym varieties. Here we can give only a very incomplete account of it. For more details we refer to the surveys [53], [39], [43], [46] and [16].

First we will discuss the recent work on the Schottky problem. Igusa proved in [62] that the Schottky locus in genus 4 is irreducible. In [112] van Geemen showed that the locus of the Jacobian varieties is a component of the Schottky locus, i.e. the locus of the Schottky-Jung identities in the moduli space of principally polarized abelian varieties of dimension $g \geq 5$ (see also [114]). Donagi strengthened this result and showed moreover that the Schottky locus consists of several components [37], [38]. Fay has related material in [45]. Moreover in [44] Fay proved another prominent theta relation, the trisecant identity, which is valid for Jacobian varieties. Beauvillé used Prym varieties to show that the Andreotti-Mayer locus in genus 4 (i.e. the locus of principally polarized abelian fourfolds whose theta divisor is not smooth) admits one other component apart from the Jacobian locus [15]. In the second volume of [81] Mumford gave a characterization of the hyperelliptic locus in the Siegel upper half space and outlined the link to the KdV-equation. Shiota showed in [104] that Jacobians are characterized by certain differential equations satisfied by their theta functions. Using these, Arbarello and De Concini were able to write some equations of the locus of Jacobians [4]. Muñoz Porras gave a geometric characterization for a Jacobian variety, valid in arbitrary characteristic (see [82], [83]).

Let $\mathcal{R}_{g+1}$ denote the moduli space of étale double coverings of curves of genus $g+1$. The map $p_g : \mathcal{R}_{g+1} \to \mathcal{M}_g$, which associates to each element of $\mathcal{R}_{g+1}$ its Prym variety, is called the Prym map. It is shown in Beauville's thesis [14] that p_g is a proper map. To be more precise this means that considering double coverings with some mild singularities, one gets a moduli space $\overline{\mathcal{R}_{g+1}}$ containing $\mathcal{R}_{g+1}$ as a dense open subset and an extended Prym map $\overline{p}_g : \overline{\mathcal{R}_{g+1}} \to \mathcal{A}_g$, which is proper. It is known that p_g is generically injective for $g \geq 6$. Proofs via degeneration were given by Friedmann-Smith [48] and Kanev [66] and using more geometric arguments by Welters [118] and Debarre [29]. However, contrary to the usual Torelli map, p_g is never injective. This is a consequence of Donagi's tetragonal construction [36]. The tetragonal conjecture in its original form stated that the tetragonal construction is the only obstruction to the injectivity of p_g. Debarre has results supporting the conjecture [28]. Verra found a counterexample in genus 10 [115], which however has Clifford index 2. This means that the conjecture had to be modified. The map p_5 is generically finite of degree 27 [40]. The structure of p_g for $g \leq 4$ has also been worked out. Recently the degenerations of the extended Prym map have been investigated by Alexeev, Birkenhake and Hulek [2]. This allows them to determine the indeterminacy locus of the Prym map extended to a compactification.

Recillas [92] showed that Prym varieties of étale double coverings of trigonal curves are Jacobian varieties. Analyzing Mumford's investigation of the singularities of the theta divisor, Shokurov characterized the Prym varieties which are isomorphic to Jacobian varieties [106].

A generalization of Prym varieties was introduced by Tyurin in [111]. A Prym-Tyurin variety of exponent e is a principally polarized abelian variety (A, Ξ) which can be represented as a subvariety of a Jacobian (J, Θ) such that the restriction $\Theta|A$ is algebraically equivalent to $e \cdot \Xi$. Kanev gave a constuction of these varieties via correspondences [65]. By the theorem of Matsusaka-Ran Prym-Tyurin varieties of exponent 1 are Jacobians (see [19]), by a theorem of Welters [117] Prym-Tyurin varieties of exponent 2 are usual Pryms. In particular, every principally polarized abelian variety of dimension 2 and 3 is a Prym-Tyurin variety of exponent 1, every principally polarized abelian variety of dimension 4 and 5 is a Prym-Tyurin variety of exponent 2. It is an open problem whether every principally polarized abelian variety of dimension 6 is a Prym-Tyurin variety of exponent 3.

7 Compactifications of bounded symmetric domains

Let D/Γ be the quotient of a bounded symmetric domain by an arithmetic group. Generalizing Satake's compactification [98] of the moduli space $\mathcal{A}_g$ of principally polarized abelian varieties, Bailey and Borel [8] found a compactification of D/Γ. These varieties have however complicated singularities in general. One wants to construct a smooth (or almost smooth) compactification of D/Γ. By blowing up along the boundary, Igusa [63] constructed a partial desingularization of the Satake compactification. The ideas of Igusa together with work of Hirzebruch on Hilbert modular surfaces [54] were the starting point for Mumford's general theory of toroidal compactifications of D/Γ which is sketched in [M8]. Details may be found in Mumford's seminars [72] and [5]. This approach is related to one proposed by Satake [99]. Namikawa showed in [86] that Igusa's compactification is a toroidal compactification in the sense of Mumford. Chai [23] and Faltings-Chai [42] extended the toroidal idea for compactifying $\mathcal{A}_g$ to schemes of general characteristic. This is relevant for the construction of a height function on $\mathcal{A}_g$.

In general there are many toroidal compactifications. They depend on several choices. One would like to find a compactification which is meaningful for moduli. In other words one would like a variety which represents a functor in terms of abelian varieties and their degenerations. Recently there has been some progress in this direction. Alexeev and Nakamura [3], [1] showed that the toroidal compactification of $\mathcal{A}_g$ which is given by the second Voronoi decomposition represents a good functor.

8 The Kodaira dimension of Siegel modular varieties

The quotient $\mathcal{A}_g$ of the Siegel upper half-space of rank g by the full modular group is a coarse moduli space for principally polarized abelian varieties of dimension g. Mumford proved in [M9] with a very elegant geometric argument that $\mathcal{A}_g$ is of general type if $g \geq 7$. This improves previous work of Tai (who in [108] proved this for $g \geq 9$) and Freitag (in [47] for $g \geq 8$). $\mathcal{A}_g$ is unirational for $g = 5$ [37], [78] and $g = 4$ [25] and rational for $g = 3$ [67] and $g = 2$ [64]. The Kodaira dimension of $\mathcal{A}_6$ is still unknown. Hulek [56] proved some general type results for moduli spaces of principally polarized abelian varieties with level structure.

If X is any complex quasi-projective variety of general type, one says that X has a canonical model (in the sense of Mori and Reid, not in the sense of Shimura as in Section 2 above), if the canonical ring $\oplus_{n \geq 0} H^0(\tilde{X}, \mathcal{O}(nK_{\tilde{X}}))$ is of finite type for some smooth projective model $\tilde{X}$ of X. Recently Shepherd-Barron [103] showed that the moduli space $\mathcal{A}_g$ has a canonical model for $g \geq 11$. To be more precise, the first Voronoi compactification of $\mathcal{A}_g$ is the canonical model for $g \geq 12$.

The Kodaira dimension has also been computed for several moduli spaces of non-principally polarized abelian varieties (see [109]). There are many general type results for moduli spaces of abelian surfaces of type $(1, n)$ with or without level structure. The main method for constructing sections of the canonical bundle of the moduli spaces is the lifting method, which can be described roughly as follows: Jacobi forms can be considered as forms on a boundary component of the moduli space. In many cases they can be extended or lifted to the whole moduli space. The idea goes back to Maaß and has been refined by Gritsenko [50], Borcherds [20] and others. Let us mention only some results. For more details see the excellent survey [58]. The Kodaira dimension of the moduli space $\mathcal{A}_d$ of abelian surfaces of type $(1, d)$ is positive for $d \geq 29, d \neq 30, 32, 35, 36, 40, 42, 48$ or 60 [51]. The moduli space $\mathcal{A}_p^{lev}$ of abelian surfaces of type $(1, p)$ with canonical level structure is of general type for any prime number $p \geq 37$ [59], [52]. Improving a result of Sankaran [95], Erdenberger [41] recently showed that $\mathcal{A}_p$ is of general type for $p = 37, 43, 53, 61, 67$ and any prime $p > 71$. The following general result is due to Borisov [21]: There are only finitely many subgroups H of $Sp(4, \mathbb{Z})$ such that the quotient of the Siegel upper half-space of rank 2 by H is not of general type.

The first author would like to thank F. O. Schreyer, H. Lange and R. Smith for their assistence in preparing (the first draft of) this report. The second author would like to that K. Hulek and E. Viehweg for some valuable hints.

Mumford's papers in this volume

[M1] On the Equations Defining Abelian Varieties I, *Inv. Math.* **1** (1966), 287-354; II, *Inv. Math.* **3** (1967), 75-135; III, *Inv. Math.* **3** (1967), 215-244.

[M2] Families of abelian varieties, *Proc. Symp. Pure Math.* AMS (1966), 347-351.

[M3] A note on Shimura's paper "discontinuous groups and abelian varieties", *Math. Ann.* **181**, (1969), 345-351.
[M4] Theta Characteristics of an Algebraic Curve, *Ann. scient. Éc. Norm. Sup.* **4** (1967), 181-192.
[M5] An Analytic Construction of Degenerating Abelian Varieties over Complete Rings, *Compos. Math.* **24** (1972), 283-316.
[M6] (with G. Horrocks) A Rank 2 Vector Bundle on $\mathbb{P}^4$ with 15000 Symmetries, *Topology* **12** (1973), 63-81.
[M7] Prym Varieties I, in: Contributions to Analysis, Volume dedicated to Lipman Bers, Academic Press (1974), 325-350.
[M8] A new Approach to Compactifying Locally Symmetric Varieties, *Proc. of the conference on discrete subgroups of Lie groups and applications to moduli, Bombay 1973*, Oxford Univ. Press (1975), 211-224.
[M9] On the Kodaira Dimension of the Siegel Modular Variety, *Springer Lect. Notes in Math.* **997** (1983), 348-375.

References

[1] V. Alexeev, Complete moduli in the presence of semiabelian group action, *Ann. of Math.* **155** (2002), 611-708.
[2] V. Alexeev, Ch. Birkenhake and K. Hulek, Degenerations of Prym varieties, *Journ. Reine Angew. Math.* **553** (2002), 73-116.
[3] V. Alexeev and I. Nakamura, On Mumford's construction of degenerating abelian varieties, *Tohoku Math. J.* **51** (1999), 399-420.
[4] E. Arbarello and C. De Concini, On a set of equations characterizing Riemann matrices, *Ann. of Math.* **120** (1984), 119-140.
[5] A. Ash, D. Mumford, M. Rapoport, R. S. Tai, Smooth Compactification of locally symmetric varieties, Math. Sci. Press, 1975.
[6] M.F. Atiyah, Riemann surfaces and spin structure, *Ann. scient. Éc. Norm. Sup.* **4** (1971), 47-62.
[7] M.F. Atiyah and E. Rees, Vector Bundles on Projective 3-Space, *Invent. Math.* **35** (1976), 131-153.
[8] W.L. Baily and A. Borel, Compactification of arithmetic quotients of bounded symmetric domains, *Ann. of Math.* **84** (1966), 442-528.
[9] W. Barth, Kummer surfaces associated with the Horrocks-Mumford bundle, *Journées de Géométrie Algébrique d'Angers 1979*, Sijthoff and Noordhoff (1980), 29-48.
[10] W. Barth, K. Hulek and R. Moore, Shioda's modular surface $S(5)$ and the Horrocks-Mumford bundle, *Proc. of the conf. on vector bundles on algebraic varieties 1984*, Oxford Univ. Press, Bombay, (1987), 35-106.
[11] W. Barth, K. Hulek and R. Moore, Degenerations of Horrocks-Mumford surfaces, *Math. Ann.* **277** (1987), 735-755.
[12] W. Barth and R. Moore, Geometry in the space of Horrock-Mumford surfaces, *Topology* **28** (1989), 231-245.
[13] W. Barth and A. van de Ven, A decomposability criterion for algebraic 2-bundles on projective spaces, *Invent. Math.* **25** (1974), 91-196.
[14] A. Beauville, Variétés de Prym et jacobiennes intermédiaires, *Ann. scient. Éc. Norm. Sup.* **10** (1977), 309-391.

[15] A. Beauville, Prym varieties and the Schottky problem, *Invent. Math.* **41** (1977), 149-196.
[16] A. Beauville, Prym varieties: A Survey, *Proc. of Symp. in Pure Math.* **49** part 1 (1989), 607-620.
[17] A. Beauville and D. Debarre, Une relation entre deux approches du problème de Schottky, *Invent. Math.* **86** (1986), 195–207.
[18] Ch. Birkenhake and H. Lange, Cubic theta relations, *J. Reine Angew. Math.* **407** (1990), 167–177.
[19] Ch. Birkenhake and H. Lange, Complex Abelian Varieties, Grundlehren **302**, Springer Verlag 1992.
[20] R.E. Borcherds, Automorphic forms with singularities on Grassmannians, *Invent. Math.* **132** (1998), 491-562.
[21] L.A. Borisov, A finiteness theorem for subgroups of $Sp(4,\mathbb{Z})$, *Journ. Math. Sci. (New York)* **94** (1999), 1073-1099.
[22] J.L. Brylinski, 1-motifs et formes automorphes (Théorie arithmétique des domaines de Siegel), *Publ. Math. Univ. Paris 7* **15** (1983), 43-106.
[23] C.-L. Chai, Compactification of Siegel moduli schemes, *London Math. Soc. Lect. Notes* **107**, Cambridge Univ. Press, 1985.
[24] H. Clemens, Applications of the theory of Prym varieties, *Proc. of ICM Vancouver* (1974), 415-421.
[25] H. Clemens, Double Solids, *Adv. Math.* **47** (1983), 107-230.
[26] H. Clemens and P. Griffiths, The intermediate Jacobian of the cubic threefold, *Ann. of Math.* **95** (1972), 281-356.
[27] A. Comessatti, Sulle superficie di Jacobi semplicemente singolari, *Mem. Soc. ital. delle Szienze* **21** (1919), 45-71.
[28] O. Debarre, Sur les variétés de Prym des courbes tetragonales, *Ann. scient. Éc. Norm. Sup.* **21** (1988), 545-560.
[29] O. Debarre, Sur le problème de Torelli pour les variétés de Prym, *Amer. J. Math.* **111** (1989), 111-134.
[30] O. Debarre, K. Hulek and J. Spandaw, Very ample linear systems on abelian varieties, *Math. Ann.* **300** (1994), 181-202.
[31] W. Decker and F. Schreyer, Pullbacks of the Horrocks-Mumford bundle, *Journ. Reine Angew. Math.* **382** (1987), 215-220.
[32] W. Decker and F. Schreyer, On the uniqueness of the Horrocks-Mumford bundle, *Math. Ann.* **273** (1986), 415-443.
[33] P. Deligne, Travaux de Shimura, Sém. Bourbaki, Exposé 389, *Springer Lect. Notes in Math.* **244** (1971), 123-165.
[34] P. Deligne, Variétés de Shimura: Interprétation modulaire, et techniques de construction de modèles canoniques, *Proc. Symp. Pure Math.* **33** part 2 (1979), 247-290.
[35] P. Deligne (Notes by J. Milne), Hodge cycles on abelian varieties, *Springer Lect. Notes in Math.* **900** (1982), 9-100.
[36] R. Donagi, The tetragonal construction, *Bull. AMS* **4** (1981), 181-185.
[37] R. Donagi, The unitrationality of $\mathcal{A}_5$, *Ann. of Math.* **119** (1984), 269-307.
[38] R. Donagi, Non-Jacobians in the Schottky loci, *Ann. of Math.* **126** (1987), 193-217.
[39] R. Donagi, The Schottky problem, in: Moduli Problems, *Springer Lect. Notes in Math.* **1337** (1989), 84-137.
[40] R. Donagi, R.C. Smith, The structure of the Prym map, *Acta Math.* **146** (1981), 25-102.

[41] C. Erdenberger, The Kodaira Dimension of Certain Moduli Spaces of Abelian Surfaces, Preprint (2002), Math.AG/0305225.

[42] G. Faltings and C.-L. Chai, Degenerations of abelian varieties, Springer Verlag, 1990.

[43] H. Farkas, Schottky-Jung theory, *Proc. Symp. in Pure Math.* **49** Part 1, (1989), 459-483.

[44] J. Fay, Theta Functions on Riemann Surfaces, *Springer Lect. Notes in Math.* **352**, 1973.

[45] J. Fay, On the even-order vanishing of Jacobians theta functions, *Duke Math. J.* **51** (1984), 109-132.

[46] J. Fay, Schottky relations on $\frac{1}{2}(C-C)$, *Proc. Symp. in Pure Math.* **49** Part 1, (1989), 485-502.

[47] E. Freitag, Siegelsche Modulfunktionen, Springer Verlag, 1983.

[48] R. Friedman and R. Smith, The generic Torelli theorem for the Prym map, *Invent. Math.* **67** (1982), 473-490.

[49] D. Gieseker, Mumford's work on the moduli of curves and vector bundles, This volume, pp ??.

[50] V. Gritsenko, Arithmetical lifting and its applications, *Proc. of the Paris seminars on Number Theory, 1992/92*, Cambr. Univ. Press (1995), 103-126.

[51] V. Gritsenko, Irrationality of the moduli spaces of polarized abelian surfaces, *Proc. of the Egloffstein Conf. on Abelian Varieties 1993*, de Gruyter (1995), 63-81.

[52] V. Gritsenko and K. Hulek, Irrationality of the moduli spaces of polarized abelian surfaces, *Proc. of the Egloffstein Conf. on Abelian Varieties 1993*, de Gruyter (1995), 83-84.

[53] R.C. Gunning, Riemann surfaces and their associated Wirtinger varieties, *Bull. AMS* **11** (1984), 287-316.

[54] F. Hirzebruch, Hilbert modular surfaces, *L'enseignement Math.* **19** (1973), 183-281.

[55] G. Horrocks, A construction for locally free sheaves, *Topology* **7** (1968), 117-120.

[56] K. Hulek, Nef divisors on moduli spaces of abelian varieties, In: Complex analysis and algebraic geometry, de Gruyter, (2000), 255-274.

[57] K. Hulek, C. Kahn and S. Weintraub, Moduli spaces of abelian surfaces: Compactification, degenerations and theta functions, de Gruyter, 1993.

[58] K. Hulek and G.K. Sankaran, The Geometry of Siegel Modular varieties, *Adv. Studies in Pure Math.* **35** (2002), 89-156.

[59] K. Hulek and G.K. Sankaran, The Kodaira dimension of certain moduli spaces of abelian surfaces, *Comp. Math.* **90** (1994), 1-35.

[60] K. Hulek and A. van de Ven, The Horrocks-Mumford bundle and the Ferrand construction, *Manuscr. Math.* **50** (1986), 313-335.

[61] J.-I. Igusa, On the graded ring of theta-constants, *Am. Journ. Math.* **86** (1964), 219-246 and **88** (1966), 221-236.

[62] J.-I. Igusa, On the irreducibility of Schottky's divisor, *J. Fac. Sci. Univ. Tokyo* **28** (1982), 531-595.

[63] J.-I. Igusa, A desingularization problem in the theory of Siegel modular functions, *Math. Ann.* **168** (1967), 228-260.

[64] J.-I. Igusa, Arithmetic variety of moduli for genus 2, *Ann. of Math.* **72** (1960), 612-649.

[65] V. Kanev, Principal polarizations of Prym-Tyurin varieties, *Comp. Math.* **64** (1987), 243-270.
[66] V. Kanev, The global Torelli theorem for Prym varieties at a generic point, *Math. USSR Izvestiza* **20** (1983), 235-258.
[67] P. Katsylo, Rationality of the moduli variety of curves of genus 3, *Comm. Math. Helv.* **71** (1996), 507-524.
[68] G. Kempf, Projective coordinate rings of abelian varieties, in: *Algebraic Analysis, Geometry and Number Theory*, The Johns Hopkins Press (1989), 225-236.
[69] G. Kempf, Linear systems on abelian varieties, *Am. Journ. Math.* **111** (1989), 65-93.
[70] G. Kempf, Notes on the inversion of integrals II, *Proc. AMS* **108** (1990), 59-67.
[71] G. Kempf, Abelian varieties and theta functions, Springer Verlag, 1991.
[72] G. Kempf, F. Knudson, D. Mumford, B. Saint-Donat, Toroidal embedddings I, *Springer Lecture Notes in Math.* **339**, 1973.
[73] S. Koizumi, Theta relations and projective normality of abelian varieties, *Am. Journ. Math.* **98** (1976), 865-889.
[74] H. Lange, Jacobian surfaces in $\mathbb{P}^4$, *Journ. Reine Angew. Math.* **372** (1986), 71-86.
[75] R. Lazarsfeld, Positivity in Algebraic Geometry, Preprint 567 p., to appear in Ergebn. der Math., Springer Verlag.
[76] B.J.J. Moonen and Y.G. Zarhin, Weil classes on abelian varieties, *Journ. Reine Angew. Math.* **496** (1998), 83-92.
[77] H. Morikawa, On theta functions and abelian varieties over valuation fields of rank one, *Nagoya Math. J.* **20** (1962), 1-27, and **21** (1962), 231-250.
[78] S. Mori and S. Mukai, The uniruledness of the moduli space of curves of genus 11, *Springer Lect. Notes in Math.* **1016** (1983), 334-353.
[79] D. Mumford, Varieties defined by quadratic equations, in: *Questions on Algebraic Varieties*, CIME, Roma (1970), 95-100.
[80] D. Mumford, Abelian varieties, Oxford Univ. Press (1970).
[81] D. Mumford, Tata Lectures on Theta I, II, III(with M. Nori and P. Norman), *Progr. in Math.* **28, 43, 97**, Birkhäuser, Boston (1983,1984,1991)
[82] J.M. Muñoz Porras, Characterization of Jacobian varieties in arbitrary characteristic, *Comp. Math.* **61** (1987), 369-381.
[83] J.M. Muñoz Porras, Geometric characterizations of Jacobians and the Schottky equations, *Proc. Symp. Pure Math.* **49** Part 1, (1989), 541-552.
[84] V.K. Murty, Exceptional Hodge classes on certain abelian varieties, *Math. Ann.* **268** (1984), 297-206.
[85] Y. Namikawa, A new compactification of the Siegel space and degeneration of abelian varieties, I, II, *Math. Ann.* **221** (1976), 97-141 and 201-241.
[86] Y. Namikawa, Toroidal compactification of Siegel spaces, *Springer Lect. Notes in Math.* **812**, 1980.
[87] A. Ohbuchi, A note on the normal generation of ample line bundles on abelian varieties, *Proc. Japan Acad.* **64** (1988), 119-120.
[88] G. Pareschi, Syzygies of abelian varieties, *J. Amer. Math. Soc.***13** (2000), 651-664.
[89] G. Pareschi and M. Popa, Regularity of abelian varieties I, *J. Amer. Math. Soc.* **16**, 285-302, II: basic results on linear series and defining equations, Preprint math.AG/0110004.

[90] H. Pohlmann, Algebraic cycles on abelian varieties of complex multiplication type, *Ann. of Math.* **88** (1968), 161-180.

[91] S. Ramanan, Ample divisors on abelian surfaces, *Proc. London Math. Soc.* **5** (1985), 231-245.

[92] S. Recillas, Jacobians of curves with a g_4^1 are Prym varieties of trigonal curves, *Bol. Soc. Mat. Mexicana* **19** (1974), 9-13.

[93] I. Reider, Vector bundles of rank 2 and linear systems on algebraic surfaces, *Ann. of Math.* **127** (1988), 309-316.

[94] B. Riemann, Collected Works, Dover edition, 1953.

[95] G. Sankaran, Moduli of Polarized Abelian Surfaces, *Math. Nachr.* **188** (1997), 321-340.

[96] N. Sasakura, A stratification theoretical method of constuction of holomorphic vector bundles, *Adv. Studies in Pure Math.* **8** (1986), 527-581.

[97] N. Sasakura, Configuration of divisors and reflexive sheaves, *R.I.M.S. Kyoto Univ.* **634** (1987), 407-513.

[98] I. Satake, On the compactification of the Siegel space, *J. Indian Math. Soc.* **20** (1956), 259-281.

[99] I. Satake, On the arithmetic of tube domains (blowing-up of the point at infinity), *Bull. AMS* **79** (1973), 1076-1094.

[100] C. Schoen, Hodge classes on self-products of a variety with an automorphism, *Comp. Math.* **65** (1988), 3-32.

[101] F. Schottky and H. Jung, Neue Sätze über Symmetralfunktionen und die abelschen Funktionen der Riemannschen Theorie, *S.-B. Preuß. Akad. Wiss.* (1909), 282-297 and 732-750.

[102] T. Sekiguchi, On the normal generation by a line bundle on an abelian variety, *Proc. of Japan Univ.* **54** (1978), 185-188.

[103] N.I. Shepherd-Barron, Canonical rings for moduli spaces of abelian varieties, Preprint, 2003.

[104] T. Shiota, Characterization of Jacobian varieties in terms of soliton equations, *Invent. Math.* **83** (1986), 333-382.

[105] T. Shioda, Algebraic cycles on abelian varieties of Fermat type, *Math. Ann.* **258** (1981), 65-80.

[106] V.V. Shokurov, Distinguishing Prymians from Jacobians, *Invent. Math.* **65** (1981), 209-219.

[107] H. Sumihiro, Elementary transformations of algebraic vector bundles, in: *Algebr. geom. and comm. algebra (vol. in honour of Nagata)* (1987), 503-516.

[108] Y.-S. Tai, On the Kodaira dimension of the moduli space of abelian varieties *Invent. Math.* **68** (1982), 425-439.

[109] Y.-S. Tai, On the Kodaira dimension of the moduli space of abelian varieties with non-principal polarization, In: *Proc. of the Egloffstein Conf on Abelian Varieties*, de Gruyter (1995), 293-302.

[110] S.G. Tankeev, Cycles on simple abelian varieties of prime dimension, Isv. Ross. Akad. Nauk SSSR **57** (1993), 192-206.

[111] A. Tyurin, Five lectures on three dimensional varieties, *Russian Math. Surveys* **27** (1972), 1-53.

[112] B. van Geemen, Siegel modular forms vanishing on the moduli space of curves, *Invent. Math.* **78** (1984), 329-349.

[113] B. van Geemen, An introduction to the Hodge conjecture for abelian varieties, *Springer Lect. Notes in Math.* **1594** (1994), 233-252.

[114] B. van Geemen and G. van der Geer, Kummer varieties and the moduli spaces of abelian varieties, *Am. Journ. Math.* **108** (1986), 615-642.

[115] A. Verra, The Prym map has degree two on plane sextics, Preprint, to appear in the proceedings of the Fano conference, Torino, 2002.

[116] A. Weil, Abelian varieties and the Hodge ring, Collected Works, vol 3, [1977c], Springer (1979), 421-429.

[117] G. Welters, Curves of twice the minimal class on principally polarized abelian varieties, *Indag. Math.* **94** (1987), 87-109.

[118] G. Welters, Recovering the curve data from a general Prym variety, *Am. Journ. Math.* **109** (1987), 165-182.

[119] W. Wirtinger, Untersuchungen über Thetafunktionen, Teubner (1895).

Invent. math. 1, 287—354 (1966)

On the Equations Defining Abelian Varieties. I*

D. MUMFORD (Cambridge, Mass.)

Contents

My aim is to set up a purely algebraic theory of theta-functions. Actually, since my methods are algebraic and not analytic, the functions themselves will not dominate the picture – although they are there. The basic idea is to construct *canonical bases* of all linear systems on all abelian varieties. The result is that one gets a very complete description first of the homogeneous coordinate ring of a single abelian variety, and second of the moduli space of all abelian varieties. We shall obtain explicit generators and relations for this moduli space. The homogeneous coordinate rings of abelian varieties are very remarkable rings. Although the abelian variety is a commutative group, these rings are acted upon (with some restriction on degrees) by a 2-step nilpotent group. Unlike the *affine* coordinate rings of linear algebraic groups, they are *not* Hopf algebras. Their structure is dominated by a symmetry of a higher order embodied in the *theta relation* of *Riemann* (a quartic relation). One might say that this is the only class of rings not "essentially" isomorphic to polynomial rings which we can describe so closely.

There are several interesting topics which I have not gone into in this paper, but which can be investigated in the same spirit: for example, the extension to inseparably polarized abelian varieties; a discussion of the transformation theory of theta-functions, especially in connection with the tower of moduli schemes; a discussion of the various standard models of the irreducible representation space for the Heisenberg commutation relations [and the adelic generalization] and the various ways in which RIEMANN's theta-function can be singled out in each of them; an analysis of special theta-functions and special abelian varieties; an analysis of degenerate theta functions and SATAKE's compactification; a tie-up between the global theory of moduli that we give, and the

* This work was partially supported by NSF grant GP 3512 and a grant from the Sloan Foundation.

infinitesimal theory of KODAIRA-SPENCER-GROTHENDIECK. All of these look like very fruitful topics. Incidentally, in § 6 there is a very annoying 8 in the main result which by all rights ought to be replaced by 4 [but I nearly despaired of getting the 8 in the course of proving what I named the "Hardest lemma" in § 6].

This paper is heavily indebted to the influence and ideas of BAILY, CARTIER, IGUSA, MAYER, SIEGEL, and WEIL. As an algebraist, I was naturally not attracted to anything called a function, and it was only because these six people all realized so clearly the significance of theta-functions that the idea got across to me. More than that, many of the key ideas are due to these people and especially to IGUSA: the reader is referred especially to the important papers [*1*, *2*, *6*, *11* and *12*]. In particular, the beautiful and far-reaching fact that the theta-null values give almost exactly the moduli space of a carefully chosen level is IGUSA'S idea.

In most of this paper, we will work over a fixed algebraically closed ground field k. At first, k will have any characteristic; later we will exclude characteristic 2. Of course we use the language of schemes. Also, if S is a finite set, $\#(S)$ denotes the cardinality of S.

A word of warning – and apology. There are several thousand formulas in this paper which allow one *or more* "sign-like ambiguities": i.e., alternate and symmetric but non-equivalent reformulations. These occur in definitions and theorems. I have made a superhuman effort to achieve consistency and even to make *correct* statements: but I still cannot guarantee the result.

§ 1. The Basic Groups Acting on Linear Systems

X will denote an abelian variety for all of this paper. All varieties that we will talk about will be abelian varieties; this will be mentioned from time to time but not invariably. When we talk of an abelian variety X, we always assume that a definite identity point $e\in X$ has been chosen: hence a definite group law on X has been chosen. Moreover, the endomorphism of an abelian variety given by multiplication by n will be denoted by $n\delta$. The inverse $-\delta$ will be denoted ι. The kernel of $n\delta$ will be denoted X_n.

Definition. If L is an invertible sheaf on X, then $H(L)$ is the subgroup of closed points $x\in X$ such that if $T_x: X\to X$ denotes translation by x, then $T_x^* L\cong L$.

We recall the basic facts about invertible sheaves on an abelian variety, and their sections (cf. [*9*], Ch. 6, § 2):

(I) L is ample if and only if $H(L)$ is finite and $\Gamma(X, L^n)$ is not (0) for all $n>0$.

(II) If L is ample and $\dim X=g$, then there is a positive integer d such that

$$\dim H^0(X, L^n)=d\cdot n^g, \quad \text{all} \quad n\geqq 1$$

$$\dim H^i(X, L^n)=0, \quad \text{all} \quad n\geqq 1, \; i\geqq 1.$$

(III) The integer d of (II) is called the degree of L, and if D is a divisor on X defining L, i.e., $L\cong o_X(D)$, then

$$(D^g)=d\cdot g!$$

(IV) Let $\hat{X}$ be the dual of X. Let $\Lambda(L): X\to\hat{X}$ be the usual homomorphism associated to L, i.e., if $x\in X_k$ then $\Lambda(L)(x)$ is the point of $\hat{X}$ corresponding to the sheaf $T_x^*(L)\otimes L^{-1}$ on X. Then

$$\text{degree}\,\Lambda(L)=d^2.$$

(V) Let $L'=\iota^* L$ be the reflection of L in the origin. Then for all integers n,

$$(n\delta)^* L\cong(L)^{\frac{n^2+n}{2}}\otimes(L')^{\frac{n^2-n}{2}}.$$

In this paper we shall be exclusively interested in invertible sheaves L such that

a) *L is ample.*

b) *If $p=\text{char}(k)$, $p\nmid$ degree (L).*

We shall refer to such sheaves as ample sheaves *of separable type.* Note that for such sheaves, by (IV), char $(k)\nmid$ degree $\Lambda(L)$, so $\Lambda(L)$ is separable. Since, by definition $H(L)$ is the kernel of $\Lambda(L)$, it follows that:

$$d^2=\text{cardinality } H(L).$$

At this point, I can define the most central concept of the entire development:

Definition. Let L be an ample invertible sheaf of separable type. Then $\mathscr{G}(L)$ is the set of pairs (x,φ), where x is a closed point of X and φ is an isomorphism:

$$L\xrightarrow[\varphi]{\sim} T_x^* L.$$

First of all: *$\mathscr{G}(L)$ is a group.* Let $(x,\varphi),(y,\psi)\in\mathscr{G}(L)$. Then the composition $T_x^*\psi\circ\varphi$:

$$L\xrightarrow{\varphi} T_x^* L\xrightarrow{T_x^*\psi} T_x^*(T_y^* L)=T_{x+y}^* L$$

is an isomorphism of L and $T_{x+y}^* L$. Define

$$(y,\psi)\circ(x,\varphi)=(x+y, T_x^*\psi\circ\varphi).$$

One checks immediately that this makes $\mathscr{G}(L)$ into a group. Secondly, the map taking (x, φ) to x puts $\mathscr{G}(L)$ into the exact sequence:

$$0 \to k^* \to \mathscr{G}(L) \to H(L) \to 0.$$

Here $\mathscr{G}(L) \to H(L)$ is surjective by the very definition of $H(L)$, and the kernel is the group of isomorphisms of L with itself: i.e., of multiplications by non-zero constants in the ground field k.

Our first objective in this section is to give a complete structure theorem for $H(L)$ and $\mathscr{G}(L)$ vis-a-vis this exact sequence. To this end, we must first examine the situation:

$$(*) \qquad \begin{array}{c} X \xrightarrow{\pi} Y \\ L \xleftarrow[\alpha]{} M. \end{array}$$

Here π stands for a separable isogeny of the abelian varieties X and Y, M stands for an invertible sheaf on Y and α stands for an isomorphism:

$$\alpha: \quad \pi^* M \xrightarrow{\sim} L.$$

Let K be the kernel of π: K is a finite subgroup of closed points of X. Moreover, if $x \in K$, then

$$L \xleftarrow[\alpha]{\sim} \pi^* M = (\pi \circ T_x)^* M = T_x^*(\pi^* M) \xrightarrow[T_x^*(\alpha)]{\sim} T_x^* L$$

defines an isomorphism of $T_x^* L$ and L, so $x \in H(L)$. Thus $K \subset H(L)$. But let the reader beware – a key point is that if $K \subset H(L)$ is an arbitrary subgroup, then there does not necessarily exist an invertible sheaf M on $Y = X/K$ such that $\pi^* M \cong L$. In fact, as GROTHENDIECK's theory of descent teaches us, it is absolutely essential to consider the pair

$$(x, T_x^*(\alpha) \circ \alpha^{-1})$$

as an element of $\mathscr{G}(L)$. If the set of these is denoted $\tilde{K}$, then one checks immediately that they form a subgroup of $\mathscr{G}(L)$. Namely:

$$\begin{aligned} &(x, T_x^*(\alpha) \circ \alpha^{-1}) \circ (y, T_y^*(\alpha) \circ \alpha^{-1}) \\ &\quad = (x+y, T_y^* \{T_x^*(\alpha) \circ \alpha^{-1}\} \circ \{T_y^*(\alpha) \circ \alpha^{-1}\}) \\ &\quad = (x+y, T_{x+y}^*(\alpha) \circ \alpha^{-1}). \end{aligned}$$

In other words, we have *split* the extension $\mathscr{G}(L) \to H(L)$ over K:

$$\begin{array}{ccccc} \mathscr{G}(L) & \longrightarrow & H(L) & \longrightarrow & 0 \\ \cup & & \cup & & \\ \tilde{K} & \xrightarrow{\sim} & K. & & \end{array}$$

Conversely, given a subgroup $K \subset H(L)$, and a lifting K of K into $\mathscr{G}(L)$, we have exactly the situation referred to by GROTHENDIECK as *descent data* for L with respect to the morphism

$$X \xrightarrow{\pi} Y = X/K.$$

His main theorem (Theorem 1.1, § 8, [3]) says that there is a $1-1$ correspondence between the set of invertible sheaves M on Y and isomorphisms $\alpha: \pi^* M \xrightarrow{\sim} L$ and the set of descent data for L with respect to π. We make the definition:

Definition. A level subgroup $\tilde{K}$ *of* $\mathcal{G}(L)$ *is a subgroup such that* $k^* \cap \tilde{K} = (0)$, i.e., $\tilde{K}$ is isomorphic to its image K in $H(L)$.

Thus, our conclusion may be rephrased:

Proposition 1. *There is a* $1-1$ *correspondence between level subgroups* $\tilde{K}$ *in* $\mathcal{G}(L)$ *and pairs* (π, α) *as in* $(*)$ *above.*

What happens to the degrees in this situation? Note first that M is certainly ample (cf. Theorem 2.6.2, Ch. 3, [5]). Say

$$d = \text{degree } L$$
$$r = \text{degree}(\pi) = \text{cardinality}(K).$$

Then if E is a Cartier divisor on Y representing M, $\pi^{-1}(E)$ is a Cartier divisor on X representing L and

$$\begin{aligned} d \cdot g! &= (\pi^{-1}(E)^g) \\ &= \pi^{-1}(E^g) \\ &= r(E^g) \\ &= r \cdot \text{degree } M \cdot g!. \end{aligned}$$

In particular, char $(k) \nmid$ degree (M), so M is of separable type too. And degree $(M) = d/r$. Moreover, it is possible to relate $\mathcal{G}(M)$ and $\mathcal{G}(L)$. If the map assigning to X and L the group $\mathcal{G}(L)$ was part of a functor, we would expect there to exist a homomorphism from $\mathcal{G}(L)$ to $\mathcal{G}(M)$ or vice versa. Actually, the situation is more complicated than that. In fact:

Proposition 2. *Given the situation* $(*)$ *above, let* $K = \ker(\pi)$ *and let* $\tilde{K}$ *be the corresponding level subgroup over* K.

i) $\pi^{-1}[H(M)] \subset H(L)$.

ii) $$\left\{\begin{matrix}\text{centralizer} \\ \text{of } \tilde{K} \text{ in } \mathcal{G}(L)\end{matrix}\right\} = \left\{ z \in \mathcal{G}(L) \,\middle|\, \begin{matrix}\text{the image of } z \text{ in } H(L) \\ \text{is in } \pi^{-1}[H(M)]\end{matrix}\right\}.$$

Call this group $\mathcal{G}(L)^*$.

iii) $$\mathcal{G}(M) \underset{\text{canonically}}{\cong} \mathcal{G}(L)^*/\tilde{K}.$$

(*In particular,* $\mathcal{G}(M)$ *is a quotient of a subgroup of* $\mathcal{G}(L)$).

Proof. Suppose $y \in Y_k$ and $M \cong T_y^* M$. Then if $x \in X_k$ and $\pi(x) = y$,

$$\begin{aligned} T_x^* L &\cong T_x^* \pi^* M \\ &\cong \pi^* T_y^* M \\ &\cong \pi^* M \\ &\cong L. \end{aligned}$$

This proves (i). More precisely, if $(y, \psi) \in \mathscr{G}(M)$, we see that

$$(x, T_x^*(\alpha) \circ \pi^* \psi \circ \alpha^{-1}) \in \mathscr{G}(L).$$

But x determines y, and the homomorphism $T_x^*(\alpha) \circ \pi^* \psi \circ \alpha^{-1}$ determines ψ. Therefore, this relation between elements of $\mathscr{G}(M)$ and $\mathscr{G}(L)$ is actually a map:

$$\mathscr{G}(L) \supset \{(x, \varphi) \in \mathscr{G}(L) \mid \pi(x) \in H(M)\} \to \mathscr{G}(M).$$

Call this subgroup $\mathscr{G}(L)^*$. This map is readily checked to be a homomorphism. Moreover, every $y \in H(M)$ is of the form $\pi(x)$, some $x \in H(L)$; and if some element (y, φ_1) is in its image, then all other elements (y, φ_2) are in the image since φ_2 is always a scalar multiple of φ_1. Therefore $\mathscr{G}(M)$ is a quotient of $\mathscr{G}(L)^*$. One checks easily that the kernel is $\tilde{K}$, and this proves (iii). Finally, (ii) is a consequence of descent theory. According to this theory, suppose M_1 and M_2 are obtained by descending (over a faithfully flat morphism π) sheaves $L_1 = \pi^* M_1$ and $L_2 = \pi^* M_2$. Then the set of homomorphisms from M_1 to M_2 is the same as the set of homomorphisms from L_1 to L_2 that "commute" with the descent data. In our case, M is obtained by descending L via the identifications:

$$L \xrightarrow[\sim]{\varphi} T_w^* L, \quad \text{all} \quad (w, \varphi) \in \tilde{K},$$

and if $y = \pi(x) \in Y_k$, then $T_y^* M$ is obtained by descending $T_x^* L$ via the identifications

$$T_x^* L \xrightarrow[\sim]{T_x^* \varphi} T_{x+w}^* L, \quad \text{all} \quad (w, \varphi) \in \tilde{K}.$$

Therefore, an isomorphism $\psi: L \to T_x^* L$ "descends" if and only if the diagram:

$$\begin{array}{ccc} L & \xrightarrow{\varphi} & T_w^* L \\ \downarrow{\scriptstyle \psi} & & \downarrow{\scriptstyle T_w^* \psi} \\ T_x^* L & \xrightarrow{T_x^* \varphi} & T_{x+w}^* L \end{array}$$

commutes, for all $(w, \varphi) \in \tilde{K}$. On the one hand, this means exactly that (x, ψ) is in the centralizer of $\tilde{K}$; on the other hand, ψ descends to an isomorphism of M and $T_y^* M$ if and only if $y \in H(M)$, i.e., $(x, \psi) \in \mathscr{G}(L)^*$. This proves (ii). *Q.E.D.*

Now consider the extension $\mathcal{G}(L)$ of $H(L)$ by k^*. Note first that k^* is contained in the *center* of $\mathcal{G}(L)$. This follows immediately from the definition of multiplication in $\mathcal{G}(L)$. The extension defines the following invariant:

$$\begin{cases} \text{Given } x, y \in H(L), \text{ let } \tilde{x}, \tilde{y} \in \mathcal{G}(L) \text{ lie over } x, y. \\ \text{Set } \; e^L(x, y) = \tilde{x} \cdot \tilde{y} \cdot \tilde{x}^{-1} \cdot \tilde{y}^{-1}. \end{cases}$$

It is immediate that this is well-defined, that $e^L(x, y)$ is an element of k^*, and that e^L is a skew-symmetric bilinear pairing from $H(L)$ to k^*: i.e.,

(a) $e^L(x_1+x_2, y) = e^L(x_1, y) \cdot e^L(x_2, y)$,

(b) $e^L(x, x) = 1$, hence $e^L(x, y) = e^L(y, x)^{-1}$.

Moreover, because k^* is divisible, for all subgroups $K \subset H(L)$, there exists a level subgroup $\tilde{K}$ over K if and only if $e^L \equiv 1$ on K. To see this, first start with a single element $x \in K$. Let $\tilde{x}' \in \mathcal{G}(L)$ lie over x. If l is the order of x, then $(\tilde{x}')^l$ must be in k^*. Let $\alpha \in k^*$ be an l-th root of $(\tilde{x}')^l$, and let $\tilde{x} = \tilde{x}' \cdot \alpha^{-1}$. Then $\tilde{x}$ lies over x and also has order l. Now writing K as a direct sum of cyclic groups, and lifting its generators x_i by this procedure, we see that since $e^L = 1$ on K, the elements $\tilde{x}_i$ generate a level subgroup $\tilde{K}$ over K. Conversely, if $\tilde{K}$ exists, then $\tilde{K}$ is commutative, so $e^L \equiv 1$ on K.

The following conditions are readily seen to be equivalent:

i) $k^* = \text{center } [\mathcal{G}(L)]$.

ii) e^L is non-degenerate: i.e., $\forall x \in H(L)$, there exists a $y \in H(L)$ such that $e^L(x, y) \neq 1$,

iii) there are subgroups K_1, K_2 in $H(L)$ such that $H(L) = K_1 \oplus K_2$, $e^L(x, y) = 1$ if $x, y \in K_1$ or if $x, y \in K_2$ and e^L is a non-degenerate pairing of K_1 and K_2 – i.e., makes

$$K_2 \cong \text{Hom}(K_1, k^*).$$

In fact, i) and ii) are trivially equivalent and are implied by iii). To go from ii) to iii), one just mimics the standard procedure for putting skew-symmetric pairings in canonical form.

Theorem 1. *The above condition on $\mathcal{G}(L)$ is, in fact, satisfied.*

Proof. Let $\tilde{K}$ be a *maximal* level subgroup in $\mathcal{G}(L)$ and let K be its image in $H(L)$. Then K is a maximal subgroup of $H(L)$ on which $e^L \equiv 1$, and the centralizer of $\tilde{K}$ in $\mathcal{G}(L)$ is just $k^* \cdot \tilde{K}$. Let $Y = X/K$ and let M be the sheaf on Y obtained by descending L. Then according to our description of $H(M)$ given above, $H(M) = \{0\}$: in other words, M has degree 1. But since degree $M = d/r$, where r = cardinality of K, this means that the cardinality of K is d. Now suppose $H_0 \subset H(L)$ is the degenerate space of e^L — the set of elements $x \in H(L)$ such that $e^L(x, y) = 1$, all $y \in H(L)$. Then $H(L)/H_0$ satisfies conditions ii) and iii); therefore

$$\#(H(L)/H_0) = l^2$$

and there exist maximal subgroups $K' \subset H(L)/H_0$ on which $e^L \equiv 1$ of cardinality l. Now if K is the inverse image of K' in $H(L)$, we see:

$$\begin{aligned} d^2 &= \#(K)^2 \\ &= \#(H_0)^2 \cdot l^2 \\ &= \#(H_0)^2 \cdot \#(H(L)/H_0) \\ &= \#(H_0) \cdot \#(H(L)) \\ &= \#(H_0) \cdot d^2. \end{aligned}$$

Therefore $H_0 = (0)$. *Q.E.D.*

Definition. Let L be an ample invertible sheaf of separable type. By Theorem 1, the elementary divisors on $H(L)$ occur in pairs. Let

$$\delta = (d_1, d_2, \ldots, d_k)$$

be the sequence of positive integers such that $d_{i+1} \mid d_i$, $d_i > 1$, and such that

$$(d_1, d_1, d_2, d_2, \ldots, d_k, d_k)$$

are the elementary divisors on $H(L)$. Then L will be said to be of type δ.

Note that char $(k) \nmid d_1$ in this sequence. Now reversing the process:

Definition. Let $\delta = (d_1, \ldots, d_k)$ be a sequence as above. Let

$$K(\delta) = \bigoplus_{i=1}^{k} \mathbf{Z}/d_i \mathbf{Z}$$

$$\widehat{K(\delta)} = \mathrm{Hom}(K(\delta), k^*)$$

$$H(\delta) = K(\delta) \oplus \widehat{K(\delta)}.$$

Let $\mathscr{G}(\delta)$, as a set, be the product

$$k^* \times K(\delta) \times \widehat{K(\delta)}.$$

Define a group law on $\mathscr{G}(\delta)$ via

$$(\alpha, x, l) \cdot (\alpha', x', l') = (\alpha \cdot \alpha' \cdot l'(x), x + x', l + l').$$

Corollary of Th. 1. *If L is of type δ, then the sequence*

$$0 \to k^* \to \mathscr{G}(L) \to H(L) \to 0$$

is isomorphic to the sequence:

$$0 \to k^* \to \mathscr{G}(\delta) \to H(\delta) \to 0.$$

Proof. As in condition (iii) preceding the theorem, let $H(L)=K_1 \oplus K_2$ and let $\tilde{K}_i$ be a level subgroup over K_i. Choose any isomorphism α between K_1 and $K(\delta)$, and then map K_2 onto $K(\delta)$ via

$$\beta: \; K_2 \underset{\text{via } e}{\cong} \operatorname{Hom}(K_1, k^*) \underset{\text{via } \alpha}{\cong} \operatorname{Hom}(K(\delta), k^*) = \widehat{K(\delta)}.$$

If we require that $(1, x, 0)$ correspond to the point of $\tilde{K}_1$ over $\alpha^{-1}(x) \in K_1$, and $(1, 0, l)$ correspond to the point $\tilde{K}_2$ over $\beta^{-1}(l) \in K_2$, then this determines an isomorphism of $\mathscr{G}(\delta)$ and $\mathscr{G}(L)$. *Q.E.D.*

So far, we have considered $\mathscr{G}(L)$ only as a group in its own right, and as the set of all possible descent data for L with respect to isogenies. What makes this game more exciting however is that $\mathscr{G}(L)$ acts on $\Gamma(X, L)$:

Definition. Let $z=(x, \varphi) \in \mathscr{G}(L)$. Then define $U_z: \Gamma(X, L) \to \Gamma(X, L)$ by

$$U_z(s) = T^*_{-x}(\varphi(s))$$

for all $s \in \Gamma(X, L)$ [i.e., $\varphi(s)$ is a section of $T^*_x L$ and $T^*_{-x}(\varphi(s))$ is a section of $L = T^*_{-x}(T^*_x L)$].

This is an action of the *group* $\mathscr{G}(L)$ since if $z=(x, \varphi)$, $w=(y, \psi)$, then

$$\begin{aligned} U_w(U_z(s)) &= T^*_{-y}\{\psi(T^*_{-x}(\varphi s))\} \\ &= T^*_{-x-y}[T^*_x(\psi(T^*_{-x}(\varphi s)))] \\ &= T^*_{-x-y}[T^*_x(\psi)(\varphi s)] \\ &= U_{(x+y,\, T^*_x(\psi)\circ\varphi)}(s). \end{aligned}$$

Also, the center k^* of $\mathscr{G}(L)$ acts on $\Gamma(X, L)$ by its natural character: i.e., $\alpha \in k^*$ acts on $\Gamma(X, L)$ as multiplication by α. Such representations are rather limited:

Proposition 3. *$\mathscr{G}(\delta)$ has a unique irreducible representation in which k^* acts by its natural character. Suppose that this representation is denoted by $V(\delta)$. Then if V is any representation of $\mathscr{G}(\delta)$ in which k^* acts in this way, V is isomorphic to the direct sum of $V(\delta)$ with itself r-times for some r. Moreover, if $\tilde{K} \subset \mathscr{G}(\delta)$ is any maximal level subgroup,*

$$r = \dim_k(V^{\tilde{K}}).$$

(Here $V^{\tilde{K}}$ is the subspace of V of $\tilde{K}$-invariants.)

We give a complete proof here since it is quite instructive. The result is an exact analog of a very general theorem of MACKEY [8].

Proof. Let V be an irreducible representation of $\mathcal{G}(\delta)$. Pick a maximal level subgroup $\tilde{K}$. Then since $\tilde{K}$ is commutative and char $(k) \nmid$ cardinality $\tilde{K}$, the action of $\tilde{K}$ can be diagonalized: i.e.,

$$V = \bigoplus_{\chi \in \operatorname{Hom}(\tilde{K}, k^*)} V_\chi,$$

where V_χ is the eigenspace with weight χ. Now let $y \in \mathcal{G}(\delta)$. Then there is a unique character χ^y of $\tilde{K}$ defined by:

$$y^{-1} \cdot z \cdot y = \chi^y(z) \cdot z, \quad \text{all} \quad z \in \tilde{K},$$

and if $s \in V_{\chi_0}$, then one checks easily that $U_y(s) \in V_{\chi_0 + \chi^y}$ (here U_y is the given representation). Moreover, note that the mapping $y \to \chi^y$ defines an isomorphism γ:

$$\mathcal{G}(\delta)/k^* \cdot \tilde{K} \xrightarrow[\gamma]{\sim} \operatorname{Hom}(\tilde{K}, k^*).$$

Therefore: (a) if $V_\chi \neq (0)$ for one χ, then $V_\chi \neq 0$ for all χ. (b) If $s \in V_{\chi_0}$, then the elements $U_y(s)$ span a subspace W of V such that $W \cap V_\chi$ is one-dimensional for all χ. It follows that if V is irreducible, $\dim V_\chi = 1$ for all χ. Now choose an arbitrary section σ:

$$\mathcal{G}(\delta) \overset{\sigma}{\underset{}{\leftrightarrows}} \mathcal{G}(\delta)/k^* \cdot \tilde{K}.$$

Moreover, choose a non-zero element $s(0) \in V_0$. Then for all $\chi \in \operatorname{Hom}(\tilde{K}, k^*)$, set $s(\chi) = U_{\sigma(\gamma^{-1}(\chi))}(s(0))$. Clearly $\{s(\chi)\}$ form a basis of V. It is easily checked that the matrices giving the representation of $\mathcal{G}(\delta)$ in terms of this basis depend only on the section σ. Therefore, all irreducible representations V are isomorphic. The last two assertions about a general representation V follow immediately from the complete reducibility of such representations. In fact, let m = order of $H(\delta)$. Then the set of elements $x \in \mathcal{G}(\delta)$ such that $x^m = 1$ form a finite subgroup $\mathcal{G}(\delta)' \subset \mathcal{G}(\delta)$. Since $k^* \mathcal{G}(\delta)' = \mathcal{G}(\delta)$, a representation V of $\mathcal{G}(\delta)$ in which k^* acts by its natural character is completely reducible if and only if it is completely reducible as a representation of $\mathcal{G}(L)'$. But char $(k) \nmid$ order $(\mathcal{G}(L)')$, so representations of $\mathcal{G}(L)'$ are completely reducible.

$$\dim(V^{\tilde{K}}) = \dim V_0 = 1$$

if V is irreducible. *Q.E.D.*

There are several natural ways to write down this unique irreducible representation explicitly. These explicit representations and the transformations between them have been closely studied by WEIL [12]. The simplest is:

Definition. Let $V(\delta)$ be the vector space of functions f on $K(\delta)$ with values in k. Let $\mathcal{G}(\delta)$ act on this by:

$$(U_{(\alpha,x,l)}f)(y)=\alpha\cdot l(y)\cdot f(x+y).$$

Here we see clearly that we have the discrete analog of the usual irreducible representation of the Heisenberg Commutation relations: in integrated form, that representation is the action a) of multiplication by unitary characters, b) of translation operators, on L^2 of a real vector space.

Theorem 2. *If L is an ample invertible sheaf of separable type, then $\Gamma(X,L)$ is an irreducible $\mathcal{G}(L)$-module.*

Proof. Let $\tilde{K}\subset\mathcal{G}(L)$ be a maximal level subgroup. Let K be its image in $H(L)$, let $Y=X/K$, and let M be the sheaf on Y such that $L\cong\pi^*M$ ($\pi: X\to Y$ being the projection). Then, as we have seen above, the maximality of $\tilde{K}$ implies that M is an ample invertible sheaf of degree 1. Therefore,

$$\dim\Gamma(Y,M)=1.$$

On the other hand, by the theory of descent, π^* maps the sections of M onto the space of sections of L *invariant under* $\tilde{K}$. Therefore

$$\dim\Gamma(X,L)^{\tilde{K}}=1. \quad Q.E.D.$$

This irreducibility has an important application: it makes possible a simple direct construction of the variety of moduli of abelian varieties. In fact, this method of constructing the moduli space is nothing but the method of theta functions applied by BAILY in the classical case. It turns out that, when suitably algebraicized, it is perfectly applicable in all characteristics, at least to separably polarized abelian varieties, in char $\neq 2$. The basic idea is this: suppose we are given

(i) an abelian variety X,

(ii) a very ample invertible sheaf L of separable type,

(iii) an isomorphism α of $\mathcal{G}(L)$ and $\mathcal{G}(\delta)$ which is the identity on the subgroups k^*.

Note that if i) and ii) are given, then there is only a *finite* set of data iii): so adding data of type (iii) means that one passes from the set of isomorphism classes of objects (X,L) to a finite covering of this set.

Definition. Given X and L as in (i), (ii) above, data of type (iii) will be called ϑ-structures on (X,L).

If X,L and a ϑ-structure are given, they determine in a canonical way *one* projective embedding of X [*n.b.* not just an equivalence class of projectively equivalent embeddings].

Step I. Since $\Gamma(X, L)$ is the unique irreducible representation of $\mathcal{G}(L)$ and $V(\delta)$ is the unique irreducible representation of $\mathcal{G}(\delta)$, among representations which restrict to the natural character on k^*, there is an isomorphism

$$\beta\colon \Gamma(X, L) \xrightarrow{\sim} V(\delta),$$

unique up to scalar multiples, such that

$$\beta\{U_x(s)\} = U_{\alpha(x)}(\beta(s))$$

all $x \in \mathcal{G}(L)$, $s \in \Gamma(X, L)$.

Step II. β induces a completely unique isomorphism

$$\boldsymbol{P}(\beta)\colon \quad \boldsymbol{P}[\Gamma(X, L)] \xrightarrow{\sim} \boldsymbol{P}[V(\delta)]$$

of projective spaces. Now pick, once and for all, a basis $\{X_i\}$ of $V(\delta)$ – say by ordering the elements $a_1, \ldots, a_m$ of $K(\delta)$ and taking X_i as the delta function at a_i. Then this defines an isomorphism:

$$\boldsymbol{P}(V(\delta)) \xrightarrow[\gamma]{\sim} \boldsymbol{P}_{m-1}.$$

Here $m = \text{cardinality}\ \{K(\delta)\}$
$= \text{degree}\ (L)$.

Step III. Since L is very ample, there is a canonical embedding

$$I\colon X \hookrightarrow \boldsymbol{P}[\Gamma(X, L)].$$

Then $J = \gamma \circ \boldsymbol{P}(\beta) \circ I$ is the canonical embedding of X in $\boldsymbol{P}_m$ which we claimed existed.

The most striking consequence of having defined a canonical embedding is that this defines a canonical *point* in $\boldsymbol{P}_{m-1}$ too: namely $J(e)$, where $e \in X$ is the identity point.

Definition. $J(e) \in \boldsymbol{P}_{m-1}$ will be called the *theta-null point* attached to X, L and the given ϑ-structure. The homogeneous coordinates of $J(e)$ will be called the *theta-null values* of X, L and the given ϑ-structure.

The ideas that we have indicated here will be fully developed in § 6.

We can make the connection of our theory with the classical theory of theta-functions more explicit, and in particular motivate our terminology "theta-null values", in the following way. The classical theta functions (when $k = C$) arise as follows: if $g = \dim X$, then C^g is the universal covering space of X. Let $\boldsymbol{o}_{\text{hol}}$ denote the trivial complex analytic invertible sheaf on C^g: i.e., the sheaf of holomorphic functions itself. Let $\pi\colon C^g \to X$ be the projection. Then one chooses "very carefully" an isomorphism

$$\lambda\colon \quad \pi^*(L_{\text{hol}}) \xrightarrow{\sim} \boldsymbol{o}_{\text{hol}}$$

of sheaves on C^g (here L_{hol} denotes the complex analytic invertible sheaf corresponding to the algebraic invertible sheaf L). Then for all

$$s \in \Gamma(X, L) = \Gamma(X, L_{\text{hol}})$$

$\Theta_s = \lambda(\pi^*(s))$ is a holomorphic function on C^g: this is the theta function corresponding to s and in this way $\Gamma(X, L)$ can be identified with a vector space of functions on C^g. We cannot imitate this procedure algebraically because $L \not\cong \mathfrak{o}_X$. However, L restricted to any finite set is isomorphic to $\mathfrak{o}_X$ on that set. In particular we can use the structure of $\mathscr{G}(L)$ to construct nearly canonical isomorphisms of L and $\mathfrak{o}_X$ *when restricted to* $H(L)$.

Definition. Let x be a closed point of X. Then let

$$L(x) = L_x \otimes_{\mathfrak{o}_x} \kappa(x),$$

when L_x is the stalk of L at x and $\kappa(x)$ is the residue field of $\mathfrak{o}_x$.

Now choose:

a) a ϑ-structure

$$\alpha: \quad \mathscr{G}(L) \xrightarrow{\sim} \mathscr{G}(\delta) \quad \text{for} \quad L,$$

b) an isomorphism

$$\lambda_0: \quad L(0) \xrightarrow{\sim} k.$$

Let the ϑ-structure induce the isomorphism $\bar{\alpha}$:

$$\begin{array}{ccccccccc} 0 & \to & k^* & \to & \mathscr{G}(L) & \to & H(L) & \to & 0 \\ & & \| & & \downarrow{\scriptstyle \alpha} & & \downarrow{\scriptstyle \bar{\alpha}} & & \\ 0 & \to & k^* & \to & \mathscr{G}(\delta) & \to & H(\delta) & \to & 0. \end{array}$$

There is a canonical section of $\mathscr{G}(\delta)$ over $H(\delta)$ obtained by mapping $(x, l) \in H(\delta)$ back to $(1, x, l) \in \mathscr{G}(\delta)$. This induces a section $\sigma: H(L) \to \mathscr{G}(L)$. In particular, for all $w \in H(L)$, this gives us an element $\sigma(w) = (w, \varphi_w) \in \mathscr{G}(L)$, where φ_w is an isomorphism:

$$\varphi_w: \quad L \to T_w^*(L).$$

Now we can use φ_w to define the composite isomorphism:

$$L(w) = T_w^* L(0) \xleftarrow[\varphi_w(0)]{\sim} L(0) \xrightarrow[\lambda_0]{\sim} k.$$

Call this λ_w. The collection of isomorphisms $\{\lambda_w\}$ is nothing but an isomorphism of the two sheaves L and $\mathfrak{o}_X$ restricted to $H(L)$.

Definition. For all $z \in H(\delta)$ and $s \in \Gamma(X, L)$, let $w = \bar{\alpha}^{-1}(z)$ and then define

$$\Theta_{[s]}(z) = \lambda_w(s|_w) = \lambda_0(\varphi_w^{-1}(T_w^* s)|_0).$$

(Here $s|_w$ denotes the image of the section s in $L(w)$.) The square brackets around s are inserted to indicate the connection with the classical notation for theta functions.

We now have a most surprising state of affairs. If a ϑ-structure on L is chosen, then we have 2 maps:

(I) $\Gamma(X,L) \xrightarrow[\beta]{\sim} \{k\text{-valued functions on } K(\delta)\} = V(\delta)$,

(II) $\Gamma(X,L) \xrightarrow[\vartheta]{} \{k\text{-valued functions on } H(\delta)\}$

both uniquely determined up to the ever-present scalar. For one thing, this means that we can combine the two, and obtain $\vartheta \circ \beta^{-1}$: a transformation taking functions on $K(\delta)$ to functions on $H(\delta)$. We will work this out shortly. But first notice the essential contrast between the two maps: the first is a purely group-theoretic affair and should reflect well the group-theoretic properties of X – as we will see in § 3. The second, like the concept of theta functions, is just a natural way of converting sections of L into functions by "evaluating" at points, so that it will respect multiplication of sections – i.e., it will extend to a homomorphism on the homogeneous coordinate ring $\bigoplus_n \Gamma(X, L^n)$ to the *ring* of functions on $H(\delta)$. The next step is to show that ϑ too has good group-theoretic properties, even if it is not defined by such properties:

Theorem 3. *Let* $s \in \Gamma(X,L)$ *and* $y \in \mathcal{G}(L)$. *Assume that*

$$\alpha(y) = (\beta, x, l), \quad \beta \in k^*, \quad x \in K(\delta), \quad l \in \widehat{K(\delta)}.$$

Then

$$\Theta_{[U_y s]}(x', l') = \beta \cdot (l' - l)(x) \cdot \Theta_{[s]}(x' - x, l' - l).$$

Proof. The linear mapping

$$s \mapsto \lambda_0(s|_0)$$

defines a linear functional $\tilde{\lambda}$: $\Gamma(X,L) \to k$. Then, if $w' \in H(L)$, $(x', l') = \bar{\alpha}(w')$, we have by definition:

$$\begin{aligned}\Theta_{[s]}(x', l') &= \lambda_{w'}(s|_{w'}) \\ &= \lambda_0(\varphi_{w'}^{-1}(T_{w'}^*(s))|_0) \\ &= \tilde{\lambda}(U_{y'}^{-1}(s))\end{aligned}$$

if $y' = (w', \varphi_{w'}) \in \mathcal{G}(L)$. But y' is the unique element of $\mathcal{G}(L)$ over w' in the section described above. Now:

$$\begin{aligned}\Theta_{[U_y(s)]}(x', l') &= \tilde{\lambda}(U_{y'}^{-1} U_y(s)) \\ &= \tilde{\lambda}(U_{y'^{-1} \cdot y}(s)).\end{aligned}$$

But let $y^{-1}\cdot y'=\gamma\cdot y''$, where $\gamma\in k^*$ and y'' is an element of $\mathscr{G}(L)$ in the distinguished section over $H(L)$. Then

$$\begin{aligned}\gamma\cdot\alpha(y'')&=\alpha(y)^{-1}\cdot\alpha(y')\\&=(\beta,x,l)^{-1}\cdot(1,x',l')\\&=(\beta^{-1}\cdot l(x)\cdot l'(x)^{-1},x'-x,l'-l).\end{aligned}$$

Therefore, $\gamma^{-1}=\beta\cdot(l'-l)(x)$, and $\alpha(y'')=(1,x'-x,l'-l)$. This implies that

$$\begin{aligned}\Theta_{[U_y(s)]}(x',l')&=\gamma^{-1}\cdot\tilde{\lambda}(U_{y''}^{-1}(s))\\&=\beta\cdot(l'-l)(x)\cdot\Theta_{[s]}(x'-x,l'-l).\quad Q.E.D.\end{aligned}$$

Definition. Let $\mathscr{G}(\delta)$ act on the vector space of functions on $H(\delta)$ by

$$\{U_{(\beta,x,l)}f\}(x',l')=\beta\cdot(l'-l)(x)\cdot f(x'-x,l'-l).$$

Corollary 1. *With respect to the above action, we have*

$$\Theta_{[U_y(s)]}=U_{\alpha(y)}(\Theta_{[s]}).$$

Corollary 2. *Either $\Theta_{[s]}=0$ for all $s\in\Gamma(X,L)$, or $\Theta_{[s]}=0$ only if $s=0$.*

Proof. This follows because the kernel of Θ is a subspace of $\Gamma(X,L)$ invariant under $\mathscr{G}(L)$.

Corollary 3. *Express the null-values of the functions $\Theta_{[s]}$ by the formula:*

$$\Theta_{[s]}(0,0)=\sum_{z\in K(\delta)}(\beta s)(z)\cdot q_L(z)$$

with a suitable k-valued function q_L on $K(\delta)$ i.e. $\{q_L(a)\}$ is the set of theta-null values of X,L; cf. p. 298. Then the transformation $\Theta\circ\beta^{-1}$ is given by:

$$\Theta_{[\beta^{-1}f]}(x,l)=\sum_{z\in K(\delta)}l(x-z)\cdot f(z-x)\cdot q_L(z).$$

Proof. Use the theorem with $x'=l'=0$, $\beta=1$, and the signs of x,l reversed, to obtain

$$\Theta_{[U_y(s)]}(0,0)=l(x)^{-1}\cdot\Theta_{[s]}(x,l).$$

Put $s=\beta^{-1}f$, use our expansion for the null-values and the Corollary comes out. *Q.E.D.*

Another consequence of the irreducibility of $\Gamma(X,L)$ under $\mathscr{G}(L)$ is that isogenies between abelian varieties induce canonical maps between corresponding linear systems. Suppose we are in the following situation, which we studied in the first part of this section:

$$(*)\qquad \begin{array}{c} X \xrightarrow{\pi} Y \\ L \underset{\alpha}{\dashleftarrow} M. \end{array}$$

Here π is a separable isogeny, L and M are ample invertible sheaves of separable type and α is an isomorphism $\alpha\colon \pi^* M \xrightarrow{\sim} L$. Then (π, α) induce a linear map:

$$(1)\qquad \pi^*\colon\ \Gamma(Y, M) \to \Gamma(X, L).$$

And, as we saw, (π, α) induces an isomorphism:

$$(2)\qquad \tau\colon\ \frac{\mathscr{G}(L)^*}{\tilde{K}} \xrightarrow{\sim} \mathscr{G}(M),$$

where

i) $\tilde{K}$ is the level subgroup over ker (π) which is the descent data on L for π associated to the descended sheaf M,

ii) $\mathscr{G}(L)^*$ is the group of elements of $\mathscr{G}(L)$ lying over the subgroup $\pi^{-1}[H(M)]$ in X,

iii) $\mathscr{G}(L)^* =$ centralizer of $\tilde{K}$.

Let δ_M and δ_L be types of M and L respectively and let

$$j_L\colon\ \mathscr{G}(L) \xrightarrow{\sim} \mathscr{G}(\delta_L)$$
$$j_M\colon\ \mathscr{G}(M) \xrightarrow{\sim} \mathscr{G}(\delta_M)$$

be ϑ-structures. Let the j's induce isomorphisms:

$$\beta_L\colon\ \Gamma(X, L) \xrightarrow{\sim} V(\delta_L)$$
$$\beta_M\colon\ \Gamma(Y, M) \xrightarrow{\sim} V(\delta_M).$$

The problem arises: what is the composite map $A = \beta_L \circ \pi^* \circ \beta_M^{-1}$:

$$V(\delta_M) \xrightarrow{\beta_M^{-1}} \Gamma(Y, M) \xrightarrow{\pi^*} \Gamma(X, L) \xrightarrow{\beta_L} V(\delta_L).$$

Theorem 4. *Assume $j_L(\tilde{K})$ is a subgroup of the form $\{1\} \times K_1 \times K_2$, where $K_1 \subset K(\delta_L)$, $K_2 \subset \widehat{K(\delta_L)}$. Let*

$$K_1^\perp = \{x \in K(\delta_L) \mid l(x) = 1, \quad \text{all} \quad l \in K_2\}$$
$$K_2^\perp = \{l \in \widehat{K(\delta_L)} \mid l(x) = 1, \quad \text{all} \quad x \in K_1\}.$$

Then $j_L(\mathscr{G}(L)^*) = k^* \times K_1^\perp \times K_2^\perp$, and $K_1^\perp / K_1$ and $K_2^\perp / K_2$ are canonically dual. Assume that there is an isomorphism

$$\sigma\colon\ K_1^\perp / K_1 \xrightarrow{\sim} K(\delta_M)$$

such that if $\hat{\sigma}\colon K_2^{\perp}/K_2 \to \widehat{K(\delta_M)}$ is dual to σ, then the following diagram commutes:

$$\begin{array}{ccc} \mathcal{G}(L)^*/\tilde{K} & \xrightarrow{\ j_L\ } & k^* \times (K_1^{\perp}/K_1) \times (K_2^{\perp}/K_2) \\ \downarrow{\scriptstyle \tau} & & \downarrow{\scriptstyle 1_{k^*} \times \sigma \times \hat{\sigma}} \\ \mathcal{G}(M) & \xrightarrow{\quad\quad j_M \quad\quad} & \mathcal{G}(\delta_M). \end{array}$$

Then there is a scalar λ such that for all $f \in V(\delta_M)$,

$$A\,f(x) = \begin{cases} 0 & \text{if } x \notin K_1^{\perp} \\ \lambda f(\sigma x) & \text{if } x \in K_1^{\perp} \end{cases}$$

all $x \in K(\delta_L)$.

Proof. Notice that π^* is injective and that its image is $\Gamma(X, L)^{\tilde{K}}$, the subspace of $\tilde{K}$-invariants. Moreover, $\Gamma(X, L)^{\tilde{K}}$ is a module over

$$\mathcal{G}' = \frac{\mathcal{G}(L)^*}{\tilde{K}}.$$

In fact, $\Gamma(X, L)^{\tilde{K}}$ as $\mathcal{G}'$-module is isomorphic to $\Gamma(Y, M)$ as $\mathcal{G}(M)$-module. Therefore, $\Gamma(X, L)^{\tilde{K}}$ is an irreducible $\mathcal{G}'$-module. This means that A is characterized, up to a scalar, by the 2 properties:

i) $\mathrm{Im}\,(A)$ = subspace of $V(\delta_L)$ of $\{1\} \times K_1 \times K_2$-invariants.

ii) If $(\alpha, x, l) \in k^* \times K_1^{\perp} \times K_2^{\perp}$, then

$$U_{(\alpha, x, l)}(A f) = A(U_{(\alpha, \sigma x, \hat{\sigma} l)} f),$$

all $f \in V(\delta_M)$.

It remains to check i) and ii) for the map A defined by our formula. As for i), the formula gives

$$\mathrm{Im}(A) = \left\{ f \in V(\delta_L) \,\middle|\, \begin{array}{l} f \equiv 0 \text{ outside } K_1^{\perp} \\ f \text{ invariant under translations by elements of } K_1 \end{array} \right\}$$

which is clearly the space of $\{1\} \times K_1 \times K_1^{\perp}$-invariants. ii) is checked in a straightforward way too. *Q.E.D.*

§ 2. Symmetric Invertible Sheaves

As in § 1, $\iota\colon X \to X$ will denote the inverse morphism on an abelian variety X. We assume in this section that $\mathrm{char}\,(k) \neq 2$.

Definition. An invertible sheaf L on an abelian variety X is *symmetric* if $\iota^* L \cong L$.

Just as isomorphisms of L with $T_x^* L$ were involved in the descent of L with respect to isogenies $X \to Y$, so isomorphisms of L and $\iota^* L$ are involved in the descent of L with respect to $X \to X/\{1, \iota\}$.

Definition. If X is an abelian variety, then the quotient of X by the involution ι will be denoted by K_X, the *Kummer variety* of X. Moreover, X^f will denote the open subset of X consisting of X minus its points of order 2, and K_X^f will be the quotient of X^f by ι.

Note that X^f is just the set of points of X where $\{1, \iota\}$ acts freely. If $\dim X \geqq 2$, then K_X^f is just the set of non-singular points of K_X and also the maximal open set over which

$$\pi: \quad X \to K_X$$

is flat.

If L is a symmetric invertible sheaf, and

$$\varphi: \quad L \xrightarrow{\sim} \iota^* L$$

is an isomorphism of L with $\iota^* L$, then for all closed points $x \in X$, φ restricts to an isomorphism:

$$\varphi(x): \quad L(x) \xrightarrow{\sim} \iota^* L(x) = L(-x).$$

We can always uniquely normalize φ by demanding that $\varphi(0)$ is the identity map from $L(0)$ to $L(0)$. Look at the composition:

$$L \xrightarrow{\varphi} \iota^* L \xrightarrow{\iota^* \varphi} \iota^*(\iota^* L) = L.$$

In general, $\iota^* \varphi \circ \varphi$ must be given by multiplication by a constant α: but if $\varphi(0)$ is the identity, then $\iota^* \varphi \circ \varphi$ acts as the identity on $L(0)$, so $\alpha = 1$.

Definition. φ is a *normalized isomorphism* of L and $\iota^* L$ if $\varphi(0) =$ identity.

Definition. Let $\varphi: L \to \iota^* L$ be the normalized isomorphism. Let $x \in X$ be a point of order 2. Define $e_*^L(x)$ to be the scalar α such that $\varphi(x)$ is multiplication by α (*n.b.* since $2x = 0$, $L(-x) = L(x)$).

First Properties.

i) $$e_*^L(x) = \pm 1, \quad e_*^L(0) = +1.$$

ii) $$e_*^{L \otimes M}(x) = e_*^L(x) \cdot e_*^M(x).$$

iii) If $\varphi: X \to Y$ is a homomorphism, and L is a symmetric invertible sheaf on Y, then

$$e_*^{\varphi^* L}(x) = e_*^L(\varphi(x))$$

for all $x \in X$ of order 2.

iv) If $\hat{X}$ is dual to X, let

$$e_2: \quad X_2 \times \hat{X}_2 \to \{\pm 1\}$$

be the canonical pairing. If $x \in X_2$, and $\alpha \in \hat{X}_2$, and if α corresponds to an invertible sheaf L of order 2 on X, then L is symmetric and

$$e_*^L(x) = e_2(x, \alpha).$$

Proofs. ii) and iii) are obvious. As for i), note that if $\varphi: L \xrightarrow{\sim} \iota^* L$ is the normalized isomorphism, then $\iota^* \varphi \circ \varphi = \text{identity}$, so for all closed points $x \in X$, $\varphi(-x) \circ \varphi(x) = \text{identity}$. To prove iv), note that for any closed point $\alpha \in \hat{X}$ corresponding to a sheaf L, $-\alpha$ corresponds to $\iota^* L$. Hence $\alpha \in \hat{X}_2$ implies $L \cong \iota^* L$. Now recall the definition of e_2 (cf. [7], pp. 188–189): let φ be an isomorphism

$$(2\delta)^* L \longrightarrow o_X.$$

Then if σ is the canonical isomorphism, the following diagram commutes up to a factor $e_2(x, \alpha)$:

$$\begin{array}{ccc} (2\delta)^* L & \xrightarrow{\varphi} & o_X \\ \sigma \downarrow & & \| \\ T_x^*(2\delta)^* L & \xrightarrow{T_x^* \varphi} & T_x^*(o_X). \end{array}$$

In particular, let $y \in X$ be such that $2y = x$. Then looking at this diagram at the point y we see that $\varphi(y)$ and $\varphi(-y)$ are maps from

$$\begin{array}{c} L(x) = L(\pm 2y) \cong (2\delta)^* L(\pm y) \to o_X(\pm y) \\ \qquad\qquad\qquad\qquad\qquad\qquad\qquad \| \wr \\ \qquad\qquad\qquad\qquad\qquad\qquad\qquad k, \end{array}$$

which differ from each other by $e_2(x, \alpha)$. On the other hand, if ψ: $L \xrightarrow{\sim} \iota^* L$ is the normalized isomorphism, then

$$\begin{array}{ccc} (2\delta)^* L & \xrightarrow{\varphi} & o_X \\ (2\delta)^* \psi \downarrow & & \| \\ \iota^*(2\delta)^* L & \xrightarrow{\iota^* \varphi} & \iota^* o_X \end{array}$$

commutes (look at the diagram at 0). Therefore $\varphi(y)$ and $\varphi(-y)$ differ by $[(2\delta)^* \psi](y)$, i.e., by $\psi(x)$. And $\psi(x)$ is by definition multiplication by $e_*^L(x)$. *Q.E.D.*

Proposition 1. *Let L be an invertible sheaf on X, and let $\pi: X \to K_X$ be the projection. Then L is of the form $\pi^* M$ for some invertible sheaf M on K_X if and only if L is symmetric and $e_*^L(x) = 1$ for all points $x \in X$ of order 2.*

Definition. A sheaf L satisfying the conditions of this Proposition will be called *totally symmetric*.

Proof of Proposition. If $L \cong \pi^* M$, then it follows immediately that L is symmetric and $e_*^L(x)=1$, all x. Conversely, assume these conditions on L and let D be a divisor on X such that $L \cong \mathbf{o}_X(D)$. Since L is symmetric, $\iota^{-1}(D)$ is linearly equivalent to D: say

$$\iota^{-1}(D)=D+(f).$$

Then:

$$\begin{aligned} D &= \iota^{-1}(\iota^{-1}(D)) \\ &= \iota^{-1} D + \iota^{-1}(f) \\ &= \iota^{-1} D + (\iota^* f), \end{aligned}$$

hence

$$(\iota^* f) \cdot f = \alpha, \quad \alpha \in k^*.$$

Replacing f by $\sqrt{\alpha} \cdot f$, we may assume $\iota^* f = f^{-1}$. Applying HILBERT's theorem 90 to the action of the group $(1, \iota^*)$ on the function field $k(X)$ of X, it follows that

$$f = \iota^* g \cdot g^{-1}$$

for some $g \in k(X)$. Now let

$$D' = D - (g).$$

Then it follows immediately that $\iota^{-1}(D') = D'$. Now if we restrict D' to the subset X^f where ι acts freely, this implies that $D' = \pi^{-1}(E)$ for some divisor E on K_X^f. But this is not automatic at points x of order 2. However, $L \cong \mathbf{o}_X(D')$ and we have assumed the existence of an isomorphism

$$\varphi: \; L \xrightarrow{\sim} \iota^* L$$

such that $\varphi(x)$ is the identity for all x of order 2. In other words, if $s_x \in L_x$ is a generator of the stalk of L at x, then $\varphi(s_x)$ differs from $\iota^*(s_x)$ by an element of $\mathbf{m}_x \cdot L_x$ ($\mathbf{m}_x$ = the maximal ideal of the local ring at x). Replacing L by $\mathbf{o}_X(D')$, this means that there is an $\alpha \in k^*$ such that if f_x is a local equation for D' at a point x of order 2, then

$$\iota^* f_x \equiv \alpha f_x \pmod{f_x \cdot \mathbf{m}_x}.$$

It is clear that $\alpha = +1$ or -1. If $\alpha = -1$, we replace D' by $D' + (h)$, where h is an element of $k(X)$ such that $\iota^* h = -h$, so that we may assume $\alpha = +1$. Now consider:

$$f'_x = \frac{\iota^* f_x \cdot f_x}{\iota^* f_x + f_x}.$$

f'_x is still a local equation for D', and $\iota^* f'_x = f'_x$. Therefore, f'_x is a local equation at $\pi(x)$ on K_X for a divisor E such that $\pi^{-1} E = D'$ at x. In this way, we see that there is a globally defined (CARTIER) divisor E on K_X such that $\pi^{-1} E = D'$, and hence $\pi^*(\mathbf{o}_{K_X}(E)) \cong \mathbf{o}_X(D') \cong L$. Q.E.D.

Note that if L is any invertible sheaf, then $L \otimes \imath^* L$ is totally symmetric: so that there are plenty of very ample totally symmetric sheaves on X. A useful remark about totally symmetric sheaves is that if L_1, L_2 are algebraically equivalent totally symmetric sheaves, then $L_1 \cong L_2$. Put in another way, no non-trivial totally symmetric sheaves are algebraically equivalent to zero. In fact, let L_α be the sheaf corresponding to the closed point $\alpha \in \hat{X}$. Then $\imath^* L_\alpha \cong L_{-\alpha}$, so L is symmetric if and only if $2\alpha = e$. Then my assertion follows from property (iv) above. Therefore, totally symmetric invertible sheaves are a very convenient class of sheaves to work with. In most of the sequel, we shall stick with this type of sheaf and the corresponding projective embeddings.

Following further the ideas in the proof of Proposition 1, we get:

Proposition 2. *Let D be a symmetric divisor, i.e., $\imath^{-1}(D) = D$, and let $L = \boldsymbol{o}_X(D)$. Then for all points x of order 2:*

$$e_*^L(x) = (-1)^{m(x) - m(0)}$$

where $m(y)$ = multiplicity of D at y, for all closed points $y \in X$.

Proof. Since D is the difference of 2 symmetric *effective* divisors, it suffices to prove the Proposition when D is itself effective. Now define φ to be the composition:

$$L = \boldsymbol{o}_X(D) = \boldsymbol{o}_X(\imath^{-1}(D)) \cong \imath^*[\boldsymbol{o}_X(D)] = \imath^* L.$$

I claim that for all points y of order 2 (including 0), $\varphi(y)$ is multiplication by $(-1)^{m(y)}$: hence $(-1)^{m(0)}$. φ is the normalized isomorphism of $\imath^* L$ and L and the result follows. To see this, let f_y be a local equation of D at y. Then $\imath^* f_y = \alpha \cdot f_y$ where α is a unit in $\boldsymbol{o}_{y,X}$. Since $\imath^* \alpha = \alpha^{-1}$, and $\imath^*$ operates trivially on $\boldsymbol{o}_y / \boldsymbol{m}_y$, it follows that $\alpha \equiv \pm 1 \pmod{\boldsymbol{m}_y}$. On the other hand, by definition, φ, at y, maps the generator f_y of the stalk L_y to the generator $\imath^* f_y$ of the stalk $(\imath^* L)_y$. Therefore $\varphi(y)$ is multiplication by the scalar α mod $\boldsymbol{m}_y$. Finally, to compute α directly, let $m = m(y)$, so that

$$f_y \in \boldsymbol{m}_y^m - \boldsymbol{m}_y^{m+1}.$$

The automorphism $\imath$ acts on $\boldsymbol{m}_y / \boldsymbol{m}_y^2$ as multiplication by -1, hence it acts on $\boldsymbol{m}_y^m / \boldsymbol{m}_y^{m+1}$ as multiplication by $(-1)^m$. Then α mod $\boldsymbol{m}_y$ is determined by the congruence:

$$\alpha \cdot f_y = \imath^* f_y \equiv (-1)^m f_y \pmod{\boldsymbol{m}_y^{m+1}}. \quad Q.E.D.$$

The analysis of symmetry in this section has proceeded along quite different lines from the study of isomorphisms of L with its translates given in the first section. Our next goal is to reinterpret the symmetry condition in general and e_*^L in particular in group-theoretic terms involving $\mathcal{G}(L)$ (actually, involving the groups $\mathcal{G}(L^n)$ also). Now assume that L is an ample symmetric sheaf of separable type.

Definition. Let $\psi: L \xrightarrow{\sim} \iota^* L$ be any isomorphism. Then if $(x, \varphi) \in \mathcal{G}(L)$, consider the composition:

$$\begin{array}{ccc} L \xrightarrow[\psi]{\sim} \iota^* L \xrightarrow{\iota^*(\varphi)} & \iota^* T_x^* L \\ & \| \\ & T_{-x}^* \iota^* L \xleftarrow{T_{-x}^* \psi} T_{-x}^* L. \end{array}$$

Set

$$\delta_{-1}((x, \varphi)) = (-x, (T_{-x}^* \psi)^{-1} \circ (\iota^* \varphi) \circ \psi).$$

Note that δ_{-1} is independent of the choice of ψ. One checks immediately that δ_{-1} is a homomorphism from $\mathcal{G}(L)$ to $\mathcal{G}(L)$ and that $\delta_{-1} \circ \delta_{-1}$ is the identity. In fact, δ_{-1} is an automorphism of $\mathcal{G}(L)$ that fits into the diagram:

$$(*)_{-1} \qquad \begin{array}{ccccccccc} 0 & \to & k^* & \to & \mathcal{G}(L) & \to & H(L) & \to & 0 \\ & & \downarrow{\scriptstyle \mathrm{id}} & & \downarrow{\scriptstyle \delta_{-1}} & & \downarrow{\scriptstyle -1} & & \\ 0 & \to & k^* & \to & \mathcal{G}(L) & \to & H(L) & \to & 0. \end{array}$$

δ_{-1} is, then, the reflection of the inverse ι in the group $\mathcal{G}(L)$. The notation δ_{-1} is motivated by the following:

Definition. If $z \in \mathcal{G}(L)$, and n is any integer, let

$$\delta_n(z) = (z)^{\frac{n^2+n}{2}} \cdot [\delta_{-1}(z)]^{\frac{n^2-n}{2}}.$$

It is a straightforward calculation to check that δ_n is a homomorphism, that $\delta_{nm} = \delta_n \circ \delta_m$ and that δ_n fits into a diagram:

$$(*)_n \qquad \begin{array}{ccccccccc} 0 & \to & k^* & \to & \mathcal{G}(L) & \to & H(L) & \to & 0 \\ & & \downarrow{\scriptstyle n^2\text{-power}} & & \downarrow{\scriptstyle \delta_n} & & \downarrow{\scriptstyle \text{mult. by } n} & & \\ 0 & \to & k^* & \to & \mathcal{G}(L) & \to & H(L) & \to & 0. \end{array}$$

We omit these calculations.

An important consequence of the existence of δ_{-1} is that we nearly have a canonical section of $\mathcal{G}(L)$ over $H(L)$. In fact, if $x \in H(L)$, then there will be exactly 2 elements $z \in \mathcal{G}(L)$ over x such that

$$(1) \qquad \delta_{-1} z = z^{-1}.$$

Starting with any $z_0 \in \mathcal{G}(L)$ over x, it follows that $\delta_{-1} z_0 = \alpha \cdot z_0^{-1}$, where $\alpha \in k^*$. The most general element over x is of the form $\beta \cdot z_0$, $\beta \in k^*$;

and

$$\delta_{-1}(\beta z_0)=\beta^2\cdot\alpha\cdot(\beta z_0)^{-1}.$$

Hence if $\beta=\alpha^{-\frac{1}{4}}$, $\pm\beta\cdot z_0$ are the elements over x satisfying formula (1).

Definition. Let $\mathcal{S}(L)=\{z\in\mathcal{G}(L)\,|\,\delta_{-1}z=z^{-1}\}$. If $(x,\varphi)\in\mathcal{S}(L)$, then φ will be called a *symmetric isomorphism* of L and $T_x^* L$. δ_{-1} and e_* are related by:

Proposition 3. *Let L be an ample symmetric invertible sheaf of separable type. Let $z\in\mathcal{G}(L)$ lie over a point $x\in H(L)$ of order 2. Then*

$$\delta_{-1}(z)=e_*^L(x)\cdot z.$$

Proof. Let $z=(x,\varphi)$, where $\varphi\colon L\xrightarrow{\sim}T_x^* L$. Let $\psi\colon L\xrightarrow{\sim}\iota^* L$ be a normalized isomorphism. Then $\delta_{-1}z=(x,(T_x^*\psi)^{-1}\circ(\iota^*\varphi)\circ\psi)$. To compare φ and $(T_x^*\psi)^{-1}\circ(\iota^*\varphi)\circ\psi$, look at the induced maps that both define from $L(0)$ to $L(x)$. In fact, the latter map gives:

$$L(0)\xrightarrow{\psi(0)}L(0)\xrightarrow{\varphi(0)}L(x)\xrightarrow{\psi(x)^{-1}}L(x),$$

which differs from $\varphi(0)$ by $e_*^L(x)=\psi(x)^{-1}$. Therefore, $(T_x^*\psi)^{-1}\circ(\iota^*\varphi)\circ\psi$ is the product of φ by the scalar $e_*^L(x)$. *Q.E.D.*

Unfortunately, this Proposition does not allow us to recapture the invariant e_*^L from δ_{-1} unless $H(L)$ contains all points of order 2. In fact, there is yet another canonical homomorphism whose existence depends on the symmetry of L. This involves the relation between $\mathcal{G}(L)$ and $\mathcal{G}(L^n)$. First, however, we define a trivial homomorphism which does *not* require the symmetry of L:

Definition. Let L be an ample invertible sheaf of separable type, and let n be an integer, $n\geqq 2$ such that char $(k)\nmid n$. Let $(x,\varphi)\in\mathcal{G}(L)$.

Define

$$\varepsilon_n(x,\varphi)=(x,\varphi^{\otimes n})$$

where $\varphi^{\otimes n}$ stands for the isomorphism

$$L^n\xrightarrow{\varphi^{\otimes n}}T_x^* L^n$$

induced by φ.

This defines a homomorphism ε_n fitting into the diagram:

$$(**)_n\qquad\begin{array}{ccccccccc}
0 & \longrightarrow & k^* & \longrightarrow & \mathcal{G}(L) & \longrightarrow & H(L) & \longrightarrow & 0\\
 & & \big\downarrow{\scriptstyle n\text{-th power}} & & \big\downarrow{\scriptstyle \varepsilon_n} & & \big\downarrow{\scriptstyle \text{inclusion}} & & \\
0 & \longrightarrow & k^* & \longrightarrow & \mathcal{G}(L^n) & \longrightarrow & H(L^n) & \longrightarrow & 0
\end{array}$$

Also note that $H(L)$ and $H(L^n)$ are related by:

Proposition 4. *If* L *is an arbitrary invertible sheaf on an abelian variety* X, *then*

$$H(L^n)=\{x \mid nx\in H(L)\}$$

and

$$H(L)=n\cdot H(L^n).$$

Proof. In fact, $H(L)$ and $H(L^n)$ are the kernels of the homomorphisms:

$$\Lambda(L)\colon\ X\to\hat{X}$$
$$\Lambda(L^n)\colon\ X\to\hat{X}$$

mentioned at the beginning of § 1. But

$$\Lambda(L^n)=n\cdot\Lambda(L)$$

(cf. [9], Ch. 6, p. 120). Therefore $\ker[\Lambda(L^n)]$ is the set of closed points $x\in X$ such that $n\cdot x\in\ker[\Lambda(L)]$: this gives the first statement. Then $H(L)$ equals $n\cdot H(L^n)$ since the group of closed points of X is divisible. *Q.E.D.*

In contrast to ε_n, the symmetry of L is involved in the existence of a non-trivial canonical homomorphism $\eta_n\colon\mathscr{G}(L^n)\to\mathscr{G}(L)$ fitting into the diagram:

$$(***)_n\qquad \begin{array}{ccccccccc} 0 & \longrightarrow & k^* & \longrightarrow & \mathscr{G}(L^n) & \longrightarrow & H(L^n) & \longrightarrow & 0\\ & & \Big\downarrow{\scriptstyle n\text{-th power}} & & \Big\downarrow{\scriptstyle \eta_n} & & \Big\downarrow{\scriptstyle \text{mult. by } n} & & \\ 0 & \longrightarrow & k^* & \longrightarrow & \mathscr{G}(L) & \longrightarrow & H(L) & \longrightarrow & 0. \end{array}$$

Definition (of η_n). Assume that L is symmetric. Start with $z=(x,\varphi)\in\mathscr{G}(L^n)$. Since L is symmetric, there is an isomorphism

$$\psi\colon\ L^{n^2}\xrightarrow{\ \sim\ }(n\delta)^*L$$

(cf. § 1, (IV)). Consider the diagram:

$$\begin{array}{ccc} L^{n^2} & \xrightarrow{\quad\varphi^{\otimes n}\quad} & T_x^*L^{n^2}\\ & & \Big\downarrow{\scriptstyle T_x^*(\psi)}\\ \Big\downarrow{\scriptstyle\psi} & & T_x^*(n\delta^*L)\\ & & \Big\Vert\\ (n\delta)^*L & \dashrightarrow & (n\delta)^*T_{nx}^*L \end{array}$$

By the previous Proposition, $nx\in H(L)$, hence there is *some* isomorphism between L and T_{nx}^*L. Then there is a unique isomorphism $\rho\colon L\xrightarrow{\sim}T_{nx}^*L$ such that

$$(n\delta)^*\rho=T_x^*\psi\circ\varphi^{\otimes n}\circ\psi^{-1}$$

i.e., such that with $(n\delta)^*\rho$ along the dotted line, the above diagram commutes. Set

$$\eta_n(x,\varphi)=(n\,x,\rho).$$

One checks easily that η_n is independent of ψ, and is a homomorphism.

Proposition 5. *Let* $\delta'_m\colon \mathcal{G}(L)\to\mathcal{G}(L)$ *and* $\delta''_m\colon \mathcal{G}(L^n)\to\mathcal{G}(L^n)$ *be the homomorphisms defined above. Then*

i) $$\delta'_{-1}\circ\eta_n=\eta_n\circ\delta''_{-1},$$

ii) $$\delta''_{-1}\circ\varepsilon_n=\varepsilon_n\circ\delta'_{-1}$$

(hence δ_m, for all m, commutes with η and ε). Moreover,

iii) $$\delta'_n=\eta_n\circ\varepsilon_n,$$

iv) $$\delta''_n=\varepsilon_n\circ\eta_n.$$

Proof. (i) and (ii) are verified easily by writing out the definitions: this will be omitted. Also, (iii) follows from (iv). Namely, if we assume (iv), then:

$$\begin{aligned}\delta'_n\circ\eta_n&=\eta_n\circ\delta''_n\\&=\eta_n\circ\varepsilon_n\circ\eta_n.\end{aligned}$$

Since η_n is surjective, (iii) must hold. Now (iv) is harder: note first that both δ''_n and $\varepsilon_n\circ\eta_n$ fit in as dotted lines in the diagram:

$$\begin{array}{ccccccccc}0&\longrightarrow&k^*&\longrightarrow&\mathcal{G}(L^n)&\longrightarrow&H(L^n)&\longrightarrow&0\\&&\big\downarrow{\scriptstyle n^2\text{-power}}&&\vdots&&\big\downarrow{\scriptstyle \text{mult. by } n}&&\\0&\longrightarrow&k^*&\longrightarrow&\mathcal{G}(L^n)&\longrightarrow&H(L)&\longrightarrow&0.\end{array}$$

Therefore, there is a homomorphism $h\colon H(L^n)\to k^*$ such that

$$\varepsilon_n(\eta_n(z))=h(\bar z)\cdot\delta''_n(z)$$

(here $z\in\mathcal{G}(L^n)$ and $\bar z$ is the image of z in $H(L^n)$). But then:

$$\begin{aligned}h(\bar z)\cdot\delta''_{-1}(\delta''_n z)&=\delta''_{-1}[h(\bar z)\cdot\delta''_n(z)]\\&=\delta''_{-1}[\varepsilon_n(\eta_n z)]\\&=\varepsilon_n(\eta_n(\delta''_{-1}z))\\&=h(\overline{\delta''_{-1}z})\cdot\delta''_n(\delta''_{-1}z)\\&=h(-\bar z)\cdot\delta''_{-n}z\\&=h(\bar z)^{-1}\cdot\delta''_{-1}(\delta''_n z),\end{aligned}$$

hence $h(\bar z)=\pm1$ for all $z\in\mathcal{G}(L^n)$. In particular, $\varepsilon_n(\eta_n(z))=\delta''_n(z)$ for all $z\in\mathcal{G}(L^n)$ such that $\bar z\in 2H(L^n)$.

Now suppose that we have a separable isogeny:

$$f: \quad Y \to X$$

and that we set $M = f^*L$. If we set $\mathscr{G}(M)^*$ equal to the subgroup of elements z whose image in $H(M)$ is in $f^{-1}[H(L)]$, then as we know from § 1,

$$\mathscr{G}(L) \cong \mathscr{G}(M)^*/\tilde{K}$$

for some level subgroup $\tilde{K}$. Similarly, $\mathscr{G}(L^n) \cong \mathscr{G}(M^n)^*/\tilde{K}_n$. It is easy to check that $\delta''_{-1}(\mathscr{G}(M^n)^*) \subset \mathscr{G}(M^n)^*, \varepsilon_n(\mathscr{G}(M)^*) \subset \mathscr{G}(M^n)^*, \eta_n(\mathscr{G}(M^n)^*) \subset \mathscr{G}(M)^*$, and that the diagram (s):

$$
\begin{array}{ccc}
\circlearrowleft \delta''_n & & \circlearrowleft \delta''_n \\
\mathscr{G}(M^n)^* & \xrightarrow{\;\alpha_n\;} & \mathscr{G}(L^n) \\
\eta_n \Big(\Big) \varepsilon_n & & \eta_n \Big(\Big) \varepsilon_n \\
\mathscr{G}(M)^* & \xrightarrow{\;\alpha\;} & \mathscr{G}(L)
\end{array}
$$

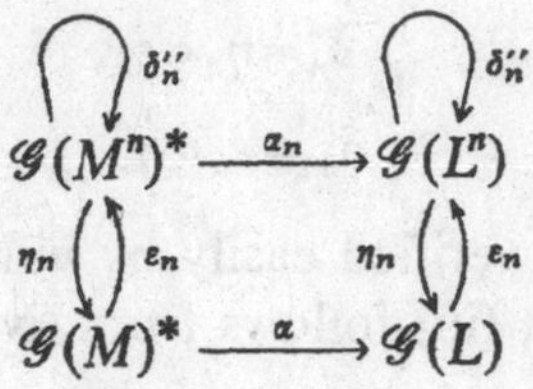

commute, (here α, α_n stand for the canonical maps). Now suppose, for $z \in \mathscr{G}(M^n)$, $\varepsilon_n(\eta_n(z))$ differs from $\delta''_n(z)$ by $\tilde{h}(\bar{z})$. Then if $z \in \mathscr{G}(M^n)^*$,

$$
\begin{aligned}
\tilde{h}(\bar{z}) &= \alpha_n(\tilde{h}(\bar{z})) \\
&= \alpha_n(\varepsilon_n(\eta_n(z)) \cdot \delta''_n(z)^{-1}) \\
&= \varepsilon_n(\eta_n(\alpha_n z)) \cdot \delta''_n(\alpha_n z)^{-1} \\
&= h(\overline{\alpha_n z}).
\end{aligned}
$$

Therefore, if every element of $f^{-1}(H(L^n))$ is divisible by two in $H(M^n)$, it follows that $\tilde{h} \equiv 1$ on $f^{-1}(H(L^n))$, hence $h \equiv 1$ on $H(L^n)$. It is easy to make this the case, however: choose $X = Y$ and $f = 2\delta$ for example. Then $M \cong L^4$, hence

$$H(M^n) = \{x \mid 4x \in H(L^n)\}$$

$$f^{-1}[H(L^n)] = \{x \mid 2x \in H(L^n)\}. \quad Q.E.D.$$

Consider, in particular, η_2. This map alone seems to contain all the useful canonical data to be extracted from the symmetry of L. On the one hand, forming $\varepsilon_2 \circ \eta_2$, we obtain $\delta_2 : \mathscr{G}(L) \to \mathscr{G}(L)$. And since $\delta_2(z) = z^3 \cdot \delta_{-1}(z)$, we can also reconstruct δ_{-1} and hence all the maps δ_n. On the other hand, unlike δ_{-1}, we can always reconstruct e_*^L from η_2:

Proposition 6. *Let* $z \in \mathscr{G}(L^2)$ *be an element of order* 2 *and let* x *be its image in* $H(L^2)$. *Then* $\eta_2(z)$ *is in* k^* *and*

$$\eta_2(z) = e_*^L(x).$$

Proof. Let $z=(x,\varphi)$, where φ is an isomorphism

$$\varphi:\ L^2 \xrightarrow{\ \sim\ } T_x^* L^2.$$

Moreover, let

$$\rho:\ L \to \iota^* L$$

be a normalized isomorphism, and let

$$\psi:\ L^4 \to (2\delta)^* L$$

be an arbitrary isomorphism. Choose a point $y \in X$ such that $2y=x$.

The first step is to consider the 2 maps:

$$\varphi(y), \rho^2(y):\ L^2(y) \xrightarrow{\ \sim\ } L^2(-y).$$

I claim that $\varphi(y)=\pm\rho^2(y)$. To see this, look at the diagram:

$$(\beta)\qquad \begin{array}{ccc} L^2 & \xrightarrow{\ \varphi\ } & T_x^* L^2 \\ {\scriptstyle \rho^2}\downarrow & & \downarrow{\scriptstyle T_x^*\rho^2} \\ & & T_x^* \iota^* L^2 \\ \iota^* L^2 & & \| \\ & \xrightarrow{\ \iota^*\varphi\ } & \iota^* T_x^* L^2 \end{array}$$

This diagram commutes: look at the induced maps at the origin

$$(\beta)_0\qquad \begin{array}{ccc} L(0)^2 & \xrightarrow{\ \varphi(x)\ } & L(x)^2 \\ {\scriptstyle \rho(0)^2}\downarrow & & \downarrow{\scriptstyle \rho(x)^2} \\ L(0)^2 & \xrightarrow{\ \varphi(x)\ } & L(x)^2. \end{array}$$

Since $\rho(0)=1$, and $\rho(x)=e_*^L(x)=\pm 1$, $(\beta)_0$ commutes, so (β) commutes also. On the other hand, look at diagram (β) at the point y. We find that

$$\begin{array}{ccc} L(y)^2 & \xrightarrow{\ \varphi(y)\ } & L(-y)^2 \\ {\scriptstyle \rho^2(y)}\downarrow & & \downarrow{\scriptstyle \rho^2(-y)} \\ L(-y)^2 & \xrightarrow{\ \varphi(-y)\ } & L(y)^2 \end{array}$$

commutes. Now since $z=(x,\varphi)$ has order 2, it follows that $T_x^*\varphi=\varphi^{-1}$. In particular, $\varphi(-y)=\varphi(y)^{-1}$. Moreover, since ρ is normalized, we know that $\iota^*\rho=\rho^{-1}$. In particular, $\rho(-y)=\rho(y)^{-1}$. Putting all this together, we conclude:

$$\rho^2(y)=\pm\varphi(y).$$

Therefore, $\rho^4(y)=\varphi^2(y)$.

What is the definition of $\eta_2(z)$? According to our recipe, in this case, $\eta_2(z)$ is the scalar α such that the diagram:

$$\begin{array}{ccc} L^4 & \xrightarrow{\varphi^2} & T_x^* L^4 \\ \downarrow{\scriptstyle\psi} & & \downarrow{\scriptstyle T_x^*\psi} \\ & & T_x^*(2\delta)^* L \\ & & \| \\ (2\delta)^* L & \xrightarrow{\alpha} & (2\delta)^* L \end{array}$$

commutes. Therefore, $\eta_2(z)$ is the composite map:

$$\begin{array}{ccccccc} L(x) & & & & & & \\ \| & & & & & & \\ [(2\delta)^* L](y) & \xrightarrow{\psi(y)^{-1}} & L^4(y) & \xrightarrow{\varphi(y)^2} & L^4(-y) & \xrightarrow{\psi(-y)} & [(2\delta)^* L](-y) \\ & & & & & & \| \\ & & & & & & L(x), \end{array}$$

(i.e., identifying the scalar $\eta_2(z)$ with multiplication by this scalar). By our first result, it follows that

$$\eta_2(z)=\psi(-y)\circ\rho(y)^4\circ\psi(y)^{-1}.$$

Thirdly, consider the diagram:

$$\begin{array}{ccc} L^4 & \xrightarrow{\rho^4} & \iota^* L^4 \\ \downarrow{\scriptstyle\psi} & & \downarrow{\scriptstyle\iota^*\psi} \\ & & \iota^*(2\delta)^* L \\ & & \| \\ (2\delta)^* L & \xrightarrow{(2\delta)^*\rho} & (2\delta)^*\iota^* L \end{array}$$

This diagram commutes: look at the induced maps at the origin

$$\begin{array}{ccc} L(0)^4 & \xrightarrow{\rho(0)^4} & L(0)^4 \\ \downarrow{\scriptstyle\psi(0)} & & \downarrow{\scriptstyle\psi(0)} \\ L(0) & \xrightarrow{\rho(0)} & L(0). \end{array}$$

Since $\rho(0)\doteq 1$, this induced diagram commutes, hence so does the full diagram of sheaves on X. Now look at the full diagram at the point y. We find that

$$\rho(x)=[(2\delta)^*\rho](y)=\psi(-y)\circ\rho(y)^4\circ\psi(y)^{-1}.$$

But, by definition, $\rho(x)=e_*^L(x)$; while we just proved that $\psi(-y)\circ\rho(y)^4\circ\psi(y)^{-1}=\eta_2(z)$. *Q.E.D.*

Corollary 1. *Let L be a symmetric invertible sheaf on an abelian variety X, and let X_2 denote the group of points of order 2 on X. Then e_*^L*

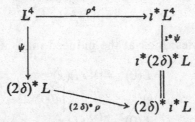

is a quadratic function from X_2 to $\{\pm 1\}$ whose associated bilinear form is $e^{(L^2)}$ (this is defined on $X_2 \times X_2$ since $X_2 \subset H(L^2)$). In other words:

$$e_*^L(x+y) = e_*^L(x) \cdot e_*^L(y) \cdot e^{(L^2)}(x, y).$$

Proof. Let z, w be elements of $\mathscr{G}(L^2)$ over x, y respectively such that $z^2 = w^2 = 1$. Then

$$\begin{aligned}(z \cdot w)^2 &= z \cdot w \cdot z \cdot w \\ &= z \cdot w \cdot z^{-1} \cdot w^{-1} \\ &= e^{(L^2)}(x, y).\end{aligned}$$

Let β be a square root of $e^{(L^2)}(x,y)$ in k^*. Then $\beta^{-1} \cdot z \cdot w$ is an element of order 2 in $\mathscr{G}(L^2)$ over $x+y$. Therefore

$$\begin{aligned}&e_*^L(x+y) \cdot e_*^L(x)^{-1} \cdot e_*^L(y)^{-1} \\ &\quad = \eta_2(\beta^{-1} \cdot z \cdot w) \cdot \eta_2(z)^{-1} \cdot \eta_2(w)^{-1} \\ &\quad = \eta_2(\beta)^{-1} \\ &\quad = \beta^{-2} \\ &\quad = e^{(L^2)}(x, y). \quad \text{Q.E.D.}\end{aligned}$$

Note that all these members (except β) are ± 1. This is the reason why the pairing $e^{(L^2)}$, which is always skew-symmetric, is also symmetric on $X_2 \times X_2$ and hence possible as the associated bilinear form to a quadratic form.

Corollary 2. *Let L be a symmetric invertible sheaf on an abelian variety X. Then L is totally symmetric if and only if*

$$\ker(\eta_2) = \{z \in \mathscr{G}(L^2) \mid z^2 = 1\}.$$

Proof. Immediate.

Corollary 3. *Let D be a symmetric divisor on an abelian variety X of dimension r. Let Σ_+ be the set of points of order 2 at which D has even multiplicity, and let Σ_- be the set where D has odd multiplicity. Then either*

case i) $\#(\Sigma_+) = 2^{2r-s-1}(2^s+1)$,
$\#(\Sigma_-) = 2^{2r-s-1}(2^s-1)$ *for some integer s, $0 \leqq s \leqq r$.*

case ii) Same as case i) *with Σ_+, Σ_- reversed.*

case iii) $\#(\Sigma_+) = \#(\Sigma_-) = 2^{2r-1}$.

Proof. This follows from Prop. 2 of this section, Cor. 1, and the elementary theory of quadratic forms over the field with 2 elements.

Corollary 4. *Let L be a totally symmetric ample invertible sheaf of separable type δ. Then $H(L)$ contains all points of order 2 on X, hence if $\delta = (d_1, \ldots, d_r)$, then $r = \dim X$ and all d_i are even.*

Proof. If L is of type $\delta=(d_1, \dots, d_r)$, then $r \leqq \dim X$ since $H(\delta)$ is isomorphic to a subgroup of X. Suppose you add some extra d's equal to 1 to make a total of $\dim X$ of them. Then by Proposition 4 and the known structure of the group of points on X of finite order, L^2 is of type 2δ. But if L is totally symmetric, then $e^{(L^2)} \equiv 1$ on points of order 2. Since $e^{(L^2)}$ is non-degenerate on $H(L^2)$, this means that every point of order 2 is twice a point of order 4 in $H(L^2)$. Therefore $H(L^2)$ contains the whole group X_4 of points of order 4 on X; hence $H(L)$ contains X_2. *Q.E.D.*

The last point that I want to deal with in this section is to give a normal form to the pair of groups $\mathscr{G}(L)$, $\mathscr{G}(L^2)$ and the maps between them.

Definition. Let $\delta=(d_1, d_2, \dots, d_k)$ be a set of elementary divisors: i.e., $d_i \in \mathbf{Z}$, $d_{i+1} | d_i$, $d_i \geqq 1$. (*N.b.*, we allow some l's at the end.) Then 2δ is the sequence of divisors $(2d_1, 2d_2, \dots, 2d_k)$. We shall always regard $K(\delta)=\bigoplus \mathbf{Z}/d_i\mathbf{Z}$ as a subgroup of $K(2\delta)=\bigoplus \mathbf{Z}/2d_i\mathbf{Z}$ under the map $(a_1, \dots, a_k) \mapsto (2a_1, \dots, 2a_k)$. Then the dual $\widehat{K(\delta)}$ is naturally a quotient of $\widehat{K(2\delta)}$. If $l \in \widehat{K(2\delta)}$, then $\bar{l}$ will denote its natural image in $\widehat{K(\delta)}$. However, if $l \in \widehat{K(\delta)}$, then there is a unique $l' \in \widehat{K(2\delta)}$ such that

$$l'(x)=l(2x)$$

for all $x \in K(2\delta)$. This defines an injection of $\widehat{K(\delta)}$ into $\widehat{K(2\delta)}$ that we denote $2*$, i.e., we set

$$l'=2*l.$$

Definition. $E_2: \mathscr{G}(\delta) \to \mathscr{G}(2\delta)$ is the map

$$E_2((\alpha, x, l))=(\alpha^2, x, 2*l).$$

Definition. $D_n: \mathscr{G}(\delta) \to \mathscr{G}(\delta)$ is the map

$$D_n((\alpha, x, l))=(\alpha^{n^2}, nx, nl).$$

Definition. $H_2: \mathscr{G}(2\delta) \to \mathscr{G}(\delta)$ is the map

$$H_2((\alpha, x, l))=(\alpha^2, 2x, \bar{l}).$$

One checks immediately that E_2, H_2 and D_n are all homomorphisms. Note also that

$$E_2 \circ H_2=D_2 \quad \text{for} \quad \mathscr{G}(2\delta)$$
$$H_2 \circ E_2=D_2 \quad \text{for} \quad \mathscr{G}(\delta)$$
$$D_n(z)=z^{\frac{n^2+n}{2}} \cdot D_{-1}(z)^{\frac{n^2-n}{2}}, \quad \text{all} \quad z \in \mathscr{G}(\delta)$$

just as was the case with ε, η and δ.

Definition. Let L be a totally symmetric ample invertible sheaf of separable type δ. A ϑ-structure

$$f: \mathscr{G}(L) \xrightarrow{\sim} \mathscr{G}(\delta)$$

will be called *symmetric* if $f \circ \delta_{-1} = D_{-1} \circ f$. A pair of ϑ-structures f_1 for L and f_2 for L^2 will be called a *symmetric ϑ-structure for* (L, L^2) or a *compatible* pair of symmetric ϑ-structures if

$$f_2 \circ \varepsilon_2 = E_2 \circ f_1$$
$$f_1 \circ \eta_2 = H_2 \circ f_2 .$$

An isomorphism

$$g: \ H(L) \xrightarrow{\sim} H(\delta)$$

will be called *symplectic* if, for all $z_1, z_2 \in H(L)$, $g(z_1) = (x_1, l_1)$, $g(z_2) = (x_2, l_2)$, then

$$e^{(L)}(z_1, z_2) = l_2(x_1) \cdot l_1(x_2)^{-1} .$$

Note that if $f: \mathscr{G}(L) \to \mathscr{G}(\delta)$ is any ϑ-structure, then the induced isomorphism of $H(L)$ and $H(\delta)$ is symplectic (check commutators!). Summarizing the situation so far, we can draw up a chart of the possible "markings" with which we can endow L, L^2 and L^4. The solid arrows indicate that a structure of one type automatically induces one of the other. (Ignore dotted arrows.)

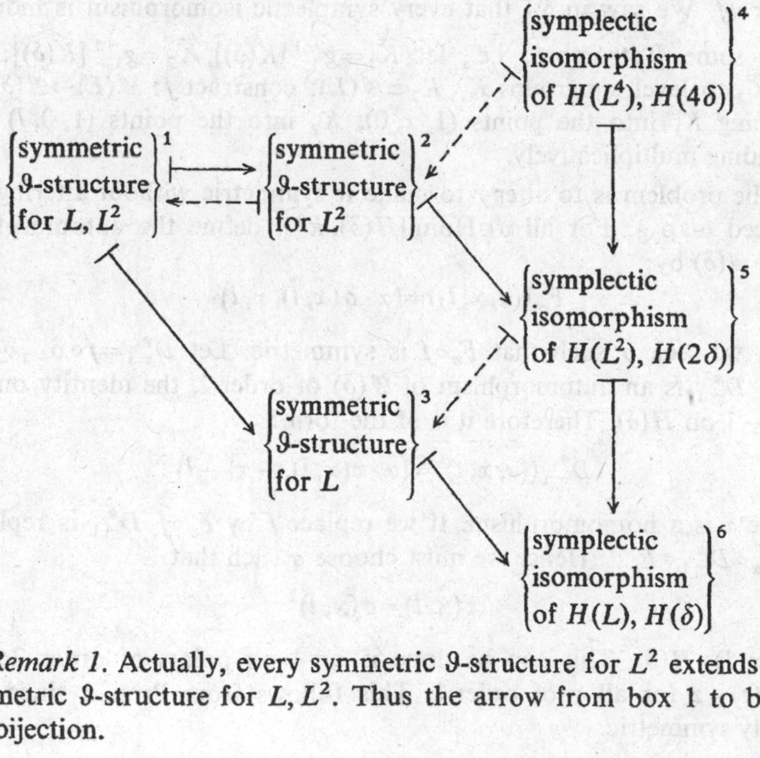

Remark 1. Actually, every symmetric ϑ-structure for L^2 extends to a symmetric ϑ-structure for L, L^2. Thus the arrow from box 1 to box 2 is a bijection.

Proof. Let f_2: $\mathscr{G}(L^2)\to\mathscr{G}(2\delta)$ be the given structure for L^2. Via η_2, $\mathscr{G}(L)$ is identified with the quotient of $\mathscr{G}(L^2)$ by its points of order 2 [Corollary 2 to Prop. 6]. Since all the d_i are even, $\mathscr{G}(\delta)$ is identified with the quotient of $\mathscr{G}(2\delta)$ by its points of order 2 via H_2. Therefore, f_2 induces a unique isomorphism f_1 of these quotients, i.e.,

$$f_1\colon\ \mathscr{G}(L)\xrightarrow{\sim}\mathscr{G}(\delta)$$

such that $H_2\circ f_2=f_1\circ\eta_2$. To check that $f_2\circ\varepsilon_2=E_2\circ f_1$, it suffices to consider this for elements of the form $\eta_2(z)$. But then

$$\begin{aligned} f_2(\varepsilon_2(\eta_2(z)))&=f_2(\delta_2(z))\\ &=f_2(z^3\cdot\delta_{-1}(z))\\ &=f_2(z)^3\cdot D_{-1}[f_2(z)]\\ &=D_2\, f_2(z)\\ &=E_2(H_2(f_2(z)))\\ &=E_2(f_1(\eta_2(z))).\quad Q.E.D. \end{aligned}$$

Remark 2. Every symplectic isomorphism $g\colon H(L)\xrightarrow{\sim}H(\delta)$ is induced by a symmetric ϑ-structure on L: i.e., the arrow from box 3 to box 6 is surjective [also that from box 2 to box 5].

Proof. We saw in § 1 that every symplectic isomorphism is induced from some ϑ-structure: i.e., let $\overline{K}_1=g^{-1}[K(\delta)]$, $\overline{K}_2=g^{-1}[\widehat{K(\delta)}]$; lift $\overline{K}_1$, $\overline{K}_2$ to level subgroups $K_1, K_2\subset\mathscr{G}(L)$; construct f: $\mathscr{G}(L)\to\mathscr{G}(\delta)$ by mapping K_1 into the points $(1, x, 0)$, K_2 into the points $(1, 0, l)$ and extending multiplicatively.

The problem is to alter f to make it symmetric without altering the induced map g. For all $\sigma\in\mathrm{Hom}\,[H(\delta), k^*]$, define the automorphism F_σ of $\mathscr{G}(\delta)$ by:

$$F_\sigma((\alpha, x, l))=(\alpha\cdot\sigma(x, l), x, l).$$

Then we seek σ such that $F_\sigma\circ f$ is symmetric. Let $D^*_{-1}=f\circ\delta_{-1}\circ f^{-1}$. Then D^*_{-1} is an automorphism of $\mathscr{G}(\delta)$ of order 2, the identity on k^*, and -1 on $H(\delta)$. Therefore it is of the form:

$$D^*_{-1}((\alpha, x, l))=(\alpha\cdot\tau(x, l), -x, -l)$$

where τ is a homomorphism. If we replace f by $F_\sigma\circ f$, D^*_{-1} is replaced by $F_\sigma\circ D^*_{-1}\circ F_\sigma^{-1}$. Hence we must choose σ such that

$$\tau(x, l)=\sigma(x, l)^2$$

all $(x, l)\in H(\delta)$. This can be done if $\tau\equiv 1$ on points of order 2; i.e., $D^*_{-1}(z)=z$ for all z of order 2. This follows from Prop. 3 since L is totally symmetric.

Remark 3. Given a symplectic isomorphism $g_2\colon H(L^2)\xrightarrow{\sim} H(2\delta)$, there is a unique symmetric ϑ-structure $f_1\colon \mathscr{G}(L)\xrightarrow{\sim}\mathscr{G}(\delta)$ such that both come from some common symmetric ϑ-structure on L^2, i.e., there is a unique dotted arrow from box 5 to box 3 giving us a commutative diagram [similarly there is a natural arrow from box 4 to box 2].

Proof. Lift g_2 to a symmetric $f_2\colon \mathscr{G}(L^2)\xrightarrow{\sim}\mathscr{G}(2\delta)$. f_2 then induces an f_1 as in Remark 1. If f_2' also lifts g_2, then $f_2'=F_\sigma\circ f_2$, where $\sigma\in\mathrm{Hom}[K(2\delta), k^*]$. Since f_2 and f_2' are symmetric, $F_\sigma\circ D_{-1}=D_{-1}\circ F_\sigma$. This implies that $\sigma(z)=\pm 1$, for all $z\in K(2\delta)$. Therefore $H_2(f_2(z))=H_2(f_2'(z))$ for all $z\in\mathscr{G}(2\delta)$. Therefore f_2 and f_2' induce the same f_1. *Q.E.D.*

We need one computation here:

Given

$$g_2\colon\ H(L^2)\to H(2\delta)\ \text{symplectic}$$

$$\Delta_1\colon\ K(2\delta)\to \text{pts of order 2 in } \widehat{K(2\delta)}$$

$$\Delta_2\colon\ \widehat{K(2\delta)}\to \text{pts of order 2 in } K(2\delta).$$

Assume

$$\Delta_1 x(y)=\Delta_1 y(x),\quad \text{all}\quad x,y\in K(2\delta),$$

$$l(\Delta_2 m)=m(\Delta_2 l),\quad \text{all}\quad l,m\in\widehat{K(2\delta)}.$$

Then:

$$g_2'=\left[1+\begin{pmatrix}0 & \Delta_1\\ \Delta_2 & 0\end{pmatrix}\right]\circ g_2$$

is also a symplectic isomorphism.

Under the correspondence of Remark 3, let $g_2\mapsto f_1, g_2'\mapsto f_1'$ where f_1, f_1' are symmetric ϑ-structures on L. Then $f_1'=F_\sigma\circ f_1$, where $\sigma\in\mathrm{Hom}(K(\delta), k^*)$ is given by:

$$\sigma((2x,\bar{l}))=\Delta_1 x(x)\cdot l(\Delta_2 l),\quad \text{all}\quad (x,l)\in K(2\delta).$$

This is a tedious but straightforward verification. With it we can prove:

Remark 4. Every symmetric ϑ-structure for L is induced by a symplectic isomorphism of $H(L^2)$, $H(2\delta)$; thus *all* arrows in our diagram are surjective.

Proof. Let $f_1\colon\mathscr{G}(L)\xrightarrow{\sim}\mathscr{G}(\delta)$ be the symmetric ϑ-structure. f_1 induces a symplectic $g_1\colon H(L)\xrightarrow{\sim} H(\delta)$, which lifts to a symplectic $g_2\colon H(L^2)\xrightarrow{\sim} H(2\delta)$. As in remark 3, g_2 induces a symmetric $f_1'\colon\mathscr{G}(L)\to\mathscr{G}(\delta)$. Then $f_1'=F_\sigma\circ f_1$ for some $\sigma\in\mathrm{Hom}(K(\delta), k^*)$. Since f_1 and f_1' are both symmetric, F_σ and D_{-1} commute, hence $\mathrm{Im}(\sigma)\subset\{\pm 1\}$. But then using the computation we have just made, it follows that we can "correct" g_2, not altering its values on $H(L)$, so that f_1' is changed to f_1. *Q.E.D.*

In conclusion, a symmetric ϑ-structure on X can be considered as something intermediate between a labeling of all the points in $H(L^2)$, or of only the points in $H(L)$. Moreover, we have proven:

Proposition 7. *Every symmetric ϑ-structure f_1 on L can be extended to a symmetric ϑ-structure (f_1, f_2) on (L, L^2). In particular, for any X, L as above, the pair of groups $\mathcal{G}(L)$, $\mathcal{G}(L^2)$ and the pairs of maps ε_2, η_2 is isomorphic to $\mathcal{G}(\delta)$, $\mathcal{G}(2\delta)$, E_2, H_2.*

§3. The Addition Formula

The most important application of the last theorem of §1 is to give an explicit description of the group law on an abelian variety X in terms of its canonical projective embedding. To express the group law algebraically, the first idea one might have would be to make the homogeneous coordinate ring:

$$R = \bigoplus_{n=0}^{\infty} \Gamma(X, L^n)$$

into a Hopf algebra, (here L is a very ample invertible sheaf on X). Thus if $\mu: X \times X \to X$ is the group law, R would be a Hopf algebra if $\mu^* L \cong p_1^* L \otimes p_2^* L$. However $\mu^*(L)$ is not isomorphic on $X \times X$ to $p_1^* L \otimes p_2^* L$ or to any other sheaf on $X \times X$ which is directly built up out of L by means of the projections of $X \times X$ onto X. So while μ^* induces a map from $\Gamma(X, L)$ to $\Gamma(X \times X, \mu^* L)$, $\Gamma(X \times X, \mu^* L)$ cannot be directly related back with $\Gamma(X, L)$. What we can do, however, is to describe the *isogeny:*

$$\xi: \quad X \times X \to X \times X$$

such that $\xi(x, y) = (x+y, x-y)$, for all closed points x, y in X. In fact, concerning ξ we have:

Proposition 1. *Let $\iota: X \to X$ be the inverse morphism on X. Then for all invertible sheaves L on X such that $\iota^* L \cong L$, if M is the invertible sheaf $p_1^* L \otimes p_2^* L$ on $X \times X$, then:*

$$\xi^*(M) \cong M^2.$$

Proof. By the see-saw principle, it suffices to check that $\xi^*(M)$ and M^2 are isomorphic when restricted to the sub-schemes $X \times \{a\}$ and $\{e\} \times X$ of $X \times X$, for all closed points $a \in X$ (here e is the identity point on X). Let

$$s_a^{(i)}: \quad X \to X \times X, \qquad i = 1, 2, \quad a \in X_k$$

be the morphisms such that $s_a^{(1)}(x)=(x,a)$, $s_a^{(2)}(x)=(a,x)$ for all closed points $x\in X$. Then:

$$\begin{aligned}(s_a^{(i)})^*(M^2)&=(s_a^{(i)})^*(p_1^*L^2\otimes p_2^*L^2)\\&=(p_1\circ s_a^{(i)})^*L^2\otimes(p_2\circ s_a^{(i)})^*L^2.\end{aligned}$$

Since $p_i\circ s_a^{(i)}=$identity on X, and if $i\neq j$, then $p_j\circ s_a^{(i)}$ maps X to the point a, we find:

$$(s_a^{(i)})^*(M^2)\cong L^2.$$

On the other hand:

$$\begin{aligned}(s_a^{(1)})^*\xi^*(M)&=(s_a^{(1)})^*\xi^*(p_1^*L\otimes p_2^*L)\\&=[p_1\circ\xi\circ s_a^{(1)}]^*L\otimes[p_2\circ\xi\circ s_a^{(2)}]^*L\\&=T_a^*L\otimes T_{-a}^*L\\&\cong L^2.\end{aligned}$$

(The last isomorphism comes from the theorem of the square.) Finally:

$$\begin{aligned}(s_e^{(2)})^*\xi^*(M)&=(s_e^{(2)})^*\xi^*(p_1^*L\otimes p_2^*L)\\&=[p_1\circ\xi\circ s_e^{(2)}]^*L\otimes[p_2\circ\xi\circ s_e^{(2)}]^*L\\&=L\otimes\iota^*L\\&\cong L^2.\end{aligned}$$

Therefore ξ^*M and M^2 have isomorphic restrictions on all the required subschemes, hence are isomorphic. *Q.E.D.*

Now, by the Künneth formula:

$$\Gamma(X\times X,M^r)\cong\Gamma(X,L^r)\otimes\Gamma(X,L^r).$$

Therefore, ξ induces a map φ:

$$\begin{array}{ccccc}\Gamma(X,L)\otimes\Gamma(X,L) & \underset{\text{Künneth}}{\cong} & \Gamma(X\times X,M) & & \\ & & \downarrow{\scriptstyle\xi^*} & & \\ & & \Gamma(X\times X,\xi^*M) & & \\ & & \wr\| & & \\ & & \Gamma(X\times X,M^2) & \underset{\text{Künneth}}{\cong} & \Gamma(X,L^2)\otimes\Gamma(X,L^2).\end{array}$$

Moreover, if L is very ample on X, then M is very ample on $X\times X$ and this map φ is sufficient to determine the morphism ξ. We want to apply

22*

Theorem 4 of § 1 to determine φ, and hence, in terms of canonical bases of $\Gamma(X,L)$ and $\Gamma(X,L^2)$, to give a canonical matrix representing φ (independent of the moduli of X!). Unfortunately, we have to make a fundamental restriction at this point:

Proposition. *If* $\dim X=g$, $\deg(\xi)=2^{2g}$; *ξ is separable if and only if* $\operatorname{char}(k)\neq 2$.

Proof. It is convenient to use the language of schemes to prove this: let $z=(x_1,x_2)$ be an R-valued point of $X\times X$, for some k-algebra R. Then z is in the kernel of ξ if and only if $x_1+x_2=x_1-x_2=e$: i.e., $x_1=x_2$ and $2x_1=e$. Therefore the map taking z to x_1 defines an isomorphism of the kernel of ξ with the kernel of

$$2\delta:\quad X\to X$$

(multiplication by 2). But in general the degree of $k\delta$ (multiplication by k) is k^{2g}, hence $\deg(\xi)=2^{2g}$. If $\operatorname{char}(k)\neq 2$, then ξ is clearly separable. If $\operatorname{char}(k)=2$, then it is well known that 2δ is not separable, hence its kernel is not reduced, hence the kernel of ξ is not reduced, hence ξ is not separable. *Q.E.D.*

From now on, we assume $\operatorname{char}(k)\neq 2$. Then if L is ample of separable type, so are L^2, M, and M^2 and the theory of § 1 is applicable.

Assume moreover that L is actually totally symmetric: hence so are L^2, M, and M^2. Now choose a symmetric ϑ-structure on (L,L^2) i.e., isomorphism f_1: $\mathscr{G}(L)\xrightarrow{\sim}\mathscr{G}(\delta)$ and f_2: $\mathscr{G}(L^2)\xrightarrow{\sim}\mathscr{G}(2\delta)$ as in Proposition 7 of § 2. Recall that we showed in § 1 that these isomorphisms induce isomorphisms:

$$\beta_1:\quad \Gamma(X,L)\xrightarrow{\sim}V(\delta)=\left\{\begin{matrix}\text{vector space of }k\text{-valued}\\ \text{functions on }K(\delta)\end{matrix}\right\}$$

$$\beta_2:\quad \Gamma(X,L^2)\xrightarrow{\sim}V(2\delta)=\left\{\begin{matrix}\text{vector space of }k\text{-valued}\\ \text{functions on }K(2\delta)\end{matrix}\right\}$$

unique up to scalar multiples. Choose some pair of β's. How about $\Gamma(X,M)$ and $\Gamma(X,M^2)$? β_1 and β_2 immediately induce:

$$\begin{array}{c}\Gamma(X\times X,M)\cong\Gamma(X,L)\otimes\Gamma(X,L)\\ \Big\downarrow \beta_1\otimes\beta_1 \quad (\text{dashed: } \beta_1^{(2)})\\ \left\{\begin{matrix}\text{fcns. on}\\ K(\delta)\end{matrix}\right\}\otimes\left\{\begin{matrix}\text{fcns. on}\\ K(\delta)\end{matrix}\right\}\cong\left\{\begin{matrix}\text{fcns. on}\\ K(\delta)\times K(\delta)\end{matrix}\right\}\end{array}$$

and similarly

$$\Gamma(X\times X, M^2) \underset{\beta_2^{(2)}}{\cong} \left\{\begin{matrix}\text{functions on}\\ K(2\delta)\times K(2\delta)\end{matrix}\right\}.$$

Actually, these maps are of the same type as β: i.e., they put the usual group action in standard form. Note:

Lemma 1. *Let X and Y be abelian varieties, and let L, M be ample invertible sheaves of separable type on X and Y. Then*

$$\mathscr{G}(p_1^* L\otimes p_2^* M)\cong \mathscr{G}(L)\times\mathscr{G}(M)/\{(\alpha,\alpha^{-1})\,|\,\alpha\in k^*\}.$$

Proof. Just note how this isomorphism is set up. Given $x\in H(L)$, $y\in H(M)$ and

$$\varphi\colon\ L \xrightarrow{\sim} T_x^* L,\qquad \psi\colon\ M\xrightarrow{\sim} T_y^* M,$$

we obtain

$$p_1^*\varphi\otimes p_2^*\psi\colon\ p_1^* L\otimes p_2^* M\xrightarrow{\sim} T_{(x,y)}^*[p_1^* L\otimes p_2^* M].$$

The lemma is now readily checked. *Q.E.D.*

Now returning to our original set-up, it follows from the lemma that the isomorphism

$$f_1\colon\ \mathscr{G}(L)\xrightarrow{\sim}\mathscr{G}(\delta)=\{(\alpha,x,l)\,|\,\alpha\in k^*, x\in K(\delta), l\in \widehat{K(\delta)}\}$$

induces an isomorphism:

$$f_1^{(2)}\colon\ \mathscr{G}(M)\xrightarrow{\sim}\mathscr{G}(\delta)^{(2)}=\{(\alpha,x_1,x_2,l_1,l_2)\,|\,\alpha\in k^*, x_i\in K(\delta), l_i\in\widehat{K(\delta)}\}.$$

Here the multiplication in $\mathscr{G}(\delta)^{(2)}$ is given by:

$$\begin{aligned}&(\alpha,x_1,x_2,l_1,l_2)\cdot(\alpha',x_1',x_2',l_1',l_2')\\ &\qquad=(\alpha\cdot\alpha'\cdot l_1'(x_1)\cdot l_2'(x_2), x_1+x_1', x_2+x_2', l_1+l_1', l_2+l_2').\end{aligned}$$

Now let $\mathscr{G}(\delta)^{(2)}$ act on the vector space of k-valued functions on $K(\delta)\times K(\delta)$ – call this vector space $V(\delta)^{(2)}$ – as follows:

$$z=(\alpha,x_1,x_2,l_1,l_2)\in\mathscr{G}(\delta)^{(2)},\qquad f\in V(\delta)^{(2)},$$
$$(U_z f)(u_1,u_2)=\alpha\, l_1(u_1)\, l_2(u_2)\, f(u_1+x_1,u_2+x_2).$$

Then $\beta_1^{(2)}$ is determined, up to scalar multiples, by the readily verified property:

$$\beta_1^{(2)}(U_z s)=U_{f_1^{(2)}(z)}(\beta_1^{(2)} s),$$

for all $z\in\mathscr{G}(M)$, $s\in\Gamma(X\times X, M)$.

Now ξ^* defines the map $\beta_2^{(2)} \circ \varphi \circ \beta_1^{(2)-1}$:

$$\begin{array}{ccc} V(\delta)^{(2)} = \left\{\begin{matrix}\text{fcns. on} \\ K(\delta)\times K(\delta)\end{matrix}\right\} & \xrightarrow[\beta_1^{(2)-1}]{\sim} & \Gamma(X\times X, M) \\ & & \downarrow \xi^* \\ & & \Gamma(X\times X, \xi^* M) \\ & & \wr\| \\ & & \Gamma(X\times X, M^2) \xrightarrow[\beta_2^{(2)}]{\sim} \left\{\begin{matrix}\text{fcns. on} \\ K(2\delta)\times K(2\delta)\end{matrix}\right\} \\ & & = V(2\delta)^{(2)}. \end{array}$$

Call this transformation Ω.

Fundamental Addition Formula. There is a scalar λ, such that for all $f \in V(\delta)^{(2)}$

$$(\Omega f)(x,y) = \begin{cases} 0 & \text{if } x+y \notin K(\delta) \\ \lambda \cdot f(x+y, x-y) & \text{if } x+y \in K(\delta) \end{cases}$$

for all $x, y \in K(2\delta)$.

Remark. By suitably normalizing our maps β, we can always assume $\lambda = 1$. In what follows, I will always assume that this has been done.

Proof. This formula is a special case of the general formula given in Theorem 4, § 1. To see this, let

$$\mathscr{G}(M^2)^* = \left\{\begin{matrix}\text{subgroup of } \mathscr{G}(M^2) \text{ of elements lying} \\ \text{over points of } X\times X \text{ in } \xi^{-1}[H(M)]\end{matrix}\right.$$

and let $\tilde{K} \subset \mathscr{G}(M^2)$ be the descent data associated to ξ. Then we have the canonical map (cf. § 1):

$$\tau: \frac{\mathscr{G}(M^2)^*}{\tilde{K}} \xrightarrow{\sim} \mathscr{G}(M).$$

First of all, $\mathscr{G}(M^2)^*$ goes over via our ϑ-structure to the subgroup of $\mathscr{G}(2\delta)^{(2)}$:

$$\mathscr{G}(2\delta)^* = \left\{(\alpha, x_1, x_2, l_1, l_2) \,\middle|\, \begin{matrix} x_i \in K(2\delta), x_1 + x_2 \in K(\delta) \\ l_i \in \widehat{K(2\delta)}, l_1 + l_2 = 2 * l \end{matrix}\right\}$$

$$\text{for some } l \in \widehat{K(\delta)}$$

In fact, we are given an isomorphism:

$$\begin{array}{c} X\times X \\ \cup \\ H(M^2) \cong K(2\delta)\times K(2\delta)\times \widehat{K(2\delta)}\times \widehat{K(2\delta)}. \end{array}$$

and ξ maps the point (x_1, x_2, l_1, l_2) in the latter group to $(x_1+x_2, x_1-x_2, l_1+l_2, l_1-l_2)$. Then $(\alpha, x_1, x_2, l_1, l_2)$ is in the group corresponding to $\mathscr{G}(M^2)^*$ if and only if $(x_1+x_2, x_1-x_2, l_1+l_2, l_1-l_2)$ corresponds to a point of $H(M)$. But this is the same as asking that $x_1+x_2 \in 2K(2\delta)$ [then x_1-x_2 is automatically also in $2K(2\delta)$] and that $l_1+l_2 \in 2\widehat{K(2\delta)}$. But $2K(2\delta)=K(\delta)$ under our identifications, and $2\widehat{K(2\delta)}$ is the group of homomorphisms $2*l$, $l \in \widehat{K(\delta)}$.

Now let T be the homomorphism defined by the diagram:

$$\begin{array}{ccc} \mathscr{G}(M^2)^* & \xrightarrow[f_2^{(2)}]{\sim} & \mathscr{G}(2\delta)^* \\ \downarrow & & \vdots \\ \mathscr{G}(M^2)^*/\tilde{K} & & \vdots\, T \\ \downarrow \tau & & \downarrow \\ \mathscr{G}(M) & \xrightarrow[f_1^{(2)}]{\sim} & \mathscr{G}(\delta)^{(2)}. \end{array}$$

The key fact is contained in the following lemma, (which uses the crucial symmetry assumption):

Lemma 2.

$$T((\alpha, x_1, x_2, l_1, l_2)) = (\alpha, x_1+x_2, x_1-x_2, l, k)$$

$$l_1+l_2=2*l,\ l_1-l_2=2*k \quad \text{for} \quad l, k \in \widehat{K(\delta)}.$$

When this is proven, our formula comes directly from Theorem 4, § 1. In fact, $\tilde{K}$ must go over via our ϑ-structure to ker (T), and this, by the lemma, equals

$$\{(1, x, x, l, l) \mid 2x=0, 2l=0\}.$$

Thus, in the notation of Theorem 4,

$$K_1 = \{(x, x) \in K(2\delta) \times K(2\delta) \mid 2x=0\},$$
$$K_1^{\perp} = \{(x, y) \in K(2\delta) \times K(2\delta) \mid x+y \in K(\delta)\}$$

$$\sigma\colon\ K_1^{\perp}/K_1 \xrightarrow{\sim} K(\delta) \times K(\delta)$$

is given by

$$\sigma((x, y)) = (x+y, x-y).$$

Then lemma 2 proves that the diagram in Theorem 4 commutes, and our formula can be read off.

Proof of lemma. T is obviously the identity on k^*, so in order to check the lemma, it will suffice to verify it for elements of the form

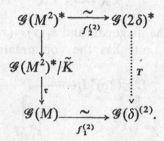

$(1, x, x, l, l)$ and of the form $(1, x, -x, l, -l)$. Namely, if $x_1, x_2 \in K(2\delta)$ and $x_1 + x_2 \in K(\delta) = 2K(2\delta)$, then there are elements $y, z \in K(2\delta)$ such that $x_1 = y + z$, $x_2 = y - z$; similarly with l_1, l_2. Therefore $\mathcal{G}(2\delta)^*$ is generated by k^* and by elements of the above form.

For elements of the form $(1, x, x, l, l)$, consider the diagram:

$$\begin{array}{ccc} X & \xrightarrow{2\delta} & X \\ {\scriptstyle\Delta}\downarrow & & \downarrow{\scriptstyle s_1} \\ X \times X & \xrightarrow{\xi} & X \times X \end{array}$$

where $\Delta(x) = (x, x)$ is the diagonal, and $s_1(x) = (x, 0)$ for closed points $x \in X$. Correspondingly, one has the commutative diagram of sheaf morphisms:

$$\begin{array}{ccc} L^4 \cong (2\delta)^* L & \xleftarrow{(2\delta)^*} & L \\ \wr\| & & \wr\| \\ \Delta^* M^2 & & s_1^* M \\ \uparrow{\scriptstyle \Delta^*} & & \uparrow{\scriptstyle s_1^*} \\ M^2 \cong \xi^* M & \xleftarrow{\xi^*} & M. \end{array}$$

On the level of H's, this gives maps between subgroups as written out here:

$$\begin{array}{ccccc} H(L^4) & & & H(L) & \\ \cup & & & \| & \\ & H(L^2) & \xrightarrow{2\delta} & H(L) & \\ & \downarrow{\scriptstyle\Delta} & & \downarrow{\scriptstyle s_1} & \\ H(M^2) & & H(M) & & \\ \cup & & \cup & & \\ & \{(x, x) \mid x \in H(L^2)\} & \xrightarrow{\xi} & H(L) \times \{0\} & \end{array}$$

Pulling these subgroups back to the $\mathcal{G}$'s, and denoting by π the projection of each $\mathcal{G}$ onto its H, we get the diagram:

$$\begin{array}{ccccc} \mathcal{G}(L^4) & & & \mathcal{G}(L) & \\ \cup & & & \| & \\ & \pi^{-1}(H(L^2)) & \xrightarrow{\widetilde{2\delta}} & \mathcal{G}(L) & \\ & \downarrow{\scriptstyle\tilde{\Delta}} & & \downarrow{\scriptstyle \tilde{s}_1} & \\ \mathcal{G}(M^2) & & \mathcal{G}(M) & & \\ \cup & & \cup & & \\ & \pi^{-1}(\{x, x \mid x \in H(L^2)\}) & \xrightarrow{\tilde{\xi}} & \pi^{-1}(H(L) \times \{0\}) & \end{array}$$

Here the tildas indicate homomorphisms naturally induced by the corresponding homomorphism of abelian varieties. $\tilde{\xi}$ is nothing but the restriction of τ to the subgroup written out above — which corresponds in $\mathcal{G}(2\delta)^*$ to the group of elements (α, x, x, l, l): Therefore this diagram is just suited to determine τ, or T, on this particular subgroup.

Now start with an element $z=(x, \varphi)\in\mathcal{G}(L^2)$. Then

$$p_1^*\varphi \otimes p_2^*\varphi:\quad M^2 \xrightarrow{\sim} T^*_{(x,x)} M^2$$

is an element of $\mathcal{G}(M^2)$ in the subgroup $\pi^{-1}(\{x, x \mid x\in H(L^2)\})$. But this isomorphism restricts on the diagonal to the isomorphism

$$\varphi^2:\quad M^4 \xrightarrow{\sim} T^*_x M^4.$$

And therefore

$$((x, x), p_1^*\varphi \otimes p_2^*\varphi)=\tilde{\Delta}[(x, \varphi^2)].$$

Moreover, if

$$\psi:\quad L \xrightarrow{\sim} T^*_{2x} L$$

is the isomorphism such that $(2\delta)^*\psi=\varphi^2$, then

$$(2x, \psi)=\widetilde{2\delta}[(x, \varphi^2)].$$

But then, referring back to the definition of η_2, this means that

$$(2x, \psi)=\eta_2[(x, \varphi)].$$

Putting all this together we get a commutative diagram:

$$\begin{array}{ccc} \mathcal{G}(L^2) & \xrightarrow{\eta_2} & \mathcal{G}(L) \\ {\scriptstyle\alpha}\downarrow & & \downarrow\uparrow{\scriptstyle\tilde{s}_1} \\ \mathcal{G}(M^2)^* & \xrightarrow{\tau} & \mathcal{G}(M) \end{array}$$

where $\alpha[(x, \varphi)]=((x, x), p_1^*\varphi\otimes p_2^*\varphi)$. Now go over to the standard groups $\mathcal{G}(\delta)$, $\mathcal{G}(2\delta)$, etc. by means of our symmetric system of isomorphisms. We get a diagram:

$$\begin{array}{ccc} \mathcal{G}(2\delta) & \xrightarrow{H_2} & \mathcal{G}(\delta) \\ {\scriptstyle A}\downarrow & & \downarrow\uparrow{\scriptstyle S} \\ \mathcal{G}(2\delta)^* & \xrightarrow{T} & \mathcal{G}(\delta)^{(2)} \end{array}$$

One checks immediately that

$$A[(\alpha, x, l)]=(\alpha^2, x, x, l, l)$$

and that

$$S[(\alpha, x, l)]=(\alpha, x, 0, l, 0).$$

Therefore, we calculate:

$$T[(1,x,x,l,l)]=S[H_2[(1,x,l)]]$$
$$=(1,2x,0,\bar{l},0)$$

and since $2*\bar{l}=2l$, this is the asserted formula.

The proof of the lemma for elements of the form $(1,x,-x,l,-l)$ is very similar, only based on the diagram:

$$\begin{array}{ccc} X & \xrightarrow{2\delta} & X \\ {\scriptstyle \Delta'}\downarrow & & \downarrow{\scriptstyle s_2} \\ X\times X & \xrightarrow{\xi} & X\times X \end{array}$$

where $\Delta'(x)=(x,-x)$, and $s_2(x)=(0,x)$ for closed points $x\in X$. This part of the proof is omitted. *Q.E.D.*

This marvelously simple formula is the basis of the whole theory which follows. The most striking thing about it is that, as promised, it does not involve the moduli of X itself — i.e., it shows, in some sense, that the *same* addition formula is valid for all abelian varieties. This is something of a cheat, however, as we have only given the map φ induced by ξ from $\Gamma(X,L)\otimes\Gamma(X,L)\to\Gamma(X,L^2)\otimes\Gamma(X,L^2)$. To have the whole story, we need also the canonical map

$$\Gamma(X,L)\otimes\Gamma(X,L)\to\Gamma(X,L^2)$$

given by tensoring a pair of sections. For L sufficiently ample, this will be surjective and hence the addition formula can be written entirely in terms of the *one* vector space $\Gamma(X,L)$, i.e., in terms of the homogeneous coordinates in one canonical projective embedding. However, the remarkable fact is that this second map is a special case of the first, so that we can pull ourselves up by our bootstraps. In fact, consider the diagram:

$$\begin{array}{ccc} X & & \\ {\scriptstyle s_1}\downarrow & \overset{\Delta}{\searrow} & \\ X\times X & \xrightarrow{\xi} & X\times X \end{array}$$

where $\Delta(x)=(x,x)$, $s_1(x)=(x,0)$ for closed points $x\in X$. Then $s_1^*M^2\cong L^2$ canonically, and if t, t' are two sections of L,

$$s_1^*[\varphi(p_1^*t\otimes p_2^*t')]=t\otimes t'\in\Gamma(X,L^2).$$

Passing over to the representation of these sections by functions on K's, this gives the diagram:

$$\begin{array}{ccc} V(\delta)\times V(\delta) & & \\ \downarrow m & & \\ V(\delta)^{(2)} & \xrightarrow{\Omega} & V(2\delta)^{(2)} \\ & & \downarrow S_1 \\ & & V(2\delta) \end{array}$$

where S_1 is obtained by carrying over s_1^*, and m takes the pair of functions f, f' into the function g of two variables

$$g(x,y)=f(x)\cdot f'(y).$$

The composition is the "multiplication" of sections of L:

Definition. If $t, t'\in\Gamma(X,L)$ and $f=\beta_1(t), f'=\beta_1(t')$, then let

$$f * f'=\beta_2(t\otimes t').$$

What is S_1? It can be given explicitly if we introduce the null-value function q_{L^2} as in § 1. Choose an isomorphism $\lambda_0^{(2)}$: $L(0)^2 \xrightarrow{\sim} k$ – actually we already did this implicitly when we identified $s_1^* M^2$ with L^2. Then there is a natural "evaluation" of sections at 0:

$$t\mapsto \lambda_0^{(2)}[t(0)], \quad t\in\Gamma(X,L^2).$$

In terms of the isomorphism β_2, this gives a unique function q_{L^2} on $K(2\delta)$ such that:

$$\lambda_0^{(2)}[t(0)]=\sum_{z\in K(2\delta)} \beta_2 t(z)\cdot q_{L^2}(z).$$

Now then, say $p_1^* t\otimes p_2^* t'$ is a given section of M^2. Then

$$s_1^*[p_1^* t\otimes p_2^* t']=\lambda_0^{(2)}[t'(0)]\cdot t$$

(if the identification of $s_1^* M^2$ with L^2 is chosen properly – otherwise a constant must be put in). In terms of the V's, this means

$$[S_1(g)](x)=\Big[\sum_{y\in K(2\delta)} f'(y)\cdot q_{L^2}(y)\Big]\cdot f(x)$$

if

$$g(x,y)=f(x)\cdot f'(y)\in V(2\delta)^{(2)}.$$

Since S_1 is linear, this means that for *any* $g\in V(2\delta)^{(2)}$,

$$[S_1(g)](x)=\sum_{y\in K(2\delta)} g(x,y)\cdot q_{L^2}(y).$$

Putting this and the addition formula together, we conclude:

Multiplication Formula. If $f, f' \in V(\delta)$, then

$$(f * f')(x) = \sum_{y \in x + K(\delta)} f(x+y) \cdot f'(x-y) \cdot q_{L^2}(y) \quad \text{for all} \quad x \in K(2\delta).$$

Another application of the fundamental addition formula is to the duplication and inverse formulas. To obtain the first, we choose symmetric ϑ-structures now on L, L^2 *and* L^4 such that those for the pair (L, L^2) and for the pair (L^2, L^4) are compatible. Then we obtain *3* isomorphisms:

$$\beta_1: \Gamma(X, L) \xrightarrow{\sim} V(\delta)$$

$$\beta_2: \Gamma(X, L^2) \xrightarrow{\sim} V(2\delta)$$

$$\beta_4: \Gamma(X, L^4) \xrightarrow{\sim} V(4\delta).$$

The endomorphism $2\delta: X \to X$ then gives the diagram:

$$\begin{array}{ccc} \Gamma(X, L) & \xrightarrow{(2\delta)^*} & \Gamma(X, (2\delta)^* L) \cong \Gamma(X, L^4) \\ \downarrow \wr & & \downarrow \wr \\ V(\delta) & \dashrightarrow & V(4\delta). \end{array}$$

Let the dotted arrow be denoted [2]: this is the duplication homomorphism. Then the result is:

Duplication Formula. If $f \in V(\delta)$, then

$$\begin{aligned} ([2] f)(x) &= 0 \quad \text{if} \quad x \notin K(2\delta) \\ &= f(2x) \quad \text{if} \quad x \in K(2\delta) \end{aligned}$$

$$\text{for all} \quad x \in K(4\delta).$$

Proof. We use the fact that $\xi \circ \xi = 2\delta \times 2\delta$. This implies that if $f_1, f_2 \in V(\delta)$, and if $f_1 \otimes f_2$ is the induced function in $V(\delta)^{(2)}$, then

$$\Omega(\Omega f_1 \otimes f_2) = [2] f_1 \otimes [2] f_2.$$

In other words,

$$(\Omega(\Omega f_1 \otimes f_2))(x, y) = [2] f_1(x) \cdot [2] f_2(y)$$

for all $x, y \in K(4\delta)$. (Here the first Ω is for the map $\xi^*: M^2 \to M^4$, and the second Ω is for the map $\xi^*: M \to M^2$.) Substituting the addition formula, the duplication formula follows immediately (at least up to ± 1: but this disappears if we correctly normalize the β's). *Q.E.D.*

The second formula requires only a symmetric ϑ-structure on L, hence

$$\beta_1: \Gamma(X, L) \xrightarrow{\sim} V(\delta).$$

Using ι, we get a diagram

$$\begin{array}{ccccc} \Gamma(X,L) & \xrightarrow{\iota^*} & \Gamma(X,\iota^* L) & \overset{\text{via normalized isomorphism}}{\cong} & \Gamma(X,L) \\ \downarrow \wr & & & & \downarrow \\ V(\delta) & & \dashrightarrow & & V(\delta). \end{array}$$

Let the dotted arrow be denoted $[-1]$. The result is:

Inverse Formula. If $f \in V(\delta)$, then

$$([-1]f)(x) = f(-x), \quad \text{for all} \quad x \in K(\delta).$$

Proof. Up to a constant, the result is quite easy: we use the fact that $\iota = p_2 \circ \xi \circ s_2$, while $1_X = p_1 \circ \xi \circ s_2$. Therefore, if $s, t \in \Gamma(X, L)$

$$s_2^* \xi^* [p_1^* s \otimes p_2^* t] \cong s \otimes \iota^* t.$$

In terms of functions, if $f, f' \in V(\delta)$, and $g(x, y) = f(x) \cdot f'(y)$ then

$$(f * [-1] f')(x) = \sum_{y \in K(2\delta)} (\Omega g)(y, x) \cdot q_{L^2}(y).$$

This gives the formula immediately. To get the correct constant (this is meaningful in this case because we are dealing with an *automorphism* of $\Gamma(X, L)$), note first that $[-1] \circ [-1] = \text{identity}$. Therefore the true formula is either $([-1]f)(x) = f(-x)$ or $-f(-x)$. Since we have used a *normalized* isomorphism of $\iota^* L$ and L to define $[-1]$, it follows that

$$\sum_{x \in K(\delta)} ([-1] f)(x) \cdot q_L(x) = \sum_{x \in K(\delta)} f(x) \cdot q_L(x).$$

In other words, $q_L(x)$ is an even function in the first case, and an odd function in the second case. But look at formula (A) below: it is invariant when you interchange u, v and replace x, y by $x, -y$. Therefore q_L is even and our formula is correct. *Q.E,D.*

To conclude this section, we want to show what restrictions these formulas place on the null-functions q, and particularly how RIEMANN's theta-relation can be deduced. One word of caution: this relation will not be exactly the same as in the classical case, because our q's are not the same as the usual theta-null values. But it can be shown that they are related by a non-singular linear transformation with roots of unity as coefficients: so the formulas are trivial modifications of each other. As before, we start by choosing an isomorphism

$$\lambda_0 : \ L(0) \xrightarrow{\sim} k.$$

This induces

$$\lambda_0^{(2)}:\ L^2(0) \xrightarrow{\sim} k$$

and

$$\lambda_0^{(4)}:\ L^4(0) \xrightarrow{\sim} k$$

($\lambda_0^{(2)}$ may as well be the same as the map we used before). Then sections of L, L^2, L^4 have "values" at 0 and we define the q's via:

$$\lambda_0[t(0)] = \sum_{z\in K(\delta)} \beta_1\, t(z)\cdot q_L(z), \qquad t\in\Gamma(X,L)$$

$$\lambda_0^{(2)}[t(0)] = \sum_{z\in K(2\delta)} \beta_2\, t(z)\cdot q_{L^2}(z), \qquad t\in\Gamma(X,L^2)$$

$$\lambda_0^{(4)}[t(0)] = \sum_{z\in K(4\delta)} \beta_4\, t(z)\cdot q_{L^4}(z), \qquad t\in\Gamma(X,L^4).$$

Also, λ_0 induces

$$\mu_0^{(i)}:\ M^i(0) \xrightarrow{\sim} k, \qquad i=1,2,4,$$

and one checks immediately that:

$$\mu_0^{(i)}[t(0)] = \sum_{x,y\in K(\delta)} \beta_i^{(2)}\, t(x,y)\cdot q_{L^i}(x)\cdot q_{L^i}(y)$$

for all $t\in\Gamma(X\times X, M^i)$, $i=1$, 2 or 4 [i.e., check it first for t of the form $p_1^* t_1 \otimes p_2^* t_2$ and then use the linearity of both sides]. But there is a constant κ such that

$$\mu_0^{(2)}[\xi^*\, t(0)] = \kappa\cdot\mu_0^{(1)}(t(0))$$

all $t\in\Gamma(X\times X, M)$. And if ξ^* is replaced by $\kappa^{-1}\cdot\xi^*$, we may as well assume $\kappa=1$. This means that for $f\in V(\delta)^{(2)}$,

$$\sum_{x,y\in K(\delta)} f(x,y)\cdot q_L(x)\cdot q_L(y) = \sum_{x,y\in K(2\delta)} \Omega f(x,y)\cdot q_{L^2}(x)\cdot q_{L^2}(y).$$

Substituting our addition formula into this, and using the fact that this is true for *any* f, we conclude immediately:

(A)
$$q_L(x)\cdot q_L(y) = \sum_{\substack{u,v\in K(2\delta)\\ \text{such that}\\ u+v=x\\ u-v=y}} q_{L^2}(u)\cdot q_{L^2}(v).$$

Now recall that according to Cor. 4, Prop. 6, § 2,

$$X_2 \subset H(L) \subset H(L^2) \subset H(L^4)$$

where X_2 is the group of points of order 2 on X. Therefore, the 2-torsion subgroups of $K(\delta)$, $K(2\delta)$, and $K(4\delta)$ are equal: call this common subgroup Z_2. Then (A) can be re-written:

(A′) For all $u, v \in K(2\delta)$ such that $u+v \in K(\delta)$,

$$q_L(u+v)\cdot q_L(u-v) = \sum_{\eta\in Z_2} q_{L^2}(u+\eta)\cdot q_{L^2}(v+\eta).$$

We can obtain an even simpler relation between q_L and q_{L^4} either by applying (A) to both of the pairs (L, L^2) and (L^2, L^4) and eliminating q_{L^2}, or by proceeding directly using the duplication formula instead of the addition formula. Following the latter procedure, first note that there is a constant κ such that

$$\lambda_0^{(4)}[(2\delta)^*\, t(0)] = \kappa\cdot\lambda_0(t(0))$$

for all $t\in\Gamma(X, L)$. If the homomorphism $(2\delta)^*\colon L\to L^4$ is chosen suitably, the κ again disappears. Then the formula goes over on the $V(\delta)$-level to:

$$\sum_{x\in K(\delta)} f(x)\cdot q_L(x) = \sum_{x\in K(4\delta)} [2]f(x)\cdot q_{L^4}(x)$$

for all $f\in V(\delta)$. Substituting the duplication formula into this, we get out:

(B) For all $x\in K(2\delta)$

$$q_L(2x) = \sum_{\eta\in Z_2} q_{L^4}(x+\eta).$$

To get a formula involving q_L alone, first choose elements $x, y\in K(2\delta)$ such that $x+y\in K(\delta)$. Then for all homomorphisms $l\colon Z_2\to\{\pm 1\}$, we obtain:

$$\begin{aligned}\sum_{\zeta\in Z_2} l(\zeta)\, q_L(x+y+\zeta)\, q_L(x-y+\zeta) &= \sum_{\substack{\zeta\in Z_2\\ u,v\in K(2\delta)\\ u+v=x+y+\zeta\\ u-v=x-y+\zeta}} l(\zeta)\, q_{L^2}(u)\, q_{L^2}(v)\\ &= \sum_{\zeta_1,\zeta_2\in Z_2} l(\zeta_1+\zeta_2)\, q_{L^2}(x+\zeta_1)\, q_{L^2}(y+\zeta_2)\\ &= \Big[\sum_{\eta\in Z_2} l(\eta)\, q_{L^2}(x+\eta)\Big]\cdot\Big[\sum_{\eta\in Z} l(\eta)\, q_{L^2}(y+\eta)\Big].\end{aligned}$$

Therefore, if $x, y, u, v\in K(2\delta)$ and $x+y$, $x+u$, and $x+v\in K(\delta)$, (so that $y+u$, $y+v$, $u+v\in K(\delta)$ too), then

$$\begin{aligned}&\Big[\sum_{\zeta\in Z_2} l(\zeta)\, q_L(x+y+\zeta)\, q_L(x-y+\zeta)\Big]\cdot\Big[\sum_{\zeta\in Z_2} l(\zeta)\, q_L(u+v+\zeta)\, q_L(u-v+\zeta)\Big]\\ &=\Big[\sum_{\zeta\in Z_2} l(\zeta)\, q_L(x+u+\zeta)\, q_L(x-u+\zeta)\Big]\cdot\Big[\sum_{\zeta\in Z_2} l(\zeta)\, q_L(y+v+\zeta)\, q_L(y-v+\zeta)\Big].\end{aligned}$$

Writing these 2 terms together and summing over l, we finally obtain:

(C)
$$\begin{aligned}&\sum_{\zeta\in Z_2} q_L(x+y+\zeta)\, q_L(x-y+\zeta)\, q_L(u+v+\zeta)\, q_L(u-v+\zeta)\\ &= \sum_{\eta\in Z_2} q_L(x+u+\eta)\, q_L(x-u+\eta)\, q_L(y+v+\eta)\, q_L(y-v+\eta).\end{aligned}$$

To complement this relation, there is the key symmetry property of q_L:

(D) $$q_L(-x)=q_L(x).$$

This follows as in the proof of the inversion formula. To understand the structure of RIEMANN's theta formula (C) a bit better, note that the variables $x\pm y$, $u\pm v$ are just 4 arbitrary elements of $K(\delta)$, and that the condition $x+u\in K(\delta)$ amounts to

$$(x+y)+(x-y)+(u+v)+(u-v)\in 2K(\delta).$$

After observing this, it is easy to rearrange the variables in (C) a bit so as to obtain:

(C′) For all $x, y, u, v, z\in K(\delta)$ such that $x+y+u+v=-2z$,

$$\sum_{\eta\in Z_2} q_L(x+\eta)\,q_L(y+\eta)\,q_L(u+\eta)\,q_L(v+\eta) = \sum_{\eta\in Z_2} q_L(x+z+\eta)\,q_L(y+z+\eta)\,q_L(u+z+\eta)\,q_L(v+z+\eta).$$

We shall give some examples of this in § 5.

The more familiar form of RIEMANN's theta formula is obtained by making a partial Fourier transformation of the function q_L. Suppose Z is a subgroup of $K(\delta)$ such that:

$$Z_2\subset Z\subset 2K(\delta).$$

Let $H=K(\delta)\times\hat{Z}$, and define a function ϑ on H by

$$\vartheta(x)=\sum_{\eta\in Z} l(\eta)\,q_L(a+\eta)$$

if $x=(a, l)$. ϑ has an obvious periodicity with respect to $Z\times\{0\}$; hence ϑ is "essentially" a function of $(K(\delta)/Z)\times\hat{Z}$. Let

$$H_2=\tfrac{1}{2}Z\times(\hat{Z})_2 = \left\{x\in H \;\middle|\; \begin{matrix} x \text{ is 2-torsion modulo} \\ Z\times\{0\}\end{matrix}\right\}.$$

(C″) For all $x, y, u, v, z\in H$ such that $x+y+u+v=-2z$,

$$\vartheta(x)\,\vartheta(y)\,\vartheta(u)\,\vartheta(v) = \frac{1}{\#(\frac{1}{2}Z)}\sum_{\xi\in H_2} A_\xi\cdot\vartheta(x+z+\xi)\,\vartheta(y+z+\xi)\,\vartheta(u+z+\xi)\,\vartheta(v+z+\xi),$$

where if $\xi=(\zeta, m)$ and $z=(e, r)$, $A_\xi=(r+m)\,(2\zeta)$.

To obtain (C″) from (C′), write the left-hand side out, assuming that

$$x=(a,l)$$
$$y=(b,k)$$
$$u=(c,p)$$
$$v=(d,q)$$
$$z=(e,r).$$

We get

$$\sum_{\eta_1,\eta_2,\eta_3,\eta_4\in Z} l(\eta_1)\,k(\eta_2)\,p(\eta_3)\,q(\eta_4)\,q_L(a+\eta_1)\,q_L(b+\eta_2)\,q_L(c+\eta_3)\,q_L(d+\eta_4).$$

Since $l+k+p+q=-2r$, this character kills Z_2. Therefore, the above sum decomposes into a sum of terms of the type

$$\sum_{\eta\in Z_2} q_L(a'+\eta)\,q_L(b'+\eta)\,q_L(c'+\eta)\,q_L(d'+\eta)$$

all of which can be rewritten using (C′). We get

$$\frac{1}{2^g}\sum_{\substack{\eta_1,\ldots,\eta_4\in Z\\ \zeta\in\frac{1}{2}Z\\ \eta_1+\eta_2+\eta_3+\eta_4=-2\zeta}} l(\eta_1)\,k(\eta_2)\,p(\eta_3)\,q(\eta_4)\cdot q_L(a+e+\zeta+\eta_1)\cdot q_L(b+e+\zeta+\eta_2)\times \times q_L(c+e+\zeta+\eta_3)\cdot q_L(d+e+\zeta+\eta_4)$$

which is the same as:

$$\frac{1}{2^g\cdot\#Z}\sum_{\substack{\eta_1,\eta_2,\eta_3,\eta_4\in Z\\ \zeta\in\frac{1}{2}Z\\ m\in\hat{Z}}} m(\eta_1+\eta_2+\eta_3+\eta_4+2\zeta)\cdot l(\eta_1)\cdot k(\eta_2)\cdot p(\eta_3)\cdot q(\eta_4)\times \times q_L(a+e+\zeta+\eta_1)\ldots q_L(d+e+\zeta+\eta_4).$$

But for fixed m, this sum is zero unless $2m+l+k+p+q=0$, i.e., $m\in r+(\hat{Z})_2$. Therefore, the sum works out to be

$$\frac{1}{2^g\cdot\#Z}\sum_{\substack{m\in(\hat{Z})_2\\ \zeta\in\frac{1}{2}Z\\ \eta_i\in Z}} (l+r+m)(\eta_1)\cdot q_L(a+e+\zeta+\eta_1)\times$$

$$\times(k+r+m)(\eta_2)\cdot q_L(b+e+\zeta+\eta_2)\cdot(p+r+m)(\eta_3)\cdot q_L(c+e+\zeta+\eta_3)\times$$

$$\times(q+r+m)(\eta_4)\cdot q_L(d+e+\zeta+\eta_4)\cdot(r+m)(2\zeta).$$

This is exactly the right-hand side of (C″).

We omit the proof that (C″)⇒(C′) which is similar.

§4. Structure of the Homogeneous Coordinate Ring

Let X be an abelian variety, and let L be a totally symmetric ample invertible sheaf of separable type. Of course, char $(k) \neq 2$. In this section we shall study

$$R = \bigoplus_{n=0}^{\infty} \Gamma(X, L^n),$$

the homogeneous coordinate ring associated to L. In particular, we would like criteria for the properties:

(a) R is generated by

$$R_1 = \Gamma(X, L).$$

(b) If $S^* R_1$ is the symmetric algebra on R_1 over k, and I is the kernel:

$$0 \to I \to S^* R_1 \to R$$

then I is generated, as $S^* R_1$-module, by the quadratic relations I_2.

In the first place, there are very general methods, based on CASTELNUOVO's lemma, that give results of this type. I can prove:

Proposition 1. *If* L *is replaced by* L^n, *where* $n \geqq \dim X + 2$, *then both* (a) *and* (b) *are true.*

This is not very useful, however, for the present theory. In the theorems that we want, we shall assume much less about the sheaf. However, as a consequence of our somewhat peculiar methods, we only obtain criteria for:

(a)′: $R_{(2^n)}$ generated by the (2^n)-th symmetric power of R_1, for all n.

(b)′: $I_{(2^n)}$ generated by I_2 times $S^{2^n-2}(R_1)$ for all n.

Although these are strange conditions, still if they are true in some case, then the following also hold:

(a)″: R_m is generated by the m-th symmetric power of R_1, for all m sufficiently large.

(b)″: I_m is generated by I_2 times $S^{m-2}(R_1)$ for all m sufficiently large.

Incidentally, (a)″ is *equivalent* to the condition that L be *very* ample.

Lemma. (a)′ ⇒ (a)″,
(b)′ ⇒ (b)″.

Proof. Assume (a)′. For all integers m, let

$$R(m) = \bigoplus_{k=0}^{\infty} R_{mk}.$$

It is well known that for all large m, $R(m)$ is generated by $R(m)_1 = R_m$. In particular, for large n, $R(2^n)$ is generated by $R_{(2^n)}$ and hence by $S^* R_1$. On the other hand, R is an $R(2^n)$-module of finite type for all n. Therefore, *a fortiori*, R is an $S^* R_1$-module of finite type. Now let $M = R/\mathrm{Im}\, S^* R_1$. Consider M as an $S^* R_1$-module. It is of finite type since R is; and $M_{(2^n)} = (0)$ for all n. But, given any graded R-module N of finite type, either $N_n \neq (0)$ for *all* large n, or $N_n = (0)$ for *all* large n. Hence (a)″ holds.

Assume (b)′. Let $J \subset I$ be the ideal generated by I_2. Then I/J is an $S^* R_1$-module of finite type such that $(I/J)_{(2^n)} = (0)$ for all n. As above, this implies that $(I/J)_m = (0)$ for all large m, hence (b)″ holds. *Q.E.D.*

Conditions (a)′ and (b)′ are exactly what we can get hold of by the theory of § 3. To set this up, first choose for every integer n a symmetric ϑ-structure on L^{2^n}:

$$f_n\colon\ \mathscr{G}(L^{2^n}) \xrightarrow{\sim} \mathscr{G}(2^n \delta),$$

such that (f_n, f_{n+1}) is always a symmetric ϑ-structure for $(L^{2^n}, L^{2^{n+1}})$. This is possible by Proposition 7 of § 2. This induces isomorphisms

$$\varphi_n\colon\ \Gamma(X, L^{2^n}) \xrightarrow{\sim} V(2^n \delta) = \left\{ \begin{array}{c} \text{vector space of} \\ \text{functions on} \\ K(2^n \delta) \end{array} \right\}$$

for all n, unique up to scalars. Moreover, for all n, let the function $q_n(x)$ on $K(2^n \delta)$ correspond to the linear functional on $\Gamma(X, L^{2^n})$ defined by evaluating sections at 0. Using only the functions q_n, we can define the following ring:

R^*: graded polynomial ring generated by $\bigoplus_{n=0}^{\infty} V(2^n \delta)$, where elements of $V(2^n \delta)$ are assigned degree 2^n, modulo the relations

$$f \cdot f' - g$$

for all $\left\{ \begin{array}{l} f, f' \in V(2^n \delta) \\ \quad g \in V(2^{n+1} \delta), \text{ some } n \text{ such that} \end{array} \right.$

$$g(x) = \sum_{y \in x + K(2^n \delta)} f(x+y) \cdot f'(x-y) \cdot q_{n+1}(y).$$

[Here, as in § 3, we identify the groups $K(2^n \delta)$ as subgroups of each other:

$$K(\delta) \subset K(2\delta) \subset \cdots \subset K(2^n \delta) \subset \cdots.]$$

It follows that

$$R^*_{2^n} = V(2^n \delta)$$

23*

and if any integer N is written as a sum

$$N=2^{n_1}+2^{n_2}+\cdots+2^{n_k}$$

with $n_1, n_2, \ldots, n_k$ distinct, then

$$R_N^* \cong V(2^{n_1}\delta)\otimes\cdots\otimes V(2^{n_k}\delta)/(\text{something}).$$

In any case, there is a canonical map

$$\Phi: \quad R^* \to R$$

defined by extending the maps φ_n to a homomorphism of all of R^*.

Lemma. i) *R^* is a finitely generated k-algebra and the kernel of Φ is finite-dimensional,*

ii) *Φ_N: $R_N^* \to R_N$ is an isomorphism if N is a power of 2,*

iii) *if L is very ample, then Φ_N is an isomorphism for all but a finite number of N.*

Proof. First of all, (ii) is obvious. Since L is ample, R_{2^n} is spanned by $R_n \otimes R_n$ for all sufficiently large n. Therefore, $R_{2^{n+1}}^*$ is spanned by $R_{2^n}^* \otimes R_{2^n}^*$ for all sufficiently large n. Therefore R^* is finitely generated. If $J=\ker(\Phi)$, then J is a finitely generated R^*-module; since J is zero in degrees 2^n, it follows that J is finite-dimensional. As for (iii), if L is very ample, R_1 generates all but a finite-dimensional piece of R. Therefore the subring of R^* generated by R_1^* already goes onto all but a finite-dimensional piece of R. *Q.E.D.*

This lemma means that if L is very ample, the abstract ring R^* defined using only the family of null-value functions q_n, is essentially isomorphic to R. We can therefore calculate in R^* to obtain results about the ring R. The first result of this type is:

Theorem 1. *If $H(L)$ contains X_4, the group of points of order 4, then $\Gamma(X, L^2)$ is spanned by $S^2\{\Gamma(X, L)\}$.*

Proof. We will show that every function on $K(2\delta)$ is a linear combination of functions $f * f'$, where f, f' are functions on $K(\delta)$. (Notation as in § 3). It will be convenient to work with the delta functions on $K(\delta)$, $K(2\delta)$: we write δ_a for the function:

$$\begin{aligned}\delta_a(a') &= 1 \quad \text{if} \quad a'=a\\ &= 0 \quad \text{if} \quad a' \neq a.\end{aligned}$$

There should be no cause for confusion with the use of δ in "level δ". We let Z_2 denote the subgroup of $K(\delta)$ of points of order 2.

First, we calculate $\delta_{a+b}*\delta_{a-b}$, where $a, b\in K(2\delta)$ and $a+b\in K(\delta)$

$$(\delta_{a+b}*\delta_{a-b})(x)=\sum_{z\in x+K(\delta)}\delta_{a+b}(x+z)\cdot\delta_{a-b}(x-z)\cdot q_1(z).$$

The term inside is 0 unless

$$x+z=a+b$$
$$x-z=a-b.$$

These imply that $x\in a+Z_2$. Conversely, if $x\in a+Z_2$, then the sum has exactly one non-zero term, viz., $q_1(a+b-x)$. Therefore:

$$\begin{aligned}\delta_{a+b}*\delta_{a-b}&=\sum_{x\in a+Z_2}q_1(a+b-x)\cdot\delta_x\\&=\sum_{\eta\in Z_2}q_1(b+\eta)\cdot\delta_{a+\eta}.\end{aligned}$$

These functions certainly span the image of $S^2[V(\delta)]$. However, it is convenient to use a slightly different basis. Let l be a homomorphism from Z_2 to $\{\pm 1\}$. Then

$$\begin{aligned}\sum_{\eta\in Z_2}l(\eta)\,\delta_{a+b+\eta}*\delta_{a-b+\eta}&=\sum_{\eta_1,\eta_2\in Z_2}l(\eta_1)\cdot q_1\big(a+b+\eta_1-(a+\eta_2)\big)\cdot\delta_{a+\eta_2}\\&=\Big[\sum_{\eta\in Z_2}l(\eta)\cdot q_1(b+\eta)\Big]\cdot\Big[\sum_{\eta\in Z_2}l(\eta)\cdot\delta_{a+\eta}\Big].\end{aligned}$$

If the image of $S^2[V(\delta)]$ contains all the functions $\sum l(\eta)\,\delta_{a+\eta}$, then it contains everything. Therefore, the Theorem is true if we can prove:

(*) For every $a\in K(2\delta)$ and every homomorphism $l\colon Z_2\to\{\pm 1\}$ there is an element $b\in a+K(\delta)$ such that

$$\sum_{\eta\in Z_2}l(\eta)\cdot q_1(b+\eta)\neq 0.$$

To prove this, set

$$F=\sum_{\eta\in Z_2}\delta_{2a+\eta}$$

and

$$G=\sum_{\eta\in Z_2}l(\eta)\,\delta_\eta.$$

Now calculate $F*G$:

$$(F*G)(x)=\sum_{z\in x+K(\delta)}F(x+z)\cdot G(x-z)\cdot q_1(z).$$

The inside term is zero unless:

$$x+z=2a+\eta_1$$
$$x-z=\eta_2,\quad \eta_1,\eta_2\in Z_2.$$

These imply $4x=4a$. Let Z_4 be the group of elements of $K(2\delta)$ of order 4. By our hypothesis on L, Z_4 is a subgroup of $K(\delta)$. Now suppose $x\in a+Z_4$. Then the sum works out to be:

$$= \sum_{\eta_2\in Z_2} F(2x+\eta_2)\cdot G(\eta_2)\cdot q_1(x+\eta_2)$$
$$= \sum_{\eta\in Z_2} l(\eta)\cdot q_1(x+\eta).$$

Therefore:

$$F*G=\sum_{x\in a+Z_4}\Big(\sum_{\eta\in Z_2} l(\eta)\cdot q_1(x+\eta)\Big)\cdot\delta_x.$$

Now assume $(*)$ is false. Since $Z_4\subset K(\delta)$, it follows that all the coefficients $\sum l(\eta)\, q_1(x+\eta)$ are zero. Therefore $F*G=0$. But $F\neq 0$ and $G\neq 0$, and the homogeneous coordinate ring of an abelian variety is well known to be an integral domain! This contradiction shows that $(*)$ must hold. *Q.E.D.*

Corollary. *If $H(L)\supset X_4$, then R^* is generated by R_1^* and* (a)$'$ *and* (a)$''$ *are true for the homogeneous coordinate ring*

$$\bigoplus_{n=0}^{\infty}\Gamma(X,L^n).$$

In particular, L is very ample.

The inequality proven about q_1 in the course of this proof will be quite useful in the sequel. To bring out the general picture I want to state the analogous fact about the values of q when expanded in characters of an arbitrary subgroup Z:

$$Z_2\subset Z\subset 2K(2\delta).$$

As above, let

$$\vartheta(a,l)=\sum_{\eta\in Z} l(\eta)\cdot q_1(a+\eta),\quad \text{if}\quad l\in\hat{Z},\quad a\in K(2\delta).$$

The result is:

a) for all $a\in K(2\delta)$, $l\in\hat{Z}$, there exists an element $\zeta\in\frac{1}{2}Z$ such that

$$\vartheta(a+\zeta,l)\neq 0,$$

b) for all $a\in K(2\delta)$, $l\in\hat{Z}$, there exists an element $m\in(\hat{Z})_2$, i.e., a character of $Z/2Z$, such that

$$\vartheta(a,l+m)\neq 0.$$

These facts are proven by an obvious generalization of the argument used to prove $(*)$.

The second result which we get by calculating in R^* is this:

Theorem 2. *Let*

$$I_2 = \operatorname{Ker}\{S^2 R_1 \to R_2\}$$
$$I_4 = \operatorname{Ker}\{S^4 R_1 \to R_4\}.$$

Assume $H(L)$ contains X_4. Then I_4 is generated by $R_2 \otimes I_2$.

Proof. First of all, note that by Theorem 1, R_2 and R_4 are quotients of $S^2 R_1$ and $S^4 R_1$, respectively. To compute $R_2 \otimes I_2$, we use the obvious remark:

Lemma. *Let V be a k-vector-space, and let $I \subset S^2 V$ be any subspace. For all elements $a, b \in V$, let*

$$x_{a,b} = \text{image of } a \cdot b \text{ in } S^2 V/I.$$

Then

$$S^4 V/S^2 V \cdot I \cong S^2[S^2 V/I] \Big/ \left\{\begin{matrix}\text{span of the elements}\\ x_{a,b} \cdot x_{c,d} - x_{a,c} \cdot x_{b,d}\end{matrix}\right\}.$$

For all $a, b \in R_1$, let $x_{a,b}$ denote the product of a and b in R_2. Then the lemma tells us that

$$S^4 R_1/R_2 \cdot I_2 = S^2 R_2 \Big/ \left\{\begin{matrix}\text{span of the elements}\\ x_{a,b} \cdot x_{c,d} - x_{a,c} \cdot x_{b,d}\end{matrix}\right\}.$$

Therefore it suffices to prove that the canonical map:

$$S^2 R_2 \Big/ \left\{\begin{matrix}\text{span of the elements}\\ x_{a,b} \cdot x_{c,d} - x_{a,c} \cdot x_{b,d}\end{matrix}\right\} \to R_4$$

is *injective*.

In order to do this, the key point is to compute in terms of the most carefully chosen bases of R_1, R_2 and R_4.

(I) In R_1, we use the functions δ_a, $a \in K(\delta)$.

(II) In R_2, we use the functions

$$Y_{a,l} = \sum_{\eta \in Z_2} l(\eta) \cdot \delta_{a+\eta}$$

where Z_2 = subgroup of $K(\delta)$ of points of order 2 and $l \in \hat{Z}_2$.

(III) In R_4, we use the functions

$$Z_{a,l} = \sum_{\eta \in Z_4} l(\eta) \cdot \delta_{a+\eta}$$

where Z_4 = subgroup of $K(\delta)$ of points of order 4 and $l \in \hat{Z}_4$.

Moreover, set

$$q_1(a,l)=\sum_{\eta\in Z_2} l(\eta)\cdot q_1(a+\eta),\quad l\in\hat{Z}_2,\quad a\in K(2\delta)$$

$$q_2(a,l)=\sum_{\eta\in Z_4} l(\eta)\cdot q_2(a+\eta),\quad l\in\hat{Z}_4,\quad a\in K(4\delta).$$

Now if $a, b\in K(2\delta)$ and $a+b\in K(\delta)$, calculate

$$\begin{aligned}\sum_{l\in\hat{Z}_2} q_1(b,l)\cdot Y_{a,l} &= \sum_{\substack{l\in\hat{Z}_2\\ \eta,\eta'\in Z_2}} l(\eta)\cdot l(\eta')\cdot q_1(b+\eta)\cdot\delta_{a+\eta'}\\ &=2^g\sum_{\eta\in Z_2} q_1(b+\eta)\cdot\delta_{a+\eta}\\ &=2^g\,\delta_{a+b}*\delta_{a-b}\end{aligned}$$

(cf. proof of Theorem 1). In S^2R_2, denote the symmetric product by $\odot$. Then the problem is to show that:

$$S^2R_2\Big/\left\{\begin{array}{l}\text{span of the elements:}\\ \qquad\sum_l q_1(b,l)\cdot Y_{a,l}\odot\sum_l q_1(d,l)\cdot Y_{c,l}-\\ -\sum_l q_1(a+e,l)\cdot Y_{-b-e,l}\odot\sum_l q_1(c+e,l)\cdot Y_{-d-e,l}\\ \text{for all } a,b,c,d,e\in K(2\delta)\text{ such that}\\ \qquad a+b+c+d=-2e\\ \qquad a+b,\,c+d\in K(\delta)\end{array}\right\}$$

is isomorphic to R_4. Let L be the span of the elements inside the braces. Then take a typical generator of L, and write

$$\begin{aligned}a&=A+B\\ c&=A-B\\ -b-e&=A+C\\ -d-e&=A-C\end{aligned}$$

with $A, B, C\in K(4\delta)$. It follows that L is the span of the elements

$$\sum_{l,l'\in\hat{Z}_2} q_1(A+C+e,l)\cdot q_1(A-C+e,l')\cdot Y_{A+B,l}\odot Y_{A-B,l'}-$$
$$-\sum_{l,l'\in\hat{Z}_2} q_1(A+B+e,l)\cdot q_1(A-B+e,l')\cdot Y_{A+C,l}\odot Y_{A-C,l'}$$

for all A, B, C, $e\in K(4\delta)$ such that

$$A+B,\,A+C,\,e\in K(2\delta)$$

and

$$B-C+e\in K(\delta).$$

Next, we replace these by another set of generators of L obtained by a partial Fourier transformation: in a typical generator, replace

$$\begin{aligned} A &\text{ by } A+\zeta \\ B &\text{ by } B+\zeta' \\ C &\text{ by } C+\zeta'' \\ e &\text{ by } e-\zeta \end{aligned}$$

where $\zeta, \zeta', \zeta'' \in Z_4$. This is $0K$, since $Z_4 \subset K(\delta)$. Let m, m', m'' be elements of $\hat{Z}_4$. Multiply by $m(\zeta) \cdot m'(\zeta') \cdot m''(\zeta'')$ and sum over ζ, ζ', ζ''. This gives us generators for L of the form:

$$\begin{aligned} &\sum_{l, l' \in \hat{Z}_2} \Big[\sum_{\zeta'' \in Z_4} m''(\zeta'') \cdot q_1(A+C+e+\zeta'', l) \cdot q_1(A-C+e-\zeta'', l') \Big] \times \\ &\qquad \times \Big[\sum_{\zeta, \zeta' \in Z_4} m(\zeta) \cdot m'(\zeta') \cdot Y_{A+B+\zeta+\zeta', l} \odot Y_{A-B+\zeta-\zeta', l'} \Big] - \\ &- \sum_{l, l' \in \hat{Z}_2} \Big[\sum_{\zeta' \in Z_4} m'(\zeta') \cdot q_1(A+B+e+\zeta', l) \cdot q_1(A-B+e-\zeta', l') \Big] \times \\ &\qquad \times \Big[\sum_{\zeta, \zeta'' \in Z_4} m(\zeta) \cdot m''(\zeta'') \cdot Y_{A+C+\zeta+\zeta'', l} \odot Y_{A-C+\zeta-\zeta'', l'} \Big]. \end{aligned}$$

This expression simplifies a great deal. The terms in the 1st sum over l and l' are 0 unless l and l' satisfy:

$$\begin{aligned} l(2x) &= (m+m')(x), \quad \text{all} \quad x \in Z_4 \\ l'(2x) &= (m-m')(x), \quad \text{all} \quad x \in Z_4 \\ m''(x) &= (l+l')(x), \quad \text{all} \quad x \in Z_2. \end{aligned}$$

In particular, the whole 1st term is 0 unless m, m' and m'' are equal on Z_2. Similarly, in the 2nd term, we get 0 unless

$$\begin{aligned} l(2x) &= (m+m'')(x), \quad \text{all} \quad x \in Z_4 \\ l'(2x) &= (m-m'')(x), \quad \text{all} \quad x \in Z_4 \\ m'(x) &= (l+l')(x), \quad \text{all} \quad x \in Z_2. \end{aligned}$$

Therefore the *whole* expression is 0 unless m, m' and m'' are equal on Z_2. We may as well set

$$\begin{aligned} m &= k+k' \\ m' &= k-k' \\ m'' &= k-k'+2k^* \end{aligned}$$

for suitable elements $k, k', k^* \in \hat{Z}_4$. Then, up to a constant, the expression reduces to:

$$[\sum_{\zeta \in Z_4} (k-k'+2k^*)(\zeta) \cdot q_1(A+C+e+\zeta, k) \cdot q_1(A-C+e-\zeta, k')] \times$$
$$\times [\sum_{\zeta \in Z_4} (k+k')(\zeta) \cdot Y_{A+B+\zeta,k} \odot Y_{A-B+\zeta,k'}] -$$
$$-[\sum_{\zeta \in Z_4} (k-k')(\zeta) \cdot q_1(A+B+e+\zeta, k+k^*) \cdot q_1(A-B+e-\zeta, k'-k^*)] \times$$
$$\times [\sum_{\zeta \in Z_4} (k+k')(\zeta) \cdot Y_{A+C+\zeta,k+k^*} \odot Y_{A-C+\zeta,k'-k^*}].$$

The 1st and the 3rd bracketed expressions can be written in terms of q_2: let $a, b \in K(4\delta)$ and let $l, l \in \hat{Z}_4$ such that $a+b \in K(2\delta)$. Then

$$\sum_{\zeta \in Z_4} (l+l')(\zeta) \cdot q_1(a+b+\zeta, l) \cdot q_1(a-b-\zeta, l')$$
$$= \sum_{\substack{\zeta \in Z_4 \\ \eta, \eta' \in Z_2}} l(\zeta+\eta) \cdot l'(\zeta+\eta') \cdot q_1(a+b+\zeta+\eta) \cdot q_1(a-b-\zeta-\eta')$$
$$= \sum_{\zeta_1, \zeta_2 \in Z_4} l(\zeta_1+\zeta_2) \cdot l'(\zeta_1-\zeta_2) \cdot q_1(a+b+\zeta_1+\zeta_2) \cdot q_1(a-b+\zeta_2-\zeta_1)$$
$$= \sum_{\zeta_1, \zeta_2 \in Z_4} l(\zeta_1+\zeta_2) \cdot l'(\zeta_1-\zeta_2) \cdot q_2(a+\zeta_2) \cdot q_2(b+\zeta_1)$$

(cf. § 3)

$$= q_2(a, l-l') \cdot q_2(b, l+l').$$

Substituting this formula, we find that L is spanned by the elements

$$q_2(A+e, k+k'+2k^*) \times$$
$$\times \begin{Bmatrix} q_2(C, k-k'+2k^*) \cdot \sum_{\zeta \in Z_4} (k+k')(\zeta) \cdot Y_{A+B+\zeta,k} \odot Y_{A-B+\zeta,k'} - \\ -q_2(B, k-k') \cdot \sum_{\zeta \in Z_4} (k+k')(\zeta) \cdot Y_{A+C+\zeta,k+k^*} \odot Y_{A-C+\zeta,k'-k^*} \end{Bmatrix}.$$

What happens to these sums when they are pushed from $S^2 R_2$ into R_4? We calculate

$$\sum_{\zeta \in Z_4} (k+k')(\zeta) \cdot Y_{A+B+\zeta,k} * Y_{A-B+\zeta,k'}$$
$$= \sum_{\substack{\zeta \in Z_4 \\ \eta, \eta' \in Z_2}} k(\zeta+\eta) \cdot k'(\zeta+\eta') \cdot \delta_{A+B+\zeta+\eta} * \delta_{A-B+\zeta+\eta'}$$
$$= \sum_{\zeta_1, \zeta_2 \in Z_4} k(\zeta_1+\zeta_2) \cdot k'(\zeta_1-\zeta_2) \cdot \delta_{A+B+\zeta_1+\zeta_2} * \delta_{A-B+\zeta_1-\zeta_2}$$
$$= \sum_{\zeta_1, \zeta_2 \in Z_4} k(\zeta_1+\zeta_2) \cdot k'(\zeta_1-\zeta_2) \cdot q_2(B+\zeta_2) \cdot \delta_{A+\zeta_1}$$

(cf. proof of Theorem 1)

$$=q_2(B, k-k')\cdot Z_{A,k+k'}.$$

Therefore, in order to check that L is big enough, it is necessary and sufficient, for every *pair* of terms

$$T_i=\sum_{\zeta\in Z_4}(k_i+k_i')(\zeta)\cdot Y_{A_i+B_i+\zeta,k_i}\odot Y_{A_i-B_i+\zeta,k_i'},\qquad i=1,2$$

[here $A_i, B_i\in K(4\delta)$, $A_i+B_i\in K(2\delta)$, $k_i, k_i'\in\hat{Z}_4$] for which $A_1=A_2$ and $k_1+k_1'=k_2+k_2'$, to show that $c_1T_1+c_2T_2$ is in L for some non-zero pair (c_1, c_2) of constants. [One must check at this point that S^2R_2 is itself spanned by these terms, but this is nearly obvious.]

Now set

$$\begin{aligned}\Sigma(A,B,C;k,k',k^*)&\\ =q_2(C,k-k'+2k^*)&\cdot\sum_{\zeta\in Z_4}(k+k')(\zeta)\cdot Y_{A+B+\zeta,k}\odot Y_{A-B+\zeta,k'}-\\ -q_2(B,k-k')&\cdot\sum_{\zeta\in Z_4}(k+k')(\zeta)\cdot Y_{A+C+\zeta,k+k^*}\odot Y_{A-C+\zeta,k'-k^*}.\end{aligned}$$

I claim it is enough to show that for every $A, B, C\in K(4\delta)$, $k, k', k^*\in\hat{Z}_4$ such that $A+B, A+C\in K(2\delta)$, the expression $\Sigma(A, B, C; k, k', k^*)$ is in L. Namely, if T_1 and T_2 are any pair of terms as above, then

$$\Sigma(A_1, B_1, B_2; k_1, k_1', k_2-k_1)$$

is clearly of the form $c_1T_1+c_2T_2$. The only problem is that c_1 and c_2 might be 0. But T_1 is mapped into $c_2Z_{A_1,k_1+k_1'}$ in R_4 and T_2 is mapped into $c_1Z_{A_1,k_1+k_1'}$ in R_4. Therefore $c_1=c_2=0$ only if T_1 and T_2 go to 0 in R_4. But since S^2R_2 is mapped *onto* R_4 by Theorem 1, for every term T_1, there is some term T_3 with $A_3=A_1$, $k_3+k_3'=k_1+k_1'$ such that T_3 is mapped to a *non-zero* multiple of $Z_{A_1,k_1+k_1'}$. Suppose all the Σ's are in L. Then

$$\Sigma(A_1, B_1, B_3; k_1, k_1', k_3-k_1)\in L$$
$$\Sigma(A_2, B_2, B_3; k_2, k_2', k_3-k_2)\in L.$$

In other words, L contains expressions of the form $d_1T_1+d_3'T_3$, with $d_1\neq 0$; and $d_2T_2+d_3''T_3$ with $d_2\neq 0$. Therefore L contains an expression $c_1T_1+c_2T_2$ with c_1 or $c_2\neq 0$.

The identity:

$$\begin{aligned}&q_2(C,k-k'+2k^*)\cdot\Sigma(A,B,D;k,k',l^*)+\\ &\qquad\qquad+q_2(B,k-k')\cdot\Sigma(A,D,C;k+l^*,k'-l^*,k^*-l^*)\\ &\qquad\qquad=q_2(D,k-k'+2l^*)\cdot\Sigma(A,B,C;k,k',k^*)\end{aligned}$$

is easy to check. [Here A, B, C, D are in $K(4\delta)$, $k, k', k^*, l^* \in \hat{Z}_4$ and $A+B$, $A+C$, $A+D \in K(2\delta)$.] We have proven, that for *all* $A, B, C \in K(4\delta)$ and $k, k', k^* \in \hat{Z}_4$ such that $A+B$, $A+C \in K(2\delta)$, then

$$q_2(A+e, k+k'+2k^*) \cdot \Sigma(A, B, C; k, k', k^*) \in L$$

for any element $e \in K(4\delta)$ such that

$$e \in B-C+K(\delta).$$

This embarrassing factor $q_2(A+e, k+k'+2k^*)$ is the crux of the problem at this point. The e here gives us very little flexibility: if it is varied by an element of Z_4 (which is permitted), the q_2 is multiplied only by a root of 1. So if one of these vanishes, all of them do. And since we have only assumed that $Z_4 \subset K(\delta)$, this is really the *only* variation possible in e. We may as well set $e = B-C$ therefore; this gives us the fact:

$$q_2(A+B-C, k+k'+2k^*) \cdot \Sigma(A, B, C; k, k', k^*) \in L.$$

But now the cocycle identity gives us also the fact:

$$\begin{aligned}
&q_2(A+B-D, k+k'+2l^*) \cdot q_2(A+D-C, k+k'+2k^*-2l^*) \times \\
&\quad \times q_2(D, k-k'+2l^*) \cdot \Sigma(A, B, C; k, k', k^*) \\
&= q_2(C, k-k'+2k^*) \cdot q_2(A+D-C, k+k'+2k^*-2l^*) \times \\
&\quad \times [q_2(A+B-D, k+k'+2l^*) \cdot \Sigma(A, B, D; k, k', l^*)] + \\
&\quad + q_2(B, k-k') \cdot q_2(A+B-D, k+k'+2l^*) \times \\
&\quad \times [q_2(A+D-C, k+k'+2k^*-2l^*) \cdot \Sigma(A, D, C, k+l^*, k'-l^*, k^*-l^*)] \\
&\in L,
\end{aligned}$$

for all $l^* \in \hat{Z}_4$, $D \in A+K(2\delta)$. The product of 3 q_2's at the beginning of this equation can be made non-zero by an appropriate choice of l^* and D! This is the content of:

Triple Vanishing Lemma. *Let A_1, A_2, A_3 be any three elements of $K(4\delta)$; let k_1, k_2, k_3 be any three elements of $\hat{Z}_4$. Then there exists*

$$\zeta \in Z_8 \qquad (= \textit{points of order } 8 \textit{ in } K(2\delta))$$

$$l \in 2\hat{Z}_4$$

such that

$$\prod_{i=1}^{3} q_2(A_i+\zeta, k_i+l) \neq 0.$$

Proof. To show this, we first establish the following general formula:

Lemma. *Let L be a totally symmetric ample invertible sheaf on X of separable type δ. Choose compatible ϑ-structures for L, L^2, L^4. Let $A_1, A_2, A_3, A_4 \in K(\delta)$, and let $q_1 \in V(\delta)$ correspond to evaluation at 0. Then*

$$\sum_{\eta \in Z_2} \delta_{A_1+\eta} * \delta_{A_2+\eta} * \delta_{A_3+\eta} * \delta_{A_4+\eta}$$

$$= \sum_{\substack{E \in K(4\delta) \\ 4E = A_1+A_2+A_3+A_4}} q_1(A_1+A_2-2E) \cdot q_1(A_1+A_3-2E) \cdot q_1(A_1+A_4-2E) \cdot \delta_E.$$

Proof. Write

$$A_1 = \alpha+\delta+\beta$$

$$A_2 = \alpha+\delta-\beta$$

$$A_3 = -\alpha+\delta+\gamma$$

$$A_4 = -\alpha+\delta-\gamma$$

where $\beta, \gamma \in K(2\delta)$, $\alpha, \delta \in K(4\delta)$. Then combining the 1st 2 and last 2 terms by the formula in Theorem 1, we find

$$\sum_{\eta \in Z_2} \delta_{A_1+\eta} * \delta_{A_2+\eta} * \delta_{A_3+\eta} * \delta_{A_4+\eta}$$

$$= \sum_{\eta, \eta_1, \eta_2 \in Z_2} q_2(\beta+\eta_1) \cdot q_2(\gamma+\eta_2) \cdot \delta_{\alpha+\delta+\eta+\eta_1} * \delta_{-\alpha+\delta+\eta+\eta_2}.$$

Combining them again by the same formula and rearranging one finds:

$$= \sum_{\substack{\eta \in Z_2 \\ \zeta_1, \zeta_2 \in Z_4 \\ \zeta_1+\zeta_2 \in Z_2}} q_2(\beta+\zeta_1+\zeta_2+\eta) \cdot q_2(\gamma+\zeta_1-\zeta_2+\eta) \cdot q_3(\alpha+\zeta_2) \cdot \delta_{\zeta_1+\delta}.$$

Using the relations among q_1, q_2, q_3 established in § 2, this reduces to:

$$= \sum_{\zeta \in Z_4} q_1(\beta+\gamma+2\zeta) \cdot q_1(\beta-\gamma+2\zeta) \cdot q_1(2\alpha+2\zeta) \cdot \delta_{\zeta+\delta}.$$

With $E = \zeta+\delta$, this is exactly what we were to prove. *Q.E.D.*

To apply the lemma in our situation, we consider the sections of L^{16}, which are described by functions on $K(16\delta)$. Let Z_8 be the group of points of order 8 in $K(16\delta)$ (or in $K(2\delta)$). For all $a \in K(16\delta)$, $l \in \hat{Z}_8$, set

$$W_{a,l} = \sum_{\eta \in Z_8} l(\eta) \cdot \delta_{a+\eta}.$$

Now suppose $A_1, A_2, A_3, A_4 \in K(16\delta)$ and $l_1, l_2, l_3, l_4 \in \hat{Z}_8$ satisfy $A_1+A_2+A_3+A_4=0$, $l_1+l_2+l_3+l_4=0$. We use the lemma to compute

$$W_{A_1,l_1} * W_{A_2,l_2} * W_{A_3,l_3} * W_{A_4,l_4}$$

$$= \sum_{\substack{\eta_1,\eta_2,\eta_3,\eta_4 \in Z_8 \\ \zeta \in Z_{32} \\ 4\zeta=\eta_1+\eta_2+\eta_3+\eta_4}} l_1(\eta_1)\cdot l_2(\eta_2)\cdot l_3(\eta_3)\cdot l_4(\eta_4)\cdot q_4(A_1+A_2+\eta_1+\eta_2-2\zeta)\times$$

$$\times q_4(A_1+A_3+\eta_1+\eta_3-2\zeta)\cdot q_4(A_1+A_4+\eta_1+\eta_4-2\zeta)\cdot \delta_\zeta$$

$$=\frac{1}{8^g}\sum_{\eta \in \hat{Z}_8}\sum_{\substack{\eta_1,\ldots,\eta_4 \in Z_8 \\ \zeta \in Z_{32}}} m(4\zeta-\eta_1-\eta_2-\eta_3-\eta_4)\cdot l_1(\eta_1)\cdot l_2(\eta_2)\times$$

$$\times l_3(\eta_3)\cdot l_4(\eta_4)\cdot q_4(A_1+A_2+\eta_1+\eta_2-2\zeta)\times$$

$$\times q_4(A_1+A_3+\eta_1+\eta_3-2\zeta)\cdot q_4(A_1+A_4+\eta_1+\eta_4-2\zeta)\cdot\delta_\zeta$$

where $g=\dim X$, (i.e., $8^g=\text{Card}\,(Z_8)$). If, as usual, we let

$$q_4(a,l)=\sum_{\eta \in Z_8} l(\eta)\cdot q_4(a+\eta),$$

then this works out to be

$$=\frac{1}{8^g}\sum_{\substack{m \in \hat{Z}_8 \\ \zeta \in Z_{32}}}\Big(\sum_{\eta_1 \in Z_8} l_1(\eta_1)^2\cdot m(\eta_1)^{+2}\Big)\cdot m(4\zeta)\cdot q_4(A_1+A_2-2\zeta, l_2-m)\times$$

$$\times q_4(A_1+A_3-2\zeta, l_3-m)\cdot q_4(A_1+A_4-2\zeta, l_4-m)\cdot\delta_\zeta$$

$$=\sum_{\substack{k \in 4\hat{Z}_8 \\ \zeta \in Z_{32}}} l_1(4\zeta)^{-1}\cdot k(4\zeta)^{-1}\cdot q_4(A_1+A_2-2\zeta, l_1+l_2+k)\times$$

$$\times q_4(A_1+A_3-2\zeta, l_1+l_3+k)\cdot q_4(A_1+A_4-2\zeta, l_1+l_4+k)\cdot\delta_\zeta.$$

Now since the homogeneous coordinate ring of X is an integral domain, it follows that for all A's and l's as above, there are elements $k \in 4\hat{Z}_8$ and $\zeta \in Z_{16}$ such that

$$\prod_{i=2}^{4} q_4(A_1+A_i+\zeta, l_1+l_i+k) \neq 0.$$

Now the triple $(A_1+A_2, A_1+A_3, A_1+A_4)$ is an arbitrary triple of elements (B_1, B_2, B_3) in $K(16\delta)$ such that $B_1+B_2+B_3 \in K(8\delta)$. Similarly the triple $(l_1+l_2, l_1+l_3, l_1+l_4)$ is an arbitrary triple of elements (k_1, k_2, k_3) in $\hat{Z}_8$ such that $k_1+k_2+k_3 \in 2\hat{Z}_8$. Therefore, for any triples

$$\begin{cases} B_1, B_2, B_3 \in K(8\delta) \\ k_1, k_2, k_3 \in 2\hat{Z}_8 \end{cases}$$

$$\prod_{i=1}^{3} q_4(B_i+\zeta, k_i+k) \neq 0$$

for some $\zeta \in Z_{16}$, $k \in 4\hat{Z}_8$. But by the results of § 3 on the q's,

$$q_2(2B, k) = q_4(B, 2*k)$$

for any $B \in K(8\delta)$, $k \in \hat{Z}_4$ [where $2*k$ is the character $(2*k)(x) = k(2x)$ as before]. This proves the triple vanishing lemma. *Q.E.D.*

Corollary. *If* $X_4 \subset H(L)$, *then* (b)′ *and* (b)″ *are true for the homogeneous coordinate ring*

$$\bigoplus_{n=0}^{\infty} \Gamma(X, L^n).$$

In particular, if $\varphi: X \to \mathbf{P}_n$ *is the embedding defined by the complete linear system* $\Gamma(X, L)$, *then* $\varphi(X)$, *as a subscheme of* $\mathbf{P}_n$, *is an intersection of quadric hypersurfaces.*

§ 5. Examples

We shall consider 4 special cases of the preceding theory:

a) dim $X=1$, $\delta=(2)$, char $\neq 2$

b) dim $X=1$, $\delta=(3)$, char $\neq 3$

c) dim $X=1$, $\delta=(4)$, char $\neq 2$

d) dim $X=2$, $\delta=(2,2)$, char $\neq 2$.

If $e \in X$ is the origin, then in the first 3 cases, we shall take L to be $\mathbf{o}_X(2e)$, $\mathbf{o}_X(3e)$, and $\mathbf{o}_X(4e)$, respectively. As is well known, these linear systems define maps:

a) $\varphi: X \to \mathbf{P}_1$ (a double covering),

b) $\varphi: X \hookrightarrow \mathbf{P}_2$ (image a non-singular cubic curve),

c) $\varphi: X \hookrightarrow \mathbf{P}_3$ (image a non-singular quartic curve, which is the complete intersection of 2 quadrics),

d) $\varphi: X \to \mathbf{P}_3$ (X a double covering of φX, φX a quartic surface with 16 nodes).

The fact that in case c), $\varphi(X)$ is a complete intersection of 2 quadrics is the simplest example of Theorem 2 of § 4. Incidentally, note that in cases b) and c), $\bigoplus_n \Gamma(X, L^n)$ is actually generated by $\Gamma(X, L)$: at least in these cases, Theorem 1 is not the best possible result.

Case a. $H(L)=X_2$, and the action of the group of translations by points of X_2 extends to $\boldsymbol{P}_1$ so as to make φ commute with these actions. This means that the group $\boldsymbol{Z}/2\boldsymbol{Z}\oplus\boldsymbol{Z}/2\boldsymbol{Z}$ acts on $\boldsymbol{P}_1$, and this action can be normalized by a suitable choice of coordinates in $\boldsymbol{P}_1$ so that it is given by the matrices:

$$\begin{pmatrix}1&0\\0&1\end{pmatrix},\quad \begin{pmatrix}1&0\\0&-1\end{pmatrix},\quad \begin{pmatrix}0&1\\1&0\end{pmatrix},\quad \begin{pmatrix}0&1\\-1&0\end{pmatrix}.$$

This normalizes the map φ among projectively equivalent maps. Since the 4 branch points must be permuted by the action of this group, they have coordinates:

$$(\lambda,-\lambda,1/\lambda,-1/\lambda).$$

Here λ is the coordinate of $\varphi(e)$, and it can be anything (except for degenerate values $0,\pm 1,\pm i,\infty$): moreover, the abelian variety X is determined by λ.

Case b. $H(L)=X_3$, and the action of the group of translations by points of X_3 extends to $\boldsymbol{P}_2$. Therefore, the group $\boldsymbol{Z}/3\boldsymbol{Z}\oplus\boldsymbol{Z}/3\boldsymbol{Z}$ acts on $\boldsymbol{P}_2$, and this action can be normalized by a suitable choice of coordinates in $\boldsymbol{P}_2$ so that it is given by the matrices:

$$\begin{pmatrix}1&0&0\\0&1&0\\0&0&1\end{pmatrix},\quad \begin{pmatrix}1&0&0\\0&\omega&0\\0&0&\omega^2\end{pmatrix},\quad \begin{pmatrix}1&0&0\\0&\omega^2&0\\0&0&\omega\end{pmatrix}$$

$$\begin{pmatrix}0&1&0\\0&0&1\\1&0&0\end{pmatrix},\quad \begin{pmatrix}0&1&0\\0&0&\omega\\\omega^2&0&0\end{pmatrix},\quad \begin{pmatrix}0&1&0\\0&0&\omega^2\\\omega&0&0\end{pmatrix}$$

$$\begin{pmatrix}0&0&1\\1&0&0\\0&1&0\end{pmatrix},\quad \begin{pmatrix}0&0&1\\\omega&0&0\\0&\omega^2&0\end{pmatrix},\quad \begin{pmatrix}0&0&1\\\omega^2&0&0\\0&\omega&0\end{pmatrix}.$$

The image cubic must be invariant under this group, and must be contained in the open set

$$\boldsymbol{P}_2-\left\{\begin{matrix}(1,0,0), & (1,1,1), & (1,1,\omega), & (1,1,\omega^2)\\ (0,1,0), & (1,\omega,\omega^2), & (1,\omega,1), & (1,\omega^2,1)\\ (0,0,1), & (1,\omega^2,\omega), & (\omega,1,1), & (\omega^2,1,1)\end{matrix}\right\}$$

where $\boldsymbol{Z}/3\boldsymbol{Z}\oplus\boldsymbol{Z}/3\boldsymbol{Z}$ acts freely. This implies that the image cubic is given by an equation:

$$X^3+Y^3+Z^3-3\mu XYZ,\quad \mu\neq 1,\omega,\omega^2,\infty.$$

Moreover, $\varphi(e)$ must be a point of inflexion of this curve, and it turns out that the 9 points of inflexion are the 9 base points of this pencil of cubics:

$$\begin{array}{l}(0,\ 1,\ -1)\\(0,\ 1,\ -\omega)\\(0,\ 1,\ -\omega^2)\\(1,\ 0,\ -1)\\(1,\ 0,\ -\omega)\\(1,\ 0,\ -\omega^2)\\(1,\ -1,\ 0)\\(1,\ -\omega,\ 0)\\(1,\ -\omega^2,\ 0).\end{array}$$

So in this case, $\varphi(e)$ is completely determined (up to a finite choice) and its coordinates do not determine X. In other words, the function q_L on $K(\delta)=\boldsymbol{Z}/3\boldsymbol{Z}$ is independent of X in this case: but since L is not totally symmetric, this restriction cannot come out of our theory.

Case c. $H(L)=X_4$, and we get, exactly as before, a group isomorphic to $\boldsymbol{Z}/4\boldsymbol{Z}\oplus\boldsymbol{Z}/4\boldsymbol{Z}$ acting on $\boldsymbol{P}_3$. Again, in suitable coordinates, it is the group generated by

$$U=\begin{pmatrix}0&1&0&0\\0&0&1&0\\0&0&0&1\\1&0&0&0\end{pmatrix} \quad\text{and}\quad V=\begin{pmatrix}1&0&0&0\\0&i&0&0\\0&0&-1&0\\0&0&0&-i\end{pmatrix}.$$

In this case, it will be easier to write everything down if we spell out the connection with our previous notation more precisely. First of all, the group generated by U, V and scalar matrices is exactly $\mathscr{G}(\delta)$, if we let U correspond to $(1;1,0)$ and let V correspond to $(1;0,\chi)$ where $\chi\in\mathrm{Hom}(\boldsymbol{Z}/4\boldsymbol{Z},k^*)$ is defined by $\chi(1)=i$. Moreover, the affine space k^4 on which U, V act is exactly $V(\delta)$ if we let $(\alpha_0,\alpha_1,\alpha_2,\alpha_3)\in k^4$ correspond to the function f:

$$f(i)=\alpha_i.$$

[One should check that the action of $\mathscr{G}(\delta)$ on $V(\delta)$ goes over to matrix multiplication.] Regard $\boldsymbol{P}_3$ as the set of hyperplanes in k^4 (á la GROTHENDIECK), so that the following are isomorphic:

$$V(\delta)\cong k^4\cong\Gamma(\boldsymbol{P}_3,\boldsymbol{o}(1))\cong\Gamma(X,L).$$

Here homogeneous coordinates X_0, X_1, X_2, X_3 in $\boldsymbol{P}_3$ correspond to the points $(1,0,0,0)$, $(0,1,0,0)$, $(0,0,1,0)$, and $(0,0,0,1)$ in k^4 and to the functions $\delta_0, \delta_1, \delta_2, \delta_3$ in $V(\delta)$. And the 4-tuple $(q_L(0), q_L(1), q_L(2), q_L(3))$ is a set of homogeneous coordinates of $\varphi(e) \in \mathbf{P}_3$. To determine the quadratic equations satisfied by the coordinates X_i on $\varphi(X)$, we refer back to the proof of Theorem 1, § 4, where the quadratic expressions in the functions δ_a were expressed in terms of a *basis* of $\Gamma(X, L^2)$: or, more precisely, in terms of functions on $K(2\delta) \cong \mathbf{Z}/8\mathbf{Z}$. We find

$$\begin{aligned}
\delta_0 * \delta_0 + \delta_2 * \delta_2 &= (q_{L^2}(0) + q_{L^2}(4)) \cdot (\delta_0 + \delta_4)\\
\delta_0 * \delta_1 + \delta_2 * \delta_3 &= (q_{L^2}(1) + q_{L^2}(5)) \cdot (\delta_1 + \delta_5)\\
2\delta_0 * \delta_2 &= (q_{L^2}(2) + q_{L^2}(6)) \cdot (\delta_2 + \delta_6)\\
\delta_0 * \delta_3 + \delta_2 * \delta_1 &= (q_{L^2}(3) + q_{L^2}(7)) \cdot (\delta_3 + \delta_7)\\
\delta_1 * \delta_1 + \delta_3 * \delta_3 &= (q_{L^2}(0) + q_{L^2}(4)) \cdot (\delta_2 + \delta_6)\\
2\delta_1 * \delta_3 &= (q_{L^2}(2) + q_{L^2}(6)) \cdot (\delta_0 + \delta_4)\\
\delta_0 * \delta_0 - \delta_2 * \delta_2 &= (q_{L^2}(0) - q_{L^2}(4)) \cdot (\delta_0 - \delta_4)\\
\delta_0 * \delta_1 - \delta_2 * \delta_3 &= (q_{L^2}(1) - q_{L^2}(5)) \cdot (\delta_1 - \delta_5)\\
\delta_0 * \delta_3 - \delta_2 * \delta_1 &= (q_{L^2}(3) - q_{L^2}(7)) \cdot (\delta_3 - \delta_7)\\
\delta_1 * \delta_1 - \delta_3 * \delta_3 &= (q_{L^2}(0) - q_{L^2}(4)) \cdot (\delta_2 - \delta_6).
\end{aligned}$$

Therefore:

$$\begin{aligned}
(q_{L^2}(2) + q_{L^2}(6)) \cdot [\delta_0 * \delta_0 + \delta_2 * \delta_2] &= (q_{L^2}(0) + q_{L^2}(4)) \cdot [2\delta_1 * \delta_3],\\
(q_{L^2}(0) + q_{L^2}(4)) \cdot [2\delta_0 * \delta_2] &= (q_{L^2}(2) + q_{L^2}(6)) \cdot [\delta_1 * \delta_1 + \delta_3 * \delta_3].
\end{aligned}$$

Note that $q_{L^2}(2) + q_{L^2}(6) = 2q_{L^2}(2)$ is not equal to 0, or else the above equations do not define an integral domain. Set $\lambda = (q_{L^2}(0) + q_{L^2}(4))/2q_{L^2}(2)$. Then $\varphi(X)$ is given by the 2 equations:

$$\begin{aligned}
X_1^2 + X_3^2 &= 2\lambda X_0 X_2\\
X_0^2 + X_2^2 &= 2\lambda X_1 X_3.
\end{aligned}$$

Conversely, for any $\lambda \neq 0, \pm 1, \pm i, \infty$, this defines an elliptic curve invariant under $\mathscr{G}(\delta)$. Now consider the coordinates of $\varphi(e)$: by symmetry, $q_L(1) = q_L(3)$, i.e., $\varphi(e)$ lies on the plane $X_1 = X_3$. Therefore it satisfies:

$$\begin{aligned}
q_L(1)^2 &= \lambda q_L(0) \cdot q_L(2)\\
q_L(0)^2 + q_L(2)^2 &= 2\lambda q_L(1)^2.
\end{aligned}$$

Therefore, λ can be calculated from the coordinates of $\varphi(e)$ by

$$\lambda = \frac{q_L(1)^2}{q_L(0) \cdot q_L(2)}$$

and the q_L's themselves also satisfy:

$$q_L(0)^3 \cdot q_L(2) + q_L(0) \cdot q_L(2)^3 = 2 \cdot q_L(1)^4.$$

This equation is the one equation to which RIEMANN's theta relations reduce in this case: it is a simple transformation of the ancient theta relation of JACOBI. Moreover, one can show that this is the only equation satisfied by the q_L's, aside from the inequalities:

$$q_L(n) \neq 0, \quad \text{any } n$$
$$q_L(0) \neq \pm q_L(2).$$

JACOBI's theta relation comes out by using the usual basis for the vector space of theta-null values:

$$\vartheta\begin{bmatrix}0\\0\end{bmatrix} = q(0) + q(2)$$
$$\vartheta\begin{bmatrix}0\\1\end{bmatrix} = q(1) + q(3) = 2q(1)$$
$$\vartheta\begin{bmatrix}1\\0\end{bmatrix} = q(0) - q(2)$$
$$\vartheta\begin{bmatrix}1\\1\end{bmatrix} = q(1) - q(3) = 0.$$

(Compare the comments at the end of § 3.) We get:

$$\vartheta\begin{bmatrix}0\\0\end{bmatrix}^4 = \vartheta\begin{bmatrix}0\\1\end{bmatrix}^4 + \vartheta\begin{bmatrix}1\\0\end{bmatrix}^4.$$

and

$$\vartheta\begin{bmatrix}0\\0\end{bmatrix}, \quad \vartheta\begin{bmatrix}0\\1\end{bmatrix}, \quad \vartheta\begin{bmatrix}1\\0\end{bmatrix} \neq 0.$$

Case d. $H(L) = X_2 \cong \mathbf{Z}/2\mathbf{Z} \oplus \mathbf{Z}/2\mathbf{Z} \oplus \mathbf{Z}/2\mathbf{Z} \oplus \mathbf{Z}/2\mathbf{Z}$. This group acts on the $\mathbf{P}_3$ ambient to $\varphi(X)$. In suitable coordinates, it is the group $\mathcal{G}(2, 2)$ generated by the matrices:

$$\begin{pmatrix} 0&1&0&0\\ 1&0&0&0\\ 0&0&0&1\\ 0&0&1&0 \end{pmatrix}, \quad \begin{pmatrix} 0&0&1&0\\ 0&0&0&1\\ 1&0&0&0\\ 0&1&0&0 \end{pmatrix},$$

$$\begin{pmatrix} 1&0&0&0\\ 0&-1&0&0\\ 0&0&1&0\\ 0&0&0&-1 \end{pmatrix}, \quad \begin{pmatrix} 1&0&0&0\\ 0&1&0&0\\ 0&0&-1&0\\ 0&0&0&-1 \end{pmatrix}.$$

Then φX is a quartic surface, invariant under this group and containing only a finite number of points fixed under non-trivial transformations in this group. This implies that φX is given by the zeroes of an equation:

$$F(x,y,z,w)=A(x^4+y^4+z^4+w^4)+B(xyzw)+C(x^2y^2+z^2w^2)+ \\ +D(x^2w^2+y^2z^2)+E(x^2z^2+y^2w^2)=0.$$

These equations define a 4-dimensional family of quartics with no base points: the generic member of this family is non-singular so it could not be equal to φX. But as soon as a surface in this family has 1 node, it acquires 16 of them: namely the images of this node under the group $\mathscr{G}(2,2)$. Such a node exists if and only if

$$\Delta(A,B,C,D,E)=0$$

where Δ is the discriminant of F. The nodes are the points $\varphi(a)$, $a\in X_2$, and they may exist anywhere at all in $\boldsymbol{P}_3$, except on a degenerate subvariety that we will not describe. Any quartic of the above type with exactly 16 nodes and no higher singularities is of the type φX for some abelian surface X.

References

[1] BAILY, W. jr.: On the theory of ϑ-functions, the moduli of abelian varieties, and the moduli of curves. Annals of Math. **75**, 342–381 (1962).

[2] CARTIER, P.: Unitary representations and theta functions. Proc. of 1965 AMS Summer Institute in Boulder, to appear.

[3] GROTHENDIECK, A.: Séminaire de géométrie algébrique. Inst. des Hautes Études Sci., 1960–61 (Mimeographed).

[4] — Séminaire: Schemas en groupes. Inst. des Hautes Études Sci., 1963–64 (Mimeographed).

[5] —, et J. DIEUDONNÉ: Éléments de la géométrie algébrique. Publ. de l'Inst. des Hautes Études Sci., No 4, 8, 11, ...

[6] IGUSA, J.-I.: On the graded ring of theta-constants. Am. J. Math. **86**, 219–246 (1964); **88**, 221–236 (1966).

[7] LANG, S.: Abelian varieties. New York: Interscience 1958.

[8] MACKEY, G.: On a theorem of Stone and von Neumann. Duke Math. J. **16**, 313–340 (1949).

[9] MUMFORD, D.: Geometric invariant theory. Berlin-Heidelberg-New York: Springer 1965.

[10] — Curves on an algebraic surface (to appear in Annals of Math. Studies).

[11] SIEGEL, C.: Moduln abelscher funktionen. Nachrichten der Akad. Göttingen, 1964.

[12] WEIL, A.: Sur certaines groupes d'operateurs unitaire. Acta Math. **111**, 145–211 (1964).

Department of Mathematics
Harvard University
Cambridge, Massachussetts

(Received January 17, 1966)

Inventiones math. 3, 75–135 (1967)

On the Equations Defining Abelian Varieties. II

D. MUMFORD (Cambridge, Mass.)

Contents

In the first part of this paper, we have analyzed a single abelian variety X. In particular, if L is an ample invertible sheaf on X, we have analyzed the vector space $\Gamma(X, L)$ and the ring

$$\bigoplus_n \Gamma(X, L^n)$$

and have shown 1) how to choose canonical bases of these vector spaces, 2) how to express this ring as a quotient of a polynomial ring by an explicit homogeneous ideal involving coefficients which are essentially the "theta-null werte" of X. In this second part, we shall apply the first part to embed both the moduli spaces of abelian varieties, and the inverse limit of these spaces over successively higher levels, as open sets in projective schemes associated to homogeneous coordinate rings defined by *explicit* homogeneous ideals. We also introduce algebraic theta functions, defined on a 2-adic vector space in terms of which our results on moduli take on a simple form.

I want to offer some explanation of why the 2-adics play such a central role in this theory. The situation is this: if you stick to abelian varieties of char. $p (p \neq 2)$, then you can build up a theory of theta functions for these over any (restricted) product

$$\prod_{l \in S}' Q_l$$

where S is any set of primes containing 2, but not containing p. In other words, Q_2 always has to be there, but you can throw in plenty of other factors if you like. Using only Q_2 seemed to have two advantages: (i) you can deal simultaneously with all characteristics except 2, and (ii) the resulting theta-functions are more concise, i.e., are defined on the smallest locally compact group which admits them. I might have written a *general* theory for some arbitrary set S – clearly this is the accepted French approach – but there seemed no point in not sticking to the simplest and most basic case. The essential features are related to the fact that multiplication by 2 does not preserve Haar measure.

§ 6. Structure of the Moduli Space

To study questions of moduli, we must first have a theory of families of the objects to be classified. Therefore, we must generalize our theory to abelian schemes:

Definition. Let S be a scheme. An *abelian scheme* $\mathscr{X}$ over S is a group scheme $\mathscr{X}$ finitely presented over S such that the projection $\pi: \mathscr{X} \to S$ is proper and smooth and the geometric fibres of π are connected.

For some of the basic facts about abelian schemes, we refer the reader to [9], Ch. 6[1].

Definition. Let L be an invertible sheaf on S. Then 1. $H(L)$ is the group of sections $\alpha: S \to \mathscr{X}$ of π such that if $T_\alpha: \mathscr{X} \to \mathscr{X}$ is translation by α, then

$$T_\alpha^* L \cong L \otimes \pi^* M$$

for some invertible sheaf M on S.

2. $H_0(L)$ is the subgroup of those α such that

$$T_\alpha^* L \cong L.$$

3. $\mathscr{G}(L)$ is the group of pairs (α, φ) where $\alpha \in H_0(L)$ and $\varphi: L \to T_\alpha^* L$ is an isomorphism.

Following a familiar procedure, we note first that H and $\mathscr{G}$ can be extended to *functors* from the category of S-schemes to the category of groups:

Definition. For all S-schemes $f: T \to S$, let $H^f(L) = H(L')$, $\mathscr{G}^f(L) = \mathscr{G}(L')$, where if F is the morphism in the diagram:

$$\begin{array}{ccc} \mathscr{X} \times_S T & \xrightarrow{F} & \mathscr{X} \\ \downarrow & & \downarrow \\ T & \xrightarrow{f} & S \end{array}$$

then $L' = F^* L$. Then $f \mapsto H^f(L)$, $f \mapsto \mathscr{G}^f(L)$ are functors in an obvious way.

Proposition 1. *Assume that L is relatively ample over S. Then the 2 functors $f \mapsto H^f$ and $f \mapsto \mathscr{G}^f$ are representable by group schemes $\underline{H}(L)$, $\underline{\mathscr{G}}(L)$ flat and of finite presentation over S. $\underline{H}(L)$ is a closed sub-group scheme of $\mathscr{X}$ itself, finite over S, and there is a canonical exact sequence:*

$$0 \to \mathbf{G}_{m,S} \to \underline{\mathscr{G}}(L) \to \underline{H}(L) \to 0.$$

[1] If any doubt should arise as to whether results in [9] are still valid if S is non-noetherian, it should be dispelled by noticing that any abelian scheme over an affine S is obtained by base extension from an abelian scheme over Spec(R), where R is a Z-algebra of finite type, [5], Ch. 4, §§ 8, 11.

Proof. First of all, $\underline{H}(L)$ is nothing but the kernel of the canonical homomorphism: $\Lambda(L)\colon \mathscr{X} \to \hat{\mathscr{X}}$. $\Lambda(L)$ is flat and finite and of finite presentation (cf. [9], p. 122), hence $\underline{H}(L)$ is flat and finite and finitely presented over S. Secondly, the morphism from the functor $\mathscr{G}^f$ to the functor H^f is represented (relatively) by G_m-bundles. In fact, if $\alpha\colon T \to \mathscr{X} \times_S T$ is an element of $H^f(T)$, and if

$$M = \pi_*(T_\alpha^*(L') \otimes L'^{-1})$$

[here L' is the sheaf on $\mathscr{X} \times_S T$ induced by L and π is the projection from $\mathscr{X} \times_S T$ to T], then

$$T_\alpha^* L' \cong L' \otimes \pi^* M,$$

and multiplication by non-zero sections of M defines the isomorphisms from L' to $T_\alpha^* L'$. Therefore the relative functor in this case is represented by the line bundle M on T corresponding to M^{-1}, minus its 0-section. *Q.E.D.*

Definition. Let $\delta = (d_1, d_2, \ldots, d_g)$ be any set of elementary divisors (d_i integers, $\geqq 1$, $d_{i+1} | d_i$). Define a functor $\mathscr{G}(\delta)$ on the category of all schemes S to the category of groups by:

$\mathscr{G}^S(\delta) =$ group of triples (α, x, l), $\alpha \in \Gamma(S, 0_S^*)$,

x is a map from the set $\pi_0(X)$ of connected components of X to $K(\delta)$, the discrete group $\oplus \mathbf{Z}/d_i\mathbf{Z}$.

$l = (l_1, \ldots, l_g)$, where l_i is a d_i-th root of 1 in $\Gamma(S, 0_S^*)$.

Multiplication is:

$$(\alpha, x, l) \cdot (\alpha', x', l') = (\alpha \cdot \alpha' \cdot l'(x), x + x', l + l')$$

where

$$l + l' = (l_1 \cdot l'_1, \ldots, l_g \cdot l'_g)$$

$$l'(x), \quad \text{on the component } Y, = \prod_{i=1}^{g} l_i'^{a_i}$$

$$\text{if } x(Y) = (a_1, \ldots, a_g).$$

[We add the l's instead of multiplying them to be consistent with our previous notation.]

It is easy to check that $S \mapsto \mathscr{G}^S(\delta)$ is represented by a group scheme $\underline{\mathscr{G}}(\delta)$, flat and of finite type over $\mathbf{Z}$, that fits into a canonical exact sequence:

$$0 \to G_m \to \underline{\mathscr{G}}(\delta) \to \underline{H}(\delta) \to 0$$

of group schemes, where

$$\underline{H}(\delta) = [\bigoplus_i \mathbf{Z}/d_i\mathbf{Z}] \oplus [\bigoplus_i \boldsymbol{\mu}_{d_i}]$$

and where $\mathbf{Z}/d_i\mathbf{Z}$ is taken as a discrete reduced group scheme and $\boldsymbol{\mu}_{d_i}$ is the usual group scheme of d_i-th roots of 1.

6*

Definition. A *ϑ-structure* on a relatively ample invertible sheaf L is an isomorphism over S of the group schemes $\underline{\mathcal{G}}(L)$ and $\underline{\mathcal{G}}(\delta)\times S$, for some δ, which is the identity on the sub-group schemes $G_{m,S}$. When this exists, δ is called the *type* of L.

In this definition, we have included some types of non-separable invertible sheaves [in fact, all ample invertible sheaves on an abelian variety with p^g points of order p have a type in the above sense] because in the present categorical approach it is no trouble. However, this was just for the fun of it and we shall now restrict ourselves to the separable case.

Fix $\delta=(d_1,\dots,d_g)$. *Assume all* d_i *even. Assume all schemes are schemes over* Spec $\mathbf{Z}[d^{-1}]$, *where*

$$d=\prod_{i=1}^{g} d_i.$$

Definition. An invertible sheaf L on $\mathcal{X}/S$ is *symmetric* if $\iota^*L\cong L$, where $\iota\colon \mathcal{X}\to\mathcal{X}$ is the inverse. It is *totally symmetric* if there is an isomorphism $\varphi\colon L\xrightarrow{\sim}\iota^*L$ which restricts to the identity on $L\otimes\underline{O}_{\mathcal{X}_2}$, where $\mathcal{X}_2\subset\mathcal{X}$ is the kernel of 2δ. It is *normalized* if $\varepsilon^*L\cong\underline{O}_S$ where $\varepsilon\colon S\to\mathcal{X}$ is the identity section.

Definition. Let $\lambda\colon \underline{\mathcal{G}}(L)\xrightarrow{\sim}\underline{\mathcal{G}}(\delta)\times S$ be a ϑ-structure for a symmetric relatively ample invertible sheaf L on $\mathcal{X}$. Let $\psi\colon L\xrightarrow{\sim}\iota^*L$ be any isomorphism. We define:

i) an automorphism δ_{-1} of the functor $f\to\mathcal{G}^f(L)$. Given $f\colon T\to S$, let $L'=L\otimes\underline{O}_T$ be the induced sheaf on $\mathcal{X}\times_S T$. Let $\alpha\colon T\to\mathcal{X}\times_S T$ be a section and

$$\varphi\colon L'\xrightarrow{\sim} T_\alpha^*(L')$$

an isomorphism, so that $(\alpha,\varphi)\in\mathcal{G}^f(L)$. Let $\psi'\colon L'\to\iota^*L'$ be the isomorphism induced by ψ. Then

$$\delta_{-1}((\alpha,\varphi))=\big(\iota\circ\alpha,(T_{\iota\circ\alpha}^*\psi')^{-1}\circ\iota^*\varphi\circ\psi'\big),$$

i.e.,

$$\begin{array}{ccc} L'\xrightarrow[\sim]{\psi'}\iota^*L'\xrightarrow[\sim]{\iota^*\varphi} & \iota^*(T_\alpha^*L') & \\ & \| & \\ & T_{\iota\circ\alpha}^*(\iota^*L') & \xleftarrow[T_{\iota\circ\alpha}\psi']{\sim} T_{\iota\circ\alpha}^*L', \end{array}$$

where $\iota\colon\mathcal{X}\times_S T\to\mathcal{X}\times_S T$ is the inverse.

ii) The automorphism δ_{-1} of the functor induces an automorphism $\underline{\delta}\colon\underline{\mathcal{G}}(L)\to\underline{\mathcal{G}}(L)$ of the scheme.

iii) Similarly, the map

$$(\alpha,x,l)\mapsto(\alpha,-x,-l)$$

where if $l=(l_1, \ldots, l_g)$, then $-l=(l_1^{-1}, \ldots, l_g^{-1})$ gives an automorphism D_{-1} of the functor $S \to \mathscr{G}^S(\delta)$.

iv) This induces an automorphism D of the group scheme $\underline{\mathscr{G}}(\delta)$.

v) Then λ is *symmetric* if $\underline{D} \circ \lambda = \lambda \circ \underline{\delta}$.

Definition. We consider triples of the following type:

i) an abelian scheme $\mathfrak{X}$ over S,

ii) a relatively ample, totally symmetric, normalized invertible sheaf L on $\mathfrak{X}$,

iii) a symmetric ϑ-structure $\lambda: \underline{\mathscr{G}}(L) \xrightarrow{\sim} \underline{\mathscr{G}}(\delta) \times S$ for L.

We shall call this triple an *abelian scheme with a δ-marking.*

Definition. For all schemes S, let $\mathscr{M}_\delta(S)$ denote the set of abelian schemes $\mathfrak{X}$ over S with δ-markings, taken modulo isomorphisms. As S varies, these sets form a functor $\mathscr{M}_\delta$ in S. This will be called the *moduli functor for abelian schemes with δ-marking.*

The object of this section will be to show that $\mathscr{M}_\delta$ is representable, and to represent it by an open subset of a definite projective variety. The next step is to study the representations of $\mathscr{G}(\delta)$.

Definition. Let V_δ be the free $\mathbf{Z}[d^{-1}]$-module of functions from $K(\delta)$ to $\mathbf{Z}[d^{-1}]$. Then for all schemes S, and invertible sheaves L on S, $V_\delta \otimes_{\mathbf{Z}} L$ is the sheaf of functions from $K(\delta)$ to L. The discrete group $\mathscr{G}^S(\delta)$ acts $\underline{O}_S$-linearly on this sheaf, exactly as $\mathscr{G}(\delta)$ acted on V_δ in § 1: e.g., if $(\alpha, x, l) \in \mathscr{G}^S(\delta)$,

$$f \in \Gamma(U, V_\delta \otimes_{\mathbf{Z}} L).$$

Assume for simplicity that U is connected, that $x(U)=(a_1, \ldots, a_g)$ and that $l=(l_1, \ldots, l_g)$. Then (α, x, l) takes f into f^*, where

$$f^*(b_1, \ldots, b_g) = \alpha \cdot \prod l_i^{b_i} \cdot f(a_1+b_1, \ldots, a_g+b_g).$$

Let $\mathbf{E}_\delta = V(V_\delta)$: this is a vector bundle of rank d over Spec $\mathbf{Z}[d^{-1}]$, and it is a direct sum of trivial line bundles $\mathbf{L}_a$ with canonical sections $[a]$: Recall that, in general, an S-valued point of $\mathbf{E}_\delta$ corresponds to a homomorphism from V_δ to $\Gamma(S, \underline{O}_S)$. Then $[a]$ corresponds to the homomorphism

$$f \mapsto f(a)$$

from $V(\delta)$ to $\mathbf{Z}[d^{-1}]$. All the actions of $\mathscr{G}^S(\delta)$ on $V_\delta \otimes \underline{O}_S$ can be put together dually in an anti-representation of $\underline{\mathscr{G}}(\delta)$ on $\mathbf{E}_\delta$, i.e., a representation in which the order of multiplication is reversed. This is clear from a functorial point of view, but it may be useful to define this anti-representation directly by putting together representations on the various subgroups:

the subgroup G_m, with S-valued points $(\alpha, 0, 0)$, acts by homotheties on E_δ,

the discrete subgroup $\oplus Z/d_i Z$, with S-valued points $(1, a, 0)$, acts by permuting the sections $[a]$: thus the point b takes $[a]$ to $[a+b]$,

the discrete but twisted subgroup $\oplus \mu_{d_i}$ with S-valued points $(1,0,l)$, acts diagonally: thus the point $(l_1, \ldots, l_g)$ takes $[a]$ to $\Pi\, l_i^a \cdot [a]$.

Proposition 2. *Let S be a scheme and let*

$$\rho\colon [\underline{\mathscr{G}}(\delta) \times S] \times_S F \to F$$

be an anti-representation over S of $\underline{\mathscr{G}}(\delta) \times S$ *on a vector bundle* F *over S of rank d. Assume that the subgroup* G_m *acts on* F *in the standard way. Then there is a line bundle* L *over S and an isomorphism*

$$F \xrightarrow{\sim} [E_\delta \times S] \otimes L$$

(where $\otimes$ *denotes tensor product of vector bundles over S) such that the action of* $\underline{\mathscr{G}}(\delta)$ *on* F *corresponds to the above action of* $\underline{\mathscr{G}}(\delta)$ *on* E_δ *tensored with the trivial action on* L. *Moreover, this isomorphism is unique up to multiplication by an element of* $\Gamma(S, \underline{O}_S^*)$.

Proof. This is nearly the same as that of Prop. 3, § 1, except that we must use a basic result on the representations of μ_d established in [4]. It is shown there that if a group scheme of the form $\oplus \mu_{d_i}$ is represented, over S, in a vector bundle F, then F is a direct sum of sub-vector bundles F_a, $a \in K(\delta)$, where the group acts on F_a by the character

$$\oplus \mu_{d_i} \to G_m$$

$$(l_1, \ldots, l_g) \mapsto \prod_{i=1}^{g} l_i^{a_i}.$$

Now we realize $\oplus \mu_{d_i}$ as the subgroup of $\underline{\mathscr{G}}(\delta)$ of triples $(1, 0, l)$; and decompose F accordingly. Exactly as in §1, the action of the $Z[d^{-1}]$-valued points $\sigma_a = (1, a, 0)$ of $\mathscr{G}(\delta)$ permutes these F_a transitively. Therefore all F_a are non-empty; since the rank of F is d, each F_a is a line bundle over S. Let $L = F_0$. We set up the required isomorphism by first identifying

$$F_0 = L \cong (L_0 \times S) \otimes L$$

then identifying F_a with $(L_a \times S) \otimes L$ by using the actions of σ_a and the first identification; and then taking the direct sum. The details work out exactly as in §1. *Q.E.D.*

Now start with an abelian scheme $\mathscr{X}/S$ with δ-marking. Let L be the given sheaf on $\mathscr{X}$, and let $\boldsymbol{F}=\boldsymbol{V}(\pi_* L)$: this is a vector bundle over S of rank d. Let $f\colon T\to S$ be a morphism of schemes. Then a T-valued point of $\boldsymbol{F}/S$, i.e., a morphism g is the diagram:

$$\begin{array}{ccc} T & \xrightarrow{\;g\;} & F \\ & {\scriptstyle f}\searrow \quad \swarrow{\scriptstyle \text{canonical morphism}} & \\ & S & \end{array}$$

is the same thing as a homomorphism γ:

$$f^*(\pi_* L)\xrightarrow{\;\gamma\;}\underline{O}_T$$

of $\underline{O}_T$-modules. I claim that $\underline{\mathscr{G}}(L)$ is anti-represented over S on the vector bundle $\boldsymbol{F}$ in a canonical way. In fact, if $f\colon T\to S$ is a morphism of schemes, a T-valued point of $\underline{\mathscr{G}}(L)/S$ is given by a section

$$\alpha\colon\ T\to\mathscr{X}\times_S T$$

and an isomorphism

$$\varphi\colon\ L'\xrightarrow{\;\sim\;}T_\alpha^* L'$$

if L' is the induced sheaf on $\mathscr{X}\times_S T$. Then φ induces

$$\begin{array}{ccc} \pi_* L' & \xrightarrow{\;\varphi^*\;} & \pi_*(T_\alpha^* L') \\ \wr\| & & \wr\| \\ f^*(\pi_* L) & & (\pi\circ T_\alpha)_*(T_\alpha^* L') \\ & & \wr\| \\ & & \pi_*(T_\alpha)_* T_\alpha^*(L') \\ & & \wr\| \\ & & \pi_* L' \\ & & \wr\| \\ & & f^*(\pi_* L) \end{array}$$

by standard canonical identifications. Let the composite isomorphism on $f^*(\pi_* L)$ be called $[\varphi]$. Then this acts on a T-valued point $\gamma\colon f^*(\pi_* L)\to\underline{O}_T$ of $\boldsymbol{F}/S$ by taking it to the new point $\gamma\circ[\varphi]$. This gives us an anti-representation of the functor $\mathscr{G}(L)$ on the functor associated to $\boldsymbol{F}$: hence an anti-representation of the scheme $\underline{\mathscr{G}}(L)$ on $\boldsymbol{F}$.

But $\mathscr{X}/S$ has a given δ-marking. Hence we also have an anti-representation over S of $\underline{\mathscr{G}}(\delta)\times S$ on $\boldsymbol{F}$. Applying Proposition 2, we get an isomorphism

$$\boldsymbol{F}\xrightarrow{\;\sim\;}(\boldsymbol{E}_\delta\times S)\otimes\boldsymbol{L},$$

for some line bundle $\boldsymbol{L}$ on S, unique up to multiplication by elements of $\Gamma(S,O_S)$. In terms of sheaves, this gives us an isomorphism

$$\pi_* L\xleftarrow{\;\sim\;}V_\delta\otimes_{\mathbf{Z}} K$$

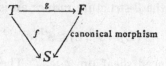

for some invertible sheaf K on S, unique up to multiplication by elements of $\Gamma(S, \underline{O}_S)$. The fact that the isomorphism of $\boldsymbol{F}$ and $(E_\delta \times S) \otimes \boldsymbol{L}$ commutes with pair of actions of $\underline{\mathscr{G}}(L)$ and $\underline{\mathscr{G}}(\delta)$ gives a corresponding fact on the sheaf level:

for all $f: T \to S$, let the ϑ-structure induce an isomorphism

$$\mathscr{G}(L') = \mathscr{G}^f(L) \cong \mathscr{G}^T(\delta)$$

where L' is the induced sheaf on $\mathscr{X} \times_S T$. Then the action of $\mathscr{G}(L')$ on $\pi_* L'$ – which is canonically isomorphic to $f^* \pi_* L$ – goes over under this isomorphism to the action of $\mathscr{G}^T(\delta)$ on $V_\delta \otimes_{\boldsymbol{Z}} f^* K$.

It follows, incidentally, that the above isomorphism of $\pi_* L$ and $V_\delta \otimes_{\boldsymbol{Z}} K$ determines the isomorphism of $\mathscr{G}(L')$ and $\mathscr{G}^T(\delta)$ for all f: hence it determines the given isomorphism of $\underline{\mathscr{G}}(L)$ and $\underline{\mathscr{G}}(\delta) \times S$. Now assume L is relatively very ample: this occurs if all d_i are divisible by 4 for example. Then we get a closed immersion over S:

$$i: \mathscr{X} \hookrightarrow \boldsymbol{P}(\pi_* L) \cong \boldsymbol{P}[V_\delta \otimes_{\boldsymbol{Z}} K] \cong \boldsymbol{P}(V_\delta) \times S,$$

which is determined by the original δ-marking on $\mathscr{X}$. *Conversely*, this immersion determines the δ-marking: the sheaf L, for example, is obtained

a) by pulling back $\underline{O}_{\boldsymbol{P}}(1)$ via $p_1 \circ i: \mathscr{X} \to \boldsymbol{P}(V_\delta)$,

b) by then normalizing this sheaf on the identity section. In other words

$$(*) \qquad L \cong (p_1 \circ i)^* [\underline{O}_{\boldsymbol{P}}(1)] \otimes (p_1 \circ i \circ \varepsilon \circ \pi)^* [\underline{O}_{\boldsymbol{P}}(-1)].$$

And the morphism from $\underline{\mathscr{G}}(\delta) \times S$ to $\underline{\mathscr{G}}(L)$ is determined as follows: let $K = (p_1 \circ i \circ \varepsilon)^* (\underline{O}_{\boldsymbol{P}}(-1))$. Then the isomorphism (*) determines the isomorphism:

$$\begin{aligned} \pi_* L &\cong \pi_* [(p_1 \circ i)^* (\underline{O}_{\boldsymbol{P}}(1))] \otimes K \\ &\cong p_{2,*} [\underline{O}_{\boldsymbol{P}}(1) \otimes \underline{O}_S] \otimes K \\ &\cong V_\delta \otimes_{\boldsymbol{Z}} K, \end{aligned}$$

which, as we just saw, determines the ϑ-structure. Summarizing, the abelian scheme $\mathscr{X}/S$ with δ-marking determines, and is determined by the closed immersion:

$$i: \mathscr{X} \hookrightarrow \boldsymbol{P}(V_\delta) \times S.$$

Recall that the group structure on $\mathscr{X}$ is determined completely by the identity section $\varepsilon: S \to \mathscr{X}$, ([9], p. 117). Therefore, the whole functor $\mathscr{M}_\delta$ is isomorphic to a subfunctor of the functor $\mathrm{Hilb}^{(1)}_{\boldsymbol{P}(V_\delta)}$ of all flat families of closed subschemes of $\boldsymbol{P}(V_\delta)$ with distinguished section (cf. [9], Ch. 0, § 5).

In particular, we have defined, from $\mathscr{X}/S$ with δ-marking, the S-valued point of $\boldsymbol{P}(V_\delta)$: $p_1 \circ i \circ \varepsilon: S \to \boldsymbol{P}(V_\delta)$. It is not hard to verify that this defines a morphism of functors:

$$\mathscr{M}_\delta \xrightarrow{t} h_{\boldsymbol{P}(V_\delta)}.$$

The next step is to describe the image. First define a closed subscheme $\overline{M}_\delta \subset \boldsymbol{P}(V_\delta)$ by RIEMANN's theta relations:

let $Q(a) \in V_\delta$ be the function which is 1 at a, and 0 elsewhere;

let $Z_2 \subset K(\delta)$ be the subgroup of points of order 2; embed $K(\delta) \subset K(2\delta)$ as before.

1. For all elements $a, b, c, d \in K(2\delta)$ such that they are all congruent modulo $K(\delta)$, and for all $l \in Z_2$, set

$$\begin{aligned}
&\Big[\sum_{\eta \in Z_2} l(\eta) \cdot Q(a+b+\eta) \cdot Q(a-b+\eta)\Big] \times \\
&\qquad \times \Big[\sum_{\eta \in Z_2} l(\eta) \cdot Q(c+d+\eta) \cdot Q(c-d+\eta)\Big] - \\
&\qquad - \Big[\sum_{\eta \in Z_2} l(\eta) \cdot Q(a+d+\eta) \cdot Q(a-d+\eta)\Big] \times \\
&\qquad \times \Big[\sum_{\eta \in Z_2} l(\eta) \cdot Q(c+b+\eta) \cdot Q(c-b+\eta)\Big] = 0;
\end{aligned}$$

2. For all $a \in K(\delta)$, set

$$Q(a) - Q(-a) = 0.$$

The main result can now be stated:

Theorem. *If $d_1, \ldots, d_g$ are all divisible by* 8, *then there is an open subset M_δ of $\overline{M}_\delta$ such that t defines an isomorphism of $\mathscr{M}_\delta$ with the functor of points of the subscheme M_δ of $\boldsymbol{P}(V_\delta)$.*

We can go even further: not only can we identify the moduli scheme M_δ, but we can write down the universal projective family of abelian varieties over M_δ obtained by the immersion i. Define

$$\bar{A}_\delta \subset \boldsymbol{P}(V_\delta) \times \overline{M}_\delta$$

by the relations:

let X_a be the same function as $Q(a)$, but in the copy of V_δ in the first factor:

1. For all a, b, c, d, l as before, set

$$\begin{aligned}
&\Big[\sum_{\eta \in Z_2} l(\eta) \cdot Q(c+d+\eta) \cdot Q(c-d+\eta)\Big] \cdot \Big[\sum_{\eta \in Z_2} l(\eta) \cdot X_{a+b+\eta} \cdot X_{a-b+\eta}\Big] - \\
&- \Big[\sum_{\eta \in Z_2} l(\eta) \cdot Q(c+b+\eta) \cdot Q(c-b+\eta)\Big] \cdot \Big[\sum_{\eta \in Z_2} l(\eta) \cdot X_{a+d+\eta} \cdot X_{a-d+\eta}\Big] = 0.
\end{aligned}$$

Thus via the projection p_2, $\bar{A}_\delta$ is a projective scheme over $\bar{M}_\delta$. Moreover, the diagonal Δ is a section

$$\varepsilon\colon \bar{M}_\delta \to \bar{A}_\delta$$

of the projective scheme $\bar{A}_\delta/\bar{M}_\delta$.

Further Theorem. *Let $A_\delta = p_1^{-1}(M_\delta)$. Then A_δ/M_δ with section Δ is an abelian scheme over M_δ and it is equal to the universal abelian scheme, with its identity section, embedded in $\mathbf{P}(V_\delta)\times M_\delta$ by the morphism i defined earlier in this section.*

The remainder of this section will be devoted to proving these theorems.

Step 1. Without passing to any subset of $\bar{M}_\delta$ at all, $\bar{A}_\delta/\bar{M}_\delta$ has a kind of ϑ-structure. In fact, let M be the sheaf induced on $\bar{A}_\delta$ by $\underline{O}_{\mathbf{P}}(1)$. Then for all morphisms

$$f\colon T \to \bar{M}_\delta,$$

let M' be the induced sheaf on $\bar{A}_\delta \times_{\bar{M}_\delta} T$. We get a canonical homomorphism:

$$\{T\text{-valued points of } \underline{\mathcal{G}}(\delta)\} \to \left\{\begin{array}{l}\text{group of } T\text{-automorphisms}\\ \lambda\colon \bar{A}_\delta \times_M T \to \bar{A}_\delta \times_M T,\\ \text{plus isomorphisms}\\ \varphi\colon M' \xrightarrow{\sim} \lambda^* M'\end{array}\right\}.$$

And whenever $\bar{A}_\delta \times_{\bar{M}_\delta} T$ is the subscheme of $\mathbf{P}(V_\delta)\times T$ induced from an abelian scheme with δ-marking, these homomorphisms give us back the ϑ-structure. To define this, assume for simplicity that T is connected, and that

$$(\alpha, x, l) \in \mathcal{G}^T(\delta),$$

$$x = (a_1, \ldots, a_g),$$

$$l = (l_1, \ldots, l_g).$$

Then we get a projective transformation μ of $\mathbf{P}(V_\delta)\times T$ by the linear map

$$V_\delta \otimes \Gamma(T, \underline{O}_T) \to V_\delta \otimes \Gamma(T, \underline{O}_T),$$

$$f \mapsto f^*,$$

$$f^*((b_1, \ldots, b_g)) = \alpha \cdot \prod l_i^{b_i} \cdot f(a_1 + b_1, \ldots, a_g + b_g)$$

(regarding $V_\delta \otimes \Gamma(T, \underline{O}_T)$ as $\Gamma(T, \underline{O}_T)$-valued functions on $K(\delta)$). In particular,

$$X_b \otimes 1 \to X_{b-x} \otimes (\alpha \cdot \prod l_i^{b_i - a_i}).$$

One checks immediately that the equations defining $\bar{A}_\delta$ are invariant under this substitution. Therefore the projective transformation μ restricts to an automorphism λ of $\bar{A}_\delta \times_{\bar{M}_\delta} T$. But the linear map defining μ also defines an isomorphism

$$\psi\colon [\underline{O}_{\boldsymbol{P}}(1) \otimes \underline{O}_T] \xrightarrow{\sim} \mu^*[\underline{O}_{\boldsymbol{P}}(1) \otimes \underline{O}_T].$$

And ψ restricts on $\bar{A}_\delta \times_{\bar{M}_\delta} T$ to an isomorphism

$$\varphi\colon M' \xrightarrow{\sim} \lambda^* M'.$$

Step II. In the first step, we have used a little bit the actual structure of the equations defining $\bar{M}_\delta$: now we shall use this structure in detail. We want to define first a canonical morphism

$$\bar{M}_{2\delta} \xrightarrow{\pi} \bar{M}_\delta$$

which will be used in Step V.

Let M (resp. M') be the invertible sheaf on $\bar{M}_\delta$ (resp. $\bar{M}_{2\delta}$) obtained by restricting $\underline{O}_{\boldsymbol{P}}(1)$ from the ambient $\boldsymbol{P}(V_\delta)$ (resp. $\boldsymbol{P}(V_{2\delta})$). By definition, we shall have

$$\pi^*(M^2) \cong M'^2,$$

and we define π by its effect on the sections of the very ample sheaf M^2:

$$(*)\qquad \pi^*[Q(a+b)\cdot Q(a-b)] = \sum_{\eta \in Z_2} Q'(a+\eta)\cdot Q'(b+\eta),$$

where Q and Q' are the sections of M and M' defined above, and $a, b \in K(2\delta)$ satisfy $a+b \in K(\delta)$. To see that this defines a morphism π, we must check that when these values for π^* are substituted, the following are zero:

$$(1)\qquad \begin{aligned} &\pi^*[Q(a)\cdot Q(b)] - \pi^*[Q(b)\cdot Q(a)] \\ &\pi^*[Q(a)\cdot Q(b)]\cdot \pi^*[Q(c)\cdot Q(d)] - \\ &\qquad - \pi^*[Q(a)\cdot Q(d)]\cdot \pi^*[Q(c)\cdot Q(b)] \\ &\pi^*[Q(-a)\cdot Q(b)] - \pi^*[Q(a)\cdot Q(b)] \end{aligned}$$

for all $a, b, c, d \in K(\delta)$,

$$(2)\qquad \begin{aligned} &\sum_{\eta \in Z_2} l(\eta)\cdot \pi^*[Q(a+b+\eta)\cdot Q(a-b+\eta)] \times \\ &\qquad \times \sum_{\eta \in Z_2} l(\eta)\cdot \pi^*[Q(c+d+\eta)\cdot Q(c-d+\eta)] - \\ &\qquad - \sum_{\eta \in Z_2} l(\eta)\cdot \pi^*[Q(a+d+\eta)\cdot Q(a-d+\eta)] \times \\ &\qquad \times \sum_{\eta \in Z_2} l(\eta)\cdot \pi^*[Q(c+b+\eta)\cdot Q(c-b+\eta)] \end{aligned}$$

for all $a, b, c, d \in K(2\delta)$, $l \in \hat{Z}_2$ such that a, b, c, d are congruent modulo $K(\delta)$.

The first 2 expressions in (1) are 0 by virtue of the relations imposed on the Q''s; the last expression in (1) and the expression (2) are identically zero by virtue of the commutativity and associativity of multiplication in the Q''s.

It is clear from the definition of π that if L is an invertible sheaf of type δ on an abelian variety and we choose symmetric ϑ-structures on (L, L^2), then the corresponding geometric points x_1, x_2 in $\overline{M}_\delta, \overline{M}_{2\delta}$ are related by $\pi(x_2)=x_1$. The main result that we will need is:

Hardest Lemma. *Let a geometric point x in $\overline{M}_{2\delta}$ be the theta-null point assigned to some abelian variety with 2δ-marking. Then π is étale in a neighborhood of x.*

Proof. Let $y=\pi(x)$. By the infinitesimal criterion for a morphism to be étale ([3], § 3, Cor. 3.2), we must check the following: let A be an artin local ring with residue field k, the field of definition of x. Assume that σ: $\mathrm{Spec}(A)\to\overline{M}_\delta$ is an A-valued point extending y: $\mathrm{Spec}(k)\to\overline{M}_\delta$. Then there exists one and only one A-valued point τ: $\mathrm{Spec}(A)\to\overline{M}_{2\delta}$ such that $\sigma=\pi\circ\tau$, and such that τ extends x.

In down-to-earth language, suppose that at x, the coordinates $Q'(a)$ have values $\bar{q}'(a)$ in k; and that at σ, the coordinates $Q(a)$ have values $q(a)$ in A. Then $q(a)$, $\bar{q}'(a)$ satisfy the Riemann theta relation and symmetry, and if $\bar{q}(a)$ is the image of $q(a)$ in k, then $\bar{q}'(a)$, $\bar{q}(a)$ are related by (*). We must show that there are elements $q'(a)\in A$ such that $q'(a)$ lifts $\bar{q}'(a)$, and is still a point of $\overline{M}_{2\delta}$ and such that $q'(a)$ and $q(a)$ are related by (*). We have one more thing to help us: the values $\bar{q}(a)$ and $\bar{q}'(a)$ come from an abelian variety.

The first thing to observe is that (*) nearly determines $q'(a)$: in fact, we get

$$(*)'\quad \sum_{\eta\in Z_2} l(\eta)\cdot q(a+b+\eta)\cdot q(a-b+\eta) = \sum_{\eta\in Z_2} l(\eta)\cdot q'(a+\eta)\cdot \sum_{\eta\in Z_2} l(\eta)\cdot q'(b+\eta).$$

Since the $\bar{q}'$'s come from an abelian variety and since $4\,|\,d_i$ for all i, we also have

For all $l\in Z_2$ and $a\in K(\delta)$, there is an element $b\in a+K(\delta)$ such that

$$\sum_{\eta\in Z_2} l(\eta)\cdot \bar{q}'(b+\eta)\neq 0$$

(cf. § 4, Proof of Theorem 1).

Set

$$U(a,b,l)=\sum_{\eta\in Z_2} l(\eta)\cdot q(a+b+\eta)\cdot q(a-b+\eta),$$

$$x(a,l)=\sum_{\eta\in Z_2} l(\eta)\cdot q'(a+\eta),$$

$$\bar{x}(a,l)=\sum_{\eta\in Z_2} l(\eta)\cdot \bar{q}'(a+\eta).$$

For each $l \in Z_2$, and $\alpha \in K(2\delta)/K(\delta)$, choose an element $a_0 \in K(2\delta)$ lifting α such that $\bar{x}(a_0, l) \neq 0$. Then $U(a_0, a_0, l)$ is a unit in A, and $x(a_0, l)$ is to satisfy:

i) $x(a_0, l)^2 = U(a_0, a_0, l)$, $x(a_0, l) \mapsto \bar{x}(a_0, l)$ in k.

Since $\mathrm{char}(k) \neq 2$, this determines one and only one $x(a_0, l)$. For any other $a \in K(2\delta)$ lifting α, set

ii) $x(a, l) = \dfrac{U(a, a_0, l)}{x(a_0, l)}$.

From the x's, we determine the q''s by summing over l. This proves the uniqueness of the values $q'(a)$, i.e., of the point τ, and shows that π is unramified at x. It remains to show that if $\{x(a, l)\}$, hence $\{q'(a)\}$ are determined by i), ii), then they satisfy all the requirements.

First of all, $x(a, l)$ lifts $\bar{x}(a, l)$ since $\bar{x}(a, l)$ satisfies (ii) with bars in it. Hence $q'(a)$ lifts $\bar{q}'(a)$.

Secondly, $x(a, l)$ and $U(a, b, l)$ satisfy $(*)'$: let $a, b \in K(2\delta)$ both lie over α. Then

$$\begin{aligned} x(a, l) \cdot x(b, l) &= \frac{U(a, a_0, l) \cdot U(b, a_0, l)}{x(a_0, l)^2} \\ &= \frac{U(a, a_0, l) \cdot U(b, a_0, l)}{U(a_0, a_0, l)} \\ &= U(a, b, l) \end{aligned}$$

by RIEMANN's theta relation for the q's. Therefore, the corresponding $q'(a)$'s satisfy $(*)$.

Thirdly, q' is even. In fact, I claim $x(-a, l) = x(a, l)$. This follows from the equation $U(-a, a_0, l) = U(a, a_0, l)$ which follows from the evenness of q.

Fourthly, we must check that q' satisfies RIEMANN's theta relation. This is really rather tricky. It is convenient to use a different form of these relations:

let $H = K(2\delta) \times \hat{Z}_4$

[*n.b.*: $\hat{Z}_4$ is the *discrete* group $\mathrm{Hom}(Z_4, A^*)$].

If $b_i = (a_i, l_i) \in H$, $i = 1, 2$, then set

$$\begin{aligned} C(b_1, b_2) &= \sum_{\substack{\eta_1, \eta_2 \in Z_4 \\ \eta_1 + \eta_2 \in Z_2}} l_1(\eta_1)\, l_2(\eta_2)\, q'(a_1 + \eta_1) \cdot q'(a_2 + \eta_2) \\ &= \frac{1}{2^g} \sum_{\eta \in Z_4} (l_1 + l_2)(\eta) \cdot x(a_1 + \eta, \bar{l}_1) \cdot x(a_2 + \eta, \bar{l}_2) \end{aligned}$$

[here $\bar{l}_i$ = restrictions of l_i to Z_2].

The relations are:

$$C(b_1,b_2)\cdot C(b_3,b_4)=C(b_1+\beta,b_2+\beta)\cdot C(b_3+\beta,b_4+\beta)$$

$$(**)\qquad \text{for all } b_1,b_2,b_3,b_4,\beta\in H \text{ such that}$$

$$b_1+b_2+b_3+b_4=-2\beta.$$

To check that these relations are equivalent to the ones we have been using involving U''s is quite straightforward and we omit it. [The method is the same as the one we will use in the next step.]

We also need a *third* form of these relations, that is, the Riemann form itself involving the x's (cf. § 3, right at the end):

$$x(a_1,l_1)\cdot x(a_2,l_2)\cdot x(a_3,l_3)\cdot x(a_4,l_4)$$

$$=\frac{1}{2^{2g}}\sum_{\substack{k\in\hat{Z}_2\\ \eta\in Z_4}} k(2\eta)\cdot x(a_1+b+\eta,l_1+k)\cdot x(a_2+b+\eta,l_2+k)\times$$

$$\times x(a_3+b+\eta,l_3+k)\cdot x(a_4+b+\eta,l_4+k)$$

$$(***)\qquad \text{for all } a_1,a_2,a_3,a_4,b\in K(2\delta)$$

$$l_1,l_2,l_3,l_4\in\ddot{Z}_2$$

such that

$$a_1+a_2+a_2+a_4=-2b,$$

$$l_1+l_2+l_3+l_4=0.$$

This is how to go back and forth between (**) and (***):

1) take expression (**), substitute l_1+2k_1 for l_1, l_3+2k_3 for l_3, and $m-k_1-k_3$ for m, [here $b_i=(a_i,l_i)$, and $\beta=(\alpha,m)$]. Sum over all choices of k_1, k_3, rearrange the right-hand side and you get (***).

2) take expression (***), substitute $a_1+\zeta$ for a_1, $a_2+\zeta$ for a_2, $a_3+\zeta'$ for a_3, $a_4+\zeta'$ for a_4, $b-\zeta-\zeta'$ for b, multiply by $(l_1+l_2)(\zeta)\cdot(l_3+l_4)(\zeta')$, and sum over all $\zeta,\zeta'\in Z_4$. Rearrange the right-hand side and you get (**).

The point to notice here is that you don't need all of the relations (**) to get a particular one of the relations (***), or vice versa.

Next, there are some relations (**) that we do get immediately. In fact, suppose $b_1+b_2\in 2H$. Then $a_1+a_2\in K(\delta)$ and $l_1=l_2$ on $\hat{Z}_2$. Using (*)′, one finds that $C(b_1,b_2)$ splits into a product:

$$C(b_1,b_2)=\frac{1}{2^{2g}}\Big[\sum_{\zeta\in Z_4}(l_1+l_2)(\zeta)\cdot q(a_1+a_2+2\zeta)\Big]\times$$

$$\times\Big[\sum_{\zeta\in Z_4}(l_1-l_2)(\zeta)\cdot q(a_1-a_2+2\zeta)\Big].$$

One gets similar equations for $C(b_3, b_4)$, $C(b_1+\beta, b_2+\beta)$ and $C(b_3+\beta, b_4+\beta)$: substituting, one gets (**) from the associative law.

Next, we can show that the relations (**) also hold if $b_1+b_3 \in 2H$. We prove this in 3 steps using the precise description of which relations (**) are needed to prove which relations (***) and vice versa.

i) relations (**) for all values of $b_1, b_2, b_3, b_4, \beta$ such that $b_1+b_2 \in 2H$ imply relations (***) for all a_i, l_i, b such that $a_1+a_2 \in K(\delta)$, $l_1=l_2$ on Z_2.

ii) relations (***) are symmetric under permutation of the variables. So we also get relations (***) when $a_1+a_3 \in K(\delta)$, $l_1=l_3$ on Z_2.

iii) Using the fact that $Z_4 \subset K(\delta)$, if $a_1+a_3 \in K(\delta)$, then $a_1+a_3+\zeta+\zeta' \in K(\delta)$ and we can go back: we get relations (**) whenever $b_1+b_3 \in 2H$.

Next, we can show that the square of relation (**) is always true:

$$C(b_1, b_2)^2 \cdot C(b_3, b_4)^2 = C(b_1+\beta, b_2+\beta)^2 \cdot C(b_3+\beta, b_4+\beta)^2.$$

In fact, for all $b_1, b_2 \in H$,

$$\begin{aligned} C(b_1, b_2)^2 &= C(b_1, b_2) \cdot C(b_1, -b_2) \\ &= C(0, b_2-b_1) \cdot C(0, -b_1-b_2) \end{aligned}$$

by relation (**) with $\beta = -b_1$

$$= C(b_1-b_2, 0) \cdot C(b_1+b_2, 0).$$

Therefore:

$$\begin{aligned} &C(b_1+\beta, b_2+\beta)^2 \cdot C(b_3+\beta, b_4+\beta)^2 \\ &\quad = C(b_1+b_2+2\beta, 0) \cdot C(b_1-b_2, 0) \cdot C(b_3+b_4+2\beta, 0) \times \\ &\quad\quad \times C(b_3-b_4, 0). \end{aligned}$$

But

$$b_1+b_2+2\beta = -b_3-b_4, \quad \text{and} \quad b_3+b_4+2\beta = -b_1-b_2: \quad \text{so we get}$$

$$\begin{aligned} &= C(b_1+b_2, 0) \cdot C(b_1-b_2, 0) \cdot C(b_3+b_4, 0) \cdot C(b_3-b_4, 0) \\ &= C(b_1, b_2)^2 \cdot C(b_3, b_4)^2. \end{aligned}$$

But the Riemann relations definitely hold when you take the image of $C(b_1, b_2)$, etc., in k. Therefore, if

$$\bar{C}(b_1, b_2) \cdot \bar{C}(b_3, b_4) \neq 0 \quad \text{in } k,$$

it follows that $C(b_1, b_2) \cdot C(b_3, b_4)$ and $C(b_1+\beta, b_2+\beta) \cdot C(b_3+\beta, b_4+\beta)$ are square roots of the same number, are units, and have the same images in k: hence they are equal.

Now suppose

$$b_1+b_2+b_3+b_4=-2\beta$$

$$b_1'+b_2'+b_3'+b_4'=-2\beta'$$

where $b_i+b_i'\in 2H$, all i, and $\beta+\beta'\in 2H$. Assume moreover that

$$C(b_1', b_2')\cdot C(b_3', b_4') \qquad \text{is a unit}.$$

Then we can also show that

$$C(b_1, b_2)\cdot C(b_3, b_4)=C(b_1+\beta, b_2+\beta)\cdot C(b_3+\beta, b_4+\beta).$$

In fact, we know this equality, when there are primes everywhere. Hence it suffices to show:

$$\begin{aligned}&C(b_1, b_2)\cdot C(b_1', b_2')\cdot C(b_3, b_4)\cdot C(b_3', b_4')\\ (\alpha)\qquad &=C(b_1+\beta, b_2+\beta)\cdot C(b_1'+\beta', b_2'+\beta')\cdot C(b_3+\beta, b_4+\beta)\times\\ &\times C(b_3'+\beta', b_4'+\beta').\end{aligned}$$

But assume $b_i+b_i'=-2\gamma_i$. Then using the relations (**) that we know to be true, we get:

$$\begin{aligned}&C(b_1, b_2)\cdot C(b_1', -b_2')\\ &\quad=C(\gamma_1-\gamma_2+b_1-b_2, \gamma_1-\gamma_2)\cdot C(\gamma_1+\gamma_2+b_1'+b_2', \gamma_1+\gamma_2)\\ &C(b_3, b_4)\cdot C(b_3', -b_4')\\ &\quad=C(\gamma_3-\gamma_4+b_3-b_4, \gamma_3-\gamma_4)\cdot C(\gamma_3+\gamma_4+b_3'+b_4', \gamma_3+\gamma_4)\\ &C(b_1+\beta, b_2+\beta)\cdot C(b_1'+\beta', -b_2'-\beta')\\ &\quad=C(b_1-b_2+\gamma_1-\gamma_2, \gamma_1-\gamma_2)\times\\ &\quad\times C(\gamma_1+\gamma_2+b_1'+b_2'+\beta'-\beta, \gamma_1+\gamma_2-\beta-\beta')\\ &C(b_3+\beta, b_4+\beta)\cdot C(b_3'+\beta', -b_4'-\beta')\\ &\quad=C(b_3-b_4+\gamma_3-\gamma_4, \gamma_3-\gamma_4)\times\\ &\quad\times C(\gamma_3+\gamma_4+b_3'+b_4'+\beta'-\beta, \gamma_3+\gamma_4-\beta-\beta').\end{aligned}$$

We can assume that $\gamma_1+\gamma_2+\gamma_3+\gamma_4=\beta+\beta'$ by altering one of the γ's by an element of order 2. Using the symmetries $C(A, B)=C(A, -B)=C(B, A)$ one checks that the product of the 1st two right-hand sides equals the product of the 2nd two. Hence the same for the left-hand sides, hence (α) is true.

It remains only to check that for *any* $b_1, b_2, b_3, b_4, \beta$ such that $b_1+b_2+b_3+b_4=-2\beta$, there exists $b_1', b_2', b_3', b_4', \beta'$ differing from the b_i's and β by $2H$, such that

$$b_1'+b_2'+b_3'+b_4'=-2\beta'$$
$$\bar{C}(b_1', b_2')\cdot\bar{C}(b_3', b_4')\neq 0.$$

Now we know that the functions $\bar{q}, \bar{q}'$ come from an abelian variety. Therefore choosing a compatible 4δ-marking on this abelian variety, there is a null-value function $\bar{q}''$ on $K(4\delta)$ with values in k, such that

$$\bar{q}'(a+b)\cdot\bar{q}'(a-b)=\sum_{\eta\in Z_2}\bar{q}''(a+\eta)\cdot\bar{q}''(b+\eta).$$

Now if $a\in K(4\delta)$, $l\in\hat{Z}_4$, set:

$$\bar{z}(a, l)=\sum_{\eta\in Z_4} l(\eta)\,\bar{q}''(a+\eta).$$

Then if $b_1=(a+b, l)$, $b_2=(a-b, m)$ are 2 elements of H, with $a, b\in K(4\delta)$, $l, m\in\hat{Z}_4$, one checks immediately that:

$$\bar{C}(b_1, b_2)=z(a, l+m)\cdot\bar{z}(b, l-m).$$

Therefore, translating the question into one involving $\bar{z}$'s, we have to check:

$$\left\{\begin{array}{ll}\text{for all } a_1, a_2, a_3, a_4\in K(4\delta), & \text{congruent mod } K(2\delta),\\ \text{for all } l_1, l_2, l_3, l_4\in\hat{Z}_4, & \text{congruent mod } 2\hat{Z}_4,\\ \text{there exists } \alpha_1, \alpha_2, \alpha_3, \alpha_4\in K(2\delta), & \text{congruent mod } K(\delta)\\ \text{and } k\in 2\ddot{Z}_4, \text{ such that} & \\ \qquad\prod_{i=1}^{4}\bar{z}(a_i+\alpha_i, l_i+m)\neq 0. & \end{array}\right.$$

But using the hypothesis that $Z_8\subset K(\delta)$, this follows from:

$$\left\{\begin{array}{l}\text{for all } a\in K(4\delta),\ l\in\hat{Z}_4, \text{ there exists } \alpha\in Z_8 \text{ such that}\\ \qquad\bar{z}(a+\alpha, l)\neq 0.\end{array}\right.$$

This is a special case of the result stated following Theorem 1, § 4. Phew!

The same analysis also gives the result:

If $\{\bar{q}'(a)_1\}$ are the coordinates of a geometric point x_1 in $\bar{M}_{2\delta}$ which corresponds to an abelian variety with 2δ-marking, then the coordinates $\{\bar{q}'(a)_2\}$ of any other geometric point x_2 such that $\pi(x_1)=\pi(x_2)$ are determined by:

$$\sum_{\eta\in Z_2} l(\eta)\,\bar{q}'(a+\eta)_2=\gamma(a, l)\cdot\sum_{\eta\in Z_2} l(\eta)\cdot\bar{q}'(a+\eta)_1$$

where γ is a quadratic character:

$$\gamma: K(2\delta)/K(\delta)\times Z_2 \to \{\pm 1\}.$$

We leave it to the reader to check this change from the $\bar{q}'(a)_1$'s to the $\bar{q}'(a)_2$'s also comes about by a suitable modification of the 2δ-marking. Therefore:

if a geometric point x of $\bar{M}_{2\delta}$ corresponds to an abelian variety with 2δ-marking, so do all geometric points in $\pi^{-1}(\pi(x))$.

Step III. Let $\mathscr{X}/S$ be an abelian scheme with δ-markings. Let $i: \mathscr{X} \hookrightarrow \boldsymbol{P}(V_\delta)\times S$ be the closed immersion defined above. Then

i) $p_1 \circ i \circ \varepsilon$ is actually a morphism j of S to $\bar{M}_\delta$,

ii) $i(\mathscr{X})$ equals the subscheme y of $\boldsymbol{P}(V_\delta)\times S$ obtained as the fibre product:

$$\begin{array}{ccc} \mathscr{Y} & \hookrightarrow & \boldsymbol{P}(V_\delta)\times S \\ \downarrow & & \downarrow {\scriptstyle id\times j} \\ \bar{A}_\delta & \longrightarrow & \boldsymbol{P}(V_\delta)\times \bar{M}_\delta . \end{array}$$

In order to establish this, it clearly suffices to take the case $S=\operatorname{Spec}(A)$, where A is an artin local ring with algebraically closed residue field. Let p be the characteristic of the residue field k. Then there are considerable technical simplifications:

a) for all positive integers n not divisible by p, A contains exactly n n-th roots of 1, which lift the n n-th roots of 1 in k,

b) if $\mathscr{X}/S$ is an abelian scheme, then for all positive integers n not divisible by p, the subscheme $\mathscr{X}_n$ of points of order n is the disjoint union of n^{2g} subschemes, each being the image of a section of $\mathscr{X}/S$: i.e., all points of order n are rational over S.

Cor. of (a), (b). Let $\mathscr{X}/S$ be an abelian scheme, and let X/k be its fibre over the residue field. Let L be a relatively ample sheaf of degree d on $\mathscr{X}$. Then $H(L)$ is a disjoint union of the images of d^2 distinct sections of $\mathscr{X}/S$, and there is a one-one correspondence between ϑ-structures for L and for the induced sheaf $\bar{L}$ on X.

In this case, we can deal entirely with the group of rational points $\mathscr{G}^S(\delta)$ instead of with the combersome group schemes. Thus $\mathscr{G}^S(\delta)$ is just the (discrete) group extension:

$$0\to A^*\to \mathscr{G}^S(\delta)\to K(\delta)\times \widehat{K(\delta)}\to 0$$

and where $\widehat{K(\delta)}$ is just a (discrete) group isomorphic to $K(\delta)$. And the representation theory boils down to:

There is a unique A-linear representation of $\mathscr{G}^S(\delta)$ on a free A-module of rank d, in which the subgroup A^* acts by homotheties.

In particular, a ϑ-structure on L induces an isomorphism:

$$\Gamma(\mathscr{X}, L) \cong V_\delta \otimes_{\mathbf{Z}} A$$

unique up to multiplication by elements of A^*.

To get down to the proof itself, we are given $\mathscr{X}/S$, plus a totally symmetric relatively ample L, plus a symmetric isomorphism f_1 of $\mathscr{G}(L)$ and $\mathscr{G}^S(\delta)$. By the results of § 2, there is a ϑ-structure

$$\bar{f}_2 \colon \mathscr{G}(\bar{L}^2) \xrightarrow{\sim} \mathscr{G}^S(2\delta)$$

such that the pair $\bar{f}_1$ and $\bar{f}_2$ of ϑ-structures on X is symmetric. Lift $\bar{f}_2$ to a ϑ-structure f_2 for L^2. Then f_1 and f_2 induce isomorphisms:

$$\Gamma(\mathscr{X}, L) \cong V_\delta \otimes_{\mathbf{Z}} A$$

$$\Gamma(\mathscr{X}, L^2) \cong V_{2\delta} \otimes_{\mathbf{Z}} A,$$

where the first is the isomorphism used to define $i \colon \mathscr{X} \hookrightarrow \mathbf{P}(V_\delta) \times S$. Also, evaluation of sections at ε defines particular A-valued functions q_L and q_{L^2} on $K(\delta)$ and $K(2\delta)$ via

$$\begin{array}{ccc} \Gamma(\mathscr{X}, L^i) \cong V_{i\delta} \otimes_{\mathbf{Z}} A & & \\ \Big\downarrow \varepsilon^* & f \;\mapsto & \displaystyle\sum_{a \in K(i\delta)} f(a)\, q_{L^i}(a). \\ A \;=\; A & & \end{array}$$

Here $q_L(a)$ is just the value of $Q(a)$ at ε: so $\{q_L(a)\}$ is a set of homogeneous coordinates for the point $i \circ \varepsilon \colon S \to \mathbf{P}(V_\delta) \times S$. This means that $\{q_L(a)\}$ are the homogeneous coordinates of $j(S)$, and these are exactly the numbers which are to satisfy Riemann's theta relations.

The only thing that must really be checked is that the pairing

$$\Gamma(\mathscr{X}, L) \times \Gamma(\mathscr{X}, L) \to \Gamma(\mathscr{X}, L^2)$$

given by multiplying sections is defined by the same multiplication formula as in § 3. But the proof given in § 3 of this formula goes over without any change to this more general case: simply replace the ground field k by the ground ring A. Once the multiplication formula is known, Riemann's relations on $\{q_L(a)\}$ follow formally, as in § 3. This proves (i).

To prove (ii), the first point is that $\mathscr{X} \subset \mathscr{Y}$. This follows immediately from the multiplication formula. In fact, on $\mathscr{X}$, the coordinates $Q(a)$ have the constant values $q_L(a)$ and the coordinates X_a induce the sections of L on $\mathscr{X}$ denoted by δ_a previously. The identity which we need comes out immediately by reducing the quadratic expressions in the δ_a's to

7*

functions on $K(2\delta)$ and using the relation between q_L and q_{L^2}. Secondly, if Y is the fibre of $\mathscr{Y}$ over k, then $X=Y$. This follows from Theorem 2, § 4, and the fact that the equations on X_a are a complete set of quadratic relations. In fact, for fixed $l\in Z_2$, $a\in K(2\delta)$, these relations assert that all the quadratic expressions

$$\{\sum_{\eta\in Z_1}^{\bar{z}} l(\eta)\, X_{a+b+\eta}\cdot X_{a-b+\eta}\,|\, b\in a+K(\delta)\}$$

are proportional. Since, in fact, on X each of these sets consists in sections of L^2 corresponding to the function $Y_{a,l}$:

$$Y_{a,l}(b)=\begin{cases}0 & \text{if } b-a\notin Z_2\\ l(b-a) & \text{if } b-a\in Z_2,\end{cases}$$

and since the $Y_{a,l}$'s are linearly *independent*, these do exhaust the quadratic relations on the X_a. Thirdly, we note the obvious:

Lemma. *Let U be a closed subscheme of V, both being noetherian schemes over* Spec(A). *Then if*

1. *U is flat over A,*
2. *the fibres $\overline{U}$ and $\overline{V}$ over* Spec(k) *are equal, then $U=V$.*

This proves (ii).

Step IV. Suppose that N is a connected subscheme of $\overline{M}_\delta$ such that if $\bar{A}_\delta$ induces a subscheme B over N, then B is smooth over N. Suppose that some geometric point x of N is in the image of $\mathscr{M}_\delta$ under t. Then the N-valued point 1_N of N is in the image of $\mathscr{M}_\delta$, i.e., there is an abelian scheme $\mathscr{Y}/N$ with δ-marking such that $i(\mathscr{Y})=B$.

By assumption, one geometric fibre $\bar{B}$ of B/N is an abelian variety. Therefore, by Theorem 6.14, Ch. 6, [9], B is an abelian scheme over N with identity section Δ. By construction, $B\subset \mathbf{P}(V_\delta)\times N$. Let $\underline{O}_{\mathbf{P}}(1)$ induce the sheaf M on B. Normalize it on the identity section, i.e., set

$$L=M\otimes p_2^*\,\Delta^*(M^{-1}).$$

I claim that the inverse ι for B is given by restricting the projective transformation j to B, where

$$j\colon\ \mathbf{P}(V_\delta)\times N\to\mathbf{P}(V_\delta)\times N$$

$$j^*(X_a)=X_{-a}.$$

But the inverse is given by j on $\bar{B}$ (cf. § 3). Since $\iota=j$ on one geometric fibre of B/N and on the section Δ of B/N, $\iota=j$ everywhere by the rigidity lemma ([9], Ch. 6, §1). Since j is a projective transformation, this shows that $\iota^*L\cong L$, i.e., L is symmetric. Now j has 2 disjoint subspaces of fixed points:

$$L_1 \text{ defined by } X_a = X_{-a}, \qquad \text{all } a \in K(\delta)$$

$$L_2 \text{ defined by } X_a = -X_{-a}, \qquad \text{all } a \in K(\delta)$$

and the identity section Δ is completely contained in L_1. To say that L is totally symmetric is the same as saying that B_2, the kernel of multiplication by 2, is completely contained in L_1. But $B_2 \subset L_1 \cup L_2$ since $j = id$ on B_2; and $\bar{B}_2$, the points of order 2 on $\bar{B}$, is contained in L_1 since the projective embedding of $\bar{B}$ is assumed to come from a δ-marking. Since S is connected, this implies that $B_2 \subset L_1$, and hence that L is totally symmetric.

Moreover, by Step I, we get a homomorphism

$$\lambda\colon\ \mathscr{G}(\delta) \times N \to \mathscr{G}(\underline{L})$$

of group schemes over N such that the actions on $p_{2,*}L$ and V_δ match up. Since λ induces an isomorphism of the group schemes $\mathscr{G}(\delta) \times \text{Spec}(k)$, $\mathscr{G}(\bar{L})$ corresponding to the situation in the geometric fibre $\bar{B}$, λ is an isomorphism everywhere[2]. Therefore, λ is a ϑ-structure. Also λ is symmetric at one point, hence it is symmetric everywhere. Therefore, B/N has a ϑ-marking which induces the given embedding in $\mathbf{P}(V_\delta) \times N$. *Q. E. D.*

Step V. Recall the main result on flattening stratifications ([*10*], Lecture 8):

Given a projective morphism $f\colon X \to Y$ of noetherian schemes, Y can be decomposed into a disjoint union of locally closed connected subschemes Y_α, such that if $g\colon Y' \to Y$ is a morphism from a connected noetherian scheme Y' to Y, and we look at the fibre product:

$$\begin{array}{ccc} X' & \xrightarrow{f'} & Y' \\ \downarrow & & \downarrow g \\ X & \xrightarrow{f} & Y, \end{array}$$

then f' is flat if and only if g factors through one of the subschemes Y_α.

Apply this to $\bar{A}_\delta/\bar{M}_\delta$: if $Z \subset \bar{M}_\delta$ is one of the pieces, and some fibre of $\bar{A}_\delta$ over Z is obtained from an abelian scheme with δ-marking, then we must show that Z is *open*. If we do this, both theorems are proven. The openness of Z follows from:

Key Lemma. *Let $X/\text{Spec}(k)$ be an abelian variety with δ-marking over an algebraically closed field k. Let this define a geometric point x of $\bar{M}_\delta$.*

[2] In fact, since λ is an isomorphism of the subgroups $G_{m,S}$ in any case, it is only a question of whether λ is an isomorphism of the quotients $\underline{H}(\delta)$, $\underline{H}(L)$. Since these are etale over S, a homomorphism is an isomorphism if it is so at one point.

Let A be any artin local ring with residue field k, and let

$$f\colon \operatorname{Spec}(A) \to \overline{M}_\delta$$

be a morphism extending x. Then the scheme over Spec(*A*) *induced by* $\bar{A}_\delta$ *is flat over* Spec(*A*).

Proof of Lemma. Let $\bar{A}_\delta$ define, over A, the projective scheme

$$\mathscr{X} \subset \boldsymbol{P}(V_\delta) \times \operatorname{Spec}(A).$$

Let L be the invertible sheaf on $\mathscr{X}$ obtained by restricting $\underline{O}_{\boldsymbol{P}}(1)$ to $\mathscr{X}$. We shall show:

(*) $\quad \Gamma(\mathscr{X}, L^{2^n})\quad$ is a free A-module, for all large n.

This certainly is enough to prove the lemma, in view of the following elementary observation:

Sublemma. *Let A be an artin local ring with residue field k, and let R be a graded A-algebra generated over* $A = R_0$ *by* R_1. *Assume that* R_1 *is a finite A-module, and that* R_n *is a free A-module for an infinite set of integers n. Then* R_n *is free for all but a finite set of n's.*

Proof. Without loss of generality, we can assume that k is infinite. Let $\bar{R} = R \otimes_A k$. In $\bar{R}$, let

$$(0) = Q_1 \cap \cdots \cap Q_l, \qquad P_i = \sqrt{Q_i}$$

be the primary decomposition of (0). Let P_1 be the irrelevant ideal

$$\bigoplus_{n \geqq 1} \bar{R}_n.$$

Since k is infinite, there is an element $g \in R_1$ such that

$$\bar{g} \notin P_2 \cup \cdots \cup P_l.$$

Let n_0 be an integer such that $Q_1 \supset P_1^{n_0}$, i.e., $Q_1 \supset R_n$ for all $n \geqq n_0$. *I* claim that R_m is a free A-module for any integer $m \geqq n_0$. Choose $f_1, \ldots, f_k$ in R_m such that $\bar{f}_1, \ldots, \bar{f}_k$ are a basis of $\bar{R}_m$. Note that for all integers p,

$$\bar{g}^p \bar{f}_1, \ldots, \bar{g}^p \bar{f}_k \in \bar{R}_{m+p}$$

are independent over k. In fact, if

$$0 = \sum \alpha_i \bar{g}^p \bar{f}_i = \bar{g}^p \cdot \left(\sum \alpha_i \bar{f}_i\right)$$

then $\bar{g} \notin P_i$, for $i \geqq 2$ implies $\sum \alpha_i \bar{f}_i \in Q_i$, $i \geqq 2$. And since the degree is large enough, $\sum \alpha_i \bar{f}_i \in Q_1$. Hence $\sum \alpha_i \bar{f}_i = 0$, which is a contradiction unless all the α's are 0. Now suppose that R_m is not a free A-module. Then

$$\sum_{i=1}^{k} a_i f_i = 0$$

for some elements $a_1, \ldots, a_k \in A$, not all 0. Let p be a positive integer such that R_{m+p} *is* a free A-module. Then since $\bar{R}_{m+p}$ has a basis of the form $\bar{g}^p \bar{f}_1, \ldots, \bar{g}^p \bar{f}_k, \bar{h}_1, \ldots, \bar{h}_l$, it follows that R_{m+p} has a basis of the form $g^p f_1, \ldots, g^p f_k, h_1, \ldots, h_l$. But

$$\sum_{i=1}^{k} a_i (f_i g^p) = 0 .$$

Therefore, the a's are all 0 which is a contradiction. *Q.E.D.*

Once we know that $\Gamma(\mathscr{X}, L^n)$ is a free A-module for all large n, then $\mathscr{X}$ is flat over Spec(A) ([5], Ch. 3, 7.9.14). To prove $(*)$, let $I_\delta \subset S^*(V_\delta) \otimes A$ be the ideal generated by the quadratic defining relations of $\bar{A}_\delta$. Then

$$\Gamma(\mathscr{X}, L^n) \cong (S^n V_\delta \otimes A)/(I_\delta)_n$$

for all large n, and it suffices to prove:

$$(**) \qquad \begin{array}{l} (I_\delta)_{2^n} \quad \text{is a direct summand of} \quad (S^{2^n} V_\delta) \otimes A \\ \text{for } all \ n = 0, 1, 2, \ldots . \end{array}$$

We shall prove $(**)$ by induction on n, assuming at each stage that it is known for $n \leqq n_0$ and *all* δ. To start things off, $(I_\delta)_1$ equals (0), for $(**)$ is always true for $n=0$. To get from $(**)$ for $n=n_0$ to $(**)$ for $n=n_0+1$, we shall simply show for all A-valued points of $\bar{M}_\delta$ as above, there is an A-valued point of $\bar{M}_{2\delta}$ with the same properties such that:

$$(***) \qquad \bigoplus_{n=0}^{\infty} \big((S^{2^n} V_\delta) \otimes A\big)/(I_\delta)_{2^n} \underset{T}{\cong} \big((S^* V_{2\delta}) \otimes A\big)/I_{2\delta} .$$

In particular,

$$\big((S^{2^{n+1}} V_\delta) \otimes A\big)/(I_\delta)_{2^{n+1}} \cong (S^{2^n} V_{2\delta} \otimes A)/(I_{2\delta})_{2^n}$$

so $(**)$ will be completely proven.

To lift the A-valued point of $\bar{M}_\delta$, we use the morphism

$$\pi\colon \bar{M}_{2\delta} \to \bar{M}_\delta$$

defined in Step II. The underlying k-valued point of f corresponds to an abelian variety with δ-marking. Choose a compatible 2δ-marking on this abelian variety. This defines a k-valued point y of $\bar{M}_{2\delta}$ over x. By Step II, π is étale at x. Therefore, there is one (and only one) A-valued point

$$g\colon \text{Spec}(A) \to \bar{M}_{2\delta}$$

lifting f with underlying k-valued point y.

Let these A-valued points be given by homogeneous coordinates $q(a)$, $a \in K(\delta)$, and $q'(b)$, $b \in K(2\delta)$, where $q(a), q'(b) \in A$. These satisfy the usual equations (cf. Step II). Then in concrete terms:

$$S^* V_\delta \otimes A/I_\delta \cong \frac{A[\ldots, X_a, \ldots]_{a \in K(\delta)}}{\left\{\begin{array}{l}(\Sigma\, l(\eta)\, q(c+d+\eta) \cdot q(c-d+\eta)) \cdot (\Sigma\, l(\eta)\, X_{a+b+\eta} X_{a-b+\eta}) \\ -(\Sigma\, l(\eta)\, q(c+b+\eta) \cdot q(c-b+\eta)) \cdot (\Sigma\, l(\eta)\, X_{a+d+\eta} X_{a-d+\eta})\end{array}\right\}}$$

$$S^* V_{2\delta} \otimes A/I_{2\delta} \cong \frac{A[\ldots, X'_a, \ldots]_{a \in K(2\delta)}}{\left\{\begin{array}{l}(\Sigma\, l(\eta)\, q'(c+d+\eta) \cdot q'(c-d+\eta)) \cdot (\Sigma\, l(\eta)\, X'_{a+b+\eta} X'_{a-b+\eta}) \\ -(\Sigma\, l(\eta)\, q'(c+b+\eta) \cdot q'(c-b+\eta)) \cdot (\Sigma\, l(\eta)\, X'_{a+d+\eta} X'_{a-d+\eta})\end{array}\right\}}.$$

We set up the isomorphism (***) by requiring that

$$T[X_{a+b} \cdot X_{a-b}] = \sum_{\eta \in Z_2} q'(b+\eta) \cdot X'_{a+\eta}$$

for all $a, b \in K(2\delta)$ such that $a+b \in K(\delta)$. This implies:

$$T\Big[\sum_{\eta \in Z_2} l(\eta)\, X_{a+b+\eta} X_{a-b+\eta}\Big] = \sum_{\eta \in Z_2} l(\eta)\, q'(b+\eta) \cdot \sum_{\eta \in Z_2} l(\eta)\, X'_{a+\eta}$$

for all $a, b \in K(2\delta)$, $l \in \hat{Z}_2$ such that $a+b \in K(\delta)$. Since

$$\Sigma\, l(\eta)\, q(a+b+\eta)\, q(a-b+\eta) = \Sigma\, l(\eta)\, q'(a+\eta) \cdot \Sigma\, l(\eta)\, q'(b+\eta),$$

this second equation implies that the quadratic relations in the ring of X's go to 0 in the ring of X''s. Moreover, since for all $a \in K(2\delta)$, $l \in \hat{Z}_2$, there is an element $b \in a + K(\delta)$ such that $\sum (l(\eta)\, q'(b+\eta))$ is a *unit*, this also shows that the map is surjective. Similarly, since the elements $\sum l(\eta) X'_{a+\eta}$ are linearly independent it is clear from this second equation that every quadratic expression in the X's which goes to 0 via T is a combination of expressions of the form:

$$\sum_{b \in a+K(\delta)} A_b \Big\{\sum_{\eta \in Z_2} l(\eta)\, X_{a+b+\eta} X_{a-b+\eta}\Big\}.$$

And, in fact, the expressions

$$\begin{aligned}&[\Sigma\, l(\eta)\, q'(b_2+\eta)] \cdot [\Sigma\, l(\eta)\, X_{a+b_1+\eta} \cdot X_{a-b_1+\eta}] \\ &\quad -[\Sigma\, l(\eta)\, q'(b_1+\eta)] \cdot [\Sigma\, l(\eta)\, X_{a+b_2+\eta} X_{a-b_2+\eta}]\end{aligned}$$

will span the kernel. But if $b_1 = b$, $b_2 = d$ and we multiply this expression by a *unit* of the form $\sum l(\eta)\, q'(c+\eta)$, $c \in a + K(\delta)$, we get the typical quadratic relation on the X's. Therefore, T sets up an isomorphism:

$$S^2 V_\delta \otimes A/(I_\delta)_2 \xrightarrow{\sim} S^1 V_{2\delta} \otimes A/(I_{2\delta})_1.$$

At this stage in the proof, we have already shown that for *all* δ, $(I_\delta)_2$ is a direct summand of $S^2 V_\delta \otimes A$.

Now try to extend T to a homomorphism of the whole ring. Let U be the free A-module $S^2 V_\delta \otimes A/(I_\delta)_2$. Let $Y_{a,b}$ be the element $X_a X_b$ in U. Then, as observed in the proof of Theorem 2, § IV,

$$\bigoplus_{n=0}^{\infty} S^{2n} V_\delta \otimes A/(I_\delta)_{2n} \cong S^* U \Big/ \begin{Bmatrix} \text{ideal generated by} \\ Y_{a,b} Y_{c,d} - Y_{a,d} Y_{c,b} \end{Bmatrix}.$$

Therefore, all we have to check is that when T is extended to $S^2 U$, the A-module generated by the expressions $T(Y_{a,b}) \cdot T(Y_{c,d}) - T(Y_{a,d}) \times T(Y_{c,b})$ is the same as the A-module of quadratic relations in the X''s, and the proof will be complete. Call the first module N_1, and the second N_2. It is easy to check that $N_1 \subset N_2$, simply by calculating out the expressions $T(Y_{a,b}) \cdot T(Y_{c,d}) - T(Y_{a,d}) \cdot T(Y_{c,b})$.

Consider the diagram

$$\begin{array}{ccccc} N_1 & \overset{i}{\hookrightarrow} & N_2 & \hookrightarrow & S^2 V_{2\delta} \otimes A \\ \downarrow & & \downarrow & & \downarrow \\ N_1 \otimes k & \xrightarrow{j} & N_2 \otimes k & \xrightarrow{l} & S^2 V_{2\delta} \otimes k. \end{array}$$

Since N_2 is a direct summand of $S^2 V_{2\delta} \otimes A$ according to the first part of this proof, it follows that l is injective. But in Theorem 2, § 4, we showed that $N_1 = N_2$ in case $A = k$. Therefore the images of $N_1 \otimes k$ and $N_2 \otimes k$ in $S^2 V_{2\delta} \otimes k$ are equal. This implies that j is surjective, hence that i is surjective, and hence $N_1 = N_2$. *Q.E.D.*

§ 7. The 2-Adic Limit

Up to this point, we have been studying pairs (X, L), consisting of abelian varieties X, and ample invertible sheaves L of separable type. To push the theory further, however, it seems almost essential to make it freer of variations within one isogeny type. The simplest way to do this is to study simultaneously a whole tower of abelian varieties. All the simplifications that occur however have to do with dividing by 2, so it doesn't seem necessary or fruitful to look at *all* isogenies: instead we look only at isogenies of degree 2^n, some n. This is economical, too, because our moduli will then be the values of functions on 2-adic vector spaces, rather than of functions on adelic spaces. As always, we assume $\mathrm{char}(k) \neq 2$.

The basic definition is:

*Definition 1. A 2-tower of abelian varieties (*or *tower) is an inverse system $\{X_\alpha\}_{\alpha \in S}$ of abelian varieties,* i.e., S is a partially ordered set such that $\forall\, \alpha_1, \alpha_2 \in S$, $\exists\, \beta \in S$, $\beta > \alpha_1, \alpha_2$, and whenever $\alpha, \beta \in S$, $\alpha > \beta$, we are given an isogeny of degree 2^n:

$$p_{\alpha,\beta}: X_\alpha \longrightarrow X_\beta$$

such that:

1) If $\alpha>\beta>\gamma$, the diagram of isogenies

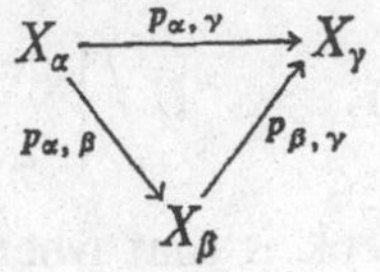

commutes,

2) If $\alpha>\beta_1, \beta_2$ and $K_i=\text{kernel}\{X_\alpha \to X_{\beta_i}\}$, then $K_1 \subseteq K_2 \Leftrightarrow \beta_1 \geqq \beta_2$.

3a) For all $\alpha\in S$, and all isogenies $X_\alpha \to Y$, of degree 2^n (some n), $\exists\beta\in S$ such that $\alpha\geqq\beta$, and one has a diagram:

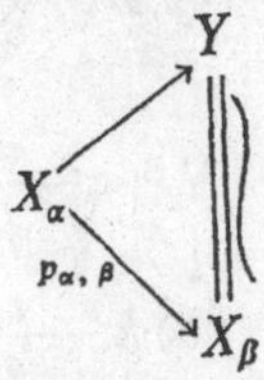

3b) For all $\alpha\in S$, and all isogenies $Y\to X_\alpha$ of degree 2^n (some n), $\exists\beta\in S$ such that $\beta\geqq\alpha$ and one has a diagram:

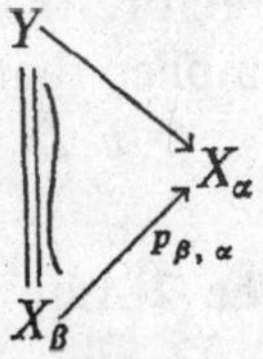

Note that, starting with *one* abelian variety X, we can generate in a canonical way a 2-tower of abelian varieties by taking coverings and quotients of degrees 2^n, starting with X.

Definition 2. If X is an abelian variety, let $\text{tor}_2(X)$ denote the group of closed points $x\in X$ of order 2^n, some n.

If $\underleftarrow{X}=\{X_\alpha\}$ is a tower of abelian varieties, we get a derived inverse system of discrete groups $\{\text{tor}_2(X_\alpha)\}$. Let

$$V(\underleftarrow{X})=\varprojlim_{\alpha\in S} \text{tor}_2(X_\alpha).$$

If X is one abelian variety, we also let $V(X)$ denote $V(\underleftarrow{X})$, where $\underleftarrow{X}$ is the 2-tower generated by X. This is the usual 2-Tate group of X. For all $\alpha\in S$, there is a canonical surjection

$$p_\alpha: V(\underleftarrow{X}) \longrightarrow \text{tor}_2(X_\alpha).$$

Denote the kernel by $T(\alpha)$. Each $T(\alpha)$ is an inverse limit of finite groups, so if we topologize $V(\underset{\leftarrow}{X})$ by taking the $T(\alpha)$'s as a basis of open neighborhoods of the origin, then $V(\underset{\leftarrow}{X})$ becomes a locally compact group, and each $T(\alpha)$ becomes a compact, open subgroup. From well-known structure theorems, we know that there are topological isomorphisms:

$$\begin{array}{ccccccccc}
0 \longrightarrow & T(\alpha) & \longrightarrow & V(\underset{\leftarrow}{X}) & \longrightarrow & \mathrm{tor}_2(X_\alpha) & \longrightarrow 0 \\
& \wr\| & & \wr\| & & \wr\| & \\
0 \longrightarrow & (\boldsymbol{Z}_2)^{2g} & \longrightarrow & (\boldsymbol{Q}_2)^{2g} & \longrightarrow & \begin{Bmatrix}\text{subgroup of} \\ \boldsymbol{Q}/\boldsymbol{Z} \text{ of elts.} \\ \text{of order } 2^n \\ \text{(some } n)\end{Bmatrix}^{2g} & \longrightarrow 0
\end{array}$$

(where $g = \dim X$). In particular,

$$S \cong \begin{Bmatrix}\text{partially ordered set of compact,} \\ \text{open subgroups } T \subset (\boldsymbol{Q}_2)^{2g}\end{Bmatrix}.$$

When $\underset{\leftarrow}{X}$ is generated from one given X, the kernel of the canonical homomorphism $V(\underset{\leftarrow}{X}) \to \mathrm{tor}_2(X)$ is usually denoted $T(X)$, and $T(X)$ is also called the 2-Tate group of X.

A final point: when $k = \boldsymbol{C}$, one can associate to a tower of abelian varieties a *common universal covering space*. In fact, for all $\alpha \geqq \beta$, $p_{\alpha\beta}$: $X_\alpha \to X_\beta$ induces an isomorphism

$$\hat{p}_{\alpha\beta}: \hat{X}_\alpha \xrightarrow{\sim} \hat{X}_\beta$$

between the universal covering spaces. Therefore let $\underset{\leftarrow}{\hat{X}}$ stand for a complex vector space canonically isomorphic to each $\hat{X}_\alpha$. In the classical theory, $\underset{\leftarrow}{\hat{X}}$ plays a role analogous to $V(\underset{\leftarrow}{X})$. For all α, there is a canonical surjection q_α: $\underset{\leftarrow}{\hat{X}} \to X_\alpha$. Let $L(\alpha)$ denote its kernel. Then $\{L(\alpha)\}$ is a family of commensurable lattices in $\underset{\leftarrow}{\hat{X}}$, and $\alpha \geqq \beta$ if and only if $L(\alpha) \subseteq L(\beta)$. Since $X_\alpha \cong \underset{\leftarrow}{\hat{X}}/L(\alpha)$, the whole tower can be generated by starting with the complex vector space $\underset{\leftarrow}{\hat{X}}$ and dividing by these lattices $L(\alpha)$.

Definition 3. A *polarized tower of abelian varieties* is a tower $\{X_\alpha\}$, $\alpha \in S$, plus a set of totally symmetric ample invertible sheaves of degree 2^n, some n: L_α on X_α, for α in a subset $S_0 \subset S$, plus isomorphisms

$$p_{\alpha,\beta}^*(L_\beta) \xrightarrow{\sim} L_\alpha$$

whenever $\alpha, \beta \in S_0, \alpha > \beta$. We require:

1) If $\alpha, \beta \in S$, and $\alpha > \beta$, then $\beta \in S_0 \Rightarrow \alpha \in S_0$.

2) If $\alpha > \beta > \gamma$, $\gamma \in S_0$, then the isomorphisms of $L_\alpha, L_\beta, L_\gamma$ – all pulled up to X_α – are to be compatible,

3) If $\alpha>\beta$, $\alpha\in S_0$, and if there exists a totally symmetric sheaf M on X_β such that $p^*_{\alpha,\beta}M\cong L_\alpha$, then $\beta\in S_0$ too, [in which case, L_β will have to be M too].

The object of this section is to generalize the theory of §1 to a polarized tower of abelian varieties. So from now on, let's suppose given one such tower $\mathscr{T}=\{X_\alpha, L_\alpha\}$. The first important observation is:

"4>2" Lemma. *For all $x\in V(\underset{\sim}{X})$, there exists an $\alpha\in S_0$ such that $p_\alpha(x)\in H(L_\alpha)$.*

Proof. Start with any $\alpha_1\in S_0$. Let 2^n be the order of the point $p_{\alpha_1}(x)$. Let $\alpha_2\in S_0$ be the element such that $\alpha_2>\alpha_1$ and such that the isogeny p_{α_2,α_1} is $2^n\,\delta$, i.e.,

$$\begin{array}{ccc} X_{\alpha_2} & & \\ & \searrow^{p_{\alpha_2,\alpha_1}} & \\ \Big\downarrow q & & X_{\alpha_1} \\ & \nearrow_{2^n\delta} & \\ X_{\alpha_1} & & \end{array}$$

Since $(2^n\,\delta)^*L_{\alpha_1}\cong(L_{\alpha_1})^{2^{2n}}$, $H((2^n\delta)^*L_{\alpha_1})$ contains all points of X_{α_1} of order 2^{2n}. Therefore $H(L_{\alpha_2})$ contains all points of X_{α_2} of order 2^{2n}. But

$$0=2^n\,p_{\alpha_1}(x)=2^n\,p_{\alpha_2,\alpha_1}(p_{\alpha_2}(x))=2^{2n}[q(p_{\alpha_2}(x))]$$

so $p_{\alpha_2}(x)$ has order 2^{2n}. *Q.E.D.*

Now let $S_0^x=\{\alpha\in S_0\,|\,p_\alpha(x)\in H(L_\alpha)\}$. Like S_0, it has the property: $\beta>\alpha$, $\alpha\in S_0^x\Rightarrow\beta\in S_0^x$. Suppose that for *one* $\alpha\in S_0^x$, you choose an isomorphism

$$\varphi_\alpha\colon L_\alpha\xrightarrow{\sim}T^*_{p_\alpha(x)}(L_\alpha).$$

Then *I* claim this determines canonically isomorphisms

$$\varphi_\beta\colon L_\beta\xrightarrow{\sim}T^*_{p_\beta(x)}(L_\beta)$$

for *all* $\beta\in S_0^x$. This is clear – first suppose $\beta>\alpha$, and then choose φ_β so that:

$$\begin{array}{ccc} L_\beta & \xrightarrow{\varphi_\beta} & T^*_{p_\beta(x)}(L_\beta) \\ \Big\| & & \Big\| \\ & & T^*_{p_\beta(x)}(p^*_{\beta,\alpha}(L_\alpha)) \\ \Big\| & & \Big\| \\ p^*_{\beta,\alpha}(L_\alpha) & \xrightarrow{p^*_{\beta,\alpha}(\varphi_\alpha)} & p^*_{\beta,\alpha}(T^*_{p_\alpha(x)}(L_\alpha)) \end{array}$$

commutes. In general, first determine φ_γ for some $\gamma\in S_0^x$ such that $\gamma>\alpha,\beta$, then go backwards to determine φ_β from φ_γ. Call such a system of φ_α's, all $\alpha\in S_0^x$, a *compatible* set of isomorphisms $L_\alpha\xrightarrow{\sim}T^*_{p_\alpha(x)}(L_\alpha)$.

Definition 4. $\mathcal{G}(\mathcal{T})$ = set of pairs $(x, \{\varphi_\alpha\})$, where $x \in V(\underline{X})$, and $\{\varphi_\alpha\}$, $\alpha \in S_0^x$, is a compatible set of isomorphisms.

This forms a group in the usual way. Given $(x, \{\varphi_\alpha\}), (y, \{\psi_\alpha\}) \in \mathcal{G}(\mathcal{T})$, choose any $\gamma \in S_0^x \cap S_0^y$. Then form the composition:

$$L_\gamma \xrightarrow[\psi_\gamma]{} T^*_{p_\gamma(y)}(L_\gamma) \xrightarrow[T^*_{p_\gamma(y)}(\varphi_\gamma)]{} T^*_{p_\gamma(x+y)}(L_\gamma).$$

Call this ρ_γ and generate with it a compatible set of isomorphisms $\{\rho_\alpha\}$, all $\alpha \in S_0^{x+y}$. Then let

$$(x, \{\varphi_\alpha\}) \circ (y, \{\psi_\alpha\}) = (x+y, \{\rho_\alpha\}).$$

Moreover, we get an exact sequence:

$$0 \longrightarrow k^* \longrightarrow \mathcal{G}(\mathcal{T}) \xrightarrow{\pi} V(\underline{X}) \longrightarrow 0$$

as usual. Notice the simplification here over the theory of §1: the finite group $H(L)$ that depended on L has been replaced by the big group $V(\underline{X})$ depending only on the tower and not on the polarization. It is not hard to interpret $\mathcal{G}(\mathcal{T})$ as a simultaneous direct and inverse limit of the $\mathcal{G}(L_\alpha)$'s, with respect to the connections induced between them whenever $\alpha > \beta$, as in Prop. 2, §1. In particular, for all $\alpha_1 \in S_0$, if we look at the subgroup:

$$\begin{array}{c} \mathcal{G}^*_{\alpha_1}(\mathcal{T}) = \pi^{-1}[p_{\alpha_1}^{-1}(H(L_{\alpha_1}))] \\ \cap \\ \mathcal{G}(\mathcal{T}) \end{array}$$

then for all $(x, \{\varphi_\alpha\}) \in \mathcal{G}^*_{\alpha_1}(\mathcal{T})$, $p_{\alpha_1}(x) \in H(L_{\alpha_1})$ so φ_{α_1} is defined, and

$$(x, \{\varphi_\alpha\}) \mapsto (p_{\alpha_1}(x), \varphi_{\alpha_1})$$

defines a surjection:

$$\mathcal{G}^*_{\alpha_1}(\mathcal{T}) \xrightarrow{q_{\alpha_1}} \mathcal{G}(L_{\alpha_1}) \longrightarrow 0.$$

Let $K(\alpha_1)$ denote the kernel of this map. We get the picture:

$$\begin{array}{ccccccccc}
 & & & & 0 & & 0 & & \\
 & & & & \downarrow & & \downarrow & & \\
 & & & & K(\alpha) & \xrightarrow{\sim} & T(\alpha) & & \\
 & & & & \downarrow & & \downarrow & & \\
 & \longrightarrow & k^* & \longrightarrow & \mathcal{G}^*_\alpha(\mathcal{T}) & \xrightarrow{\pi} & p_\alpha^{-1}(H(L_\alpha)) & \longrightarrow & 0 \\
 & & \| & & \downarrow{\scriptstyle q_\alpha} & & \downarrow{\scriptstyle p_\alpha} & & \\
0 & \longrightarrow & k^* & \longrightarrow & \mathcal{G}(L_\alpha) & \xrightarrow{\pi} & H(L_\alpha) & \longrightarrow & 0 \\
 & & & & \downarrow & & \downarrow & & \\
 & & & & 0 & & 0 & &
\end{array}$$

Lemma 2. *For all $\alpha \in S_0$, $\mathscr{G}_\alpha^*(\mathscr{T})$ is the centralizer of $K(\alpha)$ in $\mathscr{G}(\mathscr{T})$, and $k^* \cdot K(\alpha)$ is the center of $\mathscr{G}_\alpha^*(\mathscr{T})$.*

Proof. If $x \in K(\alpha)$ and $y \in \mathscr{G}_\alpha^*(\mathscr{T})$, then $y \cdot x \cdot y^{-1}$ is still in $K(\alpha)$, and it has the same image in $T(\alpha)$ as x has, since $p_\alpha^{-1}(H(L_\alpha))$ is a commutative group. Therefore $x = y \cdot x \cdot y^{-1}$, i.e., $K(\alpha) \subset$ center $[\mathscr{G}_\alpha^*(\mathscr{T})]$. Since k^* is even in the center of $\mathscr{G}(\mathscr{T})$, $k^* \cdot K(\alpha) \subset$ center $[\mathscr{G}_\alpha^*(\mathscr{T})]$ too. But if $x \in$ center $[\mathscr{G}_\alpha^*(\mathscr{T})]$, $q_\alpha(x) \in$ center $[\mathscr{G}(L_\alpha)]$ and we know k^* is the whole center of $\mathscr{G}(L_\alpha)$. Therefore $k^* \cdot K(\alpha)$ is exactly the center of $\mathscr{G}_\alpha^*(\mathscr{T})$. Now suppose $y \in \mathscr{G}(\mathscr{T})$ centralized $K(\alpha)$. y is certainly in $\mathscr{G}_\beta^*(\mathscr{T})$ for some $\beta \in S$, $(\beta > \alpha)$. Then the image $q_\beta(y)$ in $\mathscr{G}(L_\beta)$ commutes with the image $q_\beta[K(\alpha)]$ in $\mathscr{G}(L_\beta)$. Now $q_\beta[K(\alpha)]$ is a subgroup of $\mathscr{G}(L_\beta)$ lying over the subgroup $\mathrm{Ker}(p_{\beta\alpha})$ of $H(L_\beta)$. Therefore by Prop. 2, §1, $q_\beta(y)$ is an element of $\mathscr{G}(L_\beta)$ whose image in $H(L_\beta)$ is in $p_{\beta,\alpha}^{-1}[H(L_\alpha)]$, i.e., the image of y in $V(\underset{\leftarrow}{X})$ is in $p_{\alpha,\beta}^{-1}(H(L_\alpha))$, or $y \in \mathscr{G}_\alpha^*(\mathscr{T})$. *Q.E.D.*

Corollary. *$k^* =$ center of $\mathscr{G}(\mathscr{T})$.*

As in §1, we can describe the non-commutativity by a skew-symmetric form e_λ: $V(\underset{\leftarrow}{X}) \times V(\underset{\leftarrow}{X}) \to k^*$:

$$e_\lambda(\pi x, \pi y) = x \cdot y \cdot x^{-1} \cdot y^{-1}$$

all $x, y \in \mathscr{G}(\mathscr{T})$. Then the lemma tells us that:

1) $p_\alpha^{-1}(H(L_\alpha))$ is the group of elements $x \in V(\underset{\leftarrow}{X})$ such that $e_\lambda(x, y) = 1$, all $y \in T(\alpha)$, i.e., $T(\alpha)^\perp$.

2) $T(\alpha)$ is the group of elements $x \in V(\underset{\leftarrow}{X})$ such that $e_\lambda(x, y) = 1$, all $y \in p_\alpha^{-1}(H(L_\alpha))$, i.e., the degenerate subspace for the pairing e_λ on $p_\alpha^{-1}(H(L_\alpha))$.

3) For all $x \in V(\underset{\leftarrow}{X})$, there is a $y \in V(\underset{\leftarrow}{X})$ such that $e_\lambda(x, y) \neq 1$, i.e., e_λ is non-degenerate.

In particular, for all $\alpha \in S_0$, (a) $e_\lambda \equiv 1$ on $T(\alpha)$, (b) $T(\alpha)^\perp / T(\alpha) \cong H(L_\alpha)$, and (c) the pairing induced on $T(\alpha)^\perp / T(\alpha)$ by e_λ corresponds to the old pairing e_{L_α} on $H(L_\alpha)$.

Using the symmetry of the L_α's, we obtain an automorphism

$$\delta_{-1}: \mathscr{G}(\mathscr{T}) \to \mathscr{G}(\mathscr{T})$$

exactly as in §2. But now since $2 \cdot V(\underset{\leftarrow}{X}) = V(\underset{\leftarrow}{X})$, δ_{-1} is much more convenient than before. In fact, it induces a *canonical section* of $\mathscr{G}(\mathscr{T})$ over $V(\underset{\leftarrow}{X})$:

Definition 5. Let $x \in V(\underset{\leftarrow}{X})$. Let $z \in \mathscr{G}(\mathscr{T})$ satisfy $\pi(z) = x/2$, $\delta_{-1}(z) = z^{-1}$: there are exactly 2 such elements in $\mathscr{G}(\mathscr{T})$, z and $(-1) \cdot z$ (here -1 is an element of k^* and we multiply in $\mathcal{G}(\mathcal{T})$). Let

$$\sigma(x)=z^2$$

(which is independent of the choice of z).

Therefore, $\mathcal{G}(\mathcal{T})$ decomposes *as a set:*

$$\mathcal{G}(\mathcal{T})\cong k^*\times V(\underset{\sim}{X})$$

if

$$\alpha\cdot\sigma(x)\leftrightarrow(\alpha,x).$$

Let's compute what happens to the group law:

Lemma 3. *For all* $x, y\in V(\underset{\sim}{X})$, $\sigma(x)\cdot\sigma(y)=e_\lambda(x, y/2)\cdot\sigma(x+y)$.

Proof. Let $z, w\in\mathcal{G}(\mathcal{T})$ lie over $x/2, y/2$ respectively and satisfy $\delta_{-1}z=z^{-1}, \delta_{-1}w=w^{-1}$. Let

$$s=e_\lambda(-x/4, y/2)\cdot z\cdot w.$$

Then s lies over $(x+y)/2$ and satisfies:

$$\begin{aligned}\delta_{-1}s&=e_\lambda(-x/4, y/2)\cdot\delta_{-1}z\cdot\delta_{-1}w\\&=e_\lambda(-x/4, y/2)\cdot[z^{-1}\cdot w^{-1}\cdot z\cdot w]\cdot(z\cdot w)^{-1}\\&=e_\lambda(-x/4, y/2)\cdot e_\lambda(-x/2, -y/2)\cdot(z\cdot w)^{-1}\\&=s^{-1}.\end{aligned}$$

Therefore

$$\begin{aligned}e_\lambda(x, y/2)\,\sigma(x+y)&=e_\lambda(x, y/2)\cdot s^2\\&=e_\lambda(x/2, y/2)\cdot z\cdot w\cdot z\cdot w\\&=e_\lambda(x/2, y/2)\cdot z^2(z^{-1}\cdot w\cdot z\cdot w^{-1})\,w^2\\&=\sigma(x)\cdot\sigma(y).\end{aligned}$$

Q.E.D.

In other words, the group law, carried over to $k^*\times V(\underset{\sim}{X})$, is

$$(\alpha,x)\cdot(\beta,y)=(\alpha\cdot\beta\cdot e_\lambda(x, y/2), x+y).$$

Let's give a complete structure theorem for $\mathcal{G}(\mathcal{T})$ and $V(\underset{\sim}{X})$. Let

$\chi\colon \boldsymbol{Q}_2\to k^*$ be an additive character with kernel $\boldsymbol{Z}_2$.

Definition 6. A *symplectic isomorphism* of $V(\underset{\sim}{X})$ and $\boldsymbol{Q}_2^g\times\boldsymbol{Q}_2^g$ is an isomorphism $\bar{c}$ such that if $\varphi_1 x, \varphi_2 x$ are the 1st g and 2nd g components of $\bar{c}(x)$, then

$$e_\lambda(x, y)=\chi[{}^t\varphi_1 x\cdot\varphi_2 y-{}^t\varphi_1 y\cdot\varphi_2 x]$$

for all $x, y \in V(\underset{\sim}{X})$ (here t denotes the transpose vector, and $\cdot$ is multiplication of $1 \times g$ and $g \times 1$ matrices).

We leave it to the reader to check that such an isomorphism always exists: for example by constructing it inductively via some cofinal series $\alpha_1 < \alpha_2 < \cdots$ in S.

Definition 7. $\mathscr{G}_g = k^* \times Q_2^g \times Q_2^g$, with group law

$$(\alpha, x, y) \cdot (\beta, u, v) = (\alpha \cdot \beta \cdot \chi({}^t x \cdot v), x+u, y+v).$$

D_{-1}: $\mathscr{G}_g \to \mathscr{G}_g$ is the automorphism $D_{-1}((\alpha, x, y)) = (\alpha, -x, -y)$. $\Sigma: Q_2^g \times Q_2^g \to \mathscr{G}_g$ is the section

$$\Sigma((x, y)) = \left(\chi\left(\frac{{}^t x \cdot y}{2}\right), x, y\right).$$

Definition 8. A *full* ϑ*-structure* for $\mathscr{T}$ is an isomorphism $c: \mathscr{G}(\mathscr{T}) \xrightarrow{\sim} \mathscr{G}_g$, which is the identity on the subgroups k^* and such that $c \circ \delta_{-1} = D_{-1} \circ c$.

It follows immediately that a full ϑ-structure c induces a symplectic isomorphism $\bar{c}$: $V(\underset{\sim}{X}) \xrightarrow{\sim} Q_2^{2g}$, and also that $\Sigma \circ \bar{c} = c \circ \sigma$. Therefore c is even determined by $\bar{c}$. This is a simplification which was foreshadowed in the discussion at the end of § 2. In fact, given a symplectic isomorphism $\bar{c}$, we can define the unique full ϑ-structure c extending $\bar{c}$ by:

$$\begin{aligned} c(\lambda \cdot \sigma(x)) &= \lambda \cdot \Sigma(\bar{c}(x)) \\ &= \left(\lambda \cdot \chi\left(\frac{{}^t\varphi_1 x \cdot \varphi_2 x}{2}\right), \varphi_1 x, \varphi_2 x\right) \end{aligned}$$

for all $\lambda \in k^*$, $x \in V(\underset{\sim}{X})$, and one checks all the requirements easily. In particular, one full ϑ-structure always exists and we have a structure theorem for $\mathscr{G}(\mathscr{T})$ and its maps.

So far, in our polarized tower, we have considered only the *totally symmetric* L_α's that can be put in an inverse system on the X_α's. For a few more α's, however, we may be able to find a *symmetric* invertible sheaf L_α on X_α, such that for some element $\beta \in S_0$ for which $\beta \geqq \alpha$, $p_{\alpha,\beta}^* L_\alpha$ is isomorphic to the L_β that we already have: such an L_α will be said to be *compatible* with the given polarization.

If we define as before

$$\mathscr{G}_\alpha^*(\mathscr{T}) = \pi^{-1}(p_\alpha^{-1}(H(L_\alpha))),$$

then every element $(\{x_\beta\}, \{\varphi_\beta\}) \in \mathscr{G}_\alpha^*(\mathscr{T})$ is compatible with a unique isomorphism

$$\varphi_\alpha: L_\alpha \xrightarrow{\sim} T_{x_\alpha}^*(L_\alpha)$$

and we obtain a homomorphism q_α: $\mathscr{G}_\alpha^*(\mathscr{T}) \to \mathscr{G}(L_\alpha)$ just as before. Let $K(\alpha)$ be its kernel: a subgroup of $\mathscr{G}(\mathscr{T})$ isomorphic under π with $T(\alpha)$.

Lemma 4. *If L_α is a symmetric sheaf on X_α compatible with our polarization as above, then $K(\alpha)$ is the set of elements*

$$e_*^{L_\alpha}\big(p_\alpha(\tfrac{1}{2}x)\big)\cdot\sigma(x)$$

$x\in T(\alpha)$. In particular, if L_α is totally symmetric, then $K(\alpha)=\sigma[T(\alpha)]$.

Proof. To unwind the definition of $\sigma(x)$, let $\beta>\alpha$ be the element of S giving us a diagram:

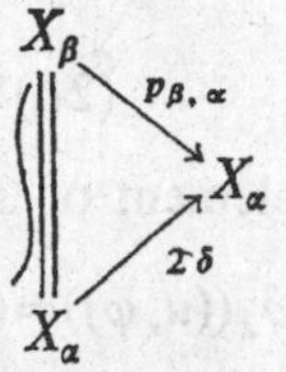

Then $p_{\beta,\alpha}^* L_\alpha$ is totally symmetric, hence isomorphic to L_β, and $u=p_\beta(\tfrac{1}{2}x)$ is in $H(L_\beta)$. Choose an isomorphism

$$\psi: L_\beta \xrightarrow{\;\sim\;} T_u^* L_\beta$$

such that $\delta_{-1}((u,\psi))=(u,\psi)^{-1}$. Let $(u,\psi)^2=(2u,\varphi_\beta)$. Then $\sigma(x)$ is represented by the elements $(2u,\varphi_\alpha)\in\mathscr{G}(L_\alpha)$. Since $x\in T(\alpha)$, $2u\in\mathrm{Ker}(p_{\beta,\alpha})$ and there is a scalar λ such that:

$$
\begin{array}{ccc}
L_\beta & \xrightarrow{\;\varphi_\beta\;} & T_{2u}^* L_\beta \\
\| & & \| \\
p_{\beta,\alpha}^* L_\alpha & & T_{2u}^* p_{\beta,\alpha}^* L_\alpha \\
& \text{mult. by }\lambda\searrow & \| \\
& & p_{\beta,\alpha}^* L_\alpha
\end{array}
$$

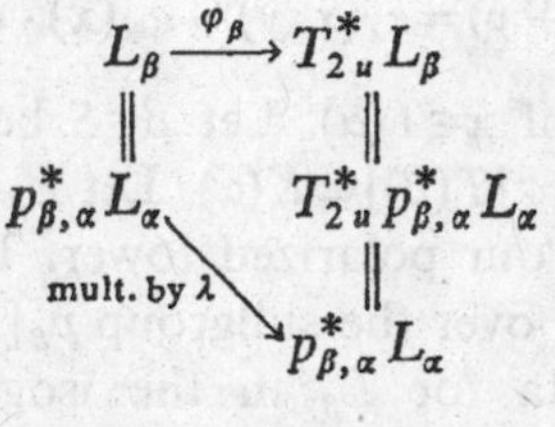

commutes. This means that $\lambda^{-1}\cdot\sigma(x)$ is compatible with the identity map from L_α to L_α, i.e., is in the kernel of q_α. The lemma boils down then to checking that $\lambda=e_*^{L_\alpha}(p_{\beta,\alpha}(u))$. Throwing out irrelevant notation, we can restate this fact as:

Lemma 5. *Let L be a symmetric invertible sheaf of separable type on an abelian variety X. Let $u\in X_4$ and let $(u,\varphi)\in\mathscr{G}((2\delta)^* L)$ satisfy $\delta_{-1}((u,\varphi))$ $=(u,\varphi)^{-1}$. Let $(u,\varphi)^2=(2u,\psi)$. Then the composite isomorphism:*

$$(2\delta)^* L \xrightarrow{\;\psi\;} T_{2u}^*(2\delta)^* L\cong(2\delta)^* T_{4u}^* L=(2\delta)^* L$$

is multiplication by $e_*^L(2u)$.

Proof. Since L is symmetric, $(2\delta)^* L\cong L^4$. Let $v=2u$ and let (v,ρ) be the element $\eta_2((u,\varphi))$ in $\mathscr{G}(L^2)$. Since $\eta_2\circ\delta_{-1}=\delta_{-1}\circ\eta_2$, $\delta_{-1}((v,\rho))=$ $(v,\rho)^{-1}$. But $e_*^{L^2}(v)=[e_*^L(v)]^2=1$, so by Prop. 3, §2, $\delta_{-1}((v,\rho))=(v,\rho)$.

Therefore $(v,\rho)^2=1$ in $\mathscr{G}(L^2)$. Therefore by Prop. 6, § 2, $\eta_2((v,\rho))=(0,e_*^L(v))$ in $\mathscr{G}(L)$. Explicitly, this means that:

$$\begin{array}{ccc}
L^4 & \xrightarrow{\rho^2} & T_v^* L^4 \\
\| & & \| \\
(2\delta)^* L & \dashrightarrow & T_v^*(2\delta)^* L \\
 & \searrow{\scriptstyle \text{mult. by } e_*^L(v)} & \| \\
 & & (2\delta)^* T_{2v}^* L \\
 & & \| \\
 & & (2\delta)^* L
\end{array}$$

commutes. But (v,ρ^2), as an element of $\mathscr{G}(L^4)$, is $\varepsilon_2((v,\rho))$. And

$$\varepsilon_2((v,\rho))=\varepsilon_2(\eta_2((u,\varphi)))=\delta_2((u,\varphi))=(u,\varphi)^3\cdot\delta_{-1}((u,\varphi))=(v,\psi).$$

Therefore the dotted arrow is ψ and the lemma is proven. *Q.E.D.*

Conversely, suppose we start with any $\alpha\in S$ such that $e\equiv 1$ on $T(\alpha)\times T(\alpha)$, i.e., $T(\alpha)$ is isotropic, and try to make a subgroup of $\mathscr{G}(\mathscr{T})$ via

$$K(\alpha)=\{e_*(\tfrac{1}{2}x)\cdot\sigma(x)\mid x\in T(\alpha)\}$$

where e_* is a function from $\frac{1}{2}T(\alpha)$ to $\{\pm 1\}$. This works if we take for e_* any function satisfying

$$e_*(x+y)=e_\lambda(x,y)^2\cdot e_*(x)\cdot e_*(y).$$

In particular, $e_*(x)=1$ if $x\in T(\alpha)$. Let $\beta\in S$ be such that $2T(\alpha)=T(\beta)$. Then $\beta\in S_0$, and $K(\beta)=\sigma[T(\beta)]\subset K(\alpha)$. Let L_β be the totally symmetric sheaf on X_β defined by our polarized tower. Then $K(\alpha)/K(\beta)$ is a level subgroup of $\mathscr{G}(L_\beta)$ lying over the subgroup $p_\beta[T(\alpha)]$ of $H(L_\beta)$. As in §1, it provides descent data for L_β in the isogeny $p_{\beta\alpha}\colon X_\beta\to X_\alpha$, since $\operatorname{Ker}(p_{\beta\alpha})=p_\beta[T(\alpha)]$. Let it define L_α on X_α. This L_α is easily seen to be symmetric, compatible with the polarization, and satisfying

$$e_*^{L_\alpha}(p_\alpha(x))=e_*(x),\qquad \text{all } x\in\tfrac{1}{2}T(\alpha).$$

Conclusion. Symmetric sheaves L_α on some X_α, compatible with the polarization, are in $1-1$ correspondence with all possible "level" subgroups $K\subset\mathscr{G}(\mathscr{T})$ such that

1) $K\cap k^*=\{1\}$,
2) for all $x\in K$, $\sigma(\pi(x))=\pm x$.

This ends our discussion of the groups that are involved in a polarized tower of abelian varieties. Next, we turn to their representations. The family of vector spaces

$$\{\Gamma(X_\alpha,L_\alpha)\},\qquad \alpha\in S_0$$

forms a direct system, and we define:

$$\Gamma(\mathcal{T}) = \varinjlim_{\alpha \in S_0} \Gamma(X_\alpha, L_\alpha).$$

Just as in §1, $\mathcal{G}(\mathcal{T})$ is represented on the vector space $\Gamma(\mathcal{T})$. Moreover, as in §1, we check that we can recover $\Gamma(X_\alpha, L_\alpha)$ from $\Gamma(\mathcal{T})$ since:

$$\Gamma(X_\alpha, L_\alpha) \cong \left\{ \begin{array}{l} \text{elements of } \Gamma(\mathcal{T}) \\ \text{invariant under } K(\alpha) \end{array} \right\}.$$

[The same holds for any symmetric L_α compatible with the polarization and the corresponding $K(\alpha)$.] What representations does a group like $\mathcal{G}(\mathcal{T})$ have? We make the following restriction:

Definition 9. An *admissible representation* $\alpha \mapsto U_\alpha$ of $\mathcal{G}(\mathcal{T})$ [resp. $\mathcal{G}_g$] in a vector space V is one in which the subgroup k^* acts via its natural character (i.e., if $\alpha \in k^*$, $U_\alpha = \alpha \cdot (id)_V$), and such that for all $x \in V$:

$$\{\sigma \in \mathcal{G}(\mathcal{T}), \text{ resp. } \mathcal{G}_g \mid U_\sigma x = \text{multiple of } x\},$$

is the inverse image in $\mathcal{G}(\mathcal{T})$ [resp. $\mathcal{G}_g$] of an *open* subset of $V(\underset{\sim}{X})$ [resp. Q_2^{2g}].

Prop. 3 of §1 generalizes easily to:

Theorem. $\mathcal{G}_g$, *and hence* $\mathcal{G}(\mathcal{T})$, *has one and only one irreducible admissible representation. All other admissible representations break up into direct sums of the irreducible one with itself.*

The proof of this is roughly as follows: choose a maximal open subgroup $U \subset Q_2^{2g}$ on which the skew-symmetric form

$$e((x_1; y_1), (x_2; y_2)) = \chi[{}^t x_1 \cdot y_2 - {}^t x_2 \cdot y_1]$$

vanishes identically, such as $\mathbf{Z}^{2g}$. Then construct eigenvectors in V for the subgroup $k^* \times U$ of $\mathcal{G}_g$. All other elements in the group permute these eigenvectors and we show that this permutation can be described simply, (independent of V). We leave the details to the reader as they are similar to those in the proof of Prop. 3, §1.

This irreducible representation can be written down like this:

Let

$$\mathcal{H}_g = \left\{ \begin{array}{l} \text{vector space of } k\text{-valued, locally constant} \\ \text{functions } f \text{ on } Q_2^g, \text{ with compact support} \end{array} \right\}.$$

$$[U_{(\alpha, x, y)} f](z) = \alpha \cdot \chi({}^t y \cdot z) \cdot f(x+z)$$

$$\text{if } f \in \mathcal{H}_g, \ (\alpha, x, y) \in \mathcal{G}_g, \ z \in Q_2^g.$$

8*

Notice that the only elements $f \in \mathcal{H}_g$ invariant under the operations $U_{(1,x,y)}$, all $x, y \in \mathbf{Z}_2^g$, are the multiples of the characteristic function:

$$D(z)=\begin{cases}0, & z \notin \mathbf{Z}_2^g\\ 1, & z \in \mathbf{Z}_2^g.\end{cases}$$

Therefore $\mathcal{H}_g$ is irreducible (i.e., in view of the Theorem, if any subspace canonically attached to the representation is *one*-dimensional, the representation must be irreducible).

On the other hand, we have:

Theorem. *$\Gamma(\mathcal{T})$ is an irreducible admissible representation for $\mathcal{G}(\mathcal{T})$.*

Proof. It is an admissible representation since every $x \in \Gamma(\mathcal{T})$ is in some $\Gamma(X_\alpha, L_\alpha)$, hence is an eigenvector for $k^* \cdot K(\alpha)$. H is irreducible since if $K(\alpha) \subset \mathcal{G}(\mathcal{T})$ corresponds to a symmetric L_α on X_α of *degree 1*, then the subspace of $K(\alpha)$-invariants is isomorphic to $\Gamma(X_\alpha, L_\alpha)$ and this is 1-dimensional. *Q.E.D.*

It follows that if we choose a ϑ-structure c: $\mathcal{G}(\mathcal{T}) \to \mathcal{G}_g$, we get a unique isomorphism:

$$\Gamma(\mathcal{T}) \xrightarrow[\sim]{\beta} \mathcal{H}_g$$

such that $\beta(U_z(s)) = U_{c(z)}(\beta(s))$, all $z \in \mathcal{G}(\mathcal{T})$, $s \in \Gamma(\mathcal{T})$. The isomorphisms β extend the isomorphisms β in §1 in the following way:

a. let $\alpha \in S_0$ and let $T(\alpha) \subset V(\underset{\sim}{X})$ be a compact open subgroup such that $\bar{c}(T(\alpha))$ in $\mathbf{Q}_2^g \times \mathbf{Q}_2^g$ is of the form $U \times V$.

b. Then $p_\alpha^{-1}(H(L_\alpha))$ is the orthogonal subgroup, i.e., $V^\perp \times U^\perp$ (if $V^\perp = \{x \in \mathbf{Q}_2^g \mid \chi({}^t x \cdot y) = 1$, all $y \in V\}$, $U^\perp$ similar). Therefore $\bar{c}$ induces isomorphisms:

$$H(L_\alpha) \cong \frac{p_\alpha^{-1}(H(L_\alpha))}{T(\alpha)} \cong (V^\perp/U) \times (U^\perp/V).$$

If we choose an isomorphism

$$(V^\perp/U) \xrightarrow{\sim} K(\delta),$$

then this gives

$$(U^\perp/V) \cong \widehat{(V^\perp/U)} \cong \widehat{K(\delta)},$$

hence

$$H(L_\alpha) \cong K(\delta) \times \widehat{K(\delta)}.$$

c. Thus we get an isomorphism:

$$\begin{aligned}\mathcal{G}(L_\alpha) &\cong \frac{\mathcal{G}_\alpha^*}{K(\alpha)}\\ &\cong k^* \times (V^\perp/U) \times (U^\perp/V)\\ &\cong k^* \times K(\delta) \times \widehat{K(\delta)} = \mathcal{G}(\delta).\end{aligned}$$

d. On the other hand, β restricts as follows:

$$\begin{array}{ccc} \Gamma(\mathcal{T}) & \xrightarrow{\beta} & \mathcal{H}_g \\ \bigcup & & \bigcup \\ \Gamma(\mathcal{T})^{K(\alpha)} & \dashrightarrow & \left\{\begin{array}{l}\text{functions } \varphi \text{ on } Q_2^g\\ \text{support in } V^\perp\\ \text{constant w.r.t.}\\ \text{translations by } U\end{array}\right\} \\ \| & & \wr\| \\ \Gamma(X_\alpha, L_\alpha) & & \{\text{functions on } V^\perp/U\} \\ & & \wr\| \\ & & \{\text{functions on } K(\delta)\} = V(\delta). \end{array}$$

The induced isomorphism of $\Gamma(X_\alpha, L_\alpha)$ and $V(\delta)$ is exactly the isomorphism β of §1 corresponding to the ϑ-structure on $\mathcal{G}(L_\alpha)$ occurring in c.

In short, β is just the union of the isomorphisms β obtained on a finite level previously.

Next choose consistent isomorphisms $(L_\alpha)_0 \otimes \kappa(0) \xrightarrow{\sim} k$, for all $\alpha \in S_0$. These induce "evaluation at 0" maps: $\Gamma(X_\alpha, L_\alpha) \to k$, for all α, which fit together into one big "evaluation at 0" map:

$$\Lambda: \Gamma(\mathcal{T}) \to k.$$

To describe this piece of information in closed form, we need to describe the dual space $\mathcal{H}_g^*$:

Dual of $\mathcal{H}_g$: $\mathcal{H}_g$ is spanned by the characteristic functions φ_U of compact, open subsets $U \subset Q_g^2$, with the obvious relations:

$$(*) \qquad \varphi_{U_1 \cup U_2} + \varphi_{U_1 \cap U_2} = \varphi_{U_1} + \varphi_{U_2}.$$

Now let $\mathcal{B}$ be the Boolean algebra of compact open subsets of Q_2^g. Then a linear functional on $\mathcal{H}_g$ is determined by its values on the φ_U's, and relations which follow from $(*)$ make this set function into a measure. In other words, if we let

$$\mathcal{H}_g^* = \left\{\begin{array}{l}\text{vector space of } k\text{-valued finitely additive}\\ \text{measures } \mu \text{ defined on the Boolean algebra } \mathcal{B}\end{array}\right\}$$

then the pairing:

$$\langle f,\mu\rangle = \int_{Q_2^g} f\cdot d\mu$$

makes $\mathscr{H}_g^*$ into the dual of $\mathscr{H}_g$.

This shows that there is a unique measure $\mu\in\mathscr{H}_g^*$ such that

$$\Lambda(s) = \int_{Q_2^g} \beta(s)\cdot d\mu$$

all $s\in\Gamma(\mathscr{T})$. This μ is "built up" out of null functions q_{L_α} in the following way: let $\alpha\in S_0$ be such that $\bar{c}(T(\alpha))=U\times V$, as above. Then for all $s\in\Gamma(X_\alpha, L_\alpha)$, $\beta(s)$ is a function on Q_2^g with support in $V^\perp$, constant on cosets of U. Then

$$\begin{aligned}\Lambda(s) &= \int_{Q_2^g} \beta(s)\cdot d\mu && \text{via new theory}\\ &= \sum_{x\in V^\perp/U} \beta(s)(x)\cdot\mu(x+U)\end{aligned}$$

while

$$\Lambda(s) = \sum_{x\in K(\delta)} \beta(s)(x)\cdot q_{L_\alpha}(x) \qquad \text{via old theory}.$$

Thus identifying $K(\delta)$ with $V^\perp/U$, we find:

$$q_{L_\alpha}(x \bmod U) = \mu(x+U), \qquad \text{all } x\in V^\perp.$$

There are formulae for the other q_{L_α}'s, but they are much more complicated.

Note that since the q_{L_α}'s are known to be even functions, this formula implies that μ is an *even* measure on Q_2^g.

§ 8. 2-Adic Theta Functions

The basic idea of theta functions is to trivialize an ample sheaf L on an abelian variety X, after pulling it back to some auxiliary space V via a map $\pi\colon V\to X$; then sections of L on X become actual k-valued functions on V. In our case, let $\mathscr{T}=\{X_\alpha, L_\alpha\}$ be a polarized 2-tower of abelian varieties, and take $V=V(\underset{\sim}{X})$. First, what is the "pull-back" of the L_α to V?

Let $x=\{x_\alpha\}\in V(\underset{\sim}{X})$. Then for all $\alpha\in S$, let $L_\alpha(x_\alpha)=(L_\alpha)_{x_\alpha}\otimes_{\mathcal{O}_{X_\alpha}}\kappa(x_\alpha)$ as usual. For all $\alpha\geqq\beta$, $p_{\alpha\beta}^*$ induces an isomorphism

$$L_\beta(x_\beta)\xrightarrow[p_{\alpha\beta}^*]{\sim} L_\alpha(x_\alpha).$$

Passing to the limit, let $L(x)$ denote the vector space you get which is canonically isomorphic to all $L_\alpha(x_\alpha)$'s. The collection $\{L(x)\}$ of 1-dimensional vector spaces represents "the sheaf induced on $V(\underset{\leftarrow}{X})$ by the L_α's". In particular, every $s \in \Gamma(\mathscr{T})$ has "values" $s(x) \in L(x)$ for each $x \in V(\underset{\leftarrow}{X})$.

Now let $z = (x, \{\varphi_\alpha\}) \in \mathscr{G}(\mathscr{T})$. For all α for which φ_α is defined, it induces an isomorphism

$$\varphi_\alpha(x_\alpha)\colon L_\alpha(y) \xrightarrow{\sim} T^*_{x_\alpha}(L_\alpha)(y) = L_\alpha(y + x_\alpha),$$

all $y \in X_\alpha$. In particular, z defines an isomorphism $z_y\colon L(y) \xrightarrow{\sim} L(x+y)$ for all $y \in V(\underset{\leftarrow}{X})$. Choose as before an isomorphism λ_0 of $L(0)$ with k.

Define λ_y to be the composite:

$$L(y) \xrightarrow[\sigma(-y)_y]{\sim} L(0) \xrightarrow[\lambda_0]{\sim} k.$$

Then $\{\lambda_y\}$ is a "trivialization" of L_α pulled back to $V(\underset{\leftarrow}{X})$. In particular, if $s \in \Gamma(\mathscr{T})$, then $\lambda_y(s(y))$ is the value of s at y. Define:

$$\vartheta_{[s]}(y) = \lambda_y(s(y)).$$

This is the *algebraic theta function associated to s*. Alternatively, it can be expressed as:

$$\vartheta_{[s]}(y) = \Lambda[U_{\sigma(-y)}(s)].$$

Property I.

$$\vartheta_{[U_w(s)]}(y) = \alpha \cdot e_\lambda(x/2, y) \cdot \vartheta_{[s]}(y - x), \quad \text{if } w = \alpha \cdot \sigma(x).$$

Proof.

$$\begin{aligned}\vartheta_{[U_w(s)]}(y) &= \Lambda[U_{\sigma(-y)}(U_w(s))]\\ &= \alpha \cdot \Lambda[U_{\sigma(-y)\cdot\sigma(x)}(s)]\\ &= \alpha \cdot e_\lambda(x/2, y) \cdot \Lambda[U_{\sigma(-y+x)}(s)]\\ &= \alpha \cdot e_\lambda(x/2, y)\, \vartheta_{[s]}(y - x).\end{aligned}$$

Corollary. *$\vartheta_{[s]}$ is a locally constant function on $V(\underset{\leftarrow}{X})$. In fact, if $s \in \Gamma(X_\alpha, L_\alpha)$, and L_α is symmetric associated to $e_*\colon \frac{1}{2}T(\alpha) \to \{\pm 1\}$, then $\vartheta_{[s]}(y) = e_*(x/2)\, e_\lambda(x/2, y) \cdot \vartheta_{[s]}(y - x)$ for all $x \in T(\alpha)$.*

Proof. Use $U_w(s) = s$, whenever $w = e_*(x/2) \cdot \sigma(x)$, $x \in T(\alpha)$.

Property II. *If $s_1, \ldots, s_n \in \Gamma(X_\alpha, L_\alpha)$, and $p(X_1, \ldots, X_n) \in k[X_1, \ldots, X_n]$ is homogeneous of degree d, then*

$$P(s_1, \ldots, s_n) = 0 \quad \text{in } \Gamma(X_\alpha, L_\alpha^d)$$

if and only if

$$P(\vartheta_{[s_1]}(y), \ldots, \vartheta_{[s_n]}(y)) = 0, \quad \text{all } y \in V(\underset{\leftarrow}{X}).$$

Equivalently, the map

$$\vartheta:\ \Gamma(\mathcal{T})\rightarrow\begin{Bmatrix}\text{vector space of locally constant}\\ k\text{-valued functions on } V(X)\end{Bmatrix}$$

extends to an injective homomorphism:

$$\vartheta:\ \bigoplus_{d=0}^{\infty}\varinjlim_{\alpha}\Gamma(X_{\alpha},L_{\alpha}^{d})\longrightarrow\begin{Bmatrix}\text{ring of locally constant}\\ k\text{-valued functions on } V(X)\end{Bmatrix}.$$

Proof. The fact that ϑ extends to a homomorphism, or that $P(s_1, \ldots, s_n)=0$ implies $P(\vartheta_{[s_1]}(y), \ldots, \vartheta_{[s_n]}(y))=0$ follows from the fact that $\vartheta_{[t]}(y)$ is defined as the *value* of t at y, (when L_α is suitably trivialized). The fact that ϑ is injective is equivalent to noting that no non-zero section $t\in\Gamma(X_\alpha,L_\alpha^d)$ can vanish on *all* points $\mathrm{tor}_2(X_\alpha)$; and this is clear since $\mathrm{tor}_2(X_\alpha)$ is Zariski-dense in X_α.   *Q.E.D.*

Property III. *If σ is an automorphism of the field k, leaving fixed $k_0\subset k$, and if (X_α,L_α) is defined over k_0, then σ acts on*

1) $\Gamma(X_\alpha,L_\alpha)$

2) $V(X)$

and for all $s\in\Gamma(X_\alpha,L_\alpha)$, $y\in V(X)$

$$\vartheta_{[\sigma(s)]}(\sigma(y))=\sigma[\vartheta_{[s]}(y)].$$

(More generally, it seems reasonable to expect that $\vartheta_{[s]}$ should "be defined" — in a suitable sense — over any ring R over which an X_α and L_α are given.)

The proof is straightforward.

Property IV. *If $\bar{c}: V(\underline{X})\xrightarrow{\sim}Q_2^g\times Q_2^g$ is a symplectic isomorphism, the transformation T in the commutative diagram:*

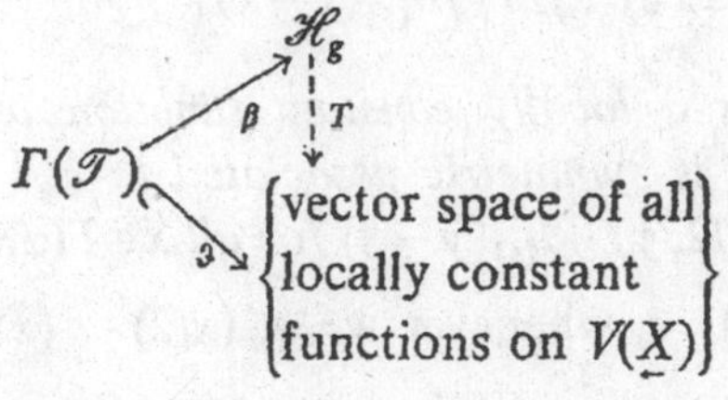

$$\begin{array}{ccc}
& & \mathcal{H}_g \\
& \nearrow^{\beta} & \Big\updownarrow T \\
\Gamma(\mathcal{T}) & & \\
& \searrow_{\vartheta} & \begin{Bmatrix}\text{vector space of all}\\ \text{locally constant}\\ \text{functions on } V(\underline{X})\end{Bmatrix}
\end{array}$$

is given by:

$$T(f)(y)=\int\limits_{Q_2^g}\chi\left({}^t\varphi_2\,y\cdot\left(\frac{\varphi_1\,y}{2}-u\right)\right)\cdot f(u-\varphi_1\,y)\cdot d\mu_u$$

where $\bar{c}(x)=(\varphi_1\,x,\ \varphi_2\,x)$, and $\mu\in\mathcal{H}_g^*$ is the theta-null measure of the last section.

Proof. If $f=\beta(s)$, then

$$\begin{aligned} T(f)(x) &= \vartheta_{[s]}(x) \\ &= \Lambda(U_{\sigma(-x)}(s)) \\ &= \int_{Q_2^g} (U_{\Sigma(-\varphi_1 x, -\varphi_2 x)} f) \cdot d\mu \\ &= \int_{Q_2^g} \chi\left(\frac{{}^t\varphi_1 x \cdot \varphi_2 x}{2}\right) \cdot \chi(-{}^t u \cdot \varphi_2 x) \cdot f(u-\varphi_1 x) \cdot d\mu_u . \end{aligned}$$

Q.E.D.

Corollary. *A basis of $\vartheta[\Gamma(\mathcal{T})]$ is given by the "classical" theta-functions:*

$$\vartheta\begin{bmatrix} a \\ b \end{bmatrix} = T\left(\delta\begin{bmatrix} a \\ b \end{bmatrix}\right)$$

where

$$\delta\begin{bmatrix} a \\ b \end{bmatrix}(x) = \begin{cases} 0 & \text{if } x \notin a + Z_2^g \\ \chi({}^t b \cdot x) & \text{if } x \in a + Z_2^g \end{cases}$$

and a, b run through coset representatives of Q_2^g/Z_2^g. Here

$$\vartheta\begin{bmatrix} a \\ b \end{bmatrix}(x) = \chi\left({}^t\left(\frac{\varphi_2 x}{2} - b\right) \cdot \varphi_1 x\right) \cdot \int_{a+\varphi_1 x + Z_2^g} \chi({}^t(b-\varphi_2 x)\cdot u)\, d\mu_u .$$

In particular,

$$\vartheta\begin{bmatrix} 0 \\ 0 \end{bmatrix}(x) = \chi\left(\frac{{}^t\varphi_2 x \cdot \varphi_1 x}{2}\right) \int_{\varphi_1 x + Z_2^g} \chi(-{}^t\varphi_2 x \cdot u)\, d\mu_u$$

is known as RIEMANN's theta function. It is the transform via T of the characteristic function $\delta\begin{bmatrix} 0 \\ 0 \end{bmatrix}$ of Z_2^g. The multiples of $\delta\begin{bmatrix} 0 \\ 0 \end{bmatrix}$ form the subspace of $\mathcal{H}_g$ invariant under the subgroup $\{1\} \times Z_2^g \times Z_2^g \subset \mathcal{G}_g$. Therefore, the multiples of $\vartheta\begin{bmatrix} 0 \\ 0 \end{bmatrix}$ are exactly the elements of $\vartheta[\Gamma(\mathcal{T})]$ invariant under the subgroup

$$K(\alpha) = c^{-1}(\{1\} \times Z_2^g \times Z_2^g) \subset \mathcal{G}(\mathcal{T}).$$

If $T(\alpha)$ is the compact open subgroup of $V(\underline{X})$:

$$T(\alpha) = \bar{c}^{-1}(Z_2^g \times Z_2^g),$$

then in fact $K(\alpha) = \{e_*(\tfrac{1}{2}x) \cdot \sigma(x) \mid x \in T(\alpha)\}$ where

$$e_*(x) = \chi(2\,{}^t\varphi_1 x \cdot \varphi_2 x), \qquad \text{all } x \in \tfrac{1}{2} T(\alpha).$$

Thus $K(\alpha)$ defines a symmetric invertible sheaf L_α on the abelian variety X_α, and $\Gamma(X_\alpha, L_\alpha)$ is just the space of $K(\alpha)$-invariants in $\Gamma(\mathcal{T})$. Since this is 1-dimensional, L_α has degree 1, and $\vartheta\begin{bmatrix}0\\0\end{bmatrix}$ is nothing but the algebraic theta function defined by the unique section of L_α (up to scalars). Another way to put it is that

$$\vartheta\begin{bmatrix}0\\0\end{bmatrix}\circ\bar{c}^{-1}$$

is the unique function f of the form $\vartheta_{[s]}\circ\bar{c}^{-1}$ to satisfy the functional equation:

$$f(x+a)=\chi\left(\frac{{}^t a_1\cdot a_2}{2}\right)\cdot\chi\left(\frac{{}^t a_1\cdot x_2}{2}-\frac{{}^t a_2\cdot x_1}{2}\right)\cdot f(x)$$

all $x\in \mathbf{Q}_2^{2g}$, $a\in \mathbf{Z}_2^{2g}$. Note that since μ is an *even* measure on $\mathbf{Q}_2^g$, $\vartheta\begin{bmatrix}0\\0\end{bmatrix}$ is an *even* function on $V(\underline{X})$:

$$\vartheta\begin{bmatrix}0\\0\end{bmatrix}(-x)=\vartheta\begin{bmatrix}0\\0\end{bmatrix}(x).$$

An important fact is that μ can be reconstructed from $\vartheta\begin{bmatrix}0\\0\end{bmatrix}$:

Lemma 1. *There is a* $1-1$ *correspondence between*

(A.) *measures* $\mu\in\mathscr{H}_g^*$ *on* $\mathbf{Q}_2^g$,

(B.) *k-valued functions* ϑ *on* $\mathbf{Q}_2^g\times\mathbf{Q}_2^g$ *such that*

$$\vartheta(x+a)=\chi\left(\frac{{}^t a_1\cdot a_2}{2}\right)\cdot\chi\left(\frac{{}^t a_1\cdot x_2}{2}-\frac{{}^t a_2\cdot x_1}{2}\right)\cdot\vartheta(x)$$

all $a\in\mathbf{Z}_2^g\times\mathbf{Z}_2^g$.

This is set up by:

$$\vartheta(x)=\chi\left(\frac{{}^t x_1\cdot x_2}{2}\right)\cdot\int_{x_1+\mathbf{Z}_2^g}\chi(-{}^t x_2\cdot u)\cdot d\mu_u$$

$$\mu(a_1+2^n\mathbf{Z}_2^g)=2^{-ng}\cdot\sum_{a_2\in 2^{-n}\mathbf{Z}_2^g/\mathbf{Z}_2^g}\chi\left(\frac{{}^t a_1\cdot a_2}{2}\right)\cdot\vartheta(a_1,a_2).$$

Proof. Left to reader.

In particular, in studying the μ's that arise from abelian varieties, it is often convenient to go back and forth between μ's and ϑ's that correspond as in this lemma. Let's consider RIEMANN's theta relation from this point of view. Before proving it, I want to set it up in both its μ and ϑ-form:

Lemma 2. *Let* μ *and* ϑ *be* even *measures/functions corresponding as in Lemma 1. The following conditions are equivalent:*

(A.) *There is a* 2nd *measure,* $\nu \in \mathscr{H}_g^*$ *related to* μ *by the identity:*

$$\mu \times \mu(U) = \nu \times \nu(\xi(U))$$

for all compact open subsets $U \subset \mathbf{Q}_2^g \times \mathbf{Q}_2^g$, *where* ξ *is the automorphism of* $\mathbf{Q}_2^g \times \mathbf{Q}_2^g$:

$$\xi(x_1, x_2) = (x_1 + x_2, x_1 - x_2).$$

(B.) *For all* $x, y, u, v \in \mathbf{Q}_2^g \times \mathbf{Q}_2^g$, *if* $r = -\frac{1}{2}(x+y+u+v)$,

$$\begin{aligned} &\vartheta(x) \cdot \vartheta(y) \cdot \vartheta(u) \cdot \vartheta(v) \\ &= 2^{-g} \cdot \sum_{\eta \in \frac{1}{2} \mathbf{Z}_2^{2g}/\mathbf{Z}_2^{2g}} \chi({}^t r_1 \cdot \eta_2 - {}^t\eta_1 \cdot r_2) \cdot \vartheta(x+r+\eta) \times \\ &\quad \times \vartheta(y+r+\eta) \cdot \vartheta(u+r+\eta) \cdot \vartheta(v+r+\eta). \end{aligned}$$

Proof. To analyze (A.), note first that it is equivalent to the existence of a measure ν such that

$$\int_{\mathbf{Q}_2^g \times \mathbf{Q}_2^g} \xi^* \left(\delta \begin{bmatrix} 2x_1 \\ -x_2 \end{bmatrix} \times \delta \begin{bmatrix} 2y_1 \\ -y_2 \end{bmatrix} \right) d\mu \times d\mu$$

$$= \int_{\mathbf{Q}_2^g \times \mathbf{Q}_2^g} \delta \begin{bmatrix} 2x_1 \\ -x_2 \end{bmatrix} \times \delta \begin{bmatrix} 2y_1 \\ -y_2 \end{bmatrix} d\nu \times d\nu$$

for all $x_1, y_1, x_2, y_2 \in \mathbf{Q}_2^g$. (Here ξ^* denotes pull-back of functions.) This is because the functions

$$\delta \begin{bmatrix} a \\ b \end{bmatrix} \times \delta \begin{bmatrix} c \\ d \end{bmatrix}$$

span the vector space of locally constant function on $\mathbf{Q}_2^{2g}$ with compact support. But

$$\begin{aligned} \xi^* \left(\delta \begin{bmatrix} 2x_1 \\ -x_2 \end{bmatrix} \times \delta \begin{bmatrix} 2y_1 \\ -y_2 \end{bmatrix} \right)(u, v) &= 0 \quad \text{if } u+v \notin 2x_1 + \mathbf{Z}_2^g \\ &\qquad \text{or if } u-v \notin 2y_1 + \mathbf{Z}_2^g \\ &= \chi(-{}^t x_2 \cdot (u+v)) \cdot \chi(-{}^t y_2 \cdot (u-v)) \\ &\qquad \text{if } u \in x_1 + y_1 + \eta + \mathbf{Z}_2^g \\ &\qquad\quad v \in x_1 - y_1 + \eta + \mathbf{Z}_2^g \\ &\qquad \text{for some } \eta \in \tfrac{1}{2} \mathbf{Z}_2^g. \end{aligned}$$

Thus

$$\xi^* \left(\delta \begin{bmatrix} 2x_1 \\ -x_2 \end{bmatrix} \times \delta \begin{bmatrix} 2y_1 \\ -y_2 \end{bmatrix} \right) = \sum_{\eta \in \frac{1}{2} \mathbf{Z}_2^g/\mathbf{Z}_2^g} \delta \begin{bmatrix} x_1 + y_1 + \eta \\ -x_2 - y_2 \end{bmatrix} \times \delta \begin{bmatrix} x_1 - y_1 + \eta \\ -x_2 + y_2 \end{bmatrix}.$$

But note that

$$\int_{Q_2^g} \delta \begin{bmatrix} x_1 \\ -x_2 \end{bmatrix} d\mu = \int_{x_1 + Z_2^g} \chi(-{}^t x_2 \cdot u)\, d\mu_u$$

$$= \chi\left(-\frac{{}^t x_1 \cdot x_2}{2}\right) \cdot \vartheta(x_1, x_2).$$

Given v, *define*

$$\Phi(x_1, x_2) = \chi({}^t x_1 \cdot x_2) \int_{2x_1 + Z_2^g} \chi(-{}^t x_2 \cdot u)\, d v_u.$$

Then equation (*) is the same as:

$$(**) \quad \sum_{\eta \in \frac{1}{2} Z_2^g / Z_2^g} \chi(-{}^t\eta \cdot x_2)\, \vartheta(x_1 + y_1 + \eta, x_2 + y_2)\, \vartheta(x_1 - y_1 + \eta, x_2 - y_2)$$
$$= \Phi(x_1, x_2) \cdot \Phi(y_1, y_2).$$

The reader can check that if there is a function Φ satisfying (**), this Φ comes from a v satisfying (*). Thus (A.) is equivalent to the existence of a Φ satisfying (**). If we let $x = (x_1, x_2)$, $y = (y_1, y_2)$, $\zeta_1 = (\eta, 0)$, then (**) becomes:

$$(**)' \quad \sum_{\zeta_1 \in \frac{1}{2}(Z_2^g \times 0)/Z_2^g \times 0} \chi(-{}^t\zeta_1 \cdot x_2)\, \vartheta(x + y + \zeta_1)\, \vartheta(x - y + \zeta_1) = \Phi(x) \cdot \Phi(y).$$

The existence of a Φ satisfying this is clearly equivalent to:

$$(***) \quad \begin{aligned} &\sum_{\zeta_1} \chi(-{}^t\zeta_1 \cdot x_2)\, \vartheta(x + y + \zeta_1)\, \vartheta(x - y + \zeta_1) \times \\ &\qquad \times \sum_{\zeta_1} \chi(-{}^t\zeta_1 \cdot u_2)\, \vartheta(u + v + \zeta_1)\, \vartheta(u - v + \zeta_1) \\ &= \sum_{\zeta_1} \chi(-{}^t\zeta_1 \cdot x_2)\, \vartheta(x + v + \zeta_1)\, \vartheta(x - v + \zeta_1) \times \\ &\qquad \times \sum_{\zeta_1} \chi(-{}^t\zeta_1 \cdot u_2)\, \vartheta(u + y + \zeta_1)\, \vartheta(u - y + \zeta_1) \end{aligned}$$

for all $x, y, u, v \in Q_2^{2g}$ (the summations being as in (**)').

In (***), replace x by $x + \zeta_2$, y by $y + \zeta_2$, u by $u + \zeta_2'$ and v by $v + \zeta_2'$, where $\zeta_2, \zeta_2' \in 0 \times (\frac{1}{2} Z_2^g)$. Multiply by $\chi({}^t\zeta_2 \cdot x_1 + y_1) \cdot \chi({}^t\zeta_2' \cdot u_1 + v_1)$ and sum over all ζ_2, ζ_2' mod $0 \times Z_2^g$. Then out comes (*B*)! Reversing this, you can get (***) out of (*B*). *Q.E.D.*

Definition. Even measures μ with property (A.) above will be called *Gaussian.*

We now intend to prove that if

$$\vartheta = \vartheta \begin{bmatrix} 0 \\ 0 \end{bmatrix},$$

and μ is the null-value measure, arising from a polarized tower $\mathcal{T}$, plus ϑ-structure c, then these conditions on μ and ϑ hold. This fact is derived by comparing

$$\mathscr{T}=\{X_\alpha, L_\alpha\}$$

with the new tower

$$\mathscr{T}^{(2)}=\{X_\alpha, L_\alpha^2\}.$$

Notice, incidentally, that the relationship between the towers $\mathscr{T}$ and $\mathscr{T}^{(2)}$ is symmetric, in that if we let

$$\mathscr{T}^{(4)}=\{X_\alpha, L_\alpha^4\}$$

then $\mathscr{T}^{(4)}$ is *isomorphic* to $\mathscr{T}$ again. In fact, for all $\alpha\in S$, let $2*\alpha$ be the new index such that $2*\alpha>\alpha$ and such that we get a diagram:

$$\begin{array}{ccc} X_{2*\alpha} & \xrightarrow[f_\alpha]{\approx} & X_\alpha \\ {\scriptstyle p_{2*\alpha,\alpha}}\big\downarrow & \swarrow {\scriptstyle 2\delta} & \\ X_\alpha & & \end{array}$$

Putting together all these isomorphisms, we find

$$L_{2*\alpha}\cong f_\alpha^*(L_\alpha^4), \quad \text{hence} \quad (X_{2*\alpha}, L_{2*\alpha})\cong(X_\alpha, L_\alpha^4)$$

so

$$\mathscr{T}=\{X_{2*\alpha}, L_{2*\alpha}\}\underset{f}{\cong}\{X_\alpha, L_\alpha^4\}=\mathscr{T}^{(4)}.$$

The natural homomorphisms ε_2 and η_2 can be defined as in § 2, and we get a diagram:

$$\begin{array}{ccccccccc}
1 & \longrightarrow & k^* & \longrightarrow & \mathscr{G}(\mathscr{T}) & \longrightarrow & V(X) & \longrightarrow & 0 \\
 & & \big\downarrow{\scriptstyle \alpha^2} & & \big\downarrow{\scriptstyle \varepsilon_2} & & \| & & \\
1 & \longrightarrow & k^* & \longrightarrow & \mathscr{G}(\mathscr{T}^{(2)}) & \longrightarrow & V(X) & \longrightarrow & 0 \\
 & & \big\downarrow{\scriptstyle \alpha^2} & {\scriptstyle \eta^2} & \big\downarrow{\scriptstyle \varepsilon_2} & & \| & & \\
1 & \longrightarrow & k^* & \longrightarrow & \mathscr{G}(\mathscr{T}^{(4)}) & \longrightarrow & V(X) & \longrightarrow & 0 \\
 & & \| & & \big\downarrow{\scriptstyle \text{via } f} & & \big\downarrow{\scriptstyle \text{mult. by } 2} & & \\
1 & \longrightarrow & k^* & \longrightarrow & \mathscr{G}(\mathscr{T}) & \longrightarrow & V(X) & \longrightarrow & 0,
\end{array}$$

and everything commutes with δ_{-1} and σ. In terms of our standard groups, this diagram goes over to:

$$\begin{array}{ccccccccc}
1 & \longrightarrow & k^* & \longrightarrow & \mathscr{G}_g & \longrightarrow & Q_2^{2g} & \longrightarrow & 0 \\
 & & \big\downarrow{\scriptstyle \alpha^2} & & \big\downarrow{\scriptstyle E_2} & & \| & & \\
1 & \longrightarrow & k^* & \longrightarrow & \mathscr{G}_g^{(2)} & \longrightarrow & Q_2^{2g} & \longrightarrow & 0 \\
 & & \big\downarrow{\scriptstyle \alpha^2} & & \big\downarrow{\scriptstyle E_2} & & \| & & \\
1 & \longrightarrow & k^* & \longrightarrow & \mathscr{G}_g^{(4)} & \longrightarrow & Q_2^{2g} & \longrightarrow & 0 \\
 & & \| & & {\scriptstyle F}\big\downarrow{\scriptstyle \wr} & & \big\downarrow{\scriptstyle 2} & & \\
1 & \longrightarrow & k^* & \longrightarrow & \mathscr{G}_g & \longrightarrow & Q_2^{2g} & \longrightarrow & 0
\end{array}$$

where $\mathscr{G}_g^{(2)}$ and $\mathscr{G}_g^{(4)}$ are both equal to $k^* \times Q_2^g \times Q_2^g$, but with group laws:

$$(\alpha, x, y) \cdot (\alpha', x', y') = (\alpha \cdot \alpha' \cdot \chi(2\,{}^t x \cdot y'), x+x', y+y')$$

resp.

$$= (\alpha \cdot \alpha' \cdot \chi(4\,{}^t x \cdot y'), x+x', y+y'$$

and

$$E_2(\alpha, x, y) = (\alpha^2, x, y),$$

$$F(\alpha, x, y) = (\alpha, 2x, 2y).$$

Thus given one symplectic isomorphism

$$V(\underset{\leftarrow}{X}) \to Q_2^g \times Q_2^g,$$

$$x \mapsto (\varphi_1 x, \varphi_2 x)$$

we get symmetric theta structures

$$c_1\colon \mathscr{G}(\mathscr{T}) \longrightarrow \mathscr{G}_g, \quad c_2\colon \mathscr{G}(\mathscr{T}^{(2)}) \xrightarrow{\sim} \mathscr{G}_g^{(2)} \text{ and } c_4\colon \mathscr{G}(\mathscr{T}^{(4)}) \xrightarrow{\sim} \mathscr{G}_g^{(4)}.$$

Via the theta function representation, we obtain injections:

$$\vartheta\colon \Gamma(\mathscr{T}) \to \begin{Bmatrix}\text{functions}\\ \text{on } V(\underset{\leftarrow}{X})\end{Bmatrix},$$

$$\vartheta^{(2)}\colon \Gamma(\mathscr{T}^{(2)}) \to \begin{Bmatrix}\text{functions}\\ \text{on } V(\underset{\leftarrow}{X})\end{Bmatrix}$$

and it is easy to check that, for all $s_1, s_2 \in \Gamma(X_\alpha, L_\alpha)$

$$\vartheta^{(2)}(s_1 \otimes s_2) = \vartheta(s_1) \cdot \vartheta(s_2),$$

i.e., tensor product of sections becomes pointwise multiplication of theta functions (compare Property II above).

Now, define actions of both $\mathscr{G}(\mathscr{T})$ and $\mathscr{G}(\mathscr{T}^{(2)})$ on the vector space of *all* k-valued functions on $V(\underset{\leftarrow}{X})$ by:

$$(U_w(\varphi))\, y = \alpha \cdot e_\lambda\left(\frac{x}{2}, y\right) \cdot \varphi(y-x)$$

if

$$w = \alpha \cdot \sigma(x) \in \mathscr{G}(\mathscr{T}),$$

$$(U_w^{(2)}(\varphi))\, y = \alpha \cdot e_\lambda(x, y) \cdot \varphi(y-x)$$

if

$$w = \alpha \cdot \sigma(x) \in \mathscr{G}(\mathscr{T}^{(2)}).$$

Then according to Property I of algebraic theta functions, Image(ϑ) is an irreducible $\mathscr{G}(\mathscr{T})$-space, and Image($\vartheta^{(2)}$) is an irreducible $\mathscr{G}(\mathscr{T}^{(2)})$-space. Moreover, Image($\vartheta$) must be generated by the various functions

$$U_w\left(\vartheta\begin{bmatrix}0\\0\end{bmatrix}\right),$$

i.e.,

$$y \mapsto e_\lambda\left(\frac{x}{2}, y\right) \cdot \vartheta \begin{bmatrix} 0 \\ 0 \end{bmatrix} (y-x)$$

and Image ($\vartheta^{(2)}$) must be generated by their various products:

$$y \mapsto e_\lambda\left(\frac{x_1+x_2}{2}, y\right) \cdot \vartheta \begin{bmatrix} 0 \\ 0 \end{bmatrix} (y-x_1) \cdot \vartheta \begin{bmatrix} 0 \\ 0 \end{bmatrix} (y-x_2).$$

Now it is a non-trivial condition that this second family of functions spans an *irreducible* $\mathscr{G}(\mathscr{T}^{(2)})$-space. In particular, let $K \subset \mathscr{G}(\mathscr{T}^{(2)})$ be the subgroup:

$$c_2^{-1}[\{(1, x, y) \mid x \in \tfrac{1}{2} \mathbf{Z}_2^g, \, y \in \mathbf{Z}_2^g\}].$$

Then in an irreducible $\mathscr{G}(\mathscr{T}^{(2)})$-space, K has a one-dimensional space of invariants. Now it is easy to check that all the functions

$$y \mapsto \vartheta \begin{bmatrix} 0 \\ 0 \end{bmatrix} (y+x) \cdot \vartheta \begin{bmatrix} 0 \\ 0 \end{bmatrix} (y-x)$$

(any $x \in V(\underset{\sim}{X})$) are invariant under $c_2^{-1}(1 \times \mathbf{Z}_2^g \times \mathbf{Z}_2^g)$. Let $V_1 \subset V(\underset{\sim}{X})$ be the subgroup $\bar{c}_2^{-1}(\mathbf{Z}_2^g \times \{0\})$. Then the functions:

$$y \mapsto \sum_{\zeta \in \frac{1}{2} V_1 / V_1} e_\lambda(y, \zeta) \cdot \vartheta \begin{bmatrix} 0 \\ 0 \end{bmatrix} (y+x+\zeta) \cdot \vartheta \begin{bmatrix} 0 \\ 0 \end{bmatrix} (y-x+\zeta)$$

are all invariant under K. It follows that they are all proportional to *one* function $\Phi(y)$. Therefore, there are constants, depending on x – call them $c(x)$ – such that

$$\sum_{\zeta \in \frac{1}{2} V_1 / V_1} e_\lambda(y, \zeta) \, \vartheta \begin{bmatrix} 0 \\ 0 \end{bmatrix} (y+x+\zeta) \cdot \vartheta \begin{bmatrix} 0 \\ 0 \end{bmatrix} (y-x+\zeta) = c(x) \cdot \Phi(y).$$

Interchanging x and y in this expression, using the evenness of $\vartheta \begin{bmatrix} 0 \\ 0 \end{bmatrix}$ and its periodicity with respect to elements of V_1, you check that the left-hand side is symmetric in x and y. Thus

$$c(x) \cdot \Phi(y) = c(y) \cdot \Phi(x), \qquad \text{all } x, y \in V(\underset{\sim}{X}).$$

Since neither c nor Φ can be identically 0, this implies that $c(x) = \alpha \cdot \Phi(x)$, for all x and some $\alpha \in k^*$. Replacing Φ by $\sqrt{\alpha} \cdot \Phi$, we get $c = \Phi$, or

$$\sum_{\zeta \in \frac{1}{2} V_1 / V_1} e_\lambda(y, \zeta) \cdot \vartheta \begin{bmatrix} 0 \\ 0 \end{bmatrix} (y+x+\zeta) \cdot \vartheta \begin{bmatrix} 0 \\ 0 \end{bmatrix} (y-x+\zeta) = \Phi(x) \cdot \Phi(y).$$

This is equation (**)′ in the proof of Lemma 2, so referring to this proof, we see that we have proven that $\vartheta \begin{bmatrix} 0 \\ 0 \end{bmatrix}$ satisfies RIEMANN's theta relation.

Because of the central significance of this result, I want to give a second proof, following the lines of the proof in § 3 in the finite case.

First, introduce the maps:

$$\begin{array}{ccc} & & \mathscr{H}_g \\ & \overset{\approx}{\nearrow} \beta & \downarrow T \\ \Gamma(\mathscr{T}) & & \\ & \searrow \vartheta & \\ & & \left\{\begin{matrix}\text{functions}\\ \text{on } V(\underset{\leftarrow}{X})\end{matrix}\right\} \end{array}, \qquad \begin{array}{ccc} & & \mathscr{H}_g \\ & \overset{\approx}{\nearrow} \beta^{(2)} & \downarrow T^{(2)} \\ \Gamma(\mathscr{T}^{(2)}) & & \\ & \searrow \vartheta^{(2)} & \\ & & \left\{\begin{matrix}\text{functions}\\ \text{on } V(\underset{\leftarrow}{X})\end{matrix}\right\} \end{array}.$$

These induce a pairing

$$\circ: \mathscr{H}_g \times \mathscr{H}_g \to \mathscr{H}_g$$

such that equivalently

$$\beta(s_1) \circ \beta(s_2) = \beta^{(2)}(s_1 \otimes s_2), \qquad s_i \in \Gamma(\mathscr{T})$$

or

$$T(f_1) \cdot T(f_2) = T^{(2)}(f_1 \circ f_2), \qquad f_i \in \mathscr{H}_g.$$

Notice that the algebraic tensor product $\mathscr{H}_g \otimes \mathscr{H}_g$ is just $\mathscr{H}_{2g}$. In particular, the map

$$f_1, f_2 \mapsto f_1 \circ f_2(0)$$

is a linear functional on $\mathscr{H}_g \otimes \mathscr{H}_g$, hence it is represented by a finitely additive measure λ on $Q_2^g \times Q_2^g$: i.e.,

$$f_1 \circ f_2(0) = \int_{Q_2^g \times Q_2^g} f_1(u) \cdot f_2(v) \cdot d\lambda_{u,v}.$$

Since for all $s_1, s_2 \in \Gamma(\mathscr{T})$, $\alpha \in \mathscr{G}(\mathscr{T})$, $U_\alpha s_1 \otimes U_\alpha s_2 = U_{\varepsilon_2(\alpha)}(s_1 \otimes s_2)$, we find that for all $f_1, f_2 \in \mathscr{H}_g$, $\alpha \in \mathscr{G}_g$, $U_\alpha f_1 \circ U_\alpha f_2 = U_{\varepsilon_2(\alpha)}(f_1 \circ f_2)$. Let $\alpha = (1, y_1, y_2)$. Then:

$$\begin{aligned} f_1 \circ f_2(y_1) &= (U_{(1,y_1,y_2)}(f_1 \circ f_2))(0) \\ &= [U_{(1,y_1,y_2)} f_1 \circ U_{(1,y_1,y_2)} f_2](0) \\ &= \int_{Q_2^g \times Q_2^g} \chi({}^t y_2 \cdot (u+v)) \cdot f_1(u+y_1) \cdot f_2(v+y_1) \cdot d\lambda_{u,v} \end{aligned}$$

for all $y_1, y_2 \in Q_2^g$. Taking combinations of these equations for various y_2's, it follows that

$$\int_{Q_2^g \times Q_2^g} g(u+v) \cdot f_1(u+y_1) \cdot f_2(v+y_2) \cdot d\lambda_{u,v} = 0$$

for all locally constant functions g such that $g(0)=0$. This shows that

$$\int h(u,v)\, d\lambda_{u,v} = 0$$

whenever $h \in \mathscr{H}_{2g}$, and $h(u,-u)=0$, all u. This implies that $\lambda_{u,v}$ is given by a measure on the set $\{(u,-u)\}$, i.e., there is a $\tilde{\lambda} \in \mathscr{H}_g^*$ such that

$$\int h(u,v)\, d\lambda_{u,v} = \int h(u,-u)\, d\tilde{\lambda}_u$$

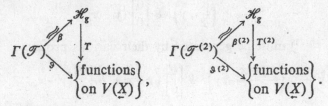

all $h \in \mathscr{H}_{2g}$. Therefore:

$$f_1 \circ f_2(y) = \int_{Q_2^g} f_1(y+u) f_2(y-u)\, d\tilde{\lambda}_u.$$

Now use the fact that $s_1 \otimes s_2(0) = s_1(0) \cdot s_2(0)$: therefore if μ and ν are the null-value measures for $\mathscr{T}$ and $\mathscr{T}^{(2)}$, we find

$$\int_{Q_2^g} f_1(u)\, d\mu_u \cdot \int_{Q_2^g} f_2(v)\, d\mu_v = \int_{Q_2^g} f_1 \circ f_2(w)\, d\nu_w$$
$$= \int_{Q_2^g \times Q_2^g} f_1(w+t) f_2(w-t)\, d\nu_w \cdot d\tilde{\lambda}_t.$$

Hence if $\xi(x, y) = (x+y, x-y)$ as usual, we find

$$\int_{Q_2^{2g}} F \cdot d(\mu \times \mu) = \int_{Q_2^{2g}} \xi^* F \cdot d(\nu \times \tilde{\lambda}) \qquad \text{for all } F \in \mathscr{H}_{2g},$$

i.e.,

$$\mu \times \mu(U) = \nu \times \lambda(\xi^{-1} U)$$

for all compact open sets $U \subset Q_2^{2g}$. Using the evenness of μ, it follows from this equation that $\nu \times \tilde{\lambda} = \tilde{\lambda} \times \nu$, hence ν and $\tilde{\lambda}$ are proportional. Changing ν by a constant, which is permissible, we may assume $\nu = \tilde{\lambda}$. Then if ν' is the measure

$$\nu'(U) = \nu(\tfrac{1}{2} U),$$

it follows that $\mu \times \mu(U) = \nu' \times \nu'(\xi U)$, all U. This is condition (A) of Lemma 2.

§ 9. The 2-Adic Moduli Space

We will now put the results of § 8 in a moduli-theoretic form, and relate these to the finite-level results of § 6. Once we have done this, we will be able to go further and determine the structure of the *boundary* of the moduli space.

The whole moduli problem for abelian varieties looks very different when viewed from an isogeny invariant point of view. The difference between polarization types disappears because any abelian variety is isogenous to a *principally polarized* abelian variety (as is easily proven by generalizing some of the results of §1 to inseparably polarized abelian varieties). The natural thing is then to view the classification in 2 steps: first one has the totality of *polarized towers*; second, within each tower one has a huge system of variously polarized abelian varieties. We will not treat this entire moduli problem since the study of all inseparably polarized abelian varieties within an isogeny type is a subject in itself. In order to a) restrict to separably polarized abelian varieties, while

b) constructing moduli schemes simultaneously in as many characteristics as possible, we shall consider only the polarized 2-tower inside each full polarized tower, and at the same time exclude only char. 2. Analogous results would be obtained if we restricted ourselves to all characteristics p, $p \nmid d$ (d a fixed *even* integer), and to isogenies within a tower of degree dividing d^N, $N \gg 0$. As far as the category of *sets* is concerned, we have the following sets and canonical maps to consider:

$$\mathcal{M}_\infty = \left\{\begin{array}{l}\text{set of polarized 2-towers } \mathcal{T}=\{X_\alpha, L_\alpha\}\\ \text{plus symplectic isomorphisms}\\ \varphi\colon V(\underset{\leftarrow}{X}) \xrightarrow{\approx} Q_2^g\times Q_2^g\\ \text{up to isomorphism}\end{array}\right\}$$

$$\Big\downarrow f_1 \quad \Big\downarrow f_2 \qquad\qquad \searrow\searrow\searrow\searrow$$

$$\mathcal{M}_{\delta_1} = \left\{\begin{array}{l}\text{set of abelian varieties } X,\\ \text{ample totally symmetric sheaves}\\ L \text{ of type } \delta_1, \text{ and symmetric}\\ \text{theta-structures } \beta\colon \mathcal{G}(L) \xrightarrow{\approx} \mathcal{G}(\delta_1),\\ \text{up to isomorphism}\end{array}\right\} \qquad \begin{array}{c}\text{---------}\\ \text{(other } \delta\text{'s)}\end{array}$$

$$\Big\downarrow g \qquad\qquad\qquad \Big\downarrow\Big\downarrow$$

$$\mathcal{M}_{\delta_1}^0 = \left\{\begin{array}{l}\text{set of abelian varieties } X, \text{ plus}\\ \text{ample totally symmetric sheaves}\\ L \text{ of type } \delta_1 \text{ up to isomorphism}\end{array}\right\} \qquad \begin{array}{c}\text{-----------}\\ \text{(other } \delta\text{'s)}\end{array}$$

$$\searrow h \qquad \swarrow \quad \swarrow$$

$$\mathcal{M}_{-\infty} = \left\{\begin{array}{l}\text{set of polarized 2-towers}\\ \mathcal{T}=\{X_\alpha, L_\alpha\}\\ \text{up to isomorphism}\end{array}\right\}.$$

Here δ_1 is any g-tuple $(2^{n_1}, 2^{n_2}, \ldots, 2^{n_g})$, $n_1 \geqq n_2 \geqq \cdots \geqq n_g \geqq 1$. The various arrows arise as follows:

(I) g takes (X, L, β) to (X, L),

(II) h takes (X, L) to the 2-tower generated by (X, L).

(III) The f_i's are given by choosing a compact, open isotropic subgroup $U \subset Q_2^{2g}$ such that $U^\perp/U \cong H(\delta_1)$, and a symmetric isomorphism

$$\beta_0\colon \frac{k^* \times \Sigma(U^\perp)}{\Sigma(U)} \xrightarrow{\approx} \mathcal{G}(\delta_1).$$

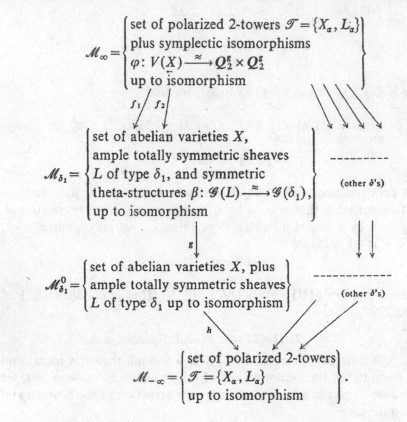

Then for all $(\mathscr{T}, \varphi)$, there is a unique level $\alpha \in S$ such that $\varphi(T(\alpha)) = U$, and $f((\mathscr{T}, \varphi))$ should be $(X_\alpha, L_\alpha, \beta)$ where β is the composite

$$\mathscr{G}(L_\alpha) \xrightarrow[\text{via } \varphi]{\approx} \frac{k^* \times \Sigma(U^\perp)}{\Sigma(U)} \xrightarrow[\beta_0]{\approx} \mathscr{G}(\delta_1).$$

It is apparent from this diagram that the various moduli sets $\mathscr{M}_\delta$ treated in § 6 are inter-related in a rather complicated way: Given δ_1, δ_2, one can choose any of an infinite number of f_1, f_2 in the diagram:

$$\begin{array}{ccc} & \mathscr{M}_\infty & \\ {}^{f_1}\swarrow & & \searrow{}^{f_2} \\ \mathscr{M}_{\delta_1} & & \mathscr{M}_{\delta_2} \end{array}$$

and relate $\mathscr{M}_{\delta_1}, \mathscr{M}_{\delta_2}$ via the, in general, many-many correspondence so obtained. Each $\mathscr{M}_\delta$ is related to itself in this way by the well-known *Hecke ring* of correspondences. The whole set-up is much easier to visualize starting from $\mathscr{M}_\infty$. Note that $G = \mathrm{Sp}(2g, \boldsymbol{Q}_2)$ acts on $\mathscr{M}_\infty$, if we let $\sigma \in G$ act as follows:

$$\sigma(\mathscr{T}, \varphi) = (\mathscr{T}, \sigma \circ \varphi).$$

Then $\mathscr{M}_{-\infty}$ is nothing but the quotient $\mathscr{M}_\infty / G$, and the $\mathscr{M}_\delta$ and $\mathscr{M}_\delta^0$'s are quotients $\mathscr{M}_\infty / \Gamma$ where $\Gamma \subset G$ is a suitable subgroup commensurable with $\mathrm{Sp}(2g, \boldsymbol{Z}_2)$. The different maps from $\mathscr{M}_\infty$ to $\mathscr{M}_\delta$ are simply the compositions of (a) action of some $\sigma \in \mathrm{Sp}(2g, \boldsymbol{Q}_2)$ and (b) the canonical map from $\mathscr{M}_\infty$ to $\mathscr{M}_\infty / \Gamma$. Clearly the most basic sets to get ahold of are $\mathscr{M}_{-\infty}$ and $\mathscr{M}_\infty$. We have seen that the $\mathscr{M}_\delta$'s "are" varieties. Thus the $\mathscr{M}_\infty$ is an infinite covering of a variety and $\mathscr{M}_{-\infty}$ is an infinite quotient. As far as I know there is no sensible object whose underlying point set is $\mathscr{M}_{-\infty}$. But $\mathscr{M}_\infty$ is an inverse limit of varieties and will turn out to be a perfectly upstanding (though non-noetherian) scheme. This moduli space (and its adelic generalizations) seem to be the most important ones for the entire moduli theory of abelian varieties.

The next step is to define the scheme an open subset of which will represent $\mathscr{M}_\infty$. We will work over the following ground ring R:

Definition 1. $R = \boldsymbol{Z}[\tfrac{1}{2}, \zeta_1, \zeta_2, \zeta_3, \ldots]$ where $\zeta_n^2 = \zeta_{n-1}$ if $n \geq 2$, $\zeta_1 = -1$. The multiplicative group generated by the ζ's is isomorphic to $\boldsymbol{Q}_2 / \boldsymbol{Z}_2$ and we define:

$$\chi \colon \boldsymbol{Q}_2 / \boldsymbol{Z}_2 \to R^*$$

via

$$\chi(m/2^n) = (\zeta_n)^m.$$

Actually, adjoining the ζ's is not essential, but it makes life easier and seems to be quite natural.

9*

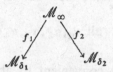

Next, to R adjoin independent transcendentals X_α, one for each $\alpha\in\boldsymbol{Q}_2^{2g}$. Then divide out by the following relations:

1) $X_{\alpha+\beta}=\chi\left(\frac{{}^t\beta_1\cdot\beta_2}{2}\right)\chi\left(\frac{{}^t\beta_1\cdot\alpha_2}{2}-\frac{{}^t\alpha_1\cdot\beta_2}{2}\right)\cdot X_\alpha$ all $\alpha\in\boldsymbol{Q}_2^{2g}$; $\beta\in\boldsymbol{Z}_2^{2g}$.

2) $X_{-\alpha}=X_\alpha$.

3) $\prod_{i=1}^{4} X_{\alpha_i}=2^{-g}\sum_{\eta\in\frac{1}{2}\boldsymbol{Z}_2^{2g}/\boldsymbol{Z}_2^{2g}}\chi({}^t\gamma_1\cdot\eta_2-{}^t\gamma_2\cdot\eta_1)\cdot\prod_{i=1}^{4}X_{\alpha_i+\gamma+\eta}$

all $\alpha_1,\alpha_2,\alpha_3,\alpha_4\in\boldsymbol{Q}_2^{2g}$, where $\gamma=-\frac{1}{2}\sum_{i=1}^{4}\alpha_i$.

In what follows, it will be convenient to abbreviate the characters in these formulas as follows:

Definition 2.

$$e(\alpha,\beta)=\chi({}^t\alpha_1\cdot\beta_2-{}^t\beta_1\cdot\alpha_2),\qquad \alpha,\beta\in\boldsymbol{Q}_2^{2g}.$$

$$e_*(\alpha)=\chi(2\,{}^t\alpha_1\cdot\alpha_2),\qquad \text{if } \alpha\in\tfrac{1}{2}\boldsymbol{Z}_2^{2g}.$$

Definition 3.

$$A=R[\ldots,X_\alpha,\ldots]\Big/\left\{\begin{matrix}\text{ideal generated by}\\ \text{relations 1, 2, 3}\end{matrix}\right\}.$$

$$\overline{M}_\infty=\operatorname{Proj}(A).$$

In order to get a preliminary idea of how big $\overline{M}_\infty$ is, introduce the subrings:

Definition 4. A_n = subring of A generated by

$$\{X_\alpha \mid 2^n\alpha\in\boldsymbol{Z}_2^{2g}\}.$$

Lemma 1. *A is integrally dependent on A_2.*

Proof. By induction, it suffices to check that X_α is integrally dependent on A_{n-1}, when $2^n\cdot\alpha\in\boldsymbol{Z}_2^{2g}$ and $n\geqq 2$. Use relation (3) with $\alpha_1=\alpha_2=\alpha_3=\alpha$, $\alpha_4=\gamma=-\alpha$. Then since $X_{-\alpha}=X_\alpha$, we find that $X_\alpha^4\in A_{n-1}$. *Q.E.D.*

Corollary. *There are* integral *affine morphisms:*

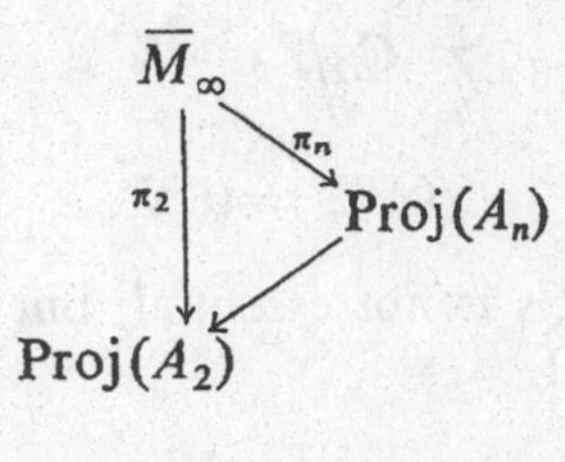

and $\overline{M}_\infty$ *is the category-theoretic* $\varprojlim$ *of the algebraic schemes* $\mathrm{Proj}(A_n)$, i.e., *for all schemes* S,

$$\mathrm{Hom}(S, \overline{M}_\infty) \cong \varprojlim \mathrm{Hom}(S, \mathrm{Proj}(A_n)).$$

Proof. Cf. EGA, Ch. 4.

The X_α's will be nothing but the values of the function $\vartheta\begin{bmatrix}0\\0\end{bmatrix}$ when we connect $\overline{M}_\infty$ to the moduli problem. It is also convenient to introduce a second set of generators of the ring A, whose values will be the values of the measure μ, in the moduli problem:

For all compact open sets $U \subset Q_2^g$, let

$$Y_U = 2^{-ng} \sum_{i=1}^{N} \sum_{\beta \in 2^{-n} Z_2^g / Z_2^g} \chi\left(\frac{{}^t\alpha_i \cdot \beta}{2}\right) X_{(\alpha_i, \beta)}$$

if

$$U = \bigcup_{i=1}^{N} [\alpha_i + 2^n Z_2^g] \quad \text{and} \quad \alpha_i \not\equiv \alpha_j \pmod{2^n Z_2^g}.$$

Using Lemmas 1 and 2, § 8, the relations on the X's go over to the following relations in the Y's:

1. $Y_{U_1} + Y_{U_2} = Y_{U_1 \cup U_2} + Y_{U_1 \cap U_2}$.

2. $Y_{-U} = Y_U$, $Y_\emptyset = 0$.

3. If we define quadratic polynomials, Z_U, for all compact open sets $U \subset Q_2^g \times Q_2^g$, by relations (1.) and $Z_{U_1 \times U_2} = Y_{U_1} Y_{U_2}$, and if $\xi(x, y) = (x+y, x-y)$ as usual, then:

$$Z_{\xi(U_1 \times U_2)} \cdot Z_{\xi(U_3 \times U_4)} = Z_{\xi(U_1 \times U_4)} \cdot Z_{\xi(U_3 \times U_2)}.$$

In particular, let $n \geqq 1$ and let

$$l\colon 2^{n-1} Z_2^g \to \{\pm 1\}$$

be a homomorphism. Define

$$Z_{\alpha,\beta} = \sum_{\eta \in 2^{n-1} Z_2^g / 2^n Z_2^g} l(\eta) \cdot Y_{\alpha+\beta+\eta+2^n Z_2^g} \cdot Y_{\alpha-\beta+\eta+2^n Z_2^g}.$$

Then the general relations imply:

$$Z_{\alpha,\beta} \cdot Z_{\gamma,\delta} = Z_{\alpha,\delta} \cdot Z_{\gamma,\beta}$$

for all $\alpha, \beta, \gamma, \delta \in Q_2^g$. Conversely, these relations, for all n's, and $l \equiv +1$, imply all the quartic relations.

Moreover, the subring A_n generated by X_α, with $\alpha \in 2^{-n} Z_2^{2g}$ is just the subring generated by

$$\{Y_U \mid U = U + 2^n Z_2^g,\ U \subset 2^{-n} Z_2^g\}$$

or by

$$\{Y_{\alpha + 2^n Z_2^g} \mid \alpha \in 2^{-n} Z_2^g\}.$$

The group $G = \mathrm{Sp}(2g, Q_2)$ acts on $\overline{M}_\infty$ in the following way:

Definition 5. For all $\sigma \in G$, let

$$(*)\quad U_\sigma(X_\alpha) = \sum_{\beta \in Z_2^{2g}/Z_2^{2g} \cap \sigma^{-1}(Z_2^{2g})} e_*(\beta/2)\, e(\beta/2, \alpha) \cdot e(\gamma/2, \alpha - \beta) \cdot X_{\sigma\alpha - \sigma\beta - \sigma\gamma},$$

where $\gamma \in Q_2^{2g}$ is some fixed element satisfying

$$e_*(\beta/2) \cdot e_*(\sigma\beta/2) = e(\gamma, \beta)$$

for all $\beta \in Z_2^{2g} \cap \sigma^{-1}(Z_2^{2g})$.

Concerning this definition, one verifies by mechanical calculation the following:

1. If the γ in the definition is varied, it must change by an element of $Z_2^{2g} + \sigma^{-1}(Z_2^{2g})$, and, if so, the operator U_σ is only changed into a constant multiple of itself:

$$U'_\sigma(X_\alpha) = c \cdot U_\sigma(X_\alpha), \qquad \text{all } \alpha.$$

We shall assume that for each σ, some fixed γ is chosen.

Note that $-I \in \mathrm{Sp}(2g, Q_2) = G$, and $U_{-I}(X_\alpha) = X_{-\alpha}$.

2. For all $\sigma \in G$, $\alpha \in Q_2^{2g}$ and $\beta \in Z_2^{2g}$,

$$U_\sigma(X_{\alpha+\beta}) = e_*(\beta/2)\, e(\beta/2, \alpha) \cdot U_\sigma(X_\alpha).$$

Therefore, if we let M be the free R-module spanned by the X_α's modulo *only* relations (1.), each U_σ defines an R-module homomorphism from M to M.

3. For all $\sigma, \tau \in G$, there is a non-zero constant $c_{\sigma,\tau}$ such that

$$U_\sigma \circ U_\tau = c_{\sigma,\tau} \cdot U_{\sigma\cdot\tau},$$

i.e., $\sigma \mapsto U_\sigma$ is a "projective" representation of G in M.

4. For all $\sigma \in G$, one computes easily that there is a sign $e_\sigma = \pm 1$ such that $U_{-I} \circ U_\sigma = e_\sigma U_\sigma \circ U_{-I}$. Unfortunately, it does not appear to be easy to show *directly* that $e_\sigma = +1$ for all σ. For example, if $\sigma(Z_2^{2g}) = Z_2^{2g}$, $e_\sigma = e_*(\gamma)$. However, using (3.), one checks that

$$\varepsilon_{\sigma\tau} = \varepsilon_\sigma \cdot \varepsilon_\tau,$$

and since G is well known to be its own commutator subgroup, this implies $\varepsilon_\sigma = +1$, all σ. But now the submodule A_1 of A of elements of degree 1 is just M modulo the span of the elements $\{X_\alpha - X_{-\alpha}\}$, and this proves that the U_σ's induce homomorphisms from A_1 to A_1. We will let U_σ denote this homomorphism too.

5. The last step is to check that all the U_σ's induce *ring* homomorphisms from A to A. Frankly, *I* balked at directly applying U_σ to the relations (3.) and seeing what comes out. But, in 2 special cases, it isn't too bad. Suppose first that σ is in the subgroup $\Gamma = \mathrm{Sp}(2g, \mathbf{Z}_2)$. Then $\sigma(\mathbf{Z}_2^{2g}) = \mathbf{Z}_2^{2g}$, and U_σ reduces to:

$$U_\sigma(X_\alpha) = e(\gamma/2, \alpha) \cdot X_{\sigma\alpha - \sigma\gamma}.$$

In this case, it's not hard to check that U_σ takes a relation of type (3.) to another relation of the same type, so that U_σ induces a map from A to A. Suppose second that σ is in the sub-semi-group:

$$H^+ = \left\{ \left(\begin{array}{c|c} A & 0 \\ \hline 0 & {}^tA^{-1} \end{array} \right) \middle| \begin{array}{l} A \in GL(g, \mathbf{Q}_2) \\ A(\mathbf{Z}_2^g) \supseteq \mathbf{Z}_2^g \end{array} \right\} \subset G.$$

For such σ, U_σ reduces to:

$$U_\sigma(X_\alpha) = \sum_{\beta \in \mathbf{Z}_2^g / A^{-1} \mathbf{Z}_2^g} e(-\beta_1/2, \alpha) \cdot X_{\sigma\alpha + \sigma\beta_1}$$

(where β_1 is the $2g$-vector $(\beta, 0)$). Now, referring back to the proof of Lemma 2, in the last section, we see that an equivalent form of the relations (3.) is:

$$Y_{\alpha,\beta} \cdot Y_{\gamma,\delta} = Y_{\alpha,\delta} \cdot Y_{\gamma,\beta}, \qquad \text{all } \alpha, \beta, \gamma, \delta \in \mathbf{Q}_2^{2g}$$

where

$$Y_{\alpha,\beta} = \sum_{\eta \in \frac{1}{2}\mathbf{Z}_2^g / \mathbf{Z}_2^g} e(-\eta_1, \alpha)\, X_{\alpha+\beta+\eta_1} \cdot X_{\alpha-\beta+\eta_1}.$$

For $\sigma \in H^+$, $Y_{\alpha,\beta}$ behaves very nicely. One computes easily that:

$$U_\sigma(Y_{\alpha,\beta}) = \sum_{\eta, \zeta \in \frac{1}{2} A \mathbf{Z}_2^g / \frac{1}{2} \mathbf{Z}_2^g} e(-\eta_1, \sigma\alpha) \cdot e(-\zeta_1, \sigma\beta) \cdot Y_{\sigma\alpha+\eta_1, \sigma\beta+\zeta_1}.$$

From this it follows immediately that $U_\sigma[Y_{\alpha,\beta} \cdot Y_{\gamma,\delta} - Y_{\alpha,\delta} \cdot Y_{\gamma,\beta}]$ is an R-linear combination of expressions $Y_{\alpha',\beta'} \cdot Y_{\gamma',\delta'} - Y_{\alpha',\delta'} \cdot Y_{\gamma',\beta'}$.

Finally, it follows from the paper of Iwahori-Matsumoto [*13*] that all the double cosets in $\Gamma \backslash G / \Gamma$ are represented by matrices:

$$\sigma = \left(\begin{array}{c|c} A & 0 \\ \hline 0 & {}^tA^{-1} \end{array} \right)$$

$$A = \begin{pmatrix} 2^{a_1} & & & 0 \\ & 2^{a_2} & & \\ & & \ddots & \\ 0 & & & 2^{a_g} \end{pmatrix}$$

$$0 \geqq a_1 \geqq a_2 \geqq \cdots \geqq a_g.$$

Since there are in H^+, $G=\Gamma\cdot H^+\cdot\Gamma$ and our 2 calculations suffice to prove that for every $\sigma\in G$, U_σ maps A to A.

Putting all this together, we conclude that Definition 5 defines a projective representation of G on A, and an action of G on the scheme $\overline{M}_\infty$. In fact, let $V_\sigma\colon \overline{M}_\infty\to\overline{M}_\infty$ denote this action, then, by definition,

$$V_\sigma^*(X_\alpha)=U_{\sigma^{-1}}(X_\alpha)$$

(the σ^{-1} makes it an action of G instead of the opposed group).

Now we can connect $\overline{M}_\infty$ to the moduli problem. For all algebraically closed fields k (char$(k)\neq 2$), there is a map:

$$\Theta\colon \mathcal{M}_\infty(k)\to\begin{Bmatrix}\text{Set of } k\text{-valued}\\ \text{points of } \overline{M}_\infty\end{Bmatrix}$$

($\mathcal{M}_\infty$ denotes the set defined at the beginning of this §), which assigns to a tower $\mathcal{T}$ and a $\varphi\colon V(\underset{\sim}{X})\overset{\approx}{\to}Q_2^{2g}$, the point with homogeneous coordinates

$$X_\alpha=\vartheta\begin{bmatrix}0\\0\end{bmatrix}(\varphi^{-1}\alpha),\quad\text{or}\quad Y_U=\mu(U)$$

(cf. Lemma 1 and 2, § 8).

Recall that G acts on $\mathcal{M}_\infty(k)$ as follows:

Definition 6. Let $(\mathcal{T},\varphi)\in\mathcal{M}_\infty(k)$ and $\sigma\in G$. Then let $U_\sigma((\mathcal{T},\varphi))$ be the pair $(\mathcal{T},\sigma\circ\varphi)$, i.e., modify the symplectic isomorphism σ to:

$$V(\underset{\sim}{X})\xrightarrow[\varphi]{\approx}Q_2^{2g}\xrightarrow[\sigma]{\approx}Q_2^{2g}.$$

Theorem 1. *Under Θ, the actions of G on $\mathcal{M}_\infty(k)$ and on the set of k-valued points of $\overline{M}_\infty$ are compatible.*

Proof. Recall that $\vartheta\begin{bmatrix}0\\0\end{bmatrix}$ is the unique element f of $\vartheta(\Gamma(\mathcal{T}))$ invariant satisfying

a) $$f(x+a)=e_*(a/2)\cdot e_\lambda(a/2,x)\cdot f(x)$$

for all $a\in\varphi^{-1}(Z_2^{2g})$, with $e_*(\varphi^{-1}(\alpha))=\chi(2^t\alpha_1\cdot\alpha_2)$, all $\alpha\in\frac{1}{2}Z_2^{2g}$. If we use the symplectic isomorphism $\sigma\circ\varphi$, the *new* function $\vartheta\begin{bmatrix}0\\0\end{bmatrix}$ will instead be the unique f in $\vartheta(\Gamma(\mathcal{T}))$ satisfying:

a') $$f(x+a)=e'_*(a/2)\cdot e_\lambda(a/2,x)\cdot f(x)$$

for all $a\in\varphi^{-1}(\sigma^{-1}(Z_2^{2g}))$, with $e'_*(\varphi^{-1}(\sigma^{-1}(\alpha)))=\chi(2^t\alpha_1\cdot\alpha_2)$ all $\alpha\in\frac{1}{2}Z_2^{2g}$. Let $\vartheta^\sigma\begin{bmatrix}0\\0\end{bmatrix}$ denote this new $\vartheta\begin{bmatrix}0\\0\end{bmatrix}$. Then

$$X_\alpha[U_\sigma(\mathcal{T},\varphi)]=\vartheta^\sigma\begin{bmatrix}0\\0\end{bmatrix}(\varphi^{-1}(\alpha)).$$

To find $\vartheta^\sigma \begin{bmatrix}0\\0\end{bmatrix}$, recall that the functions

$$\alpha \mapsto e_\lambda(\beta/2, \alpha) \cdot \vartheta \begin{bmatrix}0\\0\end{bmatrix} (\alpha - \beta), \qquad \beta \in V(\underset{\sim}{X}).$$

span $V(\underset{\sim}{X})$. Therefore, it suffices to find some linear combination of these functions satisfying a′). If we make all these into functions on $\boldsymbol{Q}_2^{2g}$ via $\sigma \circ \varphi$: $V(\underset{\sim}{X}) \xrightarrow{\sim} \boldsymbol{Q}^{2g}$, we find that

$$\vartheta^\sigma \begin{bmatrix}0\\0\end{bmatrix} (\varphi^{-1}(\sigma^{-1}(\alpha)))$$

is that linear combination f of the functions

$$g_\beta(\alpha) = e(\beta/2, \alpha) \cdot g(\sigma^{-1}\alpha - \sigma^{-1}\beta), \qquad \beta \in \boldsymbol{Q}_2^{2g}$$

$$g(\alpha) = \vartheta \begin{bmatrix}0\\0\end{bmatrix} (\varphi^{-1}\alpha)$$

such that

a″) $$f(\alpha+\beta) = e_*(\beta/2) \cdot e(\beta/2, \alpha) \cdot f(\alpha),$$

for all $\beta \in \boldsymbol{Z}_2^{2g}$. One solution of (a″) is the function:

$$g^\sigma(\alpha) = \sum_{\beta \in \boldsymbol{Z}_2^{2g}/\boldsymbol{Z}_2^{2g} \cap \sigma \boldsymbol{Z}_2^{2g}} e_*(\beta/2) \cdot e(\beta/2, \alpha) \cdot e(\gamma/2, \alpha - \beta) \times g(\sigma^{-1}\alpha - \sigma^{-1}\beta - \sigma^{-1}\gamma)$$

where γ satisfies $e_*(\beta/2) \cdot e_*(\sigma^{-1}\beta/2) = e(\gamma, \beta)$, all $\beta \in \boldsymbol{Z}_2^{2g} \cap \sigma \boldsymbol{Z}_2^{2g}$. If this function *is not* 0, it must equal

$$\vartheta^\sigma \begin{bmatrix}0\\0\end{bmatrix} \circ \varphi^{-1} \circ \sigma^{-1}.$$

But on the other hand, X_α has values $g(\alpha)$ at the geometric point x corresponding to $(\mathscr{T}, \varphi)$. Hence X_α has the values at $V_\sigma(x)$:

$$\begin{aligned} X_\alpha(V_\sigma(x)) &= V_\sigma^* X_\alpha(x) \\ &= U_{\sigma^{-1}} X_\alpha(x) \\ &= \sum_{\beta \in \boldsymbol{Z}_2^{2g}/\boldsymbol{Z}_2^{2g} \cap \sigma \boldsymbol{Z}_2^{2g}} e_*(\beta/2) \cdot e(\beta/2, \alpha) \times \\ &\qquad \times e(\gamma/2, \alpha - \beta) X_{\sigma^{-1}\alpha - \sigma^{-1}\beta - \sigma^{-1}\gamma}(x) \\ &= g^\sigma(\alpha). \end{aligned}$$

Therefore, $g^\sigma \not\equiv 0$, so $g^\sigma(\alpha)$ is also the value of X_α at the geometric point corresponding to $(\mathscr{T}, \sigma \circ \varphi)$. *Q.E.D.*

Now every moduli space is supposed to represent a functor. In our case, instead of making a big fuss over defining a *family* of 2-towers of abelian varieties, over a scheme S, it is simpler to observe that $\mathcal{M}_\infty(k)$ is an inverse limit of some of the $\mathcal{M}_\delta(k)$'s introduced in § 6, and then to *define* $\mathcal{M}_\infty(S)$ as the corresponding limit of these $\mathcal{M}_\delta(S)$'s.

Given a polarized 2-tower $\mathcal{T}=(\{X_\alpha\}, \{L_\alpha\})$ and $\varphi\colon V(\underset{\leftarrow}{X}) \xrightarrow{\sim} \mathbf{Q}_2^{2g}$, for all $n \geqq 1$, we get

a) an abelian variety $X_n = X_{\alpha_n}$, where

$$T(\alpha_n)=\varphi^{-1}(2^{-n}\mathbf{Z}_2^{2g}),$$

b) a totally symmetric ample invertible sheaf

$$L_n = L_{\alpha_n} \quad \text{on} \quad X_n, \quad \text{of type}$$

$$\delta_n=(2^{2n}, 2^{2n}, \ldots, 2^{2n}),$$

c) a symmetric ϑ-structure:

$$\lambda_n\colon \mathcal{G}(L_n) \cong \frac{\mathcal{G}^*_{\alpha_n}(\mathcal{T})}{K(\alpha_n)} \underset{via\ \varphi}{\cong} \frac{k^* \times 2^{-n}\mathbf{Z}_2^{2g}}{\{1\}\times 2^n \mathbf{Z}_2^{2g}} \cong \mathcal{G}(\delta_n).$$

This is a map

$$(\mathcal{T}, \varphi) \mapsto (X_n, L_n, \lambda_n)$$

$$\mathcal{M}_\infty(k) \to \mathcal{M}_{\delta_n}(k).$$

Note that all these X_n's are canonically isomorphic, e.g., to X_1, via diagrams:

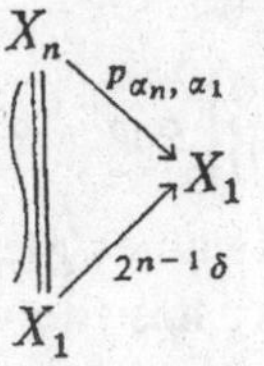

and that under these isomorphisms, L_n is just $L_1^{4^{n-1}}$.

On the other hand, between the functors $\mathcal{M}_{\delta_{n+1}}$ and $\mathcal{M}_{\delta_n}$, we have a natural transformation:

$$\pi_n\colon M_{\delta_{n+1}} \to M_{\delta_n}.$$

In fact, given $\mathcal{X}/S$, L, and λ in $\mathcal{M}_{\delta_{n+1}}(S)$, we get

a) a second ample sheaf M on $\mathcal{X}$, by descending L with respect to the isogeny

$$2\delta\colon \mathcal{X} \to \mathcal{X}$$

and the descent data $\lambda^{-1}(K_2)$, where $K_2 \subset \mathscr{G}(\delta_{n+1}) \times S$ is the subscheme representing the subfunctor of triples $(1, x, l)$, $2x = 2l = 0$.

b) Since

$$\mathscr{G}(M) \cong \frac{\text{normalizer of } \lambda^{-1}(K_2) \text{ in } \mathscr{G}(L)}{\lambda^{-1}(K_2)}$$

$$\cong \frac{\text{normalizer of } K_2 \text{ in } \mathscr{G}(\delta_{n+1})}{K_2} \cong \mathscr{G}(\delta_n)$$

we get a ϑ-structure $\mu: \mathscr{G}(M) \xrightarrow{\approx} \mathscr{G}(\delta_n)$.

It can be checked that $(\mathscr{X}/S, M, \mu) \in \mathscr{G}_{\delta_n}(S)$, so we call this $\pi_n((\mathscr{X}/S, L, \lambda))$.

Going back to k-valued points, we have a diagram:

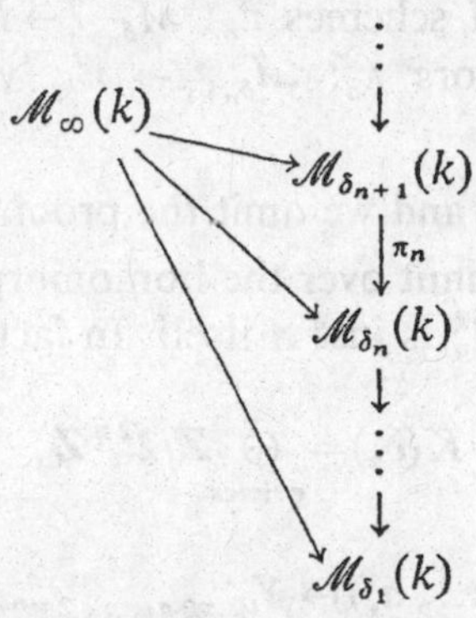

and it is clear that this induces an *isomorphism*

$$\mathscr{M}_\infty(k) \cong \varprojlim \mathscr{M}_{\delta_n}(k).$$

Definition 7. For all schemes S, let $\mathscr{M}_\infty(S) = \varprojlim \mathscr{M}_{\delta_n}(S)$.

Next, let's translate the results of § 6 on the representability of $\mathscr{M}_{\delta_n}$ into the discussion. The results there show that there is an open set:

$$M_{\delta_n} \subset \operatorname{Proj}(A_n^*) = \overline{M}_{\delta_n}$$

$$A_n^* = R[\ldots, Q^{(n)}(a), \ldots] \Big/ \left\{ \begin{matrix} \text{Modulo } Q^{(n)}(a) = Q^{(n)}(-a) \\ \text{and certain quartic relations} \end{matrix} \right\} \qquad a \in K(\delta_n)$$

such that M_{δ_n} represented $\mathscr{M}_{\delta_n}$ (over the ring R). Moreover, in Step II of the proof, a canonical morphism

$$\pi: \overline{M}_{2\delta} \to \overline{M}_\delta$$

was introduced. Iterated, this defines a morphism from $\overline{M}_{4\delta}\to\overline{M}_\delta$, e.g., from $\overline{M}_{\delta_{n+1}}$ to $\overline{M}_{\delta_n}$. If you work it out, it is just the projection:

$$\begin{array}{ccc}\overline{M}_{\delta_{n+1}} & \xrightarrow{\pi_n} & \overline{M}_{\delta_n}\\ \| & & \| \\ \mathrm{Proj}(A^*_{n+1}) & & \mathrm{Proj}(A^*_n),\end{array}$$

$$\pi_n^*(Q^{(n)}(a))=\sum_{\substack{2b=a\\ b\in K(\delta_{n+1})}} Q^{(n+1)}(b),\qquad \text{all } a\in K(\delta_n)$$

(here we identify $K(\delta_n)$ with a subgroup of $K(\delta_{n+1})$ as before). In particular, π_n is a finite morphism. Moreover, as we saw in § 6, (Step II).

(a) π_n is étale at all points of $M_{\delta_{n+1}}$ and $\pi_n^{-1}(M_{\delta_n})=M_{\delta_{n+1}}$. In fact:

(b) The morphism of schemes $\pi_n\colon M_{\delta_{n+1}}\to M_{\delta_n}$ corresponds to the transformation of functors $\pi_n\colon \mathcal{M}_{\delta_{n+1}}\to\mathcal{M}_{\delta_n}$ via the representability proven in § 6.

(This is easy to check and we omit the proof.)

Now, passing to the limit over the homomorphisms π_n^*, note that the *direct* limit of the rings A_n^* is just A itself. In fact,

$$K(\delta_n)=\bigoplus_{g\ \text{times}} \mathbf{Z}/2^{2n}\mathbf{Z},$$

and let

$$Q^{(n)}(a_1,\ldots,a_g)\mapsto Y_{(2^{-n}a_1,\ldots,2^{-n}a_n)+2^n\mathbf{Z}_2^g}$$

($a_i\in\mathbf{Z}/2^{2n}\mathbf{Z}$). It is easy to check that the relations imposed on the $Q^{(n)}(a)$'s give us exactly the defining relations on the Y_U's. Moreover, A_n is just the image of A_n^* in A. (It may very well happen that each π_n^* is injective, so that $A\cong\bigcup_n A_n^*$ and $A_n\cong A_n^*$: but *I* have no proof for this.) Geometrically, this shows that

$$\overline{M}_\infty\cong\varprojlim_n \overline{M}_{\delta_n}.$$

Definition 8. Let M_∞ be the inverse image in $\overline{M}_\infty$ of M_{δ_n} in $\overline{M}_{\delta_n}$ (independent of n by (a)).

1. $M_\infty\cong\varprojlim M_{\delta_n}$

2. $\mathcal{M}_\infty(S)\cong\varprojlim\mathcal{M}_{\delta_n}(S)$

3. The M_{δ_n}'s represent $\mathcal{M}_{\delta_n}$ (compatibly as n varies). Hence:

Theorem 2. *The scheme M_∞ represents $\mathcal{M}_\infty$.*

Recapitulating the discussion, the basic set which we are classifying is:

$$\mathcal{M}_\infty(k)=\left\{\begin{array}{l}\text{set of polarized 2-towers } \mathcal{T}=(X_\alpha, L_\alpha)\\ \text{plus symplectic isomorphisms } \varphi\colon V(\mathcal{T})\xrightarrow{\sim} Q_2^{2g}\end{array}\right\}$$

$$=\left\{\begin{array}{l}\text{set of abelian varieties } X\text{, totally}\\ \text{symmetric ample } L \text{ on } X \text{ of type } (4,4,\ldots,4),\\ \text{and symmetric } \vartheta\text{-structures } \lambda_n \text{ on } \mathcal{G}((2^n\delta)^*L),\\ \text{for all } n\text{, which are "compatible" as } n \text{ varies}\end{array}\right\}.$$

Quite clearly, we can also say:

$$\mathcal{M}_\infty(k)=\left\{\begin{array}{l}\text{set of abelian varieties } X\text{, totally symmetric}\\ \text{ample } L \text{ on } X \text{ of type } (4,4,\ldots,4)\text{, and symplectic}\\ \text{isomorphisms } \varphi\colon V(X)\xrightarrow{\approx} Q_2^{2g} \text{ such that } \varphi(T(X))=2Z_2^{2g}\end{array}\right\},$$

or:

$$\mathcal{M}_\infty(k)=\left\{\begin{array}{l}\text{set of abelian varieties } X\text{, symmetric ample } L \text{ on } X \text{ of}\\ \text{degree 1, and symplectic isomorphisms: } \varphi\colon V(X)\longrightarrow Q_2^{2g}\\ \text{such that } \varphi(T(X))=Z_2^{2g}\text{, and } e_*^L(\varphi^{-1}(\alpha))=\chi(2{}^t\alpha_1\cdot\alpha_2) \text{ all}\\ \alpha\in\tfrac{1}{2}Z_2^{2g}\end{array}\right\}.$$

In any case, the principal result, on the level of geometric points, is:

Corollary. *For all algebraically closed fields k, the map Θ is a bijection between $\mathcal{M}_\infty(k)$ and the set of k-valued points of the open set M_∞.*

In other words, the whole tower $\mathcal{T}$, plus $\varphi\colon V(X)\xrightarrow{\sim} Q_2^{2g}$, is *determined* by the theta function

$$\vartheta\begin{bmatrix}0\\0\end{bmatrix}\circ\varphi^{-1}$$

(or the measure μ) and the theta functions that arise in this way are those which satisfy some finite set of inequalities. Our next task is to determine these inequalities, and hence M_∞, *explicitly*.

Department of Mathematics
Harvard University
Cambridge, Massachussetts

(Received February 20, 1967)

Druck der Universitätsdruckerei H. Stürtz AG., Würzburg

Inventiones math. 3, 215—244 (1967)

On the Equations Defining Abelian Varieties. III*

D. MUMFORD (Cambridge, Mss.)

Contents

§ 10. Non-Degenerate Theta Functions

The third part of this paper is devoted (1) to a complete description of the boundary of the moduli space for abelian varieties described in § 9, and (2) to connecting our theory with the classical theory of theta functions. We begin by defining a theta function in a coordinate-free manner and investigating how and under what non-degeneracy restrictions we can construct a tower of abelian varieties having this as its theta function. Our goal is to find an inverse to the moduli map Θ described in § 9. Fix

o) an algebraically closed field k, char $(k)\neq 2$;

i) a $2g$-dimensional vector space V over Q_2;

ii) a skew-symmetric bi-multiplicative map:

$$e:\ V\times V\to\{2^n\text{-th roots of 1 in } k\},$$

i.e.,

$$e(\alpha,\alpha)=1$$
$$e(\alpha\cdot\beta,\gamma)=e(\alpha,\gamma)\cdot e(\beta,\gamma)$$
$$e(\alpha,\beta\cdot\gamma)=e(\alpha,\beta)\cdot e(\alpha,\gamma);$$

iii) a maximal isotropic lattice $\Lambda\subset V$ (i.e., a compact, open subgroup such that $e(\alpha,\beta)=1$, all $\alpha,\beta\in\Lambda$, maximal with this property);

iv) a quadratic character

$$e_*:\ \tfrac{1}{2}\Lambda/\Lambda\to\{\pm 1\}$$

such that

$$e_*(\alpha+\beta)\,e_*(\alpha)\,e_*(\beta)=e(\alpha,\beta)^2,$$

all $\alpha,\beta\in\tfrac{1}{2}\Lambda$.

* Part I of this paper has been published in Inventiones math. Vol. 1, pp. 287—354 and part II in Vol. 3, pp. 75—135.

We assume, however, that via a suitable isomorphism $V \cong Q_2^{2g}$, $\Lambda \cong Z_2^{2g}$, and e, e_* have the form defined in § 9. In fact, this is nearly always the case: if we write

$$e_*(\alpha)=(-1)^{Q(\alpha)}$$

where Q is a quadratic form on $\frac{1}{2}\Lambda/\Lambda$ with values in the field $F_2=\{0,1\}$, then Q has an Arf invariant $\Delta(Q)\in F_2$. It is not hard to show that (V, Λ, e, e_*) has the required form only if $\Delta(Q)=0$. We leave this point to the reader.

Definition 1. A *theta-function* Θ on V is a map $\Theta: V\to k$ satisfying

i) $\Theta(\alpha+\beta)=e_*(\beta/2)\cdot e(\beta/2,\alpha)\,\Theta(\alpha)$, all $\alpha\in V$, $\beta\in\Lambda$,

ii) $\Theta(-\alpha)=\Theta(\alpha)$, all $\alpha\in V$,

iii) $\prod_{i=1}^{4}\Theta(\alpha_i)=2^{-g}\sum_{\eta\in\frac{1}{2}\Lambda/\Lambda} e(\gamma,\eta)\cdot\prod_{i=1}^{4}\Theta(\alpha_i+\gamma+\eta)$

if $\gamma=-\frac{1}{2}\sum_1^4\alpha_i$, $\alpha_1,\dots,\alpha_4\in V$ arbitrary.

If we let

$$S_0=\{\alpha\mid\Theta(\alpha)\neq 0\}=\text{support}(\Theta),$$

then S_0 is a union of cosets of Λ. The structure of S_0 is a "fine" property of Θ, so we introduce:

Definition 2. The *coarse support* S_1 of Θ is:

$$S_1=\{\alpha\mid\Theta(\alpha+\eta)\neq 0,\text{ for some }\eta\in\tfrac{1}{2}\Lambda\}.$$

We will see in § 11 that the coarse support S_1 of a theta function is either all of V, or $\frac{1}{2}\Lambda+W$ where $W\subset V$ is a proper subvectorspace. This is the essential difference between good and bad theta functions.

Note that $S_0=-S_0$ and $S_1=-S_1$. We always assume, in what follows, that $\Theta\not\equiv 0$, i.e., $S_0\neq\phi$.

1. If $x_1\notin S_1$, $x_2, x_3, x_4\in S_0$, then $2x_1+x_2+x_3+x_4\notin S_0$.

Proof. Use the quartic relation on Θ, with $\alpha_1=2x_1+x_2+x_3+x_4$, $\alpha_2=x_2$, $\alpha_3=x_3$, $\alpha_4=x_4$, $\gamma=-x_1-x_2-x_3-x_4$. *Q.E.D.*

2. $0\in S_1$.

Proof. Assume $0\notin S_1$. Take any $y\in S_0$. Apply (1.) with $x_2=x_3=y$, $x_4=-y$ and we get a contradiction. *Q.E.D.*

3. $x, y\in S_0\Rightarrow\frac{1}{2}(x+y)\in S_1$.

Proof. Apply (1.) with $x_1=\frac{1}{2}(x+y)$, $x_2=x$, $x_3=-y$ and $x_4=-x$. *Q.E.D.*

Because of (2.), there is an $\eta_0 \in \frac{1}{2}\Lambda$ such that $\Theta(\eta_0) \neq 0$. Fix one such η_0.

4. $(0) \subseteq (S_0 + \eta_0) \subseteq (2S_0 + \Lambda) \subseteq (4S_0 + \Lambda) \subseteq \cdots$.

Proof. By (3), if $x \in S_0$, then $\frac{1}{2}(x+\eta_0) \in S_1$, so $x+\eta_0 \in 2S_0+\Lambda$. This gives the 1st inclusion. This also shows that $2x \in 4S_0+\Lambda$. Hence if $y \in 2^k S_0$, so $y = 2^k \cdot x$, $x \in S_0$, then $2^k \cdot x \in 2^{k+1} S_0 + \Lambda$. This gives the rest of the inclusions. *Q.E.D.*

Definition 3.

$$S_\infty = \bigcup_{k \geqq 1}^{\infty} [2^k S_0 + \Lambda].$$

5. S_∞ is a group.

Proof. Let $x, y \in S_\infty$. Now $x, y \in (2^l \cdot S_0 + \Lambda)$ for some $l \geqq l_0$. Then $x = 2^l \cdot x_0 + \eta$, $y = 2^l \cdot y_0 + \zeta$, $x_0, y_0 \in S_0$ and $\eta, \zeta \in \Lambda$. Therefore by (3), $\frac{1}{2}(x_0+y_0) \in S_1$, hence $2^l(x_0+y_0) \in 2^{l+1} \cdot S_0 + \Lambda$. Therefore $x+y \in (2^{l+1} S_0 + \Lambda) \subset S_\infty$. *Q.E.D.*

6. $S_\infty = W + \Lambda$, for some subvectorspace $W \subset V$.

Proof. This is easily seen to be equivalent to asserting that S_∞/Λ is a divisible subgroup of V/Λ. But if $x \in 2^k \cdot S_0 + \Lambda$, then $x = 2^k \cdot x_0 + \eta$, $x_0 \in S_0$, $\eta \in \Lambda$, hence $x - \eta \in 2\{2^{k-1} S_0\} \subset 2 \cdot S_\infty$, i.e., the image of x in S_∞/Λ is divisible by 2. *Q.E.D.*

Definition 4. A theta function is *non-degenerate* if equivalently:

(a) $S_\infty = V$.

(a′) $S_\infty \supset \frac{1}{2}\Lambda$.

(a″) For all sufficiently large n, $2^n \cdot S_0 + \Lambda \supset \frac{1}{2}\Lambda$.

(a‴) For all sufficiently large n, and $\alpha \in 2^{-n-1}\Lambda$, there is an $\eta \in 2^{-n}\Lambda$ such that $\Theta(\alpha+\eta) \neq 0$.

The next step is to form, via the function Θ, a sequence of graded rings:

Definition 5. If M is a vector space of k-valued functions on V, let

$$\mathscr{S}(M) = \bigoplus_{n=0}^{\infty} \mathscr{S}_n(M),$$

where $\mathscr{S}_0(M) = k$, $\mathscr{S}_1(M) = M$, and $\mathscr{S}_n(M)$, for $n \geqq 2$, is the vector spac of functions on V spanned by the products $f_{i_1} \cdots f_{i_n}$, ($f_{i_j} \in M$, all j). Another convenient notation is the following:

$$M^* = \begin{Bmatrix} \text{set of functions } \alpha \mapsto f(\alpha/2), \\ \text{all } f \in M \end{Bmatrix}.$$

In particular, let

$$M_{2^k} = \text{span of the functions } \Theta_{[\beta]}, \qquad \text{all } \beta \in 2^{-k}\Lambda$$

where

$$\Theta_{[\beta]}(\alpha)=e(\beta/2,\alpha)\cdot\Theta(\alpha-\beta).$$

The corresponding rings $\mathscr{S}(M_{2k})$ will be the heart of our analysis. These are only half of the rings we need, however. To define the others, choose a decomposition:

$$\Lambda=\Lambda_1\oplus\Lambda_2$$

such that $Q_2\cdot\Lambda_i=V_i$ is an isotropic subspace under e, and such that $e_*(\alpha/2)=1$ for all $\alpha\in\Lambda_1$ or Λ_2. This exists because if we choose coordinates $V\cong Q_2^{2g}$ such that Λ, e, e_* take their standard forms, then $\Lambda_1=\mathbf{Z}_g^2\times\{0\}$, $\Lambda_2=\{0\}\times\mathbf{Z}_g^2$ have these properties. In terms of Λ_1 and Λ_2, we now define a kind of "dual" theta-function ϕ. It is to satisfy the equations:

$$\sum_{\zeta\in\frac{1}{2}\Lambda_1/\Lambda_1} e(\alpha,\zeta)\cdot\Theta(\alpha+\beta+\zeta)\cdot\Theta(\alpha-\beta+\zeta)=\phi(\alpha)\cdot\phi(\beta)$$

all $\alpha,\beta\in V$. In fact, if we let $\Phi(\alpha,\beta)$ denote the left-hand side of this equation, then the quartic equations on Θ are equivalent to:

$$\Phi(\alpha,\beta)\cdot\Phi(\gamma,\delta)=\Phi(\alpha,\delta)\cdot\Phi(\gamma,\beta)$$

for all $\alpha,\beta,\gamma,\delta\in V$ (cf. proof of Lemma 2, § 8). This, plus the elementary fact $\Phi(\alpha,\beta)=\Phi(\beta,\alpha)$ implies that one and (up to scalars) only one such ϕ exists. Notice that ϕ satisfies the equations:

(i) $\phi(\alpha+\beta)=f_*(\beta)\cdot e(\beta,\alpha)\cdot\phi(\alpha)$, for all $\alpha\in V$, $\beta\in\frac{1}{2}\Lambda_1+\Lambda_2$, if $f_*(\frac{1}{2}\beta_1+\beta_2)=e(\frac{1}{2}\beta_1,\beta_2)$ $(\beta_i\in\Lambda_i)$.

(ii) $\phi(-\alpha)=\phi(\alpha)$, all $\alpha\in V$,

as well as certain quartic equations. Now let

$$M_{2k+1}=\text{span of the functions } \phi_{[\beta]},\qquad \beta\in 2^{-k-1}\cdot\Lambda$$

where

$$\phi_{[\beta]}(\alpha)=e(\beta,\alpha)\cdot\phi(\alpha-\beta).$$

Proposition 1. 1. $\mathscr{S}_2(M_{2k})\subseteq M_{2k+1}$, *equality holding if and only if for all* $\beta\in 2^{-k-1}\Lambda$, $\exists\gamma\in 2^{-k}\Lambda$ *such that* $\phi(\beta+\gamma)\neq 0$.

2. $\mathscr{S}_2(M_{2k+1})^*\subseteq M_{2k+2}$, *equality holding if and only if for all* $\beta\in 2^{-k-1}\Lambda$, $\exists\gamma\in 2^{-k}\Lambda$ *such that* $\Theta(\beta+\gamma)\neq 0$.

Proof. To compute $\mathscr{S}_2(M_{2k})$, note that it is spanned by the functions:

$$f(\alpha)=\sum_{\eta\in\frac{1}{2}\Lambda_1/\Lambda_1} e\left(\eta,\frac{\beta_1+\beta_2}{2}\right)\cdot\Theta_{[\beta_1-\eta]}(\alpha)\cdot\Theta_{[\beta_2-\eta]}(\alpha)$$

where $\beta_i \in 2^{-k}\Lambda$. But

$$f(\alpha)=e\left(\frac{\beta_1+\beta_2}{2},\alpha\right)\cdot\sum_{\eta\in\frac{1}{2}\Lambda_1/\Lambda_1} e\left(\alpha-\frac{\beta_1+\beta_2}{2},\eta\right)\times$$
$$\times\Theta(\alpha-\beta_1+\eta)\,\Theta(\alpha-\beta_2+\eta)$$
$$=e\left(\frac{\beta_1+\beta_2}{2},\alpha\right)\cdot\phi\left(\alpha-\frac{\beta_1+\beta_2}{2}\right)\cdot\phi\left(\frac{\beta_1-\beta_2}{2}\right)$$
$$=\phi_{\left[\frac{\beta_1+\beta_2}{2}\right]}(\alpha)\cdot\phi\left(\frac{\beta_1-\beta_2}{2}\right)\in M_{2k+1}.$$

We get every $\phi_{[\gamma]}$, $\gamma\in 2^{-k-1}\Lambda$, in this way, if and only if every such γ can be written:

$$\gamma=\frac{\beta_1+\beta_2}{2},\qquad \beta_i\in 2^{-k}\Lambda$$

such that

$$\phi\left(\frac{\beta_1-\beta_2}{2}\right)\neq 0.$$

This is exactly the condition in (1). To prove (2), first notice the identity:

$$(\alpha)\quad \sum_{\zeta\in\frac{1}{2}\Lambda_2/\Lambda_2} e(\alpha,\zeta)^2\cdot\phi(\alpha+\beta+\zeta)\cdot\phi(\alpha-\beta+\zeta)$$
$$=\sum_{\substack{\zeta\in\frac{1}{2}\Lambda_2/\Lambda_2\\ \eta\in\frac{1}{2}\Lambda_1/\Lambda_1}} e(\alpha,\zeta)^2\cdot e(\alpha+\beta+\zeta,\eta)\cdot\Theta(2\alpha+2\zeta+\eta)\cdot\Theta(2\beta+\eta)$$
$$=\sum_{\eta\in\frac{1}{2}\Lambda_1/\Lambda_1}\Theta(2\alpha+\eta)\cdot\Theta(2\beta+\eta)\cdot e(\alpha+\beta,\eta)\cdot\Big[\sum_{\zeta\in\frac{1}{2}\Lambda_2/\Lambda_2} e(2\zeta,\eta)\Big]$$
$$=2^g\cdot\Theta(2\alpha)\cdot\Theta(2\beta).$$

Now $\mathscr{S}_2(M_{2k+1})^*$ is spanned by the various functions:

$$f(\alpha)=\sum_{\eta\in\frac{1}{2}\Lambda_2/\Lambda_2} e(\eta,\beta_1+\beta_2)\cdot\phi_{[\beta_1-\eta]}(\alpha/2)\cdot\phi_{[\beta_2-\eta]}(\alpha/2)$$

where $\beta_i\in 2^{-k-1}\Lambda$. But this f comes out as:

$$f(\alpha)=2^g\cdot\Theta_{[\beta_1+\beta_2]}(\alpha)\cdot\Theta(\beta_1-\beta_2)\in M_{2k+2}.$$

(2) now follows just like (1). *Q.E.D.*

Corollary. *If Θ is non-degenerate, then for all $k\gg 0$,*

$$\mathscr{S}_2(M_{2k})=M_{2k+1}$$
$$\mathscr{S}_2(M_{2k+1})^*=M_{2k+2}.$$

Proof. The 2nd equality is clear, by condition (a‴) of the definition of non-degenerate. As for the first, note that by formula (α) in the proof of the Proposition,

$$2^g\,\Theta(\alpha)^2 = \sum_{\zeta\in\frac{1}{2}\Lambda_2/\Lambda_2} e(\alpha,\zeta)\cdot\phi(\alpha+\zeta)\cdot\phi(\zeta).$$

Therefore, $[\Theta(\alpha)\neq 0]\Rightarrow[\phi(\alpha+\zeta)\neq 0$, some $\zeta\in\frac{1}{2}\Lambda_2]$. Thus the non-degeneracy of Θ implies the same for ϕ, and the 1st equality follows too. *Q.E.D.*

In the following discussion, we shall assume that Θ is non-degenerate. As usual, if $R=\Sigma R_n$ is a graded ring, then $R(2)$ is the graded ring ΣR_{2n}. The Corollary shows that there exists a k_0 such that for all $k\geqq k_0$,

$$(\beta)\qquad \mathscr{S}(M_k)(2)\cong\mathscr{S}(M_{k+1}).$$

In particular, the corresponding schemes

$$X=\operatorname{Proj}(\mathscr{S}(M_k)),$$

for $k\geqq k_0$, are all canonically isomorphic. We shall prove eventually that this X is an abelian variety.

So far, we know that $\mathscr{S}(M_k)$ is finitely generated over k. Moreover, it has no nilpotents: if it did, it would have a homogeneous nilpotent element $f\in\mathscr{S}_n(M_k)$. Then $f\neq 0\Rightarrow f(\alpha)\neq 0$, some $\alpha\in V\Rightarrow f^N(\alpha)\neq 0$, all $N\Rightarrow f^N\neq 0$ in $\mathscr{S}_{nN}(M_k)$. Therefore, X is a reduced algebraic scheme over k. In fact, we can map

$$V/\Lambda\to X$$

by evaluating functions in $\mathscr{S}(M_k)$ at points of V. To be more precise, for all $\alpha\in V$, define a homogeneous prime ideal $P(\alpha)\subset\mathscr{S}(M_{2k})$ [resp. $P(\alpha)\subset\mathscr{S}(M_{2k+1})$] by:

$$P(\alpha)=\sum_n P_n(\alpha)$$

$$P_n(\alpha)=\{f\in S_n(M_{2k})\,|\,f(2^k\alpha)=0\}$$

resp.

$$=\{f\in S_n(M_{2k+1})\,|\,f(2^k\alpha)=0\}.$$

It is easy to check that for all k, if the $P(\alpha)$ in $\mathscr{S}(M_k)$ is intersected with $\mathscr{S}(M_k)$ (2), the resulting ideal is equal to the $P(\alpha)$ in $\mathscr{S}(M_{k+1})$ under the isomorphisms (β). For this reason, we omit a k in the notation $P(\alpha)$. Thus $P(\alpha)$ gives a well-defined point $\bar{P}(\alpha)\in X$. It follows easily from the definition that:

a) $\bar{P}(\alpha)$ *is a k-rational point of* X,

b) $\bar{P}(\alpha+\beta)=\bar{P}(\alpha)$, *if* $\beta\in\Lambda$.

Moreover:

c) $\{\bar{P}(\alpha)|\alpha\in V\}$ *is dense in* X.

Proof of c. Take $2k\geqq k_0$. If (c) were false, for large n, there would be a non-zero function $f\in\mathscr{S}_n(M_{2k})$ that vanished at all $\bar{P}(\alpha)$'s. But $f(\bar{P}(\alpha))=0\Leftrightarrow f(2^k\alpha)=0$, so f would vanish everywhere on V, hence $f=0$. *Q.E.D.*

One can do even more: for $\alpha\in V$, I claim that there is an automorphism T_α: $X\to X$ such that $T_\alpha(\bar{P}(\beta))=\bar{P}(\alpha+\beta)$, all $\beta\in V$. To construct T_α, let k_1 be the least integer such that $2^{k_1}\alpha\in\Lambda$. Define

$$T_\alpha^*:\quad \mathscr{S}(M_{2k})\to\mathscr{S}(M_{2k})$$

resp.:

$$\mathscr{S}(M_{2k+1})\to\mathscr{S}(M_{2k+1})$$

by:

$$T_\alpha^* f(\beta)=e(\beta,2^{k-1}\alpha)^n\cdot f(\beta+2^k\alpha),\qquad \text{all } f\in S_n(M_{2k})$$

resp.

$$=e(\beta,2^k\alpha)^n\cdot f(\beta+2^k\alpha),\qquad \text{all } f\in S_n(M_{2k+1})$$

(where we assume $k\geqq k_1$). To check that this is, indeed, an automorphism of $\mathscr{S}(M_{2k})$ [resp. $\mathscr{S}(M_{2k+1})$], it suffices to check that $T_\alpha^*\Theta_{[\gamma]}\in M_{2k}$, all $\gamma\in 2^{-k}\Lambda$; and $T_\alpha^*\phi_{[\gamma]}\in M_{2k+1}$, all $\gamma\in 2^{-k-1}\Lambda$. But, in fact, one computes:

$$(\gamma)\qquad \begin{aligned} T_\alpha^*\Theta_{[\gamma]}&=e_*(2^{k-1}\alpha)\cdot e(\gamma,2^k\alpha)\cdot\Theta_{[\gamma]}\\ T_\alpha^*\phi_{[\gamma]}&=f_*(2^k\alpha)\cdot e(\gamma,2^{k+1}\alpha)\cdot\phi_{[\gamma]}.\end{aligned}$$

Moreover, one finds that T_α^*, acting on $\mathscr{S}(M_k)$, induces the same automorphism on $\mathscr{S}(M_k)$ (2) that you get by considering the T_α^* acting on $\mathscr{S}(M_{k+1})$ and carrying it across via the isomorphisms (β) of $\mathscr{S}(M_k)$ (2) and $\mathscr{S}(M_{k+1})$. Therefore, the T_α^*'s all define one and the same automorphism T_α of X. Note that:

d) $(T_\alpha^*)^{-1}(P(\beta))=P(\alpha+\beta)$.

Proof. If $f\in\mathscr{S}_n(M_{2k})$ or $\mathscr{S}_n(M_{2k+1})$, then

$$T_\alpha^* f\in P(\beta)\ \Leftrightarrow\ T_\alpha^* f(2^k\beta)=0\ \Leftrightarrow\ f(2^k\alpha+2^k\beta)=0\ \Leftrightarrow\ f\in P(\alpha+\beta),$$

hence

d') $T_\alpha(\bar{P}(\beta))=\bar{P}(\alpha+\beta)$.

One checks also (via (γ) if you like) that:

e) $T_{\alpha_1+\alpha_2}=T_{\alpha_1}\circ T_{\alpha_2}$,

f) $T_\alpha=\text{id.}\Leftrightarrow\alpha\in\Lambda$,

so that T is a faithful action of the group V/Λ on the scheme X.

A remarkable consequence of all this is:

Proposition 2. *If* Θ *is non-degenerate, then* $\mathscr{S}(M_k)$ *is an integral domain, for all* k.

Proof. We show first that $\mathcal{S}(M_k)$ is a domain if $k \geqq k_0$. Since $\mathcal{S}(M_k)$ has no nilpotents, this is equivalent to showing that X is irreducible. Now V/Λ acts on X, so it permutes the various components of X, i.e., we have a homomorphism:

$$V/\Lambda \to S = \begin{Bmatrix} \text{gp. of permutations} \\ \text{of components of } X \end{Bmatrix}.$$

But S is a *finite* group and V/Λ is a *divisible* group. So V/Λ must map each component X_i into itself. On the other hand, the collection of points $\{\bar{P}(\alpha)\}$ forms a single orbit of the action of V/Λ on X. Therefore, all these points $\{\bar{P}(\alpha)\}$ belong to a single component of X. Since they are also dense in X, X can have only a single component. Therefore $\mathcal{S}(M_k)$ is a domain if $k \geqq k_0$.

In general, suppose some $\mathcal{S}(M_k)$ were not a domain. Then there would be homogeneous elements $f \in \mathcal{S}_n(M_k)$, $g \in \mathcal{S}_m(M_k)$ such that $f \cdot g = 0$, $f \neq 0, g \neq 0$. Now f^2 and g^2 can be considered as elements of $\mathcal{S}(M_{k+1})$. Since $f \cdot g = 0$, we still have $f^2 \cdot g^2 = 0$. Also, since $\mathcal{S}(M_k)$ has no nilpotents, $f^2 \neq 0$ and $g^2 \neq 0$. Therefore $\mathcal{S}(M_{k+1})$ is not a domain either. Continuing in this way, we find that $\mathcal{S}(M_l)$ is not a domain for all $l \geqq k$, which contradicts the first part of the proof. *Q.E.D.*

Corollary 1. *The following are equivalent:*

i) *Θ is non-degenerate,*

ii) *$S_1 = V$, i.e., for all $\alpha \in V$, $\exists \eta \in \frac{1}{2}\Lambda$ such that $\Theta(\alpha + \eta) \neq 0$.*

iii) *For all $\alpha \in \frac{1}{4}\Lambda$, $\exists \eta \in \frac{1}{2}\Lambda$ such that $\Theta(\alpha + \eta) \neq 0$.*

Proof. Clearly (ii) $\Rightarrow$ (iii) $\Rightarrow$ (i). Now assume (i) holds. If $\Theta(\alpha + \eta) = 0$, all $\eta \in \frac{1}{2}\Lambda$, then it would follow from the definition of ϕ that $\phi(\alpha + \beta) \times \phi(\beta) = 0$, all $\beta \in V$. But this means that $\phi_{[-\alpha]} \cdot \phi_{[0]} = 0$, i.e., one of the rings $\mathcal{S}(M_{2k+1})$ is not domain. This contradicts the Prop., so (ii) must hold. *Q.E.D.*

Corollary 2. $\mathcal{S}(M_k)\,(2) \cong \mathcal{S}(M_{k+1})$, *for all* $k \geqq 2$.

Proof. In view of Prop. 1, this follows from Cor. 1 provided that we check: $\forall \alpha \in V$, $\exists \eta \in \frac{1}{2}\Lambda$ such that $\phi(\alpha + \eta) \neq 0$. Looking back at the proof of the Cor. to Prop. 1, you see that this too follows from Cor. 1. *Q.E.D.*

To show that X is actually an abelian variety, we could either define the group law explicitly, using the addition formula of § 2, or else we can use only the action of V/Λ on X and combine this with general structure theorems on the automorphisms of a variety. Although the former is more elementary, we follow the latter approach as it is quicker.

X is given to us together with a projective embedding. For example, $X = \operatorname{Proj}(\mathcal{S}(M_2))$, so

$$X \subset \boldsymbol{P}(M_2).$$

Let L_2 be the invertible sheaf induced on X via this embedding. If, via the isomorphism $X \cong \operatorname{Proj}(\mathscr{S}(M_k))$, we embed X in $\boldsymbol{P}(M_k)$, the induced sheaf L_k is just:

$$L_k \cong L_2^{2^{k-2}}.$$

Let $\mathscr{P}$ denote the family of all invertible sheaves algebraically equivalent to L_2. We shall use the fact that Aut $(X, \mathscr{P})$, the group of automorphisms of the pair X, $\mathscr{P}$, is an algebraic group (MATSUSAKA [*14*], GROTHENDIECK [*15*], p. 221 – 20). For all $\alpha \in V/\Lambda$, if $2^k\alpha \in \Lambda$, then T_α is induced by an automorphism T_α^* of $\mathscr{S}(M_{2k})$; therefore $T_\alpha^*(L_{2k}) \cong L_{2k}$; therefore $T_\alpha^*(L_2)$ differs from L_2 by an invertible sheaf of finite order; therefore $T_\alpha^{-1}(\mathscr{P}) = \mathscr{P}$. In other words, the action of V/Λ on X factors through an injective homomorphism:

$$V/\Lambda \to \operatorname{Aut}(X, \mathscr{P}).$$

Let A be the Zariski-closure of V/Λ in Aut $(X, \mathscr{P})$. Then A is connected since V/Λ is divisible and dense in A (cf. proof of Prop. 2), and A is commutative since V/Λ is commutative and dense in A. Moreover, since the V/Λ-orbit of $\bar{P}_0$ is dense in X, the A-orbit of $\bar{P}_0$ must be an open dense set in X, i.e., A acts generically transitively on X. In fact, the morphism

$$\psi: A \to X$$
$$\sigma \mapsto \sigma(\bar{P}_0)$$

is an open immersion of A in X. This follows since the image $\psi(A)$ is always isomorphic to A/H, H = the stabilizer of $\bar{P}_0$; and since A is commutative and acting faithfully on X, all stabilizers are trivial.

Next, we want to compute the dimension of X. I claim that the Hilbert polynomial of (X, L_2) is given by:

Proposition 3. $\chi(L_2^n) = 4^g \cdot n^g$.

Proof. For k large,

$$\chi(L_2^{2^{2k}}) = \dim(S_{2^{2k}}(M_2)) = \dim(M_{2+2k}).$$

Now $M_{2(k+1)}$ is, by definition, the span of the $2^{2g(k+1)}$ functions $\Theta_{[\beta]}$, where β runs over cosets of $2^{-k-1}\Lambda/\Lambda$. But these functions are linearly independent. To see this, look at the automorphisms T_α^* of $\mathscr{S}(M_{2(k+1)})$, where $\alpha \in 2^{-k-1}\Lambda$. Use formulae ($\gamma$) above and note that each $\Theta_{[\gamma]}$ gives rise to a distinct set of eigenvalues for the T_α^*'s. Therefore, the $\Theta_{[\gamma]}$'s could not be dependent unless one were identically zero, and this is not the case. Therefore

$$\dim M_{2(k+1)} = 4^g \cdot (2^{2k})^g.$$

This shows that $\chi(L_2^n)$ and $4^g \cdot n^g$ agree for an infinite set of values of n. Since both are polynomials, they are always equal. *Q.E.D.*

Corollary. $\dim X = g$.

Returning to A, we find that A is a commutative g-dimensional algebraic group containing a subgroup isomorphic to $(Q_2/Z_2)^{2g}$. From well-known structure theorems on algebraic groups, the only such A's are abelian varieties. Therefore A is complete, hence $A = X$, hence:

(I) *X is an abelian variety.*

Moreover, in the course of proving this, we have also found that V/Λ is acting on X via translations, hence (comparing orders) we find:

(II) $\alpha \mapsto \bar{P}(\alpha)$ *is a group isomorphism of* V/Λ *with* $\operatorname{tor}_2(X)$.

Up to this point, identifying the various $\operatorname{Proj}(\mathscr{S}(M_k))$'s has been useful. But to go further, it is more convenient now to drop these identifications. Therefore, now let

$$X_n = \operatorname{Proj}(\mathscr{S}(M_{2^n})).$$

This is a family of isomorphic abelian varieties. However, the most natural maps between them are given by the inclusions:

$$M_{2^n} \subset M_{2^{n+2}}$$
$$\mathscr{S}(M_{2^n}) \subset \mathscr{S}(M_{2^{n+2}})$$

inducing finite morphisms:

$$X_n \xleftarrow{p} X_{n+1}.$$

To check that p is defined, we must know that $\mathscr{S}(M_{2^{n+2}})$ is integrally dependent on $\mathscr{S}(M_{2^n})$. But I claim:

$$\Theta(\gamma)^2 \cdot \Theta_{[\beta]}^2 = 2^{-g} \cdot \sum_{\eta \in \frac{1}{2}\Lambda/\Lambda} e(\eta, \gamma)\, \Theta(\eta)^2 \cdot \Theta_{[\beta+\gamma-\eta]} \cdot \Theta_{[\beta-\gamma+\eta]}.$$

[*Proof.* $\Theta(\gamma)^2 \cdot \Theta_{[\beta]}(\alpha)^2 = e(\beta, \alpha)\, \Theta(\gamma)\, \Theta(\gamma)\, \Theta(\beta-\alpha)\, \Theta(\alpha-\beta)$.

By the quartic relations on Θ, we get

$$= 2^{-g} e(\beta, \alpha) \sum_{\eta} e(-\gamma, \eta)\, \Theta(\eta)^2\, \Theta(\beta-\alpha-\gamma+\eta)\, \Theta(\alpha-\beta-\gamma+\eta)$$

$$= 2^{-g} \sum_{\eta} e(\eta, \gamma)\, \Theta(\eta)^2 \cdot \Theta_{[\beta+\gamma-\eta]}(\alpha) \cdot \Theta_{[\beta-\gamma+\eta]}(\alpha).\ \textit{Q.E.D.}]$$

Choose $\gamma \in \beta + \frac{1}{2}\Lambda$ so that $\Theta(\gamma) \neq 0$. Then if $\beta \in 2^{-n-1}\Lambda$, this equation shows that $\Theta_{[\beta]}^2 \in \mathscr{S}(M_{2^n})$. This proves that p is a finite morphism. Since X_n and X_{n+1} are abelian varieties, p must be an isogeny.

Define prime ideals:

$$P^{(k)}(\alpha) \subset \mathcal{S}(M_{2k})$$

via

$$P^{(k)}(\alpha) = \sum_n P_n^{(k)}(\alpha)$$

$$P_n^{(k)}(\alpha) = \{f \in \mathcal{S}_n(M_{2k}) \mid f(\alpha) = 0\}.$$

Then $P^{(k)}(\alpha)$ defines a k-rational point $\psi_k(\alpha) \in X_k$. We have

(a) $p(\psi_{k+1}(\alpha)) = \psi_k(\alpha)$.

(b) $\alpha \mapsto \psi_k(\alpha)$ defines an isomorphism

$$V/2^k \Lambda \xrightarrow{\approx} \mathrm{tor}_2(X_k).$$

(b) here follows from conclusion (II) above, noticing how we have reinterpreted the ideal $P(\alpha)$. In fact, if we call X the common abelian variety to which all the X_k's were previously identified, then $\bar{P}(\alpha) \in X$ corresponds exactly to $\psi_k(2^k \alpha) \in X_k$. Therefore $\psi_k(\alpha) = 0 \Leftrightarrow \bar{P}(2^{-k}\alpha) = 0 \Leftrightarrow 2^{-k}\alpha \in \Lambda$. Moreover, this shows that via these identifications, we get a morphism:

$$\begin{array}{cc}
X & \bar{P}(\alpha) \\
\wr\| & \downarrow \\
X_{k+1} & \psi_{k+1}(2^{k+1}\alpha) \\
p\downarrow & \downarrow \\
X_k & \psi_k(2^{k+1}\alpha) \\
\wr\| & \downarrow \\
X & \bar{P}(2\alpha) = 2\bar{P}(\alpha).
\end{array}$$

This map, from X to X, agrees with 2δ at all points $\bar{P}(\alpha)$. Therefore it is equal to 2δ. In particular:

(c) The degree of p is 2^{2g} and $\mathrm{Ker}(p) = \mathrm{Ker}(2\delta)$. It follows that all the X_n's generate a single 2-tower. Call this $\underset{\sim}{X} = \{X_\alpha\}_{\alpha \in S}$, and let $X_n = X_{\alpha_n}$, $\alpha_n \in S$. Moreover, these α_n's are a cofinal set in S, by (c). In view of (a)

$$\alpha \mapsto \{\psi_k(\alpha)\}$$

defines a homomorphism

$$\psi \colon V \to V(\underset{\sim}{X}),$$

and (b) implies that ψ is an isomorphism. More, (b) shows that the compact open subgroups $2^k \Lambda$ and $T(\alpha_k)$ correspond to each other under ψ.

This 2-tower is polarized too. Let L_k be the sheaf $\underline{o}(1)$ on X_k coming from its presentation as Proj $(\mathcal{S}(M_{2k}))$. Since the p's comes from gradation preserving homomorphisms of the $\mathcal{S}(M_{2k})$'s it follows that $p^*(L_k) \cong L_{k+1}$. To check that L_k is totally symmetric, we need the inverse on X_k:

Let $\iota^*(f)(\alpha)=f(-\alpha)$, all $f\in\mathcal{S}(M_{2k})$.

Then ι^* defines an involution

$$\iota: X_k \to X_k$$

such that $\iota(\psi_k(\alpha))=\psi_k(-\alpha)$.

Therefore ι agrees with the inverse of X_k on all points $\psi_k(\alpha)$, hence $\iota=$ inverse of X_k.

Since ι is induced at all by an automorphism ι^* of $\mathcal{S}(M_{2k})$, it follows that L_k is at least a symmetric sheaf. Since

$$\{\psi_k(\alpha)\,|\,\alpha\in 2^{k-1}\Lambda/2^k\Lambda\}=\text{Kernel of } 2\delta \text{ in } X_k,$$

L_k is totally symmetric if and only if ι^* is the identity in $\mathcal{S}(M_{2k})/P^{(k)}(\alpha)$, all $\alpha\in 2^{k-1}\Lambda$. This means that for all $f\in M_{2k}$, $\iota^* f-f\in P_1^{(k)}(\alpha)$, i.e., $f(\alpha)=f(-\alpha)$. But M_{2k} is spanned by $\Theta_{[\beta]}$'s, $\beta\in 2^{-k}\Lambda$, and if $\beta\in 2^{-k}\Lambda$, $\alpha\in 2^{k-1}\Lambda$, then:

$$\Theta_{[\beta]}(-\alpha)=e\left(\frac{\beta}{2},-\alpha\right)\Theta(-\alpha-\beta)=e\left(\frac{\beta}{2},\alpha\right)\Theta(\alpha-\beta)=\Theta_{[\beta]}(\alpha).$$

Therefore all the L_n's are totally symmetric and $\{X_n, L_n\}$ extends to a polarized 2-tower $\mathcal{T}=\{X_\alpha, L_\alpha\}$. We shall leave it to the reader to check the key fact that ψ is symplectic:

(d) $e_\lambda(\psi\alpha,\psi\beta)=e(\alpha,\beta)$, all $\alpha,\beta\in V$.

Recapitulating this whole section so far, we have defined an arrow:

$$\Xi:\ \left\{\begin{matrix}\text{Given a non-degenerate}\\ \text{theta function } \Theta \text{ on } V\end{matrix}\right\}\longrightarrow\left\{\begin{matrix}\text{construct a polarized}\\ \text{2-tower } \mathcal{T}=\{X_\alpha, L_\alpha\},\\ \text{plus a symplectic isomorphism}\\ \psi:\ V\xrightarrow{\approx} V(\underset{\sim}{X})\end{matrix}\right\}.$$

Now, on V we have the vector space of functions spanned by all the $\Theta_{[\beta]}$'s. On $V(\underset{\sim}{X})$, we have the vector space of all theta functions $\vartheta[\Gamma(\mathcal{T})]$ of the tower $\mathcal{T}$.

Proposition 4. *Via ψ, these vector spaces are equal:*

Span of $\Theta_{[\beta]}$'s $=\{\vartheta_{[s]}\circ\psi\,|\,s\in\Gamma(\mathcal{T})\}$.

Moreover, Θ itself is the unique function f (up to scalars) of the form $\vartheta_{[s]}\circ\psi$ satisfying the functional equation:

$$f(\alpha+\beta)=e_*(\beta/2)\cdot e(\beta/2,\alpha)\cdot f(\alpha), \qquad \textit{all } \alpha\in V,\ \beta\in\Lambda.$$

Key Corollary 1. *If $V=\mathbf{Q}_2^{2g}$, $\Lambda=\mathbf{Z}_2^{2g}$, and e, e_* have the standard forms of § 9, then Θ is exactly the theta function $\vartheta\begin{bmatrix}0\\0\end{bmatrix}\circ\psi$ associated to the* *triple* $(\underset{\leftarrow}{X}, \mathcal{T}, \psi^{-1})$ *in* § 9. *In other words,* Ξ *is an inverse to the map* Θ *of* § 9.

Proof of Prop. 4. Let $\alpha \in 2^{-k_1} \Lambda$ and let $k \geqq k_1$. Define $T_\alpha^*: \mathcal{S}(M_{2k}) \to \mathcal{S}(M_{2k})$ slightly differently from before:

$$T_\alpha^* f(\beta) = e\left(\beta, \frac{\alpha}{2}\right)^n \cdot f(\beta+\alpha), \qquad \text{all } f \in S_n(M_{2k}).$$

Note $T_\alpha^{*-1}(P^{(k)}(\beta)) = P^{(k)}(\alpha+\beta)$. Let T_α: $X_k \to X_k$ be the automorphism induced by T_α^*. Then $T_\alpha(\psi_k(\beta)) = \psi_k(\alpha+\beta)$, hence T_α is translation by the point $\psi_k(\alpha)$, i.e.,

$$T_\alpha = T_{\psi_k(\alpha)}.$$

Moreover, T_α^* also induces a compatible isomorphism:

$$g_k(\alpha): L_k \xrightarrow{\sim} T_{\psi_k(\alpha)}^* L_k.$$

For all $k \geqq k_1$, these are compatible, so the totality of pairs

$$g(\alpha) = \{(\psi_k(\alpha), g_k(\alpha) \mid k \geqq k_1\}$$

is a point of $\mathcal{G}(\mathcal{T})$.

(*) $g(\alpha) = \sigma[\psi(\alpha)]$, i.e., $g(\alpha)$ is the canonical element of $\mathcal{G}(\mathcal{T})$ over the point $\psi(\alpha)$ in $V(\underset{\sim}{X})$.

Proof of *. This requires checking 2 things: (i) $g(\alpha)$ is a symmetric element of $\mathcal{G}(\mathcal{T})$, i.e., $\delta_{-1} g(\alpha) = g(\alpha)^{-1}$, and (ii) $g(2\alpha) = g(\alpha)^2$. In terms of T_α^*, this is the same as:

(i) $\iota^* \circ T_\alpha^* = (T_\alpha^*)^{-1} \circ \iota^*$.

(ii) $T_{2\alpha}^* = T_\alpha^* \circ T_\alpha^*$.

These are both immediate. *Q.E.D.*

Next, notice that $M_{2k} \cong \Gamma(X_k, L_k)$. In fact, there is a canonical map $M_{2k} \to \Gamma(X_k, L_k)$; it is injective, since the ring $\mathcal{S}(M_{2k})$ has no nilpotents, and only nilpotent elements of $\mathcal{S}_n(M_{2k})$ define trivial sections of L_k^n; but it is easy to check that both dim M_{2k} and dim $\Gamma(X_k, L_k)$ are equal to 2^{2kg}; therefore $M_{2k} \cong \Gamma(X_k, L_k)$. Therefore,

$$\Gamma(\mathcal{T}) = \varinjlim_k \Gamma(X_k, L_k) \cong \bigcup_k M_{2k} = \left\{\begin{matrix}\text{Span of } all \text{ the} \\ \text{functions } \Theta_{[\beta]} \\ \beta \in V\end{matrix}\right\}.$$

Now let f be some linear combination of the $\Theta_{[\beta]}$. Say $f \in M_{2k_1}$. Let f define $s \in \Gamma(X_{k_1}, L_{k_1})$. I claim that:

$$(*) \qquad f(\alpha) = \vartheta_{[s]}(\psi\,\alpha), \qquad \text{all } \alpha \in V.$$

Taking a larger k_1 if necessary, we may suppose that $\alpha \in 2^{-k_1}\Lambda$. By definition, $\vartheta_{[s]}$ at $\psi\alpha$ is the "value" at the origin of X_{k_1} of the section of L_{k_1} obtained via the map:

$$\Gamma(X_{k_1}, L_{k_1}) \xrightarrow[g_{k_1}(-\alpha)]{\sim} \Gamma(X_{k_1}, T^*_{\psi_{k_1}(-\alpha)} L_{k_1}) \xrightarrow[T^*_{\psi_{k_1}(\alpha)}]{\sim} \Gamma(X_{k_1}, L_{k_1}).$$

This means that we simply apply the automorphism $(T^*_{-\alpha})^{-1}$ of M_{2k} to f, and take the value at the origin. But $T^*_{-\alpha} = T^{*-1}_{\alpha}$, and $(T^*_\alpha f)(0) = f(\alpha)$, so $(*)$ is proven. Thus the span of the $\Theta_{[\beta]}$'s is the same as the space of functions $\vartheta_{[s]} \circ \psi$, $s \in \Gamma(\mathscr{T})$.

As for the final assertion of the Proposition, on the one hand, Θ does satisfy the functional equation there; and, from the general theory of the space $\vartheta[\Gamma(\mathscr{T})]$ in § 8, we know that this functional equation has only a 1-dimensional set of solutions in $\vartheta[\Gamma(\mathscr{T})] \circ \psi$. *Q.E.D.*

Corollary 2. *All g-dimensional principally polarized abelian varieties X are isomorphic to* Proj $(\mathscr{S}(M_2))$, *where* M_2 *is the span of the* $\Theta_{[\beta]}$'s, $\beta \in \frac{1}{2}\Lambda$, *for some non-degenerate theta function* Θ *on V.*

Proof. Just take Θ to be the $\vartheta\begin{bmatrix}0\\0\end{bmatrix}$ attached to X as in § 9, and carried over to a function on V by a suitable isomorphism of V and $V(X)$. *Q.E.D.*

Corollary 3. *The open set* $M_\infty \subset \overline{M}_\infty$, *which in* § 9 *represents the moduli functor* $\mathscr{M}_\infty$, *is the open set whose geometric points represent non-degenerate theta functions, i.e.,*

$$E = \left\{ \begin{matrix} \text{set of all systems of coset representatives} \\ r\colon \tfrac{1}{4}\mathbf{Z}_2^{2g}/\tfrac{1}{2}\mathbf{Z}_2^{2g} \to \tfrac{1}{4}\mathbf{Z}_2^{2g} \end{matrix} \right\}.$$

For all $r \in E$, *let*

$$U_r = \left\{ \begin{matrix} \text{open set in } \overline{M}_\infty \text{ defined by} \\ X_\alpha \neq 0, \text{ all } \alpha \in \text{Image}(r) \end{matrix} \right\}.$$

Then

$$M_\infty = \bigcup_{r \in E} U_r.$$

§ 11. Satake's Compactification

In this section, I want to analyze the degenerate theta functions Θ on V, in the sense of § 10. In particular, they all come from lower dimensional non-degenerate theta-functions via "cusps". This will show that the whole moduli scheme $\overline{M}_\infty$ is a disjoint union of copies of the M_∞'s for dimensions g *and lower* i.e., that $\overline{M}_\infty$ is the Satake compactification of M_∞[1].

[1] *Added in Proof.* A closer study has shown that $\overline{M}_\infty$ is *not normal* along $\overline{M}_\infty - M_\infty$. Its normalization is Satake's compactification.

Return to the discussion at the beginning of § 10: let V, Λ, e, e^* be given as before. First, I want to describe a way of forming degenerate theta functions on V out of theta functions on lower dimensional spaces.

Definition 1. A *cusp* is a subspace $W \subset V$ such that $W^\perp \subset W$, i.e., if $\alpha \in V$ has the property $e(\alpha, \beta) = 1$, all $\beta \in W$, then $\alpha \in W$.

Given a cusp W, let:

$$\tilde{V} = W/W^\perp$$
$$\tilde{\Lambda} = \Lambda \cap W/\Lambda \cap W^\perp$$
$$\tilde{e} = \text{induced skew-symmetric pairing, } \tilde{V} \times \tilde{V} \to k^*.$$

Lemma. *$\tilde{\Lambda}$ is a maximal isotropic lattice in $\tilde{V}$, (for $\tilde{e}$).*

Proof. Notice that $\Lambda/\Lambda \cap W$ is a free $\mathbf{Z}_2$-module. Therefore the sequence:

$$0 \to \Lambda \cap W \to \Lambda \to \Lambda/\Lambda \cap W \to 0$$

splits, and $\Lambda = \Lambda_1 \oplus (\Lambda \cap W)$ for some sub $\mathbf{Z}_2$-Module Λ_1. Let $V_1 = \mathbf{Q}_2 \cdot \Lambda_1$, so $V = V_1 \oplus W$. Now I claim:

$$(*) \qquad (\Lambda \cap W)^\perp = \Lambda + W^\perp.$$

[In fact, let $\alpha \in V$ satisfy $e(\alpha, \beta) = 1$, all $\beta \in \Lambda \cap W$. Since V_1 and W are dual vector spaces via e, there is a $\gamma \in W^\perp$ such that $e(\alpha, \beta) = e(\gamma, \beta)$ all $\beta \in V_1$. But then $\alpha - \gamma$ is orthogonal to both V_1 and $\Lambda \cap W$, hence orthogonal to Λ, hence $\alpha - \gamma \in \Lambda$. Thus $\alpha \in W^\perp + \Lambda$.]

Now to show $\tilde{\Lambda}$ is maximal isotropic, let $\alpha \in W$ have an image $\tilde{\alpha}$ in $\tilde{V}$ perpendicular to $\tilde{\Lambda}$, i.e., $\alpha \in (W \cap \Lambda)^\perp$. By $(*)$, $\alpha = \alpha_1 + \alpha_2$, where $\alpha_1 \in \Lambda$, $\alpha_2 \in W^\perp$. But then $\alpha_1 = \alpha - \alpha_2 \in W$. Therefore $\alpha_1 \in W \cap \Lambda$ so $\tilde{\alpha} = \tilde{\alpha}_1 \in \tilde{\Lambda}$. *Q.E.D.*

Definition 2. A cusp with origin is a cusp $W \subset V$, plus an element $\eta_0 \in \frac{1}{2}\Lambda$ such that

i) $e_*(\alpha) = e(\alpha, \eta_0)^2$, all $\alpha \in W^\perp \cap (\frac{1}{2}\Lambda)$.

ii) $e_*(\eta_0) = 1$.

It is not hard to check that every cusp has at least one origin: we leave this to the reader. Given a cusp with origin, look at the map

$$\alpha \mapsto e_*(\alpha) \cdot e(\alpha, \eta_0)^2$$

where $\alpha \in \frac{1}{2}\Lambda \cap W$. If $\beta \in \frac{1}{2}\Lambda \cap W^\perp$, then

$$e_*(\alpha + \beta) \cdot e(\alpha + \beta, \eta_0)^2 = e_*(\alpha) \cdot e_*(\beta) \cdot e(\alpha, \beta)^2 \cdot e(\alpha, \eta_0)^2 \cdot e(\beta, \eta_0)^2$$
$$= e_*(\alpha) \cdot e(\alpha, \eta_0)^2.$$

16*

Thus there is a quadratic form $\tilde{e}_*: \frac{1}{2}\tilde{\Lambda}/\tilde{\Lambda} \to \{\pm 1\}$ such that

$$(*) \qquad \tilde{e}_*(\tilde{\alpha}) = e_*(\alpha) \cdot e(\alpha, \eta_0)^2, \qquad \text{all } \alpha \in \tfrac{1}{2}\Lambda \cap W.$$

It is not hard to check that the new data $(\tilde{V}, \tilde{\Lambda}, \tilde{e}, \tilde{e}_*)$ has the standard form required in § 10 (i.e., that the associated Arf-invariant is 0). We leave this to the reader also.

Now let $\tilde{\Theta}$ be a theta-function on $\tilde{V}$.

Definition 3. For all $\alpha \in V$, let

$$T_{W,\eta_0}\Theta(\alpha) = \begin{cases} 0 & \text{if } \alpha \notin \eta_0 + W + \Lambda \\ e_*\left(\frac{\eta_1}{2}\right) e\left(\frac{\eta_1}{2}, \eta_0\right) e\left(\frac{\eta_0+\eta_1}{2}, \alpha\right) \tilde{\Theta}(\tilde{\alpha}_0) & \text{if } \alpha = \eta_0 + \eta_1 + \alpha_0,\ \eta_1 \in \Lambda,\ \alpha_0 \in W. \end{cases}$$

Proposition 1. *The above $T_{W,\eta_0}\tilde{\Theta}$ is well-defined (note that the $\alpha \in V$ may be decomposed in more than way as $\alpha = \eta_0 + \eta_1 + \alpha_0$), and is a theta-function on V.*

The proof of this Proposition is a ghastly but wholly straightforward set of computations. It took me several hours to do every bit and as I was no wiser at the end – except that I knew the definition was correct – I shall omit details here. Our main result is:

Theorem. *Let Θ be any theta-function on V, and let W be the subspace of V such that $S_\infty = W + \Lambda$* (cf. § 10). *Then W is a cusp, and if η_0 is any origin for W, Θ is equal to $T_{W,\eta_0}\tilde{\Theta}$ for some non-degenerate theta-function $\tilde{\Theta}$ on $\tilde{W}$. In particular, W is characterized by:*

$$\textit{coarse support}\,(\Theta) = W + \tfrac{1}{2}\Lambda.$$

The proof of this theorem will be based on the $\Theta \leftrightarrow \mu$ correspondence, given in Lemma 1, § 8. Before taking up the proof of the Theorem, we want to give this correspondence a more intrinsic formulation. Let $V = W_1 \oplus W_2$, where W_i are maximal isotropic subspaces, such that

i) $\Lambda = \Lambda_1 \oplus \Lambda_2$, $\Lambda_i = \Lambda \cap W_i$.

ii) $e_*(\alpha/2) = 1$, all α in Λ_1 or in Λ_2.

Then

a) Define a measure μ on W_1, from a theta function Θ on V via

$$\mu(\alpha_1 + 2^n \Lambda_1) = 2^{-ng} \sum_{\alpha_2 \in 2^{-n}\Lambda_2/\Lambda_2} e\left(\alpha_1, \frac{\alpha_2}{2}\right) \cdot \Theta(\alpha_1 + \alpha_2).$$

b) Define a theta function Θ on V, from a measure μ on W_1, via

$$\Theta(\alpha_1 + \alpha_2) = e\left(\alpha_1, \frac{\alpha_2}{2}\right) \int_{\alpha_1 + \Lambda_1} e(\alpha_2, \beta) \cdot d\mu(\beta).$$

Our proof will be based on the fact that any finitely additive measure μ (on the algebra of compact open subsets of W_1) has a *support*, i.e., a smallest closed set S such that:

$$\mu(U)=0, \quad \text{all compact open } U\text{'s in } W_1-S.$$

Proof. Say S_A and S_B are closed sets such that $\mu(U)=0$ if $U\subset W_1-S_A$ or $U\subset W_1-S_B$. Then let $U\subset W_1-(S_A\cap S_B)$ be a compact open set. We must decompose U into $U_A\cup U_B$, where $U_A\subset W_1-S_A$, and $U_B\subset W_1-S_B$, and U_A and U_B are compact and open. For all $x\in U\cap S_A$, note that $x\notin S_A$, so we can find a compact, open neighborhood U_x of x such that

$$U_x\subset U\cap(W_1-S_B).$$

Since $U\cap S_A$ is compact, it can be covered by a finite set of these U_x's: say

$$U\cap S_A\subset[U_{x_1}\cup\cdots\cup U_{x_n}].$$

Let $U_B=U_{x_1}\cup\cdots\cup U_{x_n}$. By construction $U_B\subset U\cap(W_1-S_B)$ and U_B is compact and open. Let $U_A=U-U_B$. Then U_A is also compact and open and since $U_B\supset U\cap S_B$, it follows that $U_A\subset U\cap(W_1-S_B)$. By assumption on S_A and S_B, we have $\mu(U_A)=0$ and $\mu(U_B)=0$. Therefore $\mu(U)=0$. This shows that the family of sets:

$$\mathcal{S}=\{S \text{ closed in } W_1 \mid \mu(U)=0 \text{ for all compact open sets } U\subset W_1-S\}$$

is closed under finite intersections. Now let

$$S^*=\bigcap_{S\in\mathcal{S}} S.$$

I claim $S^*\in\mathcal{S}$ too. Let $U\subset W_1-S^*$ be a compact open set. Since

$$W_1-S^*=\bigcup_{S\in\mathcal{S}}(W_1-S),$$

it follows that U is covered by the open sets $U\cap(W_1-S)$, where $S\in\mathcal{S}$. Since U is compact, it can be covered by a finite number of such open sets:

$$U\subset(W_1-S_1)\cup\cdots\cup(W_1-S_n)$$

where $S_1,\ldots,S_n\in\mathcal{S}$. Now let $T\in\mathcal{S}$ be a closed set contained in all these S_i. Then $U\subset W_1-T$. But $T\in\mathcal{S}$ means that this implies $\mu(U)=0$. So $\mu(U)=0$ whenever $U\subset W_1-S^*$, i.e., $S^*\in\mathcal{S}$ too. *Q.E.D.*

Proposition. *Let μ be a non-zero even Gaussian measure on W_1* (i.e., *μ has the property* (*A*) *of Lemma* 1, § 8). *Then the support S of μ is a sub-vector space of W_1.*

Proof. Notice that if μ_1, μ_2 are 2 measures on W_1, and $\mu_1 \times \mu_2$ is the induced measure on $W_1 \times W_1$, then

$$\text{Support}(\mu_1 \times \mu_2) = \text{Support}(\mu_1) \times \text{Support}(\mu_2).$$

Let $\xi\colon W_1 \times W_1 \to W_1 \times W_1$ be the map $\xi((x, y)) = (x+y, x-y)$. By definition, a Gaussian measure μ is associated to a second measure ν such that

$$\xi_*(\mu \times \mu) = \nu \times \nu .$$

Therefore, if $S' = \text{Support}(\nu)$, it follows that $\xi(S \times S) = S' \times S'$. In particular

$$\begin{aligned} \alpha \in S &\Leftrightarrow (\alpha, \alpha) \in S \times S \\ &\Leftrightarrow (2\alpha, 0) = \xi((\alpha, \alpha)) \in S' \times S'. \end{aligned}$$

Since S is non-empty, $0 \in S'$, and $\alpha \in S \Leftrightarrow 2\alpha \in S'$, i.e., $S' = 2S$. Therefore $0 \in S$ too, and we find:

$$\begin{aligned} \alpha \in S &\Leftrightarrow (\alpha, 0) \in S \times S \\ &\Leftrightarrow (\alpha, \alpha) = \xi((\alpha, 0)) \in S' \times S' \\ &\Leftrightarrow \alpha \in S'. \end{aligned}$$

Therefore $S = S'$ also. Finally,

$$\begin{aligned} \alpha, \beta \in S &\Rightarrow (\alpha, \beta) \in S \times S \\ &\Rightarrow (\alpha+\beta, \alpha-\beta) \in S' \times S' \\ &\Rightarrow \alpha+\beta, \alpha-\beta \in S' = S. \end{aligned}$$

Thus S is a closed subgroup of W_1, such that $S = 2S$. Therefore S is a subvectorspace over $\mathbf{Q}_2$. *Q.E.D.*

Corollary. *For all $\gamma_2 \in W_2$, all theta functions Θ on V,*

$$\textit{Support}(\Theta) \subset \{\alpha \mid e(\alpha, \gamma_2) = 1\} \Rightarrow \Theta(\alpha + \lambda\gamma_2) = e\left(\alpha, \frac{\lambda\gamma_2}{2}\right)\Theta(\alpha),$$

$$\text{all } \lambda \in \mathbf{Q}_2 .$$

Proof. The assumption on the support of Θ implies (cf. (a) above) that $\mu(\alpha_1 + 2^n \Lambda_1) = 0$ if $e(\alpha_1, \gamma_2) \neq 1$. Therefore,

$$\text{Support}(\mu) \subset \{\alpha_1 \in W_1 \mid e(\alpha_1, \gamma_2) = 1\}.$$

Since this support is a vector space,

$$\text{Support}(\mu) \subset W_1 \cap (\mathbf{Q}_2 \cdot \gamma_2)^{\perp}.$$

Let H denote the hyperplane $W_1 \cap (Q_2 \cdot \gamma_2)^{\perp}$. Then

$$\Theta(\alpha_1+\alpha_2)=e\left(\alpha_1,\frac{\alpha_2}{2}\right)\int_{(\alpha_1+\Lambda_1)\cap H} e(\alpha_2,\beta)\cdot d\mu(\beta).$$

Thus

$$\Theta(\alpha_1+\alpha_2+\lambda\gamma_2)=e\left(\alpha_1,\frac{\alpha_2+\lambda\gamma_2}{2}\right)\int_{(\alpha_1+\Lambda_1)\cap H} e(\alpha_2+\lambda\gamma_2,\beta)\cdot d\mu(\beta)$$

and since $e(\lambda\gamma_2,\beta)=1$ when $\beta\in H$, this comes out

$$=e\left(\alpha_1,\frac{\lambda\gamma_2}{2}\right)\cdot\left\{e\left(\alpha_1,\frac{\alpha_2}{2}\right)\int_{(\alpha_1+\Lambda_1)\cap H} e(\alpha_2,\beta)\cdot d\mu(\beta)\right\}$$

$$=e\left(\alpha_1,\frac{\lambda\gamma_2}{2}\right)\cdot\Theta(\alpha_1+\alpha_2). \qquad Q.E.D.$$

In fact, I claim that the same Corollary holds *for all* $\gamma\in V$, not just *for* $\gamma\in W_2$. This can be seen by noting that for any $\gamma\in V$, there is a symplectic automorphism T: $V\to V$ such that $T(\Lambda)=\Lambda$, i.e., $T\in \mathrm{Sp}(V,\Lambda)$, such that $T^{-1}(\gamma)\in W_2$. Going back to the action of the symplectic group introduced in § 9, we see that:

$$\left\{\begin{array}{l}\text{If } \Theta \text{ is a theta-function, then so is } \Theta', \text{ where}\\ \qquad\qquad \Theta'(\alpha)=e(\eta/2,\alpha)\,\Theta(T\alpha-T\eta)\\ \text{where } \eta\in\frac{1}{2}\Lambda \text{ satisfies}\\ \qquad\qquad e_*(\alpha/2)\cdot e_*(T\alpha/2)=e(\eta,\alpha), \qquad \text{all } \alpha\in\Lambda.\end{array}\right.$$

Now assume $\mathrm{Supp}(\Theta)\subset\{\alpha\mid e(\alpha,\gamma)=1\}$. Then

$$\begin{aligned}\mathrm{Supp}(\Theta')&=\eta+T^{-1}(\mathrm{Supp}(\Theta))\\ &\subset\eta+\{\alpha\mid e(\alpha,T^{-1}\gamma)=1\}\\ &\subset\{\alpha\mid e(\alpha,2^n T^{-1}\gamma)=1\} \qquad (\text{if } n\gg 0).\end{aligned}$$

Therefore, by the Corollary

$$\Theta'(\alpha+\lambda T^{-1}\gamma)=e\left(\alpha,\frac{\lambda T^{-1}\gamma}{2}\right)\Theta'(\alpha), \qquad \text{all } \lambda\in Q_2,$$

from which

$$\Theta(\alpha+\lambda\gamma)=e\left(\alpha,\frac{\lambda\gamma}{2}\right)\cdot\Theta(\alpha)$$

follows immediately. We are now ready for the Proof itself:

Proof of Theorem. We know that the support of Θ meets $\frac{1}{2}\Lambda$ (cf. § 10): choose $\eta_0\in\mathrm{Supp}(\Theta)\cap\frac{1}{2}\Lambda$. Then:

$$\mathrm{Supp}(\Theta)+\eta_0\subseteq W+\Lambda$$

(§ 10, assertion (4.) at the beginning). Therefore, if $\gamma \in W^\perp \cap (2\Lambda)$ it follows that $e(\alpha, \gamma)=1$, all $\alpha \in \mathrm{Supp}(\Theta)$. But then by Corollary above – as generalized –

$$\Theta(\alpha+\lambda\cdot\gamma)=e\left(\alpha, \frac{\lambda\gamma}{2}\right)\cdot\Theta(\alpha), \qquad \text{all } \lambda\in \mathbf{Q}_2.$$

This shows that

$$(*) \qquad \Theta(\alpha+\gamma)=e\left(\alpha, \frac{\gamma}{2}\right)\cdot\Theta(\alpha), \qquad \text{all } \gamma\in W^\perp.$$

In particular, $\Theta(\eta_0+\gamma)\neq 0$, all $\gamma\in W^\perp$, hence $W^\perp+\eta_0 \subseteq W+\Lambda+\eta_0$. Therefore $W^\perp\subseteq W$, i.e., W is a cusp.

Now suppose we take an arbitrary point α in the Support of Θ. We know that α can be written as:

$$\alpha=\eta_0+\eta_1+\alpha_0, \qquad \eta_1\in\Lambda,\ \alpha_0\in W.$$

But then:

$$\begin{aligned}\Theta(\alpha)&=e_*\left(\frac{\eta_1}{2}\right)\cdot e\left(\frac{\eta_1}{2}, \eta_0+\alpha_0\right)\cdot\Theta(\eta_0+\alpha_0)\\ &=e_*\left(\frac{\eta_1}{2}\right)\cdot e\left(\frac{\eta_1}{2}, \eta_0\right)\cdot e\left(\frac{\eta_0+\eta_1}{2}, \alpha\right)\cdot\left[e\left(\alpha, \frac{\eta_0}{2}\right)\cdot\Theta(\eta_0+\alpha)\right].\end{aligned}$$

Define a function $\tilde\Theta$ on W by

$$\tilde\Theta(\alpha)=e\left(\alpha, \frac{\eta_0}{2}\right)\cdot\Theta(\alpha+\eta_0).$$

If $\gamma\in W^\perp$, we compute (using (*)):

$$\begin{aligned}\tilde\Theta(\alpha+\gamma)&=e\left(\alpha+\gamma, \frac{\eta_0}{2}\right)\cdot\Theta(\alpha+\eta_0+\gamma)\\ &=e\left(\gamma, \frac{\eta_0}{2}\right)\cdot e\left(\alpha+\eta_0, \frac{\gamma}{2}\right)\cdot e\left(\alpha, \frac{\eta_0}{2}\right)\cdot\Theta(\alpha+\eta_0)\\ &=\tilde\Theta(\alpha).\end{aligned}$$

This shows that $\tilde\Theta$ is, in reality, a function on $\tilde V=W/W^\perp$, and that Θ is exactly the function $T_{W,\eta_0}\tilde\Theta$ obtained from $\tilde\Theta$ via Definition 3.

To check that η_0 is an origin for W, look at (*) when $\gamma^\perp\in W\cap\Lambda$. Then:

$$e\left(\alpha, \frac{\gamma}{2}\right)\cdot\Theta(\alpha)=\Theta(\alpha+\gamma)=e_*\left(\frac{\gamma}{2}\right)\cdot e\left(\frac{\gamma}{2}, \alpha\right)\cdot\Theta(\alpha)$$

hence

$$e_*\left(\frac{\gamma}{2}\right)=e(\alpha, \gamma) \qquad \text{if } \Theta(\alpha)\neq 0.$$

So

$$e_*\left(\frac{\gamma}{2}\right)=e(\eta_0,\gamma), \qquad \text{all } \gamma\in W^\perp\cap\Lambda.$$

Moreover, using

$$\Theta(\eta_0)=\Theta(-\eta_0+2\eta_0)=e_*(\eta_0)\,\Theta(-\eta_0)$$

and

$$\Theta(-\eta_0)=\Theta(\eta_0)\neq 0,$$

we conclude that $e_*(\eta_0)=1$ too.

The fact that $\tilde{\Theta}$ is again a theta-function is simply a matter of applying the calculations of Prop. 1 in reverse and is quite straightforward. We omit this. The final point is that $\tilde{\Theta}$ is non-degenerate. But since $S_\infty\supseteq W$, we know that for all $\alpha\in W$, $\alpha=2^k\beta+\eta_1$, where $\Theta(\beta)\neq 0$, $\eta_1\in\Lambda$. Then $\beta=\eta_0+\eta_2+\beta_0$, $\eta_2\in\Lambda$, $\beta_0\in W$, and $\tilde{\Theta}(\beta_0)\neq 0$. Since

$$\alpha-2^k\beta_0=\eta_1+2^k\eta_0+2^k\eta_2\in W\cap\Lambda,$$

this shows that for all $\alpha\in W$, $\alpha=2^k\beta_0+\eta_3$, where $\tilde{\Theta}(\beta_0)\neq 0$, $\eta_3\in W\cap\Lambda$. This means exactly that the S_∞ for $\tilde{\Theta}$ is all of $\tilde{V}$, i.e., $\tilde{\Theta}$ is non-degenerate. *Q.E.D.*

The main Theorem can now be reformulated to give a Satake-like decomposition of $\overline{M}_\infty$. More precisely, for each integer $g\geqq 0$, let

$\overline{M}_\infty(g)$ = the Proj defined in § 9, Def. 3 with indices $\alpha\in Q_2^{2g}$.

$M_\infty(g)$ = the open set in $\overline{M}_\infty(g)$ whose geometric points are the non-degenerate theta functions.

If $h<g$, we define a vast number of closed immersions

$$i_W\colon \overline{M}_\infty(h)\to\overline{M}_\infty(g)$$

as follows: let $W\subseteq Q_2^{2g}$ be a cusp such that $2h=\dim(W/W^\perp)$. For each such W, choose an origin $\eta_0\in\frac{1}{2}\mathbf{Z}_2^{2g}$, and a symplectic isomorphism:

$$\phi\colon Q_2^{2h}\xrightarrow{\approx} W/W^\perp$$

such that

$$\phi(\mathbf{Z}_2^{2h})=W\cap\Lambda/W^\perp\cap\Lambda,$$
$$\chi(\tfrac{1}{2}{}^t a_1\cdot a_2)=\tilde{e}_*(\tfrac{1}{2}\phi(a)), \qquad \text{all } a\in\mathbf{Z}_2^{2h}.$$

Then i_W is defined by the homomorphism of the homogeneous coordinate ring:

$$i_W^*(X_\alpha^{(g)})=\begin{cases}0 & \text{if } \alpha\notin\eta_0+W+\mathbf{Z}_2^{2g}\\ e_*\left(\frac{\eta_1}{2}\right)e\left(\frac{\eta_1}{2},\eta_0\right)e\left(\frac{\eta_0+\eta_1}{2},\alpha\right)\cdot X^{(h)}_{\phi^{-1}(\alpha_0)} & \\ \qquad\qquad \text{if } \alpha=\eta_0+\alpha_0+\eta_1,\ \alpha_0\in W,\ \eta_1\in\mathbf{Z}_2^{2g}. & \end{cases}$$

(Here $X_\alpha^{(g)}, X_\alpha^{(h)}$ are the coordinates used to define $\overline{M}_\infty(g)$, $\overline{M}_\infty(h)$ respectively). Then we get the restatement:

Main Theorem.

$$\overline{M}_\infty(g) = \left\{ \begin{array}{l} \textit{disjoint union of the locally} \\ \textit{closed subschemes } i_W(M_\infty(h)) \end{array} \right\},$$

the union being taken over all cusps $W \subseteq Q_2^{2g}$.

§ 12. Analytic Theta Functions

In this section, we work over the field C of complex numbers. We have 2 purposes: (a) to sketch an approach to the classical theory of Θ-functions, analogous to our theory of algebraic Θ-functions, and (b) to use this to compute our algebraic Θ-functions via the classical ones, when $k = C$.

We will make use of the following lemma:

Lemma 1. *Let X be a compact Kähler manifold. Then the operator*

$$\frac{1}{2\pi i}\partial\bar{\partial}$$

defines a surjection:

$$\left\{ \begin{array}{l} C^\infty \textit{ real} \\ \textit{functions on } X \end{array} \right\} \to \left\{ \begin{array}{l} \textit{real closed } C^\infty \ (1,1)\textit{-forms } \Omega \textit{ on } X, \\ \textit{with } 0 \textit{ cohomology class} \end{array} \right\}$$

with kernel consisting only of constants.

Corollary. *Let* $\mathbf{L}$ *be an analytic line bundle on X. Let* $c_1(\mathbf{L}) \in H^2(X, C)$ *be its first chern class. Then for all real closed* C^∞ *(1, 1)-forms* Ω *whose cohomology class equals* $c_1(\mathbf{L})$*, there is one and (up to a constant) only one Hermitian structure* $\| \ \|$ *on* $\mathbf{L}$ *whose associated curvature form is* Ω.

The lemma is standard and we omit the proof. The Corollary can be proven by choosing one Hermitian structure $\| \ \|_0$ on $\mathbf{L}$: let Ω_0 be its curvature form. Then any other Hermitian structure on $\mathbf{L}$ is given by $\rho \cdot \| \ \|_0$, where ρ is a positive real C^∞ function on X: and its curvature form Ω is

$$\Omega = \frac{1}{2\pi i}\partial\bar{\partial}\log\rho + \Omega_0.$$

Now use the Lemma and everything comes out. *Q.E.D.*

In particular, when X is an abelian variety, an analytic line bundle $\mathbf{L}$ on X has one and (up to a constant) only one Hermitian structure $\| \ \|$ whose curvature form Ω is a translation-invariant (1, 1)-form. In what follows, we will always put this Hermitian structure on line bundles on abelian varieties. In this case, Ω is determined by its value at the origin.

Now let $\hat{X}$ be the universal covering space of X. $\hat{X}$ is a complex vector space, and if

$$p: \hat{X} \longrightarrow X$$

is the canonical homomorphism, dp induces a canonical identification between $\hat{X}$ and the tangent space of X at the origin (or at any other point). Therefore, any translation-invariant real 2-form Ω on X defines and is defined by a real-linear skew-symmetric form:

$$E: \hat{X} \times \hat{X} \longrightarrow \boldsymbol{R}.$$

E is a (1, 1)-form if and only if $E(ix, iy) = E(x, y)$, all $x, y \in X$. Moreover, let $\Lambda = \text{kernel}(p)$. Λ is a lattice in X, canonically isomorphic to $H_1(X, \boldsymbol{Z})$. Since the first chern class of a line bundle is integral, if E represents $c_1(\boldsymbol{L})$, then E must take integral values on $\Lambda \times \Lambda$:

$$E(\Lambda \times \Lambda) \subseteq \boldsymbol{Z}.$$

If we lift $\boldsymbol{L}$ to $\hat{X}$, we have a situation in which the following lemma applies:

Lemma 2. *Let Y be a complex vector space, and let $\boldsymbol{L}_1, \boldsymbol{L}_2$ be 2 analytic-Hermitian line bundles on Y. Then a holomorphic-unitary isomorphism $\phi: \boldsymbol{L}_1 \xrightarrow{\sim} \boldsymbol{L}_2$ exists if and only if the curvature forms of $\boldsymbol{L}_1, \boldsymbol{L}_2$ are equal; if so, ϕ is unique up to a scalar of absolute value 1.*

Proof. Standard methods.

In particular, let $Y = \hat{X}$, and let $\boldsymbol{M} = p^*(\boldsymbol{L})$ be induced from an abelian variety. Give $\boldsymbol{L}$ and hence $\boldsymbol{M}$ the Hermitian structure with constant curvature form E. The above lemma has 2 applications:

(I) Construction of a nilpotent group $\mathscr{G}$: If $x \in X$, and T_x denotes translation by x, then the lemma shows that $\boldsymbol{M}$ and $T_x^* \boldsymbol{M}$ are holomorphic-unitary isomorphic. If

$$\mathscr{G}(\boldsymbol{M}) = \{(x, \Phi) \mid \Phi \text{ a holo.-unit. isom. of } \boldsymbol{M} \text{ with } T_x^* \boldsymbol{M}\},$$

then $\mathscr{G}(\boldsymbol{M})$ is, as before, a group lying in an exact sequence:

$$1 \longrightarrow \boldsymbol{C}_1^* \longrightarrow \mathscr{G}(\boldsymbol{M}) \longrightarrow X \longrightarrow 0$$

($\boldsymbol{C}_1^*$ = complex numbers of absolute value 1).

(II) Construction of canonical "trivialization" of $\boldsymbol{M}$: Let $\boldsymbol{1}$ denote the trivial analytic line bundle over X with canonical section 1. To put a Hermitian structure on $\boldsymbol{1}$, we may set $\|1\| =$ any positive real C^∞-function. For example, let

$$\|1\|(x) = e^{-\pi/2H(x,x)}$$

where H is a Hermitian form on X. The corresponding curvature form $E: \hat{X} \times \hat{X} \to \boldsymbol{R}$ is easily checked to equal Im (H). But

$$H \mapsto E = \operatorname{Im}(H)$$

sets up an isomorphism:

$$\begin{Bmatrix} \text{hermitian} \\ \text{forms on } X \end{Bmatrix} \xrightarrow{\sim} \begin{Bmatrix} \text{real skew-symmetric forms } E \text{ on } X \\ \text{such that } E(i\,x, i\,y) = E(x, y) \end{Bmatrix},$$

so for each L on X with translation-invariant curvature form, we have a unique Hermitian structure on $\mathbf{1}$ of the above type so that $\mathbf{1} \cong \boldsymbol{L}$. In particular, we get a canonical

$$\mathbf{1} \cong \boldsymbol{M}.$$

We can now develop a theory along similar lines to our algebraic theory. For example, if H is positive definite, then let:

$\mathscr{H}$ = Hilbert space of L^2-holomorphic sections of $\boldsymbol{M}$ over $\hat{X}$.

Then $\mathscr{G}(\boldsymbol{M})$ has a natural unitary representation on $\mathscr{H}$, it is irreducible, and it turns out to be the only irreducible unitary representation of $\mathscr{G}(\boldsymbol{M})$ in which $\boldsymbol{C}_1^* \subset \mathscr{G}(\boldsymbol{M})$ acts by its natural character. This is the situation described by CARTIER [*2*], and studied by CARTIER and many others, e.g., MACKEY, FOCK, WEIL etc. Exactly as in § 1, $\mathscr{G}(\boldsymbol{M})$ governs the "descent" of the Hermitian bundle $\boldsymbol{M}$ to the abelian variety X, (or to other ones $X' = [\hat{X}/\text{another lattice}]$), and the "descent" of holomorphic sections of $\boldsymbol{M}$ to holomorphic sections of its descended form. Thus we get:

Proposition 1. *There is a* $1-1$ *correspondence between*

1. *Hermitian-analytic line bundles* $\boldsymbol{L}'$ *on* X *such that* $p^* \boldsymbol{L}' \cong \boldsymbol{M}$,

2. *subgroups* $K \subset \mathscr{G}(\boldsymbol{M})$, *such that* $K \cap \boldsymbol{C}_1^* = \{1\}$ *whose image in* $\hat{X}$ *is* $\Lambda = \ker(p: \hat{X} \to X)$.

Moreover, the holomorphic sections of $\boldsymbol{M}$ *of the form* $p^*(s')$, $s' \in \Gamma(X, \boldsymbol{L}')$, *are exactly those sections* s *which are invariant under* K, i.e.,

$$s = T_{-x}^*(\phi(s)), \qquad all\ (x, \phi) \in K.$$

Proof. Straightforward.

Finally, via the canonical trivialization of $\boldsymbol{M}$, holomorphic sections of $\boldsymbol{M}$ correspond to holomorphic functions on $\hat{X}$: thus each section $s \in \Gamma(X, \boldsymbol{L})$ defines a holomorphic function on $\hat{X}$. These are the classical theta-functions.

As far as moduli are concerned, the simplest and most basic result is the following: we set out to classify triples consisting of —

1. a complex vector space Y, of dimension 2;

2. an analytic, Hermitian line bundle $\boldsymbol{M}$ on Y, with curvature form $E = \operatorname{Im} H$, H positive definite.

3. Parametrized lattices in Y, i.e., monomorphisms

$$\alpha\colon \mathbf{Z}^{2g} \to Y$$

such that

$$E(\alpha x, \alpha y) = {}^t x_1 \cdot y_2 - {}^t x_2 \cdot y_1$$

if

$$x = (x_1, x_2), \quad y = (y_1, y_2).$$

Such triples arise if we start with a principally polarized abelian variety $(X, \boldsymbol{L})$, together with a symplectic isomorphism:

$$\beta\colon \mathbf{Z}^{2g} \xrightarrow{\sim} H_1(X, \mathbf{Z}).$$

Namely, let $Y = \hat{X}$, $\boldsymbol{M} = p^* \boldsymbol{L}$ with canonical Hermitian structure, and let β define α via the natural maps $H_1(X, \mathbf{Z}) \cong \operatorname{Ker}(p\colon \hat{X} \to X) \subset \hat{X}$. Conversely, the triple $(Y, \boldsymbol{M}, \alpha)$ determines X and β, and $\boldsymbol{L}$ up to replacing $\boldsymbol{L}$ by $T_x^* \boldsymbol{L}$, some $x \in X$.

Let $\mathfrak{H} =$ SIEGEL's $g \times g$ upper half-plane. Then the moduli result is:

Proposition 2. *There is a natural bijection between the set of isomorphism classes of triples* $(Y, \boldsymbol{M}, \alpha)$ *and* $\mathfrak{H}$. *In this bijection,* $\tau \in \mathfrak{H}$ *corresponds to*

$$Y = \mathbf{C}^g,$$

$$\boldsymbol{M} = \mathbf{1} \quad \textit{with hermitian structure} \quad \|1\|(x) = e^{-\frac{\pi}{2}{}^t x \cdot B \cdot \bar{x}},$$

$$\alpha((x_1, x_2)) = x_1 + \tau \cdot x_2$$

where $B = (\operatorname{Im} \tau)^{-1}$.

The final topic I want to discuss is the relation between the classical and algebraic theories. Let's start with:

X = abelian variety;

L = symmetric, ample, degree 1 sheaf on X. [Assume for simplicity that L is so chosen among its translates $T_x^* L$, $x \in X_2$, that its unique section is *even*; equivalently, that the Arf invariant of Q, where $e_*^L(x) = (-1)^{Q(x)}$, is 0.]

Let

$\boldsymbol{L}$ = line bundle on X whose holomorphic sections are L;

$\hat{X}$ = universal covering space of X;

$V_2(X)$ = 2-Tate group of X.

Also, let Λ_2 = inverse image in $\hat{X}$ of $\operatorname{tor}_2(X)$, i.e.,

$$\bigcup_n 2^{-n}\cdot\Lambda, \quad \text{if} \quad \Lambda=\operatorname{Ker}(p\colon \ddot{X}\to X).$$

Then we have canonical maps:

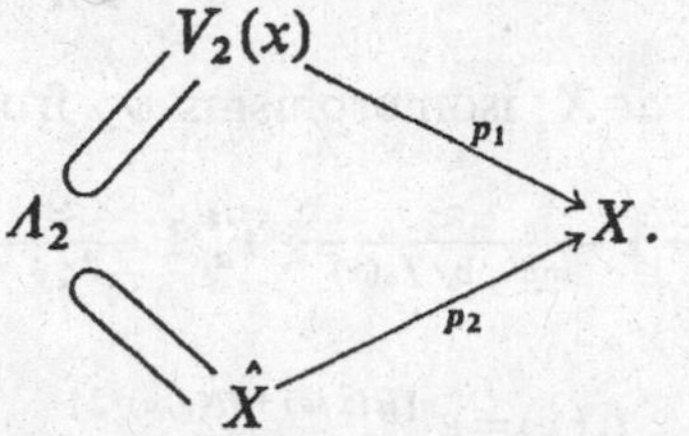

Note that Λ_2 is dense in both $V_2(X)$ and X. We have "trivialized" $\boldsymbol{L}$ when it is pulled up to $V_2(X)$ or to X, in § 8 and just above. Thus we have 2 distinct trivializations of $\boldsymbol{L}$ on Λ_2. The main result is that these differ by an elementary factor:

Theorem 3. *Let* $\mathbf{1}$ *denote the trivial complex line bundle on* Λ_2. *Then the following diagram commutes:*

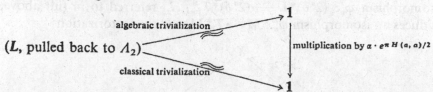

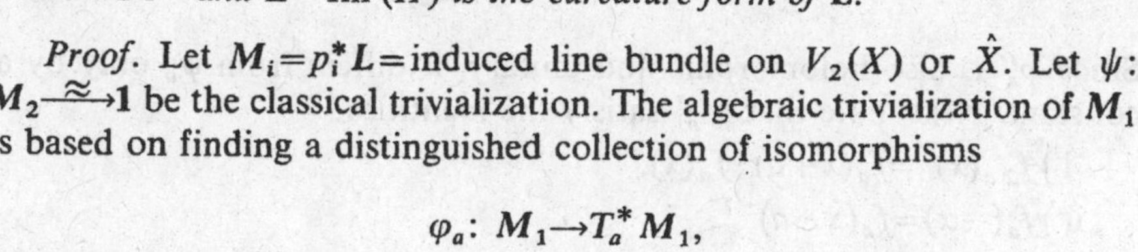

where $\alpha\in\mathbf{C}^*$ *and* $E=\operatorname{Im}(H)$ *is the curvature form of* $\boldsymbol{L}$.

Proof. Let $M_i=p_i^*L$ = induced line bundle on $V_2(X)$ or $\hat{X}$. Let ψ: $M_2\xrightarrow{\approx}\mathbf{1}$ be the classical trivialization. The algebraic trivialization of M_1 is based on finding a distinguished collection of isomorphisms

$$\varphi_a\colon M_1\to T_a^*M_1,$$

all $a\in V_2(X)$. In fact, let $\imath$ = inverse map in all our groups, and let ρ: $M_i\xrightarrow{\sim}\imath^*M_i$ be the isomorphism induced by the symmetry of $\boldsymbol{L}$. Then, for all elements $2a\in V_2(X)$, φ_{2a} is characterized by the existence of φ_a satisfying:

i) $\varphi_{2a}=T_a^*\varphi_a\circ\varphi_a$,

ii) $\imath^*\varphi_a\circ\rho=T_{-a}^*[\rho\circ\varphi_a^{-1}]$,

iii) φ_a is induced by an algebraic isomorphism

$$\varphi_a'\colon (2^n\delta)^*L\xrightarrow{\sim}(2^n\delta)^*(T_{p_1(a)}^*L)$$

for some n, i.e., via the factorization:

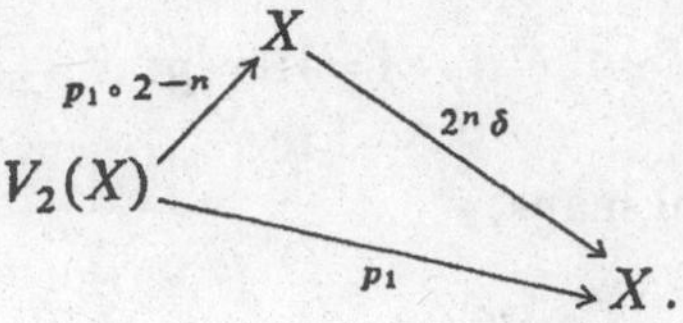

But introduce, for all $a \in X$, isomorphisms ψ_a from M_2 to $T_a^* M_2$ via:

$$M_2 \xrightarrow[\psi]{\approx} \mathbf{1} \xrightarrow[\text{mult. by } f_a(x)]{\approx} T_a^* \mathbf{1} \xleftarrow[T_a^*\psi]{\approx} T_a^* M$$

where

$$f_a(x) = e^{\pi[H(x,a)+H(a,a)/2]}.$$

Also introduce

$$\rho' \colon M_2 \xrightarrow[\psi]{\approx} \mathbf{1} \xrightarrow[\text{canonical identification}]{} \iota^* \mathbf{1} \xleftarrow[\iota^*\psi]{\approx} \iota^* M.$$

One checks easily that ψ_a and ρ' are holomorphic and unitary isomorphisms. Therefore ρ and ρ' can differ only by a constant: and since both are the identity at $0 \in X$, $\rho = \rho'$. Moreover, if $a \in 2^{-n} \Lambda$, then the algebraic isomorphism φ_a': $(2^n\delta)^* L \xrightarrow{\sim} (2^n\delta)^* T^*_{p_2(a)} L$, referred to in (iii) above, induces an isomorphism φ_a'': $M_2 \to T_a^* M_2$ via the factorization

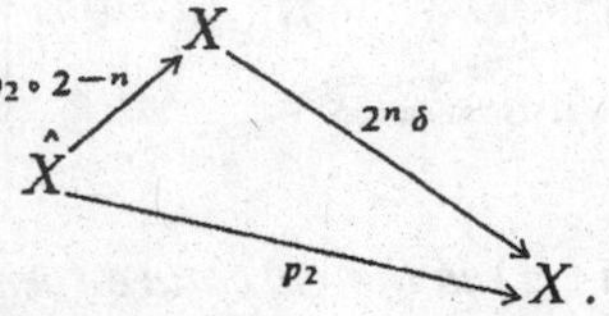

Since φ_a'' is also holomorphic and unitary, it differs from ψ_a only by a constant. Next, note that $\{f_a\}$ satisfy the identities:

i') $f_{2a}(x) = f_a(x+a) \cdot f_a(x)$,

ii') $f_a(-x) = f_a(x-a)^{-1}$.

These translate readily into the identities on the $\{\psi_a\}$:

i'') $\psi_{2a} = T_a^* \psi_a \circ \psi_a$.

ii'') $\iota^* \psi_a \circ \rho = T_{-a}^* [\rho \circ \psi_a^{-1}]$.

Finally, i'', ii'', plus the fact that φ_a' induces ψ_a, shows that ψ_a and φ_a induce the *same isomorphism* of L on Λ_2, with $T_a^*(L$ on $\Lambda_2)$, all $a \in \Lambda_2$.

Finally, to compare the 2 trivializations, start with the unit section 1 of $\mathbf{1}$ on Λ_2. This goes over, via the algebraic trivialization, to *a* section s of L on Λ_2 such that, for all $a \in \Lambda_2$,

$$s(a) = \phi_a(0)[s(0)]$$

(i.e., $\phi_a(0)$ is the induced isomorphism from the fibre L_0 or $(M_1)_0$ to the fibre $L_{p_1(a)}$ or .$(M_1)_a$) But under the classical trivialization ψ, $\psi_a(0)$ corresponds to the isomorphism of fibres:

$$\begin{array}{ccc} \mathbf{1}_0 & \xrightarrow{\text{mult. by } e^{\pi/2 H(a,a)}} & \mathbf{1}_0 \\ \| & & \| \\ \mathbf{C} & & \mathbf{C}. \end{array}$$

Therefore, the section s goes over, under the classical trivialization, to a section of $\mathbf{1}$ which, if it has value α at 0, has value

$$\alpha \cdot e^{\pi/2\, H(a,a)}$$

at a. All in all, the section 1 of $\mathbf{1}$ has gone into the section

$$g(a) = \alpha \cdot e^{\pi/2\, H(a,a)}$$

of $\mathbf{1}$. *Q.E.D.*

Corollary. *If the unique section s of* $\mathbf{L}$ *(up to scalars) defines*

a) *the holomorphic function* Θ_h *on* $\hat{X}$ *via the classical trivialization,*

b) *the* 2-adic *theta-function* Θ_a *on* $V_2(X)$ *via the algebraic trivialization, then*

$$\Theta_h(x) = \alpha \cdot e^{\frac{\pi}{2} H(x,x)} \cdot \Theta_a(x)$$

all $x \in \Lambda_2$.

To calculate Θ_h and hence Θ_a by analytic means, we must know the "descent data"

$$K \subset \mathcal{G}(M_2)$$

that defines L on X. Let $e_*: \frac{1}{2}\Lambda/\Lambda \to \{\pm 1\}$ be the quadratic character defined by L. Then, as we saw in § 8, the descent data for the pull-back M_1 of L is the group:

$$\{(x, \phi) \mid x \in \Lambda \cdot \mathbf{Z}_2,\ \phi = e_*(\tfrac{1}{2}x) \cdot \phi_x\}.$$

In view of the proof of the theorem, this implies that

$$K = \{(x, \psi) \mid x \in \Lambda,\ \psi = e_*(\tfrac{1}{2}x) \cdot \psi_x\}.$$

(Notation as in proof of Theorem). Now a K-invariant section s of M_2 is one which satisfies $T_\alpha^*(s) = \phi(s)$, all $(a, \phi) \in K$. Going back to the definition of ψ_a, one sees that if $f = \psi(s)$ is the function on $\hat{X}$ corresponding to s, then f is K-invariant if and only if

$$(*) \qquad f(x+a) = e_*(\tfrac{1}{2}a)\, f_a(x) \cdot f(x)$$

all $x \in \hat{X}$, $a \in \Lambda$. From this it follows that Θ_h must be the unique holomorphic function satisfying (*).

To go further and write down this Θ_h as an infinite series, it is convenient to introduce coordinates. Let

$$i\colon \mathbf{Z}^{2g} \xrightarrow{\approx} \Lambda \qquad \text{be a symplectic isomorphism}.$$

Coordinatize $\hat{X}$ via

$$\hat{X} \cong \mathbf{C}^g$$

so that $i((n_1, 0)) = n_1$, and let τ be the $g \times g$ matrix defined by

$$i((0, n_2)) = \tau \cdot n_2.$$

Because of our assumption on e_*^L, hence on e_*, if we choose coordinates correctly, we can assume that

$$e_*\left[\tfrac{1}{2}\, i(n_1, n_2)\right] = (-1)^{{}^t n_1 \cdot n_2}.$$

As we saw in Prop. 2, if we now express:

$$H(z, z) = {}^t z \cdot B \cdot \bar{z}$$

then $B = (\operatorname{Im} \tau)^{-1}$. Finally, set

$$\Theta_h(z) = e^{\frac{\pi}{2} {}^t z \cdot B \cdot z} \cdot \sum_{n \in \mathbf{Z}^g} e^{2\pi i [\frac{1}{2} {}^t n \cdot \tau \cdot n + {}^t n \cdot z]}.$$

It is easy to check that this is a holomorphic function satisfying (*). Therefore, this is the sought-for theta-function. Combining this with the Corollary, we find

$$\Theta_a(z) = e^{\frac{\pi}{2} {}^t z \cdot B \cdot (z - \bar{z})} \cdot \sum_{n \in \mathbf{Z}^g} e^{2\pi i [\frac{1}{2} {}^t n \cdot \tau \cdot n + {}^t n \cdot z]} \qquad \text{all } z \in \bigcup_k 2^{-k} \Lambda.$$

If

$$z = i((\alpha_1, \alpha_2)), \qquad \alpha_i \in \bigcup_k 2^{-k} \cdot (\mathbf{Z}^g),$$

then after rearranging, one finds

$$\Theta_a(\alpha_1, \alpha_2) = e^{-\pi i {}^t \alpha_1 \cdot \alpha_2} \cdot \sum_{n \in \alpha_2 + \mathbf{Z}^g} e^{2\pi i [\frac{1}{2} {}^t n \cdot \tau \cdot n + {}^t n \cdot \alpha_1]}.$$

The function so defined clearly extends to a locally constant function defined for all $\alpha_1, \alpha_2 \in \mathbf{Q}^{2g}$: it is the sought-for algebraic theta function defined in § 8. Comparing this with the formula in Lemma 1, § 8, expressing Θ_a in terms of the finitely additive measure μ on $\mathbf{Q}_2^g$, we also get an analytic description for μ:

$$\begin{cases} \mu \text{ is countably additive}, \\ \mu = \sum_{x \in D} e^{\pi i {}^t x \cdot \tau \cdot x} \cdot \delta_x, \\ \delta_x = \text{delta measure at } x, \\ D = \bigcup_k 2^{-k} \mathbf{Z}^g. \end{cases}$$

References

[1] BAILY, jr., W.: On the theory of Θ-functions, the moduli of abelian varieties, and the moduli of curves. Annals of Math. **75**, 342–381 (1962).

[2] CARTIER, P.: Quantum mechanical commutation relations and theta functions. Proc. of Symposia in Pure Math., vol. 9. Am. Math. Soc. 1966.

[3] GROTHENDIECK, A.: Séminaire de géométrie algébrique. Inst. des Hautes Études Sci. 1960/61 (Mimeographed).

[4] — Séminaire: Schémas en groupes. Inst. des Hautes Études Sci. 1963/64 (Mimeographed).

[5] —, et J. DIEUDONNE: Éléments de la géométrie algébrique. Publ. de l'Inst. des Hautes Études Sci., No. 4, 8, 11, etc.

[6] IGUSA, J.-I.: On the graded ring of theta-constants. Am. J. Math. **86**, 219–246 (1964); **88**, 221–236 (1966).

[7] LANG, S.: Abelian Varieties. New York: Interscience 1958.

[8] MACKEY, G.: On a theorem of Stone and von Neumann. Duke Math. J. **16**, 313–330 (1949).

[9] MUMFORD, D.: Geometric invariant theory. Berlin-Heidelberg-New York: Springer 1965.

[10] — Curves on an algebraic surface. Annals of Math. Studies, No. 59 (1966).

[11] SIEGEL, C.: Moduln Abelscher Funktionen. Nachrichten der Akad., Göttingen 1964.

[12] WEIL, A.: Sur certaines groupes d'opérateurs unitaires. Acta Math. **111**, 145–211 (1964).

[13] IWAHORI, N., and H. MATSUMOTO: On some Bruhat decomposition and the structure of the Hecke rings of *p*-adic Chevalley Groups. Publ. de l'Inst. des Hautes Etudes Sci., No. 25.

[14] MATSUSAKA, T.: Polarized varieties and fields of moduli. Am. J. of Math. **80**, 45–82 (1958).

[15] GROTHENDIECK, A.: Føndements de la geometrie algebrique. Secretariat math., Paris (Mimeographed).

Department of Mathematics
Harvard University
Cambridge, Massachussetts

(Received February 20, 1967)

Families of Abelian Varieties

BY

DAVID MUMFORD

In previous lectures, Kuga, Shimura, and Satake have considered various families of abelian varieties parametrized by the quotients of bounded symmetric domains by arithmetic subgroups. In particular, Shimura characterized certain of these families by means of the structure of the ring of endomorphisms—the "PEL-types." My purpose here is to show that an even larger class of Kuga's families can be characterized by intrinsic properties of the abelian varieties occurring in them. The properties in question involve the Kählerian geometry of the abelian varieties, but, assuming a famous conjecture of Hodge, they are equivalent to purely "algebro-geometric" properties of the abelian varieties. The results of this lecture are partly joint work with J. Tate.

1. **The Hodge group of a complex torus.** To give a complex torus A of dimension g is the same thing as giving

(i) a $2g$-dimensional rational vector space V;

(ii) a complex structure on $V_{\boldsymbol{R}} = V \otimes_{\boldsymbol{Q}} \boldsymbol{R}$;

(iii) a lattice $L \subset V$.

Here $V = H_1(A, \boldsymbol{Q})$, $L = H_1(A, \boldsymbol{Z})$, and the complex structure on $V_{\boldsymbol{R}}$ is induced by the natural isomorphism between $V_{\boldsymbol{R}}$ and the universal covering space of A. If we are only interested in the type of A up to isogenies, we can omit L. The datum (ii) is equivalent to either of the following objects:

(ii') an endomorphism $J: V_{\boldsymbol{R}} \to V_{\boldsymbol{R}}$ such that $J^2 = -I$,

(ii'') a homomorphism of algebraic groups,

$$\phi: T \to \mathrm{GL}(V)$$

defined over $\boldsymbol{R}$ where T is the compact 1-dimensional torus over $\boldsymbol{R}$, i.e.,

$$T_{\boldsymbol{R}} = \{z \in \boldsymbol{C} \mid |z| = 1\};$$

and such that ϕ, as a representation of G_m, has weights $+1$ and -1, each with multiplicity g.

Starting with a complex structure on $V_{\boldsymbol{R}}$, we get data (ii') and (ii'') as follows:

J = multiplication by i.

$\phi(e^{i\theta})$ = the element of $\mathrm{GL}(V)_{\boldsymbol{R}}$ given by multiplying in the complex structure on $V_{\boldsymbol{R}}$ by $e^{i\theta}$.

esp: $J = \phi(i)$.

Definition. The *Hodge group* of A, written $Hg(A)$, is the smallest algebraic subgroup of $GL(V)$ defined over Q and containing $\phi(T)$.

Since T is connected, it follows immediately that $Hg(A)$ is a connected algebraic group. A few more definitions:

Definition. Let A be a complex torus, and let

$$H^k(A, C) \cong \sum_{p+q=k} H^{p,q}(A)$$

be the Kähler decomposition of the cohomology of A. Then the *Hodge ring* of A is

$$H_0^*(A) = H^*(A, Q) \cap \sum_{p=0}^{\dim A} H^{p,p}(A).$$

Hodge's conjecture asserts that $H_0^*(A)$ is the subring of $H^*(A, Q)$ given by the Q-linear combinations of the fundamental classes of algebraic subvarieties of A.

Note that: if the complex torus A equals V_R/L, then there is a canonical isomorphism:

$$H^i(A \times \cdots \times A, Q) \cong \wedge^i(V^* \oplus \cdots \oplus V^*).$$

Therefore, there is a natural representation of $Hg(A)$ on $H^*(A^k, Q)$, defined over Q.

Proposition 1. *For all k, the Hodge ring of A^k is the ring invariants of $Hg(A)$ in $H^*(A, Q)$.*

Using this Proposition, it is easy to give examples of abelian varieties A such that their Hodge ring is not generated by elements of degree 2 [cf. §3 for the existence of abelian varieties with various Hodge groups].

2. The structure of the Hodge group of an abelian variety. The result is the following:

Theorem. *If A is an abelian variety, then*

(i) *$Hg(A)$ is a connected reductive group,*

(ii) *$\phi(-1)$ is the center of G, and centralizer $[\phi(i)]$ = centralizer $[\phi(T)]$: call this group Z,*

(iii) *Z_R^0 is a maximal compact subgroup of Hg_R^0 and Hg_R^0/Z_R^0 is a bounded symmetric domain.*

Corollary (of (i)). *$Hg(A)$ is the largest subgroup of $GL(V)$ which leaves invariant the Hodge rings of A^k for all k. Hence the Hodge group $Hg(A)$ as a subgroup of $GL(V)$ and the collection of Hodge rings $H_0^*(A^k)$ as subrings of*

$$\wedge^*[V^* \oplus \cdots \oplus V^*]$$

are "equivalent" invariants of the abelian variety A: i.e., each can be computed from the other by linear algebra.

DEFINITION. The *Hodge type* of an abelian variety A of dimension g consists in the set of "equivalent" diagrams

$$T \xrightarrow{\phi} Hg(A) \subset \mathrm{GL}(2g)$$

obtained by identifying $\mathrm{GL}[H_1(A, Q)]$ with $\mathrm{GL}(2g)$ rationally over Q; where two diagrams

$$T \xrightarrow[\phi_1]{} H_1 \subset \mathrm{GL}(2g),$$

$$T \xrightarrow[\phi_2]{} H_2 \subset \mathrm{GL}(2g),$$

are considered equivalent if there are elements $\alpha \in \mathrm{GL}(2g)_Q$, $\beta \in (H_1)_R$ such that

$$H_2 = \alpha H_1 \alpha^{-1},$$
$$\phi_2(\lambda) = \alpha\beta\phi_1(\lambda)\beta^{-1}\alpha^{-1}.$$

One should notice that once the Q-rational subgroup $H \subset \mathrm{GL}(2g)$ is given, there are only a finite number of Hodge types (H, ϕ) extending H. This follows easily from the conjugacy of maximal compact tori in H via points of H_R, and from the restriction on the weights of $\phi(T)$ in this representation.

DEFINITION. Let (H, ϕ) and (H', ϕ') be two Hodge types. Then (H, ϕ) is a refinement of (H', ϕ') if these types are represented by diagrams

$$T \xrightarrow{\phi} H \subset \mathrm{GL}(2g),$$

$$T \xrightarrow{\phi'} H' \subset \mathrm{GL}(2g),$$

where $H \subset H'$ and $\phi' = \phi$.

3. **The families.** Now suppose that a Hodge type (H, ϕ) is given. We will see that the set of all abelian varieties of this Hodge type, plus the limits which have finer Hodge types will be a family over a bounded symmetric domain, such that the action of certain arithmetic groups on the domain lifts to an action on the family.

DEFINITION. *A Hermitian symmetric pair* $(\mathscr{G}, J)$ is a real connected Lie group $\mathscr{G}$ with compact center, and an element $J \in \mathfrak{G}$, its Lie algebra, such that

(i) ad J has three eigenspaces in $\mathfrak{G}_C : \mathfrak{K}_C$ (the complexification of a real subspace $\mathfrak{K}$), $\mathfrak{p}_+$, and $\mathfrak{p}_-$ with eigenvalues $0, +2i, -2i$,.

(ii) $\mathfrak{K}$ is the Lie algebra of a maximal compact subgroup $\mathscr{K}$ in $\mathscr{G}$.

DEFINITION. Let $(\mathscr{G}, J)$ be a Hermitian symmetric pair. A faithful representation

$$\rho : \mathscr{G} \to \mathrm{GL}(2g)_R$$

is *of abelian type* if

(i) $\rho(\mathscr{G})$ is contained in $\mathrm{Sp}(2g)_R$ and is an algebraic subgroup defined over Q,

(ii) $d\rho(J)$ is conjugate under $Sp(2g)_{\mathbf{R}}$ to

$$\begin{pmatrix} 0 & I_g \\ -I_g & 0 \end{pmatrix},$$

which is the "complex structure" in $Sp(2g)_{\mathbf{R}}$.

(ii) is equivalent to asserting that

$$(B, -B\, d\rho(J))$$

form a "symplectic pair" in Kuga's sense, where

$$B = \begin{pmatrix} 0 & I_g \\ -I_g & 0 \end{pmatrix};$$

also, (ii) is condition (H_2) of Satake.

(iii) J is not contained in the Lie algebra of any *normal* subgroup $\mathscr{G}_0 \subset \mathscr{G}$ such that $\rho(\mathscr{G}_0)$ is defined over $\mathbf{Q}$.

An immediate consequence of this definition is that if we exponentiate J in $\mathscr{G}$ we obtain a homomorphism:

$$\phi: T \to \mathscr{G} \qquad \text{where} \qquad \phi(e^{i\theta}) = \exp(\theta J).$$

In fact, $(\rho(\mathscr{G}), \rho \circ \phi)$ is a Hodge type, and every Hodge type arises from an abelian representation of a symmetric pair.

Now suppose $\mathscr{G}, J$ and ρ are given. Let $\phi: T \to \mathscr{G}$ denote the above homomorphism. Let K be the compact subgroup of $\mathscr{G}$ which centralizes $\phi(T)$, and let K' be the compact subgroup of $Sp(2g)_{\mathbf{R}}$ which centralizes $\rho(\phi(T))$. Then ρ induces a holomorphic map of symmetric domains

$$\mathscr{G}/K \xrightarrow{\tau} Sp(2g)_{\mathbf{R}}/K'.$$

Via τ, the standard family of abelian varieties on Siegel's upper $\frac{1}{2}$-plane induces a family over $\mathscr{G}/K$: call it

$$\begin{array}{c} \mathfrak{X}(\mathscr{G}, J, \rho) \\ \downarrow \pi \\ \mathscr{G}/K. \end{array}$$

Since ρ is defined over $\mathbf{Q}$, ρ maps all small enough arithmetic subgroups Γ of $\mathscr{G}$ into $Sp(2g)_{\mathbf{Z}}$, and hence the action of such Γ on $\mathscr{G}/K$ lifts to an action on the family $\mathfrak{X}(\mathscr{G}, J, \rho)$.

PROPOSITION 2. *The abelian variety $\pi^{-1}(x)$ in the family $\mathfrak{X}(\mathscr{G}, J, \rho)$ is isogenous to $A = \mathbf{R}^{2g}/\mathbf{Z}^{2g}$, with complex structure defined by $\rho(g\phi(i)g^{-1})$ where $g \in \mathscr{G}$ represents $x \in \mathscr{G}/K$.*

COROLLARY. *An abelian variety A is isogenous to one in the family $\mathfrak{X}(\mathscr{G}, J, \rho)$ if and only if A has Hodge type equal to or finer than $(\rho(\mathscr{G}), \rho \circ \phi)$.*

PROPOSITION 3. *The families $\mathfrak{X}(\mathcal{G}, J, \rho)$ include all the families associated by Kuga to symplectic representations $\rho: G \to \mathrm{Sp}(2g)$ of semi-simple groups G defined over Q, in the case when $G_{\mathbf{R}}$ has no compact factors.*

4. **The conjecture.** The most intriguing possibility suggested by this theory is is an arithmetic conjecture. Serre [Colloque de Clermont-Ferrand, *Groupes de Lie l-adiques attachés aux courbes elliptiques*] has defined l-adic Lie algebras acting on $H_1(A, Q_l)$, for any abelian variety A, which are essentially the Lie algebras of the Galois group of the extension obtained by adjoining all points of order l^ν to some smallest field of definition of A. Call these $\mathfrak{G}_l$. Let $\mathrm{Lg}(A)$ be the Lie algebra of $\mathrm{Hg}(A)$. It is a sub-Lie-algebra of $\mathrm{Sl}[H_1(A, Q)]$. Then one may ask whether:

$$\mathfrak{G}_l \cap \mathrm{Sl}[H_1(A, Q_l)] = \mathrm{Lg}(A) \otimes_{Q} Q_l.$$

If $\dim A = 1$, and A is defined over Q, Serre has verified this. For A of CM-type, this result is apparently proven in Shimura-Tamiyama, *Complex multiplication of abelian varieties.*

Math. Ann. 181, 345—351 (1969)

A Note of Shimura's Paper "Discontinuous Groups and Abelian Varieties"

D. MUMFORD

Recently, there has been considerable discussion of families of abelian varieties parametrized by quotients of bounded symmetric domains by arithmetic subgroups. An exposition of this material can be found in the papers of Shimura, Kuga, Satake and myself in [1]. Subsequently in vol. 168, p. 171 of this journal, Shimura analyzed closely certain families of this type and showed that the abelian varieties in these families are characterized, surprisingly, by the existence of certain non-holomorphic endomorphisms. Furthermore, he says at the end of his paper (p. 199) that it does not seem that these families fall into the class which I constructed in [1c], which have the following property: they can be grouped into finite sets of families, i.e., families over a base space which is a finite union of varieties $V_i = \Gamma_i \backslash D_i$, such that the abelian varieties A_x in any one of these coarser families are exactly characterized by the rational (p, p)-forms on their powers $A_x \times \cdots \times A_x$*.

The purpose of this note is to clarify the relationship of the different types of families of abelian varieties. First, we recall the most general type $\mathscr{X} = \{X(x) \mid x \in V\}$, as defined by Kuga. Second, we note which of these families are "of Hodge type", i.e., defined by my procedure. Thirdly, we prove that one of Kuga's families is of Hodge type if and only if it contains one abelian variety $X(x_0)$ of CM-type. As a corollary, it follows that Shimura's families in [2] are indeed of Hodge type, contrary to his guess. Lastly, we will give another example of a family of Hodge type not characterized by its ring of endomorphisms.

§ 1. Kuga's Families

Actually, we slightly generalize his definitions, for the sake of our application. We start with:

(1) a rational vector space V of dimension $2g$,

(2) a lattice $L \subset V$,

(3) a non-degenerate skew-symmetric $A : V \times V \to \mathbb{Q}$, integral on $L \times L$,

* In the terminology of [1c], note that I showed that the *Hodge group* of an abelian variety was determined by these rational (p, p)-forms. But my families are given by the various different *Hodge types*, where a Hodge type is given by a Hodge group G, plus a conjugacy class of 1-dimensional tori $T \subset G$; where, however, the conditions on T imply that only a finite number of conjugacy classes of T's can be used to form a Hodge type.

(4) an algebraic group G, defined over $\mathbb{Q}$,

(5) a faithful symplectic representation

$$\varrho : G \to \mathrm{Sp}(V, A)$$

defined over $\mathbb{Q}$,

(6) an arithmetic subgroup $\Gamma \subset G$ such that $\varrho(\Gamma)$ preserves the lattice L.

We are interested in complex structures on $V \otimes \mathbb{R}$ so as to make $V \otimes \mathbb{R}/L$ into a complex torus. We find it convenient to adopt the following method of describing possible complex structures: let $T = \{z \in \mathbb{C} \mid |z| = 1\}$, regarded as a 1-dimensional algebraic group over the reals $\mathbb{R}$. Then, all complex structures on $V \otimes \mathbb{R}$ define a homomorphism of algebraic groups over $\mathbb{R}$:

$$\varphi : T \to GL(V)$$

via $\varphi(e^{i\theta}) =$ mult. by $e^{i\theta}$ in $V \otimes \mathbb{R}$. Conversely, any such homomorphism puts a complex structure on $V \otimes \mathbb{R}$, hence makes $V \otimes \mathbb{R}/L$ into a complex torus, which we shall call X_φ. Furthermore, if φ satisfies the Riemann conditions (a) $\varphi(T) \subset \mathrm{Sp}(V, A)$, and (b) $A(x, \varphi(i) \cdot x) > 0$ all $x \in V$, $x \neq 0$, then X_φ is an abelian variety, with a canonical polarization induced by A. Now we want a family of abelian varieties $X(x)$ parametrized by points $x \in V = \Gamma \backslash G^0_{\mathbb{R}}/K^0_{\mathbb{R}}$ for a certain maximal compact subgroup $K^0_{\mathbb{R}}$ of $G^0_{\mathbb{R}}$. In particular, the identity in $G^0_{\mathbb{R}}$ defines a double coset $\Gamma \cdot e \cdot K^0_{\mathbb{R}}$ which is a base point 0 of this base space V. Our family can be regarded as a family of perturbations of the "base abelian variety" $X(0)$ over 0. So as our next piece of data, we assume an arbitrary $X(0)$ is given:

(7) a complex structure $\varphi_0 : T \to \mathrm{Sp}(V, A) \subset GL(V)$ such that $J_0 = \varphi_0(i)$ satisfies:

$$A(x, J_0 x) > 0, \quad \text{all} \quad x \neq 0 \quad \text{in } V.$$

Now, to every point $g \in G^0_{\mathbb{R}}$, let

$$X(g) = X_{\varrho(g)\varphi_0 \cdot \varrho(g)^{-1}} = \left\{ \begin{array}{l} \text{the polarized abelian variety } V \otimes \mathbb{R}/L, \text{ with complex} \\ \text{structure } \varrho(g) \cdot \varphi_0 \cdot \varrho(g)^{-1}, \text{ Riemann form } A \end{array} \right\}.$$

If $K^0_{\mathbb{R}} = \{g \in G_{\mathbb{R}} \mid \varrho(g)\varphi_0 = \varphi_0 \varrho(g)\}^0$, then $X(g)$ depends obviously only on the image of g in the coset space $G^0_{\mathbb{R}}/K^0_{\mathbb{R}}$. Moreover, the automorphism $\varrho(\gamma)$ of V sets up isomorphisms of the abelian variety $X(g)$ with $X(\gamma g)$ for all $g \in G^0_{\mathbb{R}}$. Therefore $X(g)$ depends only on the image x of g in $\Gamma \backslash G^0_{\mathbb{R}}/K^0_{\mathbb{R}}$ and we may write for simplicity $X(x)$, instead of $X(g)$.

We still want to ensure that $G^0_{\mathbb{R}}/K^0_{\mathbb{R}}$ is a bounded symmetric domain and that $\{X(x)\}$ is a holomorphic family of abelian varieties. This is guaranteed by imposing on the data (1)—(7) the following condition (which is the integrated form of Satake's original condition (H_1)):

(H_1^*) $\quad \varrho(G)$ is normalized by $\varphi_0(T)$.

It follows easily from (H_1^*) that G is reductive, that $K^0_{\mathbb{R}}$ is a maximal compact subgroup of $G^0_{\mathbb{R}}$, that $G^0_{\mathbb{R}}/K^0_{\mathbb{R}}$ is a Hermitian symmetric space, and finally that the map induced by τ:

$$G^0_{\mathbb{R}}/K^0_{\mathbb{R}} \to \mathrm{Sp}(V, A) \Big/ \Big/ \begin{pmatrix} \text{centralizer} \\ \text{of } \varphi_0 \end{pmatrix} \cong \begin{matrix} \text{Siegel's upper} \\ \text{half-plane} \end{matrix}$$

is holomorphic. The resulting family $\{X(x) \mid x \in \Gamma\backslash G^0_{\mathbb{R}}/K^0_{\mathbb{R}}\}$ glued together into a complex analytic fibre system of abelian varieties over the base space $\Gamma\backslash G^0_{\mathbb{R}}/K^0_{\mathbb{R}}$ will be denoted by $\mathscr{X}(G, \varrho, \varphi_0)$.

Notice that the base point in Kuga's families is arbitrary. For any $g_0 \in G^0_{\mathbb{R}}$, we can replace φ_0 by $\varphi_0' = \varrho(g_0) \cdot \varphi_0 \cdot \varrho(g_0)^{-1}$, and obtain the same family as before, but with a different base point $X(0)$.

§ 2. Hodge Groups and *CM*-Type

We continue to assume that the data (V, L, A) is given. Suppose

$$\varphi: T \to \mathrm{Sp}(V, A)$$

is any complex structure on $V \otimes \mathbb{R}$ satisfying the Riemann positivity condition, so that the corresponding X_φ is an abelian variety.

Definition. The *Hodge group* $\mathrm{Hg}(X_\varphi)$ of X_φ is the smallest algebraic subgroup of $\mathrm{Sp}(V, A)$ defined over $\mathbb{Q}$ and containing $\varphi(T)$. Recall that $\mathrm{Hg}(X_\varphi)$ is always reductive, with compact center, and semi-simple part of Hermitian type.

Definition. If id.: $\mathrm{Hg}(X_\varphi) \to \mathrm{Sp}(V, A)$ ist just the inclusion map, then the families

$$\mathscr{X}(\mathrm{Hg}(X_\varphi), \mathrm{id.}, \varphi)$$

are called the *families of Hodge type*.

These are exactly the families constructed in [1c].

Definition. An abelian variety X over $\mathbb{C}$ is *of CM-type* if X is isogenous to a product $X_1 \times \cdots \times X_k$ is simple abelian varieties and there are fields $K_i \subset \mathrm{Hom}(X_i, X_i) \otimes \mathbb{Q}$ such that $[K_i : \mathbb{Q}] \geqq 2 \dim X_i$ (in which case, $[K_i : \mathbb{Q}] = 2 \dim X_i$ and $K_i = \mathrm{Hom}(X_i, X_i) \otimes \mathbb{Q}$).

The following is well known ([3], § 5):

Proposition. *X is of* CM-*type if and only if* $\mathrm{Hom}(X, X) \otimes \mathbb{Q}$ *contains a commutative semi-simple $\mathbb{Q}$-algebra R such that* $[R : \mathbb{Q}] \geqq 2 \dim X$, *and if R exists,* $[R : \mathbb{Q}] = 2 \dim X$.

Yet another characterization of CM-type is:

Proposition. *X_φ is of* CM-*type if and only if* $\mathrm{Hg}(X_\varphi)$ *is a torus algebraic group.*

Proof. Any endomorphism of X_φ has a natural representation as an endomorphism of V over $\mathbb{Q}$, and it is easily seen that for any φ,

$$\mathrm{Hom}(X_\varphi, X_\varphi) \otimes \mathbb{Q} \cong \{g \in \mathrm{Hom}(V, V)_{\mathbb{Q}} \mid g\varphi = \varphi g\}.$$

Since $\mathrm{Hg}(X_\varphi)$ is generated by $\mathrm{Im}(\varphi)$ and by its conjugates over $\mathbb{Q}$, it follows that:

$$\mathrm{Hom}(X_\varphi, X_\varphi)\otimes\mathbb{Q}\cong\{g\in\mathrm{Hom}(V,V)_\mathbb{Q}\mid gg'=g'g,\quad \text{all}\quad g'\in\mathrm{Hg}(X_\varphi)\}\,. \quad (*)$$

Now suppose that X_φ is of CM-type. Let R be the commutative semi-simple $\mathbb{Q}$-algebra given by the previous proposition. Then via the above isomorphism it follows that $\mathrm{Hg}(X_\varphi)$ commutes with a maximal commutative semi-simple subalgebra R' of $\mathrm{Hom}(V,V)$. But therefore $\mathrm{Hg}(X_\varphi)\subset$ units of R', hence $\mathrm{Hg}(X_\varphi)$ is itself commutative, hence it is a torus algebraic group. Conversely, if $\mathrm{Hg}(X_\varphi)$ is a torus algebraic group, then as a subgroup of $GL(V)_\mathbb{C}$ it is diagonalizable. Therefore, the commutator of $\mathrm{Hg}(X_\varphi)$ in $\mathrm{Hom}(V,V)_\mathbb{C}$ and hence in $\mathrm{Hom}(V,V)_\mathbb{Q}$ contains maximal commutative semi-simple subalgebras R'. Since for all such R', $[R':\mathbb{Q}]=\dim V=2\dim X_\varphi$, this implies by $(*)$ that X_φ is of CM-type. QED.

§ 3. The Theorem

Theorem. a) *Every family* $\mathscr{X}(\mathrm{Hg}(X_\varphi), \mathrm{id.}, \varphi)$ *of Hodge type contains abelian varieties of* CM-*type.*

b) *If* $\mathscr{X}(G,\varrho,\varphi_0)$ *contains a member of* CM-*type, then* $\mathscr{X}(G,\varrho,\varphi_0)$ *is isomorphic to a family of Hodge type.*

Proof. It is well known that for any algebraic group G over $\mathbb{Q}$, $G_\mathbb{Q}$ is dense in $G^0_\mathbb{R}$. This implies that for such groups G, every $G_\mathbb{R}$-conjugacy class of maximal algebraic tori in G, defined over $\mathbb{R}$, contains tori defined over $\mathbb{Q}$. Namely, let $T_1\subset G$ be a maximal torus defined over $\mathbb{R}$. If $a\in(T_1)_\mathbb{R}$ is a regular element, then T_1 is the centralizer of a, and a has an open neighborhood $U\subset G^0_\mathbb{R}$ such that the centralizer of any $a'\in U$ is a conjugate of T_1. If $a'\in U\cap G_\mathbb{Q}$, then the centralizer of a' is a conjugate of T_1 defined over $\mathbb{Q}$, as required. Now apply these results to $\mathrm{Hg}(X_\varphi)$. Let $K=$ centralizer of $\varphi(T)$, regarded as an algebraic subgroup of $\mathrm{Hg}(X_\varphi)$ defined over $\mathbb{R}$. Then K has at least one maximal algebraic torus $T_1\subset K$ defined over $\mathbb{R}$ ([1 e], p. 26). Since $\varphi(T)\subset\mathrm{center}(K)$, $\varphi(T)\subset T_1$. Moreover, if T_1' is any torus in $\mathrm{Hg}(X_\varphi)$ containing T_1, then T_1' will centralize $\varphi(T)$, hence $T_1'\subset K$, hence $T_1'=T_1$: i.e., T_1 is a maximal algebraic torus of $\mathrm{Hg}(X_\varphi)$ too. By our first remark, there are elements $g\in\mathrm{Hg}(X_\varphi)_\mathbb{R}$ such that $T_2=gT_1g^{-1}$ is defined over $\mathbb{Q}$. But then $T_2\supset g\varphi(T)g^{-1}$. Therefore $T_2\supset\mathrm{Hg}(X_{g\varphi g^{-1}})$, and so $\mathrm{Hg}(X_{g\varphi g^{-1}})$ must be an algebraic torus. Therefore $X_{g\varphi g^{-1}}$ is of CM-type and this proves (a).

To prove (b), first replace φ_0 by $\varrho(g)\varphi_0\varrho(g)^{-1}$ for a suitable $g\in G^0_\mathbb{R}$ so that the abelian variety $X(0)$ is of CM-type. This does not alter the family $\mathscr{X}(G,\varrho,\varphi_0)$. Let $T_1=\mathrm{Hg}(X_{\varphi_0})$. This is an algebraic torus, defined over $\mathbb{Q}$. Since T_1 is generated by $\varphi_0(T)$ and its $\mathbb{Q}$-conjugates, all of which normalize $\varrho(G)$, it follows that T_1 normalizes $\varrho(G)$. So $G^*=\varrho(G)$. T_1 is an algebraic subgroup of $\mathrm{Sp}(V,A)$ defined over $\mathbb{Q}$, with $\varrho(G)$ as a normal subgroup. Note that G^* is still reductive since $\varrho(G)$ and T_1 are reductive. In particular, $G^*=$ (semi-simple part) $\cdot$ (central torus), hence $G^*=\varrho(G)\cdot$ (central torus). Therefore, the two collections of complex structures:

(I) $\{\varrho(g)\varphi_0\varrho(g)^{-1} \mid g \in G^0_{\mathbb{R}}\}$,

(II) $\{g\varphi_0 g^{-1} \mid g \in G^{*,0}_{\mathbb{R}}\}$

are exactly the same. Moreover, $G^0_{\mathbb{R}}$/(centr. of φ_0) is canonically isomorphic to $G^{*,0}_{\mathbb{R}}$/(centr. of φ_0). Therefore the 2 families $\mathscr{X}(G, \varrho, \varphi_0)$ and $\mathscr{X}(G^*, \text{id.}, \varphi_0)$ (where id.: $G^* \to \mathrm{Sp}(V, A)$ is the inclusion map) are isomorphic. Finally, let $\varphi = g\varphi_0 g^{-1}$ where $g \in G^{*,0}_{\mathbb{R}}$ is a generic point of G^* over a field of definition of $\varphi_0(T)$. Then $\mathrm{Hg}(X_\varphi)$ is the smallest $\mathbb{Q}$-rational subgroup of G^* containing *all* the tori $g\varphi_0^\sigma g^{-1}$, $g \in G^*$, $\sigma \in \mathrm{Aut}(\mathbb{C})$. Hence $\mathrm{Hg}(X_\varphi)$ is just the smallest subgroup of G^* (defined over any field) containing *all* the tori gT_1g^{-1}, $g \in G^*$. In particular $\mathrm{Hg}(X_\varphi)$ is a normal subgroup of G^*, defined over $\mathbb{Q}$ and containing $\varphi(T)$ and $\varphi_0(T)$. Therefore $G^* = \mathrm{Hg}(X_\varphi) \cdot G_2$ where G_2 commutes with $\mathrm{Hg}(X_\varphi)$ and hence with $\varphi(T)$ and with $\varphi_0(T)$. It follows that the sets (I) and (II) of complex structures are equal to the set:

(III) $\{g\varphi g^{-1} \mid g \in \mathrm{Hg}(X_\varphi)^0_{\mathbb{R}}\}$.

It also follows that $G^{*,0}_{\mathbb{R}}$/(centr. of φ_0) is isomorphic to $\mathrm{Hg}(X_\varphi)^0_{\mathbb{R}}$/(centr. of φ), hence the 2 families $\mathscr{X}(G^*, \text{id.}, \varphi_0)$ and $\mathscr{X}(\mathrm{Hg}(X_\varphi), \text{id.}, \varphi)$ are isomorphic. QED.

§ 4. An Example

Since families of abelian varieties which are *not* characterized by their endomorphism rings are fairly mysterious, it seems worth-while to present as an example what seems to be the *only* family of this type of 4-dimensional abelian varieties. (In dimensions 1, 2 and 3, all families are characterized by endomorphisms.)

To define this family, we need an apparently "well-known" construction for central simple algebras: if $L \supset K$ is a finite separable extension of degree n, and D is a central simple algebra over L, with $[D:L] = e^2$, then there is canonical central simple algebra $\mathrm{Cor}_{L/K}(D)$ over K, with $[\mathrm{Cor}(D):K] = e^{2n}$, and with a homomorphism

$$\mathrm{Nm}: D^* \to \mathrm{Cor}_{L/K}(D)^*$$

of units. It is simply the corestriction map in the cohomology theory of groups applied to the Brauer group. To construct it, let Ω be a separable closure of K and let $\sigma_1, \cdots, \sigma_n: L \to \Omega$ be the distinct K-isomorphisms from L to Ω. Let $D^{(i)} = D \otimes_L (\Omega, \sigma_i)$ be the central simple Ω-algebra obtained by base change with respect to σ_i. Then $\mathrm{Gal}(\Omega/K)$ acts on

$$E = D^{(1)} \otimes \cdots \otimes D^{(n)}$$

in a natural way, i.e. if $\tau: \Omega \to \Omega$ is a K-isomorphism, then $\tau \circ \sigma_i = \sigma_{\pi(i)}$ for some permutation π, and the maps

$$d \otimes a \mapsto d \otimes \tau(a)$$

induce semi-linear-isomorphisms $D^{(i)} \xrightarrow{\sim} D^{(\pi(i))}$, hence a semi-linear automorphism of E. Let $\mathrm{Cor}_{L/K}(D)$ be the subalgebra of E left fixed by this action: this will be a central simple K-algebra such that $\mathrm{Cor}_{L/K}(D) \otimes_K \Omega \cong E$. Finally define $\mathrm{Nm}(d)$ to be the element

$$(d \otimes 1) \otimes \cdots \otimes (d \otimes 1)$$

of E, which is clearly in $\mathrm{Cor}_{L/K}(D)$.

Now let K be a totally real cubic number field, and let D be a quaternion division algebra over K such that

$$\mathrm{Cor}_{K/\mathbb{Q}}(D) \quad \text{splits, i.e.} \quad \cong M_8(\mathbb{Q}), \tag{1}$$

$$D \otimes_{\mathbb{Q}} \mathbb{R} \cong \mathbb{K} + \mathbb{K} + M_2(\mathbb{R}). \tag{2}$$

Then we get a natural homomorphism:

$$\mathrm{Nm} : D^* \to GL(8, \mathbb{Q}).$$

Let $\bar{\ }$ be the standard involution of D, and let

$$G = \{x \in D^* \mid x \cdot \bar{x} = 1\}.$$

Then G is an algebraic group over $\mathbb{Q}$, which is $\mathbb{Q}$-simple, but which by (2), is a $\mathbb{Q}$-form of the $\mathbb{R}$-algebraic group $SU(2) \times SU(2) \times SL(2, \mathbb{R})$. Moreover, let $V = \mathbb{Q}^8$; then, via Nm, G has an algebraic representation in V defined over $\mathbb{Q}$, which is a $\mathbb{Q}$-form of $\mathbb{R}$-representation:

$$SU(2) \times SU(2) \times SL(2, \mathbb{R}) \to \underbrace{SO(4) \times SL(2, \mathbb{R})}_{\text{acting on } \mathbb{R}^4 \otimes \mathbb{R}^2}$$

Over $\mathbb{R}$, this representation leaves invariant a unique symplectic form (up to scalars), so there is a unique symplectic form $A : V \times V \to \mathbb{Q}$ left fixed by our $\mathbb{Q}$-representation. Let $L \subset V$ be any lattice, and $\Gamma \subset G$ any arithmetic subgroup preserving L. Finally, let

$$\varphi_0 : T \to \mathrm{Sp}(V, A)_{\mathbb{R}}$$

be the homomorphism:

$$\varphi_0 : T \to SO(4) \times SL(2, \mathbb{R}) \subset \mathrm{Sp}(V, A)_{\mathbb{R}},$$

$$e^{i\theta} \to I_4 \otimes \begin{pmatrix} \cos\theta & \sin\theta \\ -\sin\theta & \cos\theta \end{pmatrix}.$$

This gives data (1)—(7) as in § 1, and (H_1^*) is obviously satisfied. Moreover, since G is $\mathbb{Q}$-simple, $\varrho(G)$ is exactly the Hodge group of a generic conjugate $\varrho(a) \cdot \varphi_0 \cdot \varrho(a)^{-1}$, so we have a family of Hodge type. Finally, V is an absolutely irreducible representation of G, so whenever $\mathrm{Hg}(X_\varphi) = \varrho(G)$, X_φ has no non-trivial endomorphisms.

References

1. Algebraic groups and discontinuous subgroups. Proc. Symp. in Pure Math., vol. 9, Am. Math. Soc., 1966.
1 a. Article of Shimura, p. 312.
1 b. Article of Kuga, p. 338.
1 c. Article of Mumford, p. 347.
1 d. Article of Satake, p. 352.
1 e. Article of Borel and Springer, p. 26.
2. Shimura, G.: Discontinuous groups and abelian varieties. Math. Ann. **168**, 171—199 (1967).
3. —, and Y. Taniyama: Complex multiplication of abelian varieties. Publ. of Math. Soc. Japan, 1961.

Professor D. Mumford
Department of Mathematics
Harvard University
Cambridge, Mass. 02138, USA

(Received November 4, 1968)

Ann. scient. Éc. Norm. Sup.,
4e série, t. 4, 1971, p. 181 à 192.

THETA CHARACTERISTICS OF AN ALGEBRAIC CURVE

BY DAVID MUMFORD.

Let X be a non-singular complete algebraic curve over an algebraically closed ground field k of char $\neq 2$. We are interested here in vector bundles E over X such that there exists a quadratic form ([1]) :

$$Q : \quad E \to \Omega_X^1$$

which is everywhere non-degenerate, i. e. for all $x \in X$, choosing a differential ω which is non-zero at x, Q induces a k-valued quadratic form in the fibre

$$\frac{Q}{\omega}(x) : \quad E(x) \to k$$

which is to be non-degenerate. Our first result is that for such E, $\dim \Gamma(E) \bmod 2$ *is stable under deformations of* X *and* E. If E is a line bundle L, then the existence of Q just means that $L^2 \cong \Omega_X^1$. The set of such L is called classically the set of *theta-characteristics* S (X) of X (*cf.* Krazer [K]). S (X) is a principal homogeneous space over J_2, the group of line bundles L such that $L^2 \cong \mathcal{O}_X$. Now on S (X) we have the function

$$e_*(L) = \dim \Gamma(L) \bmod 2 \in \mathbf{Z}/2\mathbf{Z}$$

and on J_2 we have the well-known skew-symmetric bilinear form

$$e_2 : J_2 \times J_2 \to \{\pm 1\}$$

(*cf.* Weil [W], Lang ([L 1], p. 173 and p. 189), or Mumford ([M 1], p. 183)).

([1]) By this we mean the composition of (i) $E \to E \otimes E$, $s \mapsto s \otimes s$, and (ii) a symmetric linear homomorphism $B : E \otimes E \to \Omega_X^1$.

Our second theorem is that e_* is a quadratic function whose associated bilinear form is e_2 (considered additively); more precisely, this means

$$(\star)\qquad \begin{cases} e_*(L)+e_*(L\otimes\alpha)+e_*(L\otimes\beta)+e_*(L\otimes\alpha\otimes\beta)=\ln e_2(\alpha,\beta), \\ L\in S(X), \quad \alpha, \beta\in J_2, \end{cases}$$

where

$$\ln(-1)=1, \qquad \ln(+1)=0.$$

This is proved by applying the first result to bundles $L\otimes\mathfrak{A}$, where $L\in S(X)$ and $\mathfrak{A}$ is a quaternionic Azumaya algebra over X. Given J_2 and e_2, it is easy to check that, up to isomorphism there are exactly two pairs (S, e_*) consisting of a principal homogeneous space S under J_2 and a function e_* on S satisfying $(\star)$. These two possibilities are distinguished by the Arf invariant of e_*, or more simply by whether e_* takes the values 0 and 1 at $2^{g-1}(2^g+1)$ and $2^{g-1}(2^g-1)$ points of S respectively, or whether the opposite happens. The third result whose proof we will only sketch is that the former happens, i. e. e_* is more often 0 than 1.

All these results in the case $E = L$ were proven by Riemann over the complex ground field using his theta function (*cf.* [R], p. 212 and 487). In the general case, they follow easily by the results of ([M 2], § 2) on abstract theta functions and by Riemann's theorem that the multiplicity of the theta divisor $\Theta\subset J^{g-1}$ at a point x equals $\dim\Gamma(L_x)$, where L_x is the line bundle of degree $g-1$ corresponding to x. This last has been proven by me in all characteristics (unpublished) and is a special case of the results in the thesis of G. Kempf, soon to be published. The inspiration of this paper came from several conversations with M. Atiyah in which he asked whether there was a simple direct proof of these results not involving the theory of theta-functions. In particular, it was his suggestion to look at all vector bundles E admitting a Q rather than only at the line bundles.

1. Stability of $\dim\Gamma(E) \bmod 2$. — Given (X, E, Q), the idea is to represent $\Gamma(E)$ as the intersection of two maximal isotropic subspaces W_1, W_2 of a big even-dimensional vector space V with non-degenerate quadratic form q, where (V, q) obviously varies continuously when (X, E, Q) vary continuously. But then it is well-known that in such a case $\dim W_1\cap W_2 \bmod 2$ is invariant under continuous deformation (*cf.* Bourbaki [B], vol. 24, *Fromes sesquilinéaries*, § 6, ex. 18 *d*). We carry this out as follows : let $\mathfrak{a}$ be a cycle on X of the form $\sum_{i=1}^{N} P_i$, where $N \gg 0$

and the P_i are distinct points. Look at the commutative diagram :

$$\begin{array}{ccccccccc}
 & & & & o & & o & & \\
 & & & & \downarrow & & \downarrow & & \\
o & \longrightarrow & E(-\mathfrak{a}) & \longrightarrow & E & \longrightarrow & E/E(-\mathfrak{a}) & \longrightarrow & o \\
 & & \| & & \downarrow & & \downarrow & & \\
o & \longrightarrow & E(-\mathfrak{a}) & \longrightarrow & E(\mathfrak{a}) & \longrightarrow & E(\mathfrak{a})/E(-\mathfrak{a}) & \longrightarrow & o \\
 & & & & \downarrow & & \downarrow & & \\
 & & & & E(\mathfrak{a})/E & = & E(\mathfrak{a})/E & & \\
 & & & & \downarrow & & \downarrow & & \\
 & & & & o & & o & &
\end{array}$$

Since $\Gamma(E(-\mathfrak{a})) = (o)$ and $H^1(E(\mathfrak{a}) = (o)$ for $\mathfrak{a}$ sufficiently positive, this gives rise to the diagram :

$$\begin{array}{ccccccccc}
 & & o & & o & & & & \\
 & & \downarrow & & \downarrow & & & & \\
o & \longrightarrow & \Gamma(E) & \longrightarrow & \Gamma(E/E(-\mathfrak{a})) & \longrightarrow & H^1(E(-\mathfrak{a})) & & \\
 & & \downarrow & & \downarrow & & \| & & \\
o & \longrightarrow & \Gamma(E(\mathfrak{a})) & \longrightarrow & \Gamma(E(\mathfrak{a})/E(-\mathfrak{a})) & \longrightarrow & H^1(E(-\mathfrak{a})) & \longrightarrow & o \\
 & & \downarrow & & \downarrow & & & & \\
 & & \Gamma(E(\mathfrak{a})/E)) & = & \Gamma(E(\mathfrak{a})/E) & & & & \\
 & & & & \downarrow & & & & \\
 & & & & o & & & &
\end{array}$$

from which it follows immediately that $\Gamma(E)$ is the intersection of the subspaces :

$$W_1 = \Gamma(E(\mathfrak{a})),$$
$$W_2 = \Gamma(E/E(-\mathfrak{a}))$$

of the vector space :

$$V = \Gamma(E(\mathfrak{a})/E(-\mathfrak{a})).$$

Next note that polarizing Q defines a non-degenerate bilinear form

$$B:\ E \otimes E \to \Omega^1,$$

hence $E \cong \mathrm{Hom}(E, \Omega^1)$. It follows immediately by Serre duality that

$$\dim H^0(E) = \dim H^0(\mathrm{Hom}(E, \Omega^1)) = \dim H^1(E),$$

so $\chi(E) = o$. Therefore $\chi(E(\mathfrak{a})) = Nr$, and since $H^1(E(\mathfrak{a})) = (o)$, $\dim W_1 = Nr$, where $r = \text{rank}\ (E)$. Obviously $\dim W_2 = Nr$ and

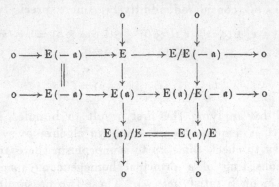

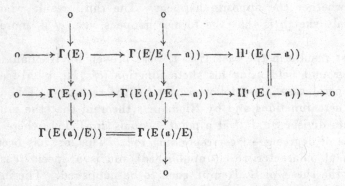

$\dim V = 2Nr$. Next define a quadratic form q on V as follows : if $a_i \in E(\mathfrak{a})_{P_i}$, $1 \leq i \leq N$, define the section $\bar{a}$ of $E(\mathfrak{a})/E(-\mathfrak{a})$, then set

$$q(\bar{a}) = \sum_{i=1}^{N} \operatorname{Res}_{P_i} Q(a_i),$$

where Q is here extended to a quadratic map $E(\mathfrak{a}) \to \Omega^1_X(2\mathfrak{a})$. If $\bar{a} \in W_1$, then the a_i all come from one global section a of $E(\mathfrak{a})$, and $q(\bar{a}) = 0$ since the sum of the residues of a rational differential on X is zero. If $\bar{a} \in W_2$, then $a_i \in E_{P_i}$, so $Q(a_i) \in (\Omega^1_X)_{P_i}$ so again $q(\bar{a}) = 0$. Thus W_1 and W_2 are isotropic subspaces of V of half the dimension, i. e. are maximal isotropic subspaces.

Now say (X, E, Q) vary in an algebraic family, i. e. we are given

(i) $\pi : \mathfrak{X} \to S$, proper smooth family of curves of genus g;

(ii) $\mathcal{E}$ on $\mathfrak{X}$: a vector bundle of rkr;

(iii) $Q : \mathcal{E} \to \Omega^1_{\mathfrak{X}/S}$ a non-degenerate quadratic form.

Theorem. — *The function* $S \to \mathbf{Z}/2\mathbf{Z}$ *defined by*

$$s \mapsto \dim \Gamma(X_s, E_s) \bmod 2$$

is constant on connected components of S.

Proof. — After an étale base change $S' \to S$, we may assume that locally $\mathfrak{X}/S$ admits N disjoint sections $\sigma_i : S \to \mathfrak{X}$. If $\mathfrak{A}$ is the relative Cartier divisor $\Sigma \sigma_i(S)$ on $\mathfrak{X}$ over S, then as above, we find three locally free sheaves on S :

$$\begin{array}{l} \mathcal{W}_1 = \pi_* \mathcal{E}(\mathfrak{A}) \hookrightarrow \pi_*[\mathcal{E}(\mathfrak{A})/\mathcal{E}(-\mathfrak{A})] = \mathcal{V} \\ \mathcal{W}_2 = \pi_*[\mathcal{E}/\mathcal{E}(-\mathfrak{A})] \hookrightarrow \pi_*[\mathcal{E}(\mathfrak{A})/\mathcal{E}(-\mathfrak{A})] = \mathcal{V} \end{array}$$

and a non-degenerate quadratic form $q : \mathcal{V} \to \mathcal{O}_S$. Moreover, for each $s \in S$, $(\mathcal{V}, \mathcal{W}_1, \mathcal{W}_2, q)$ induce, after tensoring with $k(s)$, the previous quadruple (V, W, W_2, q). In particular,

$$\Gamma(X_s, E_s) \cong [\mathcal{W}_1 \otimes_{\mathcal{O}_s} k(s)] \cap [\mathcal{W}_2 \otimes_{\mathcal{O}_s} k(s)]$$

[$\cap$ inside $\mathcal{V} \otimes_{\mathcal{O}_s} k(s)$]. But using the fact that $q \equiv 0$ on $\mathcal{W}_1$ and $\mathcal{W}_2$, by the standard constructions in the theory of quadratic forms, one finds that $\exists$ locally isomorphisms :

$$\mathcal{V} \cong \mathcal{W}_1 \oplus \mathcal{W}_1^*$$
$$\mathcal{V} \cong \mathcal{W}_2 \oplus \mathcal{W}_2^*$$

taking q into the hyperbolic quadratic form on the right. Therefore $\exists$ locally an isomorphism $\varphi : \mathcal{V} \to \mathcal{V}$ such that $\varphi(\mathcal{W}_1) = \mathcal{W}_2$ and $q \circ \varphi = q$. It follows that $\det\varphi \in \Gamma(\mathcal{O}_S^*)$ satisfies $(\det\varphi)^2 = 1$, hence $\det\varphi = +1$ or -1 on each connected component of S. But it is easy to check that for all $s \in S$,

$$(\det\varphi)(s) = (-1)^{Nr + \dim \Gamma(E_s)}$$

(*cf.* the exercise in Bourbaki referred to above).

Q. E. D.

2. Applications. — Suppose we choose one line bundle L_0 such that $L_0^2 \cong \Omega_X^1$. Then the map

$$E \mapsto E \otimes L_0$$

gives a bijection between vector bundles with non-degenerate quadratic forms $Q : E \to \mathcal{O}_X$, and vector bundles with non-degenerate quadratic forms $Q : E \to \Omega_X^1$. Now it is easy to see that locally in the étale topology, any (E, Q), where Q is $\mathcal{O}_X$-valued, is isomorphic to $(\mathcal{O}_X^r, Q_0)$, where

$$Q_0(a_1, \ldots, a_r) = \Sigma a_i^2.$$

Therefore, if $\mathbf{O}(r)$ is the full orthogonal group, the pairs (E, Q) are classified by the cohomology set

$$H^1(X_{\text{ét}}, \mathbf{O}(r)).$$

Using the structure : $\mathbf{O}(r) \cong \mathbf{Z}/2\mathbf{Z}.\ \mathbf{SO}(r)$ (the product being direct or semi-direct according as r is odd or even), and the standard isomorphisms of $\mathbf{SO}(r)$ with $\mathbf{G}_m (r = 2)$, $\mathbf{SL}(2)/(-I)$ $(r = 3)$ and $\mathbf{SL}(2) \times \mathbf{SL}(2)/(-I, -I)$ $(r = 4)$ it is easy to determine all (E, Q)'s if rank E = 1, 2, 3 or 4. We get the following bundles :

(1) $E = L$ a line bundle, $L^2 \cong \mathcal{O}_X$.

(2) $E = L \oplus L^{-1}$, L any line bundle, Q hyperbolic.

(3) If $\pi : X' \to X$ is an étale double covering, then for all line bundles L' on X' such that [2]

$$Nm_{X'/X}(L') \cong \mathcal{O}_X.$$

Let $E = \pi_* L'$. Then the norm defines $Q : E \to \mathcal{O}_X$.

(4) $E = L \otimes S^2 F$, where

F is any vector bundle of rank 2; L is a line bundle such that

$$(\bigstar) \qquad (L \otimes \Lambda^2 F)^2 \cong \mathcal{O}_X.$$

---

(2) Nm of a line bundle L' means apply Nm to transition functions defining L' for an open cover of the type $U_\alpha = \pi^{-1}(V_\alpha)$.

There is a canonical quadratic function

$$\delta: \quad S^2F \rightarrow (\Lambda^2 F)^2$$

which is the classical discriminant of a quadratic form, and δ plus the isomorphism (★) defines Q.

(5) $E = F_1 \otimes F_2$, where

F_1, F_2 are rank 2 vector bundles and

$$\Lambda^2 F_1 \otimes \Lambda^2 F_2 \cong \mathcal{O}_X.$$

The quadratic form comes from the composition

$$(F_1 \otimes F_2) \otimes (F_1 \otimes F_2) \rightarrow \Lambda^2 F_1 \otimes \Lambda^2 F_2 \overset{\sim}{\rightarrow} \mathcal{O}_X.$$

(6) $E =$ a quaternion algebra over $\mathcal{O}_X$, i. e. a locally free sheaf of rank 4, plus a multiplication $E \otimes E \rightarrow E$, making the fibres into isomorphic copies of $M_2(k)$. The quadratic form is just the *reduced* norm

$$Nm: \quad E \rightarrow \mathcal{O}_X.$$

There are other rank 4 E's whose cohomology classes are not killed by the map $H^1(o(4)) \overset{\det}{\rightarrow} H^1(\mathbf{Z}/2\mathbf{Z})$, but we will not write all these down.

In example 2, it follows easily by the Riemann-Roch theorem that

$$\dim H^0(E \otimes L_0) \equiv \deg L \pmod 2$$

so the stability is obvious in this case. In example 3, the stability is a classical theorem of Wirtinger [Wi]. In fact let $\iota: X' \rightarrow X'$ be the involution interchanging the sheets and let J, J′ be the Jacobians of X, X′ respectively, then Nm defines a homomorphism

$$Nm: \quad J' \rightarrow J$$

and it turns out that the kernel of Nm consists of exactly two components:

$$P_0 = \text{locus of the line bundles } M \otimes \iota^* M^{-1}, \quad \deg M = 0$$

and

$$P_1 = \text{locus of line bundles } M \otimes \iota^* M^{-1}. \quad \deg M = 1.$$

The result of Wirtinger is that the map

$$P_0 \cup P_1 \rightarrow \mathbf{Z}/2\mathbf{Z}$$
$$L \mapsto \dim H^0(L \otimes \pi^* L_0) \bmod 2$$

is constant on P_0 and P_1 and take different values on the two components. We can prove this as follows :

LEMMA 1. — *If* L *is a line bundle on* X' *such that* $\mathrm{Nm}\, L \cong \mathcal{O}_X$, *then* $L \cong M \otimes \iota^* M^{-1}$ *for some line bundle* M *on* X'. *Moreover*, M *can be chosen of degree* 0 *or* 1.

Proof. — In terms of divisor classes, let L be represented by a divisor $\mathfrak{a}$. Then $\mathrm{Nm} L \cong \mathcal{O}_X$ means that $\pi_* \mathfrak{a} = (f)$, for some function f on X. According to Tsen's theorem, the function field $k(X)$ is C^1 (*cf.* Lang [L 2]), hence $f = \mathrm{Nm} g$, some $g \in k(X')$. Therefore $\pi_*(\mathfrak{a} - (g)) = (f) - (\mathrm{Nm} g) = 0$. Therefore the divisor $\mathfrak{a} - (g)$ is a linear combination of the divisors $x - \iota(x)$, $x \in X'$. In other words, $\mathfrak{a} - (g)$ can be written $\mathfrak{b} - \iota^{-1}\mathfrak{b}$, for some divisor $\mathfrak{b}$. So if $M = \mathcal{O}_X(\mathfrak{b})$, L is isomorphic to $M \otimes \iota^* M^{-1}$. Finally, we may replace M by $M \otimes \pi^* N$ for any line bundle N on X without destroying the property $L \cong M \otimes \iota_* M^{-1}$. In this way, we can normalize the degree of M to be 0 or 1.

Q. E. D.

This proves that $\mathrm{Ker}(\mathrm{Nm})$ is the union of the two sets P_0, P_1 described above. Clearly P_0 and P_1 are irreducible. By our general result, the number

$$\dim H^0(L \otimes \pi^* L_0) = \dim H^0(\pi_* L \otimes L_0)$$

is constant mod 2 on P_0 and P_1. It remains to check that it has *opposite* parity on the two varieties P_0 and P_1. We can see this as follows:

Step I : For some $L \in \mathrm{Ker}\, \mathrm{Nm}$, $H^0(L \otimes \pi^* L_0) \neq (0)$.

Proof. — Choose $\omega \in \Gamma(\Omega^1_X)$ and let $\mathfrak{a} = (\omega)$.

For each point x occuring in $\mathfrak{a}$, choose an $x' \in X'$ over x, and thus find a *positive* divisor $\mathfrak{a}'$ on X' such that $\pi_* \mathfrak{a}' = \mathfrak{a}$. If $L' = \mathcal{O}_X(\mathfrak{a}')$, then $H^0(L') \neq (0)$ and $\mathrm{Nm} L' \cong \mathcal{O}_X(\mathfrak{a}) \cong \Omega^1_X$. Therefore

$$\mathrm{Nm}(L' \otimes \pi^* L_0^{-1}) \cong \mathrm{Nm}\, L' \otimes L_0^{-2} \cong \mathcal{O}_X.$$

and hence $L = L' \otimes \pi_* L_0^{-1}$ has the required properties.

Step II : If $L \in \mathrm{Ker}\, \mathrm{Nm}$ is such that $H^0(L \otimes \pi^* L_0) \neq (0)$, then for almost all $x \in X'$,

$$\dim H^0(L(x - \iota x) \otimes \pi^* L_0) = \dim H^0(L \otimes \pi^* L_0) - 1.$$

Proof. — Let $r = \dim H^0(L \otimes \pi^* L_0)$. Then provided ιx is not a common zero of all the sections of $L \otimes \pi^* L_0$,

$$\dim H^0(L(-\iota x) \otimes \pi^* L_0) = r - 1.$$

hence

$$\dim H^0(L(x - \iota x) \otimes \pi^* L_0) = r \quad \text{or} \quad r - 1.$$

If it equals r, then

$$\dim H^1(L(x - \iota x) \otimes \pi^* L_0) = \dim H^1(L(-\iota x) \otimes \pi^* L_0),$$

hence by Serre duality

$$\dim H^0(L^{-1}(\iota x - x) \otimes \pi^* L_0) = \dim H^0(L^{-1}(\iota x) \otimes \pi^* L_0),$$

hence x is a common zero of all the sections of $L^{-1}(\iota x) \otimes \pi^* L_0$, hence x is a common zero of all the sections of $L^{-1} \otimes \pi^* L_0$. But by Riemann-Roch

$$\begin{aligned}\dim H^0(L^{-1} \otimes \pi^* L_0) &= \dim H^1(L^{-1} \otimes \pi^* L_0)\\ &= \dim H^0(L \otimes \pi^* L_0) > 0\end{aligned}$$

so for almost all x, this is false.

Q. E. D.

Step III : If $L \in P_i$, then $L(x - \iota x) \in P_{1-i}$.

Proof. — Clear.

This proves all of Wirtinger's results in all char. $\neq 2$.

3. The identity between e_* and e_2. — We begin by the observation that in view of Tsen's theorem all Azumaya algebras [3] over X are *split*, i. e. equal Hom (E, E) for some vector bundle E over X (*cf.* Grothendieck [G]). In particular, this means that example 6 of quadratic bundles in paragraph 2, the quaternion bundles with reduced norm, are special cases of example 5. Now if L, $M \in J_2$, let

$$A = \mathcal{O}_X + L + M + L \otimes M$$

and make A into a quaternion bundle by fixing isomorphisms $L^2 \cong \mathcal{O}_X$, $M^2 \cong \mathcal{O}_X$ to define $l_1 . l_2$, and $m_1 . m_2$ [if $l_i \in \Gamma(U, L)$, $m_i \in \Gamma(U, M)$] and by the rule

$$l.m = -m.l, \qquad l \in \Gamma(U, L), \qquad m \in \Gamma(U, M).$$

It follows that

$$A \cong \text{Hom}(E, E)$$

for some vector bundle E of rank 2.

Lemma 2 : $e_2(L, M) = (-1)^{\deg \Lambda^2 E}$.

Proof. — Let $L = \mathcal{O}(\mathfrak{a})$, $M = \mathcal{O}(\mathfrak{b})$ for suitable divisors $\mathfrak{a}$ and $\mathfrak{b}$ with disjoint support. Then $2\mathfrak{a} = (f)$ and $2\mathfrak{b} = (g)$ for some $f, g \in k(X)$ and

(3) This means a locally free sheaf of $\mathcal{O}_X$-algebras whose fibres are isomorphic to $M_n(\mathcal{O}_X)$.

by definition $e_2(L, M) = f(\mathfrak{b})/g(\mathfrak{a})$. Let $\pi : X' \to X$ be the double covering defined by L and let $\iota : X' \to X'$ be the involution, so that $\sqrt{f} \in k(X')$. Now the vector bundle E is unique up to a substitution $E \mapsto E \otimes N$, N a line bundle on X and this substitution does not affect the truth of the lemma. We will prove the lemma by constructing one of the possible E's. By lemma 1 in the previous section, since

$$Nm(\pi^* M) \cong M^2 \cong \mathcal{O}_X,$$

it follows that there exists a line bundle P on X′ such that

$$\iota^* P \underset{\alpha}{\cong} P \otimes \pi^* M.$$

Choose α such that the composition

$$P = \iota^* \iota^* P \underset{\iota^*\alpha}{\cong} \iota^* P \otimes \iota^* \pi^* M \underset{\alpha}{\cong} (P \otimes \pi^* M) \otimes \pi^* M = P \otimes \pi^* M^2 \cong P$$

is the identity. Let $E = \pi_* P$. Then (*a*) E is a $\pi^* \mathcal{O}_{X'}$-algebra, and since $\pi_* \mathcal{O}_{X'} \cong \mathcal{O}_X + L$, we are given an action of L on E; and (*b*) the isomorphism α defines an action of M on E. It is easy to check that altogether these actions make E into an A-module. Then it follows automatically [since $M_2(k)$ has a unique module of dimension 2 over k, up to isomorphism] that $A \cong \mathrm{Hom}(E, E)$.

But now

$$\pi^* E \cong \pi^* \pi_* P \cong P + \iota^* P,$$

hence

$$\pi^* (\Lambda^2 E) \cong P \otimes \iota^* P,$$

hence

$$2 \deg \Lambda^2 E = \deg \pi^* (\Lambda^2 E) = \deg P + \deg \iota^* P = 2 \deg P,$$

or $\deg \Lambda^2 E = \deg P$. Moreover, if $P = \mathcal{O}(\mathfrak{c})$ for some divisor $\mathfrak{c}$ on X′ disjoint from $\pi^{-1}(\mathfrak{a})$ and $\pi^{-1}(\mathfrak{b})$, then the existence of α means that there is an $h \in k(X')$ such that

$$(h) + \iota^{-1} \mathfrak{c} - \mathfrak{c} = \pi^{-1} \mathfrak{b}.$$

Then

$$(Nmh) = 2\mathfrak{b} = (g),$$

so $Nmh = \lambda . g$, $\lambda \in k^*$. Therefore :

$$\begin{aligned} g(\mathfrak{a}) &= Nmh(\mathfrak{a}) = h(\pi^{-1}\mathfrak{a}) = h((\sqrt{f})) \\ &= \sqrt{f}((h)), \quad \text{by reciprocity} \\ &= \sqrt{f}(\pi^{-1}\mathfrak{b}) . \frac{\sqrt{f}(\mathfrak{c})}{\sqrt{f}(\iota^{-1}\mathfrak{c})} = f(\mathfrak{b}) . (-1)^{\deg \mathfrak{c}} \\ &= f(\mathfrak{b}) . (-1)^{\deg \Lambda^2 E}. \end{aligned}$$

Q. E. D.

Now fix L_0 such that $L_0^2 \cong \Omega_X^1$. Then

$$\begin{aligned} e_*(L_0) + e_*(L_0 \otimes L) + e_*(L_0 \otimes M) + e_*(L_0 \otimes L \otimes M) \\ \equiv \dim H^0(L_0 \otimes A) \\ = \dim H^0(L_0 \otimes \mathrm{Hom}(E, E)). \end{aligned}$$

But it is well-known that all vector bundles of given rank and degree form a connected set, i. e. any 2 can be found as fibres of a family of such vector bundles over a connected and even irreducible base S (*cf.* Seshadri [S]). Therefore by our stability theorem, if $x \in X$ is any point and $r = \deg \Lambda^2 E$, setting $E_r = \mathcal{O}_X \oplus \mathcal{O}_X(rx)$, then

$$\begin{aligned} \dim H^0(L_0 \otimes \mathrm{Hom}(E, E)) &\equiv \dim H^0(L_0 \otimes \mathrm{Hom}(E_r, E_r)) \\ &= \dim H^0(L_0 \otimes \mathcal{O}_X(rx)) + 2 \dim H_0(L_0) \\ &\quad + \dim H_0(L_0 \otimes \mathcal{O}_X(-rx)) \\ &\equiv \dim H^0(L_0 \otimes \mathcal{O}_X(rx)) - \dim H^0(L_0 \otimes \mathcal{O}_X(-rx)) \\ &= \chi(L_0 \otimes \mathcal{O}_X(rx)), \quad \text{by Serre duality} \\ &= r \\ &\equiv \ln e_2(L, M), \quad \text{by lemma 2.} \end{aligned}$$

This completes the proof of identity (★) stated in the Introduction.

4. Final comments. — The third theorem mentioned in the introduction is that e_* takes the value 0 (resp. 1) $2^{g-1}(2^g + 1)$ [resp. $2^{g-1}(2^g - 1)$] times. This can be proven as follows : by the stability theorem in paragraph 1, if two curves X_1, X_2 lie in a family over a connected base S, then to prove it for X_1 it suffices to prove it for X_2. Since the moduli space of curves of genus of g is connected (Deligne-Mumford [D-M]), it therefore suffices to prove this for one X. Now for *hyperelliptic* X the result is very elementary. In fact, let

$$\pi : \quad X \to P^1$$

be the double covering, with ramification points $P_1, \ldots, P_{2g+2} \in X$. Let $\mathfrak{a}$ be the divisor class π^{-1} (one point) on X. We have the relations

$$2P_1 \equiv \ldots \equiv 2P_{2g+2} \equiv \mathfrak{a},$$

$$P_1 + \ldots + P_{2g+2} \equiv (g+1)\mathfrak{a}.$$

$$(g-1)\mathfrak{a} \equiv K_X, \quad \text{the canonical divisor class on X.}$$

Then the elements of S(X) are represented by the divisor classes

$$\mathfrak{b}_S^{(l)} = \sum_{\alpha \in S} P_\alpha + l.\mathfrak{a}, \quad l = 0, 1, \ldots, \left[\frac{g-1}{2}\right],$$

$$S \subset \{1, 2, \ldots, 2g+2\},$$

$$\# S = g - 1 - 2l$$

and

$$\mathfrak{b}_S^{(-1)} = \sum_{\alpha \in S} P_\alpha - \mathfrak{a}, \quad S \subset \{1, 2, \ldots, 2g+2\},$$

$$\# S = g + 1,$$

where $\mathfrak{b}_S^{(-1)} = \mathfrak{b}_T^{(-1)}$ if $\{1, 2, \ldots, 2g+2\} = S \cup T$. It is easy to check that

$$\dim H^0(X, \mathcal{O}_X(\mathfrak{b}_S^{(l)})) = l + 1$$

and then adding up the number of $\mathfrak{b}_S^{(l)}$ with l odd and l even, the result follows.

A natural question to ask is what happens in char. 2? Strangely, it turns out that in this case there is a natural line bundle L such that $L^2 \cong \Omega_X^1$ In fact, for every $f \in k(X) - k(X)^2$, it follows immediately by expanding f locally as a power series that the differential df has only double zeros and double poles, i. e.

$$(df) = 2\mathfrak{a}, \quad \text{for some divisor } \mathfrak{a}.$$

Moreover, if $f_1, f_2 \in k(X) - k(X)^2$, then $\{1, f_2\}$ is a 2-basis of $k(X)$, so

$$f_1 = a^2 f_2 + b^2, \quad \text{some } a, b \in k(X).$$

Therefore $df_1 = a^2 df_2$, hence if $(df_i) = 2\mathfrak{a}_i$, we find

$$\mathfrak{a}_1 = (a) + \mathfrak{a}_2$$

Therefore the divisor *class* $\mathfrak{a}$ is independent of f, and $L = \mathcal{O}_X(\mathfrak{a})$ is a canonical square root of Ω_X^1.

REFERENCES.

[B] N. Bourbaki, *Éléments de Mathématiques*, Hermann, Paris.

[D-M] P. Deligne and D. Mumford, *Irreducibility of the space of curves*, Publ. I. H. E. S., vol. 36, 1969.

[G] A. Grothendieck, *Groupe de Brauer*, in *Dix Exposés*, North Holland, 1968.

[K] Krazer, *Lehrburch der theta-funktionen*, Teubner, 1903.

[L 1] S. LANG, *Abelian varieties*, Wiley-Interscience, 1959.

[L 2] S. LANG, *On quasi-algebraic closure* (*Annals of Math.*, vol. 55, 1952, p. 373).

[M 1] D. MUMFORD, *Abelian varieties*, Oxford University Press, 1970.

[M 2] D. MUMFORD, *On the equations defining abelian varieties* (*Inv. Math.*, vol. 1, 1966).

[R] B. RIEMANN, *Collected Works*, Dover edition, 1953.

[S] C. S. SESHADRI, *Space of unitary vector bundles on a compact Riemann surface* (*Annals of Math.*, vol. 85, 1967).

[W] A. WEIL, *Variétés abéliennes*, Hermann, Paris, 1948.

[Wi] WIRTINGER, *Untersuchungen über thetafunktionen*, Teubner, 1895.

(Manuscrit reçu le 15 novembre 1970.)

David MUMFORD,
Mathematics Institute,
University of Warwick,
Coventry,
Warwickshire, England.

COMPOSITIO MATHEMATICA, Vol. 24, Fasc. 3, 1972, pag. 239–272
Wolters-Noordhoff Publishing
Printed in the Netherlands

AN ANALYTIC CONSTRUCTION OF DEGENERATING ABELIAN VARIETIES OVER COMPLETE RINGS

by

David Mumford

This paper is a sequel to the earlier paper on the construction of degenerating curves. The basic ideas of the 2 papers are very similar and we refer the reader to the introduction of that paper for discussion and motivation. We limit ourselves here to an outline of the contents of the present paper.

Throughout this paper, A will stand for a fixed excellent integrally closed noetherian ring with quotient field K, $I \subset A$ will be an ideal such that $I = \sqrt{I}$ and A is complete with respect to the I-adic topology. We will be interested in group schemes over the base scheme:

$$S = \mathrm{Spec}(A)$$

which have particular properties over

(1) the generic point $\eta \in S$

(2) the closed subscheme $S_0 = \mathrm{Spec}\ (A/I)$.

If X is a scheme over S, X_η and X_0 will stand for the induced schemes over $\{\eta\}$ and S_0. A smooth commutative group scheme G of finite type over any base Z will be called *semi-abelian* if all its fibres G_z are connected algebraic groups without unipotent radical, i.e., each G_z is an extension of an abelian variety by a torus (= a form of $\mathbf{G}_m^r$ over $k(z)$). The *rank function* will be the map

$$z \mapsto r(z)$$

associating to each point z the dimension of the torus part of G_z; r is easily checked to be upper $\frac{1}{2}$-continuous (e.g., by looking at the cardinality of the fibres of the subscheme $\mathrm{Ker}(n_G) \subset G$, étale over Z where n is prime to the residue characteristics of Z). The semi-abelian group schemes G of constant rank are *globally over* Z extensions of an abelian scheme over Z by a torus over Z. We are interested in constructing semi-abelian group schemes G/S such that

(1) G_η is an abelian variety,

(2) G_0 has constant rank.

According to an idea of John Tate, if $r = \mathrm{rank}\ (G_0)$, then G should be canonically represented as a 'quotient' of a semi-abelian group scheme $\tilde{G}$ of constant rank r *over the whole of* S by a discrete subgroup Y of L-valued points of $\tilde{G}$ with $Y \cong \mathbf{Z}^r$. (L a finite algebraic extension of K). To simplify matters, we will consider only the case where G_0 is a split torus, i.e., $G_0 \cong \mathbf{G}_m^r \times S_0$, in which case $\tilde{G} \cong \mathbf{G}_m^r \times S$ and the points of Y are K-valued. Our plan is the following:

i) start with a set of periods $Y \subset \tilde{G}(K)$ where $\tilde{G} \cong \mathbf{G}_m^r \times S$, satisfying suitable conditions;

ii) construct a kind of compactification:

such that the action of Y by translation extends to $\tilde{P}$, and Y acts freely and discontinuously (in the *Zariski* topology) on $\tilde{P}_0$. Unlike the case of curves, $\tilde{P}$ is neither unique nor canonical!

iii) Take the I-adic completion $\tilde{\mathfrak{P}}$ of $\tilde{P}$, construct $\mathfrak{P} = \tilde{\mathfrak{P}}/Y$, algebrize $\mathfrak{P}$ to a scheme P projective over S, and take a suitable open subset $G \subset P$:

$$\tilde{G} \underset{\text{open}}{\subset} \tilde{P} \underset{\text{compl.}}{\xleftarrow{\;\;I\text{-adic}\;\;}} \tilde{\mathfrak{P}}$$

$$\downarrow \text{formal morphism}$$

$$G \underset{\text{open}}{\subset} P \underset{\text{compl.}}{\xleftarrow{\;\;I\text{-adic}\;\;}} \mathfrak{P}$$

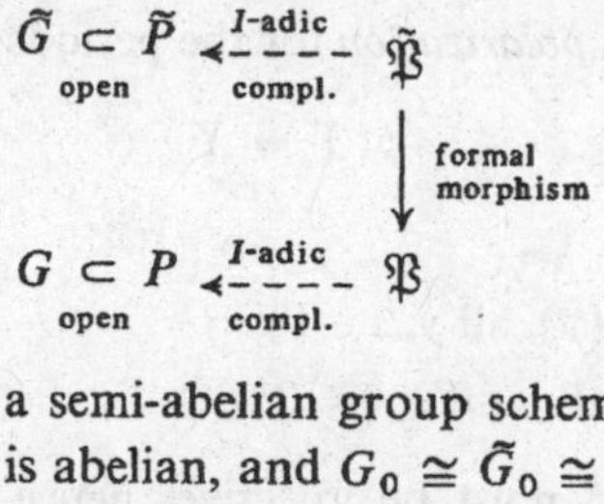

iv) prove that G is a semi-abelian group scheme over S independent of the choice of $\tilde{P}$, G_η is abelian, and $G_0 \cong \tilde{G}_0 \cong \mathbf{G}_m^r \times S_0$.

We will also show that this uniformization $\tilde{G} \to G$ is uniquely determined by G, i.e. if $\tilde{G}/Y_1 \cong \tilde{G}/Y_2$, then $Y_1 = Y_2$. However we will not discuss at all whether all semi-abelian G as in (iv) admit such a uniformization. This seems to be a fairly difficult question. In case $\dim A = 1$, and A is local, Raynaud and I have proven (independently) that all G's as in (iv) do admit uniformizations. Raynaud's method is highly analytic and mine is an application of my theory of 2-adic theta functions*. We both have reason to believe that our methods will extend to the general case, but this has not yet been done. Instead I conclude the paper with many examples. For me, one of the most enjoyable features of this research was the beauty of the examples which one works out without a great deal of extra effort. In fact, the non-uniqueness of $\tilde{P}$ gives one freedom to seek for the most elegant solutions in any particular case.

1. Periods

To begin our program, let $\tilde{G}$ be a given split torus over S. Then if X is the character group of $\tilde{G}$,for all $\alpha \in X$, the character is a canonical element

$$\mathscr{X}^{\alpha} \in \Gamma(\tilde{G}, \mathcal{O}_{\tilde{G}}).$$

Moreover $\tilde{G}$ is affine and can be described explicitly as:

$$\tilde{G} = \text{Spec}\, A[\cdots, \mathscr{X}^{\alpha}, \cdots]_{\alpha\in X} \Big/ \begin{pmatrix} \mathscr{X}^{\alpha}\cdot\mathscr{X}^{\beta} = \mathscr{X}^{\alpha+\beta} \\ \mathscr{X}^{0} = 1 \end{pmatrix}$$

Let:

$$\tilde{G}(K) = \text{group of } K\text{-valued points of } \tilde{G} \cong (K^*)^r.$$

Note that if $y \in G(K)$ and $\alpha \in X$, then the character $\mathscr{X}^{\alpha}$ takes a value on y which is an element $\mathscr{X}^{\alpha}(y) \in K^*$.

(1.1) DEFINITION: A set of *periods* is a subgroup $Y \subset \tilde{G}(K)$ isomorphic to $\mathbf{Z}^r$.

The only assumption that we will make about Y is that they admit a polarization, in the following sense:

(1.2) DEFINITION: A *polarization* for the periods Y is a homomorphism

$$\phi: Y \to X$$

such that:

i) $\mathscr{X}^{\phi(y)}(z) = \mathscr{X}^{\phi(z)}(y)$, all $y, z \in Y$,

ii) $\mathscr{X}^{\phi(y)}(y) \in I$ for all $y \in Y$, $y \neq 0$.

Note that by (ii) ϕ must be injective, hence also $[X: \phi Y] < +\infty$. Before going further, we stop to prove a basic lemma of a technical nature concerning periods and polarizations which will be very useful:

[1] On the equations defining abelian varieties, Inv. Math., Vol 1 and 3.

(1.3) Basic Lemma: *Suppose that for every $y \in Y$, $y \neq 0$, a positive integer n_y is given. Then there exists a finite set of elements $y_1, \cdots, y_k \in Y$, $y_i \neq 0$ and a finite subset $S \subset Y$ such that for all $z \in Y - S$,*

$$\mathscr{X}^{\phi(z)}(y_i) \in [\mathscr{X}^{\phi(y_i)}(y_i)]^{n_{y_i}} \cdot A \text{ for some } i.$$

Proof: Since Y is finitely generated, there is only a finite set of minimal prime ideals $\mathfrak{p} \subset A$ such that

$$\operatorname{ord}_{\mathfrak{p}} \mathscr{X}^{\phi(y)}(z) \not\equiv 0, \text{ all } y, z \in Y.$$

Let these prime ideals be $\mathfrak{p}_1, \cdots, \mathfrak{p}_n$, and let v_i be the valuation $\operatorname{ord}_{\mathfrak{p}_i}$. The axioms for a polarization tell us that

$$Q_i(y, z) = v_i[\mathscr{X}^{\phi(y)}(z)]$$

is a positive semi-definite quadratic form on Y. We can extend Q_i uniquely to a $\boldsymbol{R}$-valued quadratic form on $Y \otimes \boldsymbol{R}$. Since A is integrally closed, the assertion to be proved is equivalent to:

$$Q_j(z, y_i) \geqq n_{y_i} Q_j(y_i, y_i), \text{ for all } j.$$

For all $y \in Y$, $y \neq 0$, let

$$C_y = \left\{ z \in Y \otimes \boldsymbol{R} \;\middle|\; \begin{array}{l} Q_j(z, y) > n_y Q_j(y, y) \text{ for} \\ \text{all } j \text{ such that } y \notin \text{Null-space } (Q_j) \end{array} \right\}$$

Note that C_y is a convex open subset of $Y \otimes \boldsymbol{R}$ such that $\lambda C_y \subset C_y$ if $\lambda \in \boldsymbol{R}$, $\lambda \geqq 1$. To prove the lemma, it suffices to check that:

$$(*) \qquad C_{y_1} \cup \cdots \cup C_{y_k} \supset Y \otimes \boldsymbol{R} - (\text{a compact set})$$

for some $y_1, \cdots, y_k$. But first I claim that

$$(**) \qquad \bigcup_{N=1}^{\infty} \bigcup_{\substack{y \in Y \\ y \neq 0}} \frac{1}{N} C_y = Y \otimes \boldsymbol{R} - (0).$$

In fact, suppose $z \in Y \otimes \boldsymbol{R}$ and $z \neq 0$. If we approximate z by an element of $Y \otimes \boldsymbol{Q}$ which lies in the same Null-spaces of the Q_j that z does, and multiply this element by a large positive integer, we find a $y \in Y$ such that

$$z \in \text{Null-space } (Q_j) \Rightarrow y \in \text{Null-space } (Q_j)$$
$$z \notin \text{Null-space } (Q_j) \Rightarrow Q_j(y, z) > 0.$$

Therefore $N \cdot z \in C_y$ if $N \gg 0$, hence $z \in 1/N\, C_y$. This proves (**). By the compactness of the unit sphere in $Y \otimes \boldsymbol{R}$, it follows that there are $y_1, \cdots, y_k \in Y$, $y_i \neq 0$, and positive integers N_i such that

$$\bigcup_{i=1}^{k} \frac{1}{N_i} C_{y_i} \supset (\text{unit sphere}).$$

Then $\bigcup_{i=1}^{k} C_{y_i}$ contains all spheres of radius $\geqq \max(N_i)$ so (*) is proven.

QED

Another basic fact which is very useful is:

(1.4) LEMMA: *For any* $\alpha \in X$, *there exists an* $n \geqq 1$ *such that* $\mathscr{X}^{n\phi(y)+\alpha}(y) \in A$ *for all* $y \in Y$.

PROOF. Let $\mathfrak{p}_1, \cdots, \mathfrak{p}_n$ and $Q_1, \cdots, Q_n$ be as before. Note that using $\phi: Y \hookrightarrow X$, we can identify X with a subgroup of $Y \otimes \mathbf{R}$ so that

$$Q_i(\alpha, y) = \operatorname{ord}_{\mathfrak{p}_i} \mathscr{X}^{\alpha}(y), \qquad \alpha \in X,\ y \in Y.$$

We must show that for $n \gg 0$,

$$Q_i(ny+\alpha, y) \geqq 0, \text{ all } i, \text{ all } y.$$

But

$$nQ_i(ny+\alpha, y) = Q_i\left(ny+\frac{\alpha}{2}, ny+\frac{\alpha}{2}\right) - \tfrac{1}{4}Q_i(\alpha, \alpha).$$

Since Y projects into a lattice in $Y \otimes \mathbf{R}$/Null-space Q_i, it follows that if n is large enough, then for all y, either

a) $y \in$ Null-space (Q_i)

or

b) $ny+\alpha/2$ is arbitrarily 'big' in $Y \otimes \mathbf{R}$/Null-space Q_i. In either case, $Q_i(ny+\alpha, y) \geqq 0$.

QED

2. Relatively complete models

We now return to our basic problem: given a set of periods Y for which a polarization exists, construct *canonically* a 'quotient' of $\tilde{G}$ by Y. The main tool will be the following:

(2.1) DEFINITION. A *relatively complete model* of $\tilde{G}$ with respect to periods Y and polarization ϕ will be a collection of 5 pieces of data:

a) an integral scheme $\tilde{P}$ locally of finite type over A,

b) an open immersion $i: \tilde{G} \hookrightarrow \tilde{P}$ (we will henceforth identify $\tilde{G}$ with its image in $\tilde{P}$),

c) an invertible sheaf $\tilde{L}$ on $\tilde{P}$,

d) an action of the torus $\tilde{G}$ on $\tilde{P}$ and $\tilde{L}$ (we denote the action of an S'-valued point a of $\tilde{G}$ by

$$T_a : \tilde{P} \to \tilde{P}$$

and

$$T_a^* : \tilde{L} \to \tilde{L}),$$

e) and action of Y on $\tilde{P}$ and $\tilde{L}$ (we denote the action of $y \in Y$ by

$$S_y : \tilde{P} \to \tilde{P}$$

and

$$S_y^* : \tilde{L} \to \tilde{L}),$$

such that

(i) there exists an open $\tilde{G}$-invariant subset $U \subset \tilde{P}$ of finite type over S such that

$$\tilde{P} = \bigcup_{y \in Y} S_y(U),$$

(ii) for all valuations v on $R(\tilde{G})$, the field of rational functions on $\tilde{G}$, for which $v \geqq 0$ on A, v has a center on $\tilde{P}$ if and only if

$$* \; [\forall\, \alpha \in X, \exists y \in Y \text{ such that } v(\mathscr{X}^\alpha(y) \cdot \mathscr{X}^\alpha) \geqq 0].$$

(These valuations are allowed to have rank > 1!)

(iii) The action of $\tilde{G}$ and Y on $\tilde{P}$ extends the 'translation" action of $\tilde{G}$ and Y on $\tilde{G}$ given by the group law on $\tilde{G}$.

(iv) The actions of Y and $\tilde{G}$ on $\tilde{L}$ satisfy the identity:

$$T_a^* S_y^* = \mathscr{X}^{\phi(y)}(a) \cdot S_y^* T_a^*$$

all $y \in Y$ all S'-valued points a of $\tilde{G}$.

(v) $\tilde{L}$ is ample on $\tilde{P}$.[2]

We turn first to the construction of relatively complete models. We need one more definition:

(2.2) Definition. A *star* Σ is a finite subset of X such that $0 \in \Sigma$, $-\Sigma = \Sigma$ and Σ contains a basis of X.

Let θ be an indeterminate and consider the big graded ring:

$$\mathscr{R} = \sum_{k=0}^{\infty} \left(K[\cdots, \mathscr{X}^\alpha, \cdots] \middle/ \left(\begin{matrix} \mathscr{X}^\alpha \cdot \mathscr{X}^\beta - \mathscr{X}^{\alpha+\beta} \\ \mathscr{X}^0 - 1 \end{matrix} \right) \right) \cdot \theta^k.$$

Let Y act on $\mathscr{R}$ via operators S_y^*:

$$S_y^*(c) = c, \text{ all } c \in K,$$
$$S_y^*(\mathscr{X}^\alpha) = \mathscr{X}^\alpha(y) \cdot \mathscr{X}^\alpha, \text{ all } \alpha \in X,$$
$$S_y^*(\theta) = \mathscr{X}^{\phi(y)}(y) \cdot \mathscr{X}^{2\phi(y)} \cdot \theta.$$

[2] By definition, we take this to mean that the sections of L^n, $n \geqq 1$, form a basis of the topology of $\tilde{P}$. In EGA, II. 4.5., Grothendieck only defines ample on quasi-compact schemes, but this seems to be the best property among his equivalent defining properties for our purposes.

Let

(2.3) DEFINITION. $R_{\phi,\Sigma}$ = subring of $\mathscr{R}$ generated over A by the elements $S_y(\mathscr{X}^\alpha\theta)$, $y \in Y$, $\alpha \in \Sigma$,
i.e.,

$$R_{\phi,\Sigma} = A[\cdots, \mathscr{X}^{\phi(y)+\alpha}(y) \cdot \mathscr{X}^{2\phi(y)+\alpha} \cdot \theta, \cdots].$$

This ring is only what we want in case $\mathscr{X}^{\phi(y)+\alpha}(y) \in A$ for all $y \in Y$, $\alpha \in \Sigma$, i.e., in case

$$R_{\phi,\Sigma} \subset A[\cdots, \mathscr{X}^\alpha \cdot \theta, \cdots].$$

However, we can always make sure that this extra condition is satisfied if we replace the polarization ϕ by $n\phi$ for a large enough n. This follows from lemma (1.4). From now on, we assume that ϕ has been so chosen that $\mathscr{X}^{\phi(y)+\alpha}(y) \in A$, all $y \in Y$, $\alpha \in \Sigma$.

Now consider $\mathrm{Proj}(R_{\phi,\Sigma})$. I claim it is a relatively complete model of $\tilde{G}$! Since it is Proj of a graded ring generated by elements of degree 1, it carries a canonical ample invertible sheaf $\mathcal{O}(1)$. Moreover, the automorphisms S_y^* of $R_{\phi,\Sigma}$ induce automorphisms S_y of $\mathrm{Proj}(R_{\phi,\Sigma})$ and a compatible automorphism S_y^* of $\mathcal{O}(1)$, i.e., an action of Y on $\mathrm{Proj}(R_{\phi,\Sigma})$ and on $\mathcal{O}(1)$. To get an action of $\tilde{G}$ on $\mathrm{Proj}(R_{\phi,\Sigma})$ and on $\mathcal{O}(1)$, it suffices to define, for every A-algebra B and every B-valued point a of $\tilde{G}$, an automorphism T_a^* of $B \otimes_A R_{\phi,\Sigma}$, in a way which is functorial and compatible with compositions. In fact, let

$$T_a^*(c) = c, \text{ all } c \in A,$$
$$T_a^*(\mathscr{X}^\alpha) = \mathscr{X}^\alpha(a) \cdot \mathscr{X}^\alpha, \text{ all } \alpha \in X,$$
$$T_a^*(\theta) = \theta.$$

This clearly has all these properties and gives us our action. Since:

$$T_a^* \cdot S_y^* = R \cdot S_y^* \cdot T_a^*$$

where

$$R(c) = c, \text{ all } c \in A$$
$$R(\mathscr{X}^\alpha) = \mathscr{X}^\alpha, \text{ all } \alpha \in X$$
$$R(\theta) = \mathscr{X}^{2\phi(y)}(a) \cdot \theta,$$

it follws that the T_a's and S_y's skew-commute as required for the polarization 2ϕ.

Next $\mathrm{Proj}(R_{\phi,\Sigma})$ is covered by the affine open sets:

$$U_{\alpha,y} = \mathrm{Spec}\, A\left[\cdots, \frac{\mathscr{X}^{\phi(z)+\beta}(z)}{\mathscr{X}^{\phi(y)+\alpha}(y)} \cdot \mathscr{X}^{2\phi(z-y)+\beta-\alpha}, \cdots\right]_{\substack{\beta\in\Sigma \\ z\in Y}}.$$

Although we have here an infinite number of open sets, the action S_{y_0} of $y_0 \in Y$ carries $U_{\alpha,y}$ to $U_{\alpha,y+y_0}$, so there are only a finite number of orbits mod Y in this collection of open sets. All the affine rings are integral domains contained in $K(\cdots, \mathscr{X}^{\alpha}, \cdots)$. Moreover,

$$\begin{aligned} U_{0,0} &= \operatorname{Spec} A[\cdots, \mathscr{X}^{\phi(z)+\beta}(z) \cdot \mathscr{X}^{2\phi(z)+\beta}, \cdots]_{\substack{z\in Y\\ \beta\in\Sigma}} \\ &= \operatorname{Spec} A[\cdots, \mathscr{X}^{\beta}, \cdots]_{\beta\in\Sigma} \\ &= \operatorname{Spec} A[\cdots. \mathscr{X}^{\beta}, \cdots]_{\beta\in X} = \tilde{G}. \end{aligned}$$

Thus $\operatorname{Proj}(R_{\phi,\Sigma})$ is an integral scheme over A containing $\tilde{G}$ as a dense open subset. Next we prove:

(2.4) PROPOSITION. *$U_{\alpha,y}$ is of finite type over A.*

PROOF. Since $U_{\alpha,y}$ is isomorphic to $U_{\alpha,z}$ for any $y, z \in Y$, is suffices to check this for $U_{\alpha,0}$. The affine ring of $U_{\alpha,0}$ is generated by the infinite set of monomials:

$$M_{\beta,y} = \mathscr{X}^{\phi(y)+\beta}(y) \cdot \mathscr{X}^{2\phi(y)+\beta-\alpha}.$$

It is easy to check that:

$$(^*)\ M_{\beta,y} = [\mathscr{X}^{2\phi(y-z)-\alpha+\beta}(z)] \cdot M_{\alpha,z} \cdot M_{\beta,y-z}.$$

Now for all z, choose a positive integer n_z such that

$$\mathscr{X}^{\phi(z)}(z)^{2n_z-3} \cdot \mathscr{X}^{-\alpha+\beta}(z) \in A, \text{ all } \beta \in \Sigma.$$

Then by the basic lemma (1.3), there is a finite set $z_1, \cdots, z_k$ such that for all $y \in Y$ except for a finite set $S \subset Y$,

$$\mathscr{X}^{\phi(y)}(z_i) \in [\mathscr{X}^{\phi(z_i)}(z_i)]^{n_{z_i}} \cdot A, \text{ some } i.$$

Therefore

$$\begin{aligned} \mathscr{X}^{2\phi(y-z_i)-\alpha+\beta}(z_i) &= [\text{elt. of } A \cdot \mathscr{X}^{\phi(z_i)}(z_i)^{n_{z_i}}]^2 \cdot \mathscr{X}^{-2\phi(z_i)}(z_i) \cdot \mathscr{X}^{-\alpha+\beta}(z_i) \\ &= (\text{elt. of } A) \cdot \mathscr{X}^{\phi(z_i)}(z_i) \in I, \end{aligned}$$

so that

$$M_{\beta,y} = (\text{elt. of } I) \cdot M_{\alpha,z_i} \cdot M_{\beta,y-z_i}.$$

Thus $M_{\alpha_1 z}, \cdots, M_{\alpha,z_k}$ plus the $M_{\beta,y}$ with $y \in S$ generate the whole ring.

QED

It follows that if $U = \bigcup_{\alpha\in\Sigma} U_{\alpha,0}$, then U is an open subset of $\operatorname{Proj}(R_{\phi,\Sigma})$ of finite type over A such that

$$\bigcup_{y\in Y} S_y(U) = \operatorname{Proj}(R_{\phi,\Sigma}).$$

It remains to check the following completeness property for $\mathrm{Proj}(R_{\phi,\Sigma})$: if v is a valuation of $R(\tilde{G})$, $v \geqq 0$ on A, then

$$\begin{bmatrix} v \text{ has a center} \\ \text{on Proj}\,(R_{\phi,\Sigma}) \end{bmatrix} \Leftrightarrow \begin{bmatrix} \forall \alpha \in x, \exists y \in Y \text{ such that} \\ v(\mathscr{X}^{\alpha}(y) \cdot \mathscr{X}^{\alpha}) \geqq 0 \end{bmatrix}.$$

Clearly v has a center on $\mathrm{Proj}(R_{\phi,\Sigma})$ if and only if

$$\min_{\substack{\alpha \in \Sigma \\ y \in Y}} v(\mathscr{X}^{\phi(y)+\alpha}(y) \cdot \mathscr{X}^{2\phi(y)+\alpha}) \text{ exists.}$$

As for the other side, note that it is equivalent to any of the statements:

$$(*)'[\forall z \in Y, \exists y \in Y \text{ such that } v(\mathscr{X}^{\phi(z)}(y) \cdot \mathscr{X}^{\phi(z)}) \geqq 0]$$
$$(*)''[\forall z \in Y, \exists \alpha \in X \text{ such that } v(\mathscr{X}^{\alpha}(z) \cdot \mathscr{X}^{\phi(z)}) \geqq 0]$$
$$(*)'''[\forall z \in Y, \exists n \geqq 1 \text{ such that } v(\mathscr{X}^{\phi(z)}(z)^n \cdot \mathscr{X}^{\phi(z)}) \geqq 0].$$

Now to check the implication '$\Rightarrow$', suppose that:

$$\min_{\substack{\alpha \in \Sigma \\ y \in Y}} v(\mathscr{X}^{\phi(y)+\alpha}(y) \cdot \mathscr{X}^{2\phi(y)+\alpha}) = v(\mathscr{X}^{\phi(y_0)+\alpha_0}(y_0) \cdot \mathscr{X}^{2\phi(y_0)+\alpha_0}).$$

If $(*)''$ is false, then take the $z \in Y$ for which $v(\mathscr{X}^{\alpha}(z) \cdot \mathscr{X}^{\phi(z)}) < 0$, all $\alpha \in X$, and simply note that

$$\begin{aligned} &v(\mathscr{X}^{\phi(y_0+z)+\alpha_0}(y_0+z) \cdot \mathscr{X}^{2\phi(y_0+z)+\alpha_0}) \\ &\quad = v(\mathscr{X}^{\phi(y_0)+\alpha_0}(y_0) \cdot \mathscr{X}^{2\phi(y_0)+\alpha_0}) + v(\mathscr{X}^{\phi(z)}) + v(\mathscr{X}^{2\phi(y_0)+\phi(z)+\alpha_0}(z) \cdot \mathscr{X}^{\phi(z)}) \\ &\quad < v(\mathscr{X}^{\phi(y_0)+\alpha_0}(y_0) \cdot \mathscr{X}^{2\phi(y_0)+\alpha_0}). \end{aligned}$$

Conversely, assume that $(*)'''$ holds. For all $y \in Y$, choose n_y large enough so that

$$v(\mathscr{X}^{\phi(y)}(y)^{n_y} \cdot \mathscr{X}^{2\phi(y)}) \geqq 0$$

and

$$\mathscr{X}^{\phi(y)}(y)^{n_y - 1} \cdot \mathscr{X}^{\alpha}(y) \in A, \text{ all } \alpha \in \Sigma.$$

By the basic lemma (1.3), there exist $y_1, \cdots, y_k \in Y$ such that for all $z \notin S$, a finite set, $\mathscr{X}^{\phi(z)}(y_i) \in [\mathscr{X}^{\phi(y_i)}(y_i)]^{n_{y_i}} \cdot A$ for some i. But then

$$\begin{aligned} &v(\mathscr{X}^{\phi(z)+\alpha}(z) \cdot \mathscr{X}^{2\phi(z)+\alpha}) \\ &\quad = v(\mathscr{X}^{\phi(z-y_i)+\alpha}(z-y_i) \cdot \mathscr{X}^{2\phi(z-y_i)+\alpha}) \\ &\qquad + v(\mathscr{X}^{2\phi(z)}(y_i) \cdot \mathscr{X}^{-\phi(y_i)+\alpha}(y_i) \cdot \mathscr{X}^{2\phi(y_i)}) \\ &\quad \geqq v(\mathscr{X}^{\phi(z-y_i)+\alpha}(z-y_i) \cdot \mathscr{X}^{2\phi(z-y_i)+\alpha}) \\ &\qquad + v(\mathscr{X}^{\phi(y_i)}(y_i)^{n_{y_i}} \cdot \mathscr{X}^{2\phi(y_i)}) \\ &\qquad + v(\mathscr{X}^{\phi(y_i)}(y_i)^{n_{y_i}-1} \cdot \mathscr{X}^{\alpha}(y_i)) \\ &\quad \geqq v(\mathscr{X}^{\phi(z-y_i)+\alpha}(z-y_i) \cdot \mathscr{X}^{2\phi(z-y_i)+\alpha}) \end{aligned}$$

so that the minimum in question exists for some $z \in S$. This completes the proof of:

(2.5) THEOREM. *Let $\tilde{G}$ be a split torus over S, let $Y \subset \tilde{G}(K)$ be a set of periods and $\phi: Y \to X$ a polarization. Then if ϕ is replaced by $n\phi$ for $n \in Z$ sufficiently large, $Proj(R_{\phi,\Sigma})$ is a relatively complete model of $\tilde{G}$ over S relative to Y and 2ϕ.*

3. The Construction of the quotient

We can now forget about $\mathrm{Proj}(R_{\phi,\Sigma})$ and deal with an arbitrary relatively complete model $\tilde{P}$. The first thing I want to prove is that $\tilde{P}$ is not 'too much' bigger than $\tilde{G}$:

(3.1) PROPOSITION. *Let $y \in Y$ and let $f = \mathscr{X}^{\phi(y)}(y)$. Then in the open set $\tilde{P}_f \subset \tilde{P}$, $\mathscr{X}^{\phi(y)}$ is a unit, i.e., $\in \Gamma(\tilde{P}_f, \mathcal{O}^*_{\tilde{P}})$.*

PROOF. By the axioms for a polarization, it follows easily that for all $\alpha \in X$,

$$\mathscr{X}^{n\phi(y)+\alpha}(y) \in A \text{ if } n \gg 0.$$

Therefore $\mathscr{X}^\alpha(y) \in A_f$. In other words, the section y of $\tilde{G}$ over $\{\eta\}$ extends to a section over S_f too. For clarity, call this y'. Then we have automorphisms S_y and $T_{y'}$ of $\tilde{P}_f$ and S^*_y and $T^*_{y'}$, of $\tilde{L}_f = \tilde{L}|_{\tilde{P}_f}$. Since $S_y = T_{y'}$, on G_f, S_y equals $T_{y'}$, everywhere, so

$$S^*_y = \begin{pmatrix}\text{mult. by a unit } \lambda \\ \text{in } \tilde{P}_f\end{pmatrix} \cdot T^*_{y'}.$$

The law of skew-commutativity of S^*_y with the operators T^*_a shows that

$$\lambda(a+b) = \mathscr{X}^{\phi(y)}(b) \cdot \lambda(a)$$

for all points a, b of $\tilde{G}_\eta$ in some K-algebra B. Therefore, on $\tilde{G}_\eta$, the function $\lambda/\mathscr{X}^{\phi(y)}$ is constant along the fibres, i.e.,

$$\lambda = \zeta \cdot \mathscr{X}^{\phi(y)}, \zeta \in K.$$

But λ and $\mathscr{X}^{\phi(y)}$ are units on the bigger open set $\tilde{G}_f$, hence ζ must be a unit in A_f. Then λ and ζ^{-1} are units in $\tilde{P}_f$ so $\mathscr{X}^{\phi(y)}$ is also. *QED*

(3.2) COROLLARY. $\tilde{G}_\eta = \tilde{P}_\eta$.

Next, let's look at the closed subscheme $\tilde{P}_0$ of $\tilde{P}$.

(3.3) PROPOSITION. *Every irreducible component of $\tilde{P}_0$ is proper over A/I.*

PROOF. First apply the completeness condition (ii) to prove that if Z

is any component of $\tilde{P}_0$ and if v is any valuation of its quotient field $R(Z)$, with $v \geqq 0$ on A/I, then v has a center on Z. In fact, let v_1 be a valuation of $R(\tilde{G})$, $v \geqq 0$ on A, whose center is Z, and let v_2 be the composite of the valuations v and v_1. Since for all $z \in Y$, if $n \gg 0$, $\mathscr{X}^{\phi(z)}(z)^n \cdot \mathscr{X}^{\phi(z)}$ is regular and zero at the generic point of Z by (3.1), it follows that $v_1(\mathscr{X}^{\phi(z)}(z)^n \cdot \mathscr{X}^{\phi(z)}) > 0$, hence $v_2(\mathscr{X}^{\phi(z)}(z)^n \cdot \mathscr{X}^{\phi(z)}) > 0$. So by the completeness condition, v_2 has a center on $\tilde{P}$, hence v has a center on Z.

The Proposition now follows from the rather droll:

(3.4) LEMMA. *Let $f: X \to Y$ be a morphism locally of finite type, with X an irreducible scheme but Y arbitrary. If f satisfies the valuative criterion for properness for all valuations, then f is proper.*

PROOF. The usual valuative criterion (cf. EGA. II-7-3) would hold if we know that f was of finite type. It suffices to prove that f is quasi-compact. To prove this, we may as well replace X by X_{red}, Y by Y_{red}; then looking locally on the base, we can assume Y is affine, say Spec(A); and finally we can assume f is dominating. Then A is a subring of the function field $R(X)$ of X: let $\mathscr{X}$ be Zariski's Riemann Surface of $R(X)/A$ (cf. Zariski-Samuel, vol. II, p. 110). Set-theoretically, $\mathscr{X}$ is the set of valuations v of $R(X)$, $v \geqq 0$ on A. By the valuation criterion, every v has a center on X, so there is a natural map $\pi: \mathscr{X} \to X$ taking v to its center. Now $\mathscr{X}$ is a quasi-compact topological space and π is continuous and surjective. Therefore X is quasi-compact. *QED*

(3.5) COROLLARY. *The closure $\bar{U}_0$ of U_0 in the scheme $\tilde{P}_0$ is proper over S_0.*

PROOF. Since U_0 is of finite type over S_0, $\bar{U}_0$ has only a finite number of irreducible components, and by the Proposition, each is of finite type over S_0. *QED*

The next thing I want to prove is that Y acts freely and discontinuously on $\tilde{P}_0$ in the Zariski-topology:

(3.6) PROPOSITION. *Let $U \subset \tilde{P}$ be the open set given by the definition of a relatively complete model. Let $\bar{U}_0$ be the closure of U_0 in $\tilde{P}_0$. There is a finite subset $S \subset Y$ such that*

$$S_y(\bar{U}_0) \cap S_z(\bar{U}_0) = \phi$$

if $y - z \notin S$.

PROOF. Let $F \subset \tilde{P}$ be the closed subset which is the locus of geometric points left fixed by the action of $\tilde{G}$. The action of $\tilde{G}$ on the invertible sheaf $\tilde{L}|_F$ is a 1-dimensional representation of $\tilde{G}$ over the base scheme

F. Therefore for every *connected* subset $F' \subset F$, there is a character $\alpha \in X$ such that $\tilde{G}$ acts on $\tilde{L}|_{F'}$ via the character α. Moreover, if $y \in Y$, then $S_y(F')$ will be another connected subset of F, and by the skew-commutativity of the actions of Y and $\tilde{G}$, $\tilde{G}$ will act on $\tilde{L}|_{S_y(F')}$ through the character $\alpha+\phi(y)$. Now $F \cap \bar{U}_0$ has only a finite number of connected components: let $\alpha_1, \cdots, \alpha_n$ be the characters $\tilde{G}$ associated to the action of $\tilde{G}$ on $\tilde{L}$ on these sets. Then $\tilde{G}$ acts on $\tilde{L}$ at the points of $F \cap S_y(\bar{U}_0)$ through the characters $\alpha_1+y, \cdots, \alpha_n+y$. Now suppose $S_y(\bar{U}_0) \cap S_z(\bar{U}_0) \neq \phi$. Since this intersection is proper over Spec(A/I), by the Borel fixed point theorem (cf. A. Borel-Linear algebraic groups; Benjamin, 1959; pag. 242, th. 10.4):

$$F \cap S_y(\bar{U}_0) \cap S_z(\bar{U}_0) \neq \phi.$$

Looking at the action of $\tilde{G}$ on $\tilde{L}$ here, it follows that one of the characters α_i+y must equal one of the characters α_j+z. Let $S = \{\cdots, \alpha_i-\alpha_j, \cdots\}$. Thus $y-z \in S$.

QED

(3.7) Corollary. *Y acts freely on $\tilde{P}_0$.*

Proof. If some $y \in Y, y \neq 0$, had a fixed point x in $\tilde{P}_0$, then since y has infinite order, x would be left fixed by an infinite subgroup of Y and this would contradict the Proposition. *QED*

(3.8) Theorem. *$\tilde{P}_0$ is connected.*

Proof. Note that since A is complete in the I-adic topology, and A has no idempotents, A/I has no idempotents either, i.e., S_0 is connected. Therefore $\tilde{G}_0$ is a connected open subset of $\tilde{P}_0$ and it determines a canonical connected component of $\tilde{P}_0$. Now suppose there is a 2nd connected component. Choose a point $x \in \tilde{P}_0$ in this 2nd component and let v be a discrete rank 1 valuation of $R(\tilde{G})$, $v \geqq 0$ on A, with center x. Let $A' = \{x \in K | v(x) \geqq 0\}$. Let $S' = \text{Spec}(A')$ and $\tilde{P}' = \tilde{P} \times_S S'$. Now A' is a discrete, rank 1, valuation ring with quotient field K, so S and S' have the 'same' generic point, and $\tilde{P}, \tilde{P}'$ have the 'same' generic fibre. Let $\tilde{P}''$ be the closure in $\tilde{P}'$ of its generic fibre. $\tilde{P}''$ is an integral scheme with the same quotient field as $\tilde{P}$, namely $R(\tilde{G})$, and it is locally of finite type over the valuation ring A'. Now S' has only 2 points – its generic point and its closed point. Let $\tilde{P}_0''$ be the fibre of $\tilde{P}''$ over the closed point. There is a natural morphism

$$\tilde{P}_0'' \to \tilde{P}_0.$$

I claim that (a) $x \in$ Image, and (b) $\tilde{G}_0 \subset$ Image, hence the image meets $\geqq 2$ connected components, hence $\tilde{P}_0''$ is disconnected too. The reason for (a) is that if R_v = valuation ring of v, then we are given morphisms:

$$\begin{array}{ccc} \mathrm{Spec}\,(R_v) & \longrightarrow & \tilde{P} \\ \downarrow & & \downarrow \\ S' & \longrightarrow & S \end{array}$$

hence we get a morphism $\mathrm{Spec}(R_v) \to \tilde{P}'$, which takes the generic point of $\mathrm{Spec}(R_v)$ to the generic fibre of $\tilde{P}'$. Therefore this morphism factors through $\tilde{P}''$. The image $x' \in \tilde{P}_0''$ of the closed point lies over x. The reason for (b) is that $\tilde{G} \subset \tilde{P}$, and $\tilde{G}$ is smooth of S; so $\tilde{G} \times_s S' \subset \tilde{P}'$, and it is smooth over S'; so $\tilde{G} \times_s S' \subset \tilde{P}''$.

Now because of the completeness property (ii) of $\tilde{P}$ we know exactly which valuations of $\boldsymbol{R}(\tilde{G})$ have centers in $\tilde{P}''$ too. We have reduced the Theorem to:

(3.9) Lemma. *Let A be a discrete rank 1 valuation ring with maximal ideal (π), let G be a split torus over $S = \mathrm{Spec}(A)$, let P be an integral scheme, locally of finite type over S containing G as a dense open subset. Assume:*

1) *the generic fibres G_η, P_η are equal,*

2) *for all valuations v of $\boldsymbol{R}(G)$, $v \geqq 0$ on A and $v(\pi) > 0$, v has a center on P if and only if for all $\alpha \in X$, (the character group of G), $-nv(\pi) \leqq v(\mathscr{X}^\alpha) \leqq nv(\pi)$, if $n \gg 0$.*

Then the closed fibre P_0 of P is connected.

Proof. Introduce a basis in X, and let $\mathscr{X}_1, \cdots, \mathscr{X}_r$ be the corresponding characters. For all n, let

$$\begin{aligned} P^{(n)} &= \mathrm{Spec}\, A[\pi^n\mathscr{X}_1 \cdot \pi^n\mathscr{X}_1^{-1}, \cdots, \pi^n\mathscr{X}_r, \pi^n \cdot \mathscr{X}_r^{-1}] \\ &\cong \mathrm{Spec}\, A[U_1, V_1, \cdots, U_r, V_r]/(U_1V_1 - \pi^{2n}, \cdots, U_rV_r - \pi^{2n}). \end{aligned}$$

This scheme is a relative complete intersection in A_s^{2r} over S and is smooth over S, hence regular, outside a subset of codimension 2. Therefore $P^{(n)}$ is a normal scheme. $P^{(n)}$ and P have the same function field, so let $Z^{(n)} \subset P^{(n)} \times_A P$ be the join of this birational correspondence. By (1) and (2), all valuations v of $\boldsymbol{R}(G)$ with a center on $P^{(n)}$ also have a center on P, so $Z^{(n)} \to P^{(n)}$ satisfies the valuative criterion for properness. Since $Z^{(n)}$ is at least locally of finite type over $P^{(n)}$, by lemma (3.4), $Z^{(n)}$ is proper over $P^{(n)}$. Therefore by Zariski's connectedness theorem, all fibres of $Z^{(n)}$ over $P^{(n)}$ are connected. Now the closed fibre of $P^{(n)}$ is isomorphic to:

$$\mathrm{Spec}\,(k[U, V]/(U \cdot V)) \underset{\mathrm{Spec}\,k}{\times} \cdots \underset{\mathrm{Spec}\,(k)}{\times} \mathrm{Spec}\,(k[U, V]/(U \cdot V))$$

where $k = A/(\pi)$, which is certainly connected. Therefore the closed fibre of $Z^{(n)}$ is connected. Therefore if

$$W_n = \overline{p_2(Z_0^{(n)})} \subset P,$$

W_n is connected too. But I claim that $P_0 = \bigcup_n W_n$. In fact, every valuation v with a center x on P_0 has a center on *some* $P^{(n)}$ because of (2). Therefore x lifts to a point of $Z^{(n)}$ for some n, hence $x \in W_n$. This shows that P_0 is connected. *QED*

We are now ready to begin the construction of G. The first step is:

(3.10) THEOREM. *For every $n \geqq 1$, there exists a scheme P_n, projective over A/I^n, an ample sheaf $\mathcal{O}(1)$ on P_n, and an étale surjective morphism:*

$$\pi: \tilde{P} \times_A A/I^n \to P_n$$

such that set-theoretically, $\pi(x) = \pi(y)$ if and only if x and y are in the same Y-orbit, and such that $\mathcal{O}(1)$ on $\tilde{P} \times_A A/I^n$ is the pull-back of $\mathcal{O}(1)$ on P_n.

PROOF. First, let $k \geqq 1$ be an integer such that under the action of the subgroup $kY \subset Y$, no 2 points of any open set

$$S_y(U) \times_A A/I^n$$

are identified. Then we can form a quotient:

$$\pi' : \tilde{P} \times_A A/I^n \to P_n'$$

by the subgroup kY by the simple device of gluing these basic open sets together on bigger overlaps. Observe that since Y acts on $\mathcal{O}(1)$ we get a 'descended' form of $\mathcal{O}(1)$ on P_n'.

Choose coset representatives $y_1, \cdots, y_l \in Y$ for the cosets of kY in Y. Now notice that the restriction of π':

$$\operatorname{Res}(\pi') : \bigcup_{y \in \{y_1, \cdots, y_l\}} S_y(U) \times_A A/I \to P_n'$$

is surjective, hence so is the restriction:

$$\operatorname{Res}(\pi') : \bigcup_{y \in \{y_1, \cdots, y_l\}} S_y(\overline{U_0}) \to P_n'.$$

But the scheme on the left is a finite union of schemes proper over A/I. Therefore P_n' is proper over A/I^n. Moreover $\mathcal{O}(1)$ on P_n' pulls back to $\mathcal{O}(1)$ on the left, and $\mathcal{O}(1)$ here is ample. Therefore $\mathcal{O}(1)$ on P_n' is ample too (for instance, by Nakai's criterion, cf. S. Kleiman – A note on the Nakai-Moisezon test for ampleness of a divisor; Amer. J. Math. 87 (1965), 221–226).

Finally, the finite group Y/kY acts freely on the projective scheme P_n' and on the ample sheaf $\mathcal{O}(1)$, so a quotient $P_n = P_n'/(Y/kY)$ exists. Moreover by descent, P_n carries an ample $\mathcal{O}(1)$ too, so P_n has all the required properties. *QED*

These schemes P_n obviously fit together to form a formal scheme $\mathfrak{P}$ proper over A. Moreover the sheaves $\mathcal{O}(1)$ fit together into an ample sheaf $\mathcal{O}(1)$ on $\mathfrak{P}$. We now apply the fundamental *formal existence* theorem (EGA Ch. 3, (5.4.5)) of Grothendieck: this shows that $\mathfrak{P}$ is the formal completion of a unique scheme P, proper over A, and that $\mathcal{O}(1)$ on $\mathfrak{P}$ comes by completion from an $\mathcal{O}(1)$ on P relatively ample over A.

Now inside all of our schemes, we want to pick out a big open set:

(*I.*) $\bigcup_{y\in Y} S_y(\tilde{G}) \subset \tilde{P}$

(*II.*) $G_n =^{def} \bigcup_{y\in Y} (S_y(\tilde{G})\times_A A/I^n)/Y \subset P_n$

(*III.*) $\varinjlim_n G_n = \mathfrak{G} \subset \mathfrak{P}$.

Note that $G_n \cong \tilde{G}\times_A A/I^n$, so that $\mathfrak{G} \cong I$-adic completion of $\tilde{G}$. To pick out an open subset of P whose completion is $\mathfrak{G}$, proceed as follows:

(*I.*) Let $\tilde{B} = \tilde{P} - \bigcup_{y\in Y} S_y(\tilde{G})$ and make $\tilde{B}$ into a reduced closed subscheme of $\tilde{P}$.

(*II.*) Let $B_n = (\tilde{B}\times_A A/I^n)/Y \subset P_n$.

(*III.*) Let $\varinjlim_n B_n = \mathfrak{B} \subset \mathfrak{P}$.

Then $\mathfrak{B}$ is the formal completion of a reduced closed subscheme $B \subset P$. Finally let $G = P - B$. Then, by construction the I-adic completion of G is $\mathfrak{G}$, i.e., the I-adic completions of G and of $\tilde{G}$ are canonically isomorphic.

This G is our final goal. We will eventually prove that G is a semi-abelian group scheme.

4. *G* is semi-abelian

We begin by proving that G is smooth over S. This is best proved in a more general context:

(4.1) PROPOSITION. Given

(1) 2 *schemes, locally of finite type over* S:

$$\begin{array}{ccc} X_1 & & X_2 \\ & {\scriptstyle f_1}\searrow \quad \swarrow{\scriptstyle f_2} & \\ & S & \end{array}$$

with X_2 *proper over* S.

(2) *an étale surjective morphism of their I-adic completions*:

$$\pi: \mathscr{X}_1 \to \mathscr{X}_2$$

(3) *closed reduced subschemes* $B_1 \subset X_1$, $B_2 \subset X_2$, *assume that the following holds*:

(4) $X_1 - B_1$ *is smooth over* S, *of relative dimension* r,

(5) *if* $\mathfrak{B}_i$ = *I-adic completion of* B_i, *then we have the inclusion of formal subschemes (not just subsets)*

$$\mathfrak{B}_1 \subset \pi^{-1}(\mathfrak{B}_2),$$

then we conclude that $X_2 - B_2$ *is smooth over* S, *of relative dimension* r.

PROOF. First, let's check that $X_2 - B_2$ is flat over S. Let $\mathscr{M} \subset \mathscr{N}$ be 2 A-modules. Consider the 2 kernels:

$$0 \to \mathscr{K}_1 \to \mathscr{M} \otimes \mathcal{O}_{X_1} \to \mathscr{N} \otimes \mathcal{O}_{X_1}$$
$$0 \to \mathscr{K}_2 \to \mathscr{M} \otimes \mathcal{O}_{X_2} \to \mathscr{N} \otimes \mathcal{O}_{X_2}.$$

Since $X_1 - B_1$ is flat over S, $\mathrm{Supp}(\mathscr{K}_1) \subset B_1$, hence for all $x \in X_1$, $(\mathscr{K}_1)_x \cdot (I_{B_1})_x^n = (0)$ for some n. Taking I-adic completions, we get exact sequences:

$$0 \to \hat{\mathscr{K}}_1 \to \mathscr{M} \otimes \mathcal{O}_{\mathscr{X}_1} \to \mathscr{N} \otimes \mathcal{O}_{\mathscr{X}_1}$$
$$0 \to \hat{\mathscr{K}}_2 \to \mathscr{M} \otimes \mathcal{O}_{\mathscr{X}_2} \to \mathscr{N} \otimes \mathcal{O}_{\mathscr{X}_2}.$$

and since π is flat, it follows that $\hat{\mathscr{K}}_1 \cong \pi^* \hat{\mathscr{K}}_2$. Since for all $x \in \mathscr{X}_1$, $(\hat{\mathscr{K}}_1)_x \cdot (I_{\mathfrak{B}_1})_x^n = (0)$ for some n, and since by (5), $I_{B_1} \supset \pi^*(I_{B_2})$, it follows that $(\hat{\mathscr{K}}_2)_{\pi(x)} \cdot (I_{\mathfrak{B}_2})_{\pi(x)}^n = (0)$. This means that in an open neighborhood of $f_2^{-1}(S_0)$, $\mathscr{K}_2$ is killed by I_{B_2}, hence $\mathrm{Supp}(\mathscr{K}_2) \subset B_2$. Since X_2 is proper over S, all closed points of X_2 lie over S_0 so that $\mathrm{Supp}(\mathscr{K}_2) \subset B_2$ everywhere.

To show that $X_2 - B_2$ is actually smooth over S, it suffices to show that in addition to being flat, it is differentiably smooth: i.e., $\Omega^1_{X_2/S}$ is locally free of rank r outside B_2, and $S^n(\Omega^1_{X_2/S}) \to \mathfrak{P}^n_{X_2/S}$ is an isomorphism outside B_2 (cf. EGA, IV[4]. 16). Since we know these are true for X_1/S, we deduce, in particular, that at all points x of $f_1^{-1}(S_0)$,

a) for all $g \in I(B_1)_x$, $(\Omega^1_{X_1/S})_x \otimes_{\mathcal{O}_{x,X_1}} \mathcal{O}_{x,X_1}[1/g]$ is locally free of rank r over $\mathcal{O}[1/g]$,

b) Ker and Coker of $S^n(\Omega^1_{X_1/S})_x \to \mathfrak{P}^n_{X_1/S,x}$ are killed by powers of $I(B_1)_x$.

These 2 facts imply the corresponding facts for the formal scheme $\mathscr{X}_1$. Since π is étale, $\pi^*(\Omega^1_{\mathscr{X}_2/S}) \cong \Omega^1_{\mathscr{X}_1/S}$ and $\pi^*(\mathfrak{P}^n_{\mathscr{X}_2/S}) \cong \mathfrak{P}^n_{\mathscr{X}_1/S}$, so by assumption (5), we get the corresponding facts for $\mathscr{X}_2$. Finally, these imply that (a) and (b) hold for X_2/S at points $x \in f_2^{-1}(S_0)$. Since X_2 is proper over S, they hold everywhere on X_2. Thus X_2–B_2 is differentiably smooth over S. *QED*

(4.2) COROLLARY. *G is smooth over S.*

PROOF. Take $X_1 = \tilde{P}$, $X_2 = P$, $B_1 = \tilde{B}$, $B_2 = B$, and use the fact that $\tilde{G}$ is smooth over S. *QED*

Our next step is to show that P and hence G is irreducible:

(4.3) PROPOSITION. *P is irreducible.*

PROOF. To prove this, we may assume that P is normal. In fact, we can replace $\tilde{P}$ by its normalization and then, since A is excellent, $\tilde{\mathfrak{P}}$ is normal, hence $\mathfrak{P}$ and P are normal. Then to show that P is irreducible, it suffices to show that P is connected. But we know that $\tilde{P}_0$ is connected. Therefore $P_0 = \tilde{P}_0/Y$ is connected. But P is proper over S and so therefore P is connected too. *QED*

Next, we take up the key problem of proving that G is independent of the choice of the polarization ϕ and the model $\tilde{P}$.

(4,4) DEFINITION. A subtorus $\tilde{H} \subset \tilde{G}$ is *integrable* if

$$\operatorname{rank}(Y \cap \tilde{H}(K)) = \dim \tilde{H}.$$

The key step in our proof of independence is that an integrable subtorus $\tilde{H} \subset \tilde{G}$ defines a closed subscheme $H \subset G$ in the following way:

a) let W_1 be the closure of $\tilde{H}$ in $\tilde{P}$ (considered as a reduced closed subscheme of $\tilde{P}$). If $Y^* = Y \cap \tilde{H}(K)$, then W_1 is Y^*-invariant.

b) let $\mathfrak{W}_1$ be the I-adic completion of W_1. It is also Y^*-invariant. Then

$$\mathfrak{W}_2 = \bigcup_{y \in Y/Y^*} S_y(\mathfrak{W}_1)$$

turns out to be a locally finite union, so this defines $\mathfrak{W}_2$ as a reduced closed subscheme of $\tilde{\mathfrak{P}}$.

c) Let $\mathfrak{W}_3 = \mathfrak{W}_2/Y \subset \mathfrak{P}$.

d) Let $W_3 \subset$ P be the reduced closed subscheme whose I-adic completion is $\mathfrak{W}_3$.

e) $H = W_3 \cap G$.

The finiteness assertion in (b) is the only non-trivial step where integrability has to be used. It results from:

(4.5) PROPOSITION. *Let $\tilde{H} \subset \tilde{G}$ be an integrable subtorus and let $Y^* = Y \cap \tilde{H}(K)$. Let W_1 be the closure of $\tilde{H}$ in a relatively complete model $\tilde{P}$ of $\tilde{G}$. Then there is a finite set $S \subset Y$ such that*

$$W_1 \cap S_y(U_0) = \phi \text{ if } y \notin S+Y^*.$$

PROOF. Let $X^* \subset X$ be the group of characters that are identically 1 on $\tilde{H}$. Then $\mathscr{X}^\alpha(y) = 1$ if $\alpha \in X^*, y \in Y^*$. In particular, $Y^* \cap \phi^{-1}(X^*) = (0)$ since if $y \in Y^* \cap \phi^{-1}(X^*), y \neq 0$, then $\mathscr{X}^{\phi(y)}(y)$ is in I and is 1. If $r = \dim \tilde{G}$, $s = \dim \tilde{H}$, then rank $Y^* = s$ and rank $\phi^{-1}(X^*) =$ rank $X^* = r-s$. Therefore $Y^* + \phi^{-1}(X^*)$ has finite index in Y. Let $k \geqq 1$ be an integer such that

$$kY \subseteq Y^* + \phi^{-1}(X^*) \subseteq Y.$$

Let $Y^{**} = \{y \in Y | ny \in Y^*, \text{ some } n \geqq 1\}$. Then Y^{**}/Y^* is a finite group killed by k and Y/Y^{**} is torsion-free. Consider the quotient torus $\tilde{G}_1 = \tilde{G}/\tilde{H} \cdot Y^{**}$. Its character group is the subgroup $X^{**} \subset X^*$ of characters which are 1 on all of Y^{**}. Note that $kX^* \subset X^{**}$. In $\tilde{G}_1$, consider the group of periods $Y_1 = Y/Y^{**}$. Define a polarization $\psi: Y_1 \to X^{**}$ as follows:

let $\bar{y} \in Y_1$,

lift $\bar{y}$ to an element $y \in Y$,

write $ky = y^* + w$, where $y^* \in Y^*$, $\phi(w) \in X^*$;

set $\psi(\bar{y}) = k \cdot \phi(w)$.

If $\bar{y}_0$ is a second element of Y_1, and $ky_0 = y_0^* + w_0$ as above, then

$$\begin{aligned}
\mathscr{X}^{\psi(\bar{y})}(\bar{y}_0) &= \mathscr{X}^{k\phi(w)}(y_0) \\
&= \mathscr{X}^{\phi(w)}(ky_0) \\
&= \mathscr{X}^{\phi(w)}(y_0^* + w_0) \\
&= \mathscr{X}^{\phi(w)}(w_0) \\
&= \mathscr{X}^{\phi(w_0)}(w) = \cdots = \mathscr{X}^{\psi(\bar{y}_0)}(\bar{y})
\end{aligned}$$

so that ψ is a bona fide polarization. We shall apply the basic lemma (1.3) to this torus and this polarization.

To find the appropriate n_y's, recall that for all $y \in Y$, $\mathscr{X}^{\phi(y)}$ is a regular function on $\tilde{P}$ outside of the locus $\mathscr{X}^{\phi(y)}(y) = 0$. In particular, since U is of finite type over A,

$$[\mathscr{X}^{\phi(y)}(y)]^n \cdot \mathscr{X}^{\phi(y)}$$

is a regular function on all of U if n is large. Increasing n by one, we can even make it a function that vanishes on U_0. Now if $\bar{y} \in Y_1$ and $y \in Y$

lies over $\bar{y}$ with $ky = y^* + w$ as above, then

$$\mathscr{X}^{\psi(\bar{y})}(\bar{y}) = \mathscr{X}^{\phi(w)}(w)$$

and

$$\mathscr{X}^{\psi(\bar{y})} = [\mathscr{X}^{\phi(w)}]^k.$$

So choose an integer $n_{\bar{y}}$ such that

(*) $\qquad [\mathscr{X}^{\psi(\bar{y})}(\bar{y})]^{n_{\bar{y}}} \cdot \mathscr{X}^{\psi(\bar{y})}$ is regular on U, zero on U_0.

Applying the lemma, we find a finite set $y_1, \cdots, y_k \in Y$ and a finite subset $S \subset Y$ such that

(**) $\qquad \begin{cases} \text{for all } z \in Y,\ z \notin S + Y^{**}, \\ \mathscr{X}^{\psi(\bar{z})}(\bar{y}_i) = (\text{elt. of } A) \cdot \mathscr{X}^{\psi(\bar{y}_i)}(\bar{y}_i)^{n_{y_i}}, \text{ for some } i. \end{cases}$

Now consider the function $\mathscr{X}^{\psi(\bar{y}_i)}$ on $S_z(U)$. Via the isomorphism S_z of $S_z(U)$ and U, it corresponds to the function

$$\mathscr{X}^{\psi(\bar{y}_i)}(z) \cdot \mathscr{X}^{\psi(\bar{y}_i)}$$

on U. Combining (*) and (**), it follows that this function is regular on U and zero on U_0. Therefore $\mathscr{X}^{\psi(\bar{y}_i)}$ is regular on $S_z(U)$ and zero on $S_z(U_0)$. But $\mathscr{X}^{\psi(\bar{y}_i)} \equiv 1$ on W_1 so this shows that $W_1 \cap S_z(U_0) = \phi$. Since z was an arbitrary element of Y outside $S + Y^{**}$, this proves the Proposition. *QED*

We are now ready to prove:

(4.6) Theorem. *Let $(\tilde{G}_i, Y_i, \phi_i, \tilde{P}_i)$, $i = 1, 2$ be 2 tori plus periods, polarizations, and relatively complete models. Let G_i, $i = 1, 2$ be the 2 schemes constructed as above. Then for all S-homomorphisms $\tilde{\alpha}: \tilde{G}_1 \to \tilde{G}_2$ such that $\tilde{\alpha}(Y_1) \subset Y_2$, there is a unique S-morphism $\alpha: G_1 \to G_2$ such that under the canonical isomorphisms of the I-adic completions of G_i and $\tilde{G}_i$, α and $\tilde{\alpha}$ are formally identical.*

Proof of Theorem. Consider the torus $\tilde{G}_1 \times_s \tilde{G}_2$; $Y_1 \times Y_2$ is a set of periods for this torus, $X_1 \times X_2$ is its character group and $\phi_1 \times \phi_2$: $Y_1 \times Y_2 \to X_1 \times X_2$ is a polarization. Moreover $P_1 \times_S P_2$ is easily seen to be a relatively complete model for $\tilde{G}_1 \times \tilde{G}_2$ relative to these periods and this polarization. Now suppose $\tilde{\alpha}: \tilde{G}_1 \to \tilde{G}_2$ is an S-homomorphism such that $\tilde{\alpha}(Y_1) \subset Y_2$. Look at the graph

$$\tilde{H} = \text{Image of } (1, \tilde{\alpha}) : \tilde{G}_1 \to G_1 \times_S \tilde{G}_2.$$

It is a subtorus of $G_1 \times_S G_2$, and because $\tilde{\alpha}(Y_1) \subset Y_2$, it is integrable. As in the last Proposition, it induces a closed subscheme of $G_1 \times_S G_2$ as follows:

$$\begin{aligned} W_1 &= \text{closure of } \tilde{H} \text{ in } \tilde{P}_1 \times_S \tilde{P}_2 \\ \mathfrak{W}_2 &= \bigcup_{y \in Y_1 \times Y_2} S_y(\mathfrak{W}_1) \\ \mathfrak{W}_3 &= \mathfrak{W}_2 / Y_1 \times Y_2 \subset \mathfrak{P}_1 \times_S \mathfrak{P}_2 \\ H &= W_3 \cap (G_1 \times_S G_2). \end{aligned}$$

I claim that H is the graph of a morphism from G_1 to G_2. First of all, we prove that the projection

$$p_1 : W_3 \to P_1$$

is smooth of relative dimension 0 outside $B_1 = P_1 - G_1$. This follows by essentially the same argument used in the proof of Proposition (4.1). In fact, $p_1 : W_1 \to \tilde{\mathfrak{P}}_1$ is smooth of rel. dim. 0 outside $\tilde{B} = \tilde{\mathfrak{P}}_1 - \tilde{G}_1$. Now locally at every point $\mathfrak{W}_2$ is the formal completion of a finite union $S_{y_1}(W_1) \cup \cdots \cup S_{y_k}(W_1)$, $y_i \in Y_1 \times Y_2$, and since this is also smooth of rel. dim. 0 outside $\tilde{B}_1$, so is $p_1 : \mathfrak{W}_2 \to \tilde{\mathfrak{P}}_1$. Here by 'smooth outside $\tilde{B}_1$', we do *not* mean merely smooth at points of $\tilde{\mathfrak{P}}_1 - \tilde{B}_1$: instead we mean smoothness in the sense of properties (a) and (b) in the proof of the proposition, viz. smoothness after localizing by the ideal $I(\tilde{B}_1)$. This property descends to smoothness for $p_1 : \mathfrak{W}_3 \to \mathfrak{P}_1$, hence for $p_1 : W_3 \to P_1$. Secondly, we prove that $W_3 \cap (P_1 \times_S B_2) \subset B_1 \times_S B_2$. This follows by the same method, by descending a stronger ideal-theoretic property on the ~-schemes. In fact, since for every finite set of y_i's:

$$[S_{y_1}(W_1) \cup \cdots \cup S_{y_k}(W_1)] \cap (\tilde{P}_1 \times_S \tilde{B}_2) \subset \tilde{B}_1 \times_S \tilde{B}_2,$$

it follows that on $\tilde{\mathfrak{P}}_1 \times \tilde{\mathfrak{P}}_2$,

$$I(\mathfrak{W}_2) + I(\tilde{\mathfrak{P}}_1 \times_S \tilde{\mathfrak{B}}_2) \supset I(\tilde{\mathfrak{B}}_1 \times_S \tilde{\mathfrak{B}}_2)^N$$

for some N. This property descends, and on algebraizing, proves that $W_3 \cap (P_1 \times_S B_2) \subset B_1 \times_S B_2$. Since $p_1 : W_3 \to P_1$ is a proper morphism, this proves that the restriction $p_1 : H \to G_1$ is also proper. Combining these 2 halves, it follows that $p_1 : H \to G_1$ is finite and étale. But finally the formal I-adic completion of $\mathfrak{H}$ is obviously the graph of the formal morphism from G_1 to G_2 defined by $\tilde{H}$. Therefore $p_1 : H \to G_1$ has degree 1 over S_0, hence because G_1 is irreducible, it has degree 1 everywhere. This proves that there is an S-morphism α: $G_1 \to G_2$ extending the formal morphism defined by $\tilde{\alpha}$. Finally, since G_1 is irreducible, such an α is clearly determined by its restriction to $\mathfrak{G}_1$. *QED*

(4.7) Corollary. *The scheme G depends only on the torus $\tilde{G}$ and the periods Y, and is independent of the polarization ϕ and the relatively complete model $\tilde{P}$.*

PROOF. Apply the theorem to 2 4-tuples $(\tilde{G}, Y, \phi_1, \tilde{P}_1)$ and $(\tilde{G}, Y, \phi_2, \tilde{P}_2)$ and to the identity $1_{\tilde{G}} : \tilde{G} \to \tilde{G}$.

(4.8) COROLLARY. *G is a group scheme over S.*

PROOF. Apply the theorem to $(\tilde{G} \times_S \tilde{G},\ Y \times Y,\ \phi \times \phi,\ \tilde{P} \times_S \tilde{P})$ and $(\tilde{G}, Y, \phi, \tilde{P})$ and the multiplication map from $\tilde{G} \times_S \tilde{G}$ to $\tilde{G}$. This yields a map $\mu: G \times_S G \to G$. Apply the theorem to the inverse from $\tilde{G}$ to $\tilde{G}$. This yields a map $i: G \to G$. Since μ and i make the formal completion $\mathfrak{G}$ of G into a formal group scheme and since G is irreducible, μ and i satisfy the same identities on G on as $\mathfrak{G}$ and therefore G is a group scheme. *QED*

(4.9) COROLLARY. *G_η is an abelian variety.*

PROOF. Since $\tilde{G}_\eta = \tilde{P}_\eta$, the generic fibre $\tilde{B}_\eta$ of $\tilde{B}$ is empty. It follows that the structure sheaf $\mathcal{O}_{\tilde{B}}$ is killed by some non-zero $\tau \in A$. Therefore the structure sheaf $\mathcal{O}_B$ is also killed by τ. Therefore $B_\eta = \phi$, hence $G_\eta = P_\eta$ is proper over K. Since G is irreducible, G_η is also irreducible, hence an abelian variety. *QED*

Before we can prove that G is, in fact, a semi-abelian group scheme, we have to prove that *every* fibre of G over S is connected. This follows from a description of the points of G of finite order:Start with $\tilde{G}, Y, \phi, \tilde{P}$ again.

Let $G^* = \bigcup_{y \in Y} S_y(\tilde{G}) \subset \tilde{P}$.

Let $\sigma_y: S \to G^*$ be the section such that $\sigma_y(\eta) = y$.

Let $Z_y^{(n)}$ be the subscheme of points z such that $nz = y$, i.e., the fibre product:

$$\begin{array}{ccc} Z_y^{(n)} & \hookrightarrow & G^* \\ \downarrow & & \downarrow{\scriptstyle n} \\ S & \underset{\sigma_y}{\hookrightarrow} & G^* \end{array}$$

Note that S_z induces an isomorphism of $Z_y^{(n)}$ with $Z_{y+nz}^{(n)}$.

(4.10) THEOREM. *The kernel $G^{(n)} \subset G$ of multiplication by n is isomorphic over S to the disjoint union of the schemes $Z_y^{(n)}$ as y runs over a set of cosets of Y/nY.*

PROOF. First consider the closure $\bar{Z}_y^{(n)}$ of $Z_y^{(n)}$ in $\tilde{P}$. By the valuation property of the relatively complete model $\tilde{P}$, it follows that all valuations of $\bar{Z}_y^{(n)}$ have centers on $\bar{Z}_y^{(n)}$. By Lemma (3.4), it follows that $\bar{Z}_y^{(n)}$ is proper over S. Let $\overline{\ }_y^{(n)}$ be its I-adic completion. We have seen above that if

$$W_1^{(n)} = \text{closure in } \tilde{P} \times_S \tilde{P} \text{ of the graph of } x \mapsto nx$$

then

$$\mathfrak{W}_2^{(n)} = \bigcup_{y \in Y} S_{(0,u)}(\mathfrak{W}_1^{(n)}) \subset \tilde{\mathfrak{P}} \times_S \tilde{\mathfrak{P}}$$

is a locally finite union. Since $\overline{Z}_y^{(n)} \times_S \sigma_y(S) \subset W_1^{(n)}$, hence $\overline{Z}_y^{(n)} \times_S \sigma_0(S) \subset S_{(0,-y)}(W_1^{(n)})$, it follows that

$$\bigcup_{y \in Y} \overline{\mathfrak{Z}}_y^{(n)} \subset \tilde{\mathfrak{P}}$$

is a locally finite union, and that

$$\bigcup_{y \in Y} \overline{\mathfrak{Z}}_y^{(n)} = \mathfrak{W}_2^{(n)} \cap (\tilde{\mathfrak{P}} \times_S \sigma_0(S)).$$

Taking quotients by Y and $Y \times Y$, we obtain a formal closed subscheme $\bigcup_{y \in Y} \overline{\mathfrak{Z}}_y^{(n)}/Y \subset \mathfrak{P}$ such that

$$\overline{\mathfrak{Z}}^{(n)} = \bigcup_{y \in Y} \overline{\mathfrak{Z}}_y^{(n)}/Y = \mathfrak{W}_3^{(n)} \cap (\mathfrak{P} \times_S \sigma_0(S))$$

where $\mathfrak{W}_3^{(n)} = \mathfrak{W}_2^{(n)}/Y \times Y$. It follows that $\overline{\mathfrak{Z}}^{(n)}$ algebrizes to a subscheme $\overline{Z}^{(n)} \subset P$ such that

$$\overline{Z}^{(n)} = W_3^{(n)} \cap (P \times_S \sigma_0(S)),$$

hence $Z^{(n)} = \overline{Z}^{(n)} \cap G = H^{(n)} \cap (G \times_S \sigma_0(s))$ where $H^{(n)} \subset G \times_S G$ is the graph of the morphism $x \mapsto nx$. Thus $Z^{(n)}$ is the kernel in G of multiplication by n. Now for every *finite* subset $Y_0 \subset Y$, we have a formal morphism as follows:

$$\alpha : \bigcup_{y \in Y_0} \overline{\mathfrak{Z}}_y^{(n)} \to \overline{\mathfrak{Z}}^{(n)}.$$

Since these formal schemes are the completions of the (algebraic) schemes $\bigcup_{y \in Y_0} \overline{Z}_y^{(n)}$ and $\overline{Z}^{(n)}$, which are *proper* over S, this formal morphism extends an (algebraic) morphism:

$$\alpha : \bigcup_{y \in Y_0} \overline{Z}_y^{(n)} \to \overline{Z}^{(n)}.$$

If $\tilde{B} = \tilde{P} - \tilde{G}$ and $B = P - G$, recall that $\tilde{\mathfrak{B}}$ is the inverse image of $\mathfrak{B}$ in the étale map $\tilde{\mathfrak{P}} \to \mathfrak{P}$. Therefore $\tilde{\mathfrak{B}} \cap \bigcup_y \overline{\mathfrak{Z}}_y^{(n)}$ is the inverse image of $\mathfrak{B} \cap \overline{\mathfrak{Z}}^{(n)}$, hence $\tilde{B} \cap \bigcup_{y \in Y_0} \overline{Z}_y^{(n)}$ is the inverse image of $B \cap \overline{Z}^{(n)}$. Therefore α restricts to a proper morphism:

$$\operatorname{res}(\alpha) : \bigcup_{y \in Y_0} Z_y^{(n)} \to Z^{(n)}.$$

Next, note that

$$\alpha : \bigcup_{y \in Y} \overline{\mathfrak{Z}}_y^{(n)} \to \overline{\mathfrak{Z}}^{(n)}$$

is étale and surjective. It follows that for each fixed $y_0 \in Y$, there is a finite set $Y_0 \subset Y$ such that α: $\bigcup_{y \in Y_0} \overline{\mathfrak{Z}}_y^{(n)} \to \overline{\mathfrak{Z}}^{(n)}$ is étale at all points of

$\overline{\mathfrak{Z}}^{(n)}_{y_0}$, and is surjective. Therefore its algebraization has the same properties. On the other hand, when we intersect with $\tilde{G}$, the union $\bigcup_{y\in Y_0} Z_y^{(n)}$ is disjoint, so it follows that

$$\operatorname{res}(\alpha) : \bigcup_{y\in Y_0} Z_Y^{(n)} \to Z^{(n)}$$

is étale for every Y_0, and surjective for Y_0 big enough.

Now it is clear that the diagram:

$$\begin{array}{ccc} Z_y^{(n)} & \xrightarrow{S_z} & Z_{y+nz}^{(n)} \\ & {\scriptstyle\alpha}\searrow \quad \swarrow{\scriptstyle\alpha} & \\ & Z^{(n)} & \end{array}$$

commutes, so, in fact, if Y_0 is a set of coset representatives of Y/nY

$$\operatorname{res}(\alpha) : \bigcup_{y\in Y_0} Z_Y^{(n)} \to Z^{(n)}$$

is already surjective. More than that: identifying formally

$$\bigcup_{y\in Y_0} Z_y^{(n)} \text{ with } \bigcup_{y\in Y} Z_Y^{(n)}/Y$$

we see that $\bigcup_{y\in Y_0} Z_y^{(n)}$ has a natural group scheme structure and it is clear that $\operatorname{res}(\alpha)$ is a homomorphism. Therefore $\operatorname{res}(\alpha)$ has degree 1 everywhere if it has degree 1 over the points $\sigma_0(S) \subset G$, and this will follow if $\operatorname{res}(\alpha)$ has degree 1 over the points $\sigma_0(S_0) \subset G_0$. But it is easy to see that *over* S_0, $Z_y^{(n)} = \phi$ unless $y \in nY$, $Z_0^{(n)}$ is the kernel of n in the part of the torus $\tilde{G}$ over S_0, and $\operatorname{res}(\alpha)$ is just the restriction to the kernel of n of the canonical isomorphism of $\tilde{G}$ and G over S_0, Thus $\operatorname{res}(\alpha)$ has degree 1. *QED*

(4.11) COROLLARY. *Let $s \in S$. Let*

$$Y_1 = \{y \in Y \mid \mathscr{X}^{\phi(y)}(y) \notin m_{s,S}\}.$$

Then Y_1 is a subgroup of Y (in fact, it is a direct summand) and the kernel of n in G_s fits into an exact sequence:

$$0 \to (\text{ker of } n \text{ in } \tilde{G}_s) \to (\text{ker of } n \text{ in } G_s) \to \frac{1}{n} Y_1/Y_1 \to 0.$$

As n increases, we obtain in the limit an exact sequence

$$0 \to (\text{torsion in } \tilde{G}_s) \to (\text{torsion in } G_s) \to Y_1 \otimes \mathbf{Q}/\mathbf{Z} \to 0.$$

PROOF. This follows immediately from the Theorem and the remark that $Z_y^{(n)}$ has a non-empty fibre over y if and only if $y \in Y_1 + nY$.

(4.12) COROLLARY. *G_s is connected, hence G is semi-abelian.*

PROOF. By Corollary 1, its torsion is p-divisible for every prime p, hence G_s is connected and without unipotent radical.

5. Examples-dim $G/S = 1$

Let's look first at the 1-dimensional case $\tilde{G} = \mathbf{G}_m \times S$ where $r = 1$.

Then $X \cong \mathbf{Z}$. Choose such an isomorphism. Then we have a distinguished generator $1 \in X$, and we denote the character $\mathscr{X}^1$ imply by $\mathscr{X}$, so that

$$\tilde{G} = \operatorname{Spec} A[\mathscr{X}, \mathscr{X}^{-1}].$$

All possible ϕ's will be positive multiples of one basic ϕ which is an isomorphism of Y and X. We assume this isomorphism chosen as our ϕ. All periods are multiples of the basic period $y_0 = \phi^{-1}(1)$: let $\tau = \mathscr{X}(y_0)$.

Note that $\tau \in I$. Identifying $\tilde{G}(K)$ with K^*, Y becomes the set of powers $\{\tau^m\}_{m \in \mathbf{Z}}$. The simplest possible Σ consists in the 3 elements $\{1, 0, -1\}$. Then

$$\begin{aligned} R_{\phi,\Sigma} &= A[\cdots, \mathscr{X}^{\phi(ky_0)+\{\varepsilon\}}(ky_0) \cdot \mathscr{X}^{2\phi(ky_0)+\{\varepsilon\}} \cdot \theta, \cdots]_{\varepsilon=0,\pm 1} \\ &= A[\cdots, \tau^{k^2+k} \cdot \mathscr{X}^{2k+1} \cdot \theta, \tau^{k^2} \cdot \mathscr{X}^{2k} \cdot \theta, \tau^{k^2-k} \cdot \mathscr{X}^{2k-1} \cdot \theta, \cdots] \end{aligned}$$

with basic automorphism:

$$\begin{aligned} S(\mathscr{X}) &= \tau \cdot \mathscr{X} \\ S(\theta) &= \tau \cdot \mathscr{X}^2 \cdot \theta \end{aligned}$$

One checks easily that

$$\begin{aligned} U_{1,0} &= \operatorname{Spec} A[\tau\mathscr{X}, \mathscr{X}^{-1}] \\ U_{0,0} &= \operatorname{Spec} A[\mathscr{X}, \mathscr{X}^{-1}] \\ U_{-1,0} &= \operatorname{Spec} A[\mathscr{X}, \tau\mathscr{X}^{-1}] \end{aligned}$$

Using the automorphism S, it follows that

$$\begin{aligned} U_{1,ky_0} &= \operatorname{Spec} A[\tau^{k+1}\mathscr{X}, \tau^{-k}\mathscr{X}^{-1}] \\ U_{0,ky_0} &= \operatorname{Spec} A[\tau^{k}\mathscr{X}, \tau^{-k}\mathscr{X}^{-1}] \\ U_{-1,ky_0} &= \operatorname{Spec} A[\tau^{k}\mathscr{X}, \tau^{-k+1}\mathscr{X}^{-1}] \end{aligned}$$

In particular $U_{1,ky_0} = U_{-1,(k-1)y_0}$. One checks immediately that the the closed fibre of $\operatorname{Proj}(R_{\phi,\Sigma})$ is an infinite union of non-singular rational curves, connected in a chain and crossing each other transversely.

We may visualize $\operatorname{Proj}(R_{\phi,\Sigma})$ like this (when A is a discrete valuation ring):

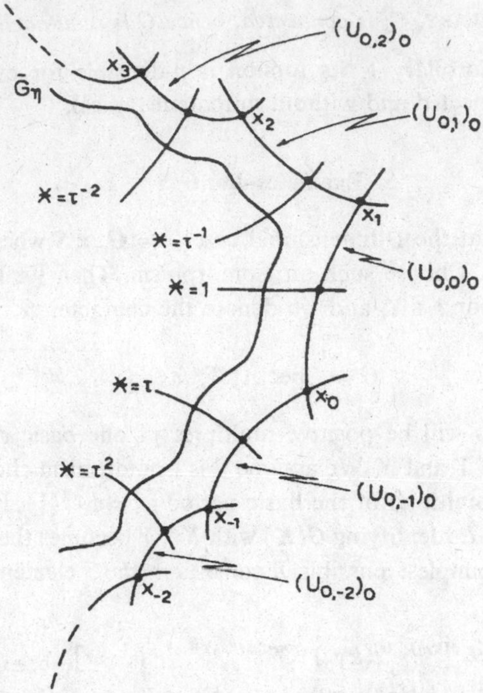

Dividing by Y, we obtain a formal scheme $\mathfrak{P}$ whose closed fibre is an irreducible rational curve with one ordinary double point. P has the same property and looks like this:

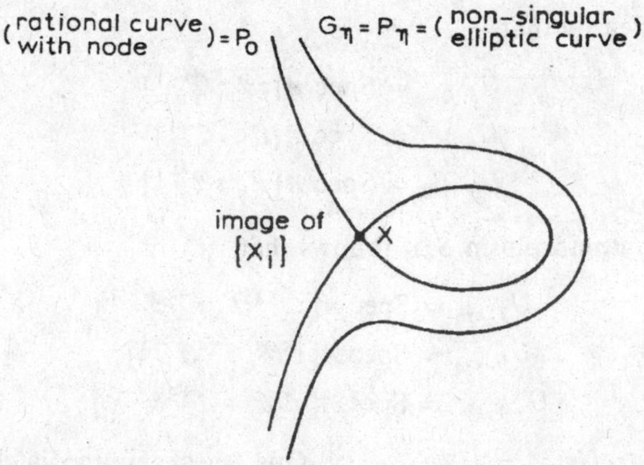

The part S which is to be thrown out will be just the single point x if A is a discrete valuation ring; in general, it will be a section of P through x over the subscheme $(\tau = 0)$ of Spec (A).

6. Examples-dim S = 1

Next, I want to look at higher dimensional tori $\tilde{G}$, but in the case where dim $A = 1$, i.e. A is a complete discrete valuation ring. Let r be the dimension of $\tilde{G}$ over S. In this case, we can describe the models $\tilde{P}$ very intuitively by a polyhedral decomposition of r-dimensional Euclidean space and we can see in this way the degree of choice in $\tilde{P}$ very clearly. We shall limit ourselves to *normal* relative complete models $\tilde{P}$ which admit an affine open covering $\{U_i\}$ such that each U_i is $\tilde{G}$-invariant. (It seems to me likely that every $\tilde{P}$ has such a covering; at least the Proj$(R_{\phi,\Sigma})$'s do). Our method is this: for every finite algebraic extension $L \supset K$, let A_L be the integral closure of A in L. A_L is again a complete discrete valuation ring. By the completeness property of $\tilde{P}$, every L-valued point ξ of $\tilde{G}$ extends uniquely to an A_L-valued point of $\tilde{P}$. We shall describe various subsets Σ of $\tilde{P}_0$ by describing those ξ whose closure meets Σ.

To begin with, the only thing that will really matter about a ξ is the *rate of growth of the various characters of* $\tilde{G}$ *on* ξ. This can be conveniently measured as follows:

(a) let $E = Hom(X, \mathbf{R})$, X = char. gp. of $\tilde{G}$ (this is a covariant functor of $\tilde{G}$),

(b) $\forall \alpha \in X$, let $l_\alpha : E \to \mathbf{R}$ be the linear functional given by evaluation at α,

(c) fix, once and for all, an embedding of the value group of A in $\mathbf{R}$. This naturally extends to an embedding of the value groups of each A_L in $\mathbf{R}$,

(d) $\forall \xi \in \tilde{G}(L)$, define $||\xi|| \in E$ by $||\xi||(\alpha) = \text{ord}\, \mathscr{X}^\alpha(\xi)$.

Note that for each L, $\{||\xi|| \mid \xi \in \tilde{G}(L)\} = N_L$ is a lattice in E; as L becomes more and more ramified over K, N_L gets denser and denser. Inside N_K, we have the smaller lattice $\{||\xi|| \mid \xi \in Y\}$ induced by the periods. We shall identify this with Y:

$$Y \subset N_K \subset \cdots \subset N_L \subset \cdots \subset E.$$

Now consider the finitely generated normal affine schemes U/S ($S = \text{Spec}\, A$), such that $U_\eta = \tilde{G}_\eta$ and such that $\tilde{G}$ acts on U extending the translation action of $\tilde{G}_\eta$ on U_η. It is easy to check that all such U are all of the type:

$$U = \text{Spec}\, A[\cdots, \pi^{r_i}\mathscr{X}^{\alpha_i}, \cdots]_{1 \leq i \leq N}$$

$$\text{where } (\pi) = \text{maximal ideal of } A$$

$$r_i \in \mathbf{Z}$$

$$\alpha_i \in X.$$

Concerning such U, we have the following:

(6.1) PROPOSITION. *There is a 1–1 correspondence between:*

(I) *normal affine schemes U, with $U_\eta = \tilde{G}_\eta$ and invariant under the action of $\tilde{G}$, and*

(II) *closed, bounded polyhedra $\Delta \subset E$ with vertices in $Q \cdot N_K$.*

This is set up by the relation: $\forall L \supset K, \forall \xi \in \tilde{G}(L)$,

$$\left\{ \begin{array}{l} \xi \textit{ extends to an} \\ A_L\textit{-valued point of } U \end{array} \right\} \Leftrightarrow \{\|\xi\| \in \Delta\}.$$

PROOF. In fact, if $\Gamma(U, \mathcal{O}_U)$ is generated by $\pi^{r_i}\mathscr{X}^{\alpha_i}$, then define Δ by

$$\Delta = \{x \in E | l_{\alpha_i}(x) \geqq -r_i \operatorname{ord}(\pi), 1 \leqq i \leqq N\}.$$

Conversely, given Δ, define

$$U = \operatorname{Spec} A[\cdots, Y_{r,\alpha}, \cdots],$$

with $Y_{r,\alpha} = \pi^r \mathscr{X}^\alpha$ for all (r, α) such that $l_\alpha + r \cdot \operatorname{ord}(\pi)$ is non-negative on Δ. We leave it to the reader to check that these set up inverse maps between the sets (I) and (II) of the Proposition.

(6.2) DEFINITION. Let $U_\Delta = \operatorname{Spec}(R_\Delta)$ be the affine scheme associated to Δ by the Proposition.

By a *closed face* of a polyhedron Δ, we mean Δ itself or a polyhedron $\Delta_1 = H \cap \Delta$, where H is a hyperplane, and Δ lies completely on side of H. By an *open face*, we mean a closed face minus all properly smaller subfaces. It is clear that every polyhedron Δ is a finite disjoint union of its open faces.

(6.3) PROPOSITION. *There is a 1–1 correspondence between:*

(I) *orbits Z of $\tilde{G}_0$ on the closed fibre $(U_\Delta)_0$ of U_Δ, and*

(II) *open faces $\sigma \subset \Delta$.*

This is set up by the relation: $\forall L \supset K, \forall \xi \in \tilde{G}(L)$,

$$(*) \qquad \left\{ \begin{array}{l} \xi \textit{ extends to a} \\ A_L\textit{-valued point meeting } Z \end{array} \right\} \Leftrightarrow \{\|\xi\| \in \sigma\}.$$

Moreover $\dim Z + \dim \sigma = r$.

PROOF. Since the union of the open faces is Δ, and the union of the orbits is $U_\Delta \times \operatorname{Spec}(\bar{k})$ it will suffice to show that for all σ there is an orbit Z satisfying (*). Let $R_\Delta = A[\cdots, \pi^{r_i}\mathscr{X}^{\alpha_i}, \cdots]_{1 \leqq i \leqq N}$. Then

$\Delta = \{x \in E | l_{\alpha_i}(x) + r_i \cdot \text{ord}\, \pi \geqq 0\}$. Let σ be defined by:

$$l_{\alpha_i}(x) + r_i \cdot \text{ord}\, \pi = 0, \; i \in I_1$$
$$l_{\alpha_i}(x) + r_i \cdot \text{ord}\, \pi > 0, \; i \in I_2$$

where $I_1 \cup I_2 = \{1, \cdots, N\}$. Write $\pi^{r_i}\mathscr{X}^{\alpha_i} = Y_i$ for simplicity. Then for all $L \supset K$, and A_L-valued points ξ such that $\|\xi\| \in \Delta$, we see that

$$\|\xi\| \in \sigma \Leftrightarrow \begin{cases} Y_i\,(\xi) \text{ a unit for } i \in I_1 \\ Y_i\,(\xi) \in \text{max. ideal of } A_L,\; i \in I_2 \end{cases}$$
$$\Leftrightarrow \xi \text{ meets } Z$$

where $Z \subset (U_\Delta)_0$ is defined by $Y_i \neq 0$, $i \in I_1$ (resp. $= 0$, $i \in I_2$). Now Z is clearly $\tilde{G}_0$-invariant. Moreover Z is easily checked to be the locus in the affine Y_i-space ($i \in I_1$) defined by:

i) $Y_i \neq 0$, all $i \in I_1$

ii) $\Pi Y_i^{s_i} = 1$ whenever $s_i \in Z$, $s_i \geqq 0$ satisfy $\sum s_i \alpha_i = 0$ (Since $\sigma \neq \emptyset$, $\sum s_i \alpha_i = 0 \Rightarrow \Sigma s_i r_i = 0!$).

If $\tilde{H} \subset \tilde{G}$ is the subtorus defined by $\mathscr{X}^{\alpha_i} = 1$, $i \in I_1$, then it follows immediately that Z is isomorphic to $\tilde{G}_0/\tilde{H}_0$ if a point of Z with coordinates $Y_i = a_i$ is associated by a point of G_0/H_0 with coordinate $\mathscr{X}^{\alpha_i} = a_i$.

(6.4) Corollary. *$(U_\Delta)_0$ is a finite union of orbits, hence each component contains a unique open orbit, and the set of components is in 1–1 correspondence with the vertices of Δ.*

(6.5) Corollary. *If Δ_1, Δ_2 are polyhedra, then U_{Δ_1} is an open subset of U_{Δ_2} if and only if Δ_1 is a closed face of Δ_2.*

Proof. By the Prop., it is certainly necessary that Δ_1 be a closed face of Δ_2. Conversely, if Δ_1 is a closed face of Δ_2, then $\Delta_1 = \Delta_2 \cap H$ where H is a hyperplane defined by $l_\alpha(X) + r \cdot \text{ord}\, \pi = 0$ and $l_\alpha(x) + r \cdot \text{ord}\, \pi \geqq 0$ on Δ_2. Then $\pi^r \cdot \mathscr{X}^\alpha \in \Gamma(\mathcal{O}_{\Delta_2})$ and U_{Δ_1} is the affine open subset of U_{Δ_2} defined by $\pi^r \cdot \mathscr{X}^\alpha \neq 0$.

(6.6) Corollary. *There is a 1–1 correspondence between:*

(I) *normal schemes $\tilde{P}$ locally of finite type over S such that*

(a) $\tilde{P}_\eta = \tilde{G}_\eta$,

(b) *the translation action of $\tilde{G}$ extends to $\tilde{P}$ and P is covered by $\tilde{G}$-invariant affine open sets, and*

(c) *for all valuations v on $\mathbf{R}(\tilde{G})$ if $v \geqq 0$ on A and if $[\forall \alpha \in X, \exists n$ such that $n \cdot v(\pi) \geqq v(\mathscr{X}^\alpha) \geqq -nv(\pi)]$ hold, then v has a centre on $\tilde{P}$, and*

(II) *Polyhedral decompositions of E, i.e. a set of polyhedron* $\Delta_\alpha \subset E$ *such that* $\bigcup \Delta_\alpha = E$, *every closed face of a* Δ_α *is a* Δ_β, $\Delta_\alpha \cap \Delta_\beta \neq \phi \Rightarrow \Delta_\alpha \cap \Delta_\beta = \Delta_\gamma$ *(some* γ*), all of whose vertices are in* $Q \cdot N_K$.

PROOF. Clear

In the 1–1 correspondence of this last corollary, certain properties of $\tilde{P}$ carry over nicely to properties of $\{\Delta_\alpha\}$:

(A) $\tilde{P} \supset \tilde{G}$ if and only one of the Δ_α's is the origin $\{0\}$.

(B) If $Y \subset \tilde{G}(K)$ is a set of periods, the action of Y on $\tilde{G}_\eta$ extends to $\tilde{P}$ if and only if the polyhedral decomposition is Y-invariant, i.e. $\forall \alpha, y$, $\Delta_\alpha + y =$ some Δ_β.

(C) $\tilde{P}$ is smooth over A at all generic points of $\tilde{P}_0$ if and only if the vertices of all Δ_α are points of N_K.

(D) $\tilde{P}$ is regular everywhere and smooth A at all generic points of $\tilde{P}_0$ if and only if the vertices of all Δ_α are in N_K, each Δ_α of dimension r is an r-simplex with volume $1/r!$ in a coordinate system in E making N_K into the integral points.

Here (A), (B) and (C) are almost immediate, and we omit the proof of (D), which is harder, because we do not need it. So far, there appears to be tremendous choice in constructing a relatively complete model $\tilde{P}$. However, the existence of an ample $\tilde{L}$ corresponding to a polarization $\phi: Y \to X$ puts very strong restrictions on the polyhedral decomposition $\{\Delta_\alpha\}$. The exact conditions on $\{\Delta_\alpha\}$ for the existence of $\tilde{L}$ are very messy, so we will only state, without proof, one partial result. First we must recall some constructions used in sphere packing problems. For all the following, cf. Rogers, *Packing and Covering*, Camb. Univ. Press, Ch. 7, § 1 and Ch. 8, § 1. Assume given a Euclidean metric on E. Start with a discrete set of points $\Sigma \subset E$ such that for some r, $E =$ union of balls around points of Σ with radius r. We may now construct 2 canonical polyhedral decompositions E:

(A) the *Voronoi decomposition*: $\forall \sigma \in \Sigma$, let

$$\Delta_\sigma = \{y \in E \mid \|y-\sigma\| \leqq \|y-\tau\|, \text{all } \tau \in \Sigma\}.$$

These are the top dimensional polyhedra. We get their faces by looking, for all $\sigma_1, \cdots, \sigma_k \in \Sigma$, at:

$$\Delta_{\sigma_1,\cdots,\sigma_k} = \left\{ y \in E \;\middle|\; \begin{array}{c} \|y-\sigma_1\| = \cdots = \|y-\sigma_k\| \leqq \|y-\tau\| \\ \text{all } \tau \in \Sigma \end{array} \right\}.$$

(B) *the Delaunay decomposition*: For all $\sigma_1, \cdots, \sigma_k$ such that

$\Delta_{\sigma_1,\cdots,\sigma_k} \neq \phi$ and such that for y in its interior,

$$\|y-\sigma_1\| = \cdots = \|y-\sigma_k\| < \|y-\tau\| \; (\tau \in \Sigma, \tau \neq \sigma_i),$$

let Δ^* = convex hull of $\sigma_1, \cdots, \sigma_k$. These polyhedra Δ^* form a decomposition of E whose vertices are exactly the points of Σ.

Now returning to our tori, let $Y \subset \tilde{G}(K)$ be a set of periods and let $\phi: Y \to X$ be a polarization. The quadratic form

$$\|y\|^2 = \text{ord}\, \mathscr{X}^{\phi(y)}(y), \; y \in Y$$

extends rationally to a Euclidean norm $\| \; \|$ on E. We find:

(6.7) PROPOSITION. *Let $\tilde{G}$, Y and ϕ be given as above. Let $\tilde{G} \subset \tilde{P}$ be given where $\tilde{P}$ is a normal scheme locally of finite type over S such that*

i) $\tilde{P}_\eta = \tilde{G}_\eta$,

ii) *$\tilde{G}$ and Y act on $\tilde{P}$,*

iii) *P has the valuative completeness property of Corollary* (6.6),

iv) *$\tilde{P}$ is covered by $\tilde{G}$-invariant affine open sets.*

Let $\tilde{P}$ correspond to the polydedral decomposition $\{\Delta_\alpha\}$ and let Σ be the 0-simplices in $\{\Delta_\alpha\}$. If $\{\Delta_\alpha\}$ is the Delaunay decomposition of E associated to Σ, with the Euclidean norm $\| \; \|$ defined by ϕ, then there exists an ample invertible sheaf $\tilde{L}$ on $\tilde{P}$, plus an action of $\tilde{G}$ and Y on $\tilde{L}$, making $\tilde{P}$ into a relatively complete model corresponding to the polarization $n \cdot \phi$ (some $n \geqq 1$).

(Proof omitted)

7. A final example

I want to give one final example which

(i) illustrates the situation where both the base and the fibre are more than one dimensional, and

(ii) is literally the keystone in the compactification of the moduli space of 2-dimensional abelian varieties.

This example was proposed by Deligne before we saw the general theory. Take as the ground ring:

$$A = k[[a, b, c]]$$

and take $\tilde{G}$ to be 2-dimensional. To express the symmetry of the situation, it is convenient not to introduce a basis of the character group X of $\tilde{G}$, but rather 3 generators $\mathscr{X}$, $\mathscr{Y}$, $\mathscr{Z}$ for X, related by one identity:

$$\mathscr{X} \cdot \mathscr{Y} \cdot \mathscr{Z} \equiv 1$$

The group of periods Y will similarly be generated by the 3 periods r, s, t with $r+s+t = 0$, given by

$$\begin{array}{lll} \mathscr{X}(r) = bc, & \mathscr{Y}(r) = c^{-1}, & \mathscr{Z}(r) = b^{-1}, \\ \mathscr{X}(s) = c^{-1}, & \mathscr{Y}(s) = ac, & \mathscr{Z}(s) = a^{-1}, \\ \mathscr{X}(t) = b^{-1}, & \mathscr{Y}(t) = a^{-1}, & \mathscr{Z}(t) = ab. \end{array}$$

The polarization ϕ will be defined by

$$\phi(r) = \mathscr{X}$$
$$\phi(s) = \mathscr{Y}$$
$$\phi(t) = \mathscr{Z}$$

and Σ will consist in the seven characters:

$$\Sigma = \{1, \mathscr{X}, \mathscr{X}^{-1}, \mathscr{Y}, \mathscr{Y}^{-1}, \mathscr{Z}, \mathscr{Z}^{-1}\}.$$

The resulting ring $R_{\phi,\Sigma}$ will have some automorphisms, in addition to those given by translation with respect to Y:

a) an automorphism σ such that $\sigma^3 = \text{id}$,

$$\sigma(a) = b, \sigma(b) = c, \sigma(c) = a$$
$$\sigma(\mathscr{X}) = \mathscr{Y}, \sigma(\mathscr{Y}) = \mathscr{Z}, \sigma(\mathscr{Z}) = \mathscr{X}$$
$$\sigma(\theta) = \theta$$

b) an involution $\tau(\tau^2 = \text{id})$,

$$\tau = \text{id on } A$$
$$\tau(\mathscr{X}) = \mathscr{X}^{-1}, \tau(\mathscr{Y}) = \mathscr{Y}^{-1}, \tau(\mathscr{Z}) = \mathscr{Z}^{-1}$$
$$\tau(\theta) = \theta.$$

Now Proj $(R_{\phi,\Sigma})$ is covered by the $U_{\alpha,y}$'s with $\alpha \neq 0$ and these are all isomorphic under these various automorphisms. We must calculate the ring of one of them. Look at $U_{\mathscr{X},0}$; its ring is generated by:

$$\mathscr{X}^{\phi(y)+\alpha}(y) \cdot \mathscr{X}^{2\phi(y)+\alpha} \cdot \mathscr{X}^{-1}, \text{ all } y \in Y, \alpha \in \Sigma.$$

One then checks easily that:

$$\begin{aligned} U_{\mathscr{X},0} &= \operatorname{Spec} A[\mathscr{Y}, c\mathscr{Y}^{-1}, \mathscr{Z}, b\mathscr{Z}^{-1}] \\ &\cong \operatorname{Spec} A[X_1, X_2, X_3, X_4]/(X_1 X_2 - c, X_3 X_4 - b). \end{aligned}$$

In particular, this is regular, so $\text{Proj}(R_{\phi,\Sigma})$, $\mathfrak{P}$ and P are all regular 5-dimensional schemes. Moreover, $(U_{\mathscr{X},0})_0$ has 4 components which are the orbits of $\tilde{G}_0$ containing the 4 sections:

$$\begin{aligned} (\mathscr{X}, \mathscr{Y}, \mathscr{Z}) &= (1, 1, 1) \\ &= (c^{-1}, c, 1) \end{aligned}$$

$$= (b^{-1}, 1, b)$$
$$= (b^{-1}c^{-1}, c, b).$$

Applying the automorphisms σ, τ and translation periods, it is not hard to construct the whole closed fibre of $\mathrm{Proj}(R_{\phi,\Sigma})$. We give the result in the following diagram (strangely isometric to the root diagram of G_2). Each triple stands for the component which is the $\tilde{G}_0$-orbit of the closed point on the section defined by that triple; 2 components meet along a curve if they are joined by a line. For each diamond, there is one point at which the 4 components corresponding to its 4 vertices meet. The components indicated by triples in square brackets are projective planes. each of them meets 3 other components along the three lines of a triangle, The components indicated by triples in round brackets are projective planes with 3 non-collinear points blown up; the 3 exceptional lines, plus the 3 lines joining pairs of blown-up points, form a hexagon along which these components meet 6 neighbouring components.

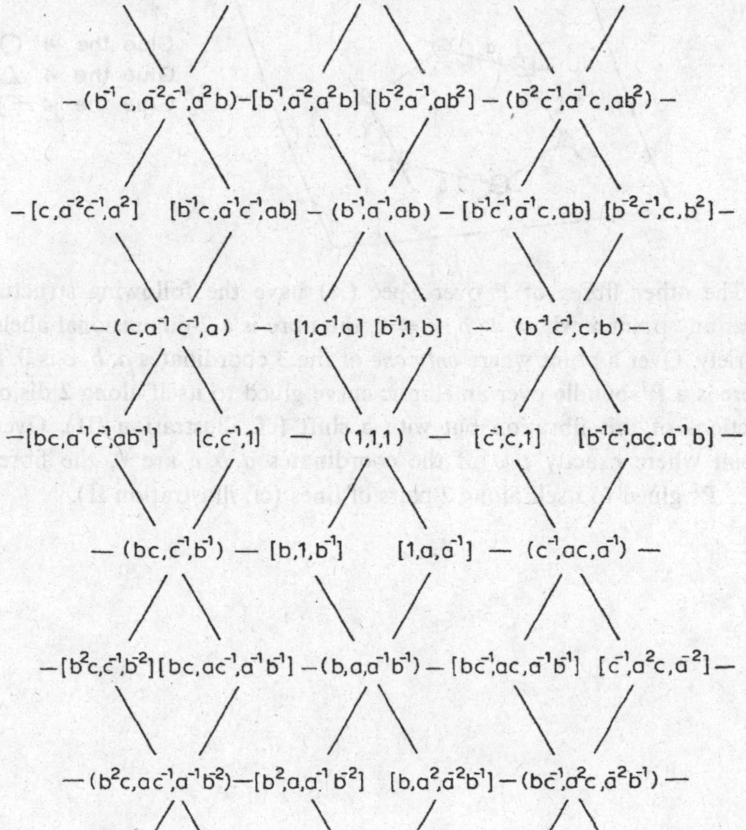

To find the structure of the closed fibre of P, we must divide by the action of Y. All the components of $\mathrm{Proj}(R_{\phi,\Sigma})_0$ collapse to 3 components: one isomorphic to the three times blown-up projective plane, which is the image of $(1, 1, 1)$; two isomorphic to projective planes which are the images, for example, of $[1, a^{-1}, a]$ and $[1, a, a^{-1}]$ respectively. They are glued as indicated below (I).

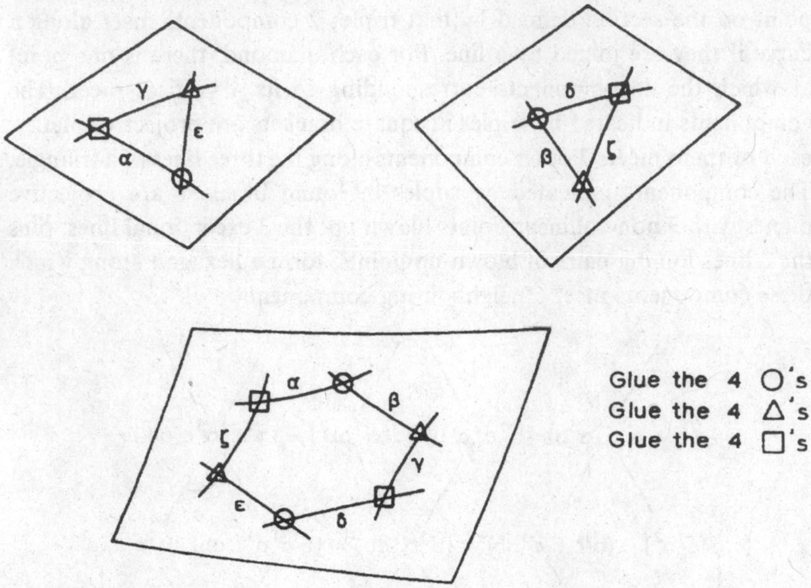

The other fibres of P over Spec (A) have the following structure: over any point x where $a \cdot b \cdot c \neq 0$, the fibre is a 2-dimensional abelian variety. Over a point where *only one* of the 3 coordinates a, b, c is 0, the fibre is a P^1-bundle over an elliptic curve glued to itself along 2 disjoint sections of this fibration but with a shift (cf. illustration III). Over a point where exactly *two* of the coordinates a, b, c are 0, the fibre is $P^1 \times P^1$ glued to itself along 2 pairs of lines (cf. illustration II).

glue via
$(o,x) - (\infty, \lambda x), \forall x$

glue via
$(x,o) - (\lambda x, \infty), \forall x$

Glue the 4 $\bigcirc$'s

glue via
$E_0 \cap \pi^{-1} a - E_\infty \cap \pi^{-1}(a+\tau)$

$\mathbb{P}^1$-bundle with
sections E_0, E_∞

——————— Elliptic curve

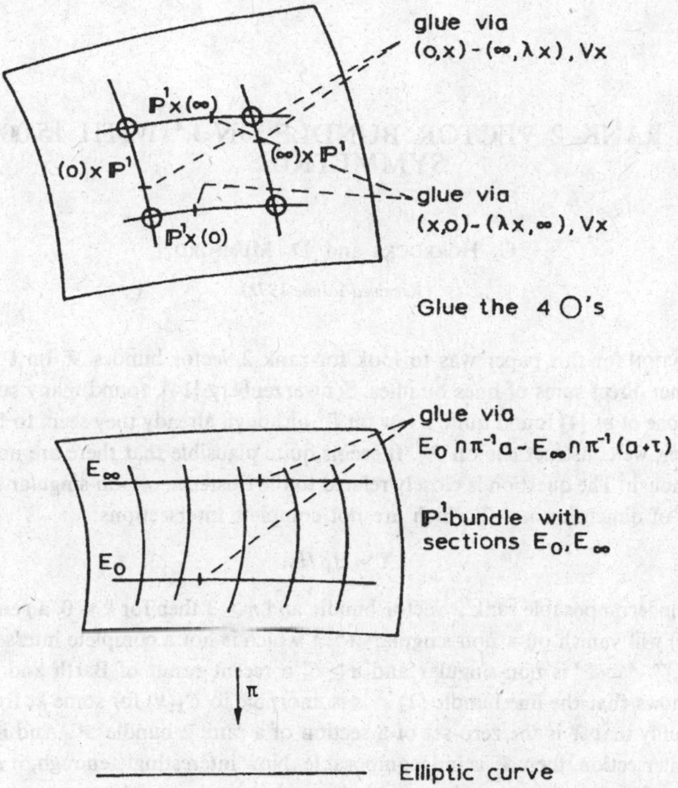

(Oblatum 16-III-1971)

Department of Mathematics,
Harvard University,
Cambridge,
Mass.,
U.S.A.

Topology Vol. 12, pp. 63–81. Pergamon Press, 1973. Printed in Great Britain

A RANK 2 VECTOR BUNDLE ON $\mathbb{P}^4$ WITH 15,000 SYMMETRIES

G. HORROCKS and D. MUMFORD

(*Received* 1 *June* 1972)

THE MOTIVATION for this paper was to look for rank 2 vector bundes $\mathscr{F}$ on $\mathbb{P}^n$ for $n \geqq 4$ which are not direct sums of lines bundles. Schwarzenberg [14], found many such bundles on $\mathbb{P}^2$ and one of us [4] found quite a few on $\mathbb{P}^3$ although already they seem to be "rarer". In this paper, we construct *one* on $\mathbb{P}^4$. It seems quite plausible that there are none on $\mathbb{P}^n$ if n is large enough. The question is closely related to the existence of non-singular subvarieties $X^{n-2} \subset \mathbb{P}^n$ of dimension $n-2$ which are not complete intersections:

$$X = H_1.H_2.$$

If $\mathscr{F}$ is an indecomposable rank 2 vector bundle and $n \geq 3$ then for $k \gg 0$, a general section $s \in \Gamma(\mathscr{F}(k))$ will vanish on a non-singular X^{n-2} which is not a complete intersection; conversely, if $X^{n-2} \subset \mathbb{P}^n$ is non-singular and $n \geq 6$, a recent result of Barth and Larsen ([1] and [9]) shows that the line bundle Ω_X^{n-2} is isomorphic to $\mathcal{O}_X(k)$ for some k, from which it follows readily that X is the zero-set of a section of a rank 2 bundle $\mathscr{F}$. And if X is not a complete intersection, then $\mathscr{F}$ is indecomposable. Now interestingly enough, it seems as far as we know that classical procedures and classical examples yield non-singular X^{n-2}'s in $\mathbb{P}^n$, which are not complete intersections, *only if* $n \leq 5$.

The vector bundle constructed here has a 4-dimensional space of sections almost all of which vanish on a non-singular $X_s \subset \mathbb{P}^4$ which is an abelian surface. We first found the bundle by establishing that such X_s's had to exist and then constructing $\mathscr{F}$ from X_s as an extension. However by then applying the general "Postnikov" construction of [3], we found a much more direct description of $\mathscr{F}$. The theory of the bundle $\mathscr{F}$ and of the surfaces X_s is united by the fact that both are acted on by the Heisenberg group H (an irreducible 2-step nilpotent subgroup of $SL_5(\mathbb{C})$ of order 125: cf. §1) which is well known from the theory of theta functions; $\mathscr{F}$ is acted on also by the normalizer N of H, of order 15,000. We have developed all our results by keeping track of the action of N at every stage and using the character table of N where necessary. This is a quick efficient method although unfortunately not very illuminating. Our main results are as follows: we construct the bundle $\mathscr{F}$ in §2 and note immediately by examining its Chern classes that it is indecomposable; in §4 we find the cohomology of $\mathscr{F}(n)$ for every n; in §5, we prove that the zero-sets X_s of its general sections are abelian and that conversely all abelian surfaces in $\mathbb{P}^4$ arise in this way; in §6, we show that as a corollary we get an explicit birational map between a certain moduli space of abelian surfaces and $\mathbb{P}^3$. We have put the character table of $SL_2(\mathbb{Z}_5)$ in an appendix for easy reference.

§1. THE HEISENBERG GROUP IN DIMENSION FIVE

The purpose of this section is to review in a special case a configuration of groups studied recently by Weil [16] (cf. also Igusa [7, Chap. 1]; Mumford [12, §1]) and closely related to the theory of theta functions and abelian varieties. Weil's construction starts from an arbitrary locally compact abelian group A, but we take $A = \mathbb{Z}_5$, the cyclic group of order 5, and proceed as follows:

Let

$$V = \mathrm{Map}(\mathbb{Z}_5, \mathbb{C}),$$

be the vector-space of complex-valued functions on $\mathbb{Z}_5$. Note that V has a natural $\mathbb{Q}$-rational structure given by the $\mathbb{Q}$-subspace $\mathrm{Map}(\mathbb{Z}_5, \mathbb{Q})$. Let $\varepsilon = e^{2\pi i/5} \in \boldsymbol{\mu}_5$, the group of 5th roots of 1. The *Heisenberg group*

$$H \subset SL_5(\mathbb{C})$$

is the subgroup generated by σ and τ, given by

$$\sigma x(i) = x(i+1)$$
$$\tau x(i) = \varepsilon^i x(i)$$

for all $x \in V$. Explicitly, H is the set of matrices

$$A_{ij} = (\varepsilon^{ai+b} \cdot \delta_{i,j+c})$$

and has order 125. An an algebraic group, H is defined over $\mathbb{Q}$, but it only splits over $\mathbb{Q}(\varepsilon)$. The Galois group Θ of $\mathbb{Q}(\varepsilon)$ over $\mathbb{Q}$ acts on H. Let $\theta \in \Theta$ be the generator given by $\theta(\varepsilon) = \varepsilon^2$ (so that $\theta^2 =$ complex conjugation). We shall sometimes use the notation $'$ to indicate the action of θ. The group H has center C equal to $\boldsymbol{\mu}_5 \cdot I_V$ and is a central extension:

$$1 \to \boldsymbol{\mu}_5 \to H \to \mathbb{Z}_5 \times \mathbb{Z}_5 \to 1, \tag{1.1}$$

where σ, τ in H are mapped to (1, 0), (0, 1). The action of Θ preserves this sequence and θ acts on $\mathbb{Z}_5 \times \mathbb{Z}_5$ by $(n, m) \to (n, 2m)$.

V is clearly an irreducible H-module, and it gives rise to three more by the action of Θ: let V_i be the representation obtained from V by composing $H \to \mathrm{Aut}\, V$ with θ^i. The trace $\langle h, V_i \rangle$ of an element $h \in H$ on V_i is given by:

$$\langle \varepsilon^r I_V, V_i \rangle = 5 \cdot \varepsilon^{2^i r}, \qquad \langle h, V_i \rangle = 0 \qquad (h \in H - C). \tag{1.2}$$

It follows that the four representations V_i are inequivalent. These plus the 25 characters of $\mathbb{Z}_5 \times \mathbb{Z}_5$ exhaust the irreducible representations of H since the sum of squares of their degrees is 125, the order of H.

Let $\phi\colon \mathbb{Z}_5 \times \mathbb{Z}_5 \to H$ be the section of (1.1) given by:

$$\phi(m, n) = \varepsilon^{2mn} \sigma^m \tau^n,$$

and define $\omega\colon \boldsymbol{\mu}_5 \times (\mathbb{Z}_5 \times \mathbb{Z}_5) \to H$ by:

$$\omega(\alpha, z) = \alpha \cdot \phi(z).$$

Then ω is bijective and the group law on H goes over to the law of composition:

$$(a, z)\cdot(a', z') = (aa'B(z, z'), z + z')$$

where

$$B(m, n; m', n') = \varepsilon^{3(mn'-m'n)}$$

is a μ_5-valued skew-symmetric form on $\mathbb{Z}_5 \times \mathbb{Z}_5$. Note that all automorphisms of H preserve the sequence (1.1) and since $B(z, z')^2 \cdot I_V$ is the commutator of (a, z) and (a', z'), they preserve the form B.

Let N be the normalizer of H in $SL_5(\mathbb{C})$. Each element of N induces by conjugation an automorphism of H, hence an automorphism of $\mathbb{Z}_5 \times \mathbb{Z}_5$ preserving B. But the group of such automorphisms is isomorphic to $SL_2(\mathbb{Z}_5)$, hence we get a homomorphism:

$$\alpha\colon N \to SL_2(\mathbb{Z}_5).$$

The kernel of α is just H itself because (a) any automorphism of H which is the identity on C and on H/C is in fact inner, and (b) since the representation V is irreducible, C is the centralizer of H in $SL_2(\mathbb{Z}_5)$. Moreover α is surjective. If $x \in SL_2(\mathbb{Z}_5)$, define $\gamma_x\colon H \to H$ by

$$\gamma_x\omega(a, z) = \omega(a, x(z)).$$

Since x preserves B, the mapping γ_x is an automorphism. The new representation of H on V obtained by composing with γ_x is equivalent to V since γ_x is the identity on C and so leaves the character fixed. So x is induced by an element of N, in fact an element of $N \cap SL_5$ $(\mathbb{Q}(\varepsilon))$. Thus α is surjective and $N \subset SL_5(\mathbb{Q}(\varepsilon))$, hence Θ acts on N and the action of N on V induces actions on each V_i.

Since $\gamma_x \cdot \gamma_y = \gamma_{xy}$ and γ_x is induced by a member of N determined up to multiplication by elements of C, it follows that N/C is a semi-direct product $(H/C)\cdot SL_2(\mathbb{Z}_5)$. Let X be the inverse image in N of the factor $SL_2(\mathbb{Z}_5)$. Then X is a central extension of $SL_2(\mathbb{Z}_5)$ by C. But the group of Schur multipliers $H^2(SL_2(\mathbb{Z}_5), \mathbb{C}^*)$ is zero [5, p. 645], hence X is a product $C\cdot SL_2(\mathbb{Z}_5)$ and the full group N is a semi-direct product $H\cdot SL_2(\mathbb{Z}_5)$.

Next, look at the dual representation V_i^* of V_i: since N acts on each V_i by unitary representations, V_i^* is isomorphic as N-module to the complex conjugate $\bar{V}_i$, i.e. to V_{i+2}.

Finally, look at the representation of N in $V_i \otimes V_i^*$. C acts trivially here so we have a representation of N/C. For all $x \in V_i$, $l \in V_i^*$, put

$$F_{x\otimes l}(h) = l(hx).$$

This gives a map

$$F\colon V_i \otimes V_i^* \to \operatorname{Map}(H, \mathbb{C})$$

which is easily seen to be injective, with image the space W_i of functions f on H such that

$$f(\alpha h) = \alpha^{2^i}\cdot f(h), \qquad \alpha \in \mu_5 I_V.$$

Moreover, for every $n \in N$, let n induce by conjugation the automorphism n^* of H. Then

$$\begin{aligned} F_{n(x\otimes l)}(h) &= F_{nx\otimes l\cdot n^{-1}}(h) \\ &= l\cdot n^{-1}(hnx) \\ &= l(n^*(h)x) \\ &= F_{x\otimes l}(n^*h) \end{aligned}$$

so F transforms the cation of n on $V_i \otimes V_i^*$ to the cation $f \mapsto f \cdot n^*$ on $\mathbb{C}$-valued functions on H. Now it is easy to check that if

$$\tilde{f}(\omega(\alpha, z)) = \alpha^{2^i} \cdot f(\omega(\alpha, 2z)),$$

then $f \mapsto \tilde{f}$ is an isomorphism of W_i and W_{i+1} commuting with the action of N: therefore the four representations $V_i \otimes V_i^*$ are all equivalent, so we may as well work with $V \otimes V^*$. This space has a decomposition $\mathbb{C} \oplus Z$ where Z is the subspace of trace zero. One sees immediately that as an H/C-module Z is the sum of the 24 non-trivial (linear) characters of H/C. Since N acts transitively on these, Z is irreducible as N/C-module. Its character ζ has values in $\mathbb{Q}$ since Z is equivalent to all its conjugates.

To summarize our conclusions, we have found groups:

$$\begin{array}{ccccccc} \text{order } 5 & & \text{order } 125 & & \text{order } 15{,}000 & & \\ C & \subset & H & \subset & N & \subset & SL_5(\mathbb{Q}(\varepsilon)). \\ \end{array} \qquad (1.3.)$$

$$\underbrace{C \subset H}_{H/C}, \quad C \cong \mu_5, \quad H/C \cong \mathbb{Z}_5 \times \mathbb{Z}_5, \quad N \cong H \cdot SL_2(\mathbb{Z}_5)$$

It is not hard to work out the explicit matrices representing elements of N. They turn out to be of two types:

$$A_{ij} = \pm \frac{\varepsilon^{ai^2+bij+cj^2+di+ej+f}}{\sqrt{5}} \qquad (a, \ldots, f \in \mathbb{Z}_5, b \neq 0)$$

and

$$A_{ij} = \pm \varepsilon^{ai^2+bi+c}\delta_{i,dj+e} \qquad (a, \ldots, e \in \mathbb{Z}_5, d \neq 0)$$

(the sign being adjusted to make the determinant $+1$). It will be necessary to identify some special elements of N of the second type for the purpose of computation. Look at the elements $\iota, \mu, \nu \in SL_5(\mathbb{Q}(\varepsilon))$ given by

$$\iota x(i) = x(-i)$$
$$\mu x(i) = -x(2i)$$
$$\nu x(i) = \varepsilon^{i^2} \cdot x(i)$$

$$\begin{aligned} &\langle \iota, V_i \rangle = 1 \\ &\langle \mu, V_i \rangle = -1, \langle \mu^2, V_i \rangle = 1 \\ &\langle \nu, V_i \rangle = \theta^i(\eta - \eta'), \langle \nu^2, V_i \rangle = \theta^i(\eta' - \eta) \end{aligned} \qquad (1.4)$$

where $\eta = \varepsilon + \varepsilon^4$, $\eta' = \varepsilon^2 + \varepsilon^3$. Conjugating σ and τ, we find

$$\iota^{-1}\sigma\iota = \sigma^{-1}, \qquad \iota^{-1}\tau\iota = \tau^{-1}$$
$$\mu^{-1}\sigma\mu = \sigma^2, \qquad \mu^{-1}\tau\mu = \tau^3$$
$$\nu^{-1}\sigma\nu = \sigma\tau^2 \bmod C, \qquad \nu^{-1}\tau\nu = \tau.$$

Thus ι, μ, $\nu \in N$ and their images $\bar{\iota}$, $\bar{\mu}$, $\bar{\nu}$ in $SL_2(\mathbb{Z}_5)$ are:

$$\bar{\iota} = \begin{pmatrix} -1 & 0 \\ 0 & -1 \end{pmatrix}$$
$$\bar{\mu} = \begin{pmatrix} 2 & 0 \\ 0 & 3 \end{pmatrix} \tag{1.5}$$
$$\bar{\nu} = \begin{pmatrix} 1 & 2 \\ 0 & 1 \end{pmatrix}$$

§2. THE BUNDLE $\mathscr{F}$

Let $\mathbb{P}$ be the projective space representing the one-dimensional subspaces of V^i. Since V is given an underlying rational vector space, $\mathbb{P}$ is to be regarded as the complexification of a scheme over $\mathbb{Q}$. In particular it is meaningful to speak of coherent sheaves and their homomorphisms as being defined over specified subfields of $\mathbb{C}$.

Write $\mathcal{O}$ for the sheaf of local rings of $\mathbb{P}$ and $\mathcal{O}(1)$ for the canonical positive invertible sheaf on $\mathbb{P}$. The general linear group acts on $\mathcal{O}(1)$ and the space of sections $\Gamma(\mathcal{O}(1))$ is canonically isomorphic to V^* the dual of V. Regard V as a sheaf over Spec $\mathbb{C}$. The external tensor product $\mathcal{O}(1) \otimes_{\mathbb{C}} V$ is a sheaf on $\mathbb{P}$ and $\Gamma(\mathcal{O}(1) \otimes_{\mathbb{C}} V)$ is isomorphic to $\mathrm{Hom}_{\mathbb{C}}(V, V)$. Let ∂ in $\Gamma(\mathcal{O}(1) \otimes_{\mathbb{C}} V)$ correspond to I_V. The Koszul complex $\mathscr{K}$ is the exterior algebra $\Lambda^*(\mathcal{O}(1) \otimes_{\mathbb{C}} V)$ with multiplication by ∂ as differential:

$$0 \to \mathcal{O} \to \mathcal{O}(1) \otimes V \to \mathcal{O}(2) \otimes \Lambda^2 V \to \mathcal{O}(3) \otimes \Lambda^3 V \to \mathcal{O}(4) \otimes \Lambda^4 V \to \mathcal{O}(5) \otimes \Lambda^5 V \to 0.$$

The quotient $\mathcal{O}(1) \otimes V/\mathcal{O}$ is isomorphic to the tangent sheaf $\mathscr{T}$ to $\mathbb{P}$ and the sheaf of cycles $\mathrm{Im}(\mathcal{O}(i) \otimes \Lambda^i V) \subset \mathcal{O}(i+1) \otimes \Lambda^{i+1} V$ is isomorphic to the ith exterior power $\Lambda^i \mathscr{T}$ of $\mathscr{T}$. For the construction of the bundle $\mathscr{F}$ the relevant part of $\mathscr{K}$ is:

$$\overbrace{\mathcal{O}(2) \otimes_{\mathbb{C}} \Lambda^2 V}^{rk\ 10} \xrightarrow{p_0} \overbrace{\Lambda^2 \mathscr{T}}^{rk\ 6} \xrightarrow{q_0} \overbrace{\mathcal{O}(3) \otimes_{\mathbb{C}} \Lambda^3 V.}^{rk\ 10} \tag{2.1}$$

Note also that $\mathscr{K}$ has a symmetric pairing

$$\mathscr{K}^i \otimes \mathscr{K}^{5-i} \to \mathcal{O}(5) \otimes_{\mathbb{C}} \Lambda^5 V \cong \mathcal{O}(5),$$

given by $x \otimes y \to (x \wedge y)_5$, and that this induces the natural pairing $\Lambda^i \mathscr{T} \otimes \Lambda^{4-i} \mathscr{T} \to \mathcal{O}(5)$ and is compatible with the action of $SL_5(\mathbb{C})$. Note that with respect to these pairings $q_0 = p_0^*(5)$.

The H-modules $\Lambda^2 V$ and $2V_1$ are isomorphic since $\langle \varepsilon I_V, \Lambda^2 V\rangle = 10\varepsilon^2$ and by (1.2) $2V_1$ is the only representation of degree 10 for which this is possible. In identifying $\Lambda^2 V$ and other such spaces as N-modules, the reader should use the general observation:

(2.2) *Let Y, Z be representation spaces for a group G and let K be a normal subgroup of G. Suppose that Y is irreducible as a K-module and that $Z \cong nY$ as K-modules, then* $\mathrm{Hom}_K(Y, Z)$ *is a G/K-module, the evaluation mapping $Y \otimes \mathrm{Hom}_K(Y, Z) \to Z$ is an isomorphism of G-modules, and Z is irreducible as a G-module if and only if* $\mathrm{Hom}_K(Y, Z)$ *is irreducible as a G/K-module.*

In the present case, put $W = \mathrm{Hom}_H(V_1, \Lambda^2 V)$. It is a representation of N/H of degree 2, and the trace of $\bar{v}$ (the image of v in N/H) is

$$\langle \bar{v}, W\rangle = (3 + \varepsilon + \varepsilon^4)/(1 + 2\varepsilon^2 + 2\varepsilon^3) = \eta'.$$

It follows that W has character χ_2 (cf. character table in Appendix).

Since N/H is perfect, W is unimodular. So W has an invariant skew symmetric pairing defined over $\mathbb{Q}$ and this form is unique up to a scale factor. Let

$$f\colon V_1 \to \Lambda^2 V \otimes W$$

be the N-homomorphism determined by this form, and let

$$g\colon \Lambda^3 V \otimes W \to V_3 (= V_1{}^*)$$

be the dual of f composed with the canonical mapping $\Lambda^3 V \otimes W \cong \Lambda^3 V \otimes W^*$. Combining these with (2.1) gives the sequence of sheaf homomorphisms

$$\mathcal{O}(2) \otimes V_1 \xrightarrow{1_{\mathcal{O}} \otimes f} \mathcal{O}(2) \otimes \Lambda^2 V \otimes W \xrightarrow{p_0 \otimes 1_W} \Lambda^2 \mathcal{T} \otimes W \xrightarrow{q_0 + 1_W}$$
$$\mathcal{O}(3) \otimes \Lambda^3 V \otimes W \xrightarrow{1_{\mathcal{O}} \otimes g} \mathcal{O}(3) \otimes V_3. \qquad (2.3)$$

This sequence is defined over $\mathbb{Q}$.

Let

$$p\colon \overbrace{\mathcal{O}(2) \otimes V_1}^{rk\ 5} \to \overbrace{\Lambda^2 \mathcal{T} \otimes W}^{rk\ 12}, \qquad q\colon \overbrace{\Lambda^2 \mathcal{T} \otimes W}^{rk\ 12} \to \overbrace{\mathcal{O}(3) \otimes V_3}^{rk\ 5}$$

be the composites of the first two and last two morphisms in 2.3. Note that $q \cong p^*(5)$. We shall prove that $qp = 0$ and p, q are locally split. From this it follows that $\mathcal{F} = \mathrm{Ker}\, q/\mathrm{Im}\, p$ is locally free of rank 2 and defined over $\mathbb{Q}$. The bundle $\mathcal{F}$ is our goal.

To prove that $qp = 0$ it is sufficient, since $\mathcal{O} \otimes V_1$ is generated by its sections, to show that $\Gamma(qp(-2)) = 0$ and to prove that p, q are locally split it is sufficient that p is — for then q is locally split by duality. The first assertion follows immediately from

LEMMA 2.4. *Let U be the symmetric square representation $S^2 W$ of degree* 3, *and let W' be the representation obtained by acting on W with the Galois automorphism θ.*

(i) $\Gamma(\Lambda^2 \mathcal{T}(-2)) \cong \Lambda^2 V \cong V_1 \otimes W$, $\Gamma(\Lambda^2 \mathcal{T}(-2) \otimes W) \cong V_1 \oplus (V_1 \otimes U)$, *and* $\Gamma(p(-2))$ *is equivalent to the inclusion* $V_1 \to V_1 \oplus (V_1 \otimes U)$.

(ii) $\Gamma(\mathcal{O}(1) \otimes V_3) \cong (V_1 \otimes U) \oplus (V_1 \otimes W')$ *and* $\Gamma(q(-2))$ *is equivalent to the homomorphism* $V_1 \oplus (V_1 \otimes U) \to (V_1 \otimes U) \oplus (V_1 \otimes W')$ *induced by the identity on* $V_1 \otimes U$.

Proof. (i) The first isomorphism follows from (2.1), and the second is just the evaluation mapping. The third isomorphism now follows from the decomposition $W \otimes W = \mathbb{C} \oplus U$. Finally, since $\Gamma(\Lambda^2 \mathcal{T}(-2)) \cong \Lambda^2 V$, the mapping $\Gamma(p(-2))$ is equivalent to f which is just the mapping induced by $\mathbb{C} \to \mathbb{C} \oplus U$.

(ii) First note that $\Gamma(\mathcal{O}(1) \otimes V_3) \cong V^* \otimes V_3$. The character of $V^* \cong V_3$ as an H-module is given by

$$\langle h, V^* \otimes V_3\rangle = 0 (h \in H - C), \quad \langle \varepsilon^i 1, V^* \otimes V_3\rangle = 25\varepsilon^{2i}.$$

So $V^* \otimes V_3 \cong 5V_1$ as an H-module. Put $X = \mathrm{Hom}_H(V_1, V^* \otimes V_3)$. The formulas (1.4) show that

$$\langle \bar{\mu}, X \rangle = -1, \langle \bar{\mu}^2, X \rangle = 1, \langle \bar{\nu}, X \rangle = \eta - \eta'.$$

The first of these shows that X must have an irreducible component X^1 of degree 3, the second that the remaining component X^2 is irreducible of degree 2, and the third that X^1, X^2 have characters χ_3, χ_2'. Since $\chi_2{}^2 = \chi_3 + \chi_1$ and $\chi_2' = \theta\chi_2$ it follows that $X^1 \cong U$ and $X^2 \cong W'$.

To prove the statement about $\Gamma q(-2)$ it is sufficient to show that $\Gamma q(-2) \neq 0$, for $V_1, V_1 \otimes U, V_1 \otimes W$ are irreducible. But $q \cong p^*(5)$ and $p \neq 0$ by (i). So $q \neq 0$, and $\Gamma q(-2) \neq 0$ since $\Gamma(\mathcal{O}(1) \otimes V_3)$ generates $\mathcal{O}(1) \otimes V_3$. *Q.E.D.*

It remains to be proven that $p(-2)$ and hence p splits locally. Let v be a non-zero element of V and write $\hat{v}$ for the corresponding point of $\mathbb{P}$. We must show that the induced map on the vector bundle fibres:

$$p(-2)_{\hat{v}}: V_1 \to \Lambda^2 T_{\hat{v}} \otimes \mathcal{O}(-2)_{\hat{v}} \otimes W$$

is injective ($T_{\hat{v}}$ = tangent space to $\mathbb{P}$ at $\hat{v}$). But via p_o

$$\Lambda^2 T_{\hat{v}} \otimes \mathcal{O}(-2)_{\hat{v}} \cong \Lambda^{2+}/v \wedge V,$$

so in view of (2.3), the injectivity of $p(-2)_{\hat{v}}$ is equivalent to:

LEMMA 2.5. *For all nonzero $v \in V$ and $t \in V_1$, the element $f(t) \notin (v \wedge V) \otimes W$.*

Proof. Let v_i be the element of V defined by $v_i(j) = \delta_{ij}$ and put

$$z_0{}^+ = v_2 \wedge v_3, z_1{}^+ = v_3 \wedge v_4, z_2{}^+ = v_4 \wedge v_0, z_3{}^+ = v_0 \wedge v_1, z_4{}^+ = v_1 \wedge v_2$$

$$z_0{}^- = v_1 \wedge v_4, z_1{}^- = v_2 \wedge v_0, z_2{}^- = v_3 \wedge v_1, z_3{}^- = v_4 \wedge v_2, z_4{}^- = v_0 \wedge v_3.$$

The linear mappings $w^+: V_1 \to \Lambda^2 V$, $w^-: V_1 \to \Lambda^2 V$ defined by $w^+(v_i) = z_i{}^+$, $w^-(v_i) = z_i{}^-$ are H-homomorphisms, and form a base for W. We wish to show that for non-zero $v \in V$, $t \in V_1$, the equations

$$w^+(t) = v \wedge y^+, w^-(t) = v \wedge y^-$$

are contradictory. But these imply that

$$w^+(t) \wedge w^-(t) = 0,$$

and if $t = \sum a_i v_i$, then one computes that

$$w^+(t) \wedge w^-(t) = \sum_{i=0}^{4} (-1)^i a_i{}^2 v_0 \wedge \cdots \wedge \hat{v}_i \wedge \cdots \wedge v_4$$

hence $t = 0$. *Q.E.D.*

This completes the proof that $\mathcal{F}$ is a locally free sheaf of rank 2. To show that $\mathcal{F}$ is indecomposable it is sufficient to verify that its total Chern class $c(\mathcal{F})$ is irreducible. By definition

$$c(\mathcal{F}) = c(\Lambda^2 \mathcal{F})^2 \cdot c(\mathcal{O}(2))^{-5} \cdot c(\mathcal{O}(3))^{-5}.$$

Let h be the positive generator for the Chow ring of $\mathbb{P}$. It follows that

$$c(\mathscr{F}) = ((1+2h)^{10}(1+h)^{-5})^2(1+2h)^{-5}(1+3h)^{-5}$$

(where the first factor comes from the resolution of $\Lambda^2\mathscr{T}$ given by the Koszul complex). Hence

$$c(\mathscr{F}) = 1 + 5h + 10h^2.$$

Applying the Riemann–Roch theorem or directly from the definition of $\mathscr{F}$, one also computes the Hilbert polynomial:

$$\chi(\mathscr{F}(n-5)) = \tfrac{1}{12}(n^2-1)(n^2-24).$$

§3. THE INVARIANT QUINTICS

This section is preliminary to the computation of $\Gamma(\mathscr{F})$ and the proof of the non-singularity of the zero set of a general section. The main results are the determination of the N/H-module $\Gamma_H(\mathcal{O}(5))$ of H-invariants of $\Gamma(\mathcal{O}(5))$ and the sheaf of ideals $\mathscr{L}$ in $\mathcal{O}$ generated by the subspace $\Gamma_H(\mathcal{O}(5))$. In the next section we show that this subspace is isomorphic to the second exterior power of $\Gamma(\mathscr{F})$, however the present section does not depend on this fact.

Write a_i for the character of $\Lambda^i V$ as an N-module, h_i for the character of $S^i V$, and a_i^*, h_i^* for the characters of the duals. Then

$$\begin{aligned} a_i^* &= \bar{a}_i = \theta^2 a_i \\ h_i^* &= \bar{h}_i = \theta^2 h_i \end{aligned} \tag{3.1}$$

since the representations are unitary. Also

$$a_i^* = a_{5-i} \tag{3.2}$$

since the representation is unimodular. As in §1, decompose $V \otimes V^*$ into $\mathbb{C} \oplus Z$ and let ζ be the character of Z. It follows that $h_1 \cdot \theta^2 h_1 = \theta h_1 \cdot \theta^3 h_1 = 1 + \zeta$.

LEMMA 3.3.
(i) $a_1 = h_1$
(ii) $a_2 = \chi_2 \cdot \theta h_1$
(iii) $a_3 = \chi_2 \cdot \theta^3 h_1$
(iv) $a_4 = \theta^2 h_1$
(v) $h_2 = \chi_3' \cdot \theta h_1$
(vi) $h_3 = (\chi_5 + \chi_2') \cdot \theta^3 h_1$
(vii) $h_4 = (\chi_4 + \chi_4^{\#} + \chi_3 + \chi_3') \cdot \theta^2 h_1$
(viii) $h_5 = (\chi_3 + \chi_3') + \zeta \cdot (\chi_3 + \chi_3' - 1)$
(ix) $h_1 \theta h_1 = (\chi_3 + \chi_2') \cdot \theta^3 h_1$.

Proof. (ii) follows from Lemma (2.4) and then (i), (iii) and (iv) follows from (3.1) and (3.2). To prove (v), note that (2.2) implies $h_2 = \chi \cdot \theta h_1$ for some character χ of N/H. A simple computation shows that $\chi(\iota) = 3$ and $\chi(v) = -\eta$. So since χ nas degree 3, $\chi = \chi_3'$. Now use the well-known formula:

$$h_i = a_1 h_{i-1} - a_2 h_{i-2} + \cdots - (-1)^i a_i h_0,$$

plus the identity (ix) proven in (2.4) and (vi), (vii) and (viii) follow by computing characters via the character table in the Appendix. *Q.E.D.*

The first of the main results of this section follows at once from part (viii) of this lemma:

THEOREM 3.5. *The character of* $\Gamma_H(\mathcal{O}(5))$ *is* $\chi_3 + \chi_3'$ *and its dimension is* 6.

Let y_i be the ith coordinate function on V, ($y_i(x) = x(i)$). The monomials

$$y_0^5, y_0^3 y_1 y_4, y_0^3 y_2 y_3, y_0^2 y_2^2 y_1, y_0^2 y_1^2 y_3, \prod_{i=0}^4 y_i,$$

are invariants of τ, and the six forms

$$S = \sum y_i^5, Q, Q', R, R', Y = 5 \prod y_i$$

obtained by summing these monomials over the powers of σ are invariants of H. Since they are linearly independent they form a base for $\Gamma_H(\mathcal{O}(5))$.

Another natural basis of $\Gamma_H(\mathcal{O}(5))$ is obtained as follows: the group H/C has six proper subgroups and these subgroups are permuted triply transitively by N. The fixed point set in $\mathbb{P}$ of the subgroup $\{C, \tau C, \tau^2 C, \tau^3 C, \tau^4 C\}$ is just the simplex of reference. The six simplexes determined in this way by the six subgroups we call the *fundamental simplexes*. Each of them determines, up to a scalar multiple, the quintic whose zero set consists of five three-dimensional faces of the simplex. The subspace of $\Gamma_H(\mathcal{O}(5))$ that these quintics span is invariant under both N and Θ. So Theorem 3.5 shows that these six quintics also form a base for $\Gamma_H(\mathcal{O}(5))$.

Now let L be the set of common zeros of the polynomials in $\Gamma_H(\mathcal{O}(5))$, and let L_i be its intersection with $y_i = 0$. But

$$S(0, y_1, \ldots, y_4) = y_1^5 + y_2^5 + y_3^5 + y_4^5$$
$$Q(0, y_1, \ldots, y_4) = y_2 y_3 (y_2^2 y_1 + y_3^2 y_4)$$
$$Q'(0, y_1, \ldots, y_4) = y_1 y_4 (y_1^2 y_3 + y_4^2 y_2)$$
$$R(0, y_1, \ldots, y_4) = y_2 y_3 (y_1^2 y_3 + y_4^2 y_2)$$
$$R'(0, y_1, \ldots, y_4) = y_1 y_4 (y_2^2 y_1 + y_3^2 y_4).$$

These equations define the set L_0 and it is straightforward to check that it consists of precisely the five lines

$$y_2 + \varepsilon^r y_3 = \varepsilon^{2r} y_1 + y_4 = 0 \qquad (*)$$

plus the 20 points

$$y_1 = y_2 = 0, y_3 = \varepsilon^r y_4$$
$$y_1 = y_3 = 0, y_2 = \varepsilon^r y_4$$
$$y_4 = y_2 = 0, y_3 = \varepsilon^r y_1$$
$$y_4 = y_3 = 0, y_2 = \varepsilon^r y_1.$$

The set of points of L_i is just $\sigma^i L_0$, the same set with $y_i = 0$ after a cyclic permutation of the coordinates. Taking the union over all i, it follows that the set L consists of the 25 skew lines:

$$y_i = y_{i+2} = \varepsilon^r y_{i+3} = \varepsilon^{2r} y_{i+1} + y_{i+4} = 0, \qquad (0 \leqq i, r \leqq 4). \tag{3.6}$$

We claim that the *scheme* $\mathcal{O}/\mathcal{L}$ is this set of 25 skew lines with reduced structure sheaf, hence is a regular scheme. If $x \in L$ lies on only one face of each of the fundamental simplices, then the ideal $\mathcal{L}_x$ is defined by six linear forms in the y_i's, so $\sqrt{\mathcal{L}_x} = \mathcal{L}_x$. On the other hand, say $x \in L$ lies on a 2-dimensional face of at least one fundamental simplex. Then x is necessarily on a 1-dimensional face or edge of this simplex (as you see by intersecting the line (3.6) with a typical 2-dimensional face $y_0 = y_1 = 0$ on the simplex of reference). But the edges of two distinct fundamental simplices do not intersect: in fact N permutes the fundamental simplices triply transitively and one may readily calculate that the edges of $\prod_i y_i = 0$ and $\prod_i (y_0 + \varepsilon^i y_1 + \varepsilon^{2i} y_2 + \varepsilon^{3i} y_3 + \varepsilon^{4i} y_4) = 0$ do not intersect. Therefore x is singular on at most one fundamental simplex. This shows that $\mathcal{L}_x$ is generated by five linear forms and one of higher degree. If these five linear forms met in a plane, L would contain more than one line in this plane, i.e. L would have two components that met. Since this is false, $\mathcal{L}_x$ is generated by the five linear forms and $\sqrt{\mathcal{L}_x} = \mathcal{L}_x$ again.

§4. THE SPACES $H^i(\mathcal{F}(n))$

Let $\psi^i(n)$ be the character of $H^i(\mathcal{F}(n))$ as a representation of N. To determine these characters we write $\mathcal{G}$ for the cokernel of p (cf. §2) and consider the exact sequences

$$\begin{gathered} 0 \to \mathcal{O}(2) \otimes V_1 \to \Lambda^2 \mathcal{T} \otimes W \to \mathcal{G} \to 0 \\ 0 \to \mathcal{F} \to \mathcal{G} \to \mathcal{O}(3) \otimes V_3 \to 0. \end{gathered} \tag{4.1}$$

Using the well-known values for the cohomology of $\mathcal{O}(n)$ and $\Lambda^2(\mathcal{T}(n))$ (which is just $\Omega^2(5+n)$, Ω^i being the shear of i-forms on $\mathbb{P}$) we get

$$\begin{aligned} & 0 \to \Gamma(\mathcal{O}(n+2)) \otimes V_1 \to \Gamma(\Lambda^2 \mathcal{T}(n)) \otimes W^n \to \Gamma(\mathcal{G}(n)) \to 0; \\ & H^1(\mathcal{G}(n)) = (0); \\ & H^2(\mathcal{G}(n)) = (0) \text{ if } n \neq -5,\ H^2(\mathcal{G}(-5)) \cong W; \\ & 0 \to \Gamma(\mathcal{F}(n)) \to \Gamma(\mathcal{G}(n)) \to \Gamma(\mathcal{O}(n+3)) \otimes V_3 \to H^1(\mathcal{F}(n)) \to 0; \\ & \psi^2(n) = 0 \text{ if } n \neq -5,\ \psi^2(-5) = \chi_2. \end{aligned} \tag{4.2}$$

Now since $\Lambda^2 \mathcal{F} \cong \mathcal{O}(5) \cong \Omega^4(10)$, Serre duality asserts that $H^i(\mathcal{F}(n))$ and $H^{4-i}(\mathcal{F}(-10-n))$ are dual. We deduce using (4.2)

$$\begin{gathered} \psi^1(n) = 0 \quad \text{if} \quad n \leqq -4, \\ \psi^3(n) = 0 \quad \text{if} \quad n \geqq -6. \end{gathered} \tag{4.3}$$

Since $\Gamma(\Lambda^2 \mathcal{T}(-3)) = (0)$ we further deduce from (4.2)

$$\psi^1(-3) = \theta^3 h_1,\ \psi^3(-7) = \theta h_1. \tag{4.4}$$

From Lemma 2.4 (ii) the image of $\Gamma(\mathcal{G}(-2))$ in $\Gamma(\mathcal{O}(1)) \otimes V_3$ has dimension 15. But the first exact sequence of (4.2) shows that the dimension of $\Gamma\mathcal{G}(-2)$ is 15. So from the second exact sequence and Serre duality we deduce that

$$\begin{aligned}\psi^0(n) &= 0 \quad \text{if} \quad n \leqq -2\\ \psi^4(n) &= 0 \quad \text{if} \quad n \geqq -8.\end{aligned} \tag{4.5}$$

We can now calculate $\psi^1(-2)$ from the exact sequences of (4.2) together with (4.5) and we find

$$\psi^1(-2) = \chi_2' \cdot \theta h_1, \psi^3(-8) = \chi_2' \cdot \theta^3 h_1. \tag{4.6}$$

Consider now the characters $\psi^i(j)$ for $i = 0$, 1 and $j = 0, -1$. The exact sequences of (4.2) together with Lemma 3.3 give

$$\begin{aligned}\psi^0(-1) - \psi^1(-1) &= -\chi_2' \cdot h_1,\\ \psi^0(0) - \psi^1(0) &= \chi_4 - \chi_2',\end{aligned} \tag{4.7}$$

and in particular it follows that

$$\psi^0(0) \geqq \chi_4, \psi^1(0) \geqq \chi_2', \tag{4.8}$$

where the inequality means that the difference between the two sides is the character of a representation. Also, since $\Gamma(\mathscr{F}(n))$ is a subspace of $\Gamma(\mathscr{G}(n))$, the exact sequences show that

$$\begin{aligned}\psi^0(-1) &\leqq (\chi_4^{\#} + \chi_4 + \chi_5)h_1,\\ \psi^0(0) &\leqq \chi_4 + \chi_5 + (\chi_4 + \chi_5 - \chi_2)\zeta.\end{aligned} \tag{4.9}$$

Since $\Lambda^2 \mathscr{F} \cong \mathcal{O}(5)$ there are homomorphisms

$$\lambda(-1)\colon \Lambda^2\Gamma(\mathscr{F}(-1)) \to \Gamma(\mathcal{O}(3)), \lambda\colon \Lambda^2\Gamma(\mathscr{F}) \to \Gamma\mathcal{O}(5).$$

LEMMA 4.10 *Let $\mathscr{S}$ be a locally free sheaf of rank 2 on $\mathbb{P}$ such that $\Gamma(\mathscr{S}(-1)) = 0$, let $\gamma\colon \Lambda^2\Gamma(\mathscr{S}) \to \Gamma\Lambda^2(\mathscr{S})$ be the canonical homomorphism, and let A be any subspace of $\Gamma(\mathscr{S})$. Then*

$$\dim \gamma(\Lambda^2 A) \geqq 2 \dim A - 3.$$

Proof. Since the Grassman cone in $\Lambda^2 A$ has dimension $2 \dim A - 3$ it is sufficient to show that the only element of this cone in $\operatorname{Ker} \gamma$ is the zero.

Suppose that $\gamma(s \wedge t) = 0$ for some s, t in A. Assume s, t are not both zero, then they generate a subsheaf $\mathscr{L}$ of $\mathscr{S}$ with rank 1. Since $\mathscr{L}$ is torsion-free its bidual is an invertible sheaf. So s, t are contained in a subsheaf isomorphic to $\mathcal{O}(r)$ for some r. As $\Gamma(\mathscr{S}(-1)) = 0$, it follows that $r \leqq 0$. Hence s, t are proportional and $s \wedge t = 0$. *Q.E.D*

First consider $\lambda(-1)$. Suppose that $\psi^0(-1) \neq 0$. Since the terms of the first inequality of (4.9) are irreducible characters with degrees at least 20, it follows that $\dim \Gamma(\mathscr{F}(-1)) \geqq 20$ and so by the lemma $\dim \Gamma(\mathcal{O}(3)) \geqq 37$. Since $\dim \Gamma\mathcal{O}(3) = 35$ it follows that

$$\psi^0(-1) = 0, \psi^1(-1) = \chi_2' \cdot h_1. \tag{4.11}$$

Now consider λ. Take A to be the subspace $\Gamma_H(\mathscr{F})$ of H-invariant sections. From Theorem 3.5, $\dim \Gamma_H(\mathcal{O}(5)) = 6$. So the lemma shows that $\dim A < 5$. Together with (4.8) and (4.9) this shows that the character of $\Gamma_H(\mathscr{F})$ is χ_4. Applying the lemma again shows that $\dim \lambda(A) \geqq 5$. So since $\Gamma_H(\mathcal{O}(5))$ has character $\chi_3 + \chi_3'$ it follows that λ is an isomorphism from $\Lambda^2\Gamma_H(\mathscr{F})$ to $\Gamma_H(\mathcal{O}(5))$.

We claim that in fact $\Gamma_H(\mathscr{F}) = \Gamma(\mathscr{F})$. If this is not true, then as a representation of

H, the space $\Gamma(\mathscr{F})$ must contain all the non-trivial characters of H/C at least once (by (4.9)), hence $\dim \Gamma(\mathscr{F}) \geqq 28$. Let $A, B \subseteq \mathbb{P}^4$ be two hyperplanes with homogeneous equations a, b and consider the exact sequence

$$0 \to \mathscr{F}(-2) \xrightarrow{(a,b)} \mathscr{F}(-1) \oplus \mathscr{F}(-1) \xrightarrow{(b,-a)} \mathscr{F} \to \mathscr{F}_{A\cdot B} \to 0.$$

We find, since $\Gamma(\mathscr{F}(-1)) = 0$,

$$\dim \Gamma(\mathscr{F}_{A\cdot B}) \geqq \dim \Gamma(\mathscr{F}) - \dim \operatorname{Ker}[(a, b) \text{ on } H^1(\mathscr{F}(-2))].$$

But from (4.2), $H^1(\mathscr{F}(-1))$ is generated by $H^0(\mathcal{O}(1)) \otimes H^1(\mathscr{F}(-2))$. Note that $h^1(\mathscr{F}(-2)) = 10$, $h^1(\mathscr{F}(-1)) = 10$ (by (4.6), (4.11)). So if a, b are sufficiently generic the image $a \cdot H^1(\mathscr{F}(-2)) + b \cdot H^1(\mathscr{F}(-2))$ in $H^1(\mathscr{F}(-1))$ has dimension at least 4. Therefore $\dim \Gamma \mathscr{F}_{A\cdot B} \geqq 28 - 10 + 4 = 22$. But let s be a non-zero section of $\mathscr{F}$. Its zero set X_s is a surface (since $\Gamma\mathscr{F}(-1) = 0$) and non-empty (since $c_2(\mathscr{F}) \neq 0$), so we can choose A, B so that $A \cdot B \cdot X_s$ is a non-empty finite set of points. Then $\mathscr{F}_{A\cdot B}/s\mathcal{O}_{A\cdot B}$ is a torsion-free rank 1 sheaf on $A \cdot B$, and hence isomorphic to $\mathscr{J} \cdot \mathcal{O}_{A\cdot B}(n)$ for some sheaf of ideals $\mathscr{J}$ defining $A \cdot B \cdot X_s$. Computing Chern classes we find $n = 5$. So, since $A \cdot B \cdot X_s$ is non-empty, $\dim \Gamma(\mathscr{F}_{A\cdot B}/s\mathcal{O}_{A\cdot B}) < 21$, and finally

$$21 \leqq \dim \Gamma(\mathscr{F}_{A\cdot B}) - 1 \leqq \dim \Gamma(\mathscr{F}_{A\cdot B}/s\mathcal{O}_{A\cdot B}) < 21,$$

which is a contradiction. So, taking account of (4.7), we have

$$\psi^0(0) = \chi_4, \psi^1(0) = \chi_2'. \tag{4.12}$$

Finally we claim $\psi^1(n) = 0$ if $n \geqq 1$. By Castelnuovo's lemma [11], it suffices to prove that $\psi^1(1) = 0$. By (4.2) the cup product $\alpha\colon \Gamma(\mathcal{O}(1)) \otimes H^1(\mathscr{F}) \to H^1(\mathscr{F}(1))$ is surjective. On the other hand N acts irreducibly on $\Gamma(\mathcal{O}(1)) \otimes H^1(\mathscr{F})$ by (2.2). Therefore either $\psi^1(1) = 0$ or α is an isomorphism. But $\Gamma(\mathcal{O}(1)) \otimes H^1(\mathscr{F}(-1)) \to H^1(\mathscr{F})$ is also surjective so for some $a \in \Gamma(\mathcal{O}(1))$, $\sigma \in H^1(\mathscr{F}(-1))$, it follows that $a \cup \sigma \neq 0$. Since $\dim \Gamma(\mathcal{O}(1)) > \dim H^1(\mathscr{F})$, it follows that $b \cup \sigma = 0$ for some other non-zero b. Therefore $\alpha(b, a \cup \sigma) = 0$ and α is not injective.

We summarize our calculations as follows:

TABLE OF $\dim H^i(\mathscr{F}(n-5))$

n	H^0	H^1	H^2	H^3	H^4
$n \geqq 6$	$\frac{(n^2-1)(n^4-24)}{12}$	0	0	0	0
5	4	2	0	0	0
4	0	10	0	0	0
3	0	10	0	0	0
2	0	5	0	0	0
1	0	0	0	0	0
0	0	0	2	0	0
−1	0	0	0	0	0
−2	0	0	0	5	0
−3	0	0	0	10	0
−4	0	0	0	10	0
−5	0	0	0	2	4
$n \leqq -6$	0	0	0	0	$\frac{(n^2-1)(n^4-24)}{12}$

§5. THE ZERO SETS X_s, $s \in \Gamma(\mathscr{F})$

THEOREM 5.1. *For almost all $s \in \Gamma(\mathscr{F})$, the zero set of s is a non-singular surface $X_s \subset \mathbb{P}$ of degree 10; when X_s is nonsingular, it is an abelian surface.*

Proof. Let Q be the projective space associated to $\Gamma(\mathscr{F})$, and let Z be the subvariety of $Q \times \mathbb{P}$ represented by pairs (s, x) $(s \in \Gamma(\mathscr{F}),\ x \in \mathbb{P})$ such that $s(x) = 0$. Since $\Lambda^2\Gamma(\mathscr{F}) \cong \Gamma_H(\mathcal{O}(5))$ the sheaf $\mathscr{F}$ is generated by $\Gamma(\mathscr{F})$ except at points of the set of 25 skew lines L whose ideal is generated by $\Gamma_H(\mathcal{O}(5))$ (see §3). It follows that Z is a fibre-bundle over $\mathbb{P} - L$, in particular it is non-singular over $\mathbb{P} - L$. Applying Sard's theorem [15] to the projection $Z \to Q$ shows that the zero variety $X_s = \{x \mid s(x) = 0\}$ of a general section s is a surface that is non-singular except possibly at points of L.

Let $x \in L$ and let e_1, e_2 be a basis of the free rank two $\mathcal{O}_x$-module $\mathscr{F}_x$. Each $s \in \psi(\mathscr{F})$ can be written

$$s = s_1 e_1 + s_2 e_2, \qquad s_i \in \mathcal{O}_x,$$

so that if $s, t \in \Gamma(\mathscr{F})$,

$$s \wedge t = (s_1 t_2 - s_2 t_1) e_1 \wedge e_2 .$$

If for every $s \in \Gamma(\mathscr{F})$, $s_1(x) = s_2(x) = 0$, then for every s and t, $s \wedge t$ would vanish to second order at x. Using again the fact that $\Lambda^2\Gamma(\mathscr{F}) \cong \Gamma_H(\mathcal{O}(5))$ and that $\Gamma_H(\mathcal{O}(5))$ generates the ideal of L, this is impossible. We may therefore choose e_1, e_2 so that e_1 is an element of $\Gamma(\mathscr{F})$. Write out a basis of $\Gamma(\mathscr{F})$ locally:

$$\begin{aligned} s &= e_1 \\ t &= f e_1 + u e_2 \\ t' &= f' e_1 + u' e_2 \qquad \text{where} \quad f, f', f'', u, u', u'' \in \mathcal{O}_x . \\ t'' &= f'' e_1 + u'' e_2 \end{aligned}$$

Then

$$\begin{aligned} s \wedge t &= u \cdot e_1 \wedge e_2 \\ s \wedge t' &= u' \cdot e_1 \wedge e_2 \\ s \wedge t'' &= u'' \cdot e_1 \wedge e_2 \end{aligned}$$

and $t^{(i)} \wedge t^{(j)}$ vanishes at x to 2nd order.

Therefore u, u', u'' must generate the ideal of L at x, i.e. their differentials are independent at x. But if $\lambda s + \mu t + \mu' t' + \mu'' t''$ is a general section in $\Gamma(\mathscr{F})$, so that $(\lambda, \mu, \mu', \mu'')$ are homogeneous coordinates in Q, then Z is described above points near x by the equations:

$$\begin{aligned} \lambda &= -(\mu f + \mu' f' + \mu'' f'') \\ 0 &= \mu u + \mu' u' + \mu'' u'' \end{aligned}$$

which are easily seen to define a non-singular subvariety of $Q \times \mathbb{P}$. Thus Z is everywhere non-singular, hence by Sard's theorem so is the set X_s of zeros of a generic section s of $\mathscr{F}$.

To prove that $X(= X_s)$ is abelian of degree 10 note that its normal bundle N in $\mathbb{P}$ is isomorphic to $\mathscr{F} \otimes \mathcal{O}_X$. So the Chern class $c(N)$ of N (in the Chow ring of X) is just the restriction to X of $1 + 5h + 10h^2$. Since $\mathbb{P}$ has Chern class $1 + 5h + 10h^2 + \cdots$, the Chern class of X is 1. So the canonical class K_X is zero and the Euler characteristic $c_2(X)$ is zero. This characterizes abelian surfaces [8, §6]. Since $c_2(N)$ is just the self intersection of X, the degree of X is 10. *Q.E.D*

THEOREM 5.2. *Every abelian surface $Z \subset \mathbb{P}$ is projectively equivalent to the zero set of some section s of $\mathscr{F}$.*

Proof. Let $\mathscr{D} = \mathcal{O}(1) \otimes \mathcal{O}_Z$. Since the Chern class of Z is 1 that of its normal bundle is the restriction of $1 + 5h + 10h^2$. As above it follows that Z has degree 10. Choose an origin on Z and let $H(\mathscr{D})$ be the subgroup

$$\{z \in Z \mid T_z{}^* \mathscr{D} \cong \mathscr{D}\}$$

where T_z is just the translation by z (cf. [13, §13]). Since $H(\mathscr{D})$ has order $(\deg Z/2)^2$ and carries a non-degenerate alternating form,

$$H(\mathscr{D}) \cong \mathbb{Z}_5 \times \mathbb{Z}_5.$$

Further the Riemann–Roch theorem for abelian varieties (*ibid.*) shows that $\dim \Gamma(\mathscr{D}) = 5$, and Lefschetz's theorem implies that Z cannot lie in a subspace of $\mathbb{P}$ (otherwise Z would be simply-connected). So the mapping

$$\phi: \Gamma\mathcal{O}(1) \to \Gamma(\mathscr{D})$$

is necessarily an isomorphism. Applying the results of [12, §1] it follows that when Z is embedded in $\mathbb{P}^4$ by the complete linear system $\Gamma(\mathscr{L})$ and a suitable isomorphism is chosen between $\mathbb{P}^4$ of §2 then Z is invariant under the action of the Heisenberg group introduced in §1. But since ϕ is an isomorphism this is just the composition of our given embedding and a projective transformation, i.e. after a projective transformation we may assume that Z is invariant under H. Actually we can go a bit further: if we choose an origin O in Z with respect to which $\mathscr{L}$ is symmetric then the map $x \mapsto -x$ for this origin extends to a projective transformation ι_0 of $\mathbb{P}$, leaving Z fixed, normalizing the action of H/C and so that $\iota_0 \cdot \eta \cdot \iota_0{}^{-1} = \eta^{-1}$ for $\eta \in H/C$. Therefore ι_0 must be induced by the element ι of N introduced in §1, and Z is invariant under H and ι.

Next, look at the natural map:

$$\psi: \Gamma_H(\mathbb{P}, \mathcal{O}(5)) \to \Gamma_H(Z, \mathscr{D}^5).$$

The group $H(\mathscr{D})$ acts on the line bundle $\mathscr{D}^5$, hence there is a line bundle $\mathscr{M}$ on $Y = Z/H(\mathscr{D})$ such that $\pi^*\mathscr{M} \cong \mathscr{D}^5$ ($\pi: Z \to Y$ the natural homomorphism). Then $\Gamma(Y, \mathscr{M}) \cong \Gamma_H(Z, \mathscr{D}^5)$ and $\deg \mathscr{M} = \deg \mathscr{D}^5/\deg \pi = 5$, so $\dim \Gamma(Y, \mathscr{M}) = 5$. In fact, under the symmetry $x \mapsto -x$, the space $\Gamma(Y, \mathscr{M})$ breaks up into the sum of an eigenspace of dimension 3 and one of dimension 2 (cf. [12, §2]; note that the action of $x \mapsto -x$ on $\Gamma(Y, \mathscr{M})$ is only well determined up to sign, so we have no obvious way of labelling one eigenspace "even" and the other "odd"). Since ι is the identity on $\Gamma_H(\mathbb{P}, \mathcal{O}(5))$, the image of ψ is contained in one of these eigenspaces. Therefore $\dim \ker(\psi) \geq 3$, *i.e.* at least three independent quintics of §3 contain Z.

Consider the map

$$\Lambda^2\Gamma(\mathscr{F}) \xrightarrow{\approx} \Gamma_H(\mathcal{O}(5)).$$

We have proven that there is a subspace $K \subset \Lambda^2\Gamma(\mathscr{F})$ of dim $\geqq 3$ consisting of elements that are mapped to zero in $\Gamma(\Lambda^2(\mathscr{F} \otimes \mathcal{O}_Z))$. It follows with a little linear algebra that there are two possibilities:

for some basis s_1, s_2, s_3, s_4 of $\Gamma(\mathscr{F})$, either

(a) $s_1 \wedge s_2, s_3 \wedge s_4, s_1 \wedge s_3 - s_2 \wedge s_4 \in K$, or

(b) $s_1 \wedge s_2, s_1 \wedge s_3$ (and a 3rd independent elt.) $\in K$.

Now if $s \wedge t \in K$, and $\bar{s}$, $\bar{t}$ are the restrictions of s, t to Z, then $\bar{s} = f \cdot \bar{t}$ for some $f \in \mathbb{C}(Z)$. Therefore in case (a),

$$\bar{s}_1 = f \cdot \bar{s}_2 \qquad \bar{s}_4 = f \cdot \bar{s}_3 .$$

Let D be the divisors of poles of f and let $\mathscr{M} = \mathcal{O}_Z(D)$. Define

$$\alpha \colon \mathscr{M} + \mathscr{M} \to \mathscr{F} \otimes \mathcal{O}_Z \quad \text{by} \quad (g_1, g_2) \mapsto g_1\bar{s}_2 + g_2\bar{s}_3 .$$

Then all four sections $\bar{s}_i$ of $\mathscr{F} \otimes \mathcal{O}_Z$ are images by α of sections of $\mathscr{M} + \mathscr{M}$ (i.e. $\alpha(1, 0)$, $\alpha(0, 1)$, $\alpha(f, 0)$ and $\alpha(0, f)$). But the s_i generate $\mathscr{F}$ everywhere except at the 25 lines L. Since Z is abelian, none of these lines is contained in Z, hence the $\bar{s}_i$ generate $\mathscr{F} \otimes \mathcal{O}_Z$ at all but a finite set of points. But α is a homomorphism of rank 2 bundles, so the support of its co-kernel is defined by the principal ideal (det α), and has codimension 1. Therefore α must be an isomorphism. But then $\mathscr{M}^2 \cong \Lambda^2 \mathscr{F} \otimes \mathcal{O}_Z \cong \mathscr{D}^5$, hence

$$4(D^2) = 25 \cdot c_1(\mathscr{D})^2 = 25 \cdot \deg Z = 250,$$

contradiction.

In case (b), either $\bar{s}_2 = f \cdot \bar{s}_1, \bar{s}_3 = g \cdot \bar{s}_1, f, g \in \mathbb{C}(Z)$ or $\bar{s}_1 = 0$. In the 1st case, as above, we get a homomorphism:

$$\alpha \colon \mathscr{M} \to \mathscr{F} \otimes \mathcal{O}_Z,$$

with three out of the four $\bar{s}_i$'s in the image $\alpha\Gamma(\mathscr{M})$. Then $\bar{s}_4$ generates the cokernel except at a finite set of points:

$$0 \to \mathcal{O}_Z \xrightarrow{\bar{s}_4} \mathscr{F} \otimes \mathcal{O}_Z/\alpha\mathscr{M} \to \underset{\text{finite support}}{\mathscr{G}} \to 0.$$

By elementary homological algebra, extensions of $\mathscr{G}$ by a line bundle split. Using this twice we find $\mathscr{G}$ must be zero, and we have

$$0 \to \mathscr{M} \xrightarrow{\alpha} \mathscr{F} \otimes \mathcal{O}_Z \to \mathcal{O}_Z \to 0,$$

hence $c_2(\mathscr{F} \otimes \mathcal{O}_Z) = Z \cdot c_2(\mathscr{F}) = 0$, which is absurd. Thus $\bar{s}_1 = 0$, i.e. $Z \subset$ zeroes of s_1. Since $\deg Z = 10 = \deg X_{s_1}$, it follows that $Z = X_{s_1}$. *Q.E.D.*

§6. CONNECTIONS WITH MODULI

The bundle $\mathscr{F}$ can be used to give an explicit representation of a certain moduli space for 2-dimensional abelian varieties. We first recall some standard results in the theory of moduli of abelian varieties:

(a) Let $g \geq 1$ be the dimension;

(b) Let $\mathfrak{H}_g$ = Siegel upper $\frac{1}{2}$-space of $g \times g$ symmetric matrices Ω, Im $\Omega > 0$,

$$\cong \mathrm{Sp}(2g, \mathbb{R})/\text{maximum compact } K;$$

(c) Fix a sequence δ of positive integers $\delta_1, \ldots, \delta_g$ such that δ_i divides δ_{i+1};

(d) Sp $(2g, \mathbb{Q})$ acts on $\mathbb{Q}^{2g}$, fixing the form

$$A(e_i, e_{g+j}) = \delta_{ij}, A(e_i, e_j) = A(e_{g+i}, e_{g+j}) = 0, \qquad 1 \leq i, j \leq g;$$

(e) Let L_δ = the sublattice $\mathbb{Z}^g \times \prod_{i=1}^g \delta_i \mathbb{Z}$ of $\mathbb{Z}^{2g}$,

$L_\delta^\perp$ = the lattice $\prod_{i=1}^g (1/\delta_i)\mathbb{Z} \times \mathbb{Z}^g$, characterized as the set of $x \in \mathbb{Q}^{2g}$ such that $A(x, y) \in \mathbb{Z}$, all $y \in L_\delta$;

(f) On $L_\delta^\perp/L_\delta$, put the multiplicative symplectic form

$$e_\delta(x, y) = e^{2\pi i A(x, y)}.$$

(g) Let $\Gamma(\delta_1, \ldots, \delta_g)_0 = \{X \in \mathrm{Sp}(2g, \mathbb{Q}) \mid X(L_\delta) = L_\delta\}$;

$\Gamma(\delta_1, \ldots, \delta_g) = \{X \in \mathrm{Sp}(2g, \mathbb{Q}) \mid X(L_\delta) = L_\delta$ and $X = \text{id. on } L_\delta^\perp/L_\delta\}$;

(h) Then the analytic quotient spaces have the significance

$$\mathfrak{U}_\delta^{(0)} \underset{\text{def}}{=} \mathfrak{H}_g/\Gamma(\delta_1, \ldots, \delta_g)_0 = \left\{\begin{array}{l}\text{moduli space of pairs } (X, \lambda), X \text{ a} \\ g\text{-dimensional abelian variety, } \lambda\text{:} \\ X \to \hat{X} \text{ a polarization such that} \\ \quad \ker(\lambda) \cong \prod_{i=1}^g (\mathbb{Z}/\delta_i\mathbb{Z})^2.\end{array}\right\}$$

$$\mathfrak{U}_\delta \underset{\text{def}}{=} \mathfrak{H}_g/\Gamma(\delta_1, \ldots, \delta_g) = \left\{\begin{array}{l}\text{moduli space of triples } (X, \lambda, \alpha), \\ (X, \lambda) \text{ as above, and} \\ \quad \alpha\colon \ker(\lambda) \xrightarrow{\cong} L_\delta^\perp/L_\delta \\ \text{a symplectic isomorphism with} \\ \text{respect to } e_\lambda \text{ and } e_\delta.\end{array}\right\}$$

(i) $\mathfrak{U}_\delta$ and $\mathfrak{U}_\delta^{(0)}$ have natural structures of quasi-projective varieties;

(j) Note that the finite "symplectic" group $\Gamma(\delta)_0/\Gamma(\delta)$ acts on $\mathfrak{U}_\delta$ and $\mathfrak{U}_\delta^{(0)}$ is the quotient $\mathfrak{U}_\delta/[\Gamma(\delta)_0/\Gamma(\delta)]$.

Now if $\lambda : X \to \hat{X}$ is a polarization, let L_λ denote one of the corresponding invertible sheaves—all such are isomorphic after a translation. The result can now be stated:

THEOREM 6.1. *Let*

$$\mathfrak{U}^*_{(5,1)} = \left\{\begin{array}{l}\textit{the Zariski-open set of points of } \mathfrak{U}_{(5,1)} \\ \textit{corresponding to triples } (X, \lambda, \alpha) \textit{ such} \\ \textit{that } L_\lambda \textit{ is very ample}\end{array}\right\}.$$

Let

$$\mathbb{P}(\Gamma(\mathscr{F}))^* = \left\{\begin{array}{l}\textit{the Zariski-open subset of } \mathbb{P}(\Gamma(\mathscr{F})) \textit{ of spaces} \\ \textit{of sections } \mathbb{C}\cdot s, \textit{ whose zero sets } X_s \textit{ are non-} \\ \textit{singular}\end{array}\right\}.$$

Then $\mathfrak{U}^*_{(5,1)} \cong \mathbb{P}(\Gamma(\mathscr{F}))^*$, *the action of* $\Gamma(5, 1)_0/\Gamma(5, 1) \cong SL_2(\mathbb{Z}_5)$ *on* $\mathfrak{U}_{(5,1)}$ *corresponding to the action of* $N/H \cong SL_2(\mathbb{Z}_5)$ *on* $\mathbb{P}(\Gamma(\mathscr{F}))$.

Proof. The idea is to set up a set-theoretic map from $\mathbb{P}(\Gamma(\mathscr{F}))^*$ to $\mathfrak{U}^*_{(5,1)}$; verify that it is a morphism and is bijective; and apply Zariski's Main Theorem. To define the map, start with a one-dimensional subspace $\mathbb{C} \cdot s \subset \Gamma(\mathscr{F})$. This determines uniquely its zero-set X_s. This variety carries a line bundle, $\mathcal{O}_{X_s}(1)$, and is invariant under the group H/C. Strictly speaking, X_s is not yet an abelian variety, since it has no distinguished origin. We can either choose any point $x \in X_s$ as origin, or if we wish to be canonical, replace X_s by its "double dual":

$$X_s' = \mathrm{Pic}^0(\mathrm{Pic}^0 X_s),$$

$$(\mathrm{Pic}^0 = \text{connected component of Grothendieck's Picard scheme}).$$

In this case, X_s is canonically a principal homogeneous space over X_s'. In both cases, $\mathcal{O}_{X_s}(1)$ induces a polarization λ on X_s (or X_s'). And the automorphisms induced by H/C are the translations by the points of $\ker(\lambda)$, so we get an isomorphism

$$\alpha\colon \ker(\lambda) \xrightarrow{\approx} H/C \underset{\text{def}}{=} \mathbb{Z}_5 \times \mathbb{Z}_5 = L^{\perp}_{(5,1)}/L_{(5,1)}.$$

This is a point of $\mathfrak{U}^*_{(5,1)}$. The fact that this is a morphism comes from checking that the above construction can be carried out universally leading to an abelian scheme $\mathfrak{X}$ over $\mathbb{P}(\Gamma(\mathscr{F}))^*$, plus a polarization $\Lambda\colon \mathfrak{X} \to \hat{\mathfrak{X}}$ plus an isomorphism of $\ker(\Lambda)$ with the constant group scheme $\mathbb{Z}_5 \times \mathbb{Z}_5$. This induces a morphism from $\mathbb{P}(\Gamma(\mathscr{F}))^*$ to $\mathfrak{U}^*_{(5,1)}$ by the universal property characterizing coarse moduli spaces (cf. [10, p. 96]). To check that this map is injective, say $\mathbb{C} \cdot s_1$ and $\mathbb{C} \cdot s_2$ lead to isomorphic triples (X, λ, α). It follows that there is an isomorphism

$$\phi\colon X_{s_1} \to X_{s_2}$$

such that $\phi^* \mathcal{O}_{X,s_2}(1)$ is algebraically equivalent to $\mathcal{O}_{X,s_1}(1)$ and such that for all $\sigma \in H/C$, if σ induces on X_{s_i} translation by $x_i \in X_{s_i}$, then $\phi T_{x_1} = T_{x_2}\phi$. But then changing ϕ by a translation, we can assume that $\phi^*(\mathcal{O}_{X,s_2}(1)) \cong \mathcal{O}_{X,s_1}(1)$, hence ϕ is the restriction to X_{s_1} of a projective transformation τ. Moreover τ satisfies $\sigma\tau \equiv \tau\sigma$ on X_{s_1}, all $\sigma \in H/C$, hence $\tau\sigma = \sigma\tau$ in $PGL_5(\mathbb{C})$. But H/C is its own centralizer so $\tau \in H/C$. Therefore $X_{s_2} = \tau(X_{s_1}) = X_{s_1}$, hence $\mathbb{C} \cdot s_1 = \mathbb{C} \cdot s_2$. Finally surjectivity follows from (5.2). *Q.E.D.*

A natural question is to analyze how the isomorphism above goes wrong outside the open sets *. We have not worked this out completely, but we state without proof two pretty facts about this:

(a) If the abelian variety X tends to $E_1 \times E_2$, E_i an elliptic curve, so that the polarization tends to $\lambda = 5\lambda_1 + \lambda_2$ ($\lambda_i\colon E_i \xrightarrow{\approx} \hat{E}_i$ the canonical isomorphism), then λ is *not* very ample. In fact L_λ has a fixed component F and defines the morphism

$$X - F \xrightarrow{p_1} E_1 \xrightarrow{\phi_1} \mathbb{P}^4$$

where ϕ_1 is the morphism defined by L_{λ_1}. Let $C_1 = \phi_1(E_1)$, an elliptic quintic curve. Then while X approaches $E_1 \times E_2$, the corresponding section s of $\mathscr{F}$ has a well-defined limit s_0

and X_{s_0} is the singular ruled surface with C_1 as cuspidal double curve equal to the union of the tangent lines to C_1. It follows that at the points $(E_1 \times E_2, 5\lambda_1 + \lambda_2, \alpha) \in \mathfrak{U}_{(5,1)}$ the correspondence with $\mathbb{P}(\Gamma(\mathscr{F}))$ given in (6.1) is still regular, but not biregular, since the image does not depend on E_2.

(b) Suppose we compactify $\mathfrak{U}_{(5,1)}$ following Igusa [6] [i.e. take his compactification of $\mathfrak{U}_{(1,1)}$ and normalize it in $\mathbb{C}(\mathfrak{U}_{(5,1)})$ via any of the canonicalmorphisms $\mathfrak{U}_{(5,1)} \to \mathfrak{U}_{(1,1)}$]. Then at some of the points at ∞ lying even on the 0-dimensional piece of Satake's compactification, the correspondence remains biregular. The corresponding X_s's depend on one parameter $\alpha \in \mathbb{C} - (0)$ and are unions of five non-singular quadric surfaces as follows:

$$X_s = Q_0 \cup Q_1 \cup Q_2 \cup Q_3 \cup Q_4$$

$$Q_i = (\text{locus } Y_i = \alpha Y_{2+i} Y_{3+i} + Y_{1+i} Y_{4+i} = 0).$$

The 10 lines $Y_i = Y_j = Y_k = 0$ $(0 \leqq i < j < k \leqq 4)$ are double lines on X_s, and the five points P_i given by $Y_j(P_i) = \delta_{ij}$ are 4-fold points of X_s. The whole configuration is readily visualized if you form a CW-complex Σ as follows:

(a) take a point $\sigma_i^{(0)}$ for each point P_i;

(b) joint $\sigma_i^{(0)}$ and $\sigma_j^{(0)}$ by a 1-simplex $\sigma_{ij}^{(1)}$ corresponding to the double line $\overline{P_i P_j}$, for each $i < j$;

(c) glue in a square $\sigma_i^{(2)}$ corresponding to Q_i filling in the loop

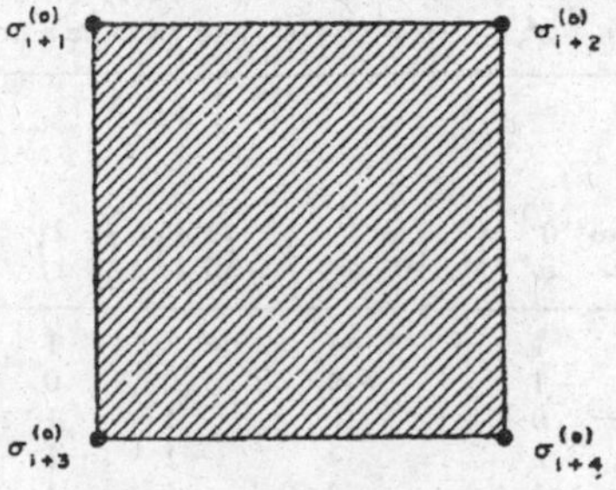

for each $0 \leqq i \leqq 4$ (read the subscripts mod 5).

Then a point or line is on a line or a quadric in $\mathbb{P}$ if and only if the corresponding 0-simplex or 1-simplex is on the corresponding 1-simplex or square in Σ. The nice thing is that the Σ you get is homeomorphic to a 2-dimensional real torus:

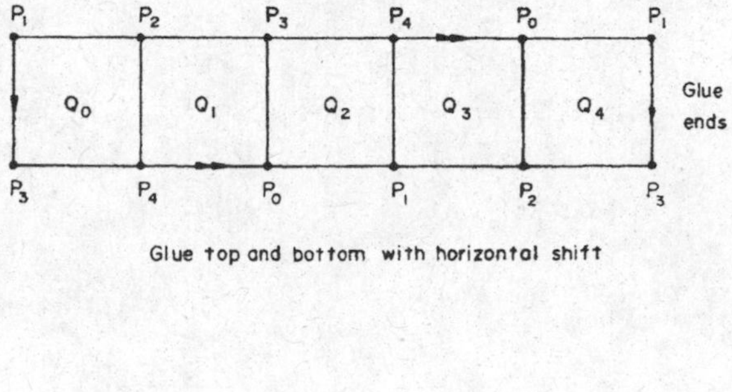

REFERENCES

1. W. Barth and M. E. Larsen: On the homotopy-groups of complex projective-algebraic varieties, *Math. Scand.* (to appear).
2. A. Borel, R. Carter, T. Springer *et al.*: *Seminar on Algebraic Groups and Related Finite Groups.* Springer Lecture Notes 131.
3. G. Horrocks: Vector bundles on the punctured spectrum of a local ring, *Proc. Lond. math. Soc.* **14** (1964), 689.
4. G. Horrocks: A construction for locally free sheaves, *Topology* 7 (1968), 117.
5. B. Huppert; *Endliche Gruppen I.* Springer, Berlin (1967).
6. J.-I. Igusa: A desingularization problem in the theory of Siegel modular functions, *Math. Annal.*
7. J.-I. Igusa: *Theta Functions.* Springer, Berlin (1972).
8. K. Kodaira: On the structure of compact, complex analytic surfaces I, *Am. J. Math.* **86** (1964), 751.
9. M. E. Larsen: On the vanishing of certain homotopy-groups of complex projective-algebraic manifolds. Preprint, Math. Inst. Copenhagen University (1971).
10. D. Mumford: *Geometric Invariant Theory.* Springer, Berlin (1965).
11. D. Mumford: Lectures on curves on an algebraic surface, *Ann. Math. Studs.* (1966).
12. D. Mumford: On the equations defining abelian varieties I, *Inv. Math.* **1** (1966), 287.
13. D. Mumford: *Abelian Varieties.* Oxford University Press (1971).
14. R. L. E. Schwarzenberger: Vector bundles on the projective plane, *Proc. Lond. math. Soc.* **11** (1961), 623.
15. S. Sternberg, *Lectures on Differential Geometry.* Prentice-Hall, N.J. (1962).
16. A. Weil: Sur certains groupes d'opérateurs unitaires, *Acta Math.* **111** (1964), 143.

University of Newcastle-upon-Tyne

Harvard University

APPENDIX

The character table of $SL_2(\mathbb{Z}_5)$ [2, p. 160]

Put $\varepsilon = \exp 2\pi\sqrt{-1}/5$, $\quad\eta = \varepsilon + \varepsilon^4$, $\quad\eta' = \varepsilon^2 + \varepsilon^3$, and let ω be a primitive root of $x^6 = 1$ over $\mathbb{Z}_5$.

			$\bar{\mu}$				$\bar{\nu}$			Symbols used in text for these representations
	$\begin{pmatrix}1&0\\0&1\end{pmatrix}$	$\begin{pmatrix}-1&0\\0&-1\end{pmatrix}$	$\begin{pmatrix}2&0\\0&3\end{pmatrix}$	$\begin{pmatrix}\omega^2&0\\0&\omega^{-2}\end{pmatrix}$	$\begin{pmatrix}\omega&0\\0&\omega^{-1}\end{pmatrix}$	$\begin{pmatrix}1&1\\0&1\end{pmatrix}$	$\begin{pmatrix}1&2\\0&1\end{pmatrix}$	$\begin{pmatrix}-1&1\\0&-1\end{pmatrix}$	$\begin{pmatrix}-1&2\\0&-1\end{pmatrix}$	
χ_1	1	1	1	1	1	1	1	1	1	I
χ_5	5	5	1	-1	-1	0	0	0	0	
χ_6	6	-6	0	0	0	1	1	-1	-1	
χ_4	4	4	0	1	1	-1	-1	-1	-1	
$\chi_4^{\#}$	4	-4	0	1	-1	-1	-1	1	1	
χ_3	3	3	-1	0	0	$-\eta$	$-\eta'$	$-\eta$	$-\eta'$	U
χ_3'	3	3	-1	0	0	$-\eta'$	$-\eta$	$-\eta'$	$-\eta$	U
χ_2	2	-2	0	-1	$+1$	η	η'	$-\eta$	$-\eta'$	W
χ_2'	2	-2	0	-1	$+1$	η'	η	$-\eta'$	$-\eta$	W'

Reprinted from:
CONTRIBUTIONS TO ANALYSIS
A Collection of Papers Dedicated to Lipman Bers

ACADEMIC PRESS, INC.

Prym Varieties I

DAVID MUMFORD
HARVARD UNIVERSITY

INTRODUCTION

This paper gives the first steps in a purely algebraic version (in all characteristics except two) of the Riemann–Prym–Wirtinger–Schottky–Jung theory of double coverings of one curve (or compact Riemann surface) over another. It also tries to incorporate some of the interesting generalizations of this theory in the thesis of Fay [4]. The basic idea is this:

$$\pi: \tilde{C} \longrightarrow C$$

is a double covering, where C and $\tilde{C}$ are nonsingular complete curves with Jacobians J and $\tilde{J}$. The involution $\iota: \tilde{C} \longrightarrow \tilde{C}$ interchanging sheets extends to $\iota: \tilde{J} \longrightarrow \tilde{J}$, and up to some points of order two, $\tilde{J}$ splits into an even part J and an odd part P, the *Prym variety*. The Prym P has a natural polarization on it, but only in two cases—where π has zero or two branch points—do we get a unique *principal* polarization on P, hence a theta divisor $\Xi \subset P$. This is discussed in the first part of this paper (Sections 1–3).

The surprise comes, however, on a closer analysis of the relations between the theta divisors $\Theta \subset J$ and $\tilde{\Theta} \subset \tilde{J}$: It turns out that they are related in a much tighter way than would be expected from looking only at the configuration of Abelian varieties and homomorphisms present. In the case of zero or two branch points this leads finally to identities relating (J, Θ) and (P, Ξ) discovered by Schottky and Jung [15] (cf. also Riemann [13] and Farkas and Rauch [3]). The point is that the existence of *any* (P, Ξ) standing in this relation to (J, Θ) means that if $g \geqslant 4$, J is not the most general Abelian variety of dimension g! Unfortunately, an efficient method of translating this into an equivalent polynomial identity on the theta nulls of J is only known at present for $g = 4$. These matters are discussed in the second part of this paper (Sections 4 and 5).

In the other direction, the curves C and $\tilde{C}$ and their geometry can be used to compute things about P. The importance of this is that it is usually quite hard to make detailed computations on the geometry of the theta divisor in a general principally polarized n-dimensional Abelian variety [which has $\frac{1}{2}n(n+1)$ moduli]; those which are Jacobians of curves of genus n (with $3n - 3$ moduli) are much better understood. However, by taking the Pryms for unramified double coverings $\tilde{C} \longrightarrow C$, genus

$C = n + 1$, we get a bigger family of principally polarized n-dimensional Abelian varieties which can be closely studied (depending on $3n$ moduli). For instance, for $n = 2, 3$ a generic principally polarized Abelian variety is a Jacobian; and according to Wirtinger, for $n = 4, 5$ a generic principally polarized Abelian variety appears to be a Prym but not of course a Jacobian. Moreover Pryms occur sometimes as the Intermediate Jacobians of unirational but not rational 3-folds (cf. Clemens and Griffiths [2], and Murre [12]). In the final part of the paper (Sections 6 and 7) with these applications in mind we compute the dimension of singular locus of the theta divisor in a Prym using results of Martens [8].

In a sequel to this paper we would like to discuss (a) how close the Schottky–Jung identities come to characterizing Jacobians among all Abelian varieties, and (b) ways of utilizing the Schottky–Jung identities in the two-branch-point case.

NOTATIONS

k	the algebraically closed ground field: always of char. $\neq 2$
$\mathbb{R}(X)$	field of rational functions on a variety X
$\mathrm{Pic}(X)$	group of divisor classes, line bundles, or invertible sheaves on a variety X
$\mathrm{Pic}^0(X)$	connected component of $0 \in \mathrm{Pic}(X)$
$\hat{X}$	another notation for $\mathrm{Pic}^0(X)$ if X is an Abelian variety (called the "dual" Abelian variety)
$\lambda_D\colon X \longrightarrow \hat{X}$	the homomorphism $x \longmapsto$ [divisor class of $T_x^{-1}D - D$], where D is a divisor on an Abelian variety X

A *polarization* of an Abelian variety X is a homomorphism $\lambda\colon X \longrightarrow \hat{X}$ such that $\lambda = \lambda_D$ for some ample D: in this case D is determined modulo $\mathrm{Pic}^0(X)$; λ is a *principle polarization* if λ is also an isomorphism, in which case $\lambda = \lambda_D$ for a positive ample D, unique up to a translation. (See my book [10] for a general reference for the facts on Abelian varieties.)

I

1. DOUBLE COVERINGS OF CURVES

The main object of our study is a morphism

$$\pi\colon \tilde{C} \longrightarrow C,$$

where C and $\tilde{C}$ are nonsingular complete curves and π is of degree two, i.e., π is surjective and via π^*, $\mathbb{R}(\tilde{C})$ is a quadratic extension of $\mathbb{R}(C)$. In fact, in this case C has an open covering by affines $U_\alpha = \mathrm{Spec}\ R_\alpha$ such that $\pi^{-1}(U_\alpha) = \mathrm{Spec}\ S_\alpha$, where S_α is an R_α-algebra of the form

$$S_\alpha \cong R_\alpha[t_\alpha]/(t_\alpha{}^2 - \beta_\alpha), \qquad \beta_\alpha \in R_\alpha.$$

Or, sheaf-theoretically, we may put this in the equivalent form $\tilde{C} = \mathbf{Spec}(\mathscr{S})$, where $\mathscr{S}$ is a sheaf of $\mathcal{O}_C$ algebras of the form

$$\mathscr{S} \cong \mathcal{O}_C \oplus L$$

with L an invertible sheaf of $\mathcal{O}_C$ modules. Multiplication is given by

$$(a + l) \cdot (b + m) = (a \cdot b + \phi(l \otimes m), a \cdot m + b \cdot l),$$

a, b sections of $\mathcal{O}_C$, l, m sections of L, for some

$$\phi\colon\ L^2 \xrightarrow{\approx} \mathcal{O}_C\Big(-\sum_{i=1}^{m} P_i\Big) \subset \mathcal{O}_C.$$

Then the zeros of β_α, or equivalently the points P_i where $\phi(L^2) \neq \mathcal{O}_C$, are the branch points of π. Since $\tilde{C}$ is nonsingular, they are all simple zeros (equivalently, $\sum P_i$ has no multiple points); and because

$$\#\ \text{branch points} = -\deg L^2 = 2(-\deg L),$$

there are an even number $m = 2n$ of them.

Let J and $\tilde{J}$ be the Jacobians of C and $\tilde{C}$: By definition, we take this to mean

$$J = \text{Pic}^0(C), \quad \tilde{J} = \text{Pic}^0(\tilde{C}).$$

Now fix base points $x_0 \in C$, and $\tilde{x}_0 \in \tilde{C}$ such that $\pi(\tilde{x}_0) = x_0$. Then we get the Albanese mappings:

$$\iota\colon\ C \longrightarrow J \quad \text{via} \quad x \longmapsto \text{divisor class } (x - x_0)$$

and

$$\tilde{\iota}\colon\ \tilde{C} \longrightarrow \tilde{J} \quad \text{via} \quad x \longmapsto \text{divisor class } (\tilde{x} - \tilde{x}_0).$$

Moreover, define

$$\text{Nm}\colon\ \tilde{J} \longrightarrow J$$

by either (a) the restriction of the map Nm,

$$\begin{array}{ccc} \tilde{J} \subset H^1(\tilde{C}, \mathcal{O}_{\tilde{C}}^*) \cong H^1(C, (\pi_* \mathcal{O}_{\tilde{C}})^*) \\ \big\downarrow \text{Nm} \\ J \hookrightarrow H^1(C, \mathcal{O}_C^*) \end{array}$$

or (b) the induced map on divisor classes given on divisors by $\mathfrak{A} \longmapsto \pi(\mathfrak{A})$ ($\mathfrak{A}$ a divisor on $\tilde{C}$). Then we get a commutative diagram:

$$\begin{array}{ccc} \tilde{C} & \xrightarrow{\tilde{\iota}} & \tilde{J} \\ \pi \big\downarrow & & \big\downarrow \text{Nm} \\ C & \xrightarrow{\iota} & J. \end{array}$$

Now this diagram defines a second by applying the functor Pic^0:

$$\begin{array}{ccc} \tilde{J} = \text{Pic}^0(\tilde{C}) & \xleftarrow{\tilde{\iota}^*} & \text{Pic}^0(\tilde{J}) \underset{\text{def}}{=} \hat{\tilde{J}} \\ \big\uparrow \pi^* & & \big\uparrow \text{Nm}^* \text{ or } \widehat{\text{Nm}} \\ J = \text{Pic}^0(C) & \xleftarrow{\iota^*} & \text{Pic}^0(J) \underset{\text{def}}{=} \hat{J} \end{array}$$

where $\hat{\tilde{J}}$ and $\hat{J}$ are the "duals" of $\tilde{J}$ and J, respectively. By the standard theory of Jacobians, ι^* and $\tilde{\iota}^*$ are isomorphisms and, in fact, if $\Theta \subset J$, $\tilde{\Theta} \subset \tilde{J}$ are the theta divisors, then

$$(\iota^*)^{-1} = -\lambda_\Theta, \qquad (\tilde{\iota}^*)^{-1} = -\lambda_{\tilde{\Theta}}$$

(where for any divisor D on an Abelian variety X, $\lambda_D \colon X \longrightarrow \hat{X}$ is the homomorphism given by $x \longrightarrow$ [divisor class $T_x^{-1}(D) - D$]). Thus the principally polarized Abelian varieties (J, Θ) and $(\tilde{J}, \tilde{\Theta})$ are related by two maps:

$$\pi^*\colon\ J \longrightarrow \tilde{J}, \qquad \mathrm{Nm}\colon\ \tilde{J} \longrightarrow J$$

and the main result is that these have two properties:

(i) π^* and Nm are dual to each other:

$$\widehat{\mathrm{Nm}} = \lambda_{\tilde{\Theta}} \cdot \pi^* \cdot \lambda_\Theta^{-1}, \qquad \hat{\pi}^* = \lambda_\Theta \cdot \mathrm{Nm} \cdot \lambda_{\tilde{\Theta}}^{-1}.$$

(ii) $\mathrm{Nm} \cdot \pi^*\colon J \longrightarrow J$ is multiplication by two.

Proof of (ii). If $\mathfrak{A}$ is a divisor class of degree zero on C, and α is the corresponding point of J, then $\pi^{-1}(\mathfrak{A})$ represents $\pi^*\alpha \in \tilde{J}$ and $\pi(\pi^{-1}\mathfrak{A})$ represents $\mathrm{Nm}(\pi^*\alpha)$. But $\pi(\pi^{-1}\mathfrak{A}) = 2\mathfrak{A}$. Q.E.D.

Rather than studying in detail the implications of (i) and (ii) in this special case, it seems easier at this point to study such a situation in general, and afterward to specialize the study to the case of Jacobians.

2. A CONFIGURATION OF ABELIAN VARIETIES

Suppose (X, θ_X) and (Y, θ_Y) are two principally polarized Abelian varieties: Thus θ_X and θ_Y are positive divisors on X and Y, given only up to translations, however, such that λ_{θ_X} and λ_{θ_Y} are isomorphisms. (It is well known then that θ_X and θ_Y are ample and are the only positive divisors D such that $\lambda_D = \lambda_{\theta_X}$ or λ_{θ_Y}.)

DATA I. Suppose $\phi\colon X \longrightarrow Y$ is a homomorphism and assume that

$$\phi^*(\theta_Y) \ \text{ algebraically equivalent to } \ 2\theta_X \tag{2.1}$$

i.e., $\phi^*\theta_Y - 2\theta_X \in \mathrm{Pic}^0(X)$. This is *equivalent* to saying

$$\lambda_{\phi^*(\theta_Y)} = 2\lambda_{\theta_X}, \tag{2.2}$$

hence (since $\lambda_{\phi^* D} = \hat{\phi} \cdot \lambda_D \cdot \phi$), it is equivalent to having the following diagram commute:

$$\begin{array}{ccc} Y & \xrightarrow{\lambda_{\theta_Y}} & Y \\ \phi \uparrow & & \downarrow \hat{\phi} \\ X & \xrightarrow[2\lambda_{\theta_X}]{} & X \end{array} \tag{2.3}$$

Thus if we define $\psi: Y \longrightarrow X$ to be the dual $\lambda_{\theta_X}^{-1} \cdot \hat{\phi} \cdot \lambda_{\theta_Y}$ of ϕ, we get $\psi \cdot \phi =$ mult. by two: exactly the situation of Section 1.

I claim that all triples $((X, \theta_X), (Y, \theta_Y), \phi)$ satisfying (2.1)—call these Data I—and only such triples arise in the following way.

DATA II.

(i) (X, θ_X) is a principally polarized Abelian variety.

(ii) P and $\rho: P \longrightarrow \hat{P}$ is some Abelian variety and a polarization of P.

(iii) $H_0 \subset H_1 \subset X_2$ are subgroups of points of order two, and $\psi: H_1/H_0 \longrightarrow \ker \rho$ is an isomorphism.

These data should satisfy:

(iv) With respect to the skew-symmetric multiplicative pairings induced by the Riemann forms of θ_X and ρ

$$e_{2,X}: \ X_2 \times X_2 \longrightarrow \{\pm 1\}, \qquad e_\rho: \ \ker \rho \times \ker \rho \longrightarrow \{\pm 1\},$$

we have the following:

(a) $e_{2,X}(\alpha, \beta) = 1$, all $\alpha, \beta \in H_0$.

(b) $H_1 = H_0^{\perp}$, where $H_0^{\perp} = \{\alpha \in X_2 \mid e_{2,X}(\alpha, \beta) = 1, \text{ all } \beta \in H_0\}$.

(c) $e_\rho(\psi\alpha, \psi\beta) = e_{2,X}(\alpha, \beta)$, all $\alpha, \beta \in H_1$.

In this case we set $Y = X \times P/H$, where

$$H = \{(\alpha, \psi\alpha) \mid \alpha \in H_1\}$$

and let ϕ be the composition of canonical maps:

$$X \longrightarrow X \times P \longrightarrow Y.$$

Moreover, if $\sigma: X \times P \longrightarrow Y$ is the canonical map, then the polarization λ_{θ_Y} is determined by the requirement that the diagram

$$\begin{array}{ccc} X \times P & \xrightarrow{2\lambda_{\theta_X} \times \rho} & \hat{X} \times \hat{P} \\ \sigma \downarrow & & \uparrow \hat{\sigma} \\ Y & \xrightarrow{\lambda_{\theta_Y}} & \hat{Y} \end{array}$$

commutes.

In other words, we find that whenever one has such a ϕ, then up to a small group H of points of order two, Y and its polarization split into a product of two natural blocks, one being X and the other we call P—which in the case of curves will be the "Prym variety." Moreover, to tie the two types of data together, I claim that:

(v) $H_0 = \ker \phi$.

(vi) There is an involution ι on Y such that

$$P = \operatorname{Im}(1_Y - \iota) = \ker(1_Y + \iota)^0, \qquad \phi(X) = \operatorname{Im}(1_Y + \iota) = \ker(1_Y - \iota)^0.$$

In fact, if $\psi = \lambda_{\theta_X}^{-1} \cdot \hat{\phi} \cdot \lambda_{\theta_Y}: Y \longrightarrow X$, then

$$\ker(1_Y - \iota) = X \times P_2/H \cong \phi(X) \times (\mathbb{Z}/2\mathbb{Z})^{2b-2c}, \qquad \ker \psi = X_2 \times P/H \cong P \times (\mathbb{Z}/2\mathbb{Z})^{a-c}$$

[for a,b,c see (viii)].

(vii) If $\sigma: X \times P \longrightarrow Y$ is the canonical map and $\tau: Y \longrightarrow X \times P$ is the map $\tau(x) = (\psi x, x - \iota x)$, then

$$\sigma \cdot \tau = 2_Y, \qquad \tau \cdot \sigma = 2_{X \times P}.$$

(viii) If $\dim X = a$, $\dim Y = a + b$, $\# \ker \phi = 2^{a-c}$, then $\dim P = b$, $\# H_0 = 2^{a-c}$, $\# H_1 = 2^{a+c}$, $\# \ker \rho = 2^{2c}$, and $0 \leqslant c \leqslant \min(a, b)$.

Much of the verification of the equivalence here is straightforward, so we will run through only the first part.

Start with Data I. Define

$$P = \lambda_{\theta_Y}^{-1}(\ker \hat{\phi})^0,$$

and ν the number of components of $\lambda_{\theta_Y}^{-1}(\ker \hat{\phi})$. Via ϕ and the inclusion of P in Y, we get $\sigma: X \times P \longrightarrow Y$. Let $H = \ker \sigma$. Note that

$$(x, y) \in H \Longrightarrow \phi(x) + y = 0 \Longrightarrow \hat{\phi}(\lambda_{\theta_Y}(\phi(x))) = 0 \Longrightarrow 2x = 0 \Longrightarrow 2y = 0;$$

hence $H \subset X_2 \times P_2$. Since $H \cap (0) \times P_2 = (0) \times (0)$, there is a subgroup $H_1 \subset X_2$ and a homomorphism $\psi: H_1 \longrightarrow P_2$ such that

$$H = \{(\alpha, \psi\alpha) \mid \alpha \in H_1\}.$$

Also, if $H_0 = \ker \phi$, then $H_0 \subset H_1$, and ψ factors as $H_1/H_0 \hookrightarrow P_2$. Moreover, for all $y \in Y$, let

$$x = \lambda_{\theta_X}^{-1}(\hat{\phi}(\lambda_{\theta_Y}(y))).$$

Then

$$2y = \phi(x) + (2y - \phi(x))$$

and

$$\hat{\phi}(\lambda_{\theta_Y}(2y - \phi(x))) = 2\lambda_{\theta_X}(x) - \hat{\phi}(\lambda_{\theta_Y}(\phi(x))) = 0;$$

hence $\nu \cdot (2y - \phi(x)) \in P$. Therefore

$$2\nu \cdot y \in \phi(x) + P \subset \operatorname{Im} \sigma$$

and since Y is a divisible group, this implies that σ is surjective. Next, the polarization λ_{θ_Y} of Y "pulls back" to a polarization of $X \times P$ given by the composition:

$$X \times P \xrightarrow{\ \sigma\ } Y \xrightarrow{\ \lambda_{\theta_Y}\ } \hat{Y} \xrightarrow{\ \hat{\sigma}\ } \hat{X} \times \hat{P},$$

which may be considered as being given by a 2×2 matrix

$$\begin{pmatrix} \alpha & \beta \\ \gamma & \delta \end{pmatrix}, \qquad \begin{matrix} \alpha: X \longrightarrow \hat{X}, & \beta: P \longrightarrow \hat{X}, \\ \gamma: X \longrightarrow \hat{P}, & \delta: P \longrightarrow \hat{P}. \end{matrix}$$

Also, because any polarization is symmetric, $\gamma = \hat{\beta}$. But by the very definition of P, the coefficient β is zero. So $\gamma = 0$, too, and the polarization splits. Note that by assumption (2.1) on Data I, $\alpha = 2\lambda_{\theta_X}$. Define ρ to be δ. Next, the fact that the polarization $(2\gamma_{\theta_X}, \rho)$ of $X \times P$ is a pullback of a principal polarization with respect to the isogeny σ is equivalent to the condition that $\ker \sigma$, as a subgroup of $\ker(2\lambda_{\theta_X}, \rho)$, is maximal isotropic for the skew-symmetric form of this polarization (cf. Mumford [10, Section 23]). Hence

$$H \subset X_2 \times \ker \rho$$

and if $(\alpha, \psi\alpha), (\beta, \psi\beta) \in H$, then

$$e_{2,X}(\alpha, \beta) \cdot e_\rho(\psi\alpha, \psi\beta) = 1.$$

This means that $\psi(H_1/H_0) \subset \ker \rho$ and ψ is "symplectic" in the sense of (iv)c. Moreover, counting orders, the maximality of H implies

$$(\# H_1)^2 = (\# H)^2 = \#(X_2 \times \ker \rho);$$

hence

$$\# H_1{}^\perp = \frac{\# X_2}{\# H_1} = \frac{\# H_1}{\# \ker \rho} \leqslant \frac{\# H_1}{\# \operatorname{Im} \psi} \leqslant \# H_0.$$

Since $\ker \rho \subseteq H_1{}^\perp$, this implies that ψ maps H_1 onto $\ker \rho$ and that $H_0 = H_1{}^\perp$, hence $H_1 = H_0{}^\perp$. Thus we have Data II.

We leave it to the reader to check now that one can go backward from Data II to Data I and that for corresponding data, (v)–(viii) hold.

3. DEFINITION OF THE PRYM VARIETY

Returning to a covering $\pi: \tilde{C} \longrightarrow C$ and their Jacobians related by $\pi^*: J \longrightarrow \tilde{J}$, we see that $\tilde{J} \cong J \times P/H$. In this case there is an involution $\imath: \tilde{C} \longrightarrow \tilde{C}$ interchanging the two sheets above any point, which induces an involution $\imath: \tilde{J} \longrightarrow \tilde{J}$. Since for any divisor $\mathfrak{A}$ on $\tilde{C}$,

$$\pi^{-1}(\pi\mathfrak{A}) = \mathfrak{A} + \imath(\mathfrak{A}),$$

it follows that

$$\pi^*(\operatorname{Nm} x) = x + \imath(x), \qquad \text{all } \; x \in \tilde{J}.$$

And since Nm is surjective, this also shows that

$$\imath(\pi^* y) = \pi^* y, \qquad \text{all } \; y \in J.$$

Therefore $\imath = +1$ on π^*J and $\imath = -1$ on ker Nm. Thus $\imath$ is precisely the involution introduced in (vi) of Section 2, and we find that

$$P \underset{\text{def}}{=} (\ker \operatorname{Nm})^0 = \ker(1_{\tilde{J}} + \imath)^0 = \operatorname{Im}(1_{\tilde{J}} - \imath),$$

i.e., P is the "odd" part of $\tilde{J}$, which we call the *Prym variety of* $\tilde{C}$ *over* C.

Let g = genus of C and let $2n$ = # of branch points. Then by Hurwitz's formula

$$\text{genus } \tilde{g} \text{ of } \tilde{C} = 2g + n - 1.$$

Therefore

$$\dim J = g, \qquad \dim \tilde{J} = 2g + n - 1, \qquad \dim P = g + n - 1.$$

To apply fully the theory of Section 2, we need only compute $\ker(\pi^*) \cong \{\text{div. classes } \mathfrak{A} \text{ on } C \mid \pi^{-1}\mathfrak{A} \equiv 0\}$. But

$$\pi^{-1}\mathfrak{A} \equiv 0 \Longrightarrow 2\mathfrak{A} = \pi\pi^{-1}\mathfrak{A} \equiv 0,$$

i.e., $\ker(\pi^*) \subset J_2$. If $\mathfrak{A}$ is any such divisor class, then $\mathfrak{A}$ defines an unramified double covering $\pi_{\mathfrak{A}}\colon C_{\mathfrak{A}} \longrightarrow C$ by "Kummer theory," i.e., $C_{\mathfrak{A}}$ is the normalization of C in $\mathbb{R}(C)(\sqrt{f})$, where $2\mathfrak{A} = (f)$, or

$$C_{\mathfrak{A}} = \mathbf{Spec}(\mathcal{O}_C \oplus \mathcal{O}_C(\mathfrak{A})), \qquad \text{mult. given by } \mathcal{O}_C(\mathfrak{A}) \times \mathcal{O}_C(\mathfrak{A}) \longrightarrow \mathcal{O}_C(2\mathfrak{A}) \cong \mathcal{O}_C.$$

Then

$$\pi^{-1}\mathfrak{A} \equiv 0 \iff \text{the double covering } C_{\mathfrak{A}} \times_C \tilde{C} \text{ of } \tilde{C} \text{ splits into two copies of } \tilde{C}$$

$$\iff \text{there is a morphism } f\colon \begin{array}{ccc} \tilde{C} & \xrightarrow{f} & C_{\mathfrak{A}} \\ & {\scriptstyle\pi}\searrow \quad \swarrow{\scriptstyle\pi_{\mathfrak{A}}} & \\ & C & \end{array}$$

and hence

$$\pi^{-1}\mathfrak{A} \equiv 0,\ \mathfrak{A} \not\equiv 0 \iff \text{there is an isomorphism } f\colon \begin{array}{ccc} \tilde{C} & \xrightarrow{f} & C_{\mathfrak{A}} \\ & {\scriptstyle\pi}\searrow \quad \swarrow{\scriptstyle\pi_{\mathfrak{A}}} & \\ & C & \end{array}$$

This proves

Lemma. If π is ramified, $\ker \pi^* = (0)$. If π is unramified, hence $\tilde{C} = C_{\mathfrak{A}}$ for some $\mathfrak{A}$, then $\ker \pi^* = \{0, \mathfrak{A}\}$.

Combining this with the results of Section 2, we deduce the following.

Corollary 1. If π is ramified, we get a symplectic injection $\psi\colon J_2 \hookrightarrow P_2$ such that
(a) $\operatorname{Im}\psi = \ker\rho$, where $\rho\colon P \longrightarrow \hat{P}$ is the polarization of P, and
(b) $\tilde{J} \cong J \times P/\{(\alpha, \psi\alpha) \mid \alpha \in J_2\}$.
If π is unramified, we get subgroups

$$\begin{array}{ccccccc} (0) & \subset & H_0 & \subset & H_1 & \subset & J_2 \\ & & \| & & \| & & \\ & & \{0, \mathfrak{A}\} & & \{\mathfrak{B} \mid e_2(\mathfrak{A}, \mathfrak{B}) = +1\} & & \\ & & \text{order } 2 & & \text{order } 2^{2g-1} & & \end{array}$$

and a symplectic isomorphism $\psi\colon H_1/H_0 \xrightarrow{\approx} P_2 = \ker\rho$ such that

$$\tilde{J} \cong J \times P/\{(\alpha, \psi\alpha) \mid \alpha \in H_1\}.$$

Corollary 2. If π is unramified or has only two branch points, then $\ker \rho = P_2$, hence $\rho = 2\lambda_{\Xi}$, where

$$\lambda_{\Xi}\colon\ P \xrightarrow{\approx} \hat{P}$$

is a principal polarization. Moreover, in these cases

$$\phi(J) = \{x \in \tilde{J} \mid \iota x = x\}.$$

II

4. RELATIONS BETWEEN THETA DIVISORS

The question arises: In the class of all positive divisors algebraically equivalent to $2\theta_X$, which ones arise as $\phi^{-1}(\theta_{Y,y})$, where $\theta_{Y,y} = T_y(\theta_Y)$ is a translate of θ_Y by y and ϕ^{-1} means its pullback as actual divisor, when defined? This class of divisors is the (disjoint) union of the linear systems $|\theta_X + \theta_{X,x}|$, $x \in X$. In particular, one can ask whether it ever happens that

$$\phi^{-1}\theta_{Y,y} = \theta_{X,x_1} + \theta_{X,x_2}$$

for some $y \in Y$, $x_1, x_2 \in X$. The situation seems to be that this does not occur in general, that it does occur for Jacobians, and that this special occurrence is the ultimate source of the "Schottky relations" satisfied by the theta nulls of Jacobians.

Let us see first what we can say about the situation in general. Since

$$\phi^{-1}(\theta_{Y,\phi(x)}) = \phi^{-1}(\theta_Y)_x$$

for all $x \in X$, we may as well restrict our attention to the divisors $\phi^{-1}(\theta_{Y,y})$ for $y \in P$. All these divisors are linearly equivalent, since

$$[\text{the div. class}\quad \phi^{-1}(\theta_{Y,y}) - \phi^{-1}(\theta_Y)\quad \text{in}\quad X] = \hat{\phi}(\lambda_{\theta_Y}(y))$$

and this is 0 if $y \in P$. Moreover, if we replace θ_X and θ_Y by suitable translates, we can then assume that θ_X and θ_Y are symmetric divisors (invariant under -1_X and -1_Y) and that†

$$\phi^{-1}(\theta_{Y,y}) \in |2\theta_X|, \qquad \text{all} \qquad y \in P.$$

Therefore we get a morphism (we change the sign of y to simplify the proposition that follows):

$$\begin{aligned} \delta\colon\ P - \{y \mid \phi(X) \subset \theta_{Y,-y}\} &\longrightarrow |2\theta_X| \\ y &\longmapsto \phi^{-1}(\theta_{Y,-y}). \end{aligned}$$

† In fact, first take any symmetric θ_X and θ_Y. Then $\phi^{-1}(\theta_Y) = 2\theta_X + D$, where $2D \equiv 0$, hence $e_*^{\theta_Y}(\phi(x)) = e_2(D, x)$ for all $x \in X_2$. This is a homomorphism from $\phi(X_2)$ to $\{\pm 1\}$: Extend it to a homomorphism $f\colon Y_2 \to \{\pm 1\}$ and represent f by $f(x) = e_2(y, x)$, for some $y \in Y_2$. Then $\theta_{Y,y}$ is still symmetric and $e_2^{\theta_{Y,y}}(\phi(x)) = 1$, all $x \in X_2$, hence $\phi^{-1}(\theta_{Y,y})$ is totally symmetric, i.e., $\in |2\theta_X|$.

Moreover, because the polarization θ_Y on Y, pulled back to $X \times P$, splits into a product, it follows that we can write

$$\sigma^*(\mathcal{O}_Y(\theta_Y)) = p_1{}^*(\mathcal{O}_X(2\theta_X)) \otimes p_2{}^*(L_\rho),$$

where L_ρ is a symmetric invertible sheaf on P representing the polarization ρ. I claim the following:

Proposition. δ is essentially the morphism of P to projective space defined by the section of L_ρ. More precisely

$$\{y \mid \phi(X) \subset \theta_{Y,-y}\} = \{y \mid s(y) = 0 \quad \text{for all} \quad s \in \Gamma(L_\rho)\}$$

—call this set B_ρ (for base points)—and there is an isomorphism

$$i\colon \ \mathbb{P}(\Gamma(L_\rho)) \hookrightarrow |2\theta_X|$$

of $\mathbb{P}(\Gamma(L_\rho))$ with a linear subspace of $|2\theta_X|$ such that the diagram

$$\begin{array}{ccc} & & \mathbb{P}(\Gamma(L_\rho)) \\ & \overset{\phi_\rho}{\nearrow} & \\ P - B_\rho & & \Big\downarrow i \\ & \underset{\delta}{\searrow} & \\ & & |2\theta_X| \end{array}$$

commutes, where ϕ_ρ is the canonical morphism defined by sections of L_ρ.

Proof. We abbreviate $\mathcal{O}_X(\theta_X)$ to L_X and $\mathcal{O}_Y(\theta_Y)$ to L_Y. Now, according to the general theory of Mumford [9, Section 1] (see also Mumford [10, Section 23]), the isomorphism

$$\sigma^* L_Y \cong p_1{}^* L_X{}^2 \otimes p_2{}^* L_\rho$$

defines a lifting of the group H:

$$\begin{array}{ccccccccc} 1 & \longrightarrow & \mathbb{G}_m & \longrightarrow & \mathcal{G}(p_1{}^*L_X{}^2 \otimes p_2{}^*L_\rho) & \longrightarrow & X_2 \times \ker \rho & \longrightarrow & 0 \\ & & & & \cup & & \cup & & \\ & & & & H^* & \xrightarrow{\approx} & H & & \end{array}$$

and the pullback $\sigma^*(s_0)$ of the unique section $s_0 \in \Gamma(L_Y)$ (unique up to scalars) is the unique element of $\Gamma(L_X{}^2) \otimes \Gamma(L_\rho)$ fixed by H^*. But for any such H^*, it is easy to describe the element fixed by H^*: in fact

$$\mathcal{G}(p_1{}^*L_X{}^2 \otimes p_2{}^*L_\rho) \cong \mathcal{G}(L_X{}^2) \times \mathcal{G}(L_\rho)/\{(\lambda, \lambda^{-1}) \mid \lambda \in \mathbb{G}_m\}$$

and any such H^* contains a subgroup $H_0{}^*$:

$$\begin{array}{ccccccccc} 1 & \longrightarrow & \mathbb{G}_m & \longrightarrow & \mathcal{G}(L_X{}^2) & \longrightarrow & X_2 & \longrightarrow & 0 \\ & & & & \cup & & \cup & & \\ & & & & H_0{}^* & \xrightarrow{\approx} & H_0 & & \end{array}$$

Then if $Z(H_0{}^*)$ is the centralizer of $H_0{}^*$, we get a Heisenberg group:

$$1 \longrightarrow \mathbb{G}_m \longrightarrow Z(H_0{}^*)/H_0{}^* \longrightarrow H_1/H_0 \longrightarrow 0$$

and H^* itself is defined by an isomorphism ψ^*:

$$\begin{array}{ccccccccc} 1 & \longrightarrow & \mathbb{G}_m & \longrightarrow & Z(H_0^*)/H_0^* & \longrightarrow & H_1/H_0 & \longrightarrow & 0 \\ & & \big\downarrow \lambda^{-1} & & \big\downarrow \psi^* & & \big\downarrow \psi & & \\ 1 & \longrightarrow & \mathbb{G}_m & \longrightarrow & \mathcal{G}(L_\rho) & \longrightarrow & \ker \rho & \longrightarrow & 0 \end{array}$$

by this connecting link

$$H^* = \{(x, \psi^* x) \mid x \in Z(H_0^*)/H_0^*\}.$$

Now the subspace $\Gamma(L_X{}^2)^{H_0^*}$ of $H_0{}^*$ invariants is the unique irreducible representation of $Z(H_0{}^*)/H_0{}^*$ on which $\mathbb{G}_m$ acts identically, and the dual $\mathrm{Hom}(\Gamma(L_\rho), k)$ is the unique irreducible representation of $\mathcal{G}(L_\rho)$ on which $\mathbb{G}_m$ acts by $\lambda \longmapsto \lambda^{-1} \cdot (\text{identity})$. Therefore ψ^* defines an isomorphism of these representations:

$$\chi\colon\ \Gamma(L_X{}^2)^{H_0^*} \xrightarrow{\ \approx\ } \mathrm{Hom}(\Gamma(L_\rho), k).$$

If $\beta_1, \ldots, \beta_d$ is a basis of $\Gamma(L_\rho)$ and $\alpha_1, \ldots, \alpha_d$ is the basis of $\Gamma(L_X{}^2)^{H_0^*}$ such that $\chi(\alpha_i)(\beta_j) = \delta_{ij}$, then it is immediate that $\sum \alpha_i \otimes \beta_i \in \Gamma(L_X{}^2) \otimes \Gamma(L_\rho)$ is H^* invariant. Thus

$$\sigma^*(s_0) = \sum_{i=1}^{d} p_1{}^*\alpha_i \otimes p_2{}^*\beta_i;$$

hence for all $y \in P$

$$\phi^{-1}(\theta_{Y,-y}) = \text{zero set of } \ \mathrm{res}_{X \times \{y\}}(\sigma^* s_0) = \text{zero set of } \sum_{i=1}^{d} \beta_i(y) \cdot \alpha_i. \qquad (4.1)$$

Thus, first of all

$$\begin{aligned} \phi^{-1}(\theta_{Y,-y}) = X &\Longleftrightarrow \sum \beta_i(y) \cdot \alpha_i \equiv 0 \Longleftrightarrow \beta_i(y) = 0, \quad \text{all } i \\ &\Longleftrightarrow y \text{ is a base point of } \Gamma(L_\rho), \end{aligned}$$

and second if $l \in \mathrm{Hom}(\Gamma(L_\rho), k)$ is "homogeneous coordinates" for a point of $\mathbb{P}(\Gamma(L_\rho))$, then set $i(l) =$ the divisor $(\sum l(\beta_i) \cdot \alpha_i = 0)$. Then (4.1) implies that $\delta = i \cdot \phi_\rho$. Q.E.D.

Now starting from the other direction, $|2\theta_X|$ contains the reducible divisors $\theta_{X,x} + \theta_{X,-x}$, $x \in X$. Therefore we get a morphism:

$$\begin{aligned} \phi_X'\colon\ X &\longrightarrow |2\theta_X| \\ x &\longmapsto \theta_{X,x} + \theta_{X,-x}. \end{aligned}$$

I claim the following

Proposition (Wirtinger). There is a nondegenerate inner product $B\colon \Gamma(L_X{}^2) \otimes \Gamma(L_X{}^2) \longrightarrow k$ (which is symmetric or skew-symmetric depending on whether $\mathrm{mult}_0\, \theta_X$ is even or odd) such that if B induces the isomorphism B',

$$\mathbb{P}(\Gamma(L_X{}^2)) \xrightarrow{\ \approx\ } \mathbb{P}(\Gamma(L_X{}^2)^*) = |2\theta_X|,$$

then the diagram

$$\begin{array}{ccc} & & \mathbb{P}(\Gamma(L_X{}^2)) \\ & \overset{\phi_X}{\nearrow} & \\ X & & \Big\downarrow{\scriptstyle B'}\ \wr\wr \\ & \underset{\phi_X'}{\searrow} & \\ & & |2\theta_X| \end{array}$$

commutes, where ϕ_X is the canonical morphism defined by sections of $L_X{}^2$.

Proof. In this case we use the morphism

$$\begin{aligned} \xi\colon X \times X &\longrightarrow X \times X \\ (x,y) &\longmapsto (x+y,\, x-y), \end{aligned}$$

and the isomorphism

$$\xi^*(p_1{}^*L_X \otimes p_2{}^*L_X) \cong p_1{}^*L_X{}^2 \otimes p_2{}^*L_X{}^2$$

(cf. Mumford 9, Section 2). Let $\{s_\alpha\}$ be a basis of $\Gamma(L_X{}^2)$: Then we can write

$$\xi^*(p_1{}^*\theta_X \otimes p_2{}^*\theta_X) = \sum_{\alpha,\beta} c_{\alpha\beta}\, p_1{}^*s_\alpha \times p_2{}^*s_\beta$$

for some matrix $c_{\alpha\beta} \in k$; or, more transparently,

$$\theta_X(u+v)\theta_X(u-v) = \sum c_{\alpha\beta}\, s_\alpha(u) \cdot s_\beta(v), \qquad \forall u, v \in X. \tag{4.2}$$

As a section of L_X, θ_X is even or odd depending on $\operatorname{mult}_0 \theta_X$ and hence interchanging u and v in this formula, we find $c_{\alpha\beta}$ is symmetric or skew-symmetric in these two cases. Moreover, the element $\xi^*(p_1{}^*\theta_X \otimes p_2{}^*\theta_X)$ is invariant under the action of $\Delta(X_2) = \{(x,x) \mid x \in X_2\}$ on $p_1{}^*L_X{}^2 \otimes p_2{}^*L_X{}^2$ [via a suitable lifting of $\Delta(X_2)$ into $\mathcal{G}(p_1{}^*L_X{}^2 \otimes p_2{}^*L_X{}^2)$]. And since X_2 acts irreducibly on $\Gamma(L_X{}^2)$, this element cannot lie in any proper subspace $W_1 \otimes W_2$ of $\Gamma(L_X{}^2) \otimes \Gamma(L_X{}^2)$. This implies that $\det c_{\alpha\beta} \neq 0$, hence $c_{\alpha\beta}$ defines a form B. Finally, for each fixed v the formula (4.2) implies

$$u \in \operatorname{support}(\theta_{X,v} + \theta_{X,-v}) \Longleftrightarrow u \in \operatorname{zeros}\left(\sum c_{\alpha\beta}\, s_\beta(v) s_\alpha\right)$$

which gives us immediately

$$\phi_X'(v) = B'(\phi_X(v)). \quad \text{Q.E.D.}$$

Corollary 1. In the abstract situation (X, θ_X), (Y, θ_Y), ϕ, we get a diagram:

$$\begin{array}{ccccc} P - B_\rho & \xrightarrow{\ \phi_\rho\ } & \mathbb{P}(\Gamma(L_\rho)) & & \\ & & & \overset{i}{\searrow} & \\ & & & & |2\theta_X|. \\ & & & \underset{\approx}{\overset{B'}{\nearrow}} & \\ X & \xrightarrow{\ \phi_X\ } & \mathbb{P}(\Gamma(L_X{}^2)) & & \end{array}$$

Then for all $y \in P - B_\rho$, $x \in X$

$$\phi^{-1}(\theta_{Y,y}) = \theta_{X,x} + \theta_{X,-x} \Longleftrightarrow i(\phi_\rho(y)) = B'(\phi_X(x)).$$

The most important case here is when $\ker \rho = P_2$, so that there is a theta divisor θ_P on P with $\rho = 2\lambda_{\theta_P}$ and $L_\rho = L_P{}^2$, where $L_P = \mathcal{O}_P(\theta_P)$. Then Corollary 1 becomes the following.

Corollary 2. In the abstract situation (X, θ_X), (Y, θ_Y), ϕ, when $\rho = 2\lambda_{\theta_P}$, we get a diagram

$$\begin{array}{ccccc} P & \xrightarrow{\ \phi_P\ } & \mathbb{P}(\Gamma(L_P{}^2)) & & \\ & & & \overset{i}{\searrow} & \\ & & & & |2\theta_X| \\ & & & \underset{B'}{\overset{\approx}{\nearrow}} & \\ X & \xrightarrow{\ \phi_X\ } & \mathbb{P}(\Gamma(L_X{}^2)) & & \end{array}$$

and for all $y \in P$, $x \in X$

$$\phi^{-1}(\theta_{Y,y}) = \theta_{X,x} + \theta_{X,-x} \Longleftrightarrow i(\phi_P(y)) = B'(\phi_X(x)).$$

5. THE SPLITTING OF $\phi^{-1}(\theta_{Y,y})$ FOR JACOBIANS

Now return to the double covering $\pi\colon \tilde{C} \longrightarrow C$. Recall the geometric meaning of the theta divisors $\Theta \subset J$, $\tilde{\Theta} \subset \tilde{J}$:

(1) Let J_k be the variety of invertible sheaves on C of degree k, and $\tilde{J}_k$ be the variety of invertible sheaves on $\tilde{C}$ of degree k [so that if we choose a *base point* on J_k or $\tilde{J}_k$, $J_k \cong J$ and $\tilde{J}_k \cong \tilde{J}$, but without such a choice $J_k(\tilde{J}_k)$ is merely a principal homogeneous space over J $(\tilde{J})$]. Note that π^* induces: $\pi^*\colon J_k \longrightarrow \tilde{J}_{2k}$ because $\deg \pi^* L = 2 \cdot \deg L$. Moreover, note that there is a canonical group structure on the big schemes:

$$\coprod_{k \in \mathbb{Z}} J_k \qquad \text{and} \qquad \coprod_{k \in \mathbb{Z}} \tilde{J}_k.$$

(2) Then we can find Θ *canonically* in J_{g-1} by

$$\Theta = \{L \in J_{g-1} \mid \Gamma(L) \neq (0)\} \subset J_{g-1}$$

and similarly

$$\tilde{\Theta} = \{L \in \tilde{J}_{\tilde{g}-1} \mid \Gamma(L) \neq (0)\} \subset \tilde{J}_{\tilde{g}-1} = \tilde{J}_{2g+n-2}.$$

(3) The various translates of the theta divisors in J and $\tilde{J}$ are given by Θ_{-y}, and $\tilde{\Theta}_{-\tilde{y}}$ for $y \in J_{g-1}$ and $\tilde{y} \in \tilde{J}_{\tilde{g}-1}$. To ask whether $\phi^{-1}(\theta_{Y,y})$ splits in this case is therefore the same as asking for points $y \in \tilde{J}_n$ if

$$(\pi^*)^{-1}(\tilde{\Theta}_{-y}) = \Theta_{x_1} + \Theta_{x_2}$$

for some $x_1, x_2 \in J$.

The double covering π gives us a unique divisor class $\mathfrak{A}$ such that:

$$2\mathfrak{A} \equiv \sum_{i=1}^{2n} P_1, \qquad P_i = \text{branch points},$$

$$\pi^{-1}\mathfrak{A} \equiv \sum_{i=1}^{2n} Q_i, \qquad Q_i = \pi^{-1}(P_i).$$

In fact, if $\mathbb{R}(\tilde{C}) = \mathbb{R}(C)(\sqrt{f})$, then $(f) = \sum_{i=1}^{2n} P_i - 2\mathfrak{A}$ on C and $(\sqrt{f}) = \sum_{i=1}^{2n} Q_i - \pi^{-1}(\mathfrak{A})$ on $\tilde{C}$ for some divisor $\mathfrak{A}$. Sheaf-theoretically, if $\tilde{C} = \mathbf{Spec}(\mathcal{O}_C \oplus L)$ as in Section 1, $L = \mathcal{O}_C(-\mathfrak{A})$. We then have the following.

Proposition. Let $x_1, \ldots, x_d$ be any d closed points on $\tilde{C}$ such that $\pi x_i \neq \pi x_j$, all $i \neq j$. Then for all invertible sheaves L of degree $g-1$ on C

$$\Gamma\left(\tilde{C}, \pi^* L\left(\sum_{i=1}^{d} x_i\right)\right) \neq (0) \Longleftrightarrow \Gamma(C, L) \neq (0) \quad \text{or} \quad \Gamma\left(C, L\left(\sum_{i=1}^{d} \pi x_i - \mathfrak{A}\right)\right) \neq (0).$$

Proof. Note that $\pi_*(\mathcal{O}_{\tilde{C}}) = \mathcal{O}_C \oplus \mathcal{O}_C(-\mathfrak{A})$, where $\mathcal{O}_C$ is the subsheaf of functions even under the involution $\imath$, and $\mathcal{O}_C(-\mathfrak{A})$ is the subsheaf of odd functions. Therefore

$$\pi_*(\pi^* L(\textstyle\sum x_i + \sum \imath x_i)) \cong L(\sum \pi x_i) \otimes \pi_* \mathcal{O}_{\tilde{C}} \cong L(\sum \pi x_i) \oplus L(\sum \pi x_i - \mathfrak{A}).$$

This sheaf has subsheaves as follows:

$$\begin{array}{ccccc} \pi_*(\pi^* L(\sum x_i + \sum \imath x_i)) & \cong & L(\sum \pi x_i) & \oplus & L(\sum \pi x_i - \mathfrak{A}) \\ \cup & & & & \\ \pi_*(\pi^* L(\sum x_i)) & & \cup & & \cup \\ \cup & & & & \\ \pi_*(\pi^* L) & \cong & L & \oplus & L(-\mathfrak{A}) \end{array}$$

but the middle sheaf does not break up into even and odd pieces, because $x_i \neq \imath x_j$ for any i, j. In fact at every point πx_i the middle sheaf is generated by $L \oplus L(-\mathfrak{A})$, plus a section (s_1, s_2) with nonzero images $\bar{s}_1 \in L(\pi x_i)/L$ and $\bar{s}_2 \in L(\pi x_i - \mathfrak{A})/L(-\mathfrak{A})$. It follows that the middle sheaf fits into an exact sequence:

$$0 \longrightarrow L \longrightarrow \pi_*(\pi^* L(\textstyle\sum x_i)) \longrightarrow L(\sum \pi x_i - \mathfrak{A}) \longrightarrow 0.$$

This gives:

$$\begin{aligned} 0 \longrightarrow \Gamma(C, L) &\longrightarrow \Gamma(\tilde{C}, \pi^* L(\textstyle\sum x_i)) \\ &\longrightarrow \Gamma(C, L(\textstyle\sum \pi x_i - \mathfrak{A})) \xrightarrow{\ \delta\ } H^1(C, L) \longrightarrow \cdots \end{aligned}$$

which gives the implication "$\Longrightarrow$" of the lemma immediately. As for "$\Longleftarrow$," the only problem would be if

$$\begin{aligned} &\Gamma(C, L) = (0) \\ &\delta\colon\ \Gamma(C, L(\textstyle\sum \pi x_i - \mathfrak{A})) \longrightarrow H^1(C, L) \quad \text{injective} \\ &\qquad\qquad \not\parallel \\ &\qquad\qquad (0). \end{aligned}$$

But $\deg L = g - 1$, so $\chi(L) = 0$ and this is impossible. Q.E.D.

Corollary 1. Let $x_1, \ldots, x_n$ be any n closed points on $\tilde{C}$ such that $\pi x_i \neq \pi x_j$, all $i \neq j$. If†

$$y = \sum_{i=1}^{n} x_i \in \tilde{J}_n \quad \text{and} \quad x = \sum_{i=1}^{n} \pi x_i - \mathfrak{A} \in J_0,$$

† To simplify notation, we are identifying divisor classes of degree k with point of J_k, e.g., writing x for $t(x)$, etc.

are compatible halves in P and J, then

$$(\pi^*)^{-1}(\tilde{\Theta})_{0,-z}) = \Theta_{0,w} + \Theta_{0,-w}.$$

Proof. Note that

$$\tilde{\Theta}_{0,-z} = (\tilde{\Theta}_{-\tilde{\zeta}})_{-\sum x_i + \delta + \pi^* w} = \tilde{\Theta}_{\pi^*(w-\zeta)-\sum x_i};$$

hence

$$\begin{aligned}(\pi^*)^{-1}(\tilde{\Theta}_{0,-z}) &= (\pi^*)^{-1}\tilde{\Theta}_{\pi^*(w-\zeta)-\sum x_i} = ((\pi^*)^{-1}\tilde{\Theta}_{-\sum x_i})_{w-\zeta} = (\Theta + \Theta_{\mathfrak{A}-\sum \pi x_i})_{w-\zeta} \\ &= \Theta_{0,w} + \Theta_{0,w+\mathfrak{A}-\sum \pi x_i} = \Theta_{0,w} + \Theta_{0,-w}. \quad \text{Q.E.D.}\end{aligned}$$

In case there are zero or two branch points, we can (a) work out more precisely what pairs (z, w) are compatible and (*b*) combine the result with Corollary 2 in Section 4 to obtain the following.

Corollary 3 (Schottky–Jung). If $\tilde{C}$ is *unramified* over C, so that $2\mathfrak{A} = 0$ as divisor class, then choose a divisor class $\mathfrak{B}$ on C such that $2\mathfrak{B} = \mathfrak{A}$. Choose any theta characteristic ζ on C and take $\tilde{\zeta} = \pi^{-1}(\zeta + \mathfrak{B})$ as theta characteristic on $\tilde{C}$. [Note that $2\tilde{\zeta} = \pi^{-1}(2\zeta + 2\mathfrak{B}) = \pi^{-1}(K + \mathfrak{A}) = \tilde{K}$ and $\operatorname{Nm}\tilde{\zeta} = 2\zeta + 2\mathfrak{B} = K + \mathfrak{A}$ as required.] These define $\Theta_0 \subset J$, $\tilde{\Theta}_0 \subset \tilde{J}$ and we have the following.

(i) $(\pi^*)^{-1}(\tilde{\Theta}_0) = \Theta_{0,\mathfrak{B}} + \Theta_{0,-\mathfrak{B}}$.

(ii) If $\Xi \subset P$ is a symmetric theta divisor on the Prym P, we get a canonical diagram

$$\begin{array}{ccccc} P & \xrightarrow{\phi_P} & \mathbb{P}(\Gamma(L_P{}^2)) & & \\ & & & \searrow^{i} & \\ & & & & |2\Theta_0| \\ & & & \nearrow_{B'}^{\approx} & \\ J & \xrightarrow{\phi_J} & \mathbb{P}(\Gamma(L_J{}^2)) & & \end{array}$$

where ϕ_P and ϕ_J are the Kummer maps defined by the linear systems $|2\Xi|$ and $|2\Theta_0|$, and i and B' are as in Section 4; and

$$i(\phi_P(0)) = B'(\phi_J(\mathfrak{B})).$$

Corollary 4 (Fay) (see *Note*, p. 350). If $\tilde{C}$ has two branch points over C, then choosing suitable theta characteristics on C and $\tilde{C}$, we get a symmetric theta divisor Ξ on P and a canonical diagram

$$\begin{array}{ccc} P & \xrightarrow{\phi_P} & \mathbb{P}(\Gamma(L_P{}^2)) \\ & & \downarrow{\scriptstyle j}\ \wr \\ J & \xrightarrow{\phi_J} & \mathbb{P}(\Gamma(L_J{}^2)) \end{array}$$

where $j = (B')^{-1} \cdot i$ is now an isomorphism. Then for every $x \in \tilde{C}$ there are compatible halves:

$$z = \tfrac{1}{2}(x - \iota x) \in P, \qquad w - \tfrac{1}{2}(\pi x - \mathfrak{A}) \in J$$

such that

$$j(\phi_P(z)) = \phi_J(w).$$

then

$$(\pi^*)^{-1}(\tilde{\Theta}_{-y}) = \Theta + \Theta_{-x}.$$

Proof. Set-theoretically, this is just a translation of the proposition. Since the divisor $\pi^{-1}(\tilde{\Theta}_{-y})$ is algebraically equivalent to 2Θ, there can be no multiplicities and the equality holds between divisors, too.

Note what happens if $\pi x_i = \pi x_j$ but $x_i \neq x_j$. Then

$$L(\pi x_i) \subset \pi_*\left(\pi^* L\left(\sum_{i=1}^{n} x_i\right)\right)$$

and since $\deg L(\pi x_i) = g$, $\Gamma(L(\pi x_i))$ is always nontrivial. Therefore in this case

$$\pi^*(J_{g-1}) \subset \tilde{\Theta}_{-y}.$$

To rephrase this corollary in a form parallel to the general description of Section 4, we must choose suitable symmetric representatives of Θ and $\tilde{\Theta}$ in J and $\tilde{J}$ themselves (instead of in J_{g-1} and $\tilde{J}_{2g+n-2}$). In fact, choose:

(a) Theta characteristics ζ and $\tilde{\zeta}$ on C and $\tilde{C}$, i.e., divisor classes such that $2\zeta = K$ (the canonical class on C) and $2\tilde{\zeta} = \tilde{K}$ (the canonical class on $\tilde{C}$), and moreover such that

$$Nm\,\tilde{\zeta} = K + \mathfrak{A}.$$

[To see that this is possible, let $\mathfrak{B}$ be a divisor class on C such that $2\mathfrak{B} \equiv \mathfrak{A} - \sum_{i=1}^{n} P_i$ (half of the P_i only). Then set $\tilde{\zeta} = \pi^{-1}(\zeta + \mathfrak{B}) + \sum_{i=1}^{n} Q_i$.]

(b) ζ and $\tilde{\zeta}$ define theta divisors $\Theta_0 = \Theta_{-\zeta}$ and $\tilde{\Theta}_0 = \tilde{\Theta}_{-\tilde{\zeta}}$ in J and $\tilde{J}$ which are well known to be symmetric. Moreover, I claim that because of our careful choice of $\tilde{\zeta}$, $(\pi^*)^{-1}\tilde{\Theta}_0 \in |2\Theta_0|$. This follows, in fact, from the next Corollary soon to be stated.

(c) Now write $\tilde{\zeta} = \pi^{-1}\zeta + \delta$ and note that $2\delta = \sum_{i=1}^{2n} Q_i$, $\mathrm{Nm}\,\delta = \mathfrak{A}$, and $\deg \delta = n$.

We make the following important definition.

Definition. If $x_1, \ldots, x_n$ are points of $\tilde{C}$, we wish to find elements

$$z = \tfrac{1}{2}\sum_{i=1}^{n}(x_i - \iota x_i) \in P, \qquad w = \tfrac{1}{2}\left(\sum_{i=1}^{n}\pi x_i - \mathfrak{A}\right) \in J.$$

We say that $z \in P$ and $w \in J$ are *compatible solutions* of the equations

$$2z = \sum_{i=1}^{n}(x_i - \iota x), \qquad 2w = \sum_{i=1}^{n}\pi x_i - \mathfrak{A}$$

if $z + \pi^* w = \sum_{i=1}^{n} x_i - \delta$.

(Note that such a pair z, w always exists: In fact $J \times P \longrightarrow \tilde{J}$ is surjective, so $\sum_{i=1}^{n} x_i - \delta$ can be written $z + \pi^* w$, where $z \in P$ and $w \in J$. Taking Nm and $1 - \iota$, it follows that $2z$ and $2w$ have the required values.)

We can now state the following result.

Corollary 2. If $x_1, \ldots, x_n$ are any points of $\tilde{C}$ such that $\pi x_i \neq \pi x_j$, all $i \neq j$, and

$$z = \tfrac{1}{2}\sum_{1}^{n}(x_i - \iota x_i), \qquad w = \tfrac{1}{2}\left(\sum_{1}^{n}\pi x_i - \mathfrak{A}\right)$$

As mentioned in the Introduction, one would hope that these last two corollaries can be used to find strong polynomial identities for the "theta-null werte" of Jacobians. Unfortunately, whereas for the projective embedding of any principally polarized Abelian variety (X, θ) defined by $|4\theta|$ one knows simple identities satisfied by the image of $0 \in X$ (namely Riemann's identities; cf. Mumford [9, Section 3]) for the morphism defined by $|2\theta|$ no analogous simple identities seem to be known. In classical terms, the problem is: Find identities for the set of 2^n functions of Z

$$f_a(Z) = \theta[{}^a_0](0, Z), \qquad a = [a_1, \ldots, a_n], \qquad a_i = 0 \quad \text{or} \quad 1.$$

($Z \in \mathfrak{H}_n$, Siegel's upper half-space). If $n = 3$, there appears to be a unique irreducible identity of order eight, which applied to $\phi_P(0)$ in Corollary 3 leads to the usual "Schottky relation" on the theta nulls of a curve C of genus 4.

To explain the strength of Corollary 4, for instance, it may be helpful to contrast it with the following result: If (X, θ_X) and (Y, θ_Y) are two principally polarized Abelian varieties and if $k \geqslant 4$, consider the diagram

$$\begin{array}{ccc} X & \xrightarrow{\phi_X} & \mathbb{P}(\Gamma(L_X{}^k)) \\ & & \wr\big\downarrow j \\ Y & \xrightarrow{\phi_Y} & \mathbb{P}(\Gamma(L_Y{}^k)) \end{array}$$

where ϕ_X and ϕ_Y are the canonical maps defined by $|k\theta_X|$ and $|k\theta_Y|$ and j is an isomorphism under which the translations by X_k [which extend uniquely to projective transformations on $\mathbb{P}(\Gamma(L_X{}^k))$] correspond to translations by Y_k. (For any X and Y a finite number of such j's always exist.) Then

$$j(\phi_X(X)) \cap \phi(Y) \neq \varnothing$$

implies

$$j(\phi_X(X)) = \phi_Y(Y);$$

hence $X \cong Y$. In other words, distinct Abelian varieties, projectively embedded by somewhat more ample linear systems, *never meet*!

III

6. GEOMETRIC DESCRIPTION OF SING Ξ, UNRAMIFIED CASE

We now consider only an unramified $\pi\colon \tilde{C} \longrightarrow C$. Recall that in this case

(a) genus $C = \dim J = g$, genus $\tilde{C} = \dim \tilde{J} = 2g - 1$, $\dim P = g - 1$, and $\ker \rho = P_2 \cong \{0, \mathfrak{A}\}^{\perp}/\{0, \mathfrak{A}\}$, $\mathfrak{A} \in J_2$ defining π (hence P principally polarized).

(b) $\{x \in \tilde{J} \mid \iota x = x\} = \pi^* J$, and $\{x \in \tilde{J} \mid \operatorname{Nm} x = 0\} \cong P \times \mathbb{Z}/2\mathbb{Z}$.

[Use the fact that in the notation of (i)–(viii) of Section 2, $a = g$, $b = g - 1$, and $c = g - 1$.] In fact, in a previous paper [11] we have shown by a different argument that if we look at the principal homogeneous space $\tilde{J}_{2g-2}$ instead of $\tilde{J} = \tilde{J}_0$, and at

$$\operatorname{Nm}\colon \quad \tilde{J}_{2g-2} \longrightarrow J_{2g-2},$$

then $\mathrm{Nm}^{-1}(K)\,(K \in J_{2g-2}$ the canonical divisor class) breaks into two components P^+, P^- such that:

$\forall$ invertible sheaves L_α on $\tilde{C}$, corresponding to $\alpha \in \tilde{J}_{2g-2}$,

if $\mathrm{Nm}\, L \cong \Omega_C^1$, then

$$\dim \Gamma(L_\alpha) \text{ even} \Longleftrightarrow \alpha \in P^+, \qquad \dim \Gamma(L_\alpha) \text{ odd} \Longleftrightarrow \alpha \in P^-; \tag{6.1}$$

moreover, for some α,

$$\dim \Gamma(L_\alpha) = 0 \quad \text{and} \quad \dim \Gamma(L_\alpha) = 1.$$

Translating these back to $\tilde{J}_0$ by any $\alpha \in \mathrm{Nm}^{-1}(K)$, P^+ and P^- correspond to P and its nontrivial coset in ker Nm. Now the theta divisors of C and $\tilde{C}$ live canonically in J_{g-1} and $\tilde{J}_{2g-2}$ and Riemann's theorem (see Kempf [7], and Szpiro [16]) asserts

$\forall$ invertible sheaves L_α on C (resp. $\tilde{C}$) corresponding to $\alpha \in J_{g-1}$ (resp. $\tilde{J}_{2-g2}$), (6.2)

$\dim \Gamma(L_\alpha) =$ mult. of α on Θ (resp. Θ).

Combining (6.1) and (6.2), we find the following result.

Proposition. (a) $\tilde{\Theta} \supset P^-$; (b) $\tilde{\Theta} \cdot P^+ = 2\Xi$, where $\Xi \subset P^+$ is a canonical representative of the theta divisor on P.

Proof. In fact

$$\alpha \in P^- \Longrightarrow \dim \Gamma(L_\alpha) \text{ odd} \Longrightarrow \dim \Gamma(L_\alpha) \geqslant 1 \Longrightarrow \alpha \in \tilde{\Theta}$$

and

$$\alpha \in \tilde{\Theta} \cap P^+ \Longrightarrow \dim \Gamma(L_\alpha) \text{ even and positive} \Longrightarrow \dim \Gamma(L_\alpha) \geqslant 2$$
$$\Longrightarrow \alpha \text{ singular on } \tilde{\Theta};$$

hence $\tilde{\Theta} \cdot P^+$ consists entirely in multiple components. But the principal polarization on $\tilde{J}$ restricts to twice that on P, so $\Theta \cdot P^+$ is in the algebraic equivalence class 2Ξ. It is easy to check that such a divisor can never have a component of multiplicity $\geqslant 3$ (or else the morphism it defines would not collapse an involution $x \longrightarrow x_0 - x$). Thus $\Theta \cdot P^+ = 2D$, D algebraically equivalent to Ξ, hence equal to it after a suitable translation. Q.E.D.

Corollary.

$$\operatorname{Sing} \Xi = \{x \in P^+ \mid \text{mult. at } x \text{ of } \tilde{\Theta} \geqslant 4\}$$
$$\cup \left\{ x \in P^+ \,\middle|\, \begin{array}{l} \text{mult. at } x \text{ of } \tilde{\Theta} = 2, \text{ but} \\ T_{x,P^+} \subset (\text{tangent cone to } \Theta \text{ at } x) \end{array} \right\}.$$

In order to apply this corollary, we must know how to compute the tangent cone to $\tilde{\Theta}$. In general, suppose J is any Jacobian and $\Theta \subset J_{g-1}$. If L_α on C corresponds to the point $\alpha \in J_{g-1}$, then not only is

$$\dim \Gamma(L_\alpha) = \text{mult. at } \alpha \text{ of } \Theta,$$

but if $k = \dim \Gamma(L_\alpha)$, $s_1, \ldots, s_k$ is the basis of $\Gamma(L_\alpha)$, $t_1, \ldots, t_k$ is the basis of $\Gamma(\Omega \otimes L_\alpha^{-1})$, and $s_i \otimes t_j \in \Gamma(\Omega)$ defines the differential ω_{ij} at $\alpha \in J_{g-1}$, then identifying $\Gamma(\Omega)$ to the cotangent space m_α/m_α^2 of J_{g-1} at α, Kempf [7] proves that $\det(\omega_{ij})=0$ is the tangent cone to Θ at α.

Now if L_α is a sheaf on $\tilde{C}$ such that $\operatorname{Nm} L_\alpha \cong \Omega_C$, then

$$L_\alpha \otimes \iota^* L_\alpha = \pi^* \operatorname{Nm} L_\alpha \cong \pi^*\Omega_C \cong \Omega_{\tilde{C}};$$

hence choosing such an isomorphism ϕ, we may use the pairing

$$\begin{aligned} \langle\,,\rangle\colon\ \Gamma(L_\alpha)\otimes\Gamma(L_\alpha) &\longrightarrow \Gamma(\Omega_{\tilde{C}}) \\ (s,t) &\longmapsto \phi(s\otimes\iota^*t)=\langle s,t\rangle \end{aligned}$$

instead of

$$\Gamma(L_\alpha)\otimes\Gamma(\Omega_{\tilde{C}}\otimes L_\alpha^{-1}) \xrightarrow{\ \otimes\ } \Gamma(\Omega_{\tilde{C}}).$$

Now ι induces $\iota^*\colon \Gamma(\Omega_{\tilde{C}}) \longrightarrow \Gamma(\Omega_{\tilde{C}})$, too: In fact, this is just the automorphism found by decomposing

$$\begin{aligned} \Gamma(\Omega_{\tilde{C}}) \cong \Gamma(\pi^*\Omega_C) &\cong \Gamma(\pi_*\pi^*\Omega_C) \\ &\subset \Gamma(\Omega_C)+\Gamma(\Omega_C(\mathfrak{A})) \\ &\qquad\qquad\quad \| \\ &\qquad \text{the “Prym differentials”} \end{aligned}$$

and letting $\iota^*=+1$ on $\Gamma(\Omega_C)$, $\iota^*=-1$ on $\Gamma(\Omega_C(\mathfrak{A}))$. It is easy to check that

$$\iota^*(\langle s,t\rangle)=\langle t,s\rangle;$$

hence the above pairing splits into two pairings:

$$\operatorname{Symm}^2\Gamma(L_\alpha)\longrightarrow\Gamma(\Omega_C),\qquad \Lambda^2\Gamma(L_\alpha)\longrightarrow\Gamma(\Omega_C(\mathfrak{A})).$$

Moreover, in the identification $\Gamma(\Omega_C)+\Gamma(\Omega_C(\mathfrak{A}))\cong\Gamma(\Omega_{\tilde{C}})\cong$ cotangent space $T^*_{\alpha, J_{2g-2}}$, clearly the even and odd subspaces under ι^* go over as follows: $\Gamma(\Omega_C)\cong$ cotangent space at α to the coset $\alpha+\pi^*(J_{2g-2})$, and $\Gamma(\Omega_C(\mathfrak{A}))\cong$ cotangent space at α to $P^\pm$.

Taking a basis $s_1,\ldots,s_k$ of $\Gamma(L_\alpha)$, let $\omega_{ij}=\langle s_i,s_j\rangle$. Then $\iota^*\omega_{ij}=\omega_{ji}$, hence decomposing ω_{ij}.

$$\omega_{ij}=\omega_{ij}^+ + \omega_{ij}^-,\qquad \omega_{ij}^+\in\Gamma(\Omega_C),\qquad \omega_{ij}^-\in\Gamma(\Omega_C(\mathfrak{A})),$$

It follows that ω_{ij}^+ is symmetric and ω_{ij}^- is skew-symmetric. Therefore if $\alpha\in P^+$, $\det(\omega_{ij})=0$ is the tangent cone to $\tilde{\Theta}$ at α, and $\det(\omega_{ij}^-)=0$ is the tangent cone to $\tilde{\Theta}\cdot P^+$ at α. But $\det(\omega_{ij}^-)=Pf(\omega_{ij}^-)^2$ ($Pf=$ Pfaffian), so that $Pf(\omega_{ij}^-)=0$ is the tangent cone to Ξ at α (unless it vanishes identically).

We use this to establish the following result.

Proposition. If $\operatorname{Nm} L_\alpha=\Omega_C$ and $\dim\Gamma(L_\alpha)=2$, then

$T_{\alpha,P^+}\subset$ tangent cone to $\tilde{\Theta}$ at $\alpha \Longleftrightarrow L_\alpha\cong\pi^*(\mathfrak{A})(\sum x_i)$ for some points $x_i\in\tilde{C}$ and a sheaf M on C such that $\dim\Gamma(M)=2$.

Proof. Let s, t be a basis of $\Gamma(L_\alpha)$. In the proceding notation

$$(\omega_{ij}^-)=\begin{pmatrix}0 & \langle s,t\rangle-\langle t,s\rangle\\ \langle t,s\rangle-\langle s,t\rangle & 0\end{pmatrix}.$$

So the linear form $\langle s, t\rangle - \langle t, s\rangle$ is the tangent cone to Ξ unless $\alpha \in \operatorname{Sing} \Xi$. Thus

$$\begin{aligned} T_{\alpha, P^+} \subset \text{tangent cone to } \tilde{\Theta} \text{ at } \alpha &\Longleftrightarrow \langle s, t\rangle = \langle t, s\rangle \\ &\Longleftrightarrow s \otimes \imath^* t = t \otimes \imath^* s \Longleftrightarrow \imath^*(s/t) = s/t \\ &\Longleftrightarrow s/t \in \mathbb{R}(C). \end{aligned}$$

In classical language, $s/t \in \mathbb{R}(C)$ says "the pencil defined by L_α is pulled back from a pencil on C." In modern language, let $\sum x_i$ be the base points of $\Gamma(L_\alpha)$, let $\mathfrak{B}$ be the poles of s/t on C, and let $M = \mathcal{O}_C(\mathfrak{B})$. Then $L_\alpha \cong \pi^* M(\sum x_i)$ and $1, s/t \in \Gamma(M)$; hence $\dim \Gamma(M) \geqslant 2$. Clearly $\dim \Gamma(M) = 2$ since $\dim \Gamma(L_\alpha) = 2$. Q.E.D.

7. DIM SING Ξ

Notations are as in Section 6. Recall that if C is a curve, a theta characteristic of C is a sheaf such that $L^2 \cong \Omega_C$; L is even or odd if $\dim \Gamma(L)$ is even or odd. We wish to prove the following theorem.

Theorem.
(a) C hyperelliptic $\Longrightarrow (P, \Xi)$ is a hyperelliptic Jacobian (hence $\dim \operatorname{Sing} \Xi = g - 4$) or a product of two such (hence $\dim \operatorname{Sing} \Xi = g - 3$).
(b) $g = 3$, C not hyperelliptic $\Longrightarrow (P, \Xi)$ is a two-dimensional Jacobian.
(c) $g = 4$; C not hyperelliptic $\Longrightarrow (P, \Xi)$ is a three-dimensional Jacobian, and Ξ is singular iff P is a hyperelliptic Jacobian iff $\exists$ is an even theta characteristic L with $\Gamma(L) \neq (0)$ and $L(\mathfrak{A})$ even.
(d) Assuming C not hyperelliptic and $g \geqslant 5$, then $\dim \operatorname{Sing} \Xi \leqslant g - 5$ and

$$\dim \operatorname{Sing} \Xi = g - 5 \Longrightarrow \begin{cases} C \text{ trigonal}, \\ \text{or } C \text{ double cover of an elliptic curve}, \\ \text{or } g = 5 \text{ and } \exists \text{ even theta characteristic } L \text{ with} \\ \Gamma(L) \neq (0) \text{ and } L(\mathfrak{A}) \text{ even}, \\ \text{or } g = 6 \text{ and } \exists \text{ odd theta characteristic } L \text{ with} \\ \dim \Gamma(L) \geqslant 3, \text{ and } L(\mathfrak{A}) \text{ even}. \end{cases}$$

In fact, in part (d), "$\Longleftarrow$" apparently also holds, but we will omit the proof of this. We first want to point out the following corollary.

Corollary. If $g \geqslant 5$ and C is neither trigonal, a double cover of an elliptic curve, nor of the preceding two special types of genus 5 or 6, then the polarized Abelian variety (P, Ξ) is *not* a Jacobian or a product of Jacobians.

This follows from the theorem and the fact that $\dim \operatorname{Sing} \Theta \geqslant \dim J - 4$ for polarized Jacobians (J, Θ) [1]. It would be quite interesting to find out in the special cases exactly which (P, Ξ) is a Jacobian.

Proof of Theorem. As shown in the previous section, the singularities of Ξ canonically embedded in P^+ arise from two sources.

Case 1: sheaves L_α such that $\operatorname{Nm} L_\alpha = \Omega_C$, $\dim \Gamma(L_\alpha) \geqslant 2$ and even, and $L_\alpha = \pi^* M(\sum x_i)$, where $\dim \Gamma(M) \geqslant 2$.

Case 2: sheaves L_α such that $\mathrm{Nm}\, L_\alpha = \Omega_C$, $\dim \Gamma(L_\alpha) \geqslant 4$ and even.
Note that in case 1

$$\Omega_C = \mathrm{Nm}\, L_\alpha = \mathrm{Nm}(\pi^* M(\textstyle\sum x_i)(= M^2(\sum \pi x_i),$$

so M satisfies the two conditions: (a) $\dim \Gamma(M) \geqslant 2$ and (b) $\dim \Gamma(\Omega_C \otimes M^{-2}) \geqslant 1$. Also, if there are no x_i, i.e., $L_\alpha = \pi^* M$, then $\dim \Gamma(L_\alpha)$ even implies (c) If $\Omega_C \cong M^2$, $\dim \Gamma(M) + \dim \Gamma(M \otimes \mathfrak{A})$ even.

Conversely, if M satisfies (a)–(c), choose an effective divisor $\sum \pi x_i$ in the linear system $\Gamma(\Omega_C \otimes M^{-2})$ and set $L_\alpha = \pi^* M(\sum x_i)$. This falls in case 1 unless there is at least one x_i and $\dim \Gamma(L_\alpha)$ odd. But as shown in a previous work [11, p. 187], we can then replace *one* x_i by $\iota(x_i)$ to make $\dim \Gamma(L_\alpha)$ even. So all M satisfying (a)–(c) define L_α in case 1.

It is not so easy to construct all the L_α in case 2 directly from sheaves on C. However, I claim the following.

Lemma. If $\dim \mathrm{Sing}\, \Xi \geqslant g-5$, then almost all $\alpha \in \mathrm{Sing}\, \Xi$ correspond to sheaves L_α in case 1.

Proof. Suppose $Z \subset \mathrm{Sing}\, \Xi$ were a component of dimension $\geqslant (g-5)$ such that

$$\dim \Gamma(L_\alpha) \geqslant 4, \qquad \text{all} \quad \alpha \in Z.$$

According to previous results [11, pp. 186–188], $\dim \Gamma(L_\alpha) = 4$ for almost all $\alpha \in Z$. Let $Z_0 \subset Z$ be the open subset where $\dim \Gamma(L_\alpha) = 4$. We wish to apply the following quite general result.

Proposition. Let C be any curve, $Z \subset J_d$ a subvariety, $Z_0 \subset Z$ an open set, and assume that for some k

$$\dim \Gamma(L_\alpha) = k, \qquad \text{all} \quad \alpha \in Z_0.$$

Then identifying T_{α, J_α} to $H^1(\mathcal{O}_C)$, hence to the dual of $\Gamma(\Omega_C)$, I claim

$$\mathrm{Im}[\Gamma(L_\alpha) \otimes \Gamma(\Omega_C \otimes L_\alpha^{-1}) \longrightarrow \Gamma(\Omega_C)]^\perp \supset T_{\alpha, Z}$$

for all $\alpha \in Z_0$.

This is proved for $k = 2$ in Lemma 2.5 of Saint-Donat [14] but the proof extends verbatim to all k. Applying this to our case, let

$$W_\alpha = \mathrm{Im}[\Lambda^2 \Gamma(L_\alpha) \longrightarrow \Gamma(\Omega_C \otimes \mathfrak{A})].$$

Identifying $\Gamma(\Omega_C \otimes \mathfrak{A})$ with T^*_{α, P^+}, we find $T_{\alpha, Z} \subset W_\alpha^\perp$. Since the codimension of Z in P^+ is $\leqslant 4$, it follows that $\dim W_\alpha \leqslant 4$. But $\dim \Lambda^2 \Gamma(L_\alpha) = 6$, so the kernel of $\Lambda^2 \Gamma(L_\alpha) \longrightarrow \Gamma(\Omega_C \otimes \mathfrak{A})$ has dimension at least two. Now the set of decomposable 2-forms $s \wedge t$ in $\Lambda^2 \Gamma(L_\alpha)$ forms a cone in $\Lambda^2 \Gamma(L_\alpha)$ of codimension one, so at least one such $s \wedge t$ lies in the kernel. But for any s, t we find

$$\begin{aligned} s \wedge t = 0 &\Longleftrightarrow \langle s, t \rangle = \langle t, s \rangle \\ &\Longleftrightarrow s/t \in \mathbb{R}(C) \qquad \text{(as before).} \end{aligned}$$

Therefore, exactly as in the last section, $L_\alpha \cong \pi^* M(\sum x_i)$ where $\dim \Gamma(M) \geqslant 2$. Q.E.D.

We are now ready to prove the theorem—or rather reduce it to a strengthened form of a theorem of Martens which is given in the appendix. First of all, say C is hyperelliptic: Let $p\colon C \longrightarrow \mathbb{P}^1$ be the double covering and let $\{z_1, \ldots, z_{2g+2}\}$ be the branch points. It is well known that all unramified double coverings $\pi\colon \tilde{C} \longrightarrow C$ arise as follows.

(a) Separate the z_i into two groups of even cardinality:

$$\{1, 2, \ldots, 2g+2\} = I' \cup I'', \qquad I' = 2h+2, \qquad I'' = 2k+2.$$

$I' \cap I'' = \phi$; hence $h + k + 1 = g$.

(b) Let $p'\colon C' \longrightarrow \mathbb{P}^1$ and $p''\colon C'' \longrightarrow \mathbb{P}^1$ be the hyperelliptic curves with branch points $\{z_i\}_{I'}$ and $\{z_i\}_{I''}$, respectively.

(c) Let $\tilde{C}$ be the normalization of $C \times_{\mathbb{P}^1} C'$. The $\mathbb{Z}/2\mathbb{Z} \times \mathbb{Z}/2\mathbb{Z}$ acts on $\tilde{C}$ and we get a tower of curves:

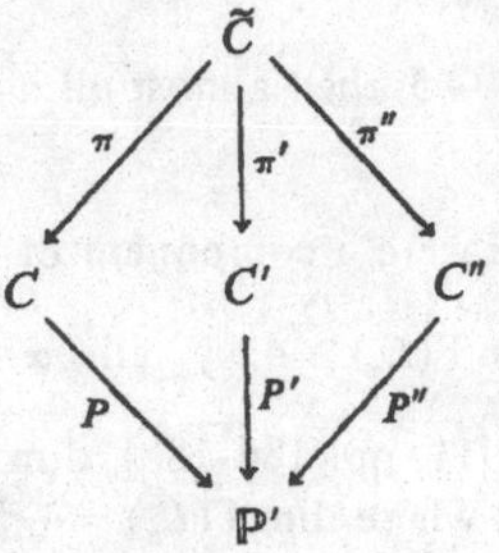

by dividing by its three subgroups of order two. Note that $\tilde{C} =$ norm. of $C \times_{\mathbb{P}^1} C'' = C' \times_{\mathbb{P}^1} C''$. I claim that in this situation

$$\operatorname{Prym}(\tilde{C}/C) \cong J' \times J''$$
$$\Xi \longleftrightarrow J' \times \Theta'' + \Theta' \times J'',$$

where J' and J'' are the Jacobians of C' and C''. (Note that if h or k is zero, one of the factors here disappears.)

Idea of Proof: Now we have $\mathbb{Z}/2\mathbb{Z} \times \mathbb{Z}/2\mathbb{Z}$ acting on $\tilde{J}$ and up to 2-isogenies, $\tilde{J}$ splits into four "eigensubvarieties"; the part invariant under the whole group will be empty, and the other three pieces will be π^*J, J', and J''. One checks that $J' \times J''$ injects into $\tilde{J}$ by the natural map $(\pi')^* \times (\pi'')^*$, and that the image is P. Finally, one checks that the $\tilde{\Theta}$ polarization on $\tilde{J}$ splits into the sum of 2Θ, $2\Theta'$, and $2\Theta''$ on the three pieces; hence Ξ splits into the sum of Θ' and Θ''. The details are left to the reader.

Now suppose C is not hyperelliptic and that $\dim \operatorname{Sing} \Xi = v \geqslant g - 5$. Almost all of these singular points must define sheaves L_α in case 1: It follows that for some d there is a v-dimensional family of pairs $\{M, \sum_{i=1}^{e} y_i\}$ where (i) M is an invertible sheaf on C; (ii) $\sum y_i$ is an effective divisor of degree e; (iii) $\deg M = d$ and $2d + e = 2g - 2$; (iv) $\dim \Gamma(M) \geqslant 2$; (v) $M^2(\sum_{i=1}^{e} y_i) \cong \Omega_C$; and (vi) if $e = 0$, then $\dim \Gamma(M) + \dim \Gamma(M \otimes \mathfrak{A})$ even.

Now for each M the set of all divisors $\sum y_i$ of this type is a projective space whose dimension equals $\dim \Gamma(\Omega_C \otimes M^{-2}) - 1$. By Clifford's theorem, we can bound this by

$$\dim \Gamma(\Omega_C \otimes M^{-2}) - 1 < \tfrac{1}{2} \deg \Omega_C \otimes M^{-2} = g - 1 - d.$$

If $d < g - 1$, by Marten's theorem (see appendix), the dimension of the set of Ms of degree d with $\dim \Gamma(M) \geqslant 2$ is bounded by $d - 3$, and if C is not trigonal, a double cover of an elliptic curve, or a nonsingular quintic, then it is bounded by $d - 4$. Therefore

$$v = (\text{dim. of possible } M\text{s}) + (\text{dim. of possible } \textstyle\sum y_i) < g - 4$$

and $v < g - 5$, except in the aforementioned special cases. Also, if $d = g - 1$, then $M^2 \cong \Omega_C$, i.e., M is one of the finite set of theta characteristics: if $g \leqslant 5$, these can give us a ($\geqslant g - 5$)-dimensional singular locus on Ξ.

Finally, let us look at the low-genus cases: If $g = 3$, the only singularities on Ξ arise from theta characteristics M. But if C is not hyperelliptic, $\dim \Gamma(M) = 0$ or 1 for all M, so Ξ is nonsingular. Thus (P, Ξ) is a principally polarized two-dimensional Abelian variety with Ξ nonsingular: Hence it is a Jacobian. If $g = 4$ and C is not hyperelliptic, again singularities on Ξ can arise only from theta characteristics. In fact, in $\mathbb{P}^3$ the canonical model of C equals $F.G$, with F a quadric, C a cubic. And if F is nonsingular, again $\dim \Gamma(M) = 0$ or 1 for all theta characteristics M. But if F is a cone, there is one even M with $\dim \Gamma(M) = 2$—namely the M defined by the divisors C.(line on F). If also $\dim \Gamma(M \otimes \mathfrak{A})$ equals zero rather than one, then Sing Ξ has a single point. Thus (P, Ξ) is a principally polarized three dimensional Abelian variety with zero or one singularity on Ξ. Now either from the fact that the moduli space over $\mathbb{Z}$ of such varieties is irreducible six dimensional, hence Jacobians are dense in it, hence by Hoyt [6] every such variety is a Jacobian or product of Jacobians; or from Harris' thesis [5], it follows that (P, Ξ) is a Jacobian. Since a three dimensional Jacobian (J, Θ) has a singular Θ if and only if J comes from a hyperelliptic curve, this proves (c). As for (d), we have proved this already modulo noting that nonsingular quintics are precisely the nonhyperelliptic curves of genus six with sheaves N such that

$$N^2 \cong \Omega_C, \qquad \dim \Gamma(M) = 3.$$

[i.e., $N = \mathcal{O}_C(1)$]. This N defines sheaves M by $M = N(-z)$, $z \in C$, hence potential singularities of Ξ by

$$L_\alpha = \pi^*(N(-z))(x_1 + x_2).$$

Then x_1 and x_2 must satisfy

$$\Omega_C \cong N^2(-2z)(\pi x_1 + \pi x_2);$$

hence $\pi x_1 = \pi x_2 = z$. Therefore

$$L_\alpha = \pi^* N(x - ix) \qquad \text{or} \qquad L_\alpha = \pi^* N.$$

But one of these will be in P^+, the other in P^-, hence Ξ will either have a whole curve of singularities parametrized by x, or exactly one singularity, and in fact

$$\dim \operatorname{Sing} \Xi = 1 \iff \pi^* N(x - \iota x) \in P^+ \iff \pi^* N \in P^-$$
$$\iff \dim \Gamma(N) + \dim \Gamma(N \otimes \mathfrak{A}) \text{ odd} \iff \dim \Gamma(N \otimes \mathfrak{A}) \text{ even.} \quad \text{Q.E.D.}$$

Precisely this final special case has turned out recently to be surprisingly interesting. The reason is that the Pryms (P, Ξ) arising from quintics $C \subset \mathbb{P}^2$ and double coverings

$\tilde{C} = \mathbf{Spec}(\mathcal{O}_C + \mathcal{O}_C(\mathfrak{A}))$ for which $\dim \Gamma(\mathcal{O}_C(1)(\mathfrak{A}))$ is *odd* include the intermediate Jacobians of cubic hypersurfaces in $\mathbb{P}^4$: By the corollary, these are *not* Jacobians and their Ξ has one singular point at which the tangent cone is in fact exactly the cubic hypersurface! Clemens and Griffiths [2] have given another proof that this intermediate Jacobian is not a Jacobian and have deduced from this that the cubic hypersurface is not rational. On the other hand, Clemens conjectures that when $\dim \Gamma(\mathcal{O}_C(1)(\mathfrak{A}))$ is *even*, then (P, Ξ) is a Jacobian.

APPENDIX: A THEOREM OF MARTENS

The purpose of this appendix is to somewhat strengthen Marten's theorem [8, Theorem 1] (see also Saint-Donat [14, Theorem 2.4]) as follows.

Theorem. If C is a nonsingular curve of genus g, and $W_d \subset J_d$, $1 \leqslant d \leqslant g-1$, is the locus of invertible sheaves of degree d *with sections*, then

$$\exists d,\quad 2 \leqslant d \leqslant g-2,\quad \text{such that}\quad \dim \operatorname{Sing} W_d \geqslant g-3$$
$$\Longleftrightarrow C \text{ is (a) hyperelliptic, or (b) trigonal, or}$$
(c) double cover of an elliptic curve, or
(d) nonsingular plane quintic.

Proof. Recall that by Kempf's results [7,16]

$$\operatorname{Sing} W_d = (\text{locus of inv. sheaves } L,\quad \dim \Gamma(L) \geqslant 2);$$

hence, in Marten's notations, $\operatorname{Sing} W_d = G_d^1$. Thus he shows that

$$\exists d,\quad 2 \leqslant d \leqslant g-2,\quad \dim \operatorname{Sing} W_d \geqslant d-2 \Longleftrightarrow C \quad \text{hyperelliptic.}$$

Excluding this case, we assume $\dim \operatorname{Sing} W_d = d-3$ for some d. If $d = 3$,

$$\operatorname{Sing} W_3 \neq \phi \Longleftrightarrow \exists L \text{ of degree three},\quad \dim \Gamma(L) \geqslant 2$$
$$\Longleftrightarrow C \quad \text{trigonal.}$$

Excluding this case, we may assume $d \geqslant 4$ (hence $g \geqslant 6$) and C not trigonal. Consider a general L of degree d with $\dim \Gamma(L) = 2$ and look at the pairing

$$\underbrace{\Gamma(L)}_{\dim 2} \otimes \underbrace{\Gamma(\Omega \otimes L^{-1})}_{\dim (g-d+1)} \xrightarrow{\phi} \Gamma(\Omega).$$

If d is the smallest d for which $\dim \operatorname{Sing} W_d = d-3$, we can assume that $\Gamma(L)$ is base-point free. Let $\alpha, \beta \in \Gamma(L)$ be a basis. Now according to Kempf's results, the pairing ϕ allows us to compute the Zariski tangent space to $\operatorname{Sing} W_d$ at any point $L \in \operatorname{Sing} W_d$ such that $\dim \Gamma(L) = 2$: namely, identify

$$T_{L, \operatorname{Sing} W_d} \subset T_{L, J_d} \cong T_{0,J} \cong H^1(\mathcal{O}_C) \cong \text{dual of } \Gamma(\Omega).$$

Then he shows that

$$\operatorname{Im} \phi = (T_{L, \operatorname{Sing} W_d})^{\perp}.$$

Therefore

$$\dim(\operatorname{Im}\phi) \leqslant g - d + 3.$$

But since α and β have no common zeros, we get an exact sequence:

$$0 \longrightarrow \alpha\otimes\beta\otimes\Gamma(\Omega\otimes L^{-2}) \longrightarrow \alpha\otimes\Gamma(\Omega\otimes L^{-1}) + \beta\otimes\Gamma(\Omega\otimes L^{-1}) \longrightarrow \operatorname{Im}\phi \longrightarrow 0; \quad \text{(A.1)}$$

hence

$$\dim \operatorname{Im}\phi = 2(g-d+1) - \dim\Gamma(\Omega\otimes L^{-2}) = g + 3 - \dim\Gamma(L^2).$$

Therefore $\dim\Gamma(L^2) \geqslant d$. In other words, the L^2s define a $(d-3)$-dimensional subset of W_{2d} of points corresponding to Ms with $\dim\Gamma(M) \geqslant d$: In Martens's notation,

$$\dim G_{2d}^{d-1} \geqslant d - 3.$$

Applying his Theorem 1 again, the only cases where this might happen are: (i) $d = 4$, $\dim \operatorname{Sing} W_4 = 1$, or (ii) $d = 5$, $g = 7$, $\dim \operatorname{Sing} W_5 = 2$.

If (i) happens, fix one L_0 of degree four, $\dim\Gamma(L_0) = 2$, $\Gamma(L_0)$ base-point free, and let L be any other. Note that $\dim\Gamma(L_0\otimes L) = 4$ in all cases where $L \not\approx L_0$ [e.g., by computing $\Gamma(L_0\otimes L)$ by an exact sequence like (A.1)]. Therefore by Riemann–Roch,

$$\dim\Gamma(\Omega\otimes L_0^{-1}) = \dim\Gamma(L_0) + 2g - 6 - g + 1 = g - 3$$

$$\dim\Gamma(\Omega\otimes L_0^{-1}\otimes L^{-1}) = \dim\Gamma(L_0\otimes L) + 2g - 10 - g + 1 = g - 5.$$

Let $P_1, \ldots, P_{g-6}$ be any $g - 6$ points on C in general position. Then

$$\Gamma\left(\Omega\otimes L_0^{-1}\otimes L^{-1}\left(-\sum_{i=1}^{g-6} P_i\right)\right) \neq (0),$$

hence if s_L is a section here, and $M = \Omega\otimes L_0^{-1}(-\sum_{i=1}^{g-6} P_i)$, we find

$$s_L \otimes \Gamma(L) \subseteq \Gamma(M)$$

for all L. Note that

$$\dim\Gamma(M) = \dim\Gamma(\Omega\otimes L_0^{-1}) - (g-6) = 3.$$

Therefore $\Gamma(M)$ defines a rational map $\pi\colon C \longrightarrow \mathbb{P}^2$ such that every base-point-free pencil $\Gamma(L)$ of degree four defines a map $C \longrightarrow \mathbb{P}^1$ which is the composition of π and a projection of $\pi(C)$ to $\mathbb{P}^1$. But if $d = \operatorname{degree}(\pi(C))$, then projecting $\pi(C)$ from a point of $\mathbb{P}^2 - \pi(C)$, or from a simple point of $\pi(C)$, gives a map of degree d, or $d - 1$, from $\pi(C)$ to $\mathbb{P}^1$. Since $\pi(C)$ has only finitely many multiple points and there are supposed to be an infinite number of Ls, we conclude that either π birational and $d \leqslant 5$ or π of degree two, $d \leqslant 3$. Since $g \geqslant 6$, C is either a nonsingular plane quintic or a double covering of an elliptic curve. Both of these do have an infinite $\operatorname{Sing} W_4$, i.e., take the line bundles $[\mathcal{O}_C\otimes\mathcal{O}_{\mathbb{P}^2}(1)](-P)$, any $P\in C$, in the first case, or π^*L, any L of degree two on the elliptic curve, in the second case.

Finally, we want to exclude (ii). Assume we have a two-dimensional family of Ls such that

$$\deg L = 5, \qquad \dim\Gamma(L) = 2, \qquad \dim\Gamma(L^2) = 5.$$

By Riemann–Roch, $\Gamma(\Omega \otimes L^{-2}) \neq (0)$; hence $L^2 \cong \Omega(-P-Q)$ for some P, Q. Since the set of all L of degree five such that $L^2 \cong \Omega(-P-Q)$ for some P and Q is irreducible and two dimensional, it follows that $\dim \Gamma(L) \geqslant 2$ for any such L. Especially if $M^2 \cong \Omega$, then $\dim \Gamma(M(-P)) \geqslant 2$ for every $P \in C$. Therefore $\dim \Gamma(M) \geqslant 3$. But for any principally polarized Abelian variety X and symmetric theta divisor $\Theta \subset X$, Θ cannot contain all points of order two on X (see Mumford [9, p. 346]). For Jacobians this means by Riemann's theorem that there is always an M with $M^2 \cong \Omega$, $\Gamma(M) = (0)$. This is a contradiction and (ii) never occurs. Q.E.D.

Note added in proof. Corollary 4 is Proposition 5.7 of Fay [4], in which there is a misprint: The lower limit of the integral in Fay should be D. To get our version, use the remarks at the top of p. 100.

REFERENCES

1. A. Andreotti and A. Mayer, On period relations for abelian integrals on algebraic curves, *Ann. Scuola Norm. Sup. Pisa* **21** (1967), 189–238.
2. H. Clemens and P. Griffiths, The intermediate Jacobian of the cubic 3-fold, *Ann. of Math.* **95** (1972), 281–356.
3. H. Farkas and H. Rauch, Period relations of Schottky type on Riemann surfaces, *Ann. of Math.* **92** (1970), 434–461.
4. J. Fay, "Theta functions on Riemann Surfaces." Springer-Verlag, Berlin and New York, 1973, Lecture Notes, Vol. 352.
5. D. Harris, A study of 3-dimensional principally polarized abelian varieties. Ph.D. Thesis, Harvard Univ., Cambridge, Massachusetts, 1972.
6. W. Hoyt, On products and algebraic families of Jacobian varieties, *Ann. of Math.* **77** (1963), 415–423.
7. G. Kempf, On the geometry of a theory of Riemann, *Ann. of Math.* **98** (1973), 178–185.
8. H. Martens, On the varieties of special divisors on a curve, *J. Reine Angew. Math.* **227** (1967), 111–120.
9. D. Mumford, On the equations defining abelian varieties, *Invent. Math.* **1** (1966).
10. D. Mumford, "Abelian Varieties." Tata Inst. Studies in Math., Oxford Univ. Press, London and New York, 1970.
11. D. Mumford, Theta characteristics of an algebraic curve, *Ann. Sci. Ecole Norm. Sup.* **4** (1971), 181–192.
12. J. Murre, Algebraic equivalence mod rational equivalence on a cubic 3-fold. *Compositio Math.* **25** (1972), 161–206.
13. B. Riemann, "Collected Works," Nochtrag IV. Dover, New York, 1953.
14. B. Saint-Donat, On Petri's analysis of the linear system of quadrics through a canonical curve, *Math. Ann.* **206** (1973), pp. 157–175.
15. F. Schottky and H. Jung, Neue Sätze über symmetralfunctionen und die Abelschen funktionen, *S.-B. Berlin Akad. Wiss.* (1909).
16. L. Szpiro, "Travaux de Kempf, Kleiman, Laksov," (Sem. Bourbaki, Exp. 417), Springer-Verlag, Berlin and New York, 1972 Lecture Notes, Vol. 317.

Reprinted from the
Proceedings of the International Colloquium on
Discrete Subgroups of Lie Groups and Applications to Moduli
Bombay, January 1973

A NEW APPROACH TO COMPACTIFYING LOCALLY SYMMETRIC VARIETIES

By DAVID MUMFORD

SUPPOSE D IS a bounded symmetric domain and $\Gamma \subset \mathrm{Aut}(D)$ is a discrete group of arithmetic type. Then Borel and Baily [2] have shown that D/Γ can be canonically embedded as a Zariski-open subset in a projective variety $\overline{D/\Gamma}$. However, Igusa [6] and others have found that the singularities of $\overline{D/\Gamma}$ are extraordinarily complicated and this presents a significant obstacle to using algebraic geometry on $\overline{D/\Gamma}$ in order to derive information on automorphic forms on D, etc. Igusa [7] for $D = \mathfrak{M}_2$ and $\mathfrak{M}_3$ ($\mathfrak{M}_n$ = Siegel's $n \times n$ upper half-space) and Γ commensurable with $Sp(2n, \mathbf{Z})$, and Hirzebruch [4] for $D = \mathfrak{M}_1 \times \mathfrak{M}_1$ and Γ commensurable with $SL(2, R)$ (R = integers in a real quadratic field) have given explicit resolutions of $\overline{D/\Gamma}$. Independently, Satake [9] and I working in collaboration with Y. Tai, M. Rapaport and A. Ash have attacked the general case, using closely related methods. The purpose of this paper is to give a very short outline of my approach. It builds in an essential way on the construction of "torus embeddings". a theory which has been published in the Springer Lecture Notes [8] (by G. Kempf, F. Knudsen, B. Saint-Donat and myself; some of which has been worked out independently by M. Demazure [3] and M. Hochster [5]). We intend to publish full details of the present research as soon as possible in a sequel "Toroidal Embeddings II" to the Notes [8]. At the present time, however, we cannot claim to have written down complete proofs of our "Main Theorem" and although I definitely believe it is true and not difficult, it is more accurate to describe the ideas below only as a suggested approach to the problem of constructing a non-singular compactification of D/Γ.

1. Let us look first at the familiar case:

$$D = \mathfrak{M}_1$$
$$\Gamma = SL(2, \mathbf{Z}).$$

We know that $D/\Gamma \cong \mathbf{C}$ via the elliptic modular function j and adding one point $j = \infty$, we get the unique non-singular compactification:

$$\begin{array}{ccc} D/\Gamma & \subset & \overline{D/\Gamma} \\ \| & & \| \\ \mathbf{C} & \subset & \mathbf{CP}^1. \end{array}$$

However, let me describe a way of glueing in the point at ∞ that will suggest generalizations:

STEP a. Factor the map $\mathfrak{M}_1 \xrightarrow{j} \mathbf{C}$ as follows

$$\begin{array}{ccccc} \mathfrak{M}_1 & \xrightarrow{\alpha} & \Delta_1^* & \xrightarrow{\beta} & \mathbf{C} \\ \| & & \| & & \\ \{\omega \mid \operatorname{Im}\omega > 0\} & & \{\zeta \mid 0 < |\zeta| < 1\} & & \end{array}$$

where $\zeta = e^{2\pi i\omega}$. If $\Gamma_0 = \left\{ \begin{pmatrix} 1 & b \\ 0 & 1 \end{pmatrix} \middle| b \in \mathbf{Z} \right\} \subset SL(2, \mathbf{Z})$, then

$$\Delta_1^* \cong \mathfrak{M}_1/\Gamma_0.$$

STEP b. *Partially* compactify Δ_1^* by adding the origin

$$\Delta_1^* \subset \Delta_1 = \{\zeta \,\big|\, |\zeta| < 1\}.$$

STEP c. Note that if

$$\mathfrak{M}_1(c) = \{\omega \,\big|\, \operatorname{Im}\omega > c\},$$

then if c is large enough, $SL(2, \mathbf{Z})$-equivalence in $\mathfrak{M}_1$ reduces, in $\mathfrak{M}_1(c)$, to Γ_0-equivalence:

$$\left.\begin{array}{l} \omega_1, \omega_2 \in \mathfrak{M}_1(c) \\ \omega_1 = \gamma(\omega_2),\ \gamma \in SL(2, \mathbf{Z}) \end{array}\right\} \Rightarrow \gamma \in \Gamma_0. \qquad (*)$$

Now $\mathfrak{M}_1(c)$ maps to Δ_b^*, where

$$\Delta_b^* = \{\zeta \,\big|\, 0 < |\zeta| < b\}$$
$$b = e^{-2\pi c},$$

and (*) says:

$$\operatorname{res}\beta : \Delta_b^* \longrightarrow \mathbf{C} \text{ is injective.}$$

Step d. This gives us the situation:

$$\begin{array}{lll} & & \mathbf{C} \\ & \overset{\text{res }\beta}{\nearrow} & \\ \Delta_b^* & & \\ & \searrow & \\ & & (\text{Interior of closure of } \Delta_b^* \text{ in } \Delta_1) = \Delta_b \end{array}$$

It is easy to see that $\mathbf{CP}^1$ is nothing but the union of $\mathbf{C}$ and Δ_b, glued on Δ_b^*.

2. Let us look next at how this procedure can be generalized to the $n \times n$ Siegel case:

$$D = \mathfrak{M}_n = \left\{ \Omega \,\middle|\, \begin{array}{l} \Omega \quad n \times n \text{ complex symmetric matrix,} \\ \text{Im } \Omega \text{ positive definite} \end{array} \right\},$$

$$\Gamma = Sp(2n, \mathbf{Z}).$$

Actually, it is usually more convenient to replace $Sp(2n, \mathbf{Z})$ by a subgroup of finite index, or else to allow V-manifold-type singularities on D/Γ and $\overline{D/\Gamma}$, because of the elements of finite order in Γ which need not act freely. We will ignore this technicality. Of course, these V-manifold singularities can also be resolved: but that involves a totally different set of problems.

Step a′: Factor $\mathfrak{M}_n \longrightarrow \mathfrak{M}_n/\Gamma$ as follows:

$$\mathfrak{M}_n \longrightarrow \mathfrak{Z}_n^0 \longrightarrow \mathfrak{M}_n/\Gamma$$

$$\Big\| $$

$$\left\{ Z \,\middle|\, \begin{array}{l} Z \;\; n \times n \text{ complex symmetric matrix, } Z_{ij} \neq 0 \\ \text{and} -\log |Z_{ij}| \text{ positive definite} \end{array} \right\}$$

where $Z_{ij} = e^{2\pi i(\Omega_{ij})}$. If $\Gamma_0 = \left\{ \begin{pmatrix} I_n & B \\ 0 & I_n \end{pmatrix} \,\middle|\, \begin{array}{l} B \in M_n(\mathbf{Z}) \\ \text{and symmetric} \end{array} \right\} \subset \Gamma,$

then

$$\mathfrak{Z}_n^0 \simeq \mathfrak{M}_n/\Gamma_0.$$

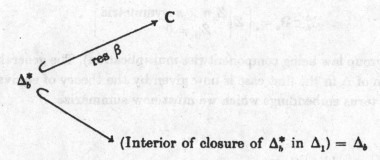

STEP b′: Note that $\mathfrak{I}_n^0$ is an open set in the algebraic torus group:

$$\mathfrak{I}_n^0 \subset \mathfrak{I}_n = \left\{ Z \left| \begin{matrix} Z\ n \times n \text{ symmetric} \\ Z_{ij} \neq 0 \end{matrix} \right. \right\}$$

(the group law being componentwise multiplication). The generalization of Δ in the first case is now given by the theory of equivariant torus embeddings which we must now summarize

Torus embedding theory :

T = an algebraic torus of dimension n, *i.e.*, $\simeq (\mathbf{C}^*)^n$,

$N = \pi_1(T)$ a free abelian group of rank n, $N_{\mathbf{R}} = N \otimes R$,

$N_{\mathbf{C}} = N \otimes \mathbf{C}$ so that

$T \simeq N_{\mathbf{C}}/N$ (via exp : $N_{\mathbf{C}} \longrightarrow T$),

$M = \mathrm{Hom}(N, \mathbf{Z}) \simeq$ [the group of characters $\mathscr{X} : T \longrightarrow \mathbf{C}^*$].

If $\alpha \in M$, write X^α: $T \longrightarrow \mathbf{C}^*$ for the associated character write $\langle x, a \rangle$ for the pairing of $M_{\mathbf{R}}$ and $N_{\mathbf{R}}$.

$\forall\ \sigma$ = closed convex rational polyhedral cone in $N_{\mathbf{R}}$ (*i.e.*, $\sigma = \{x \in N_{\mathbf{R}} \mid \langle x, \alpha_i \rangle \geqslant 0,\ 1 \leqslant i \leqslant N\}$ for some finite set of points $\alpha_i \in M$), which are "proper": $\sigma \not\supset$ pos. dim. subspace of $N_{\mathbf{R}}$

$X_\sigma \underset{\text{def}}{=} \mathrm{Spec}\,\{\mathbf{C}[\ldots, X^\alpha, \ldots]_{\alpha \in M \cap \check{\sigma}}\}$, $\check{\sigma}$ = dual of σ in $M_{\mathbf{R}}$.

Then X_σ is a normal affine variety, T is an open subset of X_σ and the action of T on itself extends to an action of T on X_σ; moreover all such embeddings $T \subset X$ arise like this. X_σ is non-singular if σ is a simplicial cone generated by a subset of a basis of N.

$\forall \{\sigma_\alpha\}$ = collection of such σ's such that every face of a σ_α is some σ_β, and every intersection $\sigma_\alpha \cap \sigma_\beta$ is a common face,

$X_{\{\sigma_\alpha\}} \underset{\text{def}}{=} \bigcup X_{\sigma_\alpha}$, where X_{σ_α} and X_{σ_β} are glued along $X_{\sigma_\alpha \cap \sigma_\beta}$ (which is, in fact, an open subset of each).

Then $X_{\{\sigma_\alpha\}}$ is an irreducible separated normal scheme, locally of finite type over $\mathbf{C}$, T is an open subset of $X_{\{\sigma_\alpha\}}$ and the action of T extends; again all such embeddings arise like this (Sumihiro [10]).

Let $N_{n,\mathbf{R}}$ = vector space of real $n \times n$ symmetric matrices,
N_n = lattice of integral ones,
C_n = cone of positive semi-definite matrices.

Then for all collections $\{\sigma_\alpha\}$, $\sigma_\alpha \subset N_{n,\mathbf{R}}$ being closed convex rational polyhedral proper cones, fitting together as above, we get

$$\mathfrak{Z}_n^0 \subset \mathfrak{Z}_n \subset X_{\{\sigma_\alpha\}}.$$

It can be shown that for all α:

$$(*)\quad \left\{\begin{array}{l} \exists\, x \in X_{\sigma_\alpha} - \bigcup_{\beta \text{ face of } \alpha} X_{\sigma_\beta} \text{ and a neighborhood} \\ U \text{ of } x \text{ in } X_{\sigma_\alpha} \text{ such that } U \subset \overline{\mathfrak{Z}_n^0} \\ \text{if and only if } \sigma_\alpha \subset C_n. \end{array}\right.$$

For this reason, we assume that $\sigma_\alpha \subset C_n$, all α, and define

$$\mathfrak{Z}_{n,\{\sigma_\alpha\}}^0 = \text{Interior of closure of } \mathfrak{Z}_n^0 \text{ in } X_{\{\sigma_\alpha\}}.$$

Then

$$\mathfrak{Z}_n^0 \subset \mathfrak{Z}_{n,\{\sigma_\alpha\}}^0$$

is the partial compactification of $\mathfrak{Z}_n^0$ which we shall use.

STEP c′. There seem to be several choices for an n-dimensional analog of $\mathfrak{M}_1(c)$ but we take :

$$\mathfrak{M}_n(c) = \left\{ \Omega \in \mathfrak{M}_n \,\middle|\, \begin{array}{l} \forall\, k \in \mathbf{Z}^n,\ k \neq (0), \\ {}^t k \cdot (\operatorname{Im} \Omega) \cdot k > c. \end{array} \right\}$$

By Siegel-Minkowski reduction theory, it can be shown that if c is large enough, $Sp(2n, \mathbf{Z})$-equivalence in $\mathfrak{M}_n$ reduces, in $\mathfrak{M}_n(c)$, not to Γ_0-equivalence but to Γ_1-equivalence; where :

$$\Gamma_1 = \left\{ \begin{pmatrix} A & B \\ 0 & {}^t A^{-1} \end{pmatrix} \,\middle|\, \begin{array}{l} A \in GL(n, \mathbf{Z}),\ B \in M_n(\mathbf{Z}) \text{ and} \\ A^{-1} B \text{ symmetric} \end{array} \right\}$$

$$= \left\{ \begin{pmatrix} A & B \\ 0 & D \end{pmatrix} \,\middle|\, \begin{array}{c} A, D \in GL(n, \mathbf{Z}) \\ B \in M_n(\mathbf{Z}) \end{array} \right\} \cap Sp(2n, \mathbf{Z}).$$

Let $\mathfrak{Z}_n^c$ = image of $\mathfrak{M}_n(c)$ in $\mathfrak{Z}_n$.

Then $\Gamma_1/\Gamma_0 \cong GL(n, \mathbf{Z})$ acts on the torus $\mathfrak{I}_n$, preserving the open subsets $\mathfrak{I}_n^0$ and $\mathfrak{I}_n^C$, and we get the situation :

$$\begin{array}{ccccc} \mathfrak{I}_n^C & \subset & \mathfrak{I}_n^0 & \subset & \mathfrak{I}_n \\ \downarrow & & \downarrow & & \\ \mathfrak{I}_n^C/(\Gamma_1/\Gamma_0) & \subset & \mathfrak{I}_n^0/(\Gamma_1/\Gamma_0) & & \\ & \searrow & \downarrow & & \\ & & \mathfrak{M}_n/\Gamma & & \end{array}$$

where the dotted arrow is injective.

STEP d′ : It is now clear how to finish up : we must assume that the collection $\{\sigma_\alpha\}$ satisfies the condition $\forall\ \gamma \in \Gamma_1/\Gamma_0$, $\forall_\alpha$, $\exists \beta$ such that $\gamma\sigma_\alpha = \sigma_\beta$ (under the natural action of $\Gamma_1 \mid \Gamma_0$ on $\mathfrak{I}_n$, hence on $N_{n,\mathbf{R}}$). Then the action of Γ_1/Γ_0 on $\mathfrak{I}_n$ extends to $X_{\{\sigma_\alpha\}}$. Define

$$\mathfrak{I}^C_{n\ \{\sigma_\alpha\}} = \text{Interior of closure of } \mathfrak{I}_n^C \text{ in } X_{\{\sigma_\alpha\}}Z$$

and consider

$$\mathfrak{I}_n^C/(\Gamma_1/\Gamma_0) \begin{array}{l} \hookrightarrow \mathfrak{M}_n/\Gamma \\ \hookrightarrow \mathfrak{I}^C_{n,\{\sigma_\alpha\}}/(\Gamma_1/\Gamma_0) \end{array}$$

and glue !

Some comments : First of all, there are many things to check in the above procedure, but we will not try to justify them here. Secondiy, this glueing alone will never give us something compact ; but what I do claim is that if you take just enough σ_α's, in the sense :

(a) $\bigcup \sigma_\alpha = \left\{ Z \in N_{n,\mathbf{R}} \;\middle|\; \begin{array}{l} Z \text{ positive semi-definite with null-} \\ \text{space defined over } \mathbf{Q} \end{array} \right\}$

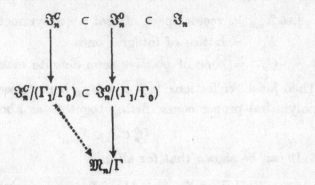

(b) Modulo Γ_1/Γ_0, the set of σ_α's is finite, then the resulting partial compactification of $\mathfrak{M}_n/\Gamma$ does cover the entire 0-dimensional boundary component; and moreover it "analytically prolongs" to a compactification $\overline{\mathfrak{M}_n/\Gamma}$ of $\mathfrak{M}_n/\Gamma$ in the following way :

(a) let

$$\mathfrak{I}^o_{n,\{\sigma_\alpha\}} = \text{Interior of closure of } \mathfrak{I}^o_n \text{ in } X_{\{\sigma_\alpha\}},$$

(b) require the existence of a map π:

$$\begin{array}{ccc} \mathfrak{I}^o_n & \subset & \mathfrak{I}^o_{n,\{\sigma_\alpha\}} \\ \downarrow & & \vdots\ \pi \\ \mathfrak{M}_n/\Gamma & \subset & \overline{\mathfrak{M}_n/\Gamma} \end{array}$$

where π is surjective and open and $\mathfrak{M}_n/\Gamma$ is open and dense in $\overline{\mathfrak{M}_n/\Gamma}$.

Thirdly, the resulting space is not necessarily non-singular. However, if the σ_α's are chosen satisfying :

(c) $\forall\alpha$, σ_α is the set of positive linear combinations of matrices $A_1, \ldots, A_k \in N_n$, which are part of a basis of the free abelian group N_n,

and if $Sp(2n, \mathbf{Z})$ is replaced by a subgroup Γ of finite index without elements of finite order, *then* we get a non-singular compactification of $\mathfrak{M}_n/\Gamma$. Even without these two conditions, the singularities are quite mild, e.g., rational (and if the σ_α's are merely simplicial cones, the singularities are V-manifolds). Fourthly, an objection may be raised that there is still a huge amount of freedom in the choice of the σ_α's, leading to a whole family of non-singular compactifications rather than one best possible one. This in fact discouraged me for several years and made me think the theory was not useful (the simplest case I know where this non-uniqueness seems really basic is $\mathfrak{M}_4$). But I believe now that this non-uniqueness is a fact of life of higher-dimensional birational geometry and that for many applications, this class of compactifications is just as usable as one canonical one would be.

3. Finally, and with many more gaps, let me sketch how I believe this procedure extends to the general case :

$$D = \text{any bounded symmetric domain,}$$
$$\Gamma = \text{an arithmetic subgroup of } \mathrm{Aut}(D).$$

Assume for simplicity that Γ has no elements of finite order.

STEP a″. For every rational boundary component F, we get groups:

$$\mathrm{Aut}(D)^0 \supset N(F) \supset U(F)$$

where

$$N(F) = \left\{ g \in \mathrm{Aut}(D^0) \middle| gF = F \right\}$$

$$U(F) = \text{center of unipotent radical of } N(F) :$$
this is just a real vector space under addition.

Let $\Gamma_0 = \Gamma \cap U(F)$: a lattice in $U(F)$ and $\Gamma_1 = \Gamma \cap N(F)$.

We factor $D \longrightarrow D/\Gamma$ via :

$$D \longrightarrow D/\Gamma_0 \longrightarrow D/\Gamma_1 \longrightarrow D/\Gamma.$$

STEP b″: To describe D relative to F suitably, embed D in $\check{D}$, its compact dual, so that the complexification $G_{\mathbf{C}}$ of $\mathrm{Aut}(D)^0$ acts on $\check{D}$. Moving D around only by $U(F)_{\mathbf{C}}$, we get an intermediate open set :

$$D \subset U(F)_{\mathbf{C}} \cdot D \subset \check{D}.$$

This gives us a description of D as a *Siegel Domain of 3rd kind* as follows : I claim

$$U(F)_{\mathbf{C}} \cdot D \cong U(F)_{\mathbf{C}} \times \mathscr{E}(F)$$

for some complex vector bundle $p : \mathscr{E}(F) \to F$ over F itself (the isomorphism being complex analytic and taking the action of $U(F)_{\mathbf{C}}$ on the left to translations in the 1st factor on the right), and that this isomorphism restricts to:

$$D \cong \left\{ (u,x) \middle| \operatorname{Im} u \in C(F) + h(x) \right\}$$

for some open convex cone $C(F) \subset U(F)$ and real analytic map $h : \mathscr{E}(F) \to U(F)$. Let $T(F) = U(F)_{\mathbf{C}}/\Gamma_0$: this is an algebraic torus group over $\mathbf{C}$. We get

$$D/\Gamma_0 \subset (U(F)_{\mathbf{C}}\cdot D)/\Gamma_0 \simeq T(F) \times \mathscr{E}(F).$$

We now choose a collection $\{\sigma_\alpha\}$ of rational polyhedral cones in $\overline{C(F)}$ and note that these define, by our general theory, an embedding

$$T(F) \subset X_{\{\sigma_\alpha\}}.$$

Define

$$(D/\Gamma_0)_{\{\sigma_\alpha\}} = \text{Interior of closure of } D/\Gamma_0 \text{ in } X_{\{\sigma_\alpha\}} \times \mathscr{E}(F).$$

Step c″. If $\gamma \in C(F)$ and $K \subset F$ is a compact set, let

$$D(\gamma, K) = \Gamma_1\cdot\left\{ (u,x) \,\middle|\, \operatorname{Im} u \in C(F) + h(x) + \gamma,\ p(x) \in K \right\}.$$

Then, I believe, *for all K, if c is large enough*, the composition

$$D(c,K)/\Gamma_1 \hookrightarrow D/\Gamma_1 \longrightarrow D/\Gamma$$

is injective.

Step d″. Assume now that the collection $\{\sigma_\alpha\}$ satisfies the conditions:

(a) $\forall\, \gamma \in \Gamma_1/\Gamma_0$, and $\forall \sigma_\alpha$, $\gamma\sigma_\alpha =$ some σ_β; and modulo this action, there are only finitely many σ_α's.

(b) $C(F) \subset \bigcup_\alpha \sigma_\alpha \subset \overline{C(F)}$.

It requires proof at this point that such $\{\sigma_\alpha\}$ exist — this seems quite likely. Define

$$(D(c,K)/\Gamma_0)_{\{\sigma_\alpha\}} = \text{Interior of closure of } D(c,K)/\Gamma_0 \text{ in } X_{\{\sigma_\alpha\}} \times \mathscr{E}(F)$$

and consider :

$$\begin{array}{ccc} & & D/\Gamma \\ & \nearrow & \\ D(c,K)/\Gamma_1 & & \\ & \searrow & \\ & & (D(c,K)/\Gamma_0)_{\{\sigma_\alpha\}}/(\Gamma_1/\Gamma_0) \end{array}$$

and glue! The whole set-up is summarized in the figure overleaf.

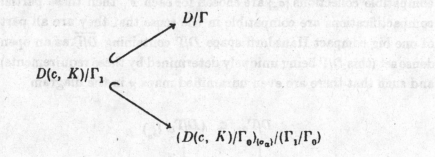

Summary of spaces

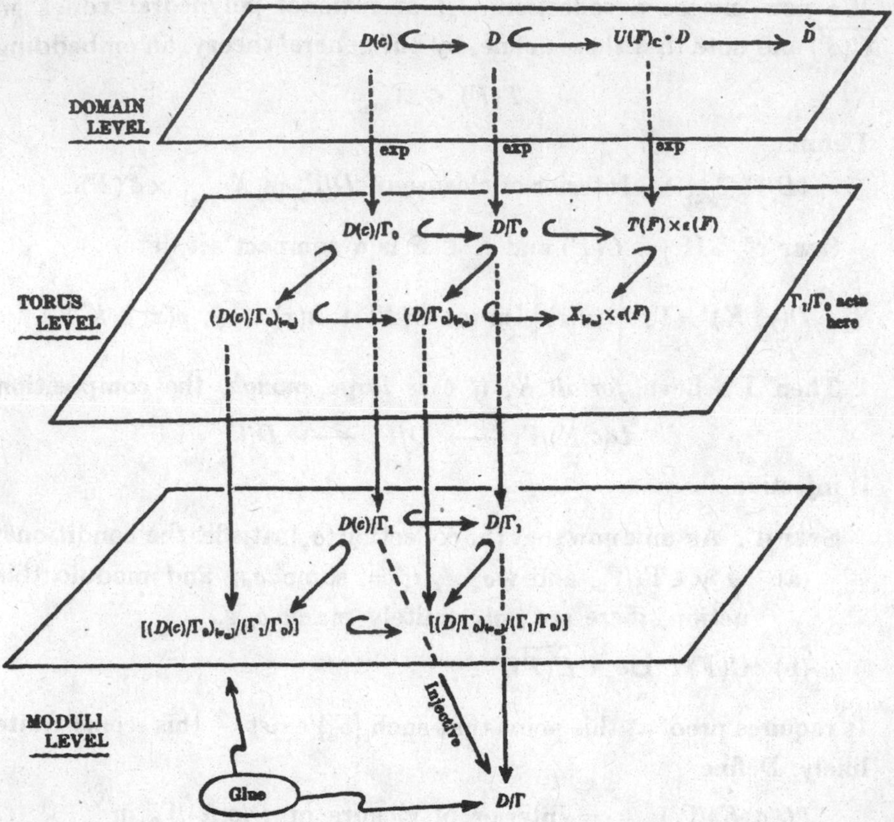

Step e′. Finally, if we let F range over the finite set of Γ-inequivalent rational boundary components, we must check that if suitably compatible collections $\{\sigma_\alpha\}$ are chosen for each F, then these partial compactifications are compatible in the sense that they are all part of one big compact Hausdorff space D/Γ containing $\overline{D/\Gamma}$ as an open dense set (this D/Γ being uniquely determined by these requirements) and such that there are even unramified maps π in the diagram

$$\begin{array}{ccc} D/\Gamma_0 & \subset & (D/\Gamma_0)_{\{\sigma_\alpha\}} \\ \downarrow & & \vdots\ \pi \\ D/\Gamma & \subset & \overline{D/\Gamma} \end{array}$$

The compatibility of the $\{\sigma_\alpha\}$'s can be expressed as follows :

Say $F_1, F_2 \subset \bar{D}$ are 2 rational boundary components and $F_1 \subset \bar{F}_2$. Then

$$U(F_1) \supset U(F_2) \quad \text{and} \quad C(F_2) \cong \text{face of } \overline{C(F_1)}.$$

Then we require that the set of cones $\sigma_\alpha^{(2)} \subset \overline{C(F_2)}$ be exactly the set $\sigma_\alpha^{(1)} \cap \overline{C(F_2)}$.

4. In order to express more clearly what our compactification depends on, and to relate it to the theory of toroidal embeddings ([8], Ch. II), it is convenient to introduce the following interesting abstract cone :

$$\Sigma = \Big(\coprod_{\substack{\text{rat. boundary}\\ \text{Comp. } F}} C(F)\Big)/\Gamma = \coprod_{\substack{\Gamma\text{-equiv. classes}\\ \text{of rat. } F}} \big(C(F)/\Gamma \cap N(F)\big)$$

where $\gamma \in \Gamma$ acts on $\coprod_F C(F)$ by the natural maps $C(F) \xrightarrow{\approx} C(\gamma F)$ for all F. To express the structure that Σ has, we use the definition.

DEFINITION. *A 'conical polyhedral complex'*† *is a topological space X, plus a finite stratification $\{S_\alpha\}$ of X, (i.e., a partition of X into disjoint locally closed pieces S_α such that each $\bar{S}_\alpha$ is a union of various*

†This definition is a slight modification of that used in [8] to allow 2 faces of the same polyhedra to be identified. Thus

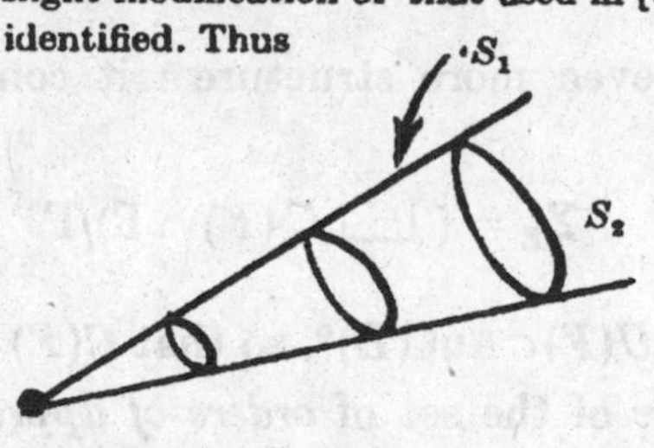

is allowed, as well as the previously allowed :

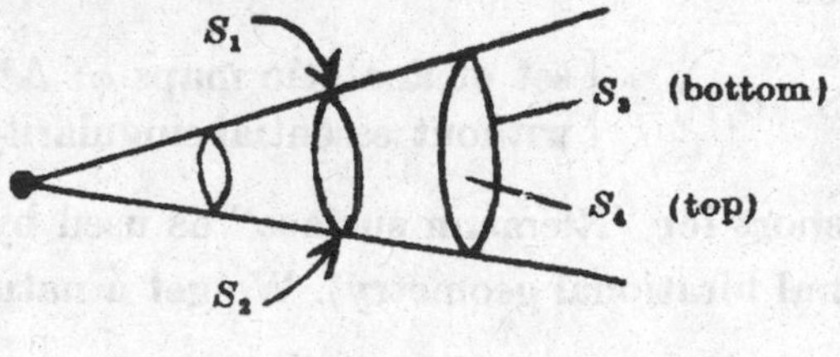

S_β's), plus for each α a finite-dimensional vector-space V_α of real-valued continuous functions on S_α such that :

(a) *if $n_\alpha = \dim V_\alpha$ and $f_1, \ldots, f_{n_\alpha}$ is a basis of V_α, then $(f_i) : S_\alpha \to \mathbf{R}^{n_\alpha}$ is a homeomorphism of S_α with an open convex polyhedral cone $C_\alpha \subset \mathbf{R}^{n_\alpha}$,*

(b) *$(f_i)^{-1}$ extends to a continuous surjective map*

$$(f_i)^{-1} : \bar{C}_\alpha \to \bar{S}_\alpha$$

mapping the open faces $C_\alpha^{(\beta)}$ of $\bar{C}_\alpha$ (= closed faces less their own faces) homeomorphically to the strata S_β in $\bar{S}_\alpha$ and inducing isomorphisms

$$\operatorname{res}_{C_\alpha^{(\beta)}} (\text{lin. fcns. } on\ \mathbf{R}^{n_\alpha}) \xrightarrow{\approx} V_\beta.$$

Now for each compatible set of decompositions $\{\sigma_{\alpha,F}\}$, Σ becomes a conical polyhedral complex : just take the S_α's to be the images of the sets ($\sigma_{\alpha,F}$—faces of $\sigma_{\alpha,F}$). In particular, this makes Σ into a topological space with piecewise-linear structure; these structures are easily seen to be independent of the choice of $\{\sigma_{\alpha,F}\}$'s. Note that conversely, the structure $\{S_\alpha, V_\alpha\}$ of conical polyhedral complex on Σ determines the $\{\sigma_{\alpha,F}\}$'s: they are just the closures of the connected components of the inverse images in the various $C(F)$'s of the strata S_α. We shall call the structures $\{S_\alpha, V_\alpha\}$ on Σ that arise from choices of $\{\sigma_{\alpha,F}\}$'s, *admissible conical polyhedral subdivisions of* Σ.

Moreover Σ has even more structure : it contains the abstract "lattice"

$$\Sigma_{\mathbf{Z}} = \Big(\coprod_F C(F) \cap \Gamma\Big)/\Gamma$$

(here regard $C(F) \subset U(F) \subset \operatorname{Aut}(D)^0$, so that $C(F) \cap \Gamma$ makes sense), which plays the role of the set of *orders of approach to* ∞ *in* D/Γ. In fact, let

$$\text{R. S.}(D/\Gamma) = \left\{ \begin{array}{l} \text{set of analytic maps } \varphi\colon \Delta^* \to D/\Gamma \\ \text{without essential singularity at } 0 \in \Delta \end{array} \right\}.$$

(R. S. is short for "Riemann surface" as used by Zariski in higher-dimensional birational geometry). We get a natural surjective map:

$$\text{ord}: \text{R.S.}\,(D/\Gamma) \to \Sigma_{\mathbf{Z}}$$

by the procedure: lift φ to $\tilde{\varphi}: H \to D$, $H = \{z \mid \operatorname{Im} z > 0\}$, such that

$$\tilde{\varphi}(z) \bmod \Gamma = \varphi(e^{2\pi i z})$$

hence for some $\gamma_0 \in \Gamma$:

$$\tilde{\varphi}(z+1) = \gamma_0 \tilde{\varphi}(z), \qquad \forall z \in H.$$

Then γ_0 can be shown to lie in $C(F)$ for some F, hence it determines an element of $\Sigma_{\mathbf{Z}}$.

We can now state the main result we hope to prove :

Main Theorem (?). *Let D be a bounded symmetric domain, $\Gamma \subset \operatorname{Aut} D^0$ an arithmetic group without elements of finite order, and Σ the piecewise-linear topological space defined by D and Γ as above. Then there is a map*

$$\left|\begin{array}{c} \textit{Admissible conical} \\ \textit{polyhedral subdivisions} \\ \{S_\alpha, V_\alpha\} \textit{ of } \Sigma \end{array}\right| \longmapsto \left|\begin{array}{l} \textit{toroidal embeddings} \\ D/\Gamma \subset \overline{D/\Gamma}\textit{, where} \\ \overline{D/\Gamma} \textit{ is a compact} \\ \textit{algebraic space} \end{array}\right|$$

such that if $\Sigma(\overline{D/\Gamma})$ is the conical polyhedral complex associated by the theory in [8] *to this toroidal embedding*†, *there is a unique isomorphism φ making the diagram*

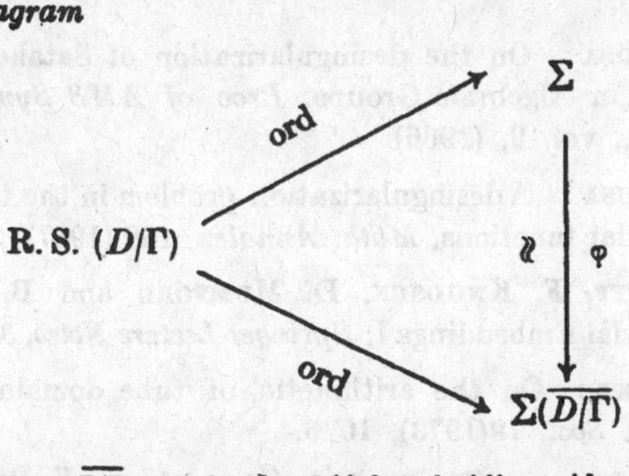

†In general, $D/\Gamma \subset \overline{D/\Gamma}$ may be a toroidal embedding *with* self-intersection, however, it is without twisting in the sense that for all strata T, the branches of $\overline{D/\Gamma} - D/\Gamma$ through T are not permuted by going around loops in T. This makes it possible to associate a complex of the type defined above to this embedding by a generalization of the procedure in ([8], Ch. II).

commute. The map $\overline{is\ a}$ functor in the $\overline{sense}$ that if $(\text{subd})_1$ *is finer than* $(\text{subd})_2$, *then* $(\overline{D/\Gamma})^{(1)}$ *dominates* $(\overline{D/\Gamma})^{(2)}$. *An integrality condition on the subdivision* $\{S_\alpha, V_\alpha\}$ *characterizes which* $\overline{D/\Gamma}$*'s are non-singular* (see p. 11 above).

I also expect that certain convexity properties of the subdivision imply $\overline{D/\Gamma}$ projective.

REFERENCES

1. M. Artin: Algebraic Spaces, *Yale Math. Monographs.*
2. A. Borel and W. Baily: Compactification of arithmetic quotients of bounded symmetric domains, *Annals of Math.*, 84 (1966), 442-528.
3. M. Demazure: Sous-groups algébriques des rang maximum du groupe de Cremona, *Annales de ENS*, 3 (1970).
4. F. Hirzebruch: Hilbert modular surfaces, *L'Enseigement Math.* (1973).
5. M. Hochster: Rings of invariants of tori, Cohen-Macauley rings generated by monomials and polytopes, *Annals of Math.* 96, (1972).
6. J.-I. Igusa: On the desingularization of Satake compactifications in Algebraic Groups, *Proc. of AMS Symposia in Pure Math.*, vol. 9, (1966).
7. J.-I. Igusa: A desingularization problem in the theory of Siegel modular functions, *Math. Annalen*, 168 (1967), 228-260.
8. G. Kempf, F. Knudsen, D. Mumford and B. Saint-Donat: Toroidal Embeddings I, *Springer Lecture Notes*, 339, (1973).
9. I. Satake: On the arithmetic of tube domains, *Bull. Amer. Math. Soc.*, 79(1973), 1076.
10. H. Sumihiro: Equivariant Completion, *Kyoto Math. J.*, 14. (1974).

Inventiones math. 42, 239 – 272 (1977)

Inventiones mathematicae

Hirzebruch's Proportionality Theorem in the Non-Compact Case

D. Mumford

Harvard University, Department of Mathematics Cambridge Massachusetts 02138, USA

Dedicated to Friedrich Hirzebruch

In the conference on Algebraic Topology [7] in 1956, F. Hirzebruch described a remarkable theorem relating the topology of a compact locally symmetric variety:

$X = D/\Gamma$,
D = bounded symmetric domain,
Γ = discrete torsion-free co-compact group of automorphisms of D

with the topology of the extremely simply rational variety $\check{D}$, the "compact dual" of D. (See §3 for full definitions.) His main result is that the Chern numbers of X are proportional to the Chern numbers of $\check{D}$, the constant of proportionality being the volume of X (in a natural metric). This is a very useful tool for analyzing the structure of X. Many of the most interesting locally symmetric varieties that arise however are not compact: they have "cusps". It seems a priori very plausible that Hirzebruch's line of reasoning should give some relation even in the non-compact case between the chern numbers of X and of $\check{D}$, with some correction terms for the cusps. The purpose of this paper is to show that this is indeed the case. We hope that the generalization that we find will have applications.

The paper is organized as follows. In §1, we make a few general definitions and observations concerning Hermitian metrics on bundles with poles and describe an instance where such metrics still enable one to calculate the Chern classes of the bundle. This section is parallel to work of Cornalba and Griffiths [6]. In §2, which is the most technical, we prove a series of estimates for a class of functions on a convex self-adjoint cone. In §3, the results of §1 and §2 are brought together, and the Proportionality Theorem is proven. One consequence is that D/Γ has the property, defined by Iitaka [8], of being of logarithmic general type. Finally, in §4, we analyze the step from logarithmic general type to general type and reprove a Theorem of Tai that D/Γ is of general type if Γ is sufficiently small.

§ 1. Singular Hermitian Metrics on Bundles

In this section we will not be concerned specifically with the locally symmetric algebraic varieties D/Γ, but with general smooth quasi-projective algebraic varieties X. When X is not compact, we want to study the order of poles of differential forms on X at infinity, and when E is moreover a vector bundle on X, we want to study Hermitian metrics on E which also "have poles at infinity". This situation has been studied by Cornalba-Griffiths [6]. The following idea of bounding various forms by local Poincaré metrics on punctured polycylinders at infinite is due to them. More precisely, we choose a smooth projective compactification $\bar{X}$:

$$X \subset \bar{X}$$

where $\bar{X}-X$ is a divisor on $\bar{X}$ with normal crossings.

Then we look at polycylinders:

$$\Delta^r \subset \bar{X} \quad \begin{pmatrix} \Delta = \text{unit disc} \\ r = \dim \bar{X} \end{pmatrix}$$

$$\text{where } \Delta^r \cap (\bar{X}-X) = \begin{Bmatrix} \text{union of coordinate hyperplanes} \\ z_1=0,\ z_2=0,\ \ldots,\ z_k=0 \end{Bmatrix}$$

hence:

$$\Delta^r \cap X = (\Delta^*)^k \times \Delta^{r-k}.$$

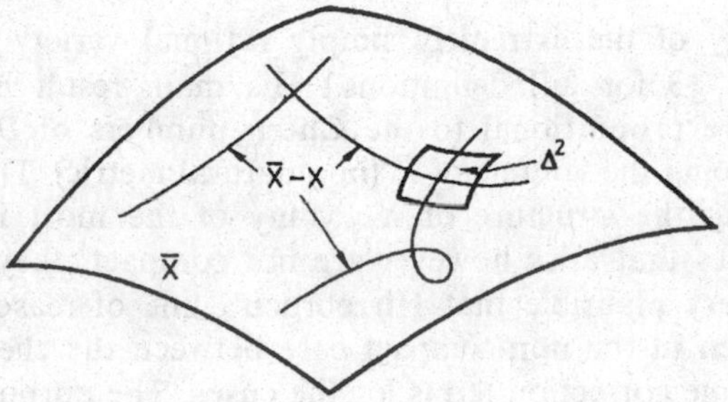

In Δ^* we have the Poincaré metric:

$$ds^2 = \frac{|dz|^2}{|z|^2(\log|z|)^2}$$

and in Δ we have the simple metric $|dz|^2$, giving us a product metric on $(\Delta^*)^k \times \Delta^{r-k}$ which we call $\omega^{(p)}$.

Definition. A complex-valued C^∞ p-form η on X is said to have *Poincaré growth* on $\bar{X}-X$ if there is a set of polycylinders $U_\alpha \subset \bar{X}$ covering $\bar{X}-X$ such that in each U_α, an estimate of the following type holds:

$$|\eta(t_1, \ldots, t_p)|^2 \leqq C_\alpha \, \omega^{(p)}_{U_\alpha}(t_1, t_1) \cdot \cdots \cdot \omega^{(p)}_{U_\alpha}(t_p, t_p)$$

(all $t_1, \ldots, t_p$ tangent vectors to $\bar{X}$ at some point of $U_\alpha \cap X$).

It is not hard to see that this property is independent of the covering U_α of $\bar{X} - X$ (but unfortunately it does depend on the compactification $\bar{X}$). Moreover, if η_1, η_2 both have Poincaré growth on $\bar{X} - X$, then so does $\eta_1 \wedge \eta_2$. This leads to the basic property:

Proposition 1.1. *A p-form η with Poincaré growth on $\bar{X} - X$ has the property that for every $C^\infty (r-p)$-form ζ on $\bar{X}$,*

$$\int_{\bar{X}-X} |\eta \wedge \zeta| < +\infty$$

hence η defines a p-current $[\eta]$ on $\bar{X}$.

Proof. Since ζ has Poincaré growth, we are reduced to checking that if η is an r-form with Poincaré growth, then

$$\int_{\bar{X}} |\eta| < +\infty.$$

In a polycylinder U_α, this amounts to the well-known fact that for all relatively compact $V \subset\subset U_\alpha$, the Poincaré metric volume of $V \cap (\Delta^{*k} \times \Delta^{r-k})$ is finite. QED

Definition. A complex-valued C^∞ p-form η on X is *good on* $\bar{X}$ if both η and $d\eta$ have Poincaré growth.

The set of all good forms η is differential graded algebra for which we have the next basic property:

Proposition 1.2. *If η is a good p-form, then*

$$d([\eta]) = [d\eta].$$

Proof.[1] By definition of $d([\eta])$, this means that for all $C^\infty (r-p-1)$-forms ζ on $\bar{X}$,

$$\int_{\bar{X}-X} d\eta \wedge \zeta = - \int_{\bar{X}-X} \eta \wedge d\zeta.$$

This comes down to asserting that if U_ε is a tube of radius ε around $\bar{X} - X$. then

$$\lim_{\varepsilon \to 0} \int_{\partial U_\varepsilon} (\eta \wedge \zeta) = 0.$$

If we take, for instance, $r=2$ and set up this integral in local coordinates x, y on $\bar{X}$ near a point where $\bar{X} - X$ has 2 branches $x=0$ and $y=0$, then this comes

[1] Compare Cornalba-Griffiths [6], p. 25

down to the assertion

$$\lim_{\varepsilon\to 0} \int_{\substack{|x|\geq\varepsilon \\ |y|=\varepsilon}} \frac{|dx|^2}{|x|^2(\log|x|)^2}\cdot\frac{|dy|}{|y|\cdot|(\log|y|)|}=0$$

which is easy to check. The general case is similar. QED

Next, let $\bar{E}$ be an analytic rank n vector bundle on $\bar{X}$, let E be the restriction of $\bar{E}$ to X and let $h: E\to\mathbb{C}$ be a Hermitian metric on E. For such h we define "good" as follows:

Definition. A Hermitian metric h on E is *good on* $\bar{X}$ if for all $x\in\bar{X}-X$ and all bases $e_1, \ldots, e_n$ of $\bar{E}$ in a neighborhood Δ^r of x in which $\bar{X}-X$ is given as above by $\prod_{i=1}^{k} z_i=0$, if $h_{ij}=h(e_i, e_j)$, then

i) $|h_{ij}|, (\det h)^{-1}\leq C\left(\sum_{i=1}^{k}\log|z_i|\right)^{2n}$, for some $C>0, n\geq 1$,

ii) the 1-forms $(\partial h\cdot h^{-1})_{ij}$ are good on $\bar{X}\cap U$.

The first point about good Hermitian metrics is that given (E, h), there is at most one extension $\bar{E}$ of E to $\bar{X}$ for which h is good. This follows from:

Proposition 1.3. *If h is good, then for all polycylinders $\Delta^r\subset\bar{X}$ in which $\bar{X}-X$ is given by $\prod_{i=1}^{k} z_i=0$,*

$$\Gamma(\Delta^r, \bar{E})=\{s\in\Gamma(\Delta^r\cap X, E)\mid h(s,s)\leq C\cdot(\textstyle\sum\log|z_i|)^{2n}, \text{ for some } C, n\}.$$

Proof. The inclusion "$\subset$" is immediate. As for "$\supset$", if $s=\sum_{i=1}^{n} a_i(z)\,e_i$ is a holomorphic section of E on $\Delta^r\cap X$, for which $h(s,s)$ is bounded as above, then it follows that

$$|a_i(z)|\leq C'(\textstyle\sum\log|z_i|)^{2m}, \quad \text{for suitable } C', m.$$

Therefore $\left(\prod_{j=1}^{k} z_i\right)\cdot a_i(z)$ is bounded on Δ^r, hence is analytic, hence $a_i(z)$ is meromorphic with simple poles on $\bar{X}-X$. But as no inequality

$$\frac{1}{|z|^2}\leq C(\log|z|)^{2n}$$

holds, $a_i(z)$ is in fact analytic. QED

The main result of this section is the following:

Theorem 1.4. *If $\bar{E}$ is a vector bundle on $\bar{X}$ and h is a good Hermitian metric on $E=\bar{E}|_X$, then the Chern forms $c_k(E,h)$ are good on $\bar{X}$ and the current $[c_k(E,h)]$ represents the cohomology class $c_k(\bar{E})\in H^{2k}(\bar{X}, \mathbb{C})$.*

Proof. Let h^* be a C^∞ Hermitian metric on $\bar{E}$. Define

$$\theta = \partial h \cdot h^{-1}, \quad \theta^* = \partial h^* \cdot h^{*-1},$$
$$K = \bar{\partial}\theta, \quad K^* = \bar{\partial}\theta^*.$$

Intrinsically, K and K^* are $\mathrm{Hom}(E, E)$-valued (1, 1)-forms and $\theta - \theta^*$ is a $\mathrm{Hom}(E, E)$-valued (1, 0)-form. According to results in Bott-Chern [5], for each k there is a universal polynomial P_k with rational coefficients in the forms K, K^* and $\theta - \theta^*$ such that *on* X:

$$c_k(E, h) - c_k(E, h^*) = d(\mathrm{Tr}\, P_k(K, K^*, \theta - \theta^*)).$$

Now K, K^* and $\theta - \theta^*$ are forms good on $\bar{X}$, hence the (k, k)-form $\mathrm{Tr}\, P_k(K, K^*, \theta - \theta^*)$ is good on $\bar{X}$. It follows that $c_k(E, h)$ is good on $\bar{X}$ and that

$$[c_k(E, h)] = d[\mathrm{tr}\, P_k] + \underbrace{[c_k(E, h^*)]}_{\text{represents } c_k(\bar{E})}. \quad \text{QED}$$

§ 2. Estimates on Cones

The results of this section are purely preliminary. We havc isolated all the inequalities needed for the general proportionality theorem which involve only the cone variables (cf. § 3, definition of Siegel Domain), and worked these out in this section.

The object of study then is a real vector space V and

$$C \subset V,$$

C an open, convex, non-degenerate ($\bar{C} \not\supset$ (pos. dim. subspace of V), or equivalently, $\exists\, l \in V$, $l > 0$ on $\bar{C}$) cone. Most of our results relate only to those C which are homogeneous and self-adjoint; for any C, we let $G \subset GL(V)$ be the group of linear maps which preserve C, and say C is *homogeneous* if G acts transitively. If, moreover, there is a positive-definite inner product $\langle\, ,\rangle$ on V for which

$$\bar{C} = \{x \in V \mid \langle x, y\rangle \geqq 0, \text{ all } y \in C\}$$

we say C is *self-adjoint*. The classification of these is well known (see [1], p. 63), as in the fact that all such arise by considering formally real Jordan algebras V and setting

$$C = \{x^2 \mid x \in V, x \text{ invertible}\}.$$

All convex non-degenerate cones C carry several canonical metrics on them. First of all, there is a canonical Finsler metric on C, which is analogous to the Caratheodory metrics on complex manifolds ([10], p. 49):

$$\forall x \in C,\ t \in T_{x,C} \cong V, \text{ let:}$$

$$\rho_x(t) = \sup_{l \in \bar{C}} \frac{|l(t)|}{l(x)}.$$

(Another canonical Finsler metric, analogous to Kobayashi's metric in the complex case, can also be defined. First introduce on the positive quadrant $\mathbb{R}_+ \times \mathbb{R}_+$ the metric $\frac{dx^2}{x^2}+\frac{dy^2}{y^2}$; then we have on any cone the definition:

$$\rho'_x(t)=\begin{Bmatrix}\text{canonical length in}\\ \text{the cone } C\cap(\mathbb{R}x+\mathbb{R}t)\end{Bmatrix} =\sqrt{\frac{1}{a_1^2}+\frac{1}{a_2^2}}$$

if $a_1>0$ and $a_2<0$ are determined by $x+a_1 t, x+a_2 t\in \bar{C}, \notin C$. We won't need this second metric however.)

The advantage of the Finsler metric is that (as in the complex case) it behaves in a monotone way when you replace C by a smaller (or bigger) cone:

Proposition 2.1. i) *If C is an open convex non-degenerate cone in V, and $a\in\bar{C}$, $x\in C$, $t\in V$, then*

$$\rho_{x+a}(t)_C \leqq \rho_x(t)_C.$$

ii) *If $C_1\subset C_2$ are 2 open convex non-degenerate cones in V, then for all $x\in C_1$, $t\in V$,*

$$\rho_x(t)_{C_1}\geqq \rho_x(t)_{C_2}.$$

(The proofs are easy.)

Now suppose C is homogeneous and self-adjoint. Then one can introduce a Riemannian metric on C as follows. Chose a base point $e\in V$ which we take as the identity for the Jordan algebra and let $\langle\ ,\ \rangle$ be an inner product on V in terms of which $G={}^tG$. Then ([1], p. 62), C is self-adjoint with respect to $\langle\ ,\ \rangle$. Moreover, $K=\mathrm{Stab}(e)$ is a maximal compact subgroup of G and if $\mathfrak{g}=\mathfrak{k}\oplus\mathfrak{p}$ is a Cartan decomposition with respect to K and $*: G\to G$ the Cartan involution, then:

1) $\langle gx, g^* y\rangle=\langle x,y\rangle$, hence $K=\exp(\mathfrak{k})$ acts by orthogonal maps while $P=\exp(\mathfrak{p})$ acts by self-adjoint maps,

2) $(gx)^{-1}=g^*(x^{-1})$ (here x^{-1} is the Jordan algebra inverse).

Now identifying $T_{e,C}$ with V, we use $\langle\ ,\ \rangle$ to define a Riemannian metric on C at e: since it is K-invariant, it globalizes to a unique G-invariant Riemannian metric on C, which we write ds_C^2. For later use, we need the following formula:

Lemma. *Take $t_1,t_2\in T_{x.C}$, and let $f(x)=\langle t_2,x^{-1}\rangle$ where x^{-1} is the Jordan algebra inverse. Then*

$$ds_C^2(t_1,t_2)=-D_{t_1}f$$

where D_{t_1} is the derivative of functions on V in the direction t_1.

Proof. Let $g\in\exp(\mathfrak{p})$ carry e to x. Then by G-invariance:

$$ds^2_{C,x}(t_1,t_2)=ds^2_{C,e}(g^{-1}t_1,g^{-1}t_2)=\langle g^{-1}t_1,g^{-1}t_2\rangle$$

and

$$D_{t_1}f(x)=D_{g^{-1}t_1}(f\circ g)(e).$$

But

$$\begin{aligned}f\circ g(x)&=\langle t_2,(gx)^{-1}\rangle\\&=\langle t_2,g^{-1}\cdot(x^{-1})\rangle\\&=\langle g^{-1}t_2,x^{-1}\rangle.\end{aligned}$$

If $\delta x\in V$ is small, then $(e+\delta x)^{-1}=e-\delta x+$(terms of lower order), hence

$$D_{g^{-1}t_1}(f\circ g)(e)=-\langle g^{-1}t_2,g^{-1}t_1\rangle.\quad \text{QED}$$

On a homogeneous cone C, any $2G$-invariant metrics are necessarily comparable, so we can deduce, from the monotone behavior of the Finsler metric ρ, a weaker monotonicity for ds_C^2:

Proposition 2.2. i) *If C is a self-adjoint homogeneous cone in V and $a\in\bar{C}$, then there is a constant $K>0$ such that*

$$ds^2_{C,x+a}(t,t)\leqq K\cdot ds^2_{C,x}(t,t)\qquad \textit{all } t\in V,\ x\in C.$$

ii) *If $C_1\subset C_2$ are 2 self-adjoint homogeneous cones in V, then there is a constant $K>0$ such that*

$$ds^2_{C_2,x}(t,t)\leqq K\cdot ds^2_{C_1,x}(t,t),\qquad \textit{all } t\in V,\ x\in C_1.$$

(The proofs are easy.)

The main estimates of this section deal with the following situation:
C a self-adjoint homogeneous cone,
$G=\mathrm{Aut}^\circ(V,C)$ ($^\circ$ means connected component),
C_n = cone of positive definite $n\times n$ Hermitian matrices,
$\rho: G\to GL(n,\mathbb{C})$ a representation,
$H: C\to C_n$ an equivariant, symmetric map with respect to ρ.

Here (ρ,H) *equivariant* means:

$$H(gx)=\rho(g)H(x)\cdot{}^t\overline{\rho(g)},\qquad \text{all } x\in C,\ g\in G, \tag{1}$$

while (ρ,H) *symmetric* means:

$$\rho(g^*)=H(e)\cdot{}^t\overline{\rho(g)}^{-1}\cdot H(e)^{-1},\qquad \text{all } g\in G. \tag{2}$$

Note that $A\mapsto H(e)\cdot{}^t\bar{A}^{-1}\cdot H(e)^{-1}$ is just the Cartan involution of $GL(n,\mathbb{C})$ with respect to the maximal compact subgroup given by the unitary group of $H(e)$: calling this $\natural$, we can rewrite the symmetry condition (2):

$$\rho(g^*)=\rho(g)^\natural. \tag{2'}$$

Condition (2) is actually independent of the choice of e: if $e' \in C$ is any other point, and $g \mapsto g^{*'}$ is the corresponding Cartan involution then one can check that (1)+(2) imply:

$$\rho(g^{*'}) = H(e')^t \overline{\rho(g)}^{-1} H(e')^{-1}. \tag{2''}$$

We will also need for applications the slightly more general situation where ρ is fixed but H depends on some extra parameters $t \in T$ *with* T *compact.* In this case, we ask that (1) and (2) hold for each H_t. Note that we can change coordinates in $\mathbb{C}^n$, to get a new pair:

$$\rho'(g) = a\rho(g)a^{-1},$$
$$H'_t(g) = aH_t(g)^t\bar{a},$$

satisfying the same identities. In this way, we can, for instance, normalize the situation so that

$$H_{t_0}(e) = I_n$$

hence:

$$\rho(K) \subset U(n),$$
$$\rho(\exp \mathfrak{p}) \subset \left\{\begin{matrix}\text{self-adjoint matrices, commuting}\\ \text{with } H_t(e), \text{ all } t\end{matrix}\right\}.$$

This normalization will not affect our estimates. The first of these is:

Proposition 2.3. *For all* $\lambda > 0$, *there is a constant* $K > 0$ *and an integer* N *such that*

$$\|H_t(x)\| \quad \textit{and} \quad |\det H_t(x)|^{-1} \leqq K \cdot \langle x, x\rangle^N, \quad \textit{all } x \in (C + \lambda \cdot e).$$

Proof. Let $A \subset \exp(\mathfrak{p})$ be a maximal $\mathbb{R}$-split torus. Then $C = K \cdot A \cdot e$ and $C + \lambda e = K \cdot (Ae + \lambda e)$. Write $x = k(a(e))$. Then

$$\begin{aligned}\|H_t(x)\| &= \|H_t(ae)\|\\ &= \|\rho(a) \cdot H_t(e) \cdot \rho(a)\|\\ &\leqq \|\rho(a)\|^2 \cdot \|H_t(e)\|.\end{aligned}$$

We may change coordinates in $\mathbb{C}^n$ by a unitary matrix so that all the matrices $\rho(a)$ are diagonalized. Write then $\rho(a) = \chi_i(a) \cdot \delta_{ij}$ where $\chi_i \colon A \to \mathbb{R}^*$ are characters. Now we may coordinatize A by

$$A \xrightarrow{\approx} A \cdot e \cong R^r_+ \subset V, \quad e \mapsto (1, \ldots, 1).$$

Let $a_1, \ldots, a_r$ be coordinates in $\mathbb{R}^r_+$. Then

$$\chi_i(a) = \prod_{j=1}^{r} a_j^{s_{ij}}, \quad s_{ij} \in \mathbb{R}.$$

Note that if $ae \in C+\lambda e$, then $ae \in \mathbb{R}^n_+ + (\lambda, \ldots, \lambda)$, i.e., $a_i \geqq \lambda$, all i. Hence

$$\begin{aligned}\|\rho(a)\|^2 &\leqq \max_{1 \leqq i \leqq n} (\chi_i(a))^2 \\ &\leqq K_1 \left(\sum_{l=1}^{r} |a_l|^2\right)^{(\max s_{ij})r} \\ &\leqq K_1 \cdot K_2 (\langle ae, ae\rangle)^{(\max s_{ij})r}\end{aligned}$$

hence

$$\|H_t(kae)\| \leqq K_1 \cdot K_2 \cdot K_3 (\langle kae, kae\rangle)^{(\max s_{ij})r}.$$

The same proof works for $|\det|^{-1}$. QED

Next, let $\xi \in V$, and let D_ξ be the derivative on V in the direction ξ. We wish to estimate the matrix-valued functions

$$(D_\xi H_t) \cdot H_t^{-1} \colon C \to M_n(\mathbb{C}).$$

To do this, we prove first:

Proposition 2.4. *For all $1 \leqq \alpha, \beta \leqq n$, let $(D_\xi H_t \cdot H_t^{-1})_{\alpha\beta}$ be the $(\alpha, \beta)^{\text{th}}$ entry in this matrix. There is a linear map*

$$C_{\alpha\beta,t} \colon V \to V$$

depending continuously on t such that

$$(D_\xi H_t \cdot H_t^{-1})_{\alpha\beta}(x) = \langle C_{\alpha\beta,t}(\xi), x^{-1}\rangle.$$

Moreover $C_{\alpha\beta,t}$ has the property:

$$\left.\begin{array}{l}\xi, \eta \in \bar{C} \\ \langle \xi, \eta\rangle = 0\end{array}\right\} \Rightarrow \langle C_{\alpha\beta,t}(\xi), \eta\rangle = 0.$$

For some reason, I can't prove this by a direct calculation, but must resort to the following trick:

Lemma. *Let $C \subset V$ be a convex open cone, let $e \in C$, and let $f \colon C \to \mathbb{C}$ be a differentiable function. Suppose that for all $W \subset V$, $\dim W = 2$, and $e \in W$, f is linear on $C \cap W$. Then f is linear.*

Proof. The hypothesis means that

$$f(ax+be) = af(x) + bf(e), \qquad \text{all } x \in C,\ a, b \in \mathbb{R}_+.$$

So we may extend f to all of V by the formula

$$f(x) = f(x+ae) - af(e), \qquad \text{provided } x+ae \in C.$$

Note that

$$f(tx)=f(tx+tae)-atf(e)$$
$$=t(f(x+ae)-af(e))$$
$$=tf(x).$$

Thus for all $x\in V$, $n\geqq 1$:

$$f(x)=\frac{f(x/n)-f(0)}{n}$$
$$\therefore f(x)=D_x f(0)$$

and the right hand side is linear in x. QED

We prove $(D_\xi H_t \cdot H_t^{-1})_{\alpha\beta}(x^{-1})$ is bilinear in ξ, x. Since it is linear in ξ, it suffices to find a basis $\{\xi_k\}$ of V such that $(D_{\xi_k} H_t \cdot H_t^{-1})_{\alpha\beta}(x^{-1})$ is linear in x. In fact, we will show that for every $\xi\in C$, $(D_\xi H_t \cdot H_t^{-1})_{\alpha\beta}(x^{-1})$ is linear in x. To do this, by the lemma, it suffices to show that for every $W\subset V$ with $\dim W=2$, $\xi\in W$, $(D_\xi H_t \cdot H_t^{-1})_{\alpha\beta}(x^{-1})$ is linear on $C\cap W$. We will do this for all α, β at once, so in verifying this we can change coordinates in $\mathbb{C}^n$.

What we do is this: we let ξ be a new base point of C and we change coordinates so that $H_{t_0}(\xi)=I_n$. This reduces us to verifying $(D_e H_t \cdot H_t^{-1})_{\alpha\beta}(x^{-1})$ is linear on $C\cap W$ when $\dim W=2$, $e\in W$. Now any such W is part of a subspace of V of the form $A\cdot e$, where $A\subset \exp(\mathfrak{p})$ is a maximal $\mathbb{R}$-split torus (of course, this $\mathfrak{p}$ corresponds to the *new* choice of e). Moreover, as above, we can diagonalize ρ on A:

$$\rho(a_1,\dots,a_n)=(\chi_i(a)\,\delta_{ij}),\qquad \chi_i(a)=\prod_{j=1}^{r} a_j^{s_{ij}}.$$

Then

$$H_t(ae)=\rho(a)^2 H_t(e),\qquad \rho(a)H_t(e)=H_t(e)\rho(a).$$

Since $e=(1,\dots,1)$, it follows:

$$D_e(H_t(ae))=\sum_{i=1}^{r}\frac{\partial}{\partial a_i}H_t(ae).$$

Using this, one calculates:

$$(D_e H_t \cdot H_t^{-1})_{\alpha\beta}(ae)=\left(\sum_{j=1}^{r}\frac{2s_{\alpha j}}{a_j}\right)\delta_{\alpha\beta}$$

and since on Ae, x^{-1} is given by $(a_1,\dots,a_r)\mapsto(a_1^{-1},\dots,a_r^{-1})$, this proves that $(D_e H_t \cdot H_t^{-1})_{\alpha\beta}$ is linear in x^{-1} on every $W\subset Ae$.

Now say $\xi,\eta\in\bar{C}$, $\langle\xi,\eta\rangle=0$. For a suitable coice of maximal torus A, we have $\xi,\eta\in\bar{A}\cdot e$. In the coordinates $(a_1,\dots,a_r)$ on $A\cdot e$, let

$$\xi=(\xi_1,\dots,\xi_r),$$
$$\eta=(\eta_1,\dots,\eta_r).$$

Since $\langle\,,\rangle$ on Ae makes A self-adjoint, it is a quadratic form of the type $\langle a\,e, b\,e\rangle = \sum \lambda_i a_i b_i$, so $\langle \xi, \eta\rangle = 0$ means that for every i, $\xi_i = 0$ or $\eta_i = 0$. A calculation like that just made shows:

$$(D_\xi H_t \cdot H_t^{-1})_{\alpha\beta}(ae) = \left(\sum_{j=1}^{r} \frac{2s_{\alpha j}\xi_j}{a_j}\right) \cdot \delta_{\alpha\beta}$$

i.e.,

$$(D_\xi \cdot H_t \cdot H_t^{-1})_{\alpha\beta}(x^{-1})|_{A \cdot e} = \left(\sum_{j=1}^{r} 2s_{\alpha j}\xi_j a_j \cdot \delta_{\alpha\beta}\right).$$

This is clearly zero if $x = \eta$. QED

For the next result, suppose δ is any vector field in the manifold of values of t. Then

Proposition 2.5. *For all vector fields δ on T,*

$$(\delta H_t \cdot H_t^{-1})(x)$$

is independent of x, i.e., depends on t alone.

Proof. By equivariance:

$$\delta H_t(gx) = \rho(g) \cdot \delta H_t(x) \cdot {}^t\overline{\rho(g)}$$

hence

$$(\delta H_t \cdot H_t^{-1})(g \cdot e) = \rho(g) \cdot (\delta H_t \cdot H_t^{-1})(e) \cdot \rho(g)^{-1}.$$

By symmetry:

$$\rho(g^*) \cdot \delta H_t(e) = \delta H_t(e) \cdot {}^t\overline{\rho(g)}^{-1}$$

hence

$$\rho(g^*) \cdot (\delta H_t \cdot H_t^{-1})(e) \cdot \rho(g^*)^{-1} = (\delta H_t \cdot H_t^{-1})(e).$$

Together, these imply the Proposition. QED

Proposition 2.4 gives us estimates on $\|D_\xi H_t \cdot H_t^{-1}\|$. To work these out, we fix a maximal flag of boundary components of C. In the notation of [1], p. 109, choosing this flag and the base point $e \in C$ is equivalent to choosing in the Jordan algebra V, a maximal set of orthogonal idempotents:

$$e = \varepsilon_1 + \dots + \varepsilon_r.$$

Let

$$C_i = \text{boundary component containing } \varepsilon_{i+1} + \dots + \varepsilon_r.$$

Then

$$\bar{C} \supsetneq \bar{C}_1 \supsetneq \bar{C}_2 \supsetneq \dots \supsetneq \bar{C}_r$$

is the flag. Also let

$$\tilde{C}=C\cup C_1\cup C_2\cup\cdots\cup C_r\cup(0)$$

and let

$$A=\sum_{i=1}^{r}\mathbb{R}\cdot\varepsilon_i.$$

Let P be the parabolic group which stabilizes the flag $\{\bar{C}_i\}$. Our estimates are based on:

Proposition 2.6. (1) *Let $\xi_1\in\tilde{C}$ and let $\xi_1'\in V$ satisfy*

$$\left.\begin{matrix}\langle\xi_1,\eta\rangle=0\\ \eta\in\bar{C}\end{matrix}\right\}\Rightarrow\langle\xi_1',\eta\rangle=0.$$

Then for every compact set $\omega\subset P$, there is a $K>0$ such that:

$$|\langle\xi_1',x^{-1}\rangle|\leqq K\sqrt{ds^2_{C,x}(\xi_1,\xi_1)},\quad \text{all } x\in\omega\cdot A\cdot e.$$

(2) *Let $\xi_1\in\tilde{C}$, $\xi_1'\in V$ be as above. Let $\xi_2\in\tilde{C}$. Then for every compact set $\omega\subset P$, there is a $K>0$ such that:*

$$|ds^2_{C,x}(\xi_1',\xi_2)|\leqq K\sqrt{ds^2_{C,x}(\xi_1,\xi_1)}\cdot\sqrt{ds^2_{C,x}(\xi_2,\xi_2)}.$$

Proof. We will use the Peirce decomposition of V defined by the idempotents ε_i:

$$V=\bigoplus_{i\leqq j}V_{ij}$$

where

$$x\in V_{ij}\Rightarrow\sum a_i\varepsilon_i\cdot x=\frac{a_i+a_j}{2}x.$$

This decomposition is orthogonal with respect to $\langle\ ,\ \rangle$ and C_k is an open cone in the subspace $\bigoplus_{k<i\leqq j}V_{ij}$. If $\xi_1\in C_k$, then note that

$$\xi_1\perp\bigoplus_{i\leqq j\leqq k}V_{ij}$$

and

$$\bigoplus_{i\leqq j\leqq k}V_{ij}\supset(\text{the boundary component of } C \text{ corresponding to } \varepsilon_1+\cdots+\varepsilon_k).$$

This boundary component is open in $\bigoplus_{i\leqq j\leqq k}V_{ij}$. Thus

$$\xi_1'\perp\bigoplus_{i\leqq j\leqq k}V_{ij}$$

or

$$\xi_1'\in\bigoplus_{\binom{i\leqq j}{k<j}}V_{ij}.$$

To prove (1), let $x = gae$, $g \in \omega$, $ae = \sum a_i \varepsilon_i$. Then

$$\begin{aligned}\langle \xi_1', x^{-1}\rangle &= \langle g^{-1}\xi_1', (ae)^{-1}\rangle \\ &= \sum_{i=1}^{r} \frac{1}{a_i} \langle (g^{-1}\xi_1')_{ii}, \varepsilon_i\rangle\end{aligned}$$

where $(g^{-1}\xi_1')_{ii}$ is the component of $(g^{-1}\xi_1')$ in V_{ii}. But ω preserves the flag, so

$$g^{-1}\xi_1' \in \bigoplus_{\substack{i \leq j \\ k<j}} V_{ij}$$

too. Thus

$$\langle \xi_1', x^{-1}\rangle = \sum_{i=k+1}^{r} \frac{1}{a_i} \langle (g^{-1}\xi_1')_{ii}, \varepsilon_i\rangle.$$

As g varies in ω, $\langle (g^{-1}\xi_1')_{ii}, \varepsilon_i\rangle$ is bounded, hence

$$|\langle \xi_1', x^{-1}\rangle| \leq K_1 \sum_{i=k+1}^{r} \frac{1}{a_i}.$$

Let $e^{(k)} = \varepsilon_{k+1} + \cdots + \varepsilon_r$ be the base point in C_k and note that, again by compactness of ω, there is a $K_2 > 0$ such that

$$g^{-1}\xi_1 \in C_k + K_2 e^{(k)}, \quad \text{all } g \in \omega,$$

hence

$$ds^2_{C,x}(K_2 e^{(k)}, K_2 e^{(k)}) \leq ds^2_{C,x}(g^{-1}\xi_1, g^{-1}\xi_1), \quad \text{all } x \in C.$$

Thus

$$\begin{aligned} ds^2_{C,gae}(\xi_1, \xi_1) &= ds^2_{C,ae}(g^{-1}\xi_1, g^{-1}\xi_1) \\ &\geq K_2^2 ds^2_{C,ae}(e^{(k)}, e^{(k)}) \\ &= K_2^2 ds^2_{C,e}(a^{-1}e^{(k)}, a^{-1}e^{(k)}) \\ &= K_2^2 \sum_{i=k+1}^{r} \frac{1}{a_i^2} \langle \varepsilon_i, \varepsilon_i\rangle \\ &\geq K_2^2 K_3 \left(\sum_{i=k+1}^{r} \frac{1}{a_i}\right)^2. \end{aligned}$$

The same procedure proves (2). If $\xi_1 \in C_k$, $\xi_2 \in C_l$, the argument goes like this:

$$\begin{aligned} ds^2_{C,gae}(\xi_1', \xi_2) &= \langle a^{-1}(g^{-1}\xi_1'), a^{-1}(g^{-1}\xi_2)\rangle \\ &= \sum_{\begin{pmatrix} i \leq j \\ l < i \\ k < j \end{pmatrix}} \frac{1}{a_i a_j} \langle (g^{-1}\xi_1')_{ij}, (g^{-1}\xi_2)_{ij}\rangle \\ &\leq K_1 \sqrt{\sum_{i=l+1}^{r} \frac{1}{a_i^2}} \cdot \sqrt{\sum_{j=k+1}^{r} \frac{1}{a_j^2}} \\ &\leq K_1 K_2 \sqrt{ds^2_{gae}(\xi_1, \xi_1)} \cdot \sqrt{ds^2_{gae}(\xi_2, \xi_2)}. \quad \text{QED} \end{aligned}$$

We can put everything we have said together as follows: Suppose $N=\dim V$ and $\xi_1,\dots,\xi_N\in\tilde{C}$ span V. Define a simplicial cone $\sigma\subset C$ by

$$\sigma=\sum_{i=1}^{N}\mathbb{R}_+\cdot\xi_i.$$

Let $l_i\colon V\to\mathbb{R}$ be a dual basis: $l_i(\xi_j)=\delta_{ij}$. Then for all ρ, H_t as above, we have the estimates:

Proposition 2.7. *For all vector fields δ,δ' to the T-space, and $a\in\bar{C}$ there is a constant $K>0$ such that for all $x\in\operatorname{Int}(\sigma+a)$:*

$$\|D_{\xi_i}H_t\cdot H_t^{-1}(x)\|\leqq\frac{K}{l_i(x)-l_i(a)},$$

$$\|\delta H_t\cdot H_t^{-1}(x)\|\leqq K,$$

$$\|D_{\xi_i}(D_{\xi_j}H_t\cdot H_t^{-1})(x)\|\leqq\frac{K}{(l_i(x)-l_i(a))\cdot(l_j(x)-l_j(a))},$$

$$\|D_{\xi_i}(\delta H_t\cdot H_t^{-1})(x)\|=0,$$

$$\|\delta(D_{\xi_i}H_t\cdot H_t^{-1})(x)\|\leqq\frac{K}{l_i(x)-l_i(a)},$$

$$\|\delta(\delta' H_t\cdot H_t^{-1})(x)\|\leqq K.$$

Proof. Combine (2.4), (2.5) and (2.6) and the formula $ds_C^2(t_1,t_2)=-D_{t_1}(\langle t_2,x^{-1}\rangle)$ to get estimates in terms of $ds_C^2(\xi_i,\xi_i)$ on sets $\omega\cdot A\cdot e$. Then apply (2.2)(i) and (ii) to the inclusion $\operatorname{Int}(\sigma+a)\to C$, plus Ash's theorem that $\operatorname{Int}(\sigma+a)\subset\omega\cdot A\cdot e$ if ω is large enough ([1], Ch. II, §4), plus the formula

$$ds_\sigma^2=\sum_{i=1}^{n}\frac{dl_i^2}{l_i^2}$$

for the canonical metric on the homogeneous convex cone σ. QED

There is one final estimate that we need. For this, we first make a definition:

Definition. A linear map $T\colon\mathbb{C}^n\to\mathbb{C}^n$ is called *ρ-upper triangular* if the following holds: For all maximal $\mathbb{R}$-split tori $A\subset G$, let $X(A)=\operatorname{Hom}(A,\mathbb{G}_m)$ be the character group of A. As is well known, there is a basis $\gamma_1,\dots,\gamma_r$ of $X(A)$ such that the weights of A acting on V are contained in $\gamma_i+\gamma_j$, $i\leqq j$ (and contain $2\gamma_i$, $1\leqq i\leqq r$). Partially order the characters by defining

$$\sum n_i\gamma_i\geqq\sum m_i\gamma_i\quad\text{if } n_i\geqq m_i,\quad\text{all } i.$$

Diagonalize $\rho(A)$:

$$\mathbb{C}^n=\bigoplus_{\lambda\in X(A)}V_\lambda.$$

Then T is ρ-upper triangular if for all A, all $\lambda_0\in X(A)$,

$$T\Big(\bigoplus_{\lambda\geqq\lambda_0}V_\lambda\Big)\subseteq\bigoplus_{\lambda\geqq\lambda_0}V_\lambda.$$

The estimate we need is:

Proposition 2.8. *If T is ρ-upper triangular, then for all $a \in C(F)$, there is a constant $K>0$ such that*

$$\|H_t(x) \cdot {}^t\bar{T} \cdot H_t(x)^{-1}\| \leqq K, \quad \text{all } x \in C(F)+a.$$

Proof. Take a as a base point of $C(F)$ and pick any maximal torus A so that $Ae = \mathbb{R}^n_+$, $a=(1,\ldots,1)$, $C(F)=K \cdot a \cdot e$. Therefore

$$C(F)+a=\{kae \mid k \in K, a=(a_1 \ldots, a_n), a_i \geqq 1, \text{ all } i\}.$$

Change coordinates in $\mathbb{C}^n$ so that

$$\rho(a)=\begin{pmatrix} \lambda_1(a) & & 0 \\ & \ddots & \\ 0 & & \lambda_n(a) \end{pmatrix}.$$

Now

$$H_t(x) \cdot {}^t\bar{T} \cdot H_t(x)^{-1} = \rho(k) H_t(e) \rho(a)^2 \cdot {}^t\overline{\rho(k)} \cdot {}^t\bar{T} \cdot {}^t\overline{\rho(k)}^{-1} \rho(a)^{-2} H_t(e)^{-1} \rho(k)^{-1}$$

so it suffices to bound $\|\rho(a)^2 \cdot {}^t(\rho(k)^{-1} \cdot T \cdot \rho(k)) \cdot \rho(a)^{-2}\|$. This means we wish to bound:

$$|\lambda_i(a)^2 \lambda_j(a)^{-2} (\rho(k)^{-1} \cdot T \cdot \rho(k))_{ji}|$$

when k ranges over K, and $a=(a_1,\ldots,a_n)$ satisfies $a_i \geqq 1$, all i. This is equivalent to:

$$(\rho(k)^{-1} \cdot T \cdot \rho(k))_{ji} \neq 0 \Rightarrow \lambda_j \geqq \lambda_i \qquad (*)$$

(weights of A being partially ordered as in the definition). But for all i define

$$W_i = \rho(k)\Big(\bigoplus_{\lambda_j \geqq \lambda_i} \mathbb{C} \cdot e_j\Big)$$

(where $e_i \in \mathbb{C}^n$ is the ith unit vector). Note that kAk^{-1} maps W_i into itself and that W_i is one of the sums of weight spaces referred to in the definition. Therefore $T(W_i) \subseteq W_i$, hence

$$(\rho(k)^{-1} T \rho(k)) e_i \in \bigoplus_{\lambda_j \geqq \lambda_i} \mathbb{C} \cdot e_j$$

which is precisely $(*)$. QED

§ 3. The Proportionality Principle

Let D be an r-dimensional bounded symmetric domain and let Γ be a neat[2] arithmetic group acting on D. Then $X=D/\Gamma$ is a smooth quasi-projective

[2] Recall that following a definition of Borel, a "neat" arithmetic subgroup Γ of an algebraic group $\mathscr{G} \subset GL(n,\mathbb{C})$ is one such that for every $x \in \Gamma$, $x \neq e$, the group generated by the eigenvalues of x is torsion-free. Every arithmetic group Γ' has neat arithmetic subgroups of finite index

variety, called a *locally symmetric variety*, or an *arithmetic variety*. In [1], Ash, Rapoport, Tai and I have introduced a family of smooth compactifications $\bar{X}$ of X such that $\bar{X}-X$ has normal crossings. We must first recall how $\bar{X}$ is described locally. At the same time, we will need various details from the whole cumbersome appartus used to manipulate D so we will rapidly sketch these too. All results stated without proof can be found in [1].

By definition, $D \cong K \backslash G$, where G is a semi-simple adjoint group and K is a maximal compact subgroup. Inside the complexification $G_{\mathbb{C}}$ of G, there is a parabolic subgroup of the form $P_+ \cdot K_{\mathbb{C}}$ (P_+ its unipotent radical which is, in fact, abelian and $K_{\mathbb{C}}$ the complexification of K) such that $K = G \cap (P_+ \cdot K_{\mathbb{C}})$ and $G \cdot (P_+ \cdot K_{\mathbb{C}})$ open in $G_{\mathbb{C}}$. This induces an open G-equivariant immersion

$$\begin{array}{ccc} D & \hookrightarrow & \check{D} \\ \| & & \| \\ K \backslash G & & P_+ \cdot K_{\mathbb{C}} \backslash G_{\mathbb{C}}. \end{array}$$

Here $\check{D}$ is a rational projective variety known as a flag space and $G_{\mathbb{C}}$ is an algebraic group acting algebraically on $\check{D}$. Let $\bar{D}$ be the closure of D in $\check{D}$. The maximal analytic submanifolds $F \subset \bar{D}-D$ are called the boundary components of D. For each F, we set

$N(F) = \{g \in G \mid gF = F\}$,

$W(F) =$ unipotent radical of $N(F)$,

$U(F) =$ center of $W(F)$, a real vector space of dimension k, say,

$V(F) = W(F)/U(F)$: known to be abelian, centralizing $U(F)$. Via "exp", we get a section and write $W(F)$ set-theoretically as $V(F) \cdot U(F)$. Also $\dim V$ is even – let it be $2l$.

Next splitting $N(F)$ into a semi-direct product of a reductive part and its unipotent radical, we decompose $N(F)$ further:

$$N(F) = \underbrace{(G_h(F) \cdot G_l(F) \cdot M(F))}_{\substack{\text{direct product mod finite} \\ \text{central group}}} \cdot V(F) \cdot U(F),$$

where

a) $G_l \cdot M \cdot V \cdot U$ acts trivially on F, G_h mod a finite center being $\mathrm{Aut}^{\circ}(F)$,

b) $G_h \cdot M \cdot V \cdot U$ commutes with $U(F)$, G_l mod a finite central group acts faithfully on $U(F)$ by inner automorphisms

c) M is compact.

Here F is said to be rational if $\Gamma \cap N(F)$ is an arithmetic subgroup of $N(F)$. Mod Γ there are only finitely many such F, and if $F_1, \ldots, F_k$ are representatives:

$$X \cup \bigcup_{i=1}^{k} (F_i / \Gamma \cap N(F_i)),$$

with suitable analytic structure is Satake-Baily-Borel's compactification of X.

Next, for each F, we define an open subset $D_F \subset \check{D}$ by

$$D_F = \bigcup_{g \in U(F)_{\mathbb{C}}} g \cdot D.$$

The embedding of D in D_F is Pjatetski-Shapiro's realization of D as "Siegel Domain of 3rd kind". In fact, there is an isomorphism:

$$D_F \cong U(F)_{\mathbb{C}} \times \mathbb{C}^l \times F$$

such that not only $N(F)$ but even the bigger group $G_h \cdot (M \cdot G_l)_{\mathbb{C}} \cdot V_{\mathbb{C}} \cdot U_{\mathbb{C}}$ acts by "semi-linear transformations":

$$(x, y, t) \mapsto (Ax + a(y, t), B_t y + b(t), g(t))$$

(A, B_t matrices, a, b vectors) and

$$D = \{(x, y, t) \mid \operatorname{Im} x + l_t(y, y) \in C(F)\}$$

where $C(F) \subset U(F)$ is a self-adjoint convex cone homogeneous under the G_l-action on $U(F)$ and $l_t: \mathbb{C}^l \times \mathbb{C}^l \to U(F)$ is a symmetric $\mathbb{R}$-bilinear form.

Moreover

$U(F) \cong$ group of automorphisms of D: $(x, y, t) \mapsto (x+a, y, t)$, $a \in U(F)$,

$U(F)_{\mathbb{C}} \cong$ group of automorphisms of $D(F)$: $(x, y, t) \mapsto (x+a, y, t)$, $a \in U(F)_{\mathbb{C}}$,

$W(F) \cong$ group of automorphisms of D: $(x, y, t) \mapsto (x + a(u, t), y + b(t), t)$ and the group $V(F)$ acts, for each t, simply transitively on the space $\mathbb{C}^l$ of possible y-values.

There is a technical lemma which we will need about this action:

Lemma. *Let $t_0 \in F$, $e_0 \in C(F)$, and let $u_0 \in U(F)_{\mathbb{C}}$ be the map $(x, y, t) \mapsto (x + ie_0, y, t)$. Let $e = (ie_0, 0, t_0)$ be a base point of D, so that* $\operatorname{Stab}_G(e) = K$, *a maximal compact of G and* $\operatorname{Stab}_{G_{\mathbb{C}}}(e) = K_{\mathbb{C}} \cdot P_+$. *Moreover,* $\operatorname{Stab}_{G_l}(e_0) = K_l$ *is a maximal compact in G_l. Since $G_l \subset \operatorname{Stab}(0, 0, t_0)$, $u_0(G_l)\, u_0^{-1} \subset \operatorname{Stab}(e)$ and we may look at*

$$\alpha: G_l \xrightarrow{\text{conj. by } u_0} \operatorname{Stab}_{G_{\mathbb{C}}}(e) \xrightarrow{\text{mod } P_+} K_{\mathbb{C}}.$$

*If * is the Cartan involution of G_l with respect to K_l, then:*

$$\alpha(g^*) = \overline{\alpha(g)}.$$

Proof. This is a straightforward calculation, for instance, using the fundamental decomposition of $\mathfrak{g} = \operatorname{Lie} G$ via $\mathfrak{sl}(2)^r \subset \mathfrak{g}$ (cf. [1], p. 182) and the description of $\operatorname{Lie}(M \cdot G_l)$ in this decomposition for the standard boundary components F_S given in [1], p. 226.

We now describe local coordinates on $\bar{X}$. Recall that $\bar{X}$ is not unique but depends on the choice of certain auxiliary simplicial decompositions. We need not recall these in detail. The chief thing is that each $\bar{X}$ is covered by a finite set of coordinate charts constructed as follows:

1) take a rational boundary component F of D,

2) take $\{\xi_1, \ldots, \xi_k\}$ a basis of $\Gamma \cap U(F)$ such that $\xi_i \in \overline{C(F)} \subset U(F)$ and in fact, $\xi_i \in C(F) \cup C_1 \cup C_2 \cup \cdots \cup C_l = \tilde{C}$, where $\overline{C(F)} \supseteq \bar{C}_1 \supseteq \bar{C}_2 \supseteq \cdots \supseteq \bar{C}_l$ is a flag of

boundary components (cf. § 2) and at least one ξ_i is in $C(F)$: say

$$\xi_1, \ldots, \xi_m \in C(F), \quad \xi_{m+1}, \ldots, \xi_k \in \overline{C(F)} - C(F),$$

3) let $l_i: U(F)_{\mathbb{C}} \to \mathbb{C}$ be dual to $\{\xi_i\}$, i.e., $l_i(\xi_j) = \delta_{ij}$,

4) consider the exponential:

$$\begin{array}{ccc} D & \subset & (U(F)_{\mathbb{C}} \times \mathbb{C}^l \times F) \\ \big\downarrow & & \big\downarrow {\scriptstyle (e^{2\pi i l_1(x)}, \ldots, e^{2\pi i l_k(x)}, y, t)} \\ D/\Gamma \cap U(F) & \subset & (\mathbb{C}^{*k} \times \mathbb{C}^l \times F) \\ \big\downarrow & & \\ X & & \end{array}$$

5) Define $(D/\Gamma \cap U(F))^{\sim}$ to be the set of $P \in \mathbb{C}^k \times \mathbb{C}^l \times F$ which have a neighborhood U such that

$$U \cap (\mathbb{C}^{*k} \times \mathbb{C}^l \times F) \subset (D/\Gamma \cap U(F)).$$

Note that

$$(D/\Gamma \cap U(F))^{\sim} \supset \bigcup_{i=1}^{m} \{(z, y, t) \mid z = (z_1, \ldots, z_k), z_i = 0\} = S(F, \{\xi_i\}).$$

6) The basic property of $\bar{X}$ is that for suitable F, $\{\xi_i\}$, the covering map p extends to a local homeomorphism

$$\bar{p}: (D/\Gamma \cap U(F))^{\sim} \to \bar{X}$$

and that every point of $\bar{X}$ is equal to $\bar{p}((z, y, t))$, where $z_i = 0$, some $1 \leqq i \leqq m$, for some such F, $\{\xi_i\}$.

We now come to the main results of this paper. Let E_0 be a G-equivariant analytic vector bundle of rank n on D. E_0 is defined by the representation

$$\sigma: K \to GL(n, \mathbb{C})$$

of the stabilizer K of the base point $e_+ \in D$ in the fibre $\mathbb{C}^n$ of E_0 over e_+. We complexify σ and extend it to $P_+ \cdot K_{\mathbb{C}}$ by letting it kill P_+. Then σ defines a $G_{\mathbb{C}}$-equivariant analytic vector bundle $\check{E}_0$ on $\check{D}$ also. In the other direction, we can divide E_0 by Γ obtaining a vector bundle E on X. Since K is compact, E_0 carries a G-invariant Hermitian metric h_0, which induces a Hermitian metric h on E. We claim:

Main Theorem 3.1. *E admits a unique extension $\bar{E}$ to $\bar{X}$ such that h is a singular Hermitian metric good on $\bar{X}$.*

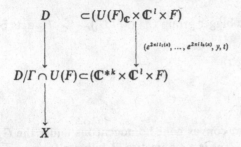

These various bundles are all linked as in this diagram:

$$\begin{array}{ccccccc} \check{E}_0 & \xrightarrow{\text{restriction}} & E_0 & \xrightarrow{\text{descent}} & E & \xrightarrow{\text{extension}} & \bar{E} \\ \downarrow & & \downarrow & & \downarrow & & \downarrow \\ D & \longleftarrow & D & \longrightarrow & X & \longrightarrow & \bar{X}. \end{array}$$

Proof. We saw in §1 that $\bar{E}$, if it existed, has as its sections the sections of E with growth $O\left(\left(\sum_{i=1}^{m} \log|z_i|\right)^{2N}\right)$ along $\bar{X}-X$. To see that the set of these sections defines an analytic vector bundle on $\bar{X}$, it suffices to check this locally, e.g., on $(D/\Gamma\cap U(F))^{\sim}$. But now the bundle $\check{E}_0$ restricts to a bundle E_F on D_F with $N(F)\cdot U(F)_{\mathbb{C}}$ acting equivariantly. Now note that the subgroup $U(F)_{\mathbb{C}}$ acts simply transitively and holomorphically on the first factor $U(F)$ of D_F in its Siegel Domain presentation. Since $\mathbb{C}^l\times F$ is contractible and Stein (F is another bounded symmetric domain), it follows that E_F has a set of n holomorphic sections $e_1, \ldots, e_n$ such that

i) e_i is $U(F)_{\mathbb{C}}$-invariant,
ii) $e_1(x), \ldots, e_n(x)$ are a basis of $E_F(x)$, all $x\in D_F$.

Dividing by $\Gamma\cap U(F)$, E_F descends to a vector bundle E'_F on $\mathbb{C}^{*k}\times\mathbb{C}^l\times F$, which is also globally trivial via the same basic sections $e_1, \ldots, e_n$. We can then extend E'_F to $\mathbb{C}^k\times\mathbb{C}^l\times F$ so as to be trivial with these basic sections. We must show that along $S(F, \{\xi_i\})$ the sheaf of sections of this extension is exactly the sheaf of sections of E'_F on $(D/\Gamma\cap U(F))$ with growth $O\left(\left(\sum_{i=1}^{k} \log|z_i|\right)^{2N}\right)$ on the coordinate hyperplanes $\bigcup_{i=1}^{k}(z_i=0)$. Equivalently, this means that $h(e_i, e_j)$ and $(\det h(e_i, e_j))^{-1}$ have this growth. To do this, it is convenient to use a 2nd basis of E'_F, which is C^∞ but not analytic. Note that $V(F)\cdot U(F)_{\mathbb{C}}$ acts simply transitively on the 1st 2 factors of D_F. So we can find $e'_1, \ldots, e'_n\in\Gamma(D_F, E_F)$ such that

i') e'_i is $V(F)\cdot U(F)_{\mathbb{C}}$-invariant,
ii') $e'_1(x), \ldots, e'_n(x)$ are a basis of $E_F(x)$, all $x\in D_F$.
iii') On $(0, 0)\times F$, $e'_i=e_i$, hence are holomorphic sections.

$\{e_i\}$ and $\{e'_i\}$ are related by an invertible $U(F)_{\mathbb{C}}$-invariant matrix S; so that $|S_{ij}|$ and $|\det S|^{-1}$ are uniformly bounded on subsets

$$(\mathbb{C}^*)^k\times\begin{pmatrix}\text{compact subset}\\ \text{of } \mathbb{C}^l\times F\end{pmatrix}\subset(\mathbb{C})^k\times\mathbb{C}^l\times F.$$

Therefore it is enough to check that $h(e'_i, e'_j)$, $(\det h(e'_i, e'_j))^{-1}$ have the required growth.

Now if $g\in G_l(F)$, note that because g normalizes $V\cdot U_{\mathbb{C}}$, $g e'_i$ is another $V\cdot U_{\mathbb{C}}$-invariant section of E'_F. Therefore

$$g\cdot(e'_i)=\sum_{j=1}^{n} r_{ij}(t)\cdot e'_j \quad \text{(here } t \text{ is the coordinate on } F\text{).}$$

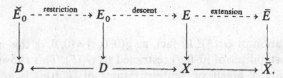

Since $g(U(F)_{\mathbb{C}} \times \mathbb{C}^l \times \{t\}) = U(F)_{\mathbb{C}} \times \mathbb{C}^l \times \{t\}$, it follows that for each t,

$$g \mapsto \rho_t(g) = (r_{ij}(t))$$

is an n-dimensional representation of $G_l(F)$. In fact, as $g(0, 0, t) = (0, 0, t)$, this is just the representation of the stabilizer of $(0, 0, t)$ restricted to $G_l(F)$. This shows that ρ_t is a holomorphic family of algebraic representations of G_l (this is not trivial because G_l has a positive dimensional center). Since G_l is reductive, we may change our basis $\{e'_i\}$ so that ρ_t is in fact independent of t.

Now consider the functions $h_{ij} = h(e'_i, e'_j)$ on D. Since h, e'_i and e'_j are $U \cdot V$-invariant, so is h_{ij}, hence in Siegel domain notation, it is a function of $u = \operatorname{Im} x + l_t(y, y)$ and t, i.e., is a function on $C(F) \times F$. For each fixed $t \in F$, and variable $u \in C(F)$,

$$H_t(u) = (h_{ij}(u, t))$$

is a map

$$C(F) \to C_n = \begin{pmatrix} \text{cone of pos. def. } n \times n \\ \text{Hermitian matrices} \end{pmatrix}.$$

I claim that (ρ, H_t) satisfy the hypothesis of §2. In fact,

$$\begin{aligned} H_t(gu)_{ij} &= h(ge'_i, ge'_j) \\ &= \sum_{k,l} r_{ik}(g)\, h(e'_k, e'_l)\, \overline{r_{jl}(g)} \\ &= (\rho(g) \cdot H_t(u) \cdot {}^t\overline{\rho(g)})_{ij}. \end{aligned}$$

Let $e = (ie_0, 0, t_0)$ be a base point of D. Since h is a K-invariant metric on E_0,

$$h(ke'_i, ke'_j)(e) = h(e'_i, e'_j)(e), \quad \text{all } k \in K.$$

Complexifying, we get too:

$$h(ke'_i, \bar{k}e'_j)(e) = h(e'_i, e'_j)(e), \quad \text{all } k \in K_{\mathbb{C}}.$$

Let $u_0 \in U_{\mathbb{C}}$ be given by $(x, y, t) \mapsto (x + ie_0, y, t)$. Then for all $g \in G_l(F)$, $u_0 g u_0^{-1}(e) = e$, so $u_0 g u_0^{-1} = k \cdot p$, where $k \in K_{\mathbb{C}}$, $p \in P_+$. By the lemma above, $u_0 g^* u_0^{-1} = \bar{k} \cdot p'$ for some other element $p' \in P_+$. Therefore:

$$\begin{aligned} h(e'_i, e'_j)(e) &= h(pe'_i, p'e'_j)(e) \quad (\text{since } P_+ \text{ acts trivially on } E_0(e)) \\ &= h(kpe'_i, \bar{k}p'e'_j)(e) \\ &= h(u_0 g u_0^{-1}(e'_i), u_0 g^* u_0^{-1}(e'_j))(e) \\ &= \sum r_{ik}(g)\, h(e'_k, e'_l)(e)\, \bar{r}_{jl}(g^*) \quad (\text{since } u_0 e'_i = e'_i, \text{ all } i) \end{aligned}$$

or

$$H_{t_0}(e_0) = \rho(g)\, H_{t_0}(e_0)\, {}^t\overline{\rho(g^*)}.$$

In fact, the same holds if t_0 is replaced by any $t \in F$ as follows easily using the $G_h(F)$-invariance of h and the fact that $G_l(F)$ and $U(F)_{\mathbb{C}}$ commute with $G_h(F)$.

Thus we have the full situation of § 2. In particular, we have available all the bounds of § 2.

As above, to describe local coordinates near boundary points of $\bar{X}$, choose a simplicial cone:

$$\sigma=\sum_{i=1}^{k}\mathbb{R}_{+}\cdot\xi_i\subset\overline{C(F)};\qquad \xi_i\in C(F)\Leftrightarrow 1\leqq i\leqq m$$

and let l_i be dual linear functionals on $U(F)_{\mathbb{C}}$. Then if (x, y, t) are Siegel domain coordinates on $D(F)$, and $z_i=e^{2\pi i\, l_i(x)}$, then (z, y, t) at points where at least one z_i is $0\,(1\leqq i\leqq m)$ are local coordinates on $\bar{X}$. Moreover each point $P\in S(F,\{\xi_i\})\subset\bar{X}$ has an open neighborhood whose intersection with X is contained in the image of

$$\{(x, y, t)\mid u=\operatorname{Im} x+l_t(y, y)\in\sigma+a,\ y\in Y',\ t\in F'\}$$

for some σ as above, $a\in C(F)$, Y' (resp. F') a relatively compact subset of Y (resp. F). Note that $\log|z_i|=-2\pi l_i(\operatorname{Im} x)$. At all points $P\in S(F,\{\xi_i\})$, we have to estimate $h(s_i', s_j')^{\pm 1}$ in terms of

$$\left(\sum\log\left|\frac{z_i}{C}\right|\right)^{2N}$$

$\left(\text{choose } C \text{ large enough so that } \left|\frac{z_i}{C}\right|<1 \text{ in a neighborhood of } P\right)$. This is the same as estimating $h(s_i', s_j')^{\pm 1}$ in terms of

$$\left(\sum l_i(u)+C\right)^{2N}$$

(choose C large enough so that $l_i(u)+C>0$ in a neighborhood of P). But if $l_i(u)+C>0$, $(\sum l_i(u)+C)^{2N}$ is comparable with $(\sum l_i(u)^2)^N$, hence with $\langle u, u\rangle^N$. This is exactly the estimate that Proposition 2.3 gives us. Next we have to estimate the connection and the curvature. Now in terms of a *holomorphic* trivialization of E_0, the connection is given by $\partial h\cdot h^{-1}$. What we have is good control of $\delta_1 h\cdot h^{-1}$ and $\delta_2(\delta_1 h\cdot h^{-1})$ for all vector fields δ_1, δ_2 in terms of the *real analytic* trivialization given by $\{e_i'\}$. Write

$$e_i=\sum_{j=1}^{n}a_{ij}\,e_j',$$
$$A=\text{matrix}\,(a_{ij}),$$
$$H^{an}=\text{matrix}\,h(e_i, e_j).$$

Then

$$H^{an}=A\cdot H\cdot{}^t\bar{A}.$$

From this you calculate:

$$\begin{pmatrix}\text{Connexion in}\\ \text{trivialization}\\ \{e_i\}\end{pmatrix}=\partial H^{an}\cdot(H^{an})^{-1}$$
$$=\partial A\cdot A^{-1}+A\cdot\partial H\cdot H^{-1}\cdot A^{-1}+A\cdot H\cdot(\partial^t\bar{A}\cdot{}^t\bar{A}^{-1})\,H^{-1}A^{-1},$$

$$
\begin{aligned}
d\begin{pmatrix}\text{connexion in}\\ \text{trivialization}\\ \{e_i\}\end{pmatrix} &= d(\partial H^{an}\cdot (H^{an})^{-1})\\
&= d(\partial A\cdot A^{-1}) + d(A\cdot \partial H\cdot H^{-1}\cdot A^{-1})\\
&\quad + d(A\cdot H\cdot (\partial{}^t\bar{A}\cdot {}^t\bar{A}^{-1})\cdot H^{-1}\cdot A^{-1})\\
&= d(\partial A\cdot A^{-1}) + dA\cdot (\partial H\cdot H^{-1})\cdot A^{-1}\\
&\quad + A\cdot d(\partial H\cdot H^{-1})\cdot A^{-1} + A\cdot \partial H\cdot H^{-1}\cdot A^{-1}\cdot dA\cdot A^{-1}\\
&\quad + dA\cdot [H\cdot (\partial{}^t\bar{A}\cdot {}^t\bar{A}^{-1}) H^{-1}]\cdot A^{-1}\\
&\quad + A(dH\cdot H^{-1})[H\cdot (\partial{}^t\bar{A}\cdot {}^t\bar{A}^{-1})\cdot H^{-1}]\cdot A^{-1}\\
&\quad + A\cdot [H\cdot d(\partial{}^t\bar{A}\cdot {}^tA^{-1})\cdot H^{-1}]\cdot A^{-1}\\
&\quad + A\cdot [H\cdot (\partial{}^t\bar{A}\cdot {}^t\bar{A}^{-1})\cdot H^{-1}]\cdot (dH\cdot H^{-1})\cdot A^{-1}\\
&\quad + A\cdot [H\cdot (\partial{}^t\bar{A}\cdot {}^t\bar{A}^{-1})\cdot H^{-1}]\cdot A^{-1}\cdot dA\cdot A^{-1}.
\end{aligned}
$$

Therefore, since A is a C^∞ metric on $\bar{X}$, to show that the connexion and its differential have Poincaré growth on $\bar{X}$ it suffices to prove that the 4 forms:

$$
\begin{aligned}
&\partial H\cdot H^{-1},\\
&d(\partial H\cdot H^{-1}),\\
&H\cdot (\partial{}^t\bar{A}\cdot {}^t\bar{A}^{-1})\cdot H^{-1},\\
&H\cdot d(\partial{}^t\bar{A}\cdot {}^t\bar{A}^{-1})\cdot H^{-1}
\end{aligned}
$$

have Poincaré growth on $\bar{X}$. To check this for the first two, note that H is a function on $C(F)\times F$, hence it suffices to bound $\delta_1 H\cdot H^{-1}$, $\delta_2(\delta_1 H\cdot H^{-1})$ for all vector fields δ_1, δ_2 on $C(F)\times F$. But the Poincaré metric is given in Siegel coordinates by:

$$
\begin{aligned}
ds^2 &= \sum \frac{|dz_i|^2}{|z_i|^2\left(\log\left|\frac{z_i}{C}\right|\right)^2} + \sum |dy_i|^2 + \sum |dt_i|^2\\
&= \sum \frac{|dl_i(x)|^2}{(l_i(\mathrm{Im}\, x)+C')^2} + \sum |dy_i|^2 + \sum |dt_i|^2
\end{aligned}
$$

(choose C large enough so that $l_i(\mathrm{Im}\, x)+C>0$ near the boundary point in question). Therefore the bounds of Proposition 2.7 imply that $\partial H\cdot H^{-1}$ and $d(\partial H\cdot H^{-1})$ have Poincaré growth.

To check the result for the last two, we need to know what sort of a function A is. Firstly, since e_i and e_i' are both $U_{\mathbb{C}}$-invariant, A is $U_{\mathbb{C}}$-invariant, i.e., is a function of y and t alone. Therefore it has derivatives only in the y and t directions, so to say these have Poincaré growth is just to say they are bounded along $\bar{X}$. Next, for all $t_0\in F$, the action of V on vector space of points (y, t_0) puts a complex structure on V. Thus it defines a splitting $V_{\mathbb{C}} = V_{t_0}^+ \oplus V_{t_0}^-$, where $V_{t_0}^\pm$ are complex subspaces and $V_{t_0}^-$ acts trivially on the points (y, t_0), while $V_{t_0}^+$ acts simply transitively. If we fix one t_0, then $V_{t_0}^+$ still acts simply transitively and holomorphically on the vector spaces of points (y, t) for t near t_0. Thus it is natural to choose our holomorphic basis e_i to be in fact $V_{t_0}^+\cdot U_{\mathbb{C}}$-invariant.

Moreover, it is easy to see that $G_l \cdot V$ normalizes $V_{t_0}^+ \cdot U_{\mathbb{C}}$ for all t_0, hence that the action of $G_l \cdot V$ in terms of the holomorphic basis is given by

$$g(e_i) = \sum_{j=1}^{n} \tilde{r}_{ij}(t) \cdot e_j$$

where

$$g \mapsto \tilde{\rho}_t(g) = (matrix\ \tilde{r}_{ij}(t))$$

is a representation of $G_l \cdot V$. Comparing ρ and $\tilde{\rho}$, we get:

$$\rho(g) = g^* A^{-1} \cdot \tilde{\rho}_t(g) \cdot A.$$

Since $e_i = e_i'$ on $(0, 0) \times F$, $\rho(g) = \tilde{\rho}_t(g)$ for all $g \in G_l$ and $A(0, t) = I_n$. Thus if $v(0, t) = (y, t)$,

$$I_n = \rho(v) = A(y, t)^{-1} \cdot \tilde{\rho}_t(v) \cdot I_n,$$

or

$$A(y, t) = \tilde{\rho}_t(v).$$

Now we use the simple:

Lemma. *Let σ be an algebraic representation of $G_l \cdot V$, and let σ_0 be the restriction of σ to G_l. Then for all $v \in V$, $\sigma(v)$ is σ_0-upper triangular.*

Proof. Let $A \subset G_l$ be a maximal $\mathbb{R}$-split torus. As in §2, there is a basis $\gamma_1, \ldots, \gamma_r$ of the character group of A such that A acts on the vector space $U(F)$ containing the cone $C(F)$ through the weights $\gamma_i + \gamma_j$. Then its action on $V(F)$ is through the weights γ_i (cf. [1], p. 224). Now if $V_i(F) \subset V(F)$ is the root space corresponding to γ_i, and if we diagonalize $\sigma(A)$:

$$\mathbb{C}^n = \bigoplus_{\lambda \in X(A)} W_\lambda,$$

then

$$V_i(F)(W_\lambda) \subset W_{\lambda + \gamma_i}$$

hence $V(F)$ acts in a σ_0-upper triangular fashion. QED

Thus $A(y, t)$ is ρ-upper triangular for all y, t, hence so are

$$A^{-1} \cdot \bar{\partial} A \quad \text{and} \quad d(A^{-1} \cdot \bar{\partial} A).$$

Applying Proposition 2.8, it follows that

$$H \cdot {}^t\overline{(A^{-1} \cdot \delta A)} \cdot H^{-1} \quad \text{and} \quad H \cdot {}^t\overline{(d(A^{-1} \cdot \bar{\partial} A))} \cdot H^{-1}$$

are bounded in a neighborhood of every point of $\bar{X}$, as required. This completes the proof of the Main Theorem.

A natural question is whether these vector bundles $\bar{E}$ are in fact pull-backs of vector bundles on less blown up compactifications of D/Γ. Thus Baily and Borel defined in [2] a "minimal" but usually highly singular compactification $(D/\Gamma)^*$ of D/Γ. Unfortunately $\bar{E}$ is only rarely a vector bundle on $(D/\Gamma)^*$ (we will see below one case where it is however). However, in [1], Ash, Rapoport, Tai and I defined not only smooth compactifications of D/Γ but also a bigger class of compactifications with toroidal singularities (cf. [9]). These are important because when you try to resolve $(D/\Gamma)^*$, often there is a $\overline{D/\Gamma}$ with relatively simple structure on the boundary but still with some toroidal singularities. It is easy to see that the construction above of $\bar{E}$ goes through equally well on all of these compactifications: it gives vector bundles on all of them such that whenever compactification a dominates compactification b, then extension a is the pull-back of extension b.

The Main Theorem, plus Hirzebruch's original proof of his proportionality theorem for compact locally symmetric varieties X, gives us easily the proportionality theorem in the general case:

Proportionality Theorem 3.2. *As above, fix:*

X = *an arithmetic variety* D/Γ, $D = K\backslash G$,

$\bar{X}$ = *a smooth compactification as in* [1],

$\check{D}$ = *compact dual of* D.

Then there is a constant K, *which in terms of a natural choice of metric on* D *is the volume of* X, *such that for all:*

$\check{E}_0$ = $G_{\mathbb{C}}$-*equivariant analytic rank* n *vector bundle on* $\check{D}$ *defined by a representation of* $\mathrm{Stab}_{G_{\mathbb{C}}}(e)$ *trivial on* P_+,

$\bar{E}$ = *corresponding vector bundle on* $\bar{X}$,

the following formula holds:

$$c^\alpha(\bar{E}) = (-1)^{\dim X} \cdot K \cdot c^\alpha(\check{E}_0), \qquad \text{all } \alpha = (\alpha_1, \ldots, \alpha_n), \ \sum \alpha_i = \dim X.$$

Proof. As above, choose a G-invariant Hermitian metric h_0 on E_0. By the Main Theorem, h_0 defined a "good" Hermitian metric h on E, hence its Chern forms $c_k(E, h)$ represent the Chern classes of $\bar{E}$. But on D, $c_k(E, h)$ are G-invariant forms, so:

$$\begin{aligned} c^\alpha(\bar{E}) &= \int_X c^\alpha(E, h) \\ &= \int_{\substack{F \text{ al Domain} \\ F \subset D}} c^\alpha(E_0, h_0) \\ &= \mathrm{vol}(F) \cdot c^\alpha(E_0, h_0)(e). \end{aligned}$$

Now if G^c is a compact form of G:

$$\mathrm{Lie}\, G = \mathfrak{k} \oplus \mathfrak{p},$$
$$\mathrm{Lie}\, G^c = \mathfrak{k} \oplus i\mathfrak{p}$$

then $\check{E}_0$ has a unique G^c-invariant Hermitian metric $\check{h}$ equal to h_0 at e. So

$$\begin{aligned} c^\alpha(\check{E}_0) &= \int_{\check{D}} c^\alpha(\check{E}_0, \check{h}) \\ &= \operatorname{vol}(\check{D}) \cdot c^\alpha(\check{E}_0, \check{h})(e). \end{aligned}$$

Then – and this is the essence of Hirzebruch's remarkable proof – a simple local calculation shows (cf. [7]):

$$c_k(E, h)(e) = (-1)^k c_k(\check{E}_0, \check{h}).$$

This proves the result.

To apply this result, it is important to describe the bundles $\bar{E}$ as closely as possible. Firstly, we can characterize their sections, precisely as a special case of the general definition of automorphic forms given by Borel [3]. Let $\rho: K \to GL(n, \mathbb{C})$ be a representation of K, and let

$$E_0 = G \times^K \mathbb{C}^n$$

be the associated G-equivariant vector bundle over $D = K \backslash G$ (i.e., E_0 = set of pairs (g, a), $\operatorname{mod}(g, a) \sim (kg, \rho(k)a)$). E_0 has a complex structure as follows: complexify ρ and extend it to $K_{\mathbb{C}} \cdot P_+$ to be trivial on P_+. Then E_0 is the restriction to D to the bundle

$$\check{E}_0 = G_{\mathbb{C}} \times^{(K_{\mathbb{C}} \cdot P_+)} \mathbb{C}^n$$

on $\check{D}$, and in the definition of $\check{E}_0$, everything is analytic. Borel introduces a measure of size on G by:

$$\|g\|_G = \operatorname{tr}(\operatorname{Ad} g^{*-1} \cdot g)$$

$* =$ Cartan involution on G w.r.t. K,

and defines holomorphic ρ-automorphic form f to be a function

$$f: G \to \mathbb{C}^n$$

such that

(1) $f(kg\gamma) = \rho(k) f(g)$, all $k \in K$, $\gamma \in \Gamma$,
(2) f induces a holomorphic section of E_0,
(3) $|f(g)| \leqq C \cdot \|g\|_G^n$, some $n \geqq 1$, $C > 0$.

Then one can show:

Proposition 3.3. *In the above notation:*

$$\Gamma(\bar{X}, \bar{E}) \cong \{\textit{vector space of holomorphic } \rho\textit{-automorphic forms}\}.$$

Sketch of Proof. The problem is to check that the bound (3) is equivalent to requiring that the corresponding section of $\bar{E}$ over X has growth $O((\sum \log |z_i|)^{2n})$

along $\bar{X}$. But $\|g\|_G$ defines a measure of size on D and on X by:

$$\forall x \in D: \ \|x\|_D = \|g\|_G \quad \text{if } x \text{ corresponds to the coset } K\cdot g,$$

$$\forall \bar{x} \in X, \text{ image of } x: \ \|\bar{x}\|_X = \min_{\gamma\in\Gamma} \|\gamma(x)\|_D = \min_{\gamma\in\Gamma} \|g\gamma\|_G.$$

Then holomorphic ρ-automorphic forms are clearly holomorphic sections s of $\bar{E}$ *over* X, such that

$$h(s,s)(x) \leqq C_1 \|x\|_X^n, \quad \text{some } n \geqq 1, \ C_1 > 0.$$

But if d_D is a G-invariant distance function on D, then it is easy to see (using $G = K\cdot A\cdot K$) that $d_D(x,e)$ and $\log\|x\|$ are bounded with respect to each other. In another paper [4], Borel has proven that if x is restricted to a Siegel set $\mathfrak{S} = \omega\cdot A_t\cdot e \subset D$, then

$$\min_{\gamma\in\Gamma} d_D(x\gamma, e) \approx d_D(x,e) \approx d_A(a(x), e)$$

(here $x = \omega(x)\,a(x)\cdot e$ and $\approx$ means the differences are bounded). Applying this to a subset of a Siegel Domain of 3rd kind of the type $\{(x,y,t) \mid y \in V', t \in F', \operatorname{Re} x \in U', \operatorname{Im} x - l_t(y,y) \in \sigma + a\}$ where $U' \subset U$, $V' \subset V$, $F' \subset F$ are compact subsets and $\sigma \subset \overline{C(F)}$ is a simplicial cone and $a \in C(F)$, we see that

$$\min_{\gamma\in\Gamma} d_D((x,y,t)\gamma, e)$$

can be bounded above and below by expressions

$$C_2 \log(\langle \operatorname{Im} x - l_t(y,y), e\rangle), \quad e \in C(F), \ C_2 > 0$$

hence $\|x\|_X$ can be bounded above and below by expressions

$$\langle \operatorname{Im} x - l_t(y,y), e\rangle^n, \quad e \in C(F), \ n \geqq 1.$$

Describing σ as $l_i \geqq 0$ as above (l_i linear functionals on U), this is of the same size as

$$\left(\sum l_i(\operatorname{Im} x) + C_3\right)^n$$

and as $z_i = e^{2\pi i\, l_i(x)}$, this is equal to

$$\left(-\sum \log(|z_i|/C_4)\right)^n. \quad \text{QED}$$

Next, there are 2 particular equivariant bundles where we can describe $\bar{E}$ more completely:

Proposition 3.4. a) *If* $E_0 = \Omega_D^1$, *the cotangent bundle, with canonical G-action, then* $\bar{E} = \Omega_{\bar{X}}^1(\log)$, *the bundle on* $\bar{X}$ *whose sections in a polycylinder* $\Delta^n \subset \bar{X}$ *such that*

$$\Delta^n \cap (\bar{X} - X) = \bigcup_{i=1}^{k} \begin{pmatrix} \text{coordinate hyperplanes} \\ z_i = 0 \end{pmatrix}$$

are given by

$$\sum_{i=1}^{k} a_i(z)\frac{dz_i}{z_i} + \sum_{i=k+1}^{n} b_i(z)dz_i.$$

b) *If E_0 is the canonical line bundle Ω^n_D, then $\bar{E}$ is the pull-back of an ample line bundle $\mathcal{O}(1)$ on the Baily-Borel compactification X^* of X. The sections of $\mathcal{O}(n)$ are the modular forms with respect to the nth power of the canonical automorphy factor given by the jacobian, hence $\mathcal{O}(n)$, $n \gg 0$, is the very ample bundle used by Baily and Borel to embed in X^* in $\mathbb{P}^N$.*

Proof. Using Siegel Domain coordinates (x, y, t) on $D(F)$, a $U(F)_{\mathbb{C}}$-invariant basis of $\Omega^1_{\underline{D}(F)}$ is given by $\{dx_i, dy_j, dt_k\}$. Therefore these span the corresponding bundle on $\bar{X}$ near the boundary F. But here

$$\{z_i = e^{2\pi i\, l_i(x)}, y_j, t_k\}$$

are coordinates and these differentials are $\left\{\dfrac{dz_i}{z_i}, dy_j, dt_k\right\}$. This proves (a).

To prove (b), recall that X^* is set-theoretically the union of X and of $F/\Gamma \cap N(F)$ for all rational boundary components F. Moreover, if $P \in F/\Gamma \cap N(F) \subset X^*$, then there exists a neighborhood $U \subset X^*$ and an open set $V \subset D$ such that V maps to $U \cap X$ and V is a $(G_l(F) \cdot V(F) \cdot U(F)) \cap \Gamma$-bundle over $U \cap X$. Now say $\{s_i\}$ is the $U(F)_{\mathbb{C}}$-invariant holomorphic basis of $\check{E}_0$ on $D(F)$ used to extend E over the F-boundary points of $\bar{X}$. If we verify that each s_i is $(G_l(F) \cdot V(F) \cdot U(F)) \cap \Gamma$-invariant it follows that E is trivial on $U \cap X$ and moreover, if $\pi\colon \bar{X} \to X^*$ is the canonical birational map, then $\{s_i\}$ are a basis of $\pi_* \bar{E}$ over U. Thus $\pi_* \bar{E}$ is a vector bundle which pulls-back to $\bar{E}$ on $\bar{X}$. Now in the case in question, E is a line bundle. s_1 can be identified with the differential form

$$(\bigwedge_i dx_i) \wedge (\bigwedge_j dy_j) \wedge (\bigwedge_k dt_k)$$

on $D(F)$, and $G_l \cdot V \cdot U$ acts on it by multiplication by the Jacobian determinant in the Siegel Domain coordinates. But Baily-Borel ([2], Prop. 3.14) showed that the Jacobian on $(G_l \cdot V \cdot U) \cap \Gamma$ was a root of unity. Since Γ is neat, it is one and $(G_l \cdot V \cdot U) \cap \Gamma$ indeed fixes s_1. The last assertion is just a restatement of Proposition 3.3 for this special case. QED

The following consequence of the proportionality principle seems to be more or less well known to experts, but does not seem to be contained in any published articles:

Corollary 3.5. *Let $L = (\Omega^n_{\check{D}})^{-1}$ be the ample line bundle on $\check{D}$, and let*

$$P(l) = \chi(L^{\otimes l})$$

be the Hilbert polynomial of $\check{D}$. Let $\pi\colon \bar{X} \to X^*$ map a smooth compactification of X onto Baily-Borel's compactification. Let $n_1 = \dim(X^* - X)$. Then there exists a *polynomial $P_1(l)$ of degree at most n_1 such that for all $l \geqq 2$:*

$$\dim \begin{bmatrix}\text{cusp forms on } D \\ \text{w.r.t. } \Gamma \text{ of weight } l\end{bmatrix} = \mathrm{vol}(X) \cdot P(l-1) + P_1(l).$$

Proof. The Riemann-Roch theorem gives us a "universal polynomial" Q such that if L is any line bundle on a smooth projective variety W, then

$$\chi(L) = Q(c_1(L); c_1(\Omega^1_W), \ldots, c_n(\Omega^1_W)).$$

Therefore if $n = \dim D$,

$$\begin{aligned}(-1)^n \mathrm{vol}(X) \cdot P(-l) &= (-1)^n \mathrm{vol}(X) \cdot \chi((\Omega^n_{\check{D}})^{\otimes l}) \\ &= (-1)^n \mathrm{vol}(X) \cdot Q(l c_1(\Omega^1_{\check{D}}); c_1(\Omega^1_{\check{D}}), \ldots, c_n(\Omega^1_{\check{D}})) \\ &= Q(l c_1(\Omega^1_{\bar{X}}(\log)); c_1(\Omega^1_{\bar{X}}(\log)), \ldots, c_n(\Omega^1_{\bar{X}}(\log)))\end{aligned}$$

by Proportionality Theorem 3.2.

Consider a typical term

$$[l c_1(\Omega^1_{\bar{X}}(\log))]^k \cdot c^\alpha(\Omega^1_{\bar{X}}(\log)), \quad |\alpha| + k = n.$$

Now by Proposition 3.4b:

$$c_1(\Omega^1_{\bar{X}}(\log)) = \pi^* H,$$

H an ample divisor on X^*. Let $n_1 = \dim(X^* - X)$. If $k > n_1$, the cycle class H^k on X^* is represented by a cycle supported on X alone, hence so is $\pi^* H_k$. Thus if $k > n_1$

$$(l \cdot \pi^* H)^k \cdot c^\alpha(\Omega^1_{\bar{X}}(\log)) = (l \cdot \pi^* H)^k \cdot c^\alpha(\Omega^1_{\bar{X}}).$$

Therefore

$$\begin{aligned}&Q(l c_1(\Omega^1_{\bar{X}}(\log)); c_1(\Omega^1_{\bar{X}}(\log)), \ldots, c_n(\Omega^1_{\bar{X}}(\log))) \\ &= Q(l c_1(\Omega^1_{\bar{X}}(\log)); c_1(\Omega^1_{\bar{X}}), \ldots, c_n(\Omega^1_{\bar{X}})) + (\text{polyn. of degree} \leqq n_1) \\ &= \chi(\Omega^n_{\bar{X}}(\log)^{\otimes l}) + (\text{polyn. of degree} \leqq n_1) \\ &= (-1)^n \chi(\Omega^n_{\bar{X}}(\log)^{\otimes(-l)} \otimes \Omega^n_{\bar{X}}) + (\text{polyn. of degree} \leqq n_1)\end{aligned}$$

by Serre duality.

Thus for suitable P_1 of degree at most n_1:

$$\mathrm{vol}(X) \cdot P(l-1) = \chi(\Omega^n_{\bar{X}}(\log)^{\otimes(l-1)} \otimes \Omega^n_{\bar{X}}) - P_1(l).$$

But since $(\Omega^n_{\bar{X}}(\log))^{\otimes N}$ is generated by its sections and maps $\bar{X}$ to X^* of the same dimension for $N \gg 0$, Kodaira Vanishing (cf. [13]) applies if $l \geqq 2$ and we have

$$h^0(\Omega^n_{\bar{X}}(\log)^{l-1} \otimes \Omega^n_{\bar{X}}) = \mathrm{vol}(X) \cdot P(l-1) + P_1(l).$$

The left-hand side is exactly the space of sections of $\Omega^n_{\bar{X}}(\log)^l$ which vanish on the boundary. By Proposition 3.3, these are exactly the cusp forms of weight l. QED

§ 4. Applications: General Type and Log General Type

The purpose of this section is to consider the application of the preceding theory to the question of when D/Γ is of general type, and to reprove as a consequence of our theory the following theorem of Y.-S. Tai ([1], Ch. IV, § 1).

Tai's Theorem 4.1. *If Γ is any arithmetic variety acting on a bounded symmetric domain D, then there is a subgroup $\Gamma_0 \subset \Gamma$ of finite index such that for all $\Gamma_1 \subset \Gamma_0$ of finite index, the variety D/Γ_1 is of general type.*

We recall that if X is any variety of dimension n, we say that X is of general type, if for one (and hence all) smooth complete varieties $\bar{X}$ birational to X, the transcendence degree of the ring

$$\bigoplus_{N=0}^{\infty} \Gamma(\bar{X}, (\Omega^n_{\bar{X}})^{\otimes N})$$

is $(n+1)$. More generally, the transcendence degree of this ring minus one is called the Kodaira dimension of X.

Recall that Iitaka [8] has recently introduced a complementary theory of "logarithmic Kodaira dimension" for arbitrary varieties Y. In fact, he first chooses a smooth blow-up Y' of Y and then a smooth compactification $\bar{Y}$ of Y' such that $\bar{Y} - Y'$ has normal crossings and defines $\Omega^1_{\bar{Y}}(\log)$ as the complex of 1-forms

$$\sum_{i=1}^{k} a_i(z)\frac{dz_i}{z_i} + \sum_{i=k+1}^{n} a_i(z)\, dz_i$$

if, locally, $\bar{Y} - Y'$ is given by $\prod_{i=1}^{k} z_i = 0$. By definition $\Omega^k_{\bar{Y}}(\log) = \Lambda^k(\Omega^1_{\bar{Y}}(\log))$. He then looks at the "logarithmic canonical ring":

$$R = \bigoplus_{N=0}^{\infty} \Gamma(\bar{Y}, \Omega^n_{\bar{Y}}(\log^{\otimes N}).$$

He shows that this ring, as well as all other vector spaces of global forms with logarithmic poles (obtained from decomposing $\Omega^1_{\bar{Y}}(\log) \otimes \cdots \otimes \Omega^1_{\bar{Y}}(\log)$ under the symmetric group and taking global sections) are independent of the choice of Y' and $\bar{Y}$. He then defines the logarithmic Kodaira dimension of Y to be the transcendence degree of R minus 1. We may restate Proposition 3.4(b) in this language as follows:

Proposition 4.2. *If Γ is a neat*[3] *arithmetic group, then D/Γ is a variety of logarithmic general type, i.e., its logarithmic Kodaira dimension equals its dimension.*

---

[3] Some hypothesis on elements of finite order is needed because $H/SL_2(\mathbb{Z}) \cong \mathbb{A}^1$ which is *not* of log general type!

Proof. In fact, by Proposition 3.4, R is just the homogeneous coordinate ring of the Baily-Borel compactification of D/Γ.

Note that D/Γ of logarithmic general type is weaker than saying D/Γ is general type.

I would like to add one comment to his theory which, in some cases, makes it easier to apply: one does not need smooth compactifications, but merely a toroidal compactification $\bar{Y}$ of Y' (cf. [9], p. 54). This means that locally $Y' \subset \bar{Y}$ is isomorphic to $(\mathbb{C}^*)^n \subset X_\sigma$, where X_σ is an affine torus embedding (i.e., $(\mathbb{C}^*)^n$ is Zariski-open in X_σ, X_σ is normal affine and translations by $\mathbb{C}^{*n}$ extend to an action of $(\mathbb{C}^*)^n$ on X_σ). On X_σ, define $\Omega^1_{X_\sigma}(\log)$ to be the sheaf generated by the $(\mathbb{C}^*)^n$-invariant 1-forms. Carrying these over, we define $\Omega^1_{\bar{Y}}(\log)$ to be the coherent sheaf of 1-forms on $\bar{Y}$, regular on Y', isomorphic locally to $\Omega^1_{X_\sigma}(\log)$. If $\bar{Y}' \to \bar{Y}$ is an "allowable" modification of toroidal embedding of Y ([9], p. 87), then $p^*(\Omega^1_{\bar{Y}}(\log)) \cong \Omega^1_{\bar{Y}'}(\log)$. In particular, there is always a smooth allowable modification $\bar{Y}'$ ([9], p. 94). So Iitaka's spaces of forms with log poles can be calculated equally well on a smooth $\bar{Y}$ or a toroidal $\bar{Y}$.

This extension is helpful in checking the analog of the above Proposition for the moduli space of curves:

Proposition 4.3. *Let $\mathfrak{M}_g^{(n)}$ be the moduli space of smooth curves of genus g with level n structure. If $n \geqq 3$, then $\mathfrak{M}_g^{(n)}$ is of log general type.*

Sketch of Proof. The proof follows the ideas of [12], § 5 very closely. Let H_g be the Hilbert scheme of e-canonically embedded smooth curves of genus g. Let $H_g^{(n)} \to H_g$ be the covering defined by the set of level n structures on these curves. Let $\bar{H}_g, \bar{\mathfrak{M}}_g$ be the compactified spaces allowing stable singular curves as well. Let $\bar{H}_g^{(n)}$, $\bar{\mathfrak{M}}_g^{(n)}$ be the normalization of $\bar{H}_g, \bar{\mathfrak{M}}_g$ in the coverings $H_g^{(n)}, \mathfrak{M}_g^{(n)}$. The group $G = PGL(\nu)$ $(\nu = (2e-1)(g-1))$ acts on $\bar{H}_g$ and on $\bar{H}_g^{(n)}$, freely on the latter, so that $\bar{\mathfrak{M}}_g \cong \bar{H}_g/G$, $\bar{\mathfrak{M}}_g^{(n)} \cong \bar{H}_g^{(n)}/G$. We have the diagram:

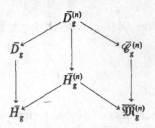

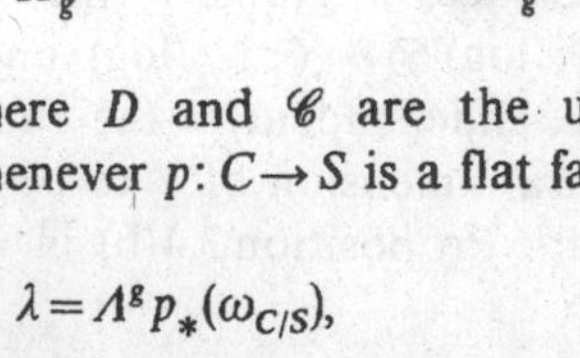

where D and $\mathscr{C}$ are the universal curves. Recall from [12] the notation: whenever $p: C \to S$ is a flat family of stable curves,

$$\lambda = \Lambda^g p_*(\omega_{C/S}),$$

$$\omega_{C/S} = \text{relative dualizing sheaf}$$

and if p is smooth over all points of S of depth zero, then $\Delta \subset S$ is the divisor of singular curves and

$$\delta = \mathcal{O}_S(\Delta).$$

Now on all 3 families above, we wish to show $\lambda^{13}\otimes\delta^{-1}$ and the sheaf of top logarithmic forms $\Omega^n(\log)$ are isomorphic line bundles. Firstly, for $p\colon \bar{D}_g\to\bar{H}_g$, we proceed like this: a) a simple modification of the proof of Theorem 5.10 [12] shows:

$$\lambda^{13}\otimes\delta^{-2}\cong\Lambda^{3g-3}(p_*(\Omega^1_{\bar{D}/\bar{H}}\otimes\omega_{\bar{D}/\bar{H}})).$$

b) Since $\bar{H}_g$ represents the functor of e-canonical stable curves, $T_{\bar{H}_g,[C]}$ is canonically isomorphic to the vector space of deformations of C. This has a subspace consisting of deformations of the e-canonical embedding where C doesn't change, and a quotient space of the deformations of C alone:

$$0\to\operatorname{Lie}G\to T_{\bar{H}_g,[C]}\to\operatorname{Ext}^1(\Omega^1_C,\mathcal{O}_C)\to 0$$

or dually:

$$0\to H^0(\Omega^1_C\otimes\omega_C)\to\Omega^1_{\bar{H}_g}\otimes\mathbb{K}(C)\to(\operatorname{Lie}G)'\to 0.$$

Therefore globally, we get

$$0\to p_*(\Omega^1_{\bar{D}/\bar{H}}\otimes\omega_{\bar{D}/\bar{H}})\to\Omega^1_{\bar{H}_g}\to(\operatorname{Lie}G)'\otimes\mathcal{O}_{\bar{H}_g}\to 0$$

hence if $m=\dim\bar{H}_g$,

$$\Omega^m_{\bar{H}_g}\cong\Lambda^{3g-3}p_*(\Omega^1_{\bar{D}/\bar{H}}\otimes\omega_{\bar{D}/\bar{H}})\cong\lambda^{13}\otimes\delta^{-2}$$

$$\therefore\ \Omega^m_{\bar{H}_g}(\log)\cong\lambda^{13}\otimes\delta^{-1}.$$

Secondly, for $p\colon\bar{D}_g^{(n)}\to\bar{H}_g^{(n)}$, $\lambda^{13}\otimes\delta^{-1}$ pulls back to the analogous sheaf on $\bar{H}_g^{(n)}$. Moreover, because $\bar{H}_g^{(n)}\to\bar{H}_g$ is ramified only along Δ which has normal crossings, $\bar{H}_g^{(n)}$ has toroidal singularities and $\Omega^m_{\bar{H}_g}(\log)$ pulls back to $\Omega^m_{\bar{H}_g^{(n)}}(\log)$. Finally both bundles descend to $\overline{\mathfrak{M}}_g^{(n)}$ and by Proposition 1.4 [11], are still isomorphic. Finally, it is proven in [12] (Th. 5.18 and 5.20: cf. diagram in § 5) that $\lambda^{13}\otimes\delta^{-1}$ is sample on $\overline{\mathfrak{M}}_g^{(n)}$. QED

In certain cases, there is a way of deducing that coverings of a variety of log general type are actually of general type. To explain this, suppose we are given a smooth quasi-projective variety Y, and a tower of connected étale Galois coverings:

$$\pi_\alpha\colon Y_\alpha\to Y,\qquad \text{group } \Gamma_\alpha.$$

We assume that any 2 covers π_α,π_β are dominated by a third one π_γ:

$$\begin{array}{ccccc} & & Y_\alpha & & \\ & \nearrow & & \searrow^{\pi_\alpha} & \\ Y_\gamma & & \xrightarrow{\ \pi_\gamma\ } & & Y, \\ & \searrow & & \nearrow_{\pi_\beta} & \\ & & Y_\beta & & \end{array}\qquad \begin{array}{ccccc} & & \Gamma_\alpha & & \\ & \nearrow & & \searrow & \\ \Gamma_\gamma & & & & (e). \\ & \searrow & & \nearrow & \\ & & \Gamma_\beta & & \end{array}$$

Let $\bar{Y}$ be a smooth compactification of Y with normal crossings at infinity. Extend the covering Y_α to a finite covering

$$\pi_\alpha\colon \bar{Y}_\alpha \to \bar{Y}$$

be defining $\bar{Y}_\alpha$ to be the normalizations of $\bar{Y}$ in the function field of Y_α. We now make the definition:

Definition. *The tower $\{\pi_\alpha\}$ is locally universally ramified over $\bar{Y}-Y$ if for all $x\in\bar{Y}-Y$, we take a nice neighborhood of x:*

$$\Delta^n\subset\bar{Y},$$

$$\Delta^n\cap(\bar{Y}-Y)=\begin{pmatrix}\text{union of coordinate hyperplanes}\\ z_1=0,\dots,z_k=0\end{pmatrix}$$

then for all m, there is an α and a commutative diagram:

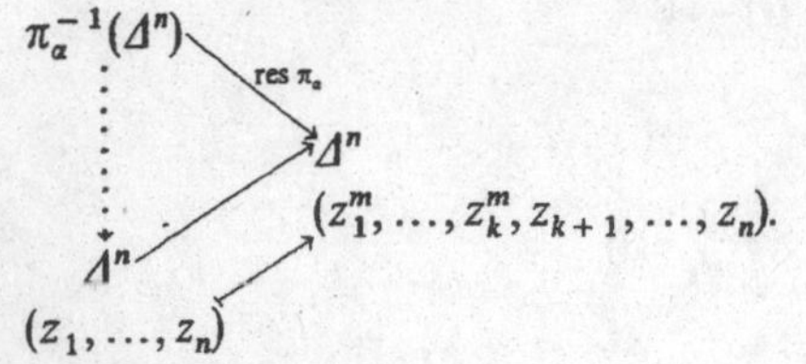

In other words, $\pi_\alpha^{-1}(\Delta^n\cap(\bar{Y}-Y))$ is cofinal in the set of all unramified coverings of $\Delta^n\cap(\bar{Y}-Y)$.

Then we assert:

Proposition 4.4. *Let $Y\subset\bar{Y}$ be as above and let $\pi_\alpha\colon\bar{Y}_\alpha\to\bar{Y}$ be a tower of coverings unramified over Y and locally universally ramified over $\bar{Y}-Y$. Then if Y is logarithmically of general type, there is an α_0 such that for all α_1 such that the covering π_{α_1} dominates π_{α_0}, Y_{α_1} is of general type.*

Proof. Let $\Delta=\bar{Y}-Y$, $n=\dim Y$ and let $\omega=\Omega^n_{\bar{Y}}(\log)$. Then we know that

a) for some N, there are differentials $\eta_0,\dots,\eta_n\in\Gamma(\bar{Y},\omega^{\otimes N})$ such that $\eta_1/\eta_0,\dots,\eta_n/\eta_0$ are a transcendence base of the function field $\mathbb{C}(Y)$,

b) $h^0(\bar{Y},\omega^{\otimes N})\geqq C_1N^{n+1}$ if $N\geqq N_0$.

From (b) it follows that

$$\Gamma(\bar{Y},\omega^{\otimes N})\to\Gamma(\Delta,\omega^{\otimes N}\otimes\mathcal{O}_\Delta)$$

has a non-zero kernel for some N: let ζ be in the kernel. Replacing η_i by $\eta_i\otimes\zeta$, we may assume that all η_i are zero on Δ. Now let's examine locally what happens to η_i when lifted to a covering of $\bar{Y}$: let $\Delta^n\subset\bar{Y}$ be a polycylinder such that $\Delta^n\cap(\bar{Y}-Y)=V\left(\prod_{i=1}^k z_i\right)$. Write out η_i:

$$\eta_1=a_i(z)\cdot\prod_{i=1}^k z_i\cdot\left(\frac{dz_1\wedge\cdots\wedge dz_n}{z_1\dots z_k}\right)^N,$$

a_i holomorphic. Let $w_i = z_i^{1/m}$, $1 \leqq i \leqq k$, $w_i = z_i$, $k+1 \leqq i \leqq n$ and let

$$\pi_m : \Delta^n \to \Delta^n$$

be the covering of the z-polycylinder by the w-polycylinder. Then

$$\frac{dw_i}{w_i} = \frac{1}{m}\frac{dz_i}{z_i}, \qquad 1 \leqq i \leqq k,$$

$$dw_i = dz_i, \qquad k+1 \leqq i \leqq n$$

hence

$$\pi_m^*(\eta_i) = m^{kN} a_i \cdot \prod_{i=1}^{k} w_i^m \cdot \left(\frac{dw_1 \wedge \cdots \wedge dw_n}{w_1 \ldots w_k} \right)^N.$$

So if $m \geqq N$, $\pi_m^*(\eta_i)$ is a holomorphic differential form on Δ^n. Now for each $x \in \bar{Y} - Y$, fix a neighborhood $U_x \subset \bar{Y}$ of this type and a covering

$$\pi_{\alpha(x)} : \bar{Y}_{\alpha(x)} \to \bar{Y}$$

which, over U_x, dominates π_N. $\bar{Y} - Y$ is covered by a finite number $\{U_{x_i}\}$ of U_x's, so we can find one cover π_{α_0} which dominates all the covers $\pi_{\alpha(x_i)}$. I claim that if π_{α_1} dominates π_{α_0}, hence $\pi_{\alpha(x_i)}$, then $\pi_{\alpha_1}^*(\eta_i)$ has no poles on a desingularization $\bar{Y}'_{\alpha_1}$ over $\bar{Y}_{\alpha_1}$. This is clear because it has an open covering by open sets V_i sitting in a diagram

$$\begin{array}{ccc} V_i & \subset & \bar{Y}'_{\alpha_1} \\ \downarrow & & \downarrow \\ U_{x_i} & & \bar{Y}_{\alpha_1} \\ \downarrow{\scriptstyle \pi_n} & & \downarrow{\scriptstyle \pi_{\alpha_1}} \\ U_{x_i} & \subset & \bar{Y} \end{array} \qquad (\pi'_{\alpha_1} : \bar{Y}'_{\alpha_1} \to \bar{Y})$$

so $\pi_m^*(\eta_i)$ regular $\Rightarrow \pi'_{\alpha_1}(\eta_i)$ regular on V_i. But now $\pi'^*_{\alpha_1}(\eta_i/\eta_0)$ are a transcendence base of the function field of $\bar{Y}_{\alpha_1}$ so $\bar{Y}_{\alpha_1}$ is of general type. QED

Let's consider the case $Y = D/\Gamma$ again. If Γ is an arithmetic group, then for every positive integer n, we have its level n subgroup $\Gamma(n)$, i.e., if

$$\Gamma = \mathscr{G}(\mathbb{Z}), \qquad \mathscr{G} \text{ algebraic group over Spec } \mathbb{Z}$$

then

$$\Gamma(n) = \operatorname{Ker}[\mathscr{G}(\mathbb{Z}) \to \mathscr{G}(\mathbb{Z}/n\mathbb{Z})].$$

It's easy to see that

$$\pi_n : D/\Gamma(n) \to D/\Gamma$$

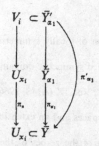

is locally universally ramified over $\overline{D/\Gamma} - D/\Gamma$. In fact, let F be a rational boundary component. Then near boundary points associated to F, the pair $D/\Gamma \subset \overline{D/\Gamma}$ is isomorphic to

$$\begin{array}{ccc} D(F)/U(F)\cap\Gamma & \subset & (D(F)/U(F)\cap\mathbb{Z})_\sigma \\ \| & & \| \\ C^{*n}\times\mathbb{C}^m\times F & & \mathbb{C}^n\times\mathbb{C}^m\times F. \end{array}$$

Thus if $U \subset \overline{D/\Gamma}$ is a small neighborhood of a point corresponding in the above chart to $(0, y, t)$, $\pi_1(U \cap (D/\Gamma))$ is isomorphic to $U(F) \cap \Gamma$. Thus we must check that for all n, there is an m such that

$$U(F)\cap\Gamma(m)\subset n\cdot U(F)\cap\Gamma.$$

But if F is rational, $U(F)$ is an algebraic subgroup of the full group $\mathcal{G}$ which is defined over $\mathbb{C}$, so this is clear. So now Proposition 4.2, Proposition 4.4 and this remark altogether imply Tai's theorem.

It is now known that this same method will show that at least some high-non-abelian levels of $\mathfrak{M}_g$ are varieties of general type too. It is not simple however to check that the Teichmuller tower is locally universally ramified at infinite. This has recently been proven by a ingenious use of dihedral level by T.-L. Brylinski.

References

1. Ash, A., Mumford, D., Rapoport, M., Tai, Y.: Smooth Compactification of locally symmetric varieties. Math. Sci. Press (53 Jordan Rd., Brookline, Mass. 02146), 1975
2. Baily, W.L., Borel, A.: Compactification of arithmetic quotients of bounded symmetric domains. Annals of Math. **84**, 744 (1966)
3. Borel, A.: Introduction to Automorphic Forms. In: Symp. in Pure Math., Vol. IX (AMS, 1966), p. 199
4. Borel, A.: Some metric properties of arithmetic quotients of symmetric spaces and an extension theorem. J. Diff. Geometry **6** (1972)
5. Bott, R., Chern, S.: Hermitian vector bundles and the equidistribution of the zeroes of their holomorphic sections. Acta math. **114**, 71 (1965)
6. Cornalba, M., Griffiths, P.: Analytic cycles and vector bundles on non-compact algebraic varieties. Inventiones math. **28**, 1 (1975)
7. Hirzebruch, F.: Automorphe Formen und der Satz von Riemann-Roch. In: Symposium Internacional de Topologia Algebraica, Unesco, 1958
8. Iitaka, S.: On logarithmic Kodaira dimension of algebraic varieties. (To appear)
9. Kempf, G., Knudsen, F., Mumford, D., Saint-Donat, B.: Toroidal Embeddings. Springer Lecture Notes 339 (1973)
10. Kobayashi, S.: Hyperbolic manifolds and holomorphic mappings. M. Dekker Inc., 1970
11. Mumford, D.: Geometric Invariant Theory. Berlin-Heidelberg-New York: Springer 1965
12. Mumford, D.: Stability of projective varieties. L'Enseignement Mathematique 1977
13. Mumford, D.: Pathogies III. Amer. J. Math. **89** (1967)

Received May 8, 1977

On the Kodaira Dimension of the Siegel Modular Variety

by David Mumford

Let $\mathcal{A}_g$ represent the quotient of Siegel's upper half-space $\mathfrak{h}_g$ of rank g by the full integral symplectic group $Sp(2g, \mathbb{Z})$: this is known as Siegel's modular variety, or as the moduli space of g-dimensional principally polarized abelian varieties (called p.p.a.v. below). $\mathcal{A}_g$ has been shown to be a variety <u>of general type</u> (i.e., Kodaira dimension = dimension) for various g's: Freitag [F1] proved this first if $24|g$; Tai [T] proved this recently for all $g \geq 9$. On the other hand, $\mathcal{A}_g$ is known to be unirational for $g \leq 5$: Donagi [D] for $g = 5$, Clemens [C] for $g = 4$, classical for $g \leq 3$. The purpose of this paper is to refine Tai's result, showing:

<u>Theorem</u>: <u>$\mathcal{A}_g$ is of general type if</u> $g \geq 7$.

Note that this leaves only the Kodaira dimension of $\mathcal{A}_6$ still to be determined. We shall use results of Freitag and Tai in a crucial way, but the idea of the proof is a direct adaption of the proof [H-M] by Harris and the author that $\mathcal{M}_g$ is of general type if $g \geq 25$, g odd. In that proof the divisor D_k of curves which are k-fold covers of $\mathbb{P}^1$, $k = \frac{g+1}{2}$, is shown to be linearly equivalent to

$$nK\text{-(ample divisor)-(effective divisor)}.$$

Here we prove the same thing except that the role of D_k is taken by the components of N_0, where

$$\mathcal{N}_k = \left[\text{locus of p.p.a.v. where dim(sing. locus of } \Theta) \geq k.\right]$$

These sets $\mathcal{N}_k$ were introduced by Andreotti and Mayer [A-M], and studied recently by Beauville [B]. I want to thank Beauville very much for stimulating discussions which led me to this result. At the same time, I would like to raise the question which seems very interesting to me: is there an explicit polynomial in theta constants, or other modular forms constructed from theta series (with quadratic forms and pluri-harmonic coefficients) whose zeroes give $\mathcal{N}_0$ with suitable multiplicities? Although important steps are taken in this direction in Andreotti-Mayer [A-M] and Beauville [B], this is not answered because the "theta nulls" $C(r,\mu,z)$ are not in general modular forms — they are theta series whose coefficients are not pluri-harmonic; esp. you cannot form a modular form out of the $\partial^2\vartheta/\partial u_k^2$'s alone without using mixed derivatives $\partial^2\vartheta/\partial u_k \partial u_\ell$ too. Finally, I want to mention the related results of Stillman [S] (based on earlier ideas of Freitag [F2]) which prove $\mathcal{A}_g$ carries holomorphic (4g-6)-forms for $g \geq 7$, $g \neq 8$. These results are directly based on the use of theta series.

§1. A partial compactification of the Siegel modular variety.

Satake's compactification $\mathcal{A}_g^*$ of $\mathcal{A}_g$ consists, set-theoretically, in the union of (g+1)-strata:

$$\mathcal{A}_g^* = \mathcal{A}_g \amalg \mathcal{A}_{g-1} \amalg \cdots \amalg \mathcal{A}_0 .$$

The Kodaira dimension of $\mathcal{A}_g$ is based on pluri-canonical differentials on a desingularization $\tilde{\mathcal{A}}_g$ of $\mathcal{A}_g^*$. However, Tai has shown that a pluri-canonical differential form with "no poles above $\mathcal{A}_g \amalg \mathcal{A}_{g-1}$", is everywhere regular, so we do not have to study the full $\tilde{\mathcal{A}}_g$. We will make this precise in a minute. The space we want to work with is a blow-up of $\mathcal{A}_g \amalg \mathcal{A}_{g-1}$ first introduced by Igusa [I] and studied by the author [M] and by Namikawa [N]. To describe this space geometrically, let us define a <u>rank 1 degeneration of a p.p.a.v.</u> as follows: it is a pair $(\overline{G},D)$ where $\overline{G}$ is a complete g-dimensional variety and D is an ample divisor (i.e., $\overline{G}$ is to be the limit of a g-dimensional abelian variety and D the limit of its theta divisor). $\overline{G}$ is constructed as follows:

1) let B^{g-1} be a (g-1)-dimensional p.p.a.v., $\Xi \subset B$ its theta divisor
2) let G be an algebraic group which is an extension of B by $\mathbb{G}_m$:
   $$0 \longrightarrow \mathbb{G}_m \longrightarrow G \longrightarrow B \longrightarrow 0.$$
3) Considering G as a $\mathbb{G}_m$-bundle over B, let $\tilde{G}$ be the associated $\mathbb{P}^1$-bundle:

$$\begin{array}{ccc} G & \subset & \tilde{G} \\ & \mathbb{G}_m \searrow \quad \pi \swarrow \mathbb{P}^1 & \\ & B & \end{array}$$

Then $\tilde{G}-G$ equals $\tilde{G}_0 \amalg \tilde{G}_\infty$, the union of 2 sections of $\tilde{G}$ over B.

4) Then $\overline{G}$ is to be the non-normal variety obtained by glueing $\tilde{G}_0, \tilde{G}_\infty$ with a translation by a point $b \in B$.

5) Note that on $\tilde{G}$

$$\tilde{G}_0 - \tilde{G}_\infty \equiv \pi^{-1}(E), \quad E \text{ algebraically equivalent to 0 on B}$$
$$\equiv \pi^{-1}(\Xi - \Xi_{b_1}), \text{ for a unique } b_1 \in B.$$

Thus

$$\tilde{G}_0 + \pi^{-1}(\Xi_{b_1}) \equiv \tilde{G}_\infty + \pi^{-1}(\Xi).$$

Let $\tilde{L} = \mathcal{O}_{\tilde{G}}(\tilde{G}_\infty + \pi^{-1}(\Xi))$. Via the Leray spectral sequence for π, we see that $h^0(\tilde{L}) = 2$ and that $\tilde{G}_0 + \pi^{-1}(\Xi_{b_1})$, $\tilde{G}_\infty + \pi^{-1}(\Xi)$ span the linear system $|\tilde{L}|$. Then $|\tilde{L}|_{\tilde{G}_0} \cong \mathcal{O}_B(\Xi)$ and $\tilde{L}|_{\tilde{G}_\infty} \cong \mathcal{O}_B(\Xi_{b_1})$, so if b is chosen to be b_1 (and only then) the line bundle $\tilde{L}$ can be descended to a line bundle L on $\overline{G}$. Choose such an L and let

$$D = \text{the unique divisor in } |L|.$$

We now define

(1.1) $\quad \overline{A}_g^{(1)} = \begin{cases} \text{coarse moduli space of p.p.a.v.}(A,\Theta) \text{ of} \\ \text{dimension g and their rank 1 degenerations} \end{cases}.$

As first shown by Igusa, this space exists, is a quasi-projective variety, and is essentially the blow-up of the open set $\mathcal{A}_g \amalg \mathcal{A}_{g-1}$ in $\mathcal{A}_g^*$ along its boundary $\mathcal{A}_{g-1}$. $\overline{\mathcal{A}}_g^{(1)}$ is the union of $\mathcal{A}_g$ and a divisor Δ parametrizing the rank 1 degenerations. Via the map

$$(\overline{G},D) \longmapsto (B,\Xi)$$

the divisor Δ is seen to be fibred:

$$\text{(1.2)} \qquad \begin{array}{c} \Delta \\ \delta \big\downarrow \\ \mathcal{A}_{g-1} \end{array} \quad \text{fibres } B/\mathrm{Aut}(B,\Xi) \; .$$

Analytically, we may consider $\overline{\mathcal{A}}_g^{(1)}$ to represent precisely the degenerations of the abelian variety $A_{\Omega(t)}$ with period matrix $\Omega(t)$ when:

$$\left.\begin{array}{c} \mathrm{Im}\, \Omega_{11} \longrightarrow \infty \\ \text{and } \Omega_{ij},\ i>1 \text{ or } j>1, \text{ have finite limits} \end{array}\right\} \text{ as } t \longrightarrow 0$$

Then $B = B_{\Omega^{(1)}}$, where $\Omega^{(1)}$ is the lower right block of the limit

$$\Omega(0) = \left(\begin{array}{c|c} i\infty & \omega \\ \hline {}^t\omega & \Omega^{(1)} \end{array} \right)$$

and b is the image of the vector $\vec{\omega} = (\Omega_{12}(0)\,\Omega_{13}(0),\cdots,\Omega_{1g}(0))$ in $B_{\Omega^{(1)}}$. To find D, we must translate $\Theta_{\Omega(t)} \subset A_{\Omega(t)}$ as $t \longrightarrow 0$.

Thus

$$\Theta_{\Omega(t)} = \left\{ \text{zeroes of } \vartheta(z,\Omega) = \sum_{n\in\mathbb{Z}^g} e^{\pi i\,{}^t n\Omega(t)n+2\pi i\,{}^t n\cdot z} \right\}.$$

Translate $\Theta_{\Omega(t)}$ by $b(t)$, the image of $(\frac{\Omega_{11}(t)}{2},0,\cdots,0)$:

$$T_{b(t)}(\Theta_{\Omega(t)}) = \left\{ \text{zeroes of } \sum e^{\pi i(n_1^2-n_1)\Omega_{11}(t)} \cdot e^{\left\{\pi i \sum_{i,j\neq 1,1} n_i n_j \Omega_{ij}(t)+2\pi i\,{}^t nz\right\}} \right\}.$$

Then $e^{\pi i(n_1^2-n_1)\Omega_n(t)} \longrightarrow 0$ unless $n_1 = 0$ or 1, hence the limit is

$$\left\{ \text{zeroes of } \sum_{n_2,\cdots,n_g\in\mathbb{Z}} e^{\left\{\pi i \sum_{i,j\geq 2} n_i n_j \Omega_{ij}(0)+2\pi i \sum_{j\leq 2} n_j z_j\right\}} \cdot \left(1+e^{2\pi i z_1}\cdot e^{2\pi i \sum_{j\leq 2} n_j\Omega_{1j}}\right) \right\}$$

(1.13)

$$= \left\{ \text{zeroes of } \vartheta(z^{(1)},\Omega^{(1)}) + e^{2\pi i z_1}\cdot \vartheta(z^{(1)}+\omega,\Omega^{(1)}) \right\}$$

where $z^{(1)} = (z_2,\cdots,z_g)$ is the analytic coordinate on $B_{\Omega^{(1)}}$. Interpreting $e^{2\pi i z_1}$ as the algebraic coordinate in the fibre $\mathbb{G}_m$ of G, and Ξ as the zeroes of $\vartheta(z^{(1)},\Omega^{(1)})$, this is immediately seen to be D if L is suitably defined.

Next, let $\overline{A}_g^{(1),0}$ be the open set in $\overline{A}_g^{(1)}$ parametrizing those pairs (A,Θ) or $(\overline{G},D)$ whose automorphism group is the minimal one, $\{\pm 1\}$. More precisely, the only non-trivial automorphism of A (or $\overline{G}$) mapping Θ (resp. D) to itself is of the form $x \longmapsto -x+a$, some a*. Then $\overline{A}_g^{(1),0}$ is locally isomorphic

*We have not normalized Θ and D to be symmetric. On the other hand, we have not fixed an origin either, so the pairs (A,Θ) and (A,Θ_c) are isomorphic by translation by c, and define the same point of $\overline{A}_g^{(1)}$.

to the universal deformation space of (A,θ) (or $(\overline{G},D)$), hence is a smooth of dimension $g(g+1)/2$. Analytically, A_g^0 is the open subset of A_g of points which are images of $\Omega \in \mathfrak{H}_g$ whose stabilizer in $Sp(2g,\mathbb{Z})$ are just $(\pm I)$. Likewise, using the analytic description of $\overline{A}_g^{(1)}$ in Ash et al [A-M-R-S], $\overline{A}_g^{(1),0}$ is the open subset of $\overline{A}_g^{(1)}$ of points which are images of points in $(\mathfrak{H}_g/U_{\mathbb{Z}})_{\{\sigma_\alpha\}}$ whose stabilizer in the normalizer of the first boundary component is just $(\pm I)$. (Compare Tai [T], §). This set includes, in particular, those $\overline{G}$ constructed from a $(B,\Xi) \in A_{g-1}^0$ and a point $b \in B$ not of order 2. We are now in a position to state one of the main results of Tai's paper [T], in the form in which we need it:

Theorem 1.4 (Tai). If $g \geq 5$, then

a) $\operatorname{codim}(\overline{A}_g^{(1)} - \overline{A}_g^{(1),0}) \geq 2$

and

b) $\Gamma(\tilde{A}_g, \mathcal{O}(nK)) = \Gamma(\overline{A}_g^{(1),0}, \mathcal{O}(nK))$, if $n \geq 1$.

This means that a pluri-canonical differential with no poles on $\overline{A}_g^{(1),0}$ is everywhere regular on a full desingularization $\tilde{A}_g$ of A_g^*.

The second result we need is the calculation of $\operatorname{Pic}(A_g^0)$. This follows from the theory of Matsushima, Borel, Wallach and others on the low cohomology groups of discrete subgroups of Lie groups. In particular, the results of Borel [Bo] imply that for any subgroup $\Gamma \subset Sp(2g,\mathbb{Z})$ of finite index:

$$H^*(\Gamma,\mathbb{Q}) \cong \mathbb{Q}[C_2, C_6, C_{10}, \ldots] \text{ , in degrees } \leq g-2 \text{ .}$$

In particular:

$$H^2(A_g,\mathbb{Q}) \cong H^2(Sp(2g,\mathbb{Z}),\mathbb{Q}) \cong \mathbb{Q} \quad \text{if } g \geq 4 \text{ .}$$

An immediate corollary* is:

Theorem 1.5 (Borel et al): $\mathrm{Pic}(\mathcal{A}_g^0) \otimes \mathbb{Q} \cong \mathbb{Q}\cdot\lambda$, if $g \geq 4$, where λ is the line bundle on $\mathcal{A}_g^0$ defined by the co-cycle $\det(C\Omega+D)$.

Corollary 1.6: $\mathrm{Pic}(\overline{\mathcal{A}}_g^{(1),0}) \otimes \mathbb{Q} \cong \mathbb{Q}\lambda + \mathbb{Q}\cdot\delta$

where δ is the divisor class of the boundary Δ.

In terms of these generators, a standard result is:

Proposition 1.7. $K_{\overline{\mathcal{A}}_g^{(1),0}} \equiv (g+1)\lambda - \delta$.

For a proof, see for instance Tai [T], §1. Another fairly standard result that we need is:

Proposition 1.8. Let (B,Ξ) be a $(g-1)$-dimension p.p.a.v. whose automorphism group is (± 1). Consider the 2-1 map

$$\phi: (B - B_2) \longrightarrow \overline{\mathcal{A}}_g^{(1),0}$$

defined by $\phi(b)$ = the pair $(\overline{G},D)$ constructed from (B,Ξ) with

---

* If $\tilde{\mathcal{A}}_g$ is a smooth compactification of $\mathcal{A}_g^0$, then use:

$$\begin{array}{ccccc}
\oplus\mathbb{Q}\delta_i & \longrightarrow & \mathrm{Pic}\tilde{\mathcal{A}}_g \otimes \mathbb{Q} & \xrightarrow{\text{res}} & \mathrm{Pic}\mathcal{A}_g^0 \otimes \mathbb{Q} \\
\| & & \downarrow & & \downarrow \\
\oplus\mathbb{Q}\delta_i & \longrightarrow & H^2(\tilde{\mathcal{A}}_g,\mathbb{Q}) & \longrightarrow & H^2(\mathcal{A}_g^0,\mathbb{Q})
\end{array}$$

plus $H^2(\mathcal{A}_g^0,\mathbb{Q}) \cong H^2(\mathcal{A}_g,\mathbb{Q}) \cong \mathbb{Q}$.

glueing via b. Then

$$\phi^*(\mathcal{O}_{\overline{A}_g}(1),0^{(\Delta)}) \cong \mathcal{O}_B(-2\Xi).$$

Proof: Let's construct over B the family of $(\overline{G},D)$'s made up with all possible b's. To do this, let P be the Poincaré bundle over $B\times B$, trivial on $e\times B$, $B\times e$. Then $P^* = P-(0\text{-section})$ serves as the universal family of G's. Let $\overline{P} \supset P$ be the associated $\mathbb{P}^1$-fibre bundle, and

$$\mathbb{P} = \overline{P}/(b_1,b_2,0) \sim (b_1,b_1+b_2,\infty).$$

Then the projection on the first factor:

$$p_1\colon \mathbb{P} \longrightarrow B$$

is the universal family of $\overline{G}$'s. The deformation theory of such a $\overline{G}$ gives an exact sequence:

$$0 \longrightarrow H^1(\overline{G},\underset{\sim}{T}^0(\mathcal{O}_{\overline{G}})) \longrightarrow T^1(\overline{G}) \longrightarrow H^0(\mathrm{Sing}\,\overline{G},\ \underset{\sim}{T}^1(\mathcal{O}_{\overline{G}}))$$
$$\| $$
$$H^0(B,N_0\otimes N_\infty)$$

where N_0,N_∞ are the normal bundles to the locus of double points of $\overline{G}$. For one $\overline{G}$, made up starting from a line bundle L over B, completed at ∞ and glued by translation by $b\in B$,

$$N_0 \otimes N_\infty \cong L\otimes T_b^*(L^{-1}).$$

Note that L must be algebraically equivalent to 0, hence $T_b^*L^{-1} \cong L^{-1}$, hence $N_0 \otimes N_\infty \cong \mathcal{O}_B$. Thus $H^0(B, N_0 \otimes N_\infty) \cong k$. This one-dimensional vector space represents the normal bundle to Δ in $\overline{\mathcal{A}}_g$ at the point $(\overline{G}, D)$. Doing this now for the whole family $\mathcal{P} \longrightarrow B$, $N_0 \otimes N_\infty$ is the line bundle on $B \times B$ given by

$$P \otimes T^*(P^{-1})$$
$$\text{where } T(x,y) = (x, x+y).$$

Then the normal bundle to Δ, pulled back to this family, is

$$p_{1,*}(P \otimes T^*(P^{-1}))$$

which is the same as the restriction of $P \otimes T^*P^{-1}$ to $B \times e$, i.e., $\delta^*(P^{-1})$, where $\delta(x) = (x,x)$. Since P, along the diagonal of $B \times B$ is $\mathcal{O}(2\Xi)$, this proves the Proposition. QED

§2. The divisor N_0 and its class in $\text{Pic}(\overline{\mathcal{A}}_g^{(1)})$.

Andreotti-Mayer [A-M] defined the important subsets N_k in $\mathcal{A}_g$:

(2.1) $$N_k = \{(A,\Theta) \mid \text{Sing } \Theta \neq \emptyset \text{ and } \dim(\text{Sing } \Theta) \geq k\}.$$

Andreotti and Mayer prove by using the Heat equation for that $N_0 \subsetneq \mathcal{A}_g$, but it is not easy to estimate the dimension of N_k in general. Nowever, we are interested only in codimension 1 and we must at least check that none of the N_k, $k \geq 1$, have codimension 1 components. This follows by an elaboration of

Andreotti-Mayer's arguments using the heat equation:

<u>Lemma 2.2</u>. <u>The codimension of</u> N_1 <u>(hence of</u> $N_2, N_3, \cdots$) <u>in</u> A_g <u>is greater than</u> 1.

<u>Proof</u>: We use the heat equation

$$(2\pi i)(1+\delta_{\alpha\beta})\frac{\partial\vartheta}{\partial\Omega_{\alpha\beta}} = \frac{\partial^2\vartheta}{\partial z_\alpha \partial z_\beta}.$$

If the lemma were false, we could find a matrix $\overline{\Omega}$, a smooth analytic hypersurface $g(\Omega) = 0$ defined in a neighborhood of $\overline{\Omega}$ and containing $\overline{\Omega}$, and a vector-valued function

$$\vec{f}(\Omega,t) \in \mathbb{C}^g$$

defined in a neighborhood of $\overline{\Omega}$ and for $|t|$ small, such that

$$\left.\begin{aligned} &\vartheta(\vec{f}(\Omega,t),\Omega) \equiv 0 \\ &\frac{\partial\vartheta}{\partial z_k}(\vec{f}(\Omega,t),\Omega) \equiv 0, \quad 1 \le k \le g \end{aligned}\right\} \text{ whenever } g(\Omega) = 0.$$

We may assume that for each Ω, $t \longmapsto \vec{f}(\Omega,t)$ is part of an algebraic curve $C_\Omega \subset A_\Omega$. Note that the lemma is obvious if $g = 2$ and if $g \ge 3$, then the codimension of the locus of non-simple abelian varieties is greater than 1. Therefore we can also assume that the abelian variety $A_{\overline{\Omega}}$ is simple. It follows that the set of differences $x-y, x,y \in C_{\overline{\Omega}}$ generates $A_{\overline{\Omega}}$, hence the set of differences $x-y, x,y \in C_\Omega$, generates A_Ω for Ω near $\overline{\Omega}$. Therefore, for no Ω near $\overline{\Omega}$ is there a vector $\vec{a}$ such that

$$\frac{\partial}{\partial t}(\vec{a}\cdot\vec{f}) = (\vec{a}\cdot\frac{\partial\vec{f}}{\partial t}) = 0, \text{ all } t.$$

We prove by induction on d that:

$(*)_d$ If $|\alpha| = d$, then $\left(\frac{\partial^{\alpha}\vartheta}{\partial z_1^{\alpha_1}\cdots\partial z_g^{\alpha_g}}\right)(\vec{f}(\Omega,t),\Omega) \equiv 0$ whenever $g(\Omega) =$

Since $(z,\bar{\Omega})$ does not vanish identically as a function of z, this is a contradiction. In fact, to prove this it will suffice to apply:

If $\eta(\Omega,z)$ satisfies the heat equation and

$$(**)\quad \left.\begin{aligned}&\eta(\vec{f}(\Omega,t),\Omega) \equiv 0\\ &\frac{\partial\eta}{\partial z_k}(\vec{f}(\Omega,t),\Omega) \equiv 0\end{aligned}\right\} \text{ whenever } g(\Omega) = 0$$

then

$$\frac{\partial^2\eta}{\partial x_k\partial z_\ell}(\vec{f}(\Omega,t),\Omega) \equiv 0 \qquad \text{whenever } g(\Omega) = 0$$

to all the partial derivatives of ϑ in turn. To prove (**), differentiate the first relation with respect to Ω. We find that if $\omega_{k\ell}$ satisfies $\sum\omega_{k\ell}\partial g/\partial\Omega_{k\ell}(\Omega) = 0$, then $\Omega+\varepsilon\omega$ is tangent to the hypersurface $g(\Omega) = 0$, hence

$$\begin{aligned}0 &= \eta(\vec{f}(\Omega+\varepsilon\omega,t),\Omega+\varepsilon\omega)\\ &= \varepsilon\left\{\sum_{k,a,b}\frac{\partial\eta}{\partial z_k}(\vec{f}(\Omega,t),\Omega)\cdot\frac{\partial f_k}{\partial\Omega_{ab}}\cdot\omega_{ab} + \sum_{a\leq b}\frac{\partial\eta}{\partial\Omega_{ab}}(\vec{f}(\Omega,t),\Omega)\cdot\omega_{ab}\right\}\\ &= \frac{\varepsilon}{4\pi i}\sum_{a,b}\frac{\partial^2\eta}{\partial z_a\partial z_b}(\vec{f}(\Omega,t),\Omega)\cdot\omega_{ab}\,.\end{aligned}$$

Therefore

$$\frac{\partial^2 \eta}{\partial z_a \partial z_b}(\vec{f}(\Omega,t),\Omega) = \phi(\Omega,t)\cdot(1+\delta_{ab})\cdot \frac{\partial g}{\partial \Omega_{ab}}(\Omega)$$

with some factor ϕ, for all Ω near $\overline{\Omega}$, all small t. Now differentiate the second relation in (**) with respect to t. We find:

$$\text{for all a,} \quad \sum_b \frac{\partial^2 \eta}{\partial z_a \partial z_b}(\vec{f}(\Omega,t),\Omega)\cdot\frac{\partial f_b}{\partial t}(\Omega,t) \equiv 0 \quad \text{whenever } g(\Omega) = 0.$$

If $\phi(\Omega,t) \equiv 0$ when $g(\Omega) = 0$, we are done. If not, we find by substitution that

$$\text{for all a,} \quad \sum_b (1+\delta_{ab})\frac{\partial g}{\partial \Omega_{ab}}(\Omega)\cdot\frac{\partial f_b}{\partial t}(\Omega,t) \equiv 0 \quad \text{whenever } g(\Omega) = 0,$$

i.e.,

$$(***) \qquad (\vec{c}(a)\cdot\frac{\partial \vec{f}}{\partial t}) = 0$$

where

$$c(a)_b = (1+\delta_{ab})\frac{\partial g}{\partial \Omega_{ab}}(\Omega).$$

For some a, $\vec{c}(a) \neq 0$ since $g(\Omega) = 0$ is a smooth hypersurface. But we saw that (***) did not occur, so thus completes the proof.

In the other direction, Beauville [B], Remark 7.7 proved*:

<u>Proposition 2.3</u> (Beauville): N_0 <u>has codimension</u> 1 <u>in</u> $\mathcal{A}_g$.

---

* The result is stated only for g = 4; however the argument works without any modification for all g.

His proof also uses an elaboration of the techniques of Andreotti-Mayer — in this case their technique for deriving "explicit" equations for the N_k. (It might be thought that this Proposition could be proven from general principles, but I don't see how, without specific information, one could have excluded the possibilities that some component of some N_k, $k > 1$, was not in the closure of $N_0 - N_1$.)

We want now to consider the closure $\overline{N}_0$ of N_0 in $\overline{A}_g^{(1)}$, and to give multiplicities to its components. To do this, we would like to use the "universal family" of pairs (A,Θ), $(\overline{G},D)$ over $\overline{A}_g^{(1)}$. However, even generically these pairs still have an automorphism group of order 2, so a universal family need not exist. However, $\overline{A}_g^{(1)}$ admits a "covering" $U_\alpha \longrightarrow \overline{A}_g^{(1)}$ such that over U_α there are flat, proper families

$$\begin{array}{c} D_\alpha \subset \overline{G}_\alpha \\ \Big\downarrow p \\ U_\alpha \end{array}$$

consisting of abelian varieties and rank 1 degenerations thereof, and such that p is locally the universal deformation space of its fibre $(\overline{G}_s, D_s)$. Outside $\Delta \cap U_\alpha$, $\overline{G}_\alpha$ will be smooth over U_α; over points of $\Delta \cap U_\alpha$, $\overline{G}_\alpha$ itself will still be smooth, but at the double points of the fibres, p will look like the universal local deformation space:

$$\begin{array}{l} \hat{\mathcal{O}}_{\overline{G}_\alpha} \cong \mathbb{C}[[z_1,z_1',z_2,\cdots,z_{g-1},t_2,\cdots,t_{g(g+1)/2}]] \\ \uparrow \\ \hat{\mathcal{O}}_{U_\alpha} \cong \mathbb{C}[[t_1,t_2,\cdots,t_{g(g+1)/2}]] \\ \quad t_1 = z_1\cdot z_1' . \end{array}$$

On $\overline{G}_\alpha$, define the subsheaf of the tangent sheaf T_{vert} to be the kernel:

$$0 \longrightarrow T_{vert} \longrightarrow T_{\overline{G}_\alpha} \longrightarrow p^*T_{U_\alpha} .$$

Note that T_{vert} is locally free of rank g (at double points of the fibres, T_{vert} is spanned by $z_1\partial/\partial z_1 - z_1'\partial/\partial z_1'$, $\partial/\partial z_2,\cdots,\partial/\partial z_g$). Using a local equation $\delta = 0$ of D_α, and interpreting sections of T as derivations, define:

$$\begin{array}{rcl} T_{vert} & \xrightarrow{\ \alpha\ } & \mathcal{O}(D_\alpha)/\mathcal{O} \\ D & \longmapsto & D\delta/\delta \qquad \text{(independent of } \delta). \end{array}$$

Let

$$\mathrm{Sing}_{vert}\, D_\alpha = \text{subscheme of } D_\alpha \text{ where } \alpha \text{ is zero.}$$

Thus $\mathrm{Sing}_{vert}\, D_\alpha$ is defined locally by g equations and has codimension at most g. Set-theoretically:

(2.4) $p(\mathrm{Sing}_{\mathrm{vert}}\, D_\alpha)$ = set of points whose fibres are of 3 types

1) fibre is (A,Θ), A abelian variety, and Θ singular

2) fibre is $(\overline{G},D)$ and D has a singularity in G

3) fibre is $(\overline{G},D)$ and the divisor $\overline{D} = D.(\overline{G}-G)$ on $\overline{G}-G$ is singular.

To see this at fibres of type $(\overline{G},D)$, at points of $\overline{G}-G$, expand δ in a power series in $z_1, z_1', z_2, \cdots, z_g$, t's: then the origin lies in $\mathrm{Sing}_{\mathrm{vert}}\, D_\alpha$ if and only if

$$\delta \in (z_1, z_1', z_i z_j \ (2 \le i,j \le g), t_i),$$

i.e., if and only if $\delta = 0$ is singular in $\mathbb{C}[[z_2,\cdots,z_g]]$. The sets $p(\mathrm{Sing}_{\mathrm{vert}}\, D_\alpha)$ patch together into a subset $\tilde{N}_0$ of $\overline{A}_g^{(1)}$. (We shall see shortly that $\tilde{N}_0 = \overline{N}_0$.)

Let us work out which $(\overline{G},D)$ arise in cases (2) and (3). Let G be the extension:

$$0 \longrightarrow \mathbb{G}_m \longrightarrow G \longrightarrow B \longrightarrow 0.$$

Then $\overline{G}-G \cong B$ and $D.(\overline{G}-G)$ is the theta divisor of B, called Ξ at the beginning of this section. Thus if $\pi\colon \Delta \longrightarrow A_{g-1}$ is the natural projection, case (3) contributes $\pi^{-1}(N_0(A_{g-1}))$ to $\tilde{N}_0$. As for case (2), if translation by $b \in B$ is used in glueing together $\overline{G}$, then a local equation of D at any point of G is of the form

$$f(x,z) = \delta_p(x) + z\cdot\delta_{p+b}(x+b)$$

Here δ_P (resp. δ_{P+b}) are local functions on B near p (resp. P+b) which define the non-zero section of $\mathcal{O}_B(\Xi)$ near P (resp. P+b), and z is a vertical coordinate on G in a local splitting $G \cong \mathbb{G}_m \times B$. (We may use the analytic equation (1.13) if we want.) Taking derivatives of f, we see that:

$f(x,z) = 0$ is singular at $x = P$, some $z \in \mathbb{C}^*$	$\Longleftrightarrow$	$P, P+b \in \Xi$ and either Ξ has the same tangent plane at P,P+b, or is singular at both pts.

Looking at points $(\overline{G},D)$ not already covered in case (3), this shows that $\tilde{N}_0$ contains the set of pairs $(\overline{G},D)$ such that $\Xi \subset B$ is smooth and Ξ, Ξ_b are tangent somewhere. If Ξ is smooth, let

$$\gamma_B: \Xi \longrightarrow \mathbb{P}^{g-2}$$

be the "Gauss map" associating to each $P \in \Xi$, the tangent plane $T_{P,\Xi}$, as a point of $\mathbb{P}(T^*_{0,B})$. Then Ξ and Ξ_b are tangent at P if and only if $\gamma_B(P) = \gamma_B(P+b)$. Thus for any principally polarized abelian variety (B,Ξ) with smooth Ξ we may define

$$\begin{aligned} c(B,\Xi) &= \text{locus of points } x-y, \text{ where } \gamma_B(x) = \gamma_B(y) \\ &= \text{locus of points } x \text{ such that } \Xi, \Xi_x \text{ are tangent somewhere.} \end{aligned}$$

Then in the description (2.4):

$$\tilde{N}_0 \cap \Delta \cong \left[\bigcup_{(\overline{G},D)} c(B,\Xi) \right] \cup \left[\delta^{-1}(N_0 \text{ for } \mathcal{A}_{g-1}) \right].$$

Next, the method of Andreotti-Mayer-Beauville extends to rank 1 degenerations, to prove that $\tilde{N}_0$ is a divisor. For abelian varieties A, their technique is to map A to $\mathbb{P}^{2^g-1}$ by $|2\Theta|$, i.e., explicitly by the theta functions

$$\vartheta_\mu(z,\Omega) = \sum_{n\in\mathbb{Z}^g} e^{2\pi i\,{}^t(n+\mu)\Omega(n+\mu)+4\pi i\,{}^t(n+\mu).z} .$$

Call this ϕ: $A \longrightarrow \mathbb{P}^{2^g-1}$.
They define a linear subspace $L_\Omega \subset \mathbb{P}^{2^g-1}$ of codimension g+1 by

$$(2.5)\qquad \begin{aligned} &\sum \vartheta_\mu(0,\Omega)\cdot X_\mu = 0 \\ &\frac{\partial^2\vartheta_\mu}{\partial z_i^2}(0,\Omega)\cdot X_\mu = 0, \quad 1 \le i \le g \end{aligned}$$

and prove

$$(2.6)\qquad \phi^{-1}(L_\Omega) = \text{Sing}\ \Theta,$$

hence

$$(A,\Theta) \in N_0 \iff L_\Omega \cap \phi(A) \neq \emptyset$$

$\iff$ Chow form of $\phi(A)$ varieties at Plücker Coord of L_Ω

Now if $\text{Im}\ \Omega_{11} \longrightarrow \infty$, the limit of $\phi(A)$ is $\phi(\overline{G})$, where ϕ is defined by the 2^g "theta functions"

$$\left.\begin{aligned} &\vartheta_\mu(z^{(1)},\Omega^{(1)}) + u^2\vartheta_\mu(z^{(1)}+\omega,\Omega^{(1)}) \\ &u\vartheta_\mu(z^{(1)}+\tfrac{1}{2}\omega,\Omega^{(1)}) \end{aligned}\right\} \qquad \mu\in \tfrac{1}{2}\mathbb{Z}^{g-1}/\mathbb{Z}^{g-1}$$

(where, as above, G is a $\mathbb{G}_m$-bundle over B, $\Omega^{(1)}$ = period matrix of B, $\overline{G}$ is glued via ω, $z^{(1)}$ is the coordinate on B, u the coordinate on $\mathbb{G}_m$). The basic theta identity on which the proof of (2.6) is based becomes

$$(2.7)\quad \begin{aligned}&[\vartheta(x+y)+uw\,\vartheta(x+y+\omega)]\cdot[\vartheta(x-y)+\tfrac{u}{w}\vartheta(x-y+\omega)] = \\ &\sum_{\mu\in\frac{1}{2}\mathbb{Z}^{g-1}/\mathbb{Z}^{g-1}} [\vartheta_\mu(x)+u^2\vartheta_\mu(x+b)]\cdot\vartheta_\mu(y)+uw\vartheta_\mu(x+\tfrac{\omega}{2})\cdot[\vartheta_\mu(y+\tfrac{\omega}{2})+\tfrac{1}{w^2}\vartheta_\mu(y-\tfrac{\omega}{2})].\end{aligned}$$

The limit of L_Ω is the linear space

$$(2.8)\quad \begin{aligned}&\sum \vartheta_\mu(0,\Omega^{(1)})\cdot X_\mu + 2\sum \vartheta_\mu(\tfrac{\omega}{2},\Omega^{(1)})\cdot Y_\mu = 0\\ &\sum \frac{\partial^2\vartheta_\mu}{\partial z_i^2}(0,\Omega^{(1)})\cdot X_\mu + 2\sum \frac{\partial^2\vartheta_\mu}{\partial z_i^2}(\tfrac{\omega}{2},\ \Omega^{(1)})\cdot Y_\mu = 0\\ &\sum \vartheta_\mu(\tfrac{\omega}{2},\Omega^{(1)})\cdot Y_\mu = 0\ .\end{aligned}$$

(The last equation comes from the 2nd derivative of (2.7) with respect to $w\,\partial/\partial w$; these equations are not the exact analogs of the (2.5) because, in passing to the limit, we have renormalized the origin.) Then it follows from (2.7) exactly as in Andreotti-Mayer-Beauville that

$$\phi^{-1}(L_\Omega) = \left(\begin{array}{c}\text{singularities of D in G plus singularities}\\ \text{of } \overline{D}\cdot(\overline{G}-G) \text{ in } \overline{G}-G\end{array}\right)$$

hence

$$\left(\begin{array}{l}\text{Chow form of } \phi(\overline{G}) \text{ is}\\ \text{zero at } L_\Omega\end{array}\right) \iff (\overline{G},D)\in \tilde{N}_0 .$$

This proves that $\tilde{N}_0$ is a divisor.

On the other hand, it is clear that for all B, $c(B,\Xi) \subsetneq B$ and for generic B,Ξ is smooth: hence $\tilde{N}_0 \cap \Delta \subsetneq \Delta$. Thus $\tilde{N}_0$ must be the closure $\overline{N}_0$ of N_0. Incidentally, this proves that $c(B,\Xi)$ is always a divisor in B. At the same time, we can now give multiplicities to the components of $\overline{N}_0$. I think the Andreotti-Mayer-Beauville equation gives artificially large multiplicities, and want, instead, to assign multiplicities via the local description of $\overline{N}_0$ in U_α as $p(\mathrm{Sing}_{\mathrm{vert}} D_\alpha)$. Let $\overline{N}_0'$ be the maximal open set of points of $\overline{N}_0$ such that for all α

$$p:\ \mathrm{Sing}_{\mathrm{vert}} D_\alpha \longrightarrow (\overline{N}_0 \cap U_\alpha)$$

is <u>finite</u> over $\overline{N}_0'$. Because N_1 has codimension at least 2, $\overline{N}_0'$ is dense in $\overline{N}_0$. Then <u>over $\overline{N}_0'$</u>

$$\dim(\mathrm{Sing}_{\mathrm{vert}} D_\alpha) = \dim N_0$$

hence

$$\mathrm{codim}(\mathrm{Sing}_{\mathrm{vert}} D_\alpha) = g+1 = \#\text{ of equations defining } \mathrm{Sing}_{\mathrm{vert}} D_\alpha$$

hence $\mathrm{Sing}_{\mathrm{vert}}\, D_\alpha$ is Cohen-Macauley. Therefore, over $\overline{N}_0'$, $p_*(\mathcal{O}_{\mathrm{Sing}_{\mathrm{vert}} D_\alpha})$ has a locally free resolution:

$$0 \longrightarrow \mathcal{E}_1 \xrightarrow{\ f\ } \mathcal{E}_0 \longrightarrow p_*(\mathcal{O}_{\mathrm{Sing}_{\mathrm{vert}} D_\alpha}) \longrightarrow 0$$

and det f gives a local equation for $\overline{N}_0' \cap D_\alpha$, and this assigns multiplicities to $\overline{N}_0$.

Next, we want to break $\overline{N}_0$ up into 2 pieces: the first piece is

$$(2.9)\qquad \vartheta_{\text{null}} = \left\{ (A,\Theta) \;\middle|\; \begin{array}{l} \text{if } \Theta \text{ is normalized to be symmetric about e,} \\ \text{then } \Theta \text{ has a singularity at a point of order 2} \end{array} \right\} .$$

It is easy to see that:

$$\vartheta_{\text{null}} \cap \Delta = \left[\bigcup_{\text{all } \overline{G}, D} 2_B(\Xi) \right] \cup \left[\delta^{-1}(\ {}_{\text{null}} \text{ for } A_{g-1}) \right]$$

where we note that (assuming Ξ is symmetric too) $c(B,\Xi)$ contains the "obvious" component:

$$2_B(\Xi) = \{2x \mid x \in \Xi\}$$

because $\gamma(-x) = \gamma(x)$, all $x \in \Xi$.

If a symmetric Θ has a singularities at a point x not of order 2, it is also singular at -x. Thus $\overline{N}_0$ breaks up:

$$\overline{N}_0 = \vartheta_{\text{null}} + 2.\overline{N}_0^*$$

where all multiplicities in the $2^{\underline{\text{nd}}}$ piece are divisible by 2. We can now state the main result of this paper:

<u>Theorem (2.10)</u>: <u>The divisor classes of</u> $\overline{N}_0$, ϑ_{null}, $\overline{N}_0^*$ <u>are given by</u>:

$$[\overline{N}_0] = \left(\frac{(g+1)!}{2} + g!\right)\lambda - \frac{(g+1)!}{12}\delta$$

$$[\vartheta_{\text{null}}] = 2^{g-2}(2^g+1)\lambda - 2^{2g-5}.\delta$$

$$[\overline{N}_0^*] = \left[\frac{(g+1)!}{4} + \frac{g!}{2} - 2^{g-3}(2^g+1)\right]\lambda - \left[\frac{(g+1)!}{24} - 2^{2g-6}\right]\delta \quad .$$

Here is a table for low degrees:

g	$[\overline{N}_0]$	$[\vartheta_{null}]$	$[N_0^*]$	slope
2	$5\lambda - \frac{1}{2}\delta$	$5\lambda - \frac{1}{2}\delta$	0	—
3	$18\lambda - 2\delta$	$18\lambda - 2\delta$	0	—
4	$84\lambda - 10\delta$	$68\lambda - 8\delta$	$8\lambda - \delta$	8
5	$480\lambda - 60\delta$	$264\lambda - 32\delta$	$108\lambda - 14\delta$	7.71
6	$3{,}240\lambda - 420\delta$	$1{,}040\lambda - 128\delta$	$1{,}100\lambda - 146\delta$	7.53
7	$25{,}200\lambda - 3{,}360\delta$	$4{,}128\lambda - 512\delta$	$10{,}536\lambda - 1{,}424\delta$	7.40

Note that the figures imply $\overline{N}_0^* = \emptyset$ for $g = 2,3$ as is well known. We also see that the divisor class of $\overline{N}_0^*$ is the same as that of the Jacobian locus for $g = 4$, confirming Beauville's results. The last column, "slope", refers to the ratio of the coefficient of λ to the coefficient of δ. As soon as this drops below the same ratio for K, $\mathcal{A}_g$ is of general type:

<u>Corollary (2.11)</u>. $\frac{(g+1)!}{12} K_{\overline{\mathcal{A}}_g}(1) \equiv [\overline{N}_0] + g!(g^2-4g-17)\lambda$.

<u>Proof</u>: Combine 1.7 and 2.10.

<u>Corollary (2.12)</u>. <u>If</u> $g \geq 7$, $\mathcal{A}_g$ <u>is of general type</u>.

<u>Proof</u>: Combine 1.4 and 2.11.

§3. Proof of the Theorem.

Now how are we going to prove the Theorem? The formula for $[\vartheta_{\text{null}}]$ is immediate, because we know the modular form that cuts out this divisor, viz.:

$$f(\Omega) = \prod_{\substack{a,b \in \frac{1}{2}\mathbb{Z}^g/\mathbb{Z} \\ {}^t(2a).(2b)\text{ even}}} \vartheta[{}^a_b](0,\Omega)$$

where

$$\vartheta[{}^a_b](0,\Omega) = \sum_{n\in\mathbb{Z}^g} e^{\pi i\,{}^t(n+a)\Omega(n+a)+2\pi i\,{}^t(n+a).b} .$$

Each is a modular form of weight 1/2 and there are $2^{g-1}(2^g+1)$ "even" pairs a,b so f has weight $2^{g-2}\cdot(2^g+1)$, and this is the coefficient of λ. On the other hand, if $\text{Im}\,\Omega_{11} \longrightarrow \infty$, we see that if $a_1 = 0$, $\lim \vartheta[{}^a_b] = 1$, while if $a_1 = \frac{1}{2}$, $\vartheta[{}^a_b]$ is divisible by

$$e^{\pi i\Omega_{11}/4}$$

hence it goes to zero. The equation of Δ is $e^{2\pi i\Omega_{11}} = 0$, and there are 2^{2g-2} "even" pairs a,b with $a_1 = \frac{1}{2}$ (take any $a_2,b_2,\cdots,a_g,b_g$, set $a_1 = \frac{1}{2}$ and make b_1 zero or one-half to force a,b to be even). Thus f goes to zero like

$$(e^{2\pi i\Omega_{11}})^{(2^{2g-5})}$$

when $\text{Im}\,\Omega_{11} \longrightarrow \infty$, hence the coefficient of δ.

It remains to prove the formula for $[\overline{N}_0]$. The value of the coefficient of λ follows from:

<u>Proposition 3.1</u>: <u>Let</u>

$$\begin{array}{ccc} \mathcal{X} & \supset & \mathcal{D} \\ \varepsilon \uparrow\downarrow p & & \\ C & & \end{array}$$

<u>be a family of p.p.a.v. over a complete curve</u> C <u>such that every theta divisor</u> $\mathcal{D}_t$ <u>has only a finite number of singularities and the generic</u> $\mathcal{D}_\eta$ <u>is smooth. Let this family define the morphism</u>

$$\varphi: C \longrightarrow \mathcal{A}_g.$$

<u>Then</u>

$$\varphi^* N_0 \equiv \left(\frac{(g+1)!}{2} + g!\right)\varphi^*\lambda + \text{torsion}.$$

(Note that such a family exists because codim $N_1 \geq 2$ and because in Satake's compactification, the whole boundary has codim ≥ 2). The coefficient of δ, on the other hand follows from:

<u>Proposition 3.2</u>: <u>Let</u> (A,Θ) <u>be a p.p.a.v. Then the divisor class of</u> $c(B,\Theta)$ <u>is given by</u>:

$$c(B,\Theta) \equiv \frac{(g+2)!}{6} \cdot \Theta$$

together with Proposition 1.8.

To prove 3.1, we use the exact sequence

$$T_{\mathfrak{X}/C} \longrightarrow \mathcal{O}_{\mathfrak{X}}(\mathfrak{D})/\mathcal{O}_{\mathfrak{X}} \longrightarrow \mathcal{O}_{\mathrm{Sing}_{\mathrm{vert}}\mathfrak{D}} \otimes \mathcal{O}_{\mathfrak{X}}(\mathfrak{D}) \longrightarrow 0$$

used to define multiplicities for N_0. It follows that $\mathrm{Sing}_{\mathrm{vert}}\mathfrak{D}$ is the scheme of zeroes of a section of

$$\Omega^1_{\mathfrak{X}/C}(\mathfrak{D}) \otimes_{\mathcal{O}_{\mathfrak{X}}} \mathcal{O}_{\mathfrak{D}}$$

hence

$$\varphi^* N_0 = p_*(c_g(\Omega^1_{\mathfrak{X}/C}(\mathfrak{D})).\mathfrak{D}) \ .$$

But if $\mathfrak{E} = p_*(\Omega^1_{\mathfrak{X}/C})$, then the bundle $\Omega^1_{\mathfrak{X}/C}$, being trivial on each fibre of $\mathfrak{X}$ over C, is isomorphic to $p^*\mathfrak{E}$. Moreover, by definition of λ,

$$\varphi^*\lambda = c_1(\mathfrak{E}).$$

Thus

$$\begin{aligned}\varphi^* N_0 &= p_*(c_g(p^*\mathfrak{E}\otimes\mathcal{O}_{\mathfrak{X}}(\mathfrak{D})).\mathfrak{D}) \\ &= p_*((\mathfrak{D}^g+\mathfrak{D}^{g-1}.c_1(p^*\mathfrak{E})).\mathfrak{D}) \\ &= p_*(\mathfrak{D}^{g+1}) + p_*(\mathfrak{D}^g)\cdot c_1(\mathfrak{E}) \quad .\end{aligned}$$

Now on each fibre $\mathfrak{D}$ is Θ and $(\Theta^g) = g!$, so the second term is $g!\varphi^*(\lambda)$. To compute the first, we apply the Grothendieck-Riemann-Roch theorem to $\mathcal{O}_{\mathfrak{X}}(\mathfrak{D})$. Note that

$$p_*(\mathcal{O}_{\mathfrak{X}}(\mathfrak{D})) \cong \mathcal{O}_C$$

$$R^i p_*(\mathcal{O}_*(\mathfrak{D})) = (0),\ i \geq 1.$$

Thus

$$\begin{aligned} 1 &= ch(p_! (\mathcal{O}_{\mathfrak{X}}(\mathfrak{D}))) \\ &= p_*(ch\,\mathcal{O}_{\mathfrak{X}}(\mathfrak{D}).Td(\Omega^1_{\mathfrak{X}/C})) \\ &= p_*(e^{\mathfrak{D}}.p^*(1-\frac{c_1(\mathbb{E})}{2})), \quad \text{mod torsion.} \end{aligned}$$

In codimension 1 on C, this says

$$0 = p_*(\frac{\mathfrak{D}^{g+1}}{(g+1)!}) - \frac{c_1(\mathbb{E})}{2}\cdot p_*(\frac{\mathfrak{D}^g}{g!})$$

or

$$p_*(\mathfrak{D}^{g+1}) = \frac{(g+1)!}{2} c_1(\mathbb{E}) \quad \text{mod torsion.}$$

This proves 3.1.

To prove 3.2, it suffices to establish the numerical equivalence of the 2 divisors. Namely, this will prove Theorem 2.10, and then Theorem 2.10 will imply Prop. 3.2 as an equality of divisor classes. Let $C \subset A$ be any curve. We shall calculate $(C.c(B,\Theta))$. Consider the map

$$C \times \Theta \xrightarrow{\ m\ } A$$

$$m(x,y) = x + y\ .$$

Then $m^{-1}(\Theta)$ is the locus of pairs (x,y) where $x+y \in \Theta$, i.e., $x = y'-y$, where $x \in C, y, y' \in \Theta$. The differential of m gives us a map

$$dm:\ p_2^* T_\Theta \otimes \mathcal{O}_{m^{-1}\Theta} \longrightarrow T_A \otimes \mathcal{O}_\Theta \longrightarrow N_{\Theta,A}$$

whose zeroes are exactly the points (x,y) such that not only is $x = y'-y$, $y,y' \in \Theta$, but also $T_{y,\Theta} = T_{y',\Theta'}$ i.e., $x \in \mathcal{D}$. Now the above dm can be thought of as a section of

$$p_2^* \Omega_\Theta^1 \otimes m^*(N_{\Theta,A}) \otimes \mathcal{O}_{m^{-1}(\Theta)}$$

hence

$$(C.\mathcal{D}) = c_{g-1}\left(p_2^* \Omega_\Theta^1 \otimes m^*(\mathcal{O}(\Theta)) \otimes \mathcal{O}_{m^{-1}\Theta}\right).$$

Let $\theta_1 = \text{pt.}\times\Theta$, $\theta_2 = m^{-1}\Theta$ be these divisor classes (mod numerical equivalence) on $C\times\Theta$. Then

$$(C.\mathcal{D}) = c_{g-1}\left(p_2^* \Omega_\Theta^1 \otimes \mathcal{O}(\theta_2)\right).\theta_2.$$

Using

$$0 \longrightarrow \mathcal{O}(-\Theta)/\mathcal{O}(-2\Theta) \longrightarrow \Omega_A^1\Big|_\Theta \longrightarrow \Omega_\Theta^1 \longrightarrow 0,$$

we see that

$$c(\Omega_\Theta^1) = (1-\Theta)^{-1}\Big|_\Theta = (1+\Theta+\Theta^2+\cdots\cdots)\Big|_\Theta .$$

Thus

$$(C.\mathcal{D}) = \theta_1^{g-1}.\theta_2 + \theta_1^{g-2}.\theta_2^2 + \cdots\cdots + \theta_2^g .$$

But now

$$(\theta_1^k\cdot\theta_2^{g-k})_{C\times\Theta} = (m(\theta_1^{k+1})\cdot \Theta^{g-k})_A$$
$$= ((C \dotplus \Theta^{k+1})\cdot\Theta^{g-k})_A$$

if $\dotplus$ is Pontryagin product. By symmetry of Θ, this is

$$= (C.(\Theta^{k+1} \dotplus \Theta^{g-k}))_A$$
$$= (C.(k+1)(g-k)(g-1)!\Theta)_A .$$

Thus

$$(C.\mathcal{D}) = (C.\Theta)(g-1)!\sum_{k=0}^{g-1}(k+1)(g-h)$$

$$= \frac{(g+2)!}{6}(C.\Theta).$$

<u>QED</u>

References

[A-M] Andreotti, A., and Mayer, A., On the period relations for abelian integrals on algebraic curves, Ann. Scuola Norm. Pisa, 21 (1971).

[A-M-R-T] Ash, A., et al, Smooth compactification of locally symmetirc varieties, Math-Sci Press, 53 Jordan Rd., Brookline, MA, 1975.

[B] Beauville, A., Prym varieties and the Schottky problem, Inv. Math., 41 (1977), p. 149.

[Bo] Borel, A., Stable real cohomology of arithmetic groups II, in Manifolds and Lie groups, Birkhauser-Boston, 1981.

[C] Clemens, H., Double solids, to appear.

[D] Donagi, R., The unirationality of $\mathcal{A}_5$, to appear.

[F1] Freitag, E., Die Kodairadimension von Körpern automorpher Funktionen, J. reine angew. Math., 296 (1977), p. 162.

[F2] Freitag, E., Der Körper der Siegelschen Modulfunktionen, Abh. Math. Sem. Hamburge, 47 (1978).

[H-M] Harris, J. and Mumford, D., On the Kodaira dimension of the moduli space of curves, to appear in Inv. Math.

[I] Igusa, J.-I., A desingularization problem in the theory of Siegel modular functions, Math. Annalen, 168 (1967), p. 228.

[M] Mumford, D., Analytic construction of degenerating abelian varieties, Comp. Math., 24 (1972), p. 239.

[N] Namikawa, A new compactification of the Siegel space and degeneration of abelian varieties, Math. Ann., 221 (1976).

[S] Stillman, M., Ph.D. Thesis, Harvard University, 1983.

[T] Tai, Y.-S., On the Kodaira dimensions of the moduli space of abelian varieties, to appear Inv. Math.

THE CLASSIFICATION OF SURFACES AND OTHER VARIETIES

COMMENTARY BY ECKART VIEHWEG

VANISHING THEOREMS AND PATHOLOGIES IN CHARACTERISTIC $p > 0$

Let X be a projective complex manifold, and let $\mathcal{L}$ be an invertible sheaf on X. For $\mathcal{L}$ ample the Kodaira Vanishing Theorem says that

$$H^i(X, \mathcal{L}^{-1}) = 0, \quad \text{for } i < \dim X.$$

Recall that the Kodaira-dimension $\kappa(\mathcal{L})$ of an invertible sheaf $\mathcal{L}$ is the maximum of the dimensions $\dim \varphi_m(X)$, where φ_m denotes the rational map given by $H^0(X, \mathcal{L}^m)$, for $m \geq 1$. If $H^0(X, \mathcal{L}^m) = 0$ for all $m > 0$, we write $\kappa(\mathcal{L}) = -\infty$. The invertible sheaf $\mathcal{L}$ is called semi-ample if, for some $m \geq 1$, the global sections generate $\mathcal{L}^m$. Mumford's Vanishing Theorem in [M3] says:

Theorem 1. *If $\mathcal{L}$ is semi-ample, and $\kappa(\mathcal{L}) \geq 2$, then*

$$H^i(X, \mathcal{L}^{-1}) = 0 \quad \text{for } i = 0 \text{ or } 1.$$

C. P. Ramanujam [78] weakened the assumption "$\mathcal{L}$ semi-ample" in Theorem 1 to "$\mathcal{L}$ numerically effective", i.e. $(\deg \mathcal{L}|_C) \geq 0$, for all curves C in X. D. Mumford later obtained Ramanujam's Vanishing Theorem on surfaces as a corollary of Bogomolov's Theorem [9] on the stability of vector bundles (Appendix to [80]).

Using analytic methods, as Mumford did in the first of the two proofs of Theorem 1 given in [M3], H. Grauert and O. Riemenschneider proved in [34] that the vanishing in Theorem 1 holds true for $i = 0, 1, \ldots, \kappa(\mathcal{L}) - 1$.

The second proof of Theorem 1 in [M3] uses two methods which reappear in the proof of several generalizations of the Kodaira Vanishing Theorem: the one-connectedness of divisors (see [78], [10] and [3], IV § 8) and the construction of cyclic coverings. The latter was used by Ramanujam in [78] and it played an essential role in the proof of the vanishing theorem for numerically effective invertible sheaves by Y. Kawamata [46] and the author [93]. The latter also allows a small normal crossing divisor as "correction term". For surfaces a similar results has been shown before by Y. Miyaoka [67].

Theorem 2. *Assume that for an effective normal crossing divisor $D = \sum_{i=1}^{r} \nu_i D_i$ and for some $N > 0$ the sheaf $\mathcal{L}^N(-D)$ is numerically effective. Then*

$$H^i(X, \mathcal{L}^{-1}([\tfrac{D}{N}])) = 0, \quad \text{for } i < \kappa(\mathcal{L}(-[\tfrac{D}{N}])).$$

Here $[\frac{D}{N}]$ denotes the integral part of the $\mathbb{Q}$-divisor $\frac{D}{N}$.

By Serre Duality, Theorem 2 is equivalent to the vanishing of $H^i(X, \omega_X \otimes \mathcal{L}(-[\frac{D}{N}]))$, for $i > \dim X - \kappa(\mathcal{L}(-[\frac{D}{N}]))$.

For $D = 0$ and $\kappa(\mathcal{L}) = \dim(X)$ one obtains for all i the surjectivity of the "adjunction map"

$$(1) \qquad H^i(X, \mathcal{L} \otimes \omega_X \otimes \mathcal{O}_X(B)) \longrightarrow H^i(B, \mathcal{L} \otimes \omega_B),$$

April 1996, updated May 2003.

where B is any effective divisor. J. Kollár has shown in [51] that the latter remains true under much weaker assumptions:

Theorem 3. *If $\mathcal{L}$ is a semi-ample invertible sheaf, B an effective divisor and if*

$$H^0(X, \mathcal{L}^M \otimes \mathcal{O}_X(-B)) \neq 0,$$

for some $M > 0$, then the adjunction map (1) is surjective, for all i.

Again, one can formulate and prove a slight generalization of Theorem 3, assuming just that $\mathcal{L}^N(-D)$ is semi-ample for some small normal crossing divisor D (see for example [29] and [30]).

In the Vanishing Theorem 2 one may allow the effective divisors D to have singularities worse than normal crossings. To this aim one considers a birational morphism $\tau : X' \to X$ with X' non-singular and $D' = \tau^* D$ a normal crossing divisor. Theorem 2 allows to show that $R^i\tau_*(\omega_{X'} \otimes \mathcal{O}_{X'}(-[\frac{D'}{N}])) = 0$, and that the sheaf $\tau_*(\omega_{X'/X} \otimes \mathcal{O}_{X'}(-[\frac{D'}{N}]))$, called the algebraic multiplier ideal sheaf, is independent of τ. One obtains (see [28]) for

$$i > \dim X - \kappa(\mathcal{L}^N(-D)) \geq \dim X - \kappa(\tau^*\mathcal{L}(-[\frac{D'}{N}]))$$

the vanishing of

$$H^i(X, \mathcal{L} \otimes \tau_*(\omega_{X'} \otimes \mathcal{O}_{X'}(-[\tfrac{D'}{N}]))). \tag{2}$$

The name "multiplier ideal" goes back to A. Nadel, who defined in [75] general multiplier ideals in the analytic context, and extended (2) to this case. In [21] J.-P. Demailly studies in a similar way how far the positivity condition can be violated along subvarieties, without effect on the vanishing of cohomology groups. Nadel's multiplier ideals, or their algebraic analogue, turned out to be a powerful tool in higher dimensional algebraic geometry, in particular for Y.-T. Siu's proof of the invariance of the plurigenera (see [87] and [88]).

There is much more to be said on vanishing theorems over a field of characteristic zero, for example: Kollár's easy proof of the Kawamata-Fujita Positivity Theorem and the Generic Vanishing Theorem of M. Green and R. Lazarsfeld. Some of those aspects are discussed in [30] and more references are given there.

Whereas Mumford's second proof of Theorem 1 is in the framework of algebraic geometry, for a long time the only known proofs of Kodaira's Vanishing Theorem and the generalizations given above were based on analytic methods. In [29] for example it is shown that Theorems 2 and 3 are both consequences of the E_1-degeneration of the Hodge to de Rham spectral sequence for logarithmic differential forms, a result shown by P. Deligne [19], using analytic methods.

G. Faltings gives in [31] the first algebraic proof of the E_1-degeneration and thereby of the vanishing theorems. K. Kato [44] in the projective case, and J.-M. Fontaine and W. Messing [32] in the proper case, obtained the E_1-degeneration for manifolds X defined over a perfect field of characteristic $p > \dim X$, provided X lifts to the ring of Witt vectors $W(k)$. Finally P. Deligne and L. Illusie [20] gave an elementary proof of this degeneration under the weaker condition that X lifts to the second Witt vectors $W_2(k)$ and that $p \geq \dim X$. The degeneration of the Hodge to de Rham spectral sequence in characteristic p implies the degeneration in characteristic zero. As explained in [30] one obtains algebraic proofs of Theorems 2 and 3. For Kodaira's original statement, one finds in [20] a short and elegant proof, due to M. Raynaud, which works in characteristic $p \geq \dim(X)$ whenever X lifts to $W_2(k)$. It is not known whether similar lifting properties imply the analogues of the Grauert-Riemenschneider Vanishing Theorem and of Theorem 2 and 3 in characteristic $p \neq 0$.

The examples of "pathologies" in characteristic $p > 0$, given by Mumford in [M1], [M2] and [M3], i.e. examples showing that neither the Vanishing Theorem 1, nor the closedness of global differential forms (and hence the E_1-degeneration), nor the symmetry of the Hodge numbers hold true in general, were complemented by counterexamples of Raynaud [79] to Kodaira's Vanishing Theorem and extended in [91]. There the reader also finds a proof, due to L. Szpiro [90], of analogues of Theorem 1 for surfaces in characteristic p. Examples of surfaces of general type in characteristic $p > 0$ with non-trivial vector fields were given by P. Russel, H. Kurke and W. E. Lang (see [62]). A list of further pathological examples can be found in [41], § 15.

Fujita's Conjecture

As mentioned in the first section, D. Mumford proved Ramanujam's Vanishing Theorem on surfaces as a corollary of Bogomolov's Theorem [9] on stability of vector bundles (published by Reid as an appendix to [80]). By a similar method I. Reider [84] found numerical conditions for an ample sheaf $\mathcal{L}$ on a surface S, which imply that $\omega_S \otimes \mathcal{L}$ is spanned by global sections, and some slightly stronger conditions for the very ampleness of $\omega_S \otimes \mathcal{L}$.

Reider's Theorem gives for surfaces an affirmative answer to Fujita's conjecture, asserting that for an ample invertible sheaf $\mathcal{L}$ on a smooth projective variety X of dimension n the sheaf $\omega_X \otimes \mathcal{L}^{n+1}$ is generated by global sections and that $\omega_X \otimes \mathcal{L}^{n+2}$ is very ample.

In the higher dimensional case, the first numerical criterion for very ampleness was obtained by Demailly [22]. His result implies that $\omega_X^2 \otimes \mathcal{L}^m$ is very ample for $m \geq 12n^n$. The bound was lowered by Siu and by Demailly himself. Their methods of proof are of analytic nature. Using tools, developed in the framework of "Mori's Program", Kollár [57] gave an explicit lower bound for numbers ν which implies the very ampleness of $(\omega_X \otimes \mathcal{L}^{n+2})^\nu$. Using algebro-geometric methods, close to those used in diophantine approximation, L. Ein, R. Lazarsfeld and M. Nakamaye [24] reproved the results of Demailly (with a slightly larger bound for m).

The first part of Fujita's conjecture, the base point freeness of $\omega_X \otimes \mathcal{L}^{n+1}$, has been verified for threefolds by Ein and Lazarsfeld [23]. Their proof, based on the Vanishing Theorem 2 and on multiplier ideals, gives an alternative approach towards Reider's result (see also [30]). Kawamata [49] proves the base point freeness of $\omega_X \otimes \mathcal{L}^5$ on four dimensional manifolds.

Without any restriction on the dimension the best numerical criterion for base point freeness is due to U. Angehrn and Y.-T. Siu [2]. They show, using analytic multiplier ideals and the Vanishing Theorem of Nadel, that $\omega_X \otimes \mathcal{L}^m$ is globally generated for $m \geq \frac{1}{2}(n^2 + n + 2)$ (slightly improved by G. Heier in [35]). Kollár and independently Ein and Lazarsfeld gave an algebraic proof of the same criterion. Lazarsfeld's forthcoming book [64] provides an excellent presentation of vanishing theorems, multiplier ideals and of the results mentioned in this section.

Classification Theory of Surfaces

The Enriques-Kodaira classification of complex surfaces has been extended to algebraic surfaces defined over an algebraically closed field k of characteristic $p \geq 0$, by D. Mumford [M5], and by E. Bombieri and D. Mumford in [M6] and [M7]. A survey on those results and methods can be found in [11].

The pathological behavior of surfaces in characteristic $p > 0$, mentioned above, has to be taken in account, and the classification table gains some additional lines in characteristic 2 and 3.

Theorem 4. *Let S be a projective surface, without exceptional curves of the first kind, defined over an algebraically closed field k of any characteristic.*

(1) *If $\kappa(S) = \kappa(\omega_S) = 2$, then $H^0(S, \omega_S^2) \neq 0$ and ω_S is semi-ample. In particular, the canonical ring*

$$R(S) = \bigoplus_{\nu \geq 0} H^0(S, \omega_S^\nu)$$

is finitely generated.

(2) *If $\kappa(S) = 1$, then either $H^0(S, \omega_S^4) \neq 0$ or $H^0(S, \omega_S^6) \neq 0$. Moreover ω_S is semi-ample and, for $n \gg 0$, ω_S^n defines an elliptic or quasi-elliptic fibration. The latter only occurs if* char $k = 2$ *or 3.*

(3) *If $\kappa(S) = 0$, then either $\omega_S^6 \cong \mathcal{O}_S$ or $\omega_S^4 \cong \mathcal{O}_S$. Moreover S belongs to one of the following classes ($q = \dim H^0(S, \Omega_S^1)$, $p_g = \dim H^0(S, \omega_S)$ and b_i are the Betti-numbers):*

b_2	b_1	$\chi(\mathcal{O}_S)$	q	p_g	*type of surface*
22	0	2	0	1	*K-3*
10	0	1	0	0	*Enriques (classical)*
10	0	1	1	1	*Enriques (non-classical) (only in char. 2)*
6	4	0	2	1	*Abelian*
2	2	0	1	0	*bi-elliptic (classical)*
2	2	0	2	1	*bi-elliptic (non-classical) (only in char. 2 or 3)*

(4) *If $\kappa(S) = -\infty$, then S is ruled. In particular, "adjunction terminates", i.e. for all invertible sheaves $\mathcal{L}$ there exists some $\nu_0 > 0$ such that $H^0(S, \mathcal{L} \otimes \omega_S^\nu) = 0$, for $\nu \geq \nu_0$.*

Some of the arguments used in [M5], [M6] and [M7] supply new proofs for the known results in characteristic 0.

Due to lack of space and knowledge I am not able even to sketch the recent developments in the theory of algebraic surfaces. [11], [77] [16] and [41] are excellent surveys, with a lot of references. A pretty complete picture of the theory of compact complex surfaces, except of the use of Mori's methods and Reider's results (discussed below) can be found in the book of Barth, Peters and Van de Ven [3] (may be there will be a second extended edition?).

Special surfaces in characteristic $p > 0$ were constructed and studied by W. E. Lang (see for example [63]).

The semi ampleness of ω_S and the finite generation of the ring $R(S)$ in Theorem 4, a) had been shown by Mumford already in an appendix to O. Zariski's paper [100]. The question, which power of ω_S defines a birational morphism or is generated by global sections was studied by E. Bombieri [10], in characteristic 0, and by T. Ekedahl [25], in positive characteristic.

Building up on Mumford's Geometric Invariant Theory [74], D. Gieseker constructed in [33] quasi-projective moduli schemes for surfaces of general type, i.e. for surfaces S with $\kappa(S) = 2$. Properties of those moduli schemes have been studied by F. Catanese (see for example [14], [15] and the survey [16]).

Iitaka's Program

Attempts to generalize the classification of surfaces, as stated in Theorem 4, to higher dimensional complex manifolds were made along two lines. The first one, started and programmed by S. Iitaka and developed by K. Ueno, E. Viehweg, T. Fujita, Y. Kawamata and J. Kollár, is explained in S. Mori's survey [71]. It builds up on:

- The Kodaira dimension $\kappa(X) = \kappa(\omega_X) \in \{-\infty, 0, 1, \ldots, \dim X\}$.
- For $1 \leq \kappa(X) \leq \dim X$ on the Iitaka map $\varphi_m : X \to \mathbb{P}^{p_m - 1}$, defined by $H^0(X, \omega_X^m)$ for the multiples m of some sufficiently large number m_0. We write $p_m = \dim H^0(X, \omega_X^m)$.
- For $\kappa(X) \leq 0$ on the Albanese morphism $\alpha_X : X \to A(X)$, where $A(X)$ is an abelian variety of dimension $q(X) = \dim H^0(X, \Omega_X^1)$.

The main results obtained in the framework of Iitaka's program can be summarized in the following three theorems (we assume from now on that the characteristic is zero).

Theorem 5. *Let X be a non-singular complex projective variety of dimension n and let $1 \leq \kappa(X) \leq n-1$. Then, for $\nu \gg 0$, the non-singular fiores $F_y = \varphi_\nu^{-1}(y)$ of the Iitaka map*

$$\varphi_\nu : X \longrightarrow \varphi_\nu(X) \subset \mathbb{P}^{p_\nu - 1}$$

have Kodaira dimension $\kappa(F_y) = 0$.

One hopes, that the statement holds true for all y in a non empty Zariski open subset of $\varphi_\nu(X)$. In Theorem 5 one uses the result of S. Iitaka [36], saying that for all y outside of a countable union of proper closed subvarieties of $\varphi_\nu(X)$ one has $\kappa(F_y) = 0$. To obtain 5, as stated one also has to use the deformation invariance of the plurigenera (Siu, [87]).

Theorem 5 reduces the study of higher dimensional varieties to the cases: $\kappa(X) = \dim(X)$, $\kappa(X) = 0$ and $\kappa(X) = -\infty$, and to families of lower dimensional varieties F with $\kappa(F) = 0$. The structure of manifolds with Kodaira dimension zero is partly described in the following theorem, which presents results of Kawamata [45], Kollár [51], [56] and Viehweg [94]. Again, the reader finds the history and complete references in [71].

Theorem 6. *Let X be a non-singular projective variety of dimension n.*

(1) *If for some $\nu > n$ one has $p_\nu = 1$ then $q \leq n$ and the Albanese morphism $\alpha_X : X \to A(X)$ is surjective with connected fibres. In particular, this holds true whenever $\kappa(X) = 0$.*
(2) *The following conditions are equivalent:*
  (a) *X is birational to an abelian variety.*
  (b) *$q = n$ and $p_\nu = 1$, for some $\nu > 2$.*
  (c) *$q = n$ and $\kappa(X) = 0$.*
(3) *If $q \geq n-2$ and if $\kappa(X) = 0$ then, up to birational equivalence, α_X is an étale fibre bundle whose fibre F is a curve or surface with $\kappa(F) = 0$.*

In dimension 3 or 4 one has (see [94] and [71]):

Theorem 7. *Let X be a non-singular projective variety, $\dim(X) \leq 4$ and $\kappa(X) = -\infty$. Let $f : X \to Y$ be the Stein factorization of the Albanese map, Y' a desingularization of Y and X' a desingularization of $X \times_Y Y'$. Then $\kappa(Y') \geq 0$, $q(Y') = q(X)$ and the induced morphism $X' \to Y'$ has a general fibre F with $\kappa(F) = -\infty$.*

The Theorems 5, 6 and 7, applied to threefolds, give a partial generalization of Theorem 4. However they say nothing about the difficult cases "$\kappa(X) = 3$", "$\kappa(X) = q(X) = 0$" and "$\kappa(X) = -\infty, q(X) = 0$".

The proofs of Theorems 6 and 7 require the study of positivity properties of the sheaves $f_*\omega_{X/Y}^\nu$ for surjective morphisms $f : X \to Y$ of manifolds. To this aim, one has to know the structure of the general fibre F and one has to study families of $(\dim X - \dim Y)$-dimensional varieties. For example, one possible proof of 6, (3) and of 7 uses results on moduli of curves and surfaces, obtained in [74] and [33]. Needless to underline that this inductive structure of the "Iitaka Program" makes it difficult to generalize Theorem 7 to higher dimensional varieties or to extend Theorem 6, (3) to the case $q(X) < n-2$. Recently the "Iitaka program" has been made more precise and extended by F. Campana [13] and S.S.Y. Lu [65].

The positivity properties of $f_*\omega^{\nu}_{X/Y}$ mentioned above have been verified only if $\kappa(F) = \dim(F)$ (Kollár) or if F has a nice "minimal model" (Kawamata). Before turning our attention to the minimal model problem, the second main line of higher dimensional classification theory, let us point out that the positivity properties of the sheaves $f_*\omega^{\nu}_{X/Y}$ can be used to construct quasi-projective moduli schemes for canonically polarized manifolds and for certain polarized manifolds of arbitrary dimension ([95], [53], [43] and [89]).

Mori's Minimal Model Program

The most powerful tools in the classification theory of threefolds came with Mori's construction of minimal models in characteristic zero.

Let us first recall the surface case. In Theorem 4 we assumed that for a surface S' with $\kappa(S') \geq 0$ there exists a surface S birational to S', with ω_S semi-ample. In order to construct such S one contracts exceptional curves E on S', i.e. curves with $\deg(\omega_{S'}|_E) = -1$. Repeating this finitely many times, one obtains a *minimal model* S of S', i.e. a surface S without exceptional curves or, equivalently, with ω_S numerically effective.

To prove Theorem 4, a) in the Appendix to [100], Mumford goes one step further. Contracting all curves C with $\deg(\omega_S|_C) = 0$, he constructs for a surface S of general type a *canonical model*, i.e. a normal surface $\tilde{S}$ with rational double points as singularities and with $\omega_{\tilde{S}}$ ample.

Unfortunately things look more complicated in the higher dimensional case (see [92] or [71]). If $\dim(X') \geq 3$ and $\kappa(X') \geq 0$, one cannot expect the existence of some manifold X birational to X', with ω_X numerically effective.

Building up on the methods he developed for his proof of the Hartshorne conjecture on the characterization of the projective space [69], Mori studied in [70] the cone of the effective curves on a threefold X' and he contracted "extremal" rational curves C with $\deg(\omega_{X'}|_C) < 0$. Inevitably such a contraction process leads to singular varieties and to proceed by induction one has to allow certain singularities.

A suitable class of singularities had been introduced by M. Reid, studying manifolds X' whose canonical ring is finitely generated. He introduced in [81] the notion of a canonical singularities. A normal variety X with such singularities is Cohen-Macaulay, as shown by Shepherd-Barron, Elkik and Flenner, and for some $r > 0$ the reflexive hull $\omega_X^{[r]}$ of $\omega_X^{\otimes r}$ is invertible. Terminal singularities are special canonical singularities.

A projective variety X is called a minimal model if it belongs to the category $\mathfrak{C}$ of $\mathbb{Q}$-Gorenstein normal varieties, with at most terminal singularities and if for some $r > 0$ the sheaf $\omega_X^{[r]}$ is invertible and numerically effective. Mori's minimal model program proposes for non uniruled manifolds X' the construction of a minimal model X in the category $\mathfrak{C}$. Unfortunately this can not be done just repeating contractions. It might happen that one has to blow up certain bad loci in order to stay in $\mathfrak{C}$. Some parts of this program have been verified in all dimensions, but for four and higher dimensional varieties it is far from being completed.

Building up on the work of Y. Kawamata, X. Benveniste, M. Reid, V. Shokurov and J. Kollár, Mori was able to finish the minimal model program in the three dimensional case [72]. The reader finds an outline of the minimal model program, of the singularities involved, of the implications and of the history of the subject in the survey articles [82], [52], [50], [98], [54], in the Lecture Notes [18] and in the contributions [73], [47] and [55] to the ICM 1990.

Let us just indicate some of the tremendous applications of Mori's result to the classification of threefolds.

Benveniste and Kawamata (for threefolds), and Kawamata (in general), have shown that for a minimal model X with $\kappa(X) = \dim X$, the sheaf $\omega_X^{[r]}$ is semi-ample. This implies that the canonical ring $R(X)$ is finitely generated and, by [81], that $\mathrm{Proj}(R(X))$ has canonical singularities. One obtains:

Theorem 8. *If X' is a threefold of general type, then $R(X')$ is finitely generated and X' has a canonical model $\tilde{X}$, i.e. a model $\tilde{X}$ with canonical singularities and with $\omega_{\tilde{X}}^{[r]}$ ample invertible, for some $r > 0$.*

For threefolds of Kodaira dimension $-\infty$, Miyaoka [68] (using Mori's result) gives the following analogue of Theorem 4 (3) for threefolds:

Theorem 9. *Let X' be a projective threefold*

(1) *X' has a minimal model X if and only if $\kappa(X') \geq 0$.*
(2) *The following three conditions are equivalent:*
  (a) *$\kappa(X') = -\infty$.*
  (b) *X' is uniruled.*
  (c) *Adjunction terminates.*

Building up on earlier work of Miyaoka, Kawamata proved in [48] the "abundance conjecture" for threefolds:

Theorem 10. *If X is a minimal threefold, then for some $r > 0$ the sheaf $\omega_X^{[r]}$ is invertible and generated by its global sections.*

Studying the blowing up and down (flips) in the construction of minimal models more closely, Kollár and Mori [60] established the invariance of the plurigenera p_ν under small deformations of threefolds, as mentioned, a result obtained in between by Siu [87] in general.

There has been a lot of progress in our understanding of 3-folds during the last years (see M. Reid's survey [83], and the references given there).

The methods developed for the minimal model program turned out to be powerful tools in the higher dimensional algebraic geometry. For example, they allow a more precise study of manifolds of Kodaira dimension $-\infty$ (see [59] and [12], the survey [76] or the book [58], and the references given there).

They were also used to define and to study "stable surfaces" in [61]. In [53] and [1] it was shown that smoothable stable surfaces of general type have a projective moduli scheme, compactifying the Gieseker moduli space. K. Karu [43] reproved their result, by showing that the existence of minimal models in dimension $n+1$ allows to define stable n-folds of general type, and to construct the corresponding projective moduli scheme.

The Lüroth Problem

Around 1971 three independent papers gave examples of unirational manifolds which are not rational. It was known that all cubic threefolds, and for some of the quartic threefolds are unirational. In [17] C. Clemens and P. Griffiths showed that the intermediate Jacobian of a rational threefold must be the product of Jacobians of curves, whereas the Jacobian of a cubic hypersurface in $\mathbb{P}^4$ does not have this property, hence it can not be rational. Studying the group $\mathrm{Bir}(X)$ of birational morphisms, V. A. Iskovskih and Yu. I. Manin showed in [40] that all quartic hypersurfaces in $\mathbb{P}^4$ are not rational.

M. Artin and D. Mumford [M8] construct in all dimensions and all characteristics $\neq 2$ unirational conic bundles X over a surface with 2-torsion in $H^3(X, \mathbb{Z})$ or in $H^3_{\text{ét}}(X, \mathbb{Z}_\ell)$.

Since those cohomology groups are shown to be birational invariants for threefolds in any characteristic and for higher dimensional manifolds in characteristic zero, X can not be rational in those cases. Their examples can not even be stably rational, i.e. for all $m \geq 0$ the manifolds $X \times \mathbb{P}^m$ are non rational.

In [5] the authors construct non-rational complex threefolds, which are stably rational. The appendix to the second edition of Manin's book [66] gives a survey of results on rationality questions, on cubic threefolds and more references.

Rational Equivalence of Zero-Cycles

Let $CH^n(X)_0$ denote the Chow group of zero-cycles of degree zero on a complex n-dimensional manifold X. One says that $CH^n(X)_0$ is finite dimensional if the natural map

$$\begin{array}{rcl} S^m X(\mathbb{C}) \times S^m X(\mathbb{C}) & \to & CH^n(X)_0 \\ (a,b) & \mapsto & a-b \end{array}$$

is surjective for some $m > 0$.

Mumford discovered in [M4] that for projective surfaces the finite dimensionality of $CH^2(X)_0$ implies that $p_g = \dim H^0(X, \omega_X) = 0$.

Theorem 11 (Mumford for $p = n = 2$, generalized by Roitman [85]). *Let X be a complex projective manifold of dimension n. If $h^{p,0} = \dim H^0(X, \Omega_X^p) > 0$ for some $p \geq 2$, then $CH^n(X)_0$ is not finite dimensional.*

In [85] Roitman showed as well that $CH^n(X)_0$ is finite dimensional if and only if $CH^n(X)_0$ is representable, i.e. if the Albanese morphism induces an isomorphism

$$\alpha_X : CH^n(X)_0 \to A(X).$$

Mumford's proof of Theorem 11 uses in an essential way the existence of a trace map for the sheaf of differential forms under finite morphisms. Unfortunately such a trace does not exist for pluricanonical forms and Mumford's arguments do not say anything about the finite dimensionality of $CH^n(X)_0$ for surfaces of general type. In fact S. Bloch proposed in [7]:

Conjecture 12. If X is a complex projective surface with $p_g = 0$, then $CH^2(X)_0$ is finite dimensional.

This conjecture was verified for surfaces X with $\kappa(X) \leq 1$ by S. Bloch, A. Kas and D. Lieberman [6]. The first examples of surfaces of general type for which Bloch's conjecture holds true were given by H. Inose and M. Mizukami [37]. A larger class of such surfaces was obtained by C. Voisin in [96].

In spite of all progresses made in the classification theory of surfaces, the conjecture 12 remains unsolved, perhaps one of the most challenging open problems for surfaces.

In [7] Bloch uses the decomposition of the diagonal, coming from the condition "$CH^n(X)_0$ finite dimensional" to reprove Mumford's Theorem 11. This method was taken up and extended in [8]. Variants of Bloch's method have been considered by several authors (see for example [42], [26] and [27]). Bloch's lecture notes [7] and the survey articles [42] and [97] explain more about motivic aspects of Mumford's Theorem and Bloch's Conjecture and about the history of the subject.

The Fake Projective Plane

D. Mumford constructs in [M9] a complex projective surface X with an ample canonical class ω_X, whose Betti-numbers $b_0, \ldots, b_4$ and Chern-numbers c_2 and c_1^2 coincide with those of $\mathbb{P}^2$. In particular one has $\chi(\mathcal{O}_X) = 1$ and $H^0(X, \omega_X) = 0$. The construction method, the

p-adic uniformization, introduced by A. Kurihara and G. A. Mustafin, and the computation of the numerical invariants have been considered by several authors in the higher dimensional case (see for example [86] and [39]).

Little is known about this surface usually called "Mumford's fake $\mathbb{P}^2$." S.-T. Yau's results [99] imply that X is the quotient of the unit ball D_2 by a discrete cocompact subgroup $\Gamma \subset SU(2,1)$, but nothing is known about Γ. The surface X is not among the ball-quotients constructed in [4] as Galois cover of $\mathbb{P}^2$, ramified along configurations of lines (loc.cit. p. 145).

In [38], the author shows that X covers an elliptic surface and he studies its bad fibres. In [39] one finds other examples of "fake $\mathbb{P}^2$'s", non-isomorphic to Mumford's surface.

Geometric properties of X itself have not been studied and, for example, Bloch's conjecture on the zero-cycles has not been verified for the fake $\mathbb{P}^2$.

Mumford's Papers on the Classification of Surfaces and Other Varieties

[M1] Pathologies of modular geometry. Amer. J. of Math **83** (1962) 339–342

[M2] Further pathologies in algebraic geometry. Amer. J. of Math., **84** (1962) 642–648

[M3] Pathologies III. Amer. J. of Math., **89** (1967) 94–104

[M4] Rational equivalences of 0-cycles on surfaces. J. of Math. of Kyoto Univ., **9** (1969) 195–204

[M5] Enriques' classification of surfaces in char. p, I. In: Global Analysis (papers in honor of K. Kodaira), Spencer and Iyanaga editors, U. of Tokyo Press, (1969) 325–339

[M6] Enriques' classification of surfaces in char. p, II (with E. Bombieri). In: Complex Analysis and Algebraic Geometry, Bailey and Shioda editors, Cambridge Univ. Press, (1977) 23–42

[M7] Enriques' classification of surfaces in char. p, III (with E. Bombieri). Invent. math., **35** (1976) 197–232

[M8] Some elementary examples of unirational varieties which are not rational (with M. Artin). J. London Math. Soc., **25** (1972) 75–95

[M9] An algebraic surface with K ample, $(K^2) = 9, p_g = q = 0$. Amer. J. of Math., **101** (1979) 233–244

Further References

[1] V. Alexeev: Boundedness and K^2 for log surfaces. Int. Journ. Math. **5** (1994) 779–810

[2] U. Angehrn, Y.-T. Siu: Effective freeness and point separation for adjoint bundles. Invent. Math. **122** (1995) 291–308.

[3] W. Barth, C. Peters, A. Van de Ven: Compact complex surfaces. Ergebnisse der Math. (3. Folge) **4** (1984) Springer, Berlin

[4] G. Barthel, F. Hirzebruch, T. Höfer: Geradenkonfigurationen und Algebraische Flächen. Aspects of Mathematics, **D4** (1987) Vieweg, Braunschweig

[5] A. Beauville, J.-L. Colliot-Thélène., J.-J. Sansuc, P. Swinnerton-Dyer: Varieétés stablement rationnelles non rationnelles. Ann. of Math. **121** (1985) 283–318

[6] S. Bloch, A. Kas, D. Lieberman: Zero cycles on surfaces with $p_q = 0$. Compositio Math. **33** (1976) 135–145

[7] S. Bloch: Lectures on algebraic cycles. Duke Univ. Math. Ser. IV (1980)

[8] S. Bloch, V. Srinivas: Remarks on correspondences and algebraic cycles. Amer. Journ. of Math. **105** (1983) 1235–1253

[9] F. A. Bogomolov: Holomorphic tensors and vector bundles on projective manifolds. Izv. Akad. Nauk SSSR Ser. Mat. **42** (1978) 1227–1287 (English translation: Math. USSR-Izv. **13** (1979) 499–555)

[10] E. Bombieri: Canonical models of surfaces of general type. Publ. math. I.H.E.S. **42** (1973) 171–219

[11] E. Bombieri, D. Husemoller: Classification and embeddings of surfaces. In: Algebraic Geometry (Arcata 1974), AMS Proc. Symp. Pure Math. **29** (1975) 329–420

[12] F. Campana: Connexit rationnelle des varits de Fano. Ann. Sci. cole Norm. Sup. **25** (1992) 539–545.

[13] F. Campana: Special Varieties and classification Theory. preprint, math.AG/0110051

[14] F. Catanese: Moduli of algebraic surfaces. In: Theory of Moduli, Proc. CIME 1985 Lecture Notes in Math. **1337** (1988) 1-83, Springer Berlin, Heidelberg, New York

[15] F. Catanese: Chow varieties, Hilbert schemes, and moduli spaces of surfaces of general type. J. Alg. Geometry **1** (1992) 561-595

[16] F. Catanese: (Some) Old and new results on algebraic surfaces. In: Proc. of the European Congress of Math. 1992, Birkhäuser, Basel (1994) 445–490

[17] C. Clemens, P. Griffiths: The intermediate Jacobian of the cubic threefold. Ann. of Math. **95** (1972) 281–356

[18] H. Clemens, J. Kollár, S. Mori: Higher dimensional complex geometry. Astérisque **166** (1988) Soc. Math. de France

[19] P. Deligne: Théorie de Hodge II. Publ. Math. I.H.E.S. **40** (1972) 5–57

[20] P. Deligne, L. Illusie: Relèvements modulo p^2 et décomposition du complexe de de Rham. Invent. math. **89** (1987) 247–270

[21] J.–P. Demailly: Une généralisation du théorème d'annulation de Kawamata-Viehweg. C. R. Acad. Sci. Paris **309** (1989) 123–126

[22] J.- P. Demailly: A numerical critierion for very ample line bundles. Journ. Diff. Geom. **37** (1993) 323–374

[23] L. Ein, R. Lazarsfeld: Global generation of pluricanonical and adjoint linear series on smooth projective threefolds. Journ. of the AMS **6** (1993) 875–903.

[24] L. Ein, R. Lazarsfeld, M. Nakamaye: Zero-estimates, intersection theory and a theorem of Demailly. Higher-dimensional complex varieties (Trento, 1994) 183–207, de Gruyter, Berlin, 1996.

[25] T. Ekedahl: Canonical models of surfaces of general type in positive characteristic. Publ. Math. I.H.E.S. **67** (1989) 97–144

[26] H. Esnault, M. Levine: Surjectivity of cycle maps. In: Journées de Géom. Alg. d'Orsay, 1992. Astérisque **218** (1993) 203–226

[27] H. Esnault, V. Srinivas, E. Viehweg: Decomposability of Chow groups implies decomposability of cohomology. In: Journées de Géom. Alg. d'Orsay, 1992. Astérisque **218** (1993) 227–242

[28] H. Esnault, E. Viehweg: Dyson's lemma for polynomials in several variables (and the theorem of Roth). Invent. Math. **78** (1984) 445–490

[29] H. Esnault, E. Viehweg: Logarithmic De Rham complexes and vanishing theorems. Invent. math. **86** (1986) 161–194

[30] H. Esnault, E. Viehweg: Lectures on vanishing theorems. DMV-Seminar **20** Birkhäuser 1992, Basel

[31] G. Faltings: p-adic Hodge theory. Journ. Amer. Math. Soc. **1** (1988) 255–299

[32] J.-M. Fontaine, W. Messing: p-adic periods and p-adic étale cohomology. In: Current Trends in Arithmetical Algebraic Geometry. Contemp. Math. **67** (1987) 179–201

[33] D. Gieseker: Global moduli for surfaces of general type. Invent. math. **43** (1977) 233–282

[34] H. Grauert, O. Riemenschneider: Verschwindungssätze für analytische Kohomologiegruppen auf komplexen Räumen. Invent. math. **11** (1970) 263–292

[35] G. Heier: Effective freeness of adjoint line bundles. Doc. Math. **7** (2002) 31–42

[36] S. Iitaka: On D-dimensions of algebraic varieties. Journ. Math. Soc. Japan **23** (1971) 356–373

[37] H. Inose, M. Mizukami: Rational equivalence of zero cycles on some surface of general type with $p_q = 0$. Math. Ann. **244** (1979) 205–217

[38] M. N. Ishida: An elliptic surface covered by Mumford's fake projektive plane. Tokoku Math. J. **40** (1988) 367–396

[39] M.-N. Ishida, F. Kato: The strong rigidity theorem for non-Archimedean uniformization. Tohoku Math. J. **50** (1998) 537–555.
[40] V. Iskovskih, Yu. Manin: Three-dimensional quartics and counter-examples to the Lüroth problem. Mat. Sb. **86** (1971) 140–166
[41] V. Iskovskih, I. R. Shafarevich: Algebraic surfaces. In: Algebraic Geometry II, EMS **35** Springer 1995, Heidelberg
[42] U. Jannsen: Motivic sheaves and filtrations on Chow groups. In: Motives. Proc. of Symp. in Pure Math. **55** (1994) Part 1, 245–302
[43] K. Karu: Minimal models and boundedness of stable varieties. J. Algebraic Geom. **9** (2000) 93–109.
[44] K. Kato: *p*-adic vanishing cycles (applications of Fontaine-Messing). In: Algebrai Geom. Adv. Studies in Pure Math. **10** (1987) 207–251
[45] Y. Kawamata: Characterization of abelian varieties. Compositio math. **43** (1981) 253–276
[46] Y. Kawamata: A generalization of Kodaira-Ramanujam's vanishing theorem. Math. Ann. **261** (1982) 43–46
[47] Y. Kawamata: Canonical and minimal models of algebraic varieties. In: Proc. of the ICM, Kyoto 1990. Mathematical Soc. of Japan (1991) 699–708
[48] Y. Kawamata: Abundance theorem for minimal threefolds. Invent. math. **108** (1992) 229–246
[49] Y. Kawamata: On Fujita's freeness conjecture for 3-folds and 4-folds. Math. Ann. **308** (1997) 491–505.
[50] Y. Kawamata, K. Matsuda, K. Matsuki: Introduction to the minimal model problem. In: Algebraic Geometry, Sendai 1985. Anvanced Studies in Pure Math. **10** (1987) 283–360
[51] J. Kollár: Higher direct images of dualizing sheaves I. Ann. of Math. **123** (1986) 11–42
[52] J. Kollár: The structure of algebraic threefolds. Bull. AMS **17** (1987) 211–273
[53] J. Kollár: Projectivity of complete moduli Journ. Diff. Geom. **32** (1990) 235–268
[54] J. Kollár: Minimal models of algebraic threefolds: Mori's program. Sém. Bourbaki **712**. In: Astérisque **177-178** (1990) 303–326
[55] J. Kollár: Flip and flop. In: Proc. of the ICM, Kyoto 1990. Mathematical Soc. of Japan (1991) 709–714
[56] J. Kollár: Effective base point freeness. Math. Ann. **296** (1993) 595–605
[57] J. Kollár: Shafarevich maps and automorphic forms. Princeton Univ. Press, 1995
[58] J. Kollár: Rational curves on algebraic varieties. Ergebnisse der Mathematik (3. Folge) **32** (1995) Springer, Berlin, Heidelberg, New York
[59] J. Kollár, Y. Miyaoka, S. Mori: Rationally connected varieties. J. Alg. Geometry. **1** (1992) 429–448
[60] J. Kollár, S. Mori: Classification of three dimensional flips. Journ. AMS **5** (1992) 533–703
[61] J. Kollár and N. I. Shepherd-Barron: Threefolds and deformations of surface singularities. Invent. math. **91** (1988) 299–338
[62] W. E. Lang: Examples of surfaces of general type with vector fields. In: Arithmetic and Geometry, Papers dedicated to I. R. Shafarevich, Vol. II, Progress in Math. **36** (1983) 167–173, Birkhäuser, Boston
[63] W. E. Lang: On Enriques surfaces in characteristic p, I. Math. Ann. **265** (1983) 45–65; II. Math. Ann. **281** (1988) 671–685
[64] R. Lazarsfeld: Positivity in Algebraic Geometry. preprint 567 pages. Ergebnisse der Mathematik, Springer, to appear
[65] S.S.Y. Lu: A refined Kodaira dimension and its canonical fibration. preprint, math.AG/0211029
[66] Yu. Manin: Cubic forms (second edition). North-Holland Math. Library **4** (1986) North-Holland, Amsterdam
[67] Y. Miyaoka: On the Mumford-Ramanujam vanishing theorem on a surface. In: Géométrie Algébrique, Angers 1979. Sijthoff and Nordhoff (1980) Oslo, 239–248
[68] Y. Miyaoka: On the Kodaira dimension of minimal threefolds. Math. Ann. **281** (1988) 325–332
[69] S. Mori: Projective manifolds with ample tangent bundle. Ann. of. Math. **110** (1979) 593–606

[70] S. Mori: Threefolds whose canonical bundles are not numerically effective. Ann. of Math. **116** (1982) 133–176

[71] S. Mori: Classification of higher-dimensional varieties. In: Algebraic Geometry (Bowdoin 1985) AMS Proc. Symp. Pure Math. **46**, Part 1 (1987) 269–331

[72] S. Mori: Flip theorem and the existence of minimal models for 3-folds. Journ. AMS **1** (1988) 117-253

[73] S. Mori: Birational Classification of algebraic threefolds. In: Proc. of the ICM, Kyoto 1990. Mathematical Soc. of Japan (1991) 235–248

[74] D. Mumford: Geometric invariant theory. Ergebnisse der Mathematik **34** (1965) Springer, Berlin, Heidelberg, New York

[75] A. Nadel: Multiplier ideal sheaves and the existence of Kähler-Einstein metrics of positive scalar curvature. Ann. of Math. **132** (1989) 549–596

[76] T. Peternell: Tangent bundles, rational curves and the geometry of manifolds of negative Kodaira dimension. In: Complex Analysis and Geometry. Plenum Press, New York (1993) 293–310

[77] C. Peters: Introduction to the theory of compact complex surfaces. In: Differential Geometry, Global Analysis and Topology (Halifax 1990) CMS Conf. Proc. **12** (1990) 129–156

[78] C. P. Ramanujam: Remarks on the Kodaira vanishing theorem. J. Indian Math. Soc. **36** (1972) 41–51

[79] M. Raynaud: Contre-exemple au "Vanishing Theorem" en caratéristique $p > 0$. In: C. P. Ramanujam–A tribute. Studies in Math. 8, Tata Institute of Fundamental Research, Bombay (1978) 273–278

[80] M. Reid: Bogomolov's theorem $c_1^2 < 4c_2$. In: Int. Symp. on Algebraic Geometry, Kyoto 1977. Kinokuniya, Tokyo. 623–642

[81] M. Reid: Canonical 3-folds. In: Géométrie Algébrique, Angers 1979. Sijthoff and Nordhoff (1980) Oslo, 273–310

[82] M. Reid: Young person's guide to canonical singularities. In: Algebraic Geometry (Bowdoin 1985) AMS Proc. Symp. Pure Math. **46**, Part 1 (1987) 345–416

[83] M. Reid: Update on 3-folds. Proc. of the ICM, Beijing 2002. Vol. II (2002) 513–524. Higher Education Press, Beijing

[84] I. Reider: Vector bundles of rank 2 and linear systems on algebraic surfaces. Ann. of Math. **127** (1988) 309–316

[85] A. Roitman: Rational equivalence of zero-cycles. Mat. Sb. **89** (1972) = Math. USSR–Sb. **18** (1974) 571–588

[86] P. Schneider: The cohomology of local systems on p-adically uniformized varieties. Math. Ann. **293** (1992) 623–650

[87] Y.-T. Siu: Extension of twisted pluricanonical sections with plurisubharmonic weight and invariance of semipositively twisted plurigenera for manifolds not necessarily of general type. In: Complex geometry. Collection of papers dedicated to Hans Grauert. Springer-Verlag, Berlin (2002) 223–277

[88] Y.-T. Siu: Some recent transcendental techniques in algebraic and complex geometry. Proc. of the ICM, Beijing 2002. Vol. I (2003) 439–448. Higher Education Press, Beijing

[89] G. Schumacher, H. Tsuji: Quasi-Projectivity of Moduli Spaces of Polarized Varieties. preprint, math.AG/0111144

[90] L. Szpiro: Sur le théorème de rigidité de Parsin et Arakelov. In: Journées de géometrie algébrique de Rennes II. Astérisque **64** (1979) 169–202

[91] L. Szpiro (Ed.): Séminaire sur les pinceaux de courbes de genre au moins deux. Astérisque **86** (1981) Paris

[92] K. Ueno: Birational geometry of algebraic threefolds. In: Géométrie Algébrique, Angers 1979. Sijthoff and Nordhoff (1980) Oslo, 311–323

[93] E. Viehweg: Vanishing theorem. Journ. reine angew. Math. **335** (1982) 1–8

[94] E. Viehweg: Weak positivity and the additivity of the Kodaira dimension for certain fibre spaces. In: Algebraic Varieties and Analytic Varieties, Advanced Studies in Pure Math. 1 (1983) 329–353

[95] E. Viehweg: Quasi-projective moduli for polarized manifolds. Ergebnisse der Mathematik (3. Folge) **30** (1995) Springer, Berlin, Heidelberg, New York

[96] C. Voisin: Sur les zéro-cycles de certaines hypersurfaces munies d'un automorphisme. Ann. Scuola Norm. di Pisa **29** (1993) 473–492

[97] C. Voisin: Transcendental methods in the study of algebraic cycles. In: Algebraic Cycles and Hodge Theory (Torino 1993). Lecture Notes in Math. **1594** (1994) Springer Verlag, Berlin, 153–223

[98] P. M. H. Wilson: Towards a birational classification of algebraic varieties. Bull. London Math. Soc. **19** (1987) 1–48

[99] S.-T. Yau: Calabi's conjecture and some new results in algebraic geometry. Proc. Nat. Acad. Sci. U.S.A. **74** (1977) 1798–1799.

[100] O. Zariski: The theorem of Riemann-Roch for high multiples of an effective divisor on an algebraic surface. Appendix by D. Mumford. Ann. of. Math. **76** (1962) 560–615

UNIVERSITÄT ESSEN, FB6 MATHEMATIK, 45117 ESSEN, GERMANY
E-mail address: viehweg@uni-essen.de

Enriques' classification of surfaces in char p: I

By David Mumford

The principal assertion in Enriques' classification of surfaces is:

Theorem. *Let F be a non-singular projective surface, without exceptional curves of the 1st kind. Let K_F be the canonical divisor class on F. Then*

(i) *if* $|12K_F| = \emptyset$, *then F is ruled,*

(ii) *if* $|12K_F| \neq \emptyset$, *then either* $12K_F \equiv 0$, *or else* $|nK_F|$ *is a linear system without base points for some n.*

As a Corollary, if we introduce the notations:

(a) $\mathcal{K} = \operatorname{tr} d_k \sum_{n=0}^{\infty} H^0(F, o_F(nK)) - 1$

(b) F *elliptic* if $\exists$ a morphism $f: F \to C$, C a curve, with almost all fibres non-singular elliptic curves,

(c) F *quasi-elliptic* if $\exists$ a morphism $f: F \to C$, C a curve, with almost all fibres singular rational curves E with $p_a(E) = 1$,[1]

(d) F of *general type* if $\exists f: F \to F_0$, f birational F_0 normal with K_{F_0} ample. Then we find that there are 4 types of surfaces F,

$\mathcal{K} = -1$	F ruled
$\mathcal{K} = 0$	$12K_F \equiv 0$
$\mathcal{K} = 1$	F elliptic, or quasi-elliptic, with K_F of positive degree
$\mathcal{K} = 2$	F of general type .

All this, plus a detailed analysis of the case $\mathcal{K} = 0$, has been proven in char 0; it is due essentially to Enriques [1], and has been worked out in detail by Kodaira [4], [5], in Šafarevič's seminar [7], and in Zariski's seminar at Harvard. The purpose of this note and its sequel is to supply some new ideas which make the proof work in char p. Some of the steps supply new proofs of parts of the theorem in char 0, which have, I believe, some interest.

[1] In view of the results of Tate [11], such surfaces can only occur if char $(k) = 2$ or 3; moreover, almost all fibres E have a single ordinary cusp.

List of Notation

F = non-singular projective surface
K = canonical divisor class on F
c_2 = 2nd Chern class of F (a number)
q = dimension of Albanese of F
B_2 = 2nd Betti number of F, dim $H^2(F, \mathbf{R})$
ρ = Base number of F
Ω_F^i = sheaf of i-forms on F,
$h^{p,q} = \dim H^q(\Omega_F^p)$
$p_g = h^{2,0} = h^{0,2}$
$p_a = h^{0,2} - h^{0,1} = \chi(o_F) - 1$

$$p_a(D) = 1 + \frac{(D\cdot D + K)}{2}, \text{ if } D \text{ is a curve on } F$$

1. $(K\cdot D) < 0$ for some effective divisor D

The situation $(K\cdot D) < 0$, some D, is known classically as the case "*adjunction terminates.*" In this case, we shall prove that F is ruled.

Step (I). *There is an ample H such that $(K\cdot H) < 0$.*

PROOF. In fact, first replacing D by a suitable component of itself, we find an *irreducible* D such that $(K\cdot D) < 0$. If $(D^2) < 0$, then since

$$(K\cdot D) + (D^2) = 2p_a(D) - 2 \geqq -2\ ,$$

it follows that $(K\cdot D) = (D^2) = -1$, i.e., D is exceptional of 1st kind. This has been excluded, so $(D^2) \geqq 0$. Let H_1 be any ample divisor on F. Then for all $n \geqq 0$, $nD + H_1$ is ample by the Nakai-Moisezon criterion of ampleness [2]. But if $n \gg 0$, $(K\cdot nD + H_1) < 0$. q.e.d.

COROLLARY OF *Step* I. $|nK| = \emptyset$, *all* $n \geqq 1$.

Step (II). *If* $(K^2) > 0$, *then F is rational, hence ruled.*

PROOF. Use the general formulas

$$12(\chi(o_F)) = (K^2) + c_2$$
$$c_2 = 2 - 4q + B_2\ .$$

In our case, $p_g = \dim |K| + 1 = 0$, hence it follows that the Picard scheme of F is reduced [6, lecture 26]. Therefore

$$\begin{aligned}\dim H^1(o_F) &= \dim(\text{tangent space to Picard scheme})\\ &= \dim(\text{Picard scheme})\\ &= q\,,\end{aligned}$$

and so

$$\begin{aligned}\chi(o_F) &= \dim H^0(o_F) - \dim H^1(o_F) + \dim H^2(o_F)\\ &= 1 - q + p_g\\ &= 1 - q\,.\end{aligned}$$

Therefore

$$12 - 12q = (K^2) + 2 - 4q + B_2\,,$$

or

$$(*) \qquad 10 = 8q + (K^2) + B_2\,.$$

But if ρ = base number of F, then $B_2 \geqq \rho > 0$ (Igusa [12]), so if $(K^2) > 0$, it follows that $q = 0$ or 1. If $q = 1$, then F admits a morphism onto an elliptic curve, hence $B_2 \geqq 2$ and (*) still cannot be satisfied. But if $q = 0$, then since $|2K| = \emptyset$, the hypotheses of Castelnuovo's criterion are met, and by Zariski [9], it follows that F is rational. q.e.d.

Step III. (Kodaira). *If* $(K^2) \leqq 0$, *then for all n there are effective divisors D on F such that*

(a) $|D + K| = \emptyset$

(b) $\dim |D| \geqq n$.

PROOF. Let H be an ample divisor such that $(K \cdot H) < 0$. For all n, note that $(nH + mK.\, H) < 0$ and hence $nH + mK$ cannot be linearly equivalent to an effective divisor, if $m \gg 0$. Let m_n be a non-negative integer such that

$$\begin{aligned}|nH + m_nK| &\neq \emptyset\\ |nH + (m_n + 1)K| &= \emptyset\,.\end{aligned}$$

Let $D_n \in |nH + m_nK|$. Write $D_n = D'_n + D''_n$, where D'_n and D''_n are positive and the components E of D'_n satisfy $(E.K) < 0$, while those of D''_n satisfy $(E \cdot K) \geqq 0$. Note that $(E \cdot K) < 0 \Rightarrow (E^2) \geqq 0$ (cf. *Step* I) so $(D'^2_n) \geqq 0$. Next, note that $|K - D'_n| = \emptyset$. In fact if not, K itself would be effective, since it would be the sum of D'_n and an effective divisor in $K - D'_n$. So by Serre duality, $H^2(o_F(D'_n)) = (0)$. Now use Riemann-Roch.

$$\begin{aligned}\dim|D'_n| &= \dim H^0(o_F(D'_n)) - 1\\ &\geqq \chi(o_F(D'_n)) - 1\\ &= \frac{(D'_n\cdot D'_n - K)}{2} + \chi(o_F) - 1\\ &\geqq -\frac{(D'_n\cdot K)}{2} + \chi(o_F) - 1\\ &\geqq -\frac{(D_n\cdot K)}{2} + \chi(o_F) - 1\\ &= -n\frac{(H\cdot K)}{2} - m_n\frac{(K^2)}{2} + \chi(o_F) - 1\\ &\geqq \frac{n}{2} + \chi(o_F) - 1\,.\end{aligned}$$

Since $|K + D'_n| = \varnothing$, D'_n has all the required properties. q.e.d.

Now comes the new idea to take care of char p.

Key Step IV. *If D is an effective divisor such that $|K+D| = \varnothing$, then the natural map*

$$\mathrm{Pic}^0_F \longrightarrow \mathrm{Pic}^0_D$$

is surjective.

PROOF. $H^0(o_F(K + D)) = (0)$ implies $H^2(o_F(-D)) = (0)$ by Serre duality. Using the exact sequence

$$0 \longrightarrow o_F(-D) \longrightarrow o_F \longrightarrow o_D \longrightarrow 0\,,$$

it follows that the natural map $H^1(o_F) \to H^1(o_D)$ is surjective. But these vector spaces are the tangent spaces at 0 to the connected and reduced schemes group Pic^0_F and Pic^0_D. (Pic^0_F is reduced since $p_g = 0$!). Hence Pic^0_F maps onto Pic^0_D. q.e.d.

Step (V). *If D is an effective divisor such that $|K + D| = \varnothing$, and if $D = \sum n_i E_i$, then*

(i) *all E_i are non-singular*

(ii) *if $n_i \geqq 2$, then*

(a) *E_i is rational or*

(b) *E_i is elliptic, $(E_i^2) = 0$ with non-trivial normal bundle or*

(c) $(E_2^i) < 0$

(iii) *the E_i are connected together without loops.*

PROOF. In fact, by Step (IV), Pic^0_D is an abelian variety. Since the natural map $\mathrm{Pic}^0_D \to \mathrm{Pic}^0_{E_i}$ is surjective, $\mathrm{Pic}^0_{E_i}$ is abelian too, so

E_i is non-singular. If (iii) were false, Pic^0_D would have subgroups of type $\mathbf{G}_m$ coming from the loops. To check (ii), use the fact that if $k_i \geqq 2$, $\mathrm{Pic}^0_D \to \mathrm{Pic}^0_{2E_i}$ is surjective, hence $\mathrm{Pic}^0_{2E_i}$ is abelian, hence $\mathrm{Pic}^0_{E_i} = \mathrm{Pic}^0_{2E_i}$. Looking at tangent spaces at 0, this implies that $H^1(o_{2E_i}) \cong H^1(o_{E_i})$. But *via* the exact sequence

$$0 \longrightarrow o_{E_i}(-E_i^2) \longrightarrow o_{2E_i} \longrightarrow o_{E_i} \longrightarrow 0 \, ,$$

this implies

$$H^1(o_{E_i}(-E_i^2)) = (0) \, .$$

But if $(E_i^2) \geqq 0$, $o_{E_i}(-(E_i^2))$ has degree $\leqq 0$, hence $H^1 \neq (0)$ except in the cases E_i rational, $o_{E_i}(-(E_i^2))$ of degree -1, 0; or E_i elliptic, $(E_i^2) = 0$ and $o_{E_i}(-(E_i^2)) \not\cong o_{E_i}$. q.e.d.

Step (VI). *There is a non-singular rational curve on F passing through every point of F.*

PROOF. Suppose to the contrary. Then there are only a finite number of non-singular rational curves on F of each degree. Moreover, on any surface there are only a finite number of non-singular curves E of each degree such that either

(a) $(E^2) < 0$, or

(b) E elliptic, $(E^2) = 0$, with *non-trivial* normal bundle.

Taking all these curves together, we get a countable set of curves, which cannot exhaust F. Let $P_1, \cdots, P_q$ be distinct points of F on none of these curves. Let D be a divisor satisfying the requirements of *Step* (III) with $\dim |D| \geqq 3q$. It follows that there is a divisor $D' \in |D|$ with double points at each P_i, $1 \leqq i \leqq q$. Let $D' = \sum n_i E_i$. If $n_i \geqq 2$, then by *Step* (V), E_i would be in the countable set of curves just described, so E_i does not contain any P_j. In other words, each P_i lies only on *simple* components of D', and since each E_i is non-singular (*Step* (V)), each P_i lies on 2 components. Since the E_i's are connected together as a tree, this shows that there are in all at least $q + 1$ E_i's containing some P_j. Note that they are all curves of genus at least 1. Therefore

$$q = \dim(\mathrm{Pic}^0_F) \geqq \dim(\mathrm{Pic}^0_{D'}) \geqq \textstyle\sum_i \dim(\mathrm{Pic}^0_{E_i}) \geqq q + 1 \, .$$

So our assumption was false. q.e.d.

Step (VII). *F is ruled.*

PROOF. This follows by any number of methods. If $q = 0$, there must be a *linear* system of positive dimension of non-singular rational curves. Apply Tsen's theorem[2] to a pencil of such curves. If $q > 0$, look at the Albanese map $\pi: F \to A$. Since π maps all rational curves on F to points, it follows from *Step* (VI) that π factors

$$F \xrightarrow{\pi'} C \hookrightarrow A$$

where C is a curve, and the fibres of π' are rational. Therefore F is ruled by Tsen's theorem. q.e.d.

2. $(K \cdot D) \geqq 0$, all effective divisors D; $(K^2) = 0$

In this section we will show that

$$\left\{\begin{matrix}(K^2) = 0 \\ (K \cdot D) \geqq 0, \text{ eff. } D\end{matrix}\right\} \Longrightarrow \left\{\begin{matrix}\text{either } 2K \equiv 0\,, \\ \text{or } \exists \text{ a pencil of curves on } F \text{ with } p_a = 1\end{matrix}\right\}$$

(Note: in char 2 or 3, the pencil may consist entirely of rational curves with cusps; Bertini's theorem tells us only that almost all the curves are irreducible, cf. [8].)

First make the definition.

DEFINITION. A curve $D = \sum n_i E_i$ on F is *of canonical type* if $(K \cdot E_i) = (D \cdot E_i) = 0$, all i. D is *indecomposable of canonical type*, if D is connected, and g.c.d. $(n_i) = 1$.

Step (I). *Either* $2K \equiv 0$ *or* F *contains at least one indecomposable curve of canonical type.*

PROOF. (Enriques; cf. also Šafarevič [7, Lemmas 9, 10, pp. 71–73)]. If $D \in |2K|$, then D is either 0 or of canonical type. In fact, let $D = \sum n_i E_i$. Then

$$0 = (K \cdot D) = \sum n_i (K \cdot E_i)\,.$$

Since $(K \cdot E_i) \geqq 0$, all i, it follows that $(K \cdot E_i) = 0$, all i; hence $(D \cdot E_i) = 2(K \cdot E_i) = 0$, too. Decomposing D, we find an indecomposable curve of canonical type. On the other hand, suppose $|2K| = \varnothing$. In this case, $q = h^{0,1}$ and $q \neq 0$ or else F would be rational by Castelnuovo; and $q \geqq 2$ would contradict the formula

$$\chi(o_F) = 1 - \dim H^1(o_F) + \dim H^2(o_F) \leqq 1 - q$$
$$12\chi(o_F) = (K^2) + (c_2) = 2 - 4q + B_2 \geqq 3 - 4q\,.$$

There remains the one really subtle case of $q = 1$. Consider the

[2] If $\pi: F \to C$ is a morphism of a surface to a curve, with fibres of $p_a = 0$, then $k(F) \cong k(C)(X)$.

Albanese morphism $\pi\colon F \to \varepsilon$, ε an elliptic curve. Let $f = \pi^{-1}(P)$, some $P \in \varepsilon$, be an irreducible fibre of π. If $p_a(f) = 1$, f is of canonical type and we are through. If $p_a(f) > 1$, then $(K\cdot f) = 2p_a(f) - 2 \geqq 2$. We then use the following argument of Enriques: for all $Q \in \varepsilon$, $Q \neq P$ consider the exact sequence

$$0 \longrightarrow o_F(2K + \pi^{-1}(Q) - f) \longrightarrow o_F(2K + \pi^{-1}(Q)) \longrightarrow o_f(2K\cdot f) \longrightarrow 0 .$$

Note that $H^2(o_F(2K + \pi^{-1}(Q) - f)) = (0)$, since if $|f - \pi^{-1}(Q) - K|$ contained an effective divisor, then $(K\cdot f) \leqq 0$. Check by Riemann-Roch that $\chi(o_F(2K + \pi^{-1}(Q) - f)) = 0$. Therefore, for all $Q \neq P$, either

(i) $|2K + \pi^{-1}(Q) - f| \neq \varnothing$, or

(ii) $H^i(o_F(2K + \pi^{-1}(Q) - f)) = (0)$, all i, hence

$$H^0(o_F(2K + \pi^{-1}(Q))) \xrightarrow{\sim} H^0(o_f(2K\cdot f)) .$$

Suppose $|2K + \pi^{-1}(Q) - f| = \varnothing$ for all Q. Then fix a non-zero $s \in H^0(o_f(2K\cdot f))$ and let A be the Cartier divisor $s = 0$ on f. For all Q, s lifts to a unique $s'_Q \in H^0(o_F(2K + \pi^{-1}(Q)))$, and let D_Q be the divisor $s'_Q = 0$. This is a 1-dimensional algebraic family of divisors, such that $D_Q\cdot f = A$ for all $Q \neq P$. Moreover, all the D_Q's are distinct since they are not even linearly equivalent. Therefore,

$$F = \text{closure of } \textstyle\bigcup_{Q\neq P} D_Q .$$

In particular, as $Q \to P$, D_Q must specialize to a divisor D_P containing the whole fibre f. Then

$$D_P = f + D_P^*$$
$$D_P \in |2K + f| .$$

So $|2K| \neq \varnothing$. So finally in all cases, $|2K + \pi^{-1}(Q) - f| \neq \varnothing$ for some $Q \in \varepsilon$ (possibly $Q = P$). But a divisor $D \in |2K + \pi^{-1}(Q) - f|$ is of canonical type just as before. q.e.d.

We would like to assert next that on any surface F and for any indecomposable D of canonical type, $\dim H^0(o_F(nD)) > 1$ and hence $|nD|$ is composite with a pencil, for some n. Unfortunately, this is false, as one sees by considering an elliptic ruled surface obtained by completing a line bundle of infinite order over an elliptic curve, the 0-section of the bundle being taken as the curve of canonical type. The rest of the proof is an exercise in avoiding this case. The following will be useful.

Lemma. *Let $D = \sum n_i E_i$ be an indecomposable curve of canonical type, and let L be an invertible sheaf on D, (regarded as a 1-dimensional scheme). If* $\deg(L \otimes o_{E_i}) = 0$, *all i, then*

$$H^0(D, L) \neq (0) \Longleftrightarrow L \cong o_F$$

and moreover $H^0(D, o_D) = k$.

Proof. Let $s \in H^0(D, L)$. It suffices to show that if $s \neq 0$, s generates L hence defines an isomorphism of o_D with L; in particular, if $L = o_D$, this shows that all non-zero elements of the algebra $H^0(D, o_D)$ are units, i.e., $H^0(D, o_D)$ is a field, hence is k. But consider the induced section of $L \otimes o_{E_i}$. Since this sheaf has degree 0, a section either generates $L \otimes o_{E_i}$ or is identically 0. If s vanishes on one E_i, it vanishes on one point of each E_j meeting E_i; since D is connected, s either generates L everywhere, or vanishes on all E_i's. Assume that s vanishes on all E_i's; let k_i = order of vanishing of s on E_i. Then $1 \leqq k_i \leqq n_i$. Whenever $k_i < n_i$, s defines a non-zero section of

$$L \otimes \left[o_F(-k_i E_i)/o_F\left(-(k_i + 1)E_i\right)\right] .$$

This section vanishes to order at least $I\left(E_i, \sum_{j \neq i} k_j E_j; P\right)$ at every $P \in E_i$ (I = intersection multiplicity). It follows that if $k_i < n_i$, then

$$\begin{aligned} (1) \qquad \left(E_i . \textstyle\sum_{j \neq i} k_j E_j\right) &\leqq \deg\left\{L \otimes o_F(-k_i E_i)/o_F\left((-k_i + 1)E_i\right)\right\} \\ &= \deg\left[o_F(-E_i)/o_F(-E_i^2)\right]^{k_i} \\ &= -k_i(E_i^2) . \end{aligned}$$

Note that if $k_i = n_i$, then since $(E_i \cdot D) = 0$,

$$(2) \qquad \left(E_i \cdot \textstyle\sum_j k_j E_j\right) = -\left(E_i \cdot \textstyle\sum_j (n_j - k_j) E_j\right) \leqq 0 .$$

So if $D_1 = \sum_j k_j E_j$, then *for all i*, using (1) or (2), according as $k_i < n_i$ or $k_i = n_i$, $(E_i \cdot D_1) \leqq 0$. On the other hand,

$$\sum n_i (E_i \cdot D_1) = (D \cdot D_1) = \sum k_i (D \cdot E_i) = 0 ,$$

so it follows that $(E_i \cdot D_1) = 0$, all i. This shows that $D - D_1$ and D_1 are of canonical type too. I claim that D and D_1 are both multiples of a 3rd divisor of canonical type, which will show that D is decomposable, unless $D = D_1$, i.e., $s \equiv 0$. To prove this, we must show that k_i/n_i is independent of i. Let $a/b = \max(k_i/n_i)$ and let $Z = aD - bD_1$. Then Z is an effective cycle and E_i occurs in Z if and only if $k_i/n_i < a/b$. Then if $k_i/n_i = a/b$, E_i is not a component of Z, and since $(Z\cdot E_i) = a(D\cdot E_i) - b(D_1\cdot E_i) = 0$, E_i does not even meet Z; so Z does not contain any E_j's meeting E_i. Since D is connected, this shows that $Z = 0$, hence k_i/n_i independent of i as needed. q.e.d.

COROLLARY 1. *If D is indecomposable of canonical type, then*

$$o_F(K + D) \otimes o_F \cong o_D .$$

PROOF. Let $\omega = o_F(K + D) \otimes o_D$: then ω is the dualizing sheaf on the Cohen-Macauley scheme D, so by Serre duality $\dim H^1(D, \omega) = \dim H^0(D, o_D) = 1$. *Via* the exact sequence

$$0 \longrightarrow o_F(K) \longrightarrow o_F(K + D) \longrightarrow \omega \longrightarrow 0 ,$$

we see that

$$\begin{aligned} \chi(\omega) &= \chi\big(o_F(K + D)\big) - \chi\big(o_F(K)\big) \\ &= \frac{(K + D\cdot D)}{2} \qquad \text{(Riemann-Roch on } F) \\ &= 0 , \end{aligned}$$

hence $\dim H^0(D, \omega) = 1$ too. Hence $\omega \cong o_D$ by the lemma. q.e.d.

COROLLARY 2. *If $D = \sum n_i E_i$ is an indecomposable divisor of canonical type, and D' is any effective divisor on F such that $(D'\cdot E_i) = 0$ all i, then*

$$D' = nD + D''$$

where $n \geqq 0$ and D'' is an effective divisor disjoint from D.

Incidentally, a complete list of all curves of canonical type can be found in Kodaira [3], [4], and the lemma could be checked case by case.

Step (II). *If $p_g = 0$ (and $(K^2) = 0$, $(K\cdot D) \geqq 0$* all eff. *Das always) and D is an indecomposable curve of canonical type, then $|nD|$ is composite with a pencil of curves of canonical type, for some n.*

PROOF. First, look at the sequences

$$0 \longrightarrow o_F\big(nK + (n-1)D\big) \longrightarrow o_F(nK + nD) \longrightarrow o_D \longrightarrow 0$$

obtained by applying Corollary 1. If $n \geqq 2$, then

$$H^2\big(o_F(nK + (n-1)D)\big) = H^2\big(o_F(nK + nD)\big) = (0) .$$

Since $H^1(o_D) \neq (0)$, we find $H^1\big(o_F(nK + nD)\big) \neq (0)$. But

$$\chi\big(o_F(nK + nD)\big) = \chi(o_F) = 0 \quad \text{or} \quad 1 \qquad \text{(cf. } \textit{Step}\text{ I)} .$$

Therefore $H^0(o_F(nK + nD)) \neq (0)$. This shows that there is a divisor $D_n \in |nK + nD|$.

D_n is of canonical type. Note that if $D = \sum n_i E_i$, then

$$(D_n \cdot E_i) = n(K \cdot E_i) + n(D \cdot E_i) = 0 .$$

So by Corollary 2, $D_n = aD + \sum k_i F_i$, where the F_i are disjoint from D. Now $(K \cdot F_i) \geqq 0$ for all i, while

$$\begin{aligned} \sum k_i(K \cdot F_i) &= (K \cdot \sum k_i F_i + aD) \\ &= (K \cdot nK + nD) \\ &= 0 \end{aligned}$$

so $(K \cdot F_i) = 0$, all i. Finally

$$(D_n \cdot F_i) = n(K \cdot F_i) + n(D \cdot F_i) ,$$

and this is 0 since D does not meet F_i.

It would seem to follow that we have produced at least one indecomposable curve of canonical type disjoint from D. But in one case this would not be so; namely, if each D_n was a multiple of D. In that case, though, K itself would be a multiple of D, hence $p_g \neq 0$ contrary to hypothesis.

Now assume that D and D' are disjoint indecomposable curves of canonical type. Then look at the sequence

$$0 \longrightarrow o_F(2K + D + D') \longrightarrow o_F(2K + 2D + 2D') \longrightarrow o_D \oplus o_{D'} \longrightarrow 0$$

(again using Corollary 1). As before, all H^2's vanish, so now

$$\dim H^1(o_F(2K + 2D + 2D')) \geqq 2 ,$$

hence calculating Euler characteristics, we find

$$\dim H^0(o_F(2K + 2D + 2D')) \geqq 2$$

so $|2K + 2D + 2D'|$ is composite with pencil of curves of canonical type. Since the intersection multiplicity $(D \cdot 2K + 2D + 2D') = 0$, it follows that one of the fibres of this pencil must be a multiple of D. q.e.d.

The final step is a new approach, not in Enriques.

Step (III). If $p_g > 0$, and D is an indecomposable curve of canonical type, then $|nD|$ is composite with a pencil of curves of canonical type, for some n.

PROOF. Let $\mathcal{F}_n$ be the quotient sheaf $o_F(nD)/o_F$. In view of the exact sequences

$$H^0(o_F(nD)) \longrightarrow H^0(\mathcal{F}_n) \longrightarrow H^1(o_F) ,$$

it will suffice to show that $\dim H^0(\mathcal{F}_n) \to \infty$ as $n \to \infty$ in order to establish *Step* (III). Let L be invertible sheaf $\mathcal{F}_1$ on the scheme D. For all n, $\mathcal{F}_{n-1}$ is a subsheaf of $\mathcal{F}_n$ with quotient L^n:

$$(*) \qquad 0 \longrightarrow \mathcal{F}_{n-1} \longrightarrow \mathcal{F}_n \longrightarrow L^n \longrightarrow 0 .$$

This proves

(A) $\dim H^0(\mathcal{F}_n)$ *is a non-decreasing function of* n.

Next, by the Riemann-Roch theorem on F, you see that $\chi(o_F(nD)) = \chi(o_F)$. Therefore $\chi(\mathcal{F}_n) = 0$. Now use the exact sequence

$$H^1(\mathcal{F}_n) \longrightarrow H^2(o_F) \longrightarrow H^2(o_F(nD)) .$$

Since D is effective, $|K - nD|$ is empty for large n, hence by Serre duality, $H^2(o_F(nD)) = (0)$. But since $p_g > 0$, $H^2(o_F) \neq (0)$, hence $H^1(\mathcal{F}_n) \neq (0)$, for large n. This proves

(B) $\dim H^0(\mathcal{F}_n) > 0$ *for* $n \gg 0$, *and* $\chi(\mathcal{F}_n) = 0$, *all* n.

Now assume that $\dim H^0(\mathcal{F}_n)$ is bounded above, and let n be the largest integer for which $\dim H^0(\mathcal{F}_{n-1}) < \dim H^0(\mathcal{F}_n)$ (there is at least one such n by (B) and the fact that $\mathcal{F}_0$ is the 0 sheaf). Using exact sequence $(*)$, it follows that L^n has a non-0 section. But since D is of canonical type, L^n has degree 0 on each component of D. So by Corollary 1, $L^n \cong o_D$. Therefore $\mathcal{F}_n$ has a section s which generates L^n everywhere as an o_D-module, hence s generates $\mathcal{F}_n$ everywhere as o_F-module. In other words, s defines an isomorphism

$$o_F/o_F(-nD) \cong \mathcal{F}_n = o_F(nD)/o_F .$$

Taking powers of s, we obtain isomorphisms

$$o_F/o_F(-nD) \cong o_F(nmD)/o_F(n(m-1)D) = \mathcal{F}_{nm}/\mathcal{F}_{n(m-1)} .$$

Now consider the diagram

$$\begin{array}{ccccccccc}
 & & & & 0 & & 0 & & \\
 & & & & \downarrow & & \downarrow & & \\
0 & \longrightarrow & o_F & \longrightarrow & o_F(n(m-1)D) & \longrightarrow & \mathcal{F}_{n(m-1)} & \longrightarrow & 0 \\
 & & \| & & \downarrow & & \downarrow & & \\
0 & \longrightarrow & o_F & \longrightarrow & o_F(nmD) & \longrightarrow & \mathcal{F}_{nm} & \longrightarrow & 0 \\
 & & & & \downarrow & & \downarrow & & \\
 & & & & \mathcal{F}_n & = & \mathcal{F}_n & & \\
 & & & & \downarrow & & \downarrow & & \\
 & & & & 0 & & 0 & &
\end{array}$$

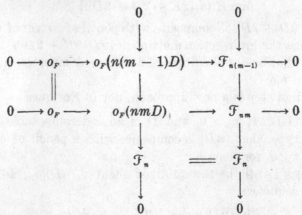

set up by means of the above isomorphisms. Taking cohomology, we get

$$\begin{array}{ccccc} H^1(\mathcal{F}_{n(m-1)}) & \longrightarrow & H^1(\mathcal{F}_{nm}) & \longrightarrow & H^1(\mathcal{F}_n) \longrightarrow 0 \\ \downarrow & & \downarrow & & \\ H^2(o_F) & = & H^2(o_F) & & \\ \downarrow & & \downarrow & & \\ 0 & & 0 & & \end{array}$$

Since $H^2(o_F) \neq (0)$, it follows that the map from $H^1(\mathcal{F}_{n(m-1)})$ to $H^1(\mathcal{F}_{nm})$ is not zero. Therefore $\dim H^1(\mathcal{F}_{nm}) > \dim H^1(\mathcal{F}_n)$. Since $\chi(\mathcal{F}_{nm}) = \chi(\mathcal{F}_n) = 0$, this shows $\dim H^0(\mathcal{F}_{nm}) > \dim H^0(\mathcal{F}_n)$, contradicting our hypothesis on n. Hence $\dim H^0(\mathcal{F}_n)$ is unbounded. q.e.d.

This establishes the existence of an elliptic or quasi-elliptic pencil on F in all cases except when $2K \equiv 0$. A propos the general question raised by this last step—namely, given a curve C such that $(C^2) = 0$, when does nC lie in a pencil, for some n?—there is one curious result that shows the situation in char p is "better" than that in char 0.

PROPOSITION. *Let C be an irreducible curve on a non-singular projective surface F. Assume* $(C^2) = 0$ *and* char $(k) \neq 0$.

$$\left\{\begin{array}{c} nC \text{ lies in a pencil} \\ \text{for some } n \end{array}\right\} \Longleftrightarrow \left\{\begin{array}{c} o_F(nC)/o_F \cong o_F/o_F(-nC) \\ \text{for some } n \end{array}\right\}.$$

The proof is left as a curiosity to the reader. To get a counter example in char 0, let F = completion of the non-trivial $\mathbf{G}_a$-bundle over an elliptic curve, let C = line at ∞.

3. All but one of the remaining steps

Note first that if $(K\cdot D) \geqq 0$, for all effective divisors D, then $(K^2) \geqq 0$ (cf. [6]). Therefore, Enriques' theorem will follow if we show

(a) $[(K\cdot D) \geqq 0$, all eff D, $(K^2) > 0] \Rightarrow |2K| \neq \emptyset$ and $|nK|$ has no base points for some n.

(b) [F elliptic or quasi-elliptic, for the morphism $f\colon F \to C$] $\Rightarrow$ $nK \equiv f^*(A)$ for some n and for some 0-cycle A on C.

(c) [F elliptic and $(K\cdot D) \geqq 0$, all eff D] $\Rightarrow |12K| \neq \emptyset$

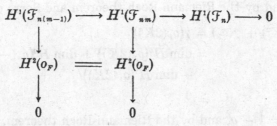

(d) [F quasi-elliptic and $(K\cdot D) \geqq 0$, all eff D] $\Rightarrow$ either F elliptic too or $|2K| \neq \emptyset$.

Proof of (a). Suppose that $|2K| = \emptyset$. Then $p_g = 0$, hence $q = h^{0,1}$, and by the Riemann-Roch theorem and Serre duality,

$$(*)\qquad \begin{aligned}(K^2) + \chi(o_F) &= \chi(o_F(2K)) \\ &= \dim H^0(o_F(2K)) + \dim H^0(o_F(-K)) \\ &\quad - \dim H^1(o_F(2K)) \\ &\leqq 0 .\end{aligned}$$

But $\chi(o_F) = 1 - q$, and by the Riemann-Roch theorem,

$$\begin{aligned}12(1-q) &= 12\chi(o_F) \\ &= (K^2) + c_2 \\ &= (K^2) + 2 - 4q + B_2 ,\end{aligned}$$

hence,

$$(**)\qquad 10 = (K^2) + 8q + B_2 .$$

Equation (**) shows that $q \leqq 1$, hence $\chi(o_F) \geqq 0$, hence by (*), $(K^2) \leqq 0$ which contradicts our hypotheses. Finally, the fact that in case (a), $|nK|$ has no base points for some n was shown in my appendix to [10].

Proof of (b). Each of the fibres of f is a connected curve of canonical type, and almost all are irreducible with multiplicity 1. A finite number of them, say $f^{-1}(x_1), \cdots, f^{-1}(x_n)$, are of the form

$$f^{-1}(x_i) = k_i D_i , \qquad k_i > 1$$

where D_i is an indecomposable curve of canonical type. Let $k =$ l.c.m. $(k_1, \cdots, k_n)$.

Now K intersects the generic fibre of f in a divisor linearly equivalent to 0. Therefore K can be represented by a divisor E_1 whose support is disjoint from the generic fibre, i.e.,

$$\operatorname{supp}(E_1) \subset f^{-1}(y_1, \cdots, y_m) .$$

Therefore for a suitable 0-cycle A on C,

$$K + f^{-1}(A) \equiv E_2 ,$$

where E_2 is effective and is a sum of components of various fibres of f. Applying Corollary 2 § 2 above, it follows that

$$E_2 = a_1 D_1 + \cdots + a_n D_n + f^{-1}(A')$$

for some integers a_i, and a 0-cycle A' on C. Therefore

$$kK \equiv f^{-1}\Big(-KA + KA' + \frac{ka_1}{k_1}x_1 + \cdots + \frac{ka_n}{k_n}x_n\Big).$$

Proof of (c). This is the step which we defer to the sequel of this paper!

Proof of (d). Assume $|2K| = \emptyset$. Then if $q = 0$, F is rational by Castelnuovo's criterion and this contradicts the hypotheses. And $q \leqq 1$ just as in the proof of (a). Therefore $q = 1$. Let

$$f: F \longrightarrow \varepsilon$$

be the Albanese map. Since the fibres of the quasi-elliptic fibration are rational, they are mapped to points under f, hence f must be the quasi-elliptic fibration. Since almost all fibres of f have a unique singular point, the set of points $x \in F$ where $df = 0$ contains a curve $\varepsilon' \subset F$ mapped generically one-to-one to ε. Now res (f): $\varepsilon' \to \varepsilon$ is flat and inseparable. Let its degree be p^n ($p = 2$ or 3 of course). Therefore

$$p^n = (\varepsilon' \cdot f^{-1}(x)), \qquad \text{all } x \in \varepsilon.$$

Now a non-singular branch can meet a cusp only with multiplicity 2 or 3, so choosing a non-singular point of ε', it follows that $(\varepsilon' \cdot f^{-1}(x)) = 2$ or 3. Therefore $n = 1$. Moreover, if ε' had a singular point y, then y would also be singular on $f^{-1}(f(y))$, so ε' would meet $f^{-1}(f(y))$ with multiplicity at least 4. Therefore ε' is non-singular and elliptic: especially it is of canonical type! So now apply the results of *Steps* II and III, § 2. We conclude that $|n\varepsilon'|$ varies in a pencil for some n, hence F is an elliptic surface.

Harvard University

References

[1] F. Enriques, Le Superficie Algebriche, Bologna, 1949.

[2] S. Kleiman, *A note on the Nakai-Moišezon test for ampleness of a divisor*, Amer. J. Math. **87** (1965), 221.

[3] K. Kodaira, "On compact analytic surfaces," in Analytic Functions, Princeton Univ. Press. 1960.

[4] ———, *On Compact complex analytic surfaces*, I, II, III, Ann. of Math.. **71**, **77**, **78** (1960, 1963), 11/563/1.

[5] ———, *On the structure of compact complex analytic surfaces*, I, II, III, IV, Amer. J. Math., 86 and 88 (1964 and 1966).

[6] D. Mumford, Lectures on curves on an algebraic surface, Annals of Math. Studies 59, Princeton, 1966.

[7] I. Šafarevič and others, *Algebraic surfaces*, Proc. of Steklov Inst. of Math., Moscow, 1965 or Am. Math. Soc., 1967.

[8] O. ZARISKI, *Proof of a theorem of Bertini,* Trans. Amer. Math. Soc., 50 (1941), 48.

[9] ———, *On Castelnuovo's criterion of rationality in the theory of algebraic surfaces,* III. J. Math. **2** (1958).

[10] ———, *The theorem of Riemann-Roch for high multiples of an effective divisor,* Ann. of Math. **76** (1962), 560.

[11] J. J. TATE, *Genus change in inseparable extensions of function fields,* Proc. Amer. Math. Soc. **3** (1952), 400.

[12] J. I. IGUSA, *Betti and Picard numbers of abstract algebraic surfaces,* Proc. Nat. Acad. Sci. U.S.A. **46** (1960), 724.

(Received October 21, 1967)

Enriques' Classification of Surfaces in Char. p, II

E. Bombieri and D. Mumford

Introduction and Preliminary Reductions

The purpose of this paper is to carry further the extension of Enriques' classification of surfaces from the case of a char. 0 groundfield to the case of a char. p groundfield. The first part of this extension was made in the paper [10] of one of the present authors. The main results of that paper are as follows[1)]: let X be a non-singular complete algebraic surface without exceptional curves over a field k of any characteristic. We may divide such X's into 4 classes:

a) $\exists$ a curve C on X with $(K_X \cdot C) < 0$

b) $\forall$ curve C on X, $(K_X \cdot C) = 0$, or equivalently, for any $l \neq$ char. p, the fundamental class $[K_X] \in H^2_{et}(X, \boldsymbol{Q}_l)$ is zero.

c) $(K_X \cdot C) \geq 0$ for all curves C and $(K_X^2) = 0$ but $(K_X \cdot H) > 0$ for all ample divisors H.

d) $(K_X \cdot C) \geq 0$ for all C and $(K_X^2) > 0$, hence $(K_X \cdot H) > 0$ for all ample H.

(Other cases are excluded by using the following well-known consequences of Hodge's Index Theorem: (1) $(K_X \cdot H) = 0$ for some ample H, $(K_X^2) \geq 0$ implies $(K_X \cdot C) = 0$ all C and (2) $(K_X \cdot C) \geq 0$ all curves C implies $(K_X^2) \geq 0$). Then in [10], it is proven that

(a) holds $\Leftrightarrow$ X is ruled, in which case $|nK_X| = \phi$, all n.

(b) holds $\Leftrightarrow$ either i) $2K_X \equiv 0$

or ii) $\exists\, \pi: X \to D$, D a curve, almost all fibres of π non-singular elliptic and hence $nK_X = \pi^*(\mathfrak{A})$, $\mathfrak{A}$ divisor on D of degree 0, $n \geq 1$ an integer.

(c) holds $\Leftrightarrow$ $\exists\, \pi: X \to D$ almost all fibres either non-singular elliptic or rational with one cusp, hence $nK_X = \pi^*(\mathfrak{A})$ where deg $(\mathfrak{A}) > 0$, $n \geq 1$.

(d) holds $\Leftrightarrow$ $|nK_X|$ is base-point free and defines a birational map from X to $\boldsymbol{P}^N$, for $n \gg 0$. Moreover, in this case $|2K_X| \neq \phi$.

Our *first goal* in this paper is to prove the following result, well known in char. 0;

Theorem 1. *In cases* (b) *and* (c), *either* $|4K_X| \neq \phi$ *or* $|6K_X| \neq \phi$. *Therefore, in case* (b), *either* $4K_X \equiv 0$ *or* $6K_X \equiv 0$, *and in case* (c), *either* $4K_X$ *or* $6K_X$ *is represented by a*

1) The notation used is summarized below in "list of notations".

positive divisor.

In particular, this shows that the 4 cases above correspond to the classification of surfaces by Kodaira-dimension κ, i.e.,

$$\kappa = \text{tr. deg.}_k \bigoplus_{n=0}^{\infty} \Gamma(X, \mathcal{O}(nK_X)) - 1.$$

Then we see that:

In case (a), $\kappa = -1$
In case (b), $\kappa = 0$
In case (c), $\kappa = 1$
In case (d), $\kappa = 2$.

Thereafter, our *next goal* in this and a subsequent 3rd paper is the further analysis of all surfaces in case (b). It turns out that these can be divided into 4 types *by their Betti numbers*. This division into 4 types is based on a rather mysterious calculation that appears again and again in all work on the classification of surfaces. This calculation is as follows:

Assume $(K_X^2)=0$. Then by the Riemann-Roch theorem on X,

$$\text{(1)} \qquad 12(\dim H^0(\mathcal{O}_X) - \dim H^1(\mathcal{O}_X) + \dim H^2(\mathcal{O}_X)) = c_{2,X} = B_0 - B_1 + B_2 - B_3 + B_4$$

hence substituting $1 = B_0 = B_4 = \dim H^0(\mathcal{O}_X)$, we find

$$\text{(2)} \qquad 10 + 12\, p_g = 8 \dim H^1(\mathcal{O}_X) + 2(2 \dim H^1(\mathcal{O}_X) - B_1) + B_2.$$

Write $\varDelta = 2 \dim H^1(\mathcal{O}_X) - B_1$. This is a "non-classical" term because when $\text{char}(k)=0$, then $\varDelta=0$. In fact, we know that for *almost all* primes l:

$$\begin{aligned}(\mathbf{Z}/l\mathbf{Z})^{B_1} &\approx H^1_{et}(X, \mathbf{Z}/l\mathbf{Z}) \\ &\cong \{x \in \text{Pic}(X) \mid lx = 0\} \\ &\cong \{x \in \text{Pic}^0(X) \mid lx = 0\} \\ &\approx (\mathbf{Z}/l\mathbf{Z})^{2q}\end{aligned}$$

hence in any characteristic $B_1 = 2q$. On the other hand,

$$H^1(\mathcal{O}_X) \cong [\text{tangent space to } \text{Pic}(X) \text{ at } 0].$$

Thus if $\text{char}(k)=0$, $\text{Pic}(X)$, like any group scheme, is reduced, hence

$$\dim H^1(\mathcal{O}_X) = q$$

and $\varDelta=0$. In general, we conclude that

$$\dim H^1(\mathcal{O}_X) \geq q$$

hence $\varDelta \geq 0$, $\varDelta$ even. We can say a bit more: if β_i are the Bockstein operators from $H^1(\mathcal{O}_X)$ to $H^2(\mathcal{O}_X)$, we know ([9], Lecture 27) that

$$\text{tangent space to } \text{Pic}^0_{\text{red}} \cong \bigcap_{i=1}^{\infty} \ker(\beta_i)$$

hence

$$\overbrace{\dim(\text{tang.sp.to Pic}^0)}^{T_P}-\overbrace{\dim(\text{tang.sp.to Pic}^0_{\text{red}})}^{T_{P,\text{red}}} = \dim H^1(\mathcal{O}_X)-\dim\bigcap_{i=1}^{\infty}\ker\beta_i$$

$$\leq \dim\bigcup_{i=1}^{\infty}\operatorname{Im}\beta_i$$

$$\leq p_g.$$

Thus

$$\Delta = 2(\dim T_P-\dim T_{P,\text{red}}) \leq 2p_g.$$

Although it is not used in what follows, it is interesting at this point to consider what happens for arbitrary *analytic* surfaces over $\boldsymbol{C}$. The equation (2) is perfectly valid and in this case Kodaira ([4], p. 755, Th. 3) has shown that $\Delta=0$ or 1 according to the parity of B_1.

Now assume that the surface X has $\kappa=0$, i.e. : $(K_X\cdot C)=0$ for all curves C. Then not only is $(K_X^2)=0$ but either $p_g=0$ or $p_g=1$ and $K_X\equiv 0$. It is then easy to list *all* solutions to equation (2) :

Table of Possible Invariants for Surfaces with $\kappa=0$

B_2	B_1	c_2	$\chi(\mathcal{O}_X)$	$\dim H^1(\mathcal{O}_X)$	p_g	Δ
22	0	24	2	0	1	0
14	2	12	1	1	1	0
10	0	12	1	0	0	0
				1	1	2
6	4	0	0	2	1	0
2	2	0	0	1	0	0
				2	1	2

(Columns B_2, B_1, c_2, $\chi(\mathcal{O}_X)$: invariants under deformation. Columns $\dim H^1(\mathcal{O}_X)$, p_g, Δ: invariants which are in general only upper semi-continuous under deformation)

Concerning these categories of surfaces, we shall prove in this paper the following results :

Theorem 5. *The surfaces with $\kappa=0$, $B_2=22$, known as K3-surfaces, have the following properties:*

i) *for all divisors D on X, $(D\cdot C)=0$ for all curves C implies $D\equiv 0$, hence*

$$\operatorname{Pic}^0(X) = (0).$$

ii) *X has no connected étale coverings, i.e.,*

$$\pi_{1,\text{alg}}(X) = (e).$$

No surfaces with $\kappa=0$, $B_2=14$ exist.

Surfaces with $\kappa=0$, $B_2=10$, $p_g=1$ cannot exist if $\operatorname{char}(k)\neq 2$.

Theorem 6. *All surfaces with $\kappa=0$, $B_2=6$ are abelian varieties.*

Moreover, the following is easy to see from the above table and the results of [15]:

Proposition. *If X is a surface with $\kappa=0$, $B_2=2$, then $B_1=2$, hence* $\mathrm{Alb}(X)$ *is an elliptic curve and the fibres of the canonical map*

$$\pi : X \to \mathrm{Alb}(X)$$

are either almost all non-singular elliptic curves, or almost all rational curves with ordinary cusps. The latter is only possible if $\mathrm{char}(k)=2$ *or* 3.

We call surfaces of this type *hyperelliptic* or *quasi-hyperelliptic* surfaces, depending on which type of fibre π has. In this paper, we shall also analyze hyperelliptic surfaces. However, the analysis of the case of quasi-hyperelliptic surfaces and the case of surfaces with $\kappa=0$, $B_2=10$, which we propose to call *Enriques surfaces* (regardless of whether $K\equiv 0$ or $K\not\equiv 0$!), we postpone to a 3rd part of the paper. Since Enriques surfaces in $\mathrm{char}(k)\neq 2$ are fairly easily seen to have the same behaviour as in char. 0, Part III of this paper will deal largely with the curious pathology of char. 2 and 3.

Finally, for use in § 2, we note that the analysis leading to the Table does not use completely the assumption $\kappa=0$: in fact, it really only uses $(K_X^2)=0$, $p_g\leq 1$. Thus the analysis also shows:

Corollary. *If X is a non-singular complete surface with $(K_X^2)=p_g=0$, then X belongs to one of the 2 following types:*

i) $B_1 = \dim H^1(\mathcal{O}_X) = 0$, *hence* $\mathrm{Pic}^0(X) = (0)$; $\chi(\mathcal{O}_X) = 1$; $B_2 = 10$

ii) $B_1 = 2$, $\dim H^1(\mathcal{O}_X) = 1$, *hence* $\mathrm{Pic}^0(X)$ *is a reduced elliptic curve* ; $\chi(\mathcal{O}_X) = 0$; $B_2 = 2$.

List of Notations

X usually a non-singular projective surface
Alb X = Albanese variety of X
Pic X = Picard scheme of X
$\mathrm{Pic}^0\, X$ = connected component of $0 \in \mathrm{Pic}(X)$
$q = \dim \mathrm{Pic}\, X = \dim \mathrm{Alb}\, X$, the "irregularity" of X
K_X = the canonical divisor class on X
$B_i = i^{\text{th}}$ Betti number of X
$h^{p,q} = \dim H^q(X, \Omega^p)$
$p_g = h^{0,2} = h^{2,0}$, the geometric genus of X
$\omega_X = \Omega_X^2$, the sheaf of 2-forms, if X is smooth
= the dualizing sheaf of Grothendieck for general Cohen-Macauley surfaces.

1. K_X of Elliptic or Quasi-elliptic Surfaces

An elliptic or quasi-elliptic surface is a fibration $f: X \to B$ of a surface X over a non-singular curve B, with $f_*\mathcal{O}_X = \mathcal{O}_B$, with almost all fibres elliptic or rational with a cusp (by a result of Tate [15], the latter situation can occur only if char$(k)=2$ or 3). Note that since the function field $k(X)$ is separable over $k(B)$, almost all fibres are generically smooth. Also every fibre of f is a curve of canonical type[1]. At finitely many points $b_1, \cdots, b_r \in B$ the fibre $f^{-1}(b_\lambda)$ is multiple, i.e.,

$$f^{-1}(b_\lambda) = m_\lambda P_\lambda$$

with $m_\lambda \geqq 2$ and P_λ indecomposable of canonical type. We have

$$R^1 f_* \mathcal{O}_X = L \oplus T$$

where L is an invertible sheaf and T is supported precisely at the points $b \in B$ at which

$$\dim H^0(f^{-1}(b), \mathcal{O}_{f^{-1}(b)}) \geqq 2.$$

To see this, note that by E. G. A. III 7. 8, the sheaf $R^1 f_* \mathcal{O}_X$ is locally free at b if and only if $\mathcal{O}_X$ is cohomologically flat at b in dimension 0.

This suggests

Definition. The fibres of f over supp T are called *wild fibres.*

Noting that if C is indecomposable of canonical type then $\dim H^0(C, \mathcal{O}_C) = 1$ (see Mumford [10], p. 332), we get

Proposition 3. *Every wild fibre is a multiple fibre.*

In the following, we consider only relatively minimal fibrations $f: X \to B$, i.e., no exceptional curve of the first kind is a component of a fibre.

Theorem 2. *Let $f: X \to B$ be a relatively minimal elliptic or quasi-elliptic fibration and let $R^1 f_* \mathcal{O}_X = L \oplus T$. Then*

$$\omega_X = f^*(L^{-1} \otimes \omega_B) \otimes \mathcal{O}(\textstyle\sum a_\lambda P_\lambda)$$

where

(i) *$m_\lambda P_\lambda$ are the multiple fibres*
(ii) $0 \leqq a_\lambda < m_\lambda$
(iii) *$a_\lambda = m_\lambda - 1$ if $m_\lambda P_\lambda$ is not wild*
(iv) $\deg(L^{-1} \otimes \omega_B) = 2p(B) - 2 + \chi(\mathcal{O}_X) + \text{length } T$

where $p(B)$ is the genus of B.

---

1) In the notation of [10], a curve $D = \sum n_i E_i$ is said to be of canonical type if $(K \cdot E_i) = (D \cdot E_i) = 0$ for all i.

Note that in the case char$(k)=0$ or in the complex analytic case there are no wild fibres, so that $a_i=m_i-1$; see Kodaira [4], p. 772, Th. 12.

Proof. For any non-multiple fibre $f^{-1}(y)$ we have

$$\mathcal{O}_{f^{-1}(y)}\otimes\omega_X\cong\omega_{f^{-1}(y)}\cong\mathcal{O}_{f^{-1}(y)},$$

hence if $y_1,\cdots,y_r$ are distinct general points of B the cohomology sequence of

$$0\to\omega_X\to\omega_X\otimes\mathcal{O}(\sum_{i=1}^{r}f^{-1}(y_i))\to\bigoplus_{i=1}^{r}\mathcal{O}_{f^{-1}(y_i)}\to 0$$

yields

$$\dim\left|\omega_X\otimes\mathcal{O}(\sum_{i=1}^{r}f^{-1}(y_i))\right|\geqq 0$$

for large enough r. If D is a divisor in the linear system above, we have

$$(D\cdot f^{-1}(y))=0$$

hence we can write

$$K_X\equiv(\text{sum of fibres})+\Delta$$

where $\Delta\geqq 0$ is contained in a union of fibres and does not contain fibres of f. Let Δ_0 be a connected component of Δ and let $C=f^{-1}(y)$ be the fibre containing Δ_0. Then Δ_0 is a rational submultiple of C, i.e., we have

$$C=mP,\qquad \Delta_0=aP$$

where P is indecomposable of canonical type and $0\leqq a<m$. This follows from

Lemma. *Let $D=\sum n_iC_i$ be an effective divisor on a surface X with each C_i irreducible. Assume that*

$$(C_i\cdot D)\leqq 0,\qquad \textit{all } i$$

and that D is connected.

Then every divisor $Z=\sum m_iC_i$ satisfies $Z^2\leqq 0$ and equality holds if and only if $D^2=0$ and $Z=\lambda D$, $\lambda\in\mathbf{Q}$.

Proof. Write $x_i=m_i/n_i$. We have

$$\begin{aligned}Z^2&=\sum x_ix_jn_in_j(C_i\cdot C_j)\\&\leqq\sum x_i^2n_i^2(C_i\cdot C_i)+\sum_{i\neq j}\frac{1}{2}(x_i^2+x_j^2)n_in_j(C_i\cdot C_j)\\&=\sum x_i^2n_i(C_i\cdot D)\leqq 0.\end{aligned}$$

If equality holds everywhere, we have either $x_i=x_j$ or $(C_i\cdot C_j)=0$ for all i,j ; since D is connected, x_i is constant, i.e., $m_i=\lambda n_i$, $\lambda\in\mathbf{Q}$. q. e. d.

Going back to the proof that $\Delta_0=aP$, if Δ_ν are the connected components of Δ, we have

$$0=K_X^2=\sum\Delta_\nu^2;$$

since each $\Delta_\nu^2\leqq 0$ by the previous lemma, we must have $\Delta_\nu^2=0$ and now the equality case of the lemma proves that Δ_ν is a rational multiple of the fibre containing it.

We have proved that

$$\omega_X = f^*\mathcal{O}_B(\mathfrak{A}) \otimes \mathcal{O}(\sum a_\lambda P_\lambda)$$

for some divisor $\mathfrak{A} \in \operatorname{div}(B)$ and integers a_λ with $0 \leq a_\lambda < m_\lambda$. We deduce that

$$f_*(\omega_X) = \mathcal{O}_B(\mathfrak{A}).$$

Now the duality theorem for a map says that

$$\begin{aligned} f_*\omega_X &= \operatorname{Hom}(R^1f_*\mathcal{O}_X, \omega_B) \\ &= L^{-1} \otimes \omega_B \end{aligned}$$

because the dual of the torsion sheaf is 0 ; this can be found in Deligne-Rapoport [2], pp. 19–20, formula (2. 2. 3). Hence

$$\omega_X = f^*(L^{-1} \otimes \omega_B) \otimes \mathcal{O}(\sum a_\lambda P_\lambda).$$

The spectral sequence of the map f yields

$$\begin{aligned} \chi(\mathcal{O}_X) &= \chi(\mathcal{O}_B) - \chi(R^1f_*\mathcal{O}_X) \\ &= \chi(\mathcal{O}_B) - \chi(L) - \text{length } T \\ &= -\deg L - \text{length } T, \end{aligned}$$

by the Riemann-Roch theorem on the curve B, and since $\deg(\omega_B) = 2p(B) - 2$ we obtain (iv) of Theorem 2.

It remains to prove (iii), and this follows from

Proposition 4. *Let m_λ, P_λ, a_λ be as in Theorem 2 and let*

$$\nu_\lambda = \operatorname{order}(\mathcal{O}_{P_\lambda} \otimes \mathcal{I}_{P_\lambda}^{-1})$$

where $\mathcal{I}_{P_\lambda}$ is the sheaf of ideals of P_λ, be the order of the normal sheaf of P_λ in X. Then we have

i) *ν_λ divides m_λ and $a_\lambda + 1$,*

ii) $\dim H^0(P_\lambda, \mathcal{O}_{(\nu_\lambda+1)P_\lambda}) \geq 2$, $\dim H^0(P_\lambda, \mathcal{O}_{\nu_\lambda P_\lambda}) = 1$,

iii) $\dim H^0(P_\lambda, \mathcal{O}_{rP_\lambda})$ *is non-decreasing with r.*

In particular, if $a_\lambda < m_\lambda - 1$ then $\nu_\lambda < m_\lambda$ and this is equivalent to the multiple fibre $m_\lambda P_\lambda$ being wild.

Proof. Let us write $m, P, a, \nu, \mathcal{I}$ for $m_\lambda, P_\lambda, a_\lambda, \nu_\lambda, \mathcal{I}_{P_\lambda}$. If $r \geq s \geq 1$, the restriction map $\mathcal{O}_{rP} \to \mathcal{O}_{sP}$ is surjective, hence $\dim H^1(P, \mathcal{O}_{rP})$ is non-decreasing with r. Since $\chi(\mathcal{O}_{rP}) = 0$, this proves that $\dim H^0(P, \mathcal{O}_{rP})$ is non-decreasing too.

We have an isomorphism

$$\mathcal{O}_P \otimes \mathcal{I}^\nu \cong \mathcal{O}_P$$

and via this isomorphism we get an exact sequence

$$0 \to \mathcal{O}_P \to \mathcal{O}_{(\nu+1)P} \xrightarrow{\text{res}} \mathcal{O}_{\nu P} \to 0$$

where res is the restriction. Since constants in $H^0(P, \mathcal{O}_{(\nu+1)P})$ are mapped into constants in $H^0(P, \mathcal{O}_{\nu P})$, the cohomology sequence shows that $\dim H^0(P, \mathcal{O}_{(\nu+1)P}) \geq 2$. Finally, ν divides both m and $a+1$, because $\mathcal{O}_P \otimes \mathcal{I}^{-m} \cong \mathcal{O}_P$ (trivial) and

$$\mathcal{O}_P \otimes \mathcal{I}^{-a-1} \cong \omega_P \cong \mathcal{O}_P$$

(Mumford [10], p. 333). q. e. d.

It is shown in Raynaud [13], Prop. 6. 3. 5, that m_λ/ν_λ is a power of the characteristic p of k. In particular the multiplicity of a wild fibre is divisible by p, and wild fibres do not occur in char. 0.

Corollary. *If* $\dim H^1(X, \mathcal{O}_X) \leqq 1$ *we have either*

$$a_\lambda + 1 = m_\lambda \quad \text{or} \quad \nu_\lambda + a_\lambda + 1 = m_\lambda.$$

Proof. Since $\chi(\mathcal{O}_{(\nu+1)P}) = 0$ and $\dim H^0(P, \mathcal{O}_{(\nu+1)P}) \geqq 2$, using duality we find that

$$\dim H^0(P, \omega_{(\nu+1)P}) \geqq 2.$$

Now the cohomology sequence of

$$0 \to \omega_X \to \mathcal{J}^{-\nu-1} \otimes \omega_X \to \omega_{(\nu+1)P} \to 0$$

yields

$$\dim H^0(X, \mathcal{J}^{-\nu-1} \otimes \omega_X) > \dim H^0(X, \omega_X),$$

since we have $\dim H^1(X, \omega_X) = \dim H^1(X, \mathcal{O}_X) \leqq 1$ by hypothesis. This increase in dimension is possible only if $\nu + a + 1 \geqq m$, or $1 + (a+1)/\nu \geqq m/\nu$. Therefore $(a+1)/\nu = m/\nu$ or $m/\nu - 1$. q. e. d.

We conclude this section with a remark on hyperelliptic or quasi-hyperelliptic surfaces.

Proposition 5. *Let* $f: X \to E$, $E = \mathrm{Alb}(X)$ *be an hyperelliptic surface. Then every fibre of f is smooth.*

Moreover if $f: X \to E$ *is quasi-hyperelliptic then every fibre of f is a rational curve with a cusp, i.e., there are no reducible fibres.*

Proof. Since $p(E) = 1$, $\chi(\mathcal{O}_X) = 0$ and $K_X \sim 0$ ($\sim$ is numerical equivalence), Theorem 2 gives

$$(\text{length } T) f^{-1}(y) + \sum a_\lambda P_\lambda \sim 0$$

therefore there are no multiple fibres. Also since the Picard number is $\rho \leqq B_2 = 2$, there are no reducible fibres. In the elliptic case the smoothness of f follows by considering the differential $f^*(\omega)$, where $\omega \in \Gamma(\Omega^1_E)$. $f^*(\omega)$ will only be zero at the points where f is not smooth and since these are finite in number,

$$c_{2,X} = [\text{number of zeroes of } f^*(\omega) \text{ counted with multiplicity}].$$

But $c_{2,X} = 0$, so $f^*\omega$ has no zeroes, so f is smooth. In any elliptic or quasi-elliptic surface, every irreducible fibre is either a) non-singular elliptic, b) rational with a node, or c) rational with a cusp. In the quasi-elliptic case, the generic fibre is of type (c) and since such a curve cannot specialize to type (a) or type (b), every irreducible fibre is rational with a cusp.

q. e. d.

2. Proof of Theorem 1

We shall prove here that if $f: X \to B$ is elliptic or quasi-elliptic, $(K_X \cdot C) \geq 0$ for all curves C and $K_X^2 = 0$, then:

$$(*) \qquad |4K_X| \neq \phi \quad \text{or} \quad |6K_X| \neq \phi.$$

In proving this result we may assume $p_g = 0$ and use Table 1 as a list of numerical invariants. Theorem 2 implies

$$p_g = \dim H^0(B, L^{-1} \otimes \omega_B)$$

and since $\chi(\mathcal{O}_X) \geq 0$, the Riemann-Roch theorem on B shows that $p_g = 0$ implies $p(B) = 0$ or 1 and if $p(B) = 1$ we must also have $T = (0)$. So if $p(B) = 1$ there are no wild fibres and $a_i = m_i - 1$ in Theorem 2. If there is a multiple fibre, it is easily seen that $|2K_X| \neq \phi$. If there are no multiple fibres at all, then

$$\omega_X = f^*(L^{-1} \otimes \omega_B)$$

and $\deg(L^{-1} \otimes \omega_B) = 0$, thus $K_X \sim 0$ and X is hyperelliptic or quasi-hyperelliptic.

Theorem 3. *If X is hyperelliptic or quasi-hyperelliptic, then there is a second structure $f: X \to \boldsymbol{P}^1$ of X as an elliptic surface over $\boldsymbol{P}^1$.*

Proof. By the results in [10], it is sufficient to show the existence of a curve C of canonical type, transversal to the Albanese fibration, $\varphi: X \to E$ with $E = \mathrm{Alb}(X)$. Let F_t be the fibre $\varphi^{-1}(t)$ of φ over $t \in E$. There exists a divisor D on X such that

$$(D^2) = 0, \quad (D \cdot F_0) > 0,$$

for example some linear combination of an ample divisor and F_0; let

$$D_t = D + F_t - F_0.$$

There is a point $t \in E$ such that $|D_t| \neq \phi$. If not, use $\chi(\mathcal{O}(D_t)) = 0$ and the Riemann-Roch theorem to prove

$$\dim H^0(X, \mathcal{O}(D_t)) = \dim H^1(X, \mathcal{O}(D_t)) = 0$$

for all t. The cohomology sequence of

$$0 \to \mathcal{O}(D_t) \to \mathcal{O}(D + F_t) \xrightarrow{r_{F_0}} \mathcal{O}_{F_0} \otimes \mathcal{O}(D) \to 0$$

then gives an isomorphism

$$r_{F_0}: H^0(X, \mathcal{O}(D + F_t)) \simeq H^0(F_0, \mathcal{O}_{F_0} \otimes \mathcal{O}(D))$$

where r_{F_0} is the restriction. Since $(D \cdot F_0) > 0$, there is a non-trivial section $\sigma \in \Gamma(\mathcal{O}_{F_0} \otimes \mathcal{O}(D))$, and let $s_t = r_{F_0}^{-1}(\sigma)$. Clearly $X = \text{closure} \bigcup_{t \neq 0} \mathrm{div}(s_t)$ and $\mathrm{div}(s_t) \cap F_0$ has support in $\mathrm{div}(\sigma)$, for all $t \neq 0$. It follows that as $t \to 0$ we must have $\mathrm{div}(s_t) \to F_0 + C \equiv D + F_0$, and $C \in |D|$, proving our assertion.

We have found a curve $C > 0$ with $(C^2) = 0$ and $(C \cdot F_0) > 0$, and we claim that C is of canonical type. In fact, since $K_X \sim 0$ and $(C^2) = 0$, our assertion will follow from the fact that X has no irreducible curve Γ with $(\Gamma^2) = -2$. Such a curve Γ

cannot be transversal to the Albanese fibering because Γ is rational, and cannot be a component of a fibre, since every fibre is irreducible by Proposition 5.

q. e. d.

In view of Theorem 3, we have only to examine the case in which $p(B)=0$. Since B is rational, the canonical bundle formula becomes

$$K_X \equiv rf^{-1}(y)+\sum_\lambda a_\lambda P_\lambda$$

where

$$r=-2+\chi(\mathcal{O}_X)+\text{length } T.$$

If H is an ample divisor on X, since $(K_X\cdot H)\geqq 0$ we have

$$r+\sum_\lambda \frac{a_\lambda}{m_\lambda}\geqq 0.$$

Moreover

$$\dim |nK_X| = nr+\sum_\lambda \left[\frac{na_\lambda}{m_\lambda}\right].$$

It is now easy to see, using $\chi(\mathcal{O}_X)\geqq 0$ and Proposition 4, Corollary that we can have only the following cases:

(A) length $T=0$, so $a_\lambda=m_\lambda-1$, $\nu_\lambda=m_\lambda$.

If $\chi(\mathcal{O}_X)=0$, then there are at least 3 multiple fibres and we can have:

a) there are 4 or more multiple fibres, i. e., $m_\lambda\geq 2$, $1\leq\lambda\leq 4$, and then $|2K_X|\neq\phi$.

b) there are 3 multiple fibres with all multiplicities $m_\lambda\geq 3$. Then $|3K_X|\neq\phi$.

c) there are 3 multiple fibres with $m_1=2$, m_2, $m_3\geq 4$. Then $|4K_X|\neq\phi$.

d) there are 3 multiple fibres with $m_1=2$, $m_2=3$, $m_3\geq 6$. Then $|6K_X|\neq\phi$.

If $\chi(\mathcal{O}_X)=1$, then there are at least 2 multiple fibres, m_1, $m_2\geq 2$, and $|2K_X|\neq\phi$.

If $\chi(\mathcal{O}_X)\geq 2$, then $|K_X|\neq\phi$.

(B) length $T=1$. If $\chi(\mathcal{O}_X)=0$, then $|K_X|=\phi$, so $\dim H^1(\mathcal{O}_X)=1$ and Prop. 4, Cor. applies. So if $f^{-1}(P_1)$ is the wild fibre, then we have $a_1=m_1-1$ or $a_1=m_1-1-\nu_1$ where $\nu_1|\text{g. c. d.}(m_1, a_1+1)$, while $a_\lambda=m_\lambda-1$, $\nu_\lambda=m_\lambda$ for $\lambda\geq 2$. Moreover there are at least 2 multiple fibres and we can have:

a′) there are 2 or more multiple fibres with $a_\lambda=m_\lambda-1$ and then $|2K_X|\neq\phi$.

b′) the wild fibre satisfies $m_1=3$, $a_1=1$, $\nu_1=1$ (hence char.$=3$) and the tame fibre satisfies $m_2\geq 3$. Then $|3K_X|\neq\phi$.

c′₁) the wild fibre satisfies $m_1=4$, $a_1=1$, $\nu_1=2$ (hence char.$=2$) and the tame fibre satisfies $m_2\geq 4$. Then $|4K_X|\neq\phi$.

c′₂) the wild fibre satisfies $m_1=\mu_1\nu_1$, where $\mu_1\geq 4$ (any positive char.). In this case, $a_1/m_1\geq 1/2$ and $|2K_X|\neq\phi$.

d′₁) the wild fibre satisfies $m_1=2\nu_1$, $a_1=\nu_1-1$, $\nu_1\geq 3$ (hence char.$=2$) and the tame fibre satisfies $m_2\geq 3$. then $|3K_X|\neq\phi$.

d′₂) the wild fibre satisfies $m_1=3\nu_1$, $a_1=2\nu_1-1$, $\nu_2\geq 2$ (hence char.$=3$). In this case $|2K_X|\neq\phi$.

If $\chi(\mathcal{O}_X)\geq 1$, then $|K_X|\neq\phi$.
(C) length $T\geq 2$, then also $|K_X|\neq\phi$.

If we specialize to the case $\kappa=0$, then we easily get the following list of possible multiple fibres for elliptic or quasi-elliptic surfaces $f: X\to P^1$ with K_X a torsion divisor:

		length T	$\chi(\mathcal{O}_X)$	a_λ/m_λ (*=wild fibre)	order K_X	char.
tame cases	i)	0	0	(1/2, 1/2, 1/2, 1/2)	2	
	ii)	0	0	(2/3, 2/3, 2/3)	3	
	iii)	0	0	(1/2, 3/4, 3/4)	4	
	iv)	0	0	(1/2, 2/3, 5/6)	6	
	v)	0	1	(1/2, 1/2)	2	
	vi)	0	2	none	1	
wild cases	vii)	1	0	(0/2*, 1/2, 1/2)	2	2
	viii)	1	0	(1/2*, 1/2)	2	2
	ix)	1	0	(1/3*, 2/3)	3	3
	x)	1	0	(1/4*, 3/4)	4	2
	xi)	1	0	(2/4*, 1/2)	2	2
	xii)	1	0	(2/6*, 2/3)	3	2
	xiii)	1	0	(3/6*, 1/2)	2	3
	xiv)	1	1	(0/2*)	1	2
	xv)	2	0	one or two wild fibres $0/p^r$	1	p

Note that each of the wild cases may be thought of as coming from the confluence of 2 tame fibres in one of the tame cases.

3. Analysis of Hyperelliptic Surfaces

In this section, we study more closely surfaces X such that:

a) $\kappa=0$

b) the Albanese mapping is $\pi: X\to E$, E elliptic

c) almost all fibres C_x of π are non-singular.

By the Table of the Introduction, it follows also that

d) $B_2=2$, $c_2=0$, $\chi(\mathcal{O}_X)=0$.

Moreover, by Proposition 5 it follows that

c′) *all* fibres C_x are non-singular elliptic.

By Theorem 3, § 2, we see:

e) There is a second elliptic pencil $\pi': X\to P^1$ on X.

We want to compare π and π' and see the effect of 2 simultaneous elliptic fibrations! Let C'_y be the fibres of π'. Then all the C'_y are finite coverings of E:

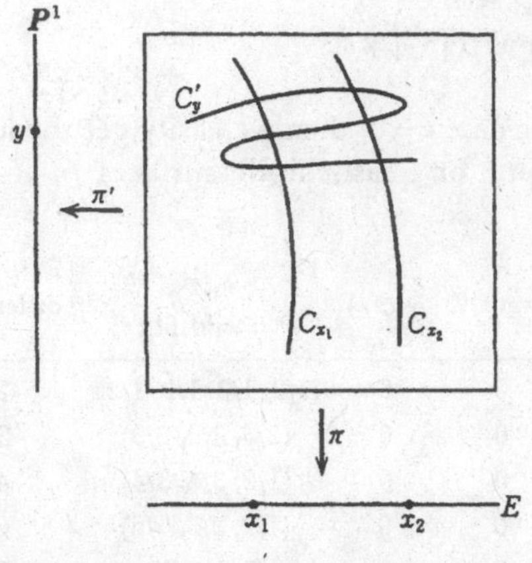

Hence all the C'_y are either non-singular elliptic or multiples of non-singular elliptic curves, and

$$p_y = \text{res } \pi : C'_y \to E$$

is an isogeny. Let $S = \{y \in \boldsymbol{P}^1 | C'_y \text{ multiple}\}$. p_y defines a pull-back on Pic^0:

$$\text{Pic}^0(C'_y) \overset{p_y^*}{\leftarrow} \text{Pic}^0(E).$$

Choosing a base point $x_0 \in E$, we can identify $\text{Pic}^0(E)$ with E by associating the sheaf $\mathcal{O}_E(x - x_0)$ with the point x. As usual, this makes E into an algebraic group with identity x_0. Now we cannot choose base points on each C'_y varying nicely with y unless $\pi' : X \to \boldsymbol{P}^1$ has a section. However, we can instead note that $\text{Pic}^0(C'_y)$ *acts* canonically on C'_y by translations: i.e., the sheaf L of degree 0 maps $u \in C'_y$ to the unique point v such that $L(u) \cong \mathcal{O}_{C'_y}(v)$. Then via the maps p_y^*, we find that E is acting by translations simultaneously on *all* the curves C'_y. If we stick to the non-multiple curves, it follows easily that this is an algebraic action of E:

$$\sigma_0 : E \times \pi'^{-1}(\boldsymbol{P}^1 - S) \to \pi'^{-1}(\boldsymbol{P}^1 - S).$$

But since X is a minimal model, any automorphism of the Zariski-open set $\pi'^{-1}(\boldsymbol{P}^1 - S)$ extends to an automorphism of X so we actually get an action:

$$\sigma : E \times X \to X.$$

To relate this action to π, say $x \in E$, $u \in C'_y$. Then x takes u to v where

$$\pi^*(\mathcal{O}_E(x - x_0)) \otimes \mathcal{O}_{C'_y}(u) \cong \mathcal{O}_{C'_y}(v).$$

Let $n = (C'_y \cdot C_x) =$ (degree of res $\pi : C'_y \to E$). Then taking $\text{Norm}_{C'_y/E}$ of the 2 sides of the above isomorphism:

$$\mathcal{O}_E(nx - nx_0 + \pi u) \cong \mathcal{O}_E(\pi v),$$

hence we get a commutative diagram

$$(*)\qquad \begin{array}{ccc} X & \xrightarrow{\text{action of } x} & X \\ \downarrow{\scriptstyle \pi} & & \downarrow{\scriptstyle \pi} \\ E & \xrightarrow[\text{by } n(x-x_0)]{\text{translation}} & E \end{array} \quad ; \quad \begin{array}{ccc} E\times X & \xrightarrow{\sigma} & X \\ \downarrow{\scriptstyle 1\times\pi} & & \downarrow{\scriptstyle \pi} \\ E\times E & \longrightarrow & E \\ x,y & \longrightarrow & nx+y \end{array}$$

We can now use this action of E to describe the whole surface X as follows : let $E_0 = C_{x_0}$ be the fibre over x_0, and let $A_n = \text{Ker}\,(n_E : E \to E)$ considered as a subgroup scheme of E. Then by $(*)$ the action of A_n on X preserves the fibres of π, hence A_n acts on E_0, and give this action the name α :

$$\alpha : A_n \to \text{Aut}(E_0) = \text{group scheme of automorphisms of } E_0.$$

Then by restriction of the action σ of E, we get a morphism :

$$\tau : E\times E_0 \to X$$

which by $(*)$ fits into a diagram :

$$\begin{array}{ccc} E\times E_0 & \xrightarrow{\tau} & X \\ & {\scriptstyle n_E\cdot p_1}\searrow \quad \swarrow{\scriptstyle \pi} & \\ & E & \end{array}$$

Note that

$$\tau(x, y) = \tau(x', y') \Leftrightarrow \sigma(x-x', y) = y'$$
$$\Leftrightarrow x-x' \in A_n \text{ and } \alpha(x-x')(y) = y'$$

hence it follows that $X \cong$ quotient $(E\times E_0/A_n)$, via the action

$$x(u, v) = (u+x, \alpha(x)(v)), \quad x \in A_n, u \in E, v \in E_0.$$

If we replace E by $E_1 = E/\text{Ker}\,\alpha$, this proves :

Theorem 4. *Every hyperelliptic surface X is of the form :*

$$X = E_1\times E_0/A, \qquad E_1, E_0 \text{ elliptic curves}$$

where A is a finite subgroupscheme of E_1, and A acts by

$$k(u, v) = (u+k, \alpha(k)(v))$$

for some injective homomorphism

$$\alpha : A \to \text{Aut}(E_0).$$

Moreover, the 2 elliptic fibrations on X are given by :

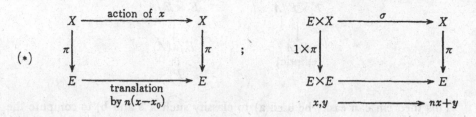

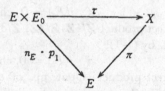

$$\begin{array}{ccc} E_1 \times E_0/A & & E_1 \times E_0/A \\ \downarrow & & \downarrow \\ E_1/A & & E_0/\alpha(A) \\ \text{(elliptic)} & & \wr\Vert \\ & & \boldsymbol{P}^1 \end{array}$$

This theorem can easily be used a) to classify such X's and b) to compute the order of K_X in Pic (X). We use the fact that choosing a base point $0 \in E_0$, $\mathrm{Aut}(E_0)$ becomes a semi-direct product:

$$\mathrm{Aut}(E_0) = E_0 \cdot \mathrm{Aut}(E_0, 0)$$

normal subgroup of translations (E_0); finite, discrete group of autos, fixing 0 ($\mathrm{Aut}(E_0, 0)$)

Note that $\alpha(A) \not\subset E_0$, or else $E_0/\alpha(A)$ would be elliptic instead of rational as required. Moreover, from the tables in Lang [5], Appendix 1, we find:

$$\begin{aligned} \mathrm{Aut}(E_0, 0) &= \{1_E, -1_E\} \cong \boldsymbol{Z}/2\boldsymbol{Z} \quad \text{if } j(E_0) \neq 0, 12^3 \\ &\cong \boldsymbol{Z}/4\boldsymbol{Z} \quad \text{if } j(E_0) = 12^3, \text{ char} \neq 2, 3 \\ &\cong \boldsymbol{Z}/6\boldsymbol{Z} \quad \text{if } j(E_0) = 0, \text{ char} \neq 2, 3 \end{aligned}$$

$\cong$ semi-direct product $\boldsymbol{Z}/4\boldsymbol{Z} \cdot \boldsymbol{Z}/3\boldsymbol{Z}$, $\boldsymbol{Z}/3\boldsymbol{Z}$ normal, $i \in \boldsymbol{Z}/4\boldsymbol{Z}$ acting by mult. by $(-1)^{2i}$

if $j(E_0) = 0$, char $= 3$

$\cong$ semi-direct product (Quat. gp. of order 8) $\cdot\, \boldsymbol{Z}/3\boldsymbol{Z}$, Quat. gp. normal, $\boldsymbol{Z}/3\boldsymbol{Z}$ permuting cyclically $i, j, k \in$ Quat. gp.

if $j(E_0) = 0$, char $= 2$

The important point here is that since A is commutative, so is $\alpha(A)$ and now even in the last 2 nasty cases, the maximal *abelian* subgroups are still $\boldsymbol{Z}/4\boldsymbol{Z}$ and $\boldsymbol{Z}/6\boldsymbol{Z}$, which in all cases are cyclic.

Let $k \in A$ be such that

$$\mathrm{Im}\, \alpha(k) \in \mathrm{Aut}(E_0)/E_0$$

generates

$$\mathrm{Im}\, \alpha(A) \subset \mathrm{Aut}(E_0)/E_0.$$

Then $\alpha(k) \notin E_0$, hence it has some fixed point. Replacing 0 by this fixed point, it follows that $\alpha(A)$ itself is a direct product:

$$\alpha(A) = A_0 \cdot \boldsymbol{Z}/n\boldsymbol{Z}$$

finite gp. scheme of translations $A_0 \subset E_0$ (A_0); cyclic gp. generated by k, $n = 2, 3, 4$ or 6 ($\boldsymbol{Z}/n\boldsymbol{Z}$)

Since A_0 and k must commute, $A_0 \subset$ (fix pt. set F of k). Again referring to Lang to check the fix point sets, we find:

a) $n = 2$, (so $k = -1_E$), then $F = \mathrm{Ker}\, 2_{E_0}$

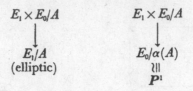

b) $n=3$, then $\#F=3$ so $F\cong \mathbf{Z}/3\mathbf{Z}$ if char $\neq 3$
$F\cong \alpha_3$ if char $=3$ (because E_0 is supersingular !)

c) $n=4$, then $\#F=2$ so $F\cong \mathbf{Z}/2\mathbf{Z}$ if char $\neq 2$
$F\cong \alpha_2$ if char $=2$ (because E_0 is supersingular !)

d) $n=6$, then $F=(e)$

We can now mechanically compile a list of all possible $\alpha(K)$' s, hence all possible X' s:

a1) $E_1\times E_0/(\mathbf{Z}/2\mathbf{Z})$; action $(x,y)\mapsto(x+a,-y)$

a2) $E_1\times E_0/(\mathbf{Z}/2\mathbf{Z})^2$; action $(x,y)\mapsto(x+a,-y)$, $(x+b,y+c)$ (here char $\neq 2$)

a3) $E_1\times E_0/(\mathbf{Z}/2\mathbf{Z})\cdot\mu_2$; action $(x,y)\mapsto(x+a,-y)$, μ_2 acts by transl. on both factors.

b1) $E_1\times E_0/(\mathbf{Z}/3\mathbf{Z})$; action $(x,y)\mapsto(x+a,\omega y)$ where $j(E_0)=0$, $\omega: E_0\to E_0$ an automorphism of order 3

b2) $E_1\times E_0/(\mathbf{Z}/3\mathbf{Z})^2$; action $(x,y)\mapsto(x+a,\omega y)$, $(x+b,y+c)$, E_0, ω as before and $\omega c=c$, order $c=3$ (here char $\neq 3$)

c1) $E_1\times E_0/(\mathbf{Z}/4\mathbf{Z})$; action $(x,y)\mapsto(x+a,iy)$, where $j(E_0)=12^3$, $i: E_0\to E_0$ an automorphism of order 4

c2) $E_1\times E_0/(\mathbf{Z}/2\mathbf{Z})\cdot(\mathbf{Z}/4\mathbf{Z})$; action $(x,y)\mapsto(x+a,iy)$, $(x+b,y+c)$, E_0, i as before and $ic=c$, order $c=2$ (here char $\neq 2$)

d) $E_1\times E_0/\mathbf{Z}/6\mathbf{Z}$; action $(x,y)\mapsto(x+a,-\omega y)$, E_0, ω as in b.

The list obtained here coincides with the classical list in characteristic 0 (see Bagnera and DeFranchis [1], Enriques and Severi [3], pp. 283–392, Šafarevič [14], p. 181). Note here that the requirements $A_0\subset E_0$ *and* $A\subset E_1$ eliminate the possibilities $n=2$, $A_0=\text{Ker } 2_{E_0}$ and $n=3$ or 4, $A_0=\alpha_3$ or α_2. A striking feature of this list are the missing cases. From a moduli point of view, even in case a1), one may ask what happens if we start with such an X in characteristic 0 and specialize to characteristic 2 in such a way that the point a goes to $0\in E_1$. One would hope for instance that the moduli spaces of these X's were proper over $\mathbf{Z}[j(E_0),j(E_1)]$ but this is not true. The answer seems to be that the X's become quasi-hyperelliptic ! This is an interesting point to investigate.

The order of K_X is easily obtained, since if ω is the 2-form on $E_1\times E_0$ with no zeros or poles, then

$$\text{order of } K_X = \text{least } n \text{ such that } A \text{ acts trivially on } \omega^{\otimes n}$$

and we find

$$\begin{aligned}\text{order of } K_X &= 2, 3, 4, 6 \text{ in cases a), b), c), d)}\\ &\quad \text{and char}(k)\neq 2,3\\ &= 1, 3, 1, 3 \text{ in cases a), b), c), d)}\\ &\quad \text{and char}(k)=2\\ &= 2, 1, 4, 2 \text{ in cases a), b), c), d)}\\ &\quad \text{and char}(k)=3\end{aligned}$$

It is interesting to check exactly which wild multiple fibres (in the sense of § 1) occur here for $\pi' : X \to P^1$. One can check that we get the following cases in the list of § 2 :

case	char. $\neq 2,3$	char. 3	char. 2
a	(i)	(i)	(xv)-one or two fibres 0/2
b	(ii)	(xv)-one fibre 0/3	(ii)
c	(iii)	(iii)	(xv)-one fibre 0/4
d	(iv)	(xiii)	(xii)

4. Proof of Theorem 5

First of all, let X be a $K3$-surface, i.e., $K_X \equiv 0$, $B_2 = 22$, $B_1 = 0$, $\chi(\mathcal{O}_X) = 2$, $H^1(\mathcal{O}_X) = (0)$ (cf. Table in Introduction). Then

i) if $\pi : Y \to X$ were a connected étale covering of degree d, one would have $K_Y \equiv \pi^* K_X \equiv 0$, hence Y would be a surface in the Table too. But

$$c_{2,Y} = \pi^{-1}(c_{2,X})$$

hence

$$\deg c_{2,Y} = 24d > 24$$

and there are no such surfaces in the table.

ii) Since $H^-(\mathcal{O}_X)$ is isomorphic to the tangent space to Pic(X), it follows that Pic^0_X is a finite discrete group. Let $L = \mathcal{O}_X(D)$ represent a point of Pic^0_X. Then $(D^2) = (D \cdot K_X) = 0$, so $\chi(L) = \chi(\mathcal{O}_X) = 2$, Therefore $H^0(L) \neq (0)$ or $H^2(L) \neq (0)$. But by Serre duality $H^2(L)$ is dual to $H^0(L^{-1})$. Thus L or L^{-1} is represented by an effective divisor E, but since it is in Pic^0, $E = 0$. So finally $L \cong \mathcal{O}_X$ and $\mathrm{Pic}^0_X = (0)$.

Secondly, let X be a surface with $K_X \equiv 0$, $B_2 = 14$, $B_1 = 2$, $\chi(\mathcal{O}_X) = 1$, dim $H^1(\mathcal{O}_X) = 1$. Since $B_1 > 0$, X has a positive dimensional Picard variety. This means that X does indeed support invertible sheaves $L = \mathcal{O}_X(D)$ such that D is numerically equivalent to zero but $D \not\equiv 0$. Then $\chi(L) = \chi(\mathcal{O}_X) = 1$, so $H^0(L) \neq (0)$ or $H^2(L) \neq (0)$. As above, Serre duality shows that $H^2(L) \neq (0) \Rightarrow H^0(L^{-1}) \neq (0)$, so L or L^{-1} is represented by an effective divisor E. E numerically equivalent to 0 implies $E = 0$, so $L \cong \mathcal{O}_X$ contrary to our assumption.

Alternatively, we could argue that because $B_1 > 0$, X has connected cyclic étale coverings $\pi : Y \to X$ of every order d prime to the characteristic. As in (i) above, $c_{2,Y} = 12d$ and if $d > 2$, no such Y appears in our table.

Arguments of the above type, using μ_p or α_p-coverings of X (cf. Mumford [11]) do not quite seem to be strong enough to prove that if X is a $K3$-surface, then $H^0(X, \Omega^1_X) = (0)$. It remains a very intriguing open question[1] whether or not

1) (added in proof) Rudakov and Šafarevič have just settled this. They show that Ω^1_X has no sections when X is a $K3$-surface. Moreover, P. Deligne has used their result to prove that all $K3$-surfaces lift to char. 0.

$H^0(X, \Omega_X^1)$ is (0) for every $K3$-surface of char. p.

Thirdly, let X be a surface with $K_X \equiv 0$, $B_2 = 10$, $B_1 = 0$, $\chi(\mathcal{O}_X) = 1$, $\dim H^1(\mathcal{O}_X) = 1$. Let $\{a_{ij}\} \in Z^1(\mathcal{O}_X)$ be a non-trivial cocycle and consider the $\boldsymbol{G}_a$-bundle

$$\pi : W \longrightarrow X$$

defined locally as $\boldsymbol{A}^1 \times U_i$, coordinate z_i on $\boldsymbol{A}^1$, and glued by

$$z_i = z_j + a_{ij}.$$

If ω is a non-zero 2-form on X with no zeroes or poles,

$$\eta = dz_i \wedge \omega$$

is a non-zero 3-form on W with no zeroes or poles, i.e., $K_W \equiv 0$. Now since $H^1(\mathcal{O}_X)$ is 1-dimensional, there is a constant $\lambda \in k$ such that $\{a_{ij}^p\}$, $\{\lambda a_{ij}\}$ are cohomologous :

$$a_{ij}^p = \lambda a_{ij} + b_i - b_j$$

Consider the global function f on W defined locally by

$$f = z_i^p - \lambda z_i - b_i.$$

Let Y be the 2-dimensional scheme $f=0$. If $\lambda \neq 0$, Y is étale over X, hence non-singular. If $\lambda = 0$, still $b_i \notin \mathcal{O}_X^p$ (or else $a_{ij} = b_i^{1/p} - b_j^{1/p}$ is cohomologous to zero), so Y is a reduced Gorenstein surface. Since $K_W \equiv 0$ and Y has trivial normal sheaf in W, in both cases $\omega_Y \cong \mathcal{O}_Y$. Thus

$$\chi(\mathcal{O}_Y) \leq \dim H^0(\mathcal{O}_Y) + \dim H^2(\mathcal{O}_Y) = \dim H^0(\mathcal{O}_Y) + \dim H^0(\omega_Y) = 2.$$

On the other hand,

$$\operatorname{res} \pi : Y \longrightarrow X$$

is finite and flat and $(\operatorname{res} \pi)_* \mathcal{O}_Y$ is filtered by the subsheaves :

$$\mathcal{O}_X \subset [\mathcal{O}_X \oplus \mathcal{O}_X \cdot z_i] \subset [\mathcal{O}_X \oplus \mathcal{O}_X \cdot z_i \oplus \mathcal{O}_X \cdot z_i^2] \subset \cdots \subset (\operatorname{res} \pi)_* \mathcal{O}_Y.$$

The quotients here are all isomorphic to $\mathcal{O}_X$, thus

$$\chi(\mathcal{O}_Y) = p \cdot \chi(\mathcal{O}_X) = p.$$

Thus $p \leq 2$ as asserted.

5. Analysis of the Case Leading to Abelian Surfaces

In this section, we prove Theorem 6, that a surface X with $K_X \equiv 0$ and $B_2 = 6$ is an abelian surface. As we see from the table in § 1, the surface X also has the properties :

a) $\dim H^1(\mathcal{O}_X) = 2$, $\dim H^2(\mathcal{O}_X) = 1$, $\chi(\mathcal{O}_X) = 0$,

b) $c_{2,X} = 0$, $B_1 = 4$, $q = 2$.

In particular, $\mathrm{Pic}^0 X$ is reduced and 2-dimensional and its dual Alb X is 2-dimensional. Let

$$\phi : X \longrightarrow \mathrm{Alb}\, X$$

be the Albanese mapping. First of all, we can see that ϕ is surjective as follows : if not, since $\phi(X)$ generates Alb X, $\phi(X)$ is a curve of genus $g \geq 2$. Consider the diagram:

$$\begin{array}{ccc} X'=X\times_{\mathrm{Alb}}\mathrm{Alb} & \xrightarrow{\psi} & \mathrm{Alb}\,X \\ \downarrow & & \downarrow{\scriptstyle n} \\ X & \xrightarrow[\phi]{} & \mathrm{Alb}\,X \end{array}$$

where n denotes multiplication by n and $p \nmid n$. Then $\psi(X')$ is an étale covering of $\phi(X)$ of degree n^{2g}. Also $\psi(X')$ is connected because $\psi(X')=n^{-1}(\phi(X))$ and $\phi(X)$ is an ample curve[1] on Alb X. Therefore, $\psi(X')$ has genus $g'>2$. Therefore, Alb X' can be mapped onto $\mathrm{Jac}(\psi(X'))$ which is an abelian variety of dimension>2: i.e., $q(X')>2$. But X' is an étale cover of X. So $K_{X'}\equiv 0$ and looking in the Table, we see that no such surface X' exists.

Therefore, ϕ is surjective, and hence of finite degree. If ϕ were separable, e.g., if char.$=0$, then we could quickly finish up as follows:

Let ω be the translation-invariant 2-form on Alb X. Then ω has no zeroes or poles and because ϕ is separable, $\phi^*\omega\neq 0$. But $\phi^*\omega$ has zeroes at all points where ϕ is not étale, and $K_X=(\phi^*\omega)$. Since $K_X\equiv 0$, $\phi^*\omega$ has no zeroes, hence ϕ is everywhere étale. But then by the Theorem of § 18 [12], X itself is an abelian surface. Unfortunately if ϕ is inseparable, this argument breaks down. However, when we are in characteristic p, we can use another trick and reduce the Theorem to the case where the ground field k is finite! In fact X lies in a smooth and proper algebraic family of surfaces defined over a finite field and all members of this family have the same invariants (e.g., because by the table in § 1, these surfaces are also characterized by saying $K_X\sim 0$ in étale cohomology and $q=2$). Therefore, if we prove that the surfaces in this family over closed points of the base are abelian, it follows that all are abelian (cf. Theorem 6.14, [8]).

Now assume the ground field k is finite. We follow a line of argument similar to that in Tate [16]. Consider the infinite sequence of surfaces:

$$\begin{array}{ccc} X_l=X\times_{\mathrm{Alb}}\mathrm{Alb} & \xrightarrow{\phi_l} & \mathrm{Alb}\,X \\ \downarrow & & \downarrow \\ X & \xrightarrow{\phi} & \mathrm{Alb}\,X \end{array}$$

for all $l\geq 2$ with $p \nmid l$. Note that $\deg\phi_l=\deg\phi$ for all: call this degree d. Note that $X_l\to X$ is étale and hence X_l is a surface of the same type as X (in fact, $K_{X_l}\equiv 0$ and $q(X_l)\geq 2$, hence by table I, $q(X_l)=2$). We can deduce quickly that ϕ and hence ϕ_l are all finite morphisms: in fact, if not, let $E\subset X$ be a curve such that $\phi(E)$ is a point $e\in\mathrm{Alb}\,X$. Then considering the Stein factorization $X\to Y\to\mathrm{Alb}$ of ϕ, we see that E can be blown down in a birational map $X\to Y$, hence $(E^2)<0$. Now

1) A suitable multiple of an ample curve C on any surface Y is a hyperplane section of Y for some projective embedding and all hyperplane sections of varieties of dimension>1 are connected.

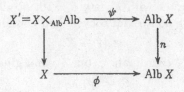

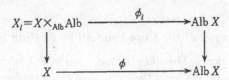

for each l, $l^{-1}(e)$ consists of l^4 points $e_i \in \text{Alb } X$, and $\phi^{-1}(e_i)$ contains a curve E_i that is contracted by ϕ_l. These curves are disjoint since $\phi_l(E_i)=e_i \neq e_j = \phi_l(E_j)$. Thus X_l has l^4 disjoint curves E_i with $(E_i^2)<0$; thus $B_2(X_l) \geq l^4$. But for all surfaces of the same type as X, $B_2=6$. This is a contradiction if $l>1$, hence ϕ is finite.

Next, fix L_0, an ample sheaf on Alb X. It follows that $L_l=\phi_l^*(L_0)$ is ample on X_l, with Hilbert polynomial

$$\chi(L_l^{\otimes n}) = d \cdot \chi(L_0^{\otimes n})$$

independent of l. By the Main Theorem of Matsusaka-Mumford [7], there is also a number N independent of l such that $L_l^{\otimes N}$ is very ample for all l. Therefore the infinite set of k-varieties X_l can all be embedded in a fixed $\boldsymbol{P}^M$ with fixed degree. Since there are only finitely many k-varieties of this degree (as k is finite), it follows that all the pairs $(X_l, L_l^{\otimes N})$ are isomorphic to finitely many of them !

Now consider the facts—

a) for any variety X and ample sheaf L, the group of automorphisms f of X such that f^*L is numerically equivalent to L is an algebraic group; esp. it has only finitely many components (Matsusaka [6]),

b) The group A_l of translations by points of order l acts on X_l since by definition, it is the fibre product $X \times_{\text{Alb}} (\text{Alb}, l)$; moreover each $g \in A_l$ carries L_l into a sheaf algebraically equivalent to L_l.

Let $(X_l, L_l^{\otimes N})$ be isomorphic to infinitely many other $(X_{l'}, L_{l'}^{\otimes N})$'s. Then $A_{l'}$ acts on X_l. Let $G_l \subset \text{Aut}(X_l)$ be the group of automorphisms f such that f^*L_l is numerically equivalent to L_l. Then $A_{l'} \subset G_l$ which implies that the order of G_l is infinite, hence G_l^0 (the connected component) is positive dimensional. But if G_l^0 contains a non-trivial *linear* subgroup, then when this acts on X_l, it would follow that X_l was a ruled surface : since $K_{X_l} \equiv 0$, this is absurd. Therefore G_l^0 is an abelian variety. On the other hand, $A_{l'} \cong (\boldsymbol{Z}/l'\boldsymbol{Z})^4$, and subgroups of fixed bounded index in $A_{l'}$ are inside G_l^0. Therefore dim $G_l^0 \geq 2$. It follows that X_l consists in only one orbit under G_l^0, hence X_l is a coset space G_l^0/H, hence X_l itself is an abelian variety. Finally X itself is now caught in the middle between 2 abelian varieties :

$$X_l \xrightarrow{\text{étale}} X \longrightarrow \text{Alb } X.$$

With a suitable origin, $X_l \to \text{Alb } X$ is then a homomorphism, hence if K is its kernel, we find :

$$X_l \times_{\text{Alb}X} X_l = \{(x, x+k) \mid x \in X_l, k \in K\}.$$

But $X_l \times_X X_l \subset X_l \times_{\text{Alb}X} X_l$ and $X_l \times_X X_l$ is (i) étale over X_l, and (ii) the graph of an equivalence relation on X_l. (i) implies that

$$X_l \times_X X_{l'} = \{(x, x+k) \mid x \in X_l, k \in K'\}$$

for some subset $K' \subset K$, and (ii) implies that K' is a subgroup. It follows that $X \cong X_l/K'$, hence X is also an abelian variety.

References

[1] Bagnera, G. and De Franchis, M. : Le nombre ρ de Picard pour les surfaces hyperelliptiques, Rend. Circ. Mat. Palermo **30** (1910).
[2] Deligne, P. and Rapoport, M. : Les schémas de modules de courbes elliptiques, in Springer Lecture Notes **349** (1973).
[3] Enriques, F. and Severi, F. : Mémoire sur les surfaces hyperelliptiques, Acta Math. **32** (1909).
[4] Kodaira, K. : On the structure of compact complex analytic surfaces I, Amer. J. Math. **86** (1964).
[5] Lang, S. : Elliptic Functions, Addison-Wesley, 1973.
[6] Matsusaka, T. : Polarized varieties and fields of moduli, Amer. J. Math. **80** (1958).
[7] Matsusaka, T. and Mumford, D. : Two fundamental theorems on deformations of polarized varieties, Amer. J. Math. **86** (1964).
[8] Mumford, D. : Geometric Invariant Theory, Springer-Verlag, 1965.
[9] —— : Lectures on curves on surfaces, Princeton Univ. Press, 1966.
[10] —— : Enriques' classification of surfaces I, in Global Analysis, Princeton Univ. Press, 1969.
[11] —— : Pathologies III, Amer. J. Math. **89** (1967).
[12] —— : Abelian Varieties, Tata Studies in Math., Oxford Univ. Press, 1970.
[13] Raynaud, M. : Spécialisation du foncteur de Picard, Publ. Math. IHES **38**, p. 27.
[14] Šafarevič, I.et al : Algebraic Surfaces, Proc. Steklov Inst. Math. **75** (1965).
[15] Tate, J. : Genus change in purely inseparable extensions of function fields, Proc. AMS **3** (1952), p. 400.
[16] —— : Endomorphisms of abelian varieties over finite fields, Inv. Math. **2** (1966).

Scuola Normale Superiore, Pisa
Harvard University

(Received January 14, 1976)

Inventiones math. 35, 197–232 (1976)

Inventiones mathematicae

Enriques' Classification of Surfaces in Char. p, III

E. Bombieri (Pisa) and D. Mumford (Cambridge)

To Jean-Pierre Serre

Introduction

This paper continues the extension to char. p of Enriques' classification of algebraic surfaces, begun in two previous papers [7, 2] and in particular deals with the special phenomena of char. 2 and 3. We have already seen that all surfaces can be divided into four classes by their "Kodaira dimension" κ:

$$\kappa = \operatorname{tr} \deg_k \bigoplus_{n=0}^{\infty} H^0(X, \mathcal{O}_X(n K_X)) - 1$$

and that

a) $\kappa = -1 \Rightarrow X$ ruled,

b) $\kappa = 0 \Rightarrow 4K_X \equiv 0$ or $6K_X \equiv 0$

c) $\kappa = 1 \Rightarrow |nK_X|$, n large, is composite with a pencil $\pi: X \to B$, making X elliptic or, in char. 2 or 3, possibly quasi-elliptic,

d) $\kappa = 2 \Rightarrow |nK_X|$, n large, defines a birational morphism onto a model X_0 with rational double points; the surface X is of "general type".

Moreover, the surfaces with $\kappa = 0$ are divided into four classes:

$B_1 = 0, \quad B_2 = 22, \quad \chi(\mathcal{O}_X) = 2,$

$B_1 = 0, \quad B_2 = 10, \quad \chi(\mathcal{O}_X) = 1,$

$B_1 = 4, \quad B_2 = 6, \quad \chi(\mathcal{O}_X) = 0,$

$B_1 = 2, \quad B_2 = 2, \quad \chi(\mathcal{O}_X) = 0.$

We have seen in Part II [2] that the third class consists in the abelian varieties of dimension 2. The first class we call K3-surfaces, the second class Enriques' surfaces. In the fourth class, the Albanese variety is an elliptic curve E and the Albanese mapping

$\pi: X \to E$

has either all fibres non-singular elliptic or rational with a cusp. These surfaces we call hyperelliptic or quasi-hyperelliptic depending on the fibres of π. In [2], we have classified the hyperelliptic surfaces too.

The goals of the present paper are to analyze Enriques' surfaces and quasi-hyperelliptic surfaces. This is largely a question of the "pathologies" of characteristic 2 and 3, since quasi-hyperelliptic surfaces only exist in this case and if char. $\neq 2$ then we also saw in [2] that Enriques surfaces had the property

$$K \not\equiv 0, \quad 2K \equiv 0.$$

Such surfaces were investigated by the unpublished thesis of M. Artin, where he showed that Enriques' classical results in char. 0 extended to all finite characteristics $p \neq 2$. However, in char. 2, we shall see that Enriques surfaces are quite varied! Here is an outline of the paper: in § 1, we shall study the formal geometry of "cuspidal fibrations", i.e., fibrations like those of quasi-elliptic surfaces. In § 2, we shall give a classification of quasi-hyperelliptic surfaces. In § 3, we look at Enriques' surfaces and we see that there are three types in char. 2 and we give examples of each type. In § 4, we shall show that every Enriques surface is elliptic or quasi-elliptic and we shall prove that $\rho = B_2$ for such surfaces.

The study of special low characteristics can be one of two types: amusing or tedious. It all depends on whether the peculiarities encountered are felt to be meaningful variations of the general picture fitting in with standard principles, such as the failure of Sard's lemma, the loss of roots of unity, etc., or are felt instead to be accidental and random, due for instance to numerological interactions between combinations of exponents and the characteristic, of interest only for the sake of having uniform theorems applicable in all characteristics. Our hope is that the geometric aspects of char. 2 and 3 studied here are not entirely of the latter kind. For example, we find quite striking the following points, which we list in the hope that they will supply some motivation for taking the plunge into char. 2 and 3:

1) that a smooth surface X fibred in curves with cusps in char. 2 defines canonically three usually distinct families of subspaces $L_P^{(i)} \subset T_{X,P}$ for all cusps P: cf. Figure 2, § 1.

2) the structure of the automorphism groupscheme of the rational line with a cusp at ∞, and the orbit of ∞ itself under this group: cf. § 2.

3) the fact that there are three types of Enriques' surfaces in char. 2, whose canonical double covering has structure group μ_2, $\mathbb{Z}/2\mathbb{Z}$, α_2, and which are at least sometimes deformable into each other: cf. § 3.

4) the intricate way in which Enriques' argument about the existence of reducible divisors in linear systems on Enriques' surfaces must be adapted in char. 2: cf. § 4.

5) the fact that isolated "pathological examples" (e.g. Igusa's surfaces) become quite natural when viewed in a broader perspective: cf. § 2.

§ 1. Remarks on the Differential Geometry of Cuspidal Fibres

In this section we study the germ (in the formal power series sense) of a general map $f\colon X \to B$, where X is a non-singular surface, B a non-singular curve, such that every fibre of f has a cusp. More precisely, if t (resp. x, y) are local coordinates on B (resp. X), then f is given by a formal power series $t = f(x, y)$, and we seek to understand

a) whether f can be put in normal form by suitable choices of the uniformizing parameters,

b) what invariants can be defined,

c) what is the differential geometry underlying the failure of Sard's lemma.

Starting with $f\colon X \to B$, let

$$\Sigma = \{P \in X \mid f \text{ is not smooth at } P\}$$

and let Σ_0 be a one-dimensional locally closed subset of Σ such that at each point $P \in \Sigma_0$, $f^{-1}(f(P))$ has an ordinary cusp at P.

Proposition 1. *We have*

(a) Σ_0 *is an étale cover of* $B^{(p^{-1})}$, $\quad p = \text{char}(k)$.

(b) $(\Sigma_0 \cdot f^{-1}(f(P)))_P = \text{char}(k)$ *for all* $P \in \Sigma_0$.

Proof. Let t be a local coordinate on B at $f(P)$ and let x, y be local coordinates on X at P. We can choose x, y such that

$$t = (\text{unit})(x^2 + y^3);$$

if char $(k) \neq 2$, then $\dfrac{\partial t}{\partial x}$ vanishes to order 1 and $\dfrac{\partial t}{\partial x} = 0$ defines a germ of a curve non-singular at P and passing through the nearby cusps, i.e. $\dfrac{\partial t}{\partial x} = 0$ is a local equation for Σ_0. If instead char $(k) = 2$, a similar argument shows that

$$\frac{\partial t}{\partial y} = (\text{unit}) \cdot \sigma^2$$

where $\sigma = 0$ is a local equation for Σ_0, and σ vanishes to exact order 1 at P, i.e. Σ_0 is non-singular. Now non-singular branches and ordinary cusps can meet only with multiplicities 2 or 3, hence

$$\dim(\mathcal{O}_{\Sigma_0, P}/\mathfrak{m}_{B, f(P)} \cdot \mathcal{O}_{\Sigma_0, P}) = (\Sigma_0 \cdot f^{-1}(f(P)))_P = 2 \quad \text{or} \quad 3.$$

On the other hand, $\Sigma_0 \to B$ factorizes as $\Sigma_0 \xrightarrow{\tau} B^{(p^{-n})} \to B$ for some n, where τ is separable and now

$$\dim(\mathcal{O}_{\Sigma_0, P}/\mathfrak{m}_{B, f(P)} \cdot \mathcal{O}_{\Sigma_0, P}) \begin{cases} = p^n & \text{generically} \\ > p^n & \text{if } \tau \text{ ramifies over } \tau(P) \end{cases}$$

This proves Proposition 1, plus the old result of Tate [13] that such a line of cusps is only possible in characteristic 2 or 3. Q.E.D.

Now it is convenient to treat separately the cases of characteristic 2 and 3, which are very different. We consider first the simpler case $\mathrm{char}(k)=3$.

We begin by finding a normal form for f. We have

Proposition 2. *Let* $\mathrm{char}(k)=3$ *and choose any local coordinate* t *on* B *at* $f(P)$. *Then in suitable power series coordinates* x, y *on* X *at* P, *we have*

$$t=x^2+y^3.$$

Proof. Choose x, y so that $x=0$ is a local equation for Σ_0. Then if $t=f(x,y)$ we have that x divides both $\frac{\partial t}{\partial x}$ and $\frac{\partial t}{\partial y}$, whence we can write t in the form

$$t=g(y)^3+x^2 r(x,y)$$

and since $t=0$ has an ordinary cusp at $(0,0)$ we have also $(dg/dy)(0)\neq 0$, $r(0,0)\neq 0$. If we change coordinates via

$$\tilde{y}=g(y), \qquad \tilde{x}=x\sqrt{r(x,y)},$$

t has the required form. Q.E.D.

By Proposition 2, we see that all such fibrations in $\mathrm{char}(k)=3$ are locally formally isomorphic. We can go a bit further and prove that they arise from smooth fibrations. Let us define

$$X^* = \textit{normalization of } X\times_B B^{(1/3)};$$

choosing x, y, t as in Proposition 2 then $s=t^{1/3}$ is a local coordinate on $B^{(1/3)}$ and $X\times_B B^{(1/3)}$ is given locally by $s^3=x^2+y^3$, i.e. $(s-y)^3=x^2$. If we put $r=\dfrac{x}{s-y}$ then $r^3=x$, so r is integral over $k[[x,y,s]]$ and r, s are local coordinates on X which is therefore non-singular along Σ_0.

Thus we have a diagram

$$\begin{array}{ccccc}
\Sigma_0 & \hookrightarrow & X^* & \xrightarrow{\ \pi\ } & X \\
\downarrow{\scriptstyle\varphi} & & \downarrow{\scriptstyle f^*} & & \downarrow{\scriptstyle f} \\
B^{(1/3)} & = & B^{(1/3)} & \xrightarrow{\ F\ } & B
\end{array}$$

where φ is étale, f^* is smooth along Σ_0, f has ordinary cusps along Σ_0, and F is the Frobenius morphism.

Since the map π is given by $x=r^3$, $y=s-r^2$ we have that if L is the line subbundle of T_{X^*} generated locally by $\frac{\partial}{\partial r}+2r\frac{\partial}{\partial s}$, then interpreting the sections of L as derivations of $\mathcal{O}_{X^*}$, we have:

$$\mathcal{O}_X=\ker(L\colon \mathcal{O}_{X^*}\to\mathcal{O}_{X^*});$$

the map π is purely inseparable of degree 3, and is said, in the theory of purely inseparable descent, to be the result of "dividing by the rank 1 distribution L".

We may make a "picture" of the map $t=x^2+y^3$ in char. 3. Notice that the line of cusps $x=0$ meets each fibre $t_0=x^2+y^3$ at its cusp with intersection multiplicity 3 and thus has the same tangent cone at this intersection. Thus we have:

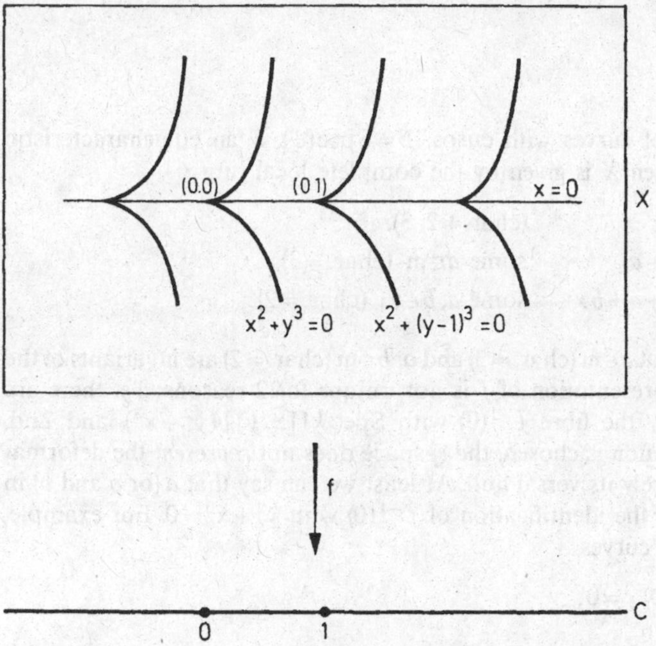

Fig. 1

The case of characteristic 2 is considerably more elaborate. The direct approach to putting the power series $t=f(x, y)$ in normal form does not work. The reason is that the local structure of the map along the line of cusps defines an invariant, which is alternatively a differential η on the line of cusps Σ_0 itself, or a differential ω on the curve $\Sigma_0^{(p)}$ which is the étale covering of the base B which parametrizes the set of cusps in the fibres (see Prop. 1): these are related by the usual p-linear isomorphism between differentials on a curve Y and those on $Y^{(p)}$. Because of this invariant, one must first choose t so that ω is in some standard form and, in fact, there are an infinite number of cases according to the multiplicity of zero of η at $t=0$. The most direct way to see why such an invariant exists is to examine the versal deformation of the cusp $y^2+x^3=0$. One calculates immediately that this definition is given in different characteristics by:

$$y^2+x^3+t_1+t_2x=0 \qquad \text{(char.} \neq 2, 3),$$
$$y^2+x^3+t_1+t_2x+t_3x^2=0 \qquad \text{(char.} = 3),$$
$$y^2+x^3+t_1+t_2x+t_3y+t_4xy=0 \qquad \text{(char.} = 2).$$

Inside the parameter space (t_1, t_2) (resp. (t_1, t_2, t_3) or $(t_1, \ldots, t_4)$) the set of points where the fibre still has a cusp is given by:

$$\begin{aligned} &t_1 = t_2 = 0 && \text{only (char.} \neq 2, 3) \\ &t_2 = t_3 = 0, && t_1 \text{ arbitrary (char.} = 3) \\ &t_3 = t_4 = 0, && t_1, t_2 \text{ arbitrary (char.} = 2). \end{aligned}$$

It follows that if

$$f: X \to B$$

is any formal family of curves with cusps, $B = \operatorname{Spec}(\mathcal{O})$, $\mathcal{O}$ an equicharacteristic complete local ring, then X is given by the complete local ring:

$$\begin{aligned} &\mathcal{O}[[x, y]]/(y^2 + x^3) && \text{(char.} \neq 2, 3), \\ &\mathcal{O}[[x, y]]/(y^2 + x^3 + a), && \text{some } a \in \mathfrak{m} \text{ (char.} = 3), \\ &\mathcal{O}[[x, y]]/(y^2 + x^3 + a + bx), && \text{some } a, b \in \mathfrak{m} \text{ (char.} = 2). \end{aligned}$$

This at first suggests that $a \in \mathfrak{m}$ (char. $= 3$) and $a, b \in \mathfrak{m}$ (char. $= 2$) are invariants of the map f. However this presentation of f is not unique for 2 reasons: 1st there are many ways to identify the fibre $f^{-1}(0)$ with $\operatorname{Spec} k[[x, y]]/(y^2 + x^3)$, and 2nd, after such an identification is chosen, the t_i-space does not *represent* the deformation functor but is merely its versal hull. At least we can say that a (or a and b) in $\mathfrak{m}/\mathfrak{m}^2$ depend only on the identification of $f^{-1}(0)$ with $y^2 + x^3 = 0$. For example, for any $\lambda \in \mathcal{O}^*$, the two curves

$$\begin{aligned} &y^2 + x^3 + \lambda^6 a + (\lambda^4 b) x = 0, \\ &y^2 + x^3 + a + b \cdot x = 0 \end{aligned}$$

are isomorphic via $y \mapsto \lambda^3 y$, $x \mapsto \lambda^2 x$. Thus, in char. 3, a can be replaced by $\lambda^6 a$ for any $\lambda \in \mathcal{O}^*$ and there is not much useful information that can be extracted from a. However, in char. 2, b^3/a^2 is at least invariant under this substitution.

Now assume $S = \operatorname{Spec} k[[t]]$, and X is regular. Then $a = a_1 t + \cdots$, $a_1 \neq 0$, hence $b^3/a^2 \in \mathfrak{m}$. We define the value of the invariant ω at the point $t = 0$ to be:

$$\omega|_{t=0} = d(b^3/a^2)|_{t=0}; \quad \text{i.e.,} \quad = b^3/a^2 \in (\mathfrak{m}/\mathfrak{m}^2).$$

Why is this independent of the choice of representation of f via a and b? Note that it only depends on the values of a and b mod $\mathfrak{m}^2$. Moreover, we claim that *all* changes in the formal isomorphism of $f^{-1}(0)$ with $y^2 + x^3 = 0$ leave b^3/a^2 mod $\mathfrak{m}^2$ fixed. This is most easily seen if we recall that the tangent space to the versal deformation space of $g = 0$ is intrinsically described as:

$$T = \operatorname{Hom}(g \cdot k[[x, y]], k[[x, y]]/(g, g_x, g_y)).$$

The automorphisms of $y^2 + x^3 = 0$ are generated by

$$\begin{aligned} &x \mapsto x + \lambda^2 x^n, \\ &y \mapsto \begin{cases} y + \lambda x^{m+1} + \lambda^2 x^{2m-1} y + \lambda^3 x^{3m}, & n = 2m \\ y + \lambda y x^m + \lambda^2 y x^{2m} + \lambda^3 y x^{3m}, & n = 2m+1. \end{cases} \end{aligned}$$

all $n \geqq 2$, which induce

$$y^2+x^3 \mapsto \begin{cases} (y^2+x^3)(1+\lambda^4 x^{4m-2}), & n=2m \\ (y^2+x^3)(1+\lambda^2 x^{2m})^3, & n=2m+1. \end{cases}$$

In our case

$$k[[x,y]]/(g, g_x, g_y) \cong k \cdot 1 + k \cdot x + k \cdot y + k \cdot xy;$$

thus these automorphisms induce the identity on T for $n \geqq 4$, and for $n=2$ and 3, they still induce the identity on the subspace

$$T_0 = \begin{bmatrix} \text{set of homomorphisms } \phi \text{ such that} \\ \phi(y^2+x^3) \in k \cdot 1 + k \cdot x \end{bmatrix}$$

where our deformations lie. This proves that $(a, b) \bmod \mathfrak{m}^2$ can only be changed to $(\lambda^6 a, \lambda^4 b) \bmod \mathfrak{m}^2$, hence $b^3/a^2 \bmod \mathfrak{m}^2$ is independent of all choices.

Still assuming $S = \operatorname{Spec} k[[t]]$, X regular, we can extend our definition of $\omega|_{t=0} \in \mathfrak{m}/\mathfrak{m}^2$ to a definition of $\omega \in \Omega^1_{\mathcal{O}/k}$. First make a base change to K = algebraic closure of $k((s))$, and then note that the substitution

$$\begin{aligned} &t+s=\tilde{t}, \\ &y+\sqrt{a(s)}+\sqrt[4]{b(s)}\cdot(x+\sqrt{b(s)})=\tilde{y}, \\ &x+\sqrt{b(s)}=\tilde{x} \end{aligned}$$

carries our surface

$$0=y^2+x^3+a(t)+b(t)\cdot x$$

into a new surface

$$0=\tilde{y}^2+\tilde{x}^3+\tilde{a}(\tilde{t})+\tilde{b}(\tilde{t})\cdot\tilde{x}$$

so that the cusp $\tilde{x}=\tilde{y}=\tilde{t}=0$ on the new surface is the image of the generic cusp on the original surface. One now calculates in a page or so that if we define ω for the original surface everywhere by:

$$\omega = \frac{\dot{b}^3}{\dot{a}^2 + b\cdot\dot{b}^2}\, dt = d(\log(\dot{a}^2 + b\cdot\dot{b}^2)) \qquad (*)$$

(Notation: $\dot{c} = dc/dt$)

then the translation $\tilde{t}=t+s$ carries the generic value of this ω to the value $\omega|_{\tilde{t}=0}$ for the new surface. Thus $(*)$ is also independent of the choice of a, b to represent f. Noting that $\Omega^1_{\mathcal{O}/k}$ is locally free of rank 1, one also has the more intrinsic formula:

$$\omega = \frac{(db)^{\otimes 3}}{da^{\otimes 2} + b\cdot(db)^{\otimes 2}}.$$

This formula is very suggestive of the usual formula for the j-invariant of the elliptic curve

$$0=y^2+x^3+a+bx$$

in char. $\neq 2, 3$. This suggests strongly that if our map f arises by specializing an elliptic fibration from char. 0, with the line of cusps "swallowing" the elliptic curves, then ω is likely to be a "ghost" of the j-invariant of that elliptic fibration. This is a good question to study.

Let's return to the algebraic setting of the beginning of this section:

$$f: X \to B,$$

where X is a smooth algebraic surface, B a smooth algebraic curve and $\Sigma_0 \subset X$ is a line of cusps. In this case, we get a closely related invariant:

$$\eta \in \Gamma(\Sigma_0, \Omega^1_{\Sigma_0}).$$

To define η, we proceed as follows:

i) let $\sigma = 0$ be a local equation of Σ_0 at a point $P \in \Sigma_0$,

ii) choose a local coordinate t near $f(P) \in B$

iii) then as we saw in the proof of Proposition 1,

$$dt = \sigma^2 \cdot \alpha$$

for some differential α on X, regular near P.

iv) Let C be Cartier's operator on differentials. Since $\sigma\, C(\alpha) = C(\sigma^2 \alpha) = C(dt) = 0$, it follows $C(\alpha) = 0$, hence $\alpha = dx$ for some function x defined near P.

v) Now $d(t + \sigma^2 x) = dt + \sigma^2 \cdot dx = 0$, so

$$t = y^2 + \sigma^2 x \tag{**}$$

for some function y defined near P.

vi) Then the fact that $t = 0$ has an ordinary cusp at P implies easily that $x, y \in \mathfrak{m}_p/\mathfrak{m}_p^2$ are independent, hence x, y are local coordinates on X near P.

vii) Let $\dfrac{\partial}{\partial y}$ be the derivation of $\mathcal{O}_X$ near P such that $\dfrac{\partial}{\partial y}(x) = 0$, $\dfrac{\partial}{\partial y}(y) = 1$. Define

$$\eta = \left.\frac{\left(\dfrac{\partial \sigma}{\partial y}\right)^2 dx}{x\left(\dfrac{\partial \sigma}{\partial y}\right)^2 + 1}\right|_{\Sigma_0} = d\left(\log\left(x\left(\frac{\partial \sigma}{\partial y}\right)^2 + 1\right)\right)\Bigg|_{\Sigma_0}$$

To see that η is independent of all choices (i.e., the choice of σ, t and x), we relate it to the previously defined ω. We recall that for any smooth curve C, there is a natural p-linear isomorphism between the differentials on C and those on $C^{(p)}$ given by

$$a \cdot db \leftrightarrow a^p \cdot d(b^p),$$

$$a, b \in \mathcal{O}_C;\ a^p, b^p \in \mathcal{O}_{C^{(p)}}.$$

If $\omega = a \cdot db$, write $\omega^{(p)} = a^p \cdot d(b^p)$. Then we have:

Proposition 3. *Let*

$$f: \hat{X} \to \hat{B}$$

be the formal completion of f near P. By Proposition 1, f sets up an isomorphism

$$B \cong (\Sigma_0)^{(2)},$$

and under this isomorphism,

$$\omega = \eta^{(2)}. \tag{***}$$

Proof. It suffices to check this at one point. Starting with (**), let $\sigma = \sigma_1 x + \sigma_2 y$ where $\sigma_i \in \mathcal{O}_{X,P}$. Note that $\sigma_1(P) \neq 0$ since $y^2 + \sigma^2 x = 0$ has an ordinary cusp at the origin. Then if we set

$$\tilde{x} = x \cdot \sqrt[3]{\frac{\sigma_1^2}{1 + x\sigma_2^2}},$$

is easily rewritten: (**)

$$0 = y^2 + \tilde{x}^3 + t\left(1 + \tilde{x}\sigma_2^2 \cdot \sqrt[3]{\frac{1 + x\sigma_2^2}{\sigma_1^2}}\right)^{-1}$$

$$= y^2 + \tilde{x}^3 + t + \frac{t\sigma_2^2}{\sigma_1^{2/3}} \cdot \tilde{x} + t\phi(y, \tilde{x}) \cdot \tilde{x}^2.$$

Therefore, to 1st order in t, this deformation is given by the homomorphism

$$y^2 + \tilde{x}^3 \mapsto t + \frac{t\sigma_2(0,0)^2}{\sigma_1(0,0)^{2/3}} \cdot \tilde{x} \in (k[[\tilde{x}, y]]/(\tilde{x}^2, y^2)).$$

Thus

$$\omega(0) = \frac{\sigma_2(0,0)^6}{\sigma_1(0,0)^2}\, dt|_{t=0}.$$

But now on Σ_0, at (0, 0), $\sigma_1(0,0)\,dx = \sigma_2(0,0)\,dy$, hence

$$\eta(0) = ((\sigma_2(0,0)^2\, dx|_{\sigma=0})|_{x=y=0}$$

$$= \left(\frac{\sigma_2(0,0)^3}{\sigma_1(0,0)}\, dy|_{\sigma=0}\right)\Bigg|_{x=y=0}$$

On Σ_0, the map f is given by $t = y^2$, hence $\eta^{(2)}(0)$ is exactly $\omega(0)$ as claimed. Q.E.D.

Corollary. *Suppose $f: X \to B$ is a quasi-elliptic surface,*

$$S = \{P \in B \mid f^{-1}(P) \text{ is reducible or multiple}\} \quad \text{and} \quad \Sigma_0 \subset X - f^{-1}(S)$$

is the line of cusps. Then the formal differential ω is actually an element $\Gamma(B - S, \Omega_B^1)$.

Proof. In this case, f defines an isomorphism of $\Sigma_0^{(2)}$ and $B - S$.

In the classification of quasi-elliptic surfaces in char. 2, this differential ω may play a role similar to the j-invariant in the classification of elliptic surfaces.

Using this invariant, we can now put cuspidal fibrations in normal forms. Recall that if C is Cartier's operator, and B is a non-singular curve, then C acts on the differentials on B and

i) $C\omega=0\Leftrightarrow\omega=df$, some f,
ii) $C\omega=\omega\Leftrightarrow\omega=df/f$, some f.

From (ii), it is quite easy to deduce:

(iii) ω is regular at $P\in B$ and $C\omega=\omega$, then $\operatorname{ord}_0\omega=2k$ for some k, and for some formal parameter t on B at P,

$$\omega=d(\log(1+t^{2k+1}))=\frac{t^{2k}}{1+t^{2k+1}}\,dt.$$

Our normal form Theorem is this:

Proposition 4. *Let $f\colon X\to S=\operatorname{Spec} k[[t]]$ be a formal map of a non-singular surface X onto S and assume that all the fibres have ordinary cusps. Let ω be the invariant on S defined above. Then for some integer k, $\operatorname{ord}_0\omega=6k$, ($k=\infty$ if $\omega\equiv 0$). If k is finite, let t' be a new parameter on S such that*

$$\omega=d(\log(1+t'^{\,6k+1})),$$

then for suitable coordinates x, y on X, f is given by:

$$t'=y^2+x^2\cdot(x+y^{2k+1}),\quad \text{if } k \text{ is finite},$$
$$t'=y^2+x^3,\quad \text{if } k \text{ is infinite}.$$

Proof. Since $\frac{\partial\sigma}{\partial x}(0)\neq 0$, we may take y and σ to be coordinates on X replacing σ by z, x by $\tau(y,z)$, the map is given by:

$$t=y^2+z^2\cdot\tau. \qquad (*)$$

It is a simple calculation to show that η is now given by

$$\eta=d\cdot\log\left(\tau\left(\frac{\partial\tau}{\partial y}\right)^2+\left(\frac{\partial\tau}{\partial x}\right)^2\right)\bigg|_{z=0}.$$

First, suppose $\omega\equiv 0$. Then $\eta\equiv 0$, hence

$$\left(\frac{\partial\tau}{\partial y}\right)^3(y,0)\cdot dy\equiv 0.$$

But this means that

$$\tau(y,z)=\tau_1(y)^2+z\cdot\tau_2(y,z).$$

Note that $\tau_2(0,0)\neq 0$. Substituting this in $(*)$ and letting $\tilde{y}=y+z\tau_1(y)$, $\tilde{z}=z\sqrt[3]{\tau_2(\eta,z)}$ the map takes the form

$$t=\tilde{y}^2+\tilde{z}^3.$$

Secondly, suppose $\omega \not\equiv 0$. Let $\omega = d \cdot \log(1+t'^{2l+1})$. To simplify notation, replace t by t' and choose new y, z, τ so that $t = y^2 + z^2 \tau$ still. Then on Σ_0, $y = \sqrt{t}$, so $\eta = d \cdot \log(1+y^{2l+1})$. It follows that

$$d \log \left(\frac{\tau \left(\frac{\partial \tau}{\partial y}\right)^2 + \left(\frac{\partial \tau}{\partial z}\right)^2}{1+y^{2l+1}} \right) \Bigg|_{z=0} \equiv 0,$$

hence

$$\frac{\tau \left(\frac{\partial \tau}{\partial y}\right)^2 + \left(\frac{\partial \tau}{\partial z}\right)^2}{1+y^{2l+1}} = \phi(y)^2 + z\psi(y,z).$$

But now expand τ:

$$\tau(y,z) = \tau_0(y)^2 + y \cdot \tau_1(y)^2 + z \cdot \tau_2(y) + z^2 \cdot \tau_3(y,z).$$

Note that $\tau_2(0) \neq 0$. Substituting, we find:

$$(\tau_0^2 + y\tau_1^2)(\tau_1^4) + \tau_2^2 = (1+y^{2l+1})\,\phi^2$$

hence

$$\tau_0\,\tau_1^2 + \tau_2 = \phi$$
$$\tau_1^3 = y^l \cdot \phi.$$

Here, if $l > 0$, then $y \mid \tau_1$, hence $\tau_1(0) = 0$, hence from the first equation $\phi(0) \neq 0$. Thus l is the exact power of y dividing τ_1^3, and so $l = 3k$. Write $\tau_1 = y^k \cdot \tilde{\tau}_1$, so that $\tilde{\tau}_1^3 = \phi = \tau_2 + \tau_0\,\tau_1^2$. Substituting into $t = y^2 + z^2\tau$, we get:

$$t = (y + \tau_0 z)^2 + z^2\,\tilde{\tau}_1^2(y^{2k+1} + z\,y^{2k}\,\tau_0 + z\tilde{\tau}_1) + z^4 \cdot t_3(y,z).$$

Letting $\tilde{y} = y + \tau_0 z$, $\tilde{z} = z\tilde{\tau}_1$, this becomes

$$t = \tilde{y}^2 + \tilde{z}^2(\tilde{y}^{2k+1} + \tilde{z}) + \tilde{z}^4 \cdot \tilde{\tau}_3(\tilde{y}, \tilde{z}),$$

which is exactly what we want except for the $\tilde{z}^4$-term. It is easy to check that a suitable substitution $\tilde{y} \mapsto \tilde{y}$, $\tilde{z} \mapsto \tilde{z} + \tilde{z}^2 f(\tilde{y} \cdot \tilde{z})$ will get rid of this $\tilde{z}^4$-term. Q.E.D.

There is a rather beautiful geometric interpretation of the invariant η. First of all, in char. 2, the tangent line to the curve of cusps Σ_0 is always *transversal* to the tangent line to each curve $f^{-1}(P)$ at its cusp, rather than being equal as in char. 3. In fact, if we write f by

$$t = y^2 + \sigma^2 \cdot x,$$

then $y = 0$ is the tangent line to the curve $f^{-1}(0)$ at the cusp $(0,0)$ and $\sigma = 0$ is the local equation of Σ_0; and in the cotangent space to X at $(0,0)$, dy and $d\sigma$ are always independent. Secondly, suppose we consider the tangent line to the fibre $f^{-1}(0)$ at points Q besides $(0,0)$:

$$T_{Q, f^{-1}(0)} \subset T_{Q, X}$$

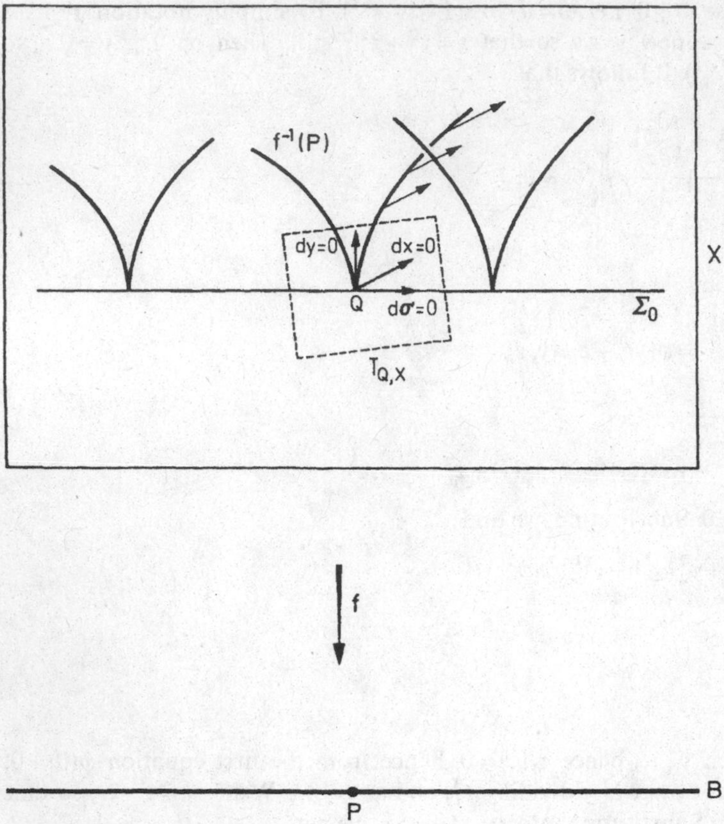

Fig. 2

and then take the limit of $T_{Q, f^{-1}(0)}$ as $Q \to (0, 0)$. $T_{Q, f^{-1}(0)}$ is the subspace of $T_{Q, X}$ defined by $dt=0$. Since $dt=\sigma^2 \cdot dx$, this is the same as the subspace defined by $dx=0$ and as $dx|_{(0,0)}$ is still non-zero, this limit is the subspace $dx|_{(0,0)}=0$. In general, this is a 3rd one-dimensional subspace of $T_{(0,0),X}$!

Thus at each point if Σ_0, the cuspidal fibration $f\colon X \to B$ determines canonically the 3 tangent directions $dy=0$, $d\sigma=0$ and $dx=0$. Here $dy=0$ is always distinct from $d\sigma=0$ and $dx=0$ (because we have an ordinary cusp) but $d\sigma=0$ and $dx=0$ may be equal. We get the following curious picture: Here is the geometric interpretation of η: look at the differential $f^*\left(\frac{dt}{t}\right)$ on X. It has a simple pole on the fibre $f^{-1}(0)$ and a double zero on the curve Σ_0. Thus at $(0, 0)$ it is indeterminate. However, take any curve $C \subset X$ on which $(0, 0)$ is a smooth point with tangent line $dx=0$ at $(0, 0)$. Then we claim

$$\lim_{\substack{Q \to (0,0) \\ Q \in C}} \left(f^* \frac{dt}{t}\right)\bigg|_Q = \eta|_{(0,0)}.$$

To see this, let $x=g(y)$ be the equation of C where $g'(0)=0$. Then

$$f^*\left(\frac{dt}{t}\right)\bigg|_{(g(y),\,y)}=\frac{\sigma(x,y)^2\,dx}{y^2+\sigma(x,y)^2\,x}\bigg|_{(g(y),\,y)}=\frac{\sigma(g(y),y)^2}{y^2+\sigma(g(y),y)^2\,g(y)}\,dx,$$

hence

$$\lim_{y\to 0} f^*\left(\frac{dt}{t}\right)\bigg|_{(g(y),\,y)}=\frac{\partial\sigma}{\partial y}(0)^2\cdot dx=\eta|_{(0,\,0)}.$$

Moreover, in this picture, note that the zeroes of the differential form η are precisely those points where the tangent lines $d\sigma=0$ and $dx=0$ are equal $\left(\text{in fact, } \frac{\partial\sigma}{\partial y}(0)=0 \text{ if and only if } d\sigma \text{ is a multiple of } dx\right)$.

As in characteristic 3, we may also reduce the study of cuspidal fibrations to that of smooth fibrations plus a "rank one distribution". In fact, as above, define

$X^*=$ *normalization of* $X\times_B B^{(1/2)}$.

Choosing t, x, y as in Proposition 4, then $s=t^{1/2}$ is a local coordinate on $B^{(1/2)}$ and $X\times_B B^{(1/2)}$ is given locally by $s^2=y^2+x^3+x^2y^{2k+1}$ (if $k=\infty$, the last term is omitted). If we put $r=\frac{s+y}{x}$, then $r^2=x+y^{2k+1}$, so r is integral over $k[[x.\ y, s]]$. As $x=r^2+y^{2k+1}$, $s=rx+y=r^3+ry^{2k+1}+y$, it follows that X^* is smooth with local coordinates (r, y), and the map $X^*\to B^{(1/2)}$ is smooth. Thus we have a diagram

$$\begin{array}{ccc} X^* & \xrightarrow{\ \pi\ } & X \\ \Big\downarrow{\scriptstyle f^*} & & \Big\downarrow \\ B^{(1/2)} & \xrightarrow{\ F\ } & B \end{array}$$

as before.

However, unlike char. 3, Σ_0 does not lift to X^* unless $k=\infty$: for finite k, the inverse image Σ_1 of Σ_0 in X^* is the curve $r^2=y^{2k+1}$ which is purely inseparable of degree 2 over $B^{(1/2)}$, i.e., is generically $B^{(1/4)}$, and which even has singularities if $k\geqq 1$, i.e., over the zeroes of η.

Since the map π is given by

$$x=r^2+y^{2k+1},$$
$$y=y$$

we have that if L is the line sub-bundle of T_{X^*} generated locally by $\partial/\partial r$, then interpreting the sections of L as derivations of $\mathcal{O}_{X^*}$, we have:

$$\mathcal{O}_X=\ker(L\colon \mathcal{O}_{X^*}\to\mathcal{O}_{X^*}).$$

This means that X is the quotient of X^* by the rank 1 distribution L.

One final remark: we can get still another definition of our invariant η from this construction. In fact, in char. 2, suppose Y is any smooth surface and

$$L_i\subset T_Y,\quad i=1,2$$

are 2 rank 1 integrable distributions (i.e., $D\in\Gamma(U,L_i)\Rightarrow D^2\in\Gamma(U,L_i)$). Let C be the curve of points P where $L_{1,P}=L_{2,P}$. Then we get an invariant differential ζ on C as follows:

i) Since the L_i are integrable, there are locally functions $x_i\in\mathcal{O}_Y$ such that $dx_i\neq 0$ and $D(x_i)=0$, all $D\in L_i$. If x_i' is another such function, $dx_i'=u_i^2\,dx_i$, u_i a unit.

ii) Along C, dx_1 is a multiple of dx_2. Define

$$\zeta=d\log(dx_1/dx_2|_C).$$

Then we claim:

Proposition 5. *Let $f: X\to B$ be a cuspidal fibration, and let $X^*=$(normalization of $X\times_B B^{(1/2)}$) as above. Let $L_1, L_2\subset T_{X^*}$ be the distributions which are tangent to the fibres of $f^*: X^*\to B^{(1/2)}$ and to the fibres of $\pi: X^*\to X$ respectively. Let $\Sigma_1\subset X^*$ be the curve where L_1, L_2 are tangent. Assume the invariant $\omega\in\Gamma(\Sigma_0,\Omega^1_{\Sigma_0})$ is not identically zero. Then $\Sigma_1=\pi_*^{-1}(\Sigma_0)$ and we get a diagram:*

$$\begin{array}{ccc} \Sigma_0^{(1/2)} & \xrightarrow{\;g\;} & \Sigma_1 \\ & {\scriptstyle F}\searrow \quad \swarrow{\scriptstyle \pi_*} & \\ & \Sigma_0 & \end{array}$$

where g is birational. Then ζ and η are related by:

$$g^*\zeta=\eta^{(1/2)}.$$

Proof. This is a simple calculation in the normal forms of Proposition 4. The details are left to the reader.

§ 2. Analysis of Quasi-Hyperelliptic Surfaces in Char. 2, 3

In this section we study quasi-hyperelliptic surfaces X. By definition, these are surfaces satisfying:

a) $K_X\sim 0$ ($\sim$ being numerical equivalence),

b) the Albanese mapping is $\pi: X\to E$, E elliptic,

c) almost all fibres C_x of π are rational with a cusp. By Table 1 of the Introduction to [2], it follows also that

d) $B_2=2$, $c_2=0$, $\chi(\mathcal{O}_X)=0$. Moreover, by Proposition 5 [2], it follows that

c') *all* fibres C_x are rational with a cusp. In Theorem 3, [2], we saw that:

e) There is a second pencil $\pi': X\to\mathbb{P}^1$ on X, this time with elliptic fibres.

We can follow now the same argument used in § 3, [2] for hyperelliptic surfaces to construct an action of E on X. Now we denote by C_0 a fixed rational curve with one ordinary cusp (all such are isomorphic) and deduce, exactly as in § 3, [2]:

Theorem 1. *Every quasi-hyperelliptic surface X is of the form:*

$$X=E_1\times C_0/K,$$

E_1 an elliptic curve, where K is a finite subgroupscheme of E_1 and K acts by

$$k(u,v)=(u+k,\alpha(k)(v))$$

for some injective homomorphism

$$\alpha\colon K\to \mathbf{Aut}(C_0).$$

Moreover the 2 fibrations on X are given by:

$$\begin{array}{cc} E_1\times C_0/K & E_1\times C_0/K \\ \Big\downarrow{\scriptstyle \text{fibres}\approx C_0} & \Big\downarrow{\scriptstyle \text{fibres elliptic}} \\ E_1/K,\ \text{elliptic} & C_0/\alpha(K)\cong \mathbb{P}^1. \end{array}$$

To use the Theorem, we must calculate $\mathbf{Aut}(C_0)$:

Proposition 6. *Let ∞ be the cusp on C_0 and let t be an affine coordinate on $C_0-(\infty)$. Then:*

i) *if* char$\neq 2, 3$, $\mathbf{Aut}(C_0)$=*reduced group of automorphisms* $t\to at+b\cong$ *semi-direct product $\mathbb{G}_m\cdot\mathbb{G}_a$, ($\mathbb{G}_a$ the normal factor).*

ii) *if* char$=3$, $\mathbf{Aut}(C_0)$=*group scheme of automorphisms* $t\to at+b+ct^3$, *where $c^3=0\cong$"semi-direct" product of 3 factors $\mathbb{G}_m\cdot A\cdot\mathbb{G}_a$, where $\mathbb{G}_m$ normalizes A and $\mathbb{G}_a$, and A normalizes $\mathbb{G}_a$ and $A\cong\alpha_3$.*

iii) *if* char$=2$, $\mathbf{Aut}(C_0)$=*group scheme of automorphisms* $t\to at+b+ct^2+dt^4$, *where $c^4=d^2=0\cong$"semi-direct" product of 3-factors $\mathbb{G}_m\cdot A\cdot\mathbb{G}_a$, where $\mathbb{G}_m$ normalizes A and $\mathbb{G}_a$ and A normalize $\mathbb{G}_a$ and A is an infinitesimal group scheme of order 8.*

Proof. In fact, $\mathbb{G}_m\cdot\mathbb{G}_a$ obviously acts on C_0, and an S-valued automorphism of C_0 is a product of an S-valued point of $\mathbb{G}_m\cdot\mathbb{G}_a$ times another S-valued automorphism which a) fixes the point $t=0$, b) acts as the identity on $\mathfrak{m}_{0,C_0}/\mathfrak{m}^2_{0,C_0}$. It suffices to show that the group of automorphisms with properties a) and b) is (e), α_3, or the above A of order 8, depending on the characteristic. Now if $L=\mathcal{O}_{C_0}((0))$, then L^3 is very ample and any automorphism is determined by its action on $\Gamma(L^3)$. But it must preserve the filtration:

$$\underset{\substack{\text{basis}\\ 1}}{\Gamma(L)}\subset\underset{\substack{\text{basis}\\ x=1/t^2}}{\Gamma(L^2)}\subset\underset{\substack{\text{basis}\\ x=1/t^2\\ y=1/t^3}}{\Gamma(L^3)},$$

and act identically on $\Gamma(L)$, $\Gamma(L^2)/\Gamma(L)$, $\Gamma(L^3)/\Gamma(L^2)$. Let it therefore act via

$$x\mapsto x+a,$$

$$y\mapsto y+bx+c.$$

But $x^3=y^2$, hence we must have

$$x^3+3x^2a+3xa^2+a^3=y^2+2byx+b^2x^2+2cy+2bcx+c^2.$$

If $p\neq 2, 3$, this shows $a=b=c=0$; if $p=3$, this shows $b=c=0$ and $a^3=0$; if $p=2$, this shows $a=b^2$ and $b^4=c^2=0$. The Proposition now follows by examining the effect of these substitutions on $t=x/y$. Q.E.D.

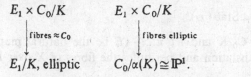

The remarkable feature of the situation of Theorem 1 is the fact that $E_1 \times C_0$ has a singular line on it, while its quotient by K is non-singular. We can give the following criterion, starting with any action of such a group K, for $E_1 \times C_0/K$ to be non-singular:

Proposition 7. *Let* $\mathrm{Stab}(\infty) \subset \mathbf{Aut}(C_0)$ *be the stabilizer of* ∞ *(as a subgroup-scheme). Then for any action of* $K \subset E_1$ *on* $E_1 \times C_0$ *as in Theorem 5,*

$$E_1 \times C_0/K \textit{ non-singular} \Leftrightarrow \alpha(K) \not\subset \mathrm{Stab}(\infty).$$

Proof. Let $E = E_1/K$, let $X = E_1 \times C_0/K$ and let $\pi: x \to E$ be the natural map. Since E_1 acts on $E_1 \times C_0/K$ by translation and permutes the fibres of X over E transitively, it follows

$$\begin{aligned} X \text{ singular anywhere} &\Leftrightarrow \text{the whole curve } E_1 \times (\infty)/K \text{ is singular on } X \\ &\Leftrightarrow \text{the generic fibre } \pi^{-1}(\eta_E) \text{ of } \pi \text{ is not regular} \\ &\quad\; (\eta_E = \text{generic point of } E). \end{aligned}$$

Now $\pi^{-1}(\eta_E)$ is a curve of arithmetic genus 1 over the field $k(\eta_E)$ which becomes isomorphic to C_0 over $\overline{k(\eta_E)}$. Let ∞ be the image of the cusp on it. Since ∞ is singular on $\pi^{-1}(\eta_E) \times \mathrm{Spec}\, \overline{k(\eta_E)}$, ∞ is a non-smooth point of $\pi^{-1}(\eta_E)$ over $k(\eta_E)$. Therefore

$$\infty \text{ a } k(\eta_E)\text{-rational point} \Rightarrow \infty \text{ not regular on } \pi^{-1}(\eta_E).$$

And if ∞ is *not* $k(\eta_E)$-rational, then ∞ is regular, or else comparing the genus of $\pi^{-1}(\eta_E)$ and its normalization it would force $p_a(\pi^{-1}(\eta_E))$ to be bigger than 1: Thus

$$\begin{aligned} \pi^{-1}(\eta_E) \text{ not regular} &\Leftrightarrow \infty \text{ is } k(\eta_E)\text{-rational} \\ &\Leftrightarrow \text{the curve } E_1 \times (\infty)/K \text{ maps by } \pi \text{ isomorphically to } E \\ &\Leftrightarrow \text{the singular line } E_1 \times (\infty)/K \text{ defines a section} \\ &\quad\; \text{of } \pi: X \to E. \end{aligned}$$

But now $E_1 \times C_0$ is recovered by fibre product:

$$\begin{array}{ccc} X & \longleftarrow & E_1 \times C_0 \\ \downarrow & & \downarrow \\ E & \longleftarrow & E_1 \end{array}$$

hence the above is equivalent to $E_1 \times (\infty)$ defining a K-equivariant section of $p_1: E_1 \times C_0 \to E_1$, or to ∞ being a fixed point by $\alpha(K)$. Q.E.D.

Since if char $\neq 2, 3$, the whole of $\mathbf{Aut}(C_0)$ stabilizes ∞, this shows again Tate's result that smooth surfaces like X can exist only if char $= 2$ or 3. Now:

$$\text{char} = 2 \Rightarrow \mathrm{Stab}(\infty) = \mathbb{G}_a \cdot A_0 \cdot \mathbb{G}_m, \qquad A_0 = \left\{ \begin{array}{c} \text{gp. of automorphisms} \\ t \mapsto t + ct^2,\ c^2 = 0 \end{array} \right\}$$

$$\text{char} = 3 \mapsto \mathrm{Stab}(\infty) = \mathbb{G}_a \cdot \mathbb{G}_m\text{: set } A_0 = (e) \text{ here for consistency.}$$

The infinitesimal orbit of ∞ can be pictured like this:

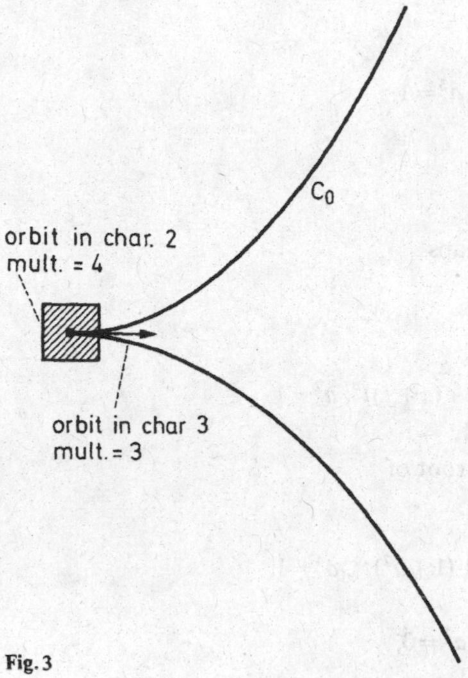

Fig. 3

Moreover, one checks that Stab(∞) is also the subgroup-*scheme* of **Aut**(C_0) of all automorphisms that lift to the normalization $\mathbb{P}^1$ of C_0. We don't know whether this can be deduced from a general principle but in our case, the automorphisms ϕ of $\mathbb{P}^1$ such that $\phi(0) \neq \infty$ are given by

$$\begin{aligned}\phi(t) &= \frac{at+b}{1-ct}, \qquad a+bc \neq 0\\ &= b+(a+bc)t+(ac+bc^2)t^2+(ac^2+bc^3)t^3+\cdots.\end{aligned}$$

If char $=3$, and ϕ is also of the form $\phi(t)=\alpha t+\beta+\gamma t^3$, with α unit, $\gamma^3=0$, it follows that $c=0$, i.e., $\phi \in$ Stab(∞). If char $=2$ and ϕ is also of the form $\phi(t)=\alpha t+\beta+\gamma t^2+\delta t^4$ with α unit, $\gamma^4=0$, $\delta^2=0$, it follows that $c^2=0$, hence $\gamma^2=\delta=0$, i.e., $\phi \in$ Stab(∞). Thus by explicit calculation we see that an S-valued point of **Aut**(C_0) lifts to $\mathbb{P}^1$ if and only if it fixes ∞.

To classify quasihyperelliptic X's, the next step is to enumerate modulo conjugacy all subgroups:

$$K \subset \mathbb{G}_m \cdot A \cdot \mathbb{G}_a$$

such that:

1) $K \not\subset \mathbb{G}_m \cdot A_0 \cdot \mathbb{G}_a$,
2) K commutative,
3) Lie K and Lie K^D at most one-dimensional ($K^D =$ Cartier dual).

This is a tedious problem but, as near as we can tell, the following is the list of all such subgroups K:

Char. 3

a) the μ_3 of maps $t \mapsto at + (1-a)t^3$, $a^3 = 1$,
b) this μ_3, plus $t \mapsto \pm t$,
c) this μ_3, plus $t \mapsto t + i$, $i \in \mathbb{Z}/3\mathbb{Z}$,
d) the α_3 of maps $t \mapsto t + at^3$, $a^3 = 0$,
e) this α_3, plus $t \mapsto \pm t$,
f) the group scheme of order 9 of maps

$$t \mapsto t + a + a^3 t^3, \quad a^9 = 0.$$

Char. 2 – for all $\lambda \in k$:

a) the μ_2 of maps $t \mapsto at + \lambda(a+1)t^2 + (a+1)t^4$, $a^2 = 1$,
b) if $\lambda = 0$, this μ_2, plus $t \mapsto \omega t$, $\omega^3 = 1$,
c) this μ_2, plus $t \mapsto t + \xi$ where ξ is a root of

$$x^4 + \lambda x^2 + x = 0,$$

d) the μ_4 of maps $t \mapsto at + (a+a^2)t^2 + (1+a^2)t^4$, $a^4 = 1$,
e) this μ_4, plus $t \mapsto t + 1$,
f) the α_2 of maps $t \mapsto t + \lambda a t^2 + a t^4$, $a^2 = 0$,
g) if $\lambda = 0$, this α_2, plus $t \mapsto \omega t$, $\omega^3 = 1$,
h) the group scheme of order 4 of maps

$$t \mapsto t + a + \lambda a^2 t^2 + a^2 t^4, \quad a^4 = 0, \ \lambda \neq 0.$$

Combined with Theorem 1, this leads immediately to a complete list of quasi-hyperelliptic surfaces. Moreover, it has the following Corollary:

Proposition 8. *If X is quasihyperelliptic, then:*
i) *if* char. $= 3$, *then* $6K_X \equiv 0$,
ii) *if* char. $= 2$, *then* $6K_X \equiv 0$ *or* $4K_X \equiv 0$.

Proof. In the notation of Theorem 1, the inverse image on $E_1 \times C_0$ of Ω_X^2 is the invertible sheaf $\omega_{E_1 \times C_0}$. But $\omega_{E_1 \times C_0} \cong \Omega_{E_1}^1 \otimes \omega_{C_0}$ and K acts trivially on $\Omega_{E_0}^1$. Therefore

(order K_X) = (least n such that $\alpha(K)$ acts trivially on ω_{C_0})

But dt spans ω_{C_0} so $\mathbf{Aut}(C_0)$ acts on ω_{C_0} through the character:

$$\mathbf{Aut}(C_0) \to \mathbf{Aut}(C_0)/\mathbb{G}_a \cdot A \cong \mathbb{G}_m.$$

Now the Proposition follows by examining all the cases in the above list and noting that the subgroups are all contained in $\mu_6 \cdot A \cdot \mathbb{G}_a$ (char. 2 or 3) or in $\mu_4 \cdot A \cdot \mathbb{G}_a$ (char. 2). Q.E.D.

Another way to describe quasihyperelliptic surfaces X is to reduce their study to that of ruled surfaces X^* with rank one distributions $L \subset T_{X^*}$, as in § 1. To do this, define:

$$X^* = \textit{normalization of } X \times_E E^{(1/p)}$$
$$\pi: X^* \to X, \qquad \text{the natural map}$$
$$L = \operatorname{Ker}[d\pi: T_{X^*} \to \pi^*(T_X)].$$

The fibres of $p: X^* \to E^{(1/p)}$ are smooth and rational, so X^* is an elliptic ruled surface. We may also described X^* in terms of the homomorphism $\alpha: K \to \mathbf{Aut}(C_0)$ as follows. Let

$$q: \mathbb{P}^1 \to C_0$$

be the normalization map, and let

$$K_0 = \alpha^{-1}(\mathrm{Stab}(\infty)).$$

Note that $K_0 \subsetneqq K$; examining the various cases in our list above, we see that we have an exact sequence

$$0 \to K_0 \to K \to (\alpha_p \text{ or } \mu_p) \to 0,$$

hence E_1/K_0 must be the elliptic curve $E^{(1/p)}$. Moreover, we have seen that the action of K_0 on C_0 lifts to an action α_0 of K_0 on $\mathbb{P}^1$. Therefore we get a diagram:

$$\begin{array}{ccccc}
E_1 \times \mathbb{P}^1 & \longrightarrow & E_1 \times \mathbb{P}^1/K_0 & & \\
\downarrow & & \downarrow & & \\
E_1 \times C_0 & \longrightarrow & E_1 \times C_0/K_0 & \longrightarrow & X \\
\downarrow & & \downarrow & & \downarrow \\
E_1 & \longrightarrow & E^{(1/p)} & \longrightarrow & E
\end{array}$$

hence:

$$E_1 \times C_0/K_0 \approx X \times_E E^{(1/p)}$$

and

$$E_1 \times \mathbb{P}^1/K_0 \cong X^*.$$

Thus X^* is the ruled surface obtained by "twisting" $E^{(1/p)} \times \mathbb{P}^1$ by the cocycle

$$\alpha_0: K_0 \to \mathbf{Aut}(\mathbb{P}^1).$$

Note that in char 3, $\alpha_0(K_0) \subset \mathbb{G}_m \cdot \mathbb{G}_a$, hence K_0 stabilizes $E_1 \times (\infty)$, hence the line of cusps $E_1 \times (\infty)/K_0$ is a section of $X^* \to E^{(1/3)}$, as we saw had to happen in § 1. On the other hand, in char 2, $\alpha_0(K_0) \subset \mathbb{G}_m \cdot A_0 \cdot \mathbb{G}_m$, so K_0 need not stabilize $E_1 \times (\infty)$ and the line of cusps need not be a section of $X^* \to E^{(1/2)}$). In fact, if ω is the invariant, defined in § 1, in $\Gamma(E, \Omega_E^1)$ defined by the cuspidal fibration $X \to E$, then we have:

$\omega \equiv 0 \Leftrightarrow$ the line of cusps is a section of $X^* \to E^{(1/2)}$

$\Leftrightarrow \alpha_0(K_0) \subset \mathbb{G}_m \cdot \mathbb{G}_a$.

$\Leftrightarrow \alpha(K)$ belongs to one of the types a, b, c, f, g, h.

However, if $\alpha(K)$ is of type *d* or *e*, then $\omega \not\equiv 0$. Since $K \subset E_1$, E_1 is an ordinary elliptic curve, hence so is E, and ω must be the unique non-zero global 1-form such that $C\omega = \omega$ (C = Cartier's operator).

§ 3. Preliminary Analysis of Enriques' Surfaces

We have defined Enriques' surfaces by the conditions:

a) $K_X \sim 0$ ($\sim$ is numerical equivalence),

b) $B_2 = 10$.

By the Basic Table in Part II, we see that such surfaces also have the properties:

c) $B_1 = 0$, $c_2 = 12$, $\chi(\mathcal{O}_X) = 1$.

Moreover either $\dim H^1(\mathcal{O}_X) = \dim H^2(\mathcal{O}_X) = 0$ hence $K_X \not\equiv 0$; or $\dim H^1(\mathcal{O}_X) = \dim H^2(\mathcal{O}_X) = 1$ hence $K_X \equiv 0$. In the first case, by Castelnuovo's Theorem, since X is not rational we find $|2K_X| \neq \emptyset$, hence in fact $2K_X \equiv 0$. We have seen in Part II, Theorem 5, that the case $K_X \equiv 0$ only occurs in char. 2. We divide Enriques' surfaces into three types according to the action of the Frobenius cohomology operation F on $H^1(\mathcal{O}_X)$:

Definition. Enriques' surfaces are called:

i) classical if $\dim H^1(\mathcal{O}_X) = 0$, hence $K_X \not\equiv 0$, $2K_X \equiv 0$;

ii) singular if $\dim H^1(\mathcal{O}_X) = 1$, hence $K_X \equiv 0$, and F is bijective on $H^1(\mathcal{O}_X)$;

iii) supersingular if $\dim H^1(\mathcal{O}_X) = 1$, hence $K_X = 0$, and F is zero on $H^1(\mathcal{O}_X)$.

The essential similarity of these three classes is brought out in the following Theorem.

Theorem 2. *Let X be an Enriques' surface. Then*[1] $\mathbf{Pic}^\tau_X$ *is a group scheme of order* 2. *Moreover:*

$$\mathbf{Pic}^\tau_X \cong \begin{cases} \mathbb{Z}/2\mathbb{Z} & \text{if } X \text{ is classical} \\ \mu_2 & \text{if } X \text{ is singular} \\ \alpha_2 & \text{if } X \text{ is supersingular.} \end{cases}$$

We give a short proof which handles the classical and non-classical cases separately and then we will sketch a uniform proof which is, perhaps, more natural.

First Proof. If X is classical, $H^1(\mathcal{O}_X) = (0)$ implies $\mathbf{Pic}^\tau_X$ has trivial tangent spaces hence is a finite discrete group. If D is a divisor numerically equivalent to 0 then

$\chi(\mathcal{O}_X(D)) = \chi(\mathcal{O}_X) = 1$,

[1] Note. $\mathbf{Pic}^\tau_X$ denotes the open subgroup of $\mathbf{Pic}_X$ parametrizing divisor classes numerically equivalent to 0.

hence $\dim H^0(\mathcal{O}_X(D))>0$ or $\dim H^2(\mathcal{O}_X(D))>0$. Therefore $|D|\neq\emptyset$ or $|K_X-D|\neq\emptyset$, hence in fact $D\equiv 0$ or $D\equiv K_X$. Therefore $\mathbf{Pic}^\tau_X\cong\mathbb{Z}/2\mathbb{Z}$. If X is not classical then the same argument shows that any such D is linearly equivalent to 0, hence $\mathbf{Pic}^\tau_X$ consists of one point. Since its tangent space is 1-dimensional, we must show that there is a non-trivial obstruction to extending a non-zero morphism $\mathrm{Spec}\, k[t]/(t^2)\to\mathbf{Pic}_X$ to a morphism on $\mathrm{Spec}\, k[t]/(t^3)$. By the theory of [5], last lecture, this follows from:

Lemma 1. *If X is a non-classical Enriques' surface, the first Bockstein operation*

$$\beta_1\colon H^1(\mathcal{O}_X)\to H^2(\mathcal{O}_X)$$

is an isomorphism.

Proof of Lemma 1. It is well-known that in char. 2, $\beta_1(x)$ is the cup-product $\beta_1(x)=x\cup x$. By Serre's duality

$$H^1(\mathcal{O}_X)\otimes H^1(\Omega^2_X)\to H^2(\Omega^2_X)$$

is an isomorphism. Since $K_X\equiv 0$, it follows that

$$H^1(\mathcal{O}_X)\otimes H^1(\mathcal{O}_X)\to H^2(\mathcal{O}_X)$$

is an isomorphism. Then β_1 is this map composed with:

$$H^1(\mathcal{O}_X)\to H^1(\mathcal{O}_X)\otimes H^1(\mathcal{O}_X)$$

where $x\mapsto x\otimes x$. Q.E.D.

This proves that $\mathbf{Pic}^\tau_X=\mu_2$ or α_2 when X is non-classical. Now the isomorphism $H^1(\mathcal{O}_X)\cong\mathrm{Lie}(\mathbf{Pic}^\tau_X)$ carries the Frobenius F on $H^1(\mathcal{O}_X)$ to p-th power on $\mathrm{Lie}(\mathbf{Pic}^\tau_X)$, so $F\neq 0$ is equivalent to p-th power $\neq 0$ on $\mathrm{Lie}(\mathbf{Pic}^\tau_X)$, hence is equivalent to $\mathbf{Pic}^\tau_X=\mu_2$. Q.E.D.

Second Proof. Let us first sketch some quite general facts about the Picard scheme of any projective variety Y. Let P be the universal invertible sheaf on $Y\times\mathbf{Pic}_Y$. Then:

(a) consider the functor which, for all morphisms $f\colon S\to\mathbf{Pic}^\tau_Y$ assigns $p_{2,*}((1\times f)^*P)$:

$$\begin{array}{ccc} Y\times S & \xrightarrow{1\times f} & Y\times\mathbf{Pic}^\tau_Y \\ {\scriptstyle p_2}\downarrow & & \downarrow \\ S & \xrightarrow{f} & \mathbf{Pic}^\tau_Y. \end{array}$$

By Grothendieck's theory of cohomology and base change (cf. EGA, Ch. 3, 2nd part; or Mumford, Abelian Varieties, § 1) there is a coherent sheaf $\mathscr{F}$ on $\mathbf{Pic}^\tau_Y$ such that

$$p_{2,*}((1\times f)^*P)=\mathrm{Hom}_{\mathcal{O}_S}(f^*\mathscr{F},\mathcal{O}_S).$$

Applying this first with $S=\mathrm{Spec}(k)$, Image(f)=point corresponding to the sheaf L on Y, it follows that $\mathrm{Supp}(\mathscr{F})=(0)$, since $H^0(Y,L)=(0)$ if $L\not\cong\mathcal{O}_Y$. Choosing next $S=\mathrm{Spec}(k[\varepsilon])$, Image$(f)=(0)$, and using the fact that if L on $Y\times\mathrm{Spec}(k[\varepsilon])$ is a non-trivial deformation of $\mathcal{O}_Y$ (that is, fits into a non-split sequence

$$0\to\mathcal{O}_Y\xrightarrow{\times\varepsilon}L\to\mathcal{O}_Y\to 0)$$

then $H^0(Y\times\mathrm{Spec}(k[\varepsilon]),L)\cong k$, it follows that for all surjective homorphisms $\mathcal{O}_{0,\mathrm{Pic}_Y}\twoheadrightarrow k[\varepsilon]$ we have

$$\mathrm{Hom}(\mathscr{F}_0\otimes_{\mathcal{O}_{0,\mathrm{Pic}_Y}}k[\varepsilon],k[\varepsilon])\cong k$$

and we deduce from this that $\mathscr{F}\cong k_0$, a sheaf with support (0) and stalk k there. We now apply this to the inclusion $f\colon H\hookrightarrow\mathbf{Pic}_Y^\tau$ of a finite group scheme into $\mathbf{Pic}_Y^\tau$. In this case

$$\mathcal{O}_{0,H}\cong k[t_1,\dots,t_n]/(t_1^{p^r},\dots,t_n^{p^{r_n}})$$

hence

$$\mathrm{Hom}_{\mathcal{O}}(k,\mathcal{O}_{0,H})\cong k.$$

This proves:

Lemma 2. *If* $\mathrm{Spec}(A)\hookrightarrow\mathbf{Pic}_Y^\tau$ *is a finite subgroup scheme, then*

$$\dim H^0(Y\times\mathrm{Spec}(A),P\otimes_{\mathcal{O}_{\mathrm{Pic}}}A)=1.$$

(b) for all finite subschemes $\mathrm{Spec}(A)\hookrightarrow\mathbf{Pic}_Y^\tau$, we may consider $P\otimes_{\mathcal{O}_{\mathrm{Pic}}}A$ as an invertible $\mathcal{O}_Y\otimes_k A$-module, hence if $n=\dim_k A$ it is also a locally free $\mathcal{O}_Y$-module of rank n. Now if $\mathrm{Spec}(A)$ is a subgroup, with points $a_1,\dots,a_m$ let translation by a_i define an automorphism $T_i\colon A\to A$, define also

$$\pi\colon A\to\mathcal{O}_{0,A}\cong k[t_1,\dots,t_n]/(t_1^{p^{r_1}},\dots,t_n^{p^{r_n}})$$

by restriction to (0), and then define $t\colon A\to k$ by

$$t(f)=\sum_{i=1}^{m}\left[\text{coeff. of}\prod_{i=1}^{n}r_i^{p^{r_i}-1}\ \text{ in }\ \pi(T_i(f))\right].$$

Then $(x,y)\mapsto t(x.y)$ is a non-degenerate quadratic form on the n-dimensional k-vector space A. Now we can pair the two locally free rank n $\mathcal{O}_Y$-sheaves $P\otimes_{\mathcal{O}_{\mathrm{Pic}}}A$ and $P^{-1}\otimes_{\mathcal{O}_{\mathrm{Pic}}}A$, via

$$(P\otimes_{\mathcal{O}_{\mathrm{Pic}}}A)\times(P^{-1}\otimes_{\mathcal{O}_{\mathrm{Pic}}}A)\xrightarrow{\text{mult.}}\mathcal{O}_{Y\times\mathrm{Pic}}\otimes_{\mathcal{O}_{\mathrm{Pic}}}A=\mathcal{O}_Y\otimes_k A\xrightarrow{1\otimes t}\mathcal{O}_Y.$$

This is similarly non-degenerate, which proves:

Lemma 3. *If* $\mathrm{Spec}(A)$ *is a finite subgroup schme of* $\mathbf{Pic}_Y^\tau$, *then*

$$P^{-1}\otimes_{\mathcal{O}_{\mathrm{Pic}}}A\cong\mathrm{Hom}_{\mathcal{O}_Y}(P\otimes_{\mathcal{O}_{\mathrm{Pic}}}A,\mathcal{O}_Y).$$

Apply this to the case in which $Y=X$, an Enriques' surface. If $\mathbf{Pic}_X$ contains a subgroup scheme $\mathrm{Spec}(A)$ of order n, then $E_A=P\otimes_{\mathcal{O}_{\mathrm{Pic}}}A$ is a locally free $\mathcal{O}_X$-

module of rank n. By Lemma 2, we have

$$\dim H^0(X, E_A) = 1,$$

and by Lemma 3 and Serre duality

$$\begin{aligned}\dim H^2(X, E_A) &= \dim H^0(X, \Omega_X^2 \otimes \mathrm{Hom}(E_A, \mathcal{O}_X)) \\ &= \dim H^0(X \times \mathrm{Spec}(A), [(\Omega_X^2 \otimes \mathcal{O}_{\mathrm{Pic}}) \otimes P^{-1}] \otimes_{\mathcal{O}_{\mathrm{Pic}}} A).\end{aligned}$$

Assume Spec(A) contains the point a of order 1 or 2 of $\mathbf{Pic}_X$ representing the sheaf Ω_X^2. Now under the automorphism $(x,p) \mapsto (x, -p)$ of $X \times \mathbf{Pic}_X$, the pull-back of P is P^{-1}; and under the automorphism $(x,p) \mapsto (x, p+a)$ the pull-back of P is $(\Omega_X^2 \otimes \mathcal{O}_{\mathrm{Pic}}) \otimes P$. Restricting the automorphism $(x,p) \mapsto (x, a-p)$ to $X \times \mathrm{Spec}(A)$, it follows that

$$\begin{aligned}&\dim H^0(X \times \mathrm{Spec}\,(A), [(\Omega_X^2 \otimes \mathcal{O}_{\mathrm{Pic}}) \otimes P^{-1}] \otimes_{\mathcal{O}_{\mathrm{Pic}}} A) \\ &\quad = \dim H^0(X \times \mathrm{Spec}(A), P \otimes_{\mathcal{O}_{\mathrm{Pic}}} A) = 1.\end{aligned}$$

Thus $\chi(E_A) \leqq 2$.

On the other hand, if $a_1, \ldots, a_m$ are the points of Spec (A) and if a_i corresponds to a sheaf L_i on X, then E_A is a direct sum of m sheaves, the i-th of these being a successive extension of copies of L_i. To see this, note that

$$A = \bigoplus_{i=1}^{m} \mathcal{O}_{a_i, \mathrm{Spec}\,(A)}$$

and choose filtrations

$$\mathcal{O}_{a_i, \mathrm{Spec}\,(A)} = I_{i,0} \supset I_{i,1} \supset \cdots \supset I_{i,s} = (0)$$

of each factor by ideals with $\dim I_{i,j}/I_{i,j+1} = 1$. Then

$$E_A = \bigoplus_{i=1}^{m} (E_A \otimes_A \mathcal{O}_{a_i, \mathrm{Spec}\,(A)})$$

and each factor is a successive extension of the sheaves

$$E_{i,j} = (E_A I_{i,j})/(E_A I_{i,j+1}).$$

If $t_{i,j} \in I_{i,j} - I_{i,j+1}$, then multiplication by $t_{i,j}$ defines

$$L_i = E_A I_{i,0}/E_A I_{i,1} \xrightarrow{\sim} E_A I_{i,j}/E_A I_{i,j+1}.$$

Since $\chi(L_i) = \chi(\mathcal{O}_X) = 1$, this proves: $\chi(E_A) = n$. Therefore $n \leqq 2$, and $\mathbf{Pic}_X^\tau$ has order 2. The rest of the argument now follows the first proof.

Combining the Theorem with the fact that for all finite commutative group schemes H, principal covering spaces

$$\pi\colon \tilde{X} \to X$$

with structure group H are classified by:

$$\mathrm{Hom}(H^D, \mathbf{Pic}_X)$$

(H^D = Cartier dual of H), cf. [8], p. 50, it follows that:

Corollary. *Let X be an Enriques surface. Then X has a canonical principal covering space*

$$\pi: \tilde{X} \to X$$

of degree 2, whose structure group is:

μ_2 *if X is classical*

$\mathbb{Z}/2\mathbb{Z}$ *if X is singular*

α_2 *if X is supersingular.*

(Recall that if char $\neq 2$, then $\mu_2 \cong \mathbb{Z}/2\mathbb{Z}$.)

We may construct this covering explicitly as follows:

Case i: X classical. Let $\{f_{ij}\} \in Z^1(\{U_i\}, \mathcal{O}_X^*)$ be a 1-cocycle representing the invertible sheaf Ω_X^2. Since $2K_X \equiv 0$, $\{f_{ij}^2\}$ is a 1-coboundary and we can write

$$f_{ij}^2 = g_i/g_j \quad \text{on } U_i \cap U_j$$

where $g_i \in \Gamma(U_i, \mathcal{O}_X^*)$. Now we define $\pi: \tilde{X} \to X$ locally by the covering

$$z_i^2 = g_i \quad \text{on } U_i \times \mathbb{A}^1$$

the glueing being given by

$$z_i/z_j = f_{ij} \quad \text{on } (U_i \times \mathbb{A}^1) \cap (U_j \times \mathbb{A}^1).$$

Case ii: X singular. Let $\{a_{ij}\} \in Z^1(\{U_i\}, \mathcal{O}_X)$ be a 1-cocycle representing an element $\eta \in H^1(\mathcal{O}_X)$ such that $F\eta = \eta$. Following Serre [11], this means that $\{a_{ij}^2 - a_{ij}\}$ is a 1-coboundary, and we can write

$$a_{ij}^2 - a_{ij} = b_i - b_j \quad \text{on } U_i \cap U_j$$

where $b_i \in \Gamma(U_i, \mathcal{O}_X)$. Now we define $\pi: \tilde{X} \to X$ locally by the Artin-Schreier covering:

$$z_i^2 - z_i = b_i \quad \text{on } U_i \times \mathbb{A}^1$$

the glueing being given by

$$z_i - z_j = a_{ij} \quad \text{on } (U_i \times \mathbb{A}^1) \cap (U_j \times \mathbb{A}^1).$$

Case iii: X supersingular. Let $\{a_{ij}\} \in Z^1(\{U_i\}, \mathcal{O}_X)$ be any 1-cocycle which is not a coboundary. Since F is zero on $H^1(\mathcal{O}_X)$, $\{a_{ij}^2\}$ is a 1-coboundary and we can write

$$a_{ij}^2 = b_i - b_j \quad \text{on } U_i \cap U_j$$

where $b_i \in \Gamma(U_i, \mathcal{O}_X)$. Now we define $\pi: \tilde{X} \to X$ locally by the inseparable covering:

$$z_i^2 = b_i \quad \text{on } U_i \times \mathbb{A}^1$$

the glueing being given by

$$z_i - z_j = a_{ij} \quad \text{on } (U_i \times \mathbb{A}^1) \cap (U_j \times \mathbb{A}^1).$$

What sort of scheme is $\tilde{X}$? If X is classical and char $\neq 2$ or if X is singular and char $= 2$, π is étale of degree 2, hence $\tilde{X}$ is a smooth surface. On the other hand, if

X is classical or supersingular and char$=2$, then π is purely inseparable. Still $\tilde{X}$ is a reduced Gorenstein surface because: a) it is codimension 1 in a smooth three-fold, and b) in the notation above, g_i (resp. b_i) are not squares because $\{f_{ij}\}$ (resp. $\{a_{ij}\}$) is not itself a 1-coboundary. Moreover, in the first set of cases, $\tilde{X}$ is certainly a $K3$-surface because $K_{\tilde{X}} \sim 0$ and $\chi(\mathcal{O}_{\tilde{X}}) = 2\chi(\mathcal{O}_X) = 2$, and X is quotient of a $K3$-surface by means of a fixed point free involution. In the second set of cases, $\tilde{X}$ must be singular. To see this, note that

$$\eta = dg_i/g_i, \qquad \text{case i, char. 2}$$

$$\eta = db_i, \qquad \text{case iii}$$

defines a global 1-form on X with no poles, and that for all $P \in \tilde{X}$:

P is singular on $\tilde{X}$ if and only if all derivatives of $z_i^2 - g_i$ (resp. $z_i^2 - b_i$) vanish at P, hence if and only if η is zero at $\pi(P)$. Since $c_2(X) = 12$, a 1-form like η will generically have 12 zeros, and it must always vanish somewhere, hence $\tilde{X}$ must be singular.

However, $\tilde{X}$ is always "$K3$-like":

Proposition 9. *The double covering $\pi: \tilde{X} \to X$ satisfies:*

$$\dim H^i(\mathcal{O}_{\tilde{X}}) = \begin{cases} 1 & i=0 \\ 0 & i=1 \\ 1 & i=2 \end{cases}$$

and $\omega_{\tilde{X}}$ (the dualizing sheaf on $\tilde{X}$) is isomorphic to $\mathcal{O}_{\tilde{X}}$.

Proof. By Grothendieck's duality theory (cf. [4]), for any finite flat morphism $g: \tilde{Y} \to Y$, write $\mathcal{A} = g_*(\mathcal{O}_{\tilde{Y}})$, so that $\tilde{Y} \cong \operatorname{Spec}(\mathcal{A})$. Then

$$\omega_{\tilde{Y}} \cong [\operatorname{Hom}_{\mathcal{O}_Y}(\mathcal{A}, \omega_Y)]^\sim$$

(i.e. regard $\operatorname{Hom}_{\mathcal{O}_Y}(\mathcal{A}, \omega_Y)$ as a sheaf of $\mathcal{A}$-modules and take the associated sheaf on $\tilde{Y}$). So to show that $\omega_{\tilde{Y}} \cong \mathcal{O}_{\tilde{Y}}$ we must show

$$\phi: \mathcal{A} \xrightarrow{\sim} \operatorname{Hom}_{\mathcal{O}_Y}(\mathcal{A}, \omega_Y) \quad \text{as } \mathcal{A}\text{-modules},$$

or equivalently we must construct an $\mathcal{O}_Y$-linear map

$$t: \mathcal{A} \to \omega_Y$$

such that $t(x \cdot y)$ is a non-degenerate quadratic form on the locally free sheaf $\mathcal{A}$ with values in the invertible sheaf ω_Y, and then define ϕ by setting $\phi(x)$ equal to the homomorphism $y \mapsto t(x \cdot y)$. Returning to Enriques' surfaces, we have:

Case i: $\pi_*(\mathcal{O}_{\tilde{X}}) \cong \mathcal{O}_X \oplus L$

where $L = \mathcal{O}_X \cdot z_i$ locally.

Since $z_i/z_j = f_{ij}$, we have $L \cong \Omega_X^2$. Now define t to be 0 on $\mathcal{O}_X$ and this isomorphism on L.

Cases ii, iii: we get an exact sequence

$$0 \to \mathcal{O}_X \xrightarrow{\alpha} \pi_*(\mathcal{O}_{\tilde{X}}) \xrightarrow{\beta} \mathcal{O}_X \to 0$$

where $\alpha(a) = a$, $\beta(a + bz_i) = b$.

Since $\Omega_X^2 \cong \mathcal{O}_X$, define t to be the map β.

This proves statement 2). Moreover, it also shows that $\chi(\mathcal{O}_{\tilde{X}})=\chi(\pi_* \mathcal{O}_{\tilde{X}})=2\chi(\mathcal{O}_X)=2$. Now $\tilde{X}$ connected and reduced implies $\dim H^0(\mathcal{O}_{\tilde{X}})=1$, so by 2) we have also

$$\dim H^2(\mathcal{O}_{\tilde{X}})=\dim H^0(\omega_{\tilde{X}})=\dim H^0(\mathcal{O}_{\tilde{X}})=1. \quad \text{Q.E.D.}$$

The same type of argument can be used to prove more generally that if $\pi\colon \tilde{Y}\to Y$ is any principal covering space with finite structure group scheme G, then, in Grothendieck's notation, $\pi^! \mathcal{O}_Y \cong \mathcal{O}_{\tilde{Y}}$, hence $\omega_{\tilde{Y}} \cong \pi^* \omega_Y$.

Finally, we would like to give some examples to show that all these types of Enriques' surfaces do exist and may even be part of the same connected family of surfaces. Enriques' original construction of his surfaces as the normalization of sextic surfaces in $\mathbb{P}^3$ passing doubly through the edges of a tetrahedron gives classical Enriques surfaces in all characteristics (cf. [6], where it was remarked that in char. 2 these surfaces carried regular 1-forms). To get all the types at once, we follow an idea of Miles Reid which adapts Serre's construction [10] to the case of Enriques' surfaces: we construct $\tilde{X}$ in $\mathbb{P}^5$ as the intersection of three quadrics and an action of μ_2, $\mathbb{Z}/2\mathbb{Z}$ or α_2 on $\mathbb{P}^5$ which restricts to a free action of this group on $\tilde{X}$. Then define X to be the orbit space.

Let $x_1, x_2, x_3, y_1, y_2, y_3$ be homogeneous coordinates on $\mathbb{P}^5$. Consider the action of

$$G=\text{group of matrices}\left\{\begin{pmatrix}1 & 0\\ a & b\end{pmatrix},\ b\in k^*,\ a\in k\right\}$$

on $\mathbb{P}^5$ given by

$$(x_i, y_i)\mapsto(x_i, a x_i + b y_i).$$

Inside G, we consider the following subgroup schemes of order 2:

a) for all $\sigma\in k^*$, $H_{(\sigma,0)}=\left\{\begin{pmatrix}1 & 0\\ 0 & 1\end{pmatrix}, \begin{pmatrix}1 & 0\\ \sigma & 1\end{pmatrix}\right\}\cong \mathbb{Z}/2\mathbb{Z}$,

b) $H_{(0,0)}=\left\{\begin{pmatrix}1 & 0\\ \varepsilon & 1\end{pmatrix}, \varepsilon^2=0\right\}\cong\alpha_2$,

c) for all $\tau\in k^*$, $H_{(0,\tau)}=\left\{\begin{pmatrix}1 & 0\\ \varepsilon & 1+\tau\varepsilon\end{pmatrix}, \varepsilon^2=0\right\}\cong\mu_2$.

Note that as either σ or τ goes to 0, $H_{(\sigma,0)}$ or $H_{(0,\tau)}$ approaches $H_{(0,0)}$. In fact, altogether they form a finite and flat group scheme over $\operatorname{Spec} k[\sigma,\tau]/(\sigma\tau)$. These group schemes act on $\mathbb{P}^5$ and the subrings of $k[x_1,\dots,y_3]$ of H-invariants of even degree are readily computed to be:

a) $(k[x_1,\dots,y_3]^{H_{(\sigma,0)}})_{\substack{\text{even}\\ \text{degree}}} = k[x_k^2, x_i x_j, y_k^2+\sigma x_k y_k, y_i x_j + y_j x_i]$

where $1\leqq i<j\leqq 3$, $1\leqq k\leqq 3$;

b) $(k[x_1,\dots,y_3]^{H_{(0,0)}})_{\substack{\text{even}\\ \text{degree}}} = k[x_k^2, x_i x_j, y_k^2, y_i x_j + y_j x_i]$

where $1\leqq i<j\leqq 3$, $1\leqq k\leqq 3$;

c) $(k[x_1, \ldots, y_3]^{H(0,\tau)})_{\substack{\text{even}\\ \text{degree}}} = k[x_k^2, x_i x_j, y_k^2, y_i x_j y_j x_i + \tau y_i y_j]$

where $1 \leqq i < j \leqq 3$, $1 \leqq k \leqq 3$.

In all three cases we take the twelve invariant quadratic forms on the right and use these to define a morphism:

$$\Phi_{(\sigma,\tau)}: \mathbb{P}^5 \to \mathbb{P}^{11}$$

for $\sigma\tau = 0$. Then $\Phi_{(\sigma,\tau)} \mathbb{P}^5 \cong \mathbb{P}^5/H_{(\sigma,\tau)}$. Define $X_{(\sigma,\tau)}$ to be a generic 2-dimensional section of this quotient:

$$X_{(\sigma,\tau)} = \Phi_{(\sigma,\tau)}(\mathbb{P}^5) \cap (L_1 = L_2 = L_3 = 0)$$

where the L_i are generic linear forms, and define $\tilde{X}_{(\sigma,\tau)}$ to be the inverse image of $X_{(\sigma,\tau)}$ in $\mathbb{P}^5$, hence

$$\tilde{X}_{(\sigma,\tau)} = \text{locus}\,(f_1 = f_2 = f_3 = 0)$$

where the f_i are the quadratic forms obtained by pull-back of the linear forms L_i. Dropping for an easier notation the subfix (σ, τ), let π be the restriction of Φ to $\tilde{X}$. Since in all cases the fixed point locus F of H is one or two planes, $\tilde{X}$ does not meet F and π: $\tilde{X} \to X$ is a principal H-bundle. Since $\Phi(\mathbb{P}^5)$ is smooth outside $\Phi(F)$, this shows that X is smooth too. On the other hand, $\tilde{X}$, being the intersection of three quadratic forms in $\mathbb{P}^5$, is Gorenstein with dualizing sheaf given by:

$$\omega_{\tilde{X}} \cong \Omega_{\mathbb{P}^5}(-\textstyle\sum \deg f_i)|_{\tilde{X}} \cong \mathcal{O}_{\tilde{X}}.$$

As we have seen, $\omega_{\tilde{X}} \cong \pi^* \omega_X$, so

$$\omega_X^{\otimes 2} \cong \text{Norm}_{\tilde{X}/X}(\omega_{\tilde{X}}) \cong \mathcal{O}_X.$$

Moreover, again because $\tilde{X}$ is a complete intersection, $H^1(\mathcal{O}_{\tilde{X}}) = (0)$; since $\dim H^2(\mathcal{O}_{\tilde{X}}) = \dim H^0(\omega_{\tilde{X}}) = \dim H^0(\mathcal{O}_{\tilde{X}}) = 1$, we find that $\chi(\mathcal{O}_{\tilde{X}}) = 2$, hence $\chi(\mathcal{O}_X) = \frac{1}{2}\chi(\mathcal{O}_{\tilde{X}}) = 1$. By the Basic Table in Part II, X must be an Enriques surface.

§ 4. Linear Systems on Enriques' Surfaces

In this section we shall continue our investigation of Enriques' surfaces and prove that they are all elliptic or quasi-elliptic. This result was proved for classical surfaces by Enriques himself; modern proofs can be found in [1, 12]. The basic argument in his proof was to show that every linear system $|D| \neq \emptyset$ with $D^2 > 0$ contains reducible curves, hence with components of lower arithmetic genus; this implies easily the existence on X of a curve of canonical type. His method of proof however breaks down in the non-classical case and we have to exploit the special features of char. 2 in order to obtain the same result.

Theorem 3. *Every Enriques' surface is elliptic or quasi-elliptic.*

Proof. In order not to repeat well-known arguments, we shall consider only the non-classical cases in char. 2. By the results in Part I, [7], it is sufficient to show the existence of a curve of canonical type on X.

Let L be an invertible sheaf on X, $L \not\cong \mathcal{O}_X$, and $|L| \neq \emptyset$. If $s \in \Gamma(L)$ and $C = \operatorname{div}(s)$ we have the exact sequence

$$0 \to \mathcal{O}_X \xrightarrow{s} L \to \mathcal{O}_C \otimes L \to 0$$

and we denote by $\Gamma(L)_0$ the vector subspace of $\Gamma(L)$ consisting of those sections s for which

$$H^1(\mathcal{O}_X) \xrightarrow{s} H^1(L) \quad \text{is the zero map,}$$

and $|L|_0$ will be the associated linear system.

Lemma 4. *Assume that there exists* $C \in |L|$ *with* $\dim H^0(\mathcal{O}_C) = 1$. *Then we have: either*

a) $\dim H^1(L) = 0$ *and* $\dim |L| = \dim |L|_0 = \frac{1}{2}(L^2)$

or

b) $\dim H^1(L) = 1$ *and* $\dim |L|_0 = \frac{1}{2}(L^2)$.

Moreover, in case b) *we have that* $D \in |L|_0$, $D > 0$ *if and only if* $\dim H^0(\mathcal{O}_D) = 2$, *hence every element of* $|L|_0$ *is reducible.*

Proof of Lemma. Since $K_X \equiv 0$, the Riemann-Roch Theorem yields

$$\dim |L| = \tfrac{1}{2}(L^2) + \dim H^1(L).$$

The cohomology sequence of

$$0 \to \mathcal{O}_X \xrightarrow{s} L \to \omega_D \to 0$$

where $D = \operatorname{div}(s)$, gives

$$H^1(\mathcal{O}_X) \xrightarrow{s} H^1(L) \to H^1(\omega_D) \to H^2(\mathcal{O}_X) \to 0.$$

Now if $\dim H^1(\omega_D) = \dim H^0(\mathcal{O}_D) = 1$, we see that $\dim H^1(L) \leqq 1$, hence we obtain either a) or b). We also get, for any D,

$$\dim H^1(L) = -1 + \dim H^0(\mathcal{O}_D) + \dim \operatorname{Im}\{H^1(\mathcal{O}_X) \xrightarrow{s} H^1(L)\}$$

and we obtain the last clause of Lemma 4. Q.E.D.

Lemma 5. *If* $L^2 > 0$, $\dim H^1(L) = 0$ *and* $|L|$ *contains an irreducible curve then*

$$\dim H^1(L^{\otimes 2}) = 0 \quad \text{and} \quad \dim |L^{\otimes 2}| = 2(L^2).$$

Proof of Lemma. Let $C = \operatorname{div}(s) \in |L|$ be irreducible. Clearly $H^1(\mathcal{O}_X) \xrightarrow{s} H^1(L^{\otimes 2})$ is the zero map, since it factors through multiplication by s alone, hence $s^2 \in \Gamma(L^{\otimes 2})_0$. On the other hand, C irreducible and $C^2 = L^2 > 0$ imply $\dim H^0(\mathcal{O}_{2C}) = 1$, and the result follows from the previous Lemma. Q.E.D.

Proposition 10. *Let* C *be an irreducible curve on* X *with* $C^2 > 0$. *Then the linear system* $|C|$ *contains a reducible divisor* D *which is not the sum of two non-singular rational curves.*

Proof of Proposition. We choose a non-trivial 1-cocycle $\{a_{ij}\} \in Z^1(\{U_i\}, \mathcal{O}_X)$ such that, as remarked in the proof of Corollary to Theorem 2,

$$a_{ij}^2 - \varepsilon_X a_{ij} = b_i - b_j \quad \text{on } U_i \cap U_j$$

with $b_i \in \Gamma(U_i, \mathcal{O}_X)$, where

$$\varepsilon_X = \begin{cases} 1 & \text{if } X \text{ is singular} \\ 0 & \text{if } X \text{ is supersingular.} \end{cases}$$

We also fix once for all such a datum $b = \{b_i\}$. Now let $\{f_{ij}\} \in Z^1(\{U_i\}, \mathcal{O}_X^*)$ be a 1-cocycle representing the class of L in Pic (X). If s is a section of L, the image of the class of $\{a_{ij}\}$ by $H^1(\mathcal{O}_X) \xrightarrow{s} H^1(L)$ is represented by the 1-cocycle $\{s_i a_{ij}\} \in Z^1(\{U_i\}, L)$, hence if $s \in \Gamma(L)_0$ then this is a 1-coboundary:

$$s_i a_{ij} = \sigma_i - f_{ij} \sigma_j \quad \text{on } U_i \cap U_j$$

with $\sigma_i \in \Gamma(U_i, \mathcal{O}_X)$. The corresponding datum $\sigma = \{\sigma_i\}$ is determined uniquely modulo sections of L. We use b and σ to construct sections of $L^{\otimes 2}$ as follows: if s, $t \in \Gamma(L)_0$ and σ, τ are associated data then $s\tau + \sigma t$ and $\sigma^2 + \varepsilon_X \sigma s + b s^2$ are sections of $L^{\otimes 2}$, as a straightforward calculation in char. 2 shows.

Now take $L = \mathcal{O}_X(C)$. By Lemma 4, if dim $H^1(L) = 1$ every element of $|C|_0$ is reducible and is not the sum of two non-singular rational curves, since by hypothesis $C^2 > 0$, and dim $H^0(\mathcal{O}_C) = 2$.

Therefore we have to consider only the case in which dim $H^1(L) = 0$. We fix a basis $s_0, \ldots, s_n$ of $\Gamma(L)$, where $n = \frac{1}{2}(L^2)$ and we fix associated data $\bar{\sigma}_0, \ldots, \bar{\sigma}_n$ which we cxtend by linearity to a well-defined datum $\bar{\sigma}$ associated to $s \in \Gamma(L)$. Any other datum σ associated to s is of type $\sigma = \bar{\sigma} + s'$, $s' \in \Gamma(L)$. Now consider in $\Gamma(L^{\otimes 2})$ sections of the form

$$\sigma^2 + \varepsilon_X \sigma s + b s^2 + s\tau + \sigma t \tag{A'}$$

and

$$s\tau + \sigma t \tag{A''}$$

There are some obvious cases in which they vanish identically: in case (A'), if $s = \sigma = 0$, or if $s = 0$, $\sigma = t$; in case (A''), if $s = \mu t$ and $\sigma = \mu \tau$ with $\mu \in k^*$, or if $s = \sigma = 0$, or $s = t = 0$, or $t = \tau = 0$. We call these the trivial relations. Writing

$$\begin{aligned} s &= \sum x_i s_i, \\ \sigma &= \sum x_i \bar{\sigma}_i + \sum u_i s_i, \\ t &= \sum y_i s_i, \\ \tau &= \sum y_i \bar{\sigma}_i + \sum (v_i + \varepsilon_X u_i) s_i \end{aligned}$$

we obtain

$$\begin{aligned} \sigma^2 &+ \varepsilon_X \sigma s + b s^2 + s\tau + \sigma t \\ &= \sum x_i^2 (\bar{\sigma}_i^2 + \varepsilon_X \bar{\sigma}_i s_i + b s_i^2) \\ &\quad + \sum_{i<j} (\varepsilon_X x_i x_j + y_i x_j + y_j x_i)(\bar{\sigma}_i s_j + \bar{\sigma}_j s_i) \\ &\quad + \sum (u_i^2 + v_i x_i + u_i y_i) s_i^2 \\ &\quad + \sum_{i<j} (v_i x_j + v_j x_i + u_i y_j + u_j y_i) s_i s_j. \end{aligned}$$

In the same way, setting $v_i' = v_i + \varepsilon_X u_i$ we find

$$\begin{aligned} s\tau + \sigma t = & \sum_{i<j} (y_i x_j + y_j x_i)(\bar{\sigma}_i s_j + \bar{\sigma}_j s_i) \\ & + \sum (v_i' x_i + u_i y_i) s_i^2 \\ & + \sum_{i<j} (v_i' x_j + v_j' x_i + u_i y_j + u_j y_i) s_i s_j . \end{aligned}$$

Now let V be a vector space over k, with basis denoted by $e_i, e_{ij}, e_i', e_{ij}', 0 \leqq i < j \leqq n$ and let M' be the algebraic cone in V consisting of points

$$\begin{aligned} & \sum x_i^2 e_i + \sum_{i<j} (\varepsilon_X x_i x_j + y_i x_j + y_j x_i) e_{ij} \\ & \quad + \sum (u_i^2 + v_i x_i + u_i y_i) e_i' + \sum_{i<j} (v_i x_j + v_j x_i + u_i y_j + u_j y_i) e_{ij}' \end{aligned}$$

and in the same way let M'' be the algebraic cone in V consisting of points

$$\begin{aligned} & \sum_{i<j} (y_i x_j + y_j x_i) e_{ij} + \sum (v_i x_i + u_i y_i) e_i' \\ & \quad + \sum_{i<j} (v_i x_j + v_j x_i + u_i y_j + u_j y_i) e_{ij}' . \end{aligned}$$

The main fact is:

(i) $M = M' \cup M''$ is closed in V,

(ii) $\dim M = 4n+2$.

If $\psi: V \to \Gamma(L^{\otimes 2})$ is the homomorphism

$$\begin{aligned} \psi(e_i) &= \bar{\sigma}_i^2 + \varepsilon_X \bar{\sigma}_i s_i + b s_i^2 , \\ \psi(e_{ij}) &= \bar{\sigma}_i s_j + \bar{\sigma}_j s_i , \\ \psi(e_i') &= s_i^2 , \\ \psi(e_{ij}') &= s_i s_j \end{aligned}$$

then its kernel is a linear subspace of V of codimension

$$\operatorname{codim} \ker(\psi) \leqq \dim \Gamma(L^{\otimes 2}) = 4n+1$$

by Lemma 5, hence by (i) and (ii) there is a point $m \in M$, $m \neq 0$, with $\psi(m) = 0$. Thus we obtain a relation of the form $(A') = 0$ or $(A'') = 0$ and this relation is non-trivial since trivial relations correspond to the origin of the cone, while $m \neq 0$.

We claim that div(s) is reducible and is not the sum of two non-singular rational curves. Suppose for example that the relation is

$$\sigma^2 + \varepsilon_X \sigma s + b s^2 + s\tau + \sigma t = 0$$

and that div(s) is irreducible; note that s is not 0, because the relation is non-trivial. Hence

$$s(\tau + \varepsilon_X \sigma + b s) = \sigma(t + \sigma)$$

and, denoting restriction to U_i by a subscript, we find either

$$\sigma_i = s_i g_i , \quad g_i \in \Gamma(U_i, \mathcal{O}_X)$$

or

$$t_i + \sigma_i = s_i\, g_i, \quad g_i \in \Gamma(U_i, \mathcal{O}_X)$$

because we assume that div (s) is irreducible. This implies easily, using $\sigma_i - f_{ij}\,\sigma_j = s_i\, a_{ij}$, that either

$$\sigma_i = s_i\, g_i \quad \text{for every } U_i,$$

or

$$t_i + \sigma_i = s_i\, g_i \quad \text{for every } U_i$$

holds. Replacing the representative σ by $t+\sigma$ if needed, we obtain $\sigma_i = s_i\, g_i$ for every U_i. This implies that

$$a_{ij} = g_i - g_j \quad \text{on } U_i \cap U_j$$

with $g_i \in \Gamma(U_i, \mathcal{O}_X)$, hence $\{a_{ij}\}$ would be a 1-coboundary, which is a contradiction.

Now assume that div $(s) = E' + E''$ where E', E'' are non-singular rational curves on X. Let $\{f'_{ij}\}$, $\{f''_{ij}\}$ be 1-cocycles in $Z^1(\{U_i\}, \mathcal{O}_X^*)$, with $f_{ij} = f'_{ij} f''_{ij}$, representing the classes of $\mathcal{O}_X(E')$ and $\mathcal{O}_X(E'')$ in Pic (X) and let s', s'' be corresponding sections with $s = s'\, s''$. Reasoning as before, we can assume that

$$\sigma_i = s'_i\, g_i \quad \text{on } U_i$$

with $g_i \in \Gamma(U_i, \mathcal{O}_X)$, and we get that

$$s''_i\, a_{ij} = g_i - f''_{ij}\, g_j \quad \text{on } U_i \cap U_j,$$

hence

$$H^1(\mathcal{O}_X) \xrightarrow{s''} H^1(\mathcal{O}_X(E'')) \quad \text{is the zero map.}$$

This contradicts the exact sequence

$$H^0(\mathcal{O}_{E''}(E'')) \to H^1(\mathcal{O}_X) \xrightarrow{s''} H^1(\mathcal{O}_X(E'')),$$

because $E''.\,E'' = -2$, hence $H^0(\mathcal{O}_{E''}(E'')) = (0)$.

The same argument applies in case of a relation of type $(A'') = 0$. Q.E.D.

We now return to the proof of Theorem 3. Let $p = \frac{1}{2}(C^2) + 1$ be the smallest genus of irreducible curves C with $C^2 > 0$, and let $D \in |C|$ be a reducible element as in the previous Proposition. We conclude our proof in the following steps:

Step 1. If E is an irreducible component of D, then $p(E) < p(C)$ hence $p(E) = 0$ or 1. In fact, let $D = E + D'$. We have, since C is irreducible and $C^2 > 0$, $DD' = CD' \geqq 0$ hence $DE \leqq C^2$. Now $DE = E^2 + ED'$ and D_{red} is connected by the degeneration principle of Enriques-Zariski, hence either $ED' > 0$ or $E^2 < 0$, and in both cases $E^2 < C^2$. Q.E.D.

Step 2. We may assume $D = \sum m_i E_i$ where $\sum m_i \geqq 3$ and the E_i are non-singular rational curves with $E_i^2 = -2$ and $E_i E_j \leqq 2$. In fact, $p(E_i) = 1$ implies that E_i is of canonical type, hence X would be elliptic or quasi-elliptic. If instead $p(E_i)=0$ for all i, then $\sum m_i \geqq 3$ by the previous Proposition. Finally, $E_i E_j \leqq 2$, otherwise $(E_i + E_j)^2 \geqq 2$, and $\dim |E_i+E_j| \geqq 1$ by the Riemann-Roch Theorem. Now $|E_i+E_j| + (D-E_i-E_j) \subseteq |C|$, and a general element of $|E_i+E_j|$ is irreducible; if $F \in |E_i+E_j|$, then $F^2 < C^2$ by the argument in Step 1, which contradicts $p(C) = \text{minimum}$.

Q.E.D.

Final Step. X contains a curve of canonical type. If $E_i E_j = 2$, then $E_i + E_j$ is of canonical type, and we are done. So assume $E_i E_j \leqq 1$ for all i, j. Consider the connected graph with vertices E_i and edges connecting E_i, E_j if $E_i E_j = 1$. If the graph contains a subgraph which is a complete Dynkin diagram $\tilde{A}_n$, $\tilde{D}_n$, $\tilde{E}_6$, $\tilde{E}_7$, $\tilde{E}_8$ (e.g. $\tilde{A}_n$ is a loop) then we find a curve of canonical type: e.g., for $\tilde{E}_8$ the curve

$$\begin{array}{cccccccc} & & ③ & & & & & \\ & & | & & & & & \\ ② — & ④ — & ⑥ — & ⑤ — & ④ — & ③ — & ② — & ① \end{array}$$

where the numbers denote the multiplicities of the components. Otherwise, the graph must itself be a Dynkin diagram, and since the associated self-intersection quadratic form is negative definite we would obtain $D^2 < 0$, a contradiction.

Q.E.D.

From the computation of the canonical bundle of an elliptic or quasi-elliptic surface in Part II, we also get:

Proposition 11. *Let $f: X \to \mathbb{P}^1$ be an elliptic or quasi-elliptic fibration on an Enriques surface. Then we have:*

(i) *if X is classical, f has 2 ordinary double fibres and*

$$R^1 f_*(\mathcal{O}_X) \cong \mathcal{O}_{\mathbb{P}^1}(-1);$$

(ii) *if X is non-classical, f has exactly one wild double fibre $2P = f^{-1}(x)$ and*

$$R^1 f_*(\mathcal{O}_X) \cong \mathcal{O}_{\mathbb{P}^1}(-2) \oplus k_x$$

where k_x is a sheaf with support x and stalk k there.

Theorem 4. *If X is an Enriques' surface then its Picard number is $\rho = B_2 = 10$.*

Proof. If X is quasi-elliptic, this comes from the following more general result:

Proposition 12. *For all quasi-elliptic surfaces we have $\rho = B_2$.*

Proof of Proposition. By the results of Section 1, there is a smooth ruled surface Y and a proper map $\pi: Y \to X$ of degree $p = \text{char}(k)$. Now use

$$\begin{array}{ccc} \mathrm{Pic}(Y) \otimes \mathbb{Z}_l & \longrightarrow\!\!\!\!\!\rightarrow & H^2_{\text{ét}}(Y, \mathbb{Z}_l) \\ {\scriptstyle \pi^*}\uparrow\downarrow{\scriptstyle \pi_*} & & {\scriptstyle \pi^*}\uparrow\downarrow{\scriptstyle \pi_*} \\ \mathrm{Pic}(X) \otimes \mathbb{Z}_l & \longrightarrow & H^2_{\text{ét}}(X, \mathbb{Z}_l) \end{array}$$

where $l \neq p$. Since $\pi_* \pi^* = \text{mult. by } p$, we obtain that $\mathrm{Pic}(X) \otimes \mathbb{Z}_l \to H^2_{\text{ét}}(X, \mathbb{Z}_l)$ is onto. Q.E.D.

Now suppose that X is elliptic. We have:

Proposition 13. *Let $X \xrightarrow{f} C$ be an elliptic surface and let $J(X)$ be the associated Jacobian surface. Then if $\rho = B_2$ for $J(X)$ we have also $\rho = B_2$ for X.*

Proof of Proposition. We use an idea of M. Artin. From the Kummer sequence

$$1 \to \mu_n \to \mathbb{G}_m \xrightarrow{n} \mathbb{G}_m \to 1$$

we get the exact sequence

$$0 \to \operatorname{Pic}(X)^{(n)} \to H^2(X, \mu_n) \to H^2(X, \mathbb{G}_m)_n \to 0$$

where $H^2(X, \mathbb{G}_m)_n$ is the subgroup of $H^2(X, \mathbb{G}_m)$ killed by multiplication by n, and where $\operatorname{Pic}(X)^{(n)} = \operatorname{Pic}(X)/n \operatorname{Pic}(X)$. This gives, for each prime $l \neq p$, that

$$B_2(X) - \rho(X) = \operatorname{corank} H^2(X, \mathbb{G}_m)(l)$$

where the symbol (l) denotes the l-torsion part.

Now the Leray spectral sequence for $f: X \to C$ shows that, denoting by $\approx$ an homomorphism with finite kernel and cokernel, we have:

$$H^2(X, \mathbb{G}_m) \approx H^1(C, R^1 f_* \mathbb{G}_m);$$

also, if η denotes the generic point of C and if $i: \eta \to C$ is the inclusion, then if J is the Jacobian of the generic fibre we have:

$$H^1(C, R^1 f_* \mathbb{G}_m) \approx H^1(C, i_* J)$$

(cfr. [3], §4 and [9]). Since the last group is the same for both fibrations $X \to C$ and $J(X) \to C$ we obtain

$$B_2(X) - \rho(X) = B_2(J(X)) - \rho(J(X))$$

and Proposition 13 is proven. Q.E.D.

Proof of Theorem 4. We have to show that $\rho(J(X)) = B_2(J(X))$ where $J(X)$ is the Jacobian fibering of the elliptic Enriques' surface X. Now the elliptic surface $J(X)$ has the same Betti numbers as X and has a section, hence no multiple fibres. The canonical bundle formula proved in Part II of this work now shows that a canonical divisor on $J(X)$ is given by the opposite of a fibre, hence all plurigenera of $J(X)$ vanish and $J(X)$ is rational by Castelnuovo's criterion. Hence $B_2(J(X)) - \rho(J(X)) = 0$. Q.E.D.

§ 5. Two Examples

Let X be a supersingular Enriques surface. We have seen in §3 of this work that X has a regular 1-form η which is locally of the form $\eta = db_i$, and also X has a nowhere vanishing regular 2-form ω. Now the formula

$$df \wedge \eta = (\vartheta f)\, \omega$$

defines a regular vector field ϑ on X. Since we are in char. 2, we also have

$$\vartheta^2 = \delta \vartheta$$

for some constant δ; replacing η by a multiple, we may assume that

$$\vartheta^2=\delta_X\,\vartheta,$$

with $\delta_X=1$ or 0.

Proposition 14. *If $\delta_X=1$ then η has exactly 12 isolated simple zeros. If instead $\delta_X=0$ then all zeros of η have even multiplicity.*

Proof. Let u, v be local formal parameters at a point P of X. Now write $\omega=\varphi\,du\wedge dv$, $\eta=db=b_u\,du+b_v\,dv$. A simple computation now shows that

$$\delta_X=(\varphi^{-1})_u\,b_v+\varphi^{-1}\,b_{uv}+(\varphi^{-1})_v\,b_u$$

where the subscripts denote partial derivatives. Taking the partial derivatives with respect to u, v we see that $(\varphi^{-1})_{uv}\,b_u=(\varphi^{-1})_{uv}\,b_v=0$, hence φ_{uv} is identically 0. Thus we must have

$$\varphi=A^2+uB^2+vC^2$$

for some power series A, B, C with $A(0,0)\neq 0$. If we introduce the new local parameters

$$\tilde{u}=u(A^2+C^2 v),$$
$$\tilde{v}=v(1+(B/A)^2 u)$$

we easily check that

$$\omega=(A^2+uB^2+vC^2)\,du\wedge dv=d\tilde{u}\wedge d\tilde{v},$$

whence we have shown that changing the local parameters u, v by multiplication with suitable units, we have the normal form $\omega=du\wedge dv$ for ω. It then follows that $\delta_X=b_{uv}$ and now we can write

$$b=A^2+uB^2+vC^2+\delta_X\,uv$$

for suitable power series A, B, C. Hence

$$\eta=(\delta_X\,v+B^2)\,du+(\delta_X\,u+C^2)\,dv.$$

If η vanishes at P, then $B(0,0)=C(0,0)=0$; if $\delta_X=0$, then η vanishes at P to even order, while if $\delta_X=1$ then η has a simple isolated zero at P, since $v+B^2$ and $u+C^2$ generate the maximal ideal of the formal local ring of X at P. Noting that $c_2(X)=12$ we get the required result. Q.E.D.

Let $Y\to X$ be the α_2-covering associated to X defined in § 3; this is locally of type $z^2=b$, hence of type

$$z^2=\delta_X\,uv+uB^2+vC^2.$$

It follows from this that if $\delta_X=1$ then Y has exactly 12 ordinary rational double points, i.e. those over the zeros of the 1-form η.

Proposition 15. *Let X be a supersingular Enriques surface with $\delta_X=1$, and let $\tilde{Y}$ be a non-singular minimal model of the associated surface Y. Then $\tilde{Y}$ is a $K3$-surface and its Picard number is $\rho(\tilde{Y})=B_2(\tilde{Y})=22$, i.e. $\tilde{Y}$ is a supersingular $K3$-surface.*

Proof. We know already that Y is $K3$-like, that is $\chi(\mathcal{O}_Y)=2$ and $\omega_Y \cong \mathcal{O}_Y$. Since Y has only 12 rational double points, $\tilde{Y}$ is a $K3$-surface. Now Theorem 4, together with the method of proof of Proposition 12, shows that $\rho(\tilde{Y}) \geqq \rho(X)+12=22=B_2(\tilde{Y})$, since the resolution of the 12 singular points brings in 12 new independent curves on $\tilde{Y}$. Q.E.D.

We have computed two examples using the construction at the end of § 3. In the first example, X is obtained from the α_2-covering $Y \to X$, where Y is the complete intersection of the three quadrics

$$x_1^2+x_1 x_2+x_2^2+x_1 y_2+x_2 y_1+y_3^2=0,$$
$$x_2^2+x_2 x_3+x_3^2+x_2 y_3+x_3 y_2+y_1^2=0,$$
$$x_3^2+x_3 x_1+x_1^2+x_3 y_1+x_1 y_3+y_2^2=0.$$

The surface Y has exactly 12 rational double points, namely: the point

$(1, 1, 1, 1, 1, 1)$,

the 9 points

$(1, t, t, t, t^2, t^2)$,
$(t, 1, t, t^2, t, t^2)$,
$(t, t, 1, t^2, t^2, t)$

where $t^3+t^2+1=0$, and the 2 points

$(1, t, t^2, 0, 0, 0)$

where $t^2+t+1=0$.

The quotient surface X by the α_2-action $(x_i, y_i) \to (x_i, \varepsilon x_i+y_i)$ where $\varepsilon^2=0$, is a smooth supersingular Enriques surface with $\delta_X=1$.

In our second example, Y is the complete intersection of the three quadrics

$$x_1^2+\bar{x}_1 x_2+y_3^2+y_1 x_2+x_1 y_2=0,$$
$$x_2^2+x_2 x_3+y_1^2+y_2 x_3+x_2 y_3=0,$$
$$x_3^2+x_1 x_3+y_2^2+y_1 x_3+x_1 y_3=0.$$

The surface Y has exactly 6 isolated singular points, namely: the point

$(1, 1, 1, 0, 0, 0)$,

the 3 points

$(t^3, t, 1, t^3, t, 1)$

where $t^3+t^2+1=0$, and the 2 points

$(t^3, t^2, t, t^2, t, 1)$

where $t^2+t+1=0$.

The quotient surface X by the α_2-action $(x_i, y_i) \to (x_i, \varepsilon x_i+y_i)$ where $\varepsilon^2=0$, is a smooth supersingular Enriques surface with $\delta_X=0$. The regular 1-form η on X has now exactly 6 double zeros (cf. Proposition 14).

References

1. Artin, M.: PhD Thesis, Harvard University, 1960
2. Bombieri, E., Mumford, D.: Enriques' classification of surfaces in char. p, II. To appear
3. Grothendieck, A.: Le groupe de Brauer III: exemples et compléments. In: Dix exposés sur la cohomologie des schémas. Amsterdam: North-Holland 1968
4. Hartshorne, R.: Residues and duality. Lecture Notes in Math. **20**. Berlin-Heidelberg-New York: Springer 1966
5. Mumford, D.: Lectures on curves on algebraic surfaces. Annals of Math. Studies 59, Princeton: University Press 1966
6. Mumford, D.: Pathologies of modular algebraic geometry. Am. J. Math. **83**, 339–342 (1961)
7. Mumford, D.: Enriques' classification of surfaces in char. p, I. In: Global Analysis. Princeton: University Press 1969
8. Raynaud, M., Spécialisation du foncteur de Picard. Publ. I.H.E.S. **38**, 27–76 (1970)
9. Raynaud, M.: Caractéristique d'Euler-Poincaré d'un faisceau et cohomologie des varétés abéliennes (d'après Ogg-Shafarévitch et Grothendieck). Sém. Bourbaki 1964–65, n. 286
10. Serre, J.-P.: Exemples de variétés projectives en car. p non relevables en car. 0. Proc. Nat. Acad. Sci. USA **47**, 108–109 (1961)
11. Serré, J-P.: Sur la topologie des variétés algébriques en caractéristique p. In: Symp. Int. de Topologia Algebraica, Mexico 1958
12. Šafarevič, I. R., Averbuh, B. G., Văinberg, Ju. R., Žižčenko, A. B., Manin, Ju. I., Moĭšezon, B. G., Tjurina, G. N., Tjurin, A. N.: Algebraic surfaces. Moskva 1965. Am. Math. Soc. Translations 1967
13. Tate, J.: Genus change in purely inseparable extensions of function fields. Proc. Am. Math. Soc. **3**, 400–406 (1952)

Received March 1, 1976

E. Bombieri
Scuola Normale Superiore
Piazza dei Cavalieri
Pisa, Italia

D. Mumford
Harvard University
Department of Mathematics
1 Oxford Street
Cambridge, Mass.
USA

PATHOLOGIES OF MODULAR ALGEBRAIC SURFACES.*

By David Mumford.

The purpose of this note is to present two counterexamples to conjectures about the geometry of algebraic surfaces, which are, on the contrary, true in characteristic zero. Thus if F is a non-singular algebraic surface defined over an algebraically closed field k, one can form the vector spaces

$$H_{0,1} = H^1(F, \Omega^0),$$
$$H_{1,0} = H^0(F, \Omega^1),$$

where Ω^i is the sheaf of regular i-forms. One also has the vector space

$$A_0 = \text{cotangent space to the Albanese Variety } A \text{ of } F \text{ at the origin } 0.$$

Among these vector spaces various maps may be defined regardless of the ground field. First, the canonical map

$$\phi: F \to A$$

induces a homomorphism

$$\phi^*: A_0 \to H_{1,0},$$

which Igusa [4] has shown to be one-one. Secondly, if an embedding of F in projective space is fixed, and hence a canonical element h in $H^1(F, \Omega^1)$, then cup product and Serre duality induces a pairing

$$\mathfrak{H}: H_{0,1} \times H_{1,0} \to k.$$

(See Kodaira [6] and Serre [9]).

Now in the classical case, the usual constructive existence proof of the Albanese variety A shows immediately that ϕ^* is onto (see Weil [12]); while the famous topological results of Lefschetz on the embeddings of varieties in projective space (see Wallace [11], and Kodaira [6]), allow one to conclude that the pairing $\mathfrak{H}$ is non-degenerate. One deduces the "fundamental equalities" of Italian surface theory:

$$\dim A_0 = \dim H_{1,0} = \dim H_{0,1},$$

* Received January 4, 1961.

proven first by Poincaré by his method of normal functions in 1910, (Poincaré [8]). A corollary of ϕ^* being onto, combined with the easy proof (valid in all characteristics—see Koizumi [7]) that regular differentials on an Abelian variety are closed, shows that the differentials in $H_{1,0}$ are all closed. This has been generalized by Hodge, and has led ultimately to the result: analytic regular differentials on a Kähler manifold are closed (see Weil [12]).

The question is what extends to the modular, or non-zero characteristic case. Igusa showed, first of all, that ϕ^* is not always onto [5]. Serre showed next that the pairing $\mathfrak{A}$ is not always non-degenerate, since it can happen that $H_{1,0} = (0)$, while $H_{0,1} \neq (0)$, (see [10]). There remain the questions:

(a) Are all differentials in $H_{1,0}$ closed?

(b) Does the pairing $\mathfrak{A}$ have the property that if x in $H_{1,0}$ is such that $\mathfrak{A}(x,y) = 0$ for all y in $H_{0,1}$, then $x = 0$?

The answer to both is no.

I.

To answer (a), we prove (assuming the characteristic $\neq 0$):

THEOREM. *Let F be a non-singular algebraic surface, ω any simple differential on F (i.e. 1-form); then there exists a non-singular surface F^* and a regular map $\phi: F^* \to F$ which is separable and algebraic, such that $\phi^*(\omega)$ is regular on F^*.*

Proof. Note first that it is enough to prove:

(#) For all P in F, there exists an open U containing P, a surface F^* and a regular map $\phi: F^* \to F$ separable and algebraic such that $\phi^{-1}(U)$ is non-singular, and $\phi^*(\omega)$ is regular on $\phi^{-1}(U)$.

If this is proven, then we can find a finite set U_i^* of open sets covering F and of surfaces F_i^*, and maps $\phi_i: F_i^* \to F$ with the above properties. Then let F^* be some non-singular model of the function field of any compositum of $k(F_i^*)$ (compatible with the identification of the common subfields $k(F)$), that dominates the models F_i^* (such exists by Abhyankar [1] and Zariski [13]). Let ϕ be the map from F^* to F. Then for all P in F^*, say $\phi(P)$ in U_i. Then $\phi_i^*(\omega)$ is regular at $\phi_i^{-1}(\phi(P))$, hence $\phi^*(\omega)$ is regular at P.

But in fact it is enough to prove:

(##) Same as (#) except for differentials $\omega = A\,dx$, x a uniformizing parameter at P.

For if at P, $\omega = A\,dx + B\,dy$, then pick $F_1{}^*$ that does the trick for $A\,dx$ and then $F_2{}^*$ that does the trick for $B\,dy$, and take a suitable non-singularization of their join as in (#).

But now if $\omega = (A_0/A_1)\,dx$, A_0 and A_1 regular at P, and $A_1 = 0$ at P, consider the extension

$$Z^p + (A_1{}^p \cdot Z) + (x) = 0, \text{ where } p \text{ is the characteristic.}$$

It follows

$$dx = -A_1{}^p dZ.$$

Now in the normalization F^* of F in this field extension, there is a unique point P^* above P. Moreover, it is simple since if x, y are uniformizing parameters at P, then Z, y are uniformizing parameters at P^*. Finally,

$$\omega = (A_0/A_1)\,dx = A_0 A_1{}^{p-1}(dx/A_1{}^p) = -A_0 A_1{}^{p-1}\,dZ.$$

Hence ω is regular at P^*, hence in an open set about P^*. QED

COROLLARY. *There exist algebraic surfaces, non-singular and with a simple regular differential ω that is not closed.*

Proof. Take in the above theorem $F = P^2$, the projective plane, and $\omega = x\,dy$. Then the differential $\phi^*(\omega)$ on the covering F^* given by the theorem satisfies the corollary. One must merely note that $d\phi^*(\omega) = \phi^*(d\omega) \neq 0$ as the covering is separable. QED

II.

To answer (b), consider the famous Enriques surface E, which is, in any characteristic, the normalization of the sextic surface E_0 in P^3:

$$0 = x^2y^2z^2 + x^2y^2 + x^2z^2 + y^2z^2 + xyzf_2(x, y, z),$$

where f_2 is a general polynomial of second degree. Its normalization is non-singular and, regardless of characteristic, can be constructed as the join of the graphs of the following set of maps:

$$\phi_1\colon E_0 \to P^1, \text{ given by } xy/z,$$
$$\phi_2\colon E_0 \to P^1, \text{ given by } xz/y,$$
$$\phi_3\colon E_0 \to P^1, \text{ given by } yz/x,$$

as the reader may with some pains verify. It follows that the surface E in characteristic 2 is a specialization of E in characteristic 0, hence has the same p_a (see Hironaka [3]), which has long been known to be 0 (for a modern treatment, see Artin's thesis [2]). On the other hand, the classical theory of adjoint surfaces tells us that canonical divisors on E, as divisors on E_0, must be cut out by $(6-4)=2$nd order forms in P^3 passing through the double lines (see Zariski [14]); in this case, we have a tetrahedron of double lines and such are contained in no quadric. Hence $p_g=0$, hence $H^{0,1}=(0)$.

Now in the remarkable case of characteristic 2, it is also the case that $H^{1,0}\neq(0)$. In fact, let t be a coordinate on P^1; then:

$$\phi_1^*(dt)=\phi_2^*(dt)=\phi_3^*(dt)=d(xyz)$$

is immediately seen to be a regular differential on E.

HARVARD UNIVERSITY.

REFERENCES.

[1] S. Abhyankar, "Local uniformization of algebraic surfaces over ground fields of characteristic $p\neq 0$," *Annals of Mathematics*, vol. 63 (1956), p. 491.

[2] M. Artin, *On Enriques' Surfaces*, Doctoral thesis, Harvard, 1960.

[3] H. Hironaka, "A note on algebraic geometry over ground rings," *Illinois Journal of Mathematics*, vol. 2 (1958), p. 355.

[4] J. I. Igusa, "A fundamental inequality in the Theory of Picard varieties," *Proceedings of the National Academy of Sciences, USA*, vol. 41 (1955), p. 317.

[5] ———, "On some problems in abstract algebraic geometry," *ibid.*, vol. 41 (1955), p. 964.

[6] K. Kodaira and D. C. Spencer, "On a theorem of Lefschetz and the lemma of Enriques-Severi-Zariski," *Proceedings of the National Academy of Sciences, USA*, vol. 39 (1953), p. 1273.

[7] S. Koizumi, "On differential forms of the first kind on algebraic varieties," *Journal of the Mathematical Society of Japan*, vol. 1 (1949), p. 273.

[8] H. Poincaré, "Sur les Courbes Tracee sur les Surfaces Algebriques," *Annales de l'Ecole Normale Supérieure*, vol. 27 (1910), p. ——.

[9] J. P. Serre, "Faisceaux Algebriques Coherents," *Annals of Mathematics*, vol. 61 (1955), p. 197.

[10] ———, "Sur la topologie des varieties algebriques en caracteristique p," *Symposium on Algebraic Topology*, Mexico, 1956.

[11] A. Wallace, *Homology Theory on Algebraic Varieties*, Pergamon Press, 1958.

[12] A. Weil, *Varietes Kahleriennes*, Paris, 1958.

[13] O. Zariski, "Simplified proof for the resolution of singularities of an algebraic surface," *Annals of Mathematics*, vol. 43 (1942), p. 583.

[14] ———, *An Introduction to the Theory of Algebraic Surfaces*, Mimeographed Lecture Notes, Harvard, 1957-8.

FURTHER PATHOLOGIES IN ALGEBRAIC GEOMETRY.*[1]

By David Mumford.

The following note is not strictly a continuation of our previous note [1]. However, we wish to present two more examples of algebro-geometric phenomena which seem to us rather startling. The first relates to characteristic p behaviour, and the second relates to the hypothesis of the completeness of the characteristic linear system of a maximal algebraic family. We will use the same notations as in [1].

I.

The first example is an illustration of a general principle that might be said to be indicated by many of the pathologies of characteristic p:

A non-singular characteristic p variety is analogous to a general non-Kähler complex manifold; in particular, a projective embedding of such a variety is not as "strong" as a Kähler metric on a complex manifold; and the Hodge-Lefschetz-Dolbeault theorems on sheaf cohomology break down in every possible way.

In this case we wish to look at the two dimensional cohomology of an algebraic surface F, non-singular, and of any characteristic but 0. The surface we shall choose will (a) be specialization of a characteristic 0 surface F', and (b) will satisfy $q = h^{0,1} = h^{1,0}$. Consequentily the second Betti number B_2 is the same, whether defined (i) as that of F' in the topological sense, (ii) as $h^{2,0} + h^{1,1} + h^{0,2}$, or (iii) following Igusa [2], as $\mathrm{Deg}(c_2) + 4q - 2$. Let ρ be the base number of F. Igusa showed that, in fact, $B_2 \geqq \rho$. However, in characteristic 0, one has the stronger result, $B_2 = h^{2,0} + h^{1,1} + h^{0,2} \geqq 2p_g + \rho$ (where $p_g = h^{2,0} = h^{0,2}$ is the geometric genus of F) as a result of the Hodge-Dolbeault theorems. Therefore the question arises whether this stronger inequality is valid in characteristic p. The answer is no.

A rather complicated example was discovered in 1961 by J. Tate and A. Ogg. Here is a very simple example: let E be a super-singular elliptic

* Received July 30, 1962.

[1] This work was supported by the Army Research Office (Durham).

curve of characteristic p (i.e. such that the rank of $End(E)$ is 4). Let $F = E \times E$. In this case, in fact:

$$\rho = B_2 = 6; \quad p_g = 1.$$

Here $p_g = 1$ since the sheaf $\Omega_F{}^2 \cong \mathcal{O}_F$; and $B_2 = 6$ by Igusa's definition, for example, since $\mathrm{Deg}(c_2) = 0$, and $q = 2$. Finally $\rho = 6$ since in general, for any two elliptic curves E_1 and E_2, one knows that the base number ρ for $E_1 \times E_2$ equals 2 plus the rank of $\mathrm{Hom}(E_1, E_2)$.

There remains one outstanding conjecture still neither proven nor disproven in characteristic p, which according to the general principle mentioned above ought to be false. This is the Regularity of the Adjoint, which may be stated as follows: if V is a non-singular projective surface and if H is a non-singular hyperplane section, then

$$H^1(\mathcal{O}_V) \to H^1(\mathcal{O}_H)$$

is injective.

II.

The second example concerns space curves in characteristic 0. Let A be any family of non-singular space curves, and let $a \in A$ represent the curve $\gamma \subset P_3$. Let T_a denote the Zariski tangent space to A at a, and let N denote the sheaf of sections of the normal bundle to γ in P_3. Then it is well-known [3] that there is a "characteristic" map:

$$T_a \to H^0(N).$$

The problem of completeness consists in asking when, for given γ, there is a family A containing γ such that the characteristic map is surjective. Kodaira [3] has shown that such a family exists if $H^1(N) = (0)$. Our example shows that if $H^1(N) \neq (0)$, then there need not be such a family.

In fact, in our example, this incompleteness holds for every curve in an open set of the corresponding Chow variety. Consequently, it is also an example where the Hilbert scheme [4] has a multiple component, i.e. is not reduced at one of its generic points. Another corollary of this example is obtained by blowing up such a space curve $\gamma \subset P_3$ to a surface E in a new three-dimensional variety V_3. Then Kodaira [5] has shown essentially that the local moduli scheme of the variety V_3 is isomorphic to the germ of the

Hilbert scheme of P_3 at the point corresponding to γ. Therefore we have constructed a non-singular projective three-dimensional variety whose local moduli scheme is nowhere reduced; in other words, any small deformation of V_3 is a variety the number of whose moduli is less than the dimension of $H^1(\Theta)$ (where Θ is the sheaf of vector fields).

The curves γ that we have in mind have degree 14 and genus 24. In the following, h will stand for the divisor class on γ induced by plane sections, and K_γ will stand for the canonical divisor on γ; also F will stand for a cubic or quartic surface in P_3, and H will stand for the (Cartier) divisor class on F induced by plane sections. The first step is partial classification of all space curves of this degree and genus, which confirms the results of M. Noether's well-known table [6].

(A) *Any non-singular space curve γ of degree 14 and genus 24 is contained in a pencil P of quartic surfaces.*

Proof. Since $\mathrm{Deg}(4h) = 56$, and $\mathrm{Deg}(K_\gamma) = 46$, the linear system $|4h|_\gamma$ is non-special,[2] and has dimension $56 - 24 = 32$. Since there is a 34-dimensional family of quartics in P_3, (A) follows.

There are 2 cases: (a) the pencil has no fixed components, and (b) the pencil has fixed components. In case (a), note first that if F' and F'' span P, then $F' \cdot F'' = \gamma + c$, where c is a conic. Now c has at most double points, hence $\gamma + c$ has at most triple points. Therefore no point x is a double point for both F' and F''. Noting that both F' and F'' are non-singular and transversal along $\gamma - c$, hence at all but a finite number of points of γ, it follows that almost every $F \in P$ is non-singular everywhere along γ.

(B) *Every algebraic family of space curves of type* (a) *has dimension less than or equal* 56.

Proof. It is enough to show that every family of pairs (γ, F) consisting of such curves γ, and quartics $F \supset \gamma$, F being non-singular along γ, has dimension at most 57. Now since all such quartics contains conics, they are not generic [7], and there is at most a $34 - 1 = 33$ dimensional family of quartics F involved in such a family of pairs. Moreover, the dimension of the set of all γ on one such F can be computed from the Riemann-Roch theorem on F:

---

[2] Here and below, $|D.|_V$ always means the linear system on V in which the Cartier divisor D varies. Also, $(D^2)_V$ always denotes the self-intersection of D, as a divisor class on D (assuming D effective).

$$\dim|\gamma|_F = \frac{\mathrm{Deg}(\gamma^2)_F}{2} + 1 + \{\dim H^1(\mathcal{O}_F(\gamma)) - \dim H^2(\mathcal{O}_F(\gamma))\}.$$

But $(\gamma^2)_F \equiv K_\gamma$ on γ, hence $\mathrm{Deg}(\gamma^2)_F = 46$. Moreover, $H^i(\mathcal{O}_F(\gamma))$ is dual to $H^{2-i}(\mathcal{O}_F(-\gamma))$ by Serre duality. This cohomology group can be computed from the exact sequence:

$$0 \to \mathcal{O}_F(-\gamma) \to \mathcal{O}_F \to \mathcal{O}_\gamma \to 0.$$

It follows that both are zero, hence $\dim|\gamma|_F = 24$. Therefore, indeed, the set of pairs (γ, F) has dimension at most $33 + 24 = 57$.[3]

Now consider case (b). Such a γ must be contained in a reducible quartic, hence in a plane, a quadric, or a cubic surface. The first two possibilities are readily checked and it happens that they contain no curves of the required degree and genus. Moreover, such a curve is contained in a *unique* cubic surface F, because $\mathrm{Deg}(\gamma) = 14 > 9 = \mathrm{Deg}(F' \cdot F'')$, for two distinct cubic surfaces F' and F''. We will say that γ is of type (b_0) if the cubic F is non-singular; otherwise, we will say that γ is of type (b_1).

(C) *Every maximal algebraic family of curves γ of type (b_0) has dimension* 56.

Proof. Let γ be a curve of type (b_0), and let F be the corresponding cubic surface. Since $K_F \equiv -H$, by the Riemann-Roch theorem on F:

$$\dim|\gamma|_F = \frac{\mathrm{Deg}(\gamma\cdot\gamma + \mathrm{H})_F}{2} + \{\dim H^1(\mathcal{O}_F(\gamma)) - \dim H^2(\mathcal{O}_F(\gamma))\}.$$

But $K_\gamma \equiv \gamma\cdot(\gamma + K_F)$, hence $46 = \mathrm{Deg}(\gamma^2)_F - \mathrm{Deg}(\gamma\cdot H)_F = \mathrm{Deg}(\gamma^2)_F - 14$, hence $\mathrm{Deg}(\gamma^2)_F = 60$. Also, $H^i(\mathcal{O}_F(\gamma))$ is dual to $H^{2-i}(\mathcal{O}_F(-H-\gamma))$, and this group can be computed from the exact sequence:

$$0 \to \mathcal{O}_F(-H-\gamma) \to \mathcal{O}_F \to \mathcal{O}_{(H+\gamma)} \to 0.$$

Since $H + \gamma$ is a reduced and connected curve, $H^0(\mathcal{O}_{H+\gamma}) = k$ (constants), and this implies $H^i(\mathcal{O}_F(-H-\gamma)) = (0)$ for $i = 1$ and 2. Putting all this together, we see that $\dim|\gamma|_F = 37$. Since there is a 19-dimensional family

[3] It may be objected that we have used the Riemann-Roch theorem, and Serre duality as though F were non-singular. But since F is non-singular along γ, the former can be proved by means of the exact sequence:

$$0 \to \mathcal{O}_F \to \mathcal{O}_F(\gamma) \to \mathcal{O}_\gamma((\gamma^2)_F) \to 0.$$

And the latter can be proven either (a) directly by resolving the singularities of F and comparing the cohomology on F and on its resolution, or (b) as a consequence of Grothendieck's general theory [8]. In the second case, one merely has to note that F is always a Cohen-Macauley variety; and since it is a *quartic* surface the canonical sheaf is simply $\mathcal{O}_F$ itself.

of cubic surfaces, (C) follows if we show that a generic γ in a maximal algebraic family is contained in a generic cubic surface. But let $\gamma \subset F$ be any curve of the family. Then recalling that the divisor class group of any non-singular cubic surface is the same as that of any other, it follows that if the set of all non-singular cubic surfaces are suitably parametrized the invertible sheaf $\mathcal{O}_F(\gamma)$ will be a specialization of an invertible sheaf L defined on the generic cubic surface F^*. And since $H^i(\mathcal{O}_F(\gamma)) = (0)$ for $i = 1$ and 2, by the upper semi-continuity of cohomology [9], we conclude that $H^i(L) = (0)$ for $i = 1$ and 2, and that all sections of $\mathcal{O}_F(\gamma)$ are specializations of sections of L. Therefore $\dim H^0(L) = 38$; and since almost all sections of $\mathcal{O}_F(\gamma)$ are non-singular, so are almost sections of L. Hence there is a non-singular $\gamma^* \subset F^*$ specializing to $\gamma \subset F$. QED.

Now suppose C is the Chow variety of non-singular curves of degree 14, and genus 24. Let $C_b \subset C$ be the locus of curves of type (b), and let $C_{b_1} \subset C_b$ be the locus of curves of type (b_1). Then it is clear that C_b and C_{b_1} are closed (possibly reducible) subvarieties of C. By (B) and (C), every component of $C - C_b$ has dimension $\leqq 56$, and every component of $C_b - C_{b_1}$ has dimension $= 56$. Therefore if C_0 equals C minus C_{b_1} and minus the closure of $C - C_b$, C_0 is an open set in the Chow variety, of dimension 56, and parametrizing *almost all* curves of type (b_0).

We shall now single out a set of components of C_0 such that, if N is the normal sheaf to a γ in one of these components, then $\dim H^0(N) = 57$. In fact, we say that $\gamma \subset F$ is of type (b'_0) if there is a line E on F such that $\gamma \equiv 4H + 2E$ on F. Then the corresponding $C'_0 \subset C_0$ which is the locus of such curves is clearly closed in C_0. But it is also open: if $\gamma^* \subset F^*$ specializes to $\gamma \subset F$, and if $\gamma \equiv 4H + 2E$ on F, then first of all, there is a line $E^* \subset F^*$ (possibly only rationally defined after a suitable base extension) which specializes to E; and secondly, since the divisor class group is discrete and constant for all non-singular cubics,

$$\gamma - 4H - 2E \equiv 0 \text{ implies } \gamma^* - 4H^* - 2E^* \equiv 0.$$

Therefore γ^* is of type (b'_0).

(D) *If* $\gamma \subset F$ *is of type* (b'_0), *then* $\dim H^0(N) = 57$.

Proof. Let N_F be the sheaf of normal vector fields to γ and *in* F, and let N_P be the sheaf of normal vector fields to F, and in P_3, which are defined along γ. Then we have the sequence:

$$0 \to N_F \to N \to N_P \to 0.$$

But if D is a non-singular divisor on a non-singular variety V, then its normal sheaf is isomorphic to $\mathcal{O}_D((D^2)_V)$. Therefore $N_F \cong \mathcal{O}_\gamma((\gamma^2)_F)$ and $N_F \cong \mathcal{O}_\gamma(3h)$. But since $K_\gamma \equiv (\gamma^2)_F + \gamma \cdot K_F \equiv (\gamma^2)_F - h$, it follows that $(\gamma^2)_F$ is a non-special divisor, of degree 60 in fact. Therefore $H^1(N_F) = (0)$ and $\dim H^0(N_F) = 60 - (24-1) = 37$. On the other hand, by the Riemann-Roch theorem for curves,

$$\begin{aligned}\dim H^0(\mathcal{O}_\gamma(3h)) &= 42 - (24-1) + \dim H^0(\mathcal{O}_\gamma(K_\gamma - 3h)) \\ &= 19 + \dim H^0(\mathcal{O}_\gamma((\gamma^2)_F - 4h)) \\ &= 19 + \dim H^0(\mathcal{O}_\gamma(2\gamma \cdot E))\end{aligned}$$

(using the hypothesis $\gamma \equiv 4H + 2E$). But now, use the exact sequence:

$$0 \to \mathcal{O}_F(-4H) \to \mathcal{O}_F(2E) \to \mathcal{O}_\gamma(2\gamma \cdot E) \to 0.$$

It is readily seen that $H^i(\mathcal{O}_F(-4H)) = (0)$ for $i = 0$ and 1, and that $\dim H^0(\mathcal{O}_F(2E)) = 1$. Putting all this information together we conclude: $\dim H^0(N) = 37 + 19 + 1 = 57$. QED.

It remains only to note:

(E) *If F is any non-singular cubic surface, and $E \subset F$ is any line, there exist non-singular curves $\gamma \in |4H + 2E|$, and they have degree* 14 *and genus* 24.

Proof. The degree and genus of such a γ are computed by the usual formulae, recalling that $\mathrm{Deg}(E^2)_F = -1$. To see that such a γ exists, it suffices, by the characteristic 0 Bertini theorem, to prove that $|4H + 2E|$ has no base points. But the only possible base points are the points of E, and we use the exact sequence:

$$0 \to \mathcal{O}_F(4H + E) \to \mathcal{O}_F(4H + 2E) \to \mathcal{O}_E(2) \to 0.$$

Since the sections of $\mathcal{O}_E(2)$ have no base points, it suffices to prove $H^1(\mathcal{O}_F(4H + E)) = (0)$. But this follows from the sequence:

$$0 \to \mathcal{O}_F(4H) \to \mathcal{O}_F(4H + E) \to \mathcal{O}_E(3) \to 0,$$

since $H^1(\mathcal{O}_F(4H)) = (0)$, and $H^1(\mathcal{O}_E(3)) = (0)$. QED.

HARVARD UNIVERSITY.

REFERENCES.

[1] D. Mumford, "Pathologies of modular algebraic surfaces," *American Journal of Mathematics*, vol. 83 (1961), p. 339.

[2] J. I. Igusa, "Betti and Picard numbers of abstract algebraic surfaces," *Proceedings of the National Academy of Sciences*, vol. 46 (1960), p. 724.

[3] K. Kodaira, "A theorem of completeness of characteristic systems for analytic families of compact submanifolds of complex manifolds," *Annals of Mathematics*, vol. 72 (1962), p. 146.

[4] A. Grothendieck, "Techniques de construction et théorèmes d'existence en géométrie algébrique, IV: Les schémas de Hilbert," *Séminaire Bourbaki*, Exposé 221, 1961.

[5] K. Kodaira, "On stability of compact submanifolds of complex manifolds," to appear in *American Journal of Mathematics*.

[6] M. Noether, *Zur Grundlegung der Theorie der Algebraischen Raumkurven*, Steiner Preisschrift, Berlin, 1882.

[7] S. Lefschetz, "On certain numerical invariants of algebraic varieties with applications to Abelian varieties," (Memoire Bordin), *Transactions of the American Mathematical Society*, vol. 22 (1921), p. 327.

[8] A. Grothendieck, "Théorèmes de dualité pour les faisceaux algebriques cohérents," *Séminaire Bourbaki*, Exposé 149, 1957.

[9] W. L. Chow and J. I. Igusa, "Cohomology theory of varieties over rings," *Proceedings of the National Academy of Sciences*, vol. 44 (1958), p. 1244.

PATHOLOGIES III.*

By D. Mumford.*

This note continues, more or less, our two previous papers [2] and [3] in which we have been presenting unpleasant facts of (algebro-geometric) life. The specific topic in this note is Kodaira's vanishing theorem, or as it is called classically, the regularity of the adjoint linear system. Except for one slight fudge (our example is a normal surface, not a non-singular one) our result is that K. V. Theorem is *false* in characteristic p. In fact, this example is presented in § 2. In § 1, we give an outline of the true result in characteristic 0, and in particular extend the theorem to singular varieties in an attempt to clarify the role played by non-singularity.

The classical form of Kodaira's Vanishing Theorem (in the surface case) may be found in Zariski's book [7], p. 144, where, in essence, it is stated:

(*) If F is a non-singular projective surface, in characteristic 0, K the canonical divisor class of F, and H a hyperplane section of F, then the linear system $|K+H|$ is regular (Picard, 1906). More generally, if H is an irreducible curve which is part of an algebraic family of curves other than an irrational pencil, then $|K+H|$ is regular (Severi, 1908).

In terms of cohomology, $|K+H|$ regular means that

$$H^1(O_F(K+H)) = H^2(O_F(K+H)) = (0).$$

Or, by Serre duality, that

$$H^0(O_F(-H)) = H^1(O_F(-H)) = (0).$$

This is the form generalized by Kodaira [1]. He showed:

(**) If V is a non-singular projective variety of dimension n, characteristic 0, and H is an ample[1] divisor class on V, then

$$H^i(O_V(-H)) = (0), \qquad i = 0, 1, \cdots, n-1.$$

1. The fact that $H^0(O_V(-H)) = (0)$ is absolutely trivial in all characteristics. Thus the reader can check:

Received February 9, 1966.

* This paper partially supported by NSF Grant GP-3512.

[1] Ample in Grothendieck's sense, of course.

PROPOSITION 1. *If V is any complete variety, and L is any invertible sheaf on V such that $\Gamma(L^n) \neq 0$ for some $n \geqq 1$, then $\Gamma(L^{-1}) = (0)$ unless $L \cong O_V$.*

On the other hand, the vanishing of H^1 already is very subtle. For H^1's I claim the following which generalizes most of Severi's assertion too:

THEOREM 2. *Let V be a complete normal variety of dimension at least 2, characteristic 0. Let L be an invertible sheaf on V such that, for large n, L^n is spanned by its sections. Let these sections define the morphism*

$$V \xrightarrow{\ \phi\ } W$$

(Recall that if $n \gg 0$, then $\phi_*(O_V) = O_W$, W is normal and the fibres of ϕ are connected). Then

$$\begin{array}{l} H^1(L^{-m}) = (0) \iff \dim W > 1. \\ \text{for all } m \geqq 1. \end{array}$$

Proof. The implication $\Rightarrow$ is nearly obvious. By definition of ϕ, there is a positive integer m and a very ample invertible sheaf M on W such that $L^m \cong \phi^*(M)$. Use Leray's spectral sequence:

$$H^p(W, R^q\phi_*(L^{-m})) \Rightarrow H^*(V, L^{-m}).$$

By definition of W, $\phi_*(O_V) = O_W$; hence $\phi_*(L^{-2m}) = M^{-2}$. The spectral sequence gives:

$$0 \to H^1(W, M^{-2}) \to H^1(V, L^{-2m}) \to \cdots.$$

But if W is a curve, and M has positive degree, then $H^1(W, M^{-2}) \neq (0)$, hence $H^1(V, L^{-2m}) \neq (0)$.

To prove the implication $\Leftarrow$ we first reduce to the case where V is non-singular and projective. To do this, introduce a projective de-singularization

$$\pi: \bar{V} \to V.$$

Let $\bar{L} = \pi^* L$. Since V is normal, $\pi_*(O_{\bar{V}}) = O_V$, hence $\pi_*(\bar{L}^{-m}) = L^{-m}$. Leray's spectral sequence again gives:

$$0 \to H^1(V, L^{-m}) \to H^1(\bar{V}, \bar{L}^{-m}) \to \cdots.$$

Therefore, $H^1(\bar{V}, \bar{L}^{-m}) = (0)$ implies $H^1(V, L^{-m}) = (0)$.

Now assume V is non-singular and projective. We can, of course, assume that the ground field is the field $\boldsymbol{C}$ of complex numbers. As above, let M be a very ample invertible sheaf on W such that $\phi^*(M) \cong L^m$, some positive m. Let $i: W \to \boldsymbol{P}_n$ be an immersion such that M is the restriction of $O(1)$ to W. Let $\boldsymbol{L}$, $\boldsymbol{M}$, and $\boldsymbol{O}_1$ be the line bundles on V, W and $\boldsymbol{P}_n$ corresponding to L, M and $O(1)$. Next, equip $\boldsymbol{O}_1$ with its standard Hermitian structure: by pull-back, this puts a Hermitian structure on $\boldsymbol{L}^m$ and $\boldsymbol{M}$, and by taking m-th roots, this puts a Hermitian structure on $\boldsymbol{L}$. Let Ω_0 be the curvature form of $\boldsymbol{O}_1$: this is well known to be *positive definite*. The curvature form of $\boldsymbol{L}^m$ is then $(i \circ \phi)^*\Omega_0$ and the curvature form Ω of $\boldsymbol{L}$ is just

$$\Omega = \frac{1}{m}(i \circ \phi)^*\Omega_0,$$

which is, therefore, *positive semi-definite.*

Now recall the fundamental inequality of Kodaira's paper. Choose a Kähler metric on V, given in local coordinates by

$$ds^2 = \sum_{\alpha,\beta=1}^{g} g_{\alpha\beta}\, dz^\alpha d\bar{z}^\beta.$$

Choose local trivializations of $\boldsymbol{L}$ and assume that the hermitian structure on $\boldsymbol{L}$ is given by

$$\| s \|^2 = a_j^{-1} \cdot | f_j |^2$$

if a section s of $\boldsymbol{L}$ is defined by the function f_j in terms of our local trivialization. By definition

$$\Omega = \frac{i}{2\pi} \bar{\partial}\partial (\log a_j) \underset{\text{locally}}{=} \frac{i}{2\pi} \sum_{\alpha,\beta} X_{\alpha\beta}\, dz^\alpha d\bar{z}^\beta$$

Now assume that an element $\phi \in H^{n-1}(V, \Omega^n \otimes L)$ is given by a harmonic $\boldsymbol{L}$-valued $(n, n-1)$-form Φ, where

$$\Phi \underset{\text{locally}}{=} \sum_{\alpha_i\text{'s}} u_{12\cdots n\alpha_1\cdots\alpha_{n-1}}\, dz^1 \wedge \cdots \wedge dz^n \wedge d\bar{z}^{\alpha_1} \wedge \cdots \wedge d\bar{z}^{\alpha_{n-1}}.$$

Then Kodaira proves that:

$$0 \geqq \int_V \frac{1}{a_j} \{ \sum_{\substack{\alpha_1,\cdots,\alpha_{n-1} \\ \gamma_1,\cdots,\gamma_{n-1} \\ \beta,\epsilon}} X_{\epsilon\alpha_1} u_{12\cdots n\beta\alpha_2\cdots\alpha_{n-1}} \cdot \bar{u}_{12\cdots n\gamma_1\cdots\gamma_{n-1}}\, g^{\epsilon\beta} \cdot g^{\gamma_1\alpha_1} \cdot \cdots \cdot g^{\gamma_{n-1}\alpha_{n-1}} \} dV.$$

(The reader may check that this integral is intrinsic.) This is easier to see in terms of the dual harmonic form

$$\Phi^\dagger \underset{\text{locally}}{=} \frac{1}{a_j} *\Phi$$

which is an L^{-1}-valued $(0,1)$-form. Express $\Phi^\dagger$:

$$\Phi^\dagger \underset{\text{locally}}{=} \sum_\alpha v_\alpha\, d\bar{z}^\alpha.$$

Then one checks that

$$v_\alpha = \frac{1}{a_j \cdot g} \sum_\epsilon \bar{u}_{12\cdots n\, 12\cdots\hat{\epsilon}\cdots n} g_{\epsilon\alpha}$$

and, after some sweat, one also can rearrange Kodaira's integral into:

$$0 \geqq \int_V g \cdot a_j \{ \sum_{\gamma,\beta} X_{\gamma\beta} g^{\gamma\beta} \cdot \sum_{\alpha,\epsilon} v_\alpha \bar{v}_\epsilon g^{\epsilon\alpha}$$
$$- \sum_{\gamma,\epsilon} X_{\gamma\epsilon} \cdot [\sum_\beta g^{\gamma\beta} v_\beta] \cdot [\sum_\alpha g^{\alpha\epsilon} \bar{v}_\alpha] \} dV.$$

To see exactly what we have here, examine the integrand at one point P of V. Choose local coordinates so that the Kähler metric is given by

$$g_{\alpha\beta}(P) = \delta_{\alpha\beta},$$

and the curvature form of L is diagonalized:

$$X_{\alpha\beta}(P) = \lambda_\alpha \cdot \delta_{\alpha\beta}.$$

Since X is positive semi-definite, $\lambda_\alpha \geqq 0$ for all α. The integrand is then:

(i) $$\sum_{\alpha \neq \beta} \lambda_\alpha \cdot |v_\beta|^2.$$

Since this is non-negative, we conclude that the integral is non-positive only if the integrand is identically 0. If $\Phi \not\equiv 0$, moreover, then $\Phi(P) \neq 0$ at a dense set of points P. From the form of (i), we conclude that at most *one* λ_α is not zero at all such points P; hence, in fact, at most one λ_α is not zero at all points. In terms of ϕ, this means that $\dim W = 1$ (i.e. use of Sard's theorem and the positive definiteness of Ω_0). Q.E.D.

Actually a completely algebro-geometric proof of this result can be given. Because this proof is so simple and because it indicates the reason why the theorem will fail in characteristic p, it seems worth giving:

Second Proof. As above, we need only prove "$\Leftarrow$." Therefore assume $\dim W > 1$. In this proof, we need to make a preliminary reduction not to the case where V is non-singular, but to the case where $L = O_V(D)$, D a reduced effective Cartier divisor on V. But by our hypothesis on L, it is clear that for large n, $L^n = O_V(D_n)$ for some reduced Cartier divisor D_n on V. Pick an affine open covering $\{U_i\}$ of V such that L is defined by the co-cycle $\{a_{ij}\}$,

7

$$a_{ij} \in \Gamma(U_i \cap U_j, O_V)$$

and D_n is defined by local equations f_i in U_i:

$$f_i \in \Gamma(U_i, O_V)$$
$$f_i = a_{ij}^n \cdot f_j.$$

Define a n-fold cyclic covering:

$$\pi : \tilde{V} \to V$$

by local equations:

$$z_i^n = f_i$$
$$z_i = a_{ij} \cdot z_j.$$

Then if $\tilde{L} = \pi^*(L)$, the equations $z_i = 0$ define a Cartier divisor $\tilde{D}$ on $\tilde{V}$ such that

$$\tilde{L} = O_{\tilde{V}}(\tilde{D}).$$

Moreover, suppose we prove that $H^1(\tilde{V}, \tilde{L}^{-1}) = (0)$. Then

$$\begin{aligned} H^1(V, \pi_*(O_{\tilde{V}}) \otimes L^{-1}) &\cong H^1(V, \pi_*(\tilde{L}^{-1})) \\ &\cong H^1(\tilde{V}, \tilde{L}^{-1}) \\ &\cong (0). \end{aligned}$$

But O_V is a direct summand of the coherent sheaf $\pi_*(O_{\tilde{V}})$ of O_V-modules:

$$O_V \underset{\beta}{\overset{\alpha}{\rightleftarrows}} \pi_*(O_{\tilde{V}})$$

where α is the canonical map π^* taking functions on V to functions on $\tilde{V}$, and $\beta = \frac{1}{n}(\text{Trace})$. Therefore $H^1(V, L^{-1}) = (0)$ also.

Now assume $L = O_V(D)$. The next step is to show that D is connected: Let H be any hyperplane section of W in the embedding defined by M. Then, since $\dim W > 1$, H is connected, hence $\phi^{-1}(H)$ is connected. But since the morphism $i \circ \phi : V \to \boldsymbol{P}_n$ is defined by the complete linear $\Gamma(L^m)$, the divisor mD equals the divisor $\phi^*(H)$ for some hyperplane section H of W. Therefore D is connected.

Now consider the exact sequence:

$$0 \to L^{-1} \to O_V \to O_D \to 0.$$

This gives:

$$H^0(O_V) \to H^0(O_D) \to H^1(L^{-1}) \to H^1(O_V) \to H^1(O_D).$$

Since D is reduced and connected, $H^0(O_D)$ consists only in constant functions, all of which lift to $H^0(O_V)$. Therefore,

$$H^1(L^{-1}) \cong \mathrm{Ker}\{H^1(O_V) \to H^1(O_D)\}.$$

Let $\mathrm{Pic}^0(V)$, $\mathrm{Pic}^0(D)$ be the connected components of the origin of the Picard schemes of V and D. Recall that $H^1(O_V)$, $H^1(O_D)$ are canonically isomorphic to the Zariski-tangent spaces to $\mathrm{Pic}^0(V)$ and $\mathrm{Pic}^0(D)$ at their origins, so that the map from $H^1(O_V)$ to $H^1(O_D)$ is just the differential of the canonical homomorphism:

$$\mathrm{Pic}^0(V) \xrightarrow{\ \alpha\ } \mathrm{Pic}^0(D)$$

(cf. [4], Lecture 24). Since the characteristic is 0, the kernel of α, as a subgroup*scheme* of $\mathrm{Pic}^0(V)$, must be reduced (cf. [4], Lecture 25). Therefore, if the differential of α has a non-trivial kernel, α itself must have a positive-dimensional kernel. Therefore, every non-trivial subgroupscheme has non-trivial points of finite order on it. Therefore, we conclude:

$$H^1(L^{-1}) \neq (0) \Rightarrow \left(\begin{array}{c}\exists\, \delta \in \mathrm{Pic}^0(V) \text{ of finite order } n,\ n > 1, \\ \text{such that } \alpha(\delta) = 0.\end{array}\right)$$

Now δ defines, in the usual way, an unramified Galois covering:

$$\begin{array}{c} V' \\ \pi \downarrow \\ V \end{array}$$

with covering group $\mathbf{Z}/n\mathbf{Z}$. $\alpha(\delta) = 0$ implies that this covering splits over D, i.e., $\pi^{-1}(D) = D_1 \cup \cdots \cup D_n$ (the D_i's disjoint and isomorphic to D). But let $L' = \pi^*L$. Then it is clear that the pair (V', L') satisfy all the requirements imposed on (V, L). Therefore the identical argument used to prove that D is conneced also proves that $D' = \pi^{-1}(D)$ is connected. This is a contradiction. Q. E. D.

Note that the only place where char $= 0$ has been used is in the step where we used the fact that $\ker(\alpha)$ must be reduced. Incidentally, this proof also can be generalized to further classes of invertible sheaves $L = O_V(D)$, which do not necessarily have the property that L^m is spanned by its sections for large m. But I don't know of any really definitive statement in this direction. One further point: it is not clear whether or not the normality of V is essential in Theorem 2. By Grothendieck's duality theorem, it follows that V must at least have the "property S2": $\forall x \in V$, depth $(O_x) \geqq 2$ (x a closed point). But I don't know whether or not singularities in codimension 1 can be allowed.

2. Now consider the question of whether or not

$$H^1(V, O_V(-D)) = (0)$$

when V is a normal variety in characteristic p, and D is an ample effective divisor. As in characteristic 0, we can analyze this via the exact sequence:

$$H^0(O_V) \to H^0(O_D) \to H^1(O_V(-D)) \to H^1(O_V) \to H^1(O_D).$$

The contribution from the left is easy to dispose of:

Proposition 3. *If V is a complete normal variety,* $\dim V \geqq 2$, *and D is an effective ample Cartier divisor on V, then $H^0(O_D)$ consists only in constants.*

Proof. In characteristic 0, this follows from Theorem 2. Now assume that the characteristic is $p > 0$. Let D be defined by local equations $f_i = 0$ with respect to a covering $\{U_i\}$ of V. Since D is ample, D is connected. Therefore, if $s \in H^0(O_D)$, s is constant on the scheme D_{red}. Assume that O_D has a non-zero section s which is zero on D_{red}. Let s be represented in U_i by a function

$$s_i \in \Gamma(U_i, O_V).$$

Then $s_i - s_j = a_{ij} f_i$, $a_{ij} \in \Gamma(U_i \cap U_j, O_V)$. It follows that for all positive ν,

$$s_i^{p^\nu} - s_j^{p^\nu} = a_{ij}^{p^\nu} \cdot f_i^{p^\nu}.$$

Therefore the collection of functions $\{s_i^{p^\nu}\}$ defines a section s_ν of O_{D_ν} where D_ν is the Cartier divisor defined by $\{f_i^{p^\nu}\}$. Notice that $s_\nu \neq 0$: for if $s_\nu = 0$, then we would have:

$$s_i^{p^\nu} = b_i \cdot f_i^{p^\nu}, \qquad b_i \in \Gamma(U_i, O_V).$$

Then (s_i/f_i) would be a rational function on V that was integrally dependent on $\Gamma(U_i, O_V)$. Since V is normal, this would imply that

$$s_i \in f_i \cdot \Gamma(U_i, O_V)$$

i.e., $s = 0$. Therefore $s_\nu \neq 0$, and by the exact sequence

$$H^0(O_V) \to H^0(O_{D_\nu}) \to H^1(O_V(-D_\nu))$$

this implies that $H^1(O_V(-D_\nu)) \neq (0)$. But

$$O_V(-D_\nu) \cong [O_V(-D)]^{\otimes p^\nu}.$$

Since we can take ν arbitrarily large, and since $O_V(D)$ is ample, this contradicts the lemma of Enriques-Severi-Zariski (cf. [5], Th. 4, p. 270). Q. E. D.

Our question is, therefore, equivalent to the injectivity of $H^1(O_V) \to H^1(O_D)$. We can further restrict this kernel by examining the frobenius cohomology operation:[2]

$$H^1(O_V) \xrightarrow{\quad F \quad} H^1(O_V)$$

defined by $F(\{a_{ij}\}) = \{a_{ij}{}^p\}$ in terms of Čech co-cycles. The point is that if we consider F on the subspace $H^1(O_V(-D))$, it factors as follows:

$$\text{(i)} \qquad \begin{array}{ccc} H^1(O_V) & \xrightarrow{\qquad F \qquad} & H^1(O_V) \\ \cup & & \cup \\ H^1(O_V(-D)) & & H^1(O_V(-D)) \\ & \searrow^{F} & \\ & & \cup \\ & & H^1(O_V(-pD)). \end{array}$$

But, for large ν, $H^1(O_V(-p^\nu D)) = (0)$ by the lemma of Enriques-Severi-Zariski. Diagram (i) implies, by induction, that

$$F^\nu\{H^1(O_V(-D))\} \subset H^1(O_V(-p^\nu D)).$$

Therefore:

PROPOSITION 4. *F is nilpotent on the image of $H^1(O_V(-D))$ in $H^1(O_V)$.*

This Proposition can also be proven by using the fact that elements of $H^1(O_V)$ idempotent for F die in p-cyclic unramified coverings of V (cf. [6], § 16) and by imitating the second proof of Theorem 2. On the other hand, the elements of $H^1(O_V)$ killed by F die in principal coverings of V with infinitesimal structure group α_p. These facts may be expressed in the language of Grothendieck cohomologies by:

$$\{\alpha \in H^1(O_V) \mid F\alpha = \alpha\} \cong \boldsymbol{H}^1_{\substack{\text{étale}\\ \text{topology}}} (V, \boldsymbol{Z}/p\boldsymbol{Z}),$$

$$\{\alpha \in H^1(O_V) \mid F\alpha = 0\} \cong \boldsymbol{H}^1_{\substack{\text{flat}\\ \text{topology}}} (V, \alpha_p).$$

[2] This fact was pointed out to me by Serre, and has also been noticed by Grauert.

But principal coverings of V with structure group α_p, even if they are varieties at all, may be non-normal and must be purely inseparable over V. This is very awkward and is what really leads to the debacle. To present the counter-example, I want first to make positive use of the preceding remarks via:

LEMMA 5. *Let $f\colon V' \to V$ be a finite surjective morphism of normal varieties coresponding to a* separable *function field extension. Let $\alpha \in H^1(V, O_V)$ be non-zero and such that $F\alpha = 0$. Then $f^*\alpha \in H^1(V', O_{V'})$ is not zero.*

Proof. Let α be represented by the Čech co-cycle $\{a_{ij}\}$ with respect to some open affine covering $\{U_i\}$ of V. Then $F\alpha = 0$ implies that there are functions $g_i \in \Gamma(U_i, O_V)$ such that

$$a_{ij}^p = g_i - g_j.$$

If $f^*\alpha = 0$, then there are also functions $h_i \in \Gamma(f^{-1}(U_i), O_{V'})$ such that

$$f^*(a_{ij}) = h_i - h_j.$$

It follows that

$$h_i^p - f^*(g_i) = h_j^p - f^*(g_j),$$

i.e., there is a constant β such that

$$f^*(g_i) = h_i^p + \beta, \text{ all } i.$$

Therefore $f^*(g_i) \in k(V')^p$. Since $k(V')$ is separable over $k(V)$, this implies that $g_i \in k(V)^p$, for all i. If $g_i = k_i^p$, $k_i \in k(V)$, then since V is normal, it follows that

$$k_i \in \Gamma(U_i, O_V).$$

Then

$$a_{ij} = k_i - k_j,$$

so $\alpha = 0$: a contradiction. Q. E. D.

Example 6. A normal complete algebraic surface V with an ample invertible sheaf L such that

$$H^1(V, L^{-1}) \neq (0).$$

Start with any normal projective algebraic surface V_0 and a non-zero element $\alpha \in H^1(V_0, O_{V_0})$ such that $F\alpha = 0$: e.g. the product of a super-singular elliptic curve with any other curve. Let H be a hyperplane section of V_0 and let $L_0 = O_{V_0}(H)$. We shall let V be the normalization of V_0 in a suitable finite separable extension field E of $k(V_0)$. If $\pi\colon V\to V_0$ is the projection, let $L=\pi^*(L_0)=O_V(\pi^{-1}(H))$: an ample sheaf on V.

Let α be represented, as in the lemma, by $\{a_{ij}\}$ and let $g_i\in\Gamma(U_i,O_{V_0})$ satisfy

$$a_{ij}^p=g_i-g_j.$$

Let $h_i=0$ be a local equation of H in U_i (replacing $\{U_i\}$ by a finer covering if necessary). Define an extension E_i of $k(V_0)$ by the separable equation:

$$z_i^p-h_i^p z_i=g_i.$$

We shall let E be a join of the extensions E_i. Then I claim that for this E and the corresponding V and L, the element $\pi^*\alpha$ is actually in the subspace:

$$H^1(V,O_V(-\pi^{-1}(H)))\subset H^1(V,O_V).$$

Cover V by the open affines $U_i^*=\pi^{-1}(U_i)$. Then $z_i\in\Gamma(U_i^*,O_V)$ and

$$\begin{aligned}\pi^*\alpha&=\text{coho. class }[a_{ij}]\\&=\text{coho. class }[a_{ij}-z_i+z_j]\end{aligned}$$

and

$$\begin{aligned}\left\{\frac{a_{ij}-z_i+z_j}{h_i}\right\}^p&=\frac{a_{ij}^p-z_i^p+z_j^p}{h_i^p}\\&=\frac{(g_i-z_i^p)-(g_j-z_j^p)}{h_i^p}\\&=-z_i+(h_j/h_i)^p z_j\\&\in\Gamma(U_i^*\cap U_j^*,O_V).\end{aligned}$$

Therefore, $a_{ij}-z_i+z_j$ is actually in $\Gamma(U_i^*\cap U_j^*,O_V(-\pi^{-1}(H)))$. Q.E.D.

REFERENCES.

[1] K. Kodaira, "On a differential-geometric method in the theory of analytic stacks," *Proceedings of the National Academy of Science, U.S.A.*, vol. 39 (1953), p. 1268.

[2] D. Mumford, "Pathologies of modular algebraic surfaces," *American Journal of Mathematics*, vol. 83 (1961), p. 339.

[3] ———, "Further pathologies in algebraic geometry," *American Journal of Mathematics*, vol. 84 (1962), p. 642.

[4] ———, "Lectures on curves on algebraic surfaces," *Annals of Mathematics Studies*, 1966.

[5] J.-P. Serre, "Faisceaux algèbriques cohèrents," *Annals of Mathematics*, vol. 61 (1955), p. 197.

[6] ———, "Sur la topologie des variétés algébriques en caractéristique *p*," in Symposio International de Topologia Algebraica, Mexico, 1958.

[7] O. Zariski, *Algebraic surfaces*, Springer-Verlag, Berlin-Heidelberg-Göttingen, 1934.

Cor of method

L ample
$h^0(L) > h^1(\mathcal{O}_X)$ $\Rightarrow h^1(L^{-1}) = (0)$

Pf. Consider $H^1(L^{-1}) \otimes H^0(L) \to H^1(\mathcal{O}_X)$

if this not zero, by hyp.

find $\alpha \otimes \beta \mapsto 0$, $\alpha \neq 0$, $\beta \neq 0$

Then via α for $0 \to L^{-1} \xrightarrow{\cup\beta} \mathcal{O}_X \to \mathcal{O}_H \to 0$

find $H^0(\mathcal{O}_H) \supsetneq k$: contradicts Prop. 3!

J. Math. Kyoto Univ.
9-2 (1969) 195-204

Rational equivalence of 0-cycles on surfaces

By

D. MUMFORD

(Communicated by Professor Nagata, November 28, 1968)

We will consider in this note 0-cycles on a complete non-singular algebraic surface F over the field C of complex numbers. We will use the language of schemes, and every scheme will be assumed separated and of finite type over C. In a very extensive set of papers, Severi set up and investigated the concept of rational equivalence (cf. [2], [3], [4], [5] among many others). It is not however very easy to find a precise definition in Severi's work, and there was a good deal of discussion on this point at the International Congress of 1954. A much more elementary approach was worked out by Chevalley in his seminar "Anneaux de Chow" [1]. For 0-cycles on F, the most elementary definition is this:

Let Σ_n = group of permutations on n letters.

Let $S^nF = F^n/\Sigma_n$, the nth symmetric power of F

$$\underset{\text{as set}}{\cong} \{A \mid A \text{ effective 0-cycle of degree } n \text{ on } F\}.$$

Definition: 2 0-cycles A_1, A_2 of degrees n_1 and n_2 are *rationally equivalent* if $n_1 = n_2$ and $\exists$ a 0-cycle B of degree m such that $A_1 + B$ and $A_2 + B$ are effective, corresponding to points $x_1, x_2 \in S^{n+m}F$, and $\exists$ a morphism $f: P^1 \to S^{n+m}F$ such that $f(0) = x_1$, $f(\infty) = x_2$.

Definiton: $A_0(F)$ = group of all zero-cycles of degree 0 on F modulo rational equivalence.

What is the structure of $A_0(F)$? Consider the following three statements:

(1) $\exists n$ such that $\forall$ 0-cycles A with $\deg(A) \geq n$,
$A \underset{\text{rat}}{\sim} A'$, A' effective.

(2) $\exists n$ such that the natural map
$$S^n F \times S^n F \to A_0(F)$$
$$(A, B) \quad \mapsto \text{class of } A - B$$
is surjective.

(3) $\exists n$ such that $\forall m$, $A \in S^m F$, $\exists$ a subvariety W:
$A \in W \subset S^m F$ of codimension $\leq n$ consisting of points rationally equivalent to A.

It is not hard to show that these are all equivalent to each other* and that, intuitively, they mean simply $A_0(F)$ *is finite-dimensional.* Severi, unfortunately, took it as almost evident that these statements were valid and it is often hard to discover which results in his papers depended on making this assumption at some point. For example in [2], p. 251, §13, he says:

"Ammettiamo che per ogni varieta k-dimensionale (virtuale) A di W, esista un numero finito di caratteri numerativi $c_1, c_2, \cdots$ tali che le relazioni $c_1 > 0, c_2 > 0, \cdots$ sieno sufficienti per affermare che A e una varieta effettiva. Con questo si vuol dire che nel sistema di equivalenza $|A|$ esittono varieta totali effettive."

The purpose of this paper is to prove that if $p_g > 0$, (1)–(3) are *false*. Now, after criticizing Severi like this, I have to admit the following: the method of disproof of (1)–(3) is due entirely to Severi: Severi created, in fact, a very excellent tool for analyzing

---

* (1)$\Longrightarrow$(2) is obvious. To show (2)$\Longrightarrow$(3), fix a base pt. $x_0 \in F$ and consider
$$Z = \{(A_1, A_2, B) \in S^n F \times S^n F \times S^m F \mid A_1 - A_2 \underset{\text{rat}}{\sim} B - m(x_0)\}.$$
By lemma 3 below, Z is countable union of closed subvarieties of $S^n F \times S^n F \times S^m F$ and by (2) one of its components Z_0 projects *onto* $S^m F$. Then for all $B \in S^m F$, take as W one of the components of $\{B' \in S^m F \mid \exists (A_1, A_2) \in S^n F \times S^n F$ such that $(A_1, A_2, B) \in Z_0$, $(A_1, A_2, B') \in Z_0\}$. To show (3)$\Longrightarrow$(1), show first that (3) implies the existence of n such that $\forall A \in S^n F$, $\forall x \in F$, $\exists A' \in S^{n-1} F$, $A \underset{\text{rat}}{\sim} A' + (x)$.

the influence of regular 2-forms on F on his systems of eqivalence. One must admit that in this case the *technique* of the Italians was superior to their vaunted intuition. This paper is certainly one of the "mémoires d'esploitation" referred to by Severi in the 1954 Congress [5], p. 541, vol. 3, but not I think part of his "prévision correspondante à mes vifs souhaits".

§1. Induced differentials

We want to study the following situation:

X = non-singular variety

G = finite group acting on X

$Y = X/G$, $\pi : X \to Y$ the canonical morphism,

$\omega \in \Gamma(X, \Omega^q_X)$ a q-form invariant under G.

If G is acting freely on X, Y will be non-singular and $\omega = \pi^*(\eta)$ for a unique $\eta \in \Gamma(Y, \Omega^q_Y)$. If G is not acting freely this is false, but the following still is true: for all non-singular varieties S and morphisms $f : S \to Y$, we get a canonical induced 2-form η_f on S. We define this as follows:

Let $\widetilde{S} = (S \times_Y X)_{red}$:

$$\begin{array}{ccc} \widetilde{S} & \xrightarrow{\widetilde{f}} & X \\ {\scriptstyle p}\downarrow & & \downarrow{\scriptstyle \pi} \\ S & \xrightarrow[f]{} & Y \end{array}$$

Then G acts on $\widetilde{S}$, and $S \cong \widetilde{S}/G$. Let $\widetilde{\omega} = \widetilde{f}^*(\omega) \in \Gamma(\widetilde{S}, \Omega^q_{\widetilde{S}})$ (here $\Omega^q_{\widetilde{S}}$ is by definition Λ^q of the Kähler differentials $\Omega^1_{\widetilde{S}}$, which is not of course locally free in general). Then $\widetilde{\omega}$ is G-invariant, and I claim there is one and only one $\eta_f \in \Gamma(S, \Omega^q_S)$ such that

(*) $$p^*(\eta_f) - \widetilde{\omega} \text{ is torsion in } \Omega^q_{\widetilde{S}}.$$

In fact, there are non-singular open dense sets $S_0 \subset S$, $\widetilde{S}_0 = p^{-1}(S_0) \subset \widetilde{S}$ such that a suitable quotient G/H is acting freely on $\widetilde{S}_0$. Therefore there is a unique $\eta_f \in \Gamma(S_0, \Omega^q_S)$ such that $p^*(\eta_f) = res_{\widetilde{S}_0}(\widetilde{\omega})$. To prove

(*) it suffices to check that this η_f is everywhere regular on S_0, and since Ω_S^q is a locally free sheaf, it suffices to check this at points of codimension 1. Let T be the normalization of $\widetilde{S}$. Then the pull-back of $\tilde{\omega}$ is regular on T at all points of codimension 1, hence the pull-back of η_f is also regular on T at all points of codimension 1. But η_f times the order n of G/H is the trace of its pull-back, and since $n \neq 0$ in C, η_f is regular at points of codimension 1.

More generally, suppose we have *any* diagram:

$$\begin{array}{ccc} \widetilde{S} & \xrightarrow{\widetilde{f}} & X \\ {\scriptstyle p}\downarrow & & \downarrow{\scriptstyle \pi} \\ S & \xrightarrow[f]{} & Y \end{array}$$

such that $\widetilde{S}$ is reduced and p is a dominating morphism. Then η_f satisfies the condition: $p^*\eta_f - \widetilde{f}^*\omega$ is torsion, and is characterized by this property. To say more we need the following lemma:

Lemma 1: *Let $h: X \to Y$ be a morphism of reduced schemes. Then $h^*: \Omega_Y^p \to \Omega_X^p$ takes torsion differentials to torsion differentials.*

Proof:* We reduce this immediately to the case X and Y varieties. By blowing up the closure of $h(X)$ and normalizing, we construct a diagram:

$$\begin{array}{ccc} X' & \longrightarrow & Y' \\ {\scriptstyle \text{dominating}}\downarrow & & \downarrow{\scriptstyle \text{birational (not necessarily proper)}} \\ X & \longrightarrow & Y \end{array}$$

with Y' non-singular. If ω is a torsion p-form on Y, then ω dies on Y', hence it dies on X'. But if η is the pull-back of ω to X, then η dies in X'. But since the characteristic is 0, the map of meromorphic p-forms

$\Omega^p_{C(X)/C} \longrightarrow \Omega^p_{C(X')/C}$ is injective, so η dies in $\Omega^p_{C(X)/C}$ too, i.e., η is torsion. *QED*

---

* This proof was given to me by H. Hironaka.

In view of the lemma, η_f has the property that for any diagram:

$$\begin{array}{ccc} \widetilde{S} & \xrightarrow{\widetilde{f}} & X \\ {\scriptstyle p}\downarrow & & \downarrow{\scriptstyle \pi} \\ S & \xrightarrow[f]{} & Y \end{array}$$

such that $\widetilde{S}$ is reduced, $p^*(\eta_f)-\widetilde{f}^*(\omega)$ is torsion. We can now prove functoriality of η_f in f: suppose we are given

$$\begin{array}{ccc} S_1 & \xrightarrow{f_1} & Y \\ & {\scriptstyle h}\searrow \quad \nearrow{\scriptstyle f_2} & \\ & S_2 & \end{array}$$

with S_1 and S_2 non-singular. Then I claim $h^*(\eta_{f_2})=\eta_{f_1}$. In fact, let $\widetilde{S_1}=(S_1\times_Y X)_{\text{red}}$, $\widetilde{S_2}=(S_2\times_Y X)_{\text{red}}$, and use the diagram:

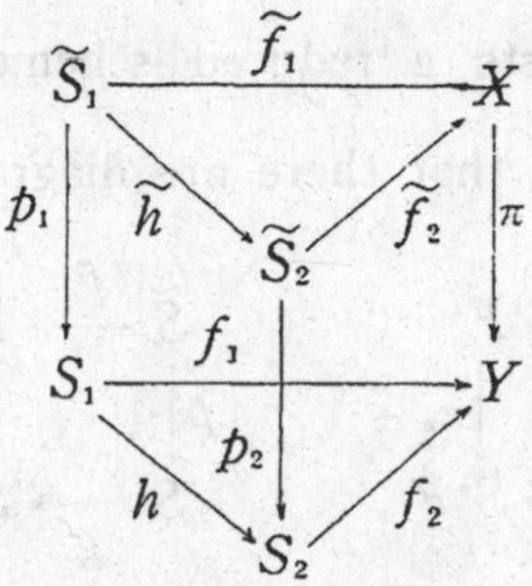

Since $p_2^*(\eta_{f_2})-\widetilde{f_2}^*(\omega)$ is torsion, it follows that

$$\widetilde{h}^*[p_2^*(\eta_{f_2})-\widetilde{f_2}^*(\omega)]=p_1^*(h^*(\eta_{f_2}))-\widetilde{f_1}^*(\omega)$$

is torsion too. Therefore $h^*(\eta_{f_2})=\eta_{f_1}$.

§2. The main theorem

We apply this machinery now to the case $X=F^n$, $G=\Sigma_n$, $Y=S^nF$. Fix for the rest of this paper a differential $\omega\in\Gamma(F,\Omega_F^2)$. Let $\omega^{(n)}=\Sigma_1^n p_i^*(\omega)\in\Gamma(F^n,\Omega_{F^n}^2)$. Then $\omega^{(n)}$ is G-invariant, and by our definitions above, for all $f:S\to S^nF$, S a non-singular variety, we obtain canonical, functorial 2-forms $\eta_f\in\Gamma(S,\Omega_S^2)$. We can now give an exact statement of our main theorem:

Theorem (*due essentially to Severi*): *For all* $f: S \to (S^n F)$ *such that all the* 0-*cycles* $f(s)$, $s \in S$, *are rationally equivalent, it follows that* $\eta_f = 0$.

Before proving the theorem, we need 2 lemmas. For the first, suppose we are given 2 morphisms

$$f: S \longrightarrow S^n F$$
$$g: S \longrightarrow S^m F.$$

Then let $f * g$ denote the composition:

$$S \xrightarrow{(f, g)} S^n F \times S^m F \xrightarrow{\pi} S^{n+m} F$$

where π is the obvious map.

Lemma 2: $\omega_{f * g} = \omega_f + \omega_g$.

Proof: There exists a reduced scheme $\widetilde{S}$ and a dominating morphism $p: \widetilde{S} \to S$ such that there are diagrams

$$\begin{array}{ccc} \widetilde{S} & \xrightarrow{\widetilde{f}} & F^n \\ p \downarrow & & \downarrow \pi_n \\ S & \xrightarrow[f]{} & S^n F \end{array} , \qquad \begin{array}{ccc} \widetilde{S} & \xrightarrow{\tilde{g}} & F^m \\ p \downarrow & & \downarrow \\ S & \xrightarrow[g]{} & S^m F \end{array} .$$

Then $p^*(\eta_f) = \widetilde{f}^*(\omega^{(n)}) + \text{torsion}$ and $p^*(\eta_g) = \tilde{g}^*(\omega^{(m)}) + \text{torsion}$. But we get also the diagram:

$$\begin{array}{ccc} \widetilde{S} & \xrightarrow{(\widetilde{f}, \tilde{g})} & F^{n+m} \\ p \downarrow & & \downarrow \pi_{n+m} \\ S & \xrightarrow[f * g]{} & S^{n+m} F \end{array} .$$

If $p_1: F^{n+m} \to F^n$, and $p_2: F^{n+m} \to F^m$ are the projections onto the first n and last m factors respectively, then $\omega^{(n+m)} = p_1^* \omega^{(n)} + p_2^* \omega^{(m)}$, and we calculate:

$$\begin{aligned} p^*(\omega_f + \omega_g) &= \widetilde{f}^*(\omega^{(n)}) + \tilde{g}^*(\omega^{(m)}) + \text{torsion} \\ &= (\widetilde{f}, \tilde{g})^*(p_1^* \omega^{(n)} + p_2^* \omega^{(m)}) + \text{torsion} \end{aligned}$$

$$=(\widetilde{f},\tilde{g})^*\omega^{(n+m)}+\text{torsion}$$

$$=p^*(\omega_{f*g})+\text{torsion}. \qquad QED$$

Lemma 3: *$S^nF\times S^nF$ contains a countable set $Z_1, Z_2, \cdots$ of closed subvarieties, such that if $(A,B)\in S^nF\times S^nF$, then*

(*) $$A\underset{\text{rat}}{\sim}B \Longleftrightarrow (A,B)\in\bigcup_{i=1}^{\infty}Z_i.$$

For each i, there is a reduced scheme W_i and a set of morphisms

$$e_i: W_i\longrightarrow Z_i$$

$$f_i: W_i\longrightarrow S^mF$$

$$g_i: W_i\times \boldsymbol{P}^1\longrightarrow S^{n+m}F$$

such that we get the equations between 0-cycles:

(**) $$\begin{aligned} g_i(w,0)&=p_1(e_i(w))+f_i(w)\\ g_i(w,\infty)&=p_2(e_i(w))+f_i(w)\end{aligned}$$

all $w\in W_i$, and e_i is surjective.

Proof: We divide up the pairs $(A,B)\in S^nF\times S^nF$ such that $A\underset{\text{rat}}{\sim}B$ into sets $S_{m,p}$, where

$(A,B)\in S_{m,p}\Longleftrightarrow \exists C\in S^mF$

$\exists$1-cycle Σn_iE_i on $S^{m+n}F$ such that

(a) $\cup E_i$ connected

(b) E_i normalized is $\boldsymbol{P}^1$

(c) $\Sigma_1^k n_i[\text{degree }(E_i)]=p$

(d) $A+C, B+C\in\cup E_i$.

Then it is easy to check using the Chow variety of $S^{m+n}F$ that $S_{m,p}$ is Zariski-closed and we may take the Z_i to be the components of the $S_{m,p}$'s. Next recall that if 2 0-cycles $A,B\in S^kF$ are joined by a chain of p rational curves $E_1,\cdots,E_p$ thus:

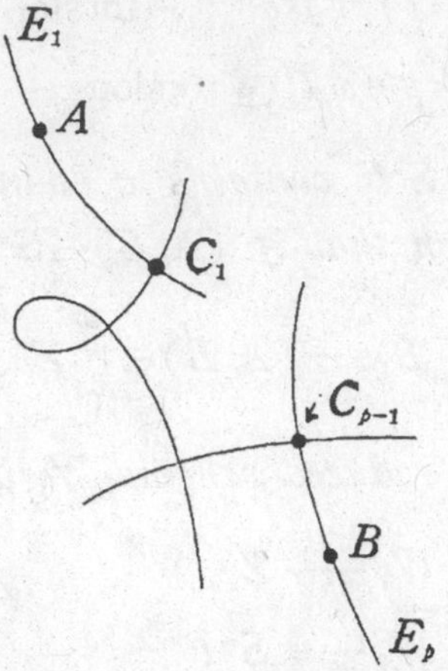

then $A+C_1+\cdots+C_{p-1}$ and $C_1+\cdots+C_{p-1}+B$ in $S^{pk}F$ are joined by a single rational curve, whose degree can be bounded by the degree of the E_i's. Therefore, there is an m and a p such that for all $(A, B)\in Z_i$, there exists $C\in S^mF$ and an irreducible rational curve $E\subset S^{m+n}F$ of degree $\leq p$ connecting $A+C$ and $B+C$. Then let

$$\mathrm{Hom}^{\leq p}(\boldsymbol{P}^1, S^{n+m}F)$$

be the scheme of all morphisms $g:\boldsymbol{P}^1\to S^{n+m}F$ such that degree $(g(\boldsymbol{P}^1))\leq p$. Define

$$W\subset Z_i\times S^mF\times \mathrm{Hom}^{\leq p}(\boldsymbol{P}^1, S^{n+m}F)$$
$$W_i=\{((A, B), C, g)\mid g(0)=A+C, g(\infty)=B+C\}.$$

QED

We can now prove the theorem. Given $f:S\to S^nF$ such that all the 0-cycles $f(s)$ are rationally equivalent, fix a base point $A_0\in \mathrm{Image}(f)$. It follows from lemma 3 that there is a non-singular variety T, a dominating morphism $e:T\to S$, and morphisms $g:T\to S^mF$, $h:T\times\boldsymbol{P}^1\to S^{n+m}F$ such that:

$$h(t, 0)=g(t)+f(e(t))$$
$$h(t, \infty)=g(t)+A_0$$

all $t\in T$. In other words, if $A_0:T\to S^nF$ is the constant map with image A_0, then

$$h|_{T\times(0)}=g_*(f\circ e)$$
$$h|_{T\times(\infty)}=g_*A_0.$$

By Lemma 2, it follows that

$$\eta_h|_{T\times(0)}=\eta_g+e^*(\eta_f)$$
$$\eta_h|_{T\times(\infty)}=\eta_g+\eta_{A_0}.$$

Now η_h is a regular 2-form on $T\times\boldsymbol{P}^1$. Since

$$\Omega^2_{T\times\boldsymbol{P}^1}\cong p_1^*(\Omega^2_T)+p_1^*(\Omega^1_T)\otimes p_2^*(\Omega^1_{\boldsymbol{P}^1})$$

and $\Omega^1_{\boldsymbol{P}^1}$ has no global sections, it follows that $\eta_h=p_1^*(\eta)$ for some $\eta\in\Gamma(T,\Omega^2_T)$. Therefore $\eta_h|_{T\times(0)}=\eta_h|_{T\times(\infty)}$, and since $\eta_{A_0}=0$, we find $e^*(\eta_f)=0$, hence $\eta_f=0$. *QED*

To apply the theorem, let

$$(S^nF)_0=\{A=\Sigma_1^n x_i\,|\,x_1,\cdots,x_n \text{ are distinct, and } \omega(x_i)\in\Lambda^2(m_{x_i}/m_{x_i}^2) \text{ is not } 0\}$$

For all $A\in(S^nF)_0$, $(S^nF)_0$ is non-singular at A, the projection $\pi_n:F^n\to S^nF$ is etale over A, and the 2-form $\omega^{(n)}$ is non-degenerate at all points over A, i.e., defines a non-degenerate skew-symmetric form in the tangent space to F^n. Therefore, there is a 2-form on $(S^nF)_0$:

$$\omega_0^{(n)}\in\Gamma((S^nF)_0,\Omega^2_{S^nF})$$

such that $\pi_n^*(\omega_0^{(n)})=\omega^{(n)}$, and $\omega_0^{(n)}$ is everywhere a non-degenerate 2-form. In particular, the maximal isotropic subspaces of $\omega_0^{(n)}(A)$ have dimension n, i.e., if

$$W\subset T_{A,S^nF}=\text{tangent space to } S^nF \text{ at } A$$

is a subspace such that $\omega_0^{(n)}|_W$ vanishes, then $\dim W\leq n$. But if S is a non-singular subvariety of $(S^nF)_0$, and $i:S\to(S^nF)_0$ is the inclusion morphism, then $\eta_i=\mathrm{res}_S(\omega_0^{(n)})$. We conclude:

Corollary: *Let F be a non-singular surface with $p_g>0$ and let $(S^nF)_0$ be defined as above. Then if $S\subset(S^nF)_0$ is a subvariety consisting of rationally equivalent 0-cycles, it follows that* $\dim S\leq n$.

This disproves (3) of the introduction, hence also (1) and (2).

References

[1] C. Chevalley, *Anneaux de Chow et applications*, mimeographed seminar, 1958, Secret. Math., Paris.

[2] F. Severi, *La base per le varieta algebriche di dimensione qualunque contenute in una data*, Mem. della R. Accad. d'Italia, **5**, (1934), p. 239.

[3] F. Severi, *Serie, sistemi d'equivalenza e corrispondenze algebriche* Roma, 1942, Edizioni Cremonese.

[4] F. Severi, *Ulteriori sviluppi della teoria delle serie d'equivalenza sulle superficie algebriche*, Comment. Pont. Acad. Sci., **6** (1942), p. 977.

[5] F. Severi, *Problems resolus et problemes nouveaux dans la theorie des systemes d'equivalence*, Proc. Int. Cong. Math., 1954, Amsterdam, vol. 3, p. 529.

DEPARTMENT OF MATHEMATICS
HARVARD UNIVERSITY

SOME ELEMENTARY EXAMPLES OF UNIRATIONAL VARIETIES WHICH ARE NOT RATIONAL

By M. ARTIN *and* D. MUMFORD†

[Received 8 November 1971]

An outstanding problem in the algebraic geometry of varieties of dimension $n \geqslant 3$ over an algebraically closed field k has been whether there exist unirational varieties which are not rational. Here V is *unirational* if it has the equivalent properties:

(a) there exists a rational surjective map $f: \mathbf{P}^n \to V$,

or, there exists an embedding $k(V) \subset k(X_1, \ldots, X_n)$;

while V *rational* means equivalently:

(b) there exists a birational map $f: \mathbf{P}^n \to V$,

or, there exists an isomorphism $k(V) \cong k(X_1, \ldots, X_n)$.

For $n = 1$, these are equivalent (Lüroth's theorem). For $n = 2$ they are equivalent in characteristic 0 (Castelnuovo's theorem) or if the map f in (a) is assumed separable (Zariski's extension of Castelnuovo's theorem). In 1959 ([**13**]), Serre clarified classical work on this problem for $n = 3$. It has been generally accepted since then that none of the examples proposed by Fano or Roth had been correctly proved irrational.

In the past year, two solutions of this problem have been found: Clemens and Griffiths ([**6**]) showed that all non-singular *cubic* hypersurfaces in $\mathbf{P}^4$ are irrational, and Iskovskikh and Manin ([**16**]) showed that all non-singular *quartic* hypersurfaces in $\mathbf{P}^4$ are irrational. Some are unirational (Segre ([**11**])).

Both of these solutions are quite deep and it seems worth while to have an elementary example as well, even if our method applies to a very special kind of variety. Ramanujam suggested using torsion in H^3 and this led us to the examples presented here. We construct varieties, of all dimensions $n \geqslant 3$ and all characteristics $p \neq 2$, which are unirational and which have 2-torsion in H^3. With the present state of resolution of singularities, we can show that such a V cannot be rational if the characteristic is 0 or if the characteristic is not 2 and $n = 3$.

† Both authors would like to thank the Mathematics Institute of the University of Warwick for its warm hospitality and generous support at the time when this research was done. We also acknowledge gratefully the support of the National Science Foundation and the Nuffield Foundation respectively.

Proc. London Math. Soc. (3) 25 (1972) 75–95

An outline of this paper is as follows. In § 1, we prove the torsion criterion for distinguishing between rational and irrational varieties. In § 2, we construct an example and prove that it has 2-torsion when $k = \mathbf{C}$ using simplicial methods. In § 3, we digress to prove a theorem on the structure of the Brauer group of a function field in two variables. We use this in § 4 to construct a whole class of examples including the particular one given in § 2, and prove that in suitable circumstances they have 2-torsion in their 2-adic étale cohomology groups.

We would like to point out that our examples belong to a general class—conic bundles over rational surfaces—which have been much studied classically, and that our theory has many points of contact with classical work: cf. Roth ([**10**], Ch. 4, §§ 4–7).

1. The criterion

Serre ([**13**]) showed that over the complex field $\mathbf{C}$ almost all cohomological properties enjoyed by non-singular projective *rational* 3-folds hold for non-singular projective *unirational* 3-folds as well. One small possible difference escaped though. To be precise, let V be a non-singular projective 3-fold over the complex field $\mathbf{C}$. Applying Poincaré duality and the universal coefficient theorem, its integral cohomology has the form in the left-hand column

$$\begin{aligned}
H^0(V) &\cong \mathbf{Z}, & &= \mathbf{Z},\\
H^1(V) &\cong \mathbf{Z}^{B_1}, & &= 0,\\
H^2(V) &\cong \mathbf{Z}^{B_2} + T_1, & &= \mathbf{Z}^{B_2},\\
H^3(V) &\cong \mathbf{Z}^{B_3} + T_2, & &= \mathbf{Z}^{B_3} + T_2,\\
H^4(V) &\cong \mathbf{Z}^{B_2} + T_2, & &= \mathbf{Z}^{B_2} + T_2,\\
H^5(V) &\cong \mathbf{Z}^{B_1} + T_1 \cong H_1(V), & &= 0,\\
H^6(V) &\cong \mathbf{Z}, & &= \mathbf{Z},
\end{aligned}$$

for suitable integers B_1, B_2, B_3 and finite groups T_1, T_2. Moreover, its complex cohomology admits the canonical decomposition given on the left:

$$\begin{aligned}
H^1(V)\otimes\mathbf{C} &\cong H^{0,1} + H^{1,0}, & &= 0,\\
H^2(V)\otimes\mathbf{C} &\cong H^{0,2} + H^{1,1} + H^{2,0}, & &= H^{1,1},\\
H^3(V)\otimes\mathbf{C} &\cong H^{0,3} + H^{1,2} + H^{2,1} + H^{3,0}, & &= H^{1,2} + H^{2,1},\\
H^4(V)\otimes\mathbf{C} &\cong H^{1,3} + H^{2,2} + H^{3,1}, & &= H^{2,2}\\
H^5(V)\otimes\mathbf{C} &\cong H^{2,3} + H^{3,2}, & &= 0.
\end{aligned}$$

Let $h^{p,q} = \dim H^{p,q}$. Serre showed that if V was unirational, then

(a) $\pi_1(V) = (0)$, hence $H_1(V) = 0$, i.e.

$$B_1 = h^{1,0} = h^{0,1} = 0 \quad \text{and} \quad T_1 = 0.$$

(b) $h^{p,0} = 0$, hence $h^{0,p} = 0$ too.

This reduces the cohomology to the form in the right-hand column. For a rational variety, the numbers B_2 and B_3 do not seem to satisfy any particularly useful further restrictions (except of course B_3 even).† However, two things are left:

(c) The Hodge decomposition on H^3 gives an abelian variety:

$$J(V) = H^3(V) \otimes \mathbf{C}/(\operatorname{Im} H^3(V, \mathbf{Z}) + H^{1,2})$$

—the *'intermediate jacobian'* of Weil ([**15**]). Via cup product, $J(V)$ carries a *canonical principal polarization* Θ, and Clemens and Griffiths have shown that for rational 3-folds:

$$(*) \qquad (J(V), \Theta) \cong \prod_i (J(C_i), \Theta_i),$$

where C_i are non-singular curves, $J(C_i)$ their jacobians, and Θ_i are the usual theta-polarizations on $J(C_i)$. On the other hand, they have shown that no non-singular cubic hypersurface in $\mathbf{P}^4$ satisfies (*), although these hypersurfaces are unirational.

(d) The torsion T_2—concerning this, we have:

Proposition 1. *The torsion subgroup $T_2 \subset H^3(V, \mathbf{Z})$ is a birational invariant of a complete non-singular complex variety V of any dimension n. In particular, $T_2 = 0$ if V is rational.*

Proof. The last assertion is of course clear since $T_2(\mathbf{P}) = 0$. Let $f\colon V' \to V$ be a morphism of smooth complete varieties which is birational. It induces maps

$$H^q(V', \mathbf{Z}) \underset{f_*}{\overset{f^*}{\rightleftarrows}} H^q(V, \mathbf{Z}),$$

the lower arrow being the Gysin map obtained via Poincaré duality. Since f is birational, $f_* f^*$ is identity.‡ Thus

$$(1.1) \qquad H^q(V', \mathbf{Z}) \approx H^q(V, \mathbf{Z}) + K^q$$

for suitable K^q. In particular,

$$T_2(V) \subset T_2(V').$$

† It is quite possible that, for rational varieties with $B_2 = 1$, B_3 can take only a few small values. But if so, this is quite likely very hard to prove.

‡ By the identity $f_*(x.f^*(y)) = f_*(x).y$, it suffices to prove that $f_* f^* 1 = 1$, where is the canonical generator of $H^0(V, \mathbf{Z})$. This is proved in [4], § 4.15.

Suppose furthermore that f is the blow-up of a smooth subvariety $Y \subset V$, say of codimension $r+1$. Then the fibres of f above Y are isomorphic to $\mathbf{P}^r$, and so the direct image $R^q f_* \mathbf{Z}$ is $\mathbf{Z}$ if $q = 0$, is the extension of $\mathbf{Z}$ by zero outside Y if $q = 2i$ $(1 \leqslant i \leqslant r)$, and is zero for other values of q. Thus the Leray spectral sequence for the map f yields an exact sequence

$$H^0(Y,\mathbf{Z}) \to H^3(V,\mathbf{Z}) \to H^3(V',\mathbf{Z}) \to H^1(Y,\mathbf{Z}) \to H^4(V,\mathbf{Z}) \to H^4(V',\mathbf{Z}).$$

By (1.1) this sequence splits, i.e.

$$H^3(V',\mathbf{Z}) \approx H^3(V,\mathbf{Z}) \oplus H^1(Y,\mathbf{Z}).$$

Since $H^1(Y,\mathbf{Z})$ is torsion-free for any Y, we have $T_2(V) \approx T_2(V')$ in this case.

Now let V, W be birationally equivalent and non-singular. According to the results of Hironaka, there is a diagram of birational morphisms

(1.2)

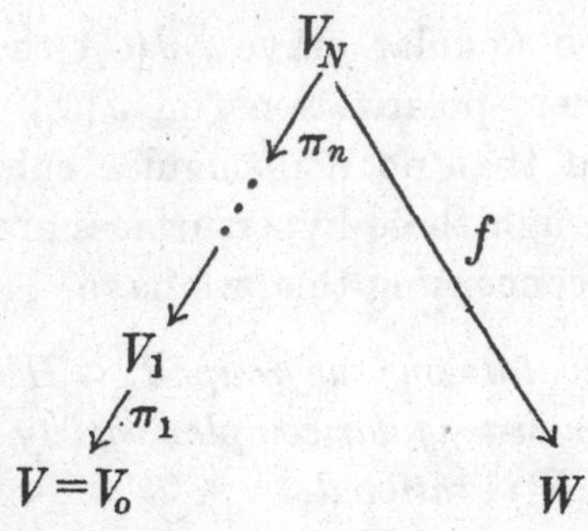

where π_i is the blow-up of a smooth $Y_{i-1} \subset V_{i-1}$. Thus the above remarks show that $T_2(W) \subset T_2(V_N) \approx T_2(V)$. By symmetry, $T_2(W) \approx T_2(V)$, as required.

Moreover, in characteristic $p \neq 0$ we have

PROPOSITION 1*. *The torsion subgroup of the étale l-adic cohomology group* $H^3(V,\mathbf{Z}_l)$ *is a birational invariant of a complete non-singular 3-fold* V *over* k, *where* k *is algebraically closed and* $l \neq \operatorname{char} k$. *In particular, this group is torsion-free if* V *is rational.*

Proof. By the results of Abhyankar ([1]), we can again find a diagram (1.2). Using the results of [2], Exposé 18, the proof goes through as before.

Note that we use the hypothesis $\dim X = 3$ only because the resolution theorem that we need is not known in higher dimension.

2. A double space with quartic branch locus

To start off, we will work over any algebraically closed field of characteristic different from 2. Let

$$A \subset \mathbf{P}^2$$

be a non-singular conic, defined by a homogeneous quadratic equation:

$$\alpha(X_0, X_1, X_2) = 0.$$

Let

$$D_1, D_2 \subset \mathbf{P}^2$$

be non-singular cubics defined by equations $\delta_1 = 0, \delta_2 = 0$ such that

(a) D_1 and D_2 meet A tangentially at six distinct points:

$$D_1 \cap A = \{P_1^{(1)}, P_2^{(1)}, P_3^{(1)}\},$$
$$D_2 \cap A = \{P_1^{(2)}, P_2^{(2)}, P_3^{(2)}\},$$

(b) D_1 meets D_2 transversally at nine distinct points $O_1, \ldots, O_9$.

It is easy to check that such cubics exist.

Next, $(D_1+D_2)\cdot A$, as a cycle on A, equals $2\mathfrak{A}$, where $\mathfrak{A} = \sum_{i=1}^2 \sum_{j=1}^3 P_j^{(i)}$. Since curves of degree 3 cut out a complete system on A, we have

$$\mathfrak{A} = B\cdot A$$

for some third cubic curve B. In homogeneous equations, this means that

$$\alpha \mid \delta_1\delta_2 - \beta^2,$$

where $\beta = 0$ is a suitable equation of B, hence

$$\delta_1\delta_2 = \beta^2 - 4\alpha\gamma$$

for some γ of degree 4.

Let

$$K \subset \mathbf{P}^3$$

be the quartic surface with homogeneous equation:

$$\alpha(X_0, X_1, X_2)X_3^2 + \beta(X_0, X_1, X_2)X_3 + \gamma(X_0, X_1, X_2) = 0.$$

If P_0 is the point $(0, 0, 0, 1)$, then P_0 is a node of K (i.e. a double point with non-singular tangent cone); projecting from P_0, K is a double cover of $\mathbf{P}^2$ ramified along the curve $\beta^2 - 4\alpha\gamma = 0$, i.e. along $D_1 \cup D_2$. Therefore K has 10 nodes in all—P_0, plus one more point P_i $(1 \leqslant i \leqslant 9)$ over each point O_i of $D_1 \cap D_2$—and no other singularities.

Next, let V_0 be the double covering of $\mathbf{P}^3$ (the 'double space') branched in K. V_0 has the weighted homogeneous equation:

$$X_4^2 = \alpha X_3^2 + \beta X_3 + \gamma; \quad \deg X_0 = \ldots = \deg X_3 = 1, \quad \deg X_4 = 2.$$

Moreover, V_0 has a node Q_i over each node P_i of K, and no other singularities. Finally, let V be the desingularization of V_0 obtained by blowing up all the Q_i to exceptional divisors E_i (cf. Figs. 1 and 2).

First of all, it is clear that V is *unirational*. In suitable affine coordinates, V_0 is just

$$X_4^2 = (X_1^2 - X_2)X_3^2 + \beta(X_1, X_2)X_3 + \gamma(X_1, X_2).$$

Consider the double covering W of this affine 3-fold defined by

$$X_5 = \sqrt{(X_1^2 - X_2)}$$

If we eliminate X_2 by the relation $X_2 = X_1^2 - X_5^2$, the new 3-fold has the equation:

$$X_4^2 = X_5^2X_3^2 + \beta(X_1, X_1^2 - X_5^2)X_3 + \gamma(X_1, X_1^2 - X_5^2). \tag{2.1}$$

This is a rational variety, via the birational map:

$$W \xrightarrow{\ f\ } \mathbf{A}^3 \quad (\text{coordinates } Y_1, Y_2, Y_3)$$

given by

$$Y_1 = X_1,$$
$$Y_2 = X_5$$
$$Y_3 = X_4 - X_5X_3.$$

In fact, to compute the fibre of $f^{-1}(a_1, a_2, a_3)$, put $X_1 = a_1$, $X_5 = a_2$, and $X_4 = a_3 + a_2X_3$ in equation (2.1). This leads to

$$a_3^2 + 2a_2a_3X_3 = \beta(a_1, a_1^2 - a_2^2)X_3 + \gamma(a_1, a_1^2 - a_2^2),$$

which almost always has a unique solution.

The really remarkable thing about V, however, is that it has 2-torsion in H^3 and H^4. We shall prove this here when $k = \mathbf{C}$, and in § 4 in general. Assuming these results, it follows from the criteria in § 1 that

(i) in any characteristic other than 2, V is unirational but not rational,

(ii) if the characteristic is zero, $V \times \mathbf{P}^n$ is an $(n+3)$-dimensional variety which is unirational but not rational.

The easiest way to compute the cohomology of V is to use the morphism:

$$f: V \to \mathbf{P}^2$$

defined outside E_0 by the composition:

$$V - E_0 \longrightarrow V_0 - \{P_0\} \longrightarrow \mathbf{P}^3 - \{Q_0\} \xrightarrow{\text{projection}} \mathbf{P}^2.$$

Let V' denote the blow-up of P_0 in V_0. Then f clearly factors:

$$V \xrightarrow{\ \pi\ } V' \xrightarrow{\ f'\ } \mathbf{P}^2.$$

If $a = (a_0, a_1, a_2)$ is a point of $\mathbf{P}^2$, the fibre $f'^{-1}(a)$ is the inverse image in V' of the line

$$X_0 : X_1 : X_2 = a_0 : a_1 : a_2$$

in $\mathbf{P}^3$, i.e. it is the conic

$$m_a:\quad X_4^2 = \alpha(a_0, a_1, a_2)X_3^2 + \beta(a_0, a_1, a_2)X_3Z + \gamma(a_0, a_1, a_2)Z^2.$$

Now $a \in D_1 \cup D_2$ if and only if $\delta_1\delta_2(a) = (\beta^2 - 4\alpha\gamma)(a) = 0$, i.e. if and only if the conic m_a is singular. Moreover, m_a can be a double line only if α, β, γ all vanish at a. This implies that $a \in A$ and that a is a double point of $D_1 \cup D_2$. There is no such a, so we conclude that

if $a \notin D_1 \cup D_2$, then $m_a \cong \mathbf{P}^1$ (a conic);

if $a \in D_1 \cup D_2$, then $m_a \cong \mathbf{P}^1 \vee \mathbf{P}^1$ (2 copies of $\mathbf{P}^1$ meeting transversely at 1 point).

Now $f'(P_i) = O_i$, so the fibres of f itself are the same as those of f' except for the fibres $f^{-1}(O_i)$; and one sees easily by calculating in local coordinates that $f^{-1}(O_i)$ is just the quadric E_i plus two lines, like this:

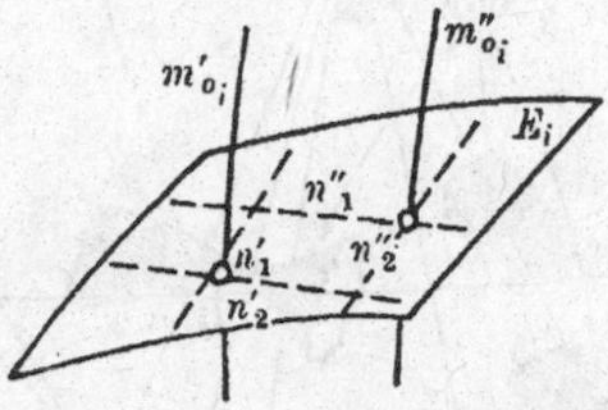

m'_{O_i}, m''_{O_i} are the proper transforms by π of two components of $f'^{-1}(O_i)$.
n'_j, n''_j are the exceptional divisors in the blow-up induced by π on surface $f'^{-1}(D_j) \subset V_0$.

FIG. 1

When $a \in D_1 \cup D_2$, let m'_a and m''_a denote the two components of m_a. The essential point now is to examine for which loops in $D_1 \cup D_2$ the two components m'_a, m''_a are interchanged when one moves continuously around them, and for which loops the two components are not interchanged. Put another way, the set of pairs

$$D'_i = \{(a, m^*) \mid a \in D_i,\ m^* \text{ a component of } m_a\}$$

is a new curve which is an unramified double covering of D_i. Which covering is it? So long as $a \notin A$, the two components can be distinguished by whether their intersection with the line $Z = 0$ is the point

$$X_4 = +X_3\sqrt{\alpha(a_0, a_1, a_2)}, \quad Z = 0$$

or

$$X_4 = -X_3\sqrt{\alpha(a_0, a_1, a_2)}, \quad Z = 0.$$

Therefore D'_i is the normalization of D_i in the field obtained by adjoining $\sqrt{\alpha}$ (or more precisely, $\sqrt{(\alpha/l^2)}$, where l is a linear form). Now note that

(a) A has intersection multiplicity 2 with D_i whenever they meet, and therefore α vanishes everywhere to even order, and D'_i is everywhere unramified over D_i;

(b) the three intersections of A and D_i are *not* collinear, and so α is *not* congruent to a square $l(X_0, X_1, X_2)^2 \bmod \delta_i$; that is, D'_i does not break up into two copies of D_i.

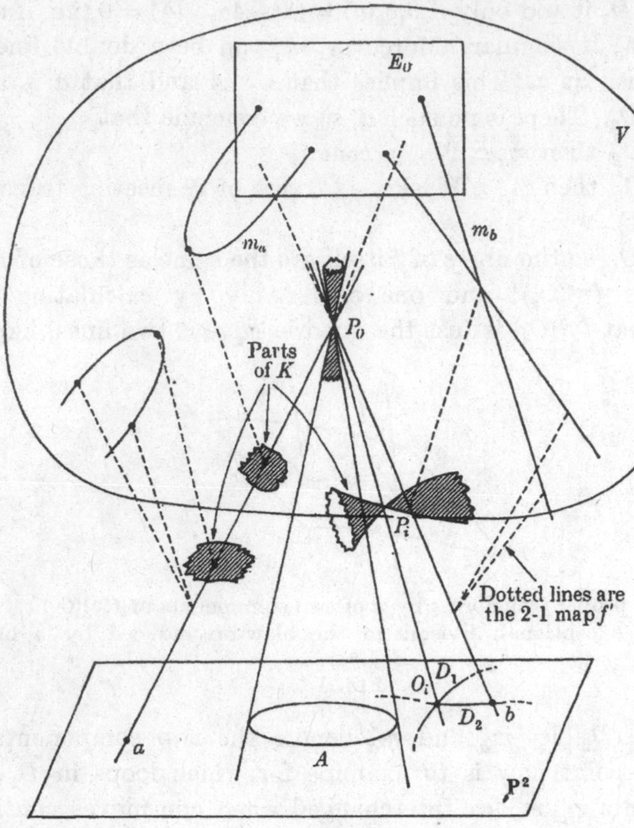

FIG. 2

This means that if we uniformize the elliptic curve D_i by the plane $\mathbf{C}$ modulo two periods ω_i', ω_i'', then for a suitable choice of periods, we get the situation:

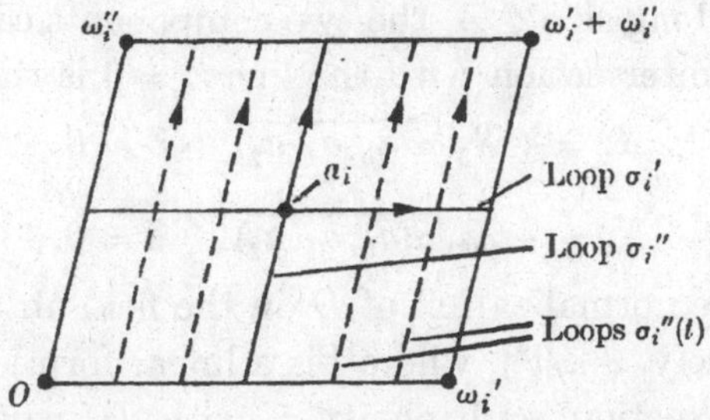

FIG. 3

(i) moving around σ_i', m_a' and m_a'' *are* interchanged,

(ii) moving around σ_i'', m_a' and m_a'' *are not* interchanged.

We are now in a position to prove that V has 2-torsion. We shall use the brutal procedure of constructing a 2-dimensional cycle α and a

3-dimensional cycle β such that

$$2\alpha = \partial\lambda,$$
$$2\beta = \partial\mu,$$
$$|\lambda| \cap |\beta| \text{ is one point } x \quad \text{and}$$
$$|\lambda|, |\beta| \text{ meet transversely at } x.$$

It follows that the cohomology classes $\bar{\alpha}, \bar{\beta}$ of α and β have order at most 2, and that their linking number is $\frac{1}{2}$; hence their order is exactly 2 ([**12**], §77).

Construction of α. Fix a base point $a_i \in D_i$ as in Fig. 3. α is to be the *algebraic* cycle

$$\alpha = m'_{a_1} - m'_{a_2}.$$

In fact, moving around the loop σ'_i it follows that

$$m'_{a_i} \sim m''_{a_i}$$

Hence

$$2m'_{a_i} \sim m'_{a_i} + m''_{a_i} = m_{a_i}.$$

But for any $b, c \in \mathbf{P}^2 - D_1 \cap D_2$,

$$b \sim c \quad \text{on} \quad \mathbf{P}^2 - D_1 \cap D_2;$$

hence

$$m_b \sim m_c \quad \text{in} \quad V.$$

Therefore

$$\begin{aligned} 2\alpha &= 2m'_{a_1} - 2m'_{a_2} \\ &\sim m_{a_1} - m_{a_2} \\ &\sim 0. \end{aligned}$$

Construction of β. Moving the cycle m''_a around the loop σ''_1, it comes back to itself. Therefore

$$\bigcup_{a \in \sigma_1''} m''_a = \beta$$

is a 3-cycle. But moving the whole loop σ''_1 around the curve D_1 as indicated by the dotted lines $\sigma''_1(t)$ in Fig. 3,† β is transformed into

$$\beta^* = \bigcup_{a \in \sigma_1''} m'_a.$$

Thus $\beta \sim \beta^*$, and $2\beta \sim \beta + \beta^* = \bigcup_{a \in \sigma_1''} m_a$. But in $\mathbf{P}^2 - D_1 \cap D_2$, $\sigma''_1 \sim 0$; therefore in V, $\beta + \beta^* \sim 0$.

Finally λ, for instance, is easily seen to be made up of

(a) chains outside $f^{-1}(D_1)$,

(b) for each $a \in \sigma'_1$, one of the two components of m_a.

† Nine of these lines will pass through points O_i. Then the definition of β should be slightly modified to include the whole curve $m''_0 + n''_1$ in the fibre $f^{-1}(O_i)$ (see Fig. 1).

If our notation is chosen suitably, we may assume that λ contains m'_a, if $a = \sigma'_1 \cap \sigma''_1$, hence $|\lambda| \cap |\beta|$ is the one point $m'_a \cap m''_a$, where $a = \sigma'_1 \cap \sigma''_1$. It is clear that the intersection is transversal.

3. The Brauer group of a function field of two variables

Let S be a complete non-singular algebraic surface over an algebraically closed field k. We propose to compute the Brauer group of its function field K in terms of the étale cohomology of S. Since our results are valid only for the part of $\mathrm{Br}\, K$ prime to the characteristic of K, we work throughout this section 'modulo p-groups'. Cohomology will mean étale cohomology ([**2**]).

If S is simply connected, the computation is particularly simple.

THEOREM 1. *Suppose that $H^1(S, \mathbf{Q}/\mathbf{Z}) = 0$. There is a canonical exact sequence*

$$0 \longrightarrow \mathrm{Br}\, S \xrightarrow{\ i\ } \mathrm{Br}\, K \xrightarrow{\ a\ } \bigoplus_{\substack{\text{curves}\\ C}} H^1(K(C), \mathbf{Q}/\mathbf{Z}) \xrightarrow{\ r\ } \bigoplus_{\substack{\text{points}\\ p}} \mu^{-1} \xrightarrow{\ s\ } \mu^{-1} \longrightarrow 0,$$

where the groups and maps are explained below.

(i) μ_n denotes the group of nth roots of unity, $\mu = \bigcup_n \mu_n$, and $\mu^{-1} = \bigcup_n \mu_n{}^{-1} = \bigcup_n \mathrm{Hom}(\mu_n, \mathbf{Q}/\mathbf{Z})$. Thus μ and μ^{-1} are non-canonically isomorphic to $\mathbf{Q}/\mathbf{Z}$.

(ii) Br S denotes the Brauer group of Azumaya algebras on S, and the map i is the restriction to the general point. Since S is a smooth surface, we have $\mathrm{Br}\, S \approx H^2(S, \mathbf{G}_m)$ ([**9**]), and this group fits into an exact sequence

$$0 \to N \otimes \mathbf{Q}/\mathbf{Z} \to H^2(S, \mu) \to \mathrm{Br}\, S \to 0,$$

where N is the Neron–Severi group of S.

(iii) The sum in the third term is taken over all irreducible curves C on S, with function field $K(C)$. Thus $H^1(K(C), \mathbf{Q}/\mathbf{Z})$ is the group of cyclic extensions of $K(C)$, or the group of cyclic *ramified* coverings of the normalization $\bar{C}$ of C.

(iv) The local ring $\mathcal{O}_{S,C}$ of S at the generic point of C is a discrete valuation ring, and so the classical theory of maximal orders ([**7**]) associates to any finite central division ring D a cyclic extension L of the residue field $K(C)$. We recall that L is obtained from a maximum order of A for D over $\mathcal{O}_{S,C}$ as $A \otimes K(C)/(\text{radical})$. This yields the map a. The division ring D is usually said to be *ramified* along the curves C for which this cyclic extension is not trivial.

(v) In the fourth term, the sum is over all closed points of S. Given a cyclic extension of $K(C)$, one may measure its ramification at a point c of $\bar{C}$. This is canonically an element of μ^{-1} ([**2**], Exposé 18 and Exposé 19 (3.3)). The map r is defined as the sum of the ramification at all points of the various $\bar{C}$ lying over p.

(vi) The map s is the sum.

We will prove the analogous result for any irreducible, regular, excellent noetherian scheme S/k whose function field K is of transcendence degree 2 over k and such that $H^1(S, \mathbf{Q}/\mathbf{Z}) = 0$. For technical reasons, we do not assume K to be finitely generated. In order to do this, it is convenient to work formally with the complement of the points p of S of codimension 2 (the residue field at such a point is necessarily the field k). By this we mean the *pro-object* of schemes

$$U = \{S - \pi\}_{\pi \in I},$$

where I denotes the filtering system of finite sets π of points of codimension 2. The cohomology of U is by definition the direct limit

$$H^q(U, F_U) = \varinjlim_{\pi} H^q(S - \pi, F)$$

for any sheaf F on S. Thus computation with this cohomology is a substitute for an obvious limit argument. The relevant morphisms are

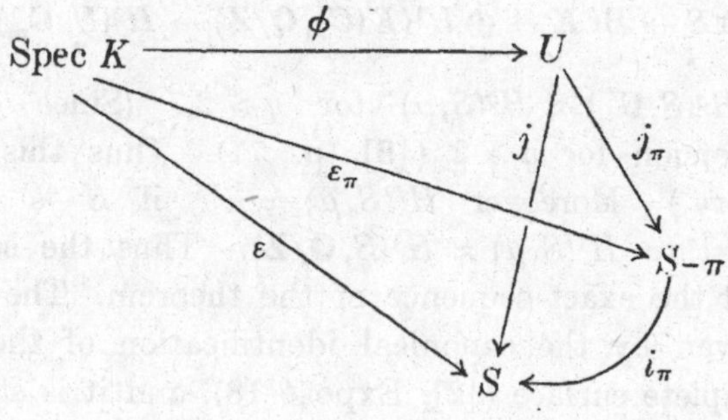

We have $R^3 i_{\pi *}\mathbf{G}_m = \oplus_{p \in \pi} \mu_p{}^{-1}$, where the subscript p denotes extension by zero outside p, and $R^q i_{\pi *}\mathbf{G}_m = 0$ if $q > 0, q \neq 3$. (To see this, note that the henselian local ring $\tilde{\mathcal{O}}_{S,p}$ of S at a point p of codimension 2 is necessarily the ring $k\{x,y\}$ of algebraic series in local parameters x, y. For, we have $k\{x,y\} \subset \tilde{\mathcal{O}}_{S,p} \subset k[[x,y]]$, the ring $k\{x,y\}$ is algebraically closed in $k[[x,y]]$, and $\tilde{\mathcal{O}}_{S,p}$ is algebraic over $k\{x,y\}$ since K has transcendence degree 2. Thus we may apply the results of [**2**] for algebraic schemes. The values of $R^q i_{\pi *}\mathbf{Z}/n\mathbf{Z}$ are given in [**2**] Exposé 16, Théorème (3.7) and the canonical twist by roots of unity is in Exposé 18, or in Exposé 19,

Théorème (3.4). Then the values for $\mathbf{G}_m$ follow from Kummer theory (Exposé 9, Théorème (3.2)).) Passing to the limit over the spectral sequences for i_π, we obtain $\operatorname{Br} S = H^2(U, \mathbf{G}_m)$, and

$$(3.1) \qquad 0 \to H^3(S, \mathbf{G}_m) \to H^3(U, \mathbf{G}_m) \to \bigoplus_p \mu^{-1} \to H^4(S, \mathbf{G}_m) \to 0.$$

Next, we have $R^q \varepsilon_{\pi *} \mathbf{G}_m = 0$ if $q > 0, q \neq 2$; and this sheaf is concentrated at the points of $S - \pi$ of codimension 2, if $q = 2$ (cohomological dimension of K ([**2**], Exposé 10)). Thus $R^q \varphi_* \mathbf{G}_m = 0$ for all $q > 0$, i.e.

$$H^q(\operatorname{Spec} K, \mathbf{G}_m) \approx H^q(U, \varphi_* \mathbf{G}_m).$$

Moreover, we have an exact sequence

$$0 \to \mathbf{G}_m \to \varphi_* \mathbf{G}_m \to \bigoplus_C \mathbf{Z}_{K(C)} \to 0,$$

where C runs over irreducible closed sets of codimension 1, and where $\mathbf{Z}_{K(C)}$ denotes the extension by zero of the constant sheaf on $\operatorname{Spec} K(C) = C \cap U$. Clearly $H^q(U, \mathbf{Z}_{K(C)}) \approx H^q(K(C), \mathbf{Z})$. Since

$$H^1(K(C), \mathbf{Z}) = 0 \quad \text{and} \quad H^3(K, \mathbf{G}_m) = 0$$

([**2**], Exposé 9 (3.6), and 10) we obtain the exact cohomology sequence

$$0 \longrightarrow H^2(U, \mathbf{G}_m) \longrightarrow H^2(K, \mathbf{G}_m)$$

$$\xrightarrow{\ r\ } \bigoplus_C H^2(K(C), \mathbf{Z}) \longrightarrow H^3(U, \mathbf{G}_m) \longrightarrow 0$$

or

$$(3.2) \qquad 0 \to \operatorname{Br} S \to \operatorname{Br} K \to \bigoplus_C H^1(K(C), \mathbf{Q}/\mathbf{Z}) \to H^3(U, \mathbf{G}_m) \to 0.$$

We have $H^q(S, \mathbf{G}_m) \approx H^q(S, \mu)$ for $q > 2$. (Since S is regular, $H^q(S, \mathbf{G}_m)$ is torsion for $q \geqslant 2$ ([**9**], p. 71). Thus this follows from Kummer theory.) Moreover $H^4(S, \mu) \approx \mu^{-1}$ if S is complete and $H^3(S, \mu)$ is dual to $H^1(S, \mu) \approx H^1(S, \mathbf{Q}/\mathbf{Z})$. Thus the sequences (3.1) and (3.2) yield the exact sequence of the theorem. The fact that s is the sum is given by the canonical identification of the fundamental class on a complete surface ([**2**], Exposé 18), and it is standard that a is the correct map.

It remains to determine the map r, and for this purpose we may pass to the henselization at a given point $p \in S$. Since then S is acyclic, the sequences (3.1) (3.2) reduce to $H^3(U, \mathbf{G}_m) \approx \mu^{-1}$ and

$$(3.3) \quad 0 \longrightarrow \operatorname{Br} K \longrightarrow \bigoplus_C H^1(K(C), \mathbf{Q}/\mathbf{Z}) \xrightarrow{\ r\ } \mu^{-1} \longrightarrow 0,$$

where now $U = S - p$. Since S is henselian, so is each C, and so there is a canonical isomorphism $H^1(K(C), \mathbf{Q}/\mathbf{Z}) \approx \mu^{-1}$. We want to check that with this identification r becomes the identity map on each summand.

If C is non-singular, this is equivalent to the transitivity assertion of [2], Exposé 19 (3.4) for the inclusions $p \subset C \subset S$, as is easily seen. In order to prove it in general, it suffices to show that the map

$$H^1(K(C), \mathbf{Q}/\mathbf{Z}) \xrightarrow{\ r\ } \mu^{-1}$$

does not change if we blow up the point p in S and rehenselize at the closed point p' of the proper transform C' of C.

Let $\pi: S' \to S$ be this blowing-up. Choose a non-singular branch D in S, tangent to C, so that its proper transform D' passes through p'. Let $\alpha \in H^1(K(C), \mathbf{Q}/\mathbf{Z})$ have image $r(\alpha)$ in μ^{-1}, and choose $\beta \in H^1(K(D), \mathbf{Q}/\mathbf{Z})$ with $r(\beta) = r(\alpha)$. By the exact sequence (3.3), there is a unique class $d \in \operatorname{Br} K$ with $a(d) = \alpha - \beta$. Consider this class on the scheme S'. The irreducible closed sets of S' of codimension 1 are the proper transforms of branches in S, and the exceptional curve E. Therefore if we denote by a prime the replacement of S by S', we have

$$a'(d) = \alpha - \beta + \varepsilon,$$

where $\varepsilon \in H^1(K(E), \mathbf{Q}/\mathbf{Z})$. Since C', D' both pass through p', ε can ramify only at p'. But E is a rational curve, and so this implies that $\varepsilon = 0$ (we are ignoring p-groups!). Therefore $r'(\alpha) = r'(\beta)$. Since D is non-singular, $r'(\beta) = r(\beta)$. Thus $r'(\alpha) = r(\alpha)$ as required.

4. Conic bundles over surfaces

Let S be a non-singular complete simply connected surface over k as in § 3, but assume now that $\operatorname{char}(k) \neq 2$. We want to specialize the results of § 3 to quaternion algebras. (By a quaternion algebra, we mean simply a rank 4 Azumaya algebra.) It is a classical result that there is a one–one correspondence between:

(a) quaternion algebras A_η over the function field K of S, and

(b) curves V_η over K, isomorphic over the algebraic closure $\bar{K}$ of K to $\mathbf{P}^1_{\bar{K}}$.

Moreover each such curve V_η is isomorphic to a conic in $\mathbf{P}^2_K$, and this conic is unique up to a projective transformation. This correspondence has been extended by Grothendieck ([9]) to show, for instance, that for any Zariski-open set $U \subset S$, there is a one–one correspondence between:

(a′) quaternion algebras A over U, and

(b′) U-schemes $\pi: V \to U$, proper and flat over U, all of whose geometric fibres are isomorphic to $\mathbf{P}^1$. Moreover, such a V can be (essentially uniquely) embedded as a bundle of conics in a $\mathbf{P}^2$-bundle over U.

The correspondence is set up as follows. Given A, define V as a functor by

$$V(S') = \{\text{left ideals } L \text{ of } A \otimes \mathcal{O}_{S'} \text{ which are locally free of rank } 2\}.$$

This is clearly a closed subscheme of the Grassmannian of 2-dimensional submodules of A, and one sees easily that it is smooth over U with fibres isomorphic to $\mathbf{P}^1$.

Next, let A_η be any quaternion algebra over K: it represents an element $d \in \operatorname{Br} K$ of order 2. By Theorem 1 of § 3, there is a finite number of curves $C_1, \ldots, C_n$ on S at which $a(d)$ is not zero. The union $C = C_1 \cup \ldots \cup C_n$ is called the *ramification curve* of the algebra D and $S - C$ is the maximal Zariski open set U in S such that A_η extends to an Azumaya algebra over U. In fact the *maximal orders* A in A_η over U are precisely the Azumaya algebras extending A_η. What happens over C however? We want to analyse the case in which C is non-singular. Choose any maximal order A in A_η over the whole of S. Since S is a smooth surface, A will be locally free of rank 4. For further details, see [**3**].

Proposition 2. *A maximal order A may be presented locally at a point $p \in C$ as the $\mathcal{O}_S$-algebra generated by elements x, y, with relations*

$$(4.1) \qquad \begin{cases} x^2 = a, \\ y^2 = bt, \\ xy = -yx, \end{cases}$$

where $t = 0$ is a local equation for C, and a, b are units in $\mathcal{O}_S$. Moreover, a is not congruent to a square (modulo t).

Conversely, when a is not congruent to a square, the algebra presented in this way is a maximal order in some (non-trivial) quaternion algebra.

Proof. We look first at a generic point of C. The local ring of X is a discrete valuation ring R with residue field $K(C)$, and we may apply the classical theory of maximal orders ([**7**]). It tells us that there is a unique prime ideal $p \subset A$ containing t, $A/p = L$ is a quadratic field extension of $K(C)$, and that $p^2 = At$. Choose $x \in A$ which reduces to a generator of L over $K(C)$, and has (reduced) trace zero, so that $x^2 = -\det x = a$ is a unit of R.

Next, note that if $y \in p$, then $\operatorname{tr} y \equiv 0 \pmod t$; for $y \to \bar{y} = \operatorname{tr} y - y$ is an anti-automorphism of A, hence maps p to p. Thus

$$\operatorname{tr} y = y + \bar{y} \in p \cap R = tR.$$

It follows that if y_0 is a non-zero element of p, then we can choose $\alpha, \beta \in Rt$ so that $y = \alpha + \beta x + y_0$ satisfies

$$\operatorname{tr} y = \operatorname{tr} xy = 0.$$

The required values are

$$(4.2)\qquad \begin{cases} \alpha = -\tfrac{1}{2}\operatorname{tr} y, \\ \beta = -\dfrac{1}{2a}\operatorname{tr} xy. \end{cases}$$

Then y, xy will form a basis for p/At over L, and by the Nakayama lemma, $\{1, x, y, xy\}$ is an R-basis for A. Note that $\overline{xy} = -xy$ since $\operatorname{tr} xy = 0$. Therefore

$$-xy = \overline{xy} = \bar{y}\bar{x} = (-y)(-x) = yx.$$

If we write $y^2 = bt$, then it is easily seen that b must be a unit of R since A is a maximal order. Thus the required presentation exists over R.

Now consider a closed point $p \in C$. Let $\mathcal{O}_X, \mathcal{O}_C$ denote the local rings of X and C there, and denote by p the kernel of the natural map $A \to A \otimes R \to L$. Then A/p is zero outside C, and hence is a free, rank 2 $\mathcal{O}_C$-module. We may choose an element $x \in A$ which reduces to a generator for A/p and has trace zero as above: $x^2 = a \in \mathcal{O}_X$. Choose $u, v \in p$ so that $\{1, x, u, v\}$ is an $\mathcal{O}_X$-basis for the free module A. By (4.2) we may adjust u, v to have trace zero.

The standard bilinear form $(\alpha, \beta) \to \operatorname{tr} \alpha\beta$ on A is non-degenerate wherever A is an Azumaya algebra, i.e. except on C. Thus its determinant δ (the discriminant of A) is of the form $\delta = \varepsilon t^r$, ε a unit. But calculation of the determinant with respect to the R-basis $\{1, x, y, xy\}$ yields $(4abt)^2$. Hence $r = 2$. Now calculate with respect to the basis $\{1, x, u, v\}$ using the fact that $\operatorname{tr} y \equiv 0 \pmod{t}$ for all $y \in p$. This gives

$$\delta = \det \begin{bmatrix} 2 & 0 & 0 & 0 \\ 0 & 2a & * & * \\ 0 & * & * & * \\ 0 & * & * & * \end{bmatrix}, \quad * \equiv 0 \pmod{t},$$

and hence

$$\varepsilon t^2 = 4a\xi t^2 + \eta t^3, \quad \text{with } \xi, \eta \in \mathcal{O}_X.$$

Therefore, a is a unit.

Since a is a unit, $A/p = \mathcal{O}_C[\sqrt{a}]$ is a semi-local dedekind domain. Therefore p/At is free of rank 1 as left A/p-module, and we choose a generator y of this module. By (4.2), we may assume that $\operatorname{tr} y = \operatorname{tr} xy = 0$, and then the above computation shows that $xy = -yx$, and $y^2 = bt$, $b \in \mathcal{O}_X$. The discriminant is $(4abt)^2 = \varepsilon t^2$. Thus b is a unit.

It remains to prove the converse, so let A have the above presentation. Then A is clearly an Azumaya algebra except on C, and if $\bar{A}$ is a maximal order containing A, then the determinant of the standard form on $\bar{A}$ is either $\equiv 0 \pmod{t^2}$ as above, or is a unit (i.e. $\bar{A}$ is an Azumaya algebra).

Comparing determinants, we see that in the former case, $A = \bar{A}$, hence A is maximal, while in the latter, the cokernel $\bar{A}/A$ must have rank 1 on C. But since a is not a square, $(A/tA)_{\text{red}}$ is irreducible and quadratic over $\mathcal{O}_C$, and so that is impossible.

Using this proposition, we can extend the correspondence between orders and conic bundles as follows.

THEOREM 2. *There is a canonical one-one correspondence between*

(a″) *maximal orders A in quaternion algebras A_η over K, whose ramification curves are non-singular,*

(b″) *non-singular S-schemes $\pi: V \to S$ proper and flat over S, all of whose geometric fibres are isomorphic to $\mathbf{P}^1$ or to $\mathbf{P}^1 \vee \mathbf{P}^1$, such that the following condition holds: for every irreducible curve C_i along which the fibres of π are reducible, the two components of $\pi^{-1}(c_i)$ ($c_i \in C_i$, the generic point) are not rational over $K(C_i)$, but define a quadratic extension of this field.*

Moreover, the quadratic extensions so defined are just those given by $a(A_\eta)$, as in § 3.

We call V the *Brauer–Severi scheme* of A.

Proof. The correspondence is set up exactly as before. Given A, we let V represent the functor of left ideals of A. We need to check the structure of this scheme above points of C. First of all, it is immediately seen from the presentation (4.1) that at any point x of S, an element of $A/m_x A$ generates a left ideal of dimension at least 2. Thus our left ideals L in $A \otimes \mathcal{O}_{S'}$ for any S' will all be principal. Next, L cannot lie entirely in the subspace spanned by $\{1, x, y\}$. Thus there is a non-zero element $u = p + qx + ry \in L$ which is unique up to scalar multiplication. Such an element u generates a left ideal of rank 2 if and only if u, xu, yu are linearly dependent. A small computation shows that this is equivalent to the equation

$$\varphi(t, p, q, r) = p^2 - aq^2 - btr^2 = 0,$$

homogeneous in p, q, r. Therefore $\varphi = 0$ defines V as a subscheme of $\mathbf{P}^2_s$. The locus defined by this equation is smooth over s except at the points $t = p = q = 0$, and at these points $\partial\varphi/\partial t \neq 0$. Thus V is non-singular. Moreover, if A is a maximal order, then a is not congruent to a square along C, hence the two components of $p^2 - aq^2$ are not rational.

Thus we have a functor from algebras A with local presentation (4.1) to non-singular S-schemes V as in the first part of (b″), and, by Proposition 2, A is a maximal order if and only if the final condition of (b″) holds.

Now the versal local deformation of $\mathbf{P}^1 \vee \mathbf{P}^1$ is the 1-parameter family

$$X_1 X_2 - t X_0^2 = 0 \tag{4.3}$$

at $t = 0$, in the plane over $k[t]$. Thus any flat, proper deformation $V \to S$ of $\mathbf{P}^1 \vee \mathbf{P}^1$ is induced locally, say for the étale topology, by a map $S \to \operatorname{Spec} k[t]$. One sees immediately that if S is our surface and V is non-singular, the locus $t = 0$ on S must be a non-singular curve C, i.e. t must be a local parameter. Thus V can be put, locally for the étale topology, into the standard form (4.3). This is isomorphic to the *Brauer–Severi scheme* of the standard algebra having presentation (4.1) with $a = b = 1$. On the other hand, any algebra with presentation (4.1) is isomorphic to this standard one in the étale neighbourhood $\operatorname{Spec}(\mathcal{O}_S[\sqrt{a}, \sqrt{b}])$. Thus to show our correspondence one–one, it suffices by 'descent', to show that the map of étale sheaves on S:

$$\operatorname{Aut} A \xrightarrow{\ \varepsilon\ } \operatorname{Aut} V$$

is an isomorphism when A is the standard algebra above. This is certainly true outside C, where A is an Azumaya algebra. Hence the map is injective. Moreover, A admits the automorphism $x \to -x$, which interchanges the line pair $\pi^{-1}(c)$ $(c \in C)$. Thus we need to consider only automorphisms fixing these pairs.

Consider the matrix representation

$$x = \begin{pmatrix} 1 & 0 \\ 0 & -1 \end{pmatrix}, \quad y = \begin{pmatrix} 0 & t \\ 1 & 0 \end{pmatrix}.$$

This identifies A with the subring of the matrix algebra $\bar{A} = M_2(\mathcal{O}_S)$ consisting of matrices

$$\begin{pmatrix} a & bt \\ c & d \end{pmatrix}, \quad a, b, c, d \in \mathcal{O}_S.$$

The inclusion $A \to \bar{A}$ induces a birational map

$$V \xrightarrow[f]{} \bar{V},$$

an isomorphism outside of C. Of course, $\bar{V} \cong \mathbf{P}^1_S$, and one sees easily that f is a morphism, and is in fact the contraction of one of the two families of lines making up $\pi^{-1}(C)$. This line family is mapped to the section v_1 of $\bar{V} \times_S C$ over C corresponding to the C-family L_1 of left ideals generated by $e_{11} = \frac{1}{2}(1 + x)$.

Clearly, any automorphism σ of V not interchanging the lines will induce an automorphism $\bar{\sigma}$ of $\bar{V}$ leaving v_1 fixed, and $\bar{\sigma}$ comes from an automorphism $\bar{\varphi}$ of $\bar{A}$ such that $\bar{\varphi}$ carries L_1 to L_1 (modulo t). But L_1 is the ideal of matrices

$$\begin{pmatrix} * & 0 \\ * & 0 \end{pmatrix} \quad \text{(modulo } t\text{)},$$

and A is exactly the subring of $\bar{A}$ of right multipliers of L_1. Thus $\bar{\varphi}$ carries A to itself, i.e. induces an automorphism φ of A, which in turn induces σ. This proves the surjectivity of ε.

As an example of Brauer–Severi schemes, note that cubic hypersurfaces $H \subset \mathbf{P}^4$ with a sufficient generic line blown up are Severi–Brauer schemes V over $\mathbf{P}^2$. In fact, it is well known ([**6**]) that there is an irreducible 2-dimensional family l_α ($\alpha \in Z$) of lines on H, and it is easy to check that for almost all these lines l_α, there is *no* plane $L \supset l_\alpha$ tangent to H along l_α, or equivalently such that $L \cdot H = 2l_\alpha + l_{\alpha'}$ (some line $l_{\alpha'}$). Thus there is only a 1-dimensional set of lines l_α ($\alpha \in Z_0 \subsetneqq Z$) such that for some plane $L \supset l_\alpha$, $L \cdot H = l_\alpha +$(double line). Pick any $\alpha \in Z - Z_0$. Let H^* be the blow up of H along l_α. Then the projection of $\mathbf{P}^4$ to $\mathbf{P}^2$ with centre l_α extends to a morphism:

$$\pi : H^* \to \mathbf{P}^2$$

and it is easily seen that all fibres $\pi^{-1}(p)$ are either irreducible conics or pairs of distinct lines. It can be checked that the ramification curve C is a non-singular quintic in $\mathbf{P}^2$ and that the generic line-pair over C does not split.

We would like to look, however, at cases where C is reducible (hence, since we are assuming C non-singular, C is also disconnected).

Proposition 3. *With notation as in Theorem* 2, *assume that C is disconnected. Then the Brauer–Severi scheme V has* 2-*torsion in* $H^4(V, \mathbf{Z}_2)$.

Here the cohomology denotes the étale 2-adic cohomology. The reader who wishes to restrict to the case $k = \mathbf{C}$ can replace this by ordinary $\mathbf{Z}$-cohomology without changing the discussion below.

Proof. The idea here is the same as in § 2. We choose one of the lines l_i, making up $\pi^{-1}(p_i)$, where $p_i \in C_i$. Then $l_1 - l_2$ represents a class in H^4 of order 2, and we just need to check that this class is not itself zero. We do this by an analysis of the spectral sequence for the map π. If $p \in S$, then the cohomology of the fibre $\pi^{-1}(p)$ is just

$$H^0(\pi^{-1}(p), \mathbf{Z}/2\mathbf{Z}) = \mathbf{Z}/2\mathbf{Z},$$

$$H^1(\pi^{-1}(p), \mathbf{Z}/2\mathbf{Z}) = (0),$$

$$H^2(\pi^{-1}(p), \mathbf{Z}/2\mathbf{Z}) = \begin{cases} \mathbf{Z}/2\mathbf{Z}, & p \notin C, \\ \mathbf{Z}/2\mathbf{Z} + \mathbf{Z}/2\mathbf{Z}, & p \in C, \end{cases}$$

where the last isomorphism is as follows: given a class in $H^2(\pi^{-1}(p), \mathbf{Z}/2\mathbf{Z})$, for $p \in C$, restrict it to each component of $\pi^{-1}(p)$ and evaluate using the isomorphism $H^2(Y, \mathbf{Z}/2\mathbf{Z}) \cong \mathbf{Z}/2\mathbf{Z}$, if Y is an irreducible curve. Therefore by the proper base change theorem:

$$\pi_*(\mathbf{Z}/2\mathbf{Z}) = (\mathbf{Z}/2\mathbf{Z})_S,$$
$$R^1\pi_*(\mathbf{Z}/2\mathbf{Z}) = (0).$$

Now locally in the étale topology, π has sections through any smooth point. Taking the fundamental class of such sections, it follows that $R^2\pi_*(\mathbf{Z}/2\mathbf{Z})$ has sections locally which, evaluated point by point, are 0 or 1 on a component l of $\pi^{-1}(p)$ according as the section misses or hits l. In particular, if $p \in C$, and $\pi^{-1}(p) = l' \cup l''$, then taking the sum of the classes of sections through l' and l'', we get a section of $R^2\pi_*(\mathbf{Z}/2\mathbf{Z})$ which is 0 outside C, and has value $(1,1)$ along C. Thus we get

$$(\mathbf{Z}/2\mathbf{Z})_C \to R^2\pi_*(\mathbf{Z}/2\mathbf{Z})$$

and it is easy to see that the cokernel is $(\mathbf{Z}/2\mathbf{Z})_S$ (we shall not need this however). To show that $\mathrm{cl}(l_1) - \mathrm{cl}(l_2) \in H^4(V, \hat{\mathbf{Z}}_2)$ is not zero, it suffices to produce an element $\zeta \in H^2(V, \mathbf{Z}/2\mathbf{Z})$ and prove that

$$[\mathrm{cl}(l_1) - \mathrm{cl}(l_2)] \cup \zeta \in H^6(V, \mathbf{Z}/2\mathbf{Z}) \cong \mathbf{Z}/2\mathbf{Z}$$

is not zero. Let i_1 and i_2 be the inclusion of l_1 and l_2 in V. Then

$$[\mathrm{cl}(l_1) - \mathrm{cl}(l_2)] \cup \zeta = i_1^*(\zeta) - i_2^*\zeta$$

(where $i_k^*(\zeta) \in H^2(l_k, \mathbf{Z}/2\mathbf{Z}) \cong \mathbf{Z}/2\mathbf{Z}$). The Leray spectral sequence for π gives

$$0 \to H^2(S, \mathbf{Z}/2\mathbf{Z}) \to H^2(V, \mathbf{Z}/2\mathbf{Z}) \to H^0(S, R^2\pi_*\mathbf{Z}/2\mathbf{Z}) \to H^3(S, \mathbf{Z}/2\mathbf{Z}).$$

But $H^3(S, \mathbf{Z}/2\mathbf{Z}) \cong (0)$ since S is simply connected. Let α_1 be a section of $(\mathbf{Z}/2\mathbf{Z})_C$ which is 1 on C_1 and 0 on the other components; let α_1' be its image in $H^0(S, R^2\pi_*\mathbf{Z}/2\mathbf{Z})$; and let $\alpha_1'' \in H^2(V, \mathbf{Z}/2\mathbf{Z})$ lift α_1'. Then

$$[\mathrm{cl}(l_1) - \mathrm{cl}(l_2)] \cup \alpha_1'' = i_1^*\alpha_1'' - i_2\alpha_1'' = 1 - 0.$$

We are now in a position to construct some examples. Choose a conic A in $\mathbf{P}^2$. If we assign arbitrarily points $p_1, \ldots, p_r$ on A, we can find non-singular curves C of degree r having a double intersection with A at each of the points $p_1, \ldots, p_r$. This is easy to see. Now choose $n \geqslant 2$ such curves $C_1, \ldots, C_n$ of degree $r_i \geqslant 3$. (The points need not be the same for the various curves.) The example of § 2 will be obtained by taking $n = 2$, $r_1 = r_2 = 3$. Let q be the rational function on $\mathbf{P}^2$ whose divisor is $A - 2L$, with L the line at infinity.

LEMMA. *If $r \geqslant 3$, the restriction of q to C is not a square in $K(C)$.*

Proof. Let $\bar{q}$ denote the restriction of q to C. Suppose that $\bar{q} = \bar{s}^2$, $\bar{s} \in K(C)$. Then $\bar{s} \in \Gamma(C, \mathcal{O}_C(C \cdot L))$. But the map

$$\Gamma(\mathbf{P}^2, \mathcal{O}_{\mathbf{P}^2}(L)) \to \Gamma(C, \mathcal{O}_C(C \cdot L))$$

is surjective, so $\bar{s}$ lifts to a function s. Let $(s) = L' - L$, where L' is another line. Then

$$\begin{aligned} A \cap C &= \text{zeros of } q \text{ on } C \\ &= \text{zeros of } s \text{ on } C \\ &= L' \cap C; \end{aligned}$$

hence $A \cap C \subset A \cap L'$ which consists of at most 2 points. But $A \cap C$ consists of $r \geqslant 3$ points. Contradiction.

Applying the lemma to our curves C_i, we obtain a quadratic extension L_i of $K(C_i)$ by adjoining $\sqrt{q}$. This extension is *everywhere unramified* on C_i, since C_i has only double intersections with the zeros and poles of q.

Let S be the result of blowing up $\mathbf{P}^2$ at the intersection points of the C_i until their proper transforms become disjoint. Denote these proper transforms by C_i as well. Now S is a rational surface, and hence has trivial Brauer group (this follows easily from the exact sequence of Theorem 1 (ii)), and is simply connected. Therefore we may apply Theorem 1 to find a unique class $d \in \operatorname{Br} K$ $(K = K(S))$ such that $a(d)$ is zero except on $C_1, \ldots, C_n$, and is the class of L_i on C_i.

This class has order 2 since each extension is quadratic. Now ordinarily we do not know which classes of order 2 in $\operatorname{Br} K$ correspond to quaternion algebras. But in this case, we can split d in the quadratic extension $K(Y) = K[\sqrt{q}]$. In fact, going back to $\mathbf{P}^2$, our element d is unramified except on C_i. The double cover Y splits the assigned classes in $H^1(K(C_i), Q/\mathbf{Z})$ by construction. Thus the pull-back of d to $K(Y)$ is everywhere unramified, i.e. lies in $\operatorname{Br} Y$. But Y is a rational surface, in fact is isomorphic to a quadric in $\mathbf{P}^3$. Hence $\operatorname{Br} Y = 0$ and so d splits in $K(Y)$. By Brauer's construction ([**14**], p. 167), a class in $\operatorname{Br} K$ which splits in a Galois extension of degree n is represented by an algebra of rank n^2. Thus d is the class of a quaternion algebra D.

Let V be the Brauer–Severi variety of a maximal order of D over $\mathcal{O}_S$. Then V is a non-singular 3-fold by Theorem 2 and $H^4(V, \mathbf{Z}_2)$ contains 2-torsion by Proposition 3. Finally, V is unirational. For, since D splits in $K(Y)$, the generic fibre of V over S becomes isomorphic to $\mathbf{P}^1$ over $K(Y)$, which is a rational function field. Thus $K(V)[\sqrt{q}]$ is rational.

REFERENCES

1. S. Abhyankar, *Resolution of singularities of embedded algebraic surfaces* (Academic Press, New York, 1966).
2. M. Artin, A. Grothendieck and J.-L. Verdier, *Séminaire de géométrie algébrique: cohomologie étale des schémas* (Inst. Hautes Etudes Sci., 1963–64, mimeographed notes).

3. M. Auslander and O. Goldman, 'The Brauer group of a commutative ring', *Trans. Amer. Math. Soc.* 97 (1960) 367–409.
4. A. Borel and A. Haefliger, 'La classe d'homologie fondamentale d'un espace analytique', *Bull. Soc. Math. France* 89 (1961) 461–513.
5. N. Bourbaki, *Algèbre commutative*, Chs. I, II (Hermann, Paris, 1961).
6. C. H. Clemens and P. A. Griffiths, 'The intermediate jacobian of the cubic threefold', *Ann. of Math.*, to appear.
7. M. Deuring, *Algebren* (Springer, Berlin, 1935).
8. G. Fano, 'Sul sistema ∞^2 di rette contenuto in una varietà cubica', *Atti R. Accad. Sci. Torino* 39 (1904) 778–92.
9. A. Grothendieck, 'Le groupe de Brauer' (*Dix exposés sur la cohomologie des schémas* (North Holland, Amsterdam, 1968)).
10. L. Roth, *Algebraic threefolds* (Springer, Berlin, 1955).
11. B. Segre, 'Variazione continua ed omotopia in geometria algebrica', *Ann. Mat. Pura Appl.* 50 (1960) 149–86.
12. H. Seifert and W. Threlfall, *Lehrbuch der Topologie* (B. G. Teubner, 1934, or Dover reprint).
13. J.-P. Serre, 'On the fundamental group of a unirational variety', *J. London Math. Soc.* 34 (1959) 481–84.
14. —— *Corps locaux* (Hermann, Paris, 1962).
15. A. Weil, 'On Picard varieties', *Amer. J. Math.* 74 (1952) 865–93.
16. V. A. Iskovskikh and Ju. I. Manin, 'Three-dimensional quartics and counterexamples to the Lüroth problem', *Mat. Sb.* 86 (1971) 140–66.

Mathematics Department
Massachusetts Institute of Technology
Cambridge, Mass. 02139

Mathematics Department
Harvard University
2 *Divinity Avenue*
Cambridge, Mass. 02138

AN ALGEBRAIC SURFACE WITH K AMPLE, $(K^2) = 9$, $p_g = q = 0$

By D. MUMFORD

Severi raised the question of whether there existed an algebraic surface X homeomorphic to $\mathbf{P}^2$ but not isomorphic to it (as a variety), and conjectured that such a surface did not exist. The essential problem in proving this is to eliminate the possibility that the canonical class K, as a member of the infinite cyclic group $H^2(X, \mathbf{Z})$ might be a *positive* multiple (in fact, 3) of the ample generator of $H^2(X, \mathbf{Z})$ instead of a negative multiple (in fact, -3) as it ought to be. That this is the problem is clear from Castelnuovo's criterion for rationality, and was analyzed and generalized to higher dimensions in the paper [3] of Hirzebruch and Kodaira where it was shown that in *odd* dimensions, $\mathbf{P}^n$ is the only variety in its homeomorphism class. Severi's question was finally answered by S. Yau [8] as a Corollary of his result that all varieties X on which K is ample carry a unique Kähler-Einstein metric. In fact, this result shows that when X is a surface on which K is ample, then the Chern numbers satisfy $c_1^2 \leq 3c_2$, with equality if and only if X is isomorphic to D_2/Γ (D_2 = unit ball in $\mathbf{C}^2$, $\Gamma \subset SU(2, 1)/(\text{center})$ a discrete torsion-free co-compact subgroup; Hirzebruch in [2] had much earlier shown that the surfaces D_2/Γ did satisfy $c_1^2 = 3c_2$). However, the question arises: how close can we come to a surface with K ample which mimics the topology of $\mathbf{P}^2$? In particular, does there exist such a surface with the *same Betti numbers as* $\mathbf{P}^2$? By the standard results on algebraic surfaces, this means that we seek a surface X such that:

$$p_g = q = 0, \quad \text{hence } \chi(\mathcal{O}_X) = 1$$

$$(c_1^2) = (K^2) = 9$$

$$B_0 = B_2 = B_4 = 1, \qquad B_1 = B_3 = 0, \qquad \text{hence } c_2 = 3$$

I shall exhibit one such surface. My method is not by complex uniformization, as used by Shavel [6] and Jenkins (unpublished) in the construction of a surface with K positive and the same Betti numbers as $\mathbf{P}^1 \times \mathbf{P}^1$, but by the p-adic uniformization introduced recently by Kurihara [1] and Mustafin [5]. After looking for such an example at some length, I would hazard the guess that there are, in fact, very few such surfaces (combining Yau's results with Weil's theorem [7] that discrete co-compact groups $\Gamma \subset SU(2, 1)$ are rigid, it follows that there are in any case only *finitely* many such surfaces). But it seems a difficult matter to find some way of enumerating all such surfaces.

1. p-adic uniformizations in general. In this section we wish to summarize and extend somewhat the results of Kurihara and Mustafin cited above, restricting ourselves however to the 2-dimensional case. Let R be a complete discrete valuation ring with fraction field K and residue field $k = R/\pi R$. We assume k is finite. The basis of the construction is a beautiful scheme $\mathfrak{X}$, locally of finite type over R, which may be described by charts as follows:

$$\mathfrak{X} = \bigcup_{A \in GL(3,K)} \operatorname{Spec} R\left[\frac{l_0}{l_1}, \frac{l_1}{l_2}, \pi\frac{l_2}{l_0}\right] - (C_0 \cup C_1 \cup C_2)$$

where $l_i = \sum_{j=0}^{2} A_{ij} x_j$, $A = (A_{ij})$

C_0 = set of curves

$$\pi = \frac{l_0}{l_1} = 0, \quad a\left(\frac{l_1}{l_2}\right)\cdot\left(\pi\frac{l_2}{l_0}\right) + b\left(\pi\frac{l_2}{l_0}\right) + c = 0$$

$a, b, c \in k$, $a \cdot c \neq 0$, plus the curves

$$\pi = \frac{l_0}{l_1} = 0, \quad \left(\pi\frac{l_2}{l_0}\right) + c = 0$$

and $$\pi = \frac{l_0}{l_1} = 0, \quad \frac{l_1}{l_2} + c = 0 \quad (c \in k^*).$$

C_1, C_2 = similar sets of curves where the role of l_0/l_1, l_1/l_2, $\pi(l_2/l_0)$ are permuted cyclically.

Here the glueing represented by the union sign is induced by the requirement that $\mathfrak{X}$ is irreducible and separated with function field

$$K\left(\frac{X_1}{X_0}, \frac{X_2}{X_0}\right),$$

which is the common fraction field of all affine rings.

The closed fibre $\mathfrak{X}_0$ of $\mathfrak{X}$ can be represented graphically by means of the Bruhat-Tits building Σ attached to PGL(3, K). In fact, the 3 sets:

Components E of $\mathfrak{X}_0$
Free rank 3 R-submodules $M \subset K\cdot X_0 \oplus K\cdot X_1 \oplus K\cdot X_2$, modulo $M \sim \pi^k\cdot M$
Vertices ν of Σ

are isomorphic. Moreover, the components of $\mathfrak{X}_0$ cross normally, and if E_i, M_i, ν_i, $i = 1, 2, 3$, correspond as above, then:

a) $E_1 \cap E_2$ is a curve $\Leftrightarrow M_1 \supsetneqq \pi^k M_2 \supsetneqq \pi M_1$, some $k \in \mathbf{Z}$

$\Leftrightarrow \nu_1, \nu_2$ are joined in Σ by a segment

b) $E_1 \cap E_2 \cap E_3$ is a triple point $\Leftrightarrow M_1 \supsetneqq \pi^k M_2 \supsetneqq \pi^l M_3 \supsetneqq \pi M_1$, some $k, l \in \mathbf{Z}$ (or same with 2, 3 interchanged)

$\Leftrightarrow \nu_1, \nu_2, \nu_3$ are the vertices of 2-simplex in Σ.

To describe $\mathfrak{X}$ in a Zariski-open neighborhood of some component E of $\mathfrak{X}_0$, we can proceed geometrically as follows: let E correspond to M and let Y_0, Y_1, Y_2 be an R-basis of M. Start with $\mathbf{P}_R^2$ based on homogeneous coordinates Y_0, Y_1, Y_2 (hence with function field $K(X_1/X_0, X_2/X_0)$ still). First, blow up all k-rational points of the closed fibre $\mathbf{P}_k^2$ of $\mathbf{P}_R^2$ (if k has q elements, there are $q^2 + q + 1$ of these). Second, blow up the proper transforms on this scheme of all k-rational lines on the original closed fibre $\mathbf{P}_k^2$ (again there are $q^2 + q + 1$ of these). Call this $\mathfrak{X}_M$ and let $E_M \subset \mathfrak{X}_M$ be the proper transform of $\mathbf{P}_k^2$. Then a suitable Zariski-neighborhood of E_M in $\mathfrak{X}_M$ is isomorphic to a neighborhood of E in $\mathfrak{X}$. In particular, all surfaces E are rational surfaces gotten by blowing up $\mathbf{P}_k^2 (q^2 + q + 1)$ times and they meet the $2(q^2 + q + 1)$ adjacent components in rational curves. These rational curves are either exceptional curves C of the first kind, in which case $(C^2) = -1$, or proper transforms of lines along which $q+1$ points have been blown up, in which case:

$$(C^2) = +1 - (q+1) = -q.$$

Thus geometrically, if $C = E_1 \cap E_2$, then $(C^2)_{E_1} = -1$ and $(C^2)_{E_2} = -q$ or vice versa; this asymmetry corresponds in (a) above to whether

$$\dim_k(M_1/\pi^k M_2) = 1 \quad \text{or} \quad \dim_k(\pi^k M_2/\pi M_1) = 1$$

and in Σ to the orientation on the segment from v_1 to v_2.

Now if $\Gamma \subset PGL(3, K)$ is a discrete torsion-free co-compact group, we define first a formal scheme $\mathfrak{X}/\Gamma$ over R by dividing the formal completion of $\mathfrak{X}$ along $\pi = 0$ by Γ (this is possible because Γ acts freely and discontinuously in the Zariski-topology on $\mathfrak{X}_0$). Secondly, one verifies that the dualizing sheaf $\omega_{\mathfrak{X}}$ is ample on each component of $\mathfrak{X}_0$, hence it descends to an invertible sheaf $\omega_{(\mathfrak{X}/\Gamma)}$ on $\mathfrak{X}/\Gamma$ with the same property: this allows one to conclude that $\mathfrak{X}/\Gamma$ can be algebraized to true projective scheme over R, which, for simplicity, we denote $\mathfrak{X}/\Gamma$. Since the generic fibre $\mathfrak{X}_\eta$ of $\mathfrak{X}$ is smooth over K, the generic fibre $(\mathfrak{X}/\Gamma)_\eta$ is also smooth over K, hence

$$\omega_{(\mathfrak{X}/\Gamma)_\eta} = \Omega^2_{(\mathfrak{X}/\Gamma)_\eta}$$

hence $(\mathfrak{X}/\Gamma)_\eta$ is a surface of general type without smooth rational curves C with $(C^2) = -1$ or -2. It is not hard to compute the invariants of $(\mathfrak{X}/\Gamma)_\eta$: to do this, note that $(\mathfrak{X}/\Gamma)_0$ consists of finite set of rational surfaces, crossing each other (possibly crossing themselves) transversally in rational double curves and triple points. Let

E_α = normalizations of the components of $(\mathfrak{X}/\Gamma)_0$, $\quad 1 \le \alpha \le \nu_2$
C_β = normalizations of the double curves of $(\mathfrak{X}/\Gamma)_0$, $\quad 1 \le \beta \le \nu_1$
P_γ = triple points of $(\mathfrak{X}/\Gamma)_0$, $\quad 1 \le \gamma \le \nu_0$

We get an exact sequence:

$$0 \to \mathcal{O}_{(\mathfrak{X}/\Gamma)_0} \to \bigoplus_{\alpha=1}^{\nu_2} \mathcal{O}_{E_\alpha} \xrightarrow{\lambda} \bigoplus_{\beta=1}^{\nu_1} \mathcal{O}_{C_\beta} \xrightarrow{\mu} \bigoplus_{\gamma=1}^{\nu_0} \mathcal{O}_{P_\gamma} \to 0$$

hence

$$\begin{aligned}\chi(\mathcal{O}_{(\mathfrak{X}/\Gamma)_\eta}) &= \chi(\mathcal{O}_{(\mathfrak{X}/\Gamma)_0})\\ &= \sum_\alpha \chi(\mathcal{O}_{E_\alpha}) - \sum_\beta \chi(\mathcal{O}_{C_\beta}) + \sum_\gamma \chi(\mathcal{O}_{P_\gamma})\\ &= \nu_2 - \nu_1 + \nu_0.\end{aligned}$$

Let N be the number of orbits when Γ acts on the vertices of Σ. Clearly $N = \nu_2$. But each E_α contains $2(q^2 + q + 1)$ double curves, each on two E_α's, so

$$\nu_1 = N(q^2 + q + 1).$$

And each double curve passes through $(q + 1)$ triple points, each on three double curves, so

$$\nu_0 = N\frac{(q^2 + q + 1)(q + 1)}{3}.$$

Thus

$$\begin{aligned}\chi(\mathcal{O}_{(\mathfrak{X}/\Gamma)_\eta}) &= N\left[1 - (q^2 + q + 1) + \frac{(q^2 + q + 1)(q + 1)}{3}\right]\\ &= N\frac{(q - 1)^2(q + 1)}{3}.\end{aligned}$$

Next:

$$\begin{aligned}(c_{1,(\mathfrak{X}/\Gamma)_\eta}{}^2) &= (c_1(\omega_{(\mathfrak{X}/\Gamma)_0})^2)\\ &= \sum_\alpha (\mathrm{res}_{E_\alpha} c_1(\omega_{(\mathfrak{X}/\Gamma)_0})^2)\\ &= \sum_\alpha \Big((c_1(\omega_{E_\alpha}) + \sum_{\beta\neq\alpha} E_\alpha \cap E_\beta)^2\Big)\end{aligned}$$

All E_α's are just B = (the blow-up of $\mathbf{P}_k{}^2$ at all $(q^2 + q + 1)$-rational points). Let $\pi: B \to \mathbf{P}_k{}^2$ be the blow-up map, let h be the divisor class of a line on $\mathbf{P}_k^2$, let $e_i \subset B$ be the exceptional divisors and let $l_j \subset B$ be the proper transforms of the lines. Then $c_1(\omega_{E_\alpha}) + \Sigma_{\beta \neq \alpha} (E_\alpha \cap E_\beta)$ corresponds on B to:

$$K_B + \sum_i e_i + \sum_j l_j \equiv (\pi^{-1}(-3h) + \Sigma e_i) + \left(\sum_i e_i\right) + \sum_j (\pi^{-1}(h) - \text{the } e_i \text{ meeting } l_j)$$
$$\equiv \pi^{-1}((q^2+q-2)h) - (q-1)\left(\sum_i e_i\right)$$

with self-intersection $3(q-1)^2(q+1)$. Thus

$$(c_{1,(\mathfrak{X}/\Gamma)_\eta}{}^2) = 3N(q-1)^2(q+1).$$

By Riemann-Roch,

$$c_{2,(\mathfrak{X}/\Gamma)_\eta} = N(q-1)^2(q+1).$$

To determine the irregularity of $(\mathfrak{X}/\Gamma)_\eta$, we can use the relative Picard scheme $Pic_{\mathfrak{X}/\Gamma}{}^0$: its closed fibre is $Pic_{(\mathfrak{X}/\Gamma)_0}{}^0$, and since $(\mathfrak{X}/\Gamma)_0$ has normal crossings and rational components, this is an algebraic torus. In particular points of finite order l, $p \nmid l$, are dense: these correspond to l-cyclic coverings of $(\mathfrak{X}/\Gamma)_0$ and such coverings lift to $(\mathfrak{X}/\Gamma)_\eta$. Therefore the points of finite order l, $p \nmid l$, of $(Pic_{\mathfrak{X}/\Gamma}{}^0)_0$ lift to points of $(Pic_{\mathfrak{X}/\Gamma}{}^0)_\eta$ and hence $Pic_{\mathfrak{X}/\Gamma}{}^0$ is flat over R. On the other hand, a line bundle on $(\mathfrak{X}/\Gamma)_0$ is a line bundle on $\mathfrak{X}_0$ with Γ action: if it is in Pic^0, it is the trivial line bundle on $\mathfrak{X}_0$ and a Γ-action is just a homomorphism from Γ to $\mathbf{G}_m$. Thus finally, using Kajdan's theorem [4] that $\Gamma/[\Gamma, \Gamma]$ is finite, we deduce

$$\begin{aligned}\text{irregularity of } (\mathfrak{X}/\Gamma)_\eta &= \dim (Pic_{\mathfrak{X}/\Gamma}{}^0)_\eta \\ &= \dim (Pic_{\mathfrak{X}/\Gamma}{}^0)_0 \\ &= \dim \operatorname{Hom}(\Gamma, \mathbf{G}_m) \\ &= rk_{\mathbf{Z}}\Gamma/[\Gamma, \Gamma] \\ &= 0.\end{aligned}$$

Thus the numbers $h^{p,q}$ of $(\mathfrak{X}/\Gamma)_\eta$ fit into the pattern:

$$\begin{array}{l|lll} q & & & \\ & M-1 & 0 & 1 \\ & 0 & M & 0 \\ & 1 & 0 & M-1 \\ \hline & & & \rightarrow p \end{array} \qquad M = N\frac{(q-1)^2(q+1)}{3}$$

In particular, if $N = 1$, $q = 2$, then $M = 1$ and $(\mathfrak{X}/\Gamma)_\eta$ is a surface of the desired type. In this case, in fact $(\mathfrak{X}/\Gamma)_0$ is one rational surface, $\mathbf{P}^2$ blown up 7 times, crossing itself in 7 rational double curves, themselves crossing in 7 triple points. The confusion arising from trying to draw the result brings vividly to mind Lewis Carroll's comment on the sandy shore—"If seven maids with seven brooms were to sweep it for half a year, do you suppose, the Walrus said, that they could get it clear?"

2. The Example. The example is based on the 7th roots of unity: fix the notation:

$$\zeta = e^{2\pi i/7}$$

$$\lambda = \zeta + \zeta^2 + \zeta^4 = \left(\frac{-1+\sqrt{-7}}{2}\right)$$

$$\bar{\lambda} = \zeta^3 + \zeta^5 + \zeta^6 = \left(\frac{-1-\sqrt{-7}}{2}\right)$$

We have the fields:

$$\begin{array}{ll} \mathbf{Q}(\zeta) & \\ \quad\big|\ \text{deg } 3 & \Big\}\quad \text{Galois group } \mathbf{Z}/3\mathbf{Z}, \text{ generator } \sigma,\ \sigma(\zeta) = \zeta^2 \\ \mathbf{Q}(\lambda) & \\ \quad\big|\ \text{deg } 2 & \Big\}\quad \text{Galois group } \mathbf{Z}/2\mathbf{Z}, \text{ generator } z \to \bar{z} \\ \mathbf{Q} & \end{array}$$

Note $\mathbf{Q}(\lambda)$ is a UFD, $2 = \lambda \cdot \bar{\lambda}$ is the prime factorization of 2 and $7 =$

$-(\sqrt{-7})^2$ is the prime factorization of 7. We set $\mathbf{Q}(\zeta) = V$ and think of it only as a 3-dimensional vector space over $\mathbf{Q}(\lambda)$. We put the Hermitian form

$$h(x, y) = \mathrm{tr}_{\mathbf{Q}(\zeta)/\mathbf{Q}(\lambda)}(x\bar{y}) = [x\bar{y} + \sigma(x\bar{y}) + \sigma^2(x\bar{y})]$$

on V. Taking $1, \zeta, \zeta^2$ as a basis of V, we find that h has the matrix

$$H = \begin{pmatrix} 3 & \bar{\lambda} & \bar{\lambda} \\ \lambda & 3 & \bar{\lambda} \\ \lambda & \lambda & 3 \end{pmatrix}$$

so that h is positive definite with determinant 7. Note that V contains the lattice $L = \mathbf{Z}[\zeta]$, with basis $1, \zeta, \zeta^2$ over $\mathbf{Z}[\lambda]$. Define

$$\Gamma_1 = \mathbf{Q}(\lambda)\text{-linear maps } \gamma: V \to V \text{ which preserve the form } h \text{ and map } L[1/2] \text{ to } L[1/2]$$

Since 2 splits in $\mathbf{Q}(\lambda)$, the λ-adic completion of $\mathbf{Q}(\lambda)$ is isomorphic to the 2-adic completion $\mathbf{Q}_2$ of $\mathbf{Q}$ (in fact, in $\mathbf{Q}_2$, we may take $\lambda = (\text{unit}) \cdot 2$, $\bar{\lambda} = \text{unit}$). So we have a canonical map $V \to$ (λ-adic completion of V) $\cong \mathbf{Q}_2 \cdot 1 \oplus \mathbf{Q}_2 \cdot \zeta \oplus \mathbf{Q}_2 \cdot \zeta^2$ and a canonical homomorphism

$$\Gamma_1 \to \mathrm{GL}(3, \mathbf{Q}_2) \to \mathrm{PGL}(3, \mathbf{Q}_2).$$

From standard results on the theory of arithmetic groups*, the image $\overline{\Gamma_1}$ of Γ_1 is discrete and co-compact. We introduce 3 elements of Γ_1: the first is σ itself; the second is

$$\tau(x) = \zeta \cdot x.$$

Note that $\sigma^3 = e$, $\tau^7 = e$ and $\sigma\tau\sigma^{-1} = \tau^2$, so together σ and τ generate

---

*Consider $U(V, h)$ as an algebraic group over $\mathbf{Q}$. Γ_1 is its $\mathbf{Z}[1/2]$-rational points. Since U is compact at the infinite place, Γ_1 is discrete and co-compact in $U(V, h)(\mathbf{Q}_2)$. But over $\mathbf{Q}_2$, $U(V, h) \cong \mathrm{GL}(3)$, so mod scalars Γ_1 defines a discrete co-compact subgroup of $\mathrm{PGL}(3, K)$.

a subgroup $\Gamma_2 \subset \Gamma_1$ of order 21. The third is a map ρ given by

$$\rho(1) = 1$$

$$\rho(\zeta) = \zeta$$

$$\rho(\zeta^2) = \lambda - \frac{\lambda^2}{\bar{\lambda}}\zeta + \frac{\lambda}{\bar{\lambda}}\zeta^2$$

It can be checked easily that $\rho \in \Gamma_1$. It can also be checked that

$$(\rho \cdot \tau)^3 = \text{multiplication by } \lambda/\bar{\lambda}.$$

Note that the scalar matrices in Γ_1 are exactly

$$\pm(\lambda/\bar{\lambda})^k \cdot I_3$$

PROPOSITION. *ρ, σ, τ and $-I$ generate Γ_1. All torsion elements in $\overline{\Gamma_1}$ are conjugate to either $\sigma^i \cdot \tau^j$ or to $(\rho \cdot \tau)^i$ (some $0 \le i \le 2$, $0 \le j \le 6$).*

Proof. Consider the action of Γ_1 on $\Sigma_0' =$ [the set of free rank $3\mathbf{Z}_2$-submodules of $\mathbf{Q}_2{}^3$]. Let M_0 be the submodule $\mathbf{Z}_2 \cdot 1 \oplus \mathbf{Z}_2 \cdot \zeta \oplus \mathbf{Z}_2 \cdot \zeta^2$ or $\mathbf{Z}_2{}^3$ for short. If an element $\alpha \in \Gamma_1$ maps M_0 to itself, then back in V, α is given by a 3×3 matrix with coefficients in $\mathbf{Z}[\lambda][1/\bar{\lambda}]$. Since α is H-unitary, its coefficients are also in $\mathbf{Z}[\lambda][1/\lambda]$, so α in fact has coefficients in $\mathbf{Z}[\lambda]$ and maps L to itself. But in L, it is easy to see that $\{\pm\zeta^i\}$ are the only elements $x \in L$ with $h(x, x) = 3$. So α permutes them. Then $\pm\tau^i \circ \alpha$ also carries the element $1 \in L$ to itself. Now the equations

$$h(x, x) = 3$$

$$h(1, x) = \bar{\lambda}$$

have only 3 solutions in L:$x = \zeta$, ζ^2 or ζ^4. So $(\pm\tau^i \cdot \alpha)$ carries ζ to ζ, ζ^2 or ζ^4. Then $(\pm\sigma^j \circ \tau^i \circ \alpha)$ fixes 1 *and* ζ and it is easy to check that such a map must be the identity. Thus $\pm\Gamma_2$ is the stabilizer of M_0.

As in the Bruhat-Tits building, call M, $M' \in \Sigma_0'$ adjacent if $M \supset M'$ and $\dim_{\mathbf{Z}/2\mathbf{Z}} M/M' = 2$ or vice versa. Then $\rho(M_0) \subset M_0$ and is adjacent to M_0. Because $M/2M \cong (\mathbf{Z}/2\mathbf{Z})^3$, there are only 7 modules $M' \subset M_0$ adjacent to M_0. One checks easily that these are the modules $\tau^i\rho(M_0)$, $0 \le i \le 6$. Thus $(\tau^i\rho)^{\pm 1}(M_0)$ is the set of all $M \in \Sigma_0'$ adjacent to M_0. Since Σ_0' is connected under adjacency, this shows that *all* elements of Σ_0' can be expressed as:

$$(\tau^{i_1}\rho)^{\epsilon_1}\cdot \ldots \cdot(\tau^{i_k}\rho)^{\epsilon_k}(M_0), \qquad 0\le i_l\le 5, \qquad \epsilon_l=\pm 1.$$

Thus the subgroup of Γ_1 generated by ρ, σ, τ and $-I$ acts as transitively as Γ_1 on Σ_0' and M_0 has the same stabilizer in both groups, so they are equal.

Now let $\alpha \in \Gamma_1$ be torsion in $\overline{\Gamma_1}$. If α is torsion in Γ_1, then α fixes e.g. the module

$$M_1=\sum_{i=1}^{\mathrm{order}(\alpha)}\alpha^i(M_0)$$

and, if $M_1=\beta(M_0)$, then $\beta^{-1}\alpha\beta$ fixes M_0. Thus $\beta^{-1}\alpha\beta\in\pm\Gamma_2$ and $\bar{\alpha}$ is conjugate to $\sigma^j\circ\tau^i$, some i, j. In general, we consider det α. Then $|\det\alpha|^2=1$ and $\det\alpha\in\mathbf{Z}[\lambda][\tfrac{1}{2}]$, hence

$$\det\alpha=\pm(\lambda/\bar{\lambda})^i.$$

Replacing α by $\pm(\lambda/\bar{\lambda})^i\alpha^{\pm1}$, we may assume $\det\alpha=\lambda/\bar{\lambda}$. Then

$$(\lambda/\bar{\lambda})^{-1}\alpha^3$$

has determinant 1 and is torsion in PGL(3), so it is torsion in GL(3). Now consider α and α^3/λ acting on $\mathbf{Q}_2{}^3$. Since $\mathbf{Z}_2[\alpha,\alpha^3/\lambda]$ is a finite ring over $\mathbf{Z}_2$, there is a free rank $3\mathbf{Z}_2$-module $M\subset\Phi_2{}^3$ such that $\alpha(M)\subset M$, $\alpha^3/\lambda(M)\subset M$. Since $\alpha^3/(\lambda/\bar{\lambda})$ is torsion and $\bar{\lambda}$ is a λ-adic unit, it follows that

$$M\supset\alpha(M)\supset\alpha^2(M)\supset\alpha^3(M)=2M.$$

As before, replacing α by a conjugate, we can assume $M=M_0$. But now it is easily checked that the 21 maps $\sigma^i\circ\tau^j$ act simply transitively on the flags

$$M_0/2M_0\supset(\text{2-dim}^l\text{ subspace})\supset(\text{1 dim}^l\text{ subspace}).$$

So conjugating α by $\sigma^i\circ\tau^j$, we can assume

$$\alpha(M_0)=(\rho\circ\tau)(M_0)$$

$$\alpha^2(M_0) = (\rho\circ\tau)^2(M_0).$$

Then $(\rho\circ\tau)^{-1}\circ\alpha$ carries M_0 into itself and fixes a flag in $M_0/2M_0$. The former implies that $(\rho\circ\tau)^{-1}\circ\alpha = \pm\sigma^i\circ\tau^j$, some i, j, and the latter implies that $i = j = 0$. Thus in $\overline{\Gamma_1}$, $\alpha = \rho\circ\tau$. Q.E.D.

It remains to choose a suitable subgroup $\Gamma \subset \Gamma_1$ of finite index such that:

a) Γ/scalars is torsion-free
b) Γ acts transitively on Σ_0', (hence Γ/scalars acts transitively on Σ_0, the vertices of the Bruhat-Tits building).

It will then follow from the results of Section 1 that the corresponding surface $(\mathfrak{X}/\Gamma)_\eta$ is a surface of the desired type. To find Γ, it is convenient to use a congruence subgroup for the prime 7. In fact, consider the maps:

$$\mathbf{Z}[\lambda, \tfrac{1}{2}] \to \mathbf{Z}[\lambda, \tfrac{1}{2}]/(\sqrt{-7}) \cong \mathbf{Z}/7\mathbf{Z}$$

$$\lambda, \bar{\lambda} \mapsto 3$$

$$L[\tfrac{1}{2}] \to \underbrace{L[\tfrac{1}{2}]/(\sqrt{-7})L[\tfrac{1}{2}]}_{\text{call this } L_0} \cong (\mathbf{Z}/7\mathbf{Z})^3$$

The induced form h_0 on L_0 is of rank 1 and has null-space $L_1 \subset L_0$ spanned by $\zeta - 1$, $\zeta^2 - 1$. Taking $1, \zeta - 1, (\zeta - 1)^2$ as a basis of L_0, it is easy to check that mod 7:

$$\sigma \mapsto \begin{pmatrix} 1 & 0 & 0 \\ 0 & 2 & 0 \\ 0 & 1 & 4 \end{pmatrix}, \qquad \tau \mapsto \begin{pmatrix} 1 & 0 & 0 \\ 1 & 1 & 0 \\ 0 & 1 & 1 \end{pmatrix}$$

$$\rho \mapsto \begin{pmatrix} 1 & 0 & 0 \\ 0 & 1 & 4 \\ 0 & 0 & 1 \end{pmatrix}, \qquad \rho\circ\tau \mapsto \begin{pmatrix} 1 & 0 & 0 \\ 1 & 5 & 4 \\ 0 & 1 & 1 \end{pmatrix}$$

In particular, considering the action of Γ_1 on L_1, we get a homomorphism

$$\pi : \Gamma_1 \to GL(2, Z/7\mathbf{Z}) \cap \{X \mid \det X = \pm 1\} \underset{\text{def}}{=} G.$$

The group G on the right has order $2^5 \cdot 3 \cdot 7$. Let H be a 2-Sylow subgroup and define $\Gamma = \pi^{-1}(H)$. Since all 21 elements $\sigma^i\tau^j$ and all 3 elements $(\rho\circ\tau)^i$ except e have non-zero images in G of orders 3 or 7, Γ is torsion-free. As the full group Γ_1 is set-theoretically $\Gamma \times \Gamma_2$, Γ acts transitively on Σ_0'. This completes the construction.

HARVARD UNIVERSITY

REFERENCES

[1] A. Kurihara, "On certain varieties uniformized by q-adic automorphic functions," to appear

[2] F. Hirzebruch, "Automorphe Formen und der Satz von Riemann-Roch," *Symp. Int. de Top. Alg.*, UNESCO, 1958

[3] F. Hirzebruch and K. Kodaira, "On the complex projective spaces," *J. Math. Pures et Appl.*, **36** (1957), p. 201.

[4] D. Kajdan, "Connection of the dual space of a group with the structure of its closed subgroups," *Funct. Anal. and its Applic.*, **1** (1967), p. 63 in English transl.

[5] G. A. Mustafin, "Non-Archimedean Uniformizations," *Mat. Sbornik*, **105** (1978), p. 207.

[6] Ira Shavel, "A class of algebraic surfaces of general type constructed from quaternion algebras," to appear, *Pac. J. Math.*

[7] A. Weil, "On discrete subgroups of Lie groups," *Ann. of Math.*, **72** (1960), p. 369.

[8] S. T. Yau, "Calabi's conjecture and some new results in algebraic geometry," *Proc. Nat. Acad. Sci.*, USA, **74** (1977), p. 1798

Printed in the United States
By Bookmasters